A GLENCOE PROGRAM

MERRILL CHEMISTRY

Student Edition
Teacher Wraparound Edition
Laboratory Manual-Student Edition
Laboratory Manual-Teacher Edition
English/Spanish Glossary
Computer Test Bank
Videodisc Correlations
Transparency Package
Problems and Solutions Manual
Solving Problems in Chemistry
Study Guide-Student Edition
Mastering Concepts in Chemistry-Software

Teacher Resource Books:
- Enrichment
- Critical Thinking/ Problem Solving
- Study Guide-Teacher Edition
- Lesson Plans
- Chemistry and Industry
- ChemActivities
- Vocabulary Review/ Concept Review
- Transparency Masters
- Reteaching
- Applying Scientific Methods in Chemistry
- Evaluation

Authors

Robert C. Smoot
Rollins Fellow in Science
McDonogh School
McDonogh, MD

Richard G. Smith
Chemistry Teacher
Bexley High School
Bexley, OH

Jack Price
Co-director
Center for Science and Mathematics Education
California State Polytechnic Institute
Pomona, CA

Contributing Author

Tom Russo
Chemistry Education Consultant
Kemtec Educational Corp.
West Chester, OH

Content Consultants

Teresa Anne McCowen
Chemistry Department
Butler University
Indianapolis, IN

Mamie W. Moy
Chemistry Department
University of Houston
Houston, TX

William M. Risen, Jr.
Chemistry Department
Brown University
Providence, RI

Safety Consultant

Joanne Neal Bowers
Plainview High School
Plainview, TX

Glencoe/McGraw-Hill
A Division of The McGraw-Hill Companies

ISBN 0-02-825527-5

Send all inquiries to: Glencoe/McGraw-Hill
8787 Orion Place, Columbus, Ohio 43240

Printed in the United States of America.

3 4 5 6 7 8 9 071/043 04 03 02 01 00

PHOTO CREDITS

3T (t)Science Source/Photo Researchers, (bl br)Mark Burnett; 11T Doug Martin; 13T (tl)Scott Cunningham, (tr)Doug Martin, (b)BLT Productions; 19T courtesy Dr. Bart Fraser-Reid; 26T Matt Meadows.

TEACHER WRAPAROUND ED[illegible]

MERRILL

CHEMISTR[illegible]

GLENCOE

McGraw-Hill

New York, New York Columbus, Ohio Woodland Hills, California Peoria, Illinois

Contents

Teacher Guide

Student Text

Course Contents and Planning Guide

You are in the best position to design a chemistry course that meets the needs of your students. The following Course Contents and Planning Guide is designed to assist you with both long-range and daily planning. If you prefer to teach chemistry in a different sequence, *Merrill Chemistry* is adaptable to your needs. You will find that many chapters can be reordered with no loss of understanding.

In the Course Contents and Planning Guide, each section of the text has been designated as Core (C), Optional (O), or Advanced (A). Core sections cover topics that are essential to a basic high school chemistry course. Optional sections are also appropriate for average chemistry students but cover topics that are not considered core material. Advanced sections cover more complex concepts that may be appropriate for honors classes or as extensions for more able students in average classes. All of the Advanced sections are labeled in the student text as *Chemistry in Depth.* All ChemActivities are considered Core material except for ChemActivities 27 and 30. ChemActivity 27 is Advanced because all of Chapter 27 is classed as Advanced. ChemActivity 30 is Optional because all of Chapter 30 is considered Optional.

C Core sections
O Optional sections
A Advanced sections (Chemistry in Depth)

Program Philosophy

The Need for New Directions in Science Education

By today's projections, seven out of every ten American jobs will be related to science, mathematics, or electronics by the year 2000. And according to the experts, if students haven't grasped the fundamentals, they probably won't go further in science and may not have a future in a global job market. Studies also reveal that students are avoiding taking "advanced" science classes.

The Time for Action Is Now!

In the past decade, educators, public policy makers, corporate America, and parents have recognized the need for reform in science education. These groups have united in a call to action to solve this national problem. *The National Science Education Standards*, recently published by the National Research Council and representing the contribution of thousands of educators and scientists, offers a comprehensive vision of a scientifically literate society. The standards not only describe what students should know, but they also offer guidelines for science teaching and assessment. If you are using, or plan to use, the standards to guide changes in your science curriculum, you can be assured that *Merrill Chemistry* aligns exceedingly well with *The National Science Education Standards*.

Merrill Chemistry Answers the Challenge!

We believe that *Merrill Chemistry* will help you bring science reform to the front lines—the classrooms of America. On every page, *Merrill Chemistry* fulfills the goals of science curriculum reform. More importantly, we believe it will help students succeed in science so that they will want to continue learning about science through high school and into adulthood.

Throughout this text, the basic learning principle of proceeding from the familiar or known information to the unknown is used. Principles of structure, matter-energy relationships, thermodynamics, and chemical equilibrium are presented several times throughout the text with varying degrees of emphasis. Thus, students develop a sense of confidence when they recognize familiar concepts presented later in greater depth. Minimum emphasis has been placed on memorization of fact. Instead, the purpose of this book is to foster understanding and the ability to predict behavior of substances under various conditions.

National Science Content Standards

Unifying Concepts and Processes	
UCP.1	Systems, order, and organization
UCP.2	Evidence, models, and explanation
UCP.3	Change, constancy, and measurement
UCP.4	Evolution and equilibrium
UCP.5	Form and function
Science as Inquiry	
A.1	Abilities necessary to do scientific inquiry
A.2	Understandings about scientific inquiry
Physical Science	
B.1	Structure of atoms
B.2	Structure and properties of matter
B.3	Chemical reactions
B.4	Motions and forces
B.5	Conservation of energy and increase in disorder
B.6	Interactions of energy and matter
Life Science	
C.1	The cell
C.2	Molecular basis of heredity
C.3	Biological evolution
C.4	Interdependence of organisms
C.5	Matter, energy, and organization in living systems
C.6	Behavior of organisms
Earth and Space Science	
D.1	Energy in the earth system
D.2	Geochemical cycles
D.3	Origin and evolution of the earth system
D.4	Origin and evolution of the universe
Science and Technology	
E.1	Abilities of technological design
E.2	Understandings about science and technology
Science in Personal and Social Perspectives	
F.1	Personal and community health
F.2	Population growth
F.3	Natural resources
F.4	Environmental quality
F.5	Natural and human-induced hazards
F.6	Science and technology in local, national, and global challenges
History and Nature of Science	
G.1	Science as a human endeavor
G.2	Nature of scientific knowledge
G.3	Historical perspectives

Science Content Standards

The National Science Content Standards for Grades 9–12 have been correlated to each major section of ***Merrill Chemistry***. You will find these correlations in the first column of the Chapter Organizer on the interleaf pages preceding each chapter. Correlations are designated according to the numbering system in the table of science content standards shown on the facing page.

Assessment Standards

The assessment standards are supported by many of the components that make up the ***Merrill Chemistry*** program. The ***Teacher Wraparound Edition*** provides many opportunities for assessing students' understanding of important concepts and ability to perform a wide range of skills. Ideas for portfolios, performance activities, written reports, and other assessment activities accompany every lesson. Other Teacher Classroom Resources that provide assessment opportunities include ***Solving Problems in Chemistry, Evaluation***, a computer ***Testbank,*** and ***Mastering Concepts in Chemistry*** software. For more details about assessment ideas and resources, see page 17T. Rubrics and performance task assessment lists can be found in *Glencoe's Professional Series* booklet ***Performance Assessment in the Science Classroom***.

Teaching Standards

Merrill Chemistry helps you make the most of every instructional moment. It offers many activities and a wide variety of effective strategies for presenting the content, processes, and skills of chemistry. The ***Enrichment, Critical Thinking/Problem Solving, Reteaching***, and ***Applying Scientific Methods in Chemistry*** booklets in the Teacher Resource Package offer additional teaching material that will help you meet the needs of all your students and aid you in teaching more effectively.

Unifying Themes in Chemistry

In every chapter, themes that run throughout the textbook help students see the big picture of chemistry. The themes as outlined on this page will help students build upon their knowledge, step by step, and see relationships to the world around them. In ***Merrill Chemistry***, themes provide a framework for integrating chemistry core concepts. Themes in chemistry provide the big ideas that link the core concepts. Although there are several possible themes around which to unify the study of chemistry, we have chosen four: Systems and Interactions, Scale and Structure, Stability and Change, and Energy.

Systems and Interactions A system can be as small as an atom's nucleus and its electrons or as large as the stars that make up a galaxy. By defining the boundaries of the system, one can study the interactions among its parts. The interactions may be the force of the attraction between the atom's positively charged nucleus and its negatively charged electrons or they may be the way two molecules come together in a reaction.

Scale and Structure A central idea in chemistry is that the physical and chemical properties of matter are a consequence of the structure and arrangement of matter's basic particles—atoms, molecules, and ions. We observe the behavior of matter on a macroscopic scale and then attempt to explain that behavior by using models of matter at the submicroscopic level, that is, at the level of individual particles and their interactions.

Stability and Change A system that is stable is constant. Often, this stability is the result of a system being in equilibrium. If a system is not stable, it will undergo change. Changes in an unstable system may be characterized as trends, as in electron configurations, cycles, as in the actions of enzymes; or irregular, as in radioactive decay.

Energy Energy is the ability of an object to change itself or its surroundings—the capacity to do work. Nearly all chemical and physical changes either take in or give up energy. The theme of energy goes hand in hand with the theme of stability and change because the availability of energy can determine whether or not a change will occur and to what degree it occurs.

Using Your *Teacher Wraparound Edition*

The *Teacher Wraparound Edition* is designed to make teaching *Merrill Chemistry* as easy as possible. Nearly all relevant teacher information is presented in wraparound margins. In addition, two pages of planning guides precede every chapter as interleaves.

Interleaves

The two interleaf pages that precede each chapter contain complete planning information for that chapter. The **Chapter Organizer** lists the objectives, text features, laboratory activities, and demonstrations for each numbered section. Correlated to each section are all the *Teacher Classroom Resources,* including worksheets and color transparencies. **National Science Content Standards** correlations to each section also are located here.

On the second interleaf page, an **Assignment Guide** coordinates all Practice Problems, Chapter Review, and supplementary problems in *Solving Problems in Chemistry* to the subsections for which they are appropriate. This correlation provides an at-a-glance guide when assigning additional practice or homework. Subsections are classified as being a part of the core curriculum, optional, or advanced.

To aid you with laboratory planning, a chart of **Laboratory Materials** needed for the Minute Labs, ChemActivities, and Demonstrations is provided. The **Glencoe Technology** box provides information on Glencoe computer software and multimedia material that can be used in the preparation and presentation of chapter content.

Wraparound Margins

The bulk of the teacher information is conveniently placed in the margin next to the student page. At the beginning of each chapter, the **CHAPTER OVERVIEW** previews the **Main Ideas** presented in the chapter. A **Theme Development** section describes how one of the themes of the text is incorporated into the chapter. **Tying to Previous Knowledge** presents strategies to help the student build on common knowledge.

INTRODUCING THE CHAPTER and the **DISCOVERY DEMO** provide suggestions for an inquiry approach to starting the chapter. **Using the Photo** assists you by providing information about the visual link between chemistry concepts and real-world situations. **Revealing Misconceptions** suggests strategies for eliciting and clearing up student misconceptions about specific ideas in the chapter.

Teaching Cycle

Each numbered section begins with **PREPARE,** which contains **Key Concepts** and **Key Terms.** In some sections, **Looking Ahead** assists you in advance preparation of laboratory or demonstration materials. The **PREPARE** section is followed by a comprehensive four-part teaching cycle designed to help students learn chemical concepts most efficiently.

1 FOCUS

The first step to teaching any set of concepts or skills is to focus the students' attention. After objectives are previewed with students, a focus idea is provided. This idea may include a **Quick Demo,** cooperative learning suggestion, analogy, audiovisual, brainstorming, or discussion question that helps motivate and focus the class so the lesson can begin.

2 TEACH

The primary aim of the *Merrill Chemistry Teacher Wraparound Edition* is to give you the tools to accomplish the task of getting chemistry concepts and problem-solving skills across to your students. Most of these tools are presented under **TEACH.** Under the heading **Concept Development,** each section provides suggestions for ways that you can present the content of the section. Many teaching suggestions are provided under a series of clearly defined headings such as **Background, Using the Illustration, Enrichment, Computer Demo,** and others. There are also more than 75 **Quick Demos. Glencoe Technology and National Geographic Society** references tell you where and how you can integrate these resources into your lessons. Also under TEACH can be found **Making Connections** and **Misconceptions.** Chapters that include problems also have **Correcting Errors** and **Practice.** You are also provided with information needed to teach each of the special features of the student edition such as **Minute Lab, Chemistry and Society, Chemistry and Technology, Frontier, Everyday Chemistry, Bridges, Chemists and Their Work, Careers in Chemistry,** and **Connections.**

3 ASSESS

To assist you in monitoring students' progress and adjusting your teaching accordingly, a **Check for Understanding** idea is provided for you in each section. For students who are having trouble, a **Reteaching** suggestion immediately follows the **Check for Understanding** hint. The **Reteaching** tip suggests a way to teach the same concepts or facts differently to adjust to students' individual learning styles. Several **Assessment** strategies in each chapter provide ideas for alternative forms of assessment. For those students in the class who do not need additional help understanding the lesson, **Extension** ideas are provided immediately following the **Reteaching** tips. These enrichments allow students to go on while others are reviewing the lesson.

4 CLOSE

Closing the lesson is the last, and one of the most important, steps in teaching the section because it helps students relate and synthesize the ideas presented in the section. The *Merrill Chemistry Teacher Wraparound Edition* gives you a variety of ways to provide effective closure to a section. The **CLOSE** options are activities, discussion questions, or demonstrations that summarize the section, bridge to the next section, or provide an application of the section.

And There's More

At the bottom of most pages in the *Merrill Chemistry Teacher Wraparound Edition* is a **Demonstration,** a **Content Background** discussion, or a **Chemistry Journal** suggestion. Chemistry Journal entries offer writing opportunities on relevant topics for your students. More than 175 demonstrations, some of which can be used as laboratory activities, permit you to make nearly every major concept visible to students. Also at the bottom of the page are **Internet** entries that will enhance your teaching and provide relevant information to you and your students. **Cultural Diversity** entries in each chapter look at the contributions of other cultures to chemistry as well as the impact of chemistry on those cultures. Answers to all **Practice Problems** and **Concept Review** and **Chapter Review** questions are presented in the margin for quick and convenient reference. There are **Program Resources** boxes at the beginning of each section and throughout the chapter to assist you in planning which of the many available resources are appropriate for your students.

As you review the *Merrill Chemistry Teacher Wraparound Edition,* you will discover that you and your students are considered to be very important. With the enormous number and diversity of teaching strategies provided by the teacher edition, you will be able to accomplish the goals of your curriculum. The materials allow and encourage adaptability and flexibility so that both student needs and curricular needs can be met.

Supplementary Materials

The *Teacher Classroom Resources* are designed to enhance the *Merrill Chemistry* program by assisting the teacher in all areas of chemistry instruction. Use the *Teacher Classroom Resources* as time-saving teaching aids in the planning, instruction, evaluation, and enrichment of your chemistry classes. Each component of the *Teacher Classroom Resources* is describe below.

Vocabulary/Concept Review—includes masters that helps students review all the key terms within each chapter and separate masters that reinforce the major concepts of the chapter.

Study Guide—contains masters that help students read and understand the major concepts of each numbered section.

Reteaching—provides masters with alternate ways of teaching more troublesome major concepts.

Enrichment—features masters that amplify and extend textbook topics or introduce more advanced topics.

Critical Thinking/Problem Solving—provides masters that require students to apply knowledge and higher-level thinking skills to analyze situations and solve problems.

ChemActivities—provides masters that have complete procedures, data forms, and answer space for all the ChemActivities in the textbook.

Transparency Masters—contains copy masters of the 50 color transparencies, 50 transparency worksheet masters, and a teaching strategy for each transparency.

Applying Scientific Methods in Chemistry—contains masters that ask students to use scientific reasoning to interpret laboratory scenarios.

Evaluation—furnishes a complete set of test masters for each chapter consisting of a multiple choice section and a section with questions and problems requiring longer answers. Each test is supplied with an alternate version of the multiple choice section. It also includes a final exam.

Chemistry and Industry—masters that describe industrial processes to show how chemistry is applied in a real-world setting.

Laboratory Manual

A well-rounded laboratory program is an integral part of the high school chemistry curriculum. The *Laboratory Manual* provides a variety of laboratory activities that serve to reinforce chemistry principles encountered in *Merrill Chemistry.* Many of the experiments employ microchemistry techniques while others use traditional equipment. A post-lab quiz is provided for each experiment.

Problems and Solutions Manual

This manual is a complete listing of all problems and questions in the text including Practice Problems, Concept Review, and Chapter Review. Each problem is restated in the manual so that you do not have to look back to the textbook and is followed by a complete solution for that problem.

Solving Problems in Chemistry

This book is keyed chapter by chapter to the textbook and provides a wealth of extra problems and questions. In addition it presents a concise review of basic concepts along with more sample problems to help students.

Color Transparencies

The *Color Transparencies* package contains 50 full-color transparencies as well as the *Transparency Masters* book, which contains a blackline version of each transparency, a student worksheet, and a teaching strategy for each transparency. The color transparencies and book are conveniently packaged in a 3-ring binder.

Study Guide - Student Edition

This is a consumable version of the *Study Guide* masters book.

Mastering Concepts in Chemistry—Software

This software has been written expressly to accompany ***Merrill Chemistry.*** It provides a colorful, animated presentation of core concepts and skills in chemistry. Individual students can also use the software for review and reinforcement of these concepts. In constructing this package, we have selected 15 groups of basic concepts that students must master if they are to progress and have success in chemistry.

Computer Test Bank

The *Computer Test Bank* provides you with a tool to generate a test from a bank of algorithms and fixed questions. Program features include selection of sections to be tested, number of questions for each section, objectives to be tested, total number of questions on a test, inclusion or exclusion of specific questions, and multiple test forms. Teachers may also add their own questions or modify existing questions. Test banks are available in IBM, Apple, and Macintosh versions.

English-Spanish Glossary

This supplement presents an English-Spanish version of the glossary in the student text.

Lesson Plans

This booklet provides a complete teaching plan for each numbered section of the textbook. Each plan is geared to the teaching cycle employed in the *Teacher Wraparound Edition* and contains a complete list of all the resources available for that section. Strategies for block scheduling also are provided.

Videodisc Correlation

The *Videodisc Correlation* book contains a complete correlation with bar codes to the Videodiscovery videodisc, *Chemistry at Work.*

Microchemistry Techniques Video

Presented by contributing author Tom Russo, this video serves as an overview of the microchemistry apparatus and techniques employed in some of the ***Merrill Chemistry*** ChemActivities in the textbook as well as in the microchemistry experiments in the *Laboratory Manual.*

CD-ROM Multimedia System

Students can better understand complex chemical processes and structures when they see them in action. In *Glencoe's CD-ROM Multimedia System,* students will be able to make choices and change variables in Explorations to challenge their critical-thinking skills. They can practice lab skills and learn core chemistry concepts with the Interactive Experiments in a virtual lab setting. They can watch 3-D animations that illustrate chemical processes at the submicroscopic level and full-color videos of chemical processes in industry and in the lab. With this flexible interactive multimedia system, many parts of the chemistry curriculum can be introduced, enriched, or remediated at any time during the course. And best of all, each part of this system is correlated to the appropriate lesson in the *Teacher Wraparound Edition.*

Videodisc Programs

Glencoe provides three videodisc programs that are correlated to ***Merrill Chemistry.*** Bar codes to all programs are displayed in the *Teacher Wraparound Edition,* right where they are needed to enhance your instruction.

The videodiscs included in the *Science and Technology Videodisc Series* contain full-motion video reports that cover a broad spectrum of topics relating to research in the fields. The video reports are derived from the *Mr. Wizard* television program and are ideal for illustrating scientific methods, laboratory techniques, and careers in chemistry.

Chemistry: Concepts and Applications videodisc program provides exciting opportunities to see chemistry come alive from the page. For example, as students read about the five different types of chemical reactions, they will be able to view them instantly on the screen.

Finally, the *National Geographic Society* videodisc series offers interactive learning tools that allow students to explore the world and all that's in it through video, photographs, sound, and text.

For you, the teacher, we have saved you time and money by filming many of the demonstrations that you would like your students to experience. If you have a limited budget, strict safety rules, and not enough time to perform the demonstrations yourself, you will find this program invaluable. Many of the submicroscopic processes of chemistry have been animated for both the CD-ROM program and the videodisc program. As an instructional advantage, while your students manipulate the CD-ROM animations frame-by-frame on their computer screens, you can show the animation on your full-size screen at the front of the classroom.

Technology in the Classroom

Accessing the Internet

The Internet is the world's largest computer network. It is a network of networks linked by high-speed data lines and wireless systems that freely exchange information. It provides access to individuals, corporations, educational institutions, government departments, and other groups.

To access the Internet, you need a computer with communication software, a modem, and a phone line. A modem is a device that enables computer data to travel to another computer via phone lines. Many computers come equipped with an internal modem. You will need a microprocessor of class 486 or greater, at least 8 megabytes of random access memory, and a modem that runs at 9600 baud or faster.

To access the Internet, you must first set up an account with an Internet service provider (server). Although these commercial providers can be expensive, depending on how many of their services you use, it is also possible in some cities to access the Internet without charge through university or local library systems.

The World Wide Web

To access a wider source of multimedia information, you and your students can access the World Wide Web (WWW) by using a web browser. Many servers now provide the use of a web browser. The first page you see when you enter a WWW site is the home page. From the home page, you can obtain complete texts and graphics from books, clips from movies, or graphical replicas of art from museums.

You can browse through books, magazines, and scientific periodicals. Go to the home pages of scientific-supply companies and order your chemicals, equipment, books, or posters. You and your students can even create your own home page on the WWW to explain to other classrooms around the country your topics of study for the week, share data collected from classroom experiments, or ask other classes for suggestions for projects.

Some Useful Free Web Sites

AskERIC The Q & A Service
http://ericir.syr.edu
This is an ongoing project that is building a digital library for educational information. You use it as you would a library.

Chemistry—Resources from ACS
gopher://acsinfo.acs.org
This gopher site is sponsored by the American Chemical Society, and it gives Internet chemistry resources.

National Renewable Energy Laboratory
http://www.nrel.gov/
The U.S. Department of Energy Lab offers information about renewable energy, including energy data, resource maps, publications, and job opportunities.

U.S. Department of Education
http://www.ed.gov/
This site provides information about various programs in the Department of Education.

Webcrawler
http://webcrawler.com
This is a search engine with useful links to high school chemistry home pages with opportunities for sharing ideas and resources.

WebElements
http://www.cchem.berkeley.edu/Table/index.html
This server provides you with a database for the Periodic Table on-line.

World Wide Web Virtual Library: Chemistry
http://www.chem.ucla.edu
This is a great starting point for access to many resources for chemistry education, including demos, free materials, conferences, and education information.

Yahoo
http://www.bham.wednet.edu
Another search engine with an extensive inventory of Web sites. Type in the key words *Chemistry Teacher* and it will list many useful resources for general chemistry.

Words of Caution

The sites referenced in the *Teacher Wraparound Edition* are not under the control of Glencoe, and therefore, Glencoe can make no representation concerning the content of these sites. Extreme care has been taken to list only reputable links by using educational and government sites whenever possible. Internet searches have been used that return only sites that do not contain content apparently intended for mature audiences.

You might wish to caution your users that any scientific information they may read on the Internet has not been reviewed and authenticated by the usual peer-review system practiced in textbook and journal publishing. However, with this caution in mind, you will soon discover the use of the computer in the classroom to be an exciting and rewarding addition to your lesson plans.

Assessment

The *Merrill Chemistry* program has been designed to provide you with a variety of assessment tools, both formal and informal, to help you develop a clearer picture of your students' progress. You will find an assortment of traditional tools for content assessment including a four-page Evaluation master for each chapter. The Computer Test Bank allows you to make customized tests from a bank of questions and problems for each chapter. In the textbook, the Concept Review questions can serve as assessment for each major section; questions and problems from the Chapter Review can also be added. Other *Merrill Chemistry* program resources that may be easily adapted as assessment tools include Solving Problems in Chemistry, as well as masters from the Study Guide, Vocabulary and Concept Review, and Reteaching booklets.

Performance Assessment

Performance assessments are becoming more common in today's schools. These assessment differ in formality and complexity, but in most cases, the teacher observes a student or group of students involved in an activity and rates the performance and/or the products that result from the activity. Background information and specific examples of performance assessment are included in Glencoe's *Alternate Assessment in the Science Classroom.*

Merrill Chemistry provides numerous opportunities to observe student behavior both in informal and formal settings. The Minute Labs, ChemActivities, and Demonstrations present many instances where you can informally observe students and evaluate their understanding of both concepts and process skills.

Group Performance Assessment

Recent research has shown that cooperative learning structures produce improved student learning outcomes for students of all ability levels. Assessments in *Merrill Chemistry* provide opportunities for cooperative learning and, as a result, many opportunities to observe group work processes and products. Glencoe's *Cooperative Learning Resource Guide* provides strategies and resources for implementing and evaluating group activities. For example, if a mixed ability, four-member laboratory work group conducts an activity, you can use a rating scale or checklist to assess the quality of both group interaction and work skills. All four members of the group are expected to review and agree on the data sheet produced by the group. In this approach, all members of the group receive the same grade on the work product. Research shows that cooperative group assessment is as valid as individual assessment. Additionally, it reduces the marking and grading work load of the teacher.

Portfolios: Putting It All Together

The purpose of a student portfolio is to present examples of the individual's work in a "non-testing" environment. A portfolio is simply a method for assembling and presenting selected examples of work products. The process of assembling the portfolio should be both integrative (of process and content) and reflective. The performance portfolio is *not* a complete collection of all worksheets and other assignments for a grading period. At its best, the portfolio should include integrated performance products that show growth in concept attainment and skill development. You can structure the portfolio development process by establishing categories and other limiting specifications. An essential component in portfolio development is the composition of a submission letter or reflective paper that lists the contents of the portfolio and discusses growth in knowledge, attitudes, and skills.

Merrill Chemistry provides opportunities for portfolio development. At least one Assessment strategy in each chapter contains a suggestion for a contribution to a student's portfolio. In addition, the results of Minute Labs, ChemActivities, and Demonstrations often afford opportunities for students to add to portfolios. Likewise, the many features to be found in *Merrill Chemistry* may spark ideas in students for additional portfolio material. These features include Chemistry and Society, Chemistry and Technology, Frontiers, Everyday Chemistry, and Bridges to Other Sciences.

Developing and Applying Thinking Processes

Science is not just a collection of facts for students to memorize. Rather, it is a process of applying observations and intuitions to situations and problems, formulating hypotheses, and drawing conclusions. This interaction of the thinking process with the content of science is the core of science and should be the focus of science study. *Merrill Chemistry* encourages the interaction between science as a process and thinking skills by offering many hands-on activities that provide excellent opportunities for students to expand their comprehension of important core concepts while practicing the basic science process skills and applying their thinking skills.

Basic Science Process Skills

Observing The most basic process is observing. Through observation—seeing, hearing, touching, smelling, tasting—the student begins to acquire information about an object or event. Observation allows a student to gather information regarding size, shape, texture, or quantity of an object or event.

Measuring is a way of quantifying information in order to classify or order it by magnitude, such as area, length, volume, or mass. Measuring can involve the use of instruments and the necessary skills to manipulate them.

Using Numbers Numbers are used to quantify information, including variables and measurements. Quantified information is useful for making comparisons in tables or graphs and for classifying data or objects.

Classifying One of the simplest ways to organize information gathered through observation is by classifying. Classifying involves the sorting of objects according to similarities or differences.

Thinking Skills

Predicting involves suggesting future events based on observations and inferences about current events. Reliable predictions are based on making accurate observations and measurements.

Recognizing Cause and Effect involves observing actions or events and making logical inferences about why they occur. Recognizing cause and effect can lead to further investigation to isolate a specific cause of a particular event.

Interpreting Data involves synthesizing information to make generalizations about the problem under study and apply those generalizations to new problems. Interpreting data may, therefore, involve many of the other process skills, such as predicting, inferring, classifying, and using numbers.

Sequencing involves arranging objects or events in a particular order and may imply a hierarchy or a chronology. Developing a sequence involves the skill of identifying relationships among objects or events. A sequence may be in the form of a numbered list or a series of objects or ideas with directional arrows.

Comparing and Contrasting Comparing is a way of identifying similarities among objects or events, whereas contrasting identifies their differences.

Inferring Inferences are logical conclusions based on observations and are made after careful evaluation of all the available facts or data. Inferences are a means of explaining or interpreting observations and are based on making judgments.

Using Space/Time Relationships This process skill involves describing the spatial relationships of objects and how those relationships change with time. For example, a student may be required to describe the motion, direction, or shape of an object.

Formulating Models A model is a way to concretely represent abstract ideas or relationships. Models can also be used to predict the outcome of future events or relationships. Models may be expressed physically in three dimensions: verbally, mathematically, or diagrammatically.

Communicating information is an important part of science. Once all the information is gathered, it is necessary to organize the observations so that the findings can be considered and shared by others. Information can be presented in tables, charts, a variety of graphs, or models.

Cultural Diversity

American classrooms reflect the rich and diverse cultural heritages of the American people. Students come from different ethnic backgrounds and different cultural experiences into a common classroom that must assist all of them in learning. The diversity itself is an important focus of the learning experience. Diversity can be repressed, creating a hostile environment; ignored, creating an indifferent environment; or appreciated, creating a receptive and productive environment.

Responding to diversity and approaching it as a part of every curriculum is challenging to any teacher, experienced or not. The goal of science is understanding. The goal of multicultural education is to promote the understanding of how people from different cultures approach and solve the basic problems all humans have in living and learning.

Merrill Chemistry addresses this issue. The **Chemists and Their Work** feature highlights individuals from diverse cultural and ethnic backgrounds as possible role models for the next generation of culturally diverse scientists. Information is provided in the margins of the *Teacher Wraparound Edition* about people and groups who have traditionally been misrepresented or omitted. The information is in the form of short articles titled **Cultural Diversity.** The intent of these features is to build, within the framework of sound chemistry instruction, awareness and appreciation for the global community in which we all live. Some features explore the interaction between cultural groups in history and today. Others look at the way other cultural groups view matter and its changes or how the groups have contributed to the shared knowledge of the nature of matter. Providing this information is helping to meet the five major goals of multicultural education:

1. promoting the strength and value of cultural diversity
2. promoting human rights and respect for those who are different from oneself
3. promoting alternative life choices for people
4. promoting social justice and equal opportunity for all people
5. promoting equity in the distribution of power among groups

Two books that provide additional information on multicultural education are:

Banks, James A. (with Cherry A. McGee Banks), *Multicultural Education: Issues and Perspectives.* Boston: Allyn and Bacon, 1989.

Selin, Helaine, *Science Across Cultures: An Annotated Bibliography of Books on Non-Western Science, Technology, and Medicine.* New York: Garland, 1992.

Bertram O. Fraser-Reid (1934-)

Bertram Fraser-Reid is a biochemist and a former professor of chemistry at Duke University. His major research efforts over the years have been in organic chemistry, notably sugars. He has worked on biological regulators for immune systems and has developed methods for producing pheromones, insect sex attractants, from glucose. These pheromones were used in place of DDT to control timber-destroying insects by disrupting their mating patterns. Fraser-Reid's work in sugars has led him to believe that many products now made from petroleum can also be made from sugar—a renewable resource.

CULTURAL DIVERSITY

Ancient Astronomers

The night skies have always intrigued humans. Early North Americans recorded some celestial events in rock writings called petroglyphs. The Anasazi, for example, observed the Crab Nebula supernova that occurred in 1054 and recorded it on a cave wall and the wall of a canyon in Arizona. The same event was also recorded by native Americans in California and New Mexico.

Through observation, the ancient astronomers were able to decipher patterns of movements. The Aztecs developed a 365-day calendar, and the Aztec sunstone, 12 feet in diameter, records the cyclic movements of the moon, Mars, and Venus. The sunstone allowed the Aztecs to regulate their religious calendar system and accurately predict events such as eclipses.

The Mayans of the Mexican isthmus developed a calendar of 365 days with an extra day added every four years—predating the Roman "leap year" solution by 500 years. The Caracól is a stone observatory at the Mayan site of Chichén Itzá. It contains alignments along its walls where observers can plot the position of various stars, planets, the sun, or moon.

Cooperative Learning

What is cooperative learning?

In cooperative learning, students work together in small groups to learn academic material and interpersonal skills. Group members learn to "sink or swim together" in that they are responsible for the group accomplishing an assigned task as well as for learning the material themselves. Cooperative learning fosters academic, personal, and social success for all students.

Establishing a Cooperative Classroom

Cooperative groups in school settings usually contain from two to five students. If students are not experienced in working in groups, you may consider grouping students in pairs and then joining the pairs later to form groups of four or five.

Generally, it is best to assign students to heterogeneous groups that contain a mixture of abilities, genders, and ethnicity. The use of heterogeneous groups exposes students to ideas different from their own and helps them to learn to work with persons different from themselves.

Initially, cooperative learning groups should work together only for a day or two. After the students are more experienced, they can do group work for longer periods of time. Some teachers change groups every week, whereas others keep groups together during the study of a chapter or unit. It is important to keep groups together long enough for each group to experience success and to change groups often enough that students have the opportunity to work with others.

It is necessary to prepare students for working in cooperative groups. Students must understand that they are responsible for themselves as well as for other group members learning the material. Before beginning, specify what interpersonal behaviors are necessary for people to work together. Discuss the basic rules for effective cooperative learning: (1) listen while others are speaking, (2) respect other people and their ideas, (3) stay on-task, and (4) be responsible for your own actions. Students can learn and practice other interpersonal skills such as speaking quietly, encouraging other group members to participate, checking for understanding, disagreeing constructively, reaching a group consensus, and criticizing ideas rather than people.

Teaching the Lesson

Merrill Chemistry contains many laboratory activities that lend themselves to cooperative learning. In addition, the *Teacher Wraparound Edition* contains many teaching strategies and activities that can be used in a group setting.

Before teaching the lesson, you must decide on the academic task and the interpersonal skills students will learn or practice in groups. Prepare the students for the academic task by teaching any material they might need to know and by giving specific instructions for the task.

Explain your criteria for evaluating group and individual learning. Specify the interpersonal skills students will be working on and how these skills will be evaluated. Explain the sharing of materials and assuming of roles.

Spend most of your time monitoring the functioning of groups. Praise group cooperation and good use of interpersonal skills. When students are having trouble with the task, answer questions, clarify the assignment, reteach, or provide background as needed. When answering questions, make certain that no students in the group can answer the question before you reply.

At the close of the lesson, reinforce student learning by having groups share their results or summarize the assignment. Answer any questions about the lesson. Use the criteria discussed at the beginning of the lesson to evaluate and give feedback on how well the academic task was mastered by the groups. Because students are responsible for their own learning and that of other group members, you can evaluate group performance during a lesson by frequently asking questions to group members picked at random. Other forms of group evaluation include having each group take a quiz together or having all students write a report and then choosing one student's paper at random to grade. Assess individual learning by your traditional method.

To assess the learning of interpersonal skills, have students evaluate how well they used interpersonal skills. Groups can list what they did well and what they could do to improve. Have groups share their analysis with the class and summarize the analysis of the whole class.

Laboratory Safety

Outlined here are some considerations on laboratory safety that are intended primarily for teachers and administrators. Safety awareness must begin with the principal, be supervised by the department head, and be important to the individual teacher.

Principals and supervisors should be familiar with the guidelines for laboratory safety and provide continual supervision to ensure compliance with those guidelines. Teachers have the ultimate responsibility for enforcing safety standards in the laboratory. They must set the proper example in the laboratory by observing all rules themselves. This behavior applies to such duties as wearing goggles and protective clothing, and not working alone in the laboratory. Planning is essential to laboratory safety, and that planning must include what to do in an emergency, as well as how to prevent accidents.

Numerous books and pamphlets are available on laboratory safety with detailed instructions on planning and preventing accidents. However, much of what they present can be summarized in the phrase, "Be prepared!" Know the rules and what common violations occur. Know where emergency equipment is stored and how to use it. Practice good laboratory housekeeping by observing these guidelines.

1. Store chemicals properly.
   a. Segregate chemicals by reaction types.
   b. Label all chemical containers; include purchase date, special precautions, and expiration date.
   c. When outdated, discard chemicals according to appropriate disposal methods.
2. Prohibit eating, drinking, and smoking in the laboratory.
3. Refuse to tolerate any behavior other than a serious, businesslike approach to laboratory work.
4. Require that eye and clothing protection be worn.
5. Expect proper use of equipment.
   a. Do not leave equipment unattended.
   b. Shield systems under pressure or vacuum.
   c. **NEVER** use open flames when a flammable solvent is in use in the same room. Use open flames only when necessary; substitute hot plates whenever possible.
   d. Instruct students in the proper handling of glass tubing.
   e. Instruct students in the proper use of pipets, cylinders, and balances.
   f. See that laboratory benches do not become catch-alls for books, jackets, and so on.
6. Dispose of wastes correctly.

Waste disposal is subject to various federal and state regulations. The agencies charged with administering disposal of wastes are the U.S. Environmental Protection Agency (EPA) and its equivalent state agency. The regulations promulgated by the EPA are found in the Code of Federal Regulations (CFR), Title 40. Since some wastes may have to be disposed of away from the school, transportation of hazardous material becomes a problem. Regulations concerning the movement of hazardous materials are established by the Department of Transportation (DOT) in CFR Title 49. Both 40 CFR and 49 CFR are published annually. Changes are published as they occur in the Federal Register. A practical guide to waste disposal for laboratories is *Prudent Practices in the Laboratory*, published by the National Research Council, Washington: National Academy Press, 1995.

Other resources on waste disposal and other aspects of laboratory management include the following: N. Irving Sax, *Dangerous Properties of Industrial Materials*, 9th ed., New York: Van Nostrand-Reinhold, 1993; the chemical catalog and reference manual published by Flinn Scientific, Inc., P.O. Box 231, Batavia, IL 60510, (312) 879 6900; a handbook, *Less Is Better*, and two pamphlets, "Hazardous Waste Management" and "Chemical Risk: A Primer," available free in single copies from the Office of Federal Regulatory Programs, ACS Department of Government Relations and Science Policy, 1155 16th St. NW, Washington, DC 20036; and the ACS Chemical Health and Safety Referral Service, at the same address, which provides referrals to literature, films, educational courses, or organizations that can provide safety information, (202) 872-4511.

DISCLAIMER

Glencoe Publishing Company makes no claims to the completeness of this discussion of laboratory safety and chemical storage. The material presented is not all-inclusive, nor does it address all of the hazards associated with handling, storage, and disposal of chemicals, or with laboratory practices and management.

Safety Symbols

Safety Symbols

These safety symbols are used to indicate possible hazards in the activities. Each activity has appropriate hazard indicators.

	**DISPOSAL ALERT** This symbol appears when care must be taken to dispose of materials properly.		**ANIMAL SAFETY** This symbol appears whenever live animals are studied and the safety of the animals and the students must be ensured.
	**BIOLOGICAL HAZARD** This symbol appears when there is danger involving bacteria, fungi, or protists.		**RADIOACTIVE SAFETY** This symbol appears when radioactive materials are used.
	**OPEN FLAME ALERT** This symbol appears when use of an open flame could cause a fire or an explosion.		**CLOTHING PROTECTION SAFETY** This symbol appears when substances used could stain or burn clothing.
	**THERMAL SAFETY** This symbol appears as a reminder to use caution when handling hot objects.		**FIRE SAFETY** This symbol appears when care should be taken around open flames.
	**SHARP OBJECT SAFETY** This symbol appears when a danger of cuts or punctures caused by the use of sharp objects exists.		**EXPLOSION SAFETY** This symbol appears when the misuse of chemicals could cause an explosion.
	**FUME SAFETY** This symbol appears when chemicals or chemical reactions could cause dangerous fumes.		**EYE SAFETY** This symbol appears when a danger to the eyes exists. Safety goggles should be worn when this symbol appears.
	**ELECTRICAL SAFETY** This symbol appears when care should be taken when using electrical equipment.		**POISON SAFETY** This symbol appears when poisonous substances are used.
	**SKIN PROTECTION SAFETY** This symbol appears when use of caustic chemicals might irritate the skin or when contact with microorganisms might transmit infection.		**CHEMICAL SAFETY** This symbol appears when chemicals used can cause burns or are poisonous if absorbed through the skin.

Chemical Storage and Disposal

The number of lawsuits against schools and the number of regulations by OSHA, EPA, and DOT are occurring at an increasing rate. All of these complications make life more difficult for chemistry teachers.

One major problem that teachers face is the prohibition by a principal of students working with any substance even remotely considered hazardous. In that circumstance, the teacher can always perform an experiment with students acting as observers, note takers, and data recorders. Even if the demonstration-type experiment is done behind a safety shield, at least the students get to see chemical phenomena really occurring.

Another problem that many teachers face is that of chemical waste disposal. One way to reduce waste is to conduct microscale experiments. Many of the ChemActivities in the textbook and experiments in the *Laboratory Manual* employ microscale equipment and techniques. Others use traditional techniques and apparatus. Thus you can tailor your laboratory chemistry program to the facilities and equipment that you have available.

A frequent question concerns what liquids may safely be disposed of in the sanitary drains. First, the teacher must be certain that the sewer flows to a wastewater treatment plant and not to a stream or other natural water course. Second, any substance from a laboratory should be flushed with at least 100 times its own volume of tap water. Third, the suggestions given below should be checked with local authorities because local regulations are often more stringent than federal requirements. The National Research Council book *Prudent Practices in the Laboratory* lists many substances that can be disposed of in the sanitary drain. In the tables are listed some positive and negative ions from the NRC lists. It is important to note that *both* the positive and negative ion of a salt must be listed in order for its drain disposal to be considered safe.

Note that, although hydrogen and hydroxide ions are listed, acids and bases should be neutralized before disposal. A good rule of thumb is that nothing of pH less than 3 or greater than 10 should be discarded without neutralizing it first.

Of the organic compounds most often found in high school laboratories, the following can be disposed of in the drain: methanol, ethanol, propanols, butanols, pentanols, ethylene glycol, glycerol, sugars, formaldehyde, formic and acetic acids, oxalic acid, sodium and potassium salts of carboxylic acids, esters with less than 5 carbon atoms, and acetone. More extensive lists are given in *Prudent Practices.*

POSITIVE IONS	NEGATIVE IONS
aluminum	borate
ammonium	bromide
bismuth	carbonate
calcium	chloride
copper	hydrogen sulfite
hydrogen	hydroxide
iron	iodide
lithium	nitrate
magnesium	phosphate
potassium	sulfate
sodium	sulfite
strontium	tetraborate
tin	thiocyanate
titanium	
zinc	

What happens to substances that do not fall into one of the categories discussed? There are three possibilities. One is to treat waste chemically to convert it to a form that is drain-disposable. A good example is iodate ion, which is too strong an oxidizing agent to go untreated. However, it is readily reduced to iodide (disposable) by acidified sodium hydrogen sulfite. Many procedures for processing laboratory waste to a discardable form are found in *Prudent Practices.*

A second possibility is the recycling of waste. Good examples here are the recovery of valuable metals such as mercury and silver. Another is the recovery of solvents through distillation.

If waste cannot be recycled or processed to disposable form, then it must be packed and shipped by a Department of Transportation-approved shipper to a landfill designated to receive chemical and hazardous waste. Since that method of disposal is very expensive, it pays to reduce such waste to a minimum. Microscale experiments, already mentioned, aid in that task. There are also numerous processes in *Prudent Practices* for reducing both the bulk and hazard of wastes. An example is the reduction of chromate and dichromate waste solutions to chromium(III) solutions, which are then made basic, precipitating chromium(III) oxide. The oxide is filtered off, dried, and crushed, resulting in a significant mass and volume saving. As hazardous wastes are accumulated awaiting shipment, it is important to observe the proper storage procedures for separating incompatible substances. Guidelines for such storage are given in the *ChemActivities* book in the *Teacher Classroom Resources.*

Most high schools qualify as Small Quantity Hazardous Waste Generators under EPA regulations. However, it is usually wise to register with the state or local environmental protection agency. That agency is a good source of information on disposal, particularly on the availability of approved packers, shippers, and landfill operators. Even though wastes are packed, shipped, and interred by licensed or approved firms, generators of waste are responsible for their waste—FOREVER! Therefore, we should examine carefully the credentials of any firm we hire to handle our waste.

Disposal of Chemicals

Local, state, and federal laws regulate the proper disposal of chemicals. Before using these disposal instructions, you should confirm them with local regulators so that they do not violate your local or state laws. *No representation, warranty, or guarantee is made by Glencoe Publishing Company or the authors as to the completeness or accuracy of these disposal procedures.*

The proper disposal of chemicals should be done only by the teacher who must be wearing a laboratory apron, goggles, and rubber gloves. The disposal procedures should be done in an operating fume hood. Even teachers should never be alone in the laboratory when disposing of chemicals. They should have the proper type of fire extinguisher and a telephone or intercom nearby.

The disposal procedures listed are for relatively small amounts of waste. If a larger amount is produced, the stoichiometry of the chemical reactions in the disposal procedure will have to be calculated. Do your disposal by reacting very small amounts of the chemical at one time. Many reactions are exothermic, so the reaction vessel will need to be in an ice water bath.

Disposal advice given in the *Teacher Wraparound Edition* is keyed to the following lettered disposal procedures.

Disposal A

These materials can be packaged and sealed in separate plastic containers and buried in a landfill approved for chemical and hazardous waste disposal. Local soil and water conditions cause local regulatory agencies to restrict what is permitted in the landfill. Contact your local regulatory agency before assuming it is acceptable to place these materials in your school's trash dumpster. Do not mix chemicals; place each one in its own container. Place the containers in a cardboard box and separate with vermiculite. Seal the closed box before disposing of it.

Disposal B

Decant the water layer into a separate beaker. Then, discard the water layer down the drain. Label the other chemical and save it for future activities, following proper storage directions.

Disposal C

Rinse the chemical down the drain with at least a 100-fold increase in the volume of water. Rinse only one chemical down the drain at a time using plenty of water to dilute it. Do not mix chemicals in the sink or drain. Dissolve any solids in a beaker before placing them down the drain. *Do not rinse these chemicals down the drain if your school's drains go into a septic system.* This disposal procedure is intended for schools whose sanitary sewers go to a waste water treatment plant. Due to local soil and water conditions, local regulations may prohibit the disposal of these chemicals in this manner. Check with the local government regulatory agency before using this procedure.

Disposal D

While being stirred, the acidic solution should be slowly added to a large beaker of cold water. Prepare a 1*M* Na_2CO_3 solution by dissolving 143 g $Na_2CO_3 \cdot 10H_2O$ to 500 cm^3 of water. Slowly add the 1*M* Na_2CO_3 solution to the diluted acid. Carbon dioxide gas will be evolved. When there is no more evolution of gas as more Na_2CO_3 solution is added, the solution can be tested with pH paper to verify that it is neutral. Rinse the neutralized solution down the drain with a large volume of water.

Disposal E

Filter the solution through filter paper. Open the filter paper and allow it to dry. Place the dried solid and filter paper in a separate, plastic container surrounded by vermiculite in a box. Seal the box and dispose of it in a landfill approved to receive chemical and hazardous waste. Dilute the filtrate by adding it slowly, while stirring, to a large beaker of cold water. Verify that it is neutral by using pH paper.

Rinse the neutral solution down the drain with a large volume of water.

Disposal F

Place the substance in a plastic container and seal it. Completely label the container and, following proper storage directions, save it for future activities.

Disposal G

For the iodine: add 18 g of sodium thiosulfate to the solution. Stir while warming the solution to 50°C. After the iodine is consumed, check with pH paper and add enough 3*M* NaOH solution (12 g NaOH in 100 cm^3 water) to neutralize the solution.

Then for the manganese or cobalt: add 6 g of sodium sulfide. After one hour of stirring, neutralize the solution with 3*M* NaOH. Verify with pH paper that the solution is neutral. Filter and place the MnS solid in a plastic bottle for disposal in an approved landfill. Treat the filtrate for excess sulfide by adding 12 g of iron(III) chloride with constant stirring. Filter the precipitate, Fe_2S_3. Place the solid in a plastic container, seal it for disposal in an approved landfill. Rinse the neutralized solution down the drain with a large volume of water.

Disposal H

Use an operating fume hood to disperse the small volume of gas produced.

Disposal I

Slowly add the substance to a large beaker of cold water while stirring. Then in an ice bath, slowly add 6*M* HCl until the solution is neutralized. Verify by using pH paper. Rinse the neutralized solution down the drain using a large volume of water.

Disposal J

This applies only to solutions used in microchemistry procedures. See the discussion on microchemistry equipment and techniques on pages 21T and 22T.

Microchemistry

Microchemistry Techniques

Chemistry teaching at the high school level is presently facing several problems at once. Concern for student safety, environmental questions, cost of materials, and the necessity of adhering to a prescribed curriculum have all worked to make the laboratory program difficult to carry out. Still, the "lab" is the most tangible, best remembered, and most visible aspect of chemistry instruction. Without direct observation of chemical phenomena and the hands-on manipulation of the substances referred to in the classroom and the textbook, chemistry can become no more than a survey of chemical theory with little real meaning to students, especially students who are visual learners.

Glencoe Publishing Company has taken the lead in answering these problems by introducing microchemistry techniques in *Merrill Chemistry*—both in the textbook in the form of ChemActivities and in the *Laboratory Manual*. When you have limited time and a limited budget, microchemistry provides a safe, inexpensive, and time-efficient means to conduct a laboratory program. It gives you, the chemistry teacher, the opportunity to keep chemistry the investigative science that it should be and that students expect it to be.

Laboratory activities and experiments that use micro-amounts of chemicals provide a way to involve students in observation and manipulation of some substances that might otherwise be regarded as hazardous. Because drastically reduced amounts of chemicals are used, safety is greatly increased. Likewise, the expense of running a laboratory chemistry program is cut. Because of the savings in both time and expense with microchemistry, students are afforded the opportunity to do much more experimental chemistry than was possible with conventional labs. As with traditional chemistry, all processes should be conducted with safety goggles and protective clothing.

Microchemistry Equipment

Microchemistry uses only two basic tools—the microplate and the plastic pipet.

The Microplate

The microplate is a plastic tray containing several shallow wells arranged in lettered rows and numbered columns. These wells are used in the same way that test tubes and beakers are used. There are three types of microplates as shown in the first photograph. A 24-well plate has 4 rows and 6 columns of wells. A 96-well microplate has 8 rows and 12 columns of smaller wells. There is also a combination microplate which consists of 4 rows of 12 wells and 2 rows of 6 wells. Most of the activities in *Merrill Chemistry* that use microplates are carried out with either 24- or 96-well plates.

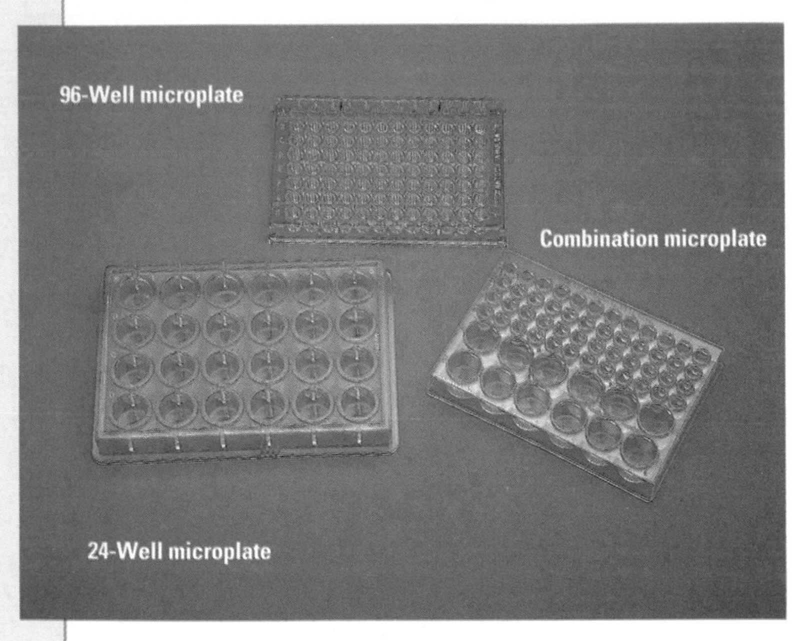

Plastic Pipets

Microchemistry uses two main types of plastic pipets—the thin-stem and microtip pipets shown in the second photograph. These pipets are used to deliver chemicals and to collect products, including gases. The pipets are made of polyethylene, which makes them soft and flexible. The microtip pipet is used when it is necessary to deliver very small droplets of liquid to a reaction. The pipets may be modified in a variety of ways to serve diverse purposes. For example, the thin-stem pipets can be used as chemical scoops, gas generators, and reaction vessels. The long, flexible tube of a thin-stem pipet makes a convenient delivery tube. Some microchemistry techniques are based on the fact that the tube of a thin-stem pipet will slide into the tube of a microtip pipet which has had its tip cut off. This arrangement produces a typical generator-collector setup.

Either pipet can be reused simply by rinsing the stem and bulb with water. The inside plastic surface of the pipet is non-wetting and therefore does not hold water or solutions the way that glass does. This means that, with proper technique, the entire contents of the pipet can be dispensed, with none of the solution left behind.

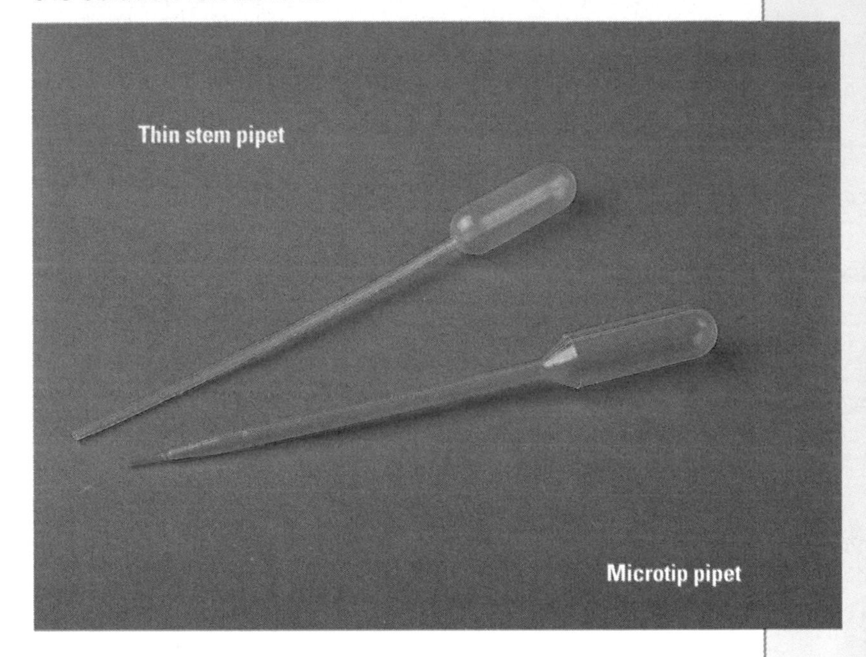

Waste Disposal in Microchemistry

Waste disposal for nearly all microchemistry activities is coded as Disposal J.

Disposal J

Collect all chemical solutions, precipitates, and rinse solutions in a polyethylene dishpan or similar container devoted to that purpose. Retain the solutions until the end of the period or day. In the fume hood, set up a hot plate with a 1-cm^3 or 2-cm^3 beaker. Pour the collected solutions into the beaker. Turn the hot plate on low and allow the beaker to heat with the hood running and the hood door nearly closed. The liquids and volatiles in the mixture will evaporate, leaving dried chemicals. Allow the beaker to cool. Continue to add solutions and waste until the beaker is 2/3 full. Treat the waste as heavy metal waste. *Dispose of the beaker and its dry contents* in an approved manner.

Supplies and Materials

These easy-to-use tables of materials can help you prepare for your chemistry classes for the year. All quantities are for one lab setup of each *ChemActivity, Minute Lab,* and *Demonstration* for the entire course. Before placing your order for supplies, determine how many classes you will be teaching and how many students you expect in each class. The amounts listed are the total required per year for all student labs and one performance of each demonstration. Therefore, if you have seven classes with ten groups of students in each class, you will need to multiply the quantities of material for each demonstration by the number of classes and the quantities of material for each student lab by the number of student groups (70).

The standard list of equipment is a set of equipment that is generally recommended for each lab bench station in the chemistry laboratory. For all lab activities in this program, it is assumed that your classroom is equipped with these items. Additional equipment required for the course is listed under **Nonconsumables.** The standard list of chemicals consists of common chemicals that are used in many lab activities and that you should always have on hand. Additional chemicals and other consumable items are listed under **Chemicals** and **Other Consumables.**

Standard Equipment List (for each station)

- apron, 1 per student
- beakers, 5: one 100-cm^3, two 250-cm^3, two 400-cm^3
- beaker tongs
- Bunsen burner and tubing
- clay triangle
- crucible and cover
- crucible tongs
- droppers, 2
- Erlenmeyer flask, 125-cm^3
- Erlenmeyer flask, 250-cm^3
- evaporating dish
- forceps
- funnel
- goggles, 1 pair per student
- graduated cylinders, 10-cm^3, 50-cm^3, 100-cm^3, 250-cm^3
- microplate, 24-well
- microplate, 96-well
- microtip pipets
- scissors
- spatula, stainless steel
- stirring rods, 2
- test-tube holder
- test-tube rack
- test tubes, 6 large
- test tubes, 6 small
- wash bottle
- watch glass
- wire gauze or screen

Classroom Eqipment List (for general use)

- balance
- barometer
- battery eliminator
- beakers, assorted small 30-cm^3, 150-cm^3
- beakers, assorted large, 600-cm^3, 1-dm^3, 2-dm^3, 4-dm^3
- burets and stands
- clamps, assorted
- corks, assorted
- dishpan, plastic
- Erlenmeyer flasks, 500-cm^3, 1-dm^3
- explosion shield
- food coloring, assorted colors
- glass tubing, 6-mm, 10-mm
- graduated cylinder, 1-dm^3
- hot plate
- iron tripod
- lighter for burner
- matches
- mortar and pestle
- overhead projector
- paper clips, metal
- ring stands and rings
- rubber bands
- rubber or tygon tubing
- rubber stoppers, assorted
- string
- stapler and staples
- tapes, electrical, masking, transparent
- thermometers (–10°C to 150°C), 3
- thermometer clamp
- thin-stem pipets
- toothpicks
- vacuum pump

Standard Chemical List (for general use)

- acetic acid, glacial
- ammonia, aqueous
- ethanol
- hydrochloric acid, 12M
- nitric acid, concentrated
- sodium chloride
- sodium hydroxide
- sulfuric acid, 18M

Nonconsumables — Chemicals

Item (Quantities required for one setup)	ChemActivities	Minute Labs	Demonstrations
spectroscope, hand	802		94, 96
spectrum tubes, hydrogen and neon (1 each)			96
spheres, 2–4 cm diameter (7)	796		
spoon, stainless steel (1)			260
straight pins (3)			354
stuffed animal, small (1)		7	
syringe, 60-cm^3 plastic (1)		429	
thumbtack (1)			68, 370
towel, cloth (1)			478
toy car, small (1)		8	
tray, cafeteria (1)			486, 488
tray, plastic (1)		558	
tubing clamp (1)			424
tubing, copper, small diameter (45 cm)			436
tubing, vacuum, rubber (1 m)			434
ultraviolet light (1)			112, 786
vial, small glass (1)			578
vegetable shredder (1)			616
voltmeter (1)	846		
wagon, small (1)		7	
washers, penny size (7)			394
wire, stiff (1)		278	624
wood squares (5)			84, 726
Chemicals			
acetic anhydride (2 cm^3)	852		
acetone (350 cm^3)			14, 256, 424, 440, 564, 630, 632, 652
aluminum chloride (1 g)			610
aluminum foil (20 g)	809	673	84, 236, 608, 696, 726
aluminum oxide packing (40 g)			368
aluminum nitrate (4 g)	836		
aluminum sulfate (0.5 g)			586
amino acid mixture (1 cm^3)			786
ammonium acetate (1 g)			610
ammonium chloride (26 g)	847	508	688
ammonium metavanadate (2 g)	816, 844		
ammonium thiocyanate (15 g)			700
barium chloride dihydrate (2.5 g)			2, 174
barium hydroxide octahydrate (33 g)			574, 700
barium nitrate (0.2 g)		228	
barium sulfate (1 g)			270
benzoic acid (0.2 g)	826		
bromcresol green indicator 0.4% (1 cm^3)		584	
bromine water (30 cm^3)	812		330, 748
bromine, liquid (12 cm^3)			260, 286, 696
bromothymol blue (0.2 g)		783	226
1-butanol (15 cm^3)	797, 823		
2-butanol (5 cm^3)	823		
tert-butanol (2-methyl-2-propanol) (5 cm^3)	823		
butyric acid (butanoic acid) (6 cm^3)			738
calcium acetate (20 g)			508
calcium carbide, pieces (7)		154	332, 750

Chemicals

Item (Quantities required for one setup)	ChemActivities	Minute Labs	Demonstrations
calcium carbonate (1 g)			46
calcium chloride (200 g)		508	94, 576
calcium fluoride (12 g)			582
calcium nitrate (2g)	816		
calcium oxide (0.1g)	840		
calcium sulfate (1 g)			270
calcium, turnings (2 g)			240
carbon, small piece (1)			154
charcoal, powdered (2 g)			642
chromium(III) nitrate (2 g)	816		
cobalt(II) chloride dihydrate (1.5 g)			440
cobalt(II) chloride hexahydrate (12 g)		558	368, 428, 564
cobalt(II) nitrate (4 g)	816, 838		
copper(I) oxide (10 g)			172
copper(II) carbonate, anhydrous (0.5 g)	802		
copper(II) chloride (17 g)			94, 674
copper(II) nitrate (2 g)	816		
copper(II) nitrate trihydrate (1 g)			6
copper(II) oxide (12 g)			172, 642
copper(II) sulfate pentahydrate (41 g)		758, 631	188
copper strip, small (1)	846		
copper, turnings (2 g)			240
cyclohexane (60 cm^3)			330, 348, 748
cyclohexene (10 cm^3)			330, 748
1,6-diaminohexane (1.1g)			768
dichloromethane (30-cm^3)			500
ethyl acetate (10-cm^3)	797		
fluorescein (0.1 g)			118
germanium, small piece (1)			154
d glucose (dextrose) (46.5 g)			520, 544, 758
glycerol (glycerin) (105 cm^3)			20, 222
glyoxal, 40% (2-cm^3)		758	
hexamethylenediamine (1 g)			4
hexane (60 cm^3)	823		176, 500
hydrogen peroxide, 3% (70 cm^3)			626, 638
hydrogen peroxide, 30% (35 cm^3)			284, 692
hydrogen peroxide, 6% (315 cm^3)		550, 558	542
iodine, crystals (7.5 g)			140, 162, 260, 286, 500
iodine/potassium iodide solution (10 cm^3)	812		
iron filings (3 g)		152, 168	
iron(II) ammonium sulfate (2 g)			174
iron(II) sulfate (34 g)			64, 172, 546, 632
iron(II) sulfate 0.1*M* (100 cm^3)	810		
iron(III) ammonium sulfate (6 g)			174
iron(III) chloride (1 g)		152	
iron(III) chloride hexahydrate (1.5 g)			368
iron(III) nitrate 0.1*M* (13 g)	810, 816, 836		
iron(III) sulfate (10 g)			172
lead(II) nitrate (8 g)			600
lead, small piece (1)			154
lead strip, small (1)	846		
lithium carbonate, anhydrous (0.5 g)	802		
lithium, small piece (2)			136, 272

Chemicals

Item (Quantities required for one setup)	ChemActivities	Minute Labs	Demonstrations
luminol (0.2 g)			626
magnesium ribbon (125 cm)	814, 846		152, 210, 240, 300, 476, 660, 684
magnesium sulfate (1 g)			270
malonic acid (4 g)			542
manganese dioxide (8 g)		550	260, 284, 692
manganese(II) nitrate (2 g)	816		
manganese(II) sulfate (1 g)			542
methanol (260 cm^3)			68, 238, 428, 738
methyl formate (3.5 cm^3)			430
methylene blue, crystals (1.1 g)	824		64, 544
nickel(II) nitrate (6 g)	810, 816		
nickel(II) chloride hexahydrate (1.5 g)			368
1-octanol (10 cm^3)	797		
oxalic acid (4 g)			546
isopentane (2-methylbutane) (6.5 cm^3)			430
isopentyl alcohol (3-methyl-1-butanol) (6 cm^3)			738
phenolphthalein, 1% solution (6 cm^3)	810		240, 632, 650, 658
polyvinyl alcohol, hydrolyzed (40 g)			776
potassium chloride (10 g)			94
potassium hexacyanoferrate(II) (0.5 g)			174
potassium hexacyanoferrate(III) (0.6 g)			626, 632
potassium hydroxide (41 g)			236, 544, 758
potassium iodate (12 g)			542
potassium iodide (8 g)			638, 650, 670, 692
potassium nitrate (22 g)	816		508
potassium nitrite (4.5 g)		584	
potassium permanganate (2 g)			2, 204, 546, 638
potassium sodium tartrate (19 g)		558, 758	
potassium thiocyanate (11 g)	816		174
potassium, small piece (2)			136, 272
1-propanol (5 cm^3)	822		
2-propanol (isopropyl alcohol) (60 cm^3)	797, 822		370, 786
salicylic acid (9 g)	852		238, 738
sebacoyl chloride (2 cm^3)			4, 768
silicon, small piece (1)			154
silver nitrate (12 g)		758	564, 602, 668, 758
soda lime, powdered (20 g)			578
sodium acetate (4 g)		612	
sodium bromide (21 g)	812		
sodium carbonate, anhydrous (3.5 g)	802		4, 610, 768
sodium fluoride (8.5 g)	812		
sodium hydrogen carbonate (270 g)	800, 847		152, 388, 462, 578
sodium hydrogen sulfite (1.5 g)			2, 174
sodium hypochlorite, 5% (60 cm^3)	812, 838, 844		286
sodium iodide (30 g)	812		
sodium nitrate (0.1 g)	826		
sodium polyacrylate (1 g)			16, 774
sodium silicate (25 cm^3)			414
sodium sulfate (0.2 g)		228	
sodium sulfate decahydrate (140 g)			226
sodium sulfate, anhydrous (9 g)	851		600
sodium sulfide (12 g)			600

Chemicals — Other Consumables

Item (Quantities required for one setup)	ChemActivities	Minute Labs	Demonstrations
sodium tetraborate (15 g)		316	776
sodium thiosulfate (5 g)	836		514
sodium, small piece (5)			136, 260, 272
strontium chloride (10 g)			94
sucrose, granulated (275 g)		783	8, 46, 52, 222, 274, 504, 506, 534
sucrose, powdered (25 g)			506
sulfur, powdered (1 g)		168	
sulfur, roll (2 g)			300
tannic acid (or strong tea) (0.5 g)			174
tin(II) chloride (1 g)		758	
tin, small piece (1)			154
trichlorotrifluoroethane (TTE) (200 cm^3)	812, 822		4, 348, 500, 768
universal indicator (10 cm^3)	832, 840, 842	612	576, 586, 606, 610
zinc dust (4 g)			140, 162
zinc granules (1.5 g)	844		310
zinc nitrate (2 g)	816		
zinc strip (2)	846		112
zinc sulfate (1 g)			652
Other Consumables			
acetate transparency film sheet (7)			38, 190, 402
Alka-Seltzer tablets (5)			552, 594
antacid tablets (variety of types) (1 each)			606
aspirin, buffered and unbuffered (2 each)	826		606, 612
bag, zip-close plastic, 1 qt (11)	832, 834	481, 584	94, 608, 688
bag, plastic, with tie, 1 qt (10)			176, 486, 552
balloons, various sizes and shapes (34)	819	79, 323, 461, 783	274, 320, 378, 452, 578
blotter paper (2 pieces)			76
bluing, liquid for laundry (10 cm^3)			408
bubble gum, sugared (5 pieces)		210	
can, soda, empty (5)		278	450, 624
candles (7)			50, 222, 304, 462
cardboard, corrugated (2 pieces)			8, 700
card stock, heavy (8)			154, 164, 442
chalk (2 pieces)	800		12
cheesecloth (1 piece)		773	
chromatography paper (2 strips)			370, 786
clay, modeling (1 pack)			100
coal, small lump (1)			408
cobalt chloride test paper (1 strip)			782
construction paper, white (1 sheet)	822		
cooking oil (30 cm^3)	797		330, 748
cotton balls (6)			100
cup, 3-oz plastic or paper (55)	824, 851, 852	210, 316	776
cup, insulated foam (1)			690
cup, porous (1)			658
dry ice (1200 g)		481	76, 478, 710
filter paper (15)	820		46, 162, 236, 238, 370, 498, 594, 684, 738
filter paper, coffee maker (1)		368	
flashcube (1)			660
flour (2 tbs.)			274

Other Consumables

Item (Quantities required for one setup)	ChemActivities	Minute Labs	Demonstrations
garbage bag, plastic with tie (1)			478
glue, white (1 bottle)		316	154, 248, 306, 400
graph paper (5)	817, 824, 837, 849		262
gumdrops, assorted colors (2 bags)	825		740, 746
household ammonia (5 cm^3)			408
kerosene (250 cm^3)			248
labels, small (6)			828, 842
lead acetate test paper (1 strip)			782
lemon (¼)	846		
lima beans (10)	849		
litmus paper, red (1 strip)			782
marble chip (1)	840		
milk, whole (60 cm^3)		11	
nail, iron, small (2)		631	644
newspaper (3)		261, 726	222
pH test paper (2 strips)			608, 612
paper bags, lunch size (2)			388
paraffin (900 g)			248, 582
peanuts (4)			782
pennies, dark and dull (17)			394, 644
pennies, new and shiny (3)			310
perfume or aftershave (1 cm^3)		385	
pipe cleaners (1 pkg.)			400
plastic bag, heavy (1)	820		
plastic film, dark (1)			152
plastic ketchup cup with lid (6)	828, 830		
plastic kitchen wrap (4 ft^2)			136, 272
polystyrene "peanuts," 1 box full (1)			14
popcorn kernels (30)			212
potato (1)			534
sand, dry (7 dm^3)			38, 190, 684
silicone caulk (1 tube)			68
soap, liquid dish (40 cm^3)		11, 154	222, 394, 692
split peas (100)	849		
spray cooking oil (1 can)			212
starch, soluble (4 g)			542
steel wool (2 rolls)		631, 758	256, 284, 628, 630, 632, 644, 652, 660, 668
straws, drinking (45)	817, 832, 842	404, 783	394, 396, 486
sugar cube (3)			506, 784
tea bag (2)			52, 504
thermite and thermite starter (1 kit)			684
tomato juice or sauce (150 cm^3)			608
topsoil (50 g)			498
vanilla flavoring (6 cm^3)		79	378
vegetable shortening, solid (25 g)		773	
vinegar (610 cm^3)		783	152, 388, 462, 578, 644
wintergreen lifesaver (1)			784
wire, copper, 16–20 ga (3 m)			230, 650, 668, 670
wire, copper, 22–26 ga (1 m)			352, 396
wood splints (8)		154, 550	152, 226, 284, 750
yeast, dry (2 packs)		783	274

Suppliers

Chemical and Equipment Suppliers

Aldrich Chemical Co., Inc.
P.O. Box 2060
Milwaukee, WI 53201

Arbor Scientific
P.O. Box 2750
Ann Arbor, MI 48106-2750

Carolina Biological Supply Co.
2700 York Road
Burlington, NC 27215

Central Scientific Co.
11222 Melrose Avenue
Franklin Park, IL 60131

Eastman Chemical Company
P.O. Box 511
Kingsport, TN 37662

Edmund Scientific Co.
103 Gloucester Pike
Barrington, NJ 08007

Fisher Scientific Co.
4901 W. LeMoyne
Chicago, IL 60651

Flinn Scientific Inc.
P.O. Box 219
Batavia, IL 60510

Frey Scientific
905 Hickory Lane
Mansfield, OH 44905

Kemtec Educational Corp.
9889 Crescent Drive
West Chester, OH 45069

LaPine Scientific Co.
6001 S. Knox Avenue
Chicago, IL 60629

McKilligan Supply Corp.
435 Main Street
Johnson City, NY 13790

Nasco
901 Janesville Avenue
Fort Atkinson, WI 53538

Pasco Scientific
10101 Foothills Blvd.
P.O. Box 619011
Roseville, CA 95661-9011

Pfaltz & Bauer, Inc.
172 E. Aurora Street
Waterbury, CT 06708

Sargent-Welch Scientific Co.
7300 N. Linder Ave.
Skokie, IL 60076

Sargent-Welch Scientific of Canada, Ltd.
285 Garyray Drive
Weston, Ontario,
Canada M9L 1P3

Science Kit and Boreal Labs
777 E. Park Drive
Tonawanda, NY 14150

Sigma Chemical Company
3050 Spruce Street
St. Louis, MO 63103

Ward's Natural Science Establishment, Inc.
P.O. Box 92912
Rochester, NY 14692

Multimedia Suppliers

BFA Educational Media
468 Park Avenue South
New York, NY 10016

Encyclopaedia Britannica Educational Corp.
425 North Michigan Avenue
Chicago, IL 60611

National Geographic Society
Educational Services
17th & M Street, N.W.
Washington, DC 20036

Optical Data Corp.
30 Technology Drive
Warren, NJ 07060

Time-Life Multimedia
100 Eisenhower Drive
Paramus, NJ 07652

Videodiscovery
1700 Westlake Ave. North
Suite 600
Seattle, WA 98109

Software Suppliers

American Chemical Society
1155 16th Street NW
Washington, DC 20036

Cross Educational Software
P.O. Box 1536
504 E. Kentucky Avenue
Ruston, LA 71270

Educational Courseware
3 Nappa Lane
Westport, CT 06880

Educational Images, Ltd.
P.O. Box 3456, West Side
Elmira, NY 14905

Educational Materials and Equipment Co. (EME)
P.O. Box 2805
Danbury, CT 06813-2805

IBM Educational Systems
Department PC
4111 Northside Parkway
Atlanta, GA 30327

Project Seraphim
Department of Chemistry
University of Wisconsin-Madison
1101 University Avenue
Madison, WI 53706

Sunburst Communications
39 Washington Avenue
Pleasantville, NY 10570

Software for the Macintosh

Beaker: An Expert System for the Organic Chemistry Student
Brooks-Cole Publishing
511 Forest Lodge Road
Pacific Grove, CA 93950
For use in Chapters 29–30

ChemAid: An Introduction to the Periodic Table of the Elements
Ventura Educational Systems
3440 Brokenhill Street
Newbury Park, CA 91320
For use in Chapters 4–7, 10–11

Doing Chemistry
American Chemical Society
Distribution Office
1155 16th Street NW
Washington, DC 20036
For use in laboratory

Elementary Data: Interactive Periodic Table for the Macintosh
Rock Ware
4251 Kipling Street
Suite 595
Wheat Ridge, CO 80033
For use in Chapters 4–6

Elements Plus
Flight Engineering
P.O. Box 661133
Miami Springs, FL 33266
For use in Chapters 4–6 and 13

MacChematics
Daedalus Scientific Software
P.O. Box 787
Oakland Gardens, NY 11364
For use in Chapter 9

MacChemistry
Fortnum Software
16742 Gothard Street
Suite 213
Huntington Beach, CA 92647
For use in Chapters 6 and 24

PeriChart 1.3
Intellimation
P.O. Box 1922
Santa Barbara, CA 93116
For use in Chapters 4–5 and 13

Physical Science Multimedia Library: Principles of Physical Science
Optical Data
30 Technology Drive
Box 4919
Warren, NJ 07060
For use in Chapters 1, 4–5, 13, and 15–19
(Also requires laser disc technology for full use.)

References

General Library References

American Chemical Society Task Force on Laboratory Waste Management Staff, *Laboratory Waste Management: A Guidebook.* Washington, DC: American Chemical Society, 1994.

Atkins, P.W., *Atoms, Electrons, and Change.* New York: Scientific American Library, 1990.

Atkins, P.W., *The Periodic Kingdom: A Journey into the Land of the Chemical Elements.* New York: Basic Books, 1995.

Bachrach, Steven M., ed. *The Internet: A Guide for Chemists.* Washington, DC: American Chemical Society, 1996.

Bard, A.J., Roger Parsons, and Joseph Jordan, *Standard Potentials in Aqueous Solution.* New York: Marcel Dekker, 1985 (an IUPAC publication).

Barrett, Jack, *Understanding Inorganic Chemistry: The Underlying Physical Principles.* New York: Prentice-Hall, 1991.

Cotton, F. Albert and Geoffrey Wilkinson, *Advanced Inorganic Chemistry.* 5th ed. New York: Wiley, 1988.

Cox, P.A., *The Elements.* New York: Oxford, 1989.

Douglas, Bodie E., et al., *Concepts and Models of Inorganic Chemistry.* 3rd ed. New York: Wiley, 1994.

Ellis, Arthur B., et al., *Teaching General Chemistry: A Materials Science Companion.* Washington, DC: American Chemical Society, 1993.

Fox, MaryeAnn and James K. Whitesell, *Organic Chemistry.* Sudbury, MA: Jones and Bartlett, 1994.

Greenwood, N.N. and A. Earnshaw, *Chemistry of the Elements.* Oxford: Pergamon Press, 1984.

Hawley, Gessner G., *The Condensed Chemical Dictionary,* 12th ed. New York: Van Nostrand-Reinhold, 1993.

Jordan, Robert B., *Reaction Mechanisms of Inorganic and Organometallic Systems.* New York: Oxford, 1991.

Kent, James A., ed. *Riegel's Handbook of Industrial Chemistry.* 9th ed. New York: Chapman & Hall. 1993.

Leigh, G.J., *Nomenclature of Inorganic Chemistry.* Oxford: Blackwell Scientific Publications, 1990 (an IUPAC publication).

Massey, A.G., *Main Group Chemistry.* Ellis Horwood, 1990.

Owen, Steven M. and Alan T. Brooker, A *Guide to Modern Inorganic Chemistry.* New York: Wiley, 1991.

Panico, R., et al., *A Guide to IUPAC Nomenclature of Organic Compounds.* Cambridge, MA: Blackwell Scientific Publications, 1994.

Prudent Practices in the Laboratory. Washington, DC: National Academy Press, 1995.

Rodgers, Glen E. *Introduction to Coordination, Solid State, and Descriptive Inorganic Chemistry.* New York: McGraw-Hill, 1994.

Snyder, Carl H. *The Extraordinary Chemistry of Ordinary Things.* 2nd ed. New York: Wiley, 1995.

Speight, James G., *The Chemistry and Technology of Petroleum.* 2nd ed. New York: Marcel Dekker, 1991.

Stevens, Malcolm P., *Polymer Chemistry.* 2nd ed. New York: Oxford, 1990.

Tabor, D., *Gases, Liquids and Solids: And Other States of Matter.* 3rd ed. New York: Cambridge University Press, 1991.

Warren, Warren S., *The Physical Basis of Chemistry.* Orlando, FL: Academic Press, 1993.

Wayne, Richard P., *Chemistry of Atmospheres.* 2nd ed. New York: Oxford, 1991.

Weast, Robert C., *Handbook of Chemistry and Physics.* Cleveland: CRC (published annually).

Windholz, Martha, ed. *The Merck Index.* 12th ed. Rahway, NJ: Merck and Co., Inc., 1996.

Winter, Mark J., *Chemical Bonding.* New York: Oxford, 1994.

References Used for Tabular Material

Bard, A.J., Roger Parsons, and Joseph Jordan, *Standard Potentials in Aqueous Solution.* New York: Marcel Dekker, 1985 (an IUPAC publication).

Chase, M.W., Jr., C.A. Davies, J.R. Downey, Jr., D.J. Friup, R.A. McDonald, and A.N. Syverud, *JANAF Thermochemical Tables.* 3rd ed. (2 volumes) Washington: American Chemical Society and American Institute of Physics for the National Bureau of Standards (now The National Institute of Standards and Technology), 1986.

Dean, John A., *Lange's Handbook of Chemistry.* 14th ed. New York: McGraw-Hill, 1992.

Emsley, John, *The Elements.* 2nd ed. New York: Oxford, 1991.

Gold, Victor, Kurt L. Loening, Alan D. McNaught, and Pamil Sehml, *Compendium of Chemical Terminology.* Oxford: Blackwell Scientific Publications, 1987 (an IUPAC publication).

Greenwood, N.N. and A. Earnshaw, *Chemistry of the Elements.* Oxford: Pergamon Press, 1984.

Kaye, G.W.C. and T.H.Laby, *Tables of Physical and Chemical Constants.* London: Longman, 1986.

Mills, Ian, Tomislav Cvitas, Klaus Homann, Nikko Kallay, and Kozo Kuchitsu, *Quantities, Units, and Symbols in Physical Chemistry.* Oxford: Blackwell Scientific Publications, 1988 (an IUPAC publication, the "Green Book").

Moses, Alfred J., *The Practicing Scientist's Handbook.* New York: Van Nostrand-Reinhold, 1978.

Porterfield, William W. *Inorganic Chemistry.* Reading, MA: Addison-Wesley, 1984.

Wagman, Donald D., William H. Evans, Vivian B. Parker, Richard H. Schumm, Iva Halow, Sylvia M. Bailey, Kenneth L. Churney, and Ralph L. Nuttall, *The NBS Tables of Chemical Thermodynamic Properties.* Washington, DC: American Chemical Society and American Institute of Physics for the National Bureau of Standards (now The National Institute of Standards and Technology), 1982.

MERRILL

CHEMISTRY

Authors

Robert C. Smoot

Richard G. Smith

Jack Price

Contributing Author

Tom Russo

GLENCOE

McGraw-Hill

New York, New York Columbus, Ohio Mission Hills, California Peoria, Illinois

MERRILL

CHEMISTRY

A GLENCOE PROGRAM

Student Edition
Teacher Wraparound Edition
Laboratory Manual-Student Edition
Laboratory Manual-Teacher Edition
English/Spanish Glossary
Computer Test Bank
Videodisc Correlations
Transparency Package
Problems and Solutions Manual
Solving Problems in Chemistry
Study Guide-Student Edition
Mastering Concepts in Chemistry-Software

Teacher Resource Books:
- **Enrichment**
- **Critical Thinking/Problem Solving**
- **Study Guide-Teacher Edition**
- **Lesson Plans**
- **Chemistry and Industry**
- **ChemActivities**
- **Vocabulary Review/Concept Review**
- **Transparency Masters**
- **Reteaching**
- **Applying Scientific Methods in Chemistry**
- **Evaluation**

REVIEWERS

DeWaine Alley
Muncie Southside High School
Muncie, IN 47302

Joanne Bowers
Plainview High School
Plainview, TX 79072

Theodore L. Boydston III
Science Curriculum Coordinator
Dade County Public Schools
Miami, FL 33156

Mae Carr
Corliss High School
Chicago, IL 60628

Victor James DeAlmeida
Alvin High School
Alvin, TX 77511

Julia Shaw Dukes
Mosley High School
Lynn Haven, FL 32444

Sandra Kay Enger
Nettleton High School
Jonesboro, AR 72401

Ronald R. Esman
Abilene High School
Abilene, TX 79603

William S. Frazer
Santa Teresa High School
San Jose, CA 95123

Janet P. Lee
Dr. Phillips High School
Orlando, FL 32819

Patricia Luco
Aldine Contemporary Education Center
Houston, TX 77088

Kathleen Munroe
Osceola High School
Kissimee, FL 34741

Erik Natti
Marblehead High School
Marblehead, MA 01945

Wylie C. Poulos
Cross Keys High School
Atlanta, GA 30319

James Eugene Sims
La Mirada High School
La Mirada, CA 90638

James F. Shannon
Pittsford-Mendon High School
Pittsford, NY 14534

Kelly Wedding
Santa Fe High School
Santa Fe, NM 87505

Earl Wilson
Cass Technical High School
Detroit, MI 49091

Cover Photograph: Erich Schrempp/Photo Researchers

Send all inquiries to:
Glencoe/McGraw-Hill
8787 Orion Place
Columbus, OH 43240

ISBN 0-02-825526-7

Printed in the United States of America.

7 8 9 071/043 04 03 02 01 00

AUTHORS

Robert C. Smoot is a chemistry teacher and Rollins Fellow Emeritus at McDonogh School, McDonogh, Maryland. He has taught chemistry at the high school level for 38 years. He has also taught courses in physics, mathematics, engineering, oceanography, electronics, and astronomy. He earned his B.S. degree in Chemical Engineering from Pennsylvania State University and his M.A. in Teaching from the Johns Hopkins University. He is a Fellow of the American Institute of Chemists and a member of the National Science Teachers Association and the American Chemical Society.

Richard G. Smith is a chemistry teacher at Bexley High School, Bexley, Ohio. He has been teaching chemistry at the high school level for 29 years. He received the outstanding teacher award from the American Chemical Society and has participated in NSF summer institutes in chemistry. Mr. Smith graduated with a B.S. degree in Education from Ohio University and earned his M.A.T. in Chemistry from Indiana University. He is a member of the American Chemical Society, National Science Teachers Association, and Phi Beta Kappa.

Jack Price is co-director of the Center for Science and Mathematics Education at California State Polytechnic University, Pomona, California. He previously taught high school chemistry and mathematics in Detroit for 13 years. He earned his B.A. degree at Eastern Michigan University and M.Ed. and Ed.D. degrees at Wayne State University, where he carried out research on organometallic compounds. He has participated in NSF summer institutes at New Mexico State University and the University of Colorado. He is also an author of a high school mathematics textbook.

CONTRIBUTING AUTHOR and MICROCHEMISTRY SPECIALIST

Tom Russo was formerly a chemistry teacher and Science Supervisor with the Millburn Township, New Jersey school system. He is currently a chemistry education consultant with Kemtec Educational Corp., West Chester, Ohio. He received a B.A. in Science Education from Jersey City State College, an M.S. in biology from Seton Hall University, and an M.S. in chemistry from Simmons College in Boston. Mr. Russo has specialized in developing microchemistry methods for classrooms. He is a Woodrow Wilson, Dreyfus Master Teacher in Chemistry, and is the author of the laboratory manual that accompanies this program.

Content Consultants:

Teresa Anne McCowen
Chemistry Department
Butler University
Indianapolis, IN 46208

Mamie W. Moy
Chemistry Department
University of Houston
Houston, TX 77204

William M. Risen, Jr.
Chemistry Department
Brown University
Providence, RI 02912

Safety Consultant:

Joanne Neal Bowers
Plainview High School
Plainview, TX 79072

CONTENTS

CONTENTS

CONTENTS

CONTENTS

17 Liquids 421

18 Gases 451

19 Gases and the Mole 477

20 Solutions and Colloids 499

21 Colligative Properties 521

CONTENTS

CONTENTS

CONTENTS

CHEMACTIVITIES

APPENDICES

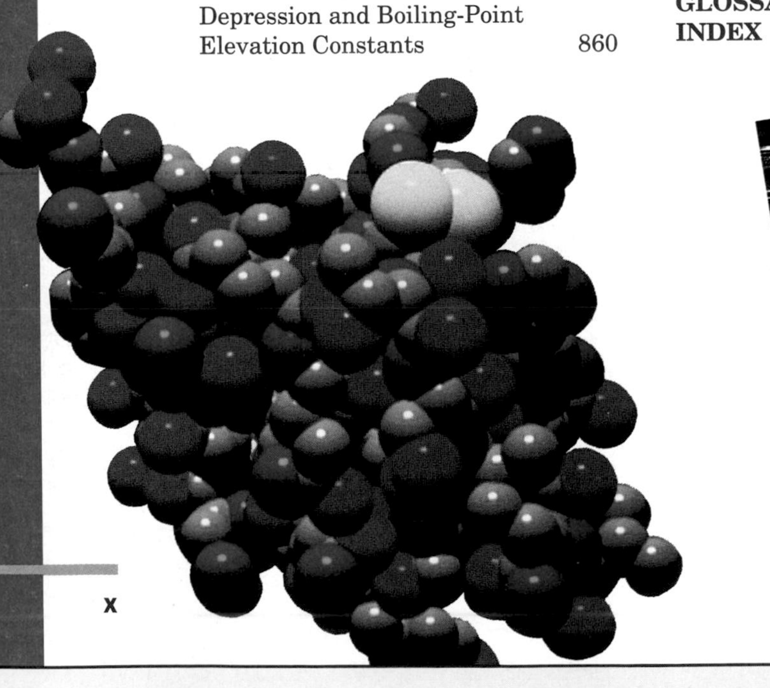

FEATURES

CHEMISTRY AND SOCIETY

What are some aspects of chemistry that have an impact on society? Develop your own viewpoint on controversial issues. Learn how to live with chemistry and use it to enrich your life.

CAREERS IN CHEMISTRY

How can you use chemistry in your career? Read the career features in your textbook. Then plan for an exciting future of discovery.

CHEMISTRY AND TECHNOLOGY

What new advances have been made by applying knowledge of chemistry? Read about these new technologies and discover more advantages of a chemical awareness.

FRONTIERS

Discover new uses for chemistry along with today's scientists. Find out how chemistry is an exciting world of discovery.

Bridges to OTHER SCIENCES

How is chemistry connected to other sciences? Chemistry provides information that forms the foundation of many of the other sciences.

CHEMISTS AND THEIR WORK

Many men and women have been dedicated to discovery in the world of chemistry. Let them be your models for a possible future in chemistry.

EVERYDAY CHEMISTRY

Every day, you are affected by chemistry. Read about familiar topics, but now learn how to explain them to your friends and family.

MINUTE LABS

Want to see it for yourself? Then try it! Just as a picture is worth a thousand words, a minute lab is worth an hour of lecture!

INTERCURRICULAR CONNECTIONS

Did you know that chemistry is connected to many other fields of study? Learn how chemistry is linked to your subjects and hobbies.

A quick tour of

MERRILL

CHEMISTRY

Have you ever wondered how a color TV produces color or what computer diskettes are made of? Have you ever watched bread dough rise or eaten a pickle? Have you ever looked at the deposits that clog up a discarded spark plug or revved a car engine to warm it up on a cold morning?

If you've done any of these things, you already have some experience in chemistry. Communications, cooking, and cars depend on chemistry for their effectiveness. In this course, you'll find out why.

The next few pages describe the many features of this book that will help you understand the exciting world of chemistry.

GETTING STARTED

Each chapter of ***Merrill CHEMISTRY*** is organized to keep you centered on the topic at hand.

CHAPTER 25

Oxidation-Reduction

The Statue of Liberty has become a symbol of freedom. Manufactured from copper sheets, it was erected in 1886 in New York harbor, where the light in the torch was considered an aid to navigation. More importantly, the statue became one of the first sights seen by immigrants seeking a better life in the United States and came to symbolize freedom and friendship.

Over the decades, the copper sheets of the Statue of Liberty, as well as the steel skeleton over which they were stretched, gradually corroded. Inspectors found evidence of decay, especially where rivets penetrated the copper sheets. In the mid-1980s, the statue was restored. The process of corrosion involves a kind of chemical reaction called oxidation-reduction. In this chapter, you will learn about these reactions and practice balancing the equations that represent them.

Preview

NTENTS

Oxidation and Reduction Processes
Types of Reactions
Oxidation and Reduction
Oxidizing and Reducing Agents
Oxidation Numbers
Assigning Oxidation Numbers
Identifying Oxidation-Reduction Reaction

EXTENSIONS AND CONNECTIONS

Bridge to Metallurgy
Ore to Metal in a Blast Furnace: How iron is ... page 629

History Connection
Politics and Chemistry—Do They Mix? page 639

Everyday Chemistry
...y: Redox reactions in print, page 642

The photos and text on the first two pages of each chapter help you focus on the chapter topic and often relate chemistry to the world around you. A detailed table of contents at the bottom of these pages gives you a quick preview of what's in each chapter.

Grasp the important ideas of chemistry

The text of ***Merrill*** **CHEMISTRY** develops ideas with descriptions, analogies, and examples. Ideas build on one another throughout the course to give you a cohesive knowledge of chemistry. Some sections, labeled "Chemistry in Depth," explore certain chemistry topics in much greater detail.

Photographs, illustrations, and **diagrams** are vital to the presentation of each topic. They will show you examples of chemistry concepts and help you visualize important processes.

Table 10.5

Electron Affinities (kilojoules per mole)

H 72.766							**He** (−21)
Li 59.8	**Be** (−241)	**B** 23	**C** 122	**N** 0	**O** 141	**F** 328	**Ne** (−29)
Na 52.9	**Mg** (−230)	**Al** 44	**Si** 120	**P** 74	**S** 200.42	**Cl** 348.7	**Ar** (−34)
K 46.36	**Ca** (−156)	**Ga** 36	**Ge** 116	**As** 77	**Se** 194.91	**Br** 324.5	**Kr** (−39)
Rb 46.88	**Sr** (−167)	**In** 34	**Sn** 121	**Sb** 101	**Te** 190.16	**I** 295.3	**Xe** (−40)
Cs 45.5	**Ba** (−52)	**Tl** 48	**Pb** 101	**Bi** 101	**Po** (170)	**At** (270)	**Rn** (−41)

Parentheses indicate a calculated rather than an experimental value.

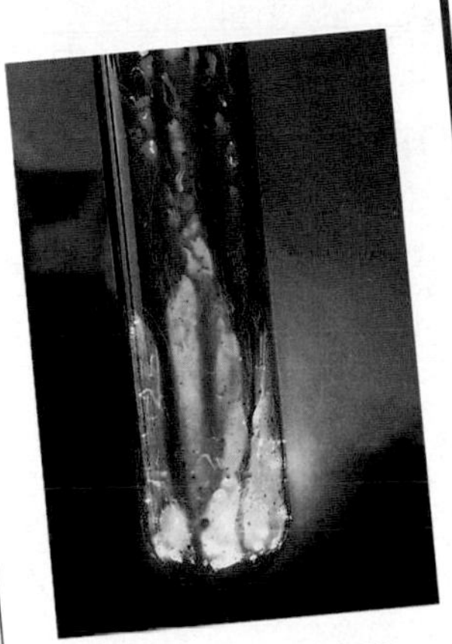

Figure 10.11 The reaction between aluminum, an element with a low electron affinity, and bromine, which has a high electron affinity, is shown here.

Electron Affinities

Now consider an atom's attraction for additional electrons. The attraction of an atom for an electron is called **electron affinity.** The same factors that affect ionization energy also affect electron affinity. In general, as electron affinity increases, an increase in ionization energy can be expected. *Metals have low electron affinities. Nonmetals have high electron affinities,* as shown in Table 10.5. Although not as regular as ionization energies, electron affinities still show periodic trends. Look at the column headed by hydrogen. The general trend as we go down the column is a decreasing tendency to gain electrons. We should expect this trend since the atoms farther down the column are larger. As a consequence, the nucleus is farther from the surface and attracts the outer electrons less strongly.

RULE OF THUMB

Rules of Thumb highlight key generalizations that you will use often when studying or solving problems.

Look at the period beginning with lithium. The general trend as we go across is a greater attraction for electrons. The increased nuclear charge of each successive nucleus accounts for the trend. These elements with high electron affinities will tend to gain electrons and form negative ions.

How do we account for the exceptions of beryllium, nitrogen, and neon in the lithium period? The more stable an atom is, the less tendency it has to gain or lose electrons. The high negative value for beryllium is associated with the stability of the full 2s sublevel. If beryllium were to gain an electron, its configuration would be less ... has a negativ... electrons mor... 2p sublevel. N... level. Becaus... attraction for...

10.2 CONCEPT REVIEW

7. Explain why magnesium has a first ionization energy higher than that of alumi... um even though, generally, ionization ...nergies increase from left to right in the ...eriodic table.
8. ...Using Table 10.4, explain why lithium has ...n oxidation number of 1+ and beryllium ...has an oxidation number of 2+.
9. What is the meaning of a neg... for the electron affinity of an...
10. Why do atoms in Group 18 (... negative electron affinities?
11. **Apply** What can you sugges... explain why antimony and b... essentially identical electron...

Concept Review questions help you check your progress as you complete each section.

Solve problems successfully

Chemistry is not just something you learn by reading. In ***Merrill*** **CHEMISTRY** you'll solve problems and apply the solutions to real-world situations. You'll discover and work with quantitative relationships that are fundamental to chemical reactions and the structure of matter.

SAMPLE PROBLEM

Enthalpy Change

Calculate the enthalpy change in the following reaction.

carbon monoxide + oxygen → carbon dioxide

Solving Process:

First, write a balanced equation. Include all the reactants and products.

$$2CO(g) + O_2(g) \rightarrow 2CO_2(g)$$

Each formula unit represents one mole. Remember that free elements have zero enthalpy by definition. Using the table of enthalpies of formation, Appendix Table A-6, the total enthalpy of the reactants is

$$\Sigma\Delta H_{f(\text{reactants})}^{\circ} = \frac{2\ \text{mol CO}}{} \left| \frac{-110.5\ \text{kJ}}{\text{mol CO}} \right. + 0\ \text{kJ} = -221.0$$

The total enthalpy of the product ($2CO_2$) is

$$\Sigma\Delta H_{f(\text{products})}^{\circ} = \frac{2\ \text{mol } CO_2}{} \left| \frac{-393.5\ \text{kJ}}{\text{mol } CO_2} \right. = -787.0\ \text{kJ}$$

The difference between the enthalpy of the reactants and the enthalpy of the product is

$$\Delta H_r^{\circ} = \Sigma\Delta H_{f(\text{products})}^{\circ} - \Sigma\Delta H_{f(\text{reactants})}^{\circ}$$

$$\Delta H_r^{\circ} = -787.0\ \text{kJ} - (-221.0\ \text{kJ}) = -566.0\ \text{kJ}$$

This difference between the enthalpy of the products and reactants (−566.0 kJ) is released as the enthalpy of re

$$2CO(g) + O_2(g) \rightarrow 2CO_2(g) + \textit{enthalpy of reaction}$$

PRACTICE PROBLEMS

5. Compute ΔH_r° for the following reaction.
$2NO(g) + O_2(g) \rightarrow 2NO_2(g)$

PROBLEM SOLVING HINT

Be careful with signs. Most enthalpies of formation are negative. Likewise, enthalpies of reaction for most spontaneous processes are negative.

STEP BY STEP PROBLEM SOLVING

As you encounter each new problem-solving situation, you are shown a carefully worked **Sample Problem** that describes the procedure and often gives you a problem-solving hint. Then, you are given immediate practice in solving that type of problem and some of its typical variations. This format lets you continue to refer to the sample problem as you begin to solve problems on your own.

You'll find solutions to the Practice Problems worked out for you in Appendix C!

Chapter 24 REVIEW

Summary

24.1 Water Equilibria

1. The solubility product constant, K_{sp}, for an ionic substance is equal to the product of the concentrations of the dissociated ions found in solution.
2. The ion product constant for water, K_w, is 1.00×10^{-14} at 25°C and is equal to the product of $[H_3O^+]$ and $[OH^-]$.
3. The pH of a solution is equal to −log $[H_3O^+]$. The pH of pure water and neutral solutions is 7. Acid solutions have a pH below 7. Basic solutions have a pH

Revie

31. Calcium car ble in water
 a. Write an dissocia
 b. Write ex the equ tion of
32. The K_{sp} of the concen rated aque
33. The conc saturated $3.4 \times 10^{-}$ sulfide.

PRACTICE and MORE PRACTICE

Each chapter ends with a **Chapter Review,** where you'll get even more problem-solving practice. In the Chapter Review, you'll answer many different types of questions, working with both words and numbers to reinforce your newly-gained knowledge of chemistry.

Experience chemistry

Is it possible for you to do some of the same experiments that led to major breakthoughs in scientific knowledge? Can you learn some of the same techniques that are used every day in industrial and research labs?

***Merrill* CHEMISTRY** supplies a number of opportunities for you to explore and experiment—to *do* chemistry. This book gives you lots of hands-on learning experience —using **Minute Labs** and **ChemActivities**—to strengthen your understanding of chemistry.

MINUTE LABS

Minute Labs appear throughout the book. They are quick, simple to set up, and often require no special equipment. Minute Labs are placed in the text margins so you can easily relate the activity to the concept it illustrates.

MINUTE LAB

Air Pressure vs. Gravity

Put on goggles and an apron. Pour water into a small fruit-juice glass until it overflows. Place a 3" × 5" index card over the mouth of the glass. Hold the card across the top of the glass while you invert the glass over a sink. Remove your hand from the card. What did you observe? Explain. Which force appears to be stronger, gravity or air pressure? What is the pressure of one standard atmosphere?

Chapter 2
ChemActivity 1

Micro

A Density Balance

ve you ever played on a teeter-totter or see-? If two children want to use one together, can they do it if one child is much heavier n the other? Both children take up the e space, but one child's mass is much ter than that of the other. The heavier must sit closer to the center of the board der for the children to play. The lighter hild is, the closer to the center the heavier must sit. The same relation between and distance can be used to determine ty.

e density of water is 1.0 g/mL. Most liq- ave densities less than that of water, other common liquid has a density of nL. Therefore, even if samples of water nother liquid have the same volume, ave different masses. The purpose of ivity is to use a balance to compare the s of some liquids

Materials

ruler with center channel
thin stem pipets (2)
ethyl alcohol
n-butyl alcohol
isopropyl alcohol
pencil
water
cooking oil
ethyl acetate
oleic acid

Proc

1. Set er mar Mov it is er li the samp balan move
2. Plac

ChemACTIVITIES

ChemActivities for each chapter are placed together at the back of the book. Doing these lab activities will get you involved with the methods and concepts of chemistry. Many of them employ micro-scale techniques, which conserve chemicals and time and are safer to do.

Discover the

Chemistry Connections

How does chemistry relate to the everyday world and to other subjects you're studying? Who are the people involved? What areas of current chemical research will provide breakthroughs in the future? How does the technology made possible by chemical research affect you? You'll find the answers to these and other questions in articles and special features throughout the text.

FRONTIERS

CHEMISTRY AND TECHNOLOGY

Bridge to PHYSICS

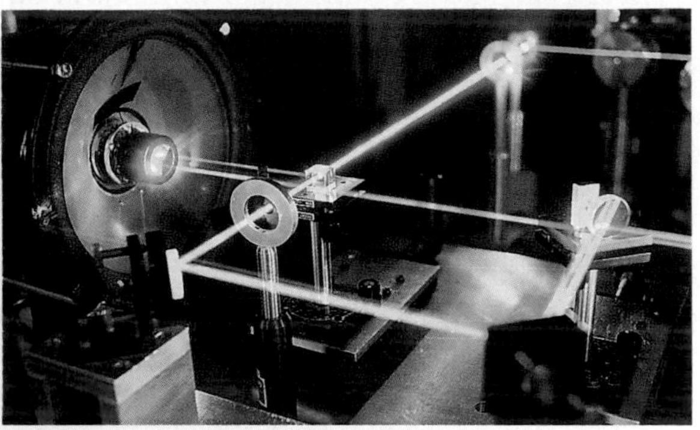

These features report on basic research and current applications of chemistry and show how chemistry relates to other sciences.

CHEMISTRY AND SOCIETY

EVERYDAY CHEMISTRY

SPORTS CONNECTION

These three features focus on the connections between chemistry and the world you live in—from daily life experiences, to art and literature, to issues of concern to society at large.

CAREERS IN CHEMISTRY

CHEMISTS AND THEIR WORK

Chemists and their careers are the focus of these brief margin features.

Read on for a message from the author of ***Merrill CHEMISTRY***

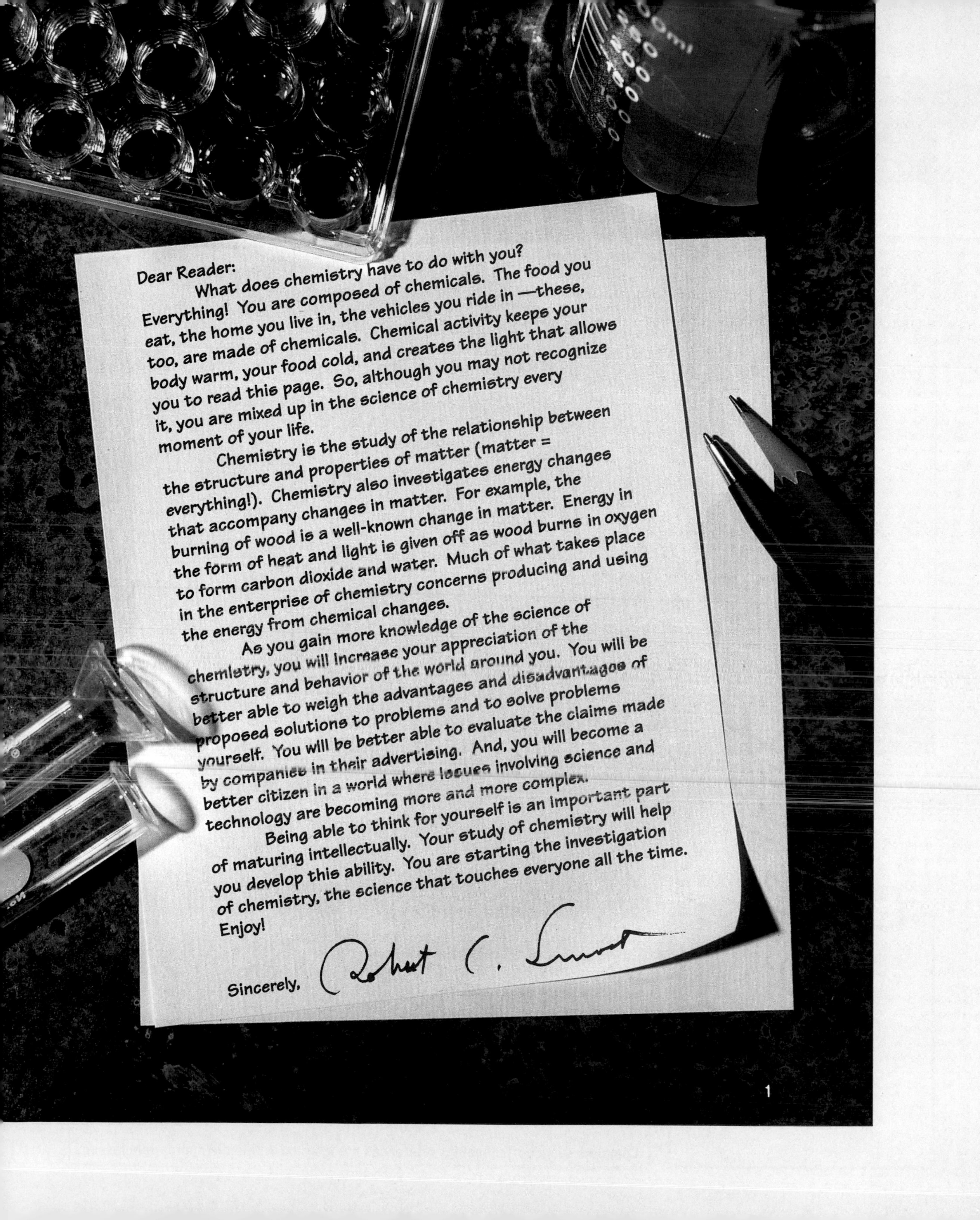

Dear Reader:

What does chemistry have to do with you? Everything! You are composed of chemicals. The food you eat, the home you live in, the vehicles you ride in —these, too, are made of chemicals. Chemical activity keeps your body warm, your food cold, and creates the light that allows you to read this page. So, although you may not recognize it, you are mixed up in the science of chemistry every moment of your life.

Chemistry is the study of the relationship between the structure and properties of matter (matter = everything!). Chemistry also investigates energy changes that accompany changes in matter. For example, the burning of wood is a well-known change in matter. Energy in the form of heat and light is given off as wood burns in oxygen to form carbon dioxide and water. Much of what takes place in the enterprise of chemistry concerns producing and using the energy from chemical changes.

As you gain more knowledge of the science of chemistry, you will increase your appreciation of the structure and behavior of the world around you. You will be better able to weigh the advantages and disadvantages of proposed solutions to problems and to solve problems yourself. You will be better able to evaluate the claims made by companies in their advertising. And, you will become a better citizen in a world where issues involving science and technology are becoming more and more complex.

Being able to think for yourself is an important part of maturing intellectually. Your study of chemistry will help you develop this ability. You are starting the investigation of chemistry, the science that touches everyone all the time.

Enjoy!

Sincerely,

CHAPTER 1

The Enterprise of Chemistry

CHAPTER ORGANIZER

Chapter Objectives	Text Features	Laboratory Options	Teacher Classroom Resources
1. Describe the role of chemists and some of the procedures that chemists use in their studies of matter and energy. **2.** Define matter, energy, and the forms of energy. **3.** Explain the law of conservation of mass-energy and its importance to chemistry. **Nat'l Science Stds: UCP.1, UCP.2, UCP.3, A.1, A.2, B.2, B.5, B.6, E.2, F.4, F.5, F.6, G.1, G.2**	**Careers in Chemistry:** Chemist, p. 9 **Chemistry and Society:** Essential Materials, p. 10 **Chemists and Their Work:** Bruce N. Ames, p. 12 **Everyday Chemistry:** Bread Making, p. 16	**Minute Lab:** Risk vs. Benefit of Inertia, p. 7 **Minute Lab:** Potential and Kinetic Energy p. 8 **Minute Lab:** Observe and Hypothesize, p. 11 **Discovery Demo:** 1-1 Wine to water to milk, p. 2 **Demonstration:** 1-2 Nylon, p. 4 **Demonstration:** 1-3 Law of Conservation of Mass, p. 6 **Demonstration:** 1-4 Sulfuric Acid and You, p. 8 **Demonstration:** 1-5 Observing Carefully, p. 12 **Demonstration:** 1-6 Recycled Plastic, p. 14 **Demonstration:** 1-7 Hard Water, p. 16 **Laboratory Manual:** 1 The Eight Solution Problem - microlab **ChemActivity 1:** A Science Experiment, p. 796	**Lesson Plans:** p.1 **Study Guide:** p.1 **Critical Thinking/Problem Solving:** The Scientific Process, p. 1 **Enrichment:** The Effects of Petroleum Spills, p. 1 **Applying Scientific Methods in Chemistry:** Using the Rules of Chemistry, p. 15 **Reteaching:** The Enterprise of Chemistry, p. 1 **ChemActivity Master 1**
Other Chapter Resources	**Vocabulary and Concept Review,** pp. 1, 31, 32 **Evaluation,** p. 1-4	**Videodisc Correlation Booklet** **Lab Partner Software**	**Test Bank**

Block Schedule

For information on block scheduling, see the Lesson Plans booklet in the Teacher Resource Package.

GLENCOE TECHNOLOGY

Chemistry: Concepts and Applications Videodisc

The Magic of Chemistry
The Law of Conservation of Mass

Chemistry: Concepts and Applications CD-ROM

The Law of Conservation of Mass

Science and Technology Videodisc Series (STVS)

Plants and Simple Organisms

Plant Chemical Repels Cockroaches
Insecticides from Desert Plants

Ecology

Wasp Biological Control
Controlling Fruit Flies
Resistance to Pesticides in Cockroaches

Complete Glencoe Technology references are inserted within the appropriate lesson.

ASSIGNMENT GUIDE

Contents	Level	Practice Problems	Chapter Review	Solving Problems in Chemistry
The People	C		10, 16	
The Ingredients	C		6, 7, 11, 15	
The Rules	C		8	
The Industry	C		12, 18	
The Process	C		13, 14, 20	
The Practice	C		21	1-9
The Reality	C			
The Future	C		19	
The Course	C		9, 17	
				Chapter Review: 1-16

C=Core, A=Advanced, O=Optional

Laboratory Materials

Minute Labs	ChemActivities	Demonstrations		
page 7 board (small) rubber band (large) stuffed animal wagon (small) **page 8** board, 2-m long bricks or books coin meter stick toy car **page 11** food coloring, four colors liquid dish soap petri dish bottom toothpicks whole milk	**page 796** masking tape spheres, 7 per group meter stick	**pages 2, 4, 6, 8, 12, 14, 16** acetone anhydrous sodium carbonate $BaCl_2{\cdot}2H_2O$ balance beakers, 100, 150, 250, 400, 600-cm^3, 1-dm^3 cardboard, 30 x 30-cm piece chalk $Cu(NO_3)_2{\cdot}3H_2O$ ethanol	Erlenmeyer flask with stopper food coloring forceps graduated cylinder 1, 6-diamino-hexane ice cubes $KMnO_4$ NaOH $NaHSO_3$ packing box of polystyrene peanuts paper towels	sebacoyl chloride sodium polyacrylate sucrose sulfuric acid TTE white teflon magnetic stir bar water, tap and distilled

CHAPTER 1
The Enterprise of Chemistry

CHAPTER OVERVIEW

Main Ideas: Chapter 1 provides students with an overview of the people and their practice of the art and science of chemistry. The relationships of matter and energy to the basic rules of science are presented early in the chapter. The students become aware that the chemical industry is a result of a scientific process. Chemistry is defined for students as they are introduced to the course. The complexities of modern science and the impact of technology in today's society are presented in a detailed case study designed to generate class discussion.

Theme Development: Systems and interactions is a theme that is developed throughout Chapter 1. Chapter 1 presents a fast-moving, interesting look at the enterprise of chemistry.

Tying To Previous Knowledge: Often the first chapter of a textbook does little to excite and motivate students in their study of the course's content. Chapter 1 of this textbook is different. It is the authors' intent to establish in each student a desire to know more about the living and growing science of chemistry. Encourage students to express their thoughts as they read the chapter.

ASSESSMENT PLANNER

Choose from the following assessment strategies.

Assess, pp. 16-17
Assessment, pp. 5, 7, 8, 10, 11, 13, 15
Portfolio, p. 13
Minute Lab, pp. 7, 8, 11
ChemActivity, pp. 796-797
Chapter Review, pp. 18-19

CHAPTER Preview

CONTENTS

2

DISCOVERY DEMO

1-1 Wine to Water to Milk

Purpose: to demonstrate that chemists use raw materials to prepare intermediates that are then used to make consumer goods

Materials: 1 g $BaCl_2 \cdot 2H_2O$, 1 g $NaHSO_3$, 0.05 g $KMnO_4$, 3 400-cm^3 beakers, 2 small test tubes, 252 cm^3 H_2O

Procedure: 1. Before class, dissolve 3 or 4 very small crystals of $KMnO_4$ in 250 cm^3 of water in a beaker. Dissolve 1 g of $NaHSO_3$ in a test tube containing 1 cm^3 water. Place 1 g $BaCl_2.2H_2O$ in a second test tube containing 1 cm^3 of water. **CAUTION:** *The solutions are very toxic.*
2. Before class, place the $NaHSO_3$ and $BaCl_2$ solutions in the empty beakers labeled *1* and *2*, respectively.
3. Show the "wine" ($KMnO_4$) solution. Pour the "wine" into beaker 1.
4. Pour the "water" into beaker 2.

The Enterprise of Chemistry

CHAPTER

1

Silkworms and spiders spin natural fibers that are stronger than nylon, stronger even than steel. Chemists have learned what there is in the chemical structure of spider silk that makes it so strong. Engineers are now able to develop a method of manufacturing synthetic spider silk.

Did you ride to school today in a vehicle? If so, it had synthetic rubber tires, a product of the chemical industry. You are probably wearing some clothing made of nylon, the same strong fiber that makes up the parachute shown on the opposite page. At lunch you may eat something that came packaged in a plastic film also produced by the chemical industry. You may be taking an antibiotic to control an infection. Everything in your life is related to chemistry. In fact, your life processes are chemical processes. You are about to begin an exciting study of the principles on which the enterprise of chemistry is based.

EXTENSIONS AND CONNECTIONS

Careers in Chemistry
Chemist, page 9

Chemistry and Society
Essential Materials: A look at how the United States is dependent on many countries for its supply of raw materials, **page 10**

Chemists and Their Work
Bruce N. Ames, page 12

Everyday Chemistry
Bread Making: How the making of bread depends on chemistry, **page 16**

Disposal: E

Results: The "wine" turns into "water" that is then changed into "milk." Avoid discussing the chemistry of this demonstration; it would be of little value at this time. Explain that the chemist has the experience to know how to change one substance into another substance having different properties.

Questions: 1. Which substance is the raw material? *the wine*

2. Which substance is the intermediate? *the water*

3. Which substance is the final consumer product? *the milk*

Follow Up Activity: Point out that chemists have given the world a substitute for cotton and wool. The development of synthetic fibers has freed the land for food production. Use this example to lead a discussion on the world's needs versus available space and resources, and the role of chemists and research.

INTRODUCING THE CHAPTER

USING THE PHOTO

The discovery and manufacturing of nylon illustrates how raw materials are used to produce intermediates that are in turn used to produce consumer goods. Nylon began with a chemist's curiosity about the properties and structure of silk and whether silk could be synthetically produced. Researchers used the silk that spiders spin for their webs as a model, both for the structure of the silk as well as how the silk is spun. Today nylon fiber is produced by forcing a polyamide polymer through tiny holes in a die fashioned after a spider's spinneret.

In discussing the photos, inform students that because chemists want to know how structure relates to properties, they must determine the structure of the spider silk before reproducing it. Explain that to a chemist "structure" means the arrangement of atoms, ions, or molecules.

REVEALING MISCONCEPTIONS

Most students believe that all real innovations, such as airplanes and computers have already been invented. Have them discuss the statement that most of them will be employed as adults in businesses that produce or use products that have yet to be invented. Encourage students to be conjective about major discoveries that may occur during their lifetimes.

GLENCOE TECHNOLOGY

Videodisc

Chemistry: Concepts and Applications

Disc 1, Side 1, Ch. 1

The Magic of Chemistry

Show this video of the Discovery Demo to introduce the concept of chemical processes.

Also available on CD-ROM.

PREPARE

Key Concepts

The major concepts presented in this chapter include inertia, potential, kinetic and radiant energy, law of conservation of mass-energy, and modeling. An overview of chemistry as a science and industry is presented.

Key Terms

matter
inertia
energy
potential energy
kinetic energy
radiant energy
law of conservation of mass
law of conservation of energy
law of conservation of mass-energy
intermediate
model
chemistry

Looking Ahead

Check the materials list for the Minute Labs to be certain you have brought suitable items from home.

1 FOCUS

Objectives

Preview with your students the objectives listed on the student page. Students can use them as a study guide for this chapter.

Activity

Write a welcome poster in invisible ink that becomes visible when sprayed with another solution. To prepare the two different inks, dissolve 1 g NH_4SCN in 100 cm^3 H_2O, and 4 g $K_4Fe(CN)_6{\cdot}3H_2O$ in 100 cm^3 H_2O. The spray that will make them visible is prepared by dissolving 6.75 g $FeCl_3{\cdot}6H_2O$ in 250 cm^3 H_2O. Prepare the poster in advance so it will be dry. The chemistry of the reactions is not important to your students at this time, but the welcome message will fascinate them. **Disposal:** F

OBJECTIVES

- describe the role of chemists and some of the procedures that chemists use in their studies of matter and energy.
- define matter, energy, and the forms of energy.
- explain the law of conservation of mass-energy and its importance to chemistry.

" 'What does chemistry mean to me?' said Mr. Averageman, as he looked at this page printed with ink made by a chemical process, and laced his shoes, made of leather tanned by a chemical process. He glanced through a pane of glass, made by a chemical process, and saw a baker's cart full of bread leavened by a chemical process, and a draper's wagon delivering a parcel of silk, made by a chemical process.

"He put on his hat, dyed by a chemical process, and stepped out upon the pavement of asphalt, compounded by a chemical process, bought a daily paper with a penny refined by a chemical process.

" 'No,' he added, 'of course not, chemistry has nothing to do with me.' "

(Herbert Newton Casson, 1869–1964)

The People

What do chemists do? Chemists solve problems that have to do with all the "stuff" of which our world is made. This "stuff" is called *matter*. Chemists have found that there is a very close relationship between the properties of matter (how it behaves) and the structure of matter (how it is put together). For instance, a diamond is very hard. It is hard because of the way its atoms are held together.

Nylon is one of the most common synthetic materials we encounter: nylon stockings, nylon fishing line, nylon bearings in the motor of a hairdryer. Nylon was developed by a chemist, Wallace H. Carothers. The pattern of Carothers' discovery is typical of what chemists do. He studied the structure and properties of silk. Once he understood how the properties of silk depended upon its structure, he could predict that a similar structure would

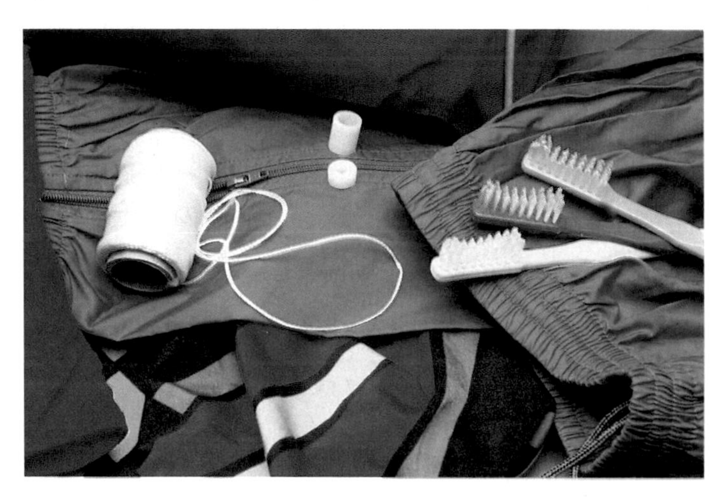

Figure 1.1 Nylon is mainly used in fibers and fabrics because of its strength, its low shrinkage, and its silky look. Carpets, swimsuits, parachutes, bristles in brushes, fishing line, bearings, gears, and even automobile body panels are made of nylon.

DEMONSTRATION

1-2 Nylon

Purpose: to duplicate the designed research of Carother's discovery of nylon

Materials: 150-cm^3 beaker, 50 cm^3 1,1,2-trichloro-1,2,2-trifluoroethane (TTE), 1 cm^3 sebacoyl chloride, 100-cm^3 beaker, 1 g hexamethylenediamine (1,6-hexanediamine 80-95% aqueous solution), 1 g anhydrous sodium carbonate, 25 cm^3 H_2O, 1 pair forceps, 100-cm^3 graduated cylinder, paper towel, 1 drop food coloring

Procedure: CAUTION: *Use a fume hood and wear goggles. Vapors are hazardous.* **1.** In the 150-cm^3 beaker place 50 cm^3 of TTE. Add 1 cm^3 of sebacoyl chloride to the TTE and stir vigorously. **2.** In the 100-cm^3 beaker, dissolve 1 g of hexamethylenediamine and 1 g of

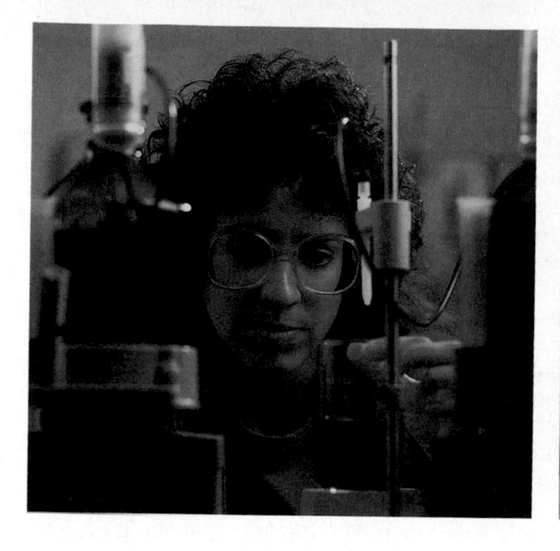

Figure 1.2 Chemists (left) develop new materials, and chemical engineers (right) design the plants and processes to produce these materials.

lead to similar properties. By reacting two chemicals, Carothers produced a material similar to silk in structure and behavior—nylon.

Once chemists like Carothers have developed a new material, they work with chemical engineers to produce the material. Together they develop a good production process. The chemical engineers then design, build, and operate the chemical plant. Chemists and chemical engineers are the principal operators of the chemical enterprise.

The things a chemist studies may be as different as the structure of the materials that transmit nerve impulses in the human brain and the bonding of rubber molecules in a car tire. In these studies, a chemist uses all of the sciences, especially mathematics and physics. The relationships between properties and structure can be organized into several basic principles, facts, and theories. A theory is simply an explanation of a phenomenon. These basic principles, facts, and theories are the foundations of chemistry. Therefore, it is not necessary for you to study the properties of *all* known materials in order to gain a knowledge of chemistry.

The Ingredients

You know that some materials are chemicals—for example, salt, copper, tin, chlorine, and petroleum. What is a chemical? Everything! All matter. Even you are a mixture of chemicals! **Matter** is anything that has the property of inertia. **Inertia** is a property of matter that shows itself as a resistance to any change in motion. This change can be in either the direction or the rate of motion, or in both. For example, suppose you are riding in a moving car. When the car is stopped suddenly, your body tends to continue moving forward. If the car makes a sharp turn, your body tends to

anhydrous sodium carbonate in 25 cm^3 H_2O. Mix well. Add a drop of food coloring to this solution.

3. Pour the second solution into the TTE solution without mixing the two.

4. Using forceps, pick up the center of the nylon film that forms at the interface of the two liquids. Slowly pull the nylon fiber from the beaker and roll it around a graduated cylinder around which a paper towel has been taped.

Disposal: A

Questions: What properties make nylon useful for clothing? *windproof, strong, light weight, resists water damage.*

Extension: Avoid a detailed discussion of the chemistry at this time. The chemistry of polymers will be presented in detail in Chapter 30. This demonstration is designed to establish a need and a desire to know more.

2 TEACH

CULTURAL DIVERSITY

A Nutty Meal

The acorn was the most important staple food of the California Indians. Acorn mush or soup was the main daily food for most. Ground acorns were used to make breads, and acorn nuts to produce coffee and tea. Acorns were also important to other native Americans such as the Apache and Northwest Coast Indians. The pilgrims survived their first Massachusetts winter in 1620 by using baskets of acorns the Indians had buried in the ground for storage.

All acorns contain tannin, a bitter substance. Humans cannot digest large quantities of tannin, so it must be removed before the acorns are eaten. Tannin is readily soluble in water. One method used by native Americans involved burying the acorns in the mud of a swamp for a year and then retrieving them. Another allowed shelled acorns to mold in baskets that were then buried in freshwater sand. The acorns were ready when they turned black.

PROGRAM RESOURCES

Critical Thinking/Problem Solving: Use the worksheet, p.1, as enrichment for the concept of The Scientific Process.

Study Guide: Use the Study Guide worksheet, p.1, for independent practice with concepts in Chapter 1.

Laboratory Manual: Use microlab 1 "The Eight Solution Problem" as enrichment for the concepts in Chapter 1.

✓ ASSESSMENT

Oral: Ask students to describe a situation, similar to the automobile acceleration in the textbook, that illustrates the property of inertia. Airplanes, trains, buses, and amusement park rides are common examples.

CONCEPT DEVELOPMENT

QUICK DEMO **Energy Transfers** Use the following demonstration as examples of energy transfers. To demonstrate how radiant energy is transformed into mechanical energy, place a radiometer in a window on a sunny day. Use a solar-powered calculator to demonstrate the photovoltaic conversion of light into electricity. A battery-powered radio demonstrates the conversion of chemical energy of the battery into electrical energy. The electrical energy is then changed into mechanical energy as the speaker coil moves the air and eventually, our eardrums. A battery-operated flash unit for a camera can be used to demonstrate the conversion of chemical energy to electrical energy to visible light and heat. Strike a match and ask your students to identify the conversions of mechanical energy (from friction) to heat and chemical energy to heat and light. **Disposal:** F

Enrichment: Ask a student to investigate the Cockcroft-Walton experiment that confirmed Einstein's mass-energy hypothesis. *In this experiment Cockcroft and Walton bombarded $^{7}_{3}Li$ nuclei with protons and obtained two alpha particles as products. From the balancing of the mass energy system before and after the nuclear change, the Einstein equivalence hypothesis was established.*

Videodisc

Newton's Apple: Physical Sciences

Newton's Laws

Chapter 3, Side B

Inertia: Newton's first law of motion

26249-29476

en: (GK) in
ergon: (GK) work or action
energy—the capacity for doing work

continue to move in its original direction. Then, your body moves toward the side of the car opposite from the direction of the turn. In both cases, your body is showing the property of inertia. All matter has the property of inertia.

Matter also has energy. In their work with the structure and properties of materials, chemists are also interested in the energy changes that take place. **Energy** is a property possessed by all matter, and, in the proper circumstances, can be made to do work. All objects possess energy. A hockey stick, an automobile, an atom, and an electron have energy. An object has two general forms of energy: potential and kinetic. **Potential energy** depends upon the position of the object with respect to some reference point. A book on a table has a greater potential energy than the same book on the floor because the table is farther from Earth. The gravitational attraction between Earth and the book gives the book greater potential energy on the table. The book could fall farther from the table to the surface of Earth than from the floor to the surface of Earth. The book could, therefore, do more work by falling from the top of the table. For the same reason, an electron close to its nucleus has less potential energy than when it is farther away. Here, however, the attraction between the electron and nucleus produces the potential energy.

Figure 1.3 The potential energy of the book on the table is greater than that of the book on the floor with respect to their distances from Earth.

Kinetic energy is the energy possessed by an object because of its motion. An airplane traveling 700 kilometers per hour has a greater kinetic energy than when it is traveling 500 kilometers per hour.

Energy can be transferred between objects in two ways: through direct contact and through electromagnetic waves. An example of direct energy transfer is the collision of two billiard balls. Kinetic energy is transferred directly from one ball to another. An example of energy transfer by electromagnetic waves

Figure 1.4 The chemical energy in gasoline is converted to mechanical energy in the engine of a car.

DEMONSTRATION

1-3 Law Of Conservation Of Mass

Purpose: to improve the understanding and retention of the law of conservation of mass

Materials: 500-cm^3 Erlenmeyer flask with stopper, 1 g $Cu(NO_3)_2 \cdot 3H_2O$, test tube, 4 g NaOH, 10 cm^3 H_2O, balance

Procedure: **1.** Place 1 g $Cu(NO_3)_2 \cdot 3H_2O$ in the 500-cm^3 Erlenmeyer flask.
2. In the test tube, dissolve 4 g of NaOH in 10 cm^3 H_2O. **CAUTION:** *$Cu(NO_3)_2$ is toxic and a strong oxidizer, and NaOH is caustic.*
3. Place the test tube in the flask with $Cu(NO_3)_2$ but do not allow the two chemicals to mix. Stopper the flask and mass the system on a balance.
4. Carefully tip the stoppered flask so the NaOH solution will mix with the $Cu(NO_3)_2$.
5. Ask a student to detect if heat is being released by the reaction by feel-

is the transfer of energy from the sun to Earth. Energy being transferred by electromagnetic waves is often called **radiant energy**.

Many other terms we use for energy are special cases or combinations of potential, kinetic, and radiant energy. Energy can be transformed from one kind to another. For instance, think about the battery-alternator system of a car. As the starter switch is turned on, the chemical energy in the battery is converted to electric energy. The electric energy is converted by the starter to mechanical energy used to start the car. When the car starts, chemical energy in the gasoline is converted into mechanical energy of the moving car. As the crankshaft gains speed, some of this mechanical energy is transferred by belt and pulley to the alternator. In the alternator, the mechanical energy is converted to electric energy. This electric energy is transferred to the battery, where it is converted to chemical energy as it recharges the battery. During this time, other energy transformations produce sound and change the temperature of engine parts.

MINUTE LAB

Risk vs. Benefit of Inertia

Place a small stuffed animal in a small wagon or cart. Place a board across a table to act as a barrier. Make a prediction of what will happen to the stuffed animal if you push the cart toward the board. Then, test your idea. What did you observe? Next, use a large rubber band as a seat belt for your stuffed animal. Again, make a prediction and then test it. Did the restraint make a difference? How would you define inertia?

The Rules

If you toss a burning match into a pile of crumpled paper, the paper will ignite. Paper is combustible. Chemists investigate the changes that matter undergoes, such as the burning of paper. You will begin studying these changes in Chapter 3. Whenever matter undergoes a change, there is an energy transfer. These changes are summarized by rules, or, as chemists refer to them, laws. Two of these laws are the *law of conservation of mass* and the *law of conservation of energy*. As you will study in Chapter 2, mass is a measure of the amount of matter in an object.

The **law of conservation of mass** states that matter is always conserved. This statement means that the total amount of matter in the universe remains constant. Matter is neither created nor destroyed. It is changed only in form. The **law of conservation of energy** states that energy is always conserved. This statement means that the total amount of energy in the universe remains the same. Energy is neither created nor destroyed. It, too, is changed only in form.

For almost two hundred years, scientists believed these two laws to be true under all circumstances. However, in the early 1900s, Albert Einstein showed that matter can be changed to energy and that energy can be changed to matter. Einstein expressed this relationship in his famous equation

$$E = mc^2$$

In this equation, E is energy, m is mass, and c is the speed of light in a vacuum (a constant).

According to Einstein's equation, mass and energy are equivalent. Thus, we see that the two conservation laws can be combined in just one law. This law is known as the **law of conservation of mass-energy**. It states that the sum of mass and

Figure 1.5 This mushroom-shaped cloud is the result of a test nuclear explosion. The large amount of energy comes from the conversion of a small amount of mass.

MINUTE LAB

Risk vs Benefit of Inertia

Purpose: To show that matter has the property of inertia, which is a resistance to any change in motion.

Materials: small wagon or cart, stuffed toy animal, small board, large rubber band.

Procedure Hints: Provide back support to prevent the animal's falling backward when the cart is accelerated. **CAUTION:** *Rubber band can break; wear goggles.* **Disposal:** F

Results: When the cart stops suddenly, the stuffed toy animal falls forward. The rubber band acts as a seat belt and keeps the animal from falling forward.

Answers To Questions: Student predictions will vary. Inertia is a property of matter that is manifest as a resistance to any change in the motion of matter.

Follow Up Activity: The Minute Lab on potential and kinetic energy in this chapter can be used to extend this lab.

ASSESSMENT

Knowledge: Give students a written definition of inertia on the chalkboard with several words missing. Have students fill in the missing words.

GLENCOE TECHNOLOGY

Videodisc

Chemistry: Concepts and Applications

Disc 1, Side 1, Ch. 3

The Law of Conservation of Mass

CD-ROM

Chemistry: Concepts and Applications

The Law of Conservation of Mass

ing the bottom of the flask. Record the observation and again mass the system. **Disposal:** I

Results: $Cu(OH)_2$ forms as a blue precipitate. The heat of the reaction may cause the blue $Cu(OH)_2$ to decompose into the black CuO. Students will observe that a chemical reaction has occurred, but there has been no change in mass.

Questions: 1. How much does the mass of the system change? *none*

2. How do you know a chemical change occurs? *heat, color change, precipitate forms*

3. State the law of conservation of mass. *Matter is neither created nor destroyed. It is changed only in form.*

Extension: Contrast this ordinary chemical reaction with a nuclear reaction when you discuss $E = mc^2$. Remind students that changes of energy to mass and mass to energy are observable only in nuclear reactions.

PROGRAM RESOURCES

Enrichment: Use the worksheet "The Effects of Petroleum Spills," p.1, to challenge more able students.

MINUTE LAB

Potential and Kinetic Energy

Purpose: to allow the student to observe that energy is conserved when potential energy is converted into kinetic energy

Materials: small toy car, 2-m long wide board, 3 bricks or 3 thick books, coin, meter stick.

Procedure Hints: Measure the distance the coin slides and graph that distance versus the number of bricks used to elevate the ramp. This activity will introduce measurement and serve as a bridge to Chapter 2. **Disposal:** F

Results: At heights of 10, 20 and 30 cm, the distances the coin slid increased.

Answers to Questions: Students will predict that the coin will slide a greater distance as the elevation of the ramp increases. The potential energy of the toy car and coin is directly related to the height of the bricks. As the potential energy increases, the speed and kinetic energies of the toy car and coin at the bottom of the ramp increase. The inertia of the coin is constant.

Follow Up Activity: Ask students how the experimental design could be altered to increase the reliability of the lab. Students may suggest that the barrier be placed at the end of the table so that the coin would fall onto the floor. This design would eliminate the effect of sliding friction on the coin.

✓ ASSESSMENT

Oral: Ask students to describe how lifting books from the floor to a desk then letting them fall to the floor can be used to illustrate kinetic and potential energy.

PROGRAM RESOURCES

Applying Scientific Methods in Chemistry: Use the worksheet, "Using the Rules of Chemistry," p.15, to help students apply and extend their lab skills.

Figure 1.6 Much of the plastic used to make consumer products, such as this telephone, is made from petroleum like that in the watch glass. The plastic pellets are an intermediate in the manufacturing process.

MINUTE LAB

Potential and Kinetic Energy

Make a ramp by raising one end of a long, wide board with a brick. Place a coin on top of a toy car. Roll the car down the hill. Allow the car to strike a barrier a short distance from the bottom and observe what happens to the coin. Increase the elevation of the ramp by using two and then three bricks. Make predictions before rolling the car with the coin down steeper hills. How do the bricks relate to potential energy? What is the observed relationship between potential and kinetic energy? Did the inertia of the coin change when it was thrown a greater distance?

energy is conserved. Mass and energy can be changed from one to the other, but their sum remains constant; it cannot be increased or decreased. Changes of energy to mass and mass to energy are observable only in nuclear reactions. Because chemical reactions do not involve nuclear changes, we may assume the original laws are correct.

The Industry

Many products of the chemical industry are for sale in drugstores (aspirin), hardware stores (vinyl siding), and other kinds of stores. These products are called consumer products. When chemists and chemical engineers begin developing a process for producing a new consumer product, they must decide where to start the process. For example, nylon is made from adipic acid and hexamethylenediamine. These two materials are called intermediates. **Intermediates** are not consumer products, but neither are they raw materials. Raw materials, such as coal, salt, air, petroleum, and metal ores, are found in nature. Intermedi-

Figure 1.7 The sulfur depot shown here is part of a sulfuric acid plant that produces large quantities of this acid. Sulfuric acid, H_2SO_4, is used in the preparation of many other chemicals.

DEMONSTRATION

1-4 Sulfuric Acid and You

Purpose: to introduce students to sulfuric acid and the need for eye protection in the laboratory

Materials: 35 g of sucrose, granulated sugar, 150-cm^3 beaker, a 30 x 30-cm piece of cardboard, 35 cm^3 of 18*M* sulfuric acid

Procedure: 1. Place 35 g of sucrose in the 150-cm^3 beaker. Place the beaker on the cardboard.
2. Add 35 cm^3 of 18*M* sulfuric acid to the beaker. Stir briefly and wait. **CAUTION:** *H_2SO_4 is corrosive; dilutions and reactions are very exothermic.*
3. Record observations. **Disposal:** A

Results: A column of black solid will rise from the beaker. Support it using a stirring rod. The smell of burned sugar will be evident. Do not allow students to touch the solid, as it is very acidic. Use caution in handling the beaker as it is very hot.

ates such as adipic acid and hexamethylenediamine are made from raw materials. If you wanted to build a plant to produce nylon (or most other consumer products), you would have to make a decision about whether to start with raw materials or to buy intermediates from another company. When you watch commercials on television, you should realize that there is a huge chemical industry that provides you with many of the comforts of modern life. Yet, much of the chemical industry consists of large companies that the general public never hears about. These companies produce intermediates for other companies.

The substance sulfuric acid is one of the major industrial chemicals of commerce. It is used as an intermediate in hundreds of other industries. In an industrial economy, sulfuric acid is so important that its annual production can be used to estimate a country's extent of industrialization. The fifty most important chemicals produced by the United States chemical industry, ranked by tonnage produced, are listed in Table 1.1.

Table 1.1

The Top 50 Chemicals	
1. Sulfuric acid	26. Ethylene oxide
2. Nitrogen	27. Hydrochloric acid
3. Oxygen	28. Toluene
4. Ethylene	29. *p*-xylene
5. Lime	30. Cumene
6. Ammonia	31. Ammonium sulfate
7. Phosphoric acid	32. Ethylene glycol
8. Sodium hydroxide	33. Acetic acid
9. Propylene	34. Phenol
10. Chlorine	35. Propylene oxide
11. Sodium carbonate	36. Butadiene
12. Methyl *tert*-butyl ether	37. Carbon black
13. Ethylene dichloride	38. Isobutylene
14. Nitric acid	39. Potash
15. Ammonium nitrate	40. Acrylonitrile
16. Benzene	41. Vinyl acetate
17. Urea	42. Titanium dioxide
18. Vinyl chloride	43. Acetone
19. Ethylbenzene	44. Butyraldehyde
20. Styrene	45. Aluminum sulfate
21. Methanol	46. Sodium silicate
22. Carbon dioxide	47. Cyclohexane
23. Xylene	48. Adipic acid
24. Formaldehyde	49. Nitrobenzene
25. Terephthalic acid	50. Bisphenol A

CAREERS IN CHEMISTRY

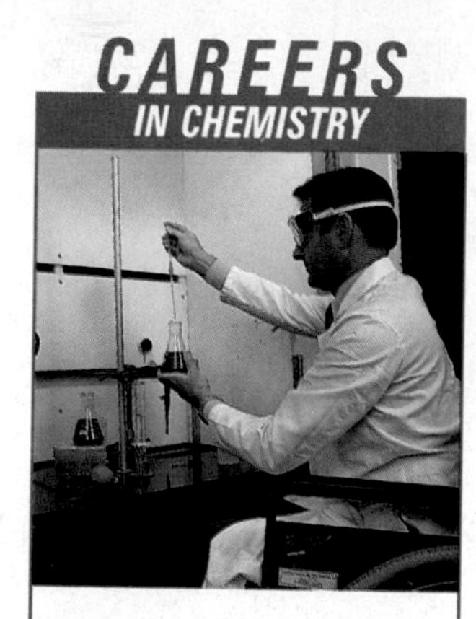

Chemist

The "nitro" fuel used in race cars, the bubble-forming agents in bubble gum and soap, the artificial sweetener in your diet cola, the aspirin you take for a headache, the fibers in your nylon jacket—all of these materials were developed by chemists. Chemists work in an enormous variety of jobs. They work in industry, government, universities, and private consulting firms.

Chemists usually specialize in a particular branch of chemistry, such as analytical, organic, or nuclear chemistry. Within a particular branch of chemistry and a particular institution, a chemist may supervise production of a product, provide marketing expertise, or carry out basic research. Throughout his or her career, a chemist may work in several different settings performing a diverse set of functions. There's one thing a chemist need never be—bored!

Questions: 1. What color change is observed? *white to black*
2. What evidence do you have that a gas is released? *"white smoke" and smell of burned sugar*
3. What element is responsible for the black color? *carbon*
4. What do you think makes up the white smoke? *water vapor and SO_2*
5. Which elements are present in sugar? *carbon, hydrogen, and oxygen*
6. What is meant when sulfuric acid is described as a dehydrating agent? *removes water*
7. How would sulfuric acid damage your skin? *by removing the water from the cells; the exothermic reaction would cause burns*

Extension: Remind students that most of their bodies are made of compounds that, like sugar, contain carbon, hydrogen, and oxygen. Stress that students must handle chemicals carefully and only when wearing proper safety equipment.

CAREERS IN CHEMISTRY
Chemist

Background: The American Chemical Society has established minimum training that a future chemist receives at a certified college or university. Courses are taken in different areas of chemistry such as inorganic, organic, analytical, physical, instrumental analysis, qualitative and quantitative analysis, and biochemistry. Mathematics includes training in calculus.

For More Information: You can write the American Chemical Society for more career information about chemists. Their address is: 1155 Sixteenth Street N.W., Washington, D.C. 20036

MAKING CONNECTIONS

Chemical Industry
One of the many products available to the consumer from the chemical industry is fiberglass. It can be used as insulation or as a fiber in fireproof cloth. The space program also uses fiberglass to make fireproof space suits. Ask your students how fiberglass is made. Have a beaker of sand, SiO_2, available to show them as a hint. Use this discussion to introduce the vocabulary: raw materials (sand), intermediates (glass fibers), and consumer products (rolls of insulation and fireproof draperies).

CONCEPT DEVELOPMENT

Enrichment: Remind students that more sulfuric acid is produced in developed countries than any other chemical. A student may want to prepare a report for the class on why so much sulfuric acid is needed. *It is used as an intermediate in hundreds of production processes.*

Using The Table: Table 1.1 can be used as a source for extra credit report topics. Students can research a chemical from the table and report to the class how it is made and used.

Computer Demo: *Sulfuric Acid,* Simulation of sulfuric acid plant operation. AP801, MC801, Project SERAPHIM

CHEMISTRY AND SOCIETY

Essential Materials

Background: Our standard of living demands products that often use rare metals that have special properties necessary for special application. For example, an automobile's catalytic converter, which is responsible for reducing air emissions, uses platinum, an imported metal.

Teaching This Feature: Remind students that several years ago all automobile bumpers were chrome plated. Chromium, an imported metal, is becoming rare and expensive. Today's car bumpers are steel covered with urethane or vinyl. The consumer is changing his or her demands as supply and cost have their effect.

Answers to
Analyzing the Issue:

1. Answers will vary. Pros may include that stockpiles would assure supply. Cons may include storage costs.

2. Answers will vary but may include development of alternate materials.

3. Answers will vary.

Explore Further: Ask interested students to research a metal selected from those listed and determine how it is used.

✓ ASSESSMENT

Oral: Henri Poincare in *La Science et l'Hypothese* wrote, "Science is built of facts the way a house is built of bricks; but an accumulation of facts is no more a science than a pile of bricks is a house." Ask students for their reaction to the quote. Guide the discussion so students realize there are organizing concepts in science. Encourage them to identify and learn these concepts as the course progresses.

CHEMISTRY AND SOCIETY

Essential Materials

The United States is dependent upon a number of countries as sources for materials used in advanced technology. The table on the right lists metals for which the United States must import more than 50% of its needs. Most of the metals in the list are probably already familiar to you. The names of others, such as germanium, indium, and tantalum, you may be seeing for the first time.

Germanium is recovered from the wastes of refining zinc ores. It also occurs to a slight extent in some coal and can be recovered from the exhaust gases in the combustion of the coal. By far, the majority of germanium is used to make semiconductor devices such as transistors, computer chips, and other solid state devices. Germanium is found in small quantities in Germany and Bolivia. In the United States, Missouri, Kansas, and Oklahoma have minute quantities of germanium available.

Indium is a very soft metal found in the ores of other metals, particularly zinc. The amount of indium in these ores is usually less than 0.001%. Very little indium is found in the United States. The major producers are Canada, Peru, Japan, and Europe, including Russia. The major use for indium is in coating bearings to give them tarnish resistance. The metal is also used in producing semiconductor devices, solder, and nuclear reactor control rods.

Tantalum is not found in useable amounts in the United States. It is imported from Canada, Thailand, Malaysia, and Brazil. Tantalum is extremely resistant to corrosion. As a consequence, it is used to make chemical-handling equipment, electronic components, dental and surgical instruments, tools, and body implants.

Many of the metals shown in the table come from Russia and the southern half of Africa. When you consider the political problems in these parts of the world, you realize that our supply could be drastically reduced without any notice. Consequently, there are industrial and government leaders who advocate stockpiling these necessary materials. Stockpiling, however, would require the investment of considerable sums of money by industry, government, or both, without any immediate likelihood of return on the investment. There are also tax laws that make it uneconomical to stockpile materials. The solution to this problem is one our elected representatives must address in the near future.

Analyzing the Issue

1. Do you think the United States should stockpile the materials it needs? Investigate the pros and cons of stockpiling critical materials and prepare a report for the class explaining your view.
2. Can you think of another solution to the problem of possible shortages of essential materials?
3. Use reference materials from the library to find out about any elements in the table with which you are not familiar. Write a short report about these elements.

Imported Metals	
Aluminum	Nickel
Beryllium	Platinum
Chromium	Silver
Cobalt	Tantalum
Germanium	Tin
Gold	Titanium
Indium	Tungsten
Manganese	Zinc
Mercury	Zirconium

CONTENT BACKGROUND

The Cost Of Production: In 1993, a farmer who sows 100 acres of corn will have a combined seed and chemical bill in excess of $9300. The agricultural chemical industry is big business. Remind students of the commercials they have seen on TV for agricultural chemicals. These costs do not include labor, land rent, fuel, maintenance, or loan payments for the machinery.

Economy of Scale: The chemical enterprise is much more pervasive than most citizens realize. Taken at its broadest, it accounts for more than 10% of our gross national product. The chemical industry alone employs more than one million people and is one of the few segments of our economy with a positive trade balance. Our chemical exports presently exceed our chemical imports by 17 billion dollars per year.

Figure 1.8 Penicillin inhibits the growth of staphylococcus bacteria, a fact discovered by Alexander Fleming in 1928. The white paper disk in the center of this petri dish has been soaked in penicillin. No bacteria colonies were able to grow in the circular region around the disk.

The Process

In 1928, Alexander Fleming returned to his laboratory at the University of London after being on vacation. He found a peculiar pattern in a dish in which he had been growing staphylococcus bacteria. The dish had become contaminated with a blue mold. He observed that the staphylococcus had failed to grow all around the mold. He then hypothesized that the mold was secreting a material deadly to staphylococcus. Fleming designed an experiment to test his hypothesis and found that he had been correct. The secreted material is what we call penicillin. Observation, hypothesis, and experiment are problem-solving processes used by all types of scientists. There are many other scientific procedures, such as classification, identification, and verification. You will come across these techniques and others as you proceed in your study of chemistry.

Sometimes accident plays a part in scientific discovery. The growth of mold in Fleming's petri dish is an example. However, there is no substitute for hard work. Even accidental discoveries must be recognized, analyzed, and expanded upon. As Sherwin B. Nuland says in his book *Doctors: The Biography of Medicine*, "... science is awash with serendipity; science is hard work when done properly, but in the hard work there is joy and in the discovery there is abundant reward. . . ."

Scientists often develop **models** in their minds to help deal with abstract ideas and objects. When you play a video game, in your mind you imagine you are in a real situation. You build a mental model of the real situation. Chemists often build mental models. In the early part of this century, the British scientist J.J. Thomson developed an idea of the atom. He pictured the atom as a big ball of positive charge with little electrons embedded in it. J.J. Thomson's model is often called the "plum pudding" model. Plum pudding is a ball of sweet bread with pieces of fruit embedded in it. Thomson compared the ball of positive charge to pudding and the electrons to fruits. In the United States, Thomson's model could be called the "chocolate chip ice cream" model.

MINUTE LAB

Observe and Hypothesize

Add whole milk to a petri dish or flat-bottomed bowl to a height of 0.5 cm. Place one drop each of four different food colorings in the milk in four different locations. Don't put a drop of coloring in the center. Dip the end of a toothpick in liquid dishwashing detergent. Touch the tip of the toothpick to the milk at the center of the dish. Then, hold the toothpick just off the bottom of the dish, but still in the milk, and observe what happens. Based on your observations and knowledge that milk contains fats, what do you hypothesize about the effect the detergent had on the fats? Can you think of an experiment to test your hypothesis?

In addition to the chemical industry, the chemical enterprise includes *all* who use chemistry in their daily occupations. Think of all the people who are trained in chemistry, but do not work directly in the chemical industry. Such people include health inspectors, engineers, science technical writers, pharmacists, physicians, and teachers. Refer to the *Careers in Chemistry* features included in this text.

The list of occupations requiring some knowledge of chemistry is growing all the time. When you look at the extent to which our lifestyles depend upon the chemical enterprise, you can understand why chemistry is called the central science.

MINUTE LAB

Observe and Hypothesize

Purpose: to develop a hypothesis based on careful observation

Materials: One plastic petri dish bottom, 15 cm^3 whole milk, drop each of four different food colorings, toothpick, drop liquid dish-washing detergent.

Procedure Hints: Have students place a drop of food coloring in each quadrant of the petri dish. **CAUTION:** *Wear aprons and goggles.* **Disposal:** A

Results: When the toothpick touches the milk, the surface tension will be temporarily destroyed by the detergent. The colors move to the outside of the dish. The fat (cream) in the milk will be emulsified by the detergent. Convection currents are established causing the colors to move from the outside toward the center.

Answers To Questions: Student answers will vary. Example: the cream in the milk uses up the detergent. An experiment would be to use a second toothpick that has no detergent on it. Another possibility is to renew the detergent toothpick and observe the increased activity.

Follow Up Activity: Have students hold the toothpick firmly on the bottom of the petri dish and observe the change in behavior of the milk. There is no convection flow unless the toothpick is held in the milk, but not touching the bottom of the dish.

✓ ASSESSMENT

Knowledge: Ask students to research the composition of the mixture called milk and record on a 3 × 5 card the names of the sugar (lactose), fat (butterfat), and protein (casein) found in milk.

Chemistry Journal

A Chemical Dictionary

Point out to students that they will encounter many new terms in their study of chemistry. Have them keep a running list of new terms and definitions in their journals.

CONCEPT DEVELOPMENT

Background: The authors believe that you and your students will enjoy reading the applications labeled The Practice, The Reality, and The Future. We do not recommend that students be tested on information contained in these special applications sections.

Discussion Question: Should cost considerations be included in the decision making process concerning chemical verses biological alternatives for killing insects? When considering alternate ways to kill insects, the resource supply, economics, and values all play an important part. When people consider the risks and the benefits, they view them from their personal perspectives rather than from society's perspective as a whole.

Activity: Have students bring in articles and advertisements from newspapers and magazines concerning chemicals that are used and released into the environment. Have students prepare a collage of the articles on a bulletin board.

CHEMISTS AND THEIR WORK
Bruce N. Ames

Background: The lethal dose for 50% of the population, LD_{50}, is usually based on tests of animals that are not closely related to human beings. The instrumentation has improved during recent years so measurements can be taken that detect parts per trillion. For example, modern analytical equipment can detect one grain of salt in a swimming pool.

Figure 1.9 A variety of chemical pesticides are used to control plant damage caused by insect populations (right). Pesticides can be applied to large areas by cropdusting planes (left).

CHEMISTS AND THEIR WORK

Bruce N. Ames (1928-)
Some pesticides have been classified by government agencies as possible carcinogens; that is, they may cause cancer. There are agricultural and industrial groups, however, that dispute the findings of the government agencies. The disagreement comes about in the attempt to predict what a substance will do to humans on the basis of what that substance has done to laboratory test animals. Dr. Bruce N. Ames, of the University of California at Berkeley, has devoted much of his professional career to developing reliable tests that predict the carcinogenic effects of chemicals on humans. He has spent much of his own time trying to bring a voice of reason to the public's fear of chemicals and cancer-causing materials.

The Practice

What did you have for lunch yesterday? Was it enough? Many people, in all countries of the world, did not get enough to eat yesterday. They are in varying stages of starvation. The chief competitor of humans for food is the insect. Insects reduce world-wide food production by 25-35%. Insects are also carriers of a large number of diseases that make us miserable: malaria, typhus, yellow fever, bubonic plague, and African sleeping sickness. The human race has been battling insects for thousands of years.

Many chemicals have been used to kill insects. Such chemicals are called pesticides. The beginning of modern pesticide chemistry was a discovery, in 1939, by the Swiss chemist Paul Müller. He produced a chemical with a very long name that came to be shortened to the initials DDT. The chemical contained many different kinds of atoms, among which were atoms of the element chlorine. The development of DDT was followed within a few years by the discovery, development, and production by chemists and the chemical industry of three other chlorine-containing pesticides: lindane, dieldrin, and chlordane.

At first, these materials appeared to be doing a great job of eliminating unwanted insects. DDT usage was the most widespread, especially against mosquitoes that spread malaria. The disease was virtually eliminated in some areas of the world. Crop production soared in areas of the world formerly plagued by large insect infestations.

But then, the insect populations started to increase again in those same areas where they had been controlled. Even greater application of pesticides did not seem to slow down the growth of the insect population very much. What had happened?

DEMONSTRATION
1-5 Observing Carefully

Purpose: to show students that observations may be prejudiced by preassumptions

Materials: piece of chalk, white teflon-coated, magnetic stir bar, 2 400-cm^3 beakers, 2 ice cubes, 200 cm^3 water, 200 cm^3 ethanol

Procedure: 1. Before class, prepare 2 beakers of "water" by adding 200 cm^3 of H_2O to one beaker and 200 cm^3 ethanol to the other.

2. In front of the class, write on the chalkboard with the piece of white chalk. Then without being observed, switch the chalk for a magnetic stir bar. Ask the students if chalk is magnetic. Then show them that the piece of "chalk" has magnetic properties by hanging it from some iron-containing metal.

3. Have the two beakers of "water"

The Reality

In every population of a species, there will be a few individuals that are resistant to a pesticide. Once the susceptible population is killed off, the resistant organisms will multiply rapidly. These resistant organisms have plenty of food and ample opportunity to reproduce. The usefulness of the pesticide is now limited.

Why not just switch to another pesticide? In some cases, that change is just what was attempted. The cycle then repeated itself, and it was necessary to try a third pesticide. Eventually, there would be no more pesticides to try. Resistant strains of insects were developing faster than chemists could devise new pesticides. Actually, other events interrupted the repetitive process just described.

All of the above chlorine-containing pesticides are persistent. That is, they remain in the environment without being destroyed by rain, sunlight, or any other natural occurrence. Chemists discovered that these persistent chemicals were being washed by rainwater into streams and creeks. These bodies of water gradually carried the pesticides to major rivers, bays, and eventually to the open sea. In the water, microscopic plants and animals called plankton absorbed these chemicals. Many small fish feed on plankton. Big fish eat small fish. Birds eat fish of all sizes. The pesticides are soluble in the fatty tissues of animals. Consequently, many fish and birds built up considerable amounts of pesticides in their bodies.

It appears that many affected bird species laid eggs with thinned shells. As a result, many chicks died before hatching. The levels of pesticides in certain fish were high enough for authorities to question whether people should eat them. It is still not known what the final consequences of this situation will be.

Another harmful effect of pesticides was observed. They killed insects indiscriminately. That is, they killed good insects as well as bad. Is there such a thing as a good insect? There certainly is. Insects pollinate most of our flowering plants. Bees produce honey and beeswax. Some silk is an insect product. Insects help keep the soil aerated. There are many insects that feed on only the insects we call pests. These predators should be encouraged, not killed. The introduction of predators to fields where their natural prey are pests is one method now in use to control insects.

As you can see, there are two sides to the use of pesticides. On the one side is the necessity of increasing crop yields to keep pace with the growth in the world's population. On the other side is the upsetting of the balance of nature.

Figure 1.10 Many insects have become resistant to pesticides. This housefly is being treated with DDT but is unaffected.

The Future

What to do? Science cannot answer that question, but it can provide some facts to be used in coming to a decision. The decision must weigh the benefits against the risks. Even that consideration is not always easy. For instance, the ability of analytical

CONCEPT DEVELOPMENT

MAKING CONNECTIONS

Agriculture

Thirteen percent of Earth's land surface is used to produce food. Natural deserts occupy one-third of Earth's land surface. Climatic changes and human activity have changed an additional 9% of Earth's land surface into desert during the last 75 years.

Computer Demo: *Six Solution Problem,* Simulation with color graphics that allows students to form and test hypotheses about identity of six solutions. AP902, MC901, PC3401, Project SERAPHIM

✓ ASSESSMENT

Portfolio: Have students write a report on DDT. Trace the history of its use from its discovery in 1939 to 1972, when it was banned in the U.S. by the EPA.

PROGRAM RESOURCES

Reteaching: Use the Reteaching worksheet, p.1, to provide students another opportunity to understand the concept of "The Enterprise of Chemistry."

on your desk. **CAUTION:** *This presents a fire risk.* Ask your students to predict if the ice cube will float or sink in the water. Then place an ice cube into the beaker of water.

4. Repeat step 3 using the beaker of ethanol. Let students discuss among themselves the discrepencies between their predictions and observations. **Disposal:** B

Results: The chalk does not have magnetic properties, but the white teflon-coated stir bar sticks to a metal chalk tray. The ice cube floats in water but sinks in ethanol.

Questions: 1. What did you observe? *See Results.*

2. Why does the piece of "chalk" stick to the metal? *It is magnetic.*

3. Why does the ice cube float in water? *Ice is less dense.*

4. Why does the ice cube sink in the second beaker of "water"? *The beaker contains a liquid less dense than ice.*

Follow Up Activity: Introduce a laboratory experiment that requires students to demonstrate the process skills of observing, hypothesizing, classifying, identifying, and verifying. Advise students that skills will improve as they gain experience in the laboratory.

MAKING CONNECTIONS

Biology

A new species of potato is being developed that has a natural resistance to pests such as aphids, leafhoppers, and some types of potato beetles. The potato plant is a cross between domestic potatoes and a wild species from the mountains of Bolivia. Crossbreeding results in plants that are covered with two types of microscopic spiky glands called trichomes. The trichomes secrete chemicals that act like the two parts of epoxy glue. When the two secretions are combined, they trap pests.

GLENCOE TECHNOLOGY

Videodisc

STVS: Plants & Simple Organisms

Disc 4, Side 2

Insecticides from Desert Plants (Ch. 15)

STVS: Ecology

Disc 6, Side 1

Wasp Biological Control (Ch. 21)

Disc 6, Side 2

Controlling Fruit Flies (Ch. 2)

Resistance to Pesticides in Cockroaches (Ch. 4)

*inter*NET CONNECTION

Information on environmental chemistry regulations and applications is available online from the EPA.

World Wide Web
http://www.epa.gov

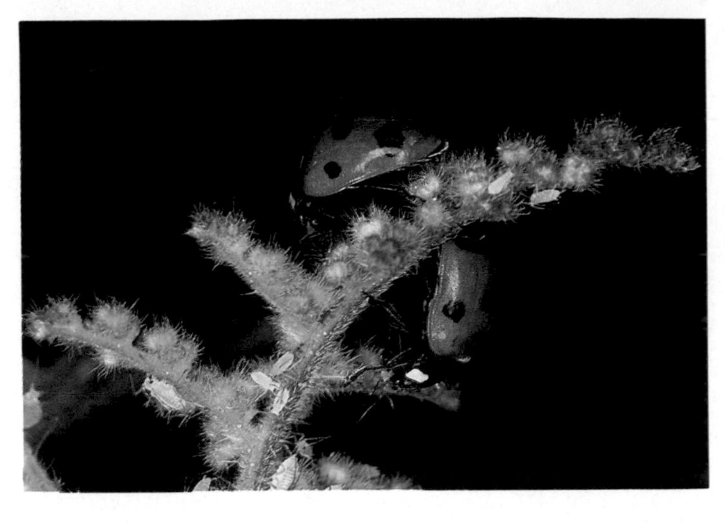

Figure 1.11 Ladybugs are used as a nonchemical means of pest control. Aphids are the ladybug's favorite food, but ladybugs also eat boll weevils, mealybugs, scale insects, corn earworms, and other insects and worms.

chemists to detect the presence of extremely small amounts of chemicals (parts per trillion) exceeds our ability to predict what effect these small quantities will have. There are some scientists who believe any amount of a harmful chemical, no matter how small, is bad. There are other scientists who believe that below a certain amount, called a threshold, a chemical can do no harm. Note that the expression is "believe." Neither viewpoint has been established by scientific methods.

There are a number of other ways to control pests. Some insects undergo one or more changes in body form during their lifetime. This process is called metamorphosis. In these insects, a chemical called a juvenile hormone prevents the insect from maturing until conditions are right. In a process that took ten

Figure 1.12 Scientists are using synthetic juvenile hormone to prevent insects from maturing and reproducing. Eventually, the population size decreases, resulting in less damage to crops. The photo shows a malformed adult American cockroach on the left and a normal adult on the right. The malformed adult, treated with juvenile hormone, cannot reproduce.

DEMONSTRATION

1-6 Recycled Plastic

Purpose: to show how chemistry can help solve the solid waste disposal problem by recycling plastics

Materials: packing box filled with polystyrene foam peanuts, 600-cm^3 beaker, 100 cm^3 acetone, 1-dm^3 beaker

Procedure: 1. Fill the large empty beaker with foam packing peanuts. Ask the class if they think the foam peanuts in the large beaker will fit into the smaller 600-cm^3 beaker.

2. Place 100 cm^3 of acetone into the 600-cm^3 beaker. **CAUTION:** *Acetone is flammable. Use a fume hood.*

3. Have a student slowly add the foam peanuts from the large beaker to the small beaker.

4. Using a fume hood, decant the acetone into a storage container.

years, chemists isolated and found the structure of an insect juvenile hormone. They then learned to make, or synthesize, the material. By manufacturing and using juvenile hormone, scientists can prevent insects from maturing and reproducing. These hormones are not persistent. Some decompose so quickly that they must be encapsulated to be released slowly if they are to be effective. They are fairly specific to particular insects so that they do not harm useful species. In some cases, by using a molecule very similar to the natural hormone, but differing slightly, the result can still be achieved without the disadvantages.

Insects do not have an internal skeleton as we do. Their skeletons are outside their bodies. The only way for insects to grow is for them to shed this skeleton when it becomes too small. This molting process is also triggered by a hormone. Chemists took 15 years to isolate and synthesize the first molting hormone. Its use can disrupt the natural molting cycle of an insect and help control insect populations.

Scientists have observed that certain plants do not seem to be affected by particular insects. They infer that these plants secrete a chemical that repels the insects. Investigation by chemists has confirmed the inference. Botanists are always trying to develop insect-resistant crops through selective breeding. Genetic engineering also offers promise of developing insect-resistant plants.

Other chemicals that have been isolated and studied structurally are sex-attractants. Female insects secrete a chemical that can be detected by males at a great distance. The males then follow the scent to mate with the females. By producing the sex-attractant and placing it in a trap, scientists have captured males and prevented them from mating with the females.

Several insect pests have been controlled by raising millions of males in captivity and then sterilizing them with nuclear radiation. The males are then released at just the right time to mate with females of the species. Being sterile, the males can produce no offspring and the population rapidly declines.

Other organisms can be used to attack insects. Fungi, viruses, bacteria, and nematodes (roundworms) can all be used to infest insect populations and produce diseases that devastate the insect population. Fortunately, most organisms used in this manner are harmless to humans.

Finally, a more recent attack on insects is to use integrated pest management, IPM. This technique makes use of several of the control measures mentioned above. Each application must be evaluated separately because of different crops, different pests, and different environments. However, pesticides are still the most effective agent in the battle against insects.

The story of pesticides is typical of major problems that chemists try to solve. Each time it appeared that the problem was solved, a negative side effect was noted. It is common in science for an investigation or experiment to raise more questions than it answers. It is often difficult to extrapolate results from the laboratory to the world.

Figure 1.13 This bag has been treated with the same sex attractant that female Japanese beetles use to attract mates. The males are lured to the trap, where they are unable to land without sliding into the bag below.

CONCEPT DEVELOPMENT

Computer Demo: *Lake Study,* No chemical knowledge is required to become involved in an investigation to identify a fish-killing pollutant, thus gaining experience using the scientific method. AP804, PC3704, Project SERAPHIM

Computer Demo: *BCTC,* An industrial pollutant simulation that shows how chemistry contributes to the solution of a societal problem. AP805, PC3705, Project SERAPHIM

5. Remove the plastic pellet from the bottom of the beaker. Place it on a paper towel and allow to dry overnight in an operating fume hood. **Disposal:** A

Results: The large volume of foam peanuts will be recycled into a small pellet of polystyrene. The dried pellet is a hard plastic.

Questions: 1. What use could be made of the recycled polystyrene pellet? *Answers will vary.*

2. Estimate the percent volume reduction that occurred. *Student answers will vary. Example: 90%*

3. What effect would this recycling effort have on available solid waste landfill space for the future? *It would extend the life of a landfill.*

Follow Up Activity: Encourage students to begin recycling plastic bottles as well as newspapers and aluminum cans. Inform them of the location of the nearest recycling center.

✓ ASSESSMENT

Performance: Have students weigh the newspapers and other paper that arrive at their residence during one week. Have students calculate the weight of paper for an entire year. Students should conduct a survey to determine the percentage of households that regularly recycle newspapers. Report the results of the survey to the school newspaper.

EVERYDAY CHEMISTRY

Bread Making

Teaching This Feature: You may wish to have a group of students prepare some bread. This is an excellent connection to the Consumer/Family Studies program. You may wish to prepare some dough in advance in a glass container so students can watch it rise in a warm room. You may wish to activate some yeast with flour and water in a closed container fitted with a delivery tube. Place the delivery tube in water, and have students observe the CO_2 bubbles in the water.

Connection to Biology: Cellular respiration produces CO_2. The gas is used to raise the dough. The students exhale CO_2 as a product of their cellular metabolism.

Answers

Exploring Further

1. Baking soda is sodium hydrogen carbonate, $NaHCO_3$. Baking powder is usually a mixture of baking soda and cream of tartar (potassium hydrogen tartrate, $KC_4H_5O_6$).
2. Carbon dioxide bubbles are produced.

3 ASSESS

CHECK FOR UNDERSTANDING

Ask students to write the answers to Concept Review problems and Review and Practice 6–10 from the end of the chapter.

RETEACHING

Activity: Have students examine a liquid crystal thermometer. Ask them to write a hypothesis about how it works based on their observations. Some students may want to design an experiment to test their ideas. They can check their hypotheses by reading about liquid crystals in Chapter 16.

Cooperative Learning: Refer to pages 28T-29T to select a teaching strategy to use with this activity.

EVERYDAY CHEMISTRY

Bread Making

From the earliest times, bread making has been an essential art of civilization. Good bread owes its existence to chemistry and chemical reactions. The major ingredients in bread are yeast, flour, water, and salt. Each is included in the recipe for a reason.

Flour contains starch and protein. Flour and water are mixed with yeast to produce a dough. As the dough is mixed, water and protein in the dough form tangled molecular chains called gluten. When the dough is kneaded, the chains align and the dough becomes smooth. The starch forms a jelly-like material with the water and gives structure to the dough.

Yeast are single-celled organisms that are related to molds. When activated by water, yeast digest starch in the flour and release carbon dioxide and alcohol. In bread making, the carbon dioxide bubbles are trapped in the dough by the gluten. As the yeast produce carbon dioxide, the dough "rises," or leavens.

Salt adds flavor and prevents the gluten from breaking down and leavening the dough too rapidly.

When the bread is baked, the trapped gas bubbles expand and cause the dough to rise even further. During baking, yeast cells are killed and the alcohol evaporates, giving off the tempting aroma of baking bread.

Exploring Further

1. Baking soda and baking powder are common household chemicals also used to leaven baked products. Find these on your shelf at home or at the grocery. List the ingredients from the containers.
2. Put a small amount of baking soda in a flat saucer. Pour a few drops of vinegar on the baking soda. What happens? This chemical reaction is similar to the one that takes place when baking cakes.

The Course

The first part of your work in chemistry will be to learn to communicate with your teacher and with each other by the use of very careful measurement and by understanding how chemists classify matter. You need to master the skills of classification and measurement to proceed with your studies in chemistry.

Second, you will begin your study of structure by examining the atom, the basic construction unit with which chemists work. Third, you will find out about how chemists represent matter and its changes with symbols. Then you will learn about putting atoms together to make molecules and other particles larger than

DEMONSTRATION

1-7 Hard Water

Purpose: to show how chemists use their knowledge of the structure and properties of matter to produce new consumer products

Materials: 250-cm^3 beaker, 0.5 g sodium polyacrylate, 100 cm^3 H_2O

Procedure: **1.** Without being observed by students, place 0.5 g sodium polyacrylate into the 250-cm^3 beaker.
2. As students observe, pour 100 cm^3 H_2O into the beaker. **Disposal:** A

Results: The water will mix with the sodium polyacrylate and gel after a few seconds.

Questions: **1.** What did you observe? *The water becomes a solid gel.*
2. Can you think of any practical uses for such a chemical? *It is used in potting soil to retain water, in paper diapers, in janitorial products that absorb water and odor (vomit).*

Figure 1.14 In chemistry, a student uses scientific problem-solving skills to learn about the structure and behavior of matter.

atoms. You will also learn about how groups of similar particles behave. Finally, you will study mixtures of different particles, both mixtures that do react and those that do not.

As you study chemistry, you will be finding out about the structure and the properties of matter. Once you have learned about the connection between a particular structure and its associated properties, you can make predictions about matter. For example, if you found a material whose atoms were held together in much the same way as those in a diamond, you could predict that material would be hard.

Chemists use this predictive ability (a scientific procedure) to make new materials with certain desirable properties. Knowing that these properties are related to certain structures, the chemist then develops a process for making new materials that have these structures. What is chemistry? **Chemistry** is the study of the structure and properties of matter. Chemistry is what chemists do! They discover the relationships between structure and properties of matter and use these to produce new materials.

CONCEPT REVIEW

1. How is inertia related to matter?
2. Distinguish between potential and kinetic energy.
3. Is the light produced by fluorescent lights potential, kinetic, or radiant energy?
4. Why do chemists need to keep developing new pesticides?
5. **Apply** Why might a chemist need to understand the law of conservation of mass-energy?

EXTENSION

In March, 1989, the tanker *Exxon Valdez* went aground in Alaska. Ten million gallons of crude oil created an oil slick larger than the state of Rhode Island. Ask a team of students to try to find out what led to the disaster and how effective were the cleanup procedures. This oil spill was carefully studied and widely reported.

Answers to
Concept Review

1. Inertia is a property of matter that shows itself as a resistance to any change in motion.
2. The potential energy of an object is based on its position with respect to some reference point; its kinetic energy is based on motion.
3. radiant
4. While a pesticide may kill many individuals of an insect species, there will always be some individuals that are resistant to the pesticide. These individuals multiply rapidly, and a new pesticide must be developed to fight the new population of insects.
5. A chemist must account for the amounts of reactants and products in a chemical reaction and the energy absorbed or released.

4 CLOSE

Discussion: Have two teams of students investigate the advantages and disadvantages respectively of expanding the nuclear power industry. Have the two teams present their reports to the class.

Follow Up Activity: NaCl from a salt shaker can be sprinkled on the gel and it will return to a liquid when stirred. Salt is a gel breaker. Sodium polyacrylate is used in oil wells to form a solid column that, with hydraulic pressure applied, will fracture the rock layers releasing the oil. Hydrochloric acid is used to break the gel and the liquid can then be pumped out of the well casing.

PROGRAM RESOURCES

Vocabulary Review/Concept Review: Use the worksheets, pp. 1, 32 to check students' understanding of the key terms and concepts of Chapter 1.

CHAPTER REVIEW 1

- Review Summary statements and Key Terms with your students.
- Complete solutions to Chapter Review Problems can be found in the Problems and Solutions Manual accompanying this text.

Answers to
Review and Practice

6. Inertia is a property of matter that shows itself as a resistance to any change in motion.
7. Potential energy is the energy of position. Kinetic energy is the energy of motion. Radiant energy is energy transferred by electromagnetic waves.
8. The law of conservation of mass states that matter is always conserved. The law of conservation of energy states that energy is always conserved. The law of conservation of mass-energy states that the sum of mass and energy is always conserved.
9. Chemistry is the study of interrelationships of structure and properties of matter.
10. Chemists develop new materials; chemical engineers produce them.

Answers to
Concept Mastery

11. Matter is anything that has the property of inertia. Energy is a property possessed by all matter which, under the proper circumstances, can be made to do work.
12. Intermediates are made from raw materials and are converted to consumer products.
13. Students' answers may include five of the following: Classification is grouping things on the basis of their similarities. Observation is noting things with the senses. Verification is checking to make sure something is true. Identification is determining the identity of something. Analysis is breaking something down into its constituent parts. Hypothesizing is proposing an explanation. Experimentation is testing a hypothesis.
14. Models are mental concepts analogous to real world systems. Many real world systems are too small, too large, or otherwise awkward to work with directly in the laboratory.
15. potential, kinetic, radiant, mechanical, thermal, nuclear, chemical, electrical

Chapter 1 REVIEW

Summary

1. Chemists investigate the relationship between structure and properties in matter. Chemical engineers design, build, and operate chemical plants after a production process has been developed.
2. Matter is anything with the property of inertia. Inertia is resistance to changes in motion.
3. Energy is the capacity to do work under the proper circumstances.
4. Potential energy is the energy an object has because of its position. Kinetic energy is the energy an object has because of its motion. Radiant energy is energy being transferred as electromagnetic waves.
5. The law of conservation of mass states that matter is always conserved. That is, it cannot be created or destroyed. The law of conservation of energy states that energy is always conserved. That is, energy cannot be created or destroyed. Both matter and energy can change form.
6. Since Einstein showed that mass can be changed to energy and vice-versa, scientists now use the law of conservation of mass-energy, which states that the sum of mass and energy cannot be increased or decreased.
7. The chemical industry converts raw materials into intermediates and consumer products.
8. Scientists tend to follow certain patterns in their work. These patterns are called scientific processes or procedures.
9. Scientists often use the results of scientific processes to build mental models to help deal with abstract ideas and objects such as atoms.
10. Chemistry is the study of how the properties of matter are a result of its structure and the use of this knowledge to make new materials.

Key Terms

matter
inertia
energy
potential energy
kinetic energy
radiant energy
law of conservation of mass
law of conservation of energy
law of conservation of mass-energy
intermediate
model
chemistry

Review and Practice

6. What is inertia?
7. Define potential, kinetic, and radiant energy.
8. What are the conservation laws of mass, energy, and mass-energy?
9. What is chemistry?
10. What is the difference between a chemist and a chemical engineer?

Concept Mastery

11. Differentiate between matter and energy.
12. How do intermediates differ from consumer products and raw materials?
13. List five scientific processes or procedures, and describe each of them.
14. What are models? Why are models important to scientists?
15. Make a list of at least five different forms of energy. Use reference materials in your school library, particularly physics texts, to help you.

Application

16. Using a dictionary, find out what aspects of nature are investigated by each of the following scientists: agronomist, astronomer, biochemist, biologist, botanist, ecologist, entomologist, geochemist, geolo-

gist, geophysicist, horticulturalist, limnologist, metallurgist, meteorologist, physicist, and zoologist.

17. Find out what kinds of careers require a knowledge of chemistry. Make use of the career education materials that your guidance counselor may have, including college catalogs, the *Dictionary of Occupational Titles*, and the *Occupational Outlook Handbook*.

18. Investigate one product from the list of top 50 chemicals, Table 1.1. Find the raw materials and intermediates from which it is made, the manufacturing process, the properties and structure of the product, and its uses.

19. In March 1989, the oil tanker *Exxon Valdez* struck a reef after leaving the port of Valdez, Alaska. A very large oil spill resulted. Describe any chemical processes that were used in the cleanup.

Critical Thinking

20. You are given a sealed box about the size of a shoe box containing an unknown object. Describe some things you might do to decide what is inside the box. Think about tests that might eliminate some classes of objects.

21. If 50 cm^3 of water are added to 50 cm^3 of ethanol, 95 cm^3 of solution result. Can you think of a model that explains this property?

Readings

Eberhart, Jonathan. "The Saying of Science." *Science News* 133, no. 5 (January 30, 1989): 72-73.

Heitz, James R. "Photoactivated Pesticides." *CHEMTECH* 18, no. 8 (August 1988): 484-488.

Lamb, James C. "Regulating Carcinogenic Pesticides." *CHEMTECH* 21, no. 1 (January 1991): 42-49.

Leng, Marguerite. "Reregistration: Its Consequences." *CHEMTECH* 21, no. 7 (July 1991): 408-413.

Levi, Primo. "The Mark of the Chemist." *Discover* 10, no. 2 (February 1989): 70-75.

Marco, Gino J., *et al.* "Silent Spring Revisited." *CHEMTECH* 18, no. 6 (June 1988): 350-353.

McClintock, J. Thomas, *et. al.* "Are Genetically Engineered Pesticides Different?" *CHEMTECH* 21, no. 8 (August 1991): 490-494.

Rhodes, Richard. *The Making of the Atomic Bomb: Part I.* New York: Simon and Schuster, 1986.

Rochow, Eugene G. "Choices." *Journal of Chemical Education* 63, no. 5 (May 1986): 400-405.

Sime, Ruth L. "The Discovery of Protactinium." *Journal of Chemical Education* 63, no. 8 (August 1986): 653-657.

Treptow, Richard S. "Conservation of Mass: Fact or Fiction?" *Journal of Chemical Education* 63, no. 2 (February 1986): 103-105.

Woods, Michael. "Nature Makes its Own Toxins." *CHEMECOLOGY* 20, no. 5 (July/August 1991): 12-13.

Answers to Application

16. agronomist: field crops and soil; astronomer: stars, planets, other bodies of the universe; biochemist: chemical processes in living things; biologist: living things; botanist: plant life; ecologist: living things and their environments; entomologist: insects; geochemist: chemical properties and changes in Earth's crust; geologist: structure and history of Earth's crust; geophysicist: physical properties of Earth's crust; horticulturist: fruits, vegetables, decorative plants: limnologist: properties and conditions of freshwater areas; metallurgist: metals; meteorologist: weather; physicist: energy-matter relationships; zoologist: animal life

17. Student answers will vary.

18. Student answers will vary.

19. Refer to the following articles for detailed information.

Lemonick, Michael D., "The Two Alaskas." *Time,* April 17, 1989, pp. 56-66.

Hackett, George; Hagen, Mary; Drew, Lisa; and Wright, Linda; "Environmental Politics." *Newsweek,* April 17, 1989, pp. 18-19.

Marshall, Eliot; Roberts, Leslie; "Valdez: The Predicted Oil Spill," *Science,* Vol. 244, No. 4900, April 1989, pp. 20-24.

Answers to Critical Thinking

20. Student answers will vary.

21. Molecules fit in the spaces between other molecules.

PROGRAM RESOURCES

Spanish/English Glossary: Provide Spanish-speaking students with the Spanish/English Glossary to assist them with understanding key terms in this text.

Evaluation: Use the Evaluation worksheets, pp.1-4, to assess students' knowledge of Chapter 1.

Computer Test Bank: Use the IBM or Apple Test Bank and Question Selection Manual to prepare your own individualized test for Chapter 1.

CHAPTER 2

Measuring and Calculating

CHAPTER ORGANIZER

Chapter Objectives	Text Features	Laboratory Options	Teacher Classroom Resources
2.1 Decision Making **1.** Compare and apply strategies for solving problems in chemistry. **2.** Realize that success in solving chemistry problems lies in knowledge and practice. **Nat'l Science Stds: UCP.2, UCP.3, A.1, A.2, G.1, G.2**	**Math Connection:** Problem Solver, p. 23 **Chemists and Their Work:** Dorothy Crowfoot Hodgkin, p. 25 **Chemistry and Society:** Chemical Research, p. 27	**Discovery Demo:** 2-1 It Floats, It Sinks, p. 20	**Lesson Plans:** pp. 2, 3 **Study Guide:** p. 3 **Applying Scientific Methods in Chemistry:** Chemistry Skills, pp. 16, 17
2.2 Numerical Problem Solving **3.** List and use the SI base units for mass, length, time, and temperature. **4.** Express and convert quantities using the common SI prefixes. **5.** Use significant digits to express the exactness of measurements. **6.** Perform calculations using density measurements. **Nat'l Science Stds: UCP.3, A.1, A.2, G.2**	**Everyday Chemistry:** Metric Measurement, p. 32 **History Connection:** Nautical Knots, p. 33 **Careers in Chemistry:** Medical Technologist, p. 34 **Bridge to Metallurgy:** Froth Flotation Separation, p. 42	**Minute Lab:** Diet Density, p. 38 **Demonstration:** 2-2 Length Conversions, p. 32 **Demonstration:** 2-3 Real Hot, p. 34 **Demonstration:** 2-4 Measurement, p. 36 **Demonstration:** 2-5 Now You See It, p. 38 **Demonstration:** 2-6 Lighter in Hot Air, p. 40 **Laboratory Manual:** 2-1 Measuring Densities of Pennies **Laboratory Manual:** 2-2 Determining Density and Specific Gravity - microlab **ChemActivity 2-1:** A Density Balance - microlab, p. 797 **ChemActivity 2-2:** A Graphical Determination of Density, p. 799	**Study Guide:** pp. 4-5 **Color Transparency 1 and Master:** SI Base Units and Prefixes **Reteaching:** Percent Error and Significant Digits, pp. 2, 3 **Critical Thinking/Problem Solving:** Graphically Speaking, p.2 **Enrichment:** Precision, Accuracy, and Significant Digits, pp. 3, 4 **ChemActivity Master 2-1** **ChemActivity Master 2-2** **Color Transparency 2 and Master:** Significant Digits
Other Chapter Resources	**Vocabulary and Concept Review** p. 2, 33, 34, **Evaluation,** pp. 5-8	**Videodisc Correlation Booklet** **Lab Partner Software**	**Test Bank**

Block Schedule

For information on block scheduling, see the Lesson Plans booklet in the Teacher Resource Package.

GLENCOE TECHNOLOGY

Mastering Concepts in Chemistry Software
- Review of the Metric System
- Using Significant Digits

Complete Glencoe Technology references are inserted within the appropriate lesson.

ASSIGNMENT GUIDE

Contents	Level	Practice Problems	Chapter Review	Solving Problems in Chemistry
2.1 Decision Making				
General Problem Solving	C		59	
Problem Solving Techniques	C		23, 24, 45, 53	
Summary of Strategies for Problem Solving	C		46	
2.2 Numerical Problem Solving				
The International System (SI)	C	4, 5	30, 55	
Mass and Weight	C		32, 47, 49, 50, 54	
Length	C			
Time	C		30	
Temperature	C		31	
Accuracy and Precision	C	6, 7	36	
Percent Error	C	8, 9		
Significant Digits	C	10	33, 34, 48	
Derived Units	C		35, 38, 51, 56	1
Density	C	11–19	25-29, 37, 39-44, 52, 53, 56-58, 60, 62	2-7
				Chapter Review: 1-3

LABORATORY MATERIALS

Minute Labs	ChemActivities	Demonstrations	
page 38 aquarium balance can of soda, diet can of soda, regular	**page 797** cooking oil ethyl acetate ethanol 2-propanol n-butyl alcohol oleic acid pencil ruler with center channel thin stem plastic pipets water **page 799** BB's balance beaker graduated cylinder, 100-mL water	**pages 20, 32, 34, 36, 38, 40** beaker, 250-cm^3 beaker tongs brown tissue liner from ditto master cellophane tape clear acetate transparency film cork from litmus tube	dropper electrical tape food coloring glycerol hot plate masking tape matches meterstick rubber stopper, size 00 sand or sawdust string water

CHAPTER 2
Measuring and Calculating

CHAPTER OVERVIEW

Main Ideas: Chapter 2 reviews basic mathematical skills and measuring principles that are important in problem solving and laboratory work. An analytical approach to problem solving is described. International metric system (SI) units are introduced. Density is presented as a real world application of measurement and calculation skills.

Theme Development: The theme of this chapter is scale and structure. The SI system is appropriate to describe matter on all levels from the submicroscopic scale of atoms and molecules to the large scale structure of stars and galaxies.

Tying To Previous Knowledge: Ask students if they know how tall they are or how much they weigh. Remind them that these and other measurements are important in their lives.

ASSESSMENT PLANNER

Choose from the following assessment strategies.

Assess, pp. 26, 41-42
Assessment, pp. 21, 25, 31, 33, 38
Portfolio, p. 25
Minute Lab, p. 38
ChemActivities, pp. 797-800
Chapter Review, pp. 43-45

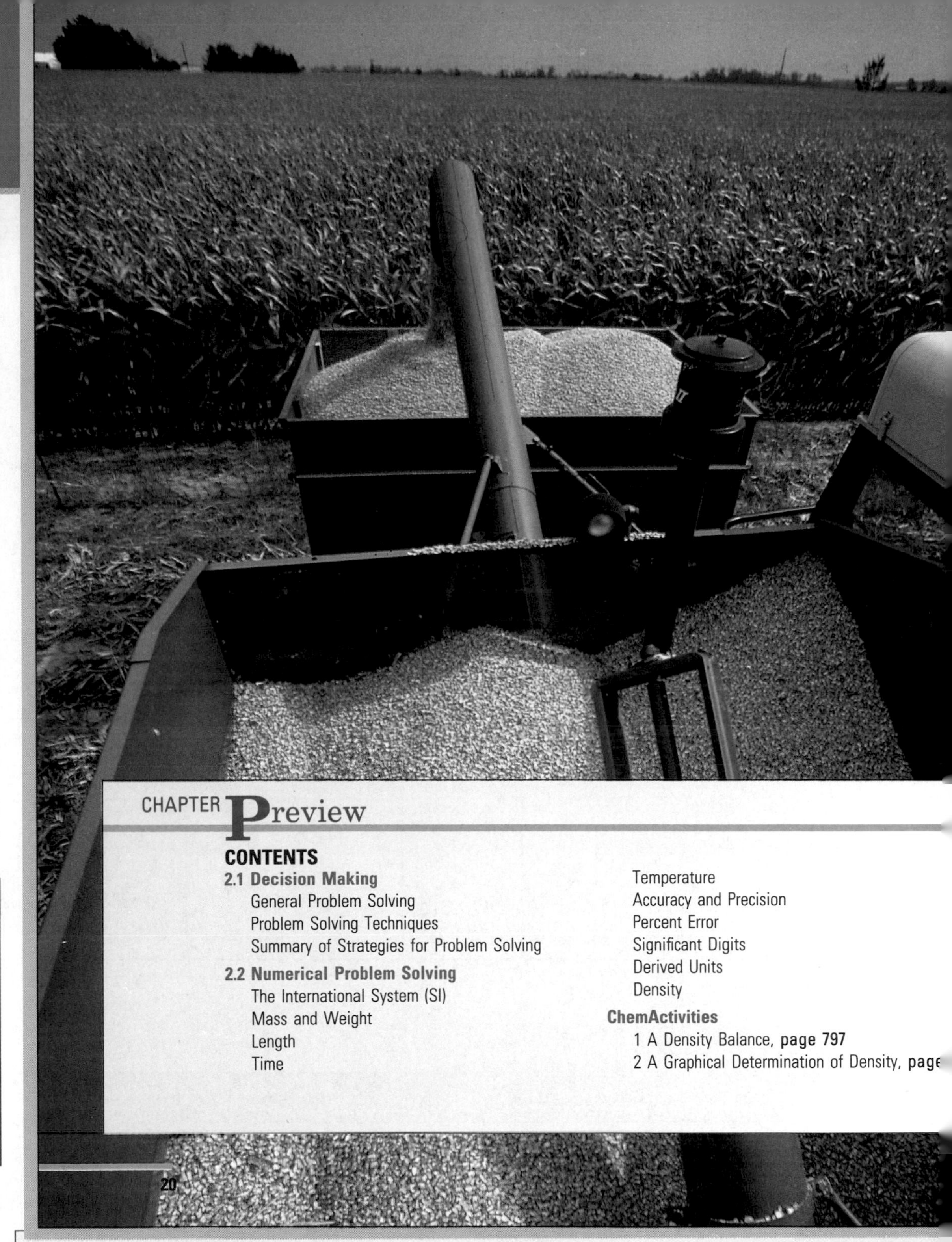

CHAPTER **P**review

CONTENTS

20

DISCOVERY DEMO

2-1 It Floats, It Sinks

Purpose: to demonstrate to students that they must use their mathematical skills in chemistry if they are to understand and explain what they observe in the laboratory

Materials: 100 cm^3 glycerol (glycerine), 250-cm^3 beaker, size-00 rubber stopper, small cork from a litmus paper tube, 100 cm^3 of water, drop of food coloring

Procedure: 1. Place 100 cm^3 of glycerol in the beaker.
2. Place the small rubber stopper, and the small cork into the glycerol.
3. Slowly add 100 cm^3 of water containing food coloring to the beaker.
Disposal: B

Results: Both the cork and the rubber stopper float on the glycerol. The cork floats in the water.

Questions: 1. Explain your observations

Measuring and Calculating

CHAPTER 2

The farmer who planted this cornfield had to solve problems to know how much seed corn and fertilizer to order and how little expensive weed killer and insecticide to buy. To make good decisions, farmers and chemists alike must use measurements and solve problems.

Why do people continue to use hazardous pesticides? The food exporting nations want to increase production in order to have more food to sell. Many developing countries must increase production in order to reduce the amount of food they import. Consequently, most nations use pesticides in agriculture. How to use pesticides efficiently and safely is a problem that must be faced by almost all countries. In this chapter, you will learn how to approach problems in a general way and how to approach chemistry problems in a more specific way.

EXTENSIONS AND CONNECTIONS

Math Connection
Problem Solver, page 23

Chemists and Their Work
Dorothy Crowfoot Hodgkin, page 25

Chemistry and Society
Chemical Research: A look at some areas in which chemistry affects society, **page 27**

Everyday Chemistry
Metric Measurement: How the slow switch to metric affects industries and consumers, **page 32**

History Connection
Nautical Knots, page 33

Careers in Chemistry
Medical Technologist, page 34

Bridge to Metallurgy
Froth Flotation Separation: How density differences help purify metal ores, **page 42**

INTRODUCING THE CHAPTER

USING THE PHOTO

Ask students how a reasonable estimate of the number of stalks of corn that are shown in the chapter opening photograph could be made. Some students will suggest counting the rows and multiplying by the number of stalks in an average row. Follow this response by asking how they would estimate the number of ears of corn. Some students will suggest multiplying the number of stalks by the number of ears of corn on each stalk. (Two ears on a stalk is a reasonable assumption.) Ask them how they would estimate the number of kernels of corn. Students will realize that this can be done by multiplying their last calculation by the number of kernels on an ear (about 200). Students will be encouraged by the fact that the math skills learned over many years will be used to solve practical, relevant problems in chemistry.

REVEALING MISCONCEPTIONS

Students should be encouraged to use programmable calculators and computers. Many students believe that the number displayed by a calculator is the correct answer. Emphasize to students that they need first to estimate the magnitude of an answer and then check the calculated answer against the estimate. They must also decide how many significant digits the calculated answer should include.

using a hypothesis. *Student answers will vary.*

2. Design an experiment that would support the hypothesis. *If students have had physical science, they realize that to calculate density ($D = m/V$) they need to measure the volume using a graduated cylinder and the mass using a balance. Use the demonstration to discuss the need for a standardized system of measurement.*

✓ ASSESSMENT

Performance: A student can make a more complicated density column. In a graduated cylinder place the following materials, in order: corn syrup (1.38 g/cm^3), rubber stopper (1.34 g/cm^3), glycerol (1.26 g/cm^3), polystyrene (1.17 g/cm^3), water (1.00 g/cm^3), corn oil (0.99 g/cm^3), and wood (0.71-0.95 g/cm^3). The student should label, exhibit, and explain the column.

2.1 Decision Making

PREPARE

Key Concepts

The major concepts presented in this section include a general approach to problem solving, general strategies that can be employed when solving chemistry problems, and a specific seven-step approach to solving problems in chemistry.

Key Terms

No new science terms will be introduced in this section.

1 FOCUS

Objectives

Preview with your students the objectives listed on the student page. Students can use them as a study guide for this major section.

Discussion

Successful problem solving requires a creative plan or strategy. Ask students to give examples of how they solved particularly difficult problems creatively. Use this discussion to emphasize that there must be a plan to solve any problem.

OBJECTIVES

- compare and apply strategies for solving problems in chemistry.
- realize that success in solving chemistry problems lies in knowledge and practice.

2.1 Decision Making

Solving problems is more than just following set procedures established by someone else. There will be times in chemistry when a set procedure is useful. However, there is no substitute for thinking through a problem on your own.

General Problem Solving

The kind of problem that you encounter most often in science—especially in chemistry—is the kind in which you are presented with a situation, various pieces of information, and are asked to apply this information to develop a solution to find the needed answer. Thus, you must have learned enough chemistry to understand the situation presented and to appreciate the importance (or unimportance) of each piece of information supplied. If you have done that, the only remaining task is to find a way to organize the information so that you can perform a calculation to give you an answer that is correct not only numerically, but also is expressed in the appropriate units.

How can you learn to solve problems in this way? There is no one answer to that question. However, there are techniques that you can apply and questions you can ask to help you solve problems in chemistry.

Problem Solving Techniques

First, let's apply these techniques to a nonscience problem.

Checcolo's in the mall has a sweater that you like very much. The price of the sweater is $75.00 including sales tax. You have been working two days a week for 5 months

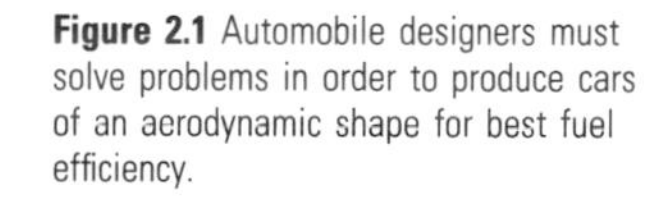

Figure 2.1 Automobile designers must solve problems in order to produce cars of an aerodynamic shape for best fuel efficiency.

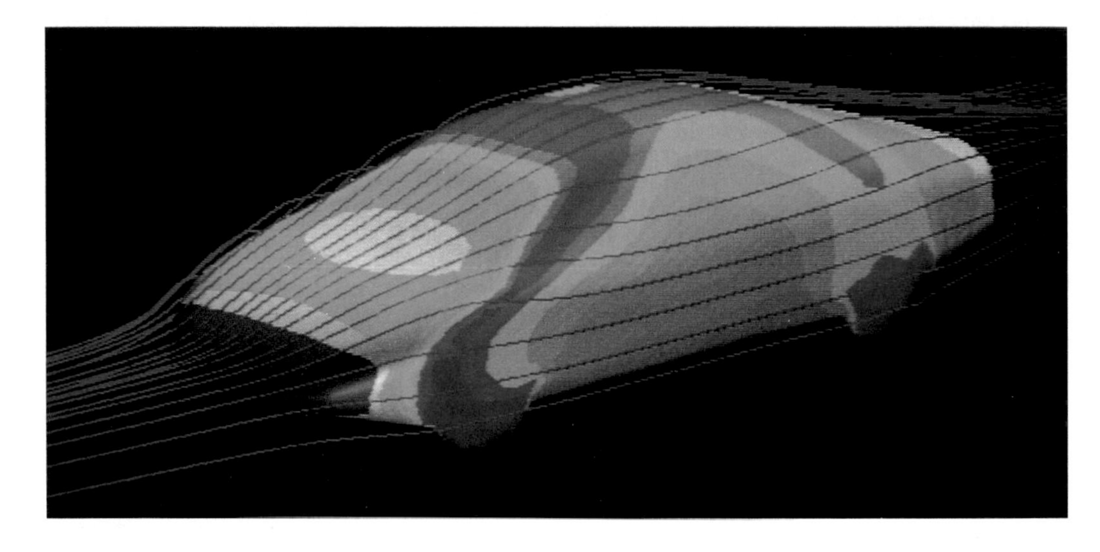

PROGRAM RESOURCES

Study Guide: Use the Study Guide worksheet, p.3, for independent practice with concepts in Section 2.1.

Lesson Plans: Use the Lesson Plan, pp. 2-3, to help prepare for Chapter 2.

Chemistry Journal

Rationalizing Beliefs

Ask students to make a list in their journals of things they have not personally seen but that they believe to exist because of indirect evidence. Have students write an analogy of their belief in each item to the belief of scientists in the existence of atoms.

Figure 2.2 How will this student determine the number of hours of work it will take to earn enough money to buy the sweater?

at the White Chateau Steak House. You are hoping for a raise to $6.75 an hour after 6 months, but you are now being paid $6.00 an hour. It takes 20 minutes to travel from your house to the restaurant. How many hours will you have to work in order to buy the sweater?

Technique 1. *Identify the known facts.* (Where am I?)
In this problem, you have the following pieces of information.

a. Price of sweater = $75.00
b. Time employed = 5 months
c. Expected raise = $6.75 per hour
d. Present wage = $6.00 per hour
e. Time from home to work = 20 minutes

Technique 2. *Define the answer required.* (Where do I want to be?)
In this problem, the answer required is hours of work.

Techniques 1 and 2 are based on careful and thorough reading of the problem. The greatest mistake you can make is to start punching numbers into a calculator without first organizing and examining the information in the problem. Read rapidly through each new problem statement in order to get the sense of the problem. Make sure you understand all the words used in the problem. If unfamiliar words, symbols, or units appear, look them up in a dictionary, glossary, or textbook. Next, read the problem again but more carefully in order to determine what facts are given. Finally, read it a third time in order to determine what sort of answer is required. Once you are confident that you fully understand the problem, you may find it helpful to write down a table showing the facts given and the answer required. It is also worthwhile to restate the problem in your own words.

MATH CONNECTION

Problem Solver
George Polya (1887–1985), a Hungarian-born American mathematician, devised a four-step plan for problem solving. His steps were (1) understand the problem, (2) devise a plan, (3) carry out the plan, and (4) look back. He wanted students to examine a problem carefully to determine what information was given and what needed to be found, to devise and carry out strategies for solving the problem, and then to check the results in terms of the problem. Polya's method is similar to the one presented in this section.

2 TEACH

CONCEPT DEVELOPMENT

Background: Before students begin to apply a strategy for solving chemistry problems, they must collect laboratory data. Give them instruction and then practice measuring volume using a graduated cylinder, pipet, and buret filled with colored water. Inform them that the volume of the liquid is determined by reading the bottom of the meniscus. Students will also need practice measuring with a balance and a metric ruler. This practice is best done before performing a laboratory experiment.

MATH CONNECTION
Problem Solver

Exploring Further: Students can read more about problem solving and discovery in George Polya's books: *How to Solve It, Mathematics and Plausible Reasoning,* and *Mathematical Discovery.*

Computer Demo: *Vernier,* A simulation that provides student practice reading the scales found on many scientific instruments, AP105, Project SERAPHIM

Computer Demo: *Mettler Balance,* A simulation that provides practice using a Mettler type H analytical balance, AP110, Project SERAPHIM

CONTENT BACKGROUND

Problem Solving: Chemistry problem solving is quantitative. This often contributes to the abstractness and perceived difficulty of the subject. Beginning the school year with a heavy mathematical emphasis does little to calm the fears of students. However, a great many chemistry concepts can be taught using only basic math skills. The authors have delayed the formal introduction of the factor-label method and scientific notation as problem-solving strategies until Chapter 8. As the laboratory program develops, students will better understand the relevance and need for a problem solving strategy.

Technique 3. *Develop possible solutions.* (What paths might take me where I want to be?)

In order to develop solutions, you need to use all relevant information. Does a fact have a bearing on the solution you are seeking? If not, it is irrelevant and may be discarded. For instance, in the present problem, the raise you hope to get, the length of time you have worked, and distance from your house have no effect on the number of hours you must work in order to buy the sweater. Therefore, these facts may be discarded.

In this problem, there are two relevant pieces of information.

a. Price of sweater = $75.00

b. Present wage = $6.00/hour

Therefore, there appear to be three possible solutions.

a. Multiply the sweater price by the wage rate.

b. Divide the sweater price by the wage rate.

c. Divide the wage rate by the sweater price.

Technique 4. *Analyze these solutions and determine the correct one.* (Which path is most likely to be the correct one?)

Now, you must consider the consequences of each solution path and choose one that produces the type of answer you're looking for. In most problems you should look at the units in which your answer will be expressed.

In the current problem, you want to get an answer that is expressed in hours (of work). Which of the three possible solutions will give you an answer in hours? If you carried out the first solution, you would be multiplying (dollars) $\times$ (dollars/hour) resulting in $(\text{dollars})^2/\text{hour}$, an absurd unit. Further examination will reveal that the second solution gives an answer in hours.

Figure 2.3 The problem this student must solve is to decide which solution and in what concentration she should add to zinc in order to produce hydrogen gas. Making this decision requires not only problem-solving skills, but also a knowledge of chemistry.

CULTURAL DIVERSITY

Weights of Wisdom

The standards for weights and measurements that different cultures have developed may combine science and art and reflect a group's way of life. Ashanti brass weights are an example. The Ashanti Kingdom, founded in the early 17th century, was part of Africa's Gold Coast. The Ashanti traded with Europeans on the coast and Muslims to the north, exchanging gold for other goods.

The Ashanti used gold dust as currency. This required the use of standard weights to measure. For this, the Ashanti created tiny brass figurines. The figures showed a great deal of skill and imagination. Many were in the form of animals or fish, and most used religious or socially significant symbolism. One showed two crocodiles with a common body. This represented the proverb "we have different mouths, but the same stomach"—an Ashanti proverb meaning that although individuals might have different desires and goals, all shared in a common concern as members of the Ashanti culture.

Today these brass weights are no longer used as standards of weight, but are collected as pieces of art.

PROGRAM RESOURCES

Applying Scientific Methods in Chemistry: Use the worksheet, "Chemistry Skills," pp.16, 17 to help students apply and extend their lab skills.

interNET CONNECTION

Students can access information on upcoming space shuttle mission payloads and archived images.

World Wide Web
http://shuttle.nasa.gov

CONTENT BACKGROUND

Industry: The following material indicates how planned research can be used to solve problems in industry. In the late 1920s Charles Kettering was director of the Delco Dayton Research Lab, a subsidiary of GM. He asked Tom Midgely, a Cornell graduate, and Alan Henning from Belgium to find a substitute for the toxic SO_2 and smelly NH_3 that were used as refrigerants. The two spent Saturday morning in

At this step, you will also discover whether you have all the information you need. You may have rejected important information or you may need to look up information in reference tables. In chemistry, you may have to solve problems in which you must look up measured properties of substances. These properties might be density, solubility, electrical conductivity, melting point, or other characteristics that you will learn about in the chemistry course.

Technique 5. *Develop the individual steps to arrive at the answer.* (Plan the trip.)
In the present problem, there is only one step, so you would write the equation

$$\text{hours of work} = \frac{\text{price of sweater}}{\text{wage rate}}$$

As you solve more complex problems in chemistry, you will learn how to develop multi-step solutions by building them up from easy single steps. You will also learn to recognize patterns that you can use again and again to solve similar types of problems. A pattern for solving a particular type of problem is called an algorithm.

Technique 6. *Solve the problem.* (Travel along the selected path.)
In this step, you perform the actual calculation.

$$\text{hours of work} = \frac{\text{price of sweater}}{\text{wage rate}} = \frac{\$75.00}{\$6.00/\text{hour}} = 12.5 \text{ hours}$$

What happens if your selected path doesn't work? First of all, don't say, "I give up! I just can't do these problems." Instead, go back through all the steps you used to arrive at a solution. Start by reading the problem again to see if something different pops out. Try to devise a new solution path. This procedure may seem tedious, but if you discover your own errors you'll become a better problem solver.

Technique 7. *Evaluate the results.* (Did I reach the place I expected?)
Once you have applied your solution and calculated an answer, it is time to take a critical look at that answer. Did your solution give you the right units? If your answer came out in dollars and you wanted hours, you would know that you did something wrong along the way. Look also at the magnitude of the answer. Is it reasonable, considering the facts of the problem? If your answer came out to be 450 hours, you should know that this answer is unreasonable. Evaluating the answer is the step most often overlooked by students, especially when they are pressed for time. However, nothing will reveal an incorrect solution more surely than a simple check to see whether the answer makes sense.

CHEMISTS AND THEIR WORK

Dorothy Crowfoot Hodgkin (1910–1994)
Dorothy Crowfoot Hodgkin, an English chemist, is one of the few women to receive a Nobel prize for chemistry. Her work was principally in the determination of molecular structures through the use of X-ray diffraction measurements. She studied at both Oxford and Cambridge Universities and later became a faculty member at Oxford. In her research, she determined the molecular structures of pepsin, penicillin, and vitamin B_{12}. Vitamin B_{12} is a very complex compound, and it was for this work that she received the Nobel prize in 1964.

MAKING CONNECTIONS

Daily Life
Ask students to determine a strategy using the seven steps presented in the text that can be used in the following situation. The 15 members of a school baseball team decide to buy caps imprinted with the school's name. Each cap costs $9.00 and each imprint costs $2.00. To raise money for the caps, the team members decide to sell candy bars at 50 cents each, of which 20 cents is profit. How many bars must each team member sell? Solution: Cost of caps: 15 x ($9.00 + $2.00) = $165.00. Number of bars that must be sold: $165.00/$.20/bar = 825. Number of bars that must be sold by each team member: 825/15 = *55*.

CHEMISTS AND THEIR WORK
Dorothy Crowfoot Hodgkin

Background: Dr. Hodgkin was born in Cairo, Egypt and educated in England. Knowledge of vitamin B_{12}'s structure allowed scientists to better understand how the body uses the vitamin to make red blood cells. Dr. Hodgkin used X-ray crystallographic analysis to determine the structure of many complex molecules. In 1964 her research resulted in the determination of the structure of insulin, which is used to treat diabetes.

Exploring Further: Ask students to report on the diseases of diabetes and pernicious anemia. Dr. Hodgkin's research aided the treatment of both.

the chemistry library and in four hours had designed a molecule of Freon. Later in the laboratory, they synthesized dichlorodifluoromethane as the first of the Freons. Freon's negative effect on the ozone layer was not discovered until studies were done in the 1970s.

✓ ASSESSMENT

Portfolio: Have students write a report on fluorocarbons and chlorofluorocarbons as an example of planned research. Make the information provided in the Content Background available to students.

3 ASSESS

CHECK FOR UNDERSTANDING

Have one team of students write a problem in which the distance traveled by a car can be determined from such factors as the speed, time, money, and fuel consumption. Have another team of students solve the problem and explain their problem solving strategy to the class. Reverse team roles for a second problem.

Cooperative Learning: Refer to pp. 28T-29T to select a teaching strategy to use with this activity.

Have students answer Concept Review problems.

RETEACHING

Write each of the seven steps of the problem solving strategy on a card. Have students arrange the mixed cards in the order that shows a planned strategy.

EXTENSION

Ask a student to report on the planned research of Charles M. Hall and his development of an inexpensive method to produce aluminum metal. In 1886, just after graduating from Oberlin College, Hall separated aluminum by electrolytic reduction. The cost fell from $254/kg in 1855 to $0.66/kg in 1909. Hall formed the Aluminum Company Of America, Alcoa.

Answers to
Concept Review

1. By evaluation, you will see if the answer is reasonable in magnitude and units.
2. Students answers may vary. One possible example is: 1. Identify the known facts. 2. Define the answer required. 3. Develop possible paths to the solution. 4. Analyze the paths and choose one. 5. Develop the individual steps to arrive at an answer.
3. Student answers may vary.

Figure 2.4 When you finish a problem, always check to see that your answer makes sense. If it doesn't, you will have a valuable clue that will help you find your error.

In the candy bar problem shown in Figure 2.4, the student reached an answer of $735.00 as the price of 7 candy bars. He should quickly realize that amount is a ridiculous price. He has obviously made a mistake in his solution.

Summary of Strategies for Problem Solving

The strategies discussed for solving chemistry problems can be summarized as follows.

1. Identify the known facts. (Where am I?)
2. Define the answer required. (Where do I want to be?)
3. Develop possible solutions. (What paths might take me where I want to be?)
4. Analyze these solutions and choose the correct one. (Which path is most likely to be the correct one?)
5. Develop the individual steps to arrive at the answer. (Plan the trip.)
6. Solve the problem. (Travel along the selected path.)
7. Evaluate the results. (Did I reach the place I expected?)

2.1 *CONCEPT REVIEW*

1. Explain why it is important to evaluate the answer to a problem.
2. List five steps that must be accomplished before beginning calculation in a chemistry problem.
3. **Apply** Explain why an understanding of the concepts of chemistry is essential in order to be successful in solving chemistry problems.

CHEMISTRY AND
SOCIETY

Chemical Research

Because the field of chemistry touches almost all aspects of our lives, the variety of research projects carried out by chemists is mind-boggling. Many current projects are in biochemistry—the chemistry of living organisms. Biochemical researchers continue an age-old quest to understand exactly how the human body operates on a molecular level. In other words, they want to know what chemical reactions are taking place in human cells. The better we understand how a properly functioning human body operates, the better able we will be to design chemical procedures for the diagnosis and treatment of illness.

Drug research occupies many chemists. Pharmaceutical companies are constantly trying to find substances that will treat illnesses presently considered incurable, such as AIDS. There is also the pursuit of improved treatments for ailments that are already controllable, but for which treatment is expensive, painful, or accompanied by high risk, such as some forms of cancer.

Chemists are also closely involved in the search for ways to improve the quality of the environment. Continued research on the chemistry of the atmosphere will lead to cleaner air and to the use of products that will not harm the protective ozone layer in the stratosphere. Investigation of the chemistry of seawater will enable us to clean up many waterways that have become badly contaminated through years of misuse and neglect. Chemists work very closely with U.S. Environmental Protection Agency technicians to find methods of cleaning up dumps and landfills which were improperly or illegally maintained.

In industry, chemists look for ways to produce better products at lower prices. For instance, use of improved plastics can greatly reduce the weight of an automobile. Less weight means greater gas mileage. Greater gas mileage means lower operating cost as well as the lowered consumption of petroleum, an important natural resource. Other new plastics are used in the manufacture of fiber for high-tech clothing that can keep us warm and dry in rain. Materials developed by chemists are used in strong, lightweight bikes and other sports equipment.

The areas of research detailed above barely scratch the surface. The opportunities available to those who study chemistry are tremendously varied and are sure to increase.

Analyzing the Issue

1. Review the annual reports of several manufacturing corporations and report on their research aims. Concentrate on one industry.
2. People often complain that the cost of prescription drugs is too high. Pharmaceutical companies answer that research and development is very expensive and they must try to recover their investment. Organize a debate on this issue. Find out what factors other than research costs affect the price of prescription drugs.

CHEMISTRY AND SOCIETY

Chemical Research

Background: The time for research to produce a finished consumer item continues to increase. The costs of that research also continue to increase. Product liability is an important consideration in today's society.

Teaching This Feature: Research benefits everyone. Emphasize this concept. Money that was spent in the early days of the space program has been returned to the citizens in many ways. Ask students to ask their parents what things they believe are a result of money invested in space exploration research.

Extension: Ask interested students to conduct a survey of local companies to determine the amount of money spent locally on research and development. Discussion topics can include using live animals and fetal tissue in research.

Answers to
Analyzing the Issue

1. Reports will vary.
2. Answers may include profit, wages, and overhead for all stages of development and sales.

4 CLOSE

Discussion: Invite a scientist from a local industry or governmental agency to speak to the class about how problems are solved through planned research. The Soil and Water Conservation Service, EPA, water treatment department, or health department can be a source of speakers. Videotape the presentation to show to later classes.

CONTENT BACKGROUND

Economics: Equipment costing 135 billion dollars, and expected to last 40 years, will be made obsolete by the year 2000. The ozone layer damaged by CFCs has caused industry research and development departments to search for alternatives. The problem is that no one knows if the substitutes will be acceptable to environmental groups. Du Pont expects to spend 1 billion dollars to fully commercialize CFC alternatives. Ask students what they think this will do to the cost of the substitute compared to the original product.

Related Reading: Haggin, Joseph, "Pressure To Market CFC Substitutes Challenges Chemical Industry," *Chemical and Engineering News,* Vol. 69 No. 36 (September 9, 1991), pp. 27-28.

2.2 Numerical Problem Solving

PREPARE

Key Concepts

The major concepts presented in this section include use of SI units to describe mass, length, time, and temperature. The roles of accuracy, precision, and significant digits in analyzing measurements are presented.

Key Terms

qualitative
quantitative
weight
mass
kilogram
balance
length
meter
time
second
temperature
kelvin
accuracy
precision
significant digit
counting number
density

Looking Ahead

Obtain three meter sticks and cover them with black electrical tape as described in Demo 2-4.

1 FOCUS

Objectives

Preview with your students the objectives listed on the student page. Students can use them as a study guide for this major section.

Activity

Have a student measure the mass of six identical objects, such as rubber stoppers, batteries, or test tubes. Have a student determine the average mass of the objects using a calculator. Ask the class how a scientist would write the calculated answer. If they don't know, use this activity to establish a need to know more about measurement and calculations involving those measurements.

OBJECTIVES

- list and use the SI base units for mass, length, time, and temperature.
- express and convert quantities using the common SI prefixes.
- use significant digits to express the exactness of measurements.
- perform calculations using density measurements.

2.2 Numerical Problem Solving

When people first began to describe the properties of materials, they talked about scruples, gills, hundredweights, minims, and drams. The same measurement did not always mean the same quantity in different countries. The king, queen, or parliament of a country ruled what a measurement should be. King Henry VIII of England decreed a *hand* to be four inches. The hand is used to this day as a unit to measure the height of horses. In Germany, a unit called the *rute* was defined as the total length of the left feet of 16 men. Sometimes measurements were changed to make a country's exports more attractive. By making a bushel a little smaller, one country could offer more bushels of wheat than another country for the same price.

The International System (SI)

Chemistry involves measuring and calculating. It is a quantitative science. When you describe a property without measurements you are characterizing the object **qualitatively**. For example, you might say the weather is hot and humid. When the property is measured and described by a number of standard units, you have characterized the object **quantitatively**. An oven temperature of 425°F is a quantitative property.

When you refer to properties as you describe materials, it is helpful to measure the property and state the result quantitatively. In order to make a measurement, you must meet three requirements.

1. Know exactly what property you are trying to measure.
2. Have some standard with which to compare whatever you are measuring.
3. Have some method of making this comparison.

Figure 2.5 SI units are used by almost all countries in the world. Some of these countries have issued stamps to help the general public become familiar with SI units of measurement.

PROGRAM RESOURCES

Study Guide: Use the Study Guide worksheets, pp. 4, 5, for independent practice with concepts in Section 2.2.

Transparency Master: Use the Transparency master and worksheet, pp. 1, 2, for guided practice in the concept "SI Base Units and Prefixes."

Color Transparency: Use Color Transparency 1 to focus on the concept of "SI Base Units and Prefixes."

Reteaching: Use the Reteaching worksheet, pp. 2, 3, to provide students another opportunity to understand the concept of "Percent Error and Significant Digits."

The standard units of measurement in science are part of a measuring system called the International System (SI). The letters are reversed in the symbol because they are taken from the French name Le Système International d'Unités. SI is used by all scientists throughout the world. It is a modern version of the metric system. The people in most countries use SI in everyday life or are in the process of converting to SI. This measurement system will be used in this text.

One important feature of SI is its simplicity. Seven base units are the foundation of the International System. These units are shown in Table 2.1. Detailed definitions of these units are found in Appendix Table A-1.

Table 2.1

SI Base Units

Quantity	Unit	Unit Symbol
Length	meter	m
Mass	kilogram	kg
Time	second	s
Electric current	ampere	A
Thermodynamic temperature	kelvin	K
Amount of substance	mole	mol
Luminous intensity	candela	cd

In SI, prefixes are added to the base units to obtain different units of a convenient size for measuring larger or smaller quantities. A *kilo*meter is one thousand meters while a *milli*meter is one thousandth of a meter. A kilometer is a convenient unit for measuring the distance between cities. Millimeters are useful for measuring the thickness of a board. The commonly used SI prefixes and their equivalents are listed in Table 2.2. These prefixes are used throughout this book and should be memorized. A complete list of SI prefixes is in Appendix Table A-2.

Table 2.2

SI Prefixes

Prefix	Symbol	Meaning	Multiplier (Numerical)	Multiplier (Exponential)
		Greater than 1		
mega	M	million	*1 000 000	1×10^{6}
kilo	k	thousand	1 000	1×10^{3}
		Less than 1		
deci	d	tenth	0.1	1×10^{-1}
centi	c	hundredth	0.01	1×10^{-2}
milli	m	thousandth	0.001	1×10^{-3}
micro	μ	millionth	0.000 001	1×10^{-6}
nano	n	billionth	0.000 000 001	1×10^{-9}
pico	p	trillionth	0.000 000 000 001	1×10^{-12}

*Spaces are used to group digits in long numbers. In some countries, a comma indicates a decimal point. Therefore, commas will not be used.

2 TEACH

CONCEPT DEVELOPMENT

Using The Table: Students are already familiar with several of the prefixes in Table 2.2. Ask them for the frequencies of their favorite AM (kilohertz) and FM (megahertz) radio stations. Students also know the magnitude of time measured as a decade, a century and a millennium. Show how the metric prefixes *deci-* (1/10), *centi-* (1/100), and *milli-* (1/1000) are related to decade (10 years), century (100 years), and millennium (1000 years), respectively. Students may be interested to note that one 53-minute chemistry class lasts one microcentury and that their brains operate in nanoseconds just as a computer does.

Reinforce: To help students learn the metric prefixes, have paired partners make flash cards and quiz each other.

Cooperative Learning: Refer to pp. 28T-29T to select a teaching strategy to use with this activity.

QUICK DEMO **SI Base Units**

As you introduce Table 2.1, use the following instruments to demonstrate each base unit. Length—meter stick, mass–standard mass, time—clock, electric current—ammeter, temperature—thermometer, amount (mole)—12 g carbon, and luminosity burning candle. These demonstrations make the abstract units more concrete for students.

CONTENT BACKGROUND

International Metric System: The metric system was an outcome of the French Revolution. The National Assembly issued a decree in 1790 establishing a committee to form a new system of measurement. This new system would be based on the meter, which was defined as one ten-millionth the distance from pole to equator along a meridian.

The seven base units of the International Metric System are presented in this section to reinforce the simplicity of metric measurement. Emphasize that SI actually makes learning chemistry less confusing by eliminating duplicate units for measurements, such as the joule and the calorie for energy, the cubic decimeter and the liter for volume, and picometer and Angstrom for small distances.

GLENCOE TECHNOLOGY

Software

Mastering Concepts in Chemistry

Unit 5, Lesson 1

Review of the Metric System

CONCEPT DEVELOPMENT

PRACTICE

Guided Practice: Have students work in class on Practice Problems 4 and 5.

Answers To
Practice Problems

4. **a.** centimeter **c.** kilogram
   b. micrometer **d.** deciliter
5. **a.** millimeter
   b microsecond
   c. centigram
   d. picosecond

Misconceptions

It is important to emphasize now and throughout the course that weight and mass are not the same. These terms should not be used interchangeably. Point out that in the laboratory, an instrument called a balance is used to determine mass. Weight is rarely measured in a chemistry laboratory. If so, weight is measured using an instrument called a scale.

Background: You might want to organize activities to coincide with National Mole Day, which occurs each year on October 23. Mole Day promotes awareness of chemistry. See the Close activity on page 215. For more information, write to National Mole Day Foundation, Inc., 1220 S. 5th Street, Prairie du Chien, Wisconsin, 53821.

PRACTICE PROBLEMS

4. Which of the units in each of the following pairs represents the larger quantity?
   a. millimeter, centimeter **c.** kilogram, centigram
   b. picometer, micrometer **d.** deciliter, milliliter
5. A decigram is 0.1 g. Give the name of each of the following quantities.
   a. 0.001 m **c.** 0.01 g
   b. 0.000 001 s **d.** 0.000 000 000 001 s

Mass and Weight

In chemistry, we will often need to know the amount of matter in an object. For instance, we may wish to measure the amount of wood in a small block. One way of measuring the block is to weigh it. Suppose we weigh such a block on a spring scale and find that its weight is one newton (N). The newton is the measurement standard for weight or force. Now suppose we take the scale and the block to the top of a high mountain. There we weigh the block again. The weight now will be slightly less than one newton. The weight has changed because the weight of an object depends on its distance from the center of Earth. **Weight** is a measure of the force of gravity between two objects. In this example, these two objects are the block and Earth. The force of gravity changes when the distance between the object and the center of Earth changes.

The weight of an object can vary from place to place. In scientific work, we need a measurement that does not change from place to place. This measurement is called mass. **Mass** is a measure of the quantity of matter. Therefore, as we saw in Chapter 1, mass is a measure of the inertia of an object.

Figure 2.6 The standard kilogram mass is the cylinder inside the glass jars shown below. Grams are more convenient units to use in chemistry. The cassette has a mass of 42 g. The mass of the headphones is 18 g.

CONTENT BACKGROUND

Common Unit Examples

UNIT EXAMPLE	LENGTH	UNIT EXAMPLE	MASS	UNIT EXAMPLE	TEMPERATURE
thickness of a dime	1 mm	nickel	5 g	normal human body temperature	37°C
diameter of a quarter	2.5 cm	newborn baby	3-4 kg	room temperature	25°C
distance between LA and New York City	4500 km	one chocolate chip	0.5 g	inside of a refrigerator	5°C
length of a new pencil	19 cm	10-speed bicycle	15 kg	cup of hot tea	55°C

Figure 2.7 A double-pan balance (left) measures mass by comparing the mass of an object placed on the pan to standard masses. The electronic balance (right) works automatically using the same principle, but provides a digital readout.

The standard for mass is a piece of metal kept at the International Bureau of Weights and Measures in Sèvres, France. This object is called the International Prototype Kilogram. Its mass is defined as one kilogram. The SI standard of mass is the **kilogram (kg).** However, the kilogram is too large a unit for convenient use in the chemical laboratory. For this reason, the gram (g), one-thousandth of a kilogram, is commonly used. The mass of a large paperclip is approximately 1.5 grams. An ordinary nickel has a mass of about 5 grams.

So far, you know what property you are going to measure—mass—and what your standard of comparison will be—the gram. Now you must compare the object with a standard by using a balance. A **balance** is an instrument used to determine the mass of an object by comparing the object's mass to known mass. To compare these masses, first place the object with unknown mass on the balance pan. Then add known masses to the beams until the masses are equal. The known masses and the unknown mass (the object) are the same distance from the center of Earth. They are, therefore, subject to the same attraction by Earth. At the top of a mountain, the unknown mass and known masses will still be equally distant from the center of Earth. Thus, Earth's attraction for each mass will still be equal. Thus, the balance will indicate no change in mass.

The process of measuring mass by comparing masses on a balance is called massing. Unfortunately, many terms that refer to weighing are often incorrectly applied to the measurement of mass. For example, standard masses used on the balance are often incorrectly called "weights." Also, using a balance to compare masses is often incorrectly called "weighing."

CONCEPT DEVELOPMENT

Background: If students use a triple beam balance in the laboratory, they sometimes wonder why it is called a balance when there aren't two pans. The triple beam balance is a mass comparison device that uses a set of different size masses. Each mass is placed in a notch a set distance from the knife edge. The distance and mass are equivalent to a standard mass that balances the unknown mass. A playground seesaw with two children of different weight is an analogy. The lighter child sits farther out on the seesaw while the heavier child sits closer to the center balance point.

Activity: When reviewing the metric system, it is helpful to have students estimate the length of common items. Have students estimate the width of the laboratory in meters; then measure it and compare the measurement with the prediction. Ask students to estimate their heights in cm. Have lab partners measure each other's height and compare the measurements and estimates. Have students estimate the lengths and widths of test tubes in mm and then measure them and compare the two values.

✓ ASSESSMENT

Knowledge: Ask students to match the words mass (balance) and weight (scale) with the measuring instrument appropriate to each.

CONTENT BACKGROUND

You will notice changes in the abbreviations and SI units in this textbook. In SI, all quantitative energy measurements are expressed in joules. Volume is expressed as the length unit cubed: m^3, dm^3, and cm^3. You will also notice that the degree symbol is not used in expressing Kelvin temperatures. Pressure is measured in pascals and kilopascals. Students will have seen the latter unit printed on automobile tire sidewalls and pressure gauges. In SI, commas are no longer used to group digits in long numbers. Rather, a space is used to group digits in threes to the right and left of the decimal point.

PROGRAM RESOURCES

Critical Thinking/Problem Solving: Use the worksheet, p.2, as enrichment for the concept of "Graphically Speaking."

EVERYDAY CHEMISTRY

Metric Measurement

Background: All tires manufactured since 1977 have sizes expressed in metric units printed on the side-walls. Tire pressures are also expressed in kilopascals.

Teaching This Feature: Show students how easy it is to use the Celsius temperature scale. When it is 0°C and below you need a coat. When it is 10°C you wear a light jacket. If the temperature is 20°C you take the jacket off, and when it is 30°C you head for the beach to cool off.

Connection To Economics: Companies wishing to do business with the United States government must submit all specifications for the bid using metric units. International trade depends on common units being used in both the importing and exporting country.

Extension: Ask students to bring in a label from an item that has both metric and English units.

Answers to
Exploring Further

1. 29.6 mL = 1 fluid ounce, 237 mL, 33.8 fluid ounces

2. Arguments for may include that the system is based on common items. Arguments against may include the difficulty of conversion among units.

CONCEPT DEVELOPMENT

Background: Thermometers are not the only instruments used to measure temperature. A thermocouple measures the potential difference across the junction of two dissimilar metals in contact. This voltage is then converted into a temperature reading. A thermistor is a solid state device whose resistance varies with temperature. The resistance is then converted into a temperature measurement. The optical pyrometer measures temperature by determining the wavelength (color) of the radiation given off by the glowing object. A thermocouple and pyrometer can be used to measure very high temperatures. Some thermistors can measure accurately to 0.01°C. A disadvantage is that the thermistor and thermocouple must be calibrated. The optical pyrometer is not very accurate.

EVERYDAY CHEMISTRY

Metric Measurement

The metric system of measurement was adopted in the United States in 1875, but we still use the English system of measurement in much of our daily lives. Recently, several government agencies have specified that they will purchase only metric goods. As a result, manufacturers must convert their operations to metric standards to continue doing business with the government. Goods to be exported to other countries must meet metric specifications, particularly if the items are to be repaired, adjusted, or used in manufacturing. The U.S. automobile industry now manufactures some engines using metric measurements. Packaged food products must have metric labeling if they are to be understood abroad. Tire pressure gauges are now calibrated in kilopascals, and the tires themselves have metric specifications. In pharmacy and medicine, the metric system has been used for many years. Dosages are given in milligrams and liquids are measured in cubic centimeters.

Complete conversion to metric measurement isn't welcomed by everyone. "Thinking metric" is hard for those who grew up with the older system. Also, conversion to metric is expensive for manufacturers, who must change all their dies and industrial tools to metric standards and retrain workers.

Exploring Further

1. Look up the factors for converting between milliliters and fluid ounces. How many milliliters are in a standard cup (8 oz.)? How many fluid ounces are in a liter?
2. The claim is sometimes made that the English system is a more "natural" system, and is therefore easier to learn. Present arguments for and against this idea.

Length

A second important measurement is that of length. **Length** is the distance covered by a straight line segment connecting two points. The standard for measuring length is defined in terms of the distance traveled by light in a unit of time (Appendix Table A-1). The standard unit of length is the **meter (m).** Length is usually measured with a ruler or similar device. A nickel has a diameter of about two centimeters. Your spiral-bound notebook is about 2.5 decimeters, or 25 centimeters, long.

Time

A third basic measurement is time. **Time** is the interval between two occurrences. Our present standard of time is defined in terms of an electron transition in an atom. In practical terms, the unit of time, the **second (s),** is 1/86 400 of an average day. The most common device for measuring time is a watch or clock. More accurate timepieces include the chronometer, the atomic clock, and the solid-state digital timer.

DEMONSTRATION

2-2 Length Conversions

Purpose: to make visible and understandable conversions of the metric prefixes mm, cm, and m

Materials: 3 identical lengths (70-75 cm) of string, meter stick

Procedure: Have three students hold the strings together to verify that they have identical lengths. Have the students measure the lengths of the strings in millimeters, centimeters, and meters, respectively. Record the measurements on the chalkboard.

Results: The digits of each measurement remained the same but the position of the decimal varied. For example, 720 mm, 72.0 cm, and 0.720 m.

Questions: 1. Is the physical length of each of the three pieces of string the same? *yes*

2. Are the measured lengths of the

Temperature

Matter is composed of small particles called atoms, ions, and molecules. These particles are in constant motion and, therefore, possess kinetic energy. The average kinetic energy of a group of particles determines the group's temperature. The **temperature** of a sample of matter is a measure of the average kinetic energy of the particles that make up the sample. If we add energy in the form of heat to an object, the kinetic energy of its particles increases. The greater the average kinetic energy of the particles, the higher the temperature of the material.

A thermometer is used to measure temperature. When the bulb is heated, the liquid inside expands and rises in the tube. When the bulb is cooled, the liquid contracts and the height of the liquid column decreases. The height of the liquid column can thus be used to measure temperature. The temperature can be read directly from the scale on the tube.

The SI unit of temperature is the **kelvin (K).** We will define this unit later in the textbook. The kelvin has a direct connection with a more familiar unit, the Celsius degree (C°).

The Celsius temperature scale is based on the fact that the freezing and boiling temperatures of pure water under normal atmospheric pressure are constant. The difference between the boiling and freezing points of water is divided into 100 equal intervals. Each interval is called a Celsius degree. The point at which water freezes is labeled zero degrees Celsius (0°C). The point at which water boils is labeled 100°C. The size of a Celsius degree is exactly equal to that of a kelvin. Thus, there are also 100 kelvins between the freezing point and the boiling point of water. Note that Celsius temperatures are expressed in °C, whereas temperature intervals and changes are expressed in C°.

Some data tables indicate that the values were measured at 25°C (Appendix Table A-10). Room temperatures are usually between 20°C and 25°C. Your average body temperature is 37°C.

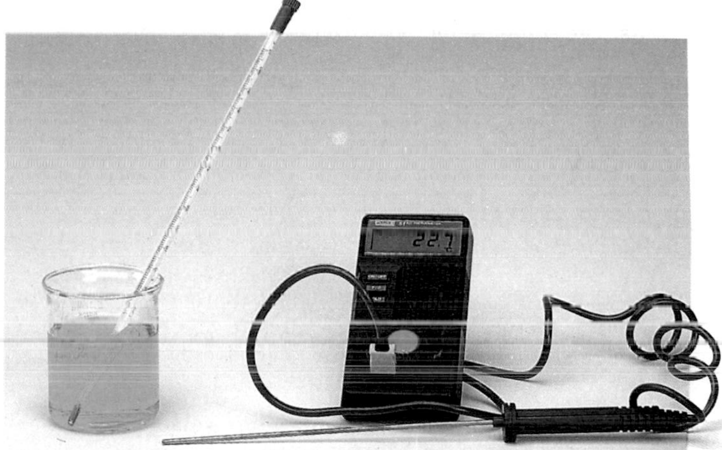

Figure 2.8 As the bulb in the mercury thermometer is warmed, the mercury expands into the capillary tube. The length of the mercury column indicates the temperature. Digital thermometers provide accurate temperature measurement with less threat of glass breakage.

HISTORY CONNECTION

Nautical Knots

The speed of ships and aircraft is generally reported in knots, (nautical miles per hour). The International Nautical Mile is a little more than 6000 feet. The term *knot* comes from the early sailing days. A line with a wooden log on it was knotted every 47 feet 3 inches and rolled onto a reel on board ship. The log was allowed to drag through the water for 28 seconds unreeling the line. The number of knots unreeled in that time was the speed of the ship. The result was recorded in the "log book." Like other units, the knot is measured today in a much more precise way.

HISTORY CONNECTION
Nautical Knots

Exploring Further: The nautical mile is 6 076.013 33 feet, a minute of arc on Earth's surface. The statute mile is 5280 feet. Twenty-eight seconds has the same ratio to an hour as 47'3" has to a nautical mile. Have students find the historical background of other interesting measurements that are mentioned in the text.

✓ ASSESSMENT

Skill: Have a student measure the outside temperature every hour and record the time and temperature on the chalkboard. The next day, show students how to construct a graph by plotting the data on the chalkboard. This activity is also an opportunity to use computer graphing software such as Lab Partner.

Misconceptions

The concepts of heat as radiant energy flowing between two objects having different temperatures, and temperature as a measure of the average kinetic energy of the particles are often confused. Use Demonstration 2-3 to help students understand that there is a difference between heat and temperature.

Extension: Discuss with students that for passive solar heat to be effective at night, there must be a large, massive heat sink to absorb the energy during the day. The initial temperature of the air coming from the heat sink is not affected by its mass. How long that initial high temperature is maintained is directly related to the mass of the heat sink.

string equivalent? *yes*

3. Do the digits in the measurements differ? *no*

4. What happens to value of the numerical part of each measurement as the size of the measurement unit increases? *The value decreases.*

PROGRAM RESOURCES

Enrichment: Use the worksheets "Precision, Accuracy, and Significant Digits," pp. 3-4, to challenge more able students.

CAREERS IN CHEMISTRY

Medical Technologist

Background: Work in medical laboratories is highly regulated by individual states. Some require that all operators in a hospital or medical lab be registered medical technologists with a college degree and one year of internship. Other states allow technicians with an associate degree from a community college to perform much of the work.

For More Information: Contact: American Society for Medical Technology, 2021 L Street, NW, Washington, DC 20036

PRACTICE

Guided Practice: Have students work in class on Practice Problems 6 and 7.

Answers to **Practice Problems**

6. Student 2 has the precise data because the largest difference in values is 0.03 g while Student 1 has a 0.59-g difference.

7. No, the balance error would subtract out.

CONCEPT DEVELOPMENT

• Remind students that when they calculate percent error, they are calculating their error relative to the total accepted value. The equation to calculate percent error is

$$\frac{|E - T|}{T} \times 100\% = \text{percent error}$$

The absolute value of the difference between the theoretical value, T, and the experimental value, E, divided by the theoretical value, multiplied by 100%, will give the percent error.

Do not use only percent error to evaluate a student's lab work. The percent error is a reflection of the techniques the student has used, but it also is directly related to the sophistication of the equipment. For example, a calorimeter used in an experiment can be expected to yield a lower percent error than a Styrofoam cup. The stage of development of the student's laboratory skills will also have an impact on percent error.

CAREERS IN CHEMISTRY

Medical Technologist

Modern medical diagnosis is heavily dependent on the results of laboratory tests made on samples of body fluids taken from a patient. Much of the work in analyzing fluids is now automated. However, the medical technologist must have a good understanding of chemistry in order to be able to calibrate and operate the equipment that does the actual analysis. The technologist must also understand the normal results to be expected from the healthy patient so that aberrant values may be double-checked by making another run, or even by obtaining a new sample for analysis. If the abnormality persists, then the physician must be notified of the impending danger to the patient.

Accuracy and Precision

Every measurement is a comparison of the physical quantity being measured with a fixed standard of measurement, such as the second or the centimeter. In describing the *reliability* of a measurement, the terms *accuracy* and *precision* are often used. **Accuracy** refers to how close a measurement is to the true or correct value for the quantity. **Precision** refers to how close a set of measurements for a quantity are to one another, regardless of whether the measurements are correct.

Suppose you check your car's odometer against mileposts on a highway. If the odometer has increased by 1.0 mile at the second milepost, you can say the odometer is accurate. If it gives this reading on every trial, you can say that the odometer is both accurate and precise.

Usually measurements that have precision are also accurate. However, it is possible to have a source of error that repeats throughout a set of measurements. Suppose, for example, that you are to find the mass of a piece of copper wire. You happen to be using a laboratory balance that is improperly adjusted and always reads 0.50 gram too low. You make five measurements of the mass of the wire and obtain the values 5.52 g, 5.54 g, 5.51 g, 5.52 g, and 5.51 g. Your values differ by at most ±0.02 g from the average value, 5.52 g. The values are, therefore, precise. They are not accurate, however, because of the 0.50-g error in the balance. The actual mass of the piece of copper wire is closer to 6.00 g.

PRACTICE PROBLEMS

6. Which student's data are precise? Explain your answer.
Student 1: 72.75 g, 73.34 g, 73.02 g, 73.25 g
Student 2: 72.01 g, 71.99 g, 72.00 g, 71.98 g

7. Using a balance that always reads 0.50 g too low, a student obtained the mass of a beaker to be 50.62 g. The student then added some sugar to the beaker and, using the same balance, obtained a total mass of 69.88 g. The student recorded the mass of the sugar as 19.26 g. Is the mass of the sugar inaccurate by 0.50 gram? Why or why not?

Percent Error

At times in your laboratory work, you will use data to determine an experimental value for a quantity that is already known. To evaluate your results, you may need to use percent error to compare your result to a value listed in an appropriate literature source. To find percent error, the absolute difference between your value and the literature value is divided by the literature value and the quotient is multiplied by 100%.

$$\text{Percent error} = \frac{|\text{your value} - \text{literature value}|}{\text{literature value}} \times 100\%$$

DEMONSTRATION

2-3 Real Hot

Purpose: to aid in understanding the difference between the concepts of heat and temperature

Materials: 250-cm^3 beaker, 150 cm^3 water, hot plate, medicine dropper, hot pad or beaker tongs

Procedure: **1.** Place about 150 cm^3 of water in the beaker, and heat to boiling on a hot plate.

2. Place a thermometer in the water and read the temperature after two minutes.

3. Fill a medicine dropper with a small amount of boiling water. **CAUTION:** *Steam and boiling water will burn skin.* Use the dropper to place a drop of the water in the palm of your hand.

Results: The drop of water in your hand has been cooled by the medicine dropper and the air. It will be

SAMPLE PROBLEM

Percent Error

A student determines the atomic mass of aluminum to be 28.9. What is her percent error?

Solving Process:

Consult the table of atomic masses on the inside back cover of your book. The atomic mass of aluminum is 27.0.

$$\text{Percent error} = \frac{|28.9 - 27.0|}{27.0} \times 100\%$$
$$= 7.0\%$$

PROBLEM SOLVING HINT

Because absolute value is used, percent error is always positive.

PRACTICE PROBLEMS

8. An experiment performed to determine the density of lead yields a value of 10.95 g/cm^3. The literature value for the density of lead is 11.342 g/cm^3. Find the percent error.
9. Find the percent error in a measurement of the boiling point of bromine if the laboratory figure is 40.6°C and the literature value is 59.35°C.

Significant Digits

Suppose we want to measure the length of a strip of metal. We have two rulers. One ruler is graduated only in whole centimeters. The other is graduated in tenths of centimeters. Which do you suppose can give the more accurate measurement? Look at Figure 2.9. The length of the strip when measured with the ruler graduated in tenths of centimeters is the more significant measurement because it is closer to the actual length of the strip. We say that this latter measurement has more significant digits than the measurement in whole centimeters.

The number of significant digits in a measurement depends on the instrument used in making the measurement. Look again at Figure 2.9. The measurement on the ruler marked in whole

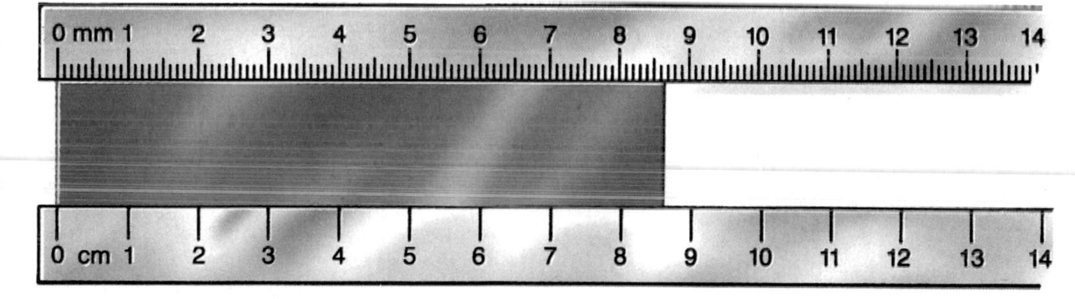

Figure 2.9 The length of the copper strip can be read as 8.63 cm on the top ruler but only as 8.6 cm on the bottom ruler. You can see that the number of significant digits in the measurement depends on the accuracy of the ruler used.

PRACTICE

Guided Practice: Have students work in class on Practice Problems 8 and 9.

Answers to

Practice Problems: Have students refer to Appendix C for complete solutions to Practice Problems.

8. 3.5%

9. 31.6%

CONCEPT DEVELOPMENT

Background: A common laboratory practice is to mass an object by determining the difference between two masses, i.e., the mass of the empty beaker subtracted from the mass of the beaker with the chemical. This procedure will produce a more accurate result, especially if the balance has not been recently calibrated.

Discussion: Additional examples of significant digits are:

2700	2 significant digits
5.10	3 significant digits
7.0200	5 significant digits
0.040 10	4 significant digits
3.00	3 significant digits

• The use of significant digits in calculations is explored in depth in Chapter 8.

Reinforce: Remind students that, in science, *significant* really means *measured.* To most students significant means important. Nonsignificant digits are important as place holders. A measurement 7000 g has four important digits, but only one measured or significant digit.

hot, but not dangerously so.

Questions: 1. What is the temperature of the boiling water? *near 100°C*

2. Would the hand be burned if the entire contents of the beaker were spilled on it? Explain your answer. *Yes, the large amount of water contains enough energy to damage skin tissue.*

PROGRAM RESOURCES

Transparency Master: Use the Transparency master and worksheet, pp.3-4, for guided practice in the concept "Significant Digits."

Color Transparency: Use Color Transparency 2 to focus on the concept of "Significant Digits."

Chemistry Journal

Accuracy and Precision

Have students write, in their own words, a comparison of *accuracy* and *precision.* They should include examples of measurements that are accurate and those that are precise.

CONCEPT DEVELOPMENT

Enrichment: Some computers can do calculations in "double precision mode." Find the meaning of that phrase. *This mode allows the computer to do calculations with many more significant digits.* Why is it necessary to sometimes have this operating mode available? *Some computers do not really round off numbers, they are truncated. In order to have the correct number of digits and the correct magnification, double precision mode is used.*

Computer Demo: *Significant Figure Drill*, AP102, PC2101, Project SERAPHIM

• Using a calculator or computer requires that students apply some rules when making calculations in order to obtain answers with reasonable numbers of digits. Calculated accuracy should at least resemble measured accuracy. A rule of thumb is that the calculated answer can be no more accurate than the least accurate piece of datum that was used in that calculation. The instrument used will determine the number of digits in the measurement, and in any calculations involving those data. It is important that students understand the role of significant digits in measurements and can perform operations using them. These operations will be explained in detail in Chapter 8. The best place to teach and emphasize significant digits is in the laboratory as students make actual measurements.

centimeters lies approximately 6/10 of the way from the 8-cm mark to the 9-cm mark. This length is recorded as 8.6 cm. On the ruler marked in tenths of a centimeter, that is, millimeters, the length lies approximately 3/10 of the way from the 8.6-cm mark to the 8.7-cm mark. This length is recorded as 8.63 cm. The measurement 8.6 cm has two significant digits. The measurement 8.63 cm has three significant digits. The last digit of each measurement is an estimate. All digits that occupy places for which actual measurement was made are referred to as **significant digits**. The places actually measured include the one uncertain, or estimated, digit.

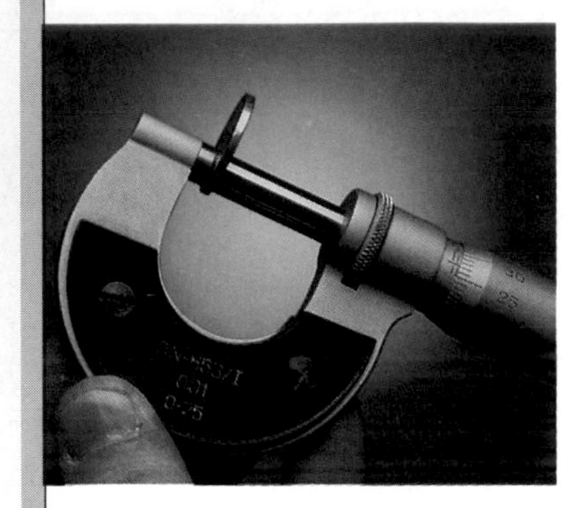

Figure 2.10 This micrometer caliper will measure lengths up to 2.5 cm to the nearest 0.001 cm. The thickness of a nickel is about 2 mm. The nickel's diameter is about 2 cm.

The exactness of measurements is an important part of experimentation. The observer and anyone reading the results of an experiment want to know how exact a measurement is. The exactness of a measurement is indicated by the number of significant digits in that measurement. The following rules are used to determine the number of significant digits in a measurement.

1. *Digits other than zero are always significant.*

96 g	2 significant digits
61.4 g	3 significant digits
0.52 g	2 significant digits

2. *One or more final zeros used after the decimal point are always significant.*

4.72 km	3 significant digits
4.7200 km	5 significant digits
82.0 m	3 significant digits

3. *Zeros between two other significant digits are always significant.*

5.029 m	4 significant digits
306 km	3 significant digits

4. *Zeros used solely for spacing the decimal point are not significant. The zeros are placeholders only.*

7000 g	1 significant digit
0.007 83 kg	3 significant digits

If the quantity 7000 grams has been measured on a balance that is accurate to the nearest gram, all four digits are significant. We must show this in a way that follows the rules for representing significant digits. To do this, we use scientific notation. Scientific notation is a type of exponential notation. It is discussed in detail in Chapter 8, on pages 194–196. For now, note that in scientific notation, you would represent 7000 g measured to the nearest gram as 7.000×10^3 g. This method shows that the mass was measured to four significant digits.

Not all numbers represent measurements. For instance, suppose there are 23 students in a chemistry class. How many significant digits are in 23? The 23 is not a measurement. We do not *measure* the number of students in a class. We *count* them. Stu-

GLENCOE TECHNOLOGY

Software

Mastering Concepts in Chemistry

Unit 5, Lesson 2

Using Significant Digits

DEMONSTRATION

2-4 Measurement

Purpose: to demonstrate that specific numbers of significant digits are associated with measurements made using different instruments

Materials: 3 meter sticks, black vinyl electrical tape, masking tape

Procedure: 1. Cover the entire front surface of one meter stick with a strip of black electrical tape. On the second meter stick, cover alternating 10-cm sections with black tape and masking tape, respectively. Cover the third meter stick's front surface with masking tape. Mark each centimeter on the masking tape.

2. Give the three meter sticks to three students. Ask each to measure the edge of the same student desk and express its length using the appropriate number of significant digits.

Results: The student using the black

dents come in natural numbers, or **counting numbers**. We cannot have 23.4 or 22.8 students. Since counted objects occur in *exact* numbers, we consider that these numbers contain an infinite number of significant digits. We can, therefore, ignore them when determining the number of significant digits in the answer to a problem. The same is true of a number such as the 2 in the equation for the area of a triangle, $A = bh/2$. It is also true in the case of a definition; for example, 1000 millimeters = 1 meter. Both the 1000 and the 1 are exact.

PRACTICE PROBLEMS

10. How many significant digits are there in each of the following quantities?

a. 20 kg **d.** 0.010 s **g.** 0.089 kg **j.** 20 cars
b. 0.0051 g **e.** 90.4°C **h.** 0.009 00 L
c. 11 m **f.** 0.004 cm **i.** 100.0°C

Derived Units

By combining the SI base units, we obtain measurement units used to express other quantities. Distance divided by time equals speed. If we multiply length by length, we get area. Area multiplied by length produces volume. The SI unit of volume is the cubic meter (m^3). However, this quantity is too large to be practical for the laboratory. Chemists often use cubic decimeters (dm^3) as the unit of volume. One cubic decimeter is given another name, the liter (L). The liter is a unit of volume. One liter equals 1000 milliliters (mL) and 1000 cubic centimeters (cm^3). From these facts, you can see that

$$1\ dm^3 = 1\ L = 1000\ mL = 1000\ cm^3$$

Much of your laboratory equipment is marked in milliliters.

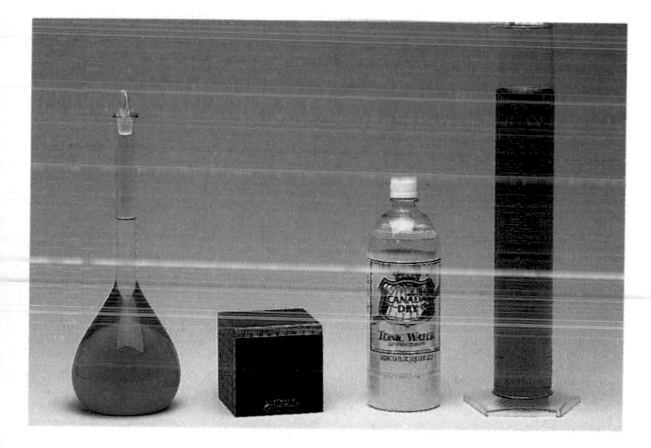

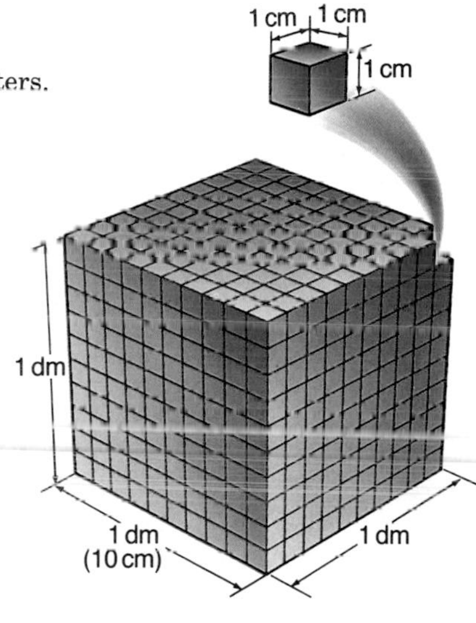

Figure 2.11 Volume is an example of a quantity that is measured in derived units. The volumes shown here all equal 1 dm^3. A cubic decimeter is the volume of a cube that is 1 dm (10 cm) on an edge. It contains $(10\ cm)^3$ or 1000 cm^3, which equals 1 L.

PRACTICE

Guided Practice: Have students work in class on Practice Problem 10.

Answers to
Practice Problems
10. a. 1 b. 2 c. 2 d. 2 e. 3 f. 1 g. 2 h. 3 i. 4 j. infinite

CONCEPT DEVELOPMENT

Background: Throughout the text, the use of cubic meter, m^3, cubic decimeter, dm^3, and cubic centimeter, cm^3, are used for expressing volumes. The liter and milliliter are not SI units. Ask students to bring in empty plastic bags from garden stores that contained peat moss or topsoil. They will see that these bags are measured in cu. ft. and dm^3. Larger bales of peat moss are measured in cu. ft. and m^3. The economics of international trade requires companies to use units that are accepted worldwide and understood by the consumer. Use graduated cylinders, large and small, to be sure students know the quantitative relationships among these units. Unfortunately, manufacturers of this equipment still mark the units as mL and L.

meter stick can estimate the desk's edge to only one significant digit, such as 0.6 m. The student using the stick marked in tenths can measure the desk to two significant digits, such as 0.64 m. The student having the stick marked in centimeters can estimate the edge to three significant digits, such as 0.645 m.

Questions: 1. How many significant digits are present in each measurement? *See Results for possible answers*

2. Why is the last significant digit estimated? *It is understood that a measurement may contain a digit that is estimated beyond the unit in which the instrument is calibrated.*

Extension: Emphasize to students that they should record as laboratory data as many significant digits as the instrument will permit. Discuss the equipment available to them in the laboratory and how many significant digits they should expect from that equipment. For example, a volume measurement made with a 10-cm^3 graduated cylinder calibrated to 0.1 cm^3 can be recorded to the nearest 0.05 cm^3.

MINUTE LAB

Diet Density

Purpose: to show students that density differences can be used to separate materials

Materials: can of regular soda, can of diet soda, water-filled aquarium, balance

Procedure Hints: Both cans must be made entirely of aluminum and should be of the same brand. **CAUTION:** *Glass could break; wear goggles.*

Answers to Questions: **1.** The can of diet soda floats and the can of regular soda sinks. **2.** The regular soda has greater mass. **3.** Sample density data: regular soda—mass 384.43 g, density 1.08 g/cm^3; diet soda—mass 372.93 g, density 1.05 g/cm^3 **4.** The calculations agree with observations. Although the density of the soda in either can is greater than that of water, the CO_2 gas trapped above the liquid also provides buoyancy. **5.** The diet drink does not contain the ten teaspoons of sugar that are in one can of regular soda.

✓ ASSESSMENT

Performance: Have students bring to class four small ripe and four small green (unripe) tomatoes. Place them in an aquarium and stir. The students will observe a separation similar to that done when machine-picked tomatoes are separated to make catsup.

PROGRAM RESOURCES

Laboratory Manual: Use macrolab 2–1 "Measuring Densities of Pennies" and microlab 2–2 "Determining Density and Specific Gravity" as enrichment for the concepts in Chapter 2.

The units used to express measurements of speed, area, and volume are called derived units. You saw that area and volume measurements are expressed using length units. The units used to express speed, such as kilometers per hour or meters per second, combine fundamental units of length and time.

MINUTE LAB

Diet Density

Density differences are used to separate ripe tomatoes from green tomatoes and coal from shale. To see how this works, fill an aquarium with water. Place a can of regular soda and a can of diet soda in the water. Note which can floats and which can sinks. Remove the cans, dry them, and mass them on a balance. Calculate the density of each type of soda using the volume printed on the can. How do your calculations explain your observations? What do you think accounts for the difference in density?

Density

Another common scientific measurement is density. **Density** is mass per unit of volume. To measure the density of a material, we must be able to measure both its mass (m) and its volume (V). Mass can be measured on a balance. The volume of a solid can be measured in different ways. For instance, the volume of a cube is the length of one edge cubed. The volume of a rectangular solid is the length multiplied by the width multiplied by the height. Width and height are simply different names for length. The volume of a liquid can be measured in a clear container graduated to indicate units of volume. You will use graduated cylinders in the laboratory to measure liquid volumes. In measurements of the densities of liquids and solids, volume is usually measured in cubic centimeters (cm^3). Density, then, is expressed in grams per cubic centimeter (g/cm^3). The units to express density are derived units. In equation form, density can be expressed as

$$\text{Density} = \frac{\text{mass}}{\text{volume}}; \quad D = \frac{m}{V}$$

People sometimes say that lead is heavier than feathers. However, a truckload of feathers is heavier than a single piece of lead buckshot. To be exact, we should say that the density of lead is greater than the density of feathers. Densities of some common materials are listed in Table 2.3. Such a table offers a convenient and accurate means of comparing the masses of equal volumes of different materials. Note that the values given for densities of gases are quit small. It is more practical to express the density of gases in g/dm

Changes in temperature change the volume of an object. Therefore, the density of a substance changes when temperature changes. Densities of most materials are given at 25°C.

Table 2.3

Densities of Some Common Materials at 25°C

Material	Density
natural gas	0.000 656 g/cm^3
air	0.001 18 g/cm^3
ethanol	0.789 48 g/cm^3
sucrose (table sugar)	1.587 g/cm^3
sodium chloride (table salt)	2.164 g/cm^3
stainless steel	8.037 g/cm^3
copper	8.94 g/cm^3
mercury	13.545 g/cm^3

DEMONSTRATION

2-5 Now You See It

Purpose: to make visible conversions of the derived metric units of mm^3, cm^3 and m^3

Materials: 3 sheets of clear acetate transparency film, cellophane tape, 500 cm^3 sand

Procedure: **1.** Obtain three sheets of clear acetate transparency film. Roll each into a cylinder and tape the sides together to form three cylinders, having diameters of about 3 cm, 6 cm and 9 cm. Stand them on end to form three concentric cylinders.
2. Fill the smallest cylinder completely with sand. Measure and record the height of the sand in millimeters.
3. Slowly lift the 3-cm diameter cylinder allowing the sand to flow into the next larger cylinder. Measure the sand's height in centimeters.
4. Slowly lift this cylinder and allow the sand to flow into the large cylinder.

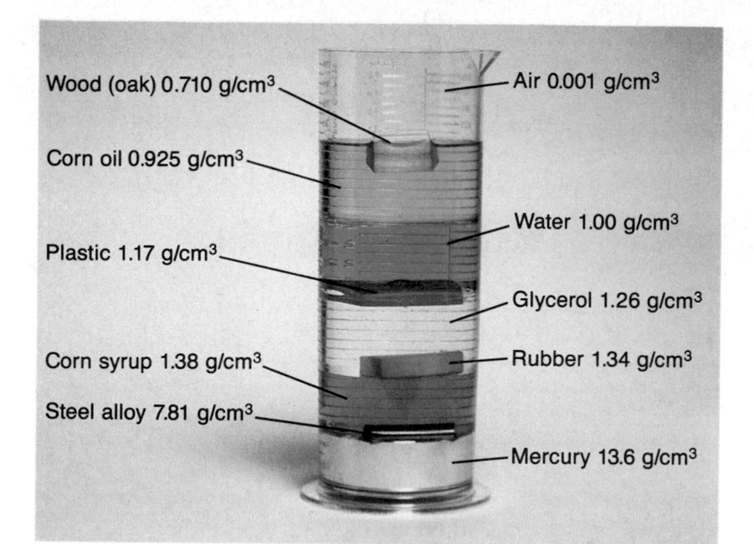

Figure 2.12 Materials of lower density will float on other materials of greater density. Because of differences in density, the materials shown here float at different levels in the container.

Correcting Errors

Review with students the units that are associated with each variable in the formula, D = m/V. Point out that mass is measured in grams or kilograms and volume is measured in cubic decimeters, cubic centimeters, or cubic meters. The units associated with a density measurement can range from gram per cubic centimeter to kilogram per cubic meter. Remind students to check their answers with those that are provided in the Appendix of their textbook.

SAMPLE PROBLEM

Calculating Density

A piece of beeswax with a volume of 8.50 cm^3 is found to have a mass of 8.06 g. What is the density of the beeswax?

Solving Process:

Density is mass per unit volume, or

$$D = \frac{m}{V}$$

Substituting the known information, we obtain

$$D = \frac{8.06\ g}{8.50\ cm^3} = 0.948\ \frac{g}{cm^3}$$

The fact that the answer is expressed in grams per cubic centimeter is a check on the work.

PROBLEM SOLVING HINT

Decide whether beeswax would be more or less dense than water (1 g/cm^3).

PRACTICE PROBLEMS

11. What is the density of a piece of concrete that has a mass of 8.76 g and a volume of 3.07 cm^3?
12. Illegal ivory is sometimes detected on the basis of density. What is the density of a sample of ivory whose volume is 14.5 cm^3 and whose mass is 26.8 g?
13. An archeologist finds that a piece of ancient pottery has a mass of 0.61 g and a volume of 0.26 cm^3. What is the density of the pottery?

PRACTICE

Guided Practice: Guide students through the Sample Problem, then have them work in class on Practice Problems 11 - 13 that follow.

Independent Practice: The homework or classroom assignment can include Chapter Review 25 and 42 from the end of the chapter.

Answers to
Practice Problems: Have students refer to Appendix C for complete solutions to Practice Problems.

11. 2.85 g/cm^3
12. 1.85 g/cm^3
13. 2.3 g/cm^3

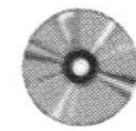

Videodisc

Newton's Apple: Physical Sciences

Buoyancy

Chapter 1, Side A

David floats and sinks

10135-12887

Fascinating fact

13648

Record the sand's height in meters.
5. Have students calculate the volume of sand using the equation $V = \pi r^2 h$
Disposal: F

Results: The volume of sand remains the same but the height in each cylinder is different. Sample data: 210 mm drops to 12 cm that drops to 0.07 m of sand. The amount of material remains the same but its value is expressed using different SI units.

Questions: 1. What happens to the height of the sand as the diameter of the cylinder is increased? *Height decreases*
2. Does the amount of sand that is in each cylinder change? *No*
3. When a measurement is expressed in different sized SI base units does the absolute value of the number change? *Yes*

CONCEPT DEVELOPMENT

- Point out that some calculators have keys other than *M* and *MR* to store values in memory. Teachers should tailor the sample problem to the kinds of calculators their students are using.

Correcting Errors

Show students the algebraic manipulation of the density formula $D = m/V$ to yield $V = m/D$ and $m = DV$.

Practice in solving for each of the three variables is given in the problems assigned in the Guided and Independent Practice. Remind students to check their answers with those that are provided in the Appendix of their textbook.

PRACTICE

Guided Practice: Guide students through the Sample Problems, then have them work in class on Practice Problems 14-17.

Independent Practice: The homework or classroom assignment can include Chapter Review 26-29 and 39-44 from the end of the chapter.

Answers to
Practice Problems: Have students refer to Appendix C for complete solutions to Practice Problems.

14. 67.7 g
15. 3.16 cm^3
16. 14.0 g
17. 1.01 m^3

PROGRAM RESOURCES

Vocabulary Review/Concept Review: Use the worksheets, pp. 2, 33-34, to check students' understanding of the Key Terms and Concepts of Chapter 2.

Figure 2.13 You can determine the concentration of acid in a car battery by using a hydrometer. The higher the float rises in the solution, the greater the solution's density. Greater density indicates a higher concentration of battery acid.

PROBLEM SOLVING HINT
Be sure your setup will result in units of mass.

SAMPLE PROBLEMS

1. Using Density to Find Volume

Cobalt is a hard magnetic metal that resembles iron in appearance. It has a density of 8.90 g/cm^3. What volume would 17.8 g of cobalt have?

Solving Process:
You know that density can be expressed as

$$D = \frac{m}{V}$$

Solving this equation for volume and substituting the known information in the resulting equation, you obtain

$$V = \frac{m}{D} = \frac{17.8\ \cancel{g}}{8.90\ \cancel{g}/cm^3} = 2.00\ cm^3$$

The answer is in cm^3, a unit of volume, which is another check on the accuracy of the problem solving approach.

2. Using Density to Find Mass

What is the mass of 19.9 cm^3 of coal that has a density of 1.50 g/cm^3?

Solving Process:
The mathematical relationship relating density, volume, and mass is

$$D = \frac{m}{V}$$

Solving the equation for m gives

$$m = DV$$

Substituting known values produces

$$m = \left(1.50\ \frac{g}{\cancel{cm^3}}\right) \times (19.9\ \cancel{cm^3}) = 29.9\ g$$

PRACTICE PROBLEMS

14. Limestone has a density of 2.72 g/cm^3. What is the mass of 24.9 cm^3 of limestone?

15. Calcium chloride is used as a deicer on roads in winter. It has a density of 2.50 g/cm^3. What is the volume of 7.91 g of this substance?

16. Ammonium magnesium chromate has a density of 1.84 g/cm^3. What is the mass of 7.62 cm^3 of this substance?

17. What is the minimum volume of a tank that can hold 795 kg of methanol whose density is 0.788 g/cm^3?

DEMONSTRATION

2-6 Lighter in Hot Air

Purpose: to demonstrate that density changes during a chemical change

Materials: 1 sheet of the brown tissue paper liner from a ditto master or thermal transparency master, matches. Note: White tissue paper will not work because it does not have enough fiber.

Procedure: 1. Fold the 8.5 x 11 inch paper in half, forming two panels that are 5.5 x 8.5 inches. Open the sheet, and fold each side in half again toward the center crease so you have four panels that are 2.75 x 8.5 inches.
2. The two end panels should be overlapped as you form a triangular prism 8.5 inches tall.
3. Turn off any fans and close the windows and door because air currents can topple the prism-shaped tube. Darken the room, and use a match to light each of the three corners at the

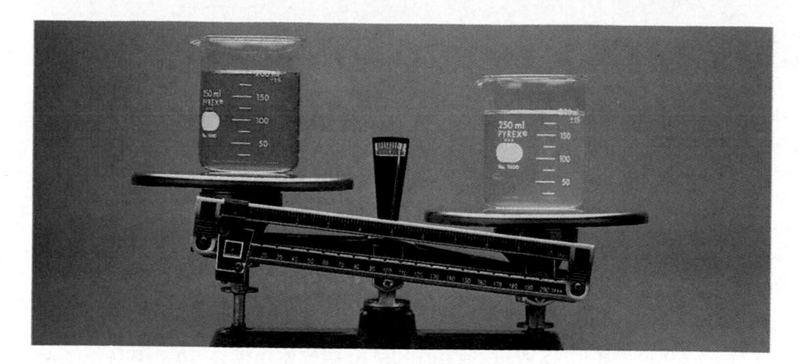

Figure 2.14 The beaker on the left contains corn oil. The beaker on the right contains water. Even though the volumes of both liquids are equal, the balance shows that their masses are not the same. Water has more mass per cubic centimeter than corn oil. That is, water is more dense than corn oil.

One type of density problem lends itself readily to the use of the memory function found on most calculators. In cases where density must be found from a mass and the dimensions of an object, the memory can be used to hold the calculated volume.

SAMPLE PROBLEM

Solving Density Problems with a Calculator

What is the density of a rectangular block of granite of mass 40.4 g and dimensions 2.00 cm by 1.09 cm by 7.04 cm?

Solving Process:

Calculate the volume of the block by multiplying length by width by height. Press keys [2] [.] [0] [0] [×] [1] [.] [0] [9] [×] [7] [.] [0] [4] [=] The result will appear as 15.3472. Don't round this result. When doing chemistry problems with a calculator, you should round off values only at the end. Press the [Min] key to store the calculated volume.

The equation for density is $D = \frac{m}{V}$. Therefore, you can complete the calculation by pressing keys [4] [0] [.] [4] [÷] [MR] [=]. The result will appear as 2.632 402 002. The data used had only three significant digits. Therefore, the answer also should be rounded to three significant digits.

$$D_{granite} = 2.63 \frac{g}{cm^3}$$

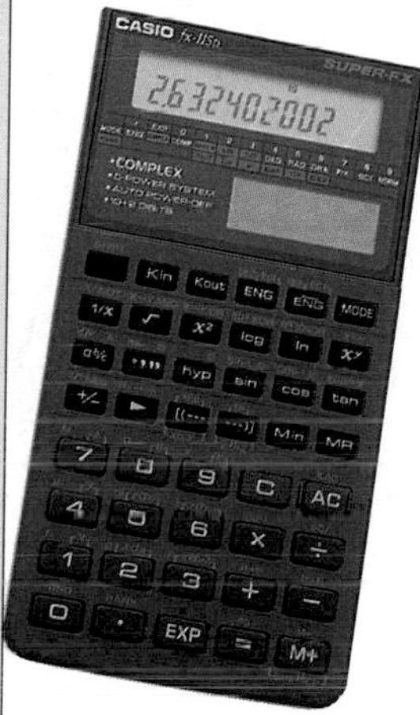

Figure 2.15 Although your calculator will give the result 2.632402002, the number must be rounded to three significant digits because the data you used to make your calculation had only three significant digits. You should report the density as 2.63 g/cm^3.

PRACTICE PROBLEMS

18. Use a calculator to determine the density of a pine board whose dimensions are 4.05 cm by 8.85 cm by 164 cm and whose mass is 2580 g.
19. Use a calculator to find the density of a 51.6 g cylindrical steel rod of diameter 0.622 cm and length 22.1 cm.

top. **CAUTION:** *Fire risk. Warn students they are not to try this demonstration at home.* **Disposal:** A

Results: A convection column of hot air is established by the burning paper. When the flame is 1 cm from the tabletop, the ash will lift off the table and float up into the air because ash is less dense than paper. You can catch the ash as it starts to fall because it will stay together in one piece.

Questions: 1. What can be said about the density of the paper and the density of its ash? *Ash is less dense.*
2. Is density a physical property? *Yes.*
3. What happens to physical properties during a chemical change? *Physical properties change.*

Extension: To gain increased student interest place the paper "rocket" on a dollar bill. The fire will get close to the bill but the ash will lift off at the last moment.

Correcting Errors

Pair students who have calculators of the same model and have them solve the Practice Problems. They will be able to help each other if problems arise in using the memory functions on their model of calculator.

Cooperative Learning: Refer to pp. 28T-29T to select a teaching strategy to use with this activity.

PRACTICE

Guided Practice: Guide students through the Sample Problem, then have them work in class on Practice Problems 18 and 19.

Independent Practice: For homework assign Chapter Review 37 from the end of the chapter.

Answers to
Practice Problems: Have students refer to Appendix C for complete solutions to Practice Problems.
18. 0.439 g/cm^3
19. 7.68 g/cm^3

3 ASSESS

CHECK FOR UNDERSTANDING

At several numbered lab stations, set up laboratory measurement problems involving the metric system. Directions are at each station. Assess each student's ability to measure using meter sticks and different metric rulers, graduated cylinders, a clock, a balance, and thermometer. Have each student determine the mass density of a rectangular solid.
Ask students to answer the Concept Review problems.

RETEACHING

Use the Review and Practice problems 23-44 at the end of the chapter.

EXTENSION

Have students observe and report where standard masses and volumes are used, and the various government seals that certify an accurate measurement. Examples: gasoline pumps, scales in grocery store's produce department.

Answers to

Concept Review

20. The first balance permits measurement within 0.1 g; the second balance permits measurement within 1 g.

21. **a.** 5000 mg **b.** 0.15 m **c.** 0.2 s **d.** 5 cm **e.** 0.06 kg **f.** 2000 µm

22. 21°C, 3 m of tape, 1.4 kg, 90 s

4 CLOSE

Discussion: Ask students to brainstorm several strategies that they use daily to solve problems. Possible answers:

list all possible solutions
look for a pattern
construct a table or graph
draw a figure or diagram
make a model
guess and check
work backwards
solve a simpler problem

BRIDGE TO METALLURGY

Froth Flotation Separation

Teaching This Feature: Students are familiar with the television portrayal of the gold miner as he pans for gold in a stream. Relate the principles used in this procedure to those used in froth flotation separation.

Connection to Economics: Machines can do the separation automatically and economically.

Extension: Ask students if they have observed any separation process that is based on density differences. Plastics at recycling centers can be separated according to type because each type has a different density.

Answers to

Exploring further

1. phosphate rock [$Ca_{10}F_2(PO_4)_6$]; coal [C]; hematite [Fe_2O_3]; galena [PbS]; argentite [Ag_2S]

2. The oil sticks because ore particles have more affinity for oil than for water.

2.2 CONCEPT REVIEW

20. Two students measure the mass of the same beaker on two different balances. The first student reports the beaker's mass to be 47.0 g. The second student reports the mass to be 47 g. Assuming the measurements were correctly reported, how can you explain the difference?

21. Express each of the following quantities in the unit named.

a. 5 grams in milligrams
b. 15 centimeters in meters
c. 200 milliseconds in seconds
d. 0.5 decimeters in centimeters
e. 60 grams in kilograms
f. 0.2 centimeters in micrometers

22. Apply Add the appropriate SI base units to the measurements in the following passage.

On a day when the thermometer read 21, we wrapped and mailed a package that required almost 3 of tape to seal. When the postal clerk put it on a balance, the dial read 1.4. We were in and out of the post office in less than 90.

Bridge to
METALLURGY

Froth Flotation Separation

Mining engineers use a process involving density differences to concentrate minerals. Some ores of copper, lead, and zinc are concentrated in this way. When minerals are removed from Earth's crust, they are mixed with dirt, rock, and other contaminants. Copper sulfide ores, for example, usually contain less than 10% copper. In the froth-flotation process, the output of the mine is first ground to a powder. Then it is mixed by a motor-driven agitator in a tank with water to which oil has been added. The oil is carefully selected to work with the mineral being separated. A froth forms when air at high pressure is blasted through the mixture. The minerals, such as the copper sulfides, have little or no attraction for the water, but they are attracted to the oil and coated by it. The oily particles stick to the air bubbles in the froth. The froth is less dense than the water because it contains a lot of air. The froth floats to the top of the tank. There, the froth can be floated off, the oil removed, and the concentrated mineral recovered. The denser dirt and rock, meanwhile, fall to the bottom of the tank. The flotation tank is flushed from time to time to remove the refuse.

Exploring Further

1. Find out the names and formulas of two ores concentrated by the froth-flotation process.
2. Why do you think the oil sticks to the ore particles, but the water does not?

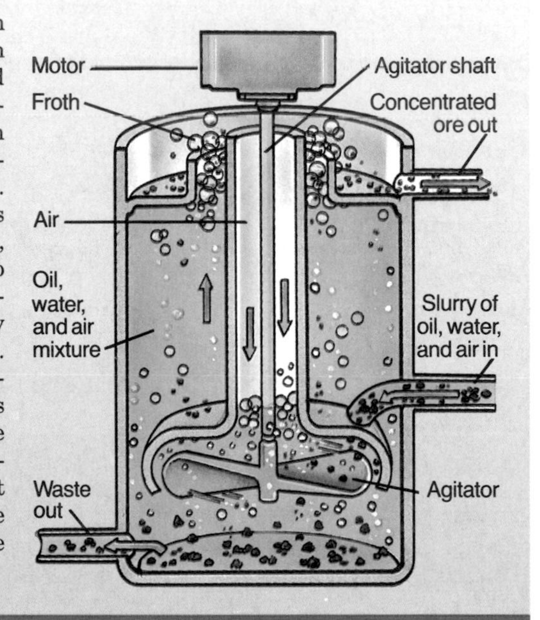

Chapter 2 REVIEW

Summary

2.1 Decision Making

1. Problem solving can be learned and uses certain techniques that are adaptable to different situations.
2. In solving problems, it is important to know exactly what information is given and what information is desired.
3. A result should always be checked against the original problem to make sure the answer makes sense.
4. Algorithms are frequently useful in solving numerical chemical problems.

2.2 Numerical Problem Solving

5. Scientific work requires a quantitative approach. In other words, to investigate a phenomenon, certain characteristics must be measured.
6. Scientists use SI measurements. This system of measurement consists of seven base units. In SI, convenient units for any measurement are obtained by using prefixes.
7. Mass is a measure of the quantity of matter. Weight is a measure of the force of gravity between two objects.
8. The SI standard unit of mass is the kilogram. For length, the unit is the meter; for time, the second; and for temperature, the kelvin.
9. Accuracy of a measurement refers to the closeness of the measurement to the true value. Precision refers to reproducibility within a group of measurements.
10. The exactness of a measurement is indicated by the number of significant digits in that measurement.
11. SI base units may be combined to form derived units such as the cubic meter, the SI standard unit of volume.
12. Density is the mass of a material in a unit volume of that material.

Key Terms

qualitative
quantitative
weight
mass
kilogram
balance
length
meter
time
second
temperature
kelvin
accuracy
precision
significant digit
counting number
density

Review and Practice

23. What is an algorithm?
24. How does determining the units of the answer to a problem help you choose a correct method of solution?
25. What is the density of a piece of cork that has a mass of 0.650 g and a volume of 2.71 cm^3?
26. Barium perchlorate has a density of 2.74 g/cm^3. What is the mass of 27.2 cm^3 of this substance?
27. Bismuth phosphate has a density of 6.32 g/cm^3. What is the mass of 25.9 cm^3 of this substance?
28. Cerium sulfate has a density of 3.17 g/cm^3. Calculate the volume of 599 g of this substance.
29. Chromium silicide has a density of 5.50 g/cm^3. Calculate the volume of 35.9 g of this substance.
30. What are the seven base units in SI?
31. How is the Celsius temperature scale defined?
32. What is the SI fundamental unit of mass?
33. List the number of significant digits for each of the following.
    a. 1 km
    b. 1.5 mL
    c. 2.15 000 cm^2
    d. 5.380 000 0 s
34. What is the difference between counting and measured numbers? Which are expressed in significant digits?

CHAPTER REVIEW 2

- Review Summary statements and Key Terms with your students.
- Complete solutions to Chapter Review Problems can be found in **Merrill Chemistry Problems and Solutions Manual.**

Answers to Review and Practice

23. An algorithm is a procedure for solving a particular type of problem.
24. If you know the units of the answer, you can choose a method of solution that will give those units in the result.
25. 0.240 g/cm^3
26. 74.5 g
27. 164 g
28. 189 cm^3
29. 6.53 cm^3
30. meter, kilogram, second, ampere, kelvin, mole, candela
31. freezing point of water = 0°C; boiling point of water = 100°C
32. kilogram
33. **a.** 1 **b.** 2 **c.** 6 **d.** 8
34. Counting numbers are whole numbers and are exact. Measured numbers are only as accurate as the device used. The exactness of a measurement is indicated by the number of significant digits in that measurement.

Answers to
Review and Practice (cont.)

35. A derived SI unit is a measurement unit that is made by combining SI base units. Example: speed (m/s), area (m^2), and volume (cm^3).
36. Accuracy is the closeness of a measurement to the true value. Precision tells how close a set of measurements (for a quantity) are to another set, regardless of whether the measurements are correct.
37. a. 3.00 g/cm^3 **b.** 2.80 g/cm^3 **c.** 0.680 g/cm^3
38. 833 cm/s
39. 1250 g
40. 36.7 g
41. 3.34 cm^3
42. 0.0012 g/cm^3
43. 32.5 cm^3
44. 319 cm^3

Answers to
Concept Mastery

45. If the calculated result differs considerably from the estimate, an error may have been made in the calculation.
46. You should evaluate the answer to make sure it is reasonable.
47. Mass does not vary with location.
48. The correct number of significant digits should be maintained so that the answer is not expressed to a greater accuracy than the measurement allows.
49. Yes, weight depends on gravitational attraction. No, mass depends on the amount of matter present.
50. No, they weigh the same, but they differ in volume and density.
51. 1.7 cm^3, 0.0017 L, 0.0017 dm^3
52. 5960 kg

35. What is a "derived" SI unit? Give two examples.

36. What is the difference between accuracy and precision?

37. Find the density in g/cm^3 of the following.
  a. concrete, if a rectangular piece 2.00 cm × 2.00 cm × 9.00 cm has a mass of 108 g
  b. granite, if a rectangular piece 5.00 cm × 10.0 cm × 23.0 cm has a mass of 3.22 kg
  c. gasoline, if 9.00 dm^3 have a mass of 6120 g

38. An automobile is traveling at the rate of 30.0 kilometers per hour. What is its rate in centimeters per second (cm/s)?

39. The density of nitrogen gas is 1.25 g/dm^3. What is the mass of 1.00 m^3 of N_2?

40. Bismuth has a density of 9.80 g/cm^3. What is the mass of 3.74 cm^3 of Bi?

41. Iron has a density of 7.87 g/cm^3. What volume would 26.3 g of Fe occupy?

42. If 1.00 km^3 of air has a mass of 1.2 billion kg, what is the density of air in g/cm^3?

43. Magnesium has a density of 1.74 g/cm^3. What is the volume of 56.6 g of Mg?

44. Tin has a density of 7.28 g/cm^3. What is the volume of 2.32 kg of Sn?

Concept Mastery

45. Why is it a good idea to estimate the answer to a problem before beginning your calculation?

46. What is the last thing you should do after calculating an answer to a chemistry problem?

47. Why is mass used instead of weight for scientific work?

48. Why is it important to use the correct number of significant digits when expressing measurements?

49. Would an astronaut's weight change on the moon? Would an astronaut's mass change on the moon? Why or why not?

50. Does a pound of feathers "weigh" more than a pound of iron? What truly is different between the two?

51. A wood block has dimensions of 2.0 cm × 1.5 cm × 5.6 mm. What is its volume in cubic centimeters? In liters? In cubic decimeters?

52. Vanadium hardens steel. Vanadium steel is used in armor plate on army tanks, for example. If its density is 5.96 g/cm^3, a cubic meter would have a mass of how many kg?

Application

53. A chemist has been given a piece of an unusual meteorite for analysis. The meteorite fell into a field 220 km from the chemist's laboratory. The meteorite had a mass of 14.9 kg. The piece sent to the chemist is irregularly shaped and is 3 cm long. A scale shows that the piece weighs 1.6 newtons. A balance shows its mass to be 163 g. The chemist decides to determine the volume of the fragment by seeing how much water it will displace when submerged. The chemist adds water to a graduated cylinder that can hold 50.0 cm^3. The level of the water in the cylinder shows that it contains 25.8 cm^3. After the meteorite piece is submerged, the water level rises to 45.9 cm^3. The density of water is 1.00 g/cm^3. What density will the chemist calculate for the meteorite sample? Organize the information just given into two lists—relevant and irrelevant data. Then, carry out the calculation.

54. A person who was on a diet measured his mass one week and found it to be 82 kilograms. Two weeks later he used the same scale again and found that his mass was 79 kilograms. (Assume that his scale has an uncertainty of 1 kilogram.)

PROGRAM RESOURCES

Evaluation: Use the Evaluation worksheets, pp.5-8, to assess students' knowledge of Chapter 2.

Computer Test Bank: Use the IBM or Apple Test Bank and Question Selection Manual to prepare your own individualized test for Chapter 2.

a. What is the person's maximum possible loss of mass?
b. What is the minimum possible loss of mass?
c. Explain the mass loss using the ± notation.

55. Name the SI unit that would be the best measure of the size of each of the following items.
a. a glass of milk
b. a tablespoon of sugar
c. the mass of an elephant
d. the surface area of a dollar bill
e. the volume of an automobile gasoline tank
f. the volume of air in a balloon

56. Aluminum, because of its light density and strength, has been used for making aircraft. It has a density of 2.70 g/cm^3. Magnesium has similar properties, and it has a density of 1.74 g/cm^3. What is the difference in mass if 1000 cm^3 of magnesium are used rather than 1000 cm^3 of aluminum?

57. The use of hydrogen for filling blimps and dirigibles led to disasters because of its flammability. Helium does not burn and is now used instead of hydrogen. Hydrogen has a density of 0.089 9 g/dm^3 and helium has a density of 0.178 g/dm^3, about twice that of hydrogen. What is the mass of helium in a 10 000-m^3 blimp?

58. The diameter of the sun is estimated at about 865 000 miles and its average density is 1.4 g/cm^3. What is its mass in kilograms? (Use $V = 4\pi r^3/3$.)

Critical Thinking

59. It is sometimes said that the calculation itself is the least important step in solving a problem while studying chemistry. Do you think that statement is valid? Explain your answer.

60. The volume of one block of wood is greater than that of a second block of wood, but the mass of the second is greater. Which has the greater density and why?

61. The definition of the liter was originally based on the volume of a certain mass of water. What is the difficulty with such a definition?

62. The mineral quartz has a density of 2.65 g/cm^3 and the mineral zircon has a density of 4.50 g/cm^3. Suppose a rock composed of quartz and zircon has a density of 3.00 g/cm^3. Find the percentage by mass and the percentage by volume of quartz in the rock.

Readings

Deavor, James P. "Chemistry: The Ultimate Liberal Art." *Journal of Chemical Education* 67, no. 10 (October 1990): 881-882.

Jakuba, Stan R. "Go Metric? Now?" *CHEMTECH* 18, no. 7 (July 1988): 424-425.

Kotz, John C., and Keith F. Purcell. *Chemistry & Chemical Reactivity*. New York: Saunders College Publishing, 1987.

Smith, Richard G., and Gary K. Himes. *Solving Problems in Chemistry*. Columbus, OH: Merrill Publishing Company, 1993.

Cumulative Review

1. List five scientific procedures.
2. What are the two general forms of energy an object can possess?
3. What is the principal aim of chemists in their work?
4. State the law of conservation of mass-energy.
5. What is the difference between a law and a theory?

Answers to
Application

53. Relevant Data
mass of sample
initial level of water in cylinder
final level of water in cylinder

Irrelevant Data
distance of field from lab
mass of meteorite
length and shape of sample
weight of sample
volume of cylinder
density of water

8.11 g/cm^3

54. a. 5 kg **b.** 1 kg **c.** 3 ± 2 kg

55. a. cm^3 **b.** g **c.** kg **d.** cm^2 **e.** dm^3 or m^3 **f.** cm^3 or dm^3

56. 960 g

57. 1 780 000 g

58. 2.0×10^{30} kg

Answers to
Critical Thinking

59. Student answers may vary. A possible answer is that the calculation will be incorrect if the problem is set up incorrectly.

60. The second block has the greater density because it has more mass in less space.

61. The volume of water varies with the temperature.

62. 81.1% by volume, 71.7% by mass

Answers to
Cumulative Review

1. observation, hypothesis, experimentation, collecting data, theorizing

2. kinetic and potential

3. Chemists study the structure and properties of matter.

4. The sum of the matter and energy of a closed system is constant and interchangeable.

5. A law states how nature behaves. A theory is an explanation of this behavior.

CHAPTER 3

Matter

CHAPTER ORGANIZER

Chapter Objectives	Text Features	Laboratory Options	Teacher Classroom Resources
3.1 Classification of Matter **1.** Describe and distinguish heterogeneous and homogeneous materials, substances, mixtures, and solutions. **2.** Describe and give examples of elements and compounds. **3.** Classify examples of matter. **Nat'l Science Stds: UCP.1, UCP.5, B.2**	**History Connection:** The Village Blacksmith, p. 51 **Frontiers:** Crystal Coolers, p. 52	**Minute Lab:** Room Enough to Spare, p. 49 **Discovery Demo:** 3-1 Separation and Testing, p. 46 **Demonstration:** 3-2 Light a Candle, p. 50 **Demonstration:** 3-3 Sweetened Tea, p. 52	**Lesson Plans:** pp. 4-6 **Study Guide:** p. 7 **Color Transparency 3 and Master:** Classification of Matter **Applying Scientific Methods in Chemistry:** Classifying Matter, p. 18
3.2 Changes in Properties **4.** Classify changes in matter as physical or chemical. **5.** Obtain information from a graph. **6.** Distinguish among extensive, intensive, physical and chemical properties. **Nat'l Science Stds: UCP.3, UCP.5, B.2, B.3**	**Bridge to Mineralogy:** Salts from the Dead Sea, p. 57	**ChemActivity 3:** Properties of Chalk and Baking Soda - microlab, p. 800 **Demonstration:** 3-4 Evidences of Chemical Change, p. 56 **Laboratory Manual:** 3-1 Using Paper Chromatography to Separate Dyes-microlab **Laboratory Manual:** 3-2 Determining Boiling Point-microlab **Laboratory Manual:** 3-3 Comparing Physical and Chemical Changes	**ChemActivity Master 3** **Study Guide:** p. 8 **Reteaching:** Matter and Its Properties, p. 4 **Color Transparency 4 and Master:** Solubility Curves
3.3 Energy **7.** Describe conditions under which heat is transferred. **8.** Convert between units used to measure energy. **9.** Describe endothermic and exothermic processes and state the function of activation energy. **10.** Perform calculations involving specific heat. **Nat'l Science Stds: UCP.1, UCP.3, B.5, B.6, D.1, F.1, F.3, F.4, F.5, F.6**	**Chemists and Their Work:** Gerty Theresa Cori, p. 63 **Everyday Chemistry:** Smog, p. 67 **Chemistry and Society:** Energy Problems, p. 70	**Demonstration:** 3-5 Photochromism, p. 64 **Demonstration:** 3-6 An Exothermic Reaction, p. 68 **Laboratory Manual:** 3-4 Measuring Specific Heat	**Study Guide:** pp. 9-10 **Enrichment:** Specific Heat, pp. 5-6 **Critical Thinking/ Problem Solving:** Changes of State in Graphic Terms, p. 3
	Vocabulary and Concept Review, pp. 3, 35-36 **Evaluation,** pp. 9-12	**Videodisc Correlation Booklet** **Lab Partner Software**	**Test Bank**

Block Schedule

For information on block scheduling, see the Lesson Plans booklet in the Teacher Resource Package.

GLENCOE TECHNOLOGY

Chemistry: Concepts and Applications Videodisc

Chemistry: Concepts and Applications CD-ROM

Complete Glencoe Technology references are inserted within the appropriate lesson.

ASSIGNMENT GUIDE

Contents	Level	Practice Problems	Chapter Review	Solving Problems in Chemistry
3.1 Classification of Matter				
Heterogeneous Materials	C		21, 38, 39, 42, 43	
Homogeneous Materials	C		22-24, 38, 43-46	
Substances	C		25-27, 38, 49, 50, 55, 67	1
3.2 Changes in Properties				
Physical and Chemical Changes	C	5	28, 42, 51-54, 58, 66, 68	2-3
Physical and Chemical Properties	C		29, 46, 56, 57, 59, 60, 68	
3.3 Energy				
Energy Transfer	C	10	31, 65, 72	
Energy and Chemical Change	C		30, 47, 59, 64, 69-71	
Measuring Energy Changes	C	11 16	32-37, 40, 41, 48, 50, 61-64, 73 76	4-8
				Chapter Review: 1-8

C=Core, A=Advanced, O=Optional

Laboratory Materials

Minute Labs	ChemActivities	Demonstrations		
page 49 beaker box of metal paper clips water	**page 800** baking soda chalk distilled water glass slides grease pencil hand lens *3M* HCl test-tube holder toothpick	**pages 46, 50, 52, 56, 64, 68** beakers, 100, 400, 600-cm^3 $CaCO_3$ candle ceramic plate cork to fit bottle drinking glass $FeSO_4$ file folder filter paper	funnel gas grill lighter H_2SO_4 hot plate ice cubes matches methanol methylene blue crystals overhead projector paper	polypropylene bottle silicone caulk string sucrose tea bags thumbtacks tongs water

CHAPTER 3

Matter

CHAPTER OVERVIEW

Main Ideas: Chapter 3 outlines a classification system that organizes matter into several broad categories. Physical and chemical properties, physical and chemical changes, and the energy related to these changes are discussed. Mathematical problem-solving skills are exercised when calculating specific heat.

Theme Development: The theme of this chapter is systems and interactions. Properties of systems consisting of homogeneous and heterogeneous mixtures are presented. The role of energy transfer into and out of systems undergoing physical changes, such as melting, and chemical changes, such as combustion, are discussed.

Tying To Previous Knowledge: Several concepts mentioned in Chapter 3 are developed fully later in the text. For example, Chapter 17 covers change of state in more depth. Try not to anticipate the material in later chapters by going into too much detail in Chapter 3. Many new terms are introduced that students will use throughout the course. You may want to remind students that an understanding of chemical concepts involves the ability to use this special vocabulary.

ASSESSMENT PLANNER

Choose from the following assessment strategies.

Assess, pp. 52-53, 61-62, 68-69
Assessment, pp. 47, 49, 51, 54, 60, 66, 72
Portfolio, p. 72
Minute Lab, p. 49
ChemActivity, pp. 800-802
Chapter Review, pp. 71-75

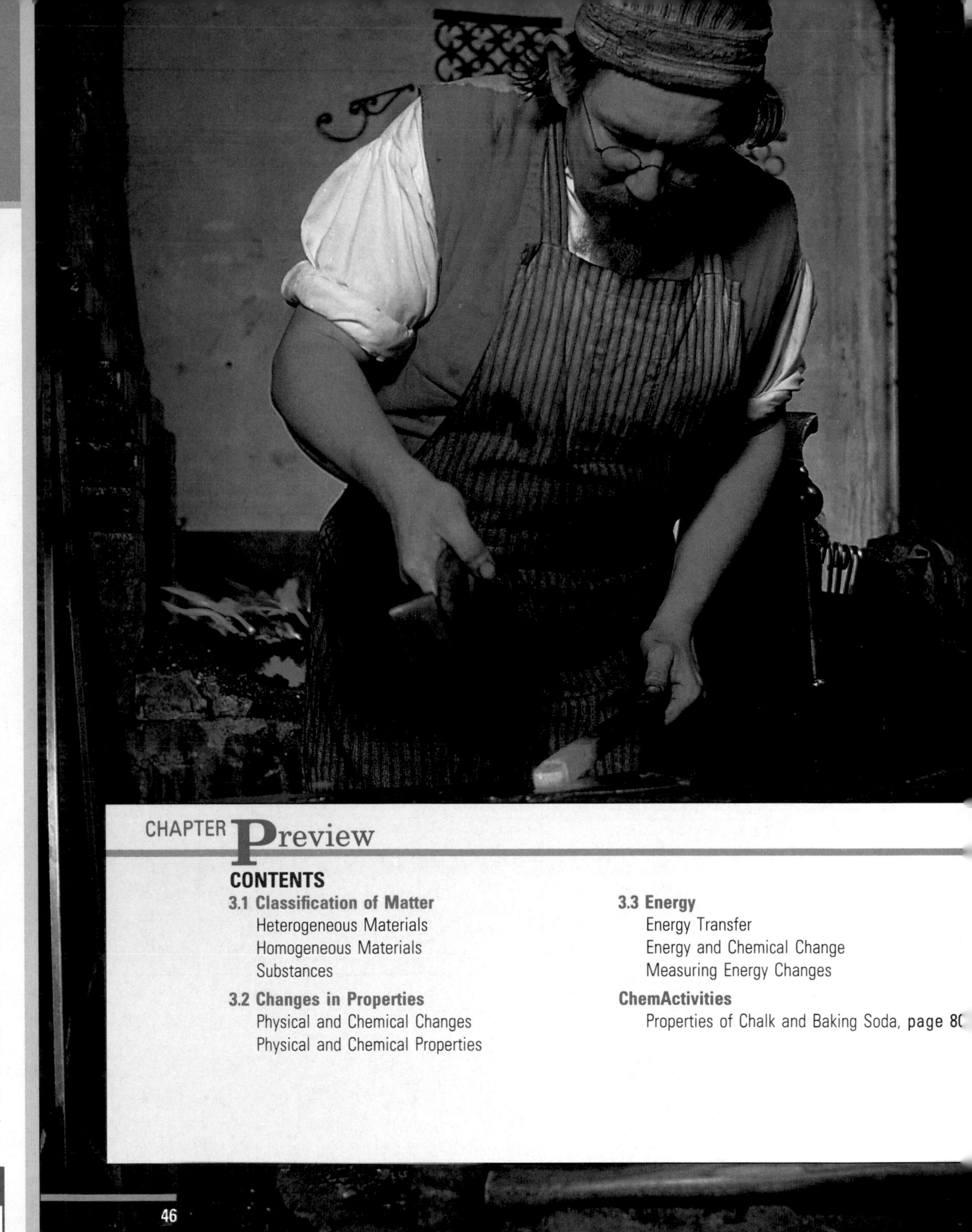

CHAPTER Preview

CONTENTS

DISCOVERY DEMO

3-1 Separation and Testing

Purpose: to make the student aware that a special vocabulary is used by scientists to describe a system and its surroundings

Materials: 1 g of sucrose, 1 g of $CaCO_3$, large test tube with stopper, 10 cm^3 of H_2O, small beaker, a few drops of 3*M* H_2SO_4 (Add 2 cm^3 18*M* H_2SO_4 to 10 cm^3 H_2O in a small test tube.), piece of filter paper, funnel

Procedure: 1. Mix 1 g of sugar and 1 g of $CaCO_3$ in the large test tube. Have a student examine the dry mixture to determine if the two chemicals are distinguishable.
2. Add 10 cm^3 of H_2O and stopper the test tube. Shake the contents to dissolve the sugar. Allow the tube to stand for a minute, then decant the liquid into a filter apparatus. Collect

Matter

CHAPTER

3

The blacksmith shapes a piece of glowing iron by hammering it on an anvil. After shaping the iron, the blacksmith may plunge it into a tub of cold water. The blacksmith's work is possible because of the properties of both iron and water.

Iron is hard and strong, but can be softened enough to shape by heating it to a temperature that is easily achieved in a forge. The blacksmith could not form tungsten or magnesium in the same way as iron. Tungsten can be heated glowing hot without softening. This property makes it useful in light bulb filaments. Magnesium has a very low density combined with high strength but, unlike iron, it cannot be heated red-hot and hammered on an open anvil as iron can. Unfortunately, hot magnesium catches fire in air. This behavior is another of magnesium's properties. In this chapter, you will see how chemists classify matter, its properties, and its changes.

EXTENSIONS AND CONNECTIONS

History Connection
The Village Blacksmith, page 51

Frontiers
Crystal Coolers: How synthetic diamonds are better than the real thing in electronics, **page 52**

Bridge to Mineralogy
Salts from the Dead Sea: Using physical properties to reap a rich harvest from a dead sea, **page 57**

Chemists and Their Work
Gerty Theresa Cori, page 63

Everyday Chemistry
Smog: Some of the causes and possible cures for smog, **page 67**

Chemistry and Society
Energy Problems: A view of some ways that chemistry can help relieve the world's energy shortage, **page 70**

INTRODUCING THE CHAPTER

USING THE PHOTO

Ask students to carefully observe and describe what they see in the photo. Discuss the effect of an increased oxygen supply on the coke in the forge. The higher temperature that the air blowing through the forge produces is used to cause the layers of iron atoms to separate as the iron expands. The hammer is used to apply a force to the iron that causes the layers of iron atoms to slide past one another. The metal becomes thinner and can be shaped on the anvil. The container of water is used to quench the iron, causing the layers of iron atoms to quickly contract. This contraction locks the layers of atoms into their new shape. The strength, hardness, and flexibility of the iron can be greatly affected by the rate of cooling.

REVEALING MISCONCEPTIONS

Show students a small glass of milk and ask if milk is a homogeneous mixture. If they answer yes, they have a common misconception about matter that appears to be the same throughout. Remind them that a chemist must consider matter from a molecular perspective. Encourage students to think at the atomic level when classifying matter. Matter mixes at the molecular level due to molecular motion. A homogenous substance is the same throughout because it consists of only one phase when viewed at the molecular level.

the filtrate in a small beaker.
3. Add a few drops of 3*M* H_2SO_4 to the solid remaining in the test tube. **CAUTION:** *Acid is corrosive.* Note the reaction.
4. Add a few drops of 3*M* H_2SO_4 to the filtrate in the beaker and contrast this reaction with that of the solid and acid. **Disposal:** C

Results: Both substances are white. Sucrose is crystalline and water soluble, and $CaCO_3$ is powder and is insoluble, in water. The $CaCO_3$ reacts with acid, liberating CO_2 gas. The sugar solution does not release any gas when treated with acid.

Questions: 1. Are the color and texture of the solids intensive or extensive physical properties? *intensive*
2. Is dissolving a physical or chemical change? *physical change*
3. Is the filtrate homogeneous or heterogeneous? *homogeneous*
4. When the acid causes the solid to bubble, is that a chemical or physical change? *chemical change.*

✓ ASSESSMENT

Oral: Ask student to describe the separation technique used. Filtration is used when there are both soluble and insoluble substances present.

CONCEPT DEVELOPMENT

MAKING CONNECTIONS

Communications Skills

Before class, place the names of 30 common elements on pieces of paper and place in a box. Ask each student to draw a name from the box. Have the student briefly research the element and write a half-page report that includes physical and chemical properties, where found, and its uses. Have a few reports read to the class each day during the study of this chapter.

QUICK DEMO **Compound Synthesis** Show the class some compounds such as $CuSO_4$, sucrose, ethanol and water. Avoid mixtures, minerals, and polymers. Name the compounds but do not give the formulas unless the class asks for them.

Obtain four small, clear glass bottles. Into the first, place a sample of Cu metal. Into the second, place sulfur. Leave the third empty and label it O_2. Add a sample of $CuSO_4$ to the fourth bottle. The combination of the elements in the first three bottles results in the formation of the compound of the fourth bottle. Discuss the difference between mixing and combining.

PROGRAM RESOURCES

Transparency Master: Use the Transparency master and worksheet, pp. 5-6, for guided practice in the concept "Classification of Matter."

Color Transparency: Use Color Transparency 3 to focus on the concept of "Classification of Matter."

Word Origins

omios: (GK) like, same
genea: (GK) origin, source

homogeneous—of the same source, uniform

The composition of homogeneous mixtures (solutions) can vary. If you put a small amount of pure salt into pure water, stir it, and let it stand, you have a solution, or homogeneous mixture. If you add a larger amount of pure salt to the same amount of pure water, you again have a solution. The composition of the second sample would differ from the first. The second sample contains more salt in an equal volume of water. In each case, the resulting material is homogeneous. A solution may also be defined as a single phase that can vary in composition. Thus, solutions such as antifreeze, seawater, and window glass vary in composition from sample to sample.

Solutions are not necessarily liquid. To a chemist, air is a homogeneous material composed of nitrogen, oxygen, and smaller quantities of other gases. Its composition varies from place to place. However, the ordinary air we breathe usually contains heterogeneous particles such as soot, spores, and pollen. Different types of window glass have different compositions, yet each type is homogeneous. Both air and glass are solutions. So is automobile radiator antifreeze. In antifreeze, the solvent is an organic liquid called 1,2-ethanediol (ethylene glycol). The solutes are various dyes, rust inhibitors, and rubber hose conditioners.

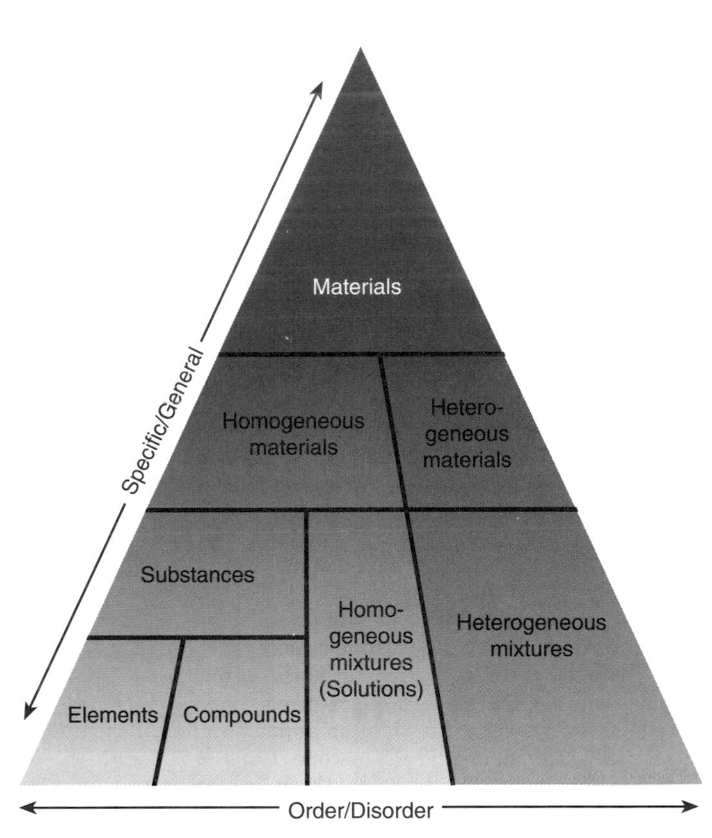

Figure 3.3 All matter can be classified from general to specific types according to composition and properties.

DEMONSTRATION

3-2 Light a Candle

Purpose: to generate a discussion about how an element differs from a compound

Materials: for each team of four students: small candle, matches, white ceramic plate, 600-cm^3 beaker, ice cube

Procedure: Have a student demonstrate the following procedures for his or her cooperative learning group. **1.** Light the candle with a match. **CAUTION:** *Fire risk.* Record observations.
2. Place a white ceramic plate in the candle flame for a few seconds. Remove and observe the underside of the plate.
3. Invert the beaker over the flame and place an ice cube on top of the beaker. Record the observations.
Disposal: F

In your laboratory work, you will be using solutions labeled with a number followed by the letter "*M*." The symbol represents the term *molarity*. The exact meaning of molarity will be studied in Chapter 8. In the meantime, you should keep in mind that molarity is used to indicate the amount of solute in a specific amount of solution. A 6*M* (six molar) solution contains 6 times as much solute as a 1*M* (one molar) solution of the same volume. Concentrated solutions have a higher ratio of solute to solvent than dilute solutions.

Percent by volume, percent by mass, molality, and normality are other methods used to express solution concentration. Normality is used by some analytical chemists and technicians.

HISTORY CONNECTION

The Village Blacksmith

At one time, every community needed a blacksmith to make hinges, hasps, weather vanes, and other pieces of necessary hardware. Even though there are now machines that manufacture these items, blacksmiths are still in great demand to shoe horses. The workers who shoe horses are called *farriers*, a term that was derived from the Latin word for iron, *ferrum*.

Substances

Pure salt, pure sugar, and pure sulfur are homogeneous materials and always have the same composition. Such materials are called **substances.** A large part of chemistry is the study of the processes by which substances change into other substances.

According to the atomic theory, matter is made of very tiny particles called atoms. Substances are divided into two classes based on the atoms they contain. Substances composed of only one kind of atom are called **elements.** Examples are sulfur, oxygen, hydrogen, nitrogen, copper, gold, and chlorine. Substances composed of more than one kind of atom are called **compounds.** The atoms in the particles of compounds are always in definite ratios. For example, water contains hydrogen atoms and oxygen atoms combined in a ratio of 2 to 1.

To summarize, all matter can be classified as heterogeneous mixture, solution, compound, or element. The development of this system of classification played a significant role in the early history of chemistry. Early chemists spent much time and energy sorting the pure substances from the mixtures.

Figure 3.4 Substances are either elements or compounds. Shown here are the elements sulfur (left), copper (top center), iodine (crystals in beaker), and bromine (red liquid in flask). Also shown are the compounds sodium dichromate (orange crystals), sodium chromate (yellow crystals), quartz (silicon dioxide, clear crystal), and galena (lead(II) sulfide, gray chunk).

HISTORY CONNECTION

The Village Blacksmith

Exploring Further: The symbol for iron, Fe, comes from the Latin word *ferrum.* The historical names for many elements are interesting, in particular those elements whose symbols are not the first one or two letters of the present name for the element.

CONCEPT DEVELOPMENT

Enrichment: Have a student investigate the structure of gelatin. In what physical state is it classified? Is it heterogeneous or homogeneous? *It is a solid, heterogeneous mixture having a dispersed phase and a continuous phase.* Also see colloids in Chapter 20.

Using The Illustration: Figure 3.3 provides the student with a visual organization chart of this section. Call their attention to the Specific/General and Order/Disorder arrows along the sides of the triangle. Inform students that, if they can draw and label the diagram from memory, they have mastered this classification system.

Results: The solid candle melts and the liquid is drawn up into the wick where it is converted into a vapor that burns in air. A black deposit appears on the white ceramic plate. The candle is slowly extinguished when covered by a beaker. Condensed water droplets appear on the inside of the beaker.

Questions: 1. Is paraffin an element or a compound? *compound*
2. Is paraffin homogeneous or heterogeneous? *homogeneous*
3. Which element do you believe is deposited on the ceramic plate? *carbon*
4. Which two elements are present in paraffin? *hydrogen and carbon* If the condensed droplets inside the beaker are water, what is the source of the oxygen in the water? *O_2 from the air*

✓ ASSESSMENT

Performance: Have several students invert different-sized beakers over burning candles. Compare the time it takes for the candle to be extinguished to the volume of air. *The larger the volume of air, the longer the candle burns.* The droplets can be shown to be condensed water by testing them with dry cobalt chloride test paper. The test paper turns pink in the presence of water.

FRONTIERS

Crystal Coolers

Background: Diamonds are crystals composed of two isotopes of carbon, carbon-12 and carbon-13. Unlike natural diamond that is composed of about 99% carbon-12 and 1% carbon-13, synthetic diamond contains almost 99.9% carbon-12. Carbon-13, retards the conduction of heat. The thermal conductivity of synthetic diamond is thus significantly greater than that of natural diamond.

Teaching This Feature: Help students to visualize thermal conductivity as the quantity of heat, q, that is transferred through a 1-meter thick piece of material with a cross sectional area of one square meter in 1 second because of a temperature difference of 1 kelvin.

Exploring Further: Have students use reference materials to identify and list other physical properties of diamonds. Possible response: *luster, transparency, index of refraction, and hardness.*

3 ASSESS

CHECK FOR UNDERSTANDING

Ask students to answer the Concept Review problems, and Review and Practice 21-27, 38-39 from the end of the chapter.

RETEACHING

Set up a lab practical and assign lab partners. At numbered lab stations, place common examples of elements, mixtures, and compounds. Have students indicate if the sample is homogeneous or heterogeneous and to classify the material as an element, compound, mixture, or solution.

Cooperative Learning: Refer to pp. 28T-29T to select a teaching strategy to use with this activity.

EXTENSION

Have interested students bring an item from home, and be prepared to tell the class whether it is an element, mixture, or compound, and whether the item is homogeneous or heterogeneous.

You have probably seen a periodic table of the elements. There may be one visible in your classroom right now. The periodic table is an indispensable tool of the chemist because of the amount of information it contains. Later in the course, you will learn more about this table and the reasons for its arrangement. Table 3.1 on pages 54–55 is a special version of the periodic table which pictures samples of nearly all the naturally-occurring elements. A complete list of the elements is found in Table 4.2 on page 84. Note that the second column of Table 4.2 also gives a symbol that is used to represent each element.

Chemists know of 88 naturally-occurring elements. The natural element with the most complex atoms is uranium. Two elements, technetium and promethium, which have simpler atoms than uranium, are not found in nature. Astatine and francium have been detected in nature. However, they are present in such small amounts that they cannot be easily separated from their ores. These four elements are not normally counted among the natural elements. These synthetic elements can be produced by scientists through nuclear reactions. Synthetic elements and nuclear reactions will be discussed later.

FRONTIERS

Crystal Coolers

This shimmering diamond was not mined from the earth. Instead, it was produced in a laboratory and is almost 99.9% pure. Because of its purity, it has properties that differ significantly from those of natural diamonds. One such property is its thermal conductivity.

A natural diamond conducts heat about six times better than copper wire at room temperature. However, synthetic diamond is almost 50% more conductive than natural diamond, making it the world's best conductor of heat.

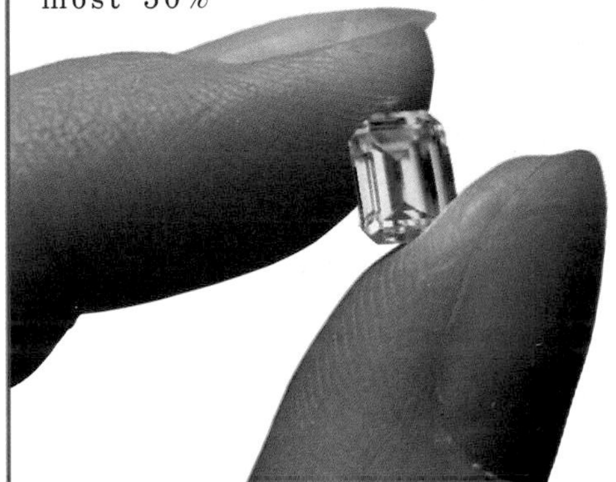

In microelectronic circuits, even minute quantities of heat may damage delicate components. To prevent damage, heat sinks are built into the circuits. Heat sinks are devices that carry away unwanted heat. Because of the synthetic diamond's high thermal conductivity, engineers will be able to replace miniature heat sinks made of natural diamond with even smaller slabs of synthetic diamond. Thus it will be possible to miniaturize circuits even further.

The high thermal conductivity and transparency of synthetic diamond may also find use in the field of laser optics. Because synthetic diamond can remove heat at a high rate, crystals of synthetic diamond can transmit intense laser light without damage. Engineers are trying to fashion mirrors and windows for lasers out of synthetic diamond that will allow them to construct more powerful and complex lasers. Perhaps in the future, these gems will make possible complex computers operated only by light.

DEMONSTRATION

3-3 Sweetened Tea

Purpose: to demonstrate that a solution is homogeneous and can be a part of a heterogeneous system

Materials: tea bag, ice cubes, water, drinking glass, 100 g sugar, 400-cm^3 beaker, hot plate

Procedure: 1. Place sugar in a glass of cold, iced tea (homogeneous) so that, after stirring, some remains undissolved on the bottom (heterogeneous) of the glass.

2. Add additional sugar to the saturated solution. Stir, and observe.

3. Transfer the mixture to the 400-cm^3 beaker and warm gently on a hot plate. Stir.

Results: The sugar initially dissolves in the cold, iced tea. The additional sugar added remains on the bottom of the glass of iced tea. When warmed, the

Figure 3.5 All organic substances contain the element carbon. Organic materials shown here include rubber, plastic, wood, and paper. Typical inorganic substances do not contain carbon. Inorganic materials shown here are glass, ceramic, metal, and water.

Chemists are interested in the reactions of elements and compounds, the analysis of compounds into their component elements, and the synthesis of compounds from elements or other compounds. These properties and processes depend upon the structure of the elements and compounds. The development of new pharmaceutical products is pursued largely on the basis of known reactions of the human body to molecules with particular structures.

There are more than 10 000 000 substances known to chemists. Obviously, no one scientist knows the characteristics of all of these substances. Thus, chemists classify them into different groups with common features. Chemists divide all substances into two large classes called *organic* and *inorganic*. **Organic substances** are compounds that contain the element carbon. **Inorganic substances** are the elements and the compounds of all elements other than carbon. These categories are defined only for convenience. Some carbon compounds are considered inorganic, for reasons we will consider later. Chemists generally concentrate their efforts in one of the two fields, *organic chemistry* or *inorganic chemistry*.

3.1 CONCEPT REVIEW

1. Classify the following materials as heterogeneous mixtures, solutions, compounds, or elements. Use a dictionary to identify any unfamiliar materials.
   a. air
   b. india ink
   c. paper
   d. table salt
   e. wood alcohol
   f. apple
   g. milk
   h. plutonium
   i. water
2. How would you determine if a piece of cloth advertised as 50% wool and 50% synthetic fiber was a heterogeneous mixture or a homogeneous mixture?
3. Distinguish between elements and compounds by contrasting examples of each.
4. **Apply** How many phases are present in a glass of soda on ice?

Answers to

Concept Review

1. a. solution (Gases in gases are homogeneous solutions rather than heterogeneous mixtures.) **b.** heterogeneous mixture **c.** heterogeneous mixture **d.** compound **e.** compound **f.** heterogeneous mixture **g.** heterogeneous mixture (When homogenized, milk is an emulsion—a colloid.) **h.** element **i.** compound

2. Examine under a microscope.

3. Copper is an element because it is a substance composed of only one kind of atom. Water is a compound because it is a substance composed of more than one kind of element, in this case, hydrogen and oxygen.

4. Four phases are present (glass, liquid, gas bubbles, ice).

4 CLOSE

Activity: Use four 500-cm^3 clear bottles. Label them C, O, H, and Zn. Place charcoal in the bottle labeled C, and mossy zinc in the one labeled Zn. Place lids or stoppers on all four bottles. Ask students which compounds could be formed by combining the elements in the bottles labelled C and O. *carbon monoxide and carbon dioxide* Ask the same question of the contents of the bottles labelled H and O. *water and hydrogen peroxide* Students will suggest zinc oxide could be made by combining the contents of bottles labeled Zn and O.

PROGRAM RESOURCES

Applying Scientific Methods in Chemistry: Use the worksheet, "Classifying Matter," p. 18, to help students apply and extend their lab skills.

amount of sugar on the bottom is reduced or disappears completely.

Questions: 1. Is a glass of unsweetened tea heterogeneous or homogeneous? *homogeneous*

2. What additional interface is introduced when excess sugar is added? *solid (sugar)/liquid (tea)*

3. What happens to the solubility of the sugar as the temperature of the tea is increased? *Solubility increases.*

Extension: Relate this demonstration to Section 3.2, Changes In Properties. Have students observe the solubility curves in Figure 3.7.

CONCEPT DEVELOPMENT

Background: Dimitri Mendeleev (1834-1907) was preoccupied with examining the physical and chemical properties of the elements. He prepared a set of index cards, one per element, on which he wrote the name, mass, and properties. He arranged and rearranged these cards, looking for regularity. Ultimately, Mendeleev recognized that there were similar properties that reoccurred on a regular, repeating basis. He arranged the cards into a table form in which the elements having similar properties were placed in vertical columns on the basis of increasing atomic mass. This periodic table was published in 1869. The table had some obvious imperfections, but the idea of periodicity of properties had been introduced. Two years later he published a much improved table. Mendeleev's "flash of genius" was evident in 1871 when he suggested that the vacant spaces in his periodic table were due to yet undiscovered elements. During the next two decades, three elements, Ga, Sc, and Ge, were discovered and verified his predictions. Moseley's work in 1913 on atomic numbers resulted in the periodic table being changed. The elements today are arranged in order of atomic number.

• This is not the time to teach the periodic table. Atomic structure will be studied in detail in Chapters 4 and 5, then a study of the periodic nature of properties based on the atomic structure will be presented in Chapters 6, 10, and 11. Focus student attention on the vocabulary of Chapter 3.

✓ ASSESSMENT

Skill: Give students a blank outline of the periodic table. Have them fill in as many chemical names and symbols as they can. Repeat this assessment for three days as the students become more familiar with the arrangement of the table.

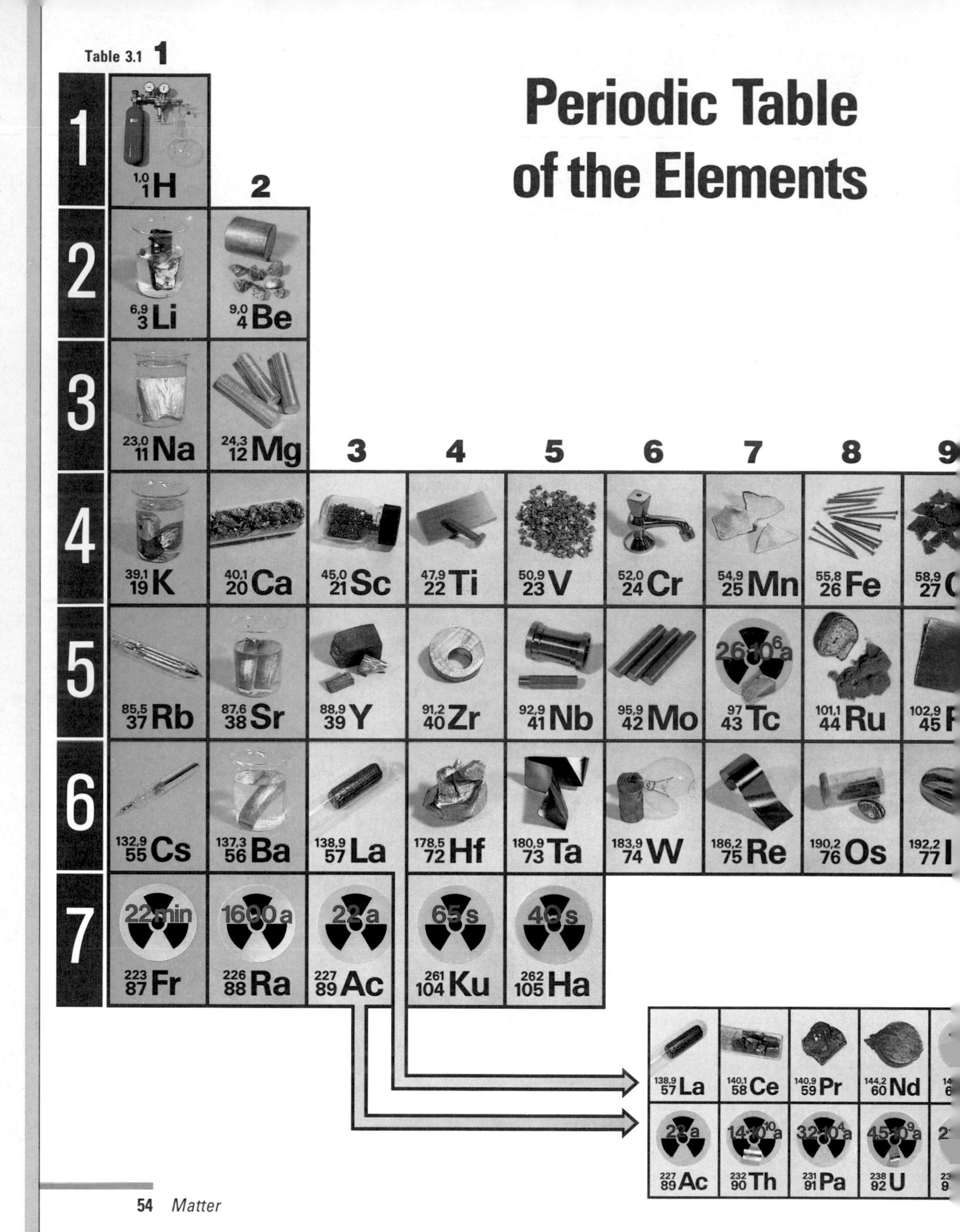

CONTENT BACKGROUND

Periodic Table: A student may notice that the periodic table pictured here is not exactly the same format as the one used in Chapter 6. For example, the names kurchatovium and hahnium were once proposed for elements 104 and 105, resulting in the symbols Ku and Ha. Since then, IUPAC has approved a systematic method for naming the elements above 103. (See the Frontiers feature on page 149.) Also, elements 106–112 are not shown here. Symbols for radioactivity designate the elements that have no stable isotopes. The half-life of the most stable isotope of each of these elements is printed in red. The symbol "a" stands for "year" (*annum*).

This table is not intended as a reference periodic table. Its sole purpose is to photographically show the physical properties of the elements. The IUPAC-approved format for the periodic table in Chapter 6 is better related to an atom's electronic structure.

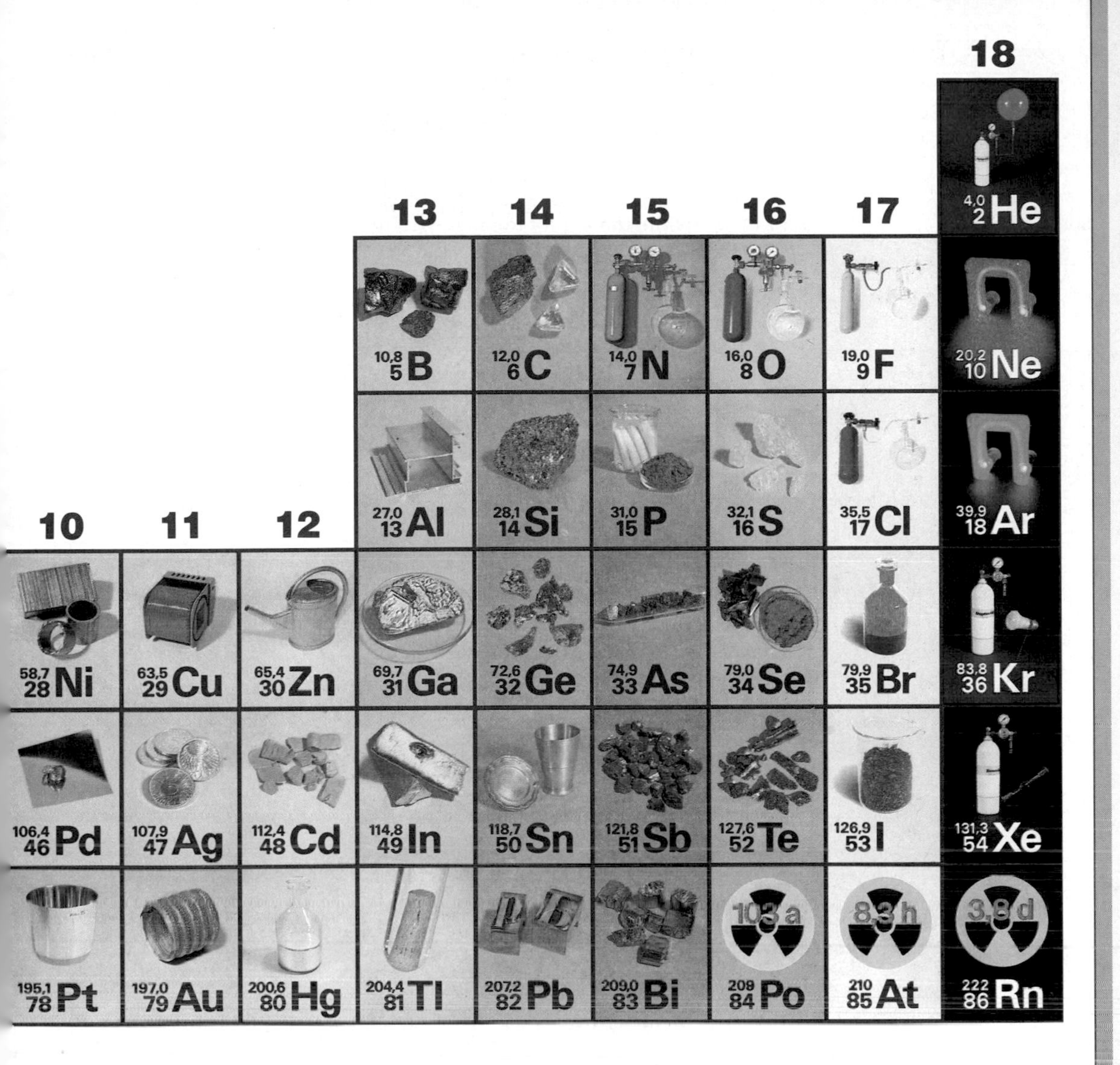

150.4 62 Sm	152.0 63 Eu	157.3 64 Gd	158.9 65 Tb	162.5 66 Dy	164.9 67 Ho	167.3 68 Er	168.9 69 Tm	173.0 70 Yb	175.0 71 Lu
244 94 Pu	243 95 Am	247 96 Cm	247 97 Bk	251 98 Cf	254 99 Es	257 100 Fm	258 101 Md	259 102 No	260 103 Lr

55

CONCEPT DEVELOPMENT

- Point out to students that most of the 112 known elements are metallic solids. Many of the nonmetals are gases at room temperature. Have students identify mercury, a metal, and bromine, a nonmetal, as the two elements that are liquids at room temperature.

Using The Table: Students need to know the properties of metals because there are so many of them. The periodic table presented here makes that visually apparent to students. Use the table to differentiate between extensive properties such as mass, volume, and length, and intensive properties such as malleability, ductility, conductivity, color, melting point, and boiling point.

Cooperative Learning: Refer to pp. 28T-29T to select a teaching strategy to use with this activity.

MAKING CONNECTIONS

Daily Life

Ask students to recognize and list elements that they have seen or know may be present at home or school. Gold, platinum, and silver jewelry, mercury in a thermometer, nitrogen and oxygen in air, copper wire and plumbing pipes, iron tools, chromium-plated faucets, argon in lightbulbs, carbon in pencil lead, aluminum beverage cans, lead fishing sinkers, and radioactive radon gas in basements can all be mentioned. Emphasize that these many elements are around and in us. Point out to students that in this introductory chemistry course they will have an opportunity to study many of these elements and their unique chemistry in detail.

3.2 Changes in Properties

PREPARE

Key Concepts

The major concepts presented in this section include physical and chemical changes and physical and chemical properties.

Key Terms

physical change
chemical change
precipitate
physical property
chemical property
extensive property
intensive property

1 FOCUS

Objectives

Preview with your students the objectives listed on the student page. Students can use them as a study guide for this major section.

Enrichment

Ask a volunteer to find out how crude oil can be obtained from shale through physical and chemical means. Possible response: *One of several methods involves preparation and then retorting. Preparation involves crushing and screening the shale into the correct size pieces. One kind of retort procedure subjects the shale to super-heated steam that vaporizes the oil. The steam also extracts ammonia, nitrogen gas, and naphtha. The vapors are scrubbed to remove the crude oil vapor from the other gases. Then the crude oil is refined.*

OBJECTIVES

- classify changes in matter as physical or chemical.
- obtain information from a graph.
- distinguish among extensive, intensive, physical, and chemical properties.

3.2 Changes in Properties

The iron that the blacksmith is shaping in the chapter opening photograph has some characteristics by which we can identify it. It is solid and gray. It also reacts with hydrochloric acid. The characteristics of solidity and color are called physical properties, while the ability to react with hydrochloric acid is called a chemical property. In this part of the chapter, you will learn about the different properties of matter and about the different kinds of changes matter can undergo.

Physical and Chemical Changes

Substances undergo changes when their conditions are changed. A change in condition could be an increase in temperature, a mechanical force, exposure to another substance, or any of a number of other alterations. If the same substance remains after the change, a **physical change** has taken place. If a new substance has appeared, a **chemical change** has occurred.

Physical Changes

Changes like pounding, pulling, or cutting do not usually change the chemical character of a substance. Pounded copper is still copper. Only its shape is changed. Cutting a piece of wood into smaller pieces, tearing paper, dissolving sugar in water, and pouring a liquid from one container to another are other examples of physical changes.

Physical changes occur when a substance melts or boils. At its melting point, a substance changes from the solid phase to the liquid phase. At its boiling point, a liquid changes to a gas. Such physical changes are called *changes of state*, because the substance is not altered except for its physical state.

Figure 3.6 Hammering copper changes only its shape. This is an example of a physical change because the substance remains copper.

GLENCOE TECHNOLOGY

Videodisc

Chemistry: Concepts and Applications

Changes of State
Disc 1, Side 2, Ch. 17

Also available on CD-ROM.

DEMONSTRATION

3-4 Evidences of Chemical Change

Purpose: to enable students to observe some evidences of chemical change

Materials: pair of crucible tongs, 10-cm x 10-cm piece of paper, matches

Procedure: 1. Hold the 10-cm x 10-cm paper over the sink with tongs. Light it on an edge, and allow it to burn with the room lights dimmed. **CAUTION:** *Keep the burning paper from other flammable materials.*
2. Discuss the changes such as color, hardness, mass, length, relative density, conductivity, that have occurred in the paper's physical properties.

Disposal: H

Results: Students will quickly realize that a chemical change produces new substances having different physical properties.

Questions: 1. Does the color change as

A knowledge of physical changes and the conditions under which they occur can be used to separate mixtures. Separating the substances in mixtures by distillation is a change-of-state operation. It is used to separate substances with different boiling points. For instance, you can separate a solution of salt in water by heating the solution to its boiling temperature. The water will then be turned to steam and escape from the container. The steam can be cooled to turn it into pure water. The salt, whose boiling point is very high compared with that of water, will remain behind. Distillation is a method that is used to produce drinking water from seawater in a few locations in the world. However, distillation is a very expensive way to produce pure water because it requires a great deal of energy, so the process is used only when necessary.

Bridge to MINERALOGY

Salts from the Dead Sea

The Dead Sea is located between the countries of Israel and Jordan. It is called the Dead Sea because almost nothing can live in its waters, which contain large quantities of dissolved salts. Far from being worthless, the Dead Sea's salts are valuable to industry and agriculture. Both Jordan and Israel are extracting these substances from the waters of the Dead Sea as a commercial enterprise. The Israeli plant was begun in the 1930s, and the Jordanian operation began in the early 1980s.

The substances extracted are sodium chloride, potassium chloride, magnesium chloride, and magnesium bromide. Of these chemicals, potassium chloride provides the most income. Two factors enable the substances in solution to be separated by fractional crystallization. One factor is the difference in the solubilities of the substances in solution. The second factor is the rate at which the solubility changes as the temperature changes. Thus, by a carefully controlled combination of evaporation and cooling, the various salts can be separated and purified.

Exploring Further

1. The water of the Dead Sea contains many dissolved materials. Would you classify it as heterogeneous or homogeneous? As a material or a substance?
2. Look up the solubilities of the four substances sodium chloride, potassium chloride, magnesium chloride, and magnesium bromide at 20°C and at 100°C. Suggest a way to separate these substances.

the paper burns? *yes*
2. Does relative density change? *yes*
3. Is a gas evolved? *yes*
4. Does the product of combustion appear to be a different substance having different properties? *yes*

Extension: This demonstration can also be used in Section 3.3 to illustrate that combustion is an exothermic chemical reaction.

PROGRAM RESOURCES

Study Guide: Use the Study Guide worksheet, p. 8, for independent practice with concepts in Section 3.2.

Reteaching: Use the Reteaching worksheet, p. 4, to provide students another opportunity to understand the concept of "Matter and Its Properties."

2 TEACH

CONCEPT DEVELOPMENT

QUICK DEMO **Salt Dissolves in Water** Help students visualize the dissolving of a salt crystal by tearing a sheet of paper into smaller and smaller pieces. Ask students which physical properties have changed (for example, mass and length) and which have not (for example color, hardness, density, and conductivity). Emphasize that changing the size of the pieces is a physical change. Point out that the two groups of properties they have just determined correspond to extensive and intensive physical properties, respectively.

BRIDGE TO MINERALOGY

Salts from the Dead Sea

Background: The Dead Sea is below sea level. The water is saline and does not support any life. In the United States the Great Salt Lake in Utah is also used as a source of minerals.

Teaching This Feature: Place samples of the chemicals listed on display on watch glasses. Students can see crystals of the various salts they are studying.

Connection to Biology: Why is the Dead Sea dead? The salt content can reach a level that is too concentrated for life to exist in the waters of the Dead Sea. Cells are dehydrated by osmosis and die.

Extension: Some students may want to grow crystals by the slow evaporation of water. Sugar crystals, or rock candy, can be grown from a saturated solution of sugar over a two-week period. Remind students not to jar the container if they want to see very large crystals.

Answers to
Exploring Further

1. *heterogeneous due to density differences, material*
2. *Slowly evaporate the water, and use fractional crystallization.*

MAKING CONNECTIONS

Daily Life

The doctor sometimes recommends gargling with salt water to relieve a sore throat. Ask students to determine whether the action of the salt water on the bacteria is chemical or physical. *The action of salt water on bacteria is physical. Too high a concentration of salt causes the cells to lose water from their cytoplasm and to dry up.* See osmosis in Chapter 21.

PRACTICE

Guided Practice: Have students interpret Figure 3.7 as a solubility temperature graph. Then have them work in class on Practice Problem 5.

Independent Practice: Your homework or classroom assignment can include Application 51-52 from the end of the chapter.

Answers to
Practice Problems

5. **a.** 38 g/100 g H_2O
b. 230 g/100 g H_2O
c. 46 g/100 g H_2O
d. 95 g/100 g H_2O

GLENCOE TECHNOLOGY

CD-ROM

Chemistry: Concepts and Applications

Exploration: *Separating Mixtures*
Interactive Experiment:
Separate the Substances

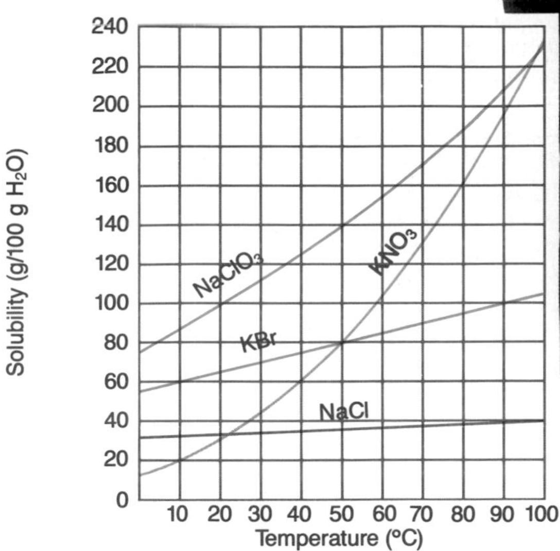

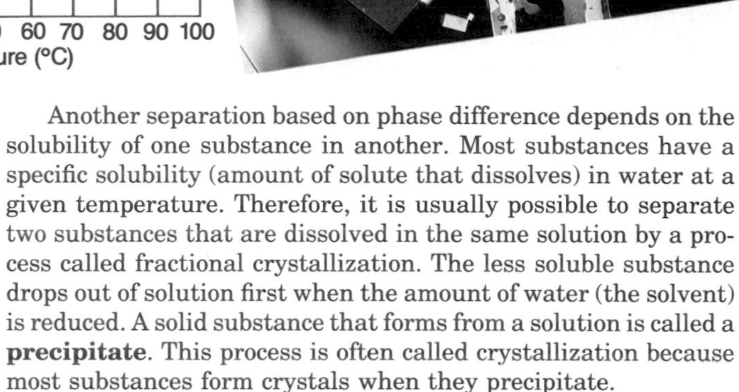

Figure 3.7 This solubility graph shows the amount of each substance that will dissolve in water at any temperature between 0°C and 100°C. When temperature is reduced or water is evaporated, crystals like those shown (above right) will form.

Another separation based on phase difference depends on the solubility of one substance in another. Most substances have a specific solubility (amount of solute that dissolves) in water at a given temperature. Therefore, it is usually possible to separate two substances that are dissolved in the same solution by a process called fractional crystallization. The less soluble substance drops out of solution first when the amount of water (the solvent) is reduced. A solid substance that forms from a solution is called a **precipitate**. This process is often called crystallization because most substances form crystals when they precipitate.

Notice in Figure 3.7 that, at 70°C, potassium bromide (KBr) is less soluble than potassium nitrate (KNO_3). If a water solution containing equal amounts of both substances is allowed to evaporate at 70°C, the potassium bromide will crystallize first. This process is used quite often in industry in the purification of many crystalline items, such as sugar, salt, and medicines.

PRACTICE PROBLEMS

5. Obtain information from the graph in Figure 3.7 to determine the solubilities of the following.
   a. NaCl at 70°C
   b. $NaClO_3$ at 100°C
   c. KNO_3 at 30°C
   d. KBr at 80°C

Chemical Changes

Suppose you are given two test tubes containing colorless liquids. One test tube contains water. The other contains nitric acid. If you place a pinch of sugar in the water, the sugar will disappear and the liquid will remain colorless. The sugar has dissolved and

PROGRAM RESOURCES

Transparency Master: Use the Transparency master and worksheet, pp. 7-8, for guided practice in the concept "Solubility Curves."

Color Transparency: Use Color Transparency 4 to focus on the concept of "Solubility Curves."

Laboratory Manual: Use microlab 3-1 "Using Paper Chromatography to Separate Dyes" as enrichment for concepts in Chapter 3.

Laboratory Manual: Use microlab 3-2 "Determining Boiling Point" as enrichment for the concepts in Chapter 3.

Sunkist®

Receipt

Caroline Ham
Sold To

21 Elliot Ave — Wenham — MA — 01984
Address — City — State — Zip

978-468-5063, cell 978-809-1291
Telephone

Amount Received — Date

Oranges

Full-Size Carton (About 40 Lbs.) — Half Carton (About 20 Lbs.)

Grapefruit

(1) $17.50

Full-Size Carton (About 40 Lbs.) — Half Carton (About 20 Lbs.)

Hamilton-Wenham Concert Choir
Group Name

Anna Evetts
Sold By

Feb 1st or 2nd
Estimated Delivery Date

Code #8126 — Litho in USA

Temp (°C)	Solubility g/100 mL H_2O
0	78
10	88
20	100
30	110
40	122
50	140
60	157
70	170
80	188
90	207
100	230

10
35 45
20
65
20
85
1.5
100

I 5 2 10
II 7 9 28
III 5 4 5 28 or
IV 3 5 12
80

7 20

you now have a sugar-water solution. A physical change has occurred. Neither the sugar nor the water has changed to a different substance. If you put a small piece of copper into the other test tube, the copper will also disappear. However, the liquid will turn blue and give off a brown gas. You now have a solution of copper(II) nitrate and some nitrogen dioxide gas, both of which are new substances. A chemical change has occurred.

Let us take another example. Sodium is a silvery, soft metal that reacts vigorously with water. Chlorine is a yellow-green gas that is highly corrosive and poisonous. However, if these two elements are brought together, they combine to produce a different substance, a white crystalline solid. This new substance is common table salt, sodium chloride, which neither reacts with water nor is poisonous. The behavior of the product is quite different from that of either reactant.

Whenever a substance undergoes a change so that one or more new substances with different characteristics are formed, a chemical change (reaction) has taken place. Burning, digestion, and fermenting are examples of chemical changes.

There are many instances in chemistry when a general statement can be made to cover most, but not all, cases of a particular phenomenon. Such a statement is called a rule of thumb. You come across rules of thumb every day. For example, one rule of thumb for driving a car is that you leave one car-length between you and the car ahead for each 10 miles per hour of speed. That statement is a rule of thumb because there are exceptions. Likewise, in chemistry, rules of thumb are generally true, but keep in mind that there may be exceptions.

Figure 3.8 When copper reacts with nitric acid, a change produces a blue solution and a brown gas (below left). Because new substances with different properties are formed, a chemical change has occurred. Sodium chloride (table salt) is the new substance resulting from the chemical change that takes place when sodium metal and chlorine gas are combined (below).

CONCEPT DEVELOPMENT

Activity: Introduce the concepts of physical and chemical change by cutting a 5-cm length of magnesium ribbon into small pieces. Put aside one piece for comparison. Place the other pieces into a porcelain crucible, cover, and heat thoroughly over a burner. **CAUTION:** *Fire risk.* While heating, raise the cover occasionally to allow oxygen to enter. Have a student describe the cut piece and then the substance in the crucible to the class.

Background: Point out to students that physical and chemical changes are always accompanied by energy changes. The energy changes occur between a system and its surroundings.

Chemical and Physical Changes

Have students make up lists of everyday chemical and physical changes. Their chemical-change list might include the use of baking soda in cooking and the burning of gasoline. The physical-change list might include the dissolving of sugar in coffee and the boiling of water. Ask students also to state any macroscopic changes that accompany chemical changes.

CONTENT BACKGROUND

Dr. Linus Pauling: A two-time Nobel prize winner, he reported an observation regarding sickle-cell anemia. In this disease, an abnormal form of hemoglobin, which is not water soluble, is present. This form of hemoglobin cannot carry oxygen in the blood and the sickle-shaped cells clog capillaries. Ask an interested student to prepare a report on sickle-cell anemia.

CULTURAL DIVERSITY

Sweet Success

Before the Civil War, Norbert Rillieux, an African American, developed an invention that revolutionized the processing of sugar. Rillieux was born into slavery in New Orleans in 1806. His father, a wealthy Cajun plantation owner and engineer, sent him to study in Paris, France. There he became an accomplished engineer, teaching and writing papers on the steam engine. He returned to Louisiana where he developed a vacuum evaporator that turned sugar cane juice into fine-grade white sugar crystals.

Juice was heated in an initial vacuum pan. The vapor from the evaporation of juice in the first pan heated the juice in the second pan. This vapor then heated the juice in the third pan, and so on. The pressure in each pan was less than the one before it, and the last pan was connected to a condenser.

Rillieux's vacuum evaporator, patented in 1846, was used extensively throughout Louisiana, and later adopted in Cuba and Mexico. Its use spread to Europe to help in the sugar beet industry. With modification, it has also been used to produce soap, gelatin, and glue.

✓ ASSESSMENT

Knowledge: Have students write the four types of evidence that a chemical change may have occurred. Precipitate, evolution of a gas, color change, and energy change should be listed from memory.

RULE OF THUMB

In this textbook, useful rules of thumb will be indicated by a logo in the margin. The following is a rule of thumb for deciding whether or not a chemical reaction has taken place. *If a precipitate, gas, color change, or energy change occurs, a chemical change has taken place.* Why is this only a rule of thumb? It is because there are other circumstances that can cause these changes. For instance, cooling a solution may cause some of the solute to precipitate. This would not be a chemical reaction.

Decomposing chemical compounds into their component elements always requires a chemical change. On the other hand, mixtures can be separated by physical means. However, it is sometimes more convenient to separate mixtures by a chemical change. For example, brass is a mixture of copper and zinc. It should be possible to separate the two elements, copper and zinc, by grinding the brass into fine particles. Then, using a microscope and a very fine pair of tweezers, we could pick out the zinc particles from among the copper particles. How much easier it is to react the brass with hydrochloric acid. The copper is unaffected, but the zinc is converted to zinc chloride, which dissolves in the acid. The copper is then recovered. The zinc chloride solution can be treated with aluminum to regenerate the zinc metal.

Physical and Chemical Properties

A description of the behavior of a substance undergoing a physical change is one type of **physical property,** while a **chemical property** describes the behavior of a substance undergoing a chemical change. For example, chlorine has the chemical property that it reacts with sodium, as described earlier. The yellow-green color of chlorine gas is a physical property. Length, color, and temperature are typical descriptive terms that are also physical properties of a substance.

Figure 3.9 A physical property of lead is that it is very malleable. It can be hammered thinner without breaking. Glass becomes pliable like taffy when it is very hot. Pliability is a physical property of glass at elevated temperatures.

CONTENT BACKGROUND

Drinking Water: The shortage of pure drinking water has become critical in many parts of the world. Energy costs determine if, when, and where water will be available. Ask students to investigate various methods used to separate drinking water from seawater. Ask students to pay particular attention to phase changes and other physical processes involved in the methods. Students may be able to find information on solar distillation, conventional evaporators, flash evaporation, electrodialysis, vapor compression, reverse osmosis, and freezing.

a

b

c

d

Figure 3.10 A mixture of sugar, sand, iron filings, and gold dust (a) is clearly heterogeneous when viewed under a microscope (b). The mixture can be separated using physical properties, such as the magnetism of iron (c). The pure, separated components of the mixture are shown in (d).

Physical properties may be divided into extensive properties and intensive properties. **Extensive properties** depend on the amount of matter present. Some of these properties are mass, length, and volume. **Intensive properties** do not depend on the amount of matter present. Intensive properties are useful for identifying substances. For example, density is an intensive property. Each sample of a substance, regardless of its size, has the same density. Other intensive properties include malleability, ductility, and conductivity. For example, copper can be hammered into thin sheets. It is more malleable than iron. Copper can also be drawn out into fine wire; it is quite ductile. Both copper and silver conduct heat and electric current well. That is, they offer little resistance to the flow of energy or electricity. A silver spoon will become hot if left in a hot pan of soup because the silver is a good conductor of heat. Most house wiring is made of copper because copper has a high electrical conductivity.

In addition to density, the intensive properties most important to the chemist are color, crystalline shape, melting point, boiling point, and refractive index (ability of material to bend light).

3 ASSESS

CHECK FOR UNDERSTANDING

Ask students to answer the Concept Review problems, and Review and Practice Problems 28-29 at the end of the chapter.

RETEACHING

QUICK DEMO **Physical Dissolving and Chemical Reacting** Partially fill a petri dish with water and place it on the overhead projector. Place a crystal of potassium permanganate in the water. Have students observe the purple streamers as you gently stir the water. Review that a solution is a homogeneous mixture and that dissolving is a physical change, therefore, solubility is a physical property. Add 1 g of sodium hydrogen sulfite to the solution and stir until the solution turns clear. Have students identify clues that indicate that a chemical change has occurred. **Disposal:** C

EXTENSION

Ask student teams to devise a procedure that will allow you to separate a mixture of copper metal filings and sodium chloride. Allow the students to test the procedure in the laboratory. Possible response: Dissolve the sodium chloride in water. Filter out the Cu metal. Wash and dry it. Evaporate the filtrate to reclaim the sodium chloride.

GLENCOE TECHNOLOGY

Videodisc

Chemistry: Concepts and Applications

Physical and Chemical Properties of Matter
Disc 1, Side 1, Ch. 4

Also available on CD-ROM.

The Unique Properties of Water
Disc 2, Side 1, Ch. 11

Also available on CD-ROM.

interNET CONNECTION

Details on current supported research are available on-line at the Department of Energy.

World Wide Web
http://www.doe.gov

Answers to
Concept Review

6. **a.** extensive **b.** intensive **c.** intensive **d.** extensive **e.** intensive

7. **a.** chemical **b.** chemical **c.** physical **d.** chemical **e.** physical **f.** chemical **g.** physical **h.** physical

8. **a.** physical **b.** chemical **c.** chemical **d.** physical **e.** physical **f.** chemical **g.** physical **h.** physical **i.** physical **j.** physical **k.** chemical

9. No, it has the chemical properties of being unreactive, nonflammable, and stable, for example.

4 CLOSE

Discussion: Ask students to imagine they are painting the walls of their bedroom. Ask them to describe what happens to the paint after it leaves the can. Are these changes chemical or physical? Possible response: The smoothing of the paint is physical. The drying of paint is a physical change as the water or thinner evaporates. Epoxy paint used on pools and appliances undergoes a chemical change as it cures.

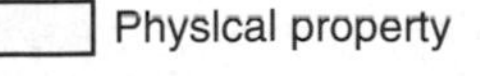

Uranium. U: atomic mass 238.029; atomic number 92; oxidation states 6+, 5+, 4+, 3+. Occurrence in the earth's crust 2×10^{-5}%; melting point 1132.3°C; boiling point 3818°C; density 19.05 g/cm³. Silver-white radioactive metal, softer than glass. Uranium is malleable, ductile, and can be polished. Half-life of the U-238 isotope is 4.51×10^9 years. Specific heat is 0.117 J/g•C°; heat of fusion 12.1 kJ/mol; heat of vaporization 460 kJ/mol. Burns in air at 150–175°C to form U_3O_8. When finely powdered, it slowly decomposes in cold water, more quickly in hot water. Burns in fluorine to form a green, volatile tetrafluoride; also burns in chlorine, bromine, and iodine. Reacts with acids with the liberation of hydrogen and the formation of salts with the 4+ oxidation state. Not attacked by alkalies.

Figure 3.11 Information concerning the physical and chemical properties of a substance can be found in a chemical handbook. The entry shown here describes both the physical and chemical (highlighted) properties of uranium.

Chemical properties describe the reaction of a substance with other materials such as air, water, acids, or a reaction within the substance itself. For example, iron reacts with air and water to form rust. Compounds can react in several ways. Thus, the organic compound paraquat has two chemical properties that make it useful as a herbicide. It is toxic to plant tissue, so it will kill certain plants. On the other hand, it is decomposed by water in soil particles, particularly clays. Thus, it does not poison the soil permanently.

For a chemist, it is just as important to find out if a particular substance *does not* react as it is to discover if it *does* react. Thus, the fact that baking soda is not flammable makes it a valuable substance to use in an emergency to extinguish a kitchen fire.

3.2 *CONCEPT REVIEW*

6. Classify the following properties as extensive or intensive.
a. mass
b. color
c. ductility
d. length
e. melting point

7. Classify each of the following changes as chemical or physical.
a. fading of dye in cloth
b. growth of a plant
c. melting of ice
d. digestion of food
e. formation of clouds in the air
f. healing of a wound
g. making of rock candy by evaporating water from a sugar solution
h. production of light by an electric lamp

8. Classify the following properties as chemical or physical. Use a dictionary to identify any unfamiliar properties.
a. color
b. reactivity
c. flammability
d. odor
e. porosity
f. stability
g. ductility
h. solubility
i. thermal expansion
j. melting point
k. catches fire in air

9. Apply Helium reacts with no known substance and forms no chemical compounds. Would it be accurate to say that helium has no chemical properties? Explain.

PROGRAM RESOURCES

Study Guide: Use the Study Guide worksheets, pp. 9-10, for independent practice with concepts in Section 3.3.

Laboratory Manual: Use macrolab 3-3 "Comparing Physical and Chemical Changes" as enrichment for the concepts in Chapter 3.

3.3 Energy

OBJECTIVES

- describe conditions under which heat is transferred.
- convert between units used to measure energy.
- describe endothermic and exothermic processes and state the function of activation energy.
- perform calculations involving specific heat.

Physical and chemical changes are always accompanied by energy changes. The energy changes occur between a system and its surroundings. Exactly what is a system? A **system** is that "piece" of the universe under consideration. Everything else is the surroundings. For example, a system may be as simple as a single molecule, or even a single atom, if that is the object under consideration. On the other hand, a system may be a laboratory setup involving several pieces of apparatus and two or three chemicals, or it could be an entire chemical refinery.

Energy Transfer

The most common form of energy change involves heat. What is heat? Consider a system of two objects, A and B. Assume that object A is at a temperature of 25°C and object B is at a temperature of 20°C. What happens when we put A into contact with B? Energy transfers from the matter at higher temperature, object A, to the matter at lower temperature, object B. The energy transferred as a result of a temperature difference is called **heat** and is represented by the letter, q. If the system remains undisturbed, energy will continue to transfer until the temperatures of objects A and B are equal.

There is another way that energy can be transferred between a system and its surroundings. The surroundings may do work on the system, or the system may do work on the surroundings. For example, if a strip of copper is the system under consideration, by hammering the strip of copper we would do work on it. On the other hand, if we consider the gases produced in an automobile engine cylinder as a system, the system does work on the engine as it expands and pushes the piston down. Quantitative measurements of energy changes are expressed in joules. The **joule** (J) is a derived SI unit rather than a base unit.

There are still many references to an older energy unit called a *calorie*. One calorie is equivalent to 4.184 joules. Undoubtedly, you have heard or read of calorie values for various foods. You may have noticed that these caloric values are given in Calories with an uppercase "C." These dietary Calories are a thousand times as large as a calorie and so they are, in fact, kilocalories. Caloric values are the amount of energy the human body can obtain by chemically breaking down food. Average caloric values for the basic food groups are listed in Table 3.2.

Table 3.2

Caloric Values

Food	joules/gram	calories/gram	Calories/gram
protein	17 000	4000	4
fat	38 000	9000	9
carbohydrate	17 000	4000	4

CHEMISTS AND THEIR WORK

Gerty Theresa Cori
(1896-1957)

A chemical change that takes place in your cells is the conversion of glucose to glycogen. Whenever you need glucose for energy, another change converts glycogen back to glucose. The research of Gerty Cori and her husband, Carl, revealed the details of these important chemical processes. Understanding how the body stores and releases energy reserves paved the way for discoveries on the causes and treatment of metabolic disorders. Cori was born in Prague, Czechoslovakia, but came to the United States with her husband in 1922 to conduct research. The Coris were awarded a Nobel Prize in 1947 for their work.

3.3 Energy

PREPARE

Key Concepts

The major concepts presented in this section include energy transfer between the system and its surroundings, endothermic and exothermic reactions, activation energy, and the measurement and calculation of specific heat using calorimeter data.

Key Terms

system, heat, joule, endothermic, exothermic, activation energy, calorimeter, specific heat

1 FOCUS

Objectives

Preview with your students the objectives listed on the student page. Students can use them as a study guide for this major section.

Demonstration

Partially fill a Pyrex test tube with metal shot. Warm the metal shot using a laboratory burner. Have students identify the test tube and metal shot as the system and the air as the surroundings. Record the temperature of the water. Slowly lower the system into a beaker of water. **CAUTION:** *Test tube may break.* Have the students determine that the system has not changed but the surroundings now include the beaker and water. Note the temperature change and discuss the energy transferred by introducing ΔT.

CHEMISTS AND THEIR WORK
Gerty Theresa Cori

Background: Cori and her husband, Carl Ferdinand, were awarded the Nobel prize in physiology. Her research described how an enzyme converts animal starch, glycogen, into blood sugar, glucose. For the first time, scientists could understand the disease diabetes mellitus.

Exploring Further: If you have a diabetic student in class, ask that student if he or she would like to share information about the disease with the other students.

CONTENT BACKGROUND

Heat Energy: Obtain a hot pack or cold pack from the school's athletic trainer. Place it so it will be visible to students as they enter the room. Ask students what it is and how it works. Use the student answers to these questions to introduce the concept of heat as a form of energy that flows between a system and its surroundings. Heat spontaneously flows from a warmer object to a cooler one. Ask the class to use their new vocabulary to describe the heat flow between the cold or hot pack and the injured area.

2 TEACH

Correcting Errors

Help students organize the information in the problems according to the problem-solving approach suggested in Chapter 2.
1. Where am I? (known facts)
2. Where do I want to be? (answer required)
3. Develop possible solutions.
4. What "bridge" exists between the known facts and the required answer?
5. Develop the individual steps to arrive at the answer.
6. Solve the problem.
7. Evaluate the results.

PRACTICE

Guided Practice: Guide students through the Sample Problem, then have them work in class on Practice Problem 10.
Independent Practice: Your homework or classroom assignment can include Review and Practice 30-31 from the end of the chapter.

Answers to
Practice Problems: Have students refer to Appendix C for complete solutions to Practice Problems.
10. a. 473 cal, **b.** 4640 J, **c.** 0.8 Cal, **d.** 0.000813 Cal or 8.13 x 10^{-4} Cal, **e.** 197 J

PROGRAM RESOURCES

Enrichment: Use the worksheet "Specific Heat," pp. 5-6, to challenge more able students.

Chemistry Journal

Endothermic and Exothermic
Ask students to make lists in their journals of all the chemical reactions they observe during a typical day and decide, if possible, whether each reaction is exothermic or endothermic.

SAMPLE PROBLEM

Energy Units

An average-sized baked potato (200 g) has an energy value of 686 000 joules. What is that value expressed in Calories?

Solving Process:
We know that 4.184 joules = 1 calorie. Therefore,

$$\frac{686\ 000\ \cancel{J}}{4.184\ \cancel{J}/\text{cal}} = 164\ 000\ \text{cal}$$

We also know that 1000 cal = 1 Cal. Therefore,

$$\frac{164\ 000\ \cancel{\text{cal}}}{1000\ \cancel{\text{cal}}/\text{Cal}} = 164\ \text{Calories}$$

PRACTICE PROBLEMS

10. Make the following conversions.
- **a.** 1980 joules to calories
- **b.** 1.11 Calories to joules
- **c.** 800 calories to Calories
- **d.** 3.40 joules to Calories
- **e.** 47.0 calories to joules

Word Origins

exo: (GK) out, outside of
endon: (GK) in, within
therme: (GK) heat

exothermic—giving out heat
endothermic—taking in heat

RULE OF THUMB

Energy and Chemical Change

Chemical changes are always accompanied by a change in energy. If energy is absorbed in a reaction, the reaction is **endothermic.** Since energy is taken in by the system, the products of the reaction are at a higher energy level than the substances that reacted. On the other hand, if energy is given off by a reaction, the reaction is **exothermic.** Since the system has lost energy, the products of the reaction have less energy than the substances that reacted. Look at Figure 3.12. It is convenient to think of an endothermic reaction as a system running uphill, while an exothermic reaction is a system running downhill. It is a rule of thumb for reacting systems that *nature tends to run downhill. Exothermic reactions tend to take place spontaneously, that is, without outside help.* The opposite is generally true of endothermic reactions. That is, they must have some external source of energy to take place.

Calcium hydroxide and phosphoric acid react with each other readily. According to our rule of thumb, we would predict that the reaction is exothermic. If it is exothermic, what would we observe about the system? Exothermic reactions give off energy, usually in the form of heat. As calcium hydroxide and phosphoric acid react, the container in which they react should get hotter if our prediction is true. Measurement of the temperature in the reaction vessel would confirm that the prediction is correct.

DEMONSTRATION

3-5 Photochromism

Purpose: to demonstrate that the photochromism of methylene blue can be used to show that chemical changes are accompanied by energy changes

Materials: 2 g $FeSO_4$, 96 cm^3 0.1*M* H_2SO_4 (Add 1 cm^3 18*M* H_2SO_4 to 95 cm^3 H_2O), 5 small crystals of methylene blue, 2 x 100-cm^3 beakers, file folder, overhead projector

Procedure: 1. The day of the demonstration dissolve 2 g of $FeSO_4$ in 96 cm^3 of 0.1*M* H_2SO_4. **CAUTION:** *Acid is very corrosive.*
2. Add 4 or 5 small crystals of methylene blue and stir to dissolve.
3. Divide the solution between 2 100-cm^3 beakers.
4. Place the opaque file folder on the stage of the overhead projector so that only half of the screen is illuminated. Place one of the beakers on

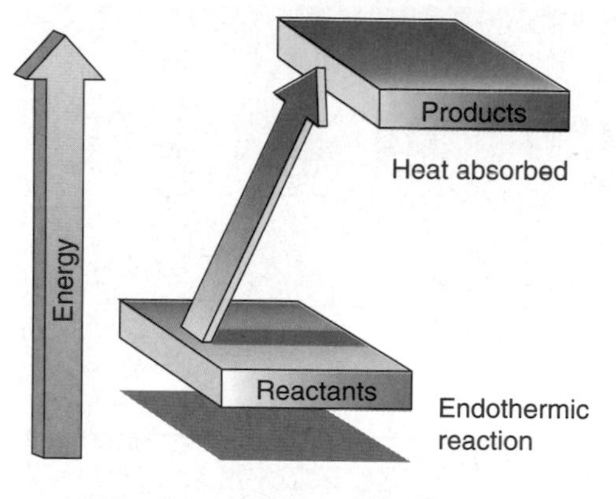

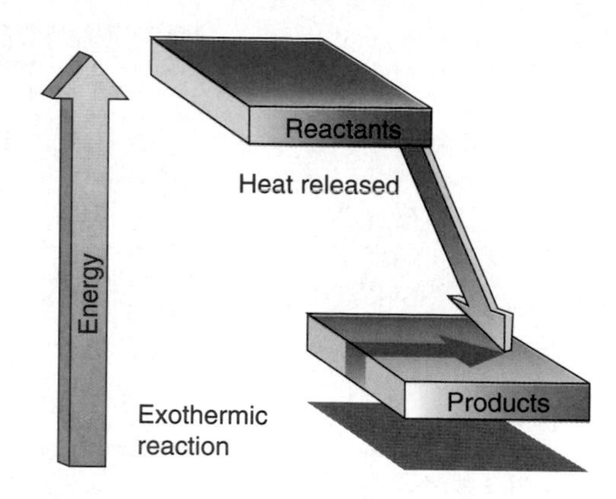

Figure 3.12 An endothermic reaction (left) absorbs energy because the products are at a higher level of energy than the reactants. An exothermic reaction (right) gives off energy because the products are lower in energy than the reactants.

Both endothermic and exothermic reactions require a certain minimum amount of energy to get started. This minimum amount of energy is called the **activation energy.** Without it, the reactant atoms or molecules will not unite to form the product and the reaction does not occur.

Chemical reactions have many possible sources for their activation energy. When you strike a match, friction produces enough heat to activate the reactants on the match head. As a result, the match ignites. Even though a small amount of energy is put in to activate the reaction, the reaction is exothermic because it produces more energy than was put in. In photography, light is the source of activation energy.

Measuring Energy Changes

A **calorimeter** is a device used to measure the energy given off or absorbed during chemical or physical changes. See Figure 3.13. There are several kinds of calorimeters. A calorimeter containing water is often used to measure the heat absorbed or released by a chemical reaction. Heat from a chemical reaction in the cup in the calorimeter causes a change in temperature of the water in the calorimeter. The temperature change of the water is used to measure the heat absorbed or released by the reaction.

To change the temperature of a substance such as water, heat must be added or removed. Some substances require little heat to cause a change in their temperature. Other substances require a great deal of heat to cause the same temperature change. For example, one gram of liquid water requires 4.184 joules of heat to cause a temperature change of one Celsius degree. It takes only 0.902 joule to raise the temperature of one gram of aluminum one Celsius degree. The heat needed to raise the temperature of one gram of a substance by one Celsius degree is called the **specific heat** (C_p) of the substance. Every substance has its own specific heat. The heat required to raise the temperature of one gram of water one Celsius degree is 4.184 joules. The specific heat of water is 4.184 J/g·C°.

calor: (L) heat
metron: (GK) measure

calorimeter—a device that measures heat

CONCEPT DEVELOPMENT

MAKING CONNECTIONS

Photography

Ask a student photo editor of the school newspaper or yearbook to report on how light affects the emulsion on photographic film. The student may want to demonstrate the developing process in the school's darkroom. What are some of the chemical reactions involved in developing exposed film? Possible response: *Exposing the film involves a light-sensitive change in AgBr. Developing the exposed film replaces the AgBr with Ag.*

$$2AgBr + C_6H_4(OH)_2 \rightarrow 2Ag + C_6H_4O_2 + 2HBr$$

Background: Use this section in the chapter to impress on your students the fact that energy changes always accompany chemical changes. Chapters 9, 22 and 27 provide detailed presentations of energy changes and activation energy. By presenting topics more than once, students have been shown to have greater retention.

Enrichment: Ask students why some food and beverages are stored in brown bottles. Possible response: *The bottles delay the decomposition of the substance by light.* Many modern food wrap films and bags are both opaque and airtight.

Extension: The synthetic dye industry develops dyes that are color fast and resist fading in sunlight. Have a student research a report on dyes and William H. Perkin. In 1856 Perkin, at the age of 18 years, made the first synthetic dye accidentally from coal tar. More than 8000 synthetic dyes are used today.

the illuminated area of the stage. If at first the solution does not go colorless, dilute the solution. Remove the beaker from the light. Have students observe that its blue color returns.

5. Place the second beaker so that half of it is in the illuminated area of the stage. **Disposal:** D

Results: Students will observe that only the half that is exposed to light turns colorless. When the folder is removed, the remainder will also become colorless. (The process can be repeated because it contains a reversible redox pair. Fe^{2+} is oxidized to Fe^{3+} while the blue form of methylene blue is reduced to the colorless form.)

Questions: 1. Is the reaction endothermic or exothermic? *endothermic*

2. What indication is there that the activation energy of this reaction is high? *Room light is not sufficient to start the reaction.*

3. What constitutes the chemical system? *the beaker and chemicals it contains*

CONCEPT DEVELOPMENT

Background: Specific heat used to be called specific heat capacity. The symbol, C_p, reminds one of the word *capacity*. Students can understand that different materials have different capacities to store energy. The sand at the beach feels hotter to a bare foot than does the boardwalk even though both are in the sun. Sand and wood have different specific heats.

ASSESSMENT

Skill: Ask students to write the equation for finding the amount of heat gained or lost by a substance: $q = (m)(\Delta T)(C_p)$ Have students explain each term and then use this formula to solve the following problem: How much heat is required to raise the temperature of 68.0 g of AlF_3 from 25.0°C to 80.0°C? See Table A-5 on page 858 for specific heat values. *Answer: 3350 J*

GLENCOE TECHNOLOGY

Videodisc

Chemistry: Concepts and Applications

Exothermic and Endothermic Reactions
Disc 3, Side 1, Ch. 2

Also available on CD-ROM.

An Exothermic Reaction
Disc 1, Side 1, Ch. 2

Show this video of Demonstration 3-6.

Also available on CD-ROM.

Activation Energy
Disc 3, Side 1, Ch. 3

Also available on CD-ROM.

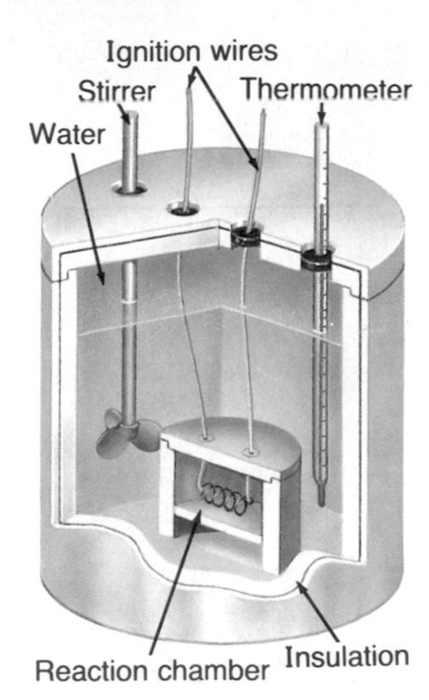

Figure 3.13 A bomb calorimeter (left) is used to measure the heat released or absorbed in the reaction taking place inside the reaction chamber. The heat released or absorbed changes the temperature of the water. The temperature change is measured with a thermometer. Energy released by chemical reactions is used to put the space shuttle (right) in orbit.

Specific heats are given in joules per gram-Celsius degree (J/g·C°). Appendix Tables A-3 and A-5 list the specific heats of some substances. Specific heats can be used in calculations involving the change in temperature of a specific mass of substance.

The law of conservation of energy means that in an insulated system, any heat lost by one quantity of matter must be gained by another. The transfer of energy takes place between two quantities of matter that are at different temperatures until the two reach the same temperature. Further, the amount of energy transferred can be calculated from the relationship:

$$\left(\begin{matrix}\textit{heat gained or}\\ \textit{lost by water}\end{matrix}\right) = \left(\begin{matrix}\textit{mass}\\ \textit{in grams}\end{matrix}\right)\left(\begin{matrix}\textit{change in}\\ \textit{temperature}\end{matrix}\right)\left(\begin{matrix}\textit{specific}\\ \textit{heat}\end{matrix}\right)$$

$$q = (m)(\Delta T)(C_p)$$

The term ΔT refers to the change in temperature. When heat is gained by water, $\Delta T = T_f - T_i$ where T_f is the final temperature and T_i is the initial temperature. When heat is lost by water, $\Delta T = T_i - T_f$.

In the calorimeter, the product of the specific heat of the water, the temperature change of the water, and the mass of the water gives the heat transferred. The heat released or absorbed by water is calculated using the following equation.

$$q = (m)(\Delta T)(4.184\ \text{J/g}\cdot\text{C°})$$

The same method can be used to calculate the transfer of heat when two dilute solutions react. The solutions are placed in a cup in the calorimeter. As the chemical reaction occurs, the temperature change is measured. Because the solutions are dilute, we can assume that the mixture has the same specific heat as water, 4.184 J/g·C°. By multiplying the specific heat by the temperature change and the mass of the solutions, we can calculate the heat produced by the reaction.

CONTENT BACKGROUND

Energy Units: The calorie gets its name from the Latin *calor*, meaning heat. It was first used in 1870 by the Frenchman Guilleman to represent a quantity of heat transferred. However, Lavoisier used the word *calorimeter* in the late eighteenth century.

There were at least three definitions of the calorie. The most common definition was the amount of heat needed to raise the temperature of 1 gram of water from 14.5°C to 15.5°C. However, because the unit is defined in terms of water, it became subject to all kinds of limitations depending upon the conditions under which it was measured. Today, the calorie is defined as exactly 4.184 joules.

The joule, adopted by the SI system in 1946, is named in honor of James Prescott Joule, an English brewer. Fifty years before Joule, Count Rumford predicted there was a rela-

EVERYDAY CHEMISTRY

Smog

Smog was originally defined as a mixture of *sm*oke and f*og*. Today, the term *smog* is applied most specifically to photochemical smog. This smog is created by the breaking down of pollutants in the air by sunlight. Photochemical smog owes many of its properties to reactions involving atmospheric oxygen, oxides of nitrogen, and hydrocarbons (compounds of hydrogen and carbon such as gasoline vapors).

Automobiles are the primary contributors of the raw materials that produce smog. Nitrogen oxides, particularly nitrogen dioxide, and hydrocarbon fumes mix with air and other pollutants. With sunlight, this mixture produces ozone, other oxides of nitrogen, and sulfur oxides.

When these products are trapped in a city surrounded by hills or mountains, the smog is particularly intense. This intensity is often increased by a temperature inversion layer in the atmosphere. The inversion results from a layer of cold air overlying the area, preventing the hot air and gases from rising and dispersing. Because blue light is scattered more by smog particles, the air often appears murky blue or brown. The color is accentuated by the brown nitrogen dioxide in the air. In addition to unhealthful air, smog also causes eye irritation and tearing probably brought on by the organic compounds formaldehyde, acrolein, and peroxyacetyl nitrate (PAN). It is also damaging to plants, affecting agriculture and forests.

Over the years, there have been many attempts to control the major source of smog, auto exhaust. Among these are engine modifications, catalytic converters, and afterburners that complete the burning of waste gases. Even if there were no photochemical reactions, auto exhaust would still produce carbon monoxide and some particles. Besides, all other forms of combustion, including smoking, also contribute to smog. Industrial operations such as refineries and service stations emit hydrocarbon vapors. Even backyard cookouts with charcoal and charcoal starter contribute particles and hydrocarbons to the atmosphere.

Obviously it is not possible to eliminate all air pollutants; but legislation and self-regulation have done much to reduce smog levels. Much, however, remains to be done.

Exploring Further

1. Find out how catalytic converters work. Why are they called "catalytic" and what do they convert?
2. Electric automobiles have been promoted as a possible solution to air pollution from internal-combustion engines. Can you think of ways in which electric cars might cause air pollution?

Notice that throughout our discussion of calorimetry, we have assumed that the calorimeter itself does not absorb any heat. We also have assumed that no heat escapes from the calorimeter. Neither of these assumptions is completely true. However, we will continue to make these assumptions in order to simplify our calculations. The error from actual losses of heat should be considered in any laboratory exercise using calorimeters.

tionship between work and heat. Rumford tried to measure it but obtained a poor result.

Joule was the first to measure the mechanical equivalent of heat accurately, a task that he completed in 1847. He felt that work and heat were interconvertible and that heat was a form of motion. His work was not immediately accepted, probably because of his lack of formal training. William Thomson, later Lord Kelvin, pointed out the value of Joule's work, and endorsements helped solidify his place in English science. It is on Joule's work that the first law of thermodynamics is based.

EVERYDAY CHEMISTRY

Smog

Background: Laws to control smog have existed since the 1960s. Photochemical smog was first recognized in Los Angeles, and its effects have been studied since the 1950s. Ozone is formed by the photochemical decomposition of nitrogen dioxide into nitric oxide and atomic oxygen. The atomic oxygen then combines with molecular oxygen in the air to produce ozone.

Teaching This Feature: You may wish to begin with a picture of pollution in a major city or weather report that lists the air pollution index.

Connection to Meteorology: Students may wish to explore more fully the role of the temperature inversion layer in smog alerts.

Extension: Ask students to explore some of the disasters in England that were caused by air pollution in industrialized cities.

Answers to
Exploring Further

1. The catalytic converter uses vanadium oxide or platinum-rhodium as a contact catalyst in the complete combustion of the exhaust gases. Students should be able to find out about the converters through the Auto Clubs in their cities. **2.** Indirectly, the electric generating plants produce pollution unless they are hydroelectric, solar, wind, or geothermal. Electricity is needed to recharge the car's batteries. Batteries can give off acidic fumes.

GLENCOE TECHNOLOGY

 **CD-ROM**

Chemistry: Concepts and Applications
Exploration: *Burning Calories*

PROGRAM RESOURCES

Critical Thinking/Problem Solving: Use the worksheet, p.3, as enrichment for the concept of "Changes of State in Graphic Terms."

Correcting Errors

To help visualize the problem, have students recall the opening photograph of the blacksmith. Ask students if the water temperature will rise more when a pony shoe is being cooled or when a large horse shoe is placed in the water. It is obvious to students that the mass of the metal affects the temperature change of the water.

PRACTICE

Guided Practice: Guide students through the Sample Problems, then have them work in class on Practice Problems 11 and 16.

Independent Practice: Your homework or classroom assignment can include the remaining Practice Problems 12 - 15, and Chapter Review 32 - 37, 40 - 41 from the end of the chapter.

Answers to **Practice Problems:** Have students refer to Appendix C for complete solutions to Practice Problems.

11. 220 000 J
12. 3170 J
13. 215 J
14. 28.6°C
15. 32.7°C
16. 1.1 J/g·C°

3 ASSESS

CHECK FOR UNDERSTANDING

Ask students to answer the Concept Review problems, and Review and Practice 30-37 from the end of the chapter.

RETEACHING

Have lab partners prepare a set of flash cards for the eight key terms in this section. Have students write the definitions on the reverse side. Have partners use the cards to drill one another until mastery is achieved.

SAMPLE PROBLEMS

1. Calculating Transfer of Heat

How much heat is lost when a solid aluminum ingot with a mass of 4110 g cools from 660.0°C to 25°C?

Solving Process:
From Appendix Table A-3, we find that the specific heat of aluminum is 0.903 J/g · C°. We also know that

$$q = (m)(\Delta T)(C_p) \quad \text{and} \quad \Delta T = 660.0°\text{C} - 25°\text{C} = 635\ \text{C}°;$$

$$\text{so } q = (4110\ \text{g})(635\ \text{C}°)\left(0.903\ \frac{\text{J}}{\text{g}\cdot\text{C}°}\right) = 2\ 360\ 000\ \text{J}$$

Note that the result is rounded to three significant digits because the data had only three significant digits.

2. Calculating Temperature

Suppose a piece of iron with a mass of 21.5 grams at a temperature of 100.0°C is dropped into an insulated container of water. The mass of the water is 132 grams and its temperature before adding the iron is 20.0°C. What will be the final temperature of the system?

Solving Process:
We know that the heat lost must equal the heat gained. Since the iron is at a higher temperature than the water, the iron will lose energy. The water will gain an equivalent amount of energy.

(a) The heat lost by the iron is

$$q = (m)(\Delta T)(C_p) = (21.5\ \text{g})(100.0°\text{C} - T_f)\left(0.449\frac{\text{J}}{\text{g}\cdot\text{C}°}\right)$$

(b) The heat gained by the water is

$$q = (m)(\Delta T)(C_p) = (132\ \text{g})(T_f - 20.0°\text{C})\left(4.184\ \frac{\text{J}}{\text{g}\cdot\text{C}°}\right)$$

(c) The heat gained must equal the heat lost

$$(132\ \text{g})(T_f - 20.0°\text{C})\left(4.184\frac{\text{J}}{\text{g}\cdot\text{C}°}\right) =$$

$$(21.5\ \text{g})(100.0°\text{C} - T_f)\left(0.449\frac{\text{J}}{\text{g}\cdot\text{C}°}\right)$$

$$T_f = 21.4°\text{C}$$

PROBLEM SOLVING HINT

Note that the iron is at 100°C and the water is at 20°C. The final temperature will fall somewhere between these two values.

The same type of calculation may be used to determine the specific heat of an unknown metal. When a warm piece of metal is dropped into a container of cool water, the energy gained by the water will be the same as that lost by the metal. We can measure the mass and the initial and final temperatures of the water, and

DEMONSTRATION

3-6 An Exothermic Reaction

Purpose: to demonstrate an exothermic reaction that requires an observable activation energy

Materials: tube of silicone caulking, piezo-electric gas grill lighter, 250-cm^3 polyethylene or polypropylene bottle, 1 cm^3 methanol, cork to fit bottle, string, thumbtack

Procedure: 1. Cut a small hole near the bottom of the side of the bottle. Use silicone caulking to cement the tip of the grill lighter into the bottle. Allow the caulking to set overnight.
2. Add 1 cm^3 of methanol to the bottle and allow a minute for it to vaporize.
3. Use a cork (not a rubber stopper) to seal the bottle. Place a thumbtack in the cork. Tie a 1-meter length of string to the thumb tack and to the neck of the bottle.

we know its specific heat. From these data, the heat gained by the water can be calculated. For the metal, we can also measure mass and initial and final temperatures. We can then compute the specific heat, C_p, using the equation, $q = (m)(\Delta T)(C_p)$. If you have an unknown metal that must be identified, measuring its specific heat is a simple task, and the result is an important physical property that aids in identifying the metal.

The values for the specific heats of metals change little over a wide range of temperatures. In fact, the specific heats of all solids and liquids are fairly constant. In contrast, the specific heats of gases vary widely with temperature.

PRACTICE PROBLEMS

11. How much heat is required to raise the temperature of 854 g H_2O from 23.5°C to 85.0°C?

12. Phosphorus trichloride, PCl_3, is a compound used in the manufacture of pesticides and gasoline additives. How much heat is required to raise the temperature of 96.7 g PCl_3 from 31.7°C to 69.2°C? The specific heat of PCl_3 is 0.874 J/g·C°.

13. Carbon tetrachloride, CCl_4, was a very popular organic solvent until it was found to be toxic. How much heat is required to raise the temperature of 10.35 g CCl_4 from 32.1°C to 56.4°C? (See Appendix Table A-5.)

14. If a piece of aluminum with mass 3.90 g and a temperature of 99.3°C is dropped into 10.0 cm^3 of water at 22.6°C, what will be the final temperature of the system? (Recall the density of water is 1.00 g/cm^3.)

15. The color of many ceramic glazes comes from cadmium compounds. If a piece of cadmium with mass 65.6 g and a temperature of 100.0°C is dropped into 25.0 cm^3 of water at 23.0°C, what will be the final temperature of the system?

16. A piece of an unknown metal with mass 23.8 g is heated to 100.0°C and dropped into 50.0 cm^3 of water at 24.0°C. The final temperature of the system is 32.5°C. What is the specific heat of the metal?

3.3 CONCEPT REVIEW

17. How do joules, calories, and Calories differ?

18. Under what conditions does heat move from one object to another?

19. Suppose you combine two solutions in a test tube and they react. How would you determine whether the reaction was endothermic or exothermic?

20. Apply A classmate argues that the burning of charcoal must be an endothermic process because getting the charcoal started requires the input of a large amount of heat. Would you agree? Explain your answer.

EXTENSION

Matter or energy is neither created nor destroyed in an ordinary chemical reaction. Ask the class where the energy comes from that is released or absorbed by the reaction. Possible response: *Any conversions of mass to energy (or vice versa) that are associated with ordinary chemical reactions are too small to be detected using present instruments. In chemical reactions (molecular rearrangements) it is the change in potential energy (relative positions) of the particles involved that manifests itself as the enthalpy of reaction (endothermic or exothermic).*

Answers to
Concept Review

17. All are units of energy. One calorie = 4.184 joules.
One Calorie = 1000 calories = 4184 joules.

18. If two objects at different temperatures are put into contact, heat will move from the one at the higher temperature to the one at the lower temperature.

19. If the two solutions are initially at the same temperature, then the reaction is endothermic if the combined solutions are at a lower temperature and exothermic if the combined solutions are at a higher temperature.

20. No, the temperature of the charcoal must first be raised to the ignition point (a physical process). Then the reaction takes place, and it is exothermic.

4. Clamp the bottle to a ring stand. **CAUTION:** *Do not point the cork at students.*

5. Dim the room lights and squeeze the handle of the lighter to generate a spark. **CAUTION:** *Use an explosion shield.*

6. Have a student feel the outside of the bottle immediately after the explosion and report the observation to the class. Note: If you plan to use the bottle next period, flush it with air.

Disposal: F

Results: The spark (activation energy) will cause the methanol-air mixture to explode and send the cork to the ceiling. The bottle will feel warm.

Questions: 1. Is the reaction inside the bottle a chemical or physical change? *chemical*

2. Is the reaction endothermic or exothermic? *exothermic*

3. What evidence do you observe that supports your answer to question 2? *Bottle is warm, light is given off.*

4. What do you call the minimum amount of energy that is needed to start the reaction? *activation energy*

Extension: Discuss how the bottle resembles an internal combustion engine. The spark between the electrodes of the lighter is like a spark plug. The methanol-air mixture is the fuel charge that is in each cylinder. The cork is like the piston that moves to provide power to the wheels.

4 CLOSE

Discussion: Ask students how a knowledge of a fuel's specific heat could be used to decide between alternative fuels. Possible response: *The amount of heat per kilogram of a fuel is a determinant in the selection of the fuel. Given equal costs for obtaining and delivering two fuels, the one with greater heat content would provide more energy per dollar.*

CHEMISTRY AND SOCIETY

Energy Problems

Background: The United States consumes a large share of the world's energy resources. As more countries become industrialized and as the standard of living improves for more people, the energy reserves will be quickly depleted. Scientists realize that we must begin to plan for the day when the world will be without fossil fuels.

Teaching This Feature: Ask students for ideas about how we can use less fossil fuel until a new technology is developed. Use of mass transportation is a common answer given.

Extension: Encourage students to check with their local utility company about ways to conserve the use of fossil fuels both at home and at school.

Answers to

Analyzing the Issue

1. While energy is not destroyed, it is converted to a less usable form. New energy sources are needed that contain energy in a more usable form. **2.** carbon, CO_2 and H_2O. **3.** Answers will vary. These sources are generally nonpolluting, but are only available in limited geographical areas.

CHEMISTRY AND SOCIETY

Energy Problems

The United States imports over three billion barrels of crude oil a year. Eventually, the world will run out of oil. The resources of natural gas and coal will also be exhausted someday. What will we do for energy sources then? Scientists are doing research in a number of areas. One area you have undoubtedly heard about is solar energy. Others are geothermal energy, wind energy, and tidal energy. All of these sources can provide some energy, but with our present technology none will be able to supply anticipated world needs.

Another area of research is the nuclear fusion reaction you will study in Chapter 28. The fusion reaction is the same process that is taking place in the sun and is the origin of most of the energy we use today. As you will learn later, there are distinct advantages of the fusion reactor over the nuclear fission reactor presently in use throughout the world. Unfortunately, scientists have not yet learned to operate a reactor powered by fusion reactions.

What part do chemists play in the solution of the energy problem? Actually, chemists are involved in almost all aspects of the problem. Let's look at some special areas where chemists can be of immediate and significant help.

Solar energy can be harnessed in several ways, but one of the most important is by the use of a device called a photovoltaic cell. This cell can take in the energy radiated by the sun and create an electric current. In order to be a useful technology, photovoltaic cells must meet two criteria. First, the cell must be capable of producing in its lifetime more energy than is consumed in producing the cell. Second, the cell must be economical and practical to manufacture. Chemists are hard at work trying to meet both of these requirements. The production of these cells involves the use of a highly purified form of the element silicon. Chemists are investigating economical, energy-efficient ways of producing the silicon.

Another attack on our energy problems is to make more efficient use of present energy resources. Here, too, chemists are making a major contribution. For instance, the process of combustion is very complex, and not fully understood at present. Chemists are active in studying combustion at the molecular and atomic level in order to find out what is really taking place. Only then can they recommend ways to make the process more efficient.

Chemists are also researching ways to produce fuels from non-traditional sources. These processes include the production of methane (natural gas) from garbage and animal waste, synthetic petroleum from water and charcoal, and the extraction of hydrocarbons from certain types of rock formations found in the western section of the United States and Canada.

Analyzing the Issue

1. In most processes, energy is neither created nor destroyed. Why, then, do we need to develop new energy sources?
2. Nearly all fuels are organic substances. What element do they contain? What substances are released into the air as a result of their combustion?
3. Research the advantages and limitations of geothermal, wind, and tidal energy sources. Debate the utility of these sources as alternatives to combustion.

PROGRAM RESOURCES

Vocabulary Review/Concept Review: Use the worksheets, p.3. and pp. 35-36, to check students' understanding of the key terms and concepts of Chapter 3.

Laboratory Manual: Use macrolab 3-4 "Measuring Specific Heat" as enrichment for the concepts in Chapter 3.

Chapter 3

REVIEW

Summary

3.1 Classification of Matter

1. A mixture is a combination of two or more substances that retain their individual properties.
2. A phase consists of a region of uniform matter. Phases are separated by boundaries called interfaces.
3. Heterogeneous matter is made of more than one phase.
4. Homogeneous matter consists of only one phase.
5. A solution is a homogeneous mixture consisting of a solute dissolved in a solvent. The component parts need not be present in specific ratios.
6. Elements are substances made of one kind of atom.
7. Compounds are substances made of more than one kind of atom. The component atoms are present in definite ratios.
8. Organic substances are compounds that contain the element carbon. Inorganic substances are the elements and the compounds that contain no carbon.

3.2 Changes in Properties

9. A substance has undergone a physical change if the same substance remains after the change. A change of state is a physical change from one state—solid, liquid, or gas—to another.
10. In a chemical change, new substances with different properties are formed. Chemical changes must be used to separate the elements of a compound.
11. Physical properties are classified as either extensive or intensive. Extensive properties, such as mass and length, depend on the amount of the substance present. Intensive properties, such as ductility and melting point, depend on the nature of the substance itself.
12. The chemical properties of a substance describe the reaction of the substance with other substances. Lack of reactivity is also a chemical property.

3.3 Energy

13. Physical and chemical changes always involve energy transfer, either in the form of work or heat, between a system and its surroundings.
14. Energy is measured in joules.
15. Energy is absorbed in an endothermic reaction and given off in an exothermic reaction.
16. The specific heat of a substance is the heat required to raise the temperature of 1 g of the substance 1 C°. The specific heat of water is 4.184 J/g·C°.
17. The energy transferred when matter changes temperature is given by $q = (m)(\Delta T)(C_p)$.

Key Terms

material
mixture
phase
heterogeneous mixture
interface
homogeneous
solution
solute
solvent
substance
element
compound
organic substance
inorganic substance
physical change
chemical change
precipitate
physical property
chemical property
extensive property
intensive property
system
heat
joule
endothermic
exothermic
activation energy
calorimeter
specific heat

Review and Practice

21. Estimate the number of phases and interfaces present in an ice-cream soda complete with whipped cream and candied cherry.

CHAPTER REVIEW 3

- Review Summary statements and Key Terms with your students.
- Complete solutions to Chapter Review Problems can be found in **Merrill Chemistry Problems and Solutions Manual.**

Answers to Review and Practice

21. The 9 phases include soda, ice cream, cherry, straw, spoon, glass, CO_2, air, cream. The 15 interfaces include soda-glass, soda-ice cream, soda-straw, soda-spoon, soda-gas, (two: in cream and "fizz"), ice cream-cream, ice cream-glass, ice cream-straw, ice cream-spoon, ice cream-gas (2), cream-gas, cream-cherry, cherry-gas.

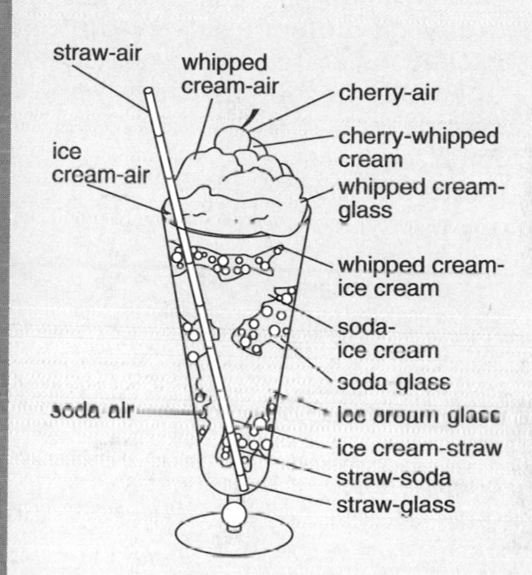

PROGRAM RESOURCES

Evaluation: Use the Evaluation worksheets, pp. 9-12, to assess students' knowledge of Chapter 3.

Computer Test Bank: Use the IBM or Apple Test and Question Selection Manual to prepare your own individualized test for Chapter 3.

Answers to
Review and Practice (cont.)

22. The solvent is present in greater quantity.
23. Water is the solvent: sugar is the solute.
24. A concentrated solution has a high ratio of solute to solvent. A dilute solution has a low ratio of solute to solvent.
25. An element consists of only one kind of atom. A compound consists of two or more elements combined in a definite ratio. An element can be trapped within a crystal of a compound.
26. There are 88 naturally-occurring elements. Two other elements, Fr and At, are detected, but not isolated; thus, 90 is an alternate answer. There are 22 synthetic elements.
27. Organic compounds contain carbon; inorganic compounds generally do not contain carbon. Exceptions are carbonic acid, carbonates, carbon oxides, and cyanides.
28. **a.** chemical
b. physical
c. physical
d. physical
e. chemical
29. **a.** physical
b. physical
c. physical
d. chemical
30. Endothermic reactions take in energy. Exothermic reactions produce energy.
31. 1 cal = 4.184 J
32. q = 89 000 J
33. q = 90 900 J
34. q = 136 J
35. q = 2100 J
36. ΔT = 376 C°
37. q = 21 000 J
38. **a.** heterogeneous mixture
b. compound
c. heterogeneous mixture
d. heterogeneous mixture
e. solution
f. element
39. phases: glass, solution, nickel, zinc ($CaCl_2$ and water do not form separate phases; the solution is a phase.)
interfaces: glass-solution, glass-nickel, glass-zinc, solution-nickel, solution-zinc, nickel-zinc
40. q = 4 000 000 000 J or 4.0×10^9 J
41. 17 000 kcal

Chapter 3
REVIEW

22. Ethanol and water will form a solution. How is it determined which one is the solute and which one is the solvent?

23. Table sugar is dissolved in a hot cup of coffee. What is the solvent of the resulting solution? What is the solute?

24. Explain the terms "concentrated" and "dilute" in terms of solvent and solute.

25. An emerald is formed when chromium atoms replace aluminum atoms in certain aluminum compounds. Differentiate between an element and a compound. How is it possible for an element to occur within a compound?

26. How many naturally-occurring elements exist? At this time, how many synthetic elements have been produced?

27. Human blood is composed of many materials, such as proteins, that are organic and other substances, such as iron and water, that are inorganic. What is the basic difference between organic substances and inorganic substances?

28. Classify the following changes as chemical or physical.
a. burning of coal
b. tearing of a piece of paper
c. kicking of a football
d. excavating of soil
e. exploding of TNT

29. Classify the following properties as chemical or physical.
a. density
b. melting point
c. length
d. flammability

30. What is the difference between endothermic reactions and exothermic reactions?

31. What is the relationship between joules and calories?

32. Heated bricks or blocks of iron were used long ago to warm beds. A 1.49-kg block of iron heated to 155°C would release how many joules of heat as it cooled to 22°C?

33. How many joules are required to heat 692 g of nickel from 22°C to 318°C?

34. How many joules are required to heat 18.2 g of tin from 14.7°C to 47.7°C?

35. Dysprosium was discovered in 1886. Its freezing point is 1400°C and its boiling point is 2600°C. If its specific heat is 0.1733 J/g·C°, how many joules are required to heat 10.0 g from its freezing point to its boiling point?

36. Copper has a specific heat of 0.384 52 J/g·C°. A 105-g sample is exposed to 15.2 kJ in an insulated container. How many degrees will the temperature of the copper sample increase?

37. Glass, which is mostly SiO_2, is not a good insulator. How much energy does a 1400-g pane of glass lose as it cools from a room temperature of 25°C to an outside temperature of 5.0°C? Use data from Appendix Table A-5.

38. Use reference materials to classify the following materials as heterogeneous mixture, solution, compound, or element.
a. paint
b. orthoclase
c. granite
d. leather
e. corn syrup
f. gold

39. Many alloys are heterogeneous mixtures. An alloy of zinc and nickel, which is a heterogeneous mixture, is placed in a beaker containing a solution of calcium chloride in water. What phases and interfaces are present?

40. A swimming pool, 10.0 m by 4.00 m, is filled to a depth of 2.50 m with water at a temperature of 20.5°C. How much energy is required to raise the temperature of the water to a more comfortable 30.0°C?

41. Fatty tissue is 15% water and 85% fat. When fat is completely broken down to carbon dioxide and water, each gram releases 9.0 kilocalories of energy. How many kilocalories are released by the loss of 2.2 kilograms of fatty tissue? (Assume that the fat is completely broken down.)

✓ **ASSESSMENT**

Portfolio: Have students make a concept map that uses the following terms and phrases. materials, homogeneous materials, heterogeneous materials, substances, homogeneous mixtures, heterogeneous mixtures. elements, compounds, solutions

Chapter 3

REVIEW

Concept Mastery

42. How does a phase differ from a state?

43. Explain how a mixture may be either heterogeneous or homogeneous. Use examples in your explanation.

44. Why is a solution always a mixture, but not every mixture a solution?

45. When would fractional crystallization be used to separate the components of a solution?

46. How is distillation used to separate a solution of two liquids? What is the main physical property that must be considered when performing a distillation?

47. In an experiment, two clear liquids are combined. A white precipitate forms and the temperature in the beaker rises.
   a. Is this reaction endothermic or exothermic?
   b. Is heat released or absorbed?
   c. Are the products higher or lower in energy than the reactants?

48. Imagine you are working in a lab and your boss gives you a sample of an unknown metal and a calorimeter. Your boss instructs you to use the calorimeter to gather data on the sample.
   a. Explain what a calorimeter measures.
   b. Describe how to use the calorimeter.
   c. What two units could be used to report your data?

49. Methane is often used as home heating fuel. When it burns, carbon dioxide and water are formed. What elements does methane contain if the oxygen in these two products comes from the air?

50. Which type of substance needs more energy to undergo a ten-degree rise in temperature, one with a high specific heat or one with a low specific heat? What factor, other than the type of substance, must you know before you can be sure of your answer to this question?

Application

51. Using Figure 3.7, determine the solubility of each of the following.
   a. NaCl at 10°C
   b. KNO_3 at 40°C
   c. KBr at 60°C
   d. $NaClO_3$ at 80°C

52. A solution contains equal amounts of KNO_3 and NaCl.
   a. If this solution is allowed to evaporate at 60°C, which compound will crystallize first?
   b. Describe the results if the evaporation had occurred at 10°C.

53. Which of the following involve a change in state of a substance?
   a. grinding beef into hamburger
   b. soldering a computer circuit board
   c. pouring milk into a glass
   d. allowing soup to cool in a bowl

54. The steps in the combustion process for a four-stroke engine are shown. Is the change that takes place in the gasoline-air mixture in the cylinder in each of steps 2 and 3 chemical or physical?

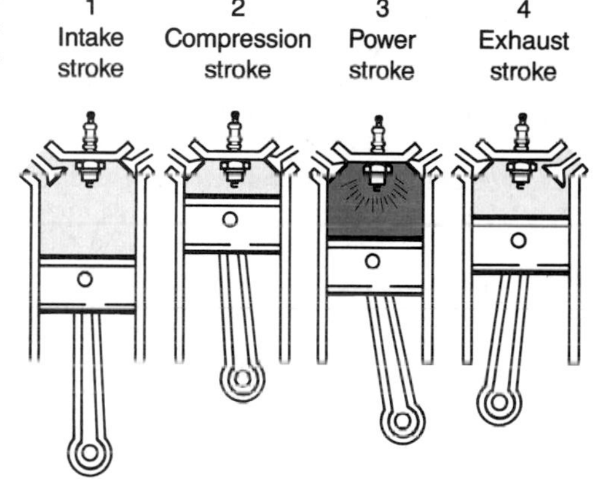

55. People often call wood, paper, milk, and paint "substances." How would you explain that paint is not a substance? How could you demonstrate the difference between paint and a true substance?

Answers to
Concept Mastery

42. A phase is matter with a uniform set of properties. A state is the form of matter in solid, liquid, or gaseous condition.

43. If a mixture is homogeneous, it is uniform throughout; for example, vinegar or salt water. If a mixture is heterogeneous, it consists of more than one phase; for example, paint.

44. All solutions are homogeneous mixtures. Some mixtures are heterogeneous and thus cannot be solutions.

45. When there are two components that have widely different solubilities at the same temperature.

46. Heat the solution until one liquid vaporizes; collect and condense the vapor. The physical property that must be considered is the boiling point of each liquid.

47. **a.** exothermic
b. released
c. lower

48. **a.** A calorimeter measures heat produced or consumed by a change.
b. Mass, then heat, the metal. Mass the water to be added to the calorimeter. Measure the initial temperatures of the water and the metal. Drop the metal into the water. Measure the final temperature of the system. Calculate C_p for the metal.
c. joules or calories

49. Methane contains carbon and hydrogen.

50. high specific heat; mass of the substance

Answers to
Application

51. **a.** 33 g/100 g water
b. 60 g/100 g water
c. 84 g/100 g water
d. 188 g/100 g water

52. **a.** NaCl will crystallize first at 60°C.
b. KNO_3 will crystallize first at 10°C.

53. b.

54. Step 2 is physical. Step 3 is chemical.

55. True substances cannot be separated physically; paint, left standing, separates and can be examined microscopically to show various phases of the mixture.

Answers to
Application (cont.)

56. Glass and plastic can both be transparent and are relatively chemically unreactive. Plastic is organic and can be more flexible than glass.
57. All have great strength, low density, and resistance to corrosion.
58. Yes, each part has a different boiling point, and air could be separated by liquefying the gases and then using fractional distillation.
59. Activation energy is needed.
60. fast cooking: slow cooking
61. $m = 6500$ g
62. $T_i = 43°C$
63. gold: lowest temperature
silver
copper
iron: highest temperature
64. 470 g

Answers to
Critical Thinking

65. $1\ J = 1\ kg \cdot m^2/s^2$
66. Sugar and salt may be separated by fractional crystallization.
67. Wöhler synthesized urea from ammonium cyanate. Organisms contain many carbon compounds.
68. a. remove iron with magnet (magnetism of iron) or dissolve the salt (solubility)
b. remove iron with magnet (magnetism of iron)
c. filter out sand (physical state)
d. physically separate (color)

56. What are the properties of glass and plastic that make them interchangeable for some uses? How are these materials different?

57. Aluminum, magnesium, and titanium have similar properties that make them useful for aircraft production. What are these properties?

58. Could distillation be used to separate air into its component parts? Explain.

59. The burning of wood is a chemical change that can be used to heat a home. The reaction proceeds by itself as long as an adequate amount of oxygen reaches the wood. However, wood does not just catch fire spontaneously. Instead, matches, paper, kindling, or other materials must be used to start the fire. Explain why these materials are needed.

60. Think of the different methods used to cook food. In what sort of cooking might you want to use cooking utensils with low specific heat? Where might utensils with high specific heat be useful?

61. A blacksmith heated an iron bar to 1445°C. The blacksmith then tempered the metal by dropping it into 42 800 cm^3 of water that had a temperature of 22°C. The final temperature of the system was 45°C. What was the mass of the bar?

62. A 752-cm^3 sample of water was placed in a 1.00-kg aluminum pan. The initial temperature of the pan was 26°C, and the final temperature of the system was 39°C. What was the initial temperature of the water?

63. Assume that samples of equal mass of the metals listed in the following table are heated in boiling water. Each sample is then dropped into a different beaker of water. All the beakers contain equal volumes of water at the same temperature. Arrange the beakers in order from lowest to highest final water temperature. Identify each beaker by the metal it contains.

Metal	Specific Heat (J/g · C°)
copper	0.384 52
gold	0.129 05
iron	0.449 4
silver	0.235 02

64. The fuel value of peanuts is 25 kJ/g. If an average adult needs 2800 kilocalories of energy a day, what mass of peanuts would meet an average adult's energy needs for the day? Assume all of the fuel value of the peanuts can be converted to useful energy.

Critical Thinking

65. The joule is an SI derived unit. Express joules in terms of SI base units.

66. A solution contains equal amounts of table salt and sugar. How might the salt be separated from the sugar?

67. Originally, organic compounds were those compounds found in or given off by living organisms, hence the name, organic. It was believed that these compounds possessed a special property called the "vital principle" that prevented humans from ever synthesizing them in the laboratory. Use reference books to find out what event changed this notion. Even today, materials produced by living things are often called organic materials. What does the use of the term *organic* tell you about the elemental composition of living organisms?

68. Physical changes are often used to separate two or more substances in a mixture. What physical change and what properties could be used to separate each of the following pairs?
a. salt and iron filings
b. iron filings and aluminum filings
c. sand and water
d. rubies and emeralds

69. The energy that our bodies need to continue life functions comes primarily from the carbohydrates that we eat. The decomposition of carbohydrates is a multistep process that includes the oxidation of glucose to form carbon dioxide and water. Predict whether this reaction is endothermic or exothermic and explain your reasoning.

70. Many athletic trainers use cold packs for injuries that occur during practice or at sporting events. In one type of cold pack, a thin lining between two chemicals breaks when the cold pack is bent in half. When the chemicals mix, the resulting solution freezes. Predict whether this reaction is endothermic or exothermic, and explain your reasoning.

71. Why are some foods, beverages, and other materials stored in brown bottles? In your answer, consider what you learned about energy changes in reactions.

72. Why are air temperatures less variable in locations near bodies of water?

73. A geologist at a mining company is trying to identify a metal sample obtained from a core sample. The unknown metal with mass 5.05 g is heated to 100.00°C and dropped into 10.0 cm^3 of water at 22.00°C. The final temperature of the system is 23.83°C. What is the specific heat of the metal? Using the data in Appendix Table A-3, determine what this metal might be.

74. Carbon has the highest melting point (3620°C) of any element and a specific heat of 0.7099 J/g·C°. Tungsten has a specific heat of 0.1320 J/g·C°. An 11.2-g sample of carbon is heated to its melting point and allowed to radiate heat to a 165.3-g sample of tungsten. The initial temperature of the tungsten is 31.0°C. How many Celsius degrees will the temperature of the tungsten sample increase?

75. A woman runs the New York Marathon in 3 hours, 42 minutes, and 18 seconds. Marathon runners need about 600 Calories per hour to supply their energy needs. How much pasta should the woman have eaten to supply her energy needs? Assume the food value of pasta to be 16 700 joules/gram.

Readings

Gordon, J. E. *The Science and Structures of Materials*. New York: Scientific American Library, 1988.

Laudise, Robert A. "Hydrothermal Synthesis of Crystals." *Chemical and Engineering News* 65, no. 39 (September 28, 1987): 30-43.

Layman, Patricia L. "Preservation of 16th Century Ship Stretches Conservation Chemistry." *Chemical and Engineering News* 65, no. 22 (June 1, 1987): 19-21.

Scientific American 255, no. 4 (October 1986). The whole issue is devoted to materials.

Cumulative Review

1. Compute the density of each of the following materials.
   a. clay, if 42.0 g occupy 19.1 cm^3
   b. cork, if 8.17 g occupy 34.0 cm^3
   c. linoleum, if 6120 g occupy 5100 cm^3
   d. ebony wood, if 201 g occupy 165 cm^3

2. What is the property possessed by all matter?

3. What is weight? How is weight measured?

4. How many significant digits are in each of the following quantities?
   a. 26.66 m
   b. 0.00402 kg
   c. 900 cm^3
   d. 100.00 cm

Answers to
Critical Thinking (cont.)

69. Exothermic; energy is released to maintain body temperature.

70. Endothermic; heat is absorbed into the products, lowering the temperature.

71. Light could provide activation energy for reactions involving the materials contained in the bottles. Brown bottles will block some of the light.

72. The high specific heat of water tends to reduce fluctuations in the water temperature, which moderates the surrounding air temperature.

73. $C_p = 0.199$ J/g · C°; the metal might be tellurium or samarium.

74. $\Delta T = 959C°$

75. 600 g pasta

Answers to
Cumulative Review

1. **a.** D = 2.20 g/cm^3
b. 0.240 g/cm^3
c. 1.2 g/cm^3
d. 1.22 g/cm^3

2. inertia

3. Weight is a measure of the force of gravity between two objects. A scale, such as a spring scale, is used to measure this force and determine weight.

4. **a.** 4
b. 3
c. 1
d. 5

CHAPTER 4

Atomic Structure

CHAPTER ORGANIZER

Chapter Objectives	Text Features	Laboratory Options	Teacher Classroom Resources
4.1 Early Atomic Models **1.** Discuss early developments in atomic theory. **2.** Explain the laws of multiple proportions and definite proportions and give examples. **3.** Determine the atomic number (Z) and mass number (A) of given isotopes of elements.	**Philosophy Connection:** Ancient Greeks, p. 78 **Chemists and Their Work:** John Dalton, p. 80; Robert Millikan, p. 82 **Frontiers:** Arsenic and Stardust, p. 86 **Photography Connection:** Unexpected Exposure, p. 88	**Discovery Demo:** 4-1 Cloud Chamber, p. 76 **Demonstration:** 4-2 Law of Definite Proportions, p. 80 **Demonstration:** 4-3 Cathode-Ray Tube, p. 82 **Demonstration:** 4-4 Geiger Counter Measures Shielding, p. 84 **Demonstration:** 4-5 Electroscope, p. 88 **Minute Lab:** Think Small, Very Small, p. 79	**Lesson Plans:** p. 7, 8 **Study Guide:** pp. 11-12 **Applying Scientific Methods in Chemistry:** Applying the Law of Definite Proportions, p. 19 **Critical Thinking/ Problem Solving:** The Nuclear Atom...What If? p. 4 **Color Transparency 5 and Master:** Rutherford's Gold Foil Experiment **Enrichment:** Thomson's Cathode-Ray Experiment, pp. 7-8
Nat'l Science Stds: UCP.1, UCP.2, A.1, A.2, B.1, B.2, B.5, G.1, G.2, G.3			
4.2 Parts of the Atom **4.** Differentiate among the major subatomic particles. **5.** Discuss the development of modern atomic theory. **6.** Calculate the average atomic mass of a mixture of isotopes of an element.	**Chemists and Their Work:** J. J. Thomson, p. 92 **Bridge to Health:** Ultraviolet Radiation, p. 93 **Chemistry and Technology:** Photoelectric Effect, p. 99	**Demonstration:** 4-6 Easy Flame Tests, p. 94 **Demonstration:** 4-7 Neon Lights p. 96 **Demonstration:** 4-8 A Massive Atom, p. 100 **ChemActivity 4:** Atomic Spectra, p. 802 **Laboratory Manual:** 4 Identifying Elements by Flame Tests - microlab	**Study Guide:** pp. 13-14 **Color Transparency 6 and Master:** Flame Tests **Color Transparency 7 and Master:** Electromagnetic Spectrum **Chemistry and Industry:** Hydrogen, pp. 3-6 **Reteaching:** Spectra of Elements, pp. 5-6 **Color Transparency 8 and Master:** Mass Spectrometer **ChemActivity Master 4**
Nat'l Science Stds: UCP.1, UCP.2, A.1, A.2, B.1, B.2, B.4, D.1, F.1, F.5, G.3			
Other Chapter Resources	**Vocabulary and Concept Review,** pp. 4, 37-38 **Evaluation,** pp. 13-16	**Videodisc Correlation Booklet** **Lab Partner Software**	**Test Bank** **Mastering Concepts in Chemistry—Software**

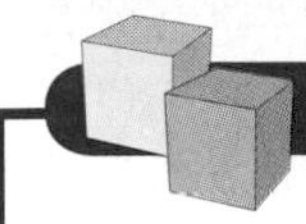

Block Schedule

For information on block scheduling, see the Lesson Plans booklet in the Teacher Resource Package.

GLENCOE TECHNOLOGY

Mastering Concepts in Chemistry Software

Basic Particles

Evidence for Alpha Particles

Thomson's Experiment

Rutherford's Gold Foil Experiment

Emission Spectra of Elements

Electrons and Energy Levels

Chemistry: Concepts and Applications Videodisc

Complete Glencoe Technology references are inserted within the appropriate lesson.

ASSIGNMENT GUIDE

Contents	Level	Practice Problems	Chapter Review	Solving Problems in Chemistry
4.1 Early Atomic Models				
Early Atomic Theory	O		19, 54, 55	
Dalton's Hypothesis	O		19-21, 41, 49	
Early Research on Atomic Particles	C		19, 22, 23, 35, 43, 61	1-2
Isotopes and Atomic Number	C	1-6	19, 25-27, 31	
The Nuclear Atom	C		19, 36, 59	
Radioactivity	C		19, 28, 56	
4.2 Parts of the Atom				
Nuclear Structure	A		29, 37	
Radiation	C		30	
The Rutherford-Bohr Atom	C		19, 38-40, 50, 53	
Planck's Hypothesis	C		19, 44	
The Hydrogen Atom and Quantum Theory	C		45-48, 51, 52	
Atomic Mass	C		24, 57, 58	
Average Atomic Mass	C	12, 13	32-34, 42, 60	3-4 Chapter Review: 1-14

C=Core, A=Advanced, O=Optional

Laboratory Materials

Minute Labs	ChemActivities	Demonstrations
page 79 balloon dropper vanilla flavoring	**page 802** hand spectroscope laboratory burner plastic bottles containing anhydrous carbonates of: copper, lithium, and sodium	**pages 76, 80, 82, 84, 88, 94, 96, 100** aluminum foil balance blotter paper chloride salts of Ca, Na, K, Cu, Sr clay cloth cloud chamber Crooke's tube dry ice DC high voltage supply electroscope ethanol fur or wool Geiger counter glass plates high intensity light laboratory burner marbles mortar and pestle packaged radioactive alpha source bar magnet plastic sandwich bag rubber or plastic rod rubber bands roofing shingles spectroscopes spectrum tubes stoppers tongs

CONCEPT DEVELOPMENT

Using The Table: Table 4.2 can be used to introduce the concept of isotopes and atomic number. Using the chalkboard or an overhead projector, draw each of the nuclei using circles to represent the protons and neutrons. The visual learner can then relate the drawings to the table and written narrative.

GLENCOE TECHNOLOGY

Software

Mastering Concepts in Chemistry

Unit 1, Lesson 2

Atomic Number, Atomic Mass Number, and Isotopes

Table 4.2

International Atomic Masses

Element	Symbol	Atomic number	Atomic mass	Element	Symbol	Atomic number	Atomic mass
Actinium	Ac	89	227.027 8*	Neon	Ne	10	20.179 7
Aluminum	Al	13	26.981 539	Neptunium	Np	93	237.048 2
Americium	Am	95	243.061 4*	Nickel	Ni	28	58.6934
Antimony	Sb	51	121.760	Niobium	Nb	41	92.906 38
Argon	Ar	18	39.948	Nitrogen	N	7	14.006 74
Arsenic	As	33	74.921 59	Nobelium	No	102	259.100 9*
Astatine	At	85	209.987 1*	Osmium	Os	76	190.2
Barium	Ba	56	137.327	Oxygen	O	8	15.999 4
Berkelium	Bk	97	247.070 3*	Palladium	Pd	46	106.42
Beryllium	Be	4	9.012 182	Phosphorus	P	15	30.973 762
Bismuth	Bi	83	208.980 37	Platinum	Pt	78	195.08
Boron	B	5	10.811	Plutonium	Pu	94	244.064 2*
Bromine	Br	35	79.904	Polonium	Po	84	208.982 4*
Cadmium	Cd	48	112.411	Potassium	K	19	39.098 3
Calcium	Ca	20	40.078	Praseodymium	Pr	59	140.907 65
Californium	Cf	98	251.079 6*	Promethium	Pm	61	144.912 8*
Carbon	C	6	12.011	Protactinium	Pa	91	231.035 88
Cerium	Ce	58	140.115	Radium	Ra	88	226.025 4
Cesium	Cs	55	132.905 43	Radon	Rn	86	222.017 6*
Chlorine	Cl	17	35.452 7	Rhenium	Re	75	186.207
Chromium	Cr	24	51.996 1	Rhodium	Rh	45	102.905 50
Cobalt	Co	27	58.933 20	Rubidium	Rb	37	85.467 8
Copper	Cu	29	63.546	Ruthenium	Ru	44	101.07
Curium	Cm	96	247.070 3*	Samarium	Sm	62	150.36
Dysprosium	Dy	66	162.50	Scandium	Sc	21	44.955 910
Einsteinium	Es	99	252.082 8*	Selenium	Se	34	78.96
Erbium	Er	68	167.26	Silicon	Si	14	28.085 5
Europium	Eu	63	151.965	Silver	Ag	47	107.868 2
Fermium	Fm	100	257.095 1*	Sodium	Na	11	22.989 768
Fluorine	F	9	18.998 403 2	Strontium	Sr	38	87.62
Francium	Fr	87	223.019 7*	Sulfur	S	16	32.066
Gadolinium	Gd	64	157.25	Tantalum	Ta	73	180.947 9
Gallium	Ga	31	69.723	Technetium	Tc	43	97.907 2*
Germanium	Ge	32	72.61	Tellurium	Te	52	127.60
Gold	Au	79	196.966 54	Terbium	Tb	65	158.925 34
Hafnium	Hf	72	178.49	Thallium	Tl	81	204.383 3
Helium	He	2	4.002 602	Thorium	Th	90	232.038 1
Holmium	Ho	67	164.930 32	Thulium	Tm	69	168.934 21
Hydrogen	H	1	1.007 94	Tin	Sn	50	118.710
Indium	In	49	114.82	Titanium	Ti	22	47.867
Iodine	I	53	126.904 47	Tungsten	W	74	183.85
Iridium	Ir	77	192.217	Unnilennium†	Une	109	266*
Iron	Fe	26	55.845	Unnilhexium†	Unh	106	263*
Krypton	Kr	36	83.80	Unniloctium†	Uno	108	265*
Lanthanum	La	57	138.905 5	Unnilseptium†	Uns	107	262*
Lawrencium	Lr	103	260.105 4*	Ununnilium†	Uun	110	269*
Lead	Pb	82	207.2	Uranium	U	92	238.028 9
Lithium	Li	3	6.941	Uranium	U	92	238.028 9
Lutetium	Lu	71	174.967	Vanadium	V	23	50.941 5
Magnesium	Mg	12	24.305 0	Xenon	Xe	54	131.29
Manganese	Mn	25	54.938 05	Ytterbium	Yb	70	173.04
Mendelevium	Md	101	258.098 6*	Yttrium	Y	39	88.905 85
Mercury	Hg	80	200.59	Zinc	Zn	30	65.39
Molybdenum	Mo	42	95.94	Zirconium	Zr	40	91.224
Neodymium	Nd	60	144.24				

*The mass of the isotope with the longest known half-life.
†See Frontiers on page 149 for an explanation of these names.

DEMONSTRATION

4-4 Geiger Counter Measures Shielding

Purpose: to determine the effects of various shielding materials

Materials: Geiger counter, squares of cloth, aluminum foil, glass panes, roofing shingles, wood squares, tongs, packaged radiation source

Procedure: 1. Mount the Geiger tube about 5 cm from the packaged radiation source. Record the meter reading. **CAUTION:** *Use packaged radiation source only. Handle it with tongs.*
2. Place a sheet of the shielding material being tested between the source and the Geiger tube.
3. Record the Geiger counter reading. Add pieces one at a time and record the readings after each addition.
4. Repeat steps 2 and 3 for each of the remaining materials. You may want to graph the data. **Disposal:** F

Table 4.3

Isotopes of Hydrogen			
Nuclide	**Protons**	**Neutrons**	**Mass Number**
protium	1	0	1
deuterium	1	1	2
tritium	1	2	3

4.3. A particular kind of atom containing a definite number of protons and neutrons is called a **nuclide.** For example, protium, hydrogen-1, is a nuclide of hydrogen.

The particles that make up the atomic nucleus are called **nucleons.** These particles are protons and neutrons. The total number of nucleons in an atom is called the **mass number** of that atom. The symbol for the mass number is A. The number of neutrons for any nuclide may be found by subtracting the atomic number from the mass number.

$$\textit{number of neutrons} = A - Z$$

In this way, you can calculate the number of neutrons in any nuclide if you know its mass number and atomic number.

SAMPLE PROBLEM

Number of Neutrons

How many electrons, neutrons, and protons are found in a copper atom of mass number 65?

Solving Process:

Using Table 4.2, you can find the atomic number of copper to be 29. Since the atomic number of an element is the number of protons in the nucleus, there are 29 protons in the nucleus of this copper atom. Atoms are electrically neutral, so an atom having 29 protons in the nucleus must also have 29 electrons. The mass number is the number of nucleons in the atom. This copper atom has 65 nucleons, of which 29 are protons. There must, therefore, be 65 − 29 = 36 neutrons in the nucleus.

PROBLEM SOLVING HINT

To check your work, add the numbers of protons and neutrons. Your total should equal the mass number.

PRACTICE PROBLEMS

1. An atom of vanadium contains 23 electrons. How many protons does it contain?
2. An atom of silver contains 47 protons. What is its atomic number?
3. An atom of sodium contains 11 electrons. What is the atomic number of this atom?

Use Table 4.2 to complete the following questions.

4. An atom contains 37 protons. What element is it?

Misconceptions

Because there are 112 known elements, students might think that there are only 112 different nuclei. Actually there are almost 2000 different nuclides.

Correcting Errors

Remind students that atomic number, *Z*, is used to identify the atom. The mass number of the atom, *A*, is the sum of the protons and neutrons after each has been assigned a mass of one atomic mass unit. The electron is not considered when determining the relative atomic mass because of its small mass.

PRACTICE

Guided Practice: Guide students through the Sample Problem, then have them work in class on Practice Problems 6a and 6b on p. 86.

Independent Practice: Your homework or classroom assignment can include the remaining Practice Problems 1 - 5, 6c and d, and Chapter Review 23 - 27 from the end of the chapter.

Answers to

Practice Problems: Have students refer to Appendix C for complete solutions to Practice Problems.

1. 23 protons
2. 47
3. 11
4. rubidium

***inter*NET CONNECTION**

Students can download and use a molecular weight calculator.

Anonymous FTP
ftp://ftpserver.sfu.ca/pub/chemcai/molecalc.sea.hqx

Results: The relative shielding effectiveness for an equal thickness of paper is 1, that of masonry is 16, iron is 50, and lead is 100.

Questions: 1. How does the thickness of the material affect the radioactive shielding? *The thicker the material, the greater the shielding.*
2. Does the density of the material affect its shielding ability? *Yes, the more dense materials are better shields.*

✓ ASSESSMENT

Performance: The inverse square law (R x 1/d^2) can be demonstrated with the Geiger counter. Have students measure and record the Geiger counter reading at various distances between the Geiger tube and the packaged radiation source. Students can use the data to construct a graph of the data.

PHOTOGRAPHY CONNECTION

Unexpected Exposure

Exploring Further: Have students make silhouettes using Sun Print® paper, which probably is available locally. A silhouette can be made by placing a small object on the blue surface of the paper. After exposing the paper to strong light for a few minutes, use water to wash off the blue coating shaded by the object. However, the blue dye exposed to light becomes insoluble and remains on the paper. Use this activity to relate how light causes chemical changes in film and photographic papers.

3 ASSESS

CHECK FOR UNDERSTANDING

Ask students to answer the Concept Review problems and Review and Practice 19–31 from the end of the chapter.

RETEACHING

Review with students the relationships between the atomic number and the mass number. Point out that the value of *Z* also is the number of electrons. *A* - *Z* yields the number of neutrons. Prepare a chart similar to Table 4-2. Select several elements and have students complete the table.

EXTENSION

Have a student investigate the use of $^{14}_{6}C$ for dating the age of once-living objects. See *Archeology Connection,* page 724, for possible responses.

PHOTOGRAPHY CONNECTION

Unexpected Exposure

Becquerel initially thought that exposure to sunlight was necessary for uranium ore to emit rays. One day when the sun did not shine he threw the uranium ore samples he was using into a drawer with a photographic plate encased in two layers of heavy black paper. A day later he developed the plate on a hunch and found the plate exposed to the same degree as when the ore was exposed to light.

Geiger and Marsden found that most of the atom is empty space. Most atoms have a diameter between 100 and 500 pm (1 pm = 1×10^{-12} m). However, the radii of the nuclei of atoms vary between 1.2×10^{-3} and 7.5×10^{-3} pm. The radius of the electron is about 2.82×10^{-3} pm. In small atoms, the distance between the nucleus and the nearest electron is about 50 pm. Thus, the nucleus occupies only about one trillionth (10^{-12}) of the volume of an atom. To help you think about this relationship, imagine the hydrogen nucleus as the size of a Ping-Pong ball. The electron is roughly the size of a tennis ball, and about 1.35 km away.

Radioactivity

In 1896, Henri Becquerel, a French physicist, found that matter containing uranium exposes sealed photographic film. This fact led another scientist, Marie Curie, and her husband, Pierre, to an important discovery. They found that rays are given off by the elements uranium and radium.

Uranium and radium can be found in nature in an ore called pitchblende. The rays from this ore have a noticeable effect on a charged electroscope. An electroscope, as shown in Figure 4.8, contains two thin, free-hanging metal leaves attached to a metal rod. When a charge is applied to the rod, the leaves become charged and repel each other. If some of this ore is placed near, but not touching, a charged electroscope, the leaves become discharged. Substances that have this effect are called radioactive substances. **Radioactivity** is the phenomenon of rays being produced spontaneously by unstable atomic nuclei.

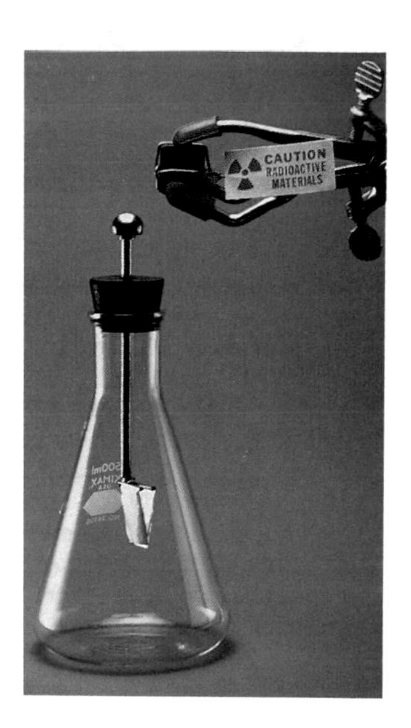

Figure 4.8 The charged electroscope on the left will become discharged if a radioactive substance is brought near the electroscope.

DEMONSTRATION

4-5 Electroscope

Purpose: to detect radioactivity with an electroscope

Materials: electroscope, rubber or plastic rod, a piece of fur or wool, pair of tongs, packaged alpha radiation source

Procedure: 1. Ask a physics teacher for an electroscope, rubber or plastic rod, and a piece of fur or wool. Rub the rod with the fur, then touch the top of the electroscope.

2. Have the students observe the foil leaves of the electroscope. Then bring the source close to the electroscope. **CAUTION:** *Use only packaged radioactive materials. Handle with tongs.* Have students observe the motion of the leaves. **Disposal:** F

Results: The two metallic leaves will spread apart as each is charged with electrons and they mutually repel one

Figure 4.9 Uranium ore is mined in both underground and surface mines. This photograph shows a view of part of a surface mine.

The rays produced by radioactive materials can be particles or energy or a mixture of both. The particles and energy are given off by the nuclei of radioactive atoms during spontaneous nuclear decay, a process covered in Chapter 28. We say the decay is spontaneous because it occurs without external influence. The amount of energy released in a nuclear change is very large. It is so large that it cannot be a result of an ordinary chemical change.

Albert Einstein, in the early 1900s, explained the origin of the energy released during nuclear changes. Einstein hypothesized that mass and energy are equivalent. This statement can be expressed in the equation

$$E = mc^2$$

where E is the energy released (in joules), m is the mass (in kilograms) of matter involved, and c is a constant, the speed of light (in meters per second).

The large amount of energy released in splitting uranium nuclei is measurable. If 1.00 g of the uranium nuclide with mass number 235 undergoes nuclear fission (splitting of the nucleus), 8.09×10^7 kJ are produced. In contrast, the energy released in normal chemical reactions is considerably less. When 1.00 g of the same uranium nuclide reacts with hydrochloric acid in a typical chemical change, only 3.25 kJ are produced.

4.1 CONCEPT REVIEW

7. Explain the difference between the law of definite proportions and the law of multiple proportions.

8. What is the difference between atomic number and mass number?

9. Find the number of electrons, neutrons, and protons in an atom of the nuclide strontium-88.

10. What contribution to atomic theory was made by Geiger and Marsden in their gold foil experiment?

11. Apply Uranium 238 undergoes a nuclear change and becomes thorium-234 and helium-4. How do you know that this change is nuclear and not chemical or physical?

another. The alpha particles remove the electrons and the leaves fall back together.

Questions: 1. Are radioactive particles visible? *No*

2. Are alpha particles charged? *Yes*

Extension: Have students research how radioactive particles affect the chemicals in photographic emulsions and how this effect is used in medicine.

Answers to
Concept Review

7. The law of definite proportions is concerned with how elements combine to form a specific substance; it says that a specific substance always contains elements in the same ratio by mass. The law of multiple proportions concerns elements that form more than one substance with each other; this law states that the ratio of masses of one element that combine with a constant mass of another element can be expressed in small whole numbers.

8. Atomic number of an element is the number of protons in the nucleus of the atom or the number of electrons in the atom. Mass number is the total number of nucleons in the nucleus of the atom (the sum of the numbers of protons and neutrons).

9. electrons: 38
neutrons: 88 - 38 = 50
protons: 38

10. The experiment showed that the atom is mostly empty space and that there is a very small core in the atom, now called the nucleus, which contains all the positive charge and almost all the mass of the atom.

11. The change is nuclear, since the products consist of different elements from the reactants.

4 CLOSE

Discussion: Heavy water is formed when two atoms of deuterium, an isotope of hydrogen, combine with an oxygen atom. Deuterium oxide, D_2O, is found naturally in very small amounts in water. Have students calculate the density of heavy water. It is used in nuclear research as a shielding material because of its increased density.

$$\text{density}_{D_2O} = \text{density}_{H_2O} \times \frac{A_{D_2O}}{A_{H_2O}}$$

$$= 1.00\ \text{g/cm}^3 \times \frac{20}{18}$$

$$= 1.11\ \text{g/cm}^3$$

4.2 Parts of the Atom

PREPARE

Key Concepts

The major concepts presented in this section include the structures in the atom's nucleus, radiation, the Rutherford-Bohr model of the atom, spectral information, Planck's hypothesis, and the hydrogen atom as viewed by quantum theory. The concept of an average atomic mass is discussed.

Key Terms

nuclear force	gamma ray
subatomic particle	spectroscopy
lepton	spectrum
hadron	electromagnetic energy
antiparticle	frequency
neutrino	hertz
quark	wavelength
baryon	quantum theory
meson	quanta
gluon	photon
alpha particle	ground state
beta particle	atomic mass

1 FOCUS

Objectives

Preview with students the objectives listed on the student page. Students can use them as a study guide for this major section.

Discussion

Ask students what particles comprise the nucleus of an atom. Many respond that protons and neutrons form the nucleus. Use Table 4.4 and Figure 4.10 to show them that the structure of the nucleus is far more complex and not yet fully understood.

CHEMISTRY IN DEPTH

The section **Nuclear Structure** on pages 90-91 is recommended for more able students.

OBJECTIVES

- differentiate among the major subatomic particles.
- discuss the development of modern atomic theory.
- calculate the average atomic mass of a mixture of isotopes of an element.

4.2 Parts of the Atom

Continued research has shown that the proton and neutron are not truly "elementary" particles. That is, they appear to be composed of yet simpler particles. All evidence about the electron, however, indicates that it is truly elementary. As you study the section on nuclear structure, refer frequently to Table 4.4 and Figure 4.10, which list subatomic particles.

CHEMISTRY IN DEPTH

Pages 90–91

Nuclear Structure

The uranium nuclide with mass number 238 contains 92 protons and 146 neutrons. These particles are bound tightly in the nucleus. Electrostatic attraction alone would not be enough to prevent the uncharged neutrons from floating away. Furthermore, because all protons have the same positive charge, electrostatic forces should cause them to fly apart. Gravitational force, which keeps us at the surface of Earth, is not strong enough to hold the nucleus together. The force that holds the protons and neutrons together in the nucleus is called the **nuclear force**. It is effective for very short distances only (about 10^{-3} pm). This distance is about the same as the diameter of the nucleus. Ideas explaining nuclear structure differ, but scientists agree on certain facts.

(1) Nucleons (protons and neutrons) have a property that corresponds to spinning on an axis.

(2) Electrons do not exist in the nucleus, yet they can be emitted from the nucleus.

The particles composing atoms are called **subatomic particles.** Nuclear scientists divide subatomic particles into two broad classes, leptons and hadrons. Current theory holds that **leptons** ("light" particles) are truly elementary particles. The electron is the best known lepton. **Hadrons,** on the other hand, appear to be made of even smaller particles. Neutrons and protons are the best known hadrons.

For every particle, a mirror-image particle called an **antiparticle** exists, or is believed to exist. Thus, there is an antielectron, called a positron, which is like an electron in every way except that it has a positive charge. Positrons are not common. They exist only until they collide with an electron. Such a collision is very likely in our world. When the collision occurs, both particles are destroyed and energy is produced. It is interesting to note that there is at least one particle that is its own antiparticle. This particle is the neutral pion.

There are several other leptons. In order to account for a certain kind of nuclear decay, a neutral particle called a **neutrino** was postulated. The neutrino has been identified and found to be essentially massless. The muon and the tau, both much more massive than the electron, make up the rest of the lepton family. A neutrino has been discovered for the muon and for the tau. There are antiparticles for all of these particles.

Table 4.4

Hadrons
Baryons
Protons
Neutrons
Other, short-lived particles
Mesons
Pions
Kaons
Other, short-lived particles

CONTENT BACKGROUND

Subatomic Particles: You may want to have students make a table listing subatomic particles. For each particle, list its mass, charge, and lifetime. Possible response: Because research in this area is so current, only a partial listing is given.

Particle	Mass	Charge	Lifetime
electron	1	±	infinite
electron's neutrino	0	0	infinite
muon	207	±	2.2×10^{-6} s
muon's neutrino	0	0	infinite
pion	273	±	2.60×10^{-8} s
pion	265	0	0.8×10^{-16} s
kaon	966	±	1.24×10^{-8} s
kaon	975	0	0.86×10^{-10} s
eta	1074	0	2.53×10^{-19} s
rho	1507	±,0	4×10^{-23} s
proton	1836	+	infinite
neutron	1839	0	0.93×10^{3} s

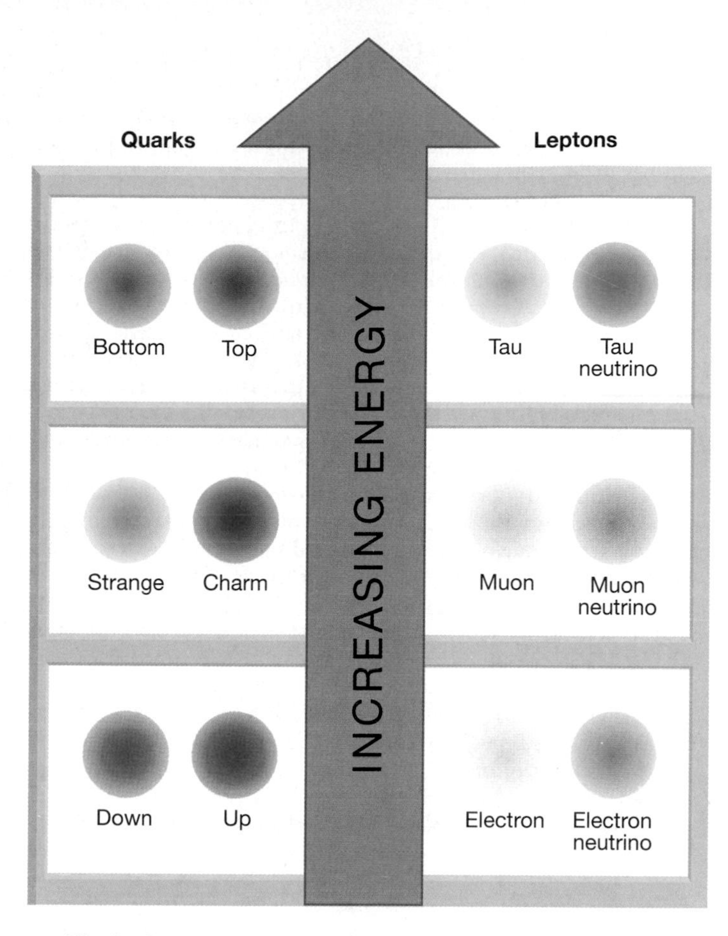

Figure 4.10 The known quarks and leptons are divided into three families. The everyday world is made from particles in the family at the bottom of the chart. Particles in the middle group are found in cosmic rays and are routinely produced in particle accelerators. Particles in the upper two families are created in high-energy collisions. This chart provides a list of known quarks and leptons. The colors, sizes, and shapes shown are not meant to represent actual particles.

The hadrons are subdivided into two groups, the mesons and the baryons. Protons and neutrons are baryons, as are a number of short-lived particles. There are several kinds of mesons.

Mesons and baryons are made of smaller particles called **quarks.** There are six kinds of quarks: "up," "down," "charm," "strange," "top," and "bottom." The names convey nothing about the properties of the quarks. They are just identification labels. Each quark comes in three "colors," red, blue, and green. Again, the color label refers only to a distinguishing property, not an appearance. Each kind of quark has an antimatter counterpart, an antiquark. **Baryons** are composed of three quarks, each of a different color. **Mesons** are composed of a quark and an antiquark of complementary color. It is believed that quarks hold together by exchanging "particles" called **gluons.** There are believed to be eight gluons, each of which is characterized by one color and one anti-color. In a similar manner, the nucleons are held together in the nucleus by the exchange of pions.

Particle	Mass	Charge	Lifetime
lambda	2183	0	2.5×10^{-10} s
sigma	2328	±,0	8.0×10^{-11} s
delta	2410	±,0	4×10^{-23} s
xi	2583	0	3.0×10^{-10} s
sigma (1385)	2710	0,-	1×10^{-10} s
omega	3287	-	1.3×10^{-10} s
tau meson	3522	±	?
D meson	3523	±,0	?
psi/J	6062	0	6×10^{-20} s
upsilon	18 590	0	?
upsilon	19 570	0	?

PROGRAM RESOURCES

Study Guide: Use the Study Guide worksheets, pp.13-14, for independent practice with concepts in Section 4.2.

CHEMISTRY IN DEPTH

Chemistry in Depth ends on this page.

2 TEACH

CONCEPT DEVELOPMENT

Background: Leptons include the less massive particles: electron, muon, and tau. Each of these particles has an antiparticle. (The electron's antiparticle is the positron.) Each particle and antiparticle has its own neutrino associated with it. Thus there are 12 leptons in all. Hadrons are the more massive particles: mesons and baryons. Mesons are each composed of a quark and an antiquark; baryons are each thought to be composed of three quarks. Each quark has an antiquark. Point out that the names of various classes of quarks have nothing to do with the physical nature of quarks.

CULTURAL DIVERSITY

Colorful Language

How many basic color words does your language have? Scientists studied over 100 languages and found that while the human eye everywhere "sees" the same color, people in different cultures classify colors differently. In an experiment, people who have a term for red all picked the same color chip as the "best" example of red, but not everyone has the same number of color terms. The number of basic color words for a culture varies from two to twelve. English has eleven: white, black, red, green, yellow, blue, brown, pink, purple, orange, and gray. Basic colors are not forms of any other color. For example, crimson is not a basic English color word because it is considered to be a shade of red. Cultures, such as the Dugum Dani of the New Guinea Highlands, with only two color terms use their own words for "black" (dark colors) and "white" (light colors). If a third term is present, it is usually red. Three-color systems are found in various cultures of Melanesia, Australia, Africa, and parts of the Americas. For example, the Toda of India, the Ndembu of the Congo, and aborigines of the Seven Rivers District in Queensland, Australia all use black, white, and red color categories.

CHEMISTS AND THEIR WORK

J. J. Thomson

Background: In 1895, Thomson believed that the cathode ray was made of tiny particles. Other scientists of his day believed the ray was a light wave. He was the first to succeed in deflecting the cathode ray using a magnetic force.

Exploring Further: Thomson is credited with the first discovery of isotopes. He detected neon-20 and neon-22. One of his students, Francis W. Aston, invented the mass spectrograph. Have a student research how this device works and report to the class.

CONCEPT DEVELOPMENT

MAKING CONNECTIONS

History

Marie Curie gave the phenomenon of radioactivity its name. Becquerel had discovered that uranium ore, called pitchblende, spontaneously emitted energy. His work led the Curies to separate the element responsible for the radiation. Curies were awarded the Nobel Prize for the discovery of the element radium.

Misconceptions

Some students think that when an atom undergoes a radioactive decay, it disappears. Emphasize that decayed atoms do not disappear, but change into different elements. Uranium eventually becomes lead.

QUICK DEMO: Background Radiation

A Geiger counter placed on the desk detects background radiation. Students should be aware that we are constantly bombarded by radiation from space and materials on Earth. Most background radiation consists of beta particles and gamma rays.

CHEMISTS AND THEIR WORK

J. J. Thomson (1856-1940)
Sir Joseph John Thomson was educated at Cambridge University in England and spent his entire professional life there as professor and director of the physics laboratories. In his own research, he is credited with the discovery of the electron and the proton, and for producing the first experimental evidence for the existence of isotopes. He was awarded the Nobel Prize in Physics in 1906. One of his greatest scientific legacies was the training of a new generation of scientists. Seven of his research assistants eventually won Nobel Prizes.

Radiation

If the structure of a nucleus is not stable, the nucleus will eject either a particle or energy until it reaches a more stable arrangement. Some nuclei are unstable as found in nature. Other nuclei can be made artificially unstable (radioactive) by bombardment in a particle accelerator.

Three forms of radiation can come from naturally radioactive nuclei. Two forms consist of particles. The third consists of energy. The particles are alpha (α) and beta (β) particles. The energy consists of gamma (γ) rays. An **alpha particle** is a helium nucleus and consists of two protons and two neutrons. A **beta particle** is a high-speed electron emitted from radioactive nuclei. Electrons do not exist as such in the nucleus but are generated at the instant of decay. **Gamma rays** are very high-energy X rays. Unstable nuclei that emit rays or particles to become stable nuclei are said to decay.

Scientists have created many radioactive nuclides that do not exist in nature on Earth. These substances are generated by bombarding stable nuclei with accelerated particles or exposing stable nuclei to neutrons in a nuclear reactor (see Chapter 28). These artificially radioactive nuclides can decay by emitting α, β, and γ rays as well as by several other methods such as positron emission. Artificially radioactive nuclides can also decay by capturing one of the electrons outside the nucleus. The electrons closest to the nucleus are sometimes called the *K* electrons. For this reason, this radioactive process is called *K*-capture. It is also sometimes called electron capture. Most antiparticles have been formed and observed during the bombardment of normal nuclei in particle accelerators. They may also be formed in other ways.

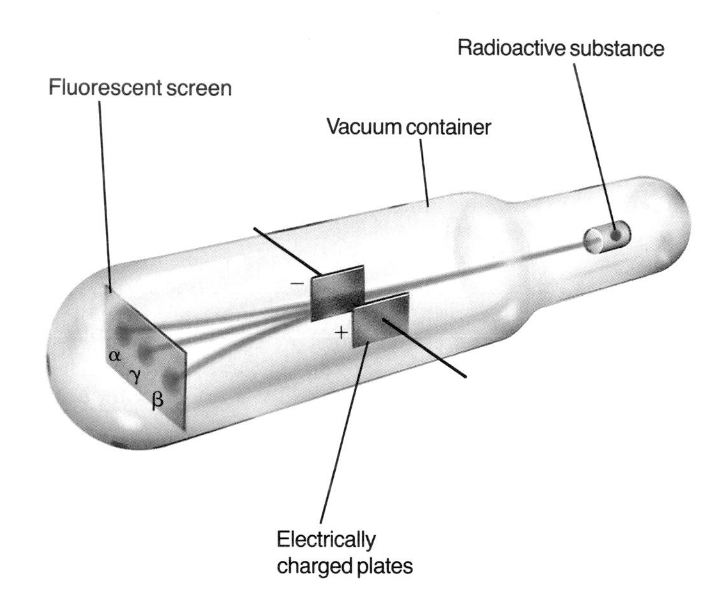

Figure 4.11 Alpha, beta, and gamma emissions behave differently in an electric field. Alpha and beta particles are deflected because of their charge.

CONTENT BACKGROUND

Natural Radioactivity: Students are aware that large amounts of heat are produced by nuclear reactors. Point out that as radium undergoes natural radioactive decay, it produces heat at a rate of 552 J/g·h. Thus, 1.0 gram of radium produces 1.00×10^{10} J of heat in its lifetime. This energy is equivalent to the heat produced by the burning of half a ton of coal. Heat from natural radioactivity can be detected deep within the ground and accounts for some of Earth's geothermal energy.

Bridge to
HEALTH

Ultraviolet Radiation

When ultraviolet light of wavelength less than 193 nm strikes an oxygen molecule, the molecule will form two oxygen atoms.

$$O_2 + h\nu \rightarrow 2O$$

These oxygen atoms, in turn, will react with other oxygen molecules to form ozone.

$$O + O_2 \rightarrow O_3$$

This process takes place in the band of the atmosphere between 15 and 30 km above the surface of Earth.

The maximum ozone concentration is reached at an altitude of approximately 25 km. This layer of ozone in the atmosphere is of vital importance to the health of living organisms on Earth because it absorbs much of the ultraviolet light in the wavelength range 280-320 nm.

This part of the ultraviolet spectrum is the energy that can produce a sunburn or skin cancer in humans.

This ultraviolet energy can also damage eyesight, and, if strong enough, can kill many microorganisms at the low end of the food chain. These organisms include many sea-living creatures called phytoplankton that produce much of the oxygen we breathe. The ultraviolet energy has a devastating effect on the DNA (deoxyribonucleic acid) molecules that control the function of living cells.

Exploring Further

1. Chlorine atoms from certain refrigerants and aerosol propellants destroy ozone in the atmosphere. What steps can be taken to prevent this from happening?
2. Explain why all atmospheric oxygen does not change to ozone.

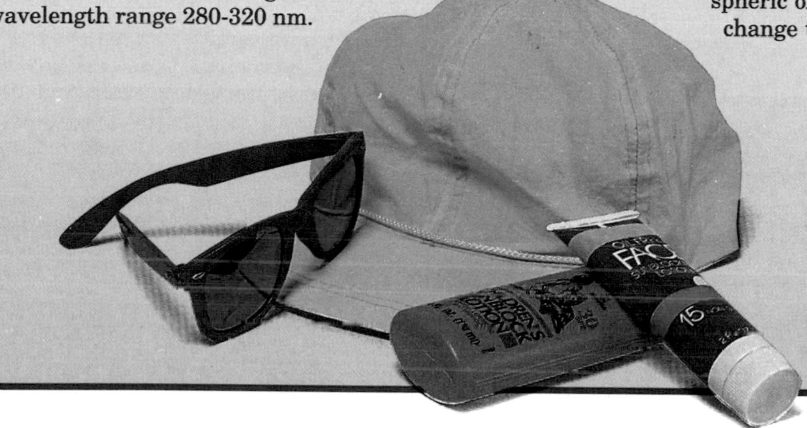

The Rutherford-Bohr Atom

Electrons are negatively charged and attracted to the positive nucleus. What prevents the electrons from being pulled into the nucleus? The discussion of this question, led by Rutherford and Bohr, resulted in a new idea. They thought of electrons as being in "orbit" around the nucleus in much the same manner as Earth is in orbit around the sun. They suggested that the relationship between the electrons and the nucleus is similar to that between the planets and the sun. The Rutherford-Bohr model of the atom is sometimes called the planetary atomic model. Thus, according to the planetary model, the hydrogen atom should be similar to a solar system consisting of a sun and one planet.

BRIDGE TO HEALTH

Ultraviolet Radiation

Background: Students are familiar with UV-blocking sunglasses. They will also probably be familiar with the number of the sun block in suntan lotions. Today people are more careful about protecting themselves from the damaging effects of ultraviolet radiation.

Teaching This Feature: Ask students why UV exposure is increasing. People have more leisure time for recreation in the sun, and they live longer which increases the exposure. Ozone depletion may be a factor in some regions.

Connection to Biology: UV light can kill microorganisms. Thirty years ago clothes dryers had ultraviolet lamps in them to kill germs as the clothes were dried. These lamps were later removed from inside the clothes dryers because the lamps were not strong enough to be effective.

Extension: Invite an optometrist to discuss the effects of ultraviolet light on the eye.

Answers to Exploring Further

1. Answers may vary. Sample answers are to use refrigerants and propellants that do not contain chlorine, recycle refrigerants, and use non-aerosol sprays or products other than sprays.
2. The chlorine atoms are regenerated in the process and go on to catalyze further reactions.

✓ ASSESSMENT

Portfolio: Have students develop a survey that can be used with a large sample of the high school's student body. The survey can include the number of students who do or do not use sun screen on their skin, and which SPF number they use; the number of students who wear UV-blocking sunglasses, and how often. Have students prepare a report that graphically presents the data they collected. Students may elect to place the survey and report in their portfolio. The report can also be submitted to the school's newspaper.

Chemistry Journal

Accelerators

Have students carry out library research to obtain more information about one type of accelerator or particle generator. They should write a journal account on their findings and include a diagram of the device.

PROGRAM RESOURCES

Transparency Master: Use the Transparency master and worksheet, pp. 11-12, for guided practice in the concept "Flame Tests."

Color Transparency: Use Color Transparency 6 to focus on the concept of "Flame Tests."

CONCEPT DEVELOPMENT

Enrichment: Although Bohr was the first to use experimental evidence to support his hypothesis, he was not the first person to propose a planetary model for the atom. For example, the Japanese physicist Hantaro Nagaoka suggested the "Saturnian" model for the atom in 1904. Ask a student to research the history of the planetary model prior to Bohr.

QUICK DEMO **An Analogy** Have the students listen carefully as you drop a textbook from three different heights. The higher the book, the louder the noise it produces. Similarly, the farther an excited electron is from the nucleus, the more energy it emits as it returns to the same state. This energy can be in the form of UV, visible, and IR light. UV light has more energy than visible light, and visible light has more energy than IR. Blue light has greater energy than red because blue light has a higher frequency.

Enrichment: A student can report to the class on the topic of nuclear magnetic resonance, NMR. This is one of the newer methods of analysis in medicine where it is called magnetic resonance imaging, MRI. Possible response: As an electron spins about the nucleus, it creates a magnetic field. If an external magnetic field is imposed, the electron spin will tend to align with the field. The energy required to "flip" the electron spin can be measured. The energy required is related to the environment of the electron and helps the chemist to deduce molecular structures. A computer provides an image from these data for the medical doctor.

In order to improve his description of atomic structure, Bohr used the experimental evidence of atoms exposed to radiant energy. When a substance is exposed to a certain intensity of light or some other form of energy, the atoms absorb some of the energy. Such atoms are said to be excited. When atoms and molecules are in an excited state, unique energy changes occur that can be used to identify the atom or molecule. Radiant energy of several different types can be emitted (given off) or absorbed (taken up) by excited atoms and molecules. The methods of studying substances that are exposed to some sort of exciting energy are called **spectroscopy.** A pattern of radiant energy studied in spectroscopy is called a **spectrum.**

Visible light is one form of radiant or **electromagnetic energy.** Other forms of electromagnetic energy are radio, infrared, ultraviolet, and X ray. This energy consists of variation in electric and magnetic fields taking place in a regular, repeating fashion. If we plot the strength of the variation against time, our graph shows "waves" of energy. The number of wave peaks that occur in a unit of time is called the **frequency** of the wave. Frequency is represented by the Greek letter *nu* (ν) and is measured in units of **hertz** (Hz). A hertz is one peak, or cycle, per second. All electromagnetic energy travels at the speed of light. The speed of light is 3.00×10^8 m/s in a vacuum and is represented by the symbol c. Another important characteristic of waves is the distance between peaks. This distance between peaks is called the **wavelength** and is represented by the Greek letter *lambda* (λ). These characteristics of waves are related by the statement

$$c = \lambda\nu$$

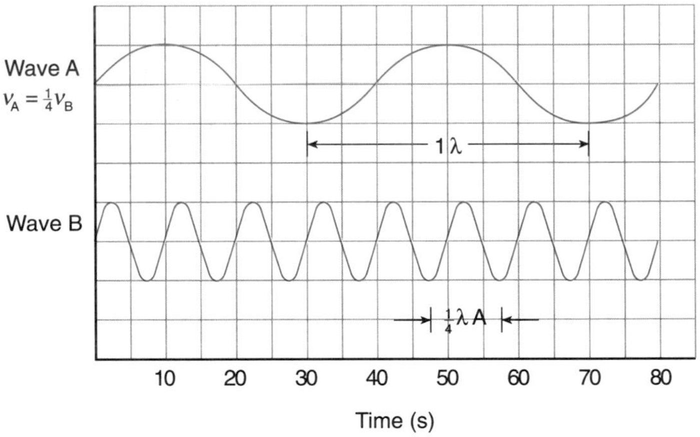

Figure 4.12 Wave A has the longer wavelength; however, wave B has the larger frequency. Short electromagnetic waves have higher frequencies than do long electromagnetic waves.

Another wave property that is of importance is the amplitude of a wave, or its maximum displacement from zero. In Figure 4.13, two waves are plotted on the same axes. Note that the amplitude of wave *A* is twice that of wave *B*, even though they have the same wavelength.

DEMONSTRATION

4-6 Easy Flame Tests

Purpose: to observe characteristic spectra of several metals

Materials: 10 g each of the chloride salts of Cu, Na, K, Ca, and Sr; 5 sealable plastic sandwich bags; laboratory burner; student spectroscopes; mortar and pestle

Procedure: **1.** Before the demonstration, grind each sample of the metal chloride salts into a fine powder using a clean mortar and pestle. Place each sample in a plastic bag and seal it.

2. For the demonstration darken the room, light a gas burner, and adjust the flame so it burns blue.

3. Shake a bag containing one of the powdered metal chloride salt crystals. Open the bag near the air vent on the burner. **CAUTION:** *Do not use toxic chemicals, such as Cd, Pb, Co, Cr, Ni, or Ba salts.* Have students use

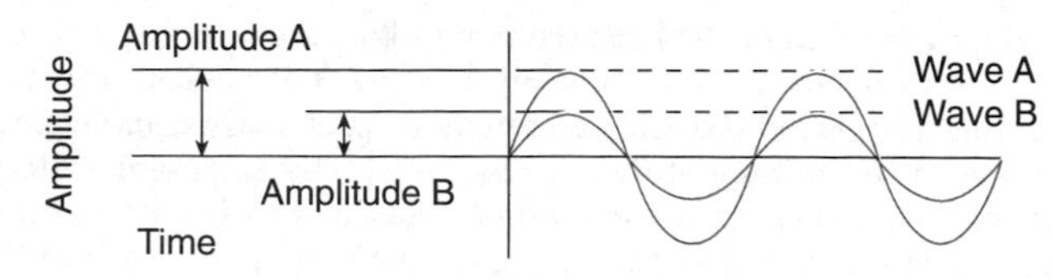

Figure 4.13 Waves A and B have the same wavelength, even though their amplitudes differ.

Excited atoms soon lose the energy they have gained. The energy emitted by gaseous atoms occurs at specific points (or lines) in a spectrum. The lines in an emission spectrum as seen in Figure 4.16 are characteristic of the element being excited. If these same atoms are exposed to light of all wavelengths, they will absorb energy. If this light is examined after it passes through the gaseous form of an element, some of the incident light will be missing. This missing light is absorbed by the gaseous atoms. The spectrum of the light leaving the gaseous atoms has lines missing. Such a spectrum is seen in Figure 4.14 and is called an absorption spectrum. These lines will be at the same wavelengths as the bright lines in the emission spectrum. The collection of lines, absorption or emission, for any element is the spectrum of that element and is unique. Spectroscopy can, therefore, be used as a means of identifying elements. Absorption and emission spectra are the fingerprints of the elements.

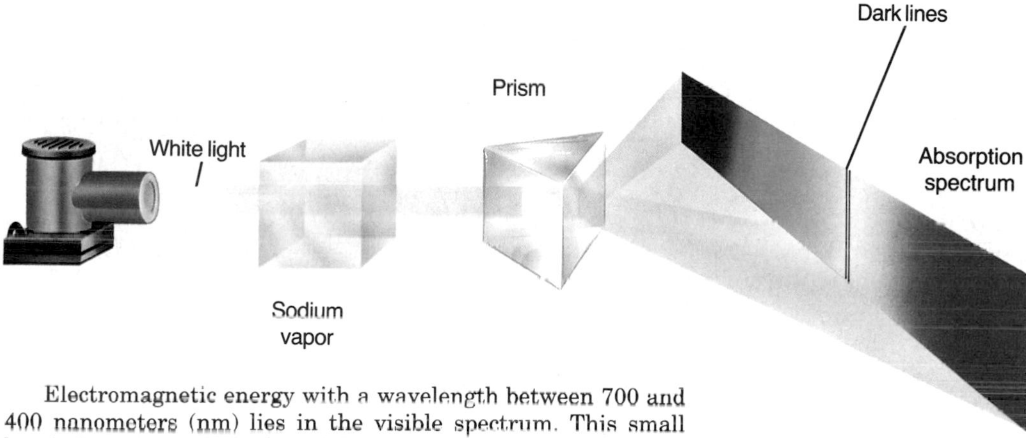

Figure 4.14 This apparatus is used to produce the absorption spectrum of sodium.

Electromagnetic energy with a wavelength between 700 and 400 nanometers (nm) lies in the visible spectrum. This small band of visible radiation has given chemists and physicists much information about the elements. Some elements (rubidium, cesium, helium, and hafnium) were discovered only after their spectra were observed. The visible spectrum may also be used for finding the concentration of substances, and for analyzing mixtures. Almost any change involving color can be measured using visible spectroscopy.

As is shown in Figure 4.15, the electromagnetic spectrum includes wavelengths much longer and wavelengths much shorter than those of visible light. Radio waves have the longest wavelengths, gamma rays the shortest. Ultraviolet radiation (UV) has

CONCEPT DEVELOPMENT

Background: Three techniques for analysis that are closely allied to spectroscopy are colorimetry, fluorimetry, and nephelometry (nef uh LAHM ih tree). You may wish to have students research and report on one of these procedures. Possible responses: *Colorimetry is the analysis of colored solutions. Fluorimetry is the analysis of fluorescent substances, i.e., those substances that absorb light in one wavelength, usually UV, and then emit light at a longer wavelength. Nephelometry is the analysis of colloids or suspensions by their light-scattering ability.*

spectroscopes to observe the flames. Repeat this step for each of the remaining salts. **Disposal:** F

Results: The flame will emit that metal's characteristic color. For example, $CuCl_2$ will emit a blue-green color. The characteristic colors of the metals used are Na - yellow, K - violet, Ca - red orange, and Sr - crimson.

Questions: 1. What colors are observed from specific metal salts? *See Results.*
2. Why are the spectra you observed in the spectroscopes called bright line emission spectra? *The atom emits a spectrum that is composed of bright lines.*

Extension: Have a student investigate how a forensic laboratory uses spectroscopy to identify a car involved in a hit-skip accident from a paint chip found on the victim's clothes.

PROGRAM RESOURCES

Transparency Master: Use the Transparency master and worksheet, pp. 13-14, for guided practice in the concept "Electromagnetic Spectrum."

Color Transparency: Use Color Transparency 7 to focus on the concept of "Electromagnetic Spectrum."

Laboratory Manual: Use microlab 4 "Identifying Elements by Flame Test" as enrichment for Chapter 4.

CONCEPT DEVELOPMENT

Background: Planck first proposed the quantum nature of light to explain the frequency distribution of black body radiation. Einstein used the quantum hypothesis to explain the photoelectric effect, which is discussed as a special feature in this chapter. Interpretation of spectroscopic data of atomic structures further supports the quantum nature of light.

✓ ASSESSMENT

Oral: Ask a student in each cooperative learning group to quickly survey the members of each group to determine the frequency of their favorite AM and FM radio stations. Then ask the Cooperative Learning groups to use Figure 4.15 to determine an approximate wavelength for each radio station selected. Then, have them check their estimates by applying the equation $C = \lambda\nu$, where C is the velocity of light, λ is wavelength, and ν is frequency. For example, 1460 KHz AM is 1.460×10^6 Hz and has a wavelength of about 205 meters, and 107.1 MHz FM is 1.071×10^8 Hz and has a wavelength of about 2.8 meters. Have the group's reporter read their results to the class.

wavelengths between 400 and 200 nm. Like visible light, ultraviolet radiation can be used to study atomic and molecular structure. Both ultraviolet spectra and visible spectra are produced by changes in the energy states of electrons. The ultraviolet spectrum of an element or compound consists of bands rather than lines. Ultraviolet radiation has such high energy, it violently excites the electrons. The transition of the electrons from the normal state to such highly excited states causes changes in the molecule being studied. Bonds between atoms may even be broken. Visible radiations are not as destructive because they have less energy than ultraviolet radiation has. Ultraviolet and visible spectroscopy are used for the same types of analyses. In order to describe completely the electronic structure of a substance, both types of analysis must be used. Infrared spectroscopy is also useful to the chemist when studying whole molecules. This type of spectroscopy will be discussed in Chapter 12.

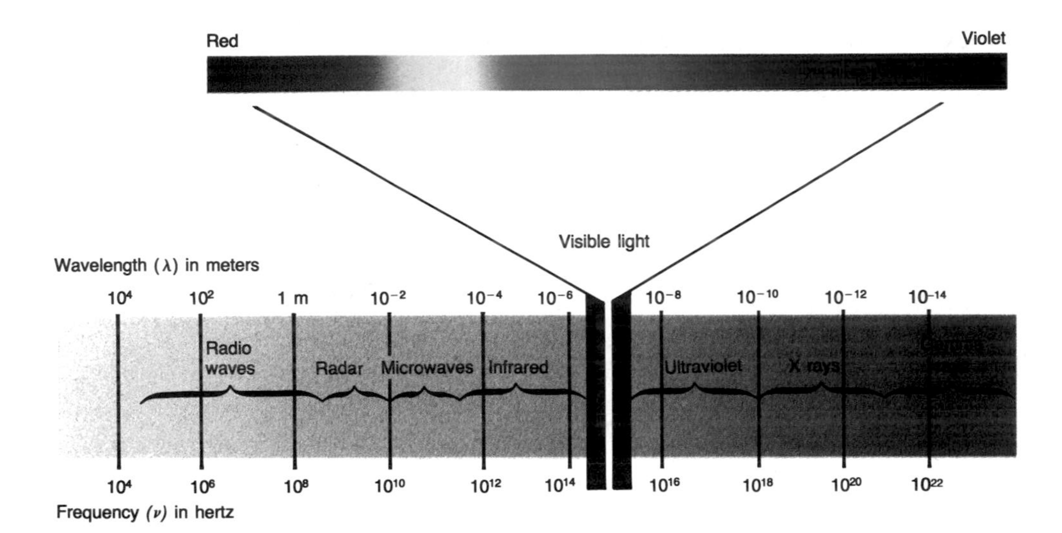

Figure 4.15 Visible light is only a small part of the electromagnetic spectrum. Note that the waves with high frequency have short wavelengths.

Planck's Hypothesis

In his attempt to explain the hydrogen spectrum, Bohr developed his planetary model of the atom. Bohr used the **quantum theory,** a theory of energy emission that had been stated by a German physicist, Max Planck. Planck assumed that energy, instead of being given off continuously, is given off in little packets, or **quanta.** Quanta of radiant energy are often called **photons.** He further stated that the amount of energy given off is directly related to the frequency of the light emitted.

Planck's idea was that one quantum of energy (light) was related to the frequency by the equation $E = h\nu$, where h is a constant. The constant is known as Planck's constant. Its value is $6.626\ 075\ 5 \times 10^{-34}$ joules per hertz.

PROGRAM RESOURCES

Chemistry and Industry: Use the worksheets, pp. 3-6, "Hydrogen: More Than Just the "H" in H_2O," as an extension to the concepts presented in Section 4.2.

DEMONSTRATION

4-7 Neon Lights

Purpose: to observe the bright line emission spectra of different elements

Materials: hydrogen and neon gas spectrum tubes, high-voltage DC power supply, student spectroscopes white light source

Procedure: 1. Hydrogen and neon gas spectrum tubes are available commercially. Your school's physics teacher may have them, as well as the high-voltage DC power supply and the student spectroscopes.

2. Have students hold the spectroscope so that the diffraction gratings are toward their eyes, and the slits are toward the white light source in a darkened room. By moving the slit slightly off the light source, a continuous spectrum will appear to one side of the spectroscope.

3. Repeat the procedure using a gas

The Hydrogen Atom and Quantum Theory

Planck's hypothesis stated that energy is given off in quanta instead of continuously. Bohr pointed out that the absorption of light by hydrogen at definite wavelengths corresponds to definite changes in the energy of the electron. He reasoned that the orbits of the electrons surrounding a nucleus must have a definite diameter. Furthermore, electrons could occupy only certain orbits. The only orbits allowed were those whose differences in energy equaled the energy absorbed when the atom was excited. Bohr thought that the electrons in an atom could absorb or emit energy only in whole numbers of photons. In other words, an electron could emit energy in one quantum or two quanta, but not in $1\frac{1}{4}$ or $3\frac{1}{2}$ quanta.

Bohr pictured the hydrogen atom as an electron circling a nucleus at a distance of about 53 pm. He also imagined that this electron could absorb a quantum of energy and move to a larger orbit. Because a quantum represents a certain amount of energy, the next orbit must be some definite distance away from the first. If still more energy is added to the electron, it moves into a still larger orbit, and it continues to move to larger orbits as more energy is added. When an electron drops from a larger orbit to a smaller one, energy is emitted. Because these orbits represent definite energy levels, a definite amount of energy is radiated.

The size of the smallest orbit an electron can occupy, the one closest to the nucleus, can be calculated. This smallest orbit is called the **ground state** of the electron. Bohr calculated the

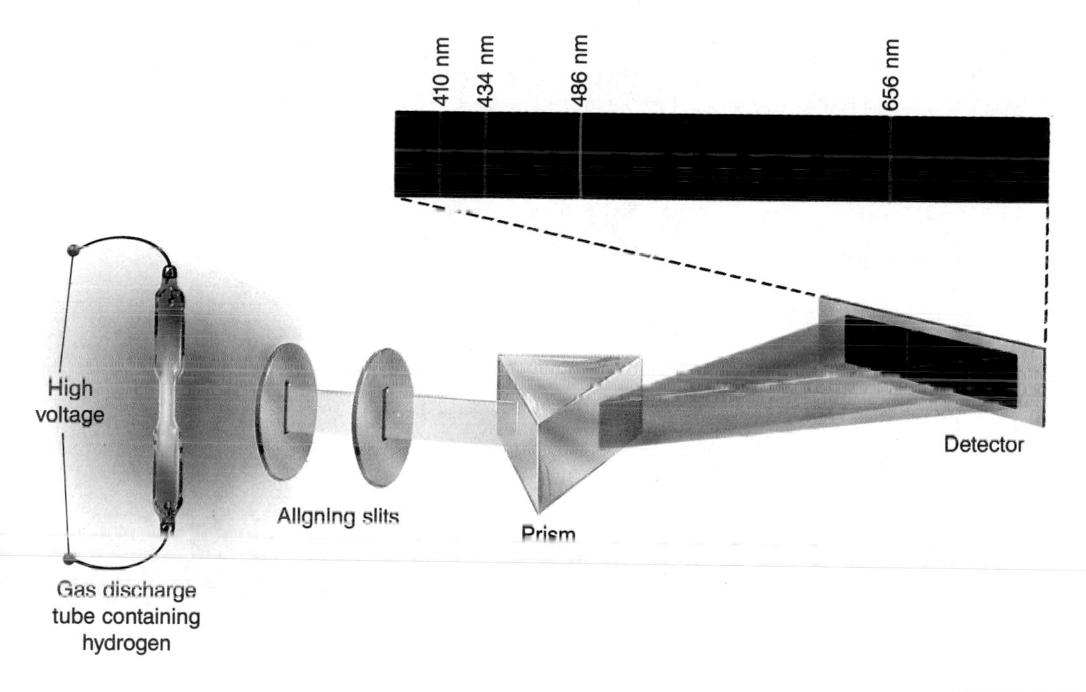

Figure 4.16 A prism spectroscope can be used to observe emission spectra. The visible region of the emission spectrum of hydrogen is a series of discrete lines.

emission tube as the light source. **Disposal:** F

Results: Students will observe the rainbow of a continuous spectrum of the white light. Students will observe a bright line emission spectrum containing green lines for neon. Usually only three lines are visible for hydrogen. Each colored line represents a different transition having its own energy.

Questions: 1. What are the colors that make up the continuous spectrum? *Roy G. Biv...red, orange, yellow, green, blue, indigo, and violet*

2. What happens when an electron drops from a high energy level to a lower one? *Energy is emitted.*

3. What source of energy is used to excite the atom's electrons? *electrical energy*

Extension: Have students investigate how astronomers use the spectra of stars to determine their composition.

CONCEPT DEVELOPMENT

Misconceptions

Students may have the impression that atoms are similar to planetary systems. Avoid illustrating atoms using Bohr models. Have students visualize electrons as clouds and their positions as probable positions, a concept that will be expanded in Chapters 5 and 6. The concept of electron clouds allows the student to better understand the chemical bond when it is presented.

GLENCOE TECHNOLOGY

Software

Mastering Concepts in Chemistry

Unit 1, Lesson 3

Description of the Atom

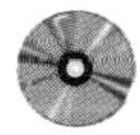

Videodisc

Chemistry: Concepts and Applications

Emission Spectra of Elements

Disc 1, Side 2, Ch. 6

Show this video of Demonstration 10-4 on pages 260–261 to reinforce the concept of emission spectra.

Also available on CD-ROM.

Videodisc

Chemistry: Concepts and Applications

Electrons and Energy Levels

Disc 1, Side 1, Ch. 9

Also available on CD-ROM.

CONCEPT DEVELOPMENT

• Emphasize to students that the modern atomic theory does not agree with Bohr's theory that stated that electrons move around the nucleus as the planets orbit the sun.

✓ ASSESSMENT

Skill: The relationships given by the equations $E = h\nu$ and $\lambda = c/\nu$ can be used by cooperative learning problem solving groups to develop answers for Application questions 44-48 on page 106. Remind students that Planck's constant is given at the bottom of page 96.

ground state of the hydrogen electron. Using the quantum theory, he calculated the frequencies for the lines that should appear in the hydrogen spectrum. His results agreed almost perfectly with the actual hydrogen spectrum. Although today we use a model of the atom that differs from the Bohr model, many aspects of his theory are still retained. The major difference is that electrons do not move around the nucleus as the planets orbit the sun. We will explore this difference in the next chapter. However, the idea of energy levels is still the basis of atomic theory. The energy level values calculated by Bohr for the hydrogen atom are still basically correct.

We can summarize the relationship between electromagnetic energy and an electron as follows: An electromagnetic wave of a certain frequency has only one possible wavelength, given by $\lambda = c/\nu$. It has only one possible amount of energy, given by $E = h\nu$. Since both c and h are constants, if frequency, wavelength, or energy is known, we can calculate the other two.

The visible and ultraviolet spectra produced by a compound can be used to determine the elements in the compound. Each line in a spectrum represents one frequency of light. Because the velocity of light is always constant, each frequency is associated with a certain energy. This energy is determined by the movement of electrons between energy levels that are specific for each element. The same set of energy levels will always produce the same spectrum.

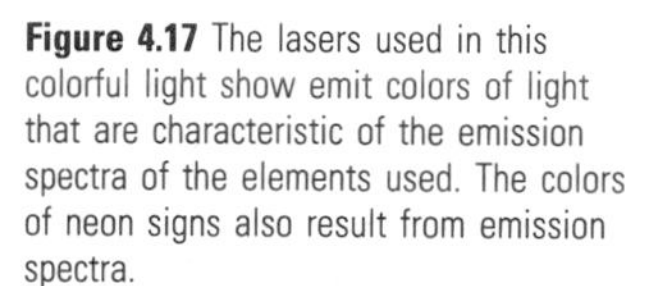

Figure 4.17 The lasers used in this colorful light show emit colors of light that are characteristic of the emission spectra of the elements used. The colors of neon signs also result from emission spectra.

CONTENT BACKGROUND

Radical Thinking: It is interesting to note that Planck's quantum proposal was so radical and the state of German physics so rigid and hierarchical that Planck actually had a brief period when he was fearful for his life.

J.J. Thomson discovered the electron by adjusting the strength of the magnetic and electric fields of his cathode-ray tube until the deflection of one field was just offset by the other. Under those conditions, the velocity of the particles was equal to the strength of the electric field divided by the strength of the magnetic field. By measuring the angle of deflection in the electric field, Thomson could compute mass-to-charge ratio for the electron.

Ernest Rutherford is the father of the nuclear atom. This distinction is the result of his interpretation of the

CHEMISTRY AND
TECHNOLOGY

Photoelectric Effect

Many stores have doors that open automatically. Some of these doors are actuated by a device whose operation depends on light. In front of the door on one side is a source of light. Opposite the source of light is a light detector. The beam of light falling on the detector causes a substance in the detector to give off electrons and an electric current to flow in a circuit. This emission of electrons is called the photoelectric effect. When you break the beam of light by stepping between the light source and the detector, the detector stops emitting electrons. Thus, the flow of electric current stops. This interruption in the flow of electric current actuates a mechanism that opens the door.

Einstein received a Nobel Prize in 1921 for his explanation of the photoelectric effect. It had been known for some time that light falling on the surface of certain substances caused electrons to be emitted. There is, however, a puzzling fact about this change. When the intensity of light (the number of photons per unit time) is reduced, the electrons emitted still have the same energy. However, fewer electrons are emitted. Einstein pointed out that Planck's hypothesis explains this observation.

A certain amount of energy is needed to remove an electron from the surface of a substance. If a photon of greater energy strikes the electron, the electron will also move away from the surface. Since it is in motion, the electron has some kinetic energy. Some of the energy of the photon is used to free the electron from the surface. The remainder of the energy becomes the kinetic energy of the electron. If light of one frequency is used, then the electrons escaping from the surface of the substance will all have the same energy.

If the light intensity is increased, but the frequency remains the same, the number of electrons being emitted will increase. If the frequency of the light is increased, the energy of the photon is increased. The amount of energy that must be used to free the electron from the atom is constant for a given substance. Thus, the electrons now leave the surface with a higher kinetic energy than they did with the lower frequency.

Planck's hypothesis, together with Einstein's explanation, confirmed the particle nature of light.

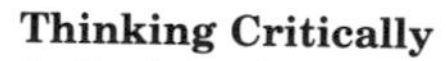

Thinking Critically

1. Explain how the photoelectric effect can be used in construction of a security system.
2. How is the photoelectric effect used in a photographic exposure meter?

results of the experiments performed between 1909 and 1913 by Hans Geiger and Ernest Marsden in Rutherford's laboratory. In his famous paper of 1911, Rutherford made no attempt to describe the arrangement of the electrons in an atom although he did make a favorable comment about the "Saturnian" model of Hantaro Nagoaka (1904).

CHEMISTRY AND TECHNOLOGY

Photoelectric Effect

Background: Classical physics before Max Planck and Albert Einstein treated light as a wave phenomenon. Refraction, reflection and diffraction all confirm that light has wave characteristics. But, the photoelectric effect confirmed that light also had particle characteristics. This duality has a paradox. Both natures of light cannot be displayed in the same experiment.

Connection to Photography: The invention of the photocell has revolutionized the automatic camera. Computer chips now measure the light intensity and set the speed and aperture based on the speed of the film being used.

Teaching This Feature: Bring to class an automatic camera, or a dusk to dawn light switch that uses a photocell. Demonstrate its use and discuss the photoelectric effect.

Answers to
Thinking Critically

1. When the beam of light is broken, the electric current stops and an alarm is activated.

2. Light from the source strikes a substance in the meter causing this substance to give off electrons. Detection of the number of electrons can give the intensity of the light.

Exploring Further: Have students think of not-yet-invented devices that would make life easier and which use the photoelectric effect.

PROGRAM RESOURCES

Reteaching: Use the Reteaching worksheets, pp. 5-6, to provide students another opportunity to understand the concept of "Spectra of Elements."

CONCEPT DEVELOPMENT

Enrichment: The mass spectrometer is used to separate isotopes for use as nuclear fuel. Have a team of students investigate this process and see how it differs from other processes used for the same purpose. Countries interested in developing a nuclear weapon have used this technique recently, even though it is considered out of date. Possible reference: Glasstone, Samuel, SOURCEBOOK FOR ATOMIC ENERGY, New York, D. Van Nostrand Co., Inc., 1958, pp. 227-230.

• Ask students if they think chemists can work conveniently with individual atoms in the laboratory. Point out that an atom may have a mass of 10^{-23} grams. Chemists generally work with moles of atoms, 6.02×10^{23} atoms, rather than individual atoms. Moles will be discussed in detail in Chapter 8.

Figure 4.18 This rhinoceros and bird have about the same mass ratio as that of a proton and an electron.

Atomic Mass

The proton and neutron are essentially equal in mass. The mass of the electron is extremely small, so almost all of the mass of an atom is located in the nucleus. Even the simplest atom, which contains only one proton and one electron, has $^{1836}/_{1837}$ of its mass in the nucleus. In other atoms that have neutrons in the nucleus, an even higher fraction of the total mass of the atom is in the nucleus.

It is possible to discuss the mass of a single atom. However, chemists use the masses of large groups of atoms. They do so because of the very small size of the particles in the atom. Chemists measure the mass of one atom in atomic mass units (u). You know that the unit gram was defined as $^{1}/_{1000}$ the mass of the International Prototype Kilogram. What is an atomic mass unit? To measure atomic masses, an atom of one element was chosen as a standard, and the other elements were compared with it. Scientists used a carbon nuclide with mass number 12 as the standard for the atomic mass scale. This carbon nuclide, carbon-12, has 6 protons and 6 neutrons in the nucleus. *One carbon-12 atom is defined as having a mass of 12 atomic mass units.* An atomic mass unit is defined to be $^{1}/_{12}$ the mass of the carbon-12 nuclide.

The most important subatomic particles you have studied thus far have the following masses.

$$\text{electron} = 9.109\ 53 \times 10^{-28}\ \text{g} = 0.000\ 549\ \text{u}$$
$$\text{proton} = 1.672\ 65 \times 10^{-24}\ \text{g} = 1.0073\ \text{u}$$
$$\text{neutron} = 1.674\ 95 \times 10^{-24}\ \text{g} = 1.0087\ \text{u}$$

Look carefully at Table 4.2. Notice that many of the elements have a mass in atomic mass units that is close to the total number of protons and neutrons in their nuclei. However, some do not.

DEMONSTRATION

4-8 A Massive Atom

Purpose: to demonstrate that the mass of the atom is concentrated in its nucleus

Materials: 12 glass marbles, modeling clay, 6 cotton balls, laboratory balance

Procedure: 1. Inform students that the marbles represent protons and neutrons. Use a small amount of clay to hold the marbles together. Mass the "nucleus" on a laboratory balance.

2. Inform students that the cotton balls simulate the atom's electrons. Add the cotton ball "electrons" to the clay-marble "nucleus". Mass the total "atom" on the balance.

Results: The mass of the nucleus is not significantly different from the mass of the atom. This simulation demonstrates that nearly all the mass of an atom is in the nucleus.

What causes the mass of chlorine or copper, for example, to be about halfway between whole numbers? The numbers in the table are based on the "average atom" of an element.

Most elements have many isotopic forms that occur naturally. It is difficult and costly to collect a large amount of a single nuclide of an element. Thus, for most calculations, the average atomic mass of the element is used.

Average Atomic Mass

Using a standard nuclide, there are two ways of determining masses for atoms of other elements. One method is by reacting the standard element with the element to be determined. Using accurate masses of the two elements and a known ratio, the atomic mass of the second element may be calculated.

Higher accuracy can be obtained by using a physical method of measurement in a device called a mass spectrometer. The mass spectrometer has many uses. Geologists, biologists, petroleum chemists, and many other research workers use the mass spectrometer as an analytical tool. Its development was based on the design of the early tubes J. J. Thomson used to find the charge/mass ratio of the electron.

Using a mass spectrometer, we can determine the relative amounts and masses of the nuclides for all isotopes of an element. The element sample, which is in gaseous form, enters a chamber where it is charged by a beam of electrons. These charged particles are then propelled by electric and magnetic fields. As in the Thomson tube, the fields bend the path of the charged particles as shown in Figure 4.19. The paths of the heavier particles are bent slightly as they pass through the fields. The paths of the lighter particles are curved more. Thus, the paths of the particles are separated by relative mass. The particles are both caught and

Figure 4.19 The mass spectrometer is used extensively as an analytical tool. Inside the spectrometer, a magnet (left) causes the positive ions to be deflected according to their mass. In the vacuum chamber (right), the process is recorded on a photographic plate or a solid-state detector.

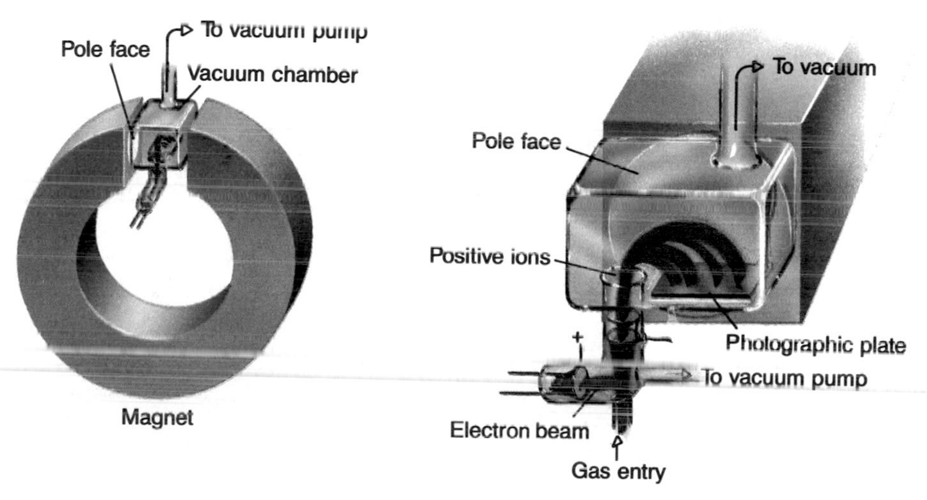

CONCEPT DEVELOPMENT

• Inform students that hydrogen appears in nature as three isotopes, having masses of 1 u (protium), 2 u (deuterium), and 3 u (tritium). Ask them to find the average atomic mass of hydrogen using the table inside the textbook cover, and predict which isotope is present in the greatest amount. *The average is 1.007 94, thus hydrogen-1 is the most common.*

• Point out to students that analytical chemists use many instruments in the laboratory. A very important one is the mass spectrometer. A description of how this instrument works is found in Figure 4.19.

Questions: 1. Using the values given in the textbook, what is the mass in grams of a nucleus having 6 protons and 6 neutrons? *$2.008\ 56 \times 10^{-23}$ g*

2. What is the mass in grams of 6 electrons? *$5.465\ 72 \times 10^{-27}$ g*

3. What is the total mass in grams of the 6 protons, 6 neutrons, and 6 electrons? *$2.009\ 11 \times 10^{-23}$ g*

4. What percent of the total mass is the nucleus? *99.9726%*

✓ ASSESSMENT

Knowledge: Ask students to determine which element is represented by the model that was made. The model used for the simulation was carbon-12. Point out to students that this nuclide was selected as the standard for the atomic mass scale because of its isotopic purity.

PROGRAM RESOURCES

Transparency Master: Use the Transparency master and worksheet, pp. 15-16, for guided practice in the concept "Mass Spectrometer."

Color Transparency: Use Color Transparency 8 to focus on the concept of "Mass Spectrometer."

CHAPTER REVIEW 4

- Review Summary statements and Key Terms with your students.
- Complete solutions to Chapter Review Problems can be found in the **Merrill Chemistry Problems and Solutions Manual.**

Answers to

Review and Practice

19. For details of scientists' contributions, see the Problems and Solutions Manual accompanying this text.

20. **(1)** All matter is composed of atoms that are indivisible.

(2) Atoms of the same element are identical; atoms of different elements are dissimilar.

(3) Atoms can unite with other atoms in simple numerical ratios to form compounds.

21. Avogadro's hypothesis: Equal volumes of gases, at the same temperature and pressure, have the same number of molecules.

22. Cathode rays are the beam of electrons in a gas discharge tube. They are called cathode rays because they begin at the cathode of the tube.

Chapter 4

REVIEW

Summary

4.1 Early Atomic Models

1. Democritus proposed the earliest recorded atomic theory. He believed that matter is composed of tiny particles, or atoms.
2. Before Dalton proposed the atomic theory, Lavoisier stated the law of conservation of mass—matter is not created or destroyed in ordinary chemical reactions. Proust's law of definite proportions stated that substances always contain elements in the same ratio by mass.
3. John Dalton used the law of conservation of mass and the law of definite proportions to state that all matter is formed of indivisible particles called atoms; all atoms of one element are the same; atoms of different elements are unlike; and atoms can unite with one another in simple whole-number ratios.
4. Modern atomic theory differs from Dalton's atomic theory due to the discovery of subatomic particles and isotopes.
5. An electron is a negatively charged particle with a very small mass. A proton is a positively charged particle with a mass 1836 times that of an electron. A neutron is an uncharged particle with a mass about the same as that of a proton.
6. The atomic number, Z, of an atom is the number of protons in its nucleus.
7. All atoms of an element contain the same number of protons in their nuclei. Atoms of the same element having different numbers of neutrons are isotopes.
8. The mass number, A, of an atom is the number of particles in its nucleus.
9. From experiments, especially those of Geiger and Marsden, Rutherford concluded that atoms consist mostly of empty space and have a small nucleus.
10. Radioactivity is particle or energy emission due to nuclear disintegration.
11. Radioactive decay is spontaneous; that is, it cannot be controlled.
12. Einstein proposed that matter and energy are equivalent.

4.2 Parts of the Atom

13. Subatomic particles are either elementary particles called leptons or complex particles called hadrons.
14. Hadrons are believed to be made of particles called quarks. Quarks and hadrons are thought to be held together by exchanging gluons and pions, respectively.
15. Naturally radioactive nuclides emit three kinds of radiation: alpha (helium nuclei), beta (electrons), and gamma (energy).
16. Many subatomic particles in addition to electrons and nucleons have been discovered. Some of these are antimatter.
17. Rutherford and Bohr pictured the atom as consisting of a central nucleus surrounded by electrons in orbits.
18. Atoms excited by energy both absorb and emit definite wavelengths of electromagnetic radiation. The unique collection of lines, absorption or emission, for any element is the spectrum of that element. Visible and ultraviolet spectroscopy are used to study atomic structure.
19. Planck stated that energy is radiated in discrete units called quanta. A photon is a quantum of light energy. The energy of a quantum of radiation varies directly as the frequency of the radiation ($E = h\nu$).
20. Bohr theorized that the energies in the hydrogen spectrum corresponded to certain quanta emitted or absorbed when the electron moved from one orbit to another. Thus, he was able to calculate the orbits for the hydrogen atom.
21. The atomic mass of an element is the weighted average mass of all its natural isotopes compared with 1/12 the mass of the carbon-12 atom. Atomic mass can be measured with the mass spectrometer.

PROGRAM RESOURCES

Evaluation: Use the Evaluation worksheets, pp. 13-16, to assess students' knowledge of Chapter 4.

Computer Test Bank: Use the IBM or Apple Test Bank and Question Selection Manual to prepare your own individualized test for Chapter 4.

Chapter 4
REVIEW

Key Terms

law of definite proportions
law of multiple proportions
anode
cathode
cathode ray
isotope
atomic number
nuclide
nucleon
mass number
radioactivity
nuclear force
subatomic particle
lepton
hadron
antiparticle
neutrino
quark
baryon
meson
gluon
alpha particle
beta particle
gamma ray
spectroscopy
spectrum
electromagnetic energy
frequency
hertz
wavelength
quantum theory
quanta
photon
ground state
atomic mass

Review and Practice

19. What did each of the following scientists contribute in forming the modern model of the atom?
a. Dalton
b. Thomson
c. Rutherford
d. Chadwick
e. Proust
f. Democritus
g. Millikan
h. Einstein
i. Moseley
j. Bohr
k. Planck
l. Avogadro
m. Lavoisier
n. Gay-Lussac
o. Becquerel
p. Geiger and Marsden

20. What are the major points in Dalton's atomic theory?

21. State Avogadro's hypothesis.

22. What are cathode rays? Why are they called cathode rays?

23. What are the differences in charge and mass among protons, neutrons, and electrons?

24. What nuclide is used as the reference standard in defining the atomic mass unit?

25. A particular atom of potassium contains 19 protons, 19 electrons, and 20 neutrons. What is the atomic number of this atom? What is its mass number? Write the symbol for this potassium nucleus.

26. How many electrons, neutrons, and protons are in atoms of chlorine with mass number 35? How many of each are in the atoms of thorium with mass number 232?

27. Yttrium was discovered in 1794. It is one of the elements used in superconductors. How many electrons, protons, and neutrons are in an atom of yttrium-88?

28. Compare the amount of energy involved in chemical changes to the amount of energy resulting from nuclear changes.

29. What are the two broad classes of subatomic particles? What are the differences between these two classes? How do protons, neutrons, and electrons fall into these classes?

30. What are the differences among the three types of natural radiation?

31. How many neutrons and protons are in each of the following nuclides?
a. carbon-14
b. phosphorus-32
c. nickel-63
d. iridium-192
e. iron-54
f. neptunium-235

32. How does a mass spectrometer separate different types of atoms?

33. Find the average atomic mass of silver if 51.83% of the silver atoms occurring in nature have mass 106.905 u and 48.17% of the atoms have mass 108.905 u.

34. Find the average atomic mass of krypton if the relative amounts are as follows.

Isotopic mass	Percentage
77.920 u	0.350
79.916 u	2.27
81.913 u	11.56
82.914 u	11.55
83.912 u	56.90
85.911 u	17.37

23. See table below. By comparing the values in the table, it can be seen that the proton and neutron have similar masses, but the electron has much less mass. Also, the charges on the electron and proton are alike in numerical value, but opposite in sign. The neutron has no charge.
24. $^{12}_{6}C$
25. 19, 39, $^{39}_{19}K$
26. Cl: 17 electrons, 18 neutrons, 17 protons; Th: 90 electrons, 142 neutrons, 90 protons
27. Y: 39 electrons, 39 protons, 49 neutrons
28. The energy involved in chemical changes is much less than the energy involved in nuclear changes because it involves no conversion of mass to energy. Mass is converted to energy in a nuclear change.
29. The two broad classes of subatomic particles are leptons and hadrons. Leptons are not made up of smaller particles, but hadrons are. An electron is a lepton. Protons and neutrons are hadrons.
30. alpha (α) particle: a helium nucleus, positive
beta (β) particle: an electron, negative
gamma (γ) rays: high-energy X rays, neutral
31. See table below.
32. A gaseous sample is ionized by a beam of electrons. These ions are then deflected by electric and magnetic fields. Lighter atoms are deflected more than are heavier atoms, and they can be separated.
33. 107.9 u
34. 83.8 u

23.

Particle	Mass	Charge
electron	$9.109\,53 \times 10^{-28}$ g = 0.000 549 u	1-
proton	$1.672\,65 \times 10^{-24}$ g = 1.0073 u	1+
neutron	$1.674\,95 \times 10^{-24}$ g = 1.0087 u	0

31.

Element	Protons	Neutrons
a. $^{14}_{6}C$	6	14 − 6 = 8
b. $^{32}_{15}P$	15	17
c. $^{63}_{28}Ni$	28	35
d. $^{192}_{77}Ir$	77	115
e. $^{54}_{26}Fe$	26	28
f. $^{235}_{93}Np$	93	142

Chapter 4
REVIEW

Concept Mastery

35. How did discovery of subatomic particles and isotopes affect Dalton's theories?
36. What was Rutherford's interpretation of the Geiger-Marsden experiment?
37. Describe the force that holds the particles of an atom's nucleus together. Does electrostatic force help or oppose the effect of this force?
38. Describe the way that excited atoms produce (a) an emission spectrum and (b) an absorption spectrum.
39. How did Bohr use spectroscopic data to formulate his model of the atom?
40. According to Bohr, what happened when an electron absorbed a photon?
41. How are the law of conservation of mass and the law of definite proportions explained by Dalton's atomic theory?
42. Why is it necessary to ionize an element sample before it can be separated into its isotopic components with a mass spectrometer?
43. Studies with gas discharge tubes had been conducted for more than 70 years before J. J. Thomson discovered the true nature of the cathode rays produced in these tubes. It was Thomson who ultimately proved that these rays consist of elementary negatively charged particles that are identical regardless of the element from which they are produced. He called these negative particles electrons. With the information that the cathode rays are actually a stream of electrons that are negatively charged, answer the following questions.
   a. To which terminal of the electric current source should the cathode of the gas discharge tube be connected?
   b. Why do electrons move toward the anode?
   c. Why do you think the cathode rays produce a colored glow in the tube?

Application

44. What is the energy of a quantum of light of frequency 4.31×10^{14} Hz?
45. A certain violet light has a wavelength of 413 nm. What is the frequency of the light? The velocity of light is equal to 3.00×10^8 m/s.
46. A certain green light has a frequency of 6.26×10^{14} Hz. What is its wavelength?
47. What is the energy content of one quantum of the light in Problem 45?
48. What is the energy content of one quantum of the light in Problem 46?
49. Hydrazine, N_2H_4, is a fuming, corrosive liquid used in rocket and jet fuels. Ammonia, NH_3, is a gas that dissolves in water to form a solution that can be used as a cleaning agent. How do hydrazine and ammonia illustrate the law of multiple proportions?
50. Explain why people need protection from ultraviolet radiation, particularly short-wavelength ultraviolet radiation.
51. What is the energy of light with wavelength 662 nm? First find the frequency in hertz of this wavelength of light.
52. Photoelectric devices rely on the ability of light to remove electrons from the surface of some substances. The energy required to release an electron from atoms on the surface of a certain substance is 3.60×10^{-19} J. What wavelength of light would be necessary to cause electrons to leave the surface of this substance?
53. How is an element's emission spectrum related to its absorption spectrum?
54. How could you demonstrate and explain the law of conservation of mass with a burning candle?
55. Design an experiment to demonstrate the law of definite proportions.

Answers to
Concept Mastery

35. Atoms are now known to consist of smaller particles; atoms are not indivisible. Not all atoms of an element have the same atomic mass; atoms of the same element do not have to be identical.
36. The atom is mostly empty space with a dense core.
37. Nuclear force is effective over very short distances and holds the nucleons in an atom together. Electrostatic force opposes the effect of this force because the protons have like charges and repel each other.
38. a. When excited electrons return to their ground state, the energy they lose is emitted in the form of light characteristic of the amount of energy lost by the electron. This energy loss results in an emission spectrum characteristic of that element.
b. If an unexcited atom is exposed to a continuous spectrum, the energies that will excite the electrons will be absorbed. The lines missing in the resulting absorption spectrum will indicate what wavelengths were absorbed.
39. Bohr used spectroscopic data to confirm that electrons have certain "allowable" energy levels, and these levels differ by quanta of energy, the value of these quanta being shown by the spectrum for the element.
40. When an electron absorbs a photon of energy, it moves to a larger orbit.
41. The law of conservation of mass is explained by the fact that atoms are indivisible and remain as they are. Dalton's theory also explains the law of definite proportions by stating that atoms can unite with other atoms in simple numerical ratios.
42. Nonionized atoms do not have a charge and will not be deflected by an electrical or magnetic field.
43. a. negative
b. The anode is positively charged.
c. The high-energy electrons in the ray collide with the atoms of the gas in the tube. The electrons in the atom are excited, then return to a lower energy level, releasing a quantum of light of a specific wavelength, and thus a specific color.

Answers to
Application

44. 2.86×10^{-19} J
45. 7.26×10^{14} Hz
46. 479 nm
47. 4.81×10^{-19} J
48. 4.15×10^{-19} J
49. If a measured mass of nitrogen reacts with hydrogen to form hydrazine, and the same mass of nitrogen reacts with hydrogen to form ammonia, the amounts of hydrogen involved will be in a ratio of 2 to 3.
50. A short wavelength means high frequency ($\lambda = c/\nu$). High frequency produces high energy ($E = h\nu$), and this energy excites electrons in the atoms in our bodies causing chemical reactions.

Chapter 4

REVIEW

Critical Thinking

56. What effect did the discovery of radioactivity and Einstein's explanation have on the law of conservation of mass?

57. Suppose calcium-40 had been used as the standard for atomic mass and had been defined as having an atomic mass of 1.00×10^2 u. How would this standard change the periodic table as we know it?

58. According to the standard stated in Problem 57, what would be the atomic mass of the atom we know as carbon-12?

59. If Geiger and Marsden had used lithium instead of gold, would they have come to the same conclusion? Explain.

60. The element boron has two naturally occurring isotopes. One has atomic mass 10.0129 u, while the other has atomic mass 11.0093 u. The average atomic mass of boron is 10.811 u. Determine the percentage occurrence of each isotope.

61. The picture tube in a television set is a cathode-ray tube. A heated cathode in the neck of the tube produces a beam of electrons. (Some color tubes emit three beams.) Attached to the tube are coils of wire that form electromagnets. The magnetic field produced by these magnets can be varied by varying the current flowing through the coils. Explain the function of these electromagnets. Use reference books to find out how an electron beam forms a television picture. In some electronic instruments, such as oscilloscopes, electrically charged plates inside the tube are used instead of electromagnets. Explain why these can produce results similar to those produced by magnets.

Readings

Crawford, Mark. "Racing after the Z Particle." *Science* 241, no. 4869 (August 26, 1988): 1031-1032.

Fisher, A. "Hunting Neutrinos." *Popular Science* 232, no. 5 (May 1988): 72-74, 115.

Rayner-Canham, Geoffrey W., and Marelene F. Rayner-Canham. "The Shell Model of the Nucleus." *The Science Teacher* 54, no. 1 (January 1987): 19-21.

Thomsen, Dietrick E. "Seeking Neutrinos Under the Ocean." *Science News* 133, no. 16 (April 16, 1988): 246.

Weisburd, Stefi. "All Charged Up for the Positron Microscope." *Science News* 133, no. 8 (February 20, 1988): 124-125.

Cumulative Review

1. Classify each object or material as a heterogeneous mixture, solution, compound, or element.

a. plastic garbage bag	**g.** milk
b. automobile	**h.** air
c. seawater	**i.** diamond
d. helium gas	**j.** soda pop
e. maple syrup	**k.** paper
f. newsprint	**l.** blood

2. Calculate the density of a material that has a mass of 7.13 grams and occupies a volume of 7.77 cm^3.

3. What is the number of significant digits in each of the following measurements?

a. 0.558 g	**d.** 0.0094 mg
b. 7.3 m	**e.** 19.000 g
c. 410 cm	**f.** 75.0 s

4. What is the difference between a physical change and a chemical change?

5. What are the seven SI base units?

6. A 50.0-g piece of iron is heated to 150°C. It is placed in 100.0 cm^3 of water at 0°C. What will be the final temperature of the water?

Answers to
Application

51. 3.00×10^{-19} J

52. 552 nm

53. The lines present in the emission spectrum will exactly match the lines missing from the absorption spectrum for any specific element.

54. Place a candle with a determined mass in a closed system containing a measured mass of oxygen. After the candle burns, mass the combustion products and the remainder of the candle. The final mass of the system equals the initial mass of the system.

55. Answers will vary. Example: Under the same conditions of temperature and pressure, combine a measured amount of NH_3 gas with HCl. After the reaction is complete, determine how much NH_3 and HCl were used. Repeat this experiment several times, varying the amounts of reactants. Use the data to confirm that the NH_3:HCl ratio for the reactants used always is the same.

Answers to
Critical Thinking

56. The law of conservation of mass had to be modified to include conservation of energy, because mass may be converted to energy. Radioactivity also accounts for some atomic mass loss from decay.

57. Each atomic mass would be 2.5 times larger than it now is. The atomic mass would no longer equal the total number of nucleons in an atom.

58. 30.0 u

59. Geiger and Marsden would have determined that the atom is mostly empty space, but the fact that the alpha particle and the Li nucleus are so similar in mass means that they would have changed their conclusion about the nucleus.

60. Percent$_1$ = 19.9%
Percent$_2$ = 80.1%

61. The electromagnets deflect the electrons in the beams so that they strike the tube at the correct location for the picture desired. The direction and amount of deflection are determined by the amount of current through the electromagnets. Electrically charged plates can produce the same results because both methods rely on attraction or repulsion of the electrons, and the strength of both can be varied.

Answers to
Cumulative Review

1. **a.** compound **b.** heterogeneous mixture **c.** solution (heterogeneous mixture if pollution is accounted for) **d.** element **e.** solution **f.** heterogeneous mixture **g.** heterogeneous mixture **h.** solution (heterogeneous mixture if you count soot, spores, etc.) **i.** element **j.** solution (unopened) **k.** heterogeneous mixture **l.** heterogeneous mixture

2. 0.918 g/cm^3

3. **a.** 3 **b.** 2 **c.** 2 **d.** 2 **e.** 5 **f.** 3

4. The same substance is present after a physical change. A chemical change produces a new substance.

5. kilogram, meter, second, ampere, mole, kelvin, candela

6. 7.7°C

CHAPTER

5 Electron Clouds and Probability

CHAPTER ORGANIZER

CHAPTER OBJECTIVES	TEXT FEATURES	LABORATORY OPTIONS	TEACHER CLASSROOM RESOURCES
5.1 Modern Atomic Structure **1.** Describe the wave-mechanical view of the hydrogen atom. **2.** Characterize the position and velocity of an electron in an atom. **3.** Describe an electron cloud. **Nat'l Science Stds: UCP.1, UCP.2, A.2, B.1, B.4, G.2, G.3**	**Careers in Chemistry:** Electron Microscopist, p. 111 **Psychology Connection:** Uncertainty and Measurement, p. 112 **Mathematics Connection:** Probability and Statistics, p. 114	**ChemActivity 5:** Felt-tip Electron Distribution, p. 803 **Discovery Demo:** 5-1 Quantum Mechanics, p. 108 **Demonstration:** 5-2 Light as a Particle, p. 112	**Study Guide:** pp. 15-16 **Lesson Plans:** pp. 9-11 **Color Transparency 9 and Master:** Electron Cloud Model **ChemActivity Master 5**
5.2 Quantum Theory **4.** Characterize the four quantum numbers. **5.** Use the Pauli exclusion principle and quantum numbers to describe an electron in an atom. **Nat'l Science Stds: UCP.1, UCP.2, A.1, A.2, B.1, B.4, B.5, D.4, G.2, G.3**	**Everyday Chemistry:** Fireworks, p. 118 **Chemists and Their Work:** Wolfgang Pauli, p. 122 **Bridge to Astronomy:** Spectra of Stars, p. 125 **Chemistry and Technology:** Thin-Layer Magnetism, p. 126	**Minute Lab:** Always an Integer, p. 120 **Demonstration:** 5-3 Energy in Light Waves, p. 118 **Demonstration:** 5-4 Shape of the Charge Cloud, p. 120 **Laboratory Manual:** 5 Measuring Electron Energy Changes	**Critical Thinking/ Problem Solving:** Orbital Investigations, p. 5 **Study Guide:** pp. 17-18 **Color Transparency 10 and Master:** Energy and Sublevels **Color Transparency 11 and Master:** Energy Level Diagram **Color Transparency 12 and Master:** Orbital Shapes **Applying Scientific Methods in Chemistry:** Energy Levels, p. 20 **Enrichment:** Infrared Spectroscopy, p. 9-10
5.3 Distributing Electrons **6.** Determine the electron configurations of the elements. **7.** Write electron dot diagrams for the elements. **Nat'l Science Stds: UCP.1, UCP.2, A.2, B.1, B.2, B.4, G.1, G.2**			**Study Guide:** p. 19 **Color Transparency 13 and Master:** The Arrow Diagram **Reteaching:** Filling an Atom with Electrons, pp. 7-8
OTHER CHAPTER RESOURCES	**Vocabulary and Concept Review,** pp. 5, 39-40 **Evaluation,** pp. 17-20	**Videodisc Correlation Booklet** **Lab Partner Software**	**Test Bank** **Mastering Concepts in Chemistry—Software**

Block Schedule

For information on block scheduling, see the Lesson Plans booklet in the Teacher Resource Package.

GLENCOE TECHNOLOGY

Mastering Concepts in Chemistry Software

Levels, Sublevels, Orbitals

Basics of Electron Distribution

Writing Electron Configurations and Dot Structures

Chemistry: Concepts and Applications Videodisc

Flame Tests

Chemistry: Concepts and Applications CD-ROM

Building Atoms

Flame Tests

Complete Glencoe Technology references are inserted within the appropriate lesson.

ASSIGNMENT GUIDE

Contents	Level	Practice Problems	Chapter Review	Solving Problems in Chemistry
5.1 Modern Atomic Structure				
The de Broglie Hypothesis	O		20, 21	
The Apparent Contradiction	O		22, 52	
Momentum	A		23, 59	
Measuring Position and Momentum	O		24, 53, 60	
Schrödinger's Work	C		25, 54	
Wave-Mechanical View of the Hydrogen Atom	C		26	
5.2 Quantum Theory				
Schrödinger's Equation	C		27, 28	
Principal Quantum Number	C	6	29, 30, 51, 55	1-8
Energy Sublevels and Orbitals	C	7,8	31-33, 51	
Shape of the Charge Cloud	C		34, 35, 50, 56	
Distribution of Electrons	C		36-38, 61	
5.3 Distributing Electrons				
Order of Filling Sublevels	C	14	39-44, 58, 63	9-11
Electron Dot Diagrams	C	15	45-47, 49, 57, 62	12-14
Electron Summary	C		64	
				Chapter Review: 1-16

C=Core, A=Advanced, O=Optional

Laboratory Materials

Minute Labs	ChemActivities	Demonstrations		
page 120 set of stairs slinky toy	**page 803** felt-tip marker, fine-point paper target	**pages 108, 112, 118, 120** beaker, 100-cm^3 cellophane flashlight fluorescein	fur hard rubber rod leaf electroscope paper rope rubber band	ruler short-wave ultraviolet light water zinc strip, 2 x 10-cm

CHAPTER 5
Electron Clouds and Probability

CHAPTER OVERVIEW

Main Ideas: The chapter presents a historical approach to the development of the wave-mechanical model of the atom based on the work of de Broglie, Heisenberg, and Schrödinger. The order of filling of the electron energy levels, sublevels, and orbitals is presented as a result of the study of quantum mechanics derived from spectra. The chapter concludes with electron dot diagrams.

Theme Development: The theme of this chapter is scale and structure. The structure of the electron cloud is best understood when students consider the very small scale of the electron cloud.

Tying To Previous Knowledge: From Chapter 4, have students recall that the atom has a small, dense, positively-charged nucleus surrounded by an electron cloud. In this chapter, students will have the opportunity to study the behavior and most probable positions of electrons in more detail.

ASSESSMENT PLANNER

Choose from the following assessment strategies.

Assess, pp. 114-115, 124-125, 130-131
Assessment, pp. 111, 113, 120, 123, 129
Portfolio, p. 123
Minute Lab, p. 120
ChemActivity, pp. 803-804
Chapter Review, pp. 132-135

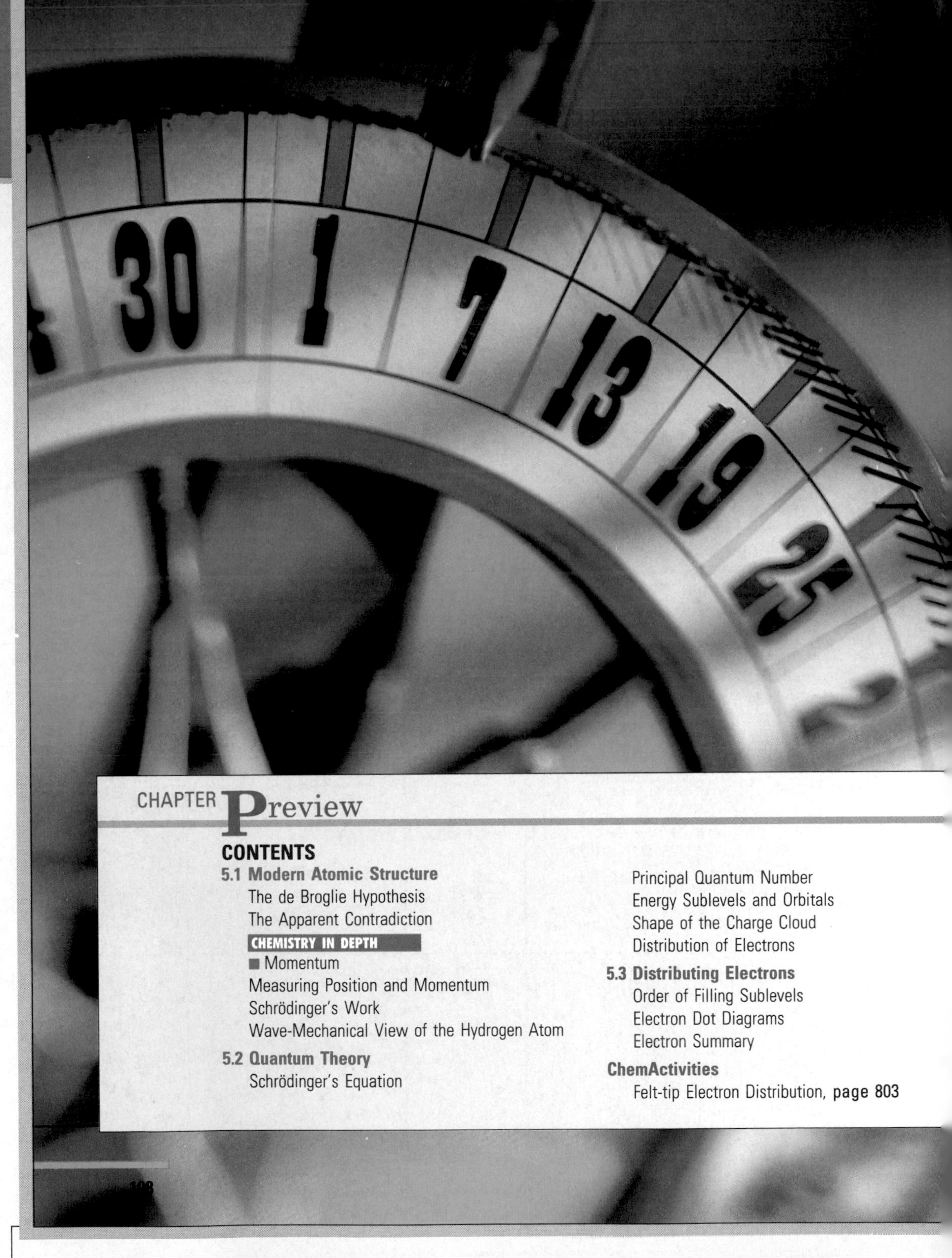

CHAPTER **Preview**

CONTENTS

DISCOVERY DEMO

5-1 Quantum Mechanics

Purpose: to simulate the difficulty of simultaneously determining the position and momentum of an electron

Materials: half-sheet of paper, ruler, rubber band

Procedure: Loosely crumple the paper into a ball and place it on a desk. Loop a rubber band over the end of a ruler and shoot it at the paper ball. **CAUTION:** *Avoid pointing the rubber band at students. Wear goggles.* **Disposal:** F

Results: When the rubber band strikes the paper ball its energy is transferred to the paper ball and the paper will move. Point out that in this model the rubber band simulates a photon of radiant energy and the paper ball simulates an electron in an atom. Use this demonstration to anticipate the discussion in the text (Measuring Position

Electron Clouds and Probability

CHAPTER

5

Often, we consider chance to be the same as luck or fate, and it seems as though we have no control over chances in our lives. Consider a wheel of fortune such as the one pictured. If you spin the wheel are the results entirely random, or can you determine what chance there is of a desired outcome?

"Chance" is a limited possibility of achieving a desired outcome. For instance, if a wheel of fortune has 30 numbers on it, a person who has chosen a particular number has one chance in 30 of winning. The probability of winning is 1/30. Chemists and physicists usually have to deal with the location of electrons in terms of the chances, or probability, of finding the electron at a particular location. Current ideas about the electronic structure of the atom are best incorporated in a mental model that will lay the groundwork for understanding chemical changes that are likely to take place.

EXTENSIONS AND CONNECTIONS

INTRODUCING THE CHAPTER

USING THE PHOTO

Use the game analogy to introduce the concept of probability and Heisenberg's uncertainty principle. The use of the word *probability* implies an ability to predict that is better than mere chance, but falls short of certainty. A student may have access to a toy wheel of fortune that can be used to demonstrate chance. Chemists are able to predict the electron's most probable position using quantum mechanics. Point out that although we cannot tell the exact position and momentum of an electron simultaneously, we can predict the most probable location of an electron in an atom.

REVEALING MISCONCEPTIONS

There are two related misconceptions that many students have as they begin their study of quantum mechanics. One misconception is that the energy levels are like planetary orbits around the sun. The other is that the energy levels are equally spaced. To see if students hold these misconceptions, have them diagram an atom. As you present the concepts contained in this chapter, emphasize that an electron's most probable position is in an electron cloud of some thickness. Point out that the great variety of spectra of different elements indicates that the distances between the clouds are not equal.

and Momentum), which describes how the collision between a photon and an electron changes either the electron's position or its momentum.

Questions: 1. Does a radar beam reflecting from an airplane cause a deflection in the airplane's flight path? *No, the energy of the photon of electromagnetic radiation is too small to deflect the mass of the airplane.* **2.** What happens to the position of the paper ball (electron) when the photon (rubber band)of light reflects off it? *The paper ball changes position.* **3.** Heisenberg stated that there is always some uncertainty about specifying the position and momentum of an electron. What is the name of that principle? *Heisenberg uncertainty principle*

Extension: Discuss with the class that classical, or Newtonian, mechanics, which describes the behavior of visible objects traveling at ordinary velocities, could not accurately predict an electron's position and momentum. Quantum mechanics successfully describes the behavior of very small particles.

5.1 Modern Atomic Structure

PREPARE

Key Concepts

The major concepts presented in this section include de Broglie's hypothesis, the wave-particle duality of nature, Heisenberg's uncertainty principle, Schrödinger's development of quantum numbers, and the wave or quantum mechanical view of the atom.

Key Terms

wave-particle duality of nature
momentum
Newtonian mechanics
quantum mechanics
Heisenberg uncertainty principle
quantum number
probability
electron cloud

1 FOCUS

Objectives

Preview with your students the objectives listed on the student page. Students can use them as a study guide for this major section.

Activity

Bring a fan to class to demonstrate that the blades rotate and appear to fill the total volume through which they turn. An electron moving very fast appears to fill the volume about the nucleus. Stop the fan to show that the blades actually occupy very little of the volume. It is the same with an electron in the atom.

OBJECTIVES

- describe the wave-mechanical view of the hydrogen atom.
- characterize the position and velocity of an electron in an atom.
- describe an electron cloud.

5.1 Modern Atomic Structure

In the attempts to refine our model of the atom, you will see that the division between matter and energy has become less clear. Radiant energy is found to have many properties of particles. Small particles of matter are found to display the characteristics of wave motion. The purpose of this section is to look more closely at this wave-particle problem.

The de Broglie Hypothesis

You saw in Chapter 4 that the frequencies predicted by Bohr for the hydrogen spectrum are "essentially" correct. Note that we did not use the word "exactly." Improved instrumentation has shown that the hydrogen spectral lines predicted by Bohr are not single lines. With a better spectroscope, you see instead that there are several lines closely spaced. To account for this "fine structure" of the hydrogen spectrum, scientists again had to refine their model of the atom.

In 1923, a French physicist, Louis de Broglie, proposed a hypothesis that led the way to the present theory of atomic structure. De Broglie knew of Planck's ideas concerning radiation being made of discrete amounts of energy called quanta. This theory seemed to give waves the properties of particles. De Broglie thought if Planck were correct, then it might be possible for particles to have some of the properties of waves.

Using Einstein's relationship between matter and energy

$$E = mc^2$$

and Planck's quantum theory

$$E = h\nu$$

de Broglie equated the two expressions for energy as follows.

$$mc^2 = h\nu$$

He then substituted $\mathbf{v}$*, a general velocity, for c, the velocity of light

$$m\mathbf{v}^2 = h\nu$$

and $\mathbf{v}/\lambda$ for ν because the frequency of a wave is equal to its velocity divided by its wavelength.

$$m\mathbf{v}^2 = \frac{h\mathbf{v}}{\lambda}$$

$$\lambda = \frac{h\mathbf{v}}{m\mathbf{v}^2} = \frac{h}{m\mathbf{v}}$$

This final expression enabled de Broglie to predict the wavelength of a particle of mass m and velocity $\mathbf{v}$. Within two years, de Broglie's hypothesis was proved correct. Scientists found by

*Symbols in boldface are vector quantities. Vectors have both magnitude and direction.

CONTENT BACKGROUND

Applying de Broglie's Hypothesis: The particle-wave characteristics of matter can be used to describe the radioactive decay of a heavy nucleus by the emission of an alpha particle. De Broglie's hypothesis states that an alpha particle has both particle and wave properties. Within the nucleus of a heavy atom, the particle properties of an alpha particle indicate that it does not have enough energy to overcome nuclear attraction and escape from the nucleus. However, its wave properties indicate that there is a finite probability for its existence outside the nucleus.

PROGRAM RESOURCES

Study Guide: Use the Study Guide worksheets, pp. 15-16, for independent practice with concepts in Section 5.1.

Lesson Plans: Use the Lesson Plan, pp. 9-11, to help you prepare for Chapter 5.

Figure 5.1 In 1993, a scanning tunneling electron microscope revealed the behavior of electrons confined inside a circular "corral" made of 48 iron atoms. You can see that the electrons act as waves, forming a ripple pattern. These wave properties of electrons occur just as de Broglie calculated in 1923.

experiment that, in some ways, an electron stream acted in the same way as a ray of light. They further showed that the wavelength of the electrons was exactly that predicted by de Broglie.

The Apparent Contradiction

Waves can act as particles, and particles can act as waves. You saw how Bohr's atom model explained light in terms of particle properties. There are also properties of light that can be explained by wave behavior.

Like light, electrons also have properties of both waves and particles. However, one cannot observe both the particle and wave properties of an electron by the same experiment. If an experiment is done to show an electron's wave properties, the electron exhibits the behavior of a wave. Another experiment, carried out to show the electron as a particle, will show that the electron exhibits the behavior of a particle. The whole idea of the two-sided nature of waves and particles is referred to as the **wave-particle duality of nature.** The duality applies to all waves and all particles. Scientists are not always interested in duality. For example, when scientists study the motion of a space shuttle, wave characteristics do not enter into their study. They are only interested in the shuttle as a particle. However, with very small particles they cannot ignore wave properties. For an electron, a study of its wave characteristics can tell as much about its behavior as a study of its particle characteristics.

CAREERS IN CHEMISTRY

Electron Microscopist

Electron microscopists use the wave properties of electrons in electron microscopes to produce images of specimens too small to be viewed by ordinary microscopes. Beams of electrons are focused by electromagnetic fields to reproduce a well-defined image of a specimen that may be too small to be investigated by ordinary microscopy. One use of electron microscopy in chemistry is in the investigation of crystal structure. Because electron microscopy is most commonly used in biology, chemistry, and engineering, microscopists must have strong backgrounds in those areas, as well as physics, mathematics, and computer science.

2 TEACH

CAREERS IN CHEMISTRY

Electron Microscopist

Background: The educational background of an electron microscopist includes courses in biology, chemistry, and physics. A knowledge of mechanical and electrical engineering subjects can be useful. An extensive background in mathematics is also very helpful. The scanning tunneling electron microscope is relatively inexpensive and it is predicted that most chemistry departments of universities will have one within a few years.

For More Information: Contact your nearest university to arrange a tour to see the electron microscope.

CONCEPT DEVELOPMENT

QUICK DEMO **Photoelectric Effect**

Use a demonstration solar cell wired to a small electric motor or a solar-powered calculator to initiate a discussion of the particle nature of light. See the Chemistry and Technology feature in Chapter 4 for a detailed explanation of the photoelectric effect proposed by Einstein.

✓ ASSESSMENT

Knowledge: Ask students to determine which nature of light, particle or wave, is demonstrated when a pencil placed in a half-filled glass of water appears bent or broken *(refraction-wave nature)* and when a solar-powered calculator is used in the light *(photoelectric effect-particle nature).* Students may need to consult a physics text.

CONTENT BACKGROUND

Electron Wave Nature: The first demonstration of an electron's wave nature occurred in the United States. In 1927, Davisson and Germer diffracted an electron beam using a crystal. Diffraction patterns have been demonstrated for fundamental particles just like those for light. The wave properties of fundamental particles are now recognized.

The equations of quantum mechanics are modifications of equations that describe waves. These equations are statistical in nature and contain variables that are described in terms of probabilities and averages. Quantum mechanics is postulated upon the wave description of particles.

CONCEPT DEVELOPMENT

MAKING CONNECTIONS

Sociology
Doors that open automatically are found in many buildings. The "electric eyes" that operate the doors utilize the photoelectric effect. A light beam shines across the door opening onto a photocell. When the beam is blocked by a person, the current stops and a mechanism is triggered that opens the door. These devices have allowed easier access for people who would otherwise find it difficult to open the door manually.

Computer Demo: *Animated Waves and Particles,* An animated demonstration of the basic concepts of particle and wave motion. AP210, Project SERAPHIM

Enrichment: Have students research experiments that have demonstrated the wave properties of electrons and those that have demonstrated the particle properties of electrons. Experiments include the diffraction of an electron beam by a crystal, the cloud chamber, bubble chamber, cyclotron, synchrotron, linear accelerator, mass spectrometer, cathode-ray tube, and Millikan's oil drop apparatus.

PSYCHOLOGY CONNECTION

Uncertainty and Measurement

Exploring Further: Students may be interested in researching other effects of the observer such as Pygmalion effect, the Halo effect, and the Hawthorne effect.

CHEMISTRY IN DEPTH

The section **Momentum** is recommended for more able students. **Chemistry in Depth** ends on this page.

CHEMISTRY IN DEPTH — Momentum

The product of the mass and velocity of an object is the **momentum** of the object. In equation form, $mv = p$, where m is the mass, v is the velocity, and p is the symbol for momentum. Note that velocity includes not only the speed but also, because it is a vector quantity, the direction of motion. Substituting momentum (p) for mv in the de Broglie equation, we can then write

$$\lambda = \frac{h}{p}$$

Note that the wavelength is inversely proportional to the momentum. The wave properties of all objects in motion are not always of interest. There is a basic difference between Newtonian mechanics and quantum mechanics. **Newtonian mechanics,** or classical mechanics, describes the behavior of visible objects at ordinary velocities. **Quantum mechanics** describes the behavior of extremely small particles at velocities near that of light.

To the chemist, the behavior of the electrons in an atom is of great interest. To be able to give a full description of an electron, you must know where it is and where it is going. In other words, you must know the electron's present position and its momentum. From the velocity and position of an electron at one time, we can calculate where the electron will be some time later.

Measuring Position and Momentum

Werner Heisenberg, a German scientist, further refined the ideas about atomic structure. He pointed out that *it is impossible to know both the exact position and the exact momentum of an object at the same time.* Let us take a closer look at Heisenberg's ideas. To locate the exact position of an electron, we must be able to "look" at it. When we look at an object large enough to see with our eyes, we actually see the light waves the object has reflected. When radar detects an object, the radar receiver is actually "seeing" the radar waves reflected by the object. In other words, for us to see an object, it must be hit by a photon of radiant energy. A photon hitting an airplane has negligible effect on the airplane. However, a collision between a photon and an electron results in a large change in the energy of the electron. Let us assume we have "seen" an electron, using some sort of radiant energy as "illumination." We have found the exact position of the electron. However, we would have little idea of the electron's velocity. The collision between it and the photons used to see it has caused its velocity to change. Thus, we would know the position of the electron, but not its velocity. On the other hand, if we measure an electron's velocity, we will change the electron's position. We would know the velocity fairly well, but not the position. Heisenberg stated that there is always some uncertainty about the position and momentum of an electron. This statement is known as the **Heisenberg uncertainty principle.**

PSYCHOLOGY CONNECTION

Uncertainty and Measurement

The effect of the observer on the observed as indicated in the Heisenberg uncertainty principle is also a problem in evaluation and measurement. This has led to the use of "unobtrusive" measures of assessment. An example of an unobtrusive measure is determining the wear patterns on the floor covering in front of certain exhibits at a museum. If the floor covering has to be replaced more often at one exhibit than at another, the first is a more popular exhibit—more feet have walked by it.

DEMONSTRATION

5-2 Light as a Particle

Purpose: to demonstrate that photons dislodge electrons in metal

Materials: leaf electroscope, short-wave ultraviolet light, hard rubber rod, piece of fur, zinc strip 2 cm x 10 cm (The first four items may be available from your school's physics teacher.)

Procedure: **1.** Steel wool the zinc strip to remove all traces of oxide. Fasten the zinc strip to the top of the electroscope being certain to have a good electrical connection.

2. Place a negative charge to the electroscope by rubbing a hard rubber rod with fur then touch the rod to the top and the zinc strip.

3. Shine the UV light on the zinc strip. **CAUTION:** *Short-wave UV light is harmful to eyes. Don't look directly at the bulb.*

The uncertainty of the position and the uncertainty of the momentum of an electron are related by Planck's constant. If Δp* is the uncertainty in the momentum and Δx is the uncertainty in the location, then

$$\Delta p \Delta x \geq h$$

Because h is constant, Δp and Δx are inversely proportional to each other. Therefore, the more certain we are of the position of the electron, the less certain we are of its momentum. Conversely, the more certain we are of its momentum, the less certain we are of its position.

*Recall the Greek letter delta (Δ) means *change in.*

Schrödinger's Work

Chemists and physicists found themselves unable to adequately describe the exact structure of the atom. Heisenberg had, in effect, stated that the exact motion of an electron was unknown and could never be determined. Notice, however, that Heisenberg's principle of uncertainty treats the electron as a particle. What happens if a moving electron is treated as a wave? The wave nature of the electron was investigated by the Austrian physicist, Erwin Schrödinger.

Schrödinger treated the electron as a wave and developed a mathematical equation to describe its wave-like behavior. Schrödinger's equation related the amplitude of the electron-wave, ψ (psi), to any point in space around the nucleus. He pointed out that there is no physical meaning to the values of ψ. Terms for the total energy and for the potential energy of the electron are also part of this equation. In computing the total energy and the potential energy, certain numbers must be used. For example, the term for the total energy is

$$2\pi^2 me^4/h^2n^2$$

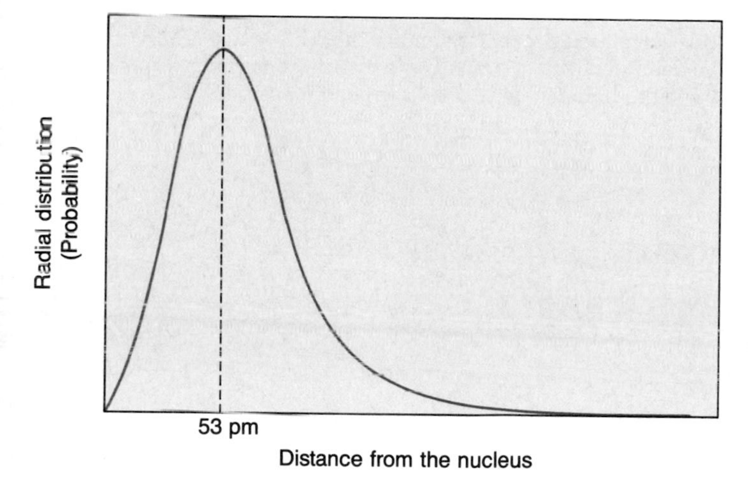

Figure 5.2 This graph shows the probability of finding the electron at a specific distance from a hydrogen nucleus. Bohr predicted the hydrogen electron to be approximately 53 pm from the nucleus. Born showed that 53 pm is merely the point of highest probability.

CONCEPT DEVELOPMENT

Discussion: Use Heisenberg's uncertainty principle to emphasize that quantum theory tells us in probability terms how the electron occupies the space outside the nucleus. For example, an electron can be as far away as 2.24 picometers from the hydrogen nucleus and as close as 0.1 pm. However, 90% of the time the electron will be between 1.42 pm and 0.29 pm from the nucleus. Probability distribution is an important concept in quantum mechanics because we do not know the exact position and momentum of an electron simultaneously.

Discussion Question: To assess the students' understanding of probability ask what the probability is of a two-month-old baby being home at 9 P.M. on Friday night. Follow that question by asking what the probability is of a twenty-year-old being home at 9 P.M. on Friday night. Point out that Schrödinger's work is based on the electron's most probable location.

Computer Demo: *Quantum Mechanics,* Graphical representations of probability distributions. AP202, Project SERAPHIM

Results: The electroscope leaves collapse as the UV light removes electrons from the zinc. The electrons on the electroscope leaves move to the zinc to replace the ones removed by the UV light.
Questions: 1. Do photons of UV light have more energy than visible light? *Yes* **2.** What is the emission of electrons from an illuminated metal called? *the photoelectric effect*

✓ ASSESSMENT

Oral: Have students positively charge the electroscope by rubbing a lucite rod with fur and touching the rod to the top and the zinc strip. Shine the UV light on the zinc strip. Ask students to explain why the electroscope remains positively charged.

PROGRAM RESOURCES

Transparency Master: Use the Transparency master and worksheet, pp. 17-18, for guided practice in the concept "Electron Cloud Model."

Color Transparency: Use Color Transparency 9 to focus on the concept "Electron Cloud Model."

MATHEMATICS CONNECTION
Probability

Exploring Further: Give extra credit to students who track a particular chemical company stock for one month. They should present their results in the form of a graph and accompany it with a probability prediction about the future price of the stock selected.

3 ASSESS

CHECK FOR UNDERSTANDING

Ask students to write the answers to Concept Review problems.

RETEACHING

When discussing that an electron has a dual nature depending upon experimental conditions, point out to students that their behavior has a dual nature. For example, a student wearing jeans and a sweatshirt at a local pizza shop behaves quite differently when wearing formal attire to a prom.

EXTENSION

Using de Broglie's wave equation, calculate the wavelength of an electron of mass 9.11×10^{-31} kg traveling at 2.19×10^{6} m/s. *Knowing that Planck's constant = 6.6261×10^{-34} J·s, but realizing that 1 J = 1 N·m, and 1 N = 1 kg·m/s^2, the constant becomes 6.6261×10^{-34} kg·m^2/s. The calculated wavelength is 3.32×10^{-10} m or 0.332 nm.*

Chemistry Journal

Electron Motion

Have students write an analogy similar to the one in the Content Background on the right. This will help them to better understand time-averaged electron motion.

MATHEMATICS CONNECTION

Probability and Statistics
The study of probability and statistics is part of the branch of mathematics called discrete, or finite, mathematics. Statistics are used in many fields besides science. For example, marketing executives study buying and population statistics carefully in deciding on new products or where to market certain items.

In this expression, m is the mass of the electron, e is the charge on the electron, h is Planck's constant, and n can take positive whole number values. The symbol n represents the first of four **quantum numbers**, which are used in describing electron behavior. They will be studied in detail in later sections. The actual wave equation involves mathematics with which you are probably not familiar, so it will not be given.

The physical significance of all this mathematics was pointed out by Max Born. He worked with the square of the absolute value of the amplitude, $|\psi|^2$. He showed that $|\psi|^2$ gave the probability of finding the electron at the point in space for which the equation was solved. This **probability** is also the ratio between the number of times the electron is in that certain position and the total number of times it is at all possible positions. The higher the probability, the more likely the electron will be found in a given position.

Wave-Mechanical View of the Hydrogen Atom

Schrödinger's wave equation is used to determine the probability of finding the hydrogen electron in any given place. The probabilities can be computed for finding the electron at different points along a given line away from the nucleus. One point will have a higher probability than any other, as shown in Figure 5.2. Note that the point of highest probability, 53 pm, is the same as that calculated by Neils Bohr. In Bohr's model, however, the electron is assumed to have a circular path and always to be found at this distance from the nucleus. To carry the process even further, computers can be used to calculate the probabilities of the location of an electron for thousands of points in space. There will be many points of equal probability. If all the points of highest probability are connected, some three-dimensional shape is formed. These shapes will be shown later in the chapter. The most probable place to find the electron will be some place on the surface of this calculated shape. Remember that this shape is only a mental model and does not actually exist. It is something we use in our minds to help locate the most probable position of the electron.

Electron cloud

Figure 5.3 The fan blades on the left appear to occupy the total volume through which they turn. An electron effectively occupies a three-dimensional space to form a cloud of negative charge (right).

CONTENT BACKGROUND

Electron Motion: To help students better understand time-averaged electron motion have them imagine a helicopter with a mounted camera is hovering over a house with the roof removed. It is night and a student in the darkened house is carrying a burning candle. Each minute the camera snaps a picture on a single piece of film. When the film is developed there will be areas on the film that contain many white spots. These areas correspond to locations where the student spent more time, for example, near the TV. This photograph then represents the probable locations of finding the student on other nights or the probable location of finding similar students in similar houses. Similarly, the shapes for *s*, *p*, and *d* orbitals contain lobes, which are areas in which there are high probabilities of finding electrons.

Figure 5.4 A spray of water from a lawn sprinkler can also be compared to an electron cloud.

There is another way of looking at this probability. The electron moves about the nucleus, passing through the points of high probability more often than through any other points. The electron is traveling at high speed. If the electron were visible to the eye, its rapid motion would cause it to appear much like a cloud. Think of an electric fan as shown in Figure 5.3. When the fan is turning, it appears to fill the complete circle through which it turns. If you try to place an object between the blades while the fan is turning, the result will show that the fan is effectively filling the entire circle. So it is with the electron. It effectively fills all the space. At any given time, it is likely to be somewhere on the surface of the shape described by the points of highest probability. The probability of finding the fan blade outside its volume is zero. However, it is possible to find the electron outside of its high probability surface. Therefore, since the volume occupied by an electron is somewhat vague, it is better to refer to it as an **electron cloud.** Let's look more closely at this electron cloud to learn more about its size and shape.

5.1 *CONCEPT REVIEW*

1. What caused de Broglie to start thinking about particles and waves?
2. Explain how measuring the position of an electron changes its velocity.
3. Why can you sometimes consider an electron as a cloud of negative charge?
4. If two particles are traveling at the same speed, does the more massive particle have a longer or shorter wavelength than the less massive particle?
5. **Apply** Why is the wave-particle duality ignored in everyday life?

Answers to
Concept Review

1. De Broglie knew of Planck's ideas concerning radiation being made of discrete amounts of energy called quanta. This theory seemed to give waves the properties of particles. De Broglie then considered if particles might have some of the properties of waves.
2. Looking at a particle to determine its exact position requires that a photon must be reflected from the particle to the observer. The velocity of the particle is changed when the particle is hit by the photon.
3. The electron moves so rapidly that its position cannot be pinpointed exactly, much like the turning blades of a fan. It is more convenient to think of the electron as a negatively-charged cloud surrounding the nucleus.
4. The more massive particle has the greater energy, corresponding to higher frequency and shorter wavelength.
5. Most of the particles encountered in everyday life are much too large to have wave characteristics of any practical importance.

4 CLOSE

Discussion: The speed of the ground state electron in hydrogen is 2.19×10^6 m/s as it travels about the nucleus. On the chalkboard calculate how many miles it would travel in 1 hour using the relationships 3600 s/h and 1 mile/1600 m. *4.93×10^6 miles.* Point out that the electron effectively fills the volume through which it travels because of its great speed.

CONTENT BACKGROUND

Probability Distribution: The probability distribution of an electron in space is described by the wave function Ψ. If a time-exposed photograph were taken of an atom, we would see where an electron spends more time and less time. The information contained in this picture is also contained in Ψ. It gives us a time-averaged view of the electron's motion. The absolute value of the square of the wave function, $|\Psi|^2$, gives us the electron's probability distribution in space. Each permitted set of quantum numbers defines an orbital. The electron motion around the nucleus in the Bohr model suggested an orbital. The true meaning of Ψ conflicts with the term orbital.

5.2 Quantum Theory

PREPARE

Key Concepts

The major concepts presented in this section include Schrödinger's equation, quantum numbers, energy sublevels, orbitals, shape of the electron cloud, and the Pauli exclusion principle.

Key Terms

principal quantum number
sublevel
orbital
degenerate
Pauli exclusion principle

1 FOCUS

Objectives

Preview with your students the objectives listed on the student page. Students can use them as a study guide for this major section.

Discussion

Write the equation for a straight line, $y = mx + b$, on the chalkboard. Ask how many variables are in the equation. *(4)* Ask the class for the name of the m variable. *(slope)* Inform the class that Schrödinger's equation has four variables *n, l, m,* and *s,* which are called quantum numbers. This is an over-simplification but it helps students to master the concept. A graph of $y = mx + b$ is a straight line. A graph of Schrödinger's equation is an electron probability plot called an orbital.

OBJECTIVES

- characterize the four quantum numbers.
- use the Pauli exclusion principle and quantum numbers to describe an electron in an atom.

5.2 Quantum Theory

Think for a moment about the chemical behavior of neon gas, atomic number 10, and sodium metal, atomic number 11. Neon gas does not react with water or with oxygen in air, yet sodium metal reacts vigorously with water and also with oxygen in the air. The chemically changed sodium atom produced in these reactions behaves more like neon gas than it does like sodium metal. The key to understanding this behavior lies in the fact that the chemical behavior of an atom is determined by the number and arrangement of electrons around the nucleus. The chemically changed sodium atom now has a number and arrangement of electrons identical to that of neon. Therefore, it is important to continue to refine our understanding of the electronic structure of the atom.

Figure 5.5 The exact location of these moving cars cannot be pinpointed. Likewise, only the most probable location of an electron can be determined.

Schrödinger's Equation

Prior to the use of computers, any solutions to the Schrödinger equation proved difficult for even the best mathematicians. It has been solved exactly only for the hydrogen atom and other one-electron systems. The key feature of the solution of the wave equation is the use of quantum numbers. These numbers represent different energy states of the electron. In Schrödinger's atomic model, changes between energy states must take place by emission or absorption of whole numbers of photons. For the hydrogen atom, solution of the wave equation gives accurate energy states. The differences between these energy states correspond to the lines observed in the hydrogen spectrum. Thus,

CONTENT BACKGROUND

Seven Energy Levels: The permitted value of the principal quantum number *n* can actually go to infinity, but in a practical sense, there are seven energy levels. The value of *n* describes an average distance from the nucleus of the electron cloud. The integers 1 to 7 are used to designate each energy level. These correspond to the periods on the periodic table.

when an electron in a hydrogen atom moves from a higher to a lower energy state, the energy change is emitted as a quantum of light. This light is one line in the emission spectrum of hydrogen.

With more complex atoms, however, more than one electron is present, and interactions between electrons make exact solution of the equation impossible. Recall that electrons all have the same charge. Thus, they tend to repel each other. In spite of this difficulty, it is possible to approximate the electronic structure of multielectron atoms. This approximation is made by first calculating the various energy states in the hydrogen atom. A quantum number is assigned different values to arrive at the energy for each state. Then, *it is assumed that the various electrons in a multielectron atom occupy these same energy states without affecting each other.* Experimental evidence has shown that these values are very close approximations to actual values.

To completely describe an electron in an atom, four quantum numbers are needed and are identified by the letters n, l, m, and s. Each electron within an atom can be described by a unique set of these four quantum numbers. We will discuss each quantum number starting with n.

Principal Quantum Number

An electron can occupy only specific energy levels. These energy levels are numbered, starting with 1 and proceeding to the higher integers. The number of the energy level, referred to as n, is called the **principal quantum number.** The principal quantum number corresponds to the energy levels 1, 2, 3, . . . n calculated for the hydrogen atom.

Electrons may be found in each energy level of an atom. The greatest number of electrons possible in any one level is $2n^2$. Figure 5.6 shows the relative energies of the various levels. It also indicates the maximum number of electrons that may be contained in each level.

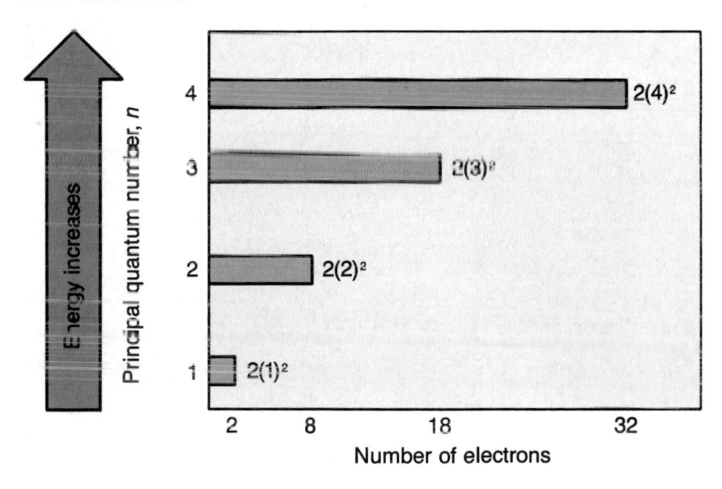

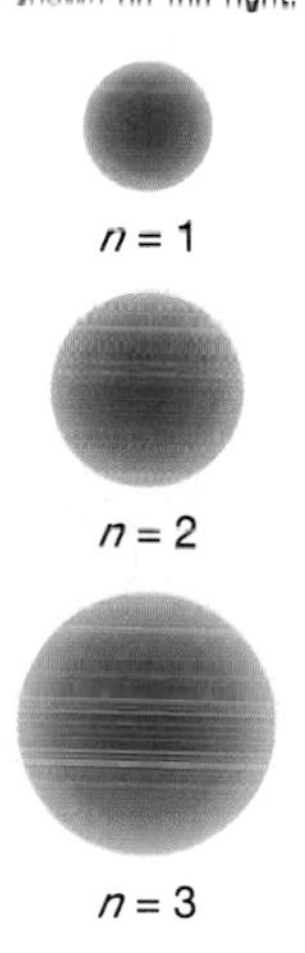

Figure 5.6 The relationship among energy, principal quantum number, and number of electrons is shown on the left. The relationship between size of electron cloud and principal quantum number is shown on the right.

2 TEACH

CULTURAL DIVERSITY

China Salt

Gunpowder probably originated from Chinese firecrackers used for religious festivals and celebrations. The first firecrackers were simply short lengths of bamboo that exploded when thrown in a fire. But by the 11th century *huo-yao*, known as the *fire drug*, had been developed and was used for firecrackers and warfare. Gunpowder was considered a drug because its main ingredients, saltpeter and sulfur, were commonly-used medicines. Gunpowder itself was used to cure skin diseases and as a desiccant and disinfectant.

Charcoal was added to the saltpeter (potassium nitrate) and sulfur. This explosive was mixed with a paste of dried leaves, tung oil, and wax, and then coated with pitch and wax to make bombs and grenades. Gunpowder was used in bamboo or paper tubes to make other weapons, including a rifle that fired iron pellets.

The Chinese also made rocket arrows that operated on the same principles as modern rockets, as early as 1126 A.D. The gun-powder was placed in a tube fused to the arrow shaft and blasted rearward, shooting the arrow into flight. A two-stage rocket called "fiery dragon emerging from water" was used as well.

Saltpeter was traded by the Chinese to the Arabs who called it China Snow, and to the Persians, who called it China Salt. The Spaniards learned about its use by translating Arabic writings. The French and English finally learned about gunpowder in the 14th century and it played a major role in European history.

Chemistry Journal

Quantized Events

Have students find and write in their journals descriptions of everyday quantized events. Examples might include steps, elevator stops, digital clocks, and push-button radios. Ask students to explain how each example is an analogy for the energy levels in the atom.

PROGRAM RESOURCES

Critical Thinking/Problem Solving: Use the worksheet, p.5, as enrichment for the concept of "Orbital Investigations."

Study Guide: Use the Study Guide worksheets, pp. 17-18, for independent practice with concepts in Section 5.2.

Correcting Errors

Remind students that there is an order in performing mathematical operations. For example, to find the product of two numbers, one of which is raised to a power, that number is raised to a power before it is multiplied. Show that $2x(2)^2 \neq (2x2)^2$.

PRACTICE

Guided Practice: Guide students through the Sample Problem, then have them work in class on Practice Problem 6.

Independent Practice: Your homework or classroom assignment can include Review and Practice 27-30.

Answers to
Practice Problem

6. The greatest number of electrons in a given energy level can be determined by finding the value of $2n^2$ where n is the energy level involved;

a. $n = 2,\ 2n^2 = 2(2)^2 = 8$
b. $n = 3,\ 2n^2 = 2(3)^2 = 18$
c. $n = 5,\ 2n^2 = 2(5)^2 = 50$
d. $n = 7,\ 2n^2 = 2(7)^2 = 98$

GLENCOE TECHNOLOGY

Software

Mastering Concepts in Chemistry
Unit 2, Lesson 1
Levels, Sublevels, Orbitals

Videodisc

Chemistry: Concepts and Applications
Flame Tests
Disc 1, Side 1, Ch. 8

Also available on CD-ROM.

PROBLEM SOLVING HINT

Be sure to find the value of n^2 before multiplying by 2.

SAMPLE PROBLEM

Capacity of Energy Levels

What is the maximum number of electrons that can occupy the first energy level in an atom? The fourth energy level?

Solving Process:

The maximum electron capacity of an energy level is given by $2n^2$ where n is the number of the energy level (principal quantum number).

For the first level, $2n^2 = 2(1)^2 = 2$.

For the fourth level, $2n^2 = 2(4)^2 = 32$.

PRACTICE PROBLEMS

6. Calculate the maximum number of electrons that can occupy the energy levels in an atom when the principal quantum number has the following values.
a. $n = 2$ **b.** $n = 3$ **c.** $n = 5$ **d.** $n = 7$

EVERYDAY CHEMISTRY

Fireworks

Much of what you see and enjoy at fireworks displays is a result of the emission spectrum of certain metal atoms. During the explosion of fireworks, a great deal of energy is released. When this energy is absorbed by metal atoms, the electrons in these atoms are raised to higher energy levels. This higher energy arrangement is not stable and the electrons quickly return to lower energy levels. The difference in energy between high and low energy arrangements is given off as brilliant light. Red lights are produced by strontium-containing compounds such as strontium nitrate. Green lights are produced by barium-containing compounds such as barium chlorate. Sodium oxalate is often included for its bright yellow light. The presence of copper sulfate will produce a bright blue-green light.

Exploring Further

1. Are the lights of fireworks more like absorption spectra or emission spectra? Explain your answer.
2. How can one Roman candle have two different explosions of two different colors?

DEMONSTRATION

5-3 Energy in Light Waves

Purpose: to demonstrate that different wavelengths of light contain different quanta of energy

Materials: 100-cm^3 beaker; 90 cm^3 water; 0.1 g fluorescein; flashlight with high-intensity halogen bulb or other halogen light source; red, green, and blue cellophane sheets used as color filters

Procedure: 1. In the beaker, dissolve the fluorescein in the water. As an alternative, use a brand of antifreeze that contains yellow-green fluorescein dye. Ask students to predict which color of light, red, green, or blue, will cause the solution to glow in the dark. Record their responses. **2.** Place a red filter between the light source and the beaker. In a darkened room, shine the filtered high-intensity light down through the fluorescein solution in the

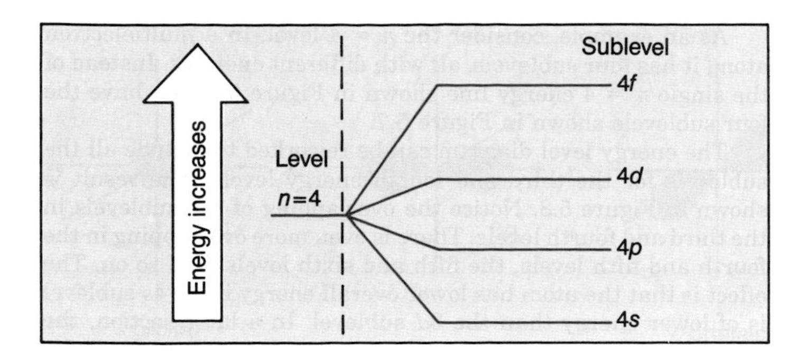

Figure 5.7 The relationship between energy and the sublevels in $n = 4$ is shown by this model.

Energy Sublevels and Orbitals

The second quantum number is l. In a hydrogen atom, there is only one energy state per level. This statement is not true for any other atom. Spectral studies have shown that in multielectron systems what may have initially appeared to be a single line in a spectrum is actually several lines closely grouped together. Therefore, an energy level is actually made of many energy states that are also closely grouped together. We refer to these states as **sublevels.** The l quantum number describes the sublevels within an energy level.

The number of sublevels for each energy level equals the value of the principal quantum number of that level. Thus, there is one sublevel in the first level, two sublevels in the second level, and three sublevels in the third level. The lowest sublevel in each level has been named *s*, and, for this sublevel, l has been assigned a value of 0. The second sublevel is named *p*, in which l has a value of 1. The third sublevel is *d*, in which l is 2. In the fourth sublevel, *f*, l has a value of 3. Thus, the first level has only an *s* sublevel. The second energy level has *s* and *p* sublevels. The third energy level has *s*, *p*, and *d* sublevels.

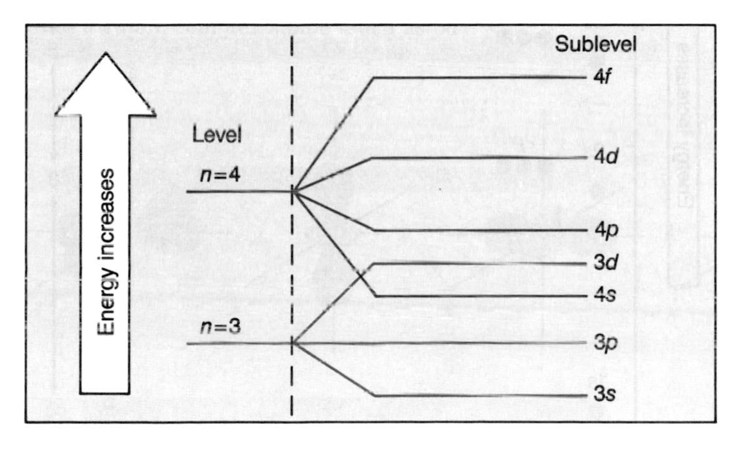

Figure 5.8 This energy level diagram shows the overlapping of sublevels that occurs between $n = 3$ and $n = 4$.

EVERYDAY CHEMISTRY

Fireworks

Background: Fireworks are also called pyrotechnics. Gunpowder packed into paper tubes is used as the lifting charge for skyrockets. As the rocket reaches its maximum altitude, a finer charge ignites and explodes. The finer charge contains the chemicals responsible for the colorful display.

Teaching This Feature: Use this opportunity to teach safety. Black powder can be purchased at sporting goods stores. Students should be cautioned against using it to make their own fireworks.

Exploring Further: Fireworks are used to save lives. Students can research how they are used in rescue operations. Police use a fusee that burns red to warn drivers of danger ahead. Railroads use a torpedo (large firecracker) that explodes when a train passes over it to warn the engineer of danger ahead. A Very pistol, used on boats and airplanes, fires a flare into the air to signal that someone is in distress. Rockets carry lifelines to ships in distress.

Answers to
Exploring Further

1. The lights are more like emission spectra because light is given off as electrons move from higher energy levels to lower energy levels.

2. Different chemicals are used. Each chemical emits its own characteristic color as the ball burns.

beaker while students view it from the side. Observe the intensity of the glow.

3. Repeat step 2 using the green and then the blue filters.

Results: The solution should glow brightly under blue light, less brightly under green light, and very little under red light.

Questions: 1. Which color of light, red, green, or blue has the longest wavelength? *red*

2. Which color light has the highest frequency? *blue*

3. The photons of which color have the most energy? *blue*

Follow Up Activity: Shine the high-intensity white light through the liquid after asking students to predict the results. White light causes the solution to glow because it contains green and blue light. Shine an ultraviolet light on the liquid and record the results.

PROGRAM RESOURCES

Transparency Master: Use the Transparency master and worksheets, pp. 19-20, for guided practice in the concept "Energy and Sublevels."

Color Transparency: Use Color Transparency 10 to focus on the concept "Energy and Sublevels."

CHEMISTRY AND TECHNOLOGY

Thin-Layer Magnetism

Teaching This Feature: You may wish to have students create an electromagnet as a starting point for the discussion. A battery connected to wire wrapped around a nail will work.

Connection to Physics: Magnetism is only one metallic property that is caused by the orientation of electrons. Electrical conductivity and transistor's behavior are two others.

Answers to Thinking Critically

1. Possible response: Thin layer deposition on light metals significantly decreases the mass of the magnet.

2. In an experiment, there are many variables. When there are a large number of interacting variables, this can lead to many hours of laboratory work. Using computer simulation, the scientist can change the variables and simulate the experiment in a fraction of the time and expense it would require in the laboratory or in nature. Only the successful simulation is verified in the laboratory.

4 CLOSE

Discussion: Ask students to demonstrate the opposite spins of two electrons in the same orbital when $s = \pm 1/2$. Have students place one hand at forehead and move it away from the head, then down and back up by the stomach and chest. With that rotation of the hand continuing, have the students rotate the other hand in the opposite direction such that it moves toward the head and down along the chest and stomach.

CHEMISTRY AND TECHNOLOGY

Thin-Layer Magnetism

Magnetism in metals results from relationships among the electrons that surround the nuclei in the atoms of the metal. In most metals, electrons are paired, and the magnetic field that is set up by one spinning electron is canceled by an opposite magnetic field set up by the oppositely-spinning electron paired with it. In some metals, however, there are unpaired electrons. These electrons lead to the magnetic properties of the metals. The degree of magnetic behavior comes from the relative numbers of unpaired electrons that are spinning in the same direction in the atoms.

The greater the number of electrons spinning in one direction, the greater the magnetic behavior of the metal. Technology has long searched for ways to make stronger magnets in smaller packages. Recent research has shown that a thin layer of iron, one atom thick, can produce a stronger magnetic field than a thicker layer of iron. In a crystal, every iron atom is surrounded by eight neighboring atoms and each of these atoms has twenty-six electrons surrounding it. The interference among electrons is great, and the ability of electrons to be induced to spin in the same direction is reduced in a bar or thick magnet. In a single layer, however, the atoms are not surrounded in three dimensions. As a result, the electrons have more freedom to move and are more easily induced to spin in the same direction.

The major problem with one-atom thin films, or monolayers, is that they cannot survive in the real world. Therefore, computer simulations were used to determine whether monolayers could be coated on another metal. Scientists had guessed that if it were possible to lay a monolayer on another metal, then that metal would become magnetic. Neither silver nor gold has unpaired electrons. However, the simulation showed that both would accept an overlay of the iron monolayer and both would then exhibit the same magnetism as the iron itself. Later actual experiments, carried out in a high vacuum to avoid contamination, showed that the computer simulation was correct.

One interesting aspect of this monolayer deposit is that its magnetic field was oriented vertically rather than horizontally. This fact has led to some interesting possibilities for magnetic tape storage of information. The usual orientation of "magnets" along a magnetic tape is horizontal. The number of magnets that can be placed along a tape limits the amount of information that can be stored. Imagine the amount that could be stored if the magnets stood on end, a one-atom end, rather than horizontally. In experiments, information density has been increased by at least 40 times.

The discovery and use of one-atom thin layers is just one more example of scientists using a prepared mind and contemporary technology to produce a technological advance from scientific curiosity.

Thinking Critically

1. What other uses might there be for thin layer magnets coated over other metals?
2. Why is the use of the computer significant in this experiment?

CONTENT BACKGROUND

Pauli Exclusion Principle: In Section 5.2, students were introduced to the Pauli exclusion principle. In Section 5.3, students are asked to put this principle into practice as the orbitals are filled in a certain order. The principle suggests that it can be derived from quantum mechanics because it mentions the four quantum numbers. Quantum mechanics does not produce the exclusion principle. Also, the fourth quantum number, spin, implies that the interaction of the pair of electrons has something to do with the exclusion principle. Actual calculations of the magnitude of this interaction show that it is negligible.

The Pauli exclusion principle is both fundamental and empirical. The principle is based on the periodic table and spectroscopic evidence rather than quantum mechanics. The Pauli exclusion principle is a mathematical

5.3 Distributing Electrons

OBJECTIVES

- determine the electron configurations of the elements.
- write electron dot diagrams for the elements.

By studying the arrangement of electrons in atoms, you will better understand chemical reactivity. Particular arrangements of electrons lead to specific reaction behaviors. It is, therefore, important to know how to find the electron arrangements of an atom in order to predict its reaction behavior. In this section, you will see how to predict the ground state electron arrangements for any atom for which you know the atomic number. In most of your subsequent work you will assume that each atom has the predicted electron arrangement. You should be aware, however, that there are several important exceptions to these predictions. A number of exceptions will be pointed out in Chapter 6.

Order of Filling Sublevels

In predicting electron arrangements, everything works well until you finish the electron configuration of argon: $1s^22s^22p^63s^23p^6$. Where will the electrons of the next element, potassium ($Z = 19$), go? You may remember that in the energy level diagram in Figure 5.8, the 4*s* level was shown below the 3*d* level. Thus, in this case, the 4*s* level fills first because that order produces an atom with lower energy. The electron configuration for potassium, therefore, is $1s^22s^22p^63s^23p^64s^1$. Calcium atoms ($Z = 20$) have the electron configuration $1s^22s^22p^63s^23p^64s^2$. Scandium, however, begins filling the 3*d* orbitals, and has a configuration of $1s^22s^22p^63s^23p^64s^23d^1$.

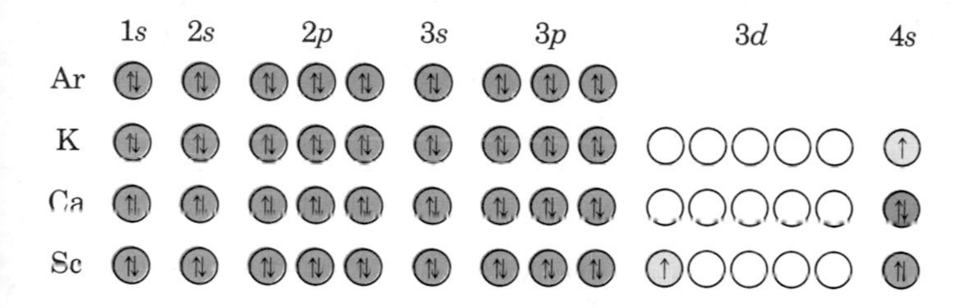

In many atoms with higher atomic numbers, the sublevels are not regularly filled. It is of little value to memorize each configuration. However, there is a rule of thumb that gives a correct configuration for most atoms in the ground state. It is a rule of thumb because it makes an assumption that is not always true. The order of increasing energy sublevels is figured for the one-electron hydrogen atom. In a multielectron atom, each electron affects the energy of the others. Consequently, there are a number of exceptions to the rule of thumb. We will explore these exceptions more fully in the next chapter. This rule of thumb is summarized in Figure 5.13. *If you follow the arrows listing the orbitals passed, you can find the electron configuration of most atoms.*

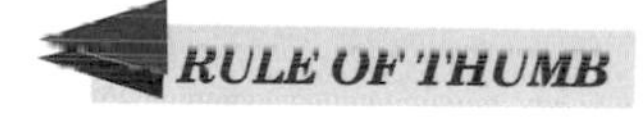

5.3 Distributing Electrons

PREPARE

Key Concepts

The major concepts in this section include predicting the order of filling sublevels (*aufbau* principle), writing electron configurations, and drawing electron dot diagrams.

Key Term

Lewis electron dot diagram

1 FOCUS

Objectives

Preview with your students the objectives listed on the student page. Students can use them as a study guide for this major section.

Discussion

Have students imagine that they are working in a new warehouse that has only empty shelves. Ask students where they would put boxes of material as they arrive. The usual response is to begin filling the bottom shelves first to keep the shelving unit stable. It also requires less energy than lifting the boxes to place them on higher shelves. In the same way, electrons first fill the sublevels closest to the nucleus. This resulting ground state atom is stable.

property of a wave function that is not required by quantum mechanics. It is an empirical restriction imposed on the wave function that makes it fit the experimentally observed facts. Although it is based on an abstract formulation, it is an important principle. Point out to students that someday one of them may produce an explanation that is not so abstract or empirical.

interNET CONNECTION

CAChe molecular modeling tutorials can be accessed.

World Wide Web
http://www.chem.vt.edu/

GLENCOE TECHNOLOGY

Software

Mastering Concepts in Chemistry
Unit 2, Lesson 2
Basics of Electron Distribution

CD-ROM

Chemistry: Concepts and Applications
Exploration: *Building Atoms*

PROGRAM RESOURCES

Study Guide: Use the Study Guide worksheet, p. 19, for independent practice with concepts in Section 5.3.

2 TEACH

CONCEPT DEVELOPMENT

Reinforce: Point out that arrow diagrams are only approximations. Cite some exceptions to the predicted order of filling such as Cr, Cu, Nb, Mo, Tc, Ru, Rh, Pd, Ag, La, Ir, Pt, Au, Ac. Their electron configurations will be discussed in Chapter 6.

Correcting Errors

Allow students to use the arrow diagram to predict electron configurations. Being familiar with this ordering will be helpful in writing electron configurations using the periodic table, which will be presented in Chapter 6.

PRACTICE

Guided Practice: Guide students through the Sample Problem, then have them work in teams on Practice Problem 14.

Coopertive Learning: Refer to pp. 28T-29T to select a teaching strategy to use with this activiy.

Independent Practice: Your homework or classroom assignment can include Review and Practice 38-44, Concept Mastery 58, Critical Thinking 63.

Answers to
Practice Problem

14. 1 $1s^1$
2 $1s^2$
3 $1s^22s^1$
4 $1s^22s^2$
5 $1s^22s^22p^1$
6 $1s^22s^22p^2$
7 $1s^22s^22p^3$
8 $1s^22s^22p^4$
9 $1s^22s^22p^5$
10 $1s^22s^22p^6$
11 $1s^22s^22p^63s^1$
12 $1s^22s^22p^63s^2$
13 $1s^22s^22p^63s^23p^1$
14 $1s^22s^22p^63s^23p^2$
15 $1s^22s^22p^63s^23p^3$
16 $1s^22s^22p^63s^23p^4$
17 $1s^22s^22p^63s^23p^5$
18 $1s^22s^22p^63s^23p^6$
19 $1s^22s^22p^63s^23p^64s^1$
20 $1s^22s^22p^63s^23p^64s^2$

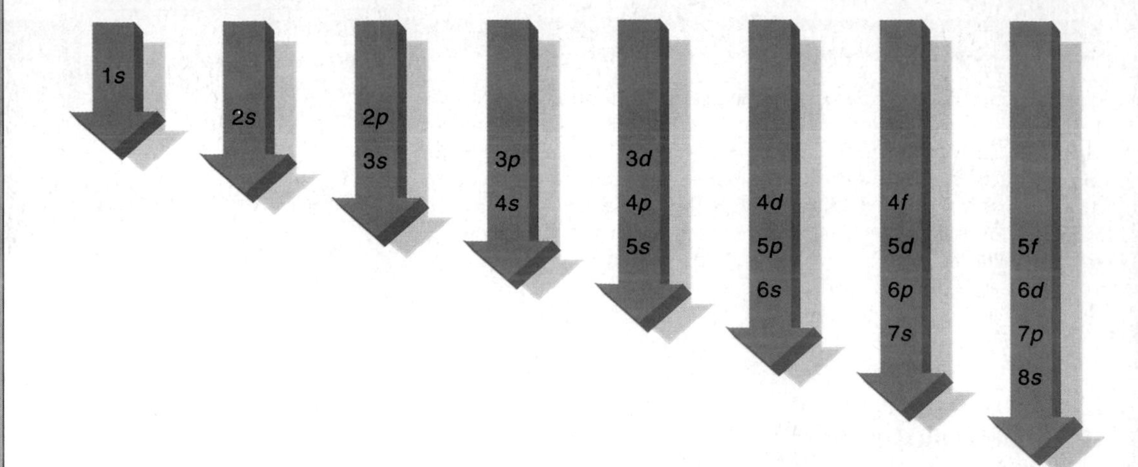

Figure 5.13 This arrow diagram provides a convenient method for writing electron configurations. Begin at top left, follow the arrow, then move right and down, always moving from the head of one arrow to the tail of the next.

SAMPLE PROBLEM

Writing Electron Configurations

Use the arrow diagram in Figure 5.13 to find the electron configuration of zirconium ($Z = 40$).

Solving Process:

Begin with the 1*s* level, move over to the 2*s*, and then over to the 2*p*. Follow the arrow down to 3*s* and over to 3*p*. Follow the arrow down to 4*s*. Move back up to the tail of the next arrow at 3*d* and follow it down through 4*p* and 5*s*. Move back to the tail of the next arrow, 4*d*, and place the remaining two electrons there. Thus, the electron configuration is

$$1s^22s^22p^63s^23p^64s^23d^{10}4p^65s^24d^2$$

A quick addition of the superscripts gives a total of 40, which is the atomic number.

PROBLEM SOLVING HINT

Add the superscripts of the electron configuration as a problem check. The total should equal *Z*.

PRACTICE PROBLEMS

14. Write the electron configurations for elements with atomic numbers $Z = 1$ through $Z = 20$.

Electron Dot Diagrams

In studying electrons in atoms in the following chapters, of primary concern will be the electrons in the outer energy level. These are the electrons most often involved in chemical reactions. It is often useful to draw these outer electrons around the symbol of an element. This notation is referred to as a **Lewis electron dot diagram,** see Table 5.2.

CONTENT BACKGROUND

Lewis Electron Dot Diagram: The Lewis electron dot diagram is presented in Chapter 5 as an application of knowing the atom's electron configuration. Inform students that being able to draw the dot diagram is an important skill that needs to be mastered because they will be using these diagrams to understand chemical bonding between atoms in later chapters.

Table 5.2

Electron Dot Diagrams for Some Elements					
Element	Configuration Ending	Dot Diagram	Element	Configuration Ending	Dot Diagram
carbon	$2s^22p^2$	·Ċ:	bromine	$4s^23d^{10}4p^5$	:B̈r:
sodium	$3s^1$	Na·	xenon	$5s^24d^{10}5p^6$	:Ẍe:
magnesium	$3s^2$	Mg:	cerium	$6s^24f^2$	Ce:
aluminum	$3s^23p^1$	Ȧl:	tungsten	$6s^24f^{14}5d^4$	W:
phosphorus	$3s^23p^3$	·Ṗ:	osmium	$6s^24f^{14}5d^6$	Os:
zinc	$4s^23d^{10}$	Zn:	uranium	$7s^25f^36d^1$	U:

The procedure for drawing electron dot diagrams follows.

Step 1. Let the symbol of the element represent the nucleus and all electrons except those in the outer level.

Step 2. Write the electron configuration of the element. From the configuration, select the electrons that are in the outer energy level. Those electrons in the outer level are the ones with the largest principal quantum number. Remember that n represents electron cloud size.

Step 3. Each "side" (top, bottom, left, right) of the symbol represents an orbital. Draw dots on the appropriate sides to represent the electrons in that orbital. It is important to remember which electrons are paired and which are not. *It is not important which side represents which orbital.*

Note the following special considerations. First, when choosing the outer electrons, ignore the highest energy electrons in the atom if they are not in the outermost energy level. Their principal quantum number is less than that of the electrons in the outermost energy level. Second, note that the s electrons are always shown as paired, or on the same side of the symbol. Also, remember from page 124 that electrons repel each other because of their negative charge. Therefore, electrons will occupy an empty degenerate orbital before pairing with other electrons. For example, four p electrons are distributed as one pair of electrons and two single electrons. No pairing takes place until the fourth electron enters the p sublevel.

SAMPLE PROBLEM

Electron Dot Diagrams

Write the electron dot diagrams for hydrogen, helium, oxygen, calcium, and cadmium.

Solving Process:

(a) Begin with the symbols for the elements required.

H, He, O, Ca, and Cd

CONCEPT DEVELOPMENT

GLENCOE TECHNOLOGY

Software

Mastering Concepts in Chemistry
Unit 2, Lesson 3
Writing Electron Configurations and Dot Structures

Computer Demo: *Isotopes,* For Apple IIe, shows electron dot diagram of atom or ion, available from: Richard G. Smith, 1416 Walker Dr. N.W. Lancaster, OH 43130, for the cost of postage. This program is also useful in Chapter 4.

Correcting Errors

After writing the electron configuration, the sublevels having the largest value of the n quantum number should be diagramed using the circles and arrows of orbital notation while observing Hund's Rule. Have students visualize a square box with the element's symbol in the center. One side show the s electrons, on the other three sides, show the p_x, p_y, and p_z electrons respectively. Only s and p electrons are shown in dot diagrams.

✓ ASSESSMENT

Skill: Ask students to use the answer to Practice Problem 14 on page 128 as the basis for writing the Lewis electron dot diagrams for the first twenty elements on the Periodic Table. Remind students that the answer to question 14 can be found on page 868. After the students have drawn the electron dot diagrams, ask them if they noticed a repeating pattern.

Chemistry Journal

An Electron's Story

Ask each student to assume he or she is an electron in the $3p$ sublevel of a phosphorus atom and to write a short journal story summarizing his or her position in the atom, any feelings of attraction within the atom, and any feelings of repulsion.

PROGRAM RESOURCES

Transparency Master: Use the Transparency master and worksheets, pp. 25-26, for guided practice in the concept "The Arrow Diagram."

Color Transparency: Use Color Transparency 13 to focus on the concept of "The Arrow Diagram."

PRACTICE

Guided Practice: Guide students through the Sample Problem, then have them work in class on Practice Problem 15a-f.

Independent Practice: Your homework or classroom assignment can include Concept Review 17, Review and Practice 43-47, Concept Mastery 56-57, Critical Thinking 62-63.

Answers to
Practice Problem

15. For dot diagrams, see Appendix C.

a. $1s^22s^22p^63s^23p^64s^23d^8$

b. $1s^22s^22p^63s^23p^6$

c. $1s^22s^22p^63s^23p^4$

d. $1s^22s^22p^63s^23p^64s^23d^{10}4p^65s^24d^9$
(actually $4d^{10}5s^1$)

e. $1s^22s^22p^63s^23p^64s^1$

f. $1s^22s^22p^63s^23p^64s^23d^{10}4p^2$

3 ASSESS

CHECK FOR UNDERSTANDING

Ask students to write the answers to Concept Review problems and Review and Practice Problems 45-47 from the end of the chapter.

RETEACHING

Refer students to Practice Problem 14. Use the electron configurations of these twenty elements as a basis for reteaching the Lewis electron dot diagram. Have students draw the orbitals of the outermost energy level using circles and arrows. Check their work. When they have mastered this, have students draw the corresponding electron dot diagrams.

(b) Write the electron configurations for each atom and determine the number of outer level electrons in the configuration.

H	$= 1s^1$	(outer electron $= 1s^1$)
He	$= 1s^2$	(outer electrons $= 1s^2$)
O	$= 1s^22s^22p^4$	(outer electrons $= 2s^22p^4$)
Ca	$= 1s^22s^22p^63s^23p^64s^2$	(outer electrons $= 4s^2$)
Cd	$= 1s^22s^22p^63s^23p^64s^23d^{10}4p^65s^24d^{10}$	(outer electrons $= 5s^2$)

(c) Use the outer configuration to determine the number of dots required for each element. Remember that each symbol represents the nucleus and inner level electrons for that atom. Then correctly place a dot for each outer electron in an orbital around the symbol.

H·, He:, :Ö·, Ca:, and Cd:

PROBLEM SOLVING HINT

To avoid misplacement of dots, think of each symbol being surrounded by a square. An orbital is represented by each side of the square.

PRACTICE PROBLEMS

15. Predict electron configurations using the arrow diagram, and draw electron dot diagrams for the following elements.
a. $Z = 28$ **c.** $Z = 16$ **e.** $Z = 19$
b. $Z = 18$ **d.** $Z = 47$ **f.** $Z = 32$

Electron Summary

We have treated the electron as a particle, a wave, and a cloud of negative charge. Which approach is correct? They all are. As pointed out earlier in the chapter, electrons can behave as particles or as waves. There are times when it is useful to consider the electron as a particle. At that time, the electron exhibits its particle characteristics. At other times, the wave properties of the electron are of the greatest importance. At still other times, it is best to consider the electrons in an atom as a cloud of negative

Figure 5.14 Scientists had known for many years that electrons behave as both particles and waves. Images produced by a scanning tunneling electron microscope have now shown that confined electrons produce an interference pattern called standing waves, which you can see as circular ripples in this photo.

CONTENT BACKGROUND

Electron Dot Diagrams: The electron dot diagrams developed by Lewis allow students to visualize the location and arrangement of the valence electrons. These outer electrons are shown as dots in paired or unpaired arrangements. One disadvantage of the electron dot diagram is that the radius of the atom is not apparent. The relative size of the atoms and ions will be presented in Chapter 7. A second disadvantage is that the electron dot diagrams are two-dimensional models of three-dimensional atoms. You may want to consider a large, permanent black marker to show the electron dots on the surface of a balloon. This will help students visualize the 3-D nature of the atom.

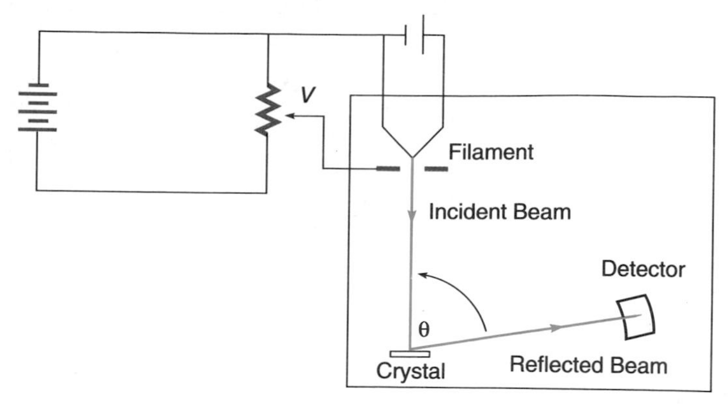

Figure 5.15 In this apparatus a beam of electrons from a hot cathode is directed at a crystal, and angles of the scattered electrons are detected. In this experiment, the electrons act as particles.

charge. Scientists do not have a single, completely satisfactory description of the structure of atoms. Consequently, they make use of more than one explanation for the properties they observe. The particular explanation that best fits each situation is the one applied to it.

You are now able to describe the electron configurations of the atoms of the elements. Your next study will be of a system of arranging elements based on their electronic structure—the periodic table.

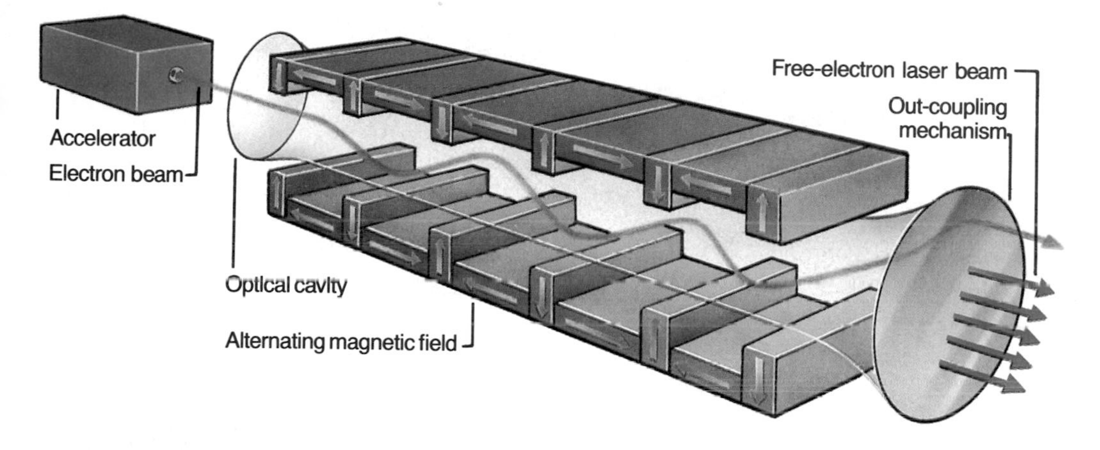

Figure 5.16 A free-electron laser shows the wave properties of an electron. An electron beam is generated, then contained by mirrors in the optical cavity. An alternating magnetic field uses this electron beam and generates a laser beam.

5.3 *CONCEPT REVIEW*

16. Write electron configurations for aluminum, hydrogen, and manganese.

17. Write the part of the electron configuration of sulfur that would be represented by dots in an electron dot diagram.

18. Draw electron dot diagrams for helium, rubidium, and silicon.

19. Apply Explain why inner-level electrons are not drawn as dots placed in orbitals when drawing a Lewis electron dot diagram of an atom.

EXTENSION

Have students write the electron configurations of sodium and chlorine. Remind students that the compound NaCl exists. Ask them for their suggestions on how the compound could form by rearranging outer electrons. Have students graphically represent their ideas for bonding between Na and Cl using the electron dot diagrams. This problem is discussed early in Chapter 6.

Answers to Concept Review

16. Aluminum, $Z = 13$: $1s^22s^22p^63s^23p^1$
Hydrogen, $Z = 1$: $1s^1$
Manganese, $Z = 25$: $1s^22s^22p^63s^23p^64s^23d^5$

17. $3s^23p^4$

18. He: Rb• •Si:

19. Usually only the outer level electrons are involved in chemical change, and these electrons are shown as dots in an electron dot diagram.

4 CLOSE

Discussion: To put the chapter into perspective have a student read aloud to the class the Electron Summary. Inform the students that in Chapter 6 they will practice and apply what they have learned in Chapter 5.

PROGRAM RESOURCES

Reteaching: Use the Reteaching worksheet, pp. 7-8, to provide students another opportunity to understand the concept of "Filling an Atom with Electrons."

Vocabulary Review/Concept Review: Use the worksheets, p. 5 and pp. 39-40, to check students' understanding of the key terms and concepts of Chapter 5.

CHAPTER REVIEW 5

- Review Summary statements and Key Terms with your students.
- Complete solutions to Chapter Review Problems can be found in **Merrill Chemistry Problems and Solutions Manual**.

Answers to
Review and Practice

20. De Broglie equated Einstein's relationship between matter and energy and Planck's quantum theory relating energy and frequency and developed an equation relating mass and wave properties, $\lambda = h/mv$

21. An electron stream acted the same way as a ray of light.

22. If the wave characteristics of a particle are being observed, the particle acts as a wave. If the particle behavior is being observed, it will act as a particle, not a wave.

23. $\lambda = h/p$

24. In order for the position of an electron to be determined, the electron absorbs energy, which causes its velocity and momentum to change.

25. Schrödinger treated the electron as a wave.

26. Even though they do not actually occupy the entire space, the blades of a fan appear to occupy the complete circle through which they turn. The electrons also effectively fill the space of the electron cloud.

Chapter 5 REVIEW

Summary

5.1 Modern Atomic Structure

1. De Broglie proposed that electrons and other particles of matter have both particle and wave properties. Two years later his hypothesis was proved correct.
2. All particles exhibit wave properties and all waves exhibit particle properties. This principle is known as the wave-particle duality of nature.
3. Chemists are interested in knowing the position and velocity of electrons in an atom.
4. Heisenberg's uncertainty principle states that it is impossible to know accurately both the position and the momentum of an electron at the same time. Measuring one quantity changes the other.
5. Schrödinger developed a mathematical equation that describes the behavior of the electron as a wave. The solution of the wave equation can be used to calculate the probability of finding an electron at a particular point around the nucleus.
6. Because of the electron's high speed, it effectively occupies all the volume defined by the path through which it moves. This volume is called the electron cloud.

5.2 Quantum Theory

7. Each electron in an atom can be described by a unique set of four quantum numbers, n, l, m, and s.
8. The principal quantum number ($n = 1, 2, 3, \ldots$) is the number of the energy level and describes the relative electron cloud size.
9. Each energy level has as many sublevels as the principal quantum number. The second quantum number ($l = 0, 1, 2, \ldots n - 1$) describes the shape of the cloud.
10. The third quantum number, m, describes the orientation in space of each orbital.
11. The fourth quantum number ($s = +\frac{1}{2}$ or $-\frac{1}{2}$) describes the spin direction of the electron.
12. Pauli's exclusion principle states that no two electrons in an atom can have the same set of quantum numbers. Each orbital may contain a maximum of one pair of electrons. Electrons in the same orbital have opposite spins.
13. Electrons normally occupy the set of orbitals that give the atom the lowest overall energy.

5.3 Distributing Electrons

14. The arrow diagram can be used to provide the correct electron configuration for most atoms.
15. The chemist is primarily concerned with the electrons in the outer energy level. Electron dot diagrams are useful in representing these outer level electrons.

Key Terms

wave-particle duality of nature
momentum
Newtonian mechanics
quantum mechanics
Heisenberg uncertainty principle
quantum number
probability
electron cloud
principal quantum number
sublevel
orbital
degenerate
Pauli exclusion principle
Lewis electron dot diagram

Review and Practice

20. How did de Broglie show the relationship between waves and particles?
21. What observation did scientists make that confirmed de Broglie's hypothesis?
22. What is meant by the wave-particle duality of nature?
23. What is the relationship between momentum and wavelength?

27. Quantum numbers represent properties of electrons in an atom.

28. The electrons interact with each other.

29. The principal quantum number, n, designates the energy level of the electron.

30. 18, $2n^2$

31. Sublevels within an energy level; n

32.

Level	Number of Sublevels	Sublevel letter(s)
1	1	s
2	2	s,p
3	3	s,p,d
4	4	s,p,d,f

Chapter 5 REVIEW

24. Explain the following statement: The more certain the position of an electron is, the less certain is its momentum.
25. What was the main assumption that Schrödinger made that enabled him to develop his equation?
26. Explain how the movement of an electron is similar to that of electric fan blades.
27. Explain what it is that quantum numbers represent.
28. Why is it difficult to apply the Schrödinger equation to multielectron atoms?
29. What does the principal quantum number, n, designate?
30. How many electrons can exist in the third energy level? Show the expression by which this value is calculated from the principal quantum number.
31. What does the second quantum number, l, designate? How many values can l have in a given energy level?
32. Copy the following table and complete it for energy levels 1-4.

Level	Number of sublevels	Sublevel letter(s)
1		
2		
3		
4		

33. What is an orbital? How many orbitals are possible at each sublevel?
34. Explain the factors that determine the size of an electron charge cloud.
35. How are degenerate orbitals alike? How do they differ from one another?
36. What are the possible values for the fourth quantum number (s)? What do these numbers signify?
37. State the Pauli exclusion principle. What does this principle tell us about two electrons occupying the same orbital?
38. The electron configuration of carbon is $1s^22s^22p^2$. Explain how this configuration is built up, electron by electron, using the Pauli exclusion principle at each step.
39. Write the electron configuration for uranium. How many electrons are in each level? Which levels are not filled?
40. What elements are composed of atoms having the following electron configurations?
    - **a.** $1s^22s^22p^63s^23p^4$
    - **b.** $1s^22s^22p^63s^23p^64s^23d^5$
    - **c.** $1s^22s^22p^63s^23p^64s^23d^{10}4p^3$
    - **d.** $1s^22s^22p^63s^23p^64s^23d^{10}4p^65s^24d^4$
41. Write the electron configurations for titanium and gallium.
42. Write the electron configurations for bismuth and ruthenium.
43. Write orbital-filling diagrams for atoms of boron, fluorine, sulfur, germanium, and krypton.
44. Why does the 4s sublevel fill with electrons before electrons are found in the 3d sublevel?
45. Draw electron dot diagrams for the elements with Z equal to 7, 15, 33, 51, and 83.
46. Selenium ($Z = 34$) was discovered by Berzelius in 1817. Write its electron configuration and its electron dot diagram.
47. Terbium (Tb), erbium (Er), ytterbium (Yb), and yttrium (Y) were all found near the town of Ytterby, Sweden, and named for it. What are the electron dot diagrams for all four of these elements?
48. Neon was first found in 1898. It occurs naturally in the air in trace amounts. Show how the four quantum numbers are used to describe the electron structure of neon.
49. An atom's electron configuration ends in $5s^24d^{10}5p^4$. Identify the element and write its electron dot diagram.

33. An orbital is the space that can be occupied by a maximum of two electrons. *s*, 1; *p*, 3; *d*, 5; *f*, 7

34. The size of an electron charge cloud is determined by the value of *n*, repulsion of other electrons, and attraction of the nucleus.

35. Degenerate orbitals are alike in size and shape; they differ in direction.

36. ± *1/2*; clockwise or counterclockwise spins.

37. No two electrons in an atom can have the same set of four quantum numbers; their spins must be opposite.

38. The first electron goes into level 1, sublevel *s*. The second electron goes into the same level and sublevel as electron 1, but it has an opposite spin. Electrons 3 and 4 both go into level 2, sublevel *s*, where they also pair up because of their opposite spins. Electron 5 goes into level 2 sublevel *p*, with orbital quantum number of −1, 0, or +1. Electron 6 goes into level 2, sublevel *p*, with orbital quantum number of −1, 0, or +1, but different from the orbital quantum number of electron 5.

39. $1s^2\ 2s^2\ 2p^6\ 3s^2\ 3p^6\ 4s^2\ 3d^{10}\ 4p^6\ 5s^2\ 4d^{10}\ 5p^6\ 6s^2\ 4f^{14}\ 5d^{10}\ 6p^6\ 7s^2\ 5f^4$

level	electrons
1	2
2	8
3	18
4	32
5	22
6	8
7	2

Levels 5, 6, and 7 are not filled.

40. **a.** S, **b.** Mn, **c.** As, **d.** Mo

41. Ti: $1s^22s^22p^63s^23p^64s^23d^2$
Ga: $1s^22s^22p^63s^23p^64s^2 3d^{10}4p^1$

42. Bi: $1s^22s^22p^63s^23p^64s^23d^{10} 4p^65s^24d^{10}5p^66s^24f^{14}5d^{10} 6p^3$
Ru: $1s^22s^22p^63s^23p^64s^23d^{10} 4p^65s^24d^6$

43. For structures, see *Merrill Chemistry Problems and Solutions Manual.*

44. The third and fourth energy levels have overlapped, and the 4*s* sublevel is lower energy than the 3*d* sublevel.

45. $Z = 7$, ·N: $Z = 51$, ·Sb:
$Z = 15$, ·P: $Z = 83$, ·Bi:
$Z = 33$, ·As:
Note that this is Column 15 (VA) of the periodic table.

46. $1s^22s^22p^63s^23p^64s^23d^{10}4p^4$ ·Se:

47. Tb:, Er:, Yb:, Y:

48.

electrons	n	l	m	s
1,2	1	0 (*s*)	0	$\pm\frac{1}{2}$
3,4	2	0 (*s*)	0	$\pm\frac{1}{2}$
5,6	2	1 (*p*)	−1	$\pm\frac{1}{2}$
7,8	2	1 (*p*)	0	$\pm\frac{1}{2}$
9,10	2	1 (*p*)	+1	$\pm\frac{1}{2}$

49. ·Te:

Answers to
Review and Practice (cont.)

50. Quantum numbers relate to charge clouds in that they explain why no two electrons can occupy the same space in the cloud; n = size of cloud, l = shape of cloud, m = orientation or orbitals and s = spin. They relate to energy by showing that there are discrete amounts of energy between levels.

51. $n = 5$, so there are 5 sublevels, with, theoretically, 2, 6, 10, 14, and 18 in them, respectively. The electrons present are arranged $5s^2\ 5p^6\ 5d^{10}\ 5f^{14}$; 18 more could be accommodated in the 5th level.

Answers to
Concept Mastery

52. wave-particle duality of nature

53. In order for any particle to be seen, it must reflect light. However, as the light strikes a small particle, it increases the energy and momentum of the particle and thus changes its position.

54. Born showed that the square of the absolute value of the amplitude of the electron wave gives the probability of finding the electron at the point in space for which the equation was solved. Electron clouds define this probability, not exact locations.

55. An electron gains energy, and a photon is absorbed.

56. The sum of all electron clouds in any sublevel (or level) is a spherical cloud.

57. Electrons that are in the 2*s* and 2*p* orbitals are shown; electrons in the 1*s* orbital are not shown.

58. For structures, see *Merrill Chemistry Problems and Solutions Manual.*

Answers to
Application

59. a. Newtonian, **b.** quantum, **c.** quantum, **d.** Newtonian

60. Traffic patterns are controlled by Newtonian, not quantum, mechanics.

Answers to
Critical Thinking

61. The Pauli exclusion principle, as a result of the work of Einstein's relating time and space.

Chapter 5 REVIEW

50. Explain how quantum numbers relate to charge clouds and energy.

51. In the fifth energy level ($n = 5$) unnilquadium (Unq, $Z = 104$) has 32 electrons. How are they arranged in sublevels? How many more electrons would the fifth energy level accommodate theoretically?

Concept Mastery

52. Why is it impossible to demonstrate the wave nature and the particle nature of electrons in the same experiment?

53. Explain why Heisenberg's uncertainty principle has a great effect on our ability to observe electrons and other very small particles.

54. What is the significance of the Schrödinger wave equation as interpreted by Max Born? How does this idea lead to the concept of electron clouds?

55. When an electron moves from the $n = 2$ to the $n = 4$ state, does it gain or lose energy? Is a photon emitted or absorbed?

56. In any energy level, the charge cloud is spherical. How, then, can orbitals have nonspherical shapes?

57. In the electron dot diagram for nitrogen, electrons from which orbitals are shown? Which electrons are not shown?

58. Niobium (Nb) and tantalum (Ta) are usually found together in nature. Both are used to make abrasives. How many paired electrons are in atoms of each?

Application

59. Determine whether each of the following is best described by Newtonian mechanics or by quantum mechanics.

- **a.** a car traveling 90 km/h
- **b.** gamma rays given off during nuclear decay
- **c.** an electron in an atom
- **d.** Earth orbiting the sun

60. Why is Heisenberg's uncertainty principle considered when a chemist examines electrons in an atom but is negligible when a traffic engineer in a helicopter observes traffic patterns at a highway intersection?

Critical Thinking

61. Classical physics would predict that only the first three quantum numbers should be necessary to describe the motion of the electron in three-dimensional space. What necessitates a fourth?

62. Draw electron dot diagrams for sodium ($Z = 11$) and chlorine ($Z = 17$). Can you explain how they might combine?

63. Shown here are orbital-filling diagrams for the elements named. Each diagram is incorrect in some way. Explain the error in each and write a correct diagram.

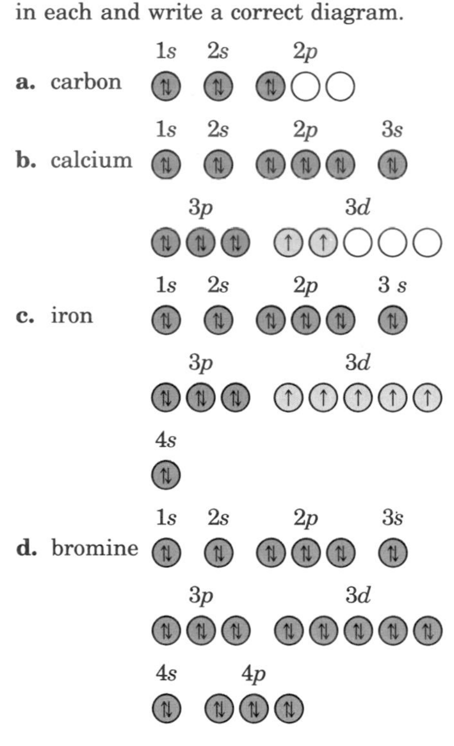

Chapter 5 REVIEW

64. Use what you have learned in this chapter to account for the complex spectra of multielectron atoms.

Readings

Freeman, Robert D. " 'New' Schemes for Applying the Aufbau Principle." *Journal of Chemical Education* 67, no. 7 (July 1990): 576.

Peterson, Ivars. "Electron Excitement in Three Dimensions." *Science News* 134, no. 16 (October 15, 1988): 252.

Peterson, Ivars. "Imitating Iron's Magnetism." *Science News* 131, no. 16 (April 18, 1987): 252-253.

Salem, Lionel. *Marvels of the Molecule*. New York: VCH Publishers, 1987.

Thomsen, Dietrick E. "A Midrash Upon Quantum Mechanics." *Science News* 132, no. 2 (July 11, 1987): 26-27.

Thomsen, Dietrick E. "Violating a Not-So-Exclusive Exclusion Principle." *Science News* 132, no. 9 (February 27, 1988): 132.

Tykodi, R. J. "The Ground State Electronic Structure for Atoms and Monatomic Ions." *Journal of Chemical Education* 64, no. 11 (November 1987): 943.

Cumulative Review

1. What is the average atomic mass of the element molybdenum if it has this isotopic composition?

Isotope	Mass	Abundance
92	91.906 808 u	15.84%
94	93.905 090 u	9.04%
95	94.905 837 u	15.72%
96	95.904 674 u	16.53%
97	96.906 023 u	9.46%
98	97.905 409 u	23.78%
100	99.907 478 u	9.63%

2. What is a nuclide? What are isotopes?

3. What is the difference between atomic number and atomic mass?

4. How much energy is needed to heat 10.0 g of tin from 25°C to 225°C? Use Appendix Table A-3.

5. What data led Bohr to formulate the planetary model of the atom?

6. What is Planck's contribution to the development of atomic theory?

7. What observations led John Dalton to formulate his atomic theory?

8. How many electrons, neutrons, and protons are in mendelevium ($A = 256$)?

9. Acetone, a common solvent, has a density of 0.788 g/cm^3. What is the volume of 2.50×10^2 grams of acetone?

10. The density of ethyl acetate is 0.898 g/cm^3. Calculate the mass of 0.588 dm^3 of ethyl acetate.

11. Distinguish between homogeneous and heterogeneous mixtures. In which category do solutions fall?

12. If you wanted to cook in a metal container that heats rapidly, would you choose an iron pan or a copper pan? Explain your answer.

PROGRAM RESOURCES

Evaluation: Use the Evaluation worksheets, pp. 17-20, to assess students' knowledge of Chapter 5.

Computer Test Bank: Use the IBM or Apple Test Bank and Question Selection Manual to prepare your own individualized test for Chapter 5.

62. Na$\cdot$, $\cdot\ddot{\text{Cl}}$:

Cl could capture the outer level electron from Na, producing Cl$^-$ and Na$^+$ ions.

63. For structures, see *Merrill Chemistry Problems and Solutions Manual.*

a. The electrons in carbon should not be paired in the *p* sublevel.

b. The Ca 4*s* sublevel should fill before the 3*d* sublevel.

c. For Fe, there are 6 electrons in the 3*d* sublevel instead of 5.

d. There are too many electrons for bromine.

64. For an electron to go from any one energy level to another, a set amount of energy is involved. This energy differs from level to level. With several electrons changing levels, many different amounts of energy are absorbed or emitted.

Answers to Cumulative Review

1. 95.89 u

2. nuclide—an atom with a specific number of protons and a specific number of neutrons.
isotopes—two different nuclides of the same element

3. atomic number = number of protons in the nucleus of an atom; atomic mass = average masses of the isotopes of an element

4. 455 J

5. the regularity of the spacing in the lines of the hydrogen atom

6. Planck proposed the quantum theory.

7. law of conservation of mass and law of definite proportions

8. 101 electrons, 155 neutrons, 101 protons.

9. 317 cm^3

10. 528 g

11. A homogeneous mixture is uniform throughout. A heterogeneous mixture is composed of different parts not uniformly dispersed.
Solutions are homogeneous.

12. Copper conducts heat better; C_p is lower for copper.

CHAPTER 6

Periodic Table

CHAPTER ORGANIZER

Chapter Objectives	Text Features	Laboratory Options	Teacher Classroom Resources
6.1 Developing the Periodic Table **1.** Describe the early attempts at classifying elements. **2.** Use the periodic table to predict the electron configurations of elements. **3.** Explain the basis for the arrangement of the modern periodic table.	**Music Connection:** Octaves, p. 139 **Chemists and Their Work:** Mendeleev, p. 140; Moseley, p. 141; Julian, p. 146 **Frontiers:** Easy as *Un-, Bi-, Tri-*, p. 149 **Chemistry and Society:** The Strontium - 90 Hazard, p. 150	**Discovery Demo:** 6-1 Group 1 (IA) Periodic Behavior, p. 136 **Demonstration:** 6-2 Metals and Nonmetals React, p. 140	**Lesson Plans:** pp. 12-13 **Study Guide:** pp. 21-22 **Color Transparency 14 and Master:** Periodic Table **Applying Scientific Methods in Chemistry:** Mendeleev's Periodic Table, p. 21 **Enrichment:** Mendeleev's Predictions, pp. 11-12 **Color Transparency 15 and Master:** Blank Periodic Table
Nat'l Science Stds: UCP.1, UCP.2, UCP.4, F.1, F.4, F.5, F.6, G.1, G.2, G.3			
6.2 Using the Periodic Table **4.** Identify metals, nonmetals, and metalloids on the periodic table **5.** Give examples of the relationship between an element's electron configuration and its placement on the periodic table. **6.** Predict the chemical stability of atoms using the octet rule.	**Chemists and Their Work:** Gertrude Belle Elion, p. 155 **Everyday Chemistry:** Beverage Cans, p. 157	**ChemActivity 6:** An Alien Periodic Table, p. 804 **Minute Lab:** *s* Comes Before *d*, p. 152 **Demonstration:** 6-3 Atomic Stability, p. 152 **Demonstration:** 6-4 Predicting a Periodic Trend, p. 154 **Minute Lab:** Bubbles That Burn, p. 154	**Study Guide:** pp. 23-24 **Color Transparency 16 and Master:** Groups of Elements **Reteaching:** Electron Configurations, p. 9 **Chemistry and Industry:** Sulfur, pp. 7-10 **Critical Thinking/ Problem Solving:** Chemical Shorthand, p. 6 **ChemActivity Master 6**
Nat'l Science Stds: UCP.1, UCP.2, UCP.5, B.2, E.1, E.2			
Other Chapter Resources	**Vocabulary and Concept Review,** pp. 6, 41-42 **Evaluation,** pp. 21-24	**Videodisc Correlation Booklet** **Lab Partner Software**	**Test Bank** **Mastering Concepts in Chemistry—Software**

Block Schedule

For information on block scheduling, see the Lesson Plans booklet in the Teacher Resource Package.

GLENCOE TECHNOLOGY

Mastering Concepts in Chemistry Software

Modern Periodic Table
Surveying the Table

Chemistry: Concepts and Applications Videodisc

Activity of Alkali Metals
A Metal and a Nonmetal React
Properties of Transition Metals
Properties of Ethyne: A Covalent Compound

Chemistry: Concepts and Applications CD-ROM

Exploring the Periodic Table
Comparing Group Traits
Determining Electron Configurations
The Periodic Table
Activity of Alkali Metals
A Metal and a Nonmetal React
Properties of Transition Metals
Properties of Ethyne: A Covalent Compound

Complete Glencoe Technology references are inserted within the appropriate lesson.

ASSIGNMENT GUIDE

Contents	Level	Practice Problems	Chapter Review	Solving Problems in Chemistry
6.1 Developing the Periodic Table				
Early Attempts at Classification: Dobereiner and Newlands	O		35, 37	
Mendeleev's Periodic Table	O		11, 12	
Modern Periodic Law	C		12	
Modern Periodic Table	C		13, 15, 17	1-5
Transition Elements	C		14, 20, 22	
The Lanthanoids and Actinoids	C		25, 30	
6.2 Using the Periodic Table				
Surveying the Table: Electron Configurations	C		21, 22, 27, 32-34, 36, 39, 41	
Relative Stability of Electron Configurations	C		26, 28, 31, 32, 38, 40, 42	
Metals and Nonmetals	C		16-21, 23, 24, 29, 33, 34	6-15
				Chapter Review: 1-13

C=Core, A=Advanced, O=Optional

LABORATORY MATERIALS

Minute Labs	ChemActivities	Demonstrations
page 152 distilled water hydrochloric acid, 6*M* iron(III) chloride iron filings test-tube rack test tubes with stoppers **page 154** beaker, 100-cm^3 burner calcium carbide chunk crucible tongs liquid dish soap stirring rod water wood splint	**page 804** copy of alien periodic table scissors	**pages 136, 140, 152, 154** baking soda beakers, 600-cm^3, 1-dm^3 burner carbon, small piece clear plastic wrap conductivity checker crucible tongs dropper dry test tube evaporating dish explosion shield germanium glue heavy card stock iodine, crystals lead, small piece lithium, small piece magnesium ribbon overhead projector plastic film, dark potassium, small piece silicon, small piece small samples of other elements sodium, small piece tin, small piece vinegar water wire screen wood splint zinc dust

CHAPTER 6

Periodic Table

CHAPTER OVERVIEW

Main Ideas: The chapter begins with the historical development of the periodic table. This approach provides the student with a rationale for the development of a classification system. The text shows students that the electron structure of atoms is the basis for the regularities and trends observed among the elements.

Theme Development: The theme of this chapter is stability and change. The theme focuses students' attention on the chemistry that best illustrates the family relationships and their periodic trends. Extensive use of the periodic trends continues in later chapters.

Tying To Previous Knowledge: Students have used the arrow diagram to predict the order of filling sublevels and to write electron configurations. Point out that students will be able to relate trends in properties of the elements to these electron configurations through the periodic table.

ASSESSMENT PLANNER

Choose from the following assessment strategies.

Assess, pp. 148-149, 156-157
Assessment, pp. 139, 142, 147, 150, 152, 153, 155, 154
Portfolio, p. 150
Minute Lab, pp. 152, 154
ChemActivity, pp. 804-805
Chapter Review, pp. 158-161

PROGRAM RESOURCES

Lesson Plans: Use the Lesson Plan, pp. 12-13 to help you prepare for Chapter 6.

Study Guide: Use the Study Guide worksheets, pp. 21-22, for independent practice with concepts in Section 6.1.

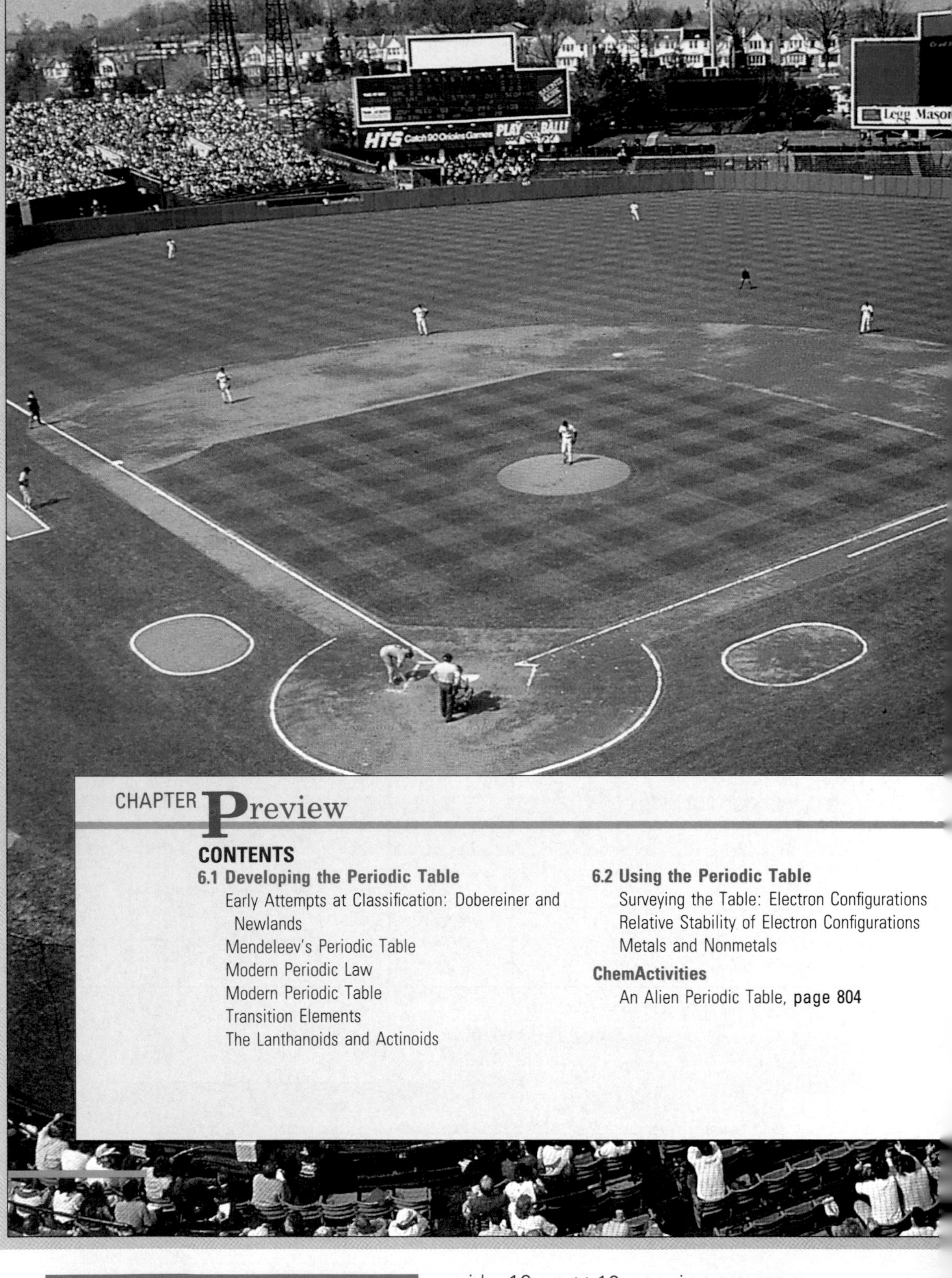

CHAPTER Preview

CONTENTS

DISCOVERY DEMO

6-1 Group 1 (IA) Periodic Behavior

Purpose: to show that chemical reactivity follows a predictable pattern related to an element's position in the periodic table

Materials: overhead projector; explosion shield; 3 600-cm^3 beakers; 300 cm^3 water; clear kitchen plastic wrap; small cubes of lithium, sodium, and potassium metals approximately 2 mm on a side; 10-cm × 10-cm wire screen

Procedure: 1. Cover the stage and lower lens of an overhead projector with clear plastic wrap. Place a 600-cm^3 beaker containing 100 cm^3 of water on the projector. **CAUTION:** *Place an explosion shield around the projector.* Turn on the projector, and darken the room. **2.** Drop a very small piece of lithium metal into the water. **CAUTION:** *Quickly cover the beaker with the wire screen to prevent the metal from flying out of*

CHAPTER 6

Periodic Table

What is the Great American Pastime? It used to be baseball, but today it may be watching the game on television. It would be a waste of time to try to find a baseball game or any other program just by switching television channels randomly at any time of day. To make wise use of time, people usually select their programs by using a published television schedule.

If the schedule is in the form of a table, you can tell at a glance what you need to know. A table is an efficient way to display large amounts of related information. Chemists (and chemistry students) need a way to organize information about the 109 elements in a manner that is as efficient for study and use as a TV schedule. In this chapter you will learn about a table that organizes information about the elements. It is called the periodic table.

EXTENSIONS AND CONNECTIONS

Music Connection

Octaves, **page 139**

Chemists and Their Work

Dmitri Ivanovich Mendeleev, page 140

Henry Gwyn-Jeffreys Moseley, page 141

Percy Lavon Julian, page 146

Gertrude Belle Elion, page 155

Frontiers

Easy as Un-, Bi-, Tri-: A look at a new system for naming new elements, **page 149**

Chemistry and Society

The Strontium-90 Hazard: An investigation of the dangers of strontium 90 in radioactive fallout, **page 150**

Everyday Chemistry

Beverage Cans: Steel versus aluminum for the can of the future, **page 157**

137

the beaker.

3. Repeat the procedure using Na and then K metal in separate beakers. The reactions of sodium and potassium are similar to that of lithium: $2Li + 2H_2O \rightarrow 2LiOH + H_2$ **Disposal:** C

4. As students observe the metals skimming across the water's surface, they will relate the speed with the metal's activity.

Results: Lithium is slow(less active), sodium is very fast (active), and potassium bursts into flame upon hitting the water (very active).

Questions: 1. Which metal reacts the fastest with water? *potassium*

2. In which metal is the outer-level electron farthest from the nucleus? *potassium*

3. How does each element's position in the column on the periodic table relate to the metal's reactivity? *Li is first, and least reactive; K is third and is the most reactive.*

INTRODUCING THE CHAPTER

USING THE PHOTO

Tables, such as those used in a television schedule, are frequently used to organize information and make predictions about related events or items. Point out to students that by knowing an element's location in the periodic table they can predict a great deal about the properties of that element.

REVEALING MISCONCEPTIONS

In Chapter 5, students used the arrow diagram to predict the electron configurations of atoms. However, nearly one out of every five elements does not follow this general pattern exactly. In this chapter students will learn to predict when and why some of these many exceptions occur.

GLENCOE TECHNOLOGY

Videodisc

Chemistry: Concepts and Applications

Activity of Alkali Metals

Disc 1, Side 1, Ch. 10

Show this video of the Discovery Demo to introduce the concept of periodic properties. Also available on CD ROM.

CONCEPT DEVELOPMENT

Enrichment: Mendeleev made predictions about five elements in addition to germanium. Ask a student to find out what these elements were and how accurate his predictions were. Possible response: Mendeleev's predictions about gallium and scandium were about as accurate as his predictions about germanium. However, his predictions about technetium, rhenium, and polonium were not as accurate.

Computer Demo: *The Periodic Table as a Database,* A graphical display of the physical and chemical properties, of the elements as they appear in the periodic table, AP207, Project SERAPHIM

CHEMISTS AND THEIR WORK
Dmitri Ivanovich Mendeleev

Background: Mendeleev was the youngest of 17 children. Their father went blind, causing the family to rely on their mother for support. Mendeleev could not gain admission to a Russian university because the education in Siberia was considered to be poor. Finally in 1855, he was accepted. In 1906, he was nominated for the Nobel Prize in chemistry but failed to win it by one vote.

Exploring Further: Mendeleev also applied his scientific knowledge to improve the quality and yield of farm crops. Why was this endeavor important in Russia? Possible response: Russia had a predominately agricultural economy.

GLENCOE TECHNOLOGY

Videodisc

Chemistry: Concepts and Applications

A Metal and a Nonmetal React
Disc 1, Side 1, Ch. 14

Show this video of Demonstration 6-2.
Also available on CD-ROM.

Figure 6.1 Mendeleev's first arrangement of the elements was published in 1869. Chemists became particularly interested in his periodic law when Mendeleev's predictions were confirmed by the discovery of gallium in 1875 and the discovery of scandium in 1879.

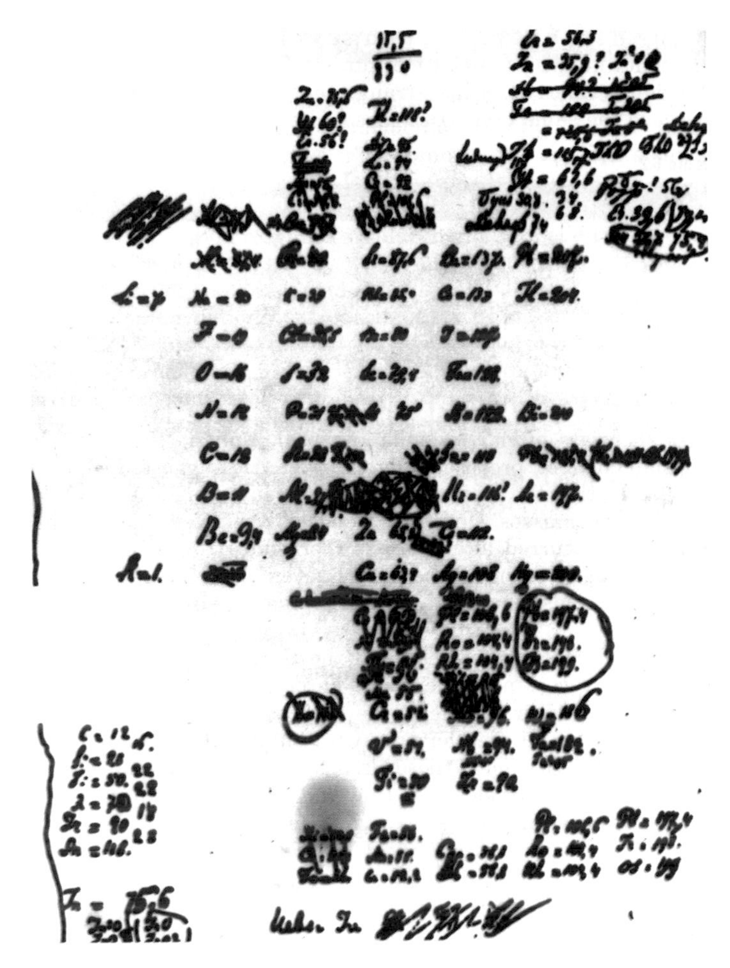

CHEMISTS AND THEIR WORK

Dmitri Ivanovich Mendeleev (1834-1907)
The periodic table is Mendeleev's lasting contribution to chemistry, but his investigations were much broader in scope throughout his scientific career. Mendeleev was born in Siberia and educated there and at the University of St. Petersburg. He also studied in France and Germany. He then returned to the University of St. Petersburg and worked as a professor until his resignation over political matters in 1890. Mendeleev was a Russian patriot who worked hard for his country. He wrote and continually updated a superb textbook for chemistry students. He received many honors and awards from foreign governments and societies, and element 101 is named in his honor.

In Mendeleev's table, the elements were arranged in order of their increasing atomic masses. The table showed that the properties of the elements are repeated in an orderly way. Such a repeating pattern is periodic. Mendeleev stated that the properties of the elements are a periodic function of their atomic masses. This statement was called the periodic law.

Modern Periodic Law

There was a problem with Mendeleev's table of elements. If the elements were arranged according to increasing atomic masses, tellurium and iodine seemed to be in the wrong columns. Their properties were different from those of other elements in the same column but were similar to those of elements in adjacent columns. Switching their positions put them in the columns where they belonged, according to their properties. If the switch

DEMONSTRATION
6-2 Metals and Nonmetals React

Purpose: to demonstrate that metals and nonmetals often combine to form a stable compound by releasing energy

Materials: 1 g zinc dust, 3 g iodine crystals, dry test tube with a stopper, evaporating dish, few drops of water, medicine dropper

Procedure: 1. Mix 1 g of zinc dust with 3 g of iodine crystals by shaking in a stoppered, dry test tube. Transfer the mixture to an evaporating dish.

2. In a fume hood, carefully add a few drops of water to the mixture using a dropper. **CAUTION:** *This demonstration should only be done in a fume hood. I_2 vapors are toxic. The dish will become very hot.*

3. The toxic, violet-colored iodine vapors can be collected in an inverted widemouth bottle clamped over the evaporating dish. Immediately cap the

were made, Mendeleev's assumption that the properties of the elements were a periodic function of their atomic masses would be wrong. Mendeleev assumed that the atomic masses of these two elements had been poorly measured, and he placed these two elements according to their properties. He thought that new mass measurements would prove his hypothesis to be correct. However, new measurements simply confirmed the original masses.

Soon, new elements were discovered, and two other pairs showed the same kind of reversal. Cobalt and nickel were known by Mendeleev, but their atomic masses had not been accurately measured. When such a determination was made, it was found that their positions in the table were also reversed. When argon was discovered, the atomic mass was found to be greater than that of potassium. If argon and potassium were put in the table on the basis of atomic masses, their positions would have been reversed.

Henry Moseley found the reason for these apparent exceptions to Mendeleev's periodic law. Moseley's X-ray experiments, in 1913, showed that the nucleus of each element has an integral positive charge, the atomic number. Iodine, nickel, and potassium have greater atomic numbers than do tellurium, cobalt, and argon, respectively. As a result of Moseley's work, the periodic law was revised. It now has as its basis the atomic numbers of the elements instead of the atomic masses. The modern statement of the **periodic law** is *the properties of the elements are a periodic function of their atomic numbers.*

Modern Periodic Table

The atomic number of an element indicates the number of protons in the nucleus of each atom of the element. Because an atom is electrically neutral, the atomic number also indicates the number of electrons surrounding the nucleus.

Certain electron arrangements are repeated periodically as atoms increase in atomic number. Elements with similar electron configurations can be placed in the same column. The elements in the column can also be listed in order of their increasing principal quantum numbers. Thus, a table of the elements like that on pages 144 and 145 can be formed. This table is the modern **periodic table** of the elements.

The periodic table is constructed in the following manner. Use the arrow diagram on page 128 to determine the order of filling the sublevels. Each s sublevel can contain two electrons, as shown in Table 6.4. Each p sublevel can contain six electrons arranged in three pairs, or orbitals. Each d sublevel can contain ten electrons in five orbitals. Each f sublevel can contain fourteen electrons in seven orbitals. Then align the elements with similar outer electron configurations. As you read about the arrangement of the periodic table, refer to Tables 6.5 through 6.9 that show electron configurations. Also refer to the diagrams that indicate the position of each element in the periodic table.

CHEMISTS AND THEIR WORK

Henry Gwyn-Jeffreys Moseley (1887-1915)

Henry Moseley was a British scientist whose work led to the use of atomic number rather than atomic mass as the basis for the periodic table. His announcement led directly to support for Mendeleev's periodic table. The concept of atomic number also enabled chemists to confirm Mendeleev's claim about missing elements. It could then be demonstrated that there were not any more missing elements.

Table 6.4

Sublevel	e^- capacity
s	2
p	6
d	10
f	14

CHEMISTS AND THEIR WORK
Henry Gwyn-Jeffreys Moseley

Background: Moseley was born in Weymouth, England. He served as a professor at the University of Manchester where he investigated radioactivity. In 1913, he reported the systematic relationship between the atomic number of the elements and the emitted X-ray spectra.

Exploring Further: Challenge students to do research to determine the relationship that exists between an emitted X ray's wavelength and the atomic number of the element being investigated. *Wavelength varies inversely as the square of the atomic number.*

CONCEPT DEVELOPMENT

QUICK DEMO **Separate and Classify**

To impress students with the idea of classification, display in random order a number of the elements and compounds from your stockroom. Students should be made aware of the difficulty of listing all the properties for each substance. A classification allows us to generalize concerning some properties. Point out that electron configurations can be used to organize elements into columns and rows. Review the *aufbau* principle (arrow diagram), the pattern in which the lowest-level electron sublevels fill first. Point out to students that this system of classification relates structure and properties.

bottle. **Disposal:** A

Results: Violet fumes are expelled from the dish as an exothermic reaction proceeds.

Questions: 1. Draw the electron dot diagrams of zinc and iodine. *Zn*:, ·Ï:

2. Use the information that one zinc atom reacts with two iodine atoms to form zinc iodide to explain how the reaction occurred. *Zinc shares one electron with each iodine atom.*

3. If metals tend to lose electrons in chemical reactions, what can be said about nonmetals as they react? *Nonmetals tend to gain electrons.*

Follow Up Activity: Use this demonstration to review and reinforce the concepts of physical and chemical properties and changes.

PROGRAM RESOURCES

Applying Scientific Methods in Chemistry: Use the worksheet, "Mendeleev's Periodic Table," p. 21, to help students apply and extend their lab skills.

CONCEPT DEVELOPMENT

✓ ASSESSMENT

Skill: Ask students to predict the appearance of the periodic table if 120 elements existed. What would it look like if 140 elements existed? If 168 elements existed?

MAKING CONNECTIONS

History
Ask students to bring in any old copies of chemistry books they can find. Compare these older periodic tables to the one in this chapter. In a 1922 high school text, only 87 elements were listed. Today's table contains 112 elements.

Activity: If possible, bring to class copies of periodic tables that are printed in different languages. Ask foreign exchange students in your school to write home to get a copy sent to you. Students will notice that the names of the elements are not in English, but the chemical symbol is the same. Chemistry is an international language. For example, the compound H_2O represents a clear, colorless, tasteless, odorless, liquid substance in all languages.

✓ ASSESSMENT

Skill: Have each student turn to the periodic table on pages 144 and 145. Point out that the rows are numbered 1-7, which corresponds to the outer energy level. Play a game in which you call out an outer configuration using only the *s* and *p* sublevels. Students write the symbol of the element that is located at that position on the periodic table. For example, the teacher calls out $3s^1$ and the student writes Na. At first, limit the game to the major group elements. Later, include the transition elements.

Table 6.5

Elements 1-10				
		1	2	
Z	Element	*s*	*s*	*p*
1	H	1		
2	He	2		
3	Li	2	1	
4	Be	2	2	
5	B	2	2	1
6	C	2	2	2
7	N	2	2	3
8	O	2	2	4
9	F	2	2	5
10	Ne	2	2	6

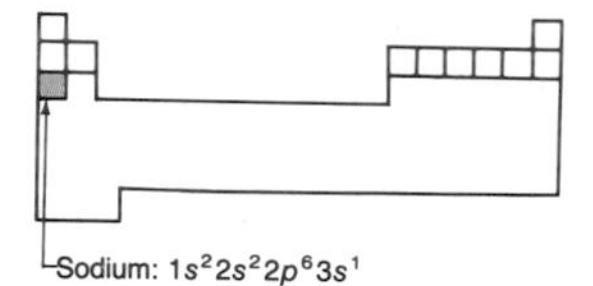

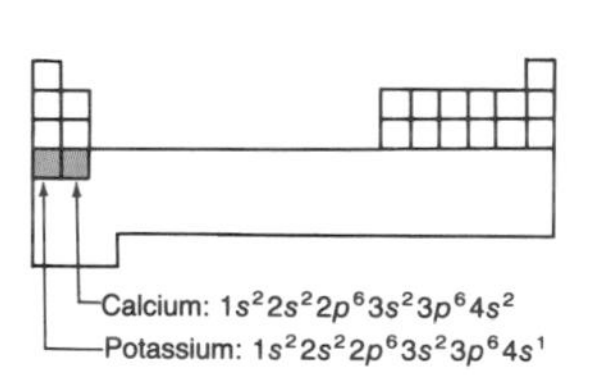

The first configuration in Table 6.5, hydrogen ($Z = 1$), consists of one electron in the 1*s* sublevel. The second configuration, helium ($Z = 2$), consists of two electrons in the 1*s* sublevel. These two electrons completely fill the 1*s* sublevel. Because the first energy level has only one sublevel (*s*), the first level is now full. The electron configurations of hydrogen and helium are not similar, so each element is in a separate column of the periodic table. Thus, the first row of the periodic table has two elements, just as the first energy level has only two electrons. The third element, lithium ($Z = 3$), has two electrons in the 1*s* sublevel and one electron in the 2*s* sublevel. Lithium is similar to hydrogen in that it has only one electron in its outermost level. Therefore, it is placed in the same column as hydrogen. The next element, beryllium ($Z = 4$), has two electrons in the 1*s* sublevel and two electrons in the 2*s* sublevel. It might seem to belong in the column with helium. However, the two electrons in helium's outermost level fill that level. Recall from Chapter 5 that the $n = 2$ level has a *p* sublevel as well as an *s* sublevel. The two electrons in the 2*s* sublevel of beryllium do not fill the second level. Therefore, beryllium starts a new column next to lithium. Boron ($Z = 5$) has a configuration composed of two 1*s* electrons, two 2*s* electrons, and one 2*p* electron, so boron heads a new column. Carbon ($Z = 6$), nitrogen ($Z = 7$), oxygen ($Z = 8$), and fluorine ($Z = 9$) atoms come next. These atoms have structures containing two, three, four, and five electrons, respectively, in the 2*p* sublevel. Each of these elements heads a new column. The atoms of neon ($Z = 10$), the tenth element, contain six 2*p* electrons. The second level ($n = 2$) is now full. Thus, neon is placed in the same column as helium.

The Group 18 (VIIIA) elements, the noble gases, have full outer energy levels. The symbol of a noble gas, in brackets, can be used to simplify writing electron configurations. For example, the electron configuration of Na could be written [Ne]$3s^1$. This notation is used in Tables 6.6 through 6.9.

Table 6.6

Elements 11-20								
		1	2		3			4
Z	Element	*s*	*s*	*p*	*s*	*p*	*d*	*s*
		2	2	6				
11	Na				1			
12	Mg				2			
13	Al				2	1		
14	Si				2	2		
15	P	[Ne]			2	3		
16	S				2	4		
17	Cl				2	5		
18	Ar				2	6		
19	K				2	6		1
20	Ca				2	6		2

CONTENT BACKGROUND

The Element Names

The names of the elements have an interesting history. As a class project, develop a table showing how each name was derived. A list of the origins can be found in *The Restless Atom* by A. Romer, Doubleday, New York (1960). Some examples include the following:

Named for discovery...

Location:	**Material:**	**Process:**
helium	lithium	neon
scandium	nitrogen	technetium
copper	calcium	promethium
gallium	radon	dysprosium

Named for Mineral...

Color:	**Property:**	**Location:**
beryllium	fluorine	strontium
boron	aluminum	yttrium
magnesium	silicon	cadmium
zirconium	manganese	terbium

Sodium atoms ($Z = 11$) have the same outer level configuration as lithium atoms, one s electron ($3s^1$). Thus, sodium is placed under lithium in the same column of the periodic table. The elements magnesium ($Z = 12$) through argon ($Z = 18$) have the same outer structures as the elements beryllium through neon have. They are also placed in the appropriate columns under atoms with similar configurations. Atoms of potassium ($Z = 19$) and calcium ($Z = 20$) have outer structures that are similar to the atoms of sodium and magnesium. Look closely at Tables 6.5 and 6.6 and the positions of these elements in the periodic table, pages 144–145. Note that each time a new energy level is started, a new row in the periodic table begins.

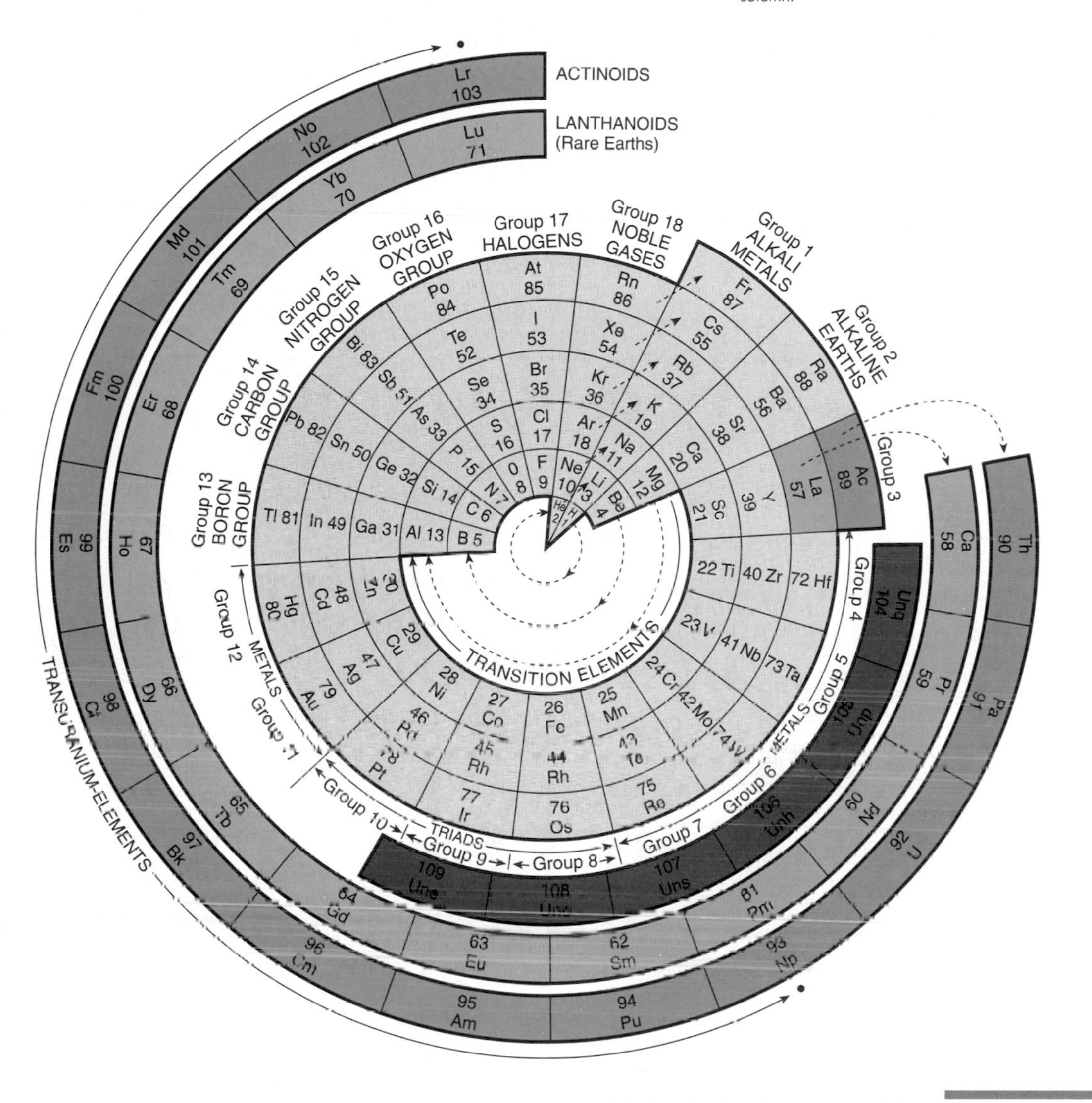

Figure 6.2 This representation of the periodic table also places elements with similar electron structures in the same column.

CULTURAL DIVERSITY

Jungle Gold

The Amazon River region of Brazil has been the site of one of the major gold discoveries of recent time. A local farmer discovered gold in 1980, and thousands of Brazilians have joined the search since then.

Brazilian prospectors, called garimpeiros, dig for gold using only hand tools such as shovels and picks. Ten thousand garimpeiros work Serra Pelada, Brazil's largest open-pit gold mine, each day. Towns with populations as large as 66,000 have appeared in the jungle. Life here is harsh: heavy rains, frequent mine cave-ins, and violence take their toll. But the gold rush continues to grow. The government depends on independent miners to produce 86 percent of Brazil's gold.

Unfortunately, the garimpeiros bring more than their shovels with them. They also bring death and disruption to the indigenous people living in the jungle, like the Yanomamo. Gerimpeiros have brought new diseases like malaria and TB, and new habits like alcoholism. Mercury, used by garimpeiros to separate out gold dust, has seeped into the rivers, contaminating the fish. The noise from airplanes and other machines has scared off the animals the Yanomamo hunt. Starvation is a very real threat, and mortality rates for Yanomamo villages are at least 14 percent - although figures are hard to obtain because the Yanomamo will not refer to anyone who has died. The clash between old and new is very clear in the Amazon jungle.

Named for heavenly body...				**Named to honor ...**	
Earth:	**Moon:**	**Planet:**	**Asteroid:**	**A person:**	**A place:**
tellurium	selenium	mercury	palladium	gadolinium	europium
		uranium	cerium	curium	francium
		plutonium		einsteinium	hafnium
				fermium	americium
Named for mythological character...				mendelevium	berkelium
titanium		vanadium		lawrencium	californium
cobalt		niobium		nobelium	
nickel		tantalum			

PROGRAM RESOURCES

Enrichment: Use the worksheet, "Mendeleev's Predictions," pp. 11-12, to challenge more able students.

CONCEPT DEVELOPMENT

Computer Demo: *Periodic Table,* A program suitable for testing student knowledge of the periodic table, AP310, Project SERAPHIM

GLENCOE TECHNOLOGY

Software

Mastering Concepts in Chemistry

Unit 3, Lesson 1

Modern Periodic Table

Unit 3, Lesson 2

Surveying the Table

CD-ROM

Chemistry: Concepts and Applications

Exploration: *Exploring the Periodic Table*

Exploration: *Comparing Group Traits*

Exploration: *Determining Electron Configurations*

Interactive Simulation: *The Periodic Table*

Periodic Table

(Based on Carbon 12 = 12.000)

Metals

Period	1* IA*	2 IIA	3 IIIB	4 IVB	5 VB	6 VIB	7 VIIB	8	9 VIIIB
1	1 H Hydrogen 1.007 94								
2	3 Li Lithium 6.941	4 Be Beryllium 9.012 182							
3	11 Na Sodium 22.989 768	12 Mg Magnesium 24.305 0							
4	19 K Potassium 39.098 3	20 Ca Calcium 40.078	21 Sc Scandium 44.955 910	22 Ti Titanium 47.88	23 V Vanadium 50.941 5	24 Cr Chromium 51.996 1	25 Mn Manganese 54.938 05	26 Fe Iron 55.847	27 Co Cobalt 58.933 20
5	37 Rb Rubidium 85.467 8	38 Sr Strontium 87.62	39 Y Yttrium 88.905 85	40 Zr Zirconium 91.224	41 Nb Niobium 92.906 38	42 Mo Molybdenum 95.94	43 Tc Technetium 97.907 2	44 Ru Ruthenium 101.07	45 Rh Rhodium 102.905 50
6	55 Cs Cesium 132.905 43	56 Ba Barium 137.327	71 Lu Lutetium 174.967	72 Hf Hafnium 178.49	73 Ta Tantalum 180.947 9	74 W Tungsten 183.85	75 Re Rhenium 186.207	76 Os Osmium 190.2	77 Ir Iridium 192.22
7	87 Fr Francium 223.019 7	88 Ra Radium 226.025 4	103 Lr Lawrencium 260.105 4	104 Unq 261	105 Unp 262	106 Unh 263	107 Uns 262	108 Uno 265	109 Une 266

Series						
LANTHANOID SERIES	57 La Lanthanum 138.905 5	58 Ce Cerium 140.115	59 Pr Praseodymium 140.907 65	60 Nd Neodymium 144.24	61 Pm Promethium 144.912 8	62 Sm Samarium 150.36
ACTINOID SERIES	89 Ac Actinium 227.027 8	90 Th Thorium 232.038 1	91 Pa Protactinium 231.035 88	92 U Uranium 238.028 9	93 Np Neptunium 237.048 2	94 Pu Plutonium 244.064 2

Gases—green, Liquids—blue, Solids—yellow, Synthetics—orange
(State is at room temperature and standard atmospheric pressure.)

interNET CONNECTION

Students can access an interactive periodic table of the elements.

World Wide Web
http://the-tech.mit.edu/Chemicool.html

Nonmetals

10	11 IB	12 IIB	13 IIIA	14 IVA	15 VA	16 VIA	17 VIIA	18 VIIIA
								2 He Helium 4.002 602
			5 B Boron 10.811	6 C Carbon 12.011	7 N Nitrogen 14.006 74	8 O Oxygen 15.999 4	9 F Fluorine 18.998 403 2	10 Ne Neon 20.179 7
			13 Al Aluminum 26.981 539	14 Si Silicon 28.085 5	15 P Phosphorus 30.973 762	16 S Sulfur 32.066	17 Cl Chlorine 35.452 7	18 Ar Argon 39.948
28 Ni Nickel 58.6934	29 Cu Copper 63.546	30 Zn Zinc 65.39	31 Ga Gallium 69.723	32 Ge Germanium 72.61	33 As Arsenic 74.921 59	34 Se Selenium 78.96	35 Br Bromine 79.904	36 Kr Krypton 83.80
46 Pd Palladium 106.42	47 Ag Silver 107.868 2	48 Cd Cadmium 112.411	49 In Indium 114.82	50 Sn Tin 118.710	51 Sb Antimony 121.757	52 Te Tellurium 127.60	53 I Iodine 126.904 47	54 Xe Xenon 131.290
78 Pt Platinum 195.08	79 Au Gold 196.966 54	80 Hg Mercury 200.59	81 Tl Thallium 204.383 3	82 Pb Lead 207.2	83 Bi Bismuth 208.980 37	84 Po Polonium 208.982 4	85 At Astatine 209.987 1	86 Rn Radon 222.017 6
110 Uun 269	111 Uuu	112 Uub						

†Metalloids lie along this heavy stairstep line.

63 Eu Europium 151.965	64 Gd Gadolinium 157.25	65 Tb Terbium 158.925 34	66 Dy Dysprosium 162.50	67 Ho Holmium 164.930 32	68 Er Erbium 167.26	69 Tm Thulium 168.934 21	70 Yb Ytterbium 173.04
95 Am Americium 243.061 4	96 Cm Curium 247.070 3	97 Bk Berkelium 247.070 3	98 Cf Californium 251.079 6	99 Es Einsteinium 252.082 8	100 Fm Fermium 257.095 1	101 Md Mendelevium 258.098 6	102 No Nobelium 259.100 9

*Currently there are two systems of labeling groups on the periodic table. A traditional system uses Roman numerals I through VIII with letters A and B. A more current system uses Arabic numerals 1 through 18, with no A and B designations. Throughout this text the current system will be used with the traditional heading following in parenthesis, for example, Group 1(IA).

CONCEPT DEVELOPMENT

Activity: Use your seating chart as a mini-periodic table. Have the class open their books to the periodic table. Assign each student in a row an element in the order it appears on the table. For example, in the first row, the first student is assigned hydrogen, the second student, lithium. Use Groups 1(IA),2(IIA), and 13-18(IIIA-VIIIA). Ask each student to write the electron configuration of the assigned element. Have each group meet together to search for a regularity. Students quickly identify that the outer electron configuration is the same. Ask students what this might suggest about the group's properties. Possible response: There are similarities.

Cooperative Learning: Refer to pp. 28T-29T to select a teaching strategy to use with this activity.

Background: Point out that the properties of an element relate to its electronic structure. The elements in both columns 1 (IA) and 11 (IB) contain only one electron in the outer level. However, column 1 (IA) metals contain eight electrons in the next-to-the-outer level and column 11 (IB) metals contain 18 electrons in this level. The column 1 (IA) metals (alkali metals) easily lose their single outer electrons and form ions, which have the stable, noble-gas configuration. Column 11 (IB) metals (coinage metals) are noncorrosive and relatively unreactive because loss of their single outer electrons does not result in the formation of appreciably more stable ions.

PROGRAM RESOURCES

Transparency Master: Use the Transparency master and worksheets, pp. 29-30, for guided practice using the "Blank Periodic Table."

Color Transparency: Use Color Transparency 15 for guided practice using the "Blank Periodic Table."

2 TEACH

MINUTE LAB

s Comes Before *d*

Purpose: to show that the properties of an element are a consequence of its outer electron structure

Materials: 1 g iron filings, 4 cm^3 6*M* hydrochloric acid, 2 test tubes, stopper, test tube rack, 1 g iron(III) chloride, 4 cm^3 distilled water

Procedure Hints: Place a white piece of paper behind the test tube containing the pale green Fe^{2+} to help students with the color determination.

Results: The iron metal filings react with HCl and are oxidized to form $FeCl_2$, which dissociates into Fe^{2+} and $2Cl^-$ ions. This solution has a pale green color. The Fe^{3+} ion has a yellow color in water solution.

Answers to Questions: **1.** Fe: $1s^22s^22p^63s^23p^64s^23d^6$ **2.** The outer $4s^2$ electrons are lost to form the Fe^{2+} ion. **3.** a $3d^6$ electron. **4.** Filled and half-filled sublevels are stable; the $3d^5$ is half filled.

✓ ASSESSMENT

Skill: Have students use Table A-3 on pages 855-857 to determine which of the elements in the first row of the transition elements, scandium through zinc, have multiple oxidation numbers. Students should prepare a written list of the ten elements and their major oxidation states.

MINUTE LAB

s Comes Before *d*

Put on an apron and goggles. Place about 1 g of iron filings into a test tube. Add 4 cm^3 of 6*M* hydrochloric acid to the test tube. **CAUTION:** *Acid is corrosive.* Place the test tube in a test-tube rack. Place approximately 1 g of iron(III) chloride, $FeCl_3$, into a second test tube. **CAUTION:** *$FeCl_3$ is a skin irritant.* Add 4 cm^3 of water to the test tube. Stopper the test tube and shake until some of the solid has dissolved. Allow the contents of both test tubes to settle, then record your observations. What is the electron configuration for iron metal? Iron loses two electrons to form the pale green substance in the first test tube. Predict what two electrons are lost. Iron loses three electrons to form the yellow iron(III) chloride in the second test tube. Which additional electron do you think was lost? Other transition elements can lose up to eight electrons. Why can iron lose only two or three electrons?

its written electron configuration end? Find Group 2 (IIA) in the periodic table. How does the written electron configuration for all elements in this group end? The same procedure can be used for Groups 13 through 18 (IIIA through VIIIA). There the endings, instead of s^1 or s^2, are p^1 through p^6 preceded by a coefficient that is the same as the number of the period.

For Groups 3 through 12 (IIIB through IIB), the endings are d^1 through d^{10}, preceded by a coefficient that is one less than the period number (for example, ... 4*s*3*d*). For these transition elements, remember that the *d* sublevel is always preceded by an *s* sublevel that is one quantum number higher. For example, iron is in the fourth period. You should expect iron to have two electrons in the 4*s* sublevel, and it does. Iron is in the sixth group, or column, of the transition elements. You should expect iron to have six electrons in the 3*d* sublevel, and it does. For the lanthanoids and actinoids, the endings are f^1 through f^{14} preceded by a coefficient that is two less than the period number (for example, ... 6*s*4*f*). The coefficient is 4 for the lanthanoids and 5 for the actinoids.

Some electron configurations are not what you might predict. To understand some of the exceptions to the arrow diagram, it is necessary to know that there appears to be a special chemical stability associated with certain electron configurations in an atom.

Relative Stability of Electron Configurations

Look at the order for filling the energy sublevels predicted by the arrow diagram on page 128:

1*s* 2*s* 2*p* 3*s* 3*p* 4*s* 3*d* **4*p* 5*s*** 4*d* **5*p* 6*s*** 4*f* 5*d* **6*p* 7*s*** 5*f* 6*d*...

Note that the outer energy level can have electrons in only the *s* and *p* sublevels. For example, if an atom has full 3*s* and 3*p* sublevels, the 4*s* sublevel will fill before electrons can enter the 3*d* sublevel. As soon as an electron enters the fourth energy level, the third energy level is no longer the outer level. In a similar

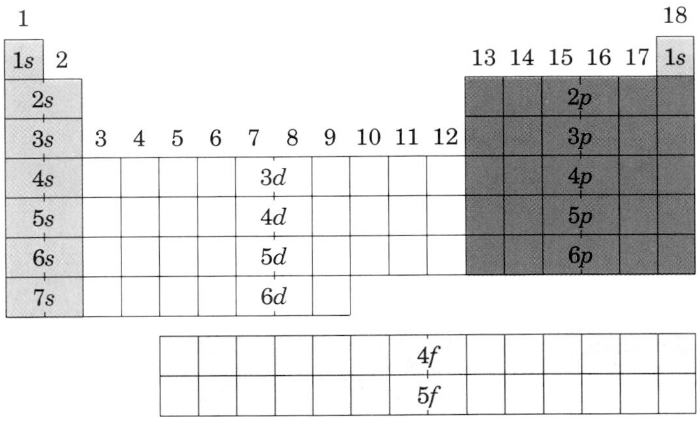

Figure 6.5 The position of an element in the periodic table can be used to determine the element's electron configuration. The outermost electron in an element is assigned to the indicated orbital.

DEMONSTRATION

6-3 Atomic Stability

Purpose: to show that an element's electron configuration is related to its chemical reactivity, and that properties are a consequence of structure

Materials: 1-dm^3 beaker with cover, 30 g baking soda, 30 cm^3 vinegar, wood splint, crucible tongs, 2-cm long piece of magnesium ribbon, laboratory burner, dark plastic film

Procedure: **1.** In the 1-dm^3 covered beaker generate CO_2 gas using baking soda and vinegar.
2. Light a wood splint in a flame. Using tongs hold the burning splint in the CO_2 atmosphere.
3. Repeat the demonstration using Mg ribbon. After the Mg is ignited hold it in the CO_2 gas. **CAUTION:** *Use a very dark colored plastic film to protect students' eyes from the bright Mg flame.*
Disposal: A

Figure 6.6 The noble gas xenon reacts with fluorine to form xenon tetrafluoride, seen here in crystalline form.

way, $4d$ is not begun until $5s$ is filled, $4f$ and $5d$ are not begun until $6s$ is filled, and $5f$ and $6d$ are not begun until $7s$ is filled. Although any energy level greater than 2 has more than the s and p sublevels, an outer energy level is considered filled when the s and p sublevels are filled. The energy level then is considered to have the maximum number of electrons that can be contained in a normal outer level.

One of the primary rules in chemistry is that atoms with full outer levels are particularly stable (less reactive). For all such elements except helium, the outer level contains eight electrons, two in the outer s sublevel and six in the outer p sublevel. These eight outer electrons are called an octet. The fact that *eight electrons in the outer level render an atom unreactive* is called the **octet rule.** Although the helium atom has only two electrons in its outer level, it, too, is one of these particularly stable elements. Helium's outer level is the first level, and it can hold only two electrons. Thus, it has a full outer level; the octet rule includes helium. Under some circumstances, it is possible to force the outer level of an atom in the third or higher period to hold more than eight electrons.

RULE OF THUMB

In addition to the outer octet, there are other electron configurations of high relative stability. *An atom having a filled or half-filled sublevel is slightly more stable (less reactive) than an atom without a filled or half-filled sublevel.* Thus, chromium is predicted to have two electrons in its $4s$ sublevel and four electrons in its $3d$ sublevel. Actually, as is shown in the diagram on page 154, chromium has one electron in its $4s$ sublevel and five electrons in its $3d$ sublevel. Note that one electron is shifted between two very closely spaced sublevels. The atom thus has two half-full sublevels instead of one full sublevel and one with no special arrangement. Copper has a similar variation from the prediction made by using the arrow diagram. Copper is predicted to have two $4s$ electrons and nine $3d$ electrons. Actually, it has one electron in its $4s$ sublevel and ten electrons in its $3d$ sublevel. One full and one half-full sublevel make an atom more stable than do one full

CONCEPT DEVELOPMENT

Using the Illustration: Many students prefer to use the periodic table rather than the arrow diagram as a visual reference when predicting the electron configurations for the elements. After your students have learned where the *s* (Groups 1,2), *p* (Groups 13-18), *d* (Groups 3-12), and *f* areas are located, they will be able to predict the order of filling of the sublevels. Each horizontal row represents a new energy level. Point out that the first d sublevel is a $3d$ and the first *f* sublevel is a $4f$. The *d* sublevel is always one less than the outer energy level ($4s$, $3d$), and the *f* sublevel is always two less than the outer energy level ($6s$, $4f$). Be sure students are aware that exceptions exist to the predicted order of filling sublevels. You may want to have them memorize Cr and Cu as examples of exceptions. Students can now begin to relate the properties of the elements that they have observed and will observe in the laboratory to the electron structure that they have predicted from the periodic table.

✓ ASSESSMENT

Knowledge: Ask students which two transition elements in the fourth period have electron configurations that are not predictable because of the stability of filled and half-filled sublevels. (Cr: $4s^1\ 3d^5$, Cu: $4s^1\ 3d^{10}$).

PROGRAM RESOURCES

Transparency Master: Use the Transparency master and worksheets, pp. 31-32, for guided practice in the concept "Groups of Elements."

Color Transparency: Use Color Transparency 16 to focus on the concept of "Groups of Elements."

Reteaching: Use the Reteaching worksheet, p. 9, to provide students another opportunity to understand the concept of "Electron Configurations."

Results: The burning splint goes out in a CO_2 atmosphere. The Mg burning becomes more violent in CO_2 than in air.

$$2Mg + CO_2 \rightarrow C + 2MgO$$

Questions: 1. Write the electron configurations of carbon and magnesium. *C: $1s^2 2s^2 2p^2$, Mg: $1s^2 2s^2 2p^6 3s^2$*

2. Which element has its electrons closer to the nucleus? *Carbon*

3. Are atoms with electrons closer to the nucleus less reactive than those atoms in which the electrons are farther away from the nucleus? *Yes*

Extension: Discuss the problem encountered by firemen when the Mg alloy frame of an executive jet is burning. Sand is the only effective agent used to extinguish a Mg fire.

3 ASSESS

CHECK FOR UNDERSTANDING

Ask students to write the answers to Concept Review problems, and Review and Practice Problems 16-24 from the end of the chapter.

RETEACHING

Randomly select elements from the periodic table and ask students to classify them as being a metal, nonmetal, or metalloid.

EXTENSION

Use a calendar to illustrate the concept of periodicity. Ask students what "property" regularly appears. Possible response: *Monday comes each week.* Use this understanding to reinforce the periodicity of outer electron configurations and the resulting metallic or nonmetallic properties.

PRACTICE

Guided Practice: Have students work in class on Practice Problem 6.

Answers to
Practice Problem: Have students refer to Appendix C for complete solutions to Practice Problems.

6. a. nonmetal
b. metal
c. metalloid
d. metal
e. metal
f. metal
g. metal

GLENCOE TECHNOLOGY

Videodisc

Chemistry: Concepts and Applications

Properties of Ethyne: A Covalent Compound
Disc 1, Side 1, Ch. 12

Show this video of the Minute Lab on page 154 to introduce the concept of reactivity of ethyne.

Also available on CD-ROM.

Figure 6.9 The metal copper, the metalloid arsenic, and the nonmetal sulfur are pictured left to right. Each element has properties characteristic of its classification.

RULE OF THUMB

As a general rule, elements with three or fewer electrons in the outer level are considered to be metals. Elements with five or more electrons in the outer level are considered to be nonmetals. There are exceptions. Some elements have properties of both metals and nonmetals. These elements are called the **metalloids.** Silicon, an element used in the manufacture of microcomputer chips, is a metalloid. On the periodic table, you will note a heavy, stairstep line toward the right side. This line is a rough dividing line between metals and nonmetals. As you might expect, the elements that lie along this line are generally metalloids.

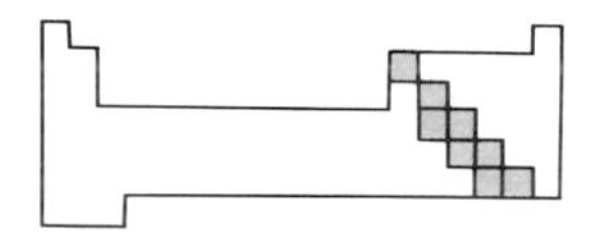

The elements of Groups 3 through 12 (IB through VIIIB), the transition metals, all have one or two electrons in the outer level. They all show metallic properties. The lanthanoids and actinoids also have two electrons in the outer level as predicted and are therefore classified as metals.

The elements of Groups 13 (IIIA) through 15 (VA) include both metals and nonmetals. At the top of the table, each of these groups contains nonmetallic elements. The metallic character of the elements increases toward the bottom of the table, and the last member of each family is distinctly metallic.

We can now look at the periodic table as a whole. Metals are located on the left and nonmetals on the right. Note again that most of the elements are metallic; that is, their atoms contain one, two, or three electrons in the outer energy level. The most unreactive atoms are those of the noble gases. Their chemical stability is explained by the octet rule.

PRACTICE PROBLEMS

6. Classify the following elements as metals, metalloids, or nonmetals.
a. oxygen
b. scandium
c. silicon
d. lithium
e. europium
f. cerium
g. mercury

Chemistry Journal

Household Elements

Ask each student to choose ten common household items or materials and, insofar as possible, list the elements that are the primary components of the items or materials. The elements should be classified as metals, nonmetals, or metalloids. Have students include the items, elements, and classsifications in their journals under the heading *Common Household Elements.*

6.2 CONCEPT REVIEW

7. Classify the following elements as metals, metalloids, or nonmetals: calcium, phosphorus, tellurium, tungsten, yttrium.
8. Find two transition elements whose actual electron configurations differ from those predicted by the arrow diagram. Explain each of these deviations.
9. State the octet rule in terms of electron pairs.
10. **Apply** The elements mercury and bromine are both liquids at room temperature. Why is mercury a metal and bromine a nonmetal?

EVERYDAY CHEMISTRY

Beverage Cans

In the past few years more than 90% of the beverage cans in the United States have been made of aluminum. The aluminum can manufacturers achieved this high percentage of the beverage can market because aluminum cans are manufactured more cheaply from single sheets of aluminum, aluminum is fairly easy to recycle, and the cans themselves have a low density, thus reducing shipping costs.

Thirty years ago most cans were made of tin plated steel, the old "tin can." They were heavy, had a tendency to leak, and gave a metallic taste to the contents. The one advantage was that iron—from which steel is made—rusts.

Steel companies have now improved production facilities and believe that they can challenge the use of aluminum cans. They are now able to produce a much thinner steel can. Because the price of aluminum has increased, the steel can is actually less expensive to make. A major problem, however, is that steel cans still need aluminum lids. A steel "flip-top" has not yet been developed.

Although a switch to steel cans might save millions of dollars annually in materials costs, assembly lines would have to be retooled and a recycling program would need to be set up. The public expects to recycle cans. More than 40 billion aluminum cans, more than half of those produced, are recycled each year. Most steel manufacturers say that it is more costly to recycle a steel can than to produce a new one. Until the steel manufacturers can solve the "flip-top" problem and cope with recycling, aluminum cans will probably continue to have an advantage.

Exploring Further

1. What properties of aluminum make it easier for cans to be made of aluminum than steel?
2. Find the mass of an aluminum can. Using its density, estimate the number of cm^3 used in the manufacture of the can. Assuming the steel manufacturers could use the same volume of steel to make a steel can, how much mass would that can have? Assume steel has approximately the same density as iron. Refer to Appendix Table A-3 for densities.

Answers to
Concept Review

7. calcium—metal
phosphorus—nonmetal
tellurium—metalloid
tungsten—metal
yttrium—metal
8. Molybdenum is predicted to have two electrons in the 5*s* sublevel and four electrons in the 4*d* sublevel. Actually, it has one electron in 5*s* and five in 4*d*, because having a half-filled 4*d* sublevel makes the atom slightly more stable. Palladium has ten 4*d* electrons and no 5*s* electrons, rather than the predicted eight 4*d* electrons and two 5*s* electrons. This is a result of the increased stability of a full 4*d* sublevel. (Other answers are possible.)
9. An atom with four pairs of electrons in its outer level is chemically stable.
10. Mercury is a metal because it contains three or fewer electrons in the outer level. Bromine is a nonmetal because it contains five or more electrons in the outer level.

EVERYDAY CHEMISTRY

Beverage Cans

Background: Many states have implemented recycling legislation that puts a deposit on aluminum cans and glass bottles. This has increased the recycling effort.

Teaching This Feature: You may wish to have students compare the mass of a steel can with the mass of an aluminum can of similar size.

Answers to
Exploring Further

1. Aluminum is more malleable and ductile than steel **2.** $V_{metal} = M_{Al}/D_{Al} = M_{Fe}/D_{Fe}$
$M_{Fe} = M_{Al} \cdot D_{Fe}/D_{Al}$
$= M_{Al} \cdot (7.9\ g/cm^3/2.7\ g/cm^3)$

4 CLOSE

Activity: As a class project, count the number of metals, metalloids, and nonmetals. Calculate the percentage of each. Draw a pie chart that shows these percentages. Review the properties of each. *78% of the elements are metals, 7% are metalloids, and 15% are nonmetals.*

PROGRAM RESOURCES

Chemistry and Industry: Use the worksheets, pp. 7-10, "Sulfur: Using Earth's Yellow Mineral" as an extension to the concepts presented in Section 6.2.

Critical Thinking/Problem Solving: Use the worksheet, p. 6, as enrichment for the concept of "Chemical Shorthand."

Vocabulary Review/Concept Review: Use the worksheets, p. 6 and pp. 41-42, to check students' understanding of the Key Terms and Concepts of Chapter 6.

CHAPTER REVIEW 6

- Review Summary statements and Key Terms with your students.
- Complete solutions to Chapter Review Problems can be found in **Merrill Chemistry Problems and Solutions Manual**.

Answers to
Review and Practice

11. Mendeleev grouped elements with similar properties in the same column on his periodic table. There were resulting blank spaces on his chart.

12. At that time, atomic masses could be determined, but atomic numbers were unknown. X-ray experiments determined the number of protons in the nucleus and thus the atomic number. The elements are now arranged in order of atomic number.

13. Each period represents an energy level, and the first energy level has only the 1*s* sublevel and can contain only two electrons.

14. 3*d*; the 4*s* sublevel is lower in energy than the 3*d*.

15. 1(IA) and 2(IIA)

16. alkali metals—Li, Na, K, Rb, Cs, Fr
alkaline earth metals—Be, Mg, Ca, Sr, Ba, Ra
chalcogens—O, S, Se, Te, Po
halogens—F, Cl, Br, I, At
noble gases—He, Ne, Ar, Kr, Xe, Rn

Chapter 6
REVIEW

Summary

6.1 Developing the Periodic Table

1. Many attempts have been made to classify the elements in a systematic manner. These attempts include Dobereiner's triads, Newlands' law of octaves, and Mendeleev's and Meyer's tables.
2. Mendeleev arranged his periodic table on the basis of the similar properties of elements. He concluded that the properties of elements were a function of atomic mass.
3. The modern periodic law states that the properties of the elements are a periodic function of their atomic numbers.
4. Today's periodic table reflects electron configurations of atoms, indicating that similarity in properties of different elements is related to similarity in their electron configurations.
5. In the transition elements, electrons are being added to a *d* sublevel. The placement of the transition elements on the periodic table reflects the overlap of energy levels.
6. In lanthanoids, electrons are being added to the 4*f* sublevel. In actinoids, electrons are filling the 5*f* sublevel.
7. All elements in a horizontal line of the table are called a period. All elements in a vertical line are called a group.

6.2 Using the Periodic Table

8. With a few exceptions, correct electron structures for atoms can be derived from examining the element's position on the periodic table.
9. According to the arrow diagram, only *s* and *p* electrons can be in the outer level of atoms, making a possible total of eight electrons in the outer level. An atom with eight electrons in the outer level is chemically stable or unreactive.
10. The most chemically stable atoms, the noble gases, have eight electrons in the outer level. Helium atoms are stable with two electrons in the outer level.
11. Filled and half-filled sublevels represent atoms in states of special chemical stability. Electrons in certain atoms are shifted between closely spaced sublevels to produce one of these more stable configurations.
12. Elements with one, two, or three electrons in the outer level usually have metallic properties. Elements with five, six, or seven outer-level electrons usually have nonmetallic properties. Elements that have both metallic and nonmetallic properties are called metalloids.

Key Terms

triad	group
law of octaves	octet rule
periodic law	family
periodic table	metal
transition element	nonmetal
lanthanoid series	metalloid
actinoid series	
period	

Review and Practice

11. What observations led Mendeleev to conclude that there were several undiscovered elements?
12. Why did Mendeleev and other chemists of his time arrange elements in the periodic table in order of atomic masses? What events changed this method? In what order are the elements arranged in the modern table?
13. Why does the first period of the periodic table contain only two elements while all other periods have eight or more elements in them?
14. To which sublevel are electrons being added across the first row of transition

17.

	Period	Group	Element	Metal, Nonmetal, or Metalloid
a.	3	14 (IVA)	silicon	metalloid
b.	5	4 (IVB)	zirconium	metal
c.	4	9 (VIIIB)	cobalt	metal
d.	3	18 (VIIIA)	argon	nonmetal
e.	5	17(VIIA)	iodine	nonmetal
f.	6	11(IB)	gold	metal

18. **a.** metal **b.** nonmetal **c.** metal **d.** metal **e.** metal **f.** metal **g.** nonmetal **h.** metal **i.** nonmetal **j.** metal **k.** metal **l.** nonmetal

19. Metals are found on the left side of the periodic table, and nonmetals are found on the right.

20. They have only two electrons in their outer energy level, which is characteristic of a metal.

21. halogens; F, Cl, Br, At; seven

22. Y, Lu, Lr; each has two *s* electrons in its outer energy level and one *d* electron in the previous energy level.

elements? Why do transition elements begin in the fourth period of the table rather than in the third period?

15. Elements whose highest-energy electrons are in *s* orbitals are found in what groups?

16. List the groups of elements that have family names. List the elements in each of these families.

17. Each of the following represents the end of an element's electron configuration. For each, tell which period and group the element is in, identify the element, and state whether the element is a metal, a nonmetal, or a metalloid.

a. $3s^2 3p^2$
b. $5s^2 4d^2$
c. $4s^2 3d^7$
d. $3s^2 3p^6$
e. $5s^2 4d^{10} 5p^5$
f. $6s^1 4f^{14} 5d^{10}$

18. Classify the following elements as metals or nonmetals.

a. manganese
b. fluorine
c. silver
d. mercury
e. cobalt
f. praseodymium
g. nitrogen
h. niobium
i. hydrogen
j. lithium
k. radium
l. carbon

19. In general, where are the metals found on the periodic table? Where are the nonmetals found?

20. Why are all of the transition elements metallic?

21. Iodine is used in many commercial chemicals and dyes. To what family does iodine belong? What are the other members of this family? How many electrons are in the outer energy level of each of these elements?

22. The existence of scandium was predicted by Mendeleev in 1871. It is the first of the transition elements. What are the other members of the group containing scandium? How are the predicted electron configurations of these elements similar?

23. List the elements generally considered to be metalloids.

24. Why are there many more metallic elements than nonmetallic elements?

Concept Mastery

25. What feature of electron configuration is unique to actinoids and lanthanoids?

26. Known elements may contain as many as 32 electrons in an energy level. If this is so, why are electrons in *s* and *p* sublevels considered so important? Why are these the only electrons considered in the octet rule?

27. What factors determine an element's chemical properties?

28. Why does the element molybdenum have an electron configuration ending $5s^1 4d^5$ rather than $5s^2 4d^4$? Explain the principle involved.

29. What are the major differences in properties among metals, nonmetals, and metalloids?

30. Why do you think the lanthanoid elements have close similarity in chemical properties?

31. If a lithium atom becomes more stable by losing an electron, does it have a stable octet? Can you say it follows the octet rule? Explain your answer.

32. Copper, silver, and gold do not follow the arrow diagram for electron configurations. Explain why they do not.

Application

33. Element X has the electron configuration $1s^2 2s^2 2p^6 3s^2 3p^6 4s^2 3d^{10} 4p^4$. Use this information to answer each of the following questions.

a. Use the periodic table to identify element X.
b. To what group and to what period does this element belong?
c. Classify the element as a metal, nonmetal, or metalloid.

Answers to
Review and Practice

23. Boron, silicon, germanium, arsenic, antimony, tellurium, polonium, and astatine are all metalloids.

24. Metals have few electrons in their outer energy levels. Because all the Groups 1(IA) through 12 (IIB) meet this requirement, there are more metals.

Answers to
Concept Mastery

25. Actinoids and lanthanoids are unique in that the *f* sublevels two levels below the outer energy level are being filled.

26. Electrons in *d* and *f* sublevels can never be in the outer level of a neutral atom. The *s* and *p* electrons are in the highest energy level in the atom and are the electrons involved in chemical reactions.

27. Electron configuration and distance of electrons from the nucleus determine an element's chemical properties.

28. An electron configuration ending in $5s^1 4d^5$ has two half-filled sublevels and is more stable than $5s^2 4d^4$, which has one complete energy sublevel.

29. metals—hard, shiny, conduct heat and electricity, lose electrons readily, malleable, ductile
nonmetals—gases or brittle solids at room temperature, dull, insulators, gain or share electrons
metalloids—properties of both metals and nonmetals, form amphoteric compounds

30. The outer two energy levels for each element have the same configurations.

31. It is stable, but does not have an octet because it has only two electrons. Yes, it has a full outer energy level.

32. If Cu ended in the predicted $4s^2 3d^9$ configuration, it would have stability with the filled *s* sublevel, but not from the unfilled $3d^9$. A configuration of $4s^1 3d^{10}$ is more stable because there is a filled sublevel (*d*) and a half-filled sublevel (*s*). Similar reasoning applies to Ag and Au.

Answers to
Application

33. a. selenium
b. Group 16 (VIA), period 4
c. nonmetal
d. gases or brittle solids at room temperature, dull, non-conductors, low melting point
e. ·Se: (electron dot structure)
f. two of: O, S, Te, Po
g. 2; krypton

PROGRAM RESOURCES

Evaluation: Use the Evaluation worksheets, pp. 21-24, to assess students' knowledge of Chapter 6.

Computer Test Bank: Use the IBM or Apple Test Bank and Question Selection Manual to prepare your own individualized test for Chapter 6.

Answers to
Application (cont.)

34.

	a		b.	d.	e.
Ba	2 (IIA)	6	metal	[Xe]$6s^2$ $1s^22s^22p^6$	Ba:
V	5 (VB)	4	metal	[Ne]$3s^23p^64s^2$ $3d^3$	V:
In	13 (IIIA)	5	metal	[Kr]$5s^24d^{10}$ $5p^2$	In·
Sn	14 (IVA)	5	metal	[Kr]$5s^24d10$ $5p^2$ $1s^22s^22p^6$	Sn·
Kr	18(VIIIA)	4	non-metal	[Ne]$3s^23p^64s^2$ $3d^{10}4p^6$	:Kr:
At	17(VIIA)	6	metal-loid	[Xe]$6s^24f^{14}$ $5d^{10}6p^5$	·At:

c. Metals—hard, shiny, conduct heat and electricity, malleable, ductile; nonmetal—gases or brittle solids at room temperature, dull, insulators; metalloids—properties of metals and nonmetals.

35. Although F is similar chemically to Cl and Br, averaging the atomic masses of F and Br would have led him to expect an element with a mass of about 50 u, not the mass of Cl. He would have found P, As, and Sb to be a triad because they are chemically similar and the average of the masses of P and Sb is very close to the mass of As. As, Sb, and Bi would not have been identified as a triad because the average mass of As and Bi is not the mass of Sb.

36. **a.** $Z = 8$ and 16
b. $Z = 3$, 11, and 37
c. $Z = 10$ and 18
d. $Z = 24$ and 42; $Z = 29$ and 47

37. Answers may vary. One possible triad is lithium, sodium, and potassium, because these alkali metals have similar properties and the atomic mass of sodium is approximately the average of the atomic masses of lithium and potassium. Another possible triad is the noble gases helium, neon, and argon.

Chapter 6 REVIEW

d. List the properties associated with the classification you chose.

e. Draw the electron dot diagram for an atom of element X.

f. List two other elements that are likely to be similar in properties to element X.

g. If this element gained electrons to achieve a stable octet, how many electrons would it gain? To which element would its outer electron arrangement then be similar?

34. The symbols of some elements are written on the periodic table shown. For each element whose symbol is shown, answer the following items.

a. To what group and what period does this element belong?

b. Classify the element as a metal, nonmetal, or metalloid.

c. List the properties this element should exhibit, based on the classification you chose.

d. Write the electron configuration of the element.

e. Draw the electron dot diagram for an atom of this element.

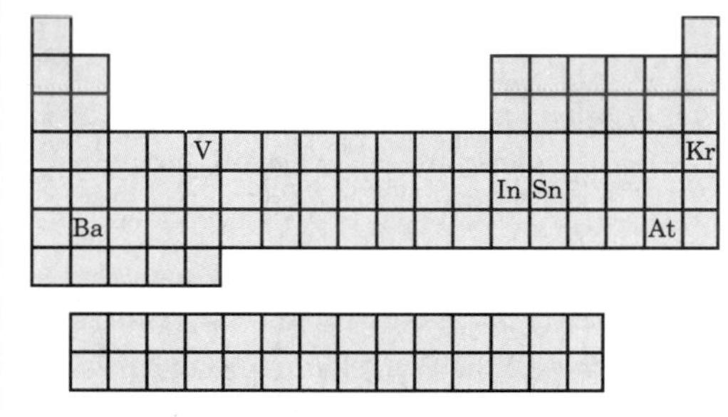

35. What conclusions do you think Dobereiner would have made if he had tried to compare F, Cl, and Br as a triad instead of Cl, Br, and I? Using the modern table, explain why F, Cl, and Br would not have fit his scheme. Would he have found P, As, and Sb to be a triad? What about As, Sb, and Bi? Explain your answers.

36. Without using a periodic table, and by working out the electron configurations, determine which elements from the following lists belong to the same group.

a. $Z = 2, 8, 16, 24$
b. $Z = 3, 11, 27, 37$
c. $Z = 4, 10, 18, 26$
d. $Z = 24, 29, 42, 47, 55$

37. Using the periodic table on pages 144 and 145, find two examples of triads other than those mentioned on page 138. Explain why each set of elements you chose is considered to be a triad.

Critical Thinking

38. The noble gases were among the last naturally-occurring elements to be discovered, even though some of them, particularly argon and neon, occur in moderate abundance on Earth. Why were the noble gases so difficult to detect and isolate?

39. In the system of numbering groups of elements with Roman numerals and letters, what is the significance of the Roman numeral in the A groups?

40. The octet rule applies to atoms within compounds as well as individual atoms. Elements unite to form a more stable configuration. How is this stability shown in the case of sodium and chlorine combining to form sodium chloride?

41. Predict the atomic mass, melting point, and boiling point for element 118. What would you expect its chemical properties to be? Use Appendix Table A-3 as needed.

42. Lithium and potassium are both elements in Group 1 (IA) and have the same electron configuration in the outer energy level. Both elements react vigorously with water, but potassium reacts more violently than does lithium. Why does potassium react more violently than lithium reacts?

160 *Periodic Table*

Answers to
Critical Thinking

38. They form no common compounds and exist naturally as gases.

39. The A number gives the number of electrons in the outer energy level of the atom.

40. Sodium loses its one outer level electron, leaving it with 10 electrons, 8 of which are now in the highest energy level. Chlorine adds the electron lost by sodium, giving chlorine 8 electrons in its highest energy level. Each now has an octet.

41. Student answers will vary. Atomic mass ≈300 u; mp ≈–30°C; bp ≈–20°C; noble gas—few chemical properties to consider; may react with oxygen or fluorine.

42. Potassium has more electrons between the nucleus and the outer energy level to shield that high-energy electron from the attraction of the nucleus. The outer electron is also farther from the nucleus.

Chapter 6
REVIEW

Readings

Ciparek, Joseph D. "Element X." *ChemMatters* 6, no. 4 (December 1987): 8-9.
Fernelius, W. Conrad. "Some Reflections on the Periodic Table and Its Use." *Journal of Chemical Education* 63, no. 3 (March 1986): 263-266.
Guenther, William B. "An Upward View of the Periodic Table." *Journal of Chemical Education* 64, no. 1 (January 1987): 9-10.
Osorio, Hanán von Marttens. "A Numerical Periodic Table and the *f* Series Chemical Elements." *Journal of Chemical Education* 67, no. 7 (July 1990): 563-565.
Peterson, Ivars. "A Superconducting Banquet From the Periodic Table." *Science News* 133, no. 10 (March 5, 1988): 148.

Cumulative Review

1. How does Schrödinger's analysis of the atom differ from that of Bohr?
2. How much energy is needed to raise the temperature of 5.24 g of ruthenium from 25°C to 202°C? See Appendix Table A-3.
3. Write the electron configurations and draw electron dot diagrams for the following elements.
   a. Al (Z = 13)
   b. S (Z = 16)
   c. Ca (Z = 20)
   d. Ti (Z = 22)
   e. V (Z = 23)
   f. Mn (Z = 25)
   g. Co (Z = 27)
   h. Ge (Z = 32)
   i. Br (Z = 35)
   j. Sr (Z = 38)
4. Explain the meaning of each symbol in the de Broglie equation.
5. Compute the average atomic mass of tungsten if its isotopes occur as follows.

Isotopic mass	Percentage
179.946 7 u	0.140%
181.948 25 u	26.41%
182.950 27 u	14.40%
183.950 97 u	30.64%
185.954 40 u	28.41%

6. How many significant digits are in the measurement 26°C?
7. What volume is occupied by 42.5 g of a substance of density 3.15 g/cm^3?
8. What are the four quantum numbers associated with each electron in an atom? What information do the first three quantum numbers give us about the electron cloud? What does the fourth tell us about the behavior of the electron?
9. An iron bar of mass 2.08 kg is heated to 150°C and is then quenched by plunging it into 50.5 kg of water at 12.0°C. What will be the final temperature of the water and iron?
10. After several years exposure to the weather, an aluminum screen door that was originally bright has become dark gray, pitted, and has a thin coating that can be scraped off to a powder. Is this change chemical or physical? Distinguish between chemical and physical changes in your answer.

Answers to Cumulative Review

1. Bohr treated electrons as particles; Schrödinger treated electrons as waves.
2. 221 J
3. a. $1s^22s^22p^63s^23p^1$ Al
   b. $1s^22s^22p^63s^23p^4$ S
   c. $1s^22s^22p^63s^23p^64s^2$ Ca:
   d. $1s^22s^22p^63s^23p^64s^23d^2$ Ti:
   e. $1s^22s^22p^63s^23p^64s^23d^3$ V:
   f. $1s^22s^22p^63s^23p^64s^23d^5$ Mn:
   g. $1s^22s^22p^63s^23p^64s^23d^7$ Co:
   h. $1s^22s^22p^63s^23p^64s^23d^{10}4p^2$ Ge
   i. $1s^22s^22p^63s^23p^64s^23d^{10}4p^5$ Br:
   j. $1s^22p^22p^63s^23p^64s^23d^{10}4p^65s^2$ Sr:
4. $\lambda = h/mv$
   λ = wavelength of a particle
   h = Planck's constant
   v = velocity of the particle
   m = mass of the particle
5. 183.8 u
6. 2 significant digits
7. 13.5 cm^3
8. principal quantum number, *n;* size; second quantum number, *l;* shape; third quantum number, *m;* orientation of orbitals; fourth quantum number, *s;* spin: clockwise or counterclockwise
9. 12.6°C
10. The change is chemical because the nature of the substance has changed.

CHAPTER 7

Chemical Formulas

CHAPTER ORGANIZER

Chapter Objectives	Text Features	Laboratory Options	Teacher Classroom Resources
7.1 Symbols and Formulas **1.** Demonstrate proficiency in writing chemical formulas. **2.** Define *oxidation number* and state oxidation numbers for common monatomic ions and charges for polyatomic ions. **Nat'l Science Stds: UCP.1, F.1, F.4, F.5**	**Chemists and Their Work:** Marie Curie, p. 165 **Everyday Chemistry:** Hazardous Materials at Home, p. 172	**Minute Lab:** Chemical Synthesis, p. 168 **ChemActivity 7:** Formulas and Oxidation Numbers, p. 806 **Discovery Demo:** 7-1 A Chemical Formula, p. 162 **Demonstration:** 7-2 Make it Neutral, p. 164 **Demonstration:** 7-3 Not Identical Twins, p. 172 **Laboratory Manual:** 7 Making Models of Compounds	**Lesson Plans:** pp. 14-15 **Study Guide:** p. 25 **Reteaching:** Chemical Formulas, p. 10 **Color Transparency 17 and Master:** Charges of Common Ions **Applying Scientific Methods in Chemistry:** Miller's Organic Compounds, pp. 22-23 **Chemistry and Industry:** Paper, pp. 11-14 **ChemActivity Master 7**
7.2 Nomenclature **3.** Demonstrate proficiency in naming chemical compounds. **4.** Distinguish between molecular and empirical formulas. **5.** Demonstrate the use of coeffients to represent the number of formula units of a substance. **Nat'l Science Stds: UCP.1, E.1, E.2**	**Careers in Chemistry:** Analytical Chemist, p. 174 **Bridge to Medicine:** Cyclopentamine for Allergies, p. 177 **Chemists and Their Work:** Todd Blumenkopf, p. 178 **Chemistry and Technology:** Making the Bone More Binding, p. 181	**Demonstration:** 7-4 The Roman Numeral, p. 174 **Demonstration:** 7-5 Descriptive Organic Chemistry, p. 176	**Color Transparency 18 and Master:** Naming Compounds **Study Guide:** p. 26 **Critical Thinking/ Problem Solving:** Ionic Clues to Atomic Identity, p. 7 **Enrichment:** Chemical Formulas, pp. 13-14
Other Chapter Resources	**Vocabulary and Concept Review,** pp. 7, 43-44 **Evaluation,** pp. 25-28	**Videodisc Correlation Booklet** **Lab Partner Software**	**Test Bank** **Mastering Concepts in Chemistry—Software**

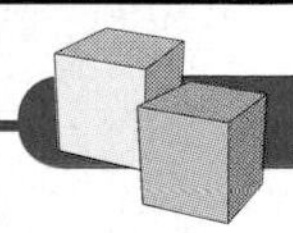

Block Schedule

For information on block scheduling, see the Lesson Plans booklet in the Teacher Resource Package.

GLENCOE TECHNOLOGY

Mastering Concepts in Chemistry Software
Symbols and Formulas
Naming Compounds

Chemistry: Concepts and Applications Videodisc
Oxidation States of Vanadium
Variable Oxidation States

Chemistry: Concepts and Applications CD-ROM
Determining a Formula
Form an Ionic Compound
Oxidation States of Vanadium
Variable Oxidation States

Complete Glencoe Technology references are inserted within the appropriate lesson.

ASSIGNMENT GUIDE

Contents	Level	Practice Problems	Chapter Review	Solving Problems in Chemistry
7.1 Symbols and Formulas				
Names of Elements	C			
Symbols for Elements	C		22, 23, 64, 71	
How Symbols are Used	C		64	
Chemical Formulas	C		31, 58, 59, 63, 71, 73, 74, 78, 79	1-2
Oxidation Number	C	1-3	24-26, 28, 29, 33, 35, 37, 39, 41, 43, 45, 46, 52, 53, 55, 56, 60-62, 66, 67, 81-83	3-6
7.2 Nomenclature				
Naming Compounds	C	8-14	27, 30, 32, 34, 36, 38, 40, 42, 44, 47, 48, 54, 57, 76, 80, 84	7-15
Molecular and Empirical Formulas	O	15	49, 50, 56, 85, 86	16
Coefficients	O	16	51, 68-70, 72, 74, 75, 77	17-18 Chapter Review: 1-14

C=Core, A=Advanced, O=Optional

Laboratory Materials

Minute Labs	ChemActivities	Demonstrations
page 168 beaker, 400 cm^3 burner cold water forceps iron filings powdered sulfur test tube	**page 806** paper pencil scissors sheet of ion models	**pages 162, 164, 172, 174, 176** battery, 9-volt with clips and wires beaker, 100-cm^3 butane lighter copper(I) oxide copper(II) oxide dropper $Fe(NH_4)_2(SO_4)_2 \cdot 6H_2O$ $FeNH_4(SO_4)_2 \cdot 12H_2O$ filter paper funnel gas outlet hexane iodine crystals iron(II) sulfate iron(III) sulfate metal paper clips overhead projector paper card stock pencil petri dish plastic sandwich bags potassium hexacyanoferr–ate(II) potassium thiocyanate ruler scissors sodium hydrogen sulfite tannic acid (or strong tea) test tubes, large with stoppers watch glasses water zinc dust

CHAPTER 7

Chemical Formulas

CHAPTER OVERVIEW

Main Ideas: Chemical symbols and oxidation numbers and their use in writing chemical formulas are presented. Nomenclature of inorganic and simple organic compounds is also presented. The difference between molecular and empirical formulas is described and the use of coefficients is introduced.

Theme Development: The theme of this chapter is systems and interactions. The current system of nomenclature and formula-writing rules of chemical compounds provides uniformity and eliminates confusion among scientists. Chemical formulas and distinctive names are used to represent compounds that are formed from the interactions of atoms.

Tying To Previous Knowledge: Students are already familiar with some chemical formulas like H_2O from earlier science courses. Likewise, they are familiar with many elemental symbols because they have studied the electron configuration of the elements using the periodic table in the last chapter.

ASSESSMENT PLANNER

Choose from the following assessment strategies.

Assess, pp. 171-172, 179-181
Assessment, pp. 166, 168, 170, 180
Portfolio, p. 180
Minute Lab, p. 168
ChemActivity, pp. 806-807
Chapter Review, pp. 182-187

CHAPTER Preview

CONTENTS

162

DISCOVERY DEMO

7-1 A Chemical Formula

Purpose: to demonstrate two elements combining in a dramatic way. The new compound is then separated and the original elements reappear

Materials: 3 g zinc dust, 2 g iodine crystals, large test tube with stopper, 21 cm^3 water, medicine dropper, funnel, filter paper, petri dish, overhead projector, 9-V battery with battery clip and wires, 2 metal paper clips.

Procedure: **1.** Place 3 g zinc dust and 2 g iodine crystals in a large test tube. Stopper, and shake to mix.
2. Place the test tube in a rack. In an operating fume hood, add 1 cm^3 of water slowly with a medicine dropper. **CAUTION:** *The reaction is exothermic; do not hold the test tube.*
3. After the reaction is complete, ask a student to carefully touch the outside of the test tube and report the obser-

Chemical Formulas

CHAPTER

7

Look at the reproduction of the music on this page. Can you make sense of the information presented? If you can read music, the strange symbols will mean something. For many other people, the cryptic entries carry no information. Musicians have evolved a system of symbols that can be used to convey a great deal of information in a small space.

Like the musician on the opposite page, chemists use a system of symbols to convey information. Early chemists developed a kind of shorthand to represent substances and chemical changes. Modern chemists continue to add to this symbolic language as they make new discoveries. In this chapter, you will learn about the symbolic representation of substances. Much of the information in this book, as well as in reference materials such as your laboratory manual, will use this same system of shorthand.

EXTENSIONS AND CONNECTIONS

Chemists and Their Work
Marie Curie, page 165
Todd Blumenkopf, page 178

Everyday Chemistry
Hazardous Materials at Home: Examine the safety of some of the materials you use every day, page 172

Careers in Chemistry
Analytical Chemist, page 174

Bridge to Medicine
Cyclopentamine for Allergies: Chemical relief for allergy sufferers, page 177

Chemistry and Technology
Making the Bone More Binding: A look at how artificial joints can be bound to existing bone, page 181

163

vation to the class. Have students observe the ZnI_2 product. **Disposal:** E

Results: Initially, the zinc and iodine combine vigorously to produce zinc iodide. There is an excess of zinc, so you should not have to worry about the production of I_2 vapor.

Questions: 1. If Zn^{2+} combines with I^-, what will be the chemical formula of zinc iodide? *ZnI_2.*
2. Does an exothermic chemical reaction release or absorb energy? *An exothermic chemical reaction releases energy.*

INTRODUCING THE CHAPTER

USING THE PHOTO

Until people trained to interpret musical symbols use their knowledge and talent to play a musical instrument, the sound cannot be appreciated. Point out to students that the chemical symbol is very similar. It is useless until a person trained to interpret the symbol combines that knowledge with chemical laboratory apparatus to make a substance that is useful and can be appreciated by others. Students know that computers (synthesizers) are useful to the musician because there are music theory rules that can be programmed into it. Chemists use computers because there are also rules in chemistry. Scientists use computer simulations before going into the laboratory. Factors such as safety, cost, and time make computer simulations a logical first step in research.

REVEALING MISCONCEPTIONS

Formula writing is a necessary skill that takes time to master. Have students do some memorization that results in greater retention and success. The quantitative aspects of formulas will be covered in Chapter 8. Chemical equation writing and balancing are covered in Chapter 9. Do not rush students into using equations if they have not mastered formula writing.

7.1 Symbols and Formulas

PREPARE

Key Concepts

The major concepts presented in this section include element symbols, chemical formulas, and oxidation numbers. Students are provided opportunities to practice writing chemical formulas.

Key Terms

chemical symbol
ion
chemical formula
oxidation number
polyatomic ion
ionic compound
molecule

Looking Ahead

Prepare the paper templates needed that represent positive and negative ions for Demonstration 7-2.

1 FOCUS

Objectives

Preview with your students the objectives listed on the student page. Students can use them as a study guide for this major section.

Discussion

Because students are familiar with the elements after studying the periodic table, ask them to imagine that they have synthesized a new element. The discoverer has the honor of naming a new element. Ask students to suggest a name for the new element. Point out that elements are often named after people, places, and observed properties.

OBJECTIVES

- demonstrate proficiency in writing chemical formulas.
- define *oxidation number* and state oxidation numbers for common monatomic ions and charges for common polyatomic ions.

7.1 Symbols and Formulas

Although there are only 109 elements, there are about ten million compounds known to chemists. It is convenient to represent elements by the use of symbols. Compounds are represented by combinations of symbols called formulas. This shorthand system improves communications among all parts of the scientific and technological community. These representations of substances are another way to help us classify these substances quickly. The ability to classify simplifies the study of the vast number of substances with which chemists work.

Names of Elements

Is there an element named for a headache? Yes, sodium. Sodium was named for one of its compounds, soda, or sodium carbonate, which was once used as a headache remedy. Soda comes from the Arabic word for headache, *suda*. The names of the elements are generally bestowed by their discoverers. However, some elements, for example gold and tin, have been known since prehistoric times, and we do not know how their names were derived.

The most common source for an element name is a property of the element. An example is nitrogen. Its name comes from the Greek words *nitron* (niter) and *genes* (to be born), meaning "niter forming." Niter was the name for naturally occurring substances that contain nitrogen. Protactinium is a radioactive element that decays to actinium. The name comes from the Greek *protos*, which means first. Protactinium comes before actinium.

After properties, the next most popular source for an element name is the place of discovery. Hafnium, for example, was discovered in Copenhagen, Denmark. The Latin name for Copenhagen was *Hafnia*.

Another source of names is the mineral in which the element was found. Lithium was found in a mineral and is named from the Greek word for stone, *lithos*. Tungsten comes from the Swedish *tung sten*, heavy stone. Boron is named for a property and a mineral. Its name is a combination of borax and carbon. Boron is

Figure 7.1 The name of the element boron (right) is taken from the mineral ***bor***ax (left) and the physical resemblance of boron to carb***on***.

DEMONSTRATION

7-2 Make it Neutral

Purpose: to illustrate the function of the oxidation number in formula writing

Materials: paper card stock (old file folders), scissors, ruler or straight edge, pencil.

Procedure: 1. Prepare several cards that have one, two, or three V-shaped notches cut into one side. These cards will represent negative ions having oxidation numbers of 1-, 2-, or 3-.
2. Prepare a second set of cards that have points sticking out from one side that will align with the V-shaped notches that have been cut into the sides of the first set of cards. Cards should have one, two, or three points protruding from the one side. These cards will represent ions that have a 1+, 2+, or 3+ oxidation numbers,

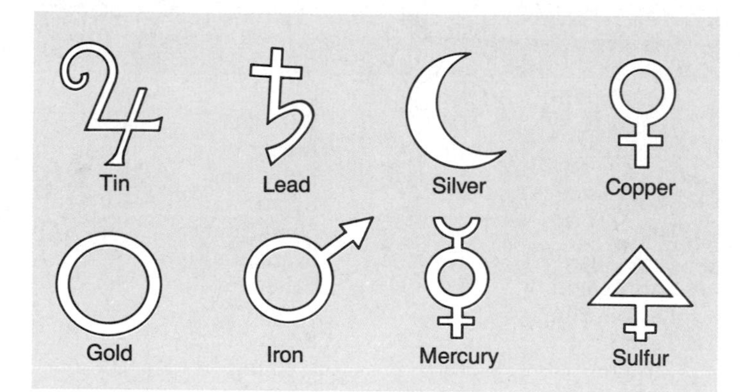

Figure 7.2 The alchemists used symbols like these to represent elements.

found in the mineral borax, and is black like carbon. The name borax is derived from the Arabic word for *glisten* because the mineral is shiny.

Finally, there are several elements named to honor a place or a person. Curium, for instance, is a radioactive element named to honor Marie and Pierre Curie, the discoverers of radium.

Symbols for Elements

The symbols for elements are derived from their names. Scientists throughout the world have agreed to represent one atom of aluminum by the symbol Al. Oxygen is given the symbol O, hydrogen H. The **chemical symbols** of the elements are shorthand representations of the elements. They take the place of the complete names of the elements.

Ancient symbols for some elements are shown in Figure 7.2. J. J. Berzelius, a Swedish chemist, is generally given credit for creating the modern symbols for elements. Berzelius proposed that all elements be given a symbol corresponding to the first letter of their names. In the case of two elements that began with the same letter, the second letter or an important letter in the name was added. In some cases, the Latin name of the element was used. Thus, the symbol for sulfur is S; selenium, Se; strontium, Sr; and sodium, Na (Latin *natrium*).

The symbols that have been agreed upon for the elements are listed in Table 7.1. Notice that they contain capital and lowercase letters. Names for elements 104 through 109 and their three-letter symbols, shown in the last column, are the result of a system adopted by the International Union of Pure and Applied Chemistry (IUPAC). In this system, Latin and Greek stems representing the atomic numbers of the elements are used for both the name and the symbol. Scientists in both the United States and Russia claim discovery of elements 104 and 105. Because these scientists cannot agree on names for these two elements, the IUPAC systematic naming was adopted for them and for all elements discovered after them.

CHEMISTS AND THEIR WORK

Marie Curie
(1867-1934)

A Nobel Prize usually caps the career of a scientist. Marie Curie won two. In 1903, she was awarded the prize in physics for her joint research with her husband Pierre on the radiation phenomena discovered by Becquerel. In 1911, she won the chemistry prize for the discovery of radium and polonium and for the study of radium compounds. Madame Curie was devoted not only to scientific research but also to the applications of the research. The use of the radioactivity of radium in the treatment of cancer is but one example of this concern.

2 TEACH

CONCEPT DEVELOPMENT

Background: Point out that no rules were followed in naming the first 103 elements. Some have names based on German, Latin, or Greek roots. Silicon is named for the Latin word *silicis*, meaning flint; rhodium is from the Greek word *rhodon*, rose, and neon is named from the German word *neos*, new. Other elements are named for geographic locations. Holmium (Stockholm), ytterbium (Ytterby, Sweden), rhenium (Rhine River, Germany), and strontium (Strontian, Scotland) are examples. The names of other elements honor scientists such as einsteinium (Albert Einstein), fermium (Enrico Fermi), and nobelium (Alfred Nobel).

CHEMISTS AND THEIR WORK
Marie Curie

Background: Marie Curie's second daughter Eve became a well-known figure in literature and wrote a biography of her mother. During World War I, Marie Curie, accompanied by her daughter Irene, took a mobile radiographic (X-ray) unit to the front lines to aid the doctors in diagnosing the injured. Marie Curie named polonium after the country of her birth, Poland.

Exploring Further: Marie Curie and her daughter Irene died of the same disease. Did radiation have anything to do with it? Possible response: Both died of leukemia, probably brought on by their years of exposure to radiation, but that is a controversial hypothesis.

respectively.

3. Inform students that the rules of the game require that all notches be filled with points. Show a card that has three notches. Ask how many cards that have one point will be required to "neutralize" the three-notched card. Students quickly see that three cards having one point will be needed. Example: NH_3, ammonia.

4. Repeat, using different cards until you are satisfied that students have gained mastery of the concept that compounds are electrically neutral.

Results: Model familiar compounds such as NaCl, H_2O, $AlCl_3$, NH_3, MgO, Al_2O_3, and $CaCl_2$.

Questions: 1. An ion with a 2+ oxidation number will react with how many negative ions that have a 1− oxidation number? *two*

2. What will be the overall charge on this compound? *0*

PROGRAM RESOURCES

Lesson Plans: Use the Lesson Plan, pp.14-15, to help you prepare for Chapter 7.

Study Guide: Use the Study Guide worksheet p. 25, for independent practice with concepts in Section 7.1.

GLENCOE TECHNOLOGY

Software

Mastering Concepts in Chemistry
Unit 4, Lesson 1
Symbols and Formulas

QUICK DEMO

An Acidic Fizz
Fill a 1-dm^3 graduated cylinder with water that has drops of universal indicator added to it. Place a piece of dry ice in the cylinder. The solid CO_2 will sink to the bottom; bubbles will rise to the top, releasing a familiar-looking fog. The color of the universal indicator will change to indicate that the water is acidic. Have a student read the label of a soft drink can. Carbonated water is listed as an ingredient. Show students the equation for the reaction: $H_2O(l) + CO_2(g) \rightarrow H_2CO_3(aq)$. Ask students to give the chemical name of each compound. **Disposal:** C

✓ ASSESSMENT

Knowledge: Ask students to look at the 36 elements in Table 7.1 that are marked with an asterisk. After students have had time to familiarize themselves with the table, give them a list of the 36 element names and have them write as many symbols as they can from memory. Reverse the procedure by giving them symbols and having them write names. Repeat this Assessment until most students are successful.

Table 7.1

Elements and Their Symbols					
Actinium	Ac	Holmium	Ho	Radon	Rn
*Aluminum	Al	*Hydrogen	H	Rhenium	Re
Americium	Am	Indium	In	Rhodium	Rh
Antimony	Sb	*Iodine	I	Rubidium	Rb
*Argon	Ar	Iridium	Ir	Ruthenium	Ru
Arsenic	As	*Iron	Fe	Samarium	Sm
Astatine	At	Krypton	Kr	Scandium	Sc
*Barium	Ba	Lanthanum	La	Selenium	Se
Berkelium	Bk	Lawrencium	Lr	*Silicon	Si
*Beryllium	Be	*Lead	Pb	*Silver	Ag
Bismuth	Bi	*Lithium	Li	*Sodium	Na
*Boron	B	Lutetium	Lu	*Strontium	Sr
*Bromine	Br	*Magnesium	Mg	*Sulfur	S
*Cadmium	Cd	*Manganese	Mn	Tantalum	Ta
*Calcium	Ca	Mendelevium	Md	Technetium	Tc
Californium	Cf	*Mercury	Hg	Tellurium	Te
*Carbon	C	Molybdenum	Mo	Terbium	Tb
Cerium	Ce	Neodymium	Nd	Thallium	Tl
Cesium	Cs	*Neon	Ne	Thorium	Th
*Chlorine	Cl	Neptunium	Np	Thulium	Tm
*Chromium	Cr	*Nickel	Ni	*Tin	Sn
*Cobalt	Co	Niobium	Nb	Titanium	Ti
*Copper	Cu	*Nitrogen	N	Tungsten	W
Curium	Cm	Nobelium	No	Unnilennium	Une
Dysprosium	Dy	Osmium	Os	Unnilhexium	Unh
Einsteinium	Es	*Oxygen	O	Unniloctium	Uno
Erbium	Er	Palladium	Pd	Unnilpentium	Unp
Europium	Eu	*Phosphorus	P	Unnilquadium	Unq
Fermium	Fm	Platinum	Pt	Unnilseptium	Uns
*Fluorine	F	Plutonium	Pu	Uranium	U
Francium	Fr	Polonium	Po	Vanadium	V
Gadolinium	Gd	*Potassium	K	Xenon	Xe
Gallium	Ga	Praseodymium	Pr	Ytterbium	Yb
Germanium	Ge	Promethium	Pm	Yttrium	Y
Gold	Au	Protactinium	Pa	*Zinc	Zn
Hafnium	Hf	Radium	Ra	Zirconium	Zr
*Helium	He				

*You should memorize those symbols marked with an asterisk as they will be used often throughout the text.

How Symbols Are Used

You saw in the last chapter that atoms can react by gaining or losing electrons. Atoms, by definition, are electrically neutral. They have equal numbers of electrons and protons. Electrons, on the other hand, are negatively charged. If an atom gains or loses electrons, it must, then, become electrically charged. An atom that has become charged is called an **ion.** Because electrons are

CONTENT BACKGROUND

History: In 1813, Jons Jakob Berzelius published a paper in which he argued that letters should be used as chemical symbols because they could be more easily written than other signs. Berzelius suggested using the first letter of the element's Latin name. He suggested that for nonmetals only the first letter be used, and for metals beginning with the same letter, the second letter be used with the first. He gave the following examples in his paper: C = carbonicum, Co = cobaltum, and Cu = cuprum.

In 1814 Berzelius wrote a paper to show how the symbols could be used to represent a "compound volume." The number of "volumes of each element" was written above the element's symbol. For example: oxidum cuprosum, CuO + SO^3 = sulfate of copper, $CuSO^4$. Although nearly all of the symbols Berzelius suggested for

negative, an atom that gains electrons becomes a negatively charged ion. An atom that loses electrons becomes a positively charged ion.

Scientists use a shorthand method of representing information about an atom or ion. Each "corner" of a symbol for an element is used to show some property of that atom or ion. Let's look at the example in Figure 7.3. The upper right-hand corner is used to show the electric charge on an ion. The lower right-hand corner is used to show the number of atoms in a formula. The upper left-hand corner is used for the mass number, or number of nucleons, in an atom. In the lower left-hand corner, the charge on the nucleus of an atom is shown. Thus, a typical sulfur nucleus or atom is represented as $^{32}_{16}S$. Fluorine would be $^{19}_{9}F$. The same system is used with subatomic particles. An electron, which has a negligible mass and a 1− charge, is represented as $_{-1}^{0}e$. An alpha particle is $^{4}_{2}He$.

Mass number | Electric charge
32 | 2−
S
16 | 2
Nuclear charge (atomic number) | Number of atoms

Figure 7.3 The diagram shows how the shorthand form for representing sulfur as $^{32}_{16}S$ is derived.

Chemical Formulas

Chemists combine symbols in chemical formulas to represent compounds. A **chemical formula** is a combination of symbols that represents the composition of a compound. A formula shows two things. It indicates the elements present in the compound and the relative number of atoms of each element in the compound. Formulas often contain numerals to indicate the ratio of elements in a compound. For example, chemists have learned from experiments that water is composed of the elements hydrogen and oxygen. It is also known that two atoms of hydrogen will react (combine chemically) with one atom of oxygen to form one molecule of water. Thus, the formula for water is written as H_2O. The small subscript, $_2$, after the H indicates that there are two atoms of hydrogen in one molecule of water. Note that there is no subscript after the oxygen. If a symbol has no subscript, it is understood that only one atom of that element is present.

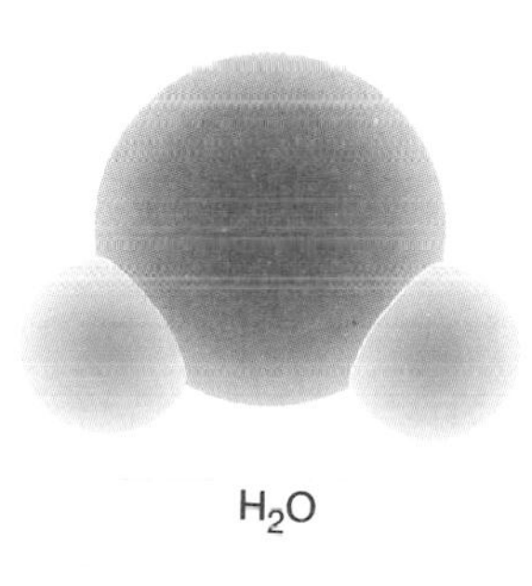

Figure 7.4 A water molecule is composed of two atoms of hydrogen and one atom of oxygen.

CULTURAL DIVERSITY

Rewriting History

Aslihan Yener, born in Turkey, moved to the United States when she was six months old. In college she planned to be a chemist, but decided to travel. She returned to Turkey and majored in art history, becoming interested in archeology. She decided to combine her talents and "to apply atomic bomb techniques to archeology . . . by analyzing the lead isotopes in mines and metals throughout the Near East."

The distinct ratios of lead isotopes to other metals in Bronze Age objects are like a fingerprint, matching the object to the mine the metal came from. Yener used this approach to analyze trace elements of lead in silver money, standard weights, and trade items.

Yener has found large Bronze Age industrial parks in the Taurus mountains—850 mines with mining tools and slag piles in one six-square mile area alone. She has also located a semisubterranean city built into the mountainside. It may have had a population 4500 years ago of 10 000, making it larger than Troy. The inhabitants pulverized tin ore, mixed it with arsenic to lower its melting temperature, and then smelted it in pottery crucibles. Evidence of a tin source in the Taurus contradicts previously-held ideas. Four thousand year old Assyrian clay tablets depict ancient Turkey as a backward area lacking tin.

elements are in use today, his method for writing chemical formulas and equations has changed. The superscript to indicate the number of atoms was replaced with a subscript. Berzelius did use coefficients in chemical equations.

By 1830 the use of symbols was widespread, but chaotic. Many scientists were using their own set of symbols. In 1834, the British Association for the Advancement of Science recommended that Berzelius's symbols be used by everyone. By the middle of the century, their use was widespread. The use of subscripts became standard practice except in France where superscripts continued to be used for another hundred years.

PROGRAM RESOURCES

Reteaching: Use the Reteaching worksheet, p. 10, to provide students another opportunity to understand the concept of "Chemical Formulas."

Laboratory Manual: Use macrolab 7 "Making Models of Compounds" as enrichment for the concepts in Chapter 7.

MINUTE LAB

Chemical Synthesis

Purpose: to show that the physical properties of a compound differ from those of the elements of which it is made

Materials: 1 g powdered sulfur, 1.75 g iron filings, test tube, laboratory burner, 400-cm^3 beaker, 300 cm^3 cold water, pair of forceps

Procedure Hints: Because SO_2 will be produced, use a fume hood if room cannot be well ventilated. **CAUTION:** *Do not allow the product to come in contact with acid; it will produce H_2S, a very poisonous gas.* Use a stained or chipped test tube since it will be destroyed. It is safe for students to handle the elements and product with their hands. Use a permanent magnet to show that the iron is magnetic before mixing and heating, but not magnetic after forming iron(II) sulfide.

Results: Initially the sulfur is yellow, and the iron is magnetic. However, the gray product is not magnetic. The equation for the reaction is Fe(cr) + S(cr) → FeS(cr).

Answers to Questions: 1. No, sulfur's yellow color has disappeared. **2.** The product is not affected by a magnet. **3.** FeS **4.** Iron(II) sulfide

✓ ASSESSMENT

Skill: Ask students to also write the formula for iron (III) sulfide (Fe_2S_3).

Chemical Synthesis

Put on an apron and goggles. Separately mass out 1.00 g of powdered sulfur and 1.75 g of iron filings. Examine each element separately, recording its physical properties of texture, color, odor, and magnetism. Then, mix the two elements on a piece of paper and pour them into a test tube. In a fume hood, use a burner to heat the test tube until the contents glow red. Immediately plunge the hot tube into a beaker of cold water. **CAUTION:** *The test tube will break.* Remove the product from the water with forceps, making sure no broken glass adheres to the product. Record its physical properties. Does the product retain the color of the sulfur? Is the product affected by a magnet? When iron reacts with sulfur, it has a 2+ oxidation number. Write the formula for the product. What is its chemical name?

Formulas for organic compounds are written according to a different set of rules. For example, the formula for acetic acid, the acid in vinegar, is written as CH_3COOH*. The actual structure of the acetic acid molecule is represented by

```
    H  O
    |  ||
 H—C—C—O—H
    |
    H
```

It shows how the atoms are joined, as well as the kind and number of atoms present. From this structural formula, you can see how the shorthand formula is derived. Other common compounds and their formulas are shown in Table 7.2.

Table 7.2

Some Common Compounds and Their Formulas

Compound	Formula	Elements
ammonia	NH_3	nitrogen, hydrogen
rust	Fe_2O_3	iron, oxygen
sucrose	$C_{12}H_{22}O_{11}$	carbon, hydrogen, oxygen
table salt	NaCl	sodium, chlorine
water	H_2O	hydrogen, oxygen

Oxidation Number

Through experiments, chemists have determined the ratios in which most elements combine. They have also learned that these ratios depend on the structure of the atoms of the elements. When an atom reacts to form an ion, the stability of an octet of electrons enables you to predict the number of electrons to be gained or lost. You can then predict the charge on the ion formed. When a single atom takes on a charge, it is called a monatomic ion. The charge on a monatomic ion is known as the **oxidation number** of the atom. Most of these charges can be verified by applying the octet rule. An ion made of more than one atom, for example, OH^-, is called a **polyatomic ion.**

Figure 7.5 Polyatomic ions, such as hydroxide (left), sulfate (center), and oxalate (right) are composed of more than one atom.

*The formula for acetic acid can also be written $HC_2H_3O_2$.

CONTENT BACKGROUND

History: Over thirty years ago, the IUPAC recommended that the system for naming elements having multiple oxidation states be changed to show a Roman numeral equivalent to the element's oxidation number. The names of chemical compounds in the literature have changed, but the labels on the bottles in the stockroom continue to be printed with both old and new nomenclature systems. Some names commonly seen on bottle labels are:

copper(I)	cuprous
copper(II)	cupric
iron(II)	ferrous
iron(III)	ferric
tin(II)	stannous
tin(IV)	stannic
mercury(I)	mercurous
mercury(II)	mercuric

Table 7.3 lists the oxidation numbers of some elements. Table 7.4 lists the charges of several polyatomic ions. You should memorize these charges. They will be important throughout your study of chemistry. You will use this information to write correct chemical formulas. Atoms and ions combine chemically in definite ratios. Oxidation numbers of elements and the charges on polyatomic ions tell us these combining ratios. The way to determine the ratio of elements in a compound is to add the charges algebraically. If the charges add up to zero, the formula for the compound has been written correctly. Not all the formulas you can write represent compounds that actually exist.

Table 7.3

Oxidation Numbers of Some Monatomic Ions*

1+	2+		
hydrogen, H^+ lithium, Li^+ potassium, K^+ silver, Ag^+ sodium, Na^+	barium, Ba^{2+} cadmium, Cd^{2+} calcium, Ca^{2+} cobalt(II), Co^{2+} copper(II), Cu^{2+} iron(II), Fe^{2+} lead(II), Pb^{2+}	magnesium, Mg^{2+} manganese(II), Mn^{2+} mercury(II), Hg^{2+} nickel(II), Ni^{2+} strontium, Sr^{2+} tin(II), Sn^{2+} zinc, Zn^{2+}	
3+	**4+**	**1−**	**2−**
aluminum, Al^{3+} chromium(III), Cr^{3+} iron(III), Fe^{3+}	lead(IV), Pb^{4+}	bromide, Br^- chloride, Cl^- fluoride, F^- hydride, H^- iodide, I^-	oxide, O^{2-} sulfide, S^{2-}

*Appendix Table A-3 lists additional monatomic ions and their charges.

Table 7.4

Charges of Common Polyatomic Ions*

1+	2+	
ammonium, NH_4^+	mercury(I), Hg_2^{2+}	
1−	**2−**	**3−**
acetate, CH_3COO^- chlorate, ClO_3^- chlorite, ClO_2^- cyanide, CN^- hydroxide, OH^- hypochlorite, ClO^- iodate, IO_3^- nitrate, NO_3^- nitrite, NO_2^- perchlorate, ClO_4^- permanganate, MnO_4^-	carbonate, CO_3^{2-} chromate, CrO_4^{2-} dichromate, $Cr_2O_7^{2-}$ oxalate, $C_2O_4^{2-}$ peroxide, O_2^{2-} silicate, SiO_3^{2-} sulfate, SO_4^{2-} sulfite, SO_3^{2-} tartrate, $C_4H_4O_6^{2-}$ tetraborate, $B_4O_7^{2-}$ thiosulfate, $S_2O_3^{2-}$	phosphate, PO_4^{3-}

*Appendix Table A-4 lists additional polyatomic ions and their charges.

CONCEPT DEVELOPMENT

Computer Demo: *Name The Ions,* AP 301, PC2201, and Valence Drill, AP305, PC2201, Project SERAPHIM.

MAKING CONNECTIONS

Daily Life
Ask students if they like to read cereal box labels while eating breakfast. Encourage all students to read labels and bring to class those that list compounds that they should recognize.

Background: In the previous two chapters, students have gained a basic knowledge about atoms. At this point, review the concepts that an atom consists of a positive nucleus surrounded by negative electrons, the nucleus of one atom attracts the electrons of another atom, and electrons can be transferred or shared. Point out that the accounting system used by chemists is oxidation number. Students often find it easier to write formulas if they treat the oxidation number from the tables as given information and use it to solve an algebraic problem. It is recommended that the charges for some ions be memorized. Daily quizzes covering a few ions at a time are helpful.

PROGRAM RESOURCES

Transparency Master: Use the Transparency master and worksheet, pp. 33-34, for guided practice in the concept of "Charges of Common Ions."

Color Transparency: Use Color Transparency 17 to focus on the concept of "Charges of Common Ions."

Applying Scientific Methods in Chemistry: Use the worksheet, "Miller's Organic Compounds," pp. 22-23, to help students apply and extend their lab skills.

Correcting Errors

Avoid using the technique that has students cross oxidation numbers to get subscripts. This shortcut results in too many errors.

PRACTICE

Guided Practice: Guide students through the Sample Problems, then have them work in class on Practice Problems 1a, 1b, 2a, and 2b that follow.

Independent Practice: Your homework or classroom assignment can include the remaining Practice Problems 1-3, and Review and Practice 25-26, 33, 35, 37, 39, 41, 43, 45-46, and 56 from the end of the chapter.

✓ ASSESSMENT

Skill: Using Tables 7.3 and 7.4 to determine the answers to Review and Practice questions 25 and 26 on page 183, have students work in cooperative learning groups.

GLENCOE TECHNOLOGY

CD-ROM

Chemistry: Concepts and Applications

Exploration: *Determining a Formula*

Interactive Experiment: *Form an Ionic Compound*

Videodisc

Chemistry: Concepts and Applications

Oxidation States of Vanadium

Disc 1, Side 2, Ch. 8

Show this video to reinforce the concept of variable oxidation states.

Also available on CD-ROM.

Common table salt, NaCl, is made from sodium, Na, and chlorine, Cl. Table 7.3 shows a 1+ charge for sodium ions and a 1− charge for chloride ions.

$$Na^{+}Cl^{-}$$

Adding these charges, you see that 1 + (1−) = 0. Therefore, the formula for salt is NaCl. This formula indicates that a one-to-one ratio exists between sodium ions and chloride ions in a crystal of salt. Sodium chloride is an **ionic compound;** that is, it is composed of ions. Note that in the formula for NaCl the positive ion is written first. This is true of all ionic compounds.

Elements also combine in another way. Neutral atoms combine to form neutral particles called **molecules.** Compounds formed from molecules are called molecular compounds. In Chapter 12, you will study ionic and molecular substances.

The element chlorine is a gas composed of molecules that are diatomic. A diatomic molecule is made of two atoms of the same element. One chlorine molecule contains two chlorine atoms. Chlorine gas is represented by the formula Cl_2. Six other common elements also occur as diatomic molecules. Hydrogen (H_2), nitrogen (N_2), oxygen (O_2), and fluorine (F_2) are diatomic gases under normal conditions. Bromine (Br_2) is a gas above 58.8°C. Iodine (I_2) is a gas above 184°C. Table 7.5 lists the diatomic elements.

The formula of a substance represents a specific amount of the substance. If a substance is composed of molecules, the formula represents one molecule. If a substance is ionic, the formula represents the ions in their lowest combining ratio.

Table 7.5

Diatomic Elements

Name	Formula
bromine	Br_2
chlorine	Cl_2
fluorine	F_2
hydrogen	H_2
iodine	I_2
nitrogen	N_2
oxygen	O_2

Figure 7.6 Iron pyrite (fool's gold) is composed of the elements iron and sulfur. The formula for this compound is FeS_2.

SAMPLE PROBLEMS

1. Writing a Simple Formula

Write a formula for a compound of calcium and bromine.

Solving Process:

Using Table 7.3, you see the oxidation states of calcium and bromide are as follows:

$$Ca^{2+} \qquad Br^{-}$$

Chemistry Journal

The Perfect Team

Ask students to write a journal entry in which they consider a sodium atom to be analogous to a sports team with one extra player and a chlorine atom to be analogous to a team with one missing player. The transfer of one electron (player) from sodium to chlorine makes both atoms (teams) more stable. Of course, the atoms become ions by losing and gaining electrons.

*inter*NET CONNECTION

A gopher-driven database of the periodic table is available on the Internet.

gopher://ucsbuxa.ucsb.edu:3001/11/.Sciences/.Chemistry/.periodic.table

Because the sum of the charges must equal zero, two Br^- are needed to balance the one Ca^{2+}. The correct formula is $CaBr_2$.

2. Writing a More Complex Formula

Write the formula for a compound made from the aluminum ion and the sulfate ion.

Solving Process:

Using Tables 7.3 and 7.4, you see that the oxidation states of aluminum and the sulfate ion are as follows:

$$Al^{3+} \qquad SO_4^{2-}$$

To make the sum of the charges equal zero, you must find the least common multiple of 3 and 2. The least common multiple is 6. Because 6/3 = 2 and 6/2 = 3, it is necessary to have two Al^{3+} and three SO_4^{2-} in the compound to maintain neutrality. Writing two aluminum ions in the formula is simple.

$$Al_2$$

For the sulfate, the entire polyatomic ion must be placed in parentheses to indicate that three sulfate ions are required.

$$(SO_4)_3$$

Thus, aluminum sulfate has the formula $Al_2(SO_4)_3$. Parentheses are used in a formula only when you are expressing multiples of a polyatomic ion. If only one sulfate ion were needed in writing a formula, parentheses would not be used. For example, the formula for the compound made from calcium, Ca^{2+}, and the sulfate ion, SO_4^{2-}, is $CaSO_4$.

PRACTICE PROBLEMS

1. Write the formula for each of the following compounds.
   - a. calcium chloride
   - b. sodium cyanide
   - c. magnesium oxide
   - d. barium oxide
   - e. sodium fluoride
   - f. aluminum nitrate
   - g. zinc iodide
   - h. cobalt(II) carbonate
2. Write the formula for the compound made from each of the following pairs.
   - a. silver and fluorine
   - b. nickel(II) and sulfur
   - c. chromium(III) and bromine
   - d. lead(II) and phosphate ion
   - e. ammonium and oxalate ions
   - f. strontium and iodine
   - g. lithium and oxygen
3. Write the formula for each of the following compounds.
   - a. potassium hydride
   - b. mercury(II) cyanide
   - c. zinc tartrate
   - d. cadmium silicate
   - e. ammonium dichromate
   - f. lead(II) nitrate
   - g. copper(II) perchlorate
   - h. sodium tetraborate

PROBLEM SOLVING HINT

Becoming proficient in formula writing will help you throughout your study of chemistry.

Figure 7.7 The elements chlorine (yellow-green), bromine (red-brown), and iodine (purple) exist as diatomic molecules.

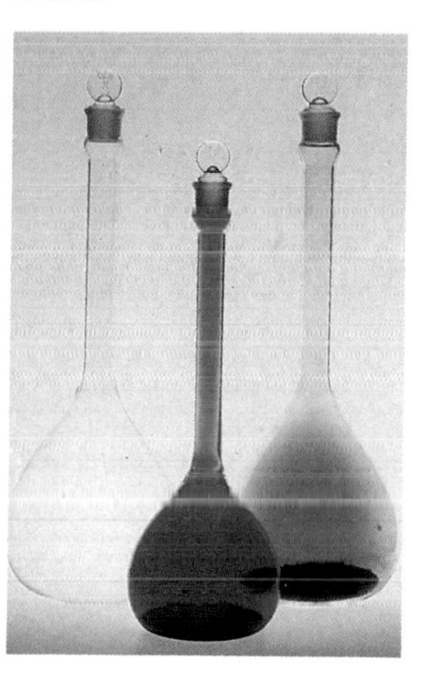

Answers to Practice Problems

1. a. $CaCl_2$, **b.** NaCN, **c.** MgO, **d.** BaO, **e.** NaF, **f.** $Al(NO_3)_3$, **g.** ZnI_2, **h.** $CoCO_3$

2. a. AgF, **b.** NiS, **c.** $CrBr_3$, **d.** $Pb_3(PO_4)_2$, **e.** $(NH_4)_2C_2O_4$, **f.** SrI_2, **g.** Li_2O

3. a. KH, **b.** $Hg(CN)_2$, **c.** $ZnC_4H_4O_6$, **d.** $CdSiO_3$, **e.** $(NH_4)_2Cr_2O_7$, **f.** $Pb(NO_3)_2$, **g.** $Cu(ClO_4)_2$, **h.** $Na_2B_4O_7$

3 ASSESS

CHECK FOR UNDERSTANDING

Ask students to answer the Concept Review problems.

RETEACHING

Use Tables 7.3 and 7.4 to have students make flash cards each with the name of an ion on one side and its symbol and oxidation number on the other side.

Cooperative Learning: Refer to pp. 28T-29T to select a teaching strategy to use with this activity.

EXTENSION

Ask a student volunteer to report on the differences between compounds known as Daltonides and those known as Bertholides. Possible response: *Daltonides are compounds of definite composition. Bertholides are compounds having variable composition. They are possible because of the occurrence of interstitial atoms in crystal lattices. Transitional metal hydrides and copper(I) sulfide are well known Bertholides.*

PROGRAM RESOURCES

Chemistry and Industry: Use the worksheets, pp.11-14, "Paper: Just What Goes Into Each Piece?" as an extension to the concepts presented in Section 7.1.

EVERYDAY CHEMISTRY

Hazardous Materials at Home

Background: This feature is written to make students aware of toxic wastes in the home.

Teaching This Feature: Ask students to survey their homes, listing chemicals that they find. Use the results of their surveys to promote discussion of the feature.

Connection to Environmental Health and Safety: Discuss with students where they can properly recycle chemicals found in the home.

Answers to
Exploring Further:

1. Student surveys will vary.

2. Many students will suggest that using less is better. Follow directions carefully and use only as much chemical as is necessary.

Answers to
Concept Review

4. a. $BaSO_4$, **b.** CaS, **c.** $Mg_3(PO_4)_2$, **d.** $SrBr_2$, **e.** $Cr(C_2H_3O_2)_3$ or $Cr(CH_3COO)_3$

5. 82; 82; 125

6. 2+; Cu^{2+}

7. Na_2SeO_4

4 CLOSE

Discussion: Encourage students to become proficient in formula writing by showing them examples of how this skill will be used throughout their study of chemistry. Show an equation of a simple and understandable chemical reaction, such as an acid being neutralized by a base: $HCl + NaOH \rightarrow NaCl$ and H_2O.

EVERYDAY CHEMISTRY

Hazardous Materials at Home

Some of the most hazardous materials, as well as those that are most polluting to the environment, are found right in your home. Hazardous materials are those that are poisonous, corrosive, or flammable. Typical poisonous substances are insecticides, some medicines, antifreeze, and rubbing alcohol. Corrosive compounds destroy tissue, metals, and other materials. Some corrosives are toilet and shower cleaners, bleach, battery acid, and oven cleaners. Flammable compounds, those that burn easily, include gasoline, lighter fluids, and some aerosols.

These and other materials are hazardous to the health and safety of people and pets in the home. For example, the common plant killer 2,4-D has been shown to cause lymphoma in dogs. The dogs pick up the chemical while in the yard and then lick it off. A single quart of oil poured into the ground can contaminate a water supply. Some propellants from aerosol cans escape into the atmosphere and contribute to the destruction of the ozone layer.

It is possible to use alternatives to some of these materials. For example, in place of the window-cleaning products that contain phosphates or ammonia, water containing vinegar can be used. Vinegar is simply a mild solution of acetic acid and is nontoxic. Products that are packaged as pump sprays or are sold as lotions can replace those that come only in harmful aerosol cans.

Many communities are starting to schedule regular collections of toxic household materials to make sure they are disposed of properly. The complete removal of hazardous materials from the home will probably not be accomplished, but where changes can be made, they should be.

Exploring Further

1. Check the shelves in your home to find out what cleaning products are there. Make a list of the active ingredients in each.

2. Aside from the alternatives given above, what other ideas do you have for reducing the use of hazardous materials?

7.1 *CONCEPT REVIEW*

4. Write formulas for the following.
- **a.** barium sulfate
- **b.** calcium sulfide
- **c.** magnesium phosphate
- **d.** strontium bromide
- **e.** chromium(III) acetate

5. How many protons are there in an atom represented by the symbol $^{207}_{82}Pb$? How many electrons? How many neutrons?

6. What is the charge on an atom of copper that has lost two electrons? How is this ion represented?

7. Apply An insecticide lists sodium selenate as one of its ingredients. What is the formula for sodium selenate?

DEMONSTRATION

7-3 Not Identical Twins

Purpose: to demonstrate that atoms of the same element having different oxidation states have different properties

Materials: 10-g display samples of iron(II) sulfate and iron(III) sulfate, copper(I) oxide and copper(II) oxide, distilled water, 4 watch glasses

Procedure: 1. Place samples of copper and iron compounds having different oxidation numbers on watch glasses.
2. Allow students to observe but not touch the samples.
3. In a small beaker, dissolve a scoop of each chemical in a similar amount of distilled water. Have students note the color of the solution and if the compound is soluble. **Disposal:** B

Results: Students will notice a color difference. The solubility behavior is

7.2 Nomenclature

There are many times when chemists prefer to describe a compound by a name rather than a formula. The rules for naming compounds are called the *nomenclature* of chemistry. Nomenclature can also refer to the names themselves. The rules have been adopted internationally under the sponsorship of IUPAC. There are several volumes of such rules, but fortunately, just a few rules are needed for the compounds discussed in this book.

OBJECTIVES

- demonstrate proficiency in naming chemical compounds.
- distinguish between molecular and empirical formulas.
- demonstrate the use of coefficients to represent the number of formula units of a substance.

Naming Compounds

Unfortunately, some compounds have both a common name and a chemical name. For example, $NaHCO_3$ is commonly known as baking soda. However, its chemical name is sodium hydrogen carbonate. We will use a systematic method of naming practically all the compounds in this book. The names of only a few compounds, particularly acids, will not be included in this system.

Naming Binary Inorganic Compounds

Compounds containing only two elements are called **binary compounds.** To name a binary compound, first write the name of the element having a positive charge. Then add the name of the negative element. The name of the negative element must be modified to end in *-ide*. For example, the compound formed by aluminum (Al^{3+}) and nitrogen (N^{3-}), with the formula AlN, is named aluminum nitride.

Table 7.6

Formulas and Names of Some Binary Compounds

Formula	Name
Al_2S_3	aluminum sulfide
$CaBr_2$	calcium bromide
H_2O	hydrogen oxide (water)
H_2Se	hydrogen selenide

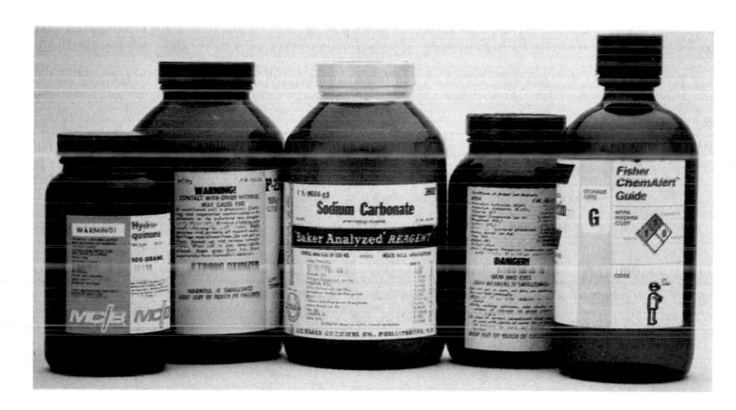

Figure 7.8 It is important to know the names and formulas of chemical compounds in order to distinguish the many different substances used in the lab and the hazards associated with them.

the same.

Questions: 1. Are the colors of the iron(II) and the iron(III) compounds the same? *No.*

2. Is the solubility of the copper(I) and copper(II) compound the same? *Yes.*

7.2 Nomenclature

PREPARE

Key Concepts

The major concepts presented in this section include naming inorganic compounds and simple organic compounds, differentiating between molecular and empirical formulas, and the use of coefficients.

Key Terms

binary compound
molecular formula
empirical formula
formula unit

1 FOCUS

Objectives

Preview with students the objectives listed on the student page. Students can use them as a study guide for this major section.

Discussion

Show the students a copy of the *Handbook of Chemistry and Physics*. Have the students observe the contents of the pages as you leaf through the book. Have students conjecture on the number of compounds in existence. Lead them to realize that there are more than ten million. Then have students discuss why rules for naming compounds must be followed by chemists in all countries. Remind students that science is an organized body of knowledge and the rules of nomenclature are a means of organization.

PROGRAM RESOURCES

Transparency Master: Use the Transparency master and worksheet, pp. 35-36, for guided practice in the concept "Naming Compounds."

Color Transparency: Use Color Transparency 18 to focus on the concept of "Naming Compounds."

Study Guide: Use the Study Guide worksheet, p. 26, for independent practice with concepts in Section 7.2.

2 TEACH

CONCEPT DEVELOPMENT

Background: Note that the text does not present the grid system for formula writing or naming compounds. This method infers that all combinations are possible which is not actually the case. Only compounds that actually do exist are used in the text.

• The term polyatomic ion is used throughout this text in place of the term radical. The term radical is only used for the free radical in organic chemistry.

MAKING CONNECTIONS

Daily Life

You can ask students to look through their kitchen and bathroom cabinets to find a product that lists its chemical ingredients. Have students use a chemical dictionary, a handbook of physics and chemistry, or the *Merck Index* to find the formula of each ingredient. Possible answers could be aspirin, $C_9H_8O_4$, ascorbic acid, $C_6H_8O_6$, or vitamin A, $C_{20}H_{29}OH$.

CAREERS IN CHEMISTRY

Analytical Chemist

Background: The analytical chemist studied instrumental analysis, quantitative and qualitative analysis in college. Some analytical chemists specialize as forensic chemists and work in police crime labs. Others specialize in environmental chemical analysis and work for government regulatory agencies such as the EPA.

For More Information: Have students write to the American Chemical Society, 1155 Sixteenth Street, N.W., Washington, DC 20036.

GLENCOE TECHNOLOGY

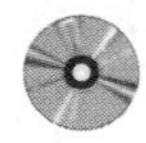

Videodisc

Chemistry: Concepts and Applications

Variable Oxidation States

Disc 1, Side 1, Ch. 17

Show this video of Demo 7-4.

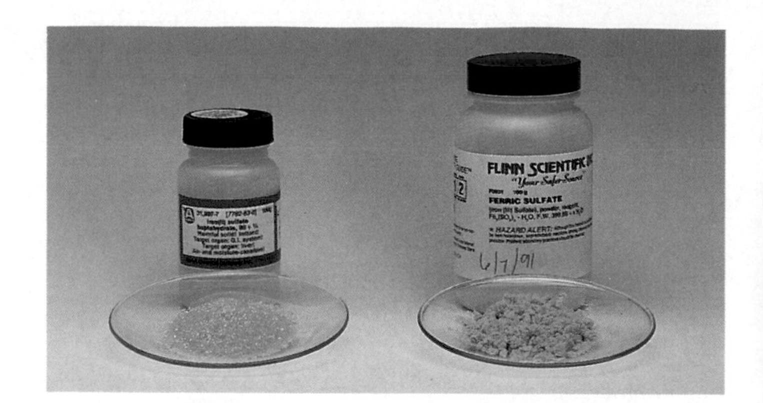

Figure 7.9 Iron exists in the 2+ and 3+ oxidation states and forms two compounds with the sulfate ion. Iron(II) sulfate (left) is blue-green and iron(III) sulfate (right) is yellow-orange. Their properties are different.

In looking at Table 7.3, you see that some elements, such as iron, have more than one possible charge. Therefore, they may form more than one compound with an element. The elements nitrogen and oxygen together form five binary compounds.

Chemists must have a way of distinguishing the names of these compounds. They tell the difference by writing the oxidation number of the positively charged element after the name of that element. The oxidation number is written as a Roman numeral in parentheses. Examples of some of these compounds are listed in Table 7.7.

Table 7.7

Formulas and Names of Some Binary Molecular Compounds Having Variable Oxidation States

Formula	Name	Formula	Name
N_2O	nitrogen(I) oxide	Cu_2S	copper(I) sulfide
NO	nitrogen(II) oxide	CuS	copper(II) sulfide
N_2O_3	nitrogen(III) oxide	SO_2	sulfur(IV) oxide
NO_2	nitrogen(IV) oxide	SO_3	sulfur(VI) oxide
N_2O_5	nitrogen(V) oxide		

CAREERS IN CHEMISTRY

Analytical Chemist

Analytical chemists are concerned chiefly with identifying the type and quantity of substances present in a material. Quality control in such fields as foods, pharmaceuticals, and safe water is one of the jobs performed by some analytical chemists. Others develop new techniques for analyzing materials. Analytical chemists may also investigate the chemical and physical properties of substances.

There are many compounds that have been named by an older system in which prefixes indicate the number of atoms present. These names have been used for so long that these more common names are usually used. Examples are listed in Table 7.8.

Table 7.8

Formulas and Common Names of Some Binary Molecular Compounds

Formula	Common Name	Formula	Common Name
CS_2	carbon disulfide	SF_6	sulfur hexafluoride
CO	carbon monoxide	SO_2	sulfur dioxide
CO_2	carbon dioxide	SO_3	sulfur trioxide
CCl_4	carbon tetrachloride		

DEMONSTRATION

7-4 The Roman Numeral

Purpose: to demonstrate the importance of the use of Roman numerals in naming compounds

Materials: 8 test tubes, 4 medicine droppers, 2 g $FeNH_4(SO_4)_2{\cdot}12H_2O$, 2 g $Fe(NH_4)_2(SO_4)_2{\cdot}6H_2O$, 0.5 g potassium thiocyanate, 0.5 g tannic acid (or very strongly brewed tea), 0.5 g sodium hydrogen sulfite, 0.5 g potassium hexacyanoferrate(II), 450 cm^3 water.

Procedure: 1. Separately prepare solutions of iron(III) ammonium sulfate (2 g $FeNH_4(SO_4)_2{\cdot}12H_2O$ in 200 cm^3 H_2O) and iron(II) ammonium sulfate (2 g $Fe(NH_4)_2(SO_4)_2{\cdot}6H_2O$ in 200 cm^3 H_2O).

2. Label 4 test tubes Fe^{2+} and 4 test tubes Fe^{3+}. Half-fill each test tube with the appropriate solution.

3. Place some crystals of each com-

Not all compounds ending in *-ide* are binary. Notice in Table 7.4 that the names of some polyatomic ions end in *-ide:* OH^- (hydroxide) and CN^- (cyanide).

Table 7.9

Numerical Prefixes	
Prefix	Number of atoms
mono-	1
di-	2
tri-	3
tetra-	4
penta-	5
hexa-	6
hepta-	7
octa-	8

SAMPLE PROBLEM

Naming Binary Compounds

Name the compound BeI_2.

Solving Process:

Be is the symbol for beryllium. I is the symbol for iodine. The positive part of the compound retains its name unchanged, beryllium. The ending of the negative part of the compound is changed to *-ide,* iodide. The name of the compound is beryllium iodide.

PRACTICE PROBLEMS

8. Name the following binary compounds.
a. BaS **c.** Mg_3N_2 **e.** ZnF_2
b. BiI_3 **d.** $PbBr_2$

9. Name the following binary compounds.
a. CaH_2 **c.** CaS **e.** $CoBr_2$
b. Na_3P **d.** TlI

Naming Other Inorganic Compounds

For naming compounds containing more than two elements, several rules apply. The simplest of these compounds is formed from one element and a polyatomic ion or from two polyatomic ions. These compounds are named in the same way as binary compounds. However, the ending of the name of the polyatomic ion is not changed. An example is $AlPO_4$, which is named aluminum phosphate. Other examples are listed in Table 7.10.

Table 7.10

Formulas and Names for Some Compounds Containing Polyatomic Ions

Formula	Name	Formula	Name
$AlAsO_4$	aluminum arsenate	$CuSO_4$	copper(II) sulfate
$(NH_4)_2SO_4$	ammonium sulfate	$Ni(OH)_2$	nickel(II) hydroxide
$Cr_2(C_2O_4)_3$	chromium(III) oxalate	$ZnCO_3$	zinc carbonate

Other rules for naming compounds and writing formulas will be discussed when the need arises. The names of the common acids, for example, do not normally follow these rules. Table 7.11 lists names and formulas for acids that you should memorize.

CONCEPT DEVELOPMENT

Using the Table: Encourage students to learn the numerical prefixes that are commonly used in chemistry. Table 7.9 organizes these prefixes for easier learning.

Correcting Errors

The subscript in a correctly-written binary formula can provide students with a clue to the oxidation number of an unfamiliar element if the oxidation number of one of the elements is known. For example, in $FeCl_2$ the oxidation number of iron is unknown, but that of the chloride ion is known to be 1–. The subscript of 2 on the chloride indicates that the iron is iron(II).

PRACTICE

Guided Practice: Guide students through the Sample Problem, then have them work in class on Practice Problem 8.

Independent Practice: Your homework or classroom assignment can include the remaining Practice Problem 9, and Review and Practice 27 and 30 from the end of the chapter.

Answers to
Practice Problems
8. a. barium sulfide, **b.** bismuth(III) iodide, **c.** magnesium nitride, **d.** lead(II) bromide, **e.** zinc fluoride
9. a. calcium hydride, **b.** sodium phosphide, **c.** calcium sulfide, **d.** thallium (I) iodide, **e.** cobalt(II) bromide

pound in a petri dish and allow the students to inspect them.
4. Prepare solutions of the following 4 compounds by adding 0.5 g of each to 10 cm^3 H_2O: potassium thiocyanate, tannic acid, sodium hydrogen sulfite, and potassium hexacyanoferrate(II). **CAUTION:** *These solutions are very toxic.*
5. Use a clean dropper to add each of these solutions to one of the four test tubes containing the ion Fe^{2+} and one of the four containing the Fe^{3+} ion.

Disposal: E
Results:

Reagent	$[Fe^{2+}]$	$[Fe^{3+}]$
KSCN	no change	deep red
$C_{76}H_{52}O_{46}$	black	blue
$NaHSO_3$	yellow	brownish orange
$K_4Fe(CN)_6 \cdot 3H_2O$	light blue	dark blue

Demonstration 7-4 illustrates that ions of different oxidation states have different chemical behaviors.

Questions: 1. Do the four reagents react the same or differently with the two forms of iron? *differently*
2. Can the two different compounds that form when iron(II) and iron(III) react with chloride both be named iron chloride? Explain your answer. *No, the same name would be applied to two different compounds having different properties, $FeCl_2$ and $FeCl_3$.*

PRACTICE

Guided Practice: Guide students through the Sample Problem on page 177, then have them work with a partner in class on Practice Problems 10-14.

Independent Practice: Your homework or classroom assignment can include Review and Practice 34, 36, 38, 40, 42, 44, 47-48, 54-55, and 57-59 from the end of the chapter.

Answers to
Practice Problems

10. a. barium chloride, **b.** sodium bromide, **c.** aluminum fluoride, **d.** lithium carbonate, **e.** potassium chloride, **f.** mercury(II) iodide, **g.** zinc nitrate, **h.** barium hydroxide

11. a. diphosphorus pentoxide
b. phosphorus pentachloride
c. sulfur hexafluoride
d. phosphorus trichloride

12. a. ammonium nitrate, **b.** acetic acid or hydrogen acetate, **c.** sodium phosphate, **d.** hydrochloric acid or hydrogen chloride, **e.** nitric acid or hydrogen nitrate, **f.** copper(II) acetate, **g.** potassium oxide, **h.** sulfuric acid or hydrogen sulfate

13. a. 4, **b.** 7, **c.** 3, **d.** 9, **e.** 8, **f.** 5

14. a. hexane, **b.** cyclobutane

CHEMISTS AND THEIR WORK
Todd Blumenkopf

Background: Most of the adaptations that were made in Blumenkopf's laboratory were designed by him. His original ideas have been copied to assist other chemists who use wheel chairs.

Exploring Further: Ask students what ways they can think of to make their chemistry laboratory more accessible to a science student who uses a wheel chair? Possible response: Construct a lab station that can be raised and lowered. Move the sink and the water and gas outlets closer to the edge of the table. Widen the entrance to the room. Lower the paper towel dispenser and safety equipment so it will be within reach.

CHEMISTS AND THEIR WORK

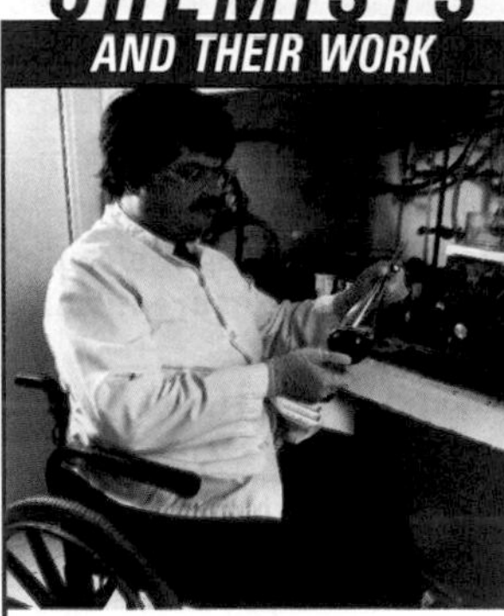

Todd Blumenkopf
(1956-)

Todd Blumenkopf is a senior research chemist at Burroughs Wellcome Co., a research-based pharmaceutical company. Blumenkopf's job involves designing and synthesizing molecules intended for therapeutic purposes. It is challenging work—more challenging when one realizes that Blumenkopf is a wheelchair user. Through adaptation of the laboratory equipment and fixtures, he has been able to carry out important work in chemistry. Blumenkopf is head of the American Chemical Society's Committee on Chemists with Disabilities and is preparing a book profiling disabled scientists.

PRACTICE PROBLEMS

10. Write the name of each of these compounds.
a. $BaCl_2$ **b.** $NaBr$ **c.** AlF_3 **d.** Li_2CO_3 **e.** KCl **f.** HgI_2 **g.** $Zn(NO_3)_2$ **h.** $Ba(OH)_2$

11. Use inorganic numerical prefixes to name each of the following compounds.
a. P_2O_5 **b.** PCl_5 **c.** SF_6 **d.** PCl_3

12. Name each of the following compounds.
a. NH_4NO_3 **b.** CH_3COOH **c.** Na_3PO_4 **d.** HCl **e.** HNO_3 **f.** $Cu(CH_3COO)_2$ **g.** K_2O **h.** H_2SO_4

13. Give the number of carbon atoms in each of the following hydrocarbons.
a. butane **b.** heptane **c.** cyclopropane **d.** nonane **e.** octane **f.** cyclopentane

14. Name these organic compounds.

a.
```
  H H H H H H
  | | | | | |
H-C-C-C-C-C-C-H
  | | | | | |
  H H H H H H
```

b.
```
  H H
  | |
H-C-C-H
  | |
H-C-C-H
  | |
  H H
```

Molecular and Empirical Formulas

The formulas for compounds that exist as molecules are called **molecular formulas.** For instance, one compound of hydrogen and oxygen is hydrogen peroxide (H_2O_2). The formula H_2O_2 is a molecular formula because one molecule of hydrogen peroxide contains two atoms of hydrogen and two atoms of oxygen. However, chemists also use another kind of formula. The atomic ratio of hydrogen to oxygen in hydrogen peroxide is one to one. Therefore, the simplest formula that would indicate the ratio between hydrogen and oxygen is HO. This simplest formula is called an **empirical formula.** As another example, both benzene (C_6H_6) and ethyne (C_2H_2) have the same empirical formula, CH. Chem-

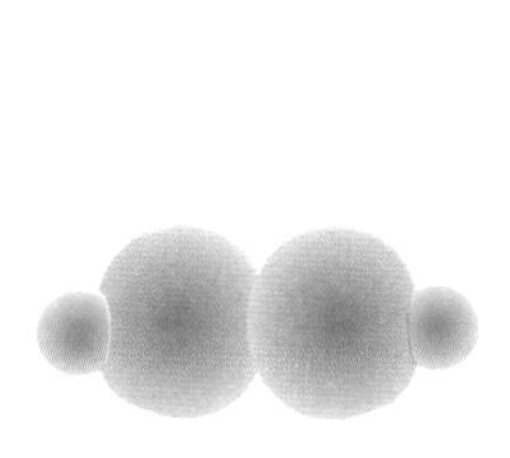

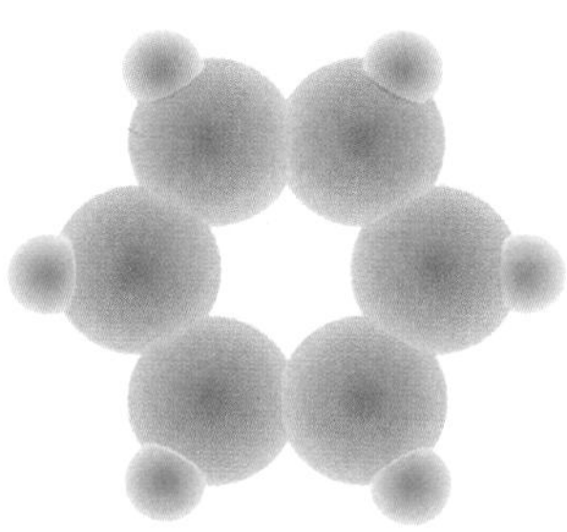

Figure 7.11 Benzene (C_6H_6) on the right and ethyne (C_2H_2) on the left have the same empirical formula (CH).

CONTENT BACKGROUND

Empirical Formulas: Students are introduced to the terms molecular formula and empirical formula in Chapter 7. In this way, students are already familiar with a working definition of the terms when they are asked to apply their mathematical problem-solving skills in Chapter 8, Section 2. In 8.2, simulated laboratory data will be presented and students will learn how to first determine the empirical formula, then use it to determine the actual molecular formula. By using two chapters to completely develop the concept, the concept density is reduced for the student.

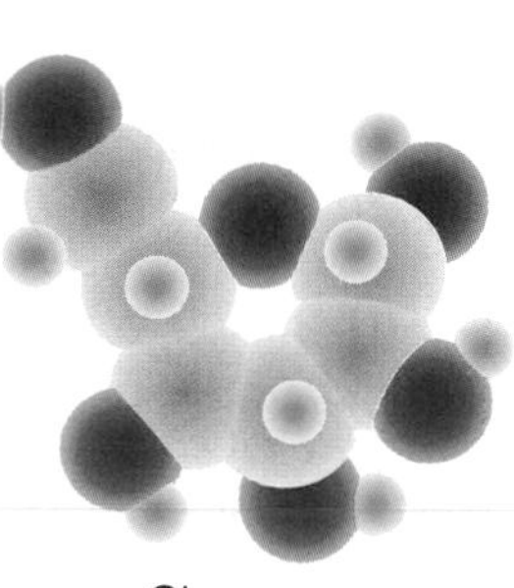

Glucose
$C_6H_{12}O_6$

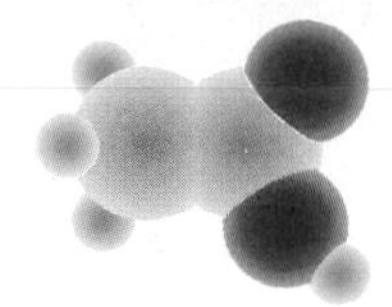

Acetic Acid
CH_3COOH

Figure 7.12 Glucose (left) and acetic acid (right) have the same empirical formula, CH_2O, but they look very different and have different molecular formulas.

ists can determine the empirical formula of an unknown substance through analysis. The empirical formula may then be used to help identify the molecular formula of the substance.

For many substances, the empirical formula is the only formula possible. The formulas for ionic compounds are almost all empirical formulas. The molecular formula of a compound is always some whole-number multiple of the empirical formula. In the above example, if we multiply CH by two, we have the formula for ethyne, C_2H_2. If we multiply by six, we have the formula for benzene, C_6H_6. The calculation of empirical formulas will be presented in Chapter 8.

Table 7.13

Molecular and Empirical Formulas for Two Common Substances

	Glucose	Acetic Acid
Empirical Formula	CH_2O	CH_2O
Molecular Formula	$C_6H_{12}O_6$	CH_3COOH
Use	Food supplement	Making plastics
Physical Properties:		
State at 25°C	Colorless solid	Colorless liquid
Molecular mass	180.16 u	60.05 u
Melting point	146°C	16.7°C
Boiling point	Decomposes	118°C

Coefficients

The formula of a compound represents a definite amount of that compound. This amount may be called a **formula unit.** One molecule of water is represented by H_2O. How do we represent two molecules of water? You use the same system as you would use in mathematics, that is, coefficients. When you wish to represent

CONCEPT DEVELOPMENT

Background: Have the students determine the ratio of boys and girls in the class. Ask them if this indicates the number of boys and girls in the class. Point out that an empirical formula represents the ratio of different atoms in a compound. The molecular formula indicates the actual number of different atoms in a molecule of the compound.

3 ASSESS

CHECK FOR UNDERSTANDING

Ask students to answer the Concept Review problems, and Review and Practice Problems 49-51 from the end of the chapter.

RETEACHING

Over 100 additional problems are available in Chapter 7, sections 5, 6 and 7, of *Solving Problems in Chemistry.*

Computer Demo, *Naming,* AP303, (see page 176) is especially good at providing a fun way to reteach nomenclature.

EXTENSION

Have a student research compounds such as sodium peroxide, Na_2O_2, and hydrogen peroxide, H_2O_2, to determine the oxidation number of each element in the compounds. Ask the student to determine what the molecular and empirical formula of each compound is. Possible response: H 1+, O 1− or O_2^{2-}, Na 1+. The empirical formulas of the compounds are HO and NaO.

PROGRAM RESOURCES

Enrichment: Use the worksheet "Chemical Formulas," pp. 13-14, to challenge more able students.

Correcting Errors

Remind students to look for the least common multiple of the subscripts in solving empirical formula problems. They can then use this number to divide all of the subscripts.

PRACTICE

Guided Practice: Guide students through the Sample Problems 1 and 2, then have them work in class on Practice Problems 15 and 16.

Independent Practice: Your homework or classroom assignment can include Review and Practice 49-51, Concept Mastery 68, Application 71-75, and Critical Thinking 86 from the end of the chapter.

Answers to
Practice Problems

15. a. NO_2, **b.** $C_3H_4O_3$, **c.** CH_3, **d.** CH_4, **e.** HgI, **f.** C_4H_9

16. a. 1, **b.** 3, **c.** 2, **d.** 4, **e.** 6, **f.** 5, **g.** 3, **h.** 3

Answers to
Concept Review

17. a. barium tartrate, **b.** calcium silicate, **c.** magnesium sulfide, **d.** sodium iodate, **e.** cobalt(II) sulfate

18. a. mercury(II) nitrate, **b.** aluminum chlorate, **c.** cobalt(II) sulfide, **d.** calcium tartrate, **e.** magnesium silicate

✓ ASSESSMENT

Portfolio: Suggest to students that they should include in their portfolio all of the homework sheets that contain the answers to exercises in which they have written chemical formulas or the names of chemical formulas. Students will need to do at least one hundred if they are to achieve mastery of this essential skill. The practice problems and chapter review provide this needed reinforcement.

two x, you write $2x$. When you wish to represent two molecules of water, you write $2H_2O$. Each formula unit of table salt consists of a sodium ion and a chloride ion, NaCl. Three formula units of sodium chloride would be written as 3NaCl.

PROBLEM SOLVING HINT

If the subscripts in a formula have no common divisor, then the formula and the empirical formula are the same.

SAMPLE PROBLEMS

1. Empirical Formula

What is the empirical formula for a compound with the formula $Cs_2C_4H_4O_6$?

Solving Process:

All of the subscripts are divisible by two. Therefore, the empirical formula is $CsC_2H_2O_3$.

2. Coefficients

What information is conveyed by the formula $3(NH_4)_2SiF_6$?

Solving Process:

The coefficient of 3 means that the quantity of the substance represented is three formula units. The name of the substance is ammonium, NH_4^+ (Table 7.4), hexafluorosilicate, SiF_6^{2-} (Appendix Table A-4). The entire representation, then, is three formula units of ammonium hexafluorosilicate.

PRACTICE PROBLEMS

15. Write the empirical formula for each of the following compounds.
a. N_2O_4 **b.** $C_6H_8O_6$ **c.** C_2H_6 **d.** CH_4 **e.** Hg_2I_2 **f.** C_8H_{18}

16. State the number of formula units that is represented by each of the following.
a. Ag_2CO_3 **b.** 3HBr **c.** $2Fe(NO_3)_2$ **d.** $4AlBr_3$ **e.** $6Ba_3(PO_4)_2$ **f.** $5SnBr_4$ **g.** $3H_3PO_4$ **h.** $3CH_3COOH$

7.2 CONCEPT REVIEW

17. Name the following compounds.
a. $BaC_4H_4O_6$ **b.** $CaSiO_3$ **c.** MgS **d.** $NaIO_3$ **e.** $CoSO_4$

18. Name the following compounds.
a. $Hg(NO_3)_2$ **b.** $Al(ClO_3)_3$ **c.** CoS **d.** $CaC_4H_4O_6$ **e.** $MgSiO_3$

19. Determine the empirical formula for ammonium oxalate, $(NH_4)_2C_2O_4$.

20. What does the expression $4Ni(IO_3)_2$ mean?

21. Apply A throw-away cigarette lighter lists only one substance in its contents, $CH_3CH_2CH_2CH_3$. What is the name of the substance?

Chemistry Journal

Ingredients

Ask students to look through their kitchen and bathroom cabinets to find several products that list their component ingredients. Hair-care products, toothpaste, mouthwash, and detergents are good choices. Have students write down the information for two or three products and bring it to class. Give them class time to look up the formulas for the compounds in the *Handbook of Chemistry and Physics or the Merck Index*, record the information in their journals, and write it on the chalkboard. Lead a class discussion about the results of the survey.

CHEMISTRY AND

TECHNOLOGY

Making the Bone More Binding

The implanting of artificial joints has been done for decades. Still the binding of bone to implants continues to create problems for doctors and technicians, as well as for the patients. Manufacturers have attempted many different mechanical methods for easing the problem of wobbling artificial joints. Now they believe there may be a chemical solution to the problem.

Prosthetic replacement for ball of femur

If the manufacturers are correct, they may have found a way for bone cells to grow right up to the implant, making it nearly as strong as the original bone. Bioceramic coatings whose structures mimic those of actual bone may provide a solution at last. Hydroxyapatite (HA) is a crystalline form of calcium phosphate that has properties very much like a component of bone. The use of HA is not new; it has been used for ten years in dentistry to fill tooth sockets and build up jaw bones. Calcium, phosphorus, and oxygen atoms also combine in other proportions to form tricalcium phosphate (TCP), a substance that encourages the growth of bone.

In research studies, bioceramic coatings have acted as frameworks for new bone growth in animals. However, an animal's bones grow much faster than a human's, so the two may not be completely analogous with regard to bone growth. Further, the coatings are often unstable and may not remain on the artificial joint. Disintegration of the coating could result in abrasions to the bone and the implant, and cause inflammation of the joint. It is difficult to determine the real corrosion and wear rates in the artificial joints.

Manufacturers of implants most often apply the bioceramic coatings directly to the implants. Powdered HA is ionized by plasma temperatures, 10 000-15 000°C, and sprayed on the joints. When the implant is used, some calcium and phosphorus is released and used by the body to produce a strong interface between the bone and the implant coating. However, some researchers believe that it is the HA itself that may eventually cause the coating to fail. These coatings have been in use only since 1986, so there is no long-term research on their ability to hold up over time.

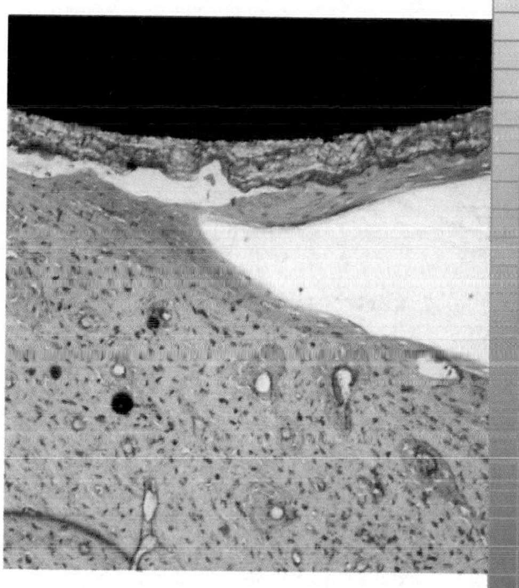

Scientists and technologists are continuing to find out more about synthetic bone, real bone, the implant, the coatings, and the interactions among all of these. They hope that this knowledge will lead to the development of better implants.

Thinking Critically

1. Find the date of the first artificial hip implant and compare it with the time of the first artificial heart. Give a possible reason for this difference.
2. What responsibility do manufacturers have to make certain that the coatings used are not injurious to the health of the patient?
3. Find out more about the structure of bone tissue.

181

19. NH_4CO_2
20. The coefficient of 4 means that there are four formula units of $Ni(IO_3)_2$. The $(IO_3)_2$ part of the substance represents two iodate ions, with a total charge of 2-. Therefore, the nickel (Ni) ion has a 2+ charge. The entire representation is thus four formula units of nickel(II) iodate.
21. butane

4 CLOSE

Discussion: Involve the class in a brief discussion about air pollution. During the discussion introduce students to the fact that sunlight plays a major role in producing photochemical smog. In the upper atmosphere, unusual compounds (dimers) can form such as N_2O_2 and N_2O_4. Ask students to give the empirical formula for each of these compounds. Have them use both prefixes and then Roman numerals to properly name them. Possible response: NO, nitrogen oxide, nitrogen (II) oxide; NO_2, nitrogen dioxide, nitrogen (IV) oxide.

CHEMISTRY AND TECHNOLOGY

Making the Bone More Binding

Background: Over half a million artificial knees and hips were implanted last year world-wide.

Connection to Medicine: A student may be able to conduct a telephone interview with a surgeon who specializes in joint replacements. The doctor or hospital may have a videotape that can be loaned that documents the procedure.

Answers to

Thinking Critically

1. The first artificial heart was implanted in 1982 several years after hip replacement surgery.
2. Student answers will vary, but most expect a good guarantee of product safety.
3. Bone tissue is made up of marrow, spongy bone, and compact bone surrounded by a membrane. The membrane supplies blood vessels and nerves to the bone.

PROGRAM RESOURCES

Vocabulary Review/Concept Review: Use the worksheets, p. 7, and pp. 43-44, to check students' understanding of the key terms and concepts of Chapter 7.

CHAPTER REVIEW 7

- Review Summary statements and Key Terms with your students.
- Complete solutions to Chapter Review Problems can be found in **Merrill Chemistry Problems and Solutions Manual.**

Answers to
Review and Practice
22. The names are taken from Latin or Greek.
23. A symbol begins with one letter. If there is already an element with that same letter, a second letter is added. The symbols with three letters, atomic numbers 104-112, use the Latin and Greek stems representing the atomic numbers of the elements.
24. Hydrogen, nitrogen, oxygen, chlorine, fluorine, bromine and iodine exist as diatomic molecules.

Chapter 7 REVIEW

Summary

7.1 Symbols and Formulas

1. A chemical symbol for an element represents one atom of that element when it appears in a formula.

2. A chemical formula is a statement in chemical symbols of the composition of one formula unit of a compound. A subscript in a formula represents the relative number of atoms of an element in the compound.

3. An ion is a charged particle that is formed when an atom gains or loses electrons.

4. A polyatomic ion is a stable, charged group of atoms.

5. The combining capacity of an atom or polyatomic ion is indicated by its oxidation number or charge.

6. In chemical compounds, atoms combine in definite ratios. Their combined charges add to zero.

7. A binary compound is composed of two elements. Its name is the name of the positive element followed by the name of the negative element modified to end in *-ide*.

8. Some elements have more than one possible oxidation number. A compound containing such an element is named by showing the oxidation number as Roman numerals in parentheses after the element.

7.2 Nomenclature

9. A compound containing a polyatomic ion is named in the same way as a binary compound. However, the ending of the name of the polyatomic ion is not changed.

10. Hydrocarbons are named using a stem that indicates the number of carbon atoms and a suffix indicating how the carbon atoms are linked.

11. A molecular formula shows the actual number of each kind of atom in a single molecule of a compound. The molecular formula is always a whole-number multiple of the empirical formula.

12. An empirical formula represents the simplest whole-number ratio between atoms in a compound.

13. A formula unit represents one molecule or the smallest number of particles giving the ratio of the elements in the compound.

14. A coefficient written before a formula indicates the number of formula units of a substance.

Key Terms

chemical symbol
ion
chemical formula
oxidation number
polyatomic ion
ionic compound
molecule
binary compound
molecular formula
empirical formula
formula unit

Review and Practice

Use tables from this chapter and Appendix Tables A-3 and A-4 when answering the following questions.

22. Why are some chemical symbols so different from the names of the elements that they represent?

23. Why do some symbols have one letter, some have two letters, and some have three letters?

24. What elements exist as diatomic molecules?

PROGRAM RESOURCES

Evaluation: Use the Evaluation worksheets, pp. 25-28, to assess students' knowledge of Chapter 7.

Computer Test Bank: Use the IBM or Apple Test Bank and Question Selection Manual to prepare your own individualized test for Chapter 7.

Chapter 7

REVIEW

25. Write the formula for each of the following compounds.
a. zinc tartrate
b. silver nitrate
c. sodium oxalate
d. barium nitrate

26. Write the formula for each of the following compounds.
a. magnesium hydroxide
b. cadmium hydroxide
c. silver acetate
d. magnesium nitrate
e. aluminum sulfate
f. potassium cyanide

27. Write the name for each of the following compounds.
a. AgBr
b. $Ca_3(PO_4)_2$
c. LiH
d. $RaCl_2$

28. Using Appendix Table A-3, find all possible formulas for binary compounds of chromium and oxygen.

29. Classify each of the following ions as either monatomic or polyatomic.
a. Al^{3+}
b. Hg_2^{2+}
c. $C_2O_4^{2-}$
d. H^-
e. O_2^{2-}
f. MnO_4^-

30. Which of the following is not a binary compound?
a. potassium chloride
b. magnesium hydroxide
c. calcium bromide
d. carbon dioxide

31. Why is it necessary to use Roman numerals in writing the name for the compound $Fe(OH)_3$?

32. Copper forms two different compounds with the chloride ion, CuCl and $CuCl_2$. In writing their names, how do we distinguish between them?

33. Write the formula for each of the following compounds.
a. iron(II) sulfate
b. manganese(II) nitrate
c. chromium(III) nitrate
d. silver perchlorate

34. Write the name for each of the following compounds.
a. $NaNO_3$
b. K_2SO_4
c. $Fe(NO_3)_3$
d. NH_4CH_3COO

35. Write the formula for each of the following compounds.
a. tin(IV) chloride
b. magnesium iodate
c. ammonium sulfide
d. cobalt(II) sulfate

36. Write the name for each of the following compounds.
a. $Pb(CN)_2$
b. MgH_2
c. MnO
d. HCl

37. Write the formula for each of the following compounds.
a. cadmium nitrate
b. cobalt(II) nitrate
c. nickel(II) nitrate
d. magnesium perchlorate
e. nitric acid

38. Write the name for each of the following compounds.
a. $Sr(NO_3)_2$
b. CaC_2O_4
c. $H_2C_2O_4$
d. HgF_2

39. Write the formula for each of the following compounds.
a. cobalt(II) hydroxide
b. magnesium carbonate
c. calcium oxide
d. lithium acetate

40. Write the name for each of the following compounds.
a. SrI_2
b. NH_4F
c. Ag_2S
d. H_2SO_4

41. Write the formula for each of the following compounds.
a. nickel(II) phosphate
b. calcium peroxide
c. cobalt(II) perchlorate
d. barium perchlorate

42. Write the name for each of the following compounds.
a. $Cr_2(SO_4)_3$
b. $Hg(IO_3)_2$
c. $CdBr_2$
d. CH_3COOH

Answers to Review and Practice

25. a. $ZnC_4H_4O_6$
b. $AgNO_3$
c. $Na_2C_2O_4$
d. $Ba(NO_3)_2$

26. a. $Mg(OH)_2$
b. $Cd(OH)_2$
c. $AgCH_3COO$
d. $Mg(NO_3)_2$
e. $Al_2(SO_4)_3$
f. KCN

27. a. silver bromide
b. calcium phosphate
c. lithium hydride
d. radium chloride

28. Students will probably predict CrO, CrO_2, Cr_2O_3, CrO_3, Cr_2O_6 ($Cr_2(O_2)_3$), and CrO_6($Cr(O_2)_3$). The peroxides do not exist.

29. a. monatomic **d.** monatomic
b. polyatomic **e.** polyatomic
c. polyatomic **f.** polyatomic

30. b. ($Mg(OH)_2$ contains three different elements.)

31. Iron exhibits two combining states, Fe^{2+} and Fe^{3+}, and they must be distinguished.

32. Use Roman numerals to indicate the oxidation number in each compound—copper(I) chloride and copper(II) chloride.

33. a. $FeSO_4$
b. $Mn(NO_3)_2$
c. $Cr(NO_3)_3$
d. $AgClO_4$

34. a. sodium nitrate
b. potassium sulfate
c. iron(III) nitrate
d. ammonium acetate

35. a. $SnCl_4$
b. $Mg(IO_3)_2$
c. $(NH_4)_2S$
d. $CoSO_4$

36. a. lead(II) cyanide
b. magnesium hydride
c. manganese(II) oxide
d. hydrochloric acid or hydrogen chloride

37. a. $Cd(NO_3)_2$
b. $Co(NO_3)_2$
c. $Ni(NO_3)_2$
d. $Mg(ClO_4)_2$
e. HNO_3

38. a. strontium nitrate
b. calcium oxalate
c. oxalic acid or hydrogen oxalate
d. mercury(II) fluoride

39. a. $Co(OH)_2$
b. $MgCO_3$
c. CaO
d. $LiCH_3COO$

40. a. strontium iodide
b. ammonium flouride
c. silver sulfide
d. hydrogen sulfate or sulfuric acid

41. a. $Ni_3(PO_4)_2$
b. CaO_2
c. $Co(ClO_4)_2$
d. $Ba(ClO_4)_2$

42. a. chromium(III) sulfate
b. mercury(II) iodate
c. cadmium bromide
d. ethanoic acid or acetic acid

Answers to
Application (cont.)

73. **a.** atom
b. molecule
c. ion
d. molecule
e. ion

74. **a.** Al, Cr
b. Cr_2O_3, Al_2O_3
c. Cr—2 Cr—2
O—3 O—3
Al—2 Al—2
They are equal.

75. 12 atoms of nitrogen

76. Some possible uses are production of fertilizers, paper products, paint, and steel; petroleum refining; manufacturing other chemicals, and processing metallic ores.

77. **a.** 44
b. A reaction requires 2 formula units of $KMnO_4$ and 5 formula units of H_2SO_4.

Answers to
Critical Thinking

78. hydrogen

79. Hydrogen and carbon are most abundant. The molecules would consist of chains and rings of carbon atoms with hydrogen atoms attached to them.

80. At least three carbon atoms are needed to form a ring.

81. 3– (Students may predict 3+. It should be pointed out that NH_3 is an exception to writing the element with the positive oxidation number first.)

82. The mercury(I) ion is Hg_2^{2+}, and is composed of two mercury(I) ions linked together.

83. X has positive oxidation states, and Y has negative oxidation states. The numerical values of these must be in a 2:3 ratio.

73. When stored, sodium is usually kept in kerosene because it reacts violently with water.

$$2Na + 2H_2O \rightarrow 2Na^+ + 2OH^- + H_2$$

Classify each of the following as an atom, a molecule, or an ion.

a. Na
b. H_2O
c. Na^+
d. H_2
e. OH^-

74. Chromium is probably best known as a rust-inhibiting plating on car parts and tools. Chromium can be produced from one of its compounds by the following reaction.

$$Cr_2O_3 + 2Al \rightarrow Al_2O_3 + 2Cr$$

a. List all of the substances that are written using coefficients.
b. List all of the substances that are written using subscripts.
c. List the number of atoms of each element shown on the left side of the equation. List the number of atoms of each element shown on the right side of the equation. What can you conclude from examining these two sets of numbers?

75. How many atoms of nitrogen are in one formula unit of $Pb_3[Fe(CN)_6]_2$?

76. Why is sulfuric acid considered to be such an important industrial chemical? Use references to find four specific industrial processes that use sulfuric acid.

77. Potassium permanganate may be used to get rid of many types of organic odors. A solution may be sprayed over a feedlot, or gases from breweries, paper mills, or meatpacking plants may be bubbled through a $KMnO_4$ solution.

a. How many oxygen atoms are in $11KMnO_4$?
b. In a chemistry book, you read that $2KMnO_4$ react with $5H_2SO_4$. What does this statement mean?

Critical Thinking

78. Hydrofluoric acid is such a reactive acid that it cannot even be stored in a glass container. It is used to etch glass for decorative purposes. From the formulas you know for other acids, what element must be present for a compound to be an acid?

79. Crude oil is a complex mixture mostly of hydrocarbons. What elements would you expect to be most abundant in oil? Describe the probable structure of most of the molecules present.

80. Why are there no compounds named cyclomethane or cycloethane?

81. Ammonia, whose solution is used for cleaning, has the formula NH_3. What is the oxidation number of nitrogen in this compound?

82. The mercury(I) ion is unusual. Look up the charge of this ion in the appropriate table. Suggest a possible explanation for its charge.

83. A compound made up of elements X and Y has the formula X_3Y_2. What does this tell you about the possible oxidation states of elements X and Y?

84. An elaborate system for naming organic compounds has been devised in recent years. Even so, this system must be revised from time to time. Why do you think a complex naming system exists and why must it undergo revision?

85. Cycloalkanes are sometimes found in petroleum and petroleum products.
   a. Write the molecular formulas for cyclopentane, cyclohexane, and cyclopropane.
   b. What is the empirical formula for each?
   c. What conclusion can you reach about cycloalkanes and their empirical formulas?
   d. What similar conclusion can you reach about the empirical formulas for butane, heptane, and octane?

86. A compound has the empirical formula CH_2O. Two molecules of this compound combine to form a substance with the molecular formula, $C_{12}H_{22}O_{11}$, and a molecule of water, H_2O. What is the molecular formula of the first compound?

Readings

Block, B. Peter, *et al. Inorganic Chemical Nomenclature.* Washington, D.C.: American Chemical Society, 1990.

Fletcher, John H., *et. al. Nomenclature of Organic Compounds.* Washington, D.C.: American Chemical Society, 1974.

Leigh, G. J. *Nomenclature of Inorganic Chemistry.* Oxford: Blackwell Scientific Publications (IUPAC), 1990.

Cumulative Review

1. Use the arrow diagram to predict electron configurations for niobium, radon, and tin.

2. Find the atomic mass of magnesium if the element has the following isotopic composition in nature.

Isotopic mass	Abundance
23.985 044	78.70%
24.985 839	10.13%
25.982 593	11.17%

3. Identify the group(s) or class(es) of elements that are most clearly identified with the following characteristics.
   a. have partially filled d sublevel
   b. have filled s and p sublevels
   c. have loosely held single s electron
   d. have half-filled p sublevel
   e. have partially filled f sublevel
   f. gain one electron to achieve a noble gas configuration
   g. have electron configurations ending in np^4, where n is the principal quantum number
   h. have electron configurations ending in $(n-1)d^7$, where n is the principal quantum number
   i. have an outer energy level with a principal quantum number of 4
   j. have 1, 2, or 3 electrons in the outer energy level
   k. have 5, 6, or 7 electrons in the outer energy level
   l. generally lose electrons to satisfy the octet rule
   m. is a group of elements that was unknown to Mendeleev
   n. have properties of both metallic and nonmetallic elements
   o. generally gain electrons to satisfy the octet rule
   p. have a Lewis electron dot structure with seven dots
   q. react to form positively charged ions
   r. have an incomplete $5f$ sublevel

84. There are so many compounds; new discoveries include new compounds and better understanding of structure not accommodated for in the current system.
85. a. C_5H_{10}, C_6H_{12}, C_3H_6
   b. CH_2, CH_2, CH_2
   c. CH_2 is the empirical formula for all cycloalkanes.
   d. C_2H_5, C_7H_{16}, C_4H_9, no pattern
86. $C_6H_{12}O_6$

Answers to
Cumulative Review
1. Nb: $1s^22s^22p^63s^23p^64s^23d^{10}4p^65s^24d^3$ (Actual [Kr]$5s^14d^4$)
Rn: $1s^22s^22p^63s^23p^64s^23d^{10}4p^65s^24d^{10}5p^66s^24f^{14}5d^{10}6p^6$
Sn: $1s^22s^22p^63s^23p^64s^23d^{10}4p^65s^24d^{10}5p^2$
2. 24.31 u
3. a. transition elements
   b. noble gases
   c. alkali metals
   d. Group 15(VA), nitrogen family
   e. lanthanoids and actinoids
   f. halogens
   g. chalcogens
   h. Group 9
   i. period 4
   j. metals
   k. nonmetals
   l. metals
   m. noble gases
   n. metalloids
   o. nonmetals
   p. halogens
   q. metals
   r. actinoids

CHAPTER

8 The Mole

CHAPTER ORGANIZER

CHAPTER OBJECTIVES	TEXT FEATURES	LABORATORY OPTIONS	TEACHER CLASSROOM RESOURCES
8.1 Factor-Label Method **1.** Use the factor-label method in calculations. **2.** Use scientific notation to express and evaluate large and small measurements. **Nat'l Science Stds: UCP.3, E.2, F.6**	**Chemistry and Society:** Economic Trade-Offs, p. 197	**Discovery Demo:** 8-1 How Much Water?, p. 188 **Demonstration:** 8-2 Change the Units, p. 190 **Laboratory Manual:** 8 -1 Determining an Empirical Formula **Laboratory Manual:** 8 -2 The Quantitative Study of a Reaction - microlab **ChemActivity 8-2:** Aluminum Foil Thickness, p. 809	**Lesson Plans:** pp. 16-17 **Study Guide:** pp. 27-28 **ChemActivity Master 8-2**
8.2 Formula-Based Problems **3.** Use the Avogadro constant to define the mole and to calculate molecular and molar mass. **4.** Use the molar mass to calculate the molarity of solutions, percentage composition, and empirical formulas. **5.** Determine the formulas of hydrates. **Nat'l Science Stds: UCP.1, UCP.3, B.2, F.3**	**Chemists and Their Work:** Amadeo Avogadro, p. 200 **Bridge to Medicine:** Making Medically Effective Products, p. 209 **Everyday Chemistry:** Trees, Natural Regulators, p. 211	**Minute Lab:** Bite the Bubble, p. 210. **Demonstration:** 8-3 A 0.1 Molar Solution, p. 204 **Demonstration:** 8-4 Empirical Formula, p. 210 **Demonstration:** 8-5 Drying Corn, p. 212 **ChemActivity 8-1:** Nuts and Bolts Chemistry, p. 807	**Study Guide:** pp. 29-32 **Enrichment:** Solution Concentration, pp. 15-16 **Reteaching:** Empirical Formula, p. 11 **Critical Thinking/ Problem Solving:** Just How Big Is It?, p. 8 **ChemActivity Master 8-1** **Applying Scientific Methods in Chemistry:** Percentage Composition, p. 24
OTHER CHAPTER RESOURCES	**Vocabulary and Concept Review,** pp. 8, 45-48 **Evaluation,** pp. 29-32	**Videodisc Correlation Booklet** **Lab Partner Software**	**Test Bank** **Mastering Concepts in Chemistry—Software**

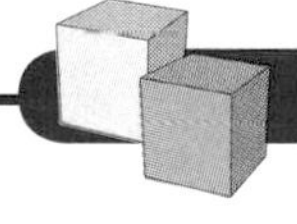

Block Schedule

For information on block scheduling, see the Lesson Plans booklet in the Teacher Resource Package.

GLENCOE TECHNOLOGY

Mastering Concepts in Chemistry Software
Using Conversion Factors, Factor-Label Method
Using Significant Digits
The Mole Concept
Molarity
Analytical Calculations Using Moles

Chemistry: Concepts and Applications Videodisc
Molar Mass
Percent Sugar in Bubblegum

Chemistry: Concepts and Applications CD-ROM
Molar Mass
Percent Sugar in Bubblegum

Science and Technology Videodisc Series (STVS)

Plants and Simple Organisms
Bitter Melon Cancer Treatment
City Trees

Complete Glencoe Technology references are inserted within the appropriate lesson.

ASSIGNMENT GUIDE

Contents	Level	Practice Problems	Chapter Review	Solving Problems in Chemistry
8.1 Factor-Label Method				
Conversion Factors	C	1-6	78, 79, 127, 130, 148	1
Factor-Label Method	C	7-20	80, 87, 126, 128, 129, 131	2-6
Handling Numbers in Science	C	21-27	76, 77, 81-86, 88, 130	7-12
8.2 Formula-Based Problems				
Molecular and Formula Mass	C	32, 33	89-92, 119, 124	13-14
The Avogadro Constant	C		118, 119, 123	
The Mole	C	34-41	93, 94, 149	15-18
Moles in Solution	C	42-49	95-100, 120, 142, 151	19-21
Percentage Composition	C	50-55	101, 105, 108-111, 132-137, 143, 144, 146, 147, 150, 152	22-24
Empirical Formulas	C	56-59	102-104, 121, 122, 125, 138-141	28-30
Molecular Formulas	A	60-64	106, 122	31-32
Hydrates	C	65-70	107, 112-117, 145	25-27, 33-34
				Chapter Review: 1-26

C=Core, A=Advanced, O=Optional

Laboratory Materials

Minute Labs	ChemActivities	Demonstrations
page 210 balance bubble gum that contains sugar (5 pieces) paper cup	**page 807** balance nuts and bolts **page 809** aluminum block aluminum foil 250-mL graduated cylinder metric ruler scissors string water	**pages 188, 190, 204, 210, 212** beakers, 100, 300, 600 cm^3 burner clay triangle clear acetate transparency film crucible $CuSO_4 \cdot 5H_2O$ dropper graduated cylinder 100 cm^3 magnesium ribbon $KMnO_4$ pipette 10 cm^3 pipette bulb popcorn kernels, 30 sand spray cooking oil test tube clamp test tube, large tongs transparent tape watch glass water wire screen

CHAPTER 8

The Mole

CHAPTER OVERVIEW

Main Ideas: Chapter 8 places a strong emphasis on the mole as the chemist's basic measuring unit. The problem-solving sections of the chapter provide practice in using the mole concept with the factor-label method. The mole concept is developed through a logical progression. Formula mass is presented first. The mole and the Avogadro constant are presented next. The remainder of the chapter provides students with applications of the mole concept, such as molarity. The calculation of percentage composition leads the student into empirical and molecular formula determination. These problems will give students a better understanding of the quantitative aspects of chemistry.

Theme Development: The theme of this chapter is systems and interactions. The mole concept is an organizing system in chemistry. Its application provides the chemist the ability to predict amounts of products and reactants.

Tying to Previous Knowledge: How much? Ask students to think of a time when the answer to that question was important to them. *How much do I get paid for doing this job?* The chemist is also interested in how much chemical will be produced or consumed in a particular reaction. The mole helps provide the answer.

ASSESSMENT PLANNER

Choose from the following assessment strategies.

Assess, pp. 196-197, 214-215
Assessment, pp. 189, 193, 195, 203, 204, 207, 210, 212
Portfolio, p. 212
Minute Lab, p. 210
ChemActivity, pp. 807-810
Chapter Review, pp. 216-221

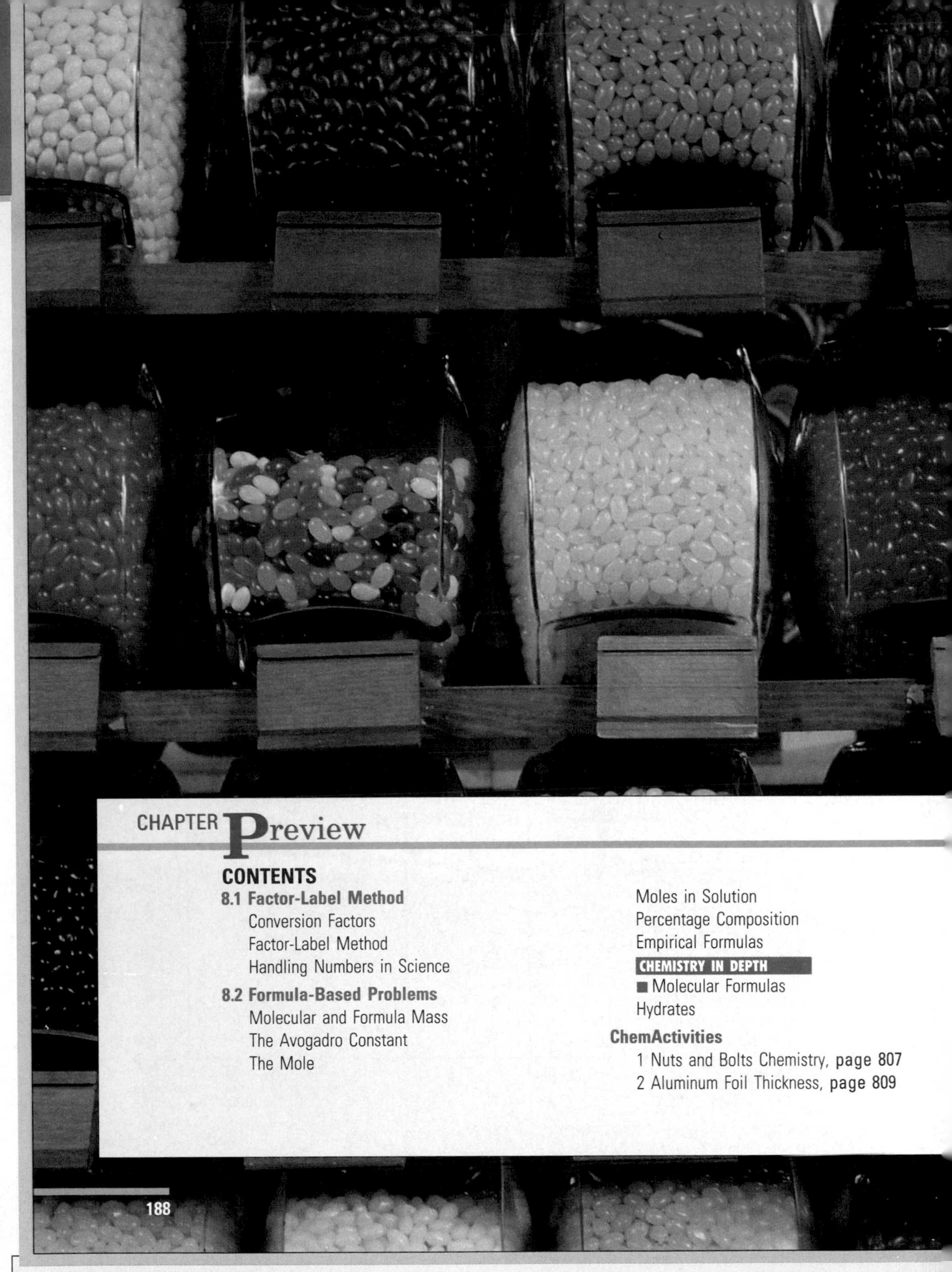

CHAPTER **Preview**

CONTENTS

188

DISCOVERY DEMO

8-1 How Much Water?

Purpose: to illustrate the need for qualitative and quantitative observation

Materials: 30 g $CuSO_4 \cdot 5H_2O$, large test tube, laboratory burner, watch glass, 2 cm^3 H_2O, medicine dropper, test tube clamp

Procedure: 1. Place about 30g of $CuSO_4 \cdot 5H_2O$ in the large test tube. Heat the test tube over a flame until the blue color disappears. Do not overheat the hydrate as it could decompose. Allow the test tube to cool, then remove the contents.
2. Place the white, anhydrous $CuSO_4$ on a watch glass and add H_2O from a dropper. **CAUTION:** *The reaction is exothermic.* **Disposal:** F

Results: Copper(II) sulfate pentahydrate loses water in an observable manner. The blue crystals become

The Mole

CHAPTER

8

How many candies in a jar? You could dump out the candies and count them, but what a tedious job! An alternative approach would be to mass a jar empty and then again with its contents. The difference in masses divided by the mass of one candy, would provide you with the number of candies in the jar.

Just as a manufacturer needs to know how many candies will fill each jar, chemists often need to know the number of atoms in a sample of a substance. Because atoms are so small, it would be impossible to count them. How do chemists know how many atoms are in a sample of a substance? In this chapter you will explore a concept that allows you to answer that question.

EXTENSIONS AND CONNECTIONS

Chemistry and Society

Economic Trade-offs: Examines factors used to make difficult decisions in our society, **page 197**

Chemists and Their Work

Amadeo Avogadro, page 200

Bridge to Medicine

Making Medically Effective Products: A look at how chemists begin the task of synthesizing naturally occurring materials for medical use, **page 209**

Everyday Chemistry

Trees, Natural Regulators: Using trees to reduce atmospheric CO_2, **page 211**

189

INTRODUCING THE CHAPTER

USING THE PHOTO

Have students conjecture how it might be possible to count objects by mass. Lead students to reason out a method of "counting" by taking the total mass and dividing it by the mass of a single item. Point out that by knowing the molecular or formula mass of a compound or the atomic mass of an element, the chemist has a count of the atoms, molecules, or formula units in a given mass of that substance. A chemist can count small particles using a balance.

REVEALING MISCONCEPTIONS

After working with the Avogadro constant, 6.022 136 7 x 10^{23}, students think this is a very exact number. It is important for them to learn that there are very few significant digits in the constant when the order of magnitude is considered.

pale blue and then white when heated. When the water is added to the white anhydrous form, the blue color returns with the evolution of heat.

Questions: 1. What do you observe? *A blue substance changes color when heated. A clear liquid condenses around the top of the test tube.*

2. Why can't you state the amount of water removed from the blue substance? *No quantitative measurements are recorded.*

3. How would you change the procedure to collect better data? *See the Follow Up Activity for a suggestion.*

✓ ASSESSMENT

Performance: Have students repeat the demonstration, but determine the mass with a balance before and after heating. Then students can determine how much water is lost upon heating.

PROGRAM RESOURCES

Lesson Plans: Use the Lesson Plan, pp.16-17 to help you prepare for Chapter 8.

Study Guide: Use the Study Guide worksheets, pp. 27-28, for independent practice with concepts in Section 8.1.

8.1 Factor-Label Method

PREPARE

Key Concepts

The major concepts presented in this section include the use of conversion factors in the factor-label method of problem solving, and the use of scientific notation to express scientific data and calculated answers.

Key Terms

factor-label method
scientific notation

1 FOCUS

Objectives

Preview with your students the objectives listed on the student page. Students can use them as a study guide for this major section.

Discussion

Ask each student to write an equality that is also a definition; such as 60 seconds = 1 minute. List these conversion factors on the chalkboard and generate problems, such as 3 hours = x seconds to illustrate their use. Point out to students that they have just been using a very powerful problem-solving technique. The technique is called the factor-label method or dimensional analysis.

OBJECTIVES

- use the factor-label method in calculations.
- use scientific notation to express and evaluate large and small measurements.

8.1 Factor-Label Method

To use mathematical relationships in solving problems in chemistry, you must first learn how to generate the relationships you need. You also must learn to handle very large and very small numbers, which are frequently used in chemistry. You will often relate quantities by using conversion factors, relationships that convert one quantity to another.

Conversion Factors

As you recall from Chapter 2, the relationships between various units that measure the same quantity can be determined from the prefixes listed in Table 2.2. For example, the relationship between the centimeter and the meter is 1 m = 100 cm. The kilogram and the gram are related by the equation 1 kg = 1000 g. Using these and similar relationships, we can form conversion factors to convert a unit to any other related unit. A conversion factor is a ratio equivalent to one.

PROBLEM SOLVING HINT

Visualize the relative sizes of the units you are converting. Since you are converting from a smaller unit to a larger unit, the numerical value of the quantity will decrease.

SAMPLE PROBLEMS

1. Determining a Conversion Factor

What conversion factor can be used to convert centimeters to meters?

Solving Process:

(a) The given quantity is in cm. The desired quantity must be in meters. We know that

$$100 \text{ cm} = 1 \text{ m}$$

(b) By dividing both sides of this equation by 100 cm, the equation becomes

$$\frac{100 \text{ cm}}{100 \text{ cm}} = \frac{1 \text{ m}}{100 \text{ cm}}; \quad 1 = \frac{1 \text{ m}}{100 \text{ cm}}$$

The fraction that is equal to one is called a conversion factor, because we multiply by it to convert one unit into another. Always make certain that the unit you wish to eliminate is properly placed in the conversion factor. If the unit to be eliminated is in the numerator of the given information, then that unit should appear in the denominator of the conversion factor. If the unit to be eliminated is in the denominator of the given information, then that unit should appear in the numerator of the conversion factor. Only then will the units divide out properly.

2. Conversion

How many cubic centimeters are there in 5 cubic decimeters of gold?

DEMONSTRATION

8-2 Change the Units

This Demonstration, used in Chapter 2, is equally pertinent to both chapters.

Purpose: to illustrate the role of a conversion factor

Materials: 3 sheets of clear acetate transparency film; transparent tape; 1 dm^3 of sand

Procedure: 1. Obtain three sheets of clear acetate transparency film. Roll each into a cylinder and tape the sides together so as to form three cylinders, having diameters of about 3 cm, 6 cm and 9 cm.
2. Stand them vertically.
3. Fill the smallest diameter cylinder completely with sand. Place the other two cylinders around it so you have three concentric cylinders.
4. Slowly remove the 3-cm cylinder

Solving Process:
The given quantity is 5 dm^3. The required quantity is expressed in cm^3. The relationship we must find is that which will take us from cm^3 to dm^3.

(a) We know that 1 dm is equal to 10 cm. The relationship between dm and cm is

$$1 \text{ dm} = 10 \text{ cm}$$

(b) We need to find a relationship between dm^3 and cm^3. If we cube both sides of the equation we get

$$1 \text{ dm}^3 = 1000 \text{ cm}^3$$

(c) Both sides of the equation can be divided by the quantity 1 dm^3. Now the equation appears as

$$1 = \frac{1000 \text{ cm}^3}{1 \text{ dm}^3}$$

The expression on the right side of the equation is the conversion factor for changing dm^3 to cm^3.

(d) Using the known quantity of 5 dm^3 that is given in the problem, we can write the equation

$$5 \text{ dm}^3 = 5 \text{ dm}^3$$

(e) We can now multiply the expression on the right side of the equation in (d) by the fraction

$$\frac{1000 \text{ cm}^3}{1 \text{ dm}^3}$$

Since this fraction equals *one,* the value of the expression on the right side of the equation is not changed. (Recall that the value of any quantity multiplied by 1 is unchanged.) The equation then becomes

$$5 \text{ dm}^3 = 5 \text{ dm}^3\left(\frac{1000 \text{ cm}^3}{1 \text{ dm}^3}\right) = \frac{(5 \text{ dm}^3)(1000 \text{ cm}^3)}{(1 \text{ dm}^3)}$$

The equation then reduces to

$$5 \text{ dm}^3 = (5)(1000 \text{ cm}^3) = 5000 \text{ cm}^3$$

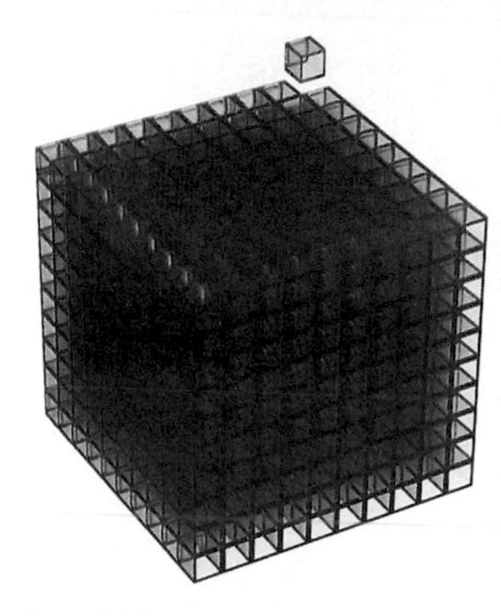

Figure 8.1 One cubic decimeter of any material contains $(10 \text{ cm})^3$ or 1000 cm^3.

PRACTICE PROBLEMS

Convert each of the following quantities.

1. 0.143 hour to seconds
2. 0.84 meter to centimeters
3. 31.5 centigrams to milligrams
4. 65.22 mg to g
5. 531 cm^3 to dm^3
6. 718 nm to cm

The principles in the procedure just described can be used to solve many kinds of problems. In order to simplify the working of chem-

allowing the sand to flow into the 6-cm cylinder at a lower level.

5. Then slowly remove the 6-cm cylinder and allow the sand to flow into the 9-cm cylinder. **Disposal:** A

Results: The amount of sand or sawdust will remain the same in each cylinder, but its height will decrease.

Questions: 1. Does the amount of sand in each cylinder change during the demonstration? *no*

2. What does change? *the height of the sand.*

3. When three hours is expressed as 180 minutes, does the interval of time change? *no*

Follow Up Activity: A 72-cm long piece of string does not change in length (amount of sand) when its length is expressed as 0.72 m or 720 mm. Have a piece of rope or string that the class can observe being measured three times, each time using a different unit.

2 TEACH

CONCEPT DEVELOPMENT

Discussion: Using unfamiliar units in a problem will help you evaluate students' mastery of the concept of conversion factors and their use. For example, if 1 wag = 12 zoom, 1000 warp = 1 bam, and 3 zoom = 1 bam; convert 8 wags to warp.

$$\frac{8 \text{ wag}}{} \left| \frac{12 \text{ zoom}}{1 \text{ wag}} \right| \frac{1 \text{ bam}}{3 \text{ zoom}} \left| \frac{1000 \text{ warp}}{1 \text{ bam}} \right. = 32\ 000 \text{ warp}$$

You will notice that the fraction bar used above is used throughout the text. It is a convenient way for students to organize information and manipulate units. It is faster for students than using parentheses.

Correcting Errors

To help students decide how a conversion factor should be expressed, such as 60 s/1 min or 1 min/60 s, solve the following problem on the chalkboard using both forms of conversion factors.
330 s = ? min

$$\frac{330 \text{ s}}{} \left| \frac{60\text{s}}{1 \text{ min}} \right. \quad \text{or} \quad \frac{330 \text{ s}}{} \left| \frac{1 \text{ min}}{60 \text{ s}} \right.$$

Point out that if the units do not divide out using the first expression of the conversion factor, use its reciprocal expression.

PRACTICE

Guided Practice: Guide students through the Sample Problems then have them work in class on Practice Problems 1-3.

Independent Practice: Your homework or classroom assignment can include the remaining Practice Problems 4-6, Review and Practice 78-80.

Answers to
Practice Problems: Have students refer to Appendix C for complete solutions to Practice Problems.

1. 515 s
2. 84 cm
3. 315 mg
4. 0.065 22 g
5. 0.531 dm^3
6. 7.18×10^{-5} cm

GLENCOE TECHNOLOGY

Software

Mastering Concepts in Chemistry

Unit 5, Lesson 3

Using Conversion Factors, Factor-Label Method

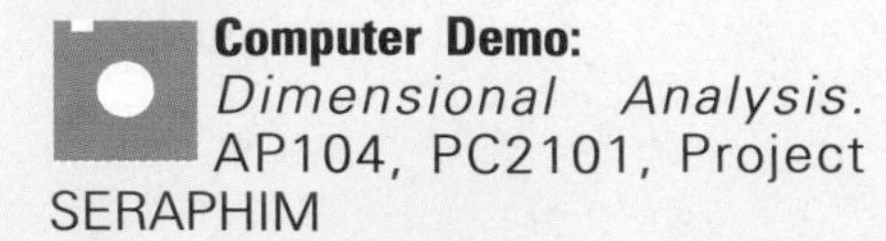

Computer Demo: *Dimensional Analysis.* AP104, PC2101, Project SERAPHIM

Correcting Errors

Help students who are unfamiliar with the metric system solve the Practice Problems by allowing them to prepare a card on which they have written common metric equalities. This will give them confidence as they learn the conversions in less time.

interNET CONNECTION

A hypercard stack for reviewing the concept of moles can be downloaded.

Anonymous FTP
ftp://archive.umich.edu/mac/misc/chemistry/moleconcept.sit.hqx

istry problems, we will use a different notation. In the last Sample Problem, we had the equation

$$5 \text{ dm}^3 = 5 \text{ dm}^3\left(\frac{1000 \text{ cm}^3}{1 \text{ dm}^3}\right)$$

That equation is equivalent to the equation

$$5 \text{ dm}^3 = \left(\frac{5 \text{ dm}^3}{1}\right)\left(\frac{1000 \text{ cm}^3}{1 \text{ dm}^3}\right)$$

Instead of enclosing every factor in parentheses, we will simply set off each factor of the equation by a vertical line. Our equation is now written as

$$5 \text{ dm}^3 = \frac{5 \text{ dm}^3 \mid 1000 \text{ cm}^3}{\mid 1 \text{ dm}^3} = 5000 \text{ cm}^3$$

Factor-Label Method

The problem-solving method discussed in the previous section is called the **factor-label method.** In effect, unit labels are treated as factors. As common factors, these labels may be divided out. The correct solution to a problem will have the correct unit label. Thus, this method aids in solving a problem and provides a check on mathematical operations. The individual conversion factors (ratios whose value is equivalent to one) may usually be written by inspection. Let us see how the method can be applied to more complex measurements.

Many measurements are simply combinations of the elementary measurements as discussed in Chapter 2. You are probably familiar with the measurement of speed. Speed is the distance covered during a single unit of time. The speed of an automobile is measured in kilometers (length) per hour (time). Most scientific measurements of speed are made in centimeters per second or meters per second.

Figure 8.2 Speed is measured in derived units composed of units of length and units of time. The speedometer reading of an automobile (left) indicates the speed, that is, the rate at which the automobile is moving at that moment (right).

CONTENT BACKGROUND

One Approach Fits All: In chemistry, units are carried along with nearly every number used. Point out that the factor-label method is a systematic approach that will enable students to see clearly the mathematical operations needed to convert a measurement from one set of units to the units sought in the problem. Emphasize to students that units are handled in division and multiplication as though they were numbers. Discourage the use of unitless ratio and proportion as a problem-solving strategy.

SAMPLE PROBLEM

Conversion of Units

Express 60.0 kilometers per hour as centimeters per second.

Solving Process:

(a) Write 60.0 kilometers per hour as a ratio. $\frac{60.0 \text{ km}}{1 \text{ h}}$

(b) We wish to convert kilometers to meters. Since 1000 m = 1 km, we can use the conversion factor

$$\frac{1000 \text{ m}}{1 \text{ km}}$$

because its value equals 1.

(c) In the same manner, we will use

$100 \text{ cm} = 1 \text{ m}$, or $\frac{100 \text{ cm}}{1 \text{ m}}$ (to convert m to cm)

$1 \text{ h} = 60 \text{ min}$, or $\frac{1 \text{ h}}{60 \text{ min}}$ (to convert h to min)

$1 \text{ min} = 60 \text{ s}$, or $\frac{1 \text{ min}}{60 \text{ s}}$ (to convert min to s)

(d) Because any number may be multiplied by 1 or its equivalent without changing its value, we can now write

km to m | m to cm | h to min | min to s

$$\frac{60.0 \text{ km}}{1 \text{ h}} = \frac{60.0 \cancel{\text{km}}}{1 \cancel{\text{h}}} \left| \frac{1000 \cancel{\text{m}}}{1 \cancel{\text{km}}} \right| \frac{100 \text{ cm}}{1 \cancel{\text{m}}} \left| \frac{1 \cancel{\text{h}}}{60 \cancel{\text{min}}} \right| \frac{1 \cancel{\text{min}}}{60 \text{ s}}$$

$= 1670 \text{ cm/s}$

Notice that ratios are arranged so that units can be divided out as factors. This procedure leaves the correct units in the answer and provides a check on the method used.

The dividing out of units in this Sample Problem will occur in almost all sample problems. Follow the same procedure in your own problem solutions.

PRACTICE PROBLEMS

Use the factor-label method to convert the following quantities.

7. 0.032 g to mg
8. 0.436 m^3 to cm^3
9. 302.1 mL to cm^3
10. 0.693 dm^3 to cm^3
11. 9.06 km/h to m/min
12. 0.307 mg/cm^3 to g/cm^3
13. 822 dm^3/s to L/min
14. 0.78 L/min to cm^3/s
15. 0.848 kg/L to mg/cm^3
16. 81.42 nm/s to cm/min
17. 7.56 mm^3/s to dm^3/min
18. 0.03 cm/s to km/h
19. 0.0775 cg/cm^3 to g/m^3
20. 0.95 kg/cm^3 to mg/mm^3

PRACTICE

Guided Practice: Guide students through the Sample Problem, then have them work in groups on Practice Problems 7-11 that follow.

Cooperative Learning: Refer to pp. 28T-29T to select a teaching strategy to use with this activity.

Independent Practice: Your homework or classroom assignment can include the remaining Practice Problems 12-20, Review and Practice 78-80 from the end of the chapter.

Answers to
Practice Problems: Have students refer to Appendix C for complete solutions to Practice Problems.

7. 32 mg
8. 4.36×10^5 cm^3
9. 302.1 cm^3
10. 693 cm^3
11. 151m/min
12. 3.07×10^{-4} g/cm^3
13. 4.93×10^4 L/min
14. 13 cm^3/s
15. 848 mg/cm^3
16. 4.885×10^{-4} cm/min
17. 4.54×10^{-4} dm^3/min
18. 0.001 km/h
19. 775 g/m^3
20. 9.5×10^2 mg/mm^3

✓ ASSESSMENT

Skill: Have students work in cooperative learning groups to solve Application questions 126-129 and 131 on page 219 using the factor-label method.

Chemistry Journal

Conversion Factors

Point out that the key to using the factor-label method is knowing a conversion factor that links the unit you have to the unit you want to obtain. Ask students to write a journal entry describing various ways they may obtain necessary conversion factors.

One method is to memorize a few important conversion factors. Another method is to look up conversion factors in a text. Sometimes, calculators include a reference card listing some frequently used conversion factors, or they display a factor when you type in a unit.

CONCEPT DEVELOPMENT

Background: Many students have purchased scientific calculators but do not know how to use the advanced features. Team students having similar calculators together and have them help one another as they work the problems in this section.

Cooperative Learning: Refer to pp. 28T-29T to select a teaching strategy to use with this activity.

Software

Mastering Concepts in Chemistry
Unit 5, Lesson 2
Using Significant Digits

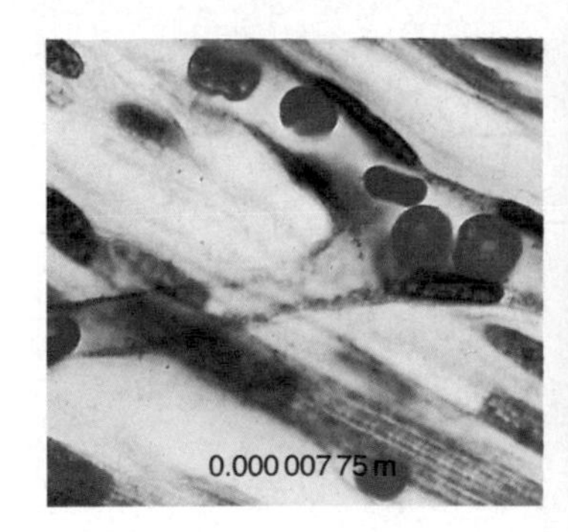

Figure 8.3 For extremely small measurements, such as the diameter of a red blood cell (left), and for very large measurements, such as the distance to the Andromeda galaxy (right), it is convenient to use scientific notation. Showing all the digits in measurements like these makes calculations difficult.

Handling Numbers in Science

In this course we will sometimes use very large numbers. For example, later in this chapter you will learn about the Avogadro constant, which is 602 214 000 000 000 000 000 000. We will also use very small numbers as in describing the distance between the particles in a sodium chloride crystal as 0.000 000 002 814 cm. In working with such numbers it is easy to make a mistake by overlooking a zero or misplacing the decimal point.

Scientific notation makes it easier to work with very large or small numbers. In **scientific notation,** all numbers are expressed as the product of a number between 1 and 10 and a whole-number power of 10.

$$M \times 10^n$$

In this expression, $1 \leq M < 10$, and n is an integer. This number is read as "M times ten to the nth". For example, 5.2×10^5 is read "five point two times ten to the fifth."

One advantage of scientific notation is that it removes any doubt about the number of significant digits in a measurement. Suppose the volume of a gas is expressed as 8000 cm^3. As written, this number contains only one significant digit. Suppose the measurement was actually made to the nearest cm^3. Then the volume 8000 cm^3 must be expressed to four significant digits. In scientific notation, we can indicate the additional significant digits by placing zeros to the right of the decimal point. Thus, 8×10^3 cm^3 has only one significant digit, while 8.000×10^3 cm^3 has four significant digits.

Once we have recorded measurements to the correct number of significant digits and expressed them in scientific notation, we can use them in calculations. To determine the number of digits that should appear in the answer to a calculation, we will use two rules.

1. *In addition and subtraction, the answer may contain only as many decimal places as the measurement having the least number of decimal places.* For example, if 345 g is added to 27.6 g, the answer must be given to the nearest whole number. In adding a column of figures such as

CONTENT BACKGROUND

Calculators and Computers: Students sometimes wonder how a computer or calculator makes computations with large numbers. A computer is basically a series of "off-on" switches (transistors). Most high speed computers operate on the binary system, which consists of two symbols, 0 and 1. It is possible to express integers using only these two symbols and the on-off positions of a series of switches as shown in the table.

A computer performs all mathematical operations by either adding or subtracting in rapid succession in a manner determined by its program. For example, if 8 is divided by 2, the computer can add 2's until the sum is 8 and then record the number of additions. It can also convert the numbers to logarithms and subtract logs in order to divide.

$$\begin{array}{r} 677.1 \text{ cm} \\ 39.24 \text{ cm} \\ 6.232 \text{ cm} \\ \hline 722.572 \text{ cm} \end{array}$$

the answer should be rounded off to the nearest tenth. The answer to the problem above is 722.6 cm.

2. *In multiplication and division, the answer may contain only as many significant digits as the measurement with the least number of significant digits.* For example, in the following problem the answer has three significant digits.

$$\underset{(3)}{(1.13 \text{ m})}\ \underset{(7)}{(5.126\ 122 \text{ m})} = 5.792\ 517\ 86 \text{ m}^2 = \underset{(3)}{5.79 \text{ m}^2}$$

The answer to the next problem has four significant digits.

$$\frac{\overset{(6)}{49.600\ 0 \text{ g}}}{\underset{(4)}{47.40 \text{ cm}^3}} = 1.046\ 413\ 5 \frac{\text{g}}{\text{cm}^3} = \overset{(4)}{1.046} \frac{\text{g}}{\text{cm}^3}$$

If you are using a calculator to obtain your numerical answer, you must be very careful. Remember, you are calculating with measured quantities, not just numbers. For example, in the problem 49.600 0 g/47.40 cm^3, the calculator may display the numerical answer 1.046413502. This answer must be rounded to the proper number of significant digits. In this case, the number of significant digits is four. Thus, the answer is 1.046 g/cm^3. Always double-check your answer against the data given. Report the answer only to the number of significant digits justified by the data. Remember, calculated accuracy cannot exceed measured accuracy.

Figure 8.4 Not all the digits that appear on a calculator are significant. In performing the calculation 49.600 g ÷ 47.40 cm^3 the calculator displays 1.046413502. The answer should be reported as 1.046 g/cm^3 because the factor 47.40 cm^3 has only four significant digits.

SAMPLE PROBLEM

Multiplying in Scientific Notation

Find the product of

$$(4.0 \times 10^{-2} \text{ cm})(3.0 \times 10^{-4} \text{ cm})(2.0 \times 10^{1} \text{ cm}).$$

Solving Process:

To multiply numbers expressed in scientific notation ($M \times 10^n$), multiply the values of M and add the exponents. The exponents do not need to be alike. When using a scientific calculator, use the [EXP] key, which stands for the "× 10" part of the number.

Multiply the values of M. $4.0 \times 3.0 \times 2.0 = 24$

Multiply the units. $\text{cm} \times \text{cm} \times \text{cm} = \text{cm}^3$

Add the exponents. $-2 + (-4) + 1 = -5$

Thus, $(4.0 \times 10^{-2})(3.0 \times 10^{-4})(2.0 \times 10^{1}) =$

$$(4.0 \times 3.0 \times 2.0) \times 10^{-2+(-4)+1} \text{ cm}^3 = 24 \times 10^{-5} \text{ cm}^3 = 2.4 \times 10^{-4} \text{ cm}^3$$

Correcting Errors

Remind students that they cannot input a number in scientific notation into their calculator as it is written in the problem. The calculator is programmed to accept the [EXP] or [EE] key instead of [×10]. Also remind students that the negative sign in an exponent that is necessary to indicate a small number, is not the same key as the subtraction symbol.

✓ ASSESSMENT

Skill: In cooperative learning groups have students trade calculators and assess their abilities with different types and brands of calculators as they solve Review and Practice problem 77 on page 216.

Integer	Binary Expression	Switches* 4 3 2 1	Integer	Binary Expression	Switches* 4 3 2 1
0	0	o o o o	5	101	o c o c
1	1	o o o c	6	110	o c c o
2	10	o o c o	7	111	o c c c
3	11	o o c c	8	1000	c o o o
4	100	o c o o	9	1001	c o o c

*o = open, c = closed

PROGRAM RESOURCES

Laboratory Manual: Use macrolab 8-1 "Determining an Empirical Formula" and microlab 8-2 "The Quantitative Study of a Reaction" as enrichment for the concepts in Chapter 8.

PRACTICE

Guided Practice: Guide students through the Sample Problems, then have them work in groups on Practice Problems 21,23,26-27.
Cooperative Learning: Refer to pp. 28T-29T to select a teaching strategy to use with this activity.
Independent Practice: Your homework or classroom assignment can include the remaining Practice Problems 22,24-25, and Review and Practice 76-77, 81-86.

Answers to
Practice Problems
21. 0.0753 m^2
22. 1.1×10^6 mm^2
23. 7.4×10^2 g/dm^3
24. 763 g
25. 9.40 kg
26. 1.33×10^{-5} cm^2
27. 3.59×10^{12} cm

3 ASSESS

CHECK FOR UNDERSTANDING

Ask students to answer the Concept Review problems, and Review and Practice 76-77 and 81-86 from the end of the chapter.

RETEACHING

Have student teams measure the length and width of the top of the lab table and then calculate the surface area. Have the team express the calculated area in m^2, cm^2, and use scientific notation to express the area in mm^2.

EXTENSION

Have an interested student show the class a series of numbers that will not fit in the display register of his or her calculator. Have the student demonstrate that scientific notation enables the numbers to be input into the calculator and acted upon by the computer logic circuits.

SAMPLE PROBLEM

Dividing in Scientific Notation

Divide. $\dfrac{(4 \times 10^3 \text{ m})(6 \times 10^{-1} \text{ m})}{(8 \times 10^2 \text{ m})}$

Solving Process:

In division the exponents are subtracted. To divide numbers in scientific notation ($M \times 10^n$), divide the values of M and subtract the sum of the exponents of the denominator from the sum of the exponents of the numerator.

$$\frac{(4 \times 6) \times 10^{3+(-1)} \text{ m}^2}{8 \times 10^2 \text{ m}} =$$

$$\frac{24}{8} \times \frac{10^2}{10^2} \times \frac{\text{m}^2}{\text{m}} = 3 \times 10^0 \text{ m}$$

$$= 3 \text{ m}$$

PRACTICE PROBLEMS

Perform the following operations. Express your answers to the correct number of significant digits. Units should be handled in the same way that numbers are.

21. (10.18 m)(0.007 40 m)
22. (302 000 mm)(3.5 mm)
23. $\dfrac{3120.14 \text{ g}}{4.2 \text{ dm}^3}$
24. 701 g + 4.24 g + 57.397 g
25. 9.66 kg − 0.256 92 kg
26. $(2.63 \times 10^{-9}$ cm$)(5.06 \times 10^3$ cm$)$
27. $\dfrac{(6.12 \times 10^2 \text{ cm})(4.92 \times 10^3 \text{ cm})}{(8.38 \times 10^{-7} \text{ cm})}$

8.1 CONCEPT REVIEW

28. Make the following conversions.
- **a.** 895 picoseconds to seconds
- **b.** 1.098 km to m
- **c.** 0.924 kilogram to milligrams
- **d.** 0.705 milligram to grams
- **e.** 491 dm^3 to liters
- **f.** 136 dm^3 to m^3
- **g.** 0.852 m^3 to cm^3

29. Perform the following operations. Express each answer in scientific notation.
- **a.** 1100 m + 596 m
- **b.** 32 144°C − 800°C
- **c.** 32 235 kg/1991 m^3
- **d.** 191 cm^2 × 0.0024 cm^2
- **e.** (32 395 kg/m^3)(32 161 m^3)/32 143 m^2

30. If the speedometer reading of a car is 55 miles per hour, what is the car's speed in km/h? in m/s?

31. **Apply** Explain which quantity, 1.0×10^3 g or 1.00×10^3 g, more nearly expresses the mass of exactly one kilogram.

CHEMISTRY AND
SOCIETY

Economic Trade-Offs

Did you know that silver is the best electrical conductor of any element? However, you won't find it in any wiring in your house. Why not? You'll know the answer if you've ever looked at the price of silver jewelry. The answer, of course, is expense. The electrical wire found in most houses is made of copper. Copper is about 10% less efficient in conducting electrical current than silver. However, it is far less expensive than silver.

The use of copper wire rather than silver wire in electrical installations is an example of what is often called an *economic trade-off*. In this circumstance, consumers have traded one set of costs for another. Even though the use of copper wire might add a small cost to an electric bill every month because it is less conductive than silver, the cost of wiring a house using silver would have been far greater.

In a technological society, company managers, political organizations, citizen groups, and individuals are constantly having to make decisions that involve economic tradeoffs.

A number of communities across the United States are now facing a decision concerning the disposition of hazardous wastes. Some of these communities are presently suffering from low employment levels. The added industry would provide much-needed jobs. However, there are in these communities large numbers of people who do not wish to have hazardous waste disposal facilities near them because they are afraid of contamination of water supplies and the local atmosphere.

Another trade-off decision made every day by millions of citizens is what type of gasoline to buy for their automobiles. Premium gasolines cost significantly more than regular gasoline. However, they also deliver better mileage.

Whenever you are considering a question involving the choice of materials or processes, it will always be wise to consider the economic consequences of your decision. In each instance, you will have to consider what the trade-off involves: what are the benefits, what are the costs, and what are the risks?

Analyzing the Issue

1. Find out what precautions are taken to prevent hazardous waste landfills from contaminating water supplies.
2. Find the price of regular and super unleaded gasolines in your area. If the regular provides 19 miles per gallon and the super provides 26 miles per gallon, which gasoline would you buy?

197

Answers to
Concept Review
Refer to *Merrill Chemistry Problems and Solutions Manual* for complete solutions.

28. **a.** 8.95×10^{-10} s
b. 1098 m
c. 9.24×10^{5} mg
d. 7.05×10^{-4} g
e. 491 L
f. 0.136 m^3
g. 8.52×10^{5} cm^3

29. **a.** 1.7×10^{3}m
b. 3.13×10^{4}°C
c. 1.619×10^{1} kg/m^3
d. 4.6×10^{-1} cm^4
e. 3.2413×10^{4} kg/m^2

30. 88 km/h
25 m/s

31. 1.00×10^{3} g indicates three significant digits, which is preferable to indicating only two significant digits.

4 CLOSE

Discussion: Have the class turn a few pages farther back in this chapter to the Sample Problems that show students how to convert molecules to moles or grams to atoms. Have them discuss how the factor label method is being used in the problems.

CHEMISTRY AND SOCIETY
Economic Trade-Offs

Background: Large concentrations of phenol fed to rats cause cancer. Currently, the EPA permits up to 3500 micrograms/dm^3 water to be discharged into a body of water that serves as a source for drinking water. Scientists have not yet established safe limits of phenol. The Federal Food & Drug Administration allows phenol concentrations of 1.4% in throat sprays.

Teaching This Feature: Use the above background information to involve students in a debate. If safe limits are unknown, should a substance be regulated? Financial and health costs can be great.

Answers to
Analyzing the Issue
Student responses will vary.

Exploring Further: Have interested students prepare a survey form dealing with the economic impact of global warming.

8.2 Formula-Based Problems

PREPARE

Key Concepts

The major concepts presented in this section include the calculation of molecular and formula mass, the Avogadro constant, the mole, molarity, percentage composition, empirical formulas, molecular formulas, and hydrate formulas.

Key Terms

molecular mass
formula mass
Avogadro constant
mole
molar mass
molarity
percentage composition
hydrate

1 FOCUS

Objectives

Preview with your students the objectives listed on the student page. Students can use them as a study guide for this major section.

Activity

In four clear, labeled, capped bottles place some copper metal, sulfur, oxygen (air), and copper(II) sulfate crystals. Place a rubber band around the three bottles containing the Cu, S, and O_2. Discuss how these elements come together to form copper(II) sulfate, $CuSO_4$. Use the Sample Problem of formula mass as an example, and have the class determine together the formula mass of copper(II) sulfate.

OBJECTIVES

- use the Avogadro constant to define the mole and to calculate molecular and molar mass.
- use the molar mass to calculate the molarity of solutions, percentage composition, and empirical formulas.
- determine the formulas of hydrates.

8.2 Formula-Based Problems

Chemical symbols and formulas (such as H and H_2O) are shorthand signs for chemical elements and compounds. The symbol of an element may represent one atom of the element. The formula of a compound may represent one molecule or one formula unit of the compound. Symbols and formulas may also represent a group of atoms or formula units. Since atoms are so very small, chemists usually deal with large groups of atoms. This section is about a group, called a mole, containing a specific number of units. Counting items in groups is a familiar thing to do. We count eggs in dozens (12), and merchants often count items in a group called a gross (144). A mole is just a particular number of atoms, ions, molecules, or formula units. It is a large number because these particles are so small, and we want these particles in a group that is big enough to obtain the mass conveniently.

Molecular and Formula Mass

A hydrogen atom has a mass of 1.67×10^{-24} gram. The mass of an oxygen atom is 16 times that of hydrogen, or 2.66×10^{-23} gram, and the mass of a carbon atom is 12 times that of hydrogen. Because these are very small numbers, it is convenient to use a mass scale defined on the atomic level as described in Chapter 4. A list of atomic masses of the elements is found inside the back cover of this book. The units associated with the numbers in the atomic mass column are atomic mass units/atom.

The atomic mass of hydrogen in atomic mass units is 1 u, and the atomic mass of oxygen is 16 u. Therefore, the total mass of a water molecule, H_2O, is 1 + 1 + 16, or 18 u. If the atomic masses of all the atoms in a molecule are added, the sum is the mass of that molecule. Such a mass is called a **molecular mass.** This name is incorrect when applied to an ionic substance. Sodium chloride, NaCl, is an ionic substance and does not exist in molecular form. A better name for the mass of ionic substances is

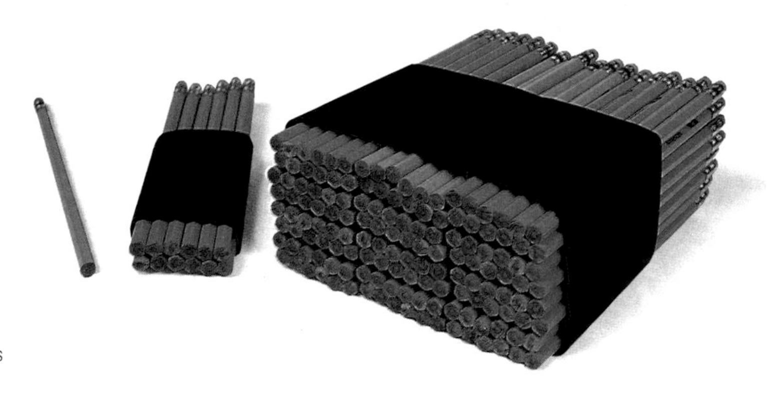

Figure 8.5 It is often convenient to count items by arranging them in a group, such as a dozen (12) or a gross (12×12).

CONTENT BACKGROUND

Cannizzaro: The Italian chemist Stanislao Cannizzaro devised a method for computing atomic masses by using Avogadro's hypothesis. Because the molar volume (22.4 dm^3) of any gaseous compound at STP contains exactly one mole of gas, that volume must also contain an integral number of moles of each of the elements that compose the compound. By comparing the masses of an element contained in a molar volume of various compounds of the element, Cannizzaro was able to compute the atomic masses of several elements. Consider the following table listing the mass of nitrogen in a molar volume of various nitrogen compounds at STP.

formula mass. The sum of the atomic masses of all atoms in the formula unit of an ionic compound is called the **formula mass.** To calculate a formula mass, add the masses of all the atoms in the formula.

In this text, atomic masses are used that will ensure three significant digits in the result. Thus, to obtain the formula mass of NaCl, the values are rounded to tenths: 23.0 + 35.5 = 58.5. For H_2, the values are rounded to hundredths: 1.01 + 1.01 = 2.02.

SAMPLE PROBLEMS

1. Molecular Mass

Find the molecular mass of 2,3-butanedione, $CH_3COCOCH_3$.

Solving Process:

Add the atomic masses of all the atoms in the $C_4H_6O_2$ formula unit.

4 C atoms	4 × 12.0 = 48.0 u
6 H atoms	6 × 1.01 = 6.06 u
2 O atoms	2 × 16.0 = 32.0 u
molecular mass of $C_4H_6O_2$	86.1 u

2. Formula Mass

Find the formula mass of calcium nitrate, $Ca(NO_3)_2$, a compound used in making explosives, fertilizers, and matches.

Solving Process:

Add the atomic masses of all the atoms in the $Ca(NO_3)_2$ formula unit. Remember that the subscript applies to the entire polyatomic ion.

1 Ca atom	1 × 40.1 = 40.1 u
2 N atoms	2 × 14.0 = 28.0 u
6 O atoms	6 × 16.0 = 96.0 u
formula mass of $Ca(NO_3)_2$	164.1 u

PRACTICE PROBLEMS

32. Calculate the molecular mass of each of the following organic compounds.
- **a.** methanol, CH_3OH
- **b.** ethane, C_2H_6
- **c.** sucrose, $C_{12}H_{22}O_{11}$
- **d.** 1-propanol, $CH_3CH_2CH_2OH$

33. Calculate the molecular or formula mass of each of the following compounds.
- **a.** tantalum carbide, TaC, used to make tools and dies
- **b.** aluminum nitride, AlN
- **c.** tetraphosphorus trisulfide, P_4S_3, used in the tips of strike-anywhere matches
- **d.** calcium phosphate, $Ca_3(PO_4)_2$
- **e.** barium hypochlorite, $Ba(ClO)_2$

Nitrogen Compound	Mass of Nitrogen
ammonia	1 × 14 g = 14 g
dinitrogen tetroxide	2 × 14 g = 28 g
nitrogen(II) oxide	1 × 14 g = 14 g
nitrogen(I) oxide	2 × 14 g = 28 g
nitrogen	2 × 14 g = 28 g

Nitrogen always occurs in multiples of 14 g. The atomic mass of nitrogen is then assumed to be 14, although it could be some sub-multiple, such as 7. Note that Cannizzaro's method demonstrates that nitrogen is diatomic. It was his demonstration of the diatomic nature of several elements that proved to be a great breakthrough in chemistry in the mid-19th century.

2 TEACH

CONCEPT DEVELOPMENT

- The terms *molecular weight* and *atomic weight* are incorrect even though they have been used for many years. Point out that weight is measured in newtons.

Background: The choice of the first atomic mass standard was arbitrary. For a long time, chemists used an atomic mass scale based on oxygen. In 1961, carbon-12 was agreed upon as the standard. This isotope is readily available in its pure form and easily measured in a mass spectrometer.

Correcting Errors

To help students arrive at the same answers that are provided in the text, point out that they should use the atomic masses expressed to three significant digits as in the tables. Three significant digits corresponds with the accuracy generally provided by high school laboratory equipment.

PRACTICE

Guided Practice: Guide students through the Sample Problems, then have them work in groups on Practice Problems 32 a-b, 33 d-e.

Cooperative Learning: Refer to pp. 28T-29T to select a teaching strategy to use with this activity.

Independent Practice: Your homework or classroom assignment can include the remaining Practice Problems 32 c-d, 33 a-c, Review and Practice 89-92 from the end of the chapter.

Answers to **Practice Problems:** Have students refer to Appendix C for complete solutions to Practice Problems.

32. a. 32.0 u, **b.** 30.1 u, **c.** 342 u, **d.** 60.1 u

33. a. 193 u, **b.** 41.0 u, **c.** 2.20×10^2 u, **d.** 3.10×10^2 u, **e.** 2.40×10^2 u

PROGRAM RESOURCES

Study Guide: Use the Study Guide worksheets, pp. 29-32, for independent practice with concepts in Section 8.2.

CHEMISTS AND THEIR WORK
Amadeo Avogadro

Background: There are three reasons why Avogadro's work was ignored during his lifetime. The first is that he failed to supply a large amount of experimental data to support his hypothesis. In fact, he was known as a careless experimenter who produced inaccurate results. The second reason is he theoretically extended his hypothesis to include metals in the vapor phase without experimental data. The third reason is that Avogadro's idea of a diatomic molecule conflicted with Berzelius's dualistic approach that was dominant in Europe at that time.

Exploring Further: Show the students that realistic atomic masses would be difficult to calculate if you assumed the reaction of hydrogen and chlorine gases to be $H + Cl \rightarrow HCl$ rather than $H_2(g) + Cl_2(g) \rightarrow 2HCl(g)$. Ask the students if the atomic masses would appear to be higher or lower than their accepted value today. Possible response: Higher

CONCEPT DEVELOPMENT

Computer Demo: *Mole Drill,* Calculation drill and practice. AP305, PC2301, Project SERAPHIM

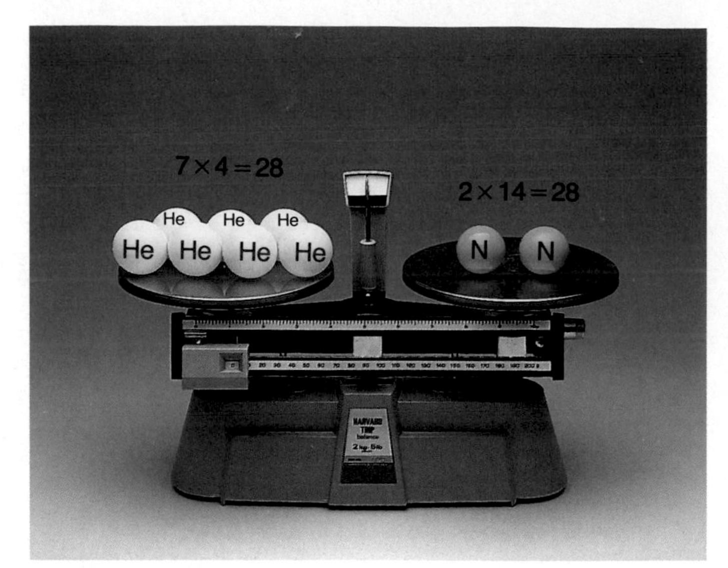

Figure 8.6 If it were possible to mass single atoms, the mass of two nitrogen atoms would equal the mass of seven helium atoms. Thus, the mass ratio of N to He is 7 to 2.

CHEMISTS AND THEIR WORK

Amadeo Avogadro
(1776-1856)

Avogadro initially studied and practiced law in his native Italy. After three years of law practice, he turned to science and spent the remainder of his life as a professor of science at the University of Turin. Avogadro is best known for the hypothesis that equal volumes of gases contain equal number of particles, which he proposed in 1811. He was the first person to distinguish between atoms and molecules. In fact, he coined the word "molecule." The ideas first proposed by Avogadro enabled chemists to determine a table of realistic atomic masses—something they could not do until they understood the difference between atoms and molecules.

The Avogadro Constant

Chemists do not deal with amounts of substances by counting out atoms or molecules. Rather, they obtain the masses of quantities of substances. Thus, it is important to obtain a relationship between mass and number of particles.

There is a problem in using the molecular masses of substances. Molecular masses are in atomic mass units. An atomic mass unit is only 1.66×10^{-24} gram. The mass of a single molecule is so small that it is impossible to measure by ordinary means in the laboratory. For everyday use in chemistry, a larger unit, such as a gram, is needed.

One helium atom has a mass of 4 u, and one nitrogen atom has a mass of 14 u. The ratio of the mass of one helium atom to one nitrogen atom is 4 to 14, or 2 to 7. Let us compare the mass of two helium atoms to that of two nitrogen atoms. The ratio would be 2×4 to 2×14, or 2 to 7. If we compare the mass of 10 atoms of each element, we will still get a 2 to 7 ratio. No matter what number of atoms we compare, equal numbers of helium and nitrogen atoms will have a mass ratio of 2 to 7. In other words, the numbers in the atomic mass table give us the relative masses of the atoms of the elements.

The laboratory unit of mass you will use is the gram. We would like to choose a number of atoms that would have a mass in grams equivalent to the mass of one atom in atomic mass units. The same number would fit all elements, because equal numbers of different atoms always have the same mass ratio. Chemists have found that 6.02×10^{23} atoms of an element have a mass in grams equivalent to the mass of one atom in atomic mass units. For example, a single atom of hydrogen has a mass of 1.0079 u;

CONTENT BACKGROUND

Avogadro Constant: Today, the most accurate determinations of the Avogadro constant are made using X-ray diffraction data on the spacing of atoms in crystals together with density data, particularly with calcite, $CaCO_3$. Another method producing reliable results is combining the charge of the electron (that is accurately measured in the Millikan oil-drop experiment) with electrolytic deposition of a massed amount of silver. Measuring radioactive decay rates and collecting the alpha particles emitted also gives fair results. One of the most famous determinations was that of Perrin, who set out to demonstrate Einstein's explanation of Brownian motion. In the process, he silenced the last critics of the atomic theory. Perrin studied the effect of gravity on the distribution of colloidal particles by height. The relationship

6.02×10^{23} atoms of hydrogen have a mass of 1.0079 g. This number, 6.02×10^{23}, is called the **Avogadro constant** in honor of a 19th century Italian scientist.

As you recall, the atomic mass unit is defined as 1/12 the mass of a carbon-12 atom. Thus, the atomic mass of a carbon-12 atom is exactly 12 u. Suppose we were to count the number of atoms in exactly 12 grams of carbon-12. The number of atoms would have the value of the Avogadro constant.

The Mole

The Avogadro constant is an accepted SI standard. We may add it to the others we have studied: the kilogram, the meter, the second, and the kelvin. The symbol used to represent the Avogadro constant is N_A. This quantity can be expressed as $6.022\,1367 \times 10^{23}$ to be more precise. This number of things is called one **mole (mol)** of the things. Chemists have chosen one mole as a standard unit for large numbers of atoms, ions, or molecules. Thus, the mole is the SI base unit representing the chemical quantity of a substance. One mole of particles (atoms, ions, molecules) has a mass in grams equivalent to that of one particle in atomic mass units. Thus, if a mole of any type of particle has a mass of 4.02 grams, then a single particle has a mass of 4.02 u. In the same manner, if a single particle has a mass of 54.03 u, then a mole of particles will have a mass of 54.03 grams. Chemists chose the value of the Avogadro constant so that the mass of N_A atoms in grams is equivalent to the mass of one atom in atomic mass units. Because of this relationship, we may associate a second set of units with the numbers in the atomic mass column: grams/mole of atoms.

Figure 8.7 Shown here are molar quantities of several substances. Clockwise from left are 32.1 g of sulfur, 342 g of sucrose (table sugar), 249 g of copper (II) sulfate pentahydrate, 58.5 g of sodium chloride (table salt), 63.5 g of copper, and 216 g of mercury(II) oxide.

he used was

$$\frac{n_1}{n_2} = e\left[\frac{mg(\rho - \rho')\,N_A\,(h_1 - h_2)}{\rho RT}\right]$$

where m was the mass of a particle of density ρ suspended in a medium of density ρ' at absolute temperature T. n_1 and n_2 are the concentrations of particles at heights h_1 and h_2. R is the universal gas constant, g is the acceleration of gravity, and N_A is the Avogadro constant. It is even possible to obtain a good value for the constant from the blue color of the sky. The sky is blue because the molecules of air are more effective at scattering blue light than red. By making quantitative measurements of sunlight and sky color and then applying the equations associated with the Rayleigh scattering law, a value for the Avogadro constant is obtained.

CONCEPT DEVELOPMENT

Background: Through the years, there have been various methods that scientists have used in determining Avogadro's number. Perrin, a French physicist, made measurements of Brownian motion in natural gum. He found the number of molecules per mole to be 6.8×10^{23}. Perrin was the first to use the term *Avogadro constant.* Rutherford later performed an experiment using helium produced by the alpha decay of certain radioactive nuclides. He calculated N_A as 6.16×10^{23} atoms per mole. Millikan's oil drop experiment gave the accepted value of 6.023×10^{23} electrons per mole.

You can use the above background information to remind students that chemistry is alive, and constantly changing. As our instruments improve, so do our measurements. Today it is possible to detect 1 grain of NaCl dissolved in a swimming pool (1 part per trillion).

- The size of a mole is nearly beyond comprehension, but computer software can be used to help students visualize this concept. See the Computer Demo for a suggestion.

GLENCOE TECHNOLOGY

Videodisc

Chemistry: Concepts and Applications
Molar Mass
Disc 2, Side 1, Ch. 6

Also available on CD-ROM.

Software

Mastering Concepts in Chemistry
Unit 6, Lesson 1
The Mole Concept
Unit 6, Lesson 2
Molarity

CONCEPT DEVELOPMENT

Discussion Question: Ask a student if he or she could drink one mole of water molecules. Show the class 1 mole of water, 18 cm^3 of water, in a graduated cylinder. Show them 1 mole of several chemicals, such as 12 g of carbon, and so on. A 14-inch (37 cm) diameter inflated beach ball contains about 25 dm^3 of air. This is about the volume of 1 mole of any gas at room temperature.

QUICK DEMO **A Mobile Mole**

As you develop the mole concept, ask interested students to build a mobile for your classroom showing the branching into particles on one side and mass on the other side. Below each hang the units. This visual reference can help students as they begin to solve mole problems.

Correcting Errors

Because the mole as a counting unit is unfamiliar to students, it may help to relate the following:

1 dozen = 12 items

1 mole = 6.02×10^{23} items

Work through "dozen" problems before you go through the Sample Problems. How many cookies are in three dozen? Use the factor-label method. Students can transfer this thinking process to the second and third Sample Problems.

N_A is a number that has been experimentally determined. In 1905, two English scientists, William Ramsay and Frederick Soddy, devised an experiment to determine the Avogadro constant using radium. Radioactive radium decays by emitting an alpha particle, which then gains two electrons and forms a helium atom. If the volume of the helium produced by a sample of radium is collected after a period of time and measured, it is possible to compute the value for N_A. Using modern instruments, the value of N_A has been found to be 6.02×10^{23} atoms. If a mole of iodine molecules is crystallized and then inspected by X-ray diffraction, the number of I_2 molecules in the crystal can be determined. Again, the number is 6.02×10^{23} molecules. The Avogadro constant has also been determined by the scattering of light and by Millikan's oil drop experiment, which was described in Chapter 4.

It is important to note that *one mole of atoms contains 6.02×10^{23} atoms. One mole of molecules contains 6.02×10^{23} molecules. One mole of formula units contains 6.02×10^{23} formula units. One mole of ions contains 6.02×10^{23} ions.* N_A, therefore, can have any of these units.

$$\frac{\text{atoms}}{\text{mole}} \text{ or } \frac{\text{molecules}}{\text{mole}} \text{ or } \frac{\text{formula units}}{\text{mole}} \text{ or } \frac{\text{ions}}{\text{mole}}$$

The mass of one mole of molecules, atoms, ions, or formula units is called the **molar mass** of that species.

SAMPLE PROBLEMS

1. Conversion of Grams to Moles

Use the factor-label method to determine how many moles are represented by 11.5 g of ethanol, C_2H_5OH.

Solving Process:

We have grams, and we wish to convert to moles. Using the atomic mass table inside the back cover, we can determine that one mole of C_2H_5OH has a mass of 46.1 g. Therefore,

$$1 \text{ mol} = 46.1 \text{ g} \quad \text{or} \quad \frac{1 \text{ mol}}{46.1 \text{ g}} = 1$$

$$\frac{11.5 \cancel{\text{g } C_2H_5OH}}{} \left| \frac{1 \text{ mol } C_2H_5OH}{46.1 \cancel{\text{g } C_2H_5OH}} \right. = 0.249 \text{ mol } C_2H_5OH$$

2. Conversion of Molecules to Moles and Grams

How many moles is 1.20×10^{25} molecules of ammonia, NH_3? What mass is this number of molecules?

Solving Process:

(a) We have molecules and wish to convert to moles. One mole equals 6.02×10^{23} molecules. Use the ratio

$$\frac{1 \text{ mol}}{6.02 \times 10^{23} \text{ molecules}}$$

CONTENT BACKGROUND

Super Computer: A supercomputer working at 8 billion instructions per second can count the world's population in 0.6 second but would require 2.4 million years to count to a number equal to the Avogadro constant. If you share this information with your students, you will need to remind them that this very large number is of very small particles.

$$\frac{1.20 \times 10^{25} \text{ molecules } NH_3}{} \left| \frac{1 \text{ mol}}{6.02 \times 10^{23} \text{ molecules}} \right.$$

$= 1.99 \times 10^1$ mol NH_3
$= 19.9$ mol NH_3

(b) Using the table inside the back cover, the molar mass of NH_3 is calculated to be 17.0 g. Use

$$\frac{17.0 \text{ g } NH_3}{1 \text{ mol } NH_3}$$

$$\frac{19.9 \text{ mol } NH_3}{} \left| \frac{17.0 \text{ g } NH_3}{1 \text{ mol } NH_3} \right. = 338 \text{ g of } NH_3$$

PRACTICE PROBLEMS

Make each of the following conversions.

34. 0.638 mole $Ba(CN)_2$ to grams
35. 50.4 grams $CaBr_2$ to moles
36. 1.26 moles NbI_5 to grams
37. 86.2 grams C_2H_4 to moles

SAMPLE PROBLEM

Conversion of Grams to Atoms

Determine the number of atoms that are in a 10.0-g sample of calcium metal.

Solving Process:
First, find the number of moles in 10.0 g Ca, and then convert to atoms. 1 molar mass of calcium is 40.1 g. Therefore, use the ratios

$$\frac{1 \text{ mol Ca}}{40.1 \text{ g Ca}} \text{ and } \frac{6.02 \times 10^{23} \text{ atoms}}{1 \text{ mol}}$$

$$\frac{10.0 \text{ g Ca}}{} \left| \frac{1 \text{ mol Ca}}{40.1 \text{ g Ca}} \right| \frac{6.02 \times 10^{23} \text{ atoms}}{1 \text{ mol}}$$

$= 1.50 \times 10^{23}$ atoms Ca

PRACTICE PROBLEMS

Make each of the following conversions.

38. 0.943 mole H_2O to molecules
39. 7.74×10^{26} formula units Al_2O_3 to moles
40. 91.9 grams NH_4IO_3 to formula units
41. 6.63×10^{23} molecules $C_6H_{12}O_6$ to moles

PROBLEM SOLVING HINT

Use the concept of the mole to relate molar mass and the mass of the sample. Use the concept of a mole again to relate the number of atoms to the molar mass.

PRACTICE

Guided Practice: Guide students through the Sample Problems, then have them work in class on Practice Problems 34-35 that follow.
Independent Practice: Your homework or classroom assignment can include the remaining Practice Problems 36-37, Review and Practice 93-94.

Answers to
Practice Problems Have students refer to Appendix C for complete solutions to Practice Problems.
34. 121 g $Ba(CN)_2$
35. 0.252 mol $CaBr_2$
36. 917 g NbI_5
37. 3.08 mol C_2H_4

Correcting Errors

Point out that students need only be concerned with two conversion factors. One involves grams and the numerical values come from the periodic table; the second utilizes Avogadro's constant.

PRACTICE

Guided Practice: Guide students through the Sample Problem, then have them work in groups on Practice Problems 38-41.
Cooperative Learning: Refer to pp. 28T-29T to select a teaching strategy to use with this activity.
Independent Practice: Your homework or classroom assignment can include Review and Practice 93-94 from the end of the chapter.

Answers to
Practice Problems: Have students refer to Appendix C for complete solutions to Practice Problems.
38. 5.68×10^{23} molecules H_2O
39. 1.29×10^3 mol Al_2O_3
40. 2.87×10^{23} formula units NH_4IO_3
41. 1.10 mol $C_6H_{12}O_6$

✓ ASSESSMENT

Knowledge: Have students construct a network tree concept map that starts with *mole* at the top branching into *particles* and *mass.* Under each of these have students place the correct units such as *atoms, ions,* and *molecules* under *particles,* and *grams* under *mass.*

CONCEPT DEVELOPMENT

- The mole concept is extended in calculating solution concentrations. Molarity is defined by the IUPAC as the ratio of the moles of solute to the volume of solution in dm^3. The change in the volume unit will not be noticed by your students since they are learning it for the first time. They will think of molarity as mol/dm^3. It is important to teach units in use throughout the world (SI) and that appear in scientific literature.

Extension: Discuss with students why we cannot continue to use dilution by rivers and oceans as a method of sewage treatment. Possible response: The concentration of potentially toxic chemicals is increasing. At some point in the future, the threshold level may be reached, resulting in catastrophic loss of life.

✓ ASSESSMENT

Performance: Place at each lab station a bottle of a different salt with a 3x5 card on which you have written the amount and concentration of a solution. Students are asked to prepare, for example, 150 cm^3 of 0.2*M* $Zn(NO_3)_2$. The lab team must have their calculations verified by you.

If you want students to go ahead and prepare solutions, have them make solutions for future lab experiments or demonstrations, thus saving you prep time in the future. Students should use distilled water. Provide appropriate precautions for handling the chemicals you have selected.

Figure 8.8 A 0.1000*M* solution of $CuSO_4 \cdot 5H_2O$ is made by dissolving 24.97 g of solid $CuSO_4 \cdot 5H_2O$ in enough water to make exactly 1000 cm^3 of solution.

Moles in Solution

Most chemical reactions take place in solution. There are several methods of expressing the relationship between the dissolved substance and the solution. The method most often used by chemists is molarity (*M*). **Molarity** is the ratio between the moles of dissolved substance and the volume of solution expressed in cubic decimeters.

$$\text{Molarity} = \frac{\text{moles of solute}}{\text{volume of solution in dm}^3}$$

A one-molar (1*M*) solution of nitric acid contains one mole of nitric acid molecules in one cubic decimeter of solution. A 0.372*M* solution of $Ba(NO_3)_2$ contains 0.372 mole of $Ba(NO_3)_2$ in 1 dm^3 of solution.

Assume that you wish to try a reaction using 0.1 mole of glucose ($C_6H_{12}O_6$). If the solution of glucose in your laboratory is 1*M*, then 0.1 mole of glucose would be contained in 0.1 dm^3, or 100 cm^3 of the glucose solution. Expressing the composition of solutions in units of molarity is a convenient way of measuring a number of particles.

Note that when the concentration of a solution is expressed in molarity, the expression uses cubic decimeters of *solution* and not cubic decimeters of solvent. For example, to make a 1.0*M* glucose solution, you do not add one cubic decimeter of water to one mole (180 g) of glucose. The final volume might not be one cubic decimeter. Rather, dissolve the glucose in water, and then add water until you reach one cubic decimeter of solution.

It is important to be able to compute the molarity of solutions if you are given their composition. It is also important to be able to compute the amount of substance you need to produce a specific solution. You will be using these calculations in your laboratory work and later in solving problems in this book.

SAMPLE PROBLEM

Molarity

What is the molarity of 2.50×10^2 cm^3 of solution containing 9.46 g CsBr, cesium bromide, a compound that is used to make optical devices such as prisms and spectrophotometer cells?

Solving Process:

The units of molarity are moles of substance per (divided by) dm^3 of solution. Thus, to obtain molarity we must divide the substance, expressed in moles, by the solution, expressed in dm^3. The table inside the back cover can be used to determine the molar mass of cesium bromide. Recall from Chapter 2 that 1 dm^3 = 1000 cm^3.

$$\frac{9.46\ \cancel{\text{g CsBr}}}{} \left| \frac{1 \text{ mol CsBr}}{213\ \cancel{\text{g CsBr}}} \right. = 0.0444 \text{ mol CsBr}$$

DEMONSTRATION

8-3 A 0.1 Molar Solution

Purpose: to demonstrate preparation of a solution of known concentration by serial dilution

Materials: 1.58 g $KMnO_4$, 280 cm^3 water, 100-cm^3 graduated cylinder or volumetric flask, 10-cm^3 pipette or graduated cylinder, pipette bulb, 3 100-cm^3 beakers

Procedure: 1. In front of your class, go through the calculations necessary to prepare 100 cm^3 of a 0.1*M* solution of $KMnO_4$ (158.04 g/mol).
2. Prepare the 0.1*M* $KMnO_4$ by dissolving 1.58 g $KMnO_4$ in enough water to make 100 cm^3 of solution.
3. Then show the class how to use this solution to prepare, by serial dilution, a 0.01*M* $KMnO_4$ by pipetting 10 cm^3 of the 0.1*M* solution into a 100-cm^3 graduated cylinder or volumetric

$$\frac{2.50 \times 10^2 \cancel{cm^3}}{} \left| \frac{1\ dm^3}{1000\ \cancel{cm^3}} \right. = 0.250\ dm^3$$

$$\frac{0.0444\ mol\ CsBr}{0.250\ dm^3} = 0.178M\ CsBr$$

To save time and to reduce the chance for error, these calculations can be combined in one continuous chain. Note that the factors are arranged to yield an answer in mol/dm³, which is molarity.

$$\frac{9.46\ \cancel{g\ CsBr}}{2.50 \times 10^2\ \cancel{cm^3}} \left| \frac{1\ mol\ CsBr}{213\ \cancel{g\ CsBr}} \right| \frac{1000\ \cancel{cm^3}}{1\ dm^3} = 0.178M\ CsBr$$

PRACTICE PROBLEMS

Compute the molarity of each of the following solutions.

42. 5.23 g $Fe(NO_3)_2$ in 100.0 cm³ of solution

43. 8.55 g NH_4I in 50.0 cm³ of solution

44. 9.94 g $CoSO_4$ in 2.50×10^2 cm³ of solution

45. 44.3 g $Pb(ClO_4)_2$ in 250.0 cm³ of solution

SAMPLE PROBLEM

Making a Solution

How would you make 5.00×10^2 cm³ of a 0.133*M* solution of $MnSeO_4$, manganese(II) selenate, and water?

Solving Process:

First, use the given quantity of solution, 5.00×10^2 cm³, to find the part of a dm³ desired. Then use the molarity to convert from solution to moles of substance. Finally, convert the moles of substance to grams.

$$\frac{5.00 \times 10^2\ \cancel{cm^3}}{} \left| \frac{1\ \cancel{dm^3}}{1000\ \cancel{cm^3}} \right| \frac{0.133\ \cancel{mol\ MnSeO_4}}{1\ \cancel{dm^3}} \left| \frac{198\ g\ MnSeO_4}{1\ \cancel{mol\ MnSeO_4}} \right. = 13.2\ g\ MnSeO_4$$

To make the solution, then, you would dissolve 13.2 g $MnSeO_4$ in sufficient water to make 5.00×10^2 cm³ of solution.

PRACTICE PROBLEMS

Describe the preparation of each of the following solutions.

46. 1.00 dm³ of 3.00*M* $NiCl_2$

47. 2.50×10^2 cm³ of 4.00*M* $CoCl_2$

48. 0.500 dm³ of 1.50*M* AgF

49. 2.50×10^2 cm³ of 0.002 00*M* $Cd(IO_3)_2$

flask and adding enough water to make 100 cm³ of solution.

4. Repeat the procedure to make 100 cm³ of the 0.001*M* solution. **Disposal:** G

Results: The change in color allows the students to verify that a dilution has been made.

Questions: 1. Which solution is the darkest in color? *The most concentrated solution is darkest.*

2. When making a dilution, what important fact must be remembered? *The total volume must remain the same.*

3. What is the definition of a 1 molar solution? *One mole of solute is dissolved in enough solvent to make one cubic decimeter of solution.*

CONCEPT DEVELOPMENT

Computer Demo: *Molarity,* A problem generator and tutor. AP502, PC2501, Project SERAPHIM

Correcting Errors

Students are sometimes confused when a solution's volume is expressed in scientific notation. Point out to them that 2.50×10^2 cm³ is not the same as 250 cm³. The volume is written in scientific notation to show properly that the volume is measured to the nearest cm³.

PRACTICE

Guided Practice: Guide students through the Sample Problems, then have them work in groups on Practice Problems 42-43, 46-47.

Cooperative Learning: Refer to pp. 28T-29T to select a teaching strategy to use with this activity.

Independent Practice: Your homework or classroom assignment can include the remaining Practice Problems 44-45, 48-49, and Review and Practice 95-100.

Answers to

Practice Problems: Have students refer to Appendix C for complete solutions to Practice Problems.

42. 0.291*M* $Fe(NO_3)_2$

43. 1.18*M* NH_4I

44. 0.257*M* $CoSO_4$

45. 0.436*M* $Pb(ClO_4)_2$

46. Dissolve 3.90×10^2 g $NiCl_2$ in enough water to make 1.00 dm³ of solution.

47. Dissolve 1.30×10^2 g $CoCl_2$ in enough water to make 2.50×10^2 cm³ of solution.

48. Dissolve 95.3 g AgF in enough water to make 0.500 dm³ of solution.

49. Dissolve 0.231 g $Cd(IO_3)_2$ in enough water to make 2.50×10^2 cm³ of solution.

PROGRAM RESOURCES

Enrichment: Use the worksheet "Solution Concentration," pp. 15-16, to challenge more able students.

MINUTE LAB

Bite the Bubble

Purpose: to have students calculate the percentage content of an ingredient in a commercial product

Materials: 5 pieces of sugared bubble gum, paper cup, balance

Results: Sample data—mass of gum before chewing and cup, 48.31g; mass of gum after chewing and cup, 22.30g; mass of dissolved sugar, 26.01g; and percentage of sugar, 64.4%.

Answers to Questions: 1. Accept reasonable answers. **2.** No, because sugar feeds the bacteria in the mouth that secrete acid that attacks the tooth enamel. **3.** The basic assumption is that only the sugar dissolves and is removed from the gum. The mass of the flavoring lost is ignored. **4.** The uncertainty due to instrumentation becomes a smaller part of the answer when the sample size increases.

✓ ASSESSMENT

Performance: Repeat the procedure and calculations using sugarless gum. Students will be surprised at the high percentage of sweetener used.

PRACTICE

Guided Practice: Guide students through the Sample Problems, then have them work in class on Practice Problems 56-57 that follow.

Independent Practice: Your homework or classroom assignment can include Review and Practice 103-104, and Application 137-141.

Answers to
Practice Problems: Have students refer to Appendix C for complete solutions to Practice Problems.

56. CeI_3
57. CH_2

PROGRAM RESOURCES

Critical Thinking/Problem Solving: Use the worksheet, p. 8, as enrichment for the concept of "Just How Big Is It?"

MINUTE LAB

Bite the Bubble

Work together in groups of five. Use a balance to determine the mass of a clean paper cup. Unwrap five pieces of bubble gum containing sugar and place them in the cup. Determine the mass of the cup and the gum. Each person in the group should chew a piece of gum to dissolve and remove the sugar from the gum. After about five minutes, collect the chewed gum in the cup that you used previously. Wash your hands after handling the chewed gum. Determine the mass of the cup and the chewed gum. From your data, determine the mass of the sugar that was dissolved from the gum. Calculate the percentage of sugar in the gum by dividing the mass of the dissolved sugar by the mass of the unchewed gum and multiplying this value by 100. What is the percentage of sugar? Do you think a dentist would recommend chewing this gum? Why? What assumption are you making about the mass of the sugar and the difference in the masses of the gum before and after chewing? How would using more pieces of gum affect the results of this Minute Lab?

2. Empirical Formula from Percentage Composition

A compound has a percentage composition of 40.0% C, 6.71% H, and 53.3% O. What is the empirical formula?

Solving Process:

To calculate the ratio of moles of these elements, we assume a convenient amount of compound, usually 100 g. Then the percentages of the elements have the same numerical value in grams. In the present example, a 100-g sample would have 40.0 g of C, 6.71 g of H, and 53.3 g of O. We then change the quantities to moles.

$$40.0\ \cancel{\text{g C}} \times \frac{1\ \text{mol C}}{12.0\ \cancel{\text{g C}}} = 3.33\ \text{mol C}$$

$$6.71\ \cancel{\text{g H}} \times \frac{1\ \text{mol H}}{1.01\ \cancel{\text{g H}}} = 6.64\ \text{mol H}$$

$$53.3\ \cancel{\text{g O}} \times \frac{1\ \text{mol O}}{16.0\ \cancel{\text{g O}}} = 3.33\ \text{mol O}$$

Dividing each result by 3.33 mol, we get 1 to 1.99 to 1. We round off to 1 to 2 to 1. The empirical formula is CH_2O.

PRACTICE PROBLEMS

56. Calculate the empirical formula of a compound that contains 1.67 g of cerium, Ce, and 4.54 g of iodine, I.

57. 2-Methylpropene is a compound used to make synthetic rubber. A sample of 2-methylpropene contains 0.556 g C and 0.0933 g H. Determine its empirical formula.

Sometimes dividing by the smallest number of moles does not yield a ratio close to a whole number. Suppose we have an empirical formula problem that produces a ratio of 2.33, or $2\frac{1}{3}$, to 1. What is the whole number ratio? If we multiply each member of the ratio by 3, the ratio becomes 7 to 3. Other common ratios are 1.33 to 1, or 4 to 3, and 1.67 to 1, or 5 to 3. When these ratios occur, another step is needed in the calculation of an empirical formula.

SAMPLE PROBLEM

Empirical Formula from Percentage Composition—Another Example

What is the empirical formula of a compound that is 66.0% Ca and 34.0% P?

Solving Process:

Assume a 100-g sample so that we have 66.0 g Ca and 34.0 g P. Convert these quantities to moles of atoms.

DEMONSTRATION

8-4 Empirical Formula

Purpose: to prepare magnesium oxide and determine its empirical formula

Materials: crucible, crucible tongs, laboratory burner, clay triangle, ring stand and ring, medicine dropper, glass stirring rod, 70-cm piece of magnesium ribbon, balance, 1 cm^3 distilled water

Procedure: 1. Record the mass of the crucible. Loosely roll the 70-cm piece of magnesium ribbon into a ball and place in the crucible. Mass the crucible and Mg ribbon.
2. Place the crucible and Mg ball on a clay triangle. Heat the crucible strongly with a laboratory burner. When the Mg begins to burn, turn off the flame. Allow the Mg to continue to burn.
3. After the crucible has cooled, powder the contents with the tip of a

$$\frac{66.0\ \text{g Ca}}{} \left| \frac{1\ \text{mol Ca}}{40.1\ \text{g Ca}} \right. = 1.65\ \text{mol Ca}$$

$$\frac{34.0\ \text{g P}}{} \left| \frac{1\ \text{mol P}}{31.0\ \text{g P}} \right. = 1.10\ \text{mol P}$$

Dividing both results by 1.10 mol, we obtain 1.50 to 1. The result is not close to a whole number. When we substitute the fractional form of 1.50, we have

$$\frac{1.5}{1} = \frac{3/2}{1}$$

We then multiply by 2/2:

$$\frac{3/2 \times 2}{1 \times 2} = \frac{3}{2}$$

Thus, the empirical formula is Ca_3P_2.

EVERYDAY CHEMISTRY

Trees, Natural Regulators

One problem facing scientists today is the buildup of carbon dioxide in the atmosphere. Nearly 20% of all carbon dioxide emissions come from electrical energy production for homes. About 0.35 kg of carbon products, including carbon dioxide, are released into the atmosphere for each kilowatt-hour (kWh) of electrical energy that is produced.

Trees absorb carbon dioxide and store it as carbon compounds in their biomass—trunks, branches, leaves, and roots. One tree can absorb about 0.6 kg of carbon dioxide for every cubic decimeter of wood in the tree. Further, through energy conservation, the tree can prevent 35 times that amount of carbon dioxide from entering the atmosphere. Planting trees around homes can save from 10-50% for cooling and 4-22% for heating.

Planting trees improves our environment in many ways. Trees provide a habitat for living creatures, beautify urban landscapes, and give shade. In the future, their role as regulators of carbon dioxide in the atmosphere will only increase our appreciation for trees.

Exploring Further

1. Estimate the volume of a tree near your home or school. How much carbon dioxide could this tree absorb?
2. What is happening to the Amazon rain forest? What effect do you think this will have on the atmosphere?
3. Find out how trees use carbon dioxide. Discuss the chemistry involved.

glass stirring rod.

4. Have a student note any odor as you slowly add 15 drops of distilled water.

5. Heat the crucible strongly for 3 to 5 minutes to dry the residue. Cool for 5 minutes. Record the mass of crucible and contents. **Disposal:** A

Results: The Mg ribbon burns in air to form MgO and Mg_3N_2. Three water molecules react with the Mg_3N_2 to form 3MgO and $2NH_3$ gas. Sample data includes 0.57 g Mg ribbon, 0.96 g MgO, and 0.39 g oxygen.

Questions: 1. Calculate the number of moles of magnesium and the number of moles of oxygen. *0.024 mol Mg, 0.024 mol O*

2. What is the simplest whole number ratio of Mg to O? *1:1*

3. Use this whole number ratio to write the empirical formula of magnesium oxide. *MgO*

EVERYDAY CHEMISTRY

Trees, Natural Regulators

Background: The destruction of tropical rain forests affects both ecology and economy.

Teaching This Feature: Ask students to debate the question after researching the issue.

Answers to **Exploring Further**

1. Answers will vary. A 30-cm diameter tree will absorb 424 kg of CO_2.

2. Millions of acres each year are being cut. CO_2 will increase.

3. $6CO_2(g) + 6H_2O(l) \rightarrow C_6H_{12}O_6(aq) + 6O_2(g)$. The carbon dioxide is converted into biomass in the presence of sunlight, chlorophyll and water.

Correcting Errors

To help students handle the more complex empirical formula problems in which the ratio of the moles of elements is not an "integer ratio, have them think of common decimal fractions, such as 0.5, 0.25, and 0.33. Ask students what factor each fraction would have to be multiplied by to have a product that is a whole number.

GLENCOE TECHNOLOGY

Videodisc

Chemistry: Concepts and Applications

Percent Sugar in Bubblegum

Disc 2, Side 1, Ch. 5

Show this video of the Minute Lab to introduce the concept of percentage composition.

Also available on CD-ROM.

Software

Mastering Concepts in Chemistry

Unit 6, Lesson 3

Analytical Calculations Using Moles

Answers to
Review and Practice (cont.)

95. a. $0.0291 M\ Cr_2(SO_4)_3$
b. $0.323 M\ NH_4Br$
c. $0.0631\ M\ CuF_2$
d. $0.0716 M\ CdCl_2$

96. a. 8.60×10^2 g $MnBr_2$
b. 521 g LiBr
c. 184 g $MgBr_2$
d. 227 g $MnSO_4$

97. 1.7 g K_2SO_4 dissolved in enough water to make 100.00 cm^3 of solution.

98. $2.50 M\ AgNO_3$

99. $0.0500 M$ NaI

100. 0.5 g KSCN dissolved in enough water to make 50.0 cm^3 of solution.

101. a. 22.3% Na
77.6% Br
b. 40.0% C
6.73% H
53.3% O
c. 84.1% Hg
15.9% F
d. 67.1% Zn
32.9% S

102. a. NaO
b. C_2H_5
c. KBr
d. $Al_2(SO_4)_3$

103. a. Rb_2O
b. UCl_5
c. $Ni_3P_2O_8$ or $Ni_3(PO_4)_2$
d. $CaSO_4$

104. a. $CrCl_3$
b. CoSe
c. $LaCl_3$
d. Ta_2O_5

105. C_6H_6

106. $P_6N_6Cl_{12}$

107. a. $Tl(NO_3) \cdot 3H_2O$
b. $Sn(NO_3)_2 \cdot 20H_2O$
c. $UO_2(HCOO)_2 \cdot H_2O$
d. $UO_2HPO_4 \cdot 4H_2O$

108. a. $AlPO_4$

109. c. $BaSiO_3$

110. b. $Bi(OH)_3$

111. $CaSO_3$ molecular formula

112. The dot indicates that there is a chemical bond between the ionic compound and water.

113. a. $Mg(NO_3)_2 \cdot 6H_2O$
b. $FeSO_4 \cdot 7H_2O$
c. $Cu(NO_3)_2 \cdot 3H_2O$
d. $SnCl_2 \cdot 2H_2O$

114. a. magnesium sulfate monohydrate
b. manganese(II) nitrate tetrahydrate
c. calcium chromate dihydrate

95. Compute the molarities for the indicated volumes of solution.
a. 5.70 g $Cr_2(SO_4)_3$ in 5.00×10^2 cm^3
b. 7.90 g NH_4Br in 2.50×10^2 cm^3
c. 6.44 g CuF_2 in 1.00×10^3 cm^3
d. 13.1 g $CdCl_2$ in 1.00×10^3 cm^3

96. Describe the preparation of each of the following solutions.
a. 1.00 dm^3 of $4.00M$ $MnBr_2$
b. 1.00 dm^3 of $6.00M$ LiBr
c. 5.00×10^2 cm^3 of $2.00M$ $MgBr_2$
d. 1.00×10^3 cm^3 of $1.50M$ $MnSO_4$

97. How would you prepare 100.0 cm^3 of $0.10M$ potassium sulfate, K_2SO_4?

98. What is the molarity of a solution of silver nitrate that contains 42.5 g $AgNO_3$ in 100.0 mL of solution?

99. What is the molarity of a solution of sodium iodide that contains 3.75 g NaI in 0.500 L of solution?

100. How would you prepare 50.0 cm^3 of $0.1M$ potassium thiocyanate, KSCN?

101. Find the percentage composition of each element in the following compounds.
a. NaBr **c.** HgF_2
b. CH_3COOH **d.** ZnS

102. Give the empirical formula of each.
a. Na_2O_2 **c.** KBr
b. C_4H_{10} **d.** $Al_2(SO_4)_3$

103. Find the empirical formulas of compounds with the following composition.
a. 63.0 g Rb, 5.90 g O
b. 0.159 g U, 0.119 g Cl
c. 7.22 g Ni, 2.53 g P, 5.25 g O
d. 0.295 g Ca, 0.236 g S, 0.469 g O

104. Find the empirical formulas of compounds with the following analyses.
a. 32.8% Cr, 67.2% Cl
b. 42.7% Co, 57.3% Se
c. 56.6% La, 43.4% Cl
d. 81.9% Ta, 18.1% O

105. The percentage composition of a compound is 92.3% C and 7.7% H. If the molecular mass is 78 u, what is the molecular formula?

106. Find the molecular formula of a compound with percentage composition 26.7% P, 12.1% N, and 61.2% Cl and molecular mass 695 u.

107. Find the formulas for the hydrates with the following analyses.
a. 5.262 g $Tl(NO_3)_3$ and 0.728 g H_2O
b. 2.94 g $Sn(NO_3)_2$ and 4.37 g H_2O
c. 8.351 g $UO_2(HCOO)_2$ and 0.421 g H_2O
d. 17.02 g UO_2HPO_4 and 3.42 g H_2O

108. A compound is found to have a percentage composition of 22.1% aluminum, 25.4% phosphorus, and 52.5% oxygen. Which of the following compounds could it be?
a. $AlPO_4$ **c.** $Al_4(P_2O_7)_3$
b. $Al(PO_3)_3$ **d.** $AlPO_3$

109. A compound is found to have a percentage composition of 64.3% barium, 13.2% silicon, and 22.5% oxygen. Which of the following compounds could it be?
a. Ba_2SiO_4 **b.** $Ba_3Si_2O_7$ **c.** $BaSiO_3$

110. A compound is found to have a percentage composition of 80.4% bismuth, 18.5% oxygen, and 1.16% hydrogen. Which of the following compounds could it be?
a. $Bi(OH)_5$ **b.** $Bi(OH)_3$ **c.** H_3BiO_4

111. A compound is found to contain 33.3% calcium, 40.0% oxygen, and 26.7% sulfur. Its formula mass is 120 u. What is the formula for the compound?

112. What is the meaning of the dot in the formula $CuSO_4 \cdot 5H_2O$?

113. Write formulas for the following hydrates.
a. magnesium nitrate hexahydrate
b. iron(II) sulfate heptahydrate
c. copper(II) nitrate trihydrate
d. tin(II) chloride dihydrate

114. Name each of the following hydrates.
a. $MgSO_4 \cdot H_2O$
b. $Mn(NO_3)_2 \cdot 4H_2O$
c. $CaCrO_4 \cdot 2H_2O$
d. $Na_2SO_4 \cdot 10H_2O$

Chapter 8
REVIEW

115. What is the formula for a hydrate that is 90.7% SrC_2O_4 and 9.30% H_2O?

116. What is the formula for a hydrate that is 76.9% $La_2(CO_3)_3$ and 23.9% H_2O?

117. What is the formula for a hydrate that is 86.7% Mo_2S_5 and 13.3% H_2O?

Concept Mastery

118. Why do chemists use 6.02×10^{23} as the number of things in a mole?

119. Why is it not practical to measure the mass of substances in atomic mass units?

120. Explain the difference between the terms mole and molarity.

121. A chemist finds that a certain compound has an empirical formula CH_3O. What does the chemist know about the compound on the basis of this formula?

122. Explain the difference between empirical formula and molecular formula.

123. Why do we not make the Avogadro constant some convenient round number such as 1×10^{25}?

124. Why is it incorrect to apply the term *molecular mass* to ionic substances?

125. The empirical formulas of acetylene and benzene are the same. How could we tell which was which?

Application

126. In a routine hospital lab test, a student is found to have a cholesterol level of 2.3 g/dm^3. Cholesterol levels are usually expressed in milligrams per deciliter of blood. Express the student's cholesterol level in mg/dL.

127. The speed of light is 2.9979×10^8 m/s. How far does light travel in a nanosecond?

128. A Tollycraft 61 Motor Yacht travels at 23.7 knots at full speed. A knot is 1.852 kilometers per hour. What is the full speed of the yacht in meters per second?

129. The magnitude (brightness) of the star Nova Cygni varies over a period of 0.140 day. Express this period in seconds.

130. How many significant digits are in the quantity, 5.67×10^3 m? What does this expression tell you about how the measurement was made? Write the quantity without using scientific notation. Why is scientific notation necessary to express this measurement appropriately?

131. A spaceship from another planet travels at a speed of 4.27 googs/mulm. There are 256 googs in a meter and 8000 mulm in one hour. What is the spaceship speed in meters per second?

132. During the late 19th century cobalt blue glassware was popular. Cobalt blue is considered to be the most durable of all blue pigments. It is a mixture of compounds including the compound $Co(AlO_2)_2$. What is the percentage composition of this compound?

133. The compound mannitol, $C_6H_8(OH)_6$, is used as a sweetener in some dietetic foods. What is the percentage composition of this compound?

134. The compound 2,3,7,8-tetrachlorodioxin, also known as TCDD or, simply, dioxin, has been implicated as a cause of a number of physical ailments. Its formula is $C_{12}H_4Cl_4O_2$. What is the percentage of chlorine in the compound?

135. Caffeine, $C_8H_{10}N_4O_2$, and theophylline, $C_7H_8N_4O_2$, are both found in tea leaves, and both are used as medicines affecting hormone action in the human body. Compare the percentage of nitrogen found in the two compounds.

136. Indoleacetic acid, $C_{10}H_9NO_2$, and gibberellic acid, $C_{19}H_{22}O_6$, are compounds that occur naturally in plants and affect plant development. Indoleacetic acid increases growth rates of plants, while

d. sodium sulfate decahydrate

115. $SrC_2O_4 \cdot H_2O$

116. $La_2(CO_3)_3 \cdot 8H_2O$

117. $Mo_2S_5 \cdot 3H_2O$

Answers to
Concept Mastery

118. It is used so that one mole of particles will have a mass in grams numerically equal to the mass of one particle in atomic mass units.

119. The unit is too small; gradations on balances would be impossible to make.

120. mole—the number of particles equal to the Avogadro constant

molarity—moles of solute per dm^3 of solution

121. The carbon/hydrogen/oxygen ratio is 1:3:1.

122. Empirical—simplest ratio of atoms

Molecular—actual ratio of atoms per molecule

123. The mass of one mole in grams would not equal the mass of one particle in atomic or formula units.

124. Ionic substances do not form molecules.

125. Use physical properties, and measure the molecular masses.

Answers to
Application

126. 2.3×10^2 mg/dL

127. 0.299 79 m/ns

128. 12.2 m/s

129. 1.21×10^4 s

130. 3; it was made to the nearest 10; 5670; the last zero is not significant.

131. 0.0371 m/s

132. 33.3% Co; 30.5% Al; 36.2% O

133. 39.6% C; 7.75% H; 52.7% O

134. 44.1% Cl

135. 28.9% N in caffeine
31.1% N in theophylline

136. 68.6% C in indoleacetic acid
65.9% C in gibberellic acid
18.3% O in indoleacetic acid
27.7% O in gibberellic acid—higher % O

CHAPTER 9
Chemical Reactions

CHAPTER OVERVIEW

Main Ideas: A chemical equation represents chemical change. The system undergoing a chemical change will be characterized by a new set of properties. In a balanced equation, equal numbers of each kind of atom are present on both sides of the equation. Reactions can be classified into five general types: single and double displacement, decomposition, combustion, and synthesis. Products may be predicted once students learn the characteristics of the five general types.

Theme Development: Stability and change as a theme of the textbook is developed through a presentation of how chemical changes are classified and represented by balanced chemical equations.

Tying To Previous Knowledge: Point out that the ability to write and balance chemical equations depends upon students' abilities to name and write chemical formulas, which they learned in Chapter 7, and to analyze quantitative relationships involving chemical formulas, which they learned in Chapter 8.

ASSESSMENT PLANNER

Choose from the following assessment strategies.

Assess, pp. 230-231, 241-242
Assessment, pp. 229, 236, 239, 242, 244
Portfolio, p. 242
Minute Lab, p. 229
ChemActivity, pp. 810-812
Chapter Review, pp. 243-247

500 DOBSON UNITS
400
300
200
100

CHAPTER Preview

CONTENTS

DISCOVERY DEMO

9-1 A Dramatic Chemical Reaction

Purpose: to illustrate writing and balancing chemical equations

Materials: 220 cm^3 H_2O, 400-cm^3 beaker, 150-cm^3 beaker, 25 cm^3 liquid detergent, 5 cm^3 glycerine, 5 g sucrose, 1-m length rubber tubing, small funnel, meter stick, candle, matches, newspaper, masking tape

Procedure: 1. Prepare a soap bubble solution by adding 160 cm^3 H_2O, 25 cm^3 liquid detergent, and 5 cm^3 glycerine to the 400-cm^3 beaker.
2. In a separate beaker dissolve 5 g of sucrose in 60 cm^3 H_2O. Gently mix the two solutions together.
3. Connect rubber tubing to the gas outlet and the other end to a small funnel. Invert the funnel into a beaker of the soap mixture. Turn on the gas. Make a bubble using the gas.

Chemical Reactions

CHAPTER 9

What is destroying the ozone layer surrounding our planet? Can the thinning of the ozone layer, shown on the opposite page, be fixed? Both industrial and environmental scientists are beginning to work together to answer these important questions and to find a solution to these problems.

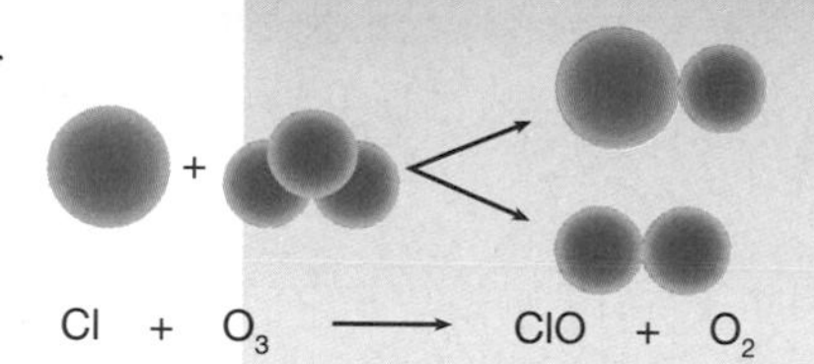

The depletion of the ozone content of the stratosphere has become a major concern to citizens throughout the world during the last few years. (See page 470 for more information about this problem.) The photograph opposite is the result of an assay of this ozone from the Nimbus 7 satellite. You can see that the area in the center has less ozone than other areas. In order to deal with this problem, chemists have had to investigate the many chemical reactions taking place in the stratosphere. Here you see the symbolic representation of one of these reactions. In this chapter, you will study how to represent reactions by equations and how to calculate the quantities of substances involved in those reactions.

EXTENSIONS AND CONNECTIONS

Chemistry and Technology

Developing Self-Fertilizing Plants: New developments in the chemical reaction that converts atmospheric nitrogen into a form that plants can use, page 232

Careers in Chemistry

Chemical Engineer, page 234

Chemists and Their Work

St. Elmo Brady, page 236

Frontiers

Gas Stations: A look at a new type of automotive fuel, page 238

Bridge to Biology

Soaking Up the Rays: How chlorophyll in plants converts sunlight to chemical energy, page 242

INTRODUCING THE CHAPTER

USING THE PHOTO

The large photo, that has been computer-enhanced from satellite data, shows a view of the South Pole. The large area in the center shows the "ozone hole" at a peak. Normally, the sparse ozone in the stratosphere shields Earth from much of the sun's ultraviolet radiation. However, ozone becomes depleted at times, allowing more ultraviolet radiation to pass through—particularly wavelengths harmful to living organisms. Scientists believe that atmospheric pollution, particularly by organic compounds called chlorofluorocarbons, or CFCs, is a major contributor to the depletion of the ozone layer. CFCs were used for many years as propellants in aerosol cans, in air conditioners (Freon), as solvents, and as the foaming agent in plastic foam. Part of the mechanism for this depletion is shown in the small diagram. Have students study this pictorial representation and describe the process taking place. Ask them to comment on the advantages and disadvantages of representing chemical processes in this way. Have students try to represent the process using symbols and formulas rather than pictures.

REVEALING MISCONCEPTIONS

As students study and master this chapter, they may think their knowledge of the classification of reactions is complete. Caution them that there are other types of reactions, such as oxidation-reduction that will be studied later.

4. Darken the room. Dislodge bubbles by turning the funnel sideways and gently shaking. As the bubble rises (natural gas) or sinks (propane gas), ignite it with a candle taped to the end of a meter stick. It is best to have an assistant hold the meter stick with the attached candle above or below the funnel before you shake it free. **CAUTION:** *Fire hazard. Do not do this demonstration near flammable materials.* Place a piece of newspaper on the floor to catch the dripping wax from the candle. **Disposal:** F

Results: There will be a flare-up as the bubble bursts and the entrapped gas burns with a luminous yellow flame.

Questions: 1. The combustion you observed is a chemical reaction between what two reactants? *methane (propane) and oxygen*

2. What products are produced? *carbon dioxide, light, heat, water vapor*

3. Is the combustion of this hydrocarbon fuel endothermic or exothermic? *exothermic*

Extension: Have students write descriptions of the reaction on the chalkboard. Then show students that a balanced chemical equation represents a more concise form of the same observation.

$$CH_4(g) + 2O_2(g) \rightarrow CO_2(g) + 2H_2O(g)$$

or

$$C_3H_8(g) + 5O_2(g) \rightarrow 3CO_2(g) + 4H_2O(g)$$

9.1 Chemical Equations

PREPARE

Key Concepts

The major concepts presented in this section include writing balanced chemical equations and classifying five chemical changes: single displacement, double displacement, decomposition, synthesis, and combustion.

Key Terms

chemical reaction
reactant
product
single displacement
double displacement
decomposition
synthesis
combustion

1 FOCUS

Objectives

Preview with your students the objectives listed on the student page. Students can use them as a study guide for this major section.

Activity

Bring a brownie mix to class. Have a student read the printed directions to the class. As the directions are read, ask each student to pictorially represent the directions on a piece of paper. Have several students show their art work to the class. Have these students repeat the directions using their art work as a guide. Point out to students that chemists graphically represent their chemical recipes with balanced equations that can easily be read by chemists all over the world in any language.

GLENCOE TECHNOLOGY

Videodisc

Chemistry: Concepts and Applications

Combustion of Methane

Disc 1, Side 2, Ch. 1

Show this video of the Discovery Demo.
Also available on CD-ROM.

OBJECTIVES

- write chemical equations to represent reactions.
- use coefficients to balance chemical equations.
- differentiate among five general types of chemical reactions.

9.1 Chemical Equations

You already know how to use symbols to represent elements and formulas to represent compounds. Now you can use that shorthand to describe the chemical changes substances undergo. That is, you can describe chemical reactions. Consider the following statement: "Two molecules of acetylene gas will react with five molecules of oxygen gas to produce four molecules of carbon dioxide gas and two molecules of water vapor." It is much easier to write this as

$$2C_2H_2(g) + 5O_2(g) \rightarrow 4CO_2(g) + 2H_2O(g)$$

Representing Chemical Changes

The formulas of compounds are used in certain combinations to represent the chemical changes that occur in a chemical reaction. A **chemical reaction** is the process by which one or more substances are changed into one or more different substances. A chemical reaction can be represented by an equation. A correct chemical equation shows what changes take place. It also shows the relative amounts of the various elements and compounds that take part in these changes. The starting substances in a chemical reaction are the **reactants**. The substances that are formed by the chemical reaction are the **products**.

$$\underbrace{2C_2H_2(g) + 5O_2(g)}_{\text{reactants}} \underbrace{\rightarrow}_{\text{yield}} \underbrace{4CO_2(g) + 2H_2O(g)}_{\text{products}}$$

The letters in parentheses indicate the physical state of each substance involved. The symbol (g) after a formula means that the substance is a gas. Liquids are indicated by the symbol (l), and solids by the symbol (cr). The symbol (cr) for solid indicates that the solid is crystalline. We can see that in the reaction described above, C_2H_2, O_2, CO_2, and H_2O are all gases.

Figure 9.1 Acetylene gas is burned in the presence of oxygen in a welder's torch.

CONTENT BACKGROUND

Language of Chemistry: Beginning chemistry students in all countries think that the chemical equations are being written in their native language. However, the symbols used in chemistry represent an international language understood by all.

GLENCOE TECHNOLOGY

Videodisc

STVS: Chemistry

Disc 2, Side 1

Hole in the Ozone (Ch. 21)

Since many chemical reactions take place in water solution, a substance dissolved in water is shown by the symbol (aq). This symbol comes from the word *aqueous*. For example, if a water solution of sulfurous acid is warmed, it decomposes. The products of this reaction are water and sulfur dioxide, a gas.

$$H_2SO_3(aq) \rightarrow H_2O(l) + SO_2(g)$$

Word Origins

aqua: (L) water

aqueous—made from, with, or by water

Writing Balanced Equations

A chemical reaction can be represented by a chemical equation. To write an equation that accurately represents the reaction, you must correctly perform three steps.

Step 1. *Determine the reactants and the products.* For example, when propane gas burns in air, the reactants are propane (C_3H_8) and oxygen (O_2). The products formed are carbon dioxide (CO_2) and water (H_2O).

Step 2. *Assemble the parts of the chemical equation.* Write the formulas for the reactants on one side of the equation, usually on the left, and connect them with plus signs. Write the formulas for the products on the right side of the equation. Connect the two sides using an arrow to show the direction of the reaction. Thus,

C_3H_8 + O_2	$\longrightarrow$	CO_2 + H_2O
propane + oxygen	yield	carbon dioxide + water
reactants	*yield*	*products*

The symbols and formulas must be correct. If not, Step 3 will be useless. Use Appendix Tables A-3 and A-4 and Tables 7.3 and 7.4 in Chapter 7 when trying to write correct formulas. We will omit the symbols that indicate physical state while we go through the procedure for writing balanced equations.

Step 3. *Balance the equation.* Balancing means showing an equal number of atoms for each element on both sides of the equation. In chemical reactions, no mass is lost or gained. The same amount of matter is present before and after the reactions. So, the same number and kinds of atoms must be present on both sides of an equation. The equation below is not balanced.

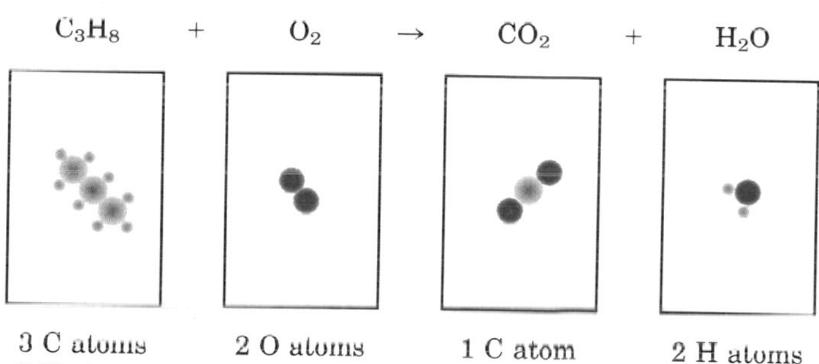

Figure 9.2 The equation $C_3H_8 + O_2 \rightarrow CO_2 + H_2O$ is not balanced. Equal numbers of carbon, hydrogen, and oxygen atoms are not present on both sides of the arrow.

2 TEACH

CONCEPT DEVELOPMENT

GLENCOE TECHNOLOGY

Software

Mastering Concepts in Chemistry

Unit 7, Lesson 1

Chemical Equations

QUICK DEMO **A Model Science**

Representing balanced chemical equations can best be done using models. If you do not have a commercial kit, obtain 18 plastic foam balls. Use latex paint to color 4 balls black to represent C and 4 balls blue to represent H. Paint 10 balls red to represent O. Use pipe cleaners or toothpicks to build two molecules of C_2H_2 and five molecules of O_2. Explain the color coding to students. Have a student rearrange the balls, using all of them, to form CO_2 and H_2O molecules. The student will quickly produce four CO_2 and two H_2O molecules. Point out that the student has balanced the equation for the reaction of $2C_2H_2 + 5O_2$. Have a student explain why it is necessary to have the same number and kind of atom on both sides of the equation.

MAKING CONNECTIONS

Aerospace

To purify air in space vehicles, lithium peroxide, Li_2O_2, reacts with CO_2 to form lithium carbonate, Li_2CO_3, and oxygen gas, O_2. Have students balance the equation for this reaction.

$2Li_2O_2 + 2CO_2 \rightarrow 2Li_2CO_3 + O_2$

interNET CONNECTION

Hypermedia text reviews of basic chemistry concepts are available for use on the Internet.

World Wide Web
http://www.scimedia.com

PROGRAM RESOURCES

Study Guide: Use the Study Guide worksheet, p. 33, for independent practice with concepts in Section 9.1.

Lesson Plans: Use the Lesson Plan, p. 18-19, to help you prepare for Chapter 9.

PRACTICE

Guided Practice: Guide students through the Sample Problem, then have them work in groups on Practice Problems 1, 2, 8, and 9.
Cooperative Learning: Refer to pp. 28T-29T to select a teaching strategy to use with this activity.
Independent Practice: Your homework or classroom assignment can include the remaining Practice Problems 3-7, 10-12, Review and Practice 43, 45-46, 58, Concept Mastery 64-65, 67-68, and Application 69-70 from the end of the chapter

Answers to
Practice Problems

1. balanced
2. $Al(NO_3)_3 + 3NaOH \rightarrow Al(OH)_3 + 3NaNO_3$
3. $2KNO_3 \rightarrow 2KNO_2 + O_2$
4. $2Fe + 3H_2SO_4 \rightarrow Fe_2(SO_4)_3 + 3H_2$
5. $3O_2 + CS_2 \rightarrow CO_2 + 2SO_2$
6. balanced
7. $3Mg + N_2 \rightarrow Mg_3N_2$
8. $CuCO_3 \rightarrow CuO + CO_2$
9. $2Na + 2H_2O \rightarrow 2NaOH + H_2$
10. $2Cu + S \rightarrow Cu_2S$
11. $2AgNO_3 + H_2SO_4 \rightarrow Ag_2SO_4 + 2HNO_3$
12. $2C_2H_6 + 7O_2 \rightarrow 4CO_2 + 6H_2O$

GLENCOE TECHNOLOGY

Videodisc

Chemistry: Concepts and Applications

Five Types of Chemical Reactions
Disc 1, Side 2, Ch. 2

Also available on CD-ROM.

Types of Chemical Reactions
Disc 1, Side 2, Ch. 3

Also available on CD-ROM.

PRACTICE PROBLEMS

Write balanced equations for each of the following reactions.

1. $Cu + H_2O \rightarrow CuO + H_2$
2. $Al(NO_3)_3 + NaOH \rightarrow Al(OH)_3 + NaNO_3$
3. $KNO_3 \rightarrow KNO_2 + O_2$
4. $Fe + H_2SO_4 \rightarrow Fe_2(SO_4)_3 + H_2$
5. $O_2 + CS_2 \rightarrow CO_2 + SO_2$
6. $Cu + Cl_2 \rightarrow CuCl_2$
7. $Mg + N_2 \rightarrow Mg_3N_2$

Substitute symbols and formulas for names, and then write a balanced equation for each of the following reactions.

8. When copper(II) carbonate is heated, it forms copper(II) oxide and carbon dioxide gas.
9. Sodium reacts with water to produce sodium hydroxide and hydrogen gas.
10. Copper combines with sulfur to form copper(I) sulfide.
11. Silver nitrate reacts with sulfuric acid to produce silver sulfate and nitric acid.
12. Ethane, a component of natural gas, burns in oxygen to produce carbon dioxide and water.

MINUTE LAB

Double Displacement
Wear an apron and goggles. Mass 0.16 g barium nitrate and place it in a test tube with 5 cm^3 H_2O. Stopper and shake until dissolved. Mass 0.15 g sodium sulfate and place it in a second test tube containing 5 cm^3 H_2O. Stopper the test tube and shake until dissolved. Observe the tubes and record your observations. Mix the contents of the two test tubes. What did you observe? Use the solubility rules in Appendix Table A-7 to determine the name of the precipitate. Write the balanced chemical equation for the chemical reaction.

Classifying Chemical Changes

There are hundreds of different kinds of chemical reactions. Many of these can be divided into five general types. Of these five, three are often used to make, or synthesize, new compounds.

Single Displacement. In this type of reaction, one element displaces another in a compound. For example, in the reaction

$$Cl_2(g) + 2KBr(aq) \rightarrow 2KCl(aq) + Br_2(l)$$

chlorine displaces bromine from potassium bromide.

In the reaction

$$2Al(cr) + Fe_2O_3(cr) \rightarrow 2Fe(cr) + Al_2O_3(cr)$$

aluminum displaces iron from iron(III) oxide. A **single displacement** reaction is recognized and predicted by its general form:

element + compound → element + compound

Double Displacement. There are hundreds of reactions in which the positive and negative portions of two compounds are interchanged.

$$PbCl_2(cr) + Li_2SO_4(aq) \rightarrow PbSO_4(cr) + 2LiCl(aq)$$
$$ZnBr_2(aq) + 2AgNO_3(aq) \rightarrow Zn(NO_3)_2(aq) + 2AgBr(cr)$$
$$BaCl_2(aq) + 2KIO_3(aq) \rightarrow Ba(IO_3)_2(cr) + 2KCl(aq)$$

CONTENT BACKGROUND

Typical Decomposition Reactions: Remind students that decomposition reactions require the input of energy in some form. The following reactions illustrate general patterns of decompositions.
metal carbonate → metal oxide + CO_2
transition metal oxide → metal + O_2
metal hydroxide → metal oxide + H_2O
metal chlorate → metal chloride + O_2
oxy-acid → nonmetal oxide + H_2O

Chemistry Journal

Using an Analogy
In their journals, have students compare reaction types with dance activities. Synthesis—two persons come together to dance; decomposition—the dance is over, and the two people part; single displacement—another person cuts in and displaces one of the two dancers; double displacement—two couples switch partners.

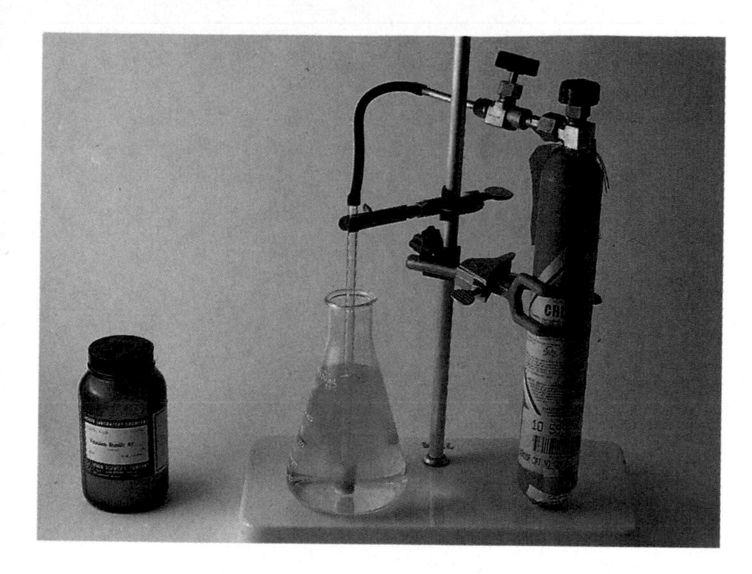

Figure 9.6 In the single displacement reaction (left), the element chlorine displaces bromine in a solution of potassium bromide, KBr. In the double displacement reaction (right), the positive and negative portions of cadmium nitrate and sodium sulfide are exchanged to form cadmium sulfide (orange) and sodium nitrate.

The form for a **double displacement** reaction is

compound + compound → compound + compound

Decomposition. Many substances will break up, or decompose, into simpler substances when energy is supplied.

$$CdCO_3(cr) \rightarrow CdO(cr) + CO_2(g)$$
$$Pb(OH)_2(cr) \rightarrow PbO(cr) + H_2O(g)$$
$$N_2O_4(g) \rightarrow 2NO_2(g)$$
$$PCl_5(cr) \rightarrow PCl_3(cr) + Cl_2(g)$$
$$H_2CO_3(aq) \rightarrow H_2O(l) + CO_2(g)$$
$$2KClO_3(cr) \rightarrow 2KCl(cr) + 3O_2(g)$$
$$2Ag_2O(cr) \rightarrow 4Ag(cr) + O_2(g)$$

Energy may be supplied in the form of heat, light, mechanical shock, or electricity. The general form for a **decomposition** reaction is

compound → two or more elements or compounds

Synthesis. In synthesis reactions, two or more substances combine to form one new substance.

$$NH_3(g) + HCl(g) \rightarrow NH_4Cl(cr)$$
$$CaO(cr) + SiO_2(cr) \rightarrow CaSiO_3(cr)$$
$$2H_2(g) + O_2(g) \rightarrow 2H_2O(g)$$

From the name, you might expect that synthesis reactions would be the most common method of preparing new compounds. However, these reactions are rarely as practical as one of the three preceding methods. The general form of a **synthesis** reaction is

element or compound + element or compound → compound

Figure 9.7 When mercury(II) oxide is heated, it decomposes to form liquid mercury and oxygen gas. Note the droplets of mercury on the sides of the test tube.

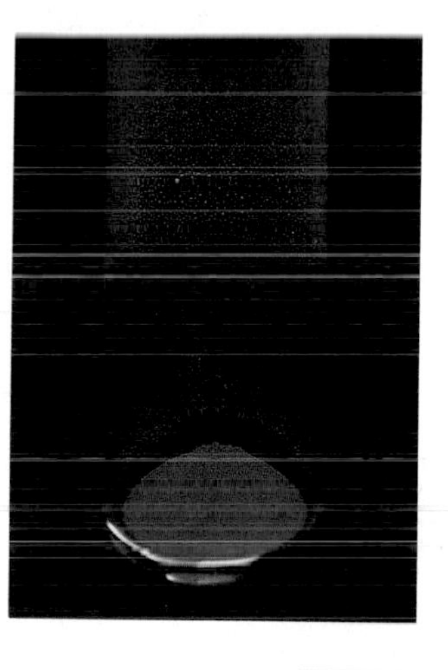

MINUTE LAB

Double Displacement

Purpose: to illustrate a double displacement reaction involving a color change

Materials: 0.16 g $Ba(NO_3)_2$, 10 cm^3 water, 2 test tubes with stoppers, 0.15 g Na_2SO_4, Petri dish

Procedure Hints: To better see the reaction, mix the two colorless solutions into a Petri dish on an overhead projector. **Disposal:** A

Results: A precipitate of $BaSO_4$ forms when the two solutions are mixed.

Answers to Questions: 1. Two clear, colorless solutions mixed to form a white milky solution. When left to stand, the white solid settled to the bottom of the test tube. **2.** Barium sulfate is the name of the precipitate. **3.** $Ba(NO_3)_2(aq) + Na_2SO_4(aq) \rightarrow BaSO_4(cr) + 2NaNO_3(aq)$

✓ ASSESSMENT

Performance: Have student laboratory groups use Appendix Table A-7 to select two other chemicals that they predict will react to form a precipitate. If those chemicals are in stock and may be safely dissolved and mixed, have the group dissolve micro amounts of them and test their hypothesis using a few drops of each. Students should write a balanced chemical equation if a precipitate forms.

MAKING CONNECTIONS

Language Arts

Have a student read the dictionary definition of *decompose* to the class. Ask the students to write a paragraph relating decomposition reactions to a possible solution to the problem of solid waste disposal caused by the closing of landfills.

PROGRAM RESOURCES

Enrichment: Use the worksheet "Classifying Organic Reactions," pp. 17-18, to challenge more able students.

Laboratory Manual: Use microlab 9-1, "Types of Chemical Reactions" as enrichment for the concepts in Chapter 9.

Laboratory Manual: Use macrolab 9-2, "Types of Chemical Reactions" as enrichment for the concepts in Chapter 9.

CONCEPT DEVELOPMENT

• Students should receive sufficient drill in predicting the products of a reaction. This skill is used throughout the course, especially in the laboratory where students will be writing balanced equations to represent their observations.

Extension: Challenge your students to find out what is a metathesis reaction. *double displacement*

CONCEPT DEVELOPMENT

Correcting Errors

To help students predict the reaction type and products, review the general pattern of each type of reaction.

PRACTICE

Guided Practice: Have students work in class on Practice Problems 13-15, 18-22.

Independent Practice: Your homework or classroom assignment can include Practice Problems 16-17, Review and Practice 47, 49, and Application 71-72 from the end of the chapter.

Answers to
Practice Problems

13. single displacement
14. decomposition
15. synthesis
16. combustion
17. single displacement
18. double displacement
19. $2Al + 3Cu(NO_3)_2 \rightarrow 3Cu + 2Al(NO_3)_3$
20. $2Hg + O_2 \rightarrow 2HgO$
21. $H_2SO_4 + 2KOH \rightarrow K_2SO_4 + 2H_2O$
22. $2C_5H_{10} + 15O_2 \rightarrow 10CO_2 + 10H_2O$

3 ASSESS

CHECK FOR UNDERSTANDING

Ask students to answer the Concept Review problems, and Review and Practice 43, 45-47, 49, and 58 from the end of the chapter.

RETEACHING

Have students use molecular model building kits or candy gum drops and toothpicks to construct the reactant molecules and then disassemble the reactants and construct the product molecules.

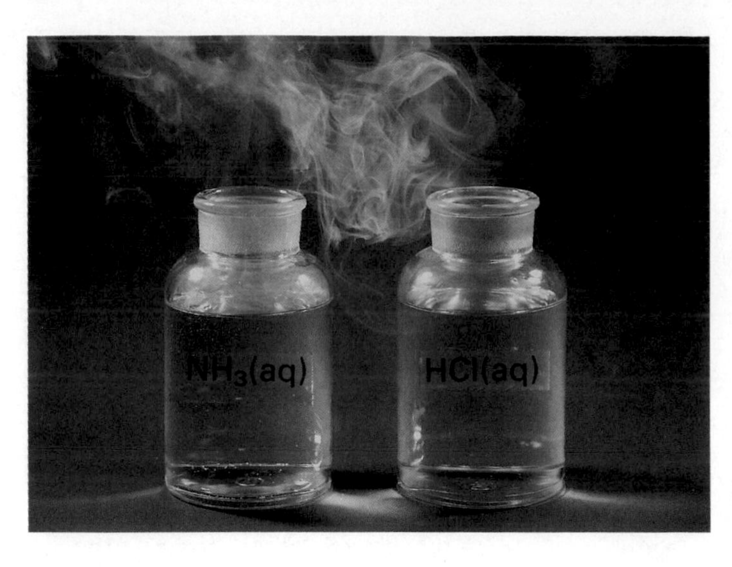

Figure 9.8 A cloud of ammonium chloride is formed when HCl and NH_3 combine in a synthesis reaction.

Combustion. Almost all organic compounds and some inorganic compounds will burn in air. Many will ignite quite readily and are considered *flammable*. When a compound burns in air, it is actually reacting with the oxygen in the air. As a result, chemists refer to combustion reactions as oxidation reactions. The products of the oxidation of a hydrocarbon under normal conditions are carbon dioxide and water vapor. As an example, consider the combustion of methane (CH_4), the main component of natural gas, and butane (C_4H_{10}), a fuel used in camping stoves.

$$CH_4 + 2O_2 \rightarrow CO_2 + 2H_2O$$
$$2C_4H_{10} + 13O_2 \rightarrow 8CO_2 + 10H_2O$$

The general form for **combustion** reactions is

hydrocarbon + oxygen → carbon dioxide + water

Not all reactions take one of these five general forms. Other classes of reactions will be considered later. Until then, we will deal chiefly with reactions of single or double displacement, decomposition, synthesis, or combustion (oxidation).

Figure 9.9 The butane in this camp stove serves as a source of energy when it combines with oxygen in a combustion reaction.

DEMONSTRATION

9-3 Reaction Types

Purpose: to review three reaction types and predict their products

Materials: 1 g Cu wire, 10 cm^3 15.8*M* HNO_3, 12 g NaOH, 250-cm^3 beaker, 500 g ice, 10 cm^3 distilled water, hot plate, stirring rod

Procedure: 1. Place 1 g sample of Cu wire in the 250-cm^3 beaker. In a fume hood, add 10 cm^3 of 15.8*M* HNO_3 acid. **CAUTION:** *Acid is corrosive. The fumes are poisonous.* Swirl the contents in the beaker. After the fumes no longer evolve, place the beaker in an ice water bath. Have students observe the reaction.

$$Cu + 4HNO_3 \rightarrow Cu(NO_3)_2 + 2NO_2 + 2H_2O$$

2. While stirring, slowly add the 6*M* NaOH (12 g NaOH dissolved in 40 cm^3 H_2O), 5 cm^3 at a time, until a total of 40 cm^3 has been added to the 250-

PRACTICE PROBLEMS

Classify each of the following reactions as single or double displacement, decomposition, synthesis, or combustion.

13. $CuO + H_2 \rightarrow Cu + H_2O$
14. $2H_2O_2 \rightarrow 2H_2O + O_2$
15. $2Ag + S \rightarrow Ag_2S$
16. $C_4H_8 + 6O_2 \rightarrow 4CO_2 + 4H_2O$
17. $2K + 2H_2O \rightarrow 2KOH + H_2$
18. $HCl + NaOH \rightarrow H_2O + NaCl$

Predict the products formed in each of the following reactions.

19. the single displacement reaction between aluminum, Al, and copper(II) nitrate, $Cu(NO_3)_2$
20. the synthesis of mercury(II) oxide from its elements
21. the double displacement reaction of sulfuric acid, H_2SO_4, and potassium hydroxide, KOH
22. the combustion of cyclopentane, C_5H_{10}

9.1 CONCEPT REVIEW

23. The illustrations below show three different types of reactions. What type of reaction is shown in each of the following?

a. 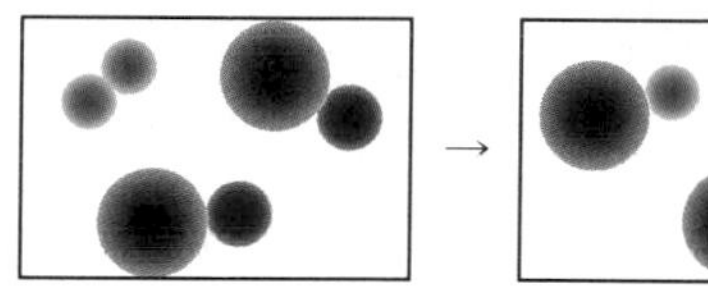 → 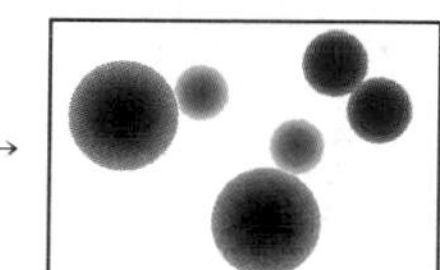

b. 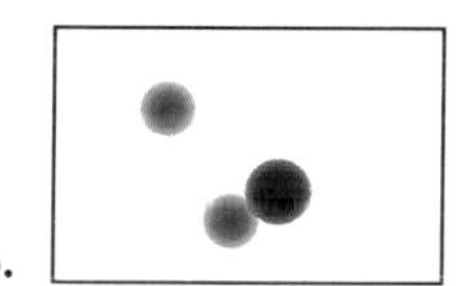 → 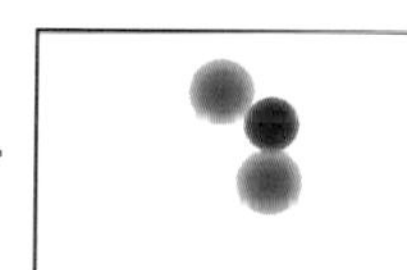

c. 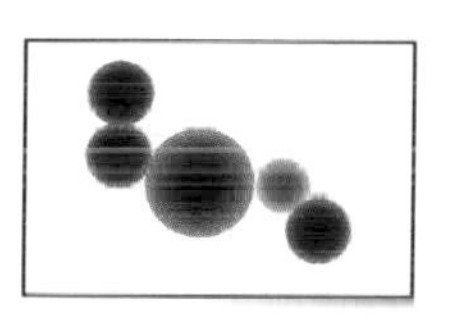 → 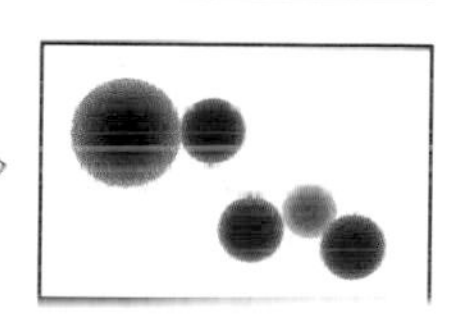

24. Write a balanced chemical equation to represent each of the following reactions.
   a. antimony and chlorine gas react to form antimony trichloride
   b. the decomposition of silver oxide to form metallic silver and oxygen gas

25. Four students try to balance the equation for the following reaction.

$$Cu_2S + O_2 \rightarrow Cu_2O + SO_2$$

   a. The students obtain the following results. Tell what is wrong with each equation and why.

$Cu_2S + 2O_2 \rightarrow Cu_2O_2 + SO_2$

$Cu_2S + O_2 \rightarrow Cu_2O + SO$

$Cu_2S + 2O_2 \rightarrow Cu_2O + SO_2$

$4Cu_2S + 6O_2 \rightarrow 4Cu_2O + 4SO_2$

   b. Write the correct balanced equation for the reaction.

26. Classify the following reactions as single displacement, double displacement, decomposition, synthesis, or combustion. Tell why you classified each reaction as you did.
   a. $2Al(cr) + 3Cl_2(g) \rightarrow 2AlCl_3(cr)$
   b. $Zn(cr) + 2HCl(aq) \rightarrow ZnCl_2(aq) + H_2(g)$
   c. $Mg(ClO_3)_2(cr) \rightarrow MgCl_2(cr) + 3O_2(g)$
   d. $3BaCl_2(aq) + 2H_3PO_4(aq) \rightarrow Ba_3(PO_4)_2(cr) + 6HCl(aq)$

27. **Apply** Why aren't mixtures representable in chemical equations?

EXTENSION

Ask students to find and know at least four of the rules for predicting the products of decomposition reactions. See **Merrill Chemistry Solving Problems in Chemistry**, Chapter 9.

Answers to
Concept Review

23. a. single displacement, **b.** synthesis, **c.** decomposition

24. a. $2Sb(cr) + 3Cl_2(g) \rightarrow 2SbCl_3(cr)$, **b.** $2Ag_2O(cr) \rightarrow 4Ag(cr) + O_2(g)$

25. a. $Cu_2S + 2O_2 \rightarrow Cu_2O_2 + SO_2$ product is Cu_2O_2 instead of Cu_2O, $Cu_2S + O_2 \rightarrow Cu_2O + SO$ product is SO instead of SO_2, $Cu_2S + 2O_2 \rightarrow Cu_2O + SO_2$ 4 O atoms on left, 3 O atoms on right, $4Cu_2S + 6O_2 \rightarrow 4Cu_2O + 4SO_2$; all coefficients can be divided by 2

b. $2Cu_2S + 3O_2 \rightarrow 2Cu_2O + 2SO_2$

26. a. synthesis; element + element → compound, **b.** single displacement; element + compound → element + compound, **c.** decomposition; compound → compound + element, **d.** double displacement; compound + compound → compound + compound

27. Mixtures are neither pure elements nor pure compounds, which are the only substances representable in a chemical reaction.

cm^3 beaker in the ice bath. **CAUTION:** *NaOH is caustic.* Have students predict the products of this double displacement reaction.

3. Add 100 cm^3 of distilled H_2O to the blue paste, $Cu(OH)_2$, and heat to boiling for 3 minutes. Keep a stirring rod in the beaker to prevent bumping. **CAUTION:** *This solution is hot and caustic.* **Disposal:** E

Results: The first reaction is a redox reaction. Explain that the five types of reactions presented in Chapter 9 are not all inclusive.

Question: 1. Write the balanced chemical equations steps from 2-3.

2. $Cu(NO_3)_2 + 2NaOH \rightarrow Cu(OH)_2 + 2NaNO_3$

3. $Cu(OH)_2 \rightarrow CuO + H_2O$

4 CLOSE

Computer Demo: *Animation,* Simulates the mechanism of methane reacting with chlorine to form chloroform and hydrogen chloride, AP603, Project SERAPHIM.

CHEMISTRY AND TECHNOLOGY

Developing Self-Fertilizing Plants

Background: Scientists have not been able to infect wheat with *Rhizobium,* the bacteria present in the nodules of legumes. However, *Azospirillum,* a bacterium that lives freely in the ground, was successfully introduced. Unlike *Rhizobium,* that enters the cells, *Azospirillum* lives between the cells. The actual development of field applications may take ten years.

Teaching This Feature: You may need to review the nitrogen cycle for the class. A biology teacher may have an overhead transparency or chart that would be useful for a pictorial review of the cycle.

Answers To

Thinking Critically

1. *Azospirillum* and *Rhizobium;* nitrogen-containing organic matter is decomposed by bacteria to ammonia. Ammonia is changed by other bacteria into the nitrite ion, and then the nitrate ion.

2. This needed research will result in increased crop yields necessary to feed a growing population. Because the fertilizer is natural, there will be less need for commercially-prepared fertilizers resulting in a savings of energy and reducing water run-off pollution.

CHEMISTRY AND

TECHNOLOGY

Developing Self-Fertilizing Plants

Nitrogen compounds are major nutrients for plants. Because approximately 78% of the air is nitrogen, you would think that there would be no problem providing plants with this nutrient. However, the major food crops of the world—rice, wheat, and corn—cannot utilize the nitrogen in the air directly. Through many industrial and natural chemical processes, nitrogen is converted into fertilizers in forms that plants can use.

The conversion of N_2 from the atmosphere into a form useful to plants is an important step in the nitrogen cycle, the interconversion of inorganic and organic nitrogen in nature. Both chemical and biological processes are involved. Commercial fertilizer production accounts for the conversion of about 3×10^7 metric tons of nitrogen each year. Other chemical processes, involving lightning, ultraviolet radiation, and combustion, account for another 4×10^7 metric tons annually. Biological sources, however, produce nearly twice as much usable nitrogen as these two sources combined. Much of this production is carried out by certain plants, such as peas and beans, whose seeds are in pods. These plants, called legumes, can use nitrogen directly from the air.

Legumes do not actually convert nitrogen themselves. Bacteria that live in nodules on the roots of the legumes can convert molecular nitrogen into ammonia, nitrites, and nitrates—compounds that plants can use. This process is called nitrogen fixation. In effect, these plants are self-fertilizing.

For years, farmers have grown legumes to help replenish nitrogen-poor soil. Farmers rotate legumes with grain crops or plow under the legumes as "green manure." In this way, the soil is enriched with the nitrogen products manufactured by the bacteria.

Researchers at the University of Sydney in Australia believe they have found a way to "infect" crops such as wheat with nitrogen-fixing bacteria. They found that low concentrations of the herbicide 2,4-D soften wheat cell walls and cause nodule-like swellings to form on the roots of wheat plants. Into these nodules, the researchers have been able to introduce a nitrogen-fixing bacterium that usually lives free in the soil. Comparison of treated wheat with untreated wheat showed that nitrogen was being fixed in the treated wheat. If the bacteria are able to remain for long periods of time in the nodules, farmers growing these altered crops will not have to rely on industrially produced fertilizers. These scientists may have come up with a means to improve production of major food crops and reduce environmental pollution without adding more synthetic fertilizer to our environment.

Thinking Critically

1. There are two kinds of bacteria that assist in nitrogen-fixing. One lives in root nodules of plants and the other is found free in the soil. Try to find out what these bacteria are and what chemical reactions are involved in nitrogen fixation.
2. Why is the Australian research important to the world's food shortage?

9.2 Stoichiometry

Stoichiometry is a word for that part of chemistry that deals with the amount of substances involved in chemical reactions, both as reactants and as products. Balanced equations represent the relationship between the number of particles that react and the number of particles produced. Because each type of particle has its own formula mass, there must be definite relationships between the masses that react and the masses of the products.

OBJECTIVES

- determine the mass of a reactant or product based on the mass of another reactant or product in a reaction.
- calculate the actual yield of a product as a percentage of the theoretical yield.
- determine the heat of reaction for a chemical reaction in which a specified amount of a substance is involved.

Mass-Mass Relationships

Balanced equations and the mass-mass relationships they convey help us answer several types of questions. How much of one reactant is needed to combine with a given amount of another reactant? How much product is produced from a specific amount of reactant? For instance, how much silver chloride can be produced from 17.0 g of silver nitrate? There are other, similar, questions. Fortunately, all these questions can be answered using the same procedure. The coefficients of a balanced equation give the relative amounts (in moles) of reactants and products. Calculations to find the masses of materials involved in reactions are called **mass-mass problems.**

SAMPLE PROBLEM

Mass-Mass

How many grams of silver chloride can be produced from the reaction of 17.0 g of silver nitrate with excess sodium chloride solution?

Solving Process:

Step 1. We are given silver nitrate and asked to find silver chloride. What do we know that would connect two different substances? A chemical equation would relate them in a quantitative way. Therefore, we write a balanced equation to represent the reaction that occurs. Silver nitrate is reacting with sodium chloride. Since we have compound plus compound, we predict that a double displacement reaction will occur.

$$AgNO_3(aq) + NaCl(aq) \rightarrow AgCl(cr) + NaNO_3(aq)$$

Step 2. We have thought of equations as written in terms of individual atoms and formula units. We can also consider the coefficients in the equation as indicating the number of *moles* of formula units that take part in the reaction. The equation above indicates that one formula unit of silver nitrate will produce one formula unit of silver chloride. It also indicates that one mole of silver nitrate formula units will produce one mole of silver chloride formula units. The problem states that

Figure 9.10 A white precipitate of silver chloride, AgCl, forms when solutions of silver nitrate, $AgNO_3$, and sodium chloride, NaCl, are mixed.

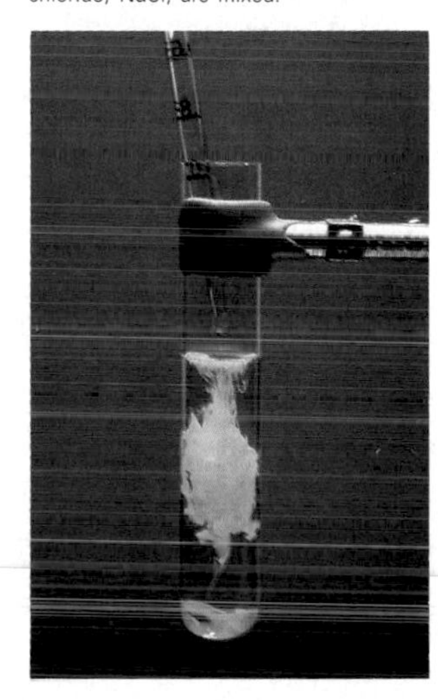

9.2 Stoichiometry

PREPARE

Key Concepts

The major concepts presented in this section include mass-mass, percentage yield, and mass-energy quantitative relationships. Problem solving strategies common to each type of problem are presented utilizing the mole as a unifying factor.

Key Terms

stoichiometry
mass-mass problem
percentage yield
mass-energy problem

1 FOCUS

Objectives

Preview with your students the objectives listed on the student page. Students can use them as a study guide for this major section.

Demonstration

Hold a potato chip with tongs, and ignite it using a burning match. Darken the room as the students observe the relatively large amount of heat and flame produced by the burning carbohydrate and oil. Discuss with students that a chemist wants to know "how much" about each part of the chemical reaction. How much oxygen is used and how much heat is produced are common questions. What percentage of the available energy was released as the chip burned is another question chemists may ask. In this section, students will learn how to answer these questions quantitatively.

CONTENT BACKGROUND

Coefficients: It is important to stress the coefficients of a balanced equation as being the ratio of moles for the mass-mass problem. Doing mass-mass problems in one step may appear at first to be more difficult for students to master. Emphasize that the factor-label method helps to eliminate mistakes in problem solving. The proportion method of solving these stoichiometry problems is actually a form of factor-label logic but the units are not expressed. As a result, students memorize rather than understand the solving process and the chemistry involved. Point out that students will later be able to use the same technique to master mass-volume and volume-volume problems presented in Chapter 19.

PROGRAM RESOURCES

Study Guide: Use the Study Guide worksheet, p. 34, for independent practice with concepts in Section 9.2.

2 TEACH

CAREERS IN CHEMISTRY
Chemical Engineer

Background: The starting salary of a chemical engineer is higher than that of a chemist. The chemical engineer enrolls in only a few chemistry courses in college. Most of the course work is in engineering with an emphasis in mathematics.

For More Information: Contact: The American Chemical Society, 1155 Sixteenth Street, NW, Washington, DC 20036

CONCEPT DEVELOPMENT

Computer Demo: *Stoichiometry,* Assists students in finding their errors, AP306, PC2301, Project SERAPHIM

GLENCOE TECHNOLOGY

Software

Mastering Concepts in Chemistry
Unit 7, Lesson 2
Mass Stoichiometry

CD-ROM

Chemistry: Concepts and Applications
Exploration: *Predicting Mass of Product*

CAREERS IN CHEMISTRY

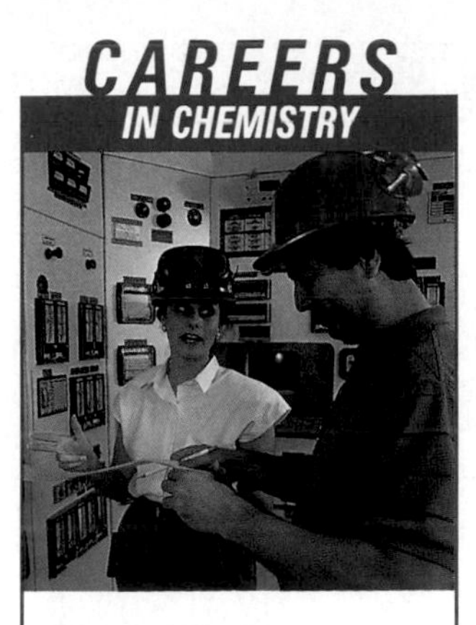

Chemical Engineer
The most important functions of chemical engineers are designing, constructing, and operating chemical plants. However, there are in the chemical enterprise many other aspects of interest to the chemical engineer. Chemical engineers need to be concerned with the economics of the chemical marketplace, and many engineers enter the field of sales. Quality control is another phase of chemical production under the supervision of chemical engineers, as is the maintenance of conditions suitable for preservation of our environment. If you are interested in chemistry but like to work with practical matters instead of more theoretical areas, you will find chemical engineering a fascinating career.

an excess of sodium chloride is used, which means that all the silver nitrate will react and there will be some sodium chloride left over.

Let us review where we are. We are given grams of silver nitrate. We are asked to find grams of silver chloride. Silver nitrate and silver chloride are related by the equation. The equation tells us that one mole of silver nitrate yields one mole of silver chloride.

Our solution begins by converting the grams of silver nitrate to moles. We use the table on the inside back cover to determine the molar mass of $AgNO_3$, which is 170 g.

$$17.0\text{ g }AgNO_3 \times \frac{1\text{ mol }AgNO_3}{170\text{ g }AgNO_3} \ldots$$

Step 3. We have now converted the silver nitrate to units that will enable us to relate it to the silver chloride. We use the equation to find the conversion factor to use in going from silver nitrate to silver chloride. The equation tells us that one mole of silver nitrate will produce one mole of silver chloride. Using this fact gives us a partial solution.

$$17.0\text{ g }AgNO_3 \times \frac{1\text{ mol }AgNO_3}{170\text{ g }AgNO_3} \times \frac{1\text{ mol AgCl}}{1\text{ mol }AgNO_3} \ldots$$

Step 4. We have now arrived at silver chloride, which is the substance we were asked to find. However, we were asked to find the answer in grams. To complete the problem, we must convert the moles of silver chloride to grams. Using the table inside the back cover, we can see that one mole of silver chloride is 144 g. The final solution then appears as:

$$17.0\text{ g }AgNO_3 \times \frac{1\text{ mol }AgNO_3}{170\text{ g }AgNO_3} \times \frac{1\text{ mol AgCl}}{1\text{ mol }AgNO_3} \times \frac{144\text{ g AgCl}}{1\text{ mol AgCl}}$$
$$= 14.4\text{ g AgCl}$$

The problem above involved finding the mass of one substance from a given mass of another substance. All mass-mass problems can be solved in this way. This problem is an example of stoichiometry.

Let us review the process.

Step 1. *Write the balanced equation.*
The first step in the solution of any mass-mass problem is to write a balanced equation for the correct reaction. In this section, we will always assume that only one reaction occurs and that all of one reactant is used. In reality, several reactions may occur, the actual reaction may not be known, or all the reactant may not react.

CONTENT BACKGROUND

History of Stoichiometry: The first stoichiometric concept that appeared in the scientific literature involved the use of the term equivalence. H.T. Richter used the concept of equivalence before Dalton proposed his atomic theory. The concept was based on experimental work. In 1814, William Hyde Wollaston extended Thomson's list of the masses of acids and bases which neutralized each other to include elements and other compounds. Wollaston invented a scale of chemical equivalents that allowed experimental and manufacturing chemists to calculate the amounts produced in chemical reactions. Although equivalents were considered to be a useful stoichiometric tool, they proved to be useless as a universal concept in chemical formulation. Elements having multiple oxidation states caused a great deal of confusion, and were ignored.

Figure 9.11 Mass must be measured carefully so that accurate data will be available for solving mass-mass problems.

Step 2. *Find the number of moles of the given substance.*
Express the mass of the given substance in moles by dividing the mass of the given substance by its molar mass.

$$\frac{\textit{grams of given substance}}{} \bigg| \frac{\textit{1 mole}}{\textit{molar mass of given substance}} \ldots$$

Step 3. *Inspect the balanced equation to determine the ratio of moles of required substance to moles of given substance.*
For example, look at the following equation.

$$2H_2(g) + O_2(g) \rightarrow 2H_2O(g)$$

In this reaction, 1 mole of oxygen reacts with 2 moles of hydrogen. From the same equation, 1 mole of oxygen will produce 2 moles of water. Also, 2 moles of water are produced by 2 moles of hydrogen. Once the equation is balanced, only the reactants and products directly involved in the problem should be in your calculations. Multiply the moles of given substance by the ratio:

$$\ldots \frac{\textit{moles of required substance}}{\textit{moles of given substance}} \ldots$$

Step 4. *Express the moles of required substance in terms of grams, then convert moles to grams.*

$$\ldots \frac{\textit{molar mass of required substance}}{\textit{1 mole}} = \textit{grams of required substance}$$

Notice that, as you work through a problem of this kind, you first convert grams of given substance to moles, and then convert moles of required substance back to grams.

(*start with grams given*) → (*convert grams to moles*) → (*use mole ratio*) → (*convert moles to grams*) → (*end with grams required*)

This method is used because the balanced equation relates the number of moles of given substance to the number of moles of required substance. Now try the following two problems.

CONCEPT DEVELOPMENT

QUICK DEMO **Quick Silver**
Illustrate the Sample Problem by placing a few drops of dilute silver nitrate (0.1*M*) in a Petri dish on an overhead projector. Add a few drops of dilute NaCl solution to the $AgNO_3$. **Disposal:** E

- Stoichiometry is the quantitative study of chemical reactions. Remind students that mass-mass problems only involve multiplication and division processes. The factor-label method makes the mathematics easy to visualize.
- Ask students to verbalize the mass-mass problem-solving process. When a student can tell you how to solve the typical problem, step-by-step, then the student has mastery of the concept and the skill.

MAKING CONNECTIONS

Laboratory Safety
When chemists work with unstable chemicals that could detonate, they limit the force of the explosion by limiting the amount of chemical present at any one time. If a larger amount is necessary, then safety equipment such as explosion shields and ear protection are used. Some noble gas compounds decompose when exposed to ultraviolet light. If very small amounts of the substances are used, the chemist is relatively safe.

Chemistry Journal

Stoichiometry Map
The four steps required to solve a stoichiometry problem are outlined on these two pages. Review the steps with students and have them make a map or algorithm in their journals to represent the process.

PROGRAM RESOURCES

Transparency Master: Use the Transparency master and worksheet, pp. 39-40, for guided practice in the concept "Solving Mass-Mass Problems."

Color Transparency: Use Color Transparency 20 to focus on the concept of "Solving Mass-Mass Problems."

Reteaching: Use the Reteaching worksheet, pp. 12-13, to provide students another opportunity to understand the concept of "Stoichiometry."

CHEMISTS AND THEIR WORK

St. Elmo Brady

Background: Brady is unique in that he established chemistry departments in four different universities.

Exploring Further: Tuskegee Institute is also associated with two other eminent African American educators, Dr. Booker T. Washington and Dr. George Washington Carver. Ask students to prepare reports on the contributions of these two men.

Correcting Errors

To help students set up mass-mass problems correctly, remind them first to change the given quantities of substances into moles and then to use the mole to mole ratio from the balanced chemical equation.

✓ ASSESSMENT

Oral: After students have studied the Sample Problems on pages 233-237, ask individual students in cooperative learning groups to use Practice Problems 30 and 31 on page 237 to demonstrate to the group members the 4-step sequence that is followed when working a stoichiometry problem.

CHEMISTS AND THEIR WORK

St. Elmo Brady
(1884-1966)

St. Elmo Brady was the first African-American student to receive a doctorate in chemistry. He spent the rest of his life making certain that other African-American students would have the same opportunity. After finishing his degree at the University of Illinois in 1916, he went to Tuskegee Institute to develop what became the chemistry department there. He moved in 1920 to Howard University and accomplished the same task. In 1927, he moved to Fisk University and served 25 years building both undergraduate and graduate programs. After retirement, when he was nearly 70 years old, Brady was again asked to build a chemistry department, this time at Tougaloo, Mississippi. As a result of his efforts, thousands of students of all races have been able to study chemistry.

SAMPLE PROBLEMS

1. Mass-Mass

How many grams of Cu_2S could be produced from 9.90 g of CuCl reacting with an excess of H_2S gas?

Solving Process:

(1) We must write the balanced equation. CuCl and H_2S are reactants, Cu_2S is one product.

$$2CuCl(aq) + H_2S(g) \rightarrow Cu_2S(cr) + 2HCl(aq)$$

If we use the wrong reaction or do not balance the equation properly, we cannot get a correct answer.

(2) Find the number of moles of the given substance.

$$\begin{aligned} 1Cu \quad 1 \times 63.5 &= 63.5\text{ g} \\ 1Cl \quad 1 \times 35.5 &= 35.5\text{ g} \\ \text{molar mass of CuCl} &= \overline{99.0\text{ g}} \end{aligned}$$

$$\frac{9.90\text{ g CuCl}}{} \left| \frac{1\text{ mol CuCl}}{99.0\text{ g CuCl}} \right. \cdots$$

(3) Determine the mole ratio of the required substance to the given substance. Notice that, although H_2S and HCl are part of the reaction, we do not consider them in this problem. They are not part of the calculations.

$$2CuCl + H_2S \rightarrow Cu_2S + 2HCl$$

$$\frac{9.90\text{ g CuCl}}{} \left| \frac{1\text{ mol CuCl}}{99.0\text{ g CuCl}} \right| \frac{1\text{ mol }Cu_2S}{2\text{ mol CuCl}} \cdots$$

(4) We have now arrived at moles of Cu_2S. We were asked to find grams of Cu_2S, so we must convert moles of Cu_2S into grams of Cu_2S. We use the table inside the back cover.

$$\begin{aligned} 2Cu \quad 2 \times 63.5 &= 127\text{ g} \\ 1S \quad 1 \times 32.1 &= 32.1\text{ g} \\ \text{molar mass of }Cu_2S &= \overline{159\text{ g}} \end{aligned}$$

$$\frac{9.90\text{ }\cancel{\text{g CuCl}}}{} \left| \frac{1\text{ }\cancel{\text{mol CuCl}}}{99.0\text{ }\cancel{\text{g CuCl}}} \right| \frac{1\text{ }\cancel{\text{mol }Cu_2S}}{2\text{ }\cancel{\text{mol CuCl}}} \left| \frac{159\text{ g }Cu_2S}{1\text{ }\cancel{\text{mol }Cu_2S}} \right.$$

$$= 7.95\text{ g of }Cu_2S$$

Thus, we predict 9.90 g of CuCl will react to produce 7.95 g of Cu_2S. If the problem is set up correctly, all the factor units will divide out except the required units.

2. Mass-Mass: Another Example

How many grams of calcium hydroxide will be needed to react completely with 10.0 g of phosphoric acid? Note that we are asked to find the mass of one reactant that will react with a given mass of another reactant, H_3PO_4.

DEMONSTRATION

9-4 Aluminum to Alum Stoichiometry

Purpose: to prepare potassium aluminum sulfate from aluminum and demonstrate the mass-mass relationship in this chemical reaction

Materials: ice water bath, 30-cm × 5-cm (1.0 g) piece aluminum foil, 4.0 g KOH, two 250-cm^3 beakers, 12 cm^3 conc. sulfuric acid, balance, filter paper, funnel, ring stand and ring, stirring rod, hot plate, water

Procedure: 1. Dissolve 4.0 g KOH in 50 cm^3 of water in a 250-cm^3 beaker. Tear up the Al foil into very small pieces and place them in the beaker in a fume hood for 15 minutes.

2. While waiting, place 13 cm^3 of water into a second beaker that is in an ice water bath, and very slowly, stirring, add 12 cm^3 of conc. sulfuric acid. **CAUTION:** *Always add acid to water.*

Solving Process:

(1) Write a balanced equation.

$$3Ca(OH)_2 + 2H_3PO_4 \rightarrow Ca_3(PO_4)_2 + 6H_2O$$

(2) Change 10.0 g H_3PO_4 to moles of H_3PO_4.

$$\frac{10.0 \text{ g } H_3PO_4}{} \left| \frac{1 \text{ mol } H_3PO_4}{98.0 \text{ g } H_3PO_4} \right. \ldots$$

(3) From the equation, 2 moles of H_3PO_4 will require 3 moles $Ca(OH)_2$.

$$\frac{10.0 \text{ g } H_3PO_4}{} \left| \frac{1 \text{ mol } H_3PO_4}{98.0 \text{ g } H_3PO_4} \right| \frac{3 \text{ mol } Ca(OH)_2}{2 \text{ mol } H_3PO_4} \ldots$$

(4) Change moles of calcium hydroxide into grams (mass) of calcium hydroxide.

$$\frac{10.0 \cancel{\text{ g } H_3PO_4}}{} \left| \frac{1 \cancel{\text{ mol } H_3PO_4}}{98.0 \cancel{\text{ g } H_3PO_4}} \right| \frac{3 \cancel{\text{ mol } Ca(OH)_2}}{2 \cancel{\text{ mol } H_3PO_4}} \left| \frac{74.1 \text{ g } Ca(OH)_2}{1 \cancel{\text{ mol } Ca(OH)_2}} \right.$$

$$= 11.3 \text{ g } Ca(OH)_2$$

PROBLEM SOLVING HINT

Use the unit(s) in your answer to check your work.

PRACTICE PROBLEMS

28. Glucose is used as a source of energy by the human body. The overall reaction in the body is

$$C_6H_{12}O_6 + 6O_2 \rightarrow 6CO_2 + 6H_2O$$

Calculate the number of grams of oxygen needed to oxidize 12.5 g of glucose to carbon dioxide and water.

29. Ammonia is synthesized from hydrogen and nitrogen according to the following equation.

$$N_2(g) + 3H_2(g) \rightarrow 2NH_3(g)$$

If an excess of nitrogen is reacted with 3.41 g of hydrogen, how many grams of ammonia can be produced?

30. Assume that in the decomposition of potassium chlorate, $KClO_3$, 80.5 g of O_2 form. How many grams of potassium chloride, the other product, would be formed?

31. In a single displacement reaction, 9.23 g of aluminum react with excess hydrochloric acid. How many grams of hydrogen will be produced?

32. The compound "cisplatin", $PtCl_2(NH_3)_2$, has been found to be effective in treating some types of cancer. It can be synthesized using the following reaction.

$$K_2PtCl_4(aq) + 2NH_3(aq) \rightarrow 2KCl(aq) + PtCl_2(NH_3)_2(aq)$$

a. How much "cisplatin" can be produced from 2.50 g K_2PtCl_4 and excess NH_3?
b. How much NH_3 would be needed?

PRACTICE

Guided Practice: Guide students through the Sample Problems, then have them work in groups on Practice Problems 28 and 29.
Cooperative Learning: Refer to pp. 28T-29T to select a teaching strategy to use with this activity.
Independent Practice: Your homework or classroom assignment can include the remaining Practice Problems 30-32, and Review and Practice 44, 48, 50-53, 59-61, from the end of the chapter.

Answers to
Practice Problems: Have students refer to Appendix C for complete solutions to Practice Problems.
28. 13.3 g O_2
29. 19.1 g NH_3
30. 125 g KCl
31. 1.04 g H_2
32. a. 1.81 g $Pt(NH_3)_2Cl_2$
b. 0.205 g NH_3

3. After the Al foil has dissolved in the KOH solution to form $KAl(OH)_4$, place the beaker in an ice water bath. Allow the black residue to settle.
4. Filter the $KAl(OH)_4$ solution into the cold sulfuric acid solution.
5. Use a hot plate to heat the beaker until the white precipitate dissolves. Allow this beaker to set overnight.
6. Decant the liquid from the crystals. Rinse once with water, decant, and set the crystals aside to dry.
7. Remove the dry alum crystals, $KAl(SO_4)_2 \bullet 12H_2O$, from the beaker. Obtain their mass.

Results: $2\ Al + 2KOH + 6H_2O \rightarrow 2KAl(OH)_4 + 3H_2$

Potassium aluminum hydroxide, $KAl(OH)_4$, is a base that is neutralized by the sulfuric acid, in two steps.

(1) $2KAl(OH)_4 + H_2SO_4 \rightarrow K_2SO_4 + 2H_2O + 2Al(OH)_3$

(2) $2Al(OH)_3 + 3H_2SO_4 \rightarrow Al_2(SO_4)_3 + 6HOH$

Questions: 1. What is the formula mass of alum, $KAl(SO_4)_2 \bullet 12H_2O$? *474 g/mol*
2. How many moles of Al were consumed? *1.02 g = 0.0378 mol Al*
3. How many moles of alum were produced? *7.14 g = 0.0151 mol alum*
4. From 1 mol of Al, 1 mol of alum should be produced. Calculate the expected yield of alum in grams based on the amount of Al with which you started. *17.9 g alum*

FRONTIERS

Gas Stations

Background: The hydrogen content of Hythane is 15% by volume. The amount of Hythane needed for a 200-mile trip would fill a tank about 3.4 times larger than that of a standard automobile at a pressure of 2×10^4 kPa.

Teaching This Feature: Have students compare the reactions discussed in the feature by writing balanced chemical equations for the combustions of hydrogen, methane, and heptane, a hydrocarbon found in gasoline.

Exploring Further: Have interested students investigate the function of ethanol in gasohol and octane-enhancing additives in gasoline.

CONCEPT DEVELOPMENT

Discussion: Point out that chemical reactions do not always go to completion. Explain that many reactions occur as multi-step mechanisms. Reversible reactions may also contribute to a reaction's inefficiency in producing a product. These concepts will be formally introduced and developed later in the text.

FRONTIERS

Gas Stations

"Five dollars worth of gas, please. And put in some H_2, too!" You won't hear this at the local service station yet, but recent research into alternative automotive fuels may put a gas in the gas tank.

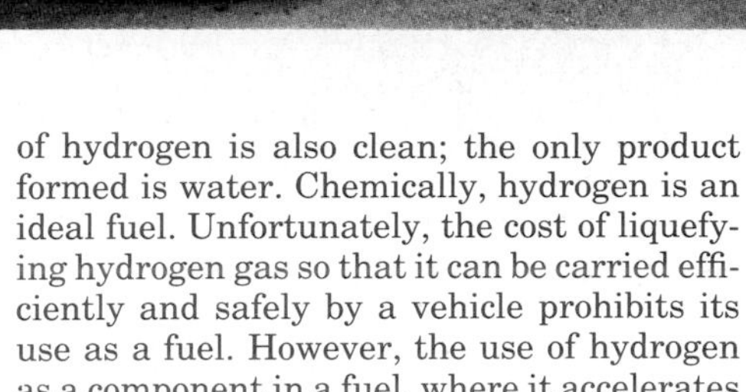

The truck shown here is modified to run on a mixture of compressed natural gas (CNG) and hydrogen gas. Because methane is the major component of natural gas, the mixture has been dubbed "Hythane" by its developer, Frank E. Lynch.

The abundance of natural gas, which is widely used for home heating, has spurred the development of compressed natural gas as an alternative automotive fuel. If the gas is compressed, a vehicle such as a delivery truck or commuter car can carry a sufficient supply of CNG conveniently for short-distance, urban use. However, because the combustion of CNG is a relatively slow reaction, incomplete combustion may take place in the engine. Unburned hydrocarbons, as well as products such as carbon monoxide, could be released to the atmosphere as pollutants.

The combustion of hydrogen, on the other hand, is rapid. That's why liquid hydrogen is used as a rocket propellant. The combustion of hydrogen is also clean; the only product formed is water. Chemically, hydrogen is an ideal fuel. Unfortunately, the cost of liquefying hydrogen gas so that it can be carried efficiently and safely by a vehicle prohibits its use as a fuel. However, the use of hydrogen as a component in a fuel, where it accelerates combustion, is feasible.

Initial tests indicate that the combustion of Hythane produces almost half as many unburned hydrocarbons but greater amounts of carbon monoxide as the combustion of CNG alone. Intrigued by these early results, researchers are testing the use of Hythane as a viable alternative to gasoline. Someday, the corner "gas" station may indeed sell gas.

Percentage Yield

The quantities we calculate in mass-mass problems are, of course, theoretical amounts. That is, we calculate the maximum possible yield for a particular product. In reality, the actual amounts of products are often much less. Chemists are frequently interested in comparing the actual amount of product from a reaction with the amount calculated by a mass-mass calculation. The comparison is usually expressed as a percentage and is called the percentage yield. The **percentage yield** of a product is the actual amount of product expressed as a percentage of the calculated theoretical yield of that product.

$$\text{Percentage yield} = \frac{\text{actual amount of product}}{\text{theoretical amount of product}} \times 100$$

DEMONSTRATION

9-5 Take A Sniff

Purpose: to illustrate the chemical reaction that is discussed in the Percentage Yield Sample Problem

Materials: 1.75 g salicylic acid, 3 cm^3 methanol, 200 cm^3 water, test tube, 250-cm^3 beaker, hot plate, 2 drops 18*M* H_2SO_4, Petri dish, filter paper

Procedure: **1.** Add 3 cm^3 of water to 1.75 g salicylic acid in the test tube.
2. Add 3 cm^3 methanol to the test tube, followed by 2 drops 18*M* H_2SO_4. **CAUTION:** *Acid is corrosive. Methanol is flammable.*
3. Heat the test tube in a very hot water bath for a few minutes.
4. Place a filter paper in the bottom of a Petri dish. Pour the contents of the test tube onto the filter paper.
5. Have students identify the smell as oil of wintergreen, an ester.

Figure 9.12 The actual miles per gallon in "average driving" that a new car delivers will usually vary from that indicated on the EPA sticker due to vehicle options, driving conditions, driving habits, and the vehicle's condition. In the same way, the actual yield of a reaction is usually less than the theoretical yield due to less than optimum conditions.

SAMPLE PROBLEM

Percentage Yield

Oil of wintergreen (methyl salicylate) is used in a variety of commercial products for its flavor and aroma. It is made by heating salicylic acid, $C_7H_6O_3$, with methanol, CH_3OH.

$$C_7H_6O_3 + CH_3OH \rightarrow C_8H_8O_3 + H_2O$$

A chemist starts with 1.75 g of salicylic acid and excess methanol and calculates the maximum possible yield to be 1.93 g. However, after the reaction is run, the chemist finds that the amount of methyl salicylate produced and isolated is only 1.42 g. What is the percentage yield of the process?

$$\frac{1.42\ g}{1.93\ g} \times 100 = 73.6\%$$

PRACTICE PROBLEMS

33. The actual amount of product in a reaction is 39.7 g although a mass-mass calculation predicted 65.6 g. What is the percentage yield of this product?

34. What is the percentage yield in the following reaction if 5.50 grams of hydrogen react with nitrogen to form 20.4 grams of ammonia?

$$N_2(g) + 3H_2(g) \rightarrow 2NH_3(g)$$

Correcting Errors

Remind students who have difficulty conceptualizing percentage yields that the smaller mass of product is placed in the numerator and the larger mass of product is placed in the denominator. The quotient is multiplied by 100% to convert to percent.

PRACTICE

Guided Practice: Guide students through the Sample Problem, then have them work in groups on Practice Problems 33 and 34 that follow.

Cooperative Learning: Refer to pp. 28T-29T to select a teaching strategy to use with this activity.

Independent Practice: Your homework or classroom assignment can include Review and Practice 54-56, and Critical Thinking 80 from the end of the chapter.

Answers to
Practice Problems: Have students refer to Appendix C for complete solutions to Practice Problems.
33. 60.5%
34. 66.0%

✓ ASSESSMENT

Oral: After you have guided the students through the Sample Problem on page 239, have them calculate the answers to Practice Problems 33 and 34. Ask two students to show their solutions on the chalkboard and explain to the class how they derived their answer.

Results: An organic acid, salicylic acid, reacts with an alcohol, methanol, to produce an ester called methyl salicylate. This is esterification.

Questions: 1. The Sample Problem shows that this reaction is only 73.6% efficient. Can you think of a way to improve the percentage yield? Possible response: *Heat the test tube longer. Cover the test tube so no vapor escapes, or condense the vapors as they escape.*

2. Should you expect to get yields of 100% in the laboratory? *no*

Extension: Have students compare prices of perfumes made from natural esters and those that are made from synthetic esters. The rarity and small quantities of natural esters, such as rose oil, make them very expensive. Synthetic esters are more efficiently, and, therefore, more economically, produced.

PROGRAM RESOURCES

Applying Scientific Methods in Chemistry: Use the worksheet "Percentage Yield," p. 25, to help students apply and extend their lab skills.

Laboratory Manual: Use microlab 9-3 "Stoichiometry of a Chemical Reaction" as enrichment for the concepts in Chapter 9.

Answers to
Review and Practice (cont.)

46. **a.** $Cr + 2HCl \rightarrow H_2 + CrCl_2$
b. $Ba(OH)_2 + CO_2 \rightarrow BaCO_3 + H_2O$
c. $CaCO_3 \rightarrow CaO + CO_2$
d. $CaC_2 + 2H_2O \rightarrow Ca(OH)_2 + C_2H_2$
e. $H_2 + S \rightarrow H_2S$
f. $C_5H_{12} + 8O_2 \rightarrow 5CO_2 + 6H_2O$

47. **a.** synthesis
b. double displacement
c. synthesis
d. single displacement
e. combustion

48. **a.** 1.94 g Na_2O
b. 4.73 kg $Pb(NO_3)_2$
c. 0.12 g H_2
3.0 g LiOH
d. 247 kg H_2O
658 kg O_2

49. **a.** copper, oxygen
$2CuO \rightarrow 2Cu + O_2$
b. copper(II) nitrate, silver
$Cu + 2AgNO_3 \rightarrow Cu(NO_3)_2 + 2Ag$
c. magnesium oxide
$2Mg + O_2 \rightarrow 2MgO$
d. nitric acid, silver chloride
$HCl + AgNO_3 \rightarrow HNO_3 + AgCl$
e. magnesium chloride, hydrogen
$Mg + 2HCl \rightarrow MgCl_2 + H_2$
f. iron(III) oxide
$4Fe + 3O_2 \rightarrow 2Fe_2O_3$
g. iron(II) sulfide
$Fe + S \rightarrow FeS$
h. calcium sulfate, water
$Ca(OH)_2 + H_2SO_4 \rightarrow CaSO_4 + 2H_2O$
i. zinc sulfate, hydrogen
$Zn + H_2SO_4 \rightarrow ZnSO_4 + H_2$
j. carbon dioxide, water
$2C_6H_6 + 15O_2 \rightarrow 12CO_2 + 6H_2O$
k. carbon dioxide, water
$2C_8H_{18} + 25O_2 \rightarrow 16CO_2 + 18H_2O$

50. 2.73 g $NaAlO_2$
51. 60.9 g HCl
52. 0.530 g H_2
53. 1.27 g NH_3
54. 63.8%
55. 37.3%

a. Chromium displaces hydrogen from hydrochloric acid, with chromium(II) chloride as the other product.
b. Barium hydroxide reacts with carbon dioxide to form barium carbonate and water.
c. Calcium carbonate decomposes to calcium oxide and carbon dioxide gas.
d. Calcium carbide, CaC_2, reacts with water to produce calcium hydroxide and ethyne gas (acetylene), C_2H_2.
e. Hydrogen combines with sulfur to form hydrogen sulfide.
f. Pentane burns in oxygen to produce carbon dioxide and water.

47. Classify the following reactions as single displacement, double displacement, decomposition, synthesis, or combustion.
a. $4Na(cr) + O_2(g) \rightarrow 2Na_2O(cr)$
b. $Pb(NO_3)_2(aq) + Na_2CrO_4(aq) \rightarrow 2NaNO_3(aq) + PbCrO_4(cr)$
c. $NbI_3(cr) + I_2(cr) \rightarrow NbI_5(cr)$
d. $2Li(cr) + 2H_2O(l) \rightarrow 2LiOH(aq) + H_2(g)$
e. $2C_7H_{14}(l) + 21O_2(g) \rightarrow 14CO_2(g) + 14H_2O(g)$

48. Use the appropriate equations in Question 47 to solve the following mass-mass problems.
a. What mass of sodium oxide is produced by the reaction of 1.44 g of sodium with oxygen?
b. How much lead(II) nitrate will be needed to react with sodium chromate to produce 4.62 kg of lead(II) chromate?
c. What quantity of hydrogen gas is formed when 0.85 g of lithium reacts with water? How much lithium hydroxide will be formed?
d. What mass of water is given off when 192 kg of C_7H_{14} burn completely in air? How much oxygen will be used in the reaction?

49. Predict the products in the following reactions. Then write balanced equations.
a. decomposition of copper(II) oxide
b. copper and silver nitrate (displacement; copper(II) compound formed)
c. magnesium and oxygen (synthesis)
d. hydrochloric acid and silver nitrate (double displacement)
e. magnesium plus hydrochloric acid (displacement)
f. iron and oxygen (synthesis; iron(III) compound is formed)
g. iron and sulfur (synthesis; iron(II) compound is formed)
h. calcium hydroxide and sulfuric acid (double displacement)
i. zinc and sulfuric acid (single displacement)
j. benzene (C_6H_6) and oxygen (combustion reaction)
k. octane (C_8H_{18}) and oxygen (combustion reaction)

50. How many grams of $NaAlO_2$ can be obtained from 4.46 g of $AlCl_3$ according to the following reaction?

$$AlCl_3(aq) + 4NaOH(aq) \rightarrow NaAlO_2(aq) + 3NaCl(aq) + 2H_2O(l)$$

51. How many grams of hydrochloric acid are required to react completely with 61.8 grams of calcium hydroxide?

52. How many grams of hydrogen are produced when 4.72 grams of aluminum react with excess sulfuric acid?

53. How many grams of NH_3 are needed when 2.96 g of $Cr(NO_3)_3$ react according to the reaction: $Cr(NO_3)_3 + 6NH_3 \rightarrow Cr(NH_3)_6(NO_3)_3$?

54. A chemist carried out a reaction that should produce 21.8 g of a product, according to a mass-mass calculation. However, the chemist was able to recover only 13.9 g of the product. What percentage yield did the chemist get?

55. A calculation indicates that 82.2 g of a product should be obtained from a certain reaction. If a chemist actually gets 30.7 g, what is the percentage yield?

✓ ASSESSMENT

Skill: Concept maps are visual representations of relationships among particular concepts. Concept maps make abstract information more concrete and useful. Have students construct an events chain which shows the steps used to solve a mass-mass problem. They can use the information provided on pages 234 and 235. Have students illustrate the events chain by working two problems from the Chapter Review in graphic ways that parallel the events chain. Students could consider including this work in their portfolios.

Chapter 9 REVIEW

56. Chromium(III) hydroxide will dissolve in concentrated sodium hydroxide solution according to the following equation.

$$NaOH + Cr(OH)_3 \rightarrow NaCr(OH)_4$$

This process is one step in making high purity chromium chemicals. If you begin with 66.0 g $Cr(OH)_3$ and obtain 38.4 g of $NaCr(OH)_4$, what is your percentage yield?

57. Zinc oxide can be prepared industrially by treating zinc sulfide with oxygen. The by-product is sulfur dioxide, SO_2. An engineer expects to obtain a 78% yield of zinc oxide by this process. How much zinc sulfide should the chemical plant have on hand in order to prepare 2.0×10^4 kg of zinc oxide? Start by writing a balanced equation for the reaction.

58. Balance each of the following equations.

a. $PbCrO_4 + HCl + FeSO_4 \rightarrow PbCl_2 + Cr_2(SO_4)_3 + FeCl_3 + H_2O + Fe_2(SO_4)_3$

b. $IrCl_3(aq) + NaOH(aq) \rightarrow Ir_2O_3(cr) + HCl(aq) + NaCl(aq)$

c. $MoO_3(cr) + Zn(cr) + H_2SO_4(l) \rightarrow Mo_2O_3(cr) + ZnSO_4(aq) + H_2O(l)$

d. $Cu_2S(cr) + HNO_3(aq) \rightarrow Cu(NO_3)_2(aq) + CuSO_4(aq) + NO_2(g) + H_2O(l)$

e. $Ce(IO_3)_4(aq) + H_2C_2O_4(aq) \rightarrow Ce_2(C_2O_4)_3(aq) + I_2(aq) + CO_2(g) + H_2O(l)$

f. $KBr(cr) + H_2SO_4(aq) + MnO_2(cr) \rightarrow KHSO_4(aq) + MnSO_4(aq) + H_2O(l) + Br_2(l)$

59. In the decomposition of sodium chlorate, 31.7 g of O_2 are formed. How many grams of sodium chloride are produced?

60. The action of carbon monoxide on iron(III) oxide can be represented by the equation, $Fe_2O_3(cr) + 3CO(g) \rightarrow 2Fe(cr) + 3CO_2(g)$. What is the minimum amount of carbon monoxide used if 57.5 grams of iron were produced?

61. Claude-Louis Berthollet first prepared ethyne (acetylene) by sparking carbon electrodes in hydrogen gas.

$$2C + H_2 \rightarrow C_2H_2$$

How many grams of the carbon electrode will be consumed when 59.8 grams of acetylene are produced?

62. Compute the heat change for the reaction of 1.24 grams of NO according to the following equation.

$$2NO + O_2 \rightarrow 2NO_2 + 114.14 \text{ kJ}$$

63. Compute the heat change for the production of 17.1 grams of Fe_2O_3 according to the following equation.

$$4FeO + O_2 \rightarrow 2Fe_2O_3 + 560.4 \text{ kJ}$$

Concept Mastery

64. Applying Dalton's atomic theory, explain why chemical equations must be balanced.

65. Why is it that subscripts should not be changed in order to write a balanced chemical equation?

66. Why is it necessary to use a balanced chemical equation when solving a mass-mass problem?

67. Nitric acid, HNO_3, ranks thirteenth on the list of chemicals produced in greatest amount in the United States. It has extensive use in industry. Nitric acid can be produced by the action of water on nitrogen(IV) oxide. The balanced equation for the reaction is $3NO_2(g) + H_2O(l) \rightarrow 2HNO_3(aq) + NO(g)$. Write a sentence that summarizes the information given in this equation.

68. The reaction that describes the rusting of an old car body is $4Fe + 3O_2 \rightarrow 2Fe_2O_3$. Which of the diagrams on the next page illustrates this reaction?

Answers to
Review and Practice

56. 41.9%
57. 3.1×10^4 kg ZnS
58. a. $2PbCrO_4 + 16HCl + 6FeSO_4 \rightarrow 2PbCl_2 + Cr_2(SO_4)_3 + 4FeCl_3 + 8H_2O + Fe_2(SO_4)_3$
b. $2IrCl_3 + 3NaOH \rightarrow Ir_2O_3 + 3HCl + 3NaCl$
c. $2MoO_3 + 3Zn + 3H_2SO_4 \rightarrow Mo_2O_3 + 3ZnSO_4 + 3H_2O$
d. $Cu_2S + 12HNO_3 \rightarrow Cu(NO_3)_2 + CuSO_4 + 10NO_2 + 6H_2O$
e. $2Ce(IO_3)_4 + 24H_2C_2O_4 \rightarrow Ce_2(C_2O_4)_3 + 4I_2 + 42CO_2 + 24H_2O$
f. $2KBr + 3H_2SO_4 + MnO_2 \rightarrow KHSO_4 + MnSO_4 + 2H_2O + Br_2$
59. 38.6 g NaCl
60. 43.3 g CO
61. 55.2g C
62. 2.36 kJ
63. 30.0 kJ

Answers to
Concept Mastery

64. Atoms are indivisible, combine in whole number ratios, and mass is conserved.
65. Changing the subscripts would change the compounds.
66. Balanced equations give the correct molar ratios.
67. 3 moles of nitrogen(IV) oxide gas react with 1 mole of liquid water to yield 2 moles of nitric acid in water solution and 1 mole of nitrogen(II) oxide gas.
68. a

Answers to
Application

69. $Fe_2O_3 + 3CO \rightarrow 2Fe + 3CO_2$
70. $CaCO_3 + 2HCl \rightarrow CaCl_2 + H_2O + CO_2$
71. $2C_3H_6 + 9O_2 \rightarrow 6CO_2 + 6H_2O$
72. $Al(OH)_3 + 3HCl \rightarrow AlCl_3 + 3H_2O$
73. **a.** $NH_3 + HNO_3 \rightarrow NH_4NO_3$
b. synthesis
c. 1600 kg NH_4NO_3
74. 0.329 g O_2
75. 250 g NaOH
220 g Cl_2
6.2 g H_2
76. 990 kg H_2O
77. decomposition

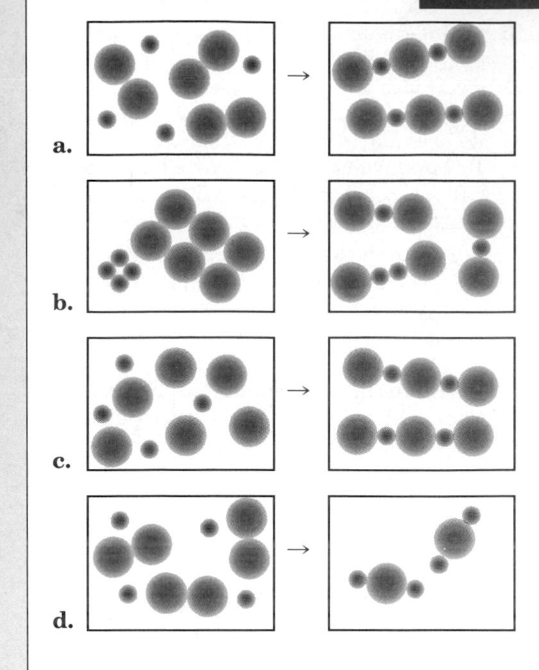

Application

69. Iron occurs in the United States in the form of oxide ores such as iron(III) oxide, Fe_2O_3. In a blast furnace the iron ore is reduced to iron metal by carbon monoxide, which becomes carbon dioxide. Write a balanced equation for the process.

70. Calcium chloride is used to melt ice or to keep it from forming on roads. This important compound can be produced by reacting calcium carbonate with hydrochloric acid. Two by-products are water and carbon dioxide. Write a balanced equation for the formation of calcium chloride by this reaction.

71. Cyclopropane, C_3H_6, was once used with oxygen as a general anesthetic. Write a balanced equation for the complete combustion of cyclopropane.

72. Aluminum hydroxide, $Al(OH)_3$, is a component in many antacids. It neutralizes the HCl in stomach acid in a double displacement reaction. Write a balanced equation for this reaction.

73. Ammonium nitrate, NH_4NO_3, is an important fertilizer and is also used in the manufacture of explosives and fireworks. It is produced by treating nitric acid, HNO_3, with ammonia gas, NH_3.
a. Write a balanced equation for the reaction.
b. What type of reaction is this?
c. If 340 kg of ammonia gas are used with all the nitric acid necessary for reaction, how much ammonium nitrate can be formed?

74. In space vehicles, air purification for the crew is partly accomplished with the use of lithium peroxide, Li_2O_2. It reacts with waste CO_2 in the air according to the reaction $2Li_2O_2 + 2CO_2 \rightarrow 2Li_2CO_3 + O_2$. How many grams of oxygen are released by the reaction of 0.905 g CO_2?

75. Chlorine is prepared industrially by passing an electric current through brine, a concentrated solution of sodium chloride.

$$2NaCl(aq) + 2H_2O(l) \rightarrow 2NaOH(aq) + Cl_2(g) + H_2(g)$$

All three products are commercially valuable. If 1500 g of a brine solution that is 24% NaCl by mass is used, how many grams of each product can be produced?

76. Borax, $Na_2B_4O_7{\cdot}10H_2O$, is an important household cleaner. Its ore, kernite, has the formula $Na_2B_4O_7{\cdot}4H_2O$. If kernite is shipped and then treated with water to form borax at the destination, how much shipping mass is saved if 2500 kg of kernite are shipped?

77. Two components of smog are NO and NO_2. NO_2 is formed in car engines and emitted into the atmosphere. When NO_2 is exposed to sunlight it forms NO and O atoms. What type of reaction is this?

Chapter 9
REVIEW

Critical Thinking

78. In the years following the French Revolution, Joseph Gay-Lussac made several important discoveries about how gases combine. For example, he noted that three volumes of hydrogen react with one volume of nitrogen to form two volumes of ammonia. Keep in mind that in those times, scientists did not know about atoms or the formulas of compounds.
    a. Write a balanced chemical equation for this reaction.
    b. How does this equation relate to Gay-Lussac's findings?
    c. What can you infer about the contents of two equal volumes of reacting gases?
    d. What word could be substituted for the word *volume* to describe this reaction on the simplest level?
    e. What type of reaction did Gay-Lussac perform?

79. A chemical engineer was trying to project the amount of materials that a chemical plant would need for the second step in the production of sulfuric acid. The engineer used the "equation" $SO_2 + O_2 \rightarrow SO_3$. What problems could result from the use of this "equation"? Answer in terms of the quantities the engineer might order.

80. What factors can you think of that might cause a yield of less than 100 percent in a reaction?

Readings

Garcia, Arcesio. "A New Method to Balance Chemical Equations." *Journal of Chemical Education* 64, no. 3 (March 1987): 247–248.

Harjadi, W. "A Simpler Method of Chemical Reaction Balancing." *Journal of Chemical Education* 63, no. 11 (November 1986): 978–979.

Cumulative Review

1. What is the molarity of a 1.00×10^3-cm^3 solution containing 0.550 g of $Ni(IO_3)_2$?
2. Butyric acid, $CH_3CH_2CH_2COOH$, is formed when butter becomes rancid. Find the percentage composition of each element in this compound.
3. Tungsten is Swedish for "heavy stone." Its symbol, W, comes from wolfram, the German name for the element. It has a density of 19.3 g/cm^3. How many atoms are in a cubic centimeter of tungsten?
4. Convert 0.633 mol $BaSeO_4$ to formula units and to grams.
5. Convert 0.0731 mol $Sr(CN)_2{\cdot}4H_2O$ to grams.
6. Find the percentage composition of each element in $BaSO_4$.
7. What is the empirical formula of a compound with the composition 63.9% Cd, 11.8% P, and 24.3% O?
8. Compute the formula mass of Sr_3N_2.
9. What is the mass of 3.00 moles of $Zn_3(PO_4)_2$?
10. How would you prepare 100.0 cm^3 of a 0.100*M* solution of $ZrCl_4$?
11. Find the percentage composition of $Co(CH_3COO)_2$.
12. The compound styrene glycol is used in the manufacture of some plastics. Its percentage composition is 69.54% carbon, 7.30% hydrogen, and 23.16% oxygen. What is its empirical formula? If the molecular mass of styrene glycol is 138.2, what is its molecular formula?

Answers to
Critical Thinking

78. **a.** $3H_2 + N_2 \rightarrow 2NH_3$
b. The ratio of moles is also the ratio of gas volumes.
c. There are equal numbers of moles.
d. molecules
e. synthesis

79. The engineer should order 2 moles of SO_2 for every 1 mole of O_2 ordered. He or she would run out of SO_2 when only half the O_2 was used up if he or she orders according to this equation.

80. The reaction may be reversible, and may not go to completion. In this case, an equilibrium is established in which both reactant(s) and products are present.

Answers to
Cumulative Review

1. 0.001 34*M* $Ni(IO_3)_2$
2. 54.5% C
   9.17% H
   36.3% O
3. 6.31×10^{22} atoms
4. 3.81×10^{23} formula units $BaSeO_4$, 177 g $BaSeO_4$
5. 15.5 g $Sr(CN)_2 \cdot 4H_2O$
6. 58.8% Ba
   13.8% S
   27.5% O
7. $P_2Cd_3O_8 = Cd_3(PO_4)_2$
8. 291 *u*
9. 1160 g $Zn_3(PO_4)_2$
10. 2.33 g $ZrCl_4$ dissolved in enough water to form 100.0 cm^3 of solution.
11. 33.3% Co
    27.1% C
    3.42% H
    36.2% O
12. C_4H_5O; empirical formula
    $C_8H_{10}O_2$; molecular formula

CHAPTER

10 Periodic Properties

CHAPTER ORGANIZER

Chapter Objectives	Text Features	Laboratory Options	Teacher Classroom Resources
10.1 Periodic Trends **1.** Use examples to explain the periodic properties of elements. **2.** State how atomic and ionic sizes change in groups and periods. **3.** Predict oxidation numbers of elements.	**History Connection:** All That Glitters..., p. 253	**ChemActivity 10:** Periodicity of Halogen Properties-microlab, p. 812 **Laboratory Manual:** 10 Periodicity and Chemical Reactivity - microlab **Discovery Demo:** 10-1 A Visible Periodic Table, p. 248 **Demonstration:** 10-2 Negative Ions Are Larger, p. 252 **Demonstration:** 10-3 Oxidation Numbers, p. 256	**Study Guide:** p. 35 **Lesson Plans:** pp. 20-21 **Critical Thinking/ Problem Solving:** Atomic Radii, p. 10 **Color Transparency 21 and Master:** Atomic and Ionic Radii **Applying Scientific Methods in Chemistry:** Elements and Their Oxidation Numbers, p. 26 **Color Transparency 22 and Master:** Size Change in Ion Formation **Color Transparency 23 and Master:** Oxidation Numbers of Groups **Enrichment:** The Periodicity and Chemical Compounds of the Transition Elements, pp. 19-20 **ChemActivity Master 10**
Nat'l Science Stds: UCP.1, UCP.2, UCP.3, UCP.5, B.1, B.4			
10.2 Reaction Tendencies **4.** Define ionization energy and electron affinity, and describe the factors that affect these properties. **5.** Use multiple ionization energies to predict oxidation numbers of elements.	**Bridge to Astronomy:** Using Ionization Energy to Make Images, p. 260 **Everyday Chemistry:** Coatings that Prevent Corrosion, p. 262 **Frontiers:** Metallic Hydrogen? Press On!, p. 264	**Minute Lab:** Smokeless Smoke, p. 261 **Demonstration:** 10-4 Halogens: Salt Formers, p. 260 **Demonstration:** 10-5 Ionization Energy Trend, p. 262	**Study Guide:** p. 36 **Reteaching:** Ionization Energy and Atomic Number, pp. 14-15 **Color Transparency 24 and Master:** Ionization Energy vs Atomic Number
Nat'l Science Stds: UCP.1, UCP.3, B.1, B.2, B.3, B.4, D.4, E.1, E.2			
Other Chapter Resources	**Vocabulary and Concept Review,** pp. 10, 53-54 **Evaluation,** pp. 37-40	**Videodisc Correlation Booklet** **Lab Partner Software**	**Test Bank** **Mastering Concepts in Chemistry—Software**

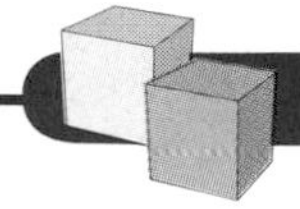

Block Schedule

For information on block scheduling, see the Lesson Plans booklet in the Teacher Resource Package.

GLENCOE TECHNOLOGY

Mastering Concepts in Chemistry Software
Surveying the Table Periodic Properties

Chemistry: Concepts and Applications Videodisc
Oxidation States of Vanadium

Chemistry: Concepts and Applications CD-ROM
Organizing Elements in a Periodic Table

Complete Glencoe Technology references are inserted within the appropriate lesson.

ASSIGNMENT GUIDE

Contents	Level	Practice Problems	Chapter Review	Solving Problems in Chemistry
10.1 Periodic Trends				
Properties and Position	C		12, 13, 24, 25, 39, 46, 49	1
Radii of Atoms	O	1	14-16, 19, 20, 21, 50	2
Radii of Ions	O	2	17-23, 37, 48, 50	3-4
Predicting Oxidation Numbers	C		23, 26-28, 35, 36, 40-42, 55	
10.2 Reaction Tendencies				
First Ionization Energy	C		29-33, 38, 47, 51-54, 56-58	5
Multiple Ionization Energies	A		43-44, 56-58	
Electron Affinities	A		34, 45, 56-58	
				Chapter Review: 1-13

C=Core, A=Advanced, O=Optional

Laboratory Materials

Minute Labs	ChemActivities	Demonstrations
page 261 matches metal pan newspaper smoke detector, battery operated	**page 812** 24-well microplate Microplate Data Form plastic micropipets white paper, one sheet TTE (trichlorotrifluoroethane) Column solutions: 2*M* solutions of sodium fluoride (NaF), sodium chloride (NaCl), sodium bromide (NaBr), sodium iodide (NaI) Row solutions: 6*M* HCl 5% sodium hypochlorite, (NaClO, bleach) bromine (Br_2) water iodine/potassium iodide (I_2/KI)	**pages 248, 252, 256, 260, 262** aluminum can, empty beaker bromine circular discs of the same size (coins) colored marker dropper elements from stockroom fume hood glass bottles with wide mouths glue graph paper HCl concentrated hot plate iodine crystals kerosene MnO_2 overhead projector paraffin wax 24 small clear containers with screw tops sodium metal, small pieces stainless steel spoon text book tin can, empty watch glasses to cover wide mouth bottles water

CHAPTER 10 *Periodic Properties*

CHAPTER OVERVIEW

Main Ideas: The periodic nature of the properties of elements as related to their electronic structure is discussed in Chapter 10. An awareness of the periodic nature of atomic and ionic radii will enable the student to better understand the trends of ionization energy and electron affinity presented later in the chapter. Chapter 10 provides the student with the prerequisite skills needed to understand the information presented in Chapters 11 and 12.

Theme Development: Scale and structure as a theme of the textbook is developed through a presentation of periodic trends of elements and their positions on the periodic table as a consequence of their atomic structures.

Tying To Previous Knowledge: Ask students why they like to wear gold chains. Possible responses: *The gold does not react with the chemicals on the skin; it has luster; and it doesn't tarnish like silver.* In this chapter students will learn why gold has different properties from silver.

ASSESSMENT PLANNER

Choose from the following assessment strategies.

Assess, pp. 256-257, 263-264
Assessment, pp. 249, 252, 253, 255, 258, 259, 261
Portfolio, p. 258
Minute Lab, p. 261
ChemActivity, pp. 812-813
Chapter Review, pp. 265-269

CHAPTER Preview

CONTENTS

DISCOVERY DEMO

10-1 A Visible Periodic Table

Purpose: to observe properties of common elements

Materials: 24 small, clear containers with plastic screw tops; elements from school's stockroom; kerosene, water, glue; paraffin wax, beaker, hot plate

Procedure: 1. Place samples of the elements in individual, clear containers. Label each container with the symbol and name. Use elements such as aluminum, antimony, bismuth, bromine, cadmium, calcium, carbon, chromium, cobalt, copper, gallium, gold foil, iodine, iron, lead, lithium, magnesium, mercury, nickel, phosphorus (red or white), potassium, selenium, silicon, silver, sodium, sulfur, tin, titanium, and zinc. **CAUTION:** *Handle these elements with care; some are carcino-*

CHAPTER 10

Periodic Properties

Have you ever been given a list and been asked to tell which item doesn't fit with the others? To answer that question, you first need to find what the other items have in common. If the list consisted of hat, gloves, flowers, shoes, *and* jeans, *the answers would be easy to find. If the list were* Curie, Stravinski, Raoult, Boyle, *and* Gibbs, *the connection might not be so obvious.*

What is the common factor in these two pictures? The pool water has been treated with chlorine to kill microorganisms. An abrasion on the skin can be treated with iodine—also to kill microorganisms. Chlorine and iodine are in the same family in the periodic table because their outer level electron configurations are the same. Therefore, we would anticipate that the two elements have similar properties. Look at the periodic table. What element might be a substitute for chlorine or iodine?

EXTENSIONS AND CONNECTIONS

History Connection

All that Glitters, page 253

Bridge to Astronomy

Using Ionization Energy to Make Images: A view of how astronomers are using ionization properties of silicon to see dim objects, **page 260**

Everyday Chemistry

Coatings that Prevent Corrosion: A look at ways rust is prevented by coating iron, **page 262**

Frontiers

Metallic Hydrogen? Press On!: Predictions made from the periodic table are being used to investigate and maybe create a strange new material, **page 264**

249

INTRODUCING THE CHAPTER

USING THE PHOTO

Elements in the same family are often used for similar purposes because they have similar properties. Students should suggest that fluorine and bromine would be considered as alternatives for iodine and chlorine, because these elements are all in the same family. Use this opportunity to point out to students how useful it is to know that the properties of the elements are periodic functions of their atomic numbers. Emphasize the relationship between atomic structure and the properties of the elements as you introduce each new topic in this chapter.

REVEALING MISCONCEPTIONS

Ask students what they think about the relative sizes of the atoms. Some may believe that because all atoms are very small, all atoms are the same size. Have the class turn in their textbooks to Table 10.1, Atomic and Ionic Radii, p. 252. These scale drawings help to dispel the misconception. Students will learn why the radii follow the periodic trends shown as they complete Chapter 10.

genic. The Group 1 elements should be submerged in kerosene and white phosphorus in water.

2. Glue the lids onto the display bottles and dip the tops in melted paraffin wax to seal. They will last for many years.

3. Have the students rotate among the lab stations and record observations about each element. Encourage the students to spend time refining their observations. **CAUTION:** *Lithium, sodium, phosphorus, potassium, and bromine should be observed only with the teacher present.* **Disposal:** F

Results: Have students compare their observations of metals and nonmetals and compare the properties of metals in a family.

Questions: 1. What can be said about the color and luster of most metals? *Most metals are gray and shiny.*

2. What color is the nonmetal sulfur? *yellow*

3. Are there more metals or nonmetals among the elements on the periodic table? *more metals*

✓ ASSESSMENT

Knowledge: Conduct a short, lab practical quiz identifying five elements during the next lab period. Choose five common elements, and cover the labels of the jars with heavy, numbered cards that correspond to numbered answer sheets.

10.1 Periodic Trends

PREPARE

Key Concepts

The major concepts presented in this section include the factors that affect the radii of atoms and ions, and how electron configuration can be used to predict oxidation numbers.

Key Terms

atomic radius
noble gas configuration
ionic radius

Looking Ahead

Have each student obtain an empty steel can and an empty aluminum can in preparation for Demonstration 10-3.

1 FOCUS

Objectives

Preview with your students the objectives on the student page. Students can use them as a study guide for this major section.

Analogy

To help students conceptualize the factors that affect an atom's radius, have them think about a rock concert. Ask students if the audience moves toward the stage. Yes, just like the larger nuclei pull in the electron cloud. At the concert, people in front can block the view of the stage just as the lower energy level electrons shield the nucleus.

interNET CONNECTION

Software tools for exploring molecules and the periodic table are available for download.

World Wide Web
http://www.cchem.berkeley.edu

OBJECTIVES

- use examples to explain the periodic properties of elements.
- state how atomic and ionic sizes change in groups and periods.
- predict oxidation numbers of elements.

10.1 Periodic Trends

The periodic table is a powerful tool of the chemist. The table is organized on the basis of the atomic structures of the elements. In this chapter, we will examine some properties whose variation depends upon electron configurations. These properties should be closely related to the positions of the elements in the periodic table. Remember, the elements do not have the properties *because* of their positions in the table. Rather, *both the position and the properties arise from the electron configurations of the atoms.*

Properties and Position

Elements in the same column have similar outer level electron configurations and the configurations change in a regular way from one column to the next as we scan across the table. Because the properties of the elements are determined by their electron configurations, we should be able to predict properties of most elements based on our knowledge of the behavior of a few.

We have already seen that as we scan across the table from left to right, we proceed from metallic elements, through metalloids, to nonmetals. When we drop down to the next period, the same pattern repeats. In other words, the properties are periodic. Similar properties occur at certain intervals of atomic number, as is stated in the modern periodic law. From the graph in Figure 10.1, you can see that density is a periodic property. The densities of the elements vary in a regular way when plotted against the atomic numbers of the elements. Metals have the highest densities and those nonmetals existing as gases at room temperature and one atmosphere of pressure have the lowest densities.

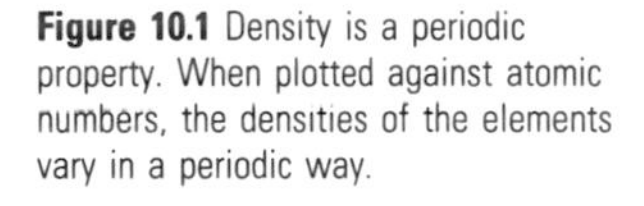

Figure 10.1 Density is a periodic property. When plotted against atomic numbers, the densities of the elements vary in a periodic way.

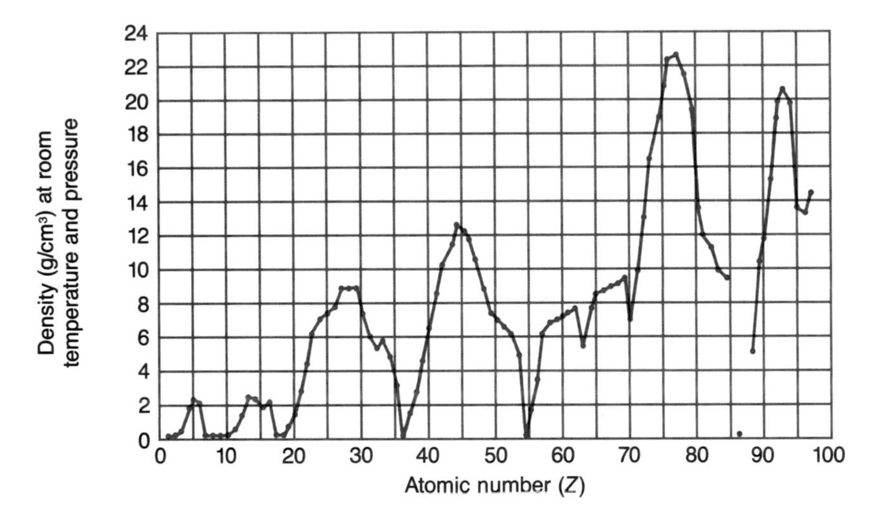

CONTENT BACKGROUND

Trends: Students can use the graphical representation of atomic and ionic radii presented in Table 10.1 to observe the trends that occur. Table 10.1 is also available as a color transparency and a blackline master. Point out that these trends in radii can be used to help explain ionization energies and chemical bonding.

Chemistry Journal

Periodic Table of Your Life

Ask students to list their own periodic activities in their journals. Have them divide these activities into groups, depending upon how frequently they occur. For example, birthdays happen yearly, visits to the dentist twice a year, and visits to grandparents weekly. Have students make a calendar of periodic activities and look for patterns.

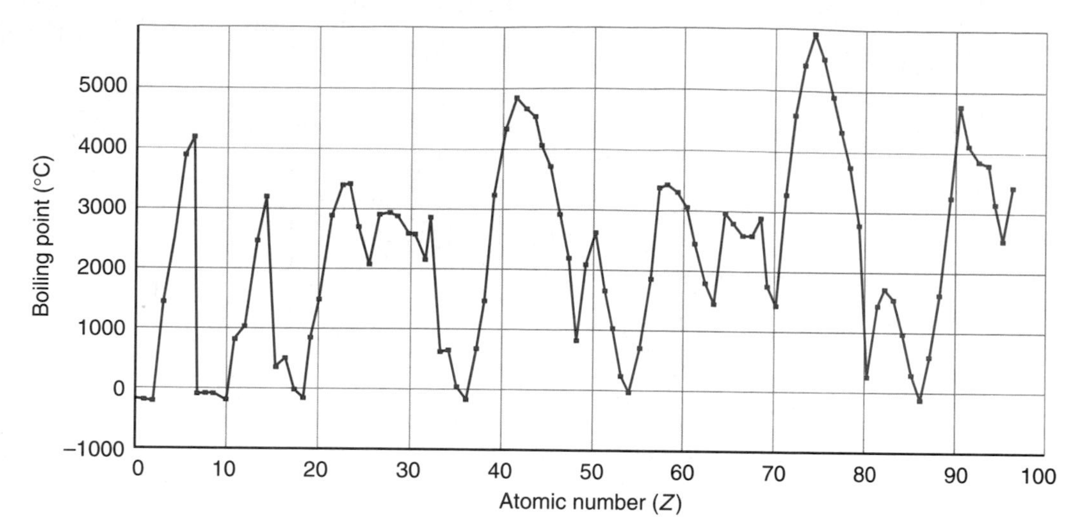

Figure 10.2 Boiling point is also a periodic property.

Radii of Atoms

As you look at the periodic table from top to bottom, each period represents a new, higher principal quantum number. *As the principal quantum number increases, the size of the electron cloud increases.* Therefore, the size of atoms in each group increases as you look down the table. Chemists discuss the size of atoms by referring to their radii. The radius of an atom without regard to surrounding atoms is the **atomic radius.** As you look across the periodic table, all the atoms in a period have the same principal quantum number. This fact might lead you to expect that all atoms in a period are of similar size. However, positive charge on the nucleus increases by one proton for each element in a period. As a result, the outer electron cloud is pulled in a little tighter. Consequently, one periodic property of atoms is that they generally decrease slightly in size from left to right across a period of the table. In summary, *atomic radii increase top to bottom and right to left in the periodic table.* In Table 10.1, you can see the general trends and the few exceptions to the rule.

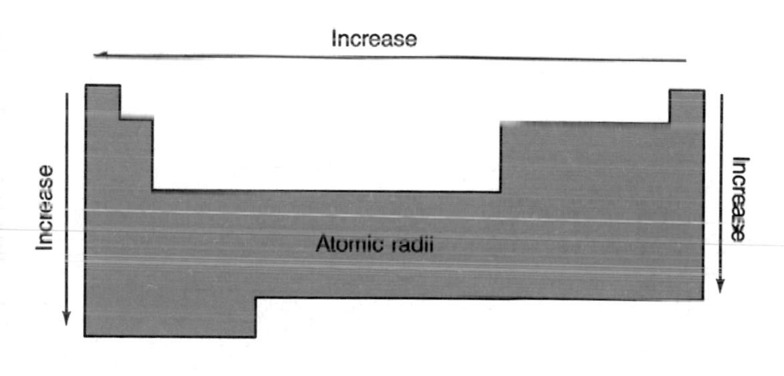

Figure 10.3 The general trend in atomic radii is an increase from right to left and top to bottom in the periodic table.

2 TEACH

CONCEPT DEVELOPMENT

Discussion: Have students explain the trends in atomic radii based on the electronic structure of atoms discussed in Chapter 6. Point out that as the energy levels of atoms increase, the atomic radius increases. At this time, you may want to introduce the shielding effect of inner-sublevel electrons in diminishing the effective nuclear attraction for outer electrons.

Background: Nuclear charge also affects atomic size. For example, the radii of period 6 elements are comparable to those of period 5 because of an increase in effective nuclear charge. The term *lanthanoid contraction* is used to describe this characteristic.

MAKING CONNECTIONS

Health

Sodium and potassium ions are essential to the proper functioning of the human nervous system. Point out that the axons of nerve cells do not conduct electricity. The sodium and potassium ions move through the cell membrane of the axons causing a change in the electric potential. This change in potential moves down the axon like a wave at a rate of 30-50 m/s.

CONTENT BACKGROUND

Ionic Radii: Students who understand this section of the chapter will not have trouble predicting oxidation numbers. Help students understand how each of the following can affect the ionic radii: the shielding effect of inner electron sublevels, the distance of outer electrons from the nucleus, and the size of the nuclear charge (protons). A brief explanation of the inverse square law ($F \propto 1/d^2$) can help students understand why an electron is removed more easily to form a positive ion when the atomic radius is large. When the distance between the electron and the nucleus is doubled, the attractive force is only one-fourth as much. This is an oversimplification, but it does help students to better understand the concept.

PROGRAM RESOURCES

Study Guide: Use the Study Guide worksheet, p. 35, for independent practice with concepts in Section 10.1.

Lesson Plans: Use the Lesson Plan, pp. 20-21, to help you prepare for Chapter 10.

Critical Thinking/Problem Solving: Use the worksheet, p. 10, as enrichment for the concept of "Atomic Radii."

Misconceptions

It is common knowledge that a diet containing too much table salt is harmful by contributing to high blood pressure. However, many people do not realize that too little salt is also harmful. The Na^+ ion is used to produce compounds, such as $NaHCO_3$ and Na_2CO_3 in the blood's buffer system. The Cl^- ion is used to produce HCl in the stomach.

CONCEPT DEVELOPMENT

Background: Atomic radii in Table 10.1, including those of the noble gases, are based on measurements made in the solid state. For more information, see *Introduction to Solid State Physics* by Charles Kittel, 6th edition, Wiley, 1986.

QUICK DEMO

Ionic Conductivity

Use a conductivity tester to show students that dry NaCl does not conduct electricity, whereas an aqueous solution of NaCl does. Heat a pure sample of $KClO_3$ salt in a crucible to melt it. The molten salt will also conduct. **CAUTION:** *Impure $KClO_3$ will explode when heated. Test a very small sample before class. Use a safety shield and goggles.* **Disposal:** A

✓ ASSESSMENT

Knowledge: Ask students to write a paragraph that describes the periodic patterns exhibited by atomic radii and ionic radii.

Table 10.1

Atomic and Ionic Radii (in picometers)

Element	Atomic radius	Ionic radius	Ion charge
H	78		
He	122		
Li	156	90	1+
Be	112	59	2+
B	85	41	3+
C	77	260	4-
N	71	132	3-
O	60	126	2-
F	69	119	1-
Ne	131		
Na	186	116	1+
Mg	160	86	2+
Al	143	67.5	3+
Si	118	54	4+
P	109	212	3-
S	103	170	2-
Cl	91	167	1-
Ar	191		
K	231	152	1+
Ca	197	114	2+
Sc	162	88.5	3+
Ti	147	74.5	4+
V	134	72	4+
Cr	128	75.5	3+
Mn	127	81	2+
Fe	126	69	3+
Co	125	79	2+
Ni	124	83	2+
Cu	128	87	2+
Zn	134	88	2+
Ga	134	76	3+
Ge	123	67	4+
As	121	72	3+
Se	117	184	2-
Br	119	182	1-
Kr	201		
Rb	248	166	1+
Sr	215	132	2+
Y	180	104	3+
Zr	160	86	4+
Nb	146	78	5+
Mo	139	73	6+
Tc	136	78.5	4+
Ru	134	76	4+
Rh	134	80.5	3+
Pd	137	100	2+
Ag	144	129	1+
Cd	151	109	2+
In	167	94	3+
Sn	141	83	4+
Sb	161	90	3+
Te	138	207	2-
I	138	206	1-
Xe	218		
Cs	262	181	1+
Ba	222	149	2+
Lu	174	100.1	3+
Hf	159	85	4+
Ta	146	78	5+
W	139	74	6+
Re	137	77	4+
Os	135	77	4+
Ir	136	76.5	4+
Pt	139	76.5	4+
Au	144	99	3+
Hg	151	116	2+
Tl	170	164	1+
Pb	175	133	2+
Bi	151	117	3+
Po	164	94	4+
At			
Rn	232		
Fr	280	180	1+
Ra	228	162	2+
La	187	117.2	3+
Ce	182	115	3+
Pr	182	99	3+
Nd	181	98.3	3+
Pm	183	97	3+
Sm	180	95.8	3+
Eu	208	94.7	3+
Gd	180	93.8	3+
Tb	177	92.3	3+
Dy	178	91.2	3+
Ho	176	90.1	3+
Er	176	89	3+
Tm	176	88	3+
Yb	193	86.8	3+
Ac	203	112	3+
Th	179	108	4+
Pa	163	78	5+
U	156	103	4+
Np	155	75	5+
Pu	162	86	4+
Am	183	97.5	3+
Cm	174	97	3+
Bk	170	96	3+

DEMONSTRATION

10-2 Negative Ions Are Larger

Purpose: to illustrate why the chloride ion is larger than the chlorine atom

Materials: overhead projector, 15 similar coins

Procedure: **1.** Arrange 7 coins on the stage of an overhead projector in a circle so that each coin is just touching its neighbors. Point out that this configuration represents the outer electrons of an atom of chlorine in which the electron-nuclear attraction is balanced against the electron-electron repulsion.

2. Beside this circle arrange the other 8 coins in a circle. Point out that this configuration represents a chloride ion in which one electron has been added. In this ion the electron-electron repulsion is greater than the electron-nuclear attraction. Have students

Sodium and chlorine are located at opposite ends of the third period. Sodium is found at the left side of the table and is a metal. Chlorine is on the right side of the table in Group 17 (VIIA) and is a nonmetal.

Both sodium and chlorine have partially filled third levels. The outer electrons that take part in reactions are separated from the positively charged nucleus by two inner energy levels. These two inner levels are filled (ten electrons). The chlorine nucleus contains seventeen protons; the sodium nucleus contains only eleven protons. The outer electrons of the chlorine atom are attracted by six more protons than are the outer electrons of the sodium atom. Therefore, the chlorine electrons are held more tightly, and the chlorine atom is smaller than the sodium atom. Sodium and chlorine atoms follow the general pattern shown in Table 10.1.

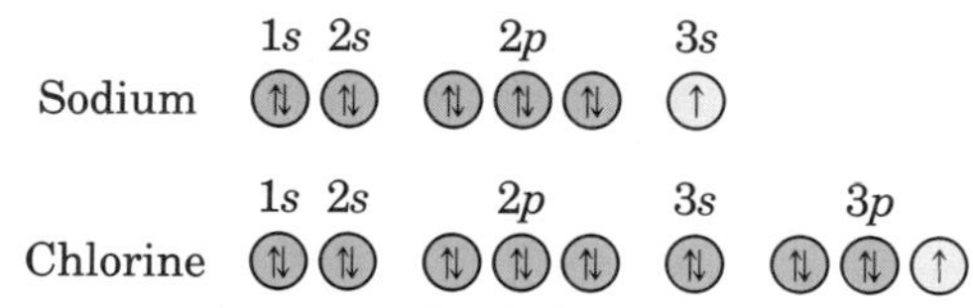

One obvious exception to the rule of thumb for size variation is the noble gas family. The sizes of the atoms for most of the elements have been determined by using a process called X-ray diffraction. In X-ray diffraction, X rays are passed through a crystal of the element. The resulting diffraction patterns reveal the arrangement, size, and spacing of the atoms. When most atoms form crystals, there is some interaction of the electrons in different atoms, pulling the atoms closer together and making the atoms smaller. In the noble gases, with a complete octet in the outer level of each atom, there is no electron interaction and the atoms remain far apart and larger.

Radii of Ions

In general, when atoms unite to form compounds, the compound is more stable than the uncombined atoms. Consider a reaction between sodium and chlorine. The sodium atom holds its single outer $3s$ electron loosely. When chlorine and sodium atoms react, the chlorine removes the outer electron from the sodium atom. The resulting sodium ion has eleven protons, but only ten electrons, so it has a 1+ charge. There are two main reasons why the sodium ion is smaller than the sodium atom. First, the positively charged nucleus is now attracting fewer electrons. Second, with the loss of the $3s^1$ electron, the ion now has two energy levels whereas the atom had three levels. The sodium ion now has a new outer level ($2s^2 2p^6$) that is the same as the outer level of the noble gas, neon. It is important to remember that **noble gas configurations** are particularly stable because the noble gases have filled outer energy levels.

HISTORY CONNECTION

All that Glitters . . .

An object submerged in water will displace its volume of water. Archimedes (287–212 B.C.) discovered the secret of density and water displacement while attempting to determine whether a king's crown was pure gold. He reasoned that a pure gold crown would displace less water than a crown of equal mass made of silver and gold. Because silver is less dense than gold, the same mass would require a greater volume than gold alone. When he placed the crown in water, he found that it displaced a greater amount than an equal mass of pure gold. The crown was not pure!

HISTORY CONNECTION

All that Glitters...

Exploring Further: You may wish to have an interested student explore displacement further. Have the student use water displacement to determine the density of a pure metal and an alloy containing that metal (Cu and brass, Cu-Zn). Provide a balance and a graduated cylinder to the student.

CONCEPT DEVELOPMENT

GLENCOE TECHNOLOGY

Software

Mastering Concepts in Chemistry

Unit 3, Lesson 2

Surveying the Table

Unit 8, Lesson 1

Periodic Properties

CD-ROM

Chemistry: Concepts and Applications

Exploration: *Organizing Elements in a Periodic Table*

Exploration: *Tracing Periodic Patterns*

Interactive Simulation: *The Periodic Table*

compare the radii of the two circles.

Disposal: none

Results: The sizes of the circles differ.

Questions: **1.** What is the name of the group that has seven outer electrons? *halogens*

2. What is the name of the group that has eight outer electrons? *noble gases*

3. Which is larger, a negative ion or its parent atom? *the negative ion*

4. Why? *In the ion, the electron-electron repulsion has a greater effect on the radius than the electron-nucleus attraction.*

ASSESSMENT

Performance: Have students repeat the activity to illustrate why positive ions are smaller than their parent atoms. This time construct the second circle with one less electron. This configuration will lead to a circle with a smaller diameter.

PROGRAM RESOURCES

Transparency Master: Use the Transparency master and worksheet, pp. 41-42, for guided practice in the concept "Atomic and Ionic Radii."

Color Transparency: Use Color Transparency 21 to focus on the concept of "Atomic and Ionic Radii."

Laboratory Manual: Use microlab 10 "Periodicity and Chemical Reactivity" as enrichment for the concepts in Chapter 10.

CONCEPT DEVELOPMENT

Computer Demo: *Valence Drill,* A fast and fun exercise, AP305, PC2201, Project SERAPHIM

CULTURAL DIVERSITY

Salt of the Earth

Adding salt supplements to food has been associated with a change from nomadic ways of life to a settled agricultural lifestyle. If raw or roasted meat or milk form a large part of a culture's diet, natural salts in the food are not lost. Where people depend mainly on cereal, vegetable, or boiled meat diets, though, supplementary salt is needed.

In many early agricultural societies in the Mediterranean and Middle Eastern areas salt was offered to the gods along with cereal gifts. Salt also was widely used as a preservative and came to symbolize fidelity and permanence of relationships in many societies. An Arab saying "there is salt between us" indicates a lasting trust between two people. The Hebrew expression "to eat the salt of the palace" means to think highly of someone, as does the English phrase that a person is "the salt of the earth".

Historically salt has been widely traded and heavily taxed. Ancient salt routes connect widely dispersed areas in Europe, Africa, and Asia. Cakes of salt have been used as money in Ethiopia and other countries in Africa, as well as in Tibet.

Salt originally was obtained primarily through evaporation of saltwater, but extensive salt deposits are found throughout the world. The Salt Range in Pakistan is one of the richest salt fields in the world and has been mined since ancient times.

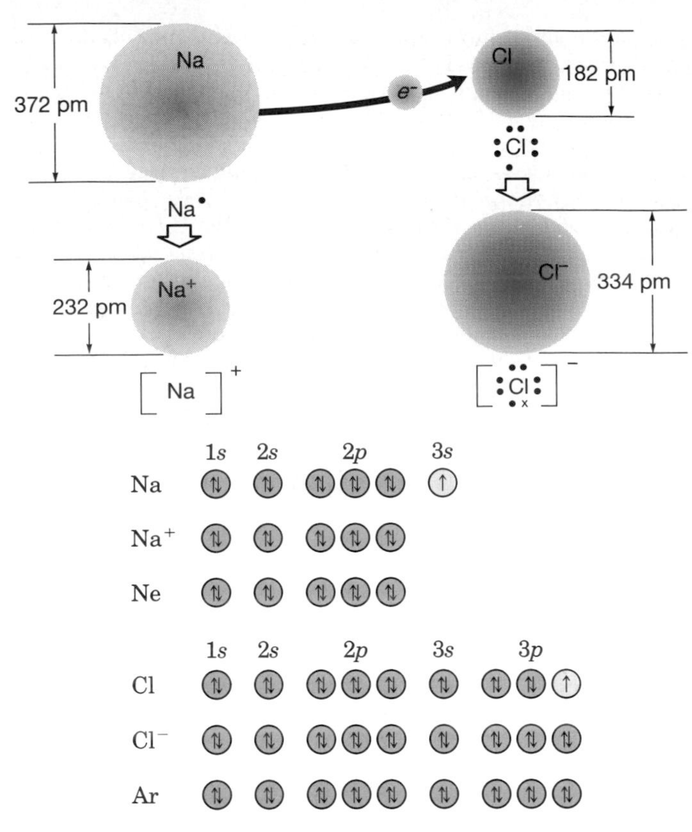

Figure 10.4 The sodium ion is smaller than the sodium atom because 11 protons are attracting only 10 electrons in the ion. The chloride ion is larger than the chlorine atom because 17 protons are attracting 18 electrons in the ion. The chlorine atom is smaller than the sodium atom because the electrons in the outer energy level of chlorine are attracted by 6 more protons than in the sodium atom.

The chlorine atom has gained an electron. The seventeen protons in its nucleus are now attracting eighteen electrons. The chloride ion thus has a 1− charge. Also, the chloride ion is larger than the chlorine atom. The ion now has the same outer level configuration as the noble gas, argon. The positively-charged sodium ions and the negatively-charged chloride ions now attract each other and form a crystal of sodium chloride, common table salt. This salt is more stable than the separate atoms or ions.

The compound formed from sodium ions and chloride ions does not consist of sodium chloride molecules. Instead, each cube of sodium chloride is a collection of equal numbers of Na^+ and Cl^- ions. When a salt is dissolved or melted, it will carry a current because the ions are free to move. However, a solid salt crystal will not carry an electric current because the individual ions are tightly bound in the crystal structure. The mobility of electric charge completes a circuit. Thus, the solution or molten substance is able to conduct a current. This property is a characteristic of compounds composed of ions.

Chemists discuss the size of ions by referring to their **ionic radii**. We have noted that the sodium ion is smaller than the sodium atom. Compared to its atom, the magnesium ion is even

CONTENT BACKGROUND

Radii: Atomic radii are measured by X-ray diffraction in the crystals of the pure metal elements or from the constants in the van der Waals equation of state for gases. In measuring ionic radii, the internuclear distances can be obtained from X-ray diffraction data. However, the problem is to decide how much of the internuclear distance to assign to each ion. By making certain assumptions about the relative sizes of ions based on theoretical size calculations, it is possible to develop a consistent set of dimensions.

smaller. In losing two outer electrons, the unbalanced nuclear positive charge is larger than the negative charge on the electron cloud. The cloud shrinks in size. The nonmetals sulfur and chlorine form ions that are larger than their respective atoms. These elements gain electrons to form ions. The elements silicon and phosphorus do not gain or lose electrons readily. Although they may form ions, they tend to form compounds by sharing their outer electrons.

We can now look at some trends that will apply to any row of the periodic table. In general, *metallic ions, on the left and in the center of the table, are formed by the loss of electrons. They are smaller than the atoms from which they are formed. Nonmetallic ions are located on the right side of the table. They are formed by the gain of electrons and are larger than the atoms from which they are formed.* The metallic ions have an outer level that resembles that of the noble gas at the end of the preceding period. Nonmetallic ions have an outer level resembling that of the noble gas to the right in the same period.

RULE OF THUMB

PRACTICE PROBLEMS

1. From each of the following pairs of particles, select the particle that is larger in radius. Use the periodic table on the inside back cover.
   a. Ar, Ne **c.** N, P **e.** Ca, Sc
   b. B, C **d.** Cl, O
2. From each of the following pairs of particles, select the particle that is smaller in radius. Use the periodic table on the inside back cover.
   a. O, O^{2-} **c.** Te, Te^{2-}
   b. Mg, Mg^{2+} **d.** Ti, Ti^{4+}

Predicting Oxidation Numbers

Those electrons that are involved in the reaction of atoms with each other are the outer and highest energy electrons. You now know about electron configurations and the stability of the noble gas structures. Thus, it is possible for you to predict what oxidation numbers atoms will have.

Consider the metals in Group 1 (IA). Each atom has one electron in its outer level. The loss of this one electron will give these metals the same configuration as a noble gas. Group 1 (IA) metals have an oxidation number of 1+. What about the element hydrogen, which also is in Group 1 (IA)? Hydrogen can lose one electron and have an oxidation number of 1+. Note also that the hydrogen atom could attain the helium configuration by gaining one electron. If this change occurred, we would say that hydrogen has a 1− oxidation number. Hydrogen does have a 1− oxidation number in some compounds. In Group 2 (IIA), we expect the loss of the

PRACTICE

Guided Practice: Remind students of the periodic trends that occur regarding atomic and ionic radii. Have students work in class on Practice Problems 1a and 1c.

Independent Practice: Your homework or classroom assignment can include the remaining Practice Problems, and Review and Practice 14-19 from the end of the chapter.

Answers to
Practice Problems
1. **a.** Ar, **b.** B, **c.** P, **d.** Cl, **e.** Ca
2. **a.** O, **b.** Mg^{2+}, **c.** Te, **d.** Ti^{4+}

✓ ASSESSMENT

Skill: Before assigning students Practice Problems 1 and 2, allow students to use Table 10.1 Atomic and Ionic Radii as you ask them to give the atomic or ionic radii of randomly chosen elements.

PROGRAM RESOURCES

Applying Scientific Methods in Chemistry: Use the worksheet, "Elements and Their Oxidation Numbers," p. 26, to help students apply and extend their lab skills.

Transparency Master: Use the Transparency master and worksheet, pp. 43-44, for guided practice in the concept "Size Change in Ion Formation."

Color Transparency: Use Color Transparency 22 to focus on the concept of "Size Change in Ion Formation."

CONCEPT DEVELOPMENT

Discussion: Review with your students the Discovery Demo in Chapter 6 which involved the reaction of lithium, sodium, and potassium with water. Use the information from the previous topics on radii and relate it to reactivity as you move down the group of alkali metals.

3 ASSESS

CHECK FOR UNDERSTANDING

Ask students to answer the Concept Review problems, and Review and Practice 12-28 from the end of the chapter.

RETEACHING

Provide each student with a blank copy of the periodic table. Have them label the chart with directional arrows as a periodic trend is discussed. Use the marked periodic table to review atomic and ionic radii as well as to predict oxidation numbers. Save the table and use it to label the trends in ionization energy and electron affinity in the next section.

EXTENSION

When doctors prescribe a low sodium diet, they often suggest KCl as a salt substitute. Ask a student to suggest how these two elements might have similar biochemistries. Possible response: Both Na and K have one outer electron and react in a similar manner with other elements.

GLENCOE TECHNOLOGY

Videodisc

Chemistry: Concepts and Applications

Oxidation States of Vanadium

Disc 1, Side 2, Ch. 8

Show this video to reinforce the concept of variable oxidation states.

Also available on CD-ROM.

Figure 10.5 Chromium shows several oxidation states, with the 3+ state being the most common. These solutions of $Cr(NO_3)_3$ (violet) and $CrCl_3$ (green) illustrate the 3+ oxidation state. Solutions of K_2CrO_4 (yellow) and $K_2Cr_2O_7$ (orange) illustrate the 6+ state.

two *s* electrons for the atom to achieve the same configuration as the prior noble gas element. That loss leads to a prediction of 2+ as the oxidation number for the alkaline earth metals. The elements in these two columns exhibit the oxidation numbers predicted for them.

Beginning with Group 3 (IIIB), we have atoms in which the highest energy electrons are not in the outer level. For instance, scandium has the configuration $1s^22s^22p^63s^23p^64s^23d^1$. Scandium's outer level is the fourth level containing two electrons. Its highest energy electron, however, is the one in the 3*d* sublevel. For the transition elements, it is possible to lose not only the outer level electrons but also some lower level electrons. The *d* electrons, because they are in an energy level one below the outer level, can be lost only after the electrons in the outer level have been lost and the outer level is empty. Further, these *d* electrons may be lost one at a time. The transition elements exhibit oxidation numbers varying from 1+ (representing loss of the outer electron or electrons) up to 8+. We would predict scandium to lose

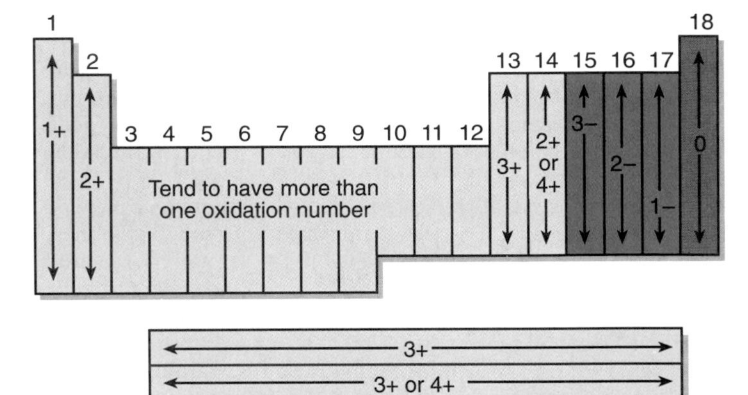

Figure 10.6 The oxidation number of an element can be predicted if one knows the element's location in the periodic table.

DEMONSTRATION

10-3 Oxidation Numbers

Purpose: to demonstrate the effects of electron loss (oxidation) by a metal

Materials: small piece of fine steel wool, 250-cm^3 beaker, 20 cm^3 acetone

Procedure: This demonstration can be performed at home by each student if the teacher completes procedures 1 and 2 in the lab. **1.** Wash a small piece of fine steel wool in acetone to remove any oil or other anti-rust coating. **CAUTION:** *Acetone is flammable and highly volatile. Do not use acetone anywhere near an open flame or glowing electric heating element.*
2. Wash the steel wool in water.
3. Place the steel wool in the beaker and add 20 cm^3 of water.
4. Place the beaker containing the water and steel wool in a warm place

Figure 10.7 When exposed to the oxygen in air, iron will change from its 2+ oxidation state (blue-green) to its more stable 3+ oxidation state (reddish brown).

the two electrons in the outer 4*s* sublevel, giving it an oxidation number of 2+. It can also then lose the 3*d* electron to give it a 3+ oxidation number. In actual practice, the element shows only the 3+ oxidation number. Titanium, which has one more 3*d* electron than scandium has, should show 2+, 3+, and 4+, and it does. These oxidation numbers represent loss of $4s^2$, $4s^2$ and $3d^1$, and $4s^2$ and $3d^2$ electrons, respectively. As we continue across the fourth row, vanadium has a maximum oxidation number of 5+, chromium 6+, and manganese 7+. Iron, which has the configuration $1s^22s^22p^63s^23p^64s^23d^6$, has only 2+ and 3+ oxidation numbers. Recall that a half-full sublevel represents a particularly stable configuration. To take iron higher than 3+ would mean removing electrons from a half-full 3*d* sublevel. However, osmium, in the same group, does have an oxidation number of 8+.

With the exception of boron, Group 13 (IIIA) elements lose three electrons and have an oxidation number of 3+. Boron combines with other atoms solely by sharing electrons. Thallium, in addition to the 3+ oxidation number, exhibits a 1+ oxidation number. If we look at its configuration, we can understand why. The thallium configuration ends $6s^24f^{14}5d^{10}6p^1$. The large energy difference between the 6*s* and the 6*p* electrons makes it possible to lose only the 6*p* electron. That loss leads to an oxidation number of 1+. If stronger reaction methods are used, thallium also has an oxidation number of 3+. For the same reason, tin and lead in Group 14 (IVA) may have either a 2+ or 4+ oxidation number.

10.1 CONCEPT REVIEW

3. Place the following atoms in order of increasing size: B, Be, Mg, N, Na.
4. In each of the following pairs of particles, which particle is larger?
   a. Al, Al^{3+} b. P, P^{3-}
5. Predict oxidation numbers for each of the following elements: Ca, Cl, Cr, K, S.
6. **Apply** Would you predict iron to be more stable in the 2+ or 3+ oxidation state? Give a reason for your answer.

Answers to
Concept Review

3. N, B, Be, Mg, Na
4. **a.** Al
b. P^{3-}
5. Ca: 2+
Cl: 1-, 1+, 3+, 5+
Cr: 2+, 3+, 6+
K: 1+
S: 2-, 4+, 6+
6. 3+ is more stable; the atom loses two electrons from the 4*s* sublevel and one from the 3*d* sublevel, leaving $3d^5$, which is more stable than $3d^6$.

4 CLOSE

Discussion: The variation in atomic and ionic radii can be explained in terms of:
1. shielding effect
2. distance of outer electrons from the nucleus
3. size of the nuclear charge.

Review the periodic trends with students and ask them to apply the three listed concepts when explaining which of the three has the greatest effect on each trend discussed. Possible response: *All three factors continually interact and one is usually not dominant.*

overnight. Do not cover the beaker.
Disposal: A

Results: The steel wool will oxidize to form red rust.

Questions: 1. Write the electron configuration of iron. *$1s^22s^22p^63s^23p^64s^23d^6$*
2. Iron is predicted to have an oxidation number of 3+ in iron(III) oxide, red rust. Which electrons are lost when iron is oxidized from Fe^0 to Fe^{3+}? *The two 4s electrons and one of the 3d electrons are lost.*
3. Is elemental iron, Fe, more stable than iron(III), Fe^{3+}? *No, Fe is less stable than Fe^{3+}.*
4. Is an atom of elemental iron smaller or larger than the Fe^{3+} ion? *Fe is larger than Fe^{3+}.*

Extension: Use this demonstration to relate the two sections of the chapter. Point out that ionization energy trends can also be used to predict oxidation numbers and reactivities more accurately .

PROGRAM RESOURCES

Transparency Master: Use the Transparency master and work sheet, pp. 45-46, for guided practice in the concept "Oxidation Numbers of Groups."

Color Transparency: Use Color Transparency 23 to focus on the concept of "Oxidation Numbers of Groups."

Enrichment: Use the worksheet "The Periodicity and Chemical Compounds of the Transition Elements," pp. 19-20, to challenge more able students.

10.2 Reaction Tendencies

PREPARE

Key Concepts

The major concepts presented in this section include the first ionization energy, multiple ionization energies, and electron affinities.

Key Terms

ionization energy
first ionization energy
shielding effect
electron affinity

1 FOCUS

Objectives

Preview with your students the objectives on the student page. Students can use them as a study guide for this major section.

Analogy

Remind students that some of them can brush their hair, and hair is pulled out by the brush. Others can brush their hair and none falls out. When an atom of one element brushes against the electron cloud of a different element, sometimes electrons are pulled out, and two ions form. The ionization energy is analogous to how strongly the hair (electron) is rooted in the scalp (nucleus).

✓ ASSESSMENT

Portfolio: This is an ideal chapter for practice in taking notes, isolating main ideas, and sorting out important facts. Encourage students to prepare a detailed set of reading notes from Chapter 10, and include these in the portfolio. They should describe periodic trends and give reasons for those trends whenever possible.

OBJECTIVES

- define ionization energy and electron affinity, and describe the factors that affect these properties.
- use multiple ionization energies to predict oxidation numbers of elements.

10.2 Reaction Tendencies

We know that when atoms form compounds, some atoms tend to give up electrons and become positive ions, while other atoms tend to gain electrons and become negative ions. The path of reactions and the properties of the products are largely dependent upon the tendencies of the reacting atoms to gain or lose electrons. We now want to examine the periodic nature of these atomic tendencies. In later chapters, we will be examining the attraction of an atom for electrons as the determining factor in the type of bond formed between atoms in a compound.

First Ionization Energy

The energy required to remove an electron from an atom is called its **ionization energy.** Our model of the atom was developed partly from determining the energy needed to remove the most loosely held electron from an atom. This energy is called the **first ionization energy** of that element. It is measured in kilojoules per mole (kJ/mol).

The first ionization energies of eighty-seven of the elements are graphed in Figure 10.8. Note that the first ionization energies, like many other properties of the elements, are periodic. In fact, the relative first ionization energies of two elements can be predicted by referring to their positions in the periodic table.

The first ionization energy tends to increase as atomic number increases in any horizontal row or period. In any column or group, there is a gradual decrease in first ionization energy as atomic

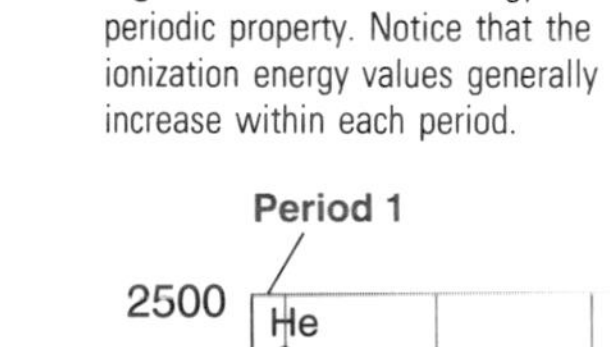

Figure 10.8 Ionization energy is a periodic property. Notice that the ionization energy values generally increase within each period.

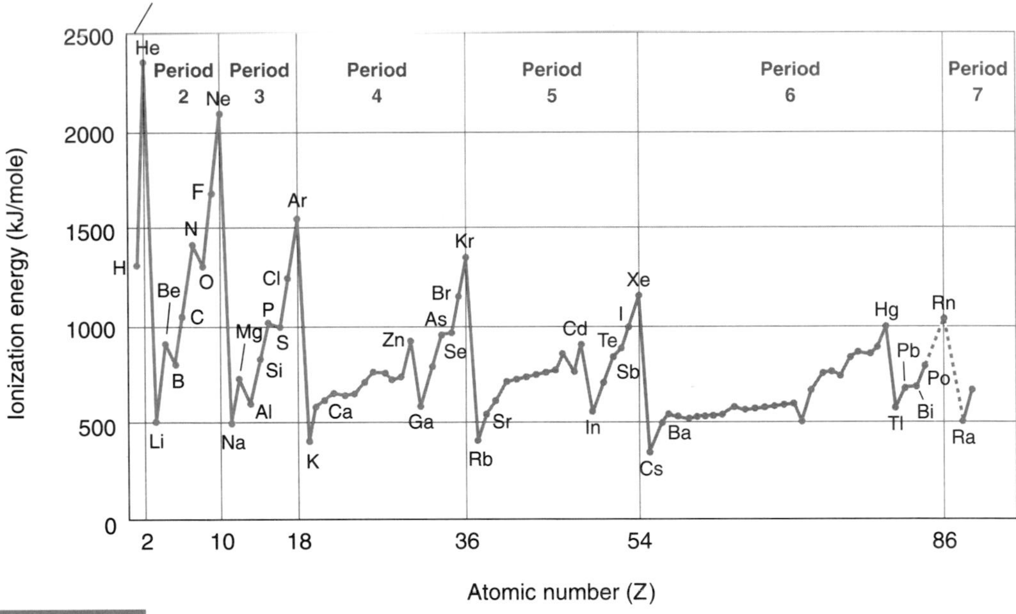

CONTENT BACKGROUND

Measuring Ionization Energy: To determine the ionization energy of a substance, the substance is vaporized in a space between two electrodes. Initially no current will flow through the vapor between the electrodes because the vapor particles have no charge. If this vapor is then bombarded with a stream of electrons, however, some of the bombarding electrons will collide with atoms and "knock off" electrons from the atoms of the vapor. The process produces positively-charged ions which then will move toward the negative electrode, producing an electric current.

To determine the necessary energy for ionization, the energy of the electron stream is increased slowly. When a current is detected, the energy of the electrons that first caused the ionization is noted. That value is the energy necessary to remove completely

Table 10.2

First Ionization Energies (kilojoules per mole)							
H 1312.0							**He** 2372.3
Li 520.2	**Be** 899.5	**B** 800.6	**C** 1086.5	**N** 1402.3	**O** 1313.9	**F** 1681.0	**Ne** 2080.7
Na 495.9	**Mg** 737.8	**Al** 577.5	**Si** 786.5	**P** 1011.8	**S** 999.6	**Cl** 1255.5	**Ar** 1520.6
K 418.8	**Ca** 589.8	**Ga** 578.8	**Ge** 761.2	**As** 947	**Se** 940.7	**Br** 1139.9	**Kr** 1350.8
Rb 403.0	**Sr** 549.5	**In** 558.2	**Sn** 708.4	**Sb** 834	**Te** 869	**I** 1008.4	**Xe** 1170.4

number increases. Note, for example, the gradual decrease in first ionization energy in the alkali metal family, lithium through cesium. The same trend is seen in the noble gas family.

RULE OF THUMB

In general, elements can be classified as metals or nonmetals on the basis of first ionization energy. *A metal is characterized by a low first ionization energy.* Metals are located at the left side of the table. *An element with a high first ionization energy is a nonmetal.* Nonmetals are found at the right side of the table.

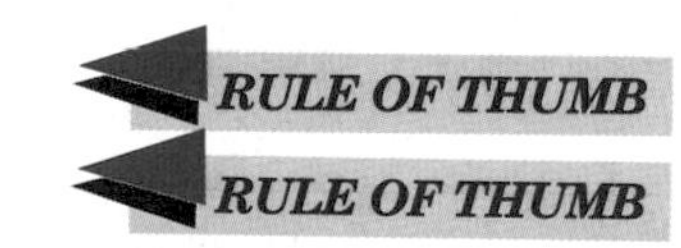

These patterns of first ionization energies provide strong evidence for the existence of energy levels in the atom. Our theories of structure are based on experimental evidence such as ionization energies and atomic spectra.

Look at Table 10.2. Notice that, in general, *first ionization energies decrease as you go down a column of the periodic table* (for instance, lithium, sodium, potassium). Both increased distance of the outer electrons from the nucleus and the **shielding effect,** in which inner electrons block the attraction of the nucleus for outer electrons, tend to lower ionization energy. Though it would seem that the increased nuclear charge of an element with a greater atomic number would increase first ionization energy, the lowering tendency caused by distance and the shielding effect is greater. Remember that the number of electrons in the outermost sublevel is the same for all elements in a particular group.

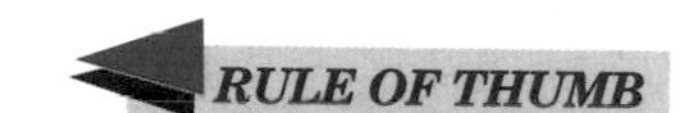

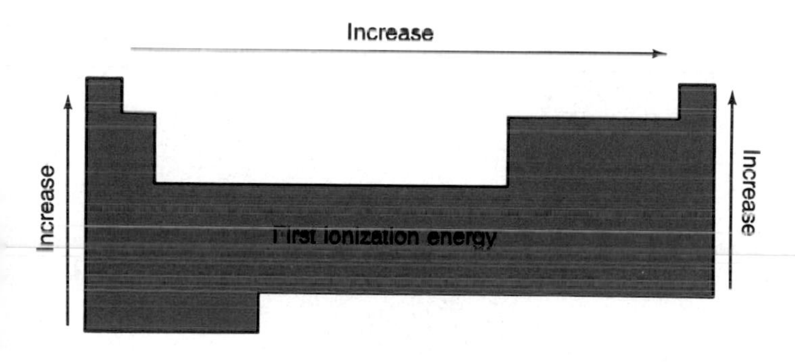

Figure 10.9 In general, the first ionization energies for the elements increase from left to right and from bottom to top of the periodic table.

2 TEACH

CONCEPT DEVELOPMENT

Using the Illustration: Point out to your students that the ionization energy falls off immediately as each maximum node is passed and then gradually builds up to the next maximum. This corresponds to the formation and filling of each new energy level. This experimental data gives evidence for:

1. the effect of increasing nuclear charge (number of protons)
2. the stability of an outer octet of electrons
3. the effect of increased radius
4. the *s* and *p* sublevels in the outer level.

The fourth factor, the effect of *s* and *p* sublevels, causes the slight drop from beryllium to boron, and from magnesium to aluminum. Electrons in the *p* sublevel have slightly more energy and therefore require less additional energy to be removed from the atom.

Computer Demo: *The Periodic Table as a Database,* A graphic display of properties, AP207, Project SERAPHIM

✓ ASSESSMENT

Oral: Ask students in cooperative learning groups to think of an analogy for shielding effect. Have the group's recorder read the analogy to the class. For example, radios work poorly in metal-walled buildings because of the shielding effect of the metal.

the most loosely held electron from an atom of the element being bombarded. This energy is the first ionization energy of that element. It is measured in kilojoules per mole (kJ/mol). This can be demonstrated using a battery-operated smoke detector. See the Minute Lab on p. 261 for details.

PROGRAM RESOURCES

Transparency Master: Use the Transparency master and worksheet, p. 47-48, for guided practice in the concept "Ionization Energy vs Atomic Number."

Color Transparency: Use Color Transparency 24 to focus on the concept of "Ionization Energy vs Atomic Number."

Study Guide: Use the Study Guide worksheet, p. 36, for independent practice with concepts in Section 10.2.

Reteaching: Use the Reteaching worksheet, p. 14-15, to provide students another opportunity to understand the concept of "Ionization Energy and Atomic Number."

In moving across a period of the periodic table, there is a general increase in first ionization energy as a result of increasing nuclear charge. However, there are some deviations from the expected trend of increasing first ionization energy. Look at the second row. There is a small decrease from beryllium ($1s^22s^2$) to boron ($1s^22s^22p^1$). In beryllium, the first ionization energy is determined by removing an s electron from a full s sublevel. In boron, it is determined by removing the lone p electron.

There is another slight decrease from nitrogen ($1s^22s^22p^3$) to oxygen ($1s^22s^22p^4$). Note that the nitrogen p sublevel is half-full. Recall from Chapter 6 that this is a state of special stability. Thus, a large amount of energy is needed to remove an electron from nitrogen's half-full p sublevel. Oxygen does not have the special stability of a half-full sublevel. Thus, oxygen has a lower first ionization energy than does nitrogen.

The patterns in ionization energy values can be explained by the same factors we discussed in Chapter 6 in connection with the periodic table. These factors are summarized in Table 10.3.

BRIDGE TO ASTRONOMY

Using Ionization Energy to Make Images

Background: The first TV cameras contained a vacuum tube that allowed the camera to form images. They required very bright TV studio lighting to operate. Camcorders use charge-coupled devices to form the image and can operate in very dim light.

Teaching This Feature: Use the school's video camera that has CCD circuitry to conduct a test. Focus the camera on a color poster. Gradually dim the lights until the room eventually is very dark. How did the camera perform? Ask students if comparing two cameras in this manner would be good before making a decision as to which one to purchase.

Extension: If your school uses a scanner to grade tests, have a student contact the manufacturer to determine if CCDs are used and how the scanner reads a test answer sheet.

Answers to
Exploring Further

1. 96.4869 kJ/mol = 1 eV/atom; 8.151 eV/atom

2. The Hubble telescope uses CCDs within the telescope for imaging. The images, however, cannot be used directly and must be passed through several ground and satellite transmitting stations before they are used on Earth.

Bridge to ASTRONOMY

Using Ionization Energy to Make Images

Astronomers are constantly trying to see dimmer objects in the universe. As additions to telescopes, they use devices that enhance their ability to see dim objects.

One such device, the charge-coupled device (CCD), makes use of the ionization properties of silicon. When a photon strikes a silicon atom, the atom becomes ionized. The number of free electrons generated varies directly with the radiant energy absorbed. These electrons are then detected electronically.

A silicon microchip can be divided into many small areas called pixels. The charge acquired by each pixel can be detected separately and the information assembled to produce images of dim objects. When cooled to temperatures as low as 100 K, the CCD can image even dimmer objects. The electronic circuitry can be designed to add images in much the same way as a time exposure is made on film. In addition, a CCD will record 70% of the photons that strike it, whereas film seldom records more than 1%. The CCD is reliable, and requires little power to operate. It is also sensitive to a wide range of wavelengths. In addition to its use in astronomy, the CCD can be used to measure and sort objects on an assembly line and read optical characters.

Exploring Further

1. Ionization energies often are reported in units of electron volts/atom. Find the conversion factor for kJ/mol to eV/atom. Using Table 10.2, write the first ionization energy of silicon in eV/atom.
2. Find out how the Hubble Space Telescope incorporates CCDs.

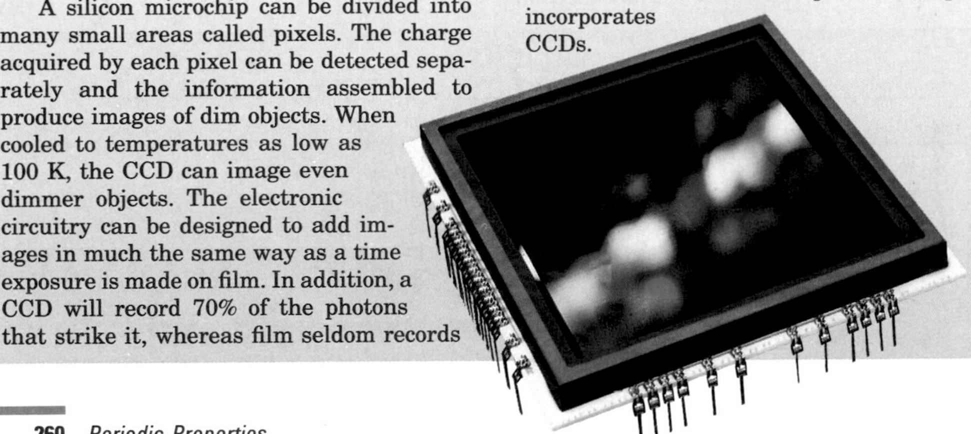

DEMONSTRATION

10-4 Halogens: Salt Formers

Purpose: to illustrate the relationship between ionization energy of nonmetals and reactivity

Materials: 20 cm^3 12*M* HCl, 6 g MnO_2, 3 very small pieces of sodium metal, medicine dropper, 1 cm^3 water, stainless steel spoon, 1 g iodine crystals, 1 cm^3 bromine, fume hood, 3 wide mouth glass bottles with watch glass covers

Procedure: 1. Halogen reactivity can be illustrated by separately reacting a very small piece (1/8 of a pea) of sodium metal in chlorine, bromine, and iodine vapors. **CAUTION:** *Do **NOT** use potassium as it will explode with bromine and iodine. Do **ALL** work in a fume hood while wearing goggles.*

2. Chlorine gas can be generated by adding 20 cm^3 12*M* HCl to 6 g MnO_2. **CAUTION:** *HCl is corrosive. Chlorine gas is very toxic. Collect the gas in a covered bottle in the fume hood.*

3. Using a small stainless steel spoon, lower the very small piece of Na metal into the chlorine gas. Carefully put a drop or two of water on the sodium to begin the reaction. Cover the bottle.

4. Repeat the procedure using bromine. **CAUTION:** *Wear rubber gloves when working with bromine in the fume hood.*

Table 10.3

Factors Affecting Ionization Energy
1. **Nuclear charge**—The larger the nuclear charge, the greater the ionization energy.
2. **Shielding effect**—The greater the shielding effect, the less the ionization energy.
3. **Radius**—The greater the distance between the nucleus and the outer electrons of an atom, the less the ionization energy.
4. **Sublevel**—An electron from a full or half-full sublevel requires additional energy to be removed.

Multiple Ionization Energies

As additional electrons are lost from an atom it is possible to measure other ionization energies of an atom. These measurements give us the same evidence for atomic structure as first ionization energies give us. For example, the second ionization energy of aluminum is about three times as large as the first ionization energy, as shown in Table 10.4. The difference can be explained by the fact that the first ionization removes a *p* electron and the second removes an *s* electron from a full *s* sublevel. The third ionization energy is about one and one-half times as large as the second ionization energy. The second and third electrons are in the same sublevel. Yet, the third electron's ionization energy is greater because the positive nuclear charge remains constant as we remove electrons. As a result, the aluminum atom's remaining electrons are more tightly held. The fourth ionization energy is about four times as large as the third. Why is the jump so large between the third and fourth ionization energies? Look at the electron configuration. Aluminum has the configuration $1s^22s^22p^63s^23p^1$. The fourth electron would come from the second energy level that is both full and closer to the nucleus. The $2s^22p^6$ level with eight electrons is stable. Thus, a large amount of energy will be required to remove that fourth electron.

Table 10.4

Ionization Energies (kilojoules per mole)

Element	1st	2nd	3rd	4th	5th	6th
H	1312.0					
He	2372.3	5220				
Li	520.2	7300	11 750			
Be	899.5	1760	14 850	20 900		
B	800.6	2420	3 660	25 020	32 660	
C	1086.5	2390	4 620	6 220	37 820	46 990
Al	577.5	1810	2 750	11 580	14 820	18 360
Ga	578.8	1980	2 970	6 170	8 680	11 390

CHEMISTRY IN DEPTH

Pages 261–263

MINUTE LAB

Smokeless Smoke

Put on an apron and goggles. Bring to class a battery-operated ionization-type smoke detector. Carefully burn a small 10 cm × 10 cm piece of newsprint in a metal pan in a sink while holding the detector about 30 cm above the flame. **CAUTION:** *Do not have the paper near other flammable objects.* How much smoke did you observe as the paper burned? Did the smoke detector sense the gases produced? Research how a smoke detector works and be prepared to discuss it in class the next day.

5. Repeat using iodine vapor in a separate bottle. Gently warm a few iodine crystals until the bottle is filled with violet vapors. **Disposal:** H

Results: Water initiates an exothermic reaction during which the sodium burns with a yellow flame producing a salt vapor that resembles thick smoke.

Questions: 1. Using Table 10.2, what is the trend of ionization energies within the halogen group? *The ionization energy decreases with increasing atomic number within the halogens.*
2. How is the ionization energy of metals and nonmetals related to reactivity? *As the ionization energy of a metal decreases, reactivity increases. As the ionization energy of a nonmetal decreases, reactivity decreases.*

Extension: Explain that metals tend to lose electrons while nonmetals tend to gain them.

MINUTE LAB

Smokeless Smoke

Purpose: to demonstrate that most smoke detectors actually detect gaseous ions rather than dense smoke

Materials: battery-powered smoke detector, matches, newspaper, metal pan.

Procedure Hints: Test your detector before class to determine the size of the burning sheet of newspaper necessary to activate the alarm. Some require 30-cm × 30-cm pieces. Burning the paper in the sink allows you to quickly extinguish it. **Disposal:** F

Results: The alarm sounds even though there is little or no visible smoke in the air.

Answers to Questions: 1. Very little smoke is observed. **2.** Yes, the alarm sounded. **3.** In a smoke detector, a radioactive source ionizes the air in a chamber containing electrodes of opposite charge. As a result of this ionization, a weak current flows between the electrodes. When smoke particles enter the chamber, the ionized air is disrupted and current stops flowing. The detector senses the "broken circuit" and sets off the alarm.

✓ ASSESSMENT

Knowledge: Ask each student to write a half-page report with a diagram describing how a smoke detector works.

CHEMISTRY IN DEPTH

The sections **Multiple Ionization Energies** and **Electron Affinities** on pages 261-263 are recommended for more able students.

CONCEPT DEVELOPMENT

Using The Table: Take time to be certain each student understands and can explain how the factors listed in Table 10.3 affect ionization energy. This understanding will be helpful when chemical bonding is discussed.

EVERYDAY CHEMISTRY

Coatings that Prevent Corrosion

Background: You may wish to review the "rules of thumb" for activity with respect to the periodic table. From these rules the students can develop a rule of thumb activity series.

Teaching This Feature: Place a piece of silver (spoon) and a piece of iron (nail) in separate beakers that are half-filled with water. After a few days, the partially submerged metals will begin to show corrosion. Show a picture of a bronze statue that has turned green.

Connection to Architecture: Ask students why an architect would want to use copper spouting on a public building. Possible response: *The green patina is attractive, and the long-lasting copper is resistant to water-induced corrosion.*

Extension: Have interested students research what other metals are self-protecting against corrosion through the formation of oxide coatings. Possible response: titanium

Answers to
Exploring Further

1. Answers will vary, but details can be found in chemistry and physics textbooks. A layer of copper is first applied before the other metal is applied.

2. Plastics are commonly used because they are light weight, do not corrode, and are relatively inexpensive.

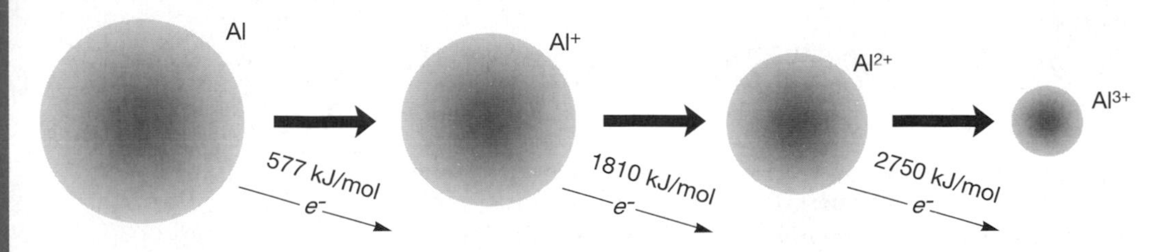

Figure 10.10 The second, third, and fourth ionization energies of aluminum are higher than the first because the same number of protons is attracting fewer electrons.

Look again at ionization energies for the aluminum atom. The first three are relatively small and close together when compared with later ionization energies. These energies help explain why aluminum has an oxidation number of 3+ in all of its compounds. This reasoning can be applied to any other atom. By knowing the ionization energies of the first six or so electrons, you can predict the most likely oxidation number or numbers by seeing where ionization energy increases greatly.

We have looked only at the aluminum atom, but the same reasoning can be used to explain similar data for other elements. This information can be applied as evidence for the theories of atomic structure.

EVERYDAY CHEMISTRY

Coatings that Prevent Corrosion

Corrosion is a popular term used to describe the reaction of some metals with their environment. For example, iron is an active metal and, assisted by some special properties of water, will unite with the oxygen of the air to form iron(III) oxide, or rust.

One possible means to prevent rust is to coat the iron or steel with a metal that does not react with its environment. Chromium plating trim on an automobile is an example of such a coating. Zinc plating is another method of protection. Under ordinary conditions, zinc reacts with water and carbon dioxide in air to produce a zinc carbonate coating that resists further corrosion. Building nails, for example, may be galvanized (coated with zinc) to prevent rusting before and after building is completed.

There are some forms of corrosion, like zinc carbonate, that are beneficial. For example, aluminum, when exposed to the air, forms a coating of aluminum oxide that prevents further corrosion from taking place. The patina on statues and sculpture and the green coating on many bronze or brass fixtures is $Cu_2CO_3(OH)_2$, also a protective covering.

Exploring Further

1. Investigate the processes by which chromium and zinc are electroplated on iron products. How do these processes differ?
2. What materials have replaced galvanized iron in buckets and downspouts? Why are these materials now used?

DEMONSTRATION

10-5 Ionization Energy Trend

Purpose: to illustrate the ionization energy trend found on the periodic table

Materials: graph paper, textbook, colored markers

Procedure: 1. Have students plot the first ionization energy of elements 2-20 on graph paper. Label each point on the graph with the element's symbol. Make an overhead transparency of the completed graph.

2. Using a marker, connect the He, Ne, and Ar peaks. Continue to connect the next two lower points, F and Cl, with a line, then N and P followed by O and S.

3. After you connect C with Si, and Be to Mg, and Ca, students will see that if the ionization energy graph is rotated about 45 degrees on the overhead,

Table 10.5

Electron Affinities (kilojoules per mole)							
H 72.766							He (−21)
Li 59.8	Be (−241)	B 23	C 122	N 0	O 141	F 328	Ne (−29)
Na 52.9	Mg (−230)	Al 44	Si 120	P 74	S 200.42	Cl 348.7	Ar (−34)
K 46.36	Ca (−156)	Ga 36	Ge 116	As 77	Se 194.91	Br 324.5	Kr (−39)
Rb 46.88	Sr (−167)	In 34	Sn 121	Sb 101	Te 190.16	I 295.3	Xe (−40)
Cs 45.5	Ba (−52)	Tl 48	Pb 101	Bi 101	Po (170)	At (270)	Rn (−41)

Parentheses indicate a calculated rather than an experimental value.

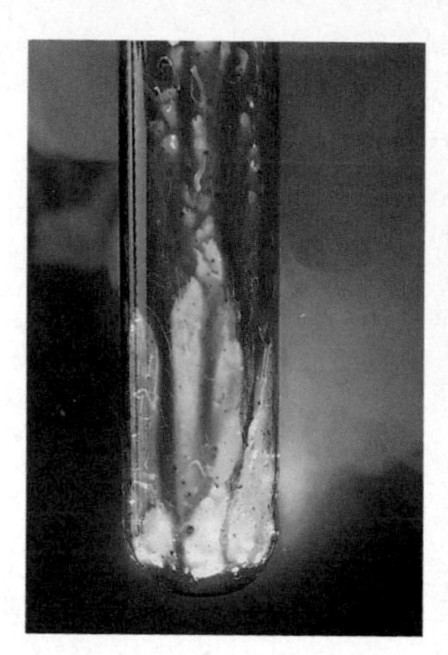

Figure 10.11 The reaction between aluminum, an element with a low electron affinity, and bromine, which has a high electron affinity, is shown here.

Electron Affinities

Now consider an atom's attraction for additional electrons. The attraction of an atom for an electron is called **electron affinity.** The same factors that affect ionization energy also affect electron affinity. In general, as electron affinity increases, an increase in ionization energy can be expected. *Metals have low electron affinities. Nonmetals have high electron affinities,* as shown in Table 10.5. Although not as regular as ionization energies, electron affinities still show periodic trends. Look at the column headed by hydrogen. The general trend as we go down the column is a decreasing tendency to gain electrons. We should expect this trend since the atoms farther down the column are larger. As a consequence, the nucleus is farther from the surface and attracts the outer electrons less strongly.

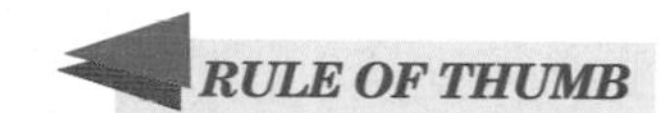

Look at the period beginning with lithium. The general trend as we go across is a greater attraction for electrons. The increased nuclear charge of each successive nucleus accounts for the trend. These elements with high electron affinities will tend to gain electrons and form negative ions.

How do we account for the exceptions of beryllium, nitrogen, and neon in the lithium period? The more stable an atom is, the less tendency it has to gain or lose electrons. The high negative value for beryllium is associated with the stability of the full 2*s* sublevel. If beryllium were to gain an electron, its configuration would be less stable than it previously was. Therefore, beryllium has a negative electron affinity. The failure of nitrogen to attract electrons more strongly is evidence of the stability of the half-full 2*p* sublevel. Neon has a stable, full octet of electrons in the outer level. Because this configuration is stable, neon will have no attraction for additional electrons. For a nonmetal, the greater the electron affinity, the greater the reactivity.

FRONTIERS

Metallic Hydrogen? Press On!

Teaching This Feature: Point out that hydrogen would gain metallic characteristics at these pressures because electrons would move into energy levels that did not formerly exist and become conducting electrons. At 2 million atmospheres the hydrogen should form a molecular metal in which electrons in the diatomic hydrogen should move into new energy levels. At 4 million atmospheres, the molecules should dissociate and form atomic metallic hydrogen in which the electrons are delocalized, that is, not associated with individual hydrogen nuclei.

3 ASSESS

CHECK FOR UNDERSTANDING

Ask students to answer Concept Review problems, and Review and Practice problems 29-34 from the end of the chapter.

RETEACHING

Provide each student with a blank copy of the periodic table. Have them label the table with directional arrows indicating trends in ionization energy and electron affinity.

EXTENSION

Ionization energies are often reported in the literature in units of electron volts per atom (eV/atom). Have a student find the conversion factor for kJ/mol to eV/atom and prepare a table of the first 10 elements with values in electron volts/atom. 1 eV/atom = 96.4869 kJ/mol. Possible response: *H, 13.60; He, 24.59; Li, 5.39; Be, 9.32; B, 8.30; C, 11.26; N, 14.53; 0, 13.62; F, 17.42; Ne, 21.57*

the diagonal lines are the groups or families of the elements.

4. Finally connect B with Al, and Li, Na, and K. **Disposal:** F

Results: The order of the lines does not agree with the order of the families on the periodic table. This anomaly will allow you to review the extraordinary stability of the filled and half-filled sublevels illustrating how the electronic structure of the atom determines its properties.

Questions: 1. Which group has the highest ionization energy for its period? *noble gases*

2. Which group has the lowest ionization energy? *alkali metals*

3. Why doesn't the graph show a smooth line? *The s and p sublevels cause deviations in the trend.*

CHEMISTRY IN DEPTH

Chemistry in Depth ends on this page.

Answers to
Concept Review

7. The first electron of Al is the lone electron in the 3*p* sublevel, whereas the first electron of Mg comes from the full 3*s* sublevel.

8. Li has an oxidation number of 1+ because the ionization energy for the second electron is relatively large, compared with that of the first electron (ratio ≈ 14:1). For beryllium, the large jump occurs between the second and third ionization energies. Therefore, beryllium has a 2+ oxidation number.

9. A negative value for the electron affinity of an atom means that electrons are repelled.

10. Atoms in Group 18 (VIIIA) have a full octet of electrons in the outer level, and have no tendency to attract electrons.

11. Bi and Sb have first ionization energies that are different by only about 15%; they are both in Group 15 (VA) of the periodic table, differing by one period; their atoms are approximately the same size; and the electron configuration in the outermost levels is identical: $5s^25p^3$ for Sb and $6s^26p^3$ for Bi.

4 CLOSE

Discussion: In the USA and Canada, 50 percent of the people drink fluoridated water. The fluoride ion is incorporated into the tooth enamel which is hardened and becomes more resistant to tooth decay. Ask students to use their knowledge of fluorine's ionization energy and electron affinity to explain the apparent stability, in this case, resistance to decay, of the compound containing fluoride. Use Figure 10.6 and Table 10.5 to obtain data. Possible response: *Fluorine's ionization energy is the third highest on the periodic table and its electron affinity is very high. Once fluorine gains an electron to become fluoride, it is very stable.*

FRONTIERS

Metallic Hydrogen? Press On!

If you look at a periodic table of elements, you see that hydrogen is a Group 1(IA) element. Hydrogen and the metals that make up the rest of the Group 1(IA) elements share one metallic characteristic: low electron affinities. This property is just about the only metallic property hydrogen has under standard conditions. However, researchers, spurred by theoretical predictions, are searching for metallic hydrogen, a strange material that may form only at pressures millions of times greater than that of Earth's atmosphere.

Hydrogen exists as a transparent molecular gas and acts as an electrical insulator, both distinctly nonmetallic characteristics. But theoretical models predict that hydrogen will start gaining greater metallic characteristics at extremely high pressures. At pressures of about 2 million times that of atmospheric pressure, hydrogen should lose its electrical resistance and become a conductor—in other words, metallic hydrogen. At twice that pressure, about the pressure at the center of Earth, hydrogen acquires its most metallic characteristic—it becomes a superconductor.

Recently, researchers created metallic hydrogen by placing a thin layer of liquid hydrogen in a device called a two-stage gas gun. In this device, a piston is driven down a gas-filled tube by a gunpowder charge. When the pressure of the gas builds up in the tube, it bursts a seal and sends a smaller piston down a narrow tube at 15 750 miles per hour. The piston hits the container of hydrogen, creating pressures of 1.8 million atmospheres and temperatures of 5000°C. Under these conditions, the liquid hydrogen begins to conduct electricity.

Astronomers believe that the core of the planet Jupiter contains large amounts of hydrogen, which is squeezed at tremendous pressures into a metal. This metallic hydrogen could conduct electricity and generate a magnetic field. Jupiter has a huge magnetic field—about ten times greater than Earth's. Metallic hydrogen may account for it.

10.2 *CONCEPT REVIEW*

7. Explain why magnesium has a first ionization energy higher than that of aluminum even though, generally, ionization energies increase from left to right in the periodic table.
8. Using Table 10.4, explain why lithium has an oxidation number of 1+ and beryllium has an oxidation number of 2+.
9. What is the meaning of a negative value for the electron affinity of an atom?
10. Why do atoms in Group 18 (VIIIA) have negative electron affinities?
11. **Apply** What can you suggest that would explain why antimony and bismuth have essentially identical electron affinities?

CONTENT BACKGROUND

More Periodic Trends: In addition to the trends in atomic and ionic size, ionization energy, electron affinity, and oxidation number, there are other trends that are important to point out to students. Molar volume and density are readily presented. Electronegativity will be covered in Chapter 12.The trends in forming acids and bases from the oxides can also be introduced here if desired. However, it is vital that students do not begin thinking of the properties as the result of a position in the table. Reiterate constantly the point that the properties and the table position are both consequences of structure.

Chapter 10
REVIEW

Summary

10.1 Periodic Trends

1. Because properties of elements are based on electron configurations, many of these properties are predictable and repeat in periodic patterns.
2. Density is a property of elements that varies periodically. In general, densities of elements increase and then decrease as we move left to right across a period of elements.
3. Within a group of elements, the atomic radii of the atoms increase with increasing atomic number.
4. Within a period, the atomic radii of the atoms generally decrease with increasing atomic number.
5. Positive ions are smaller than the atoms from which they are produced.
6. Negative ions are larger than the atoms from which they are produced.
7. When atoms form ions they tend to take on noble gas configurations in the outer energy level.
8. Metals are found on the left side and center of the periodic table. Their atoms tend to lose electrons and thus have positive oxidation numbers.
9. Nonmetals are found on the right side of the periodic table. Their atoms tend to gain electrons and thus have negative oxidation numbers.
10. Oxidation numbers can be predicted from electron configurations by making use of the special stability of the full octet as well as the stability of full and half-full sublevels.

10.2 Reaction Tendencies

11. The relative ease with which an electron can be removed from an atom is related to the type of bond the atom is likely to form with another atom.
12. First ionization energy is the energy necessary to remove the first electron from an atom, leaving a positive ion.
13. First ionization energy is a periodic property. It tends to increase from left to right across a period and decrease from top to bottom through a group.
14. Metals are characterized by low ionization energy, while nonmetals generally have high ionization energy.
15. The factors that tend to lower ionization energy as we move down through a group of elements are increased distance of electrons from the nucleus and increased shielding effect from inner electrons.
16. Electron affinity is the attraction of an atom for an additional electron.
17. Metals have low electron affinities. Nonmetals have high electron affinities.
18. Atoms with filled or half-filled sublevels in their outer energy level tend to have lower electron affinities than neighboring atoms in a period.

Key Terms

atomic radius
noble gas configuration
ionic radius
ionization energy
first ionization energy
shielding effect
electron affinity

Review and Practice

12. What determines the characteristic properties of elements?
13. Why do some properties of elements repeat in periodic patterns?
14. How do atomic radii change from the top to the bottom of a column in the periodic table?

CHAPTER REVIEW 10

- Review Summary statements and Key Terms with your students.
- Complete solutions to Chapter Review Problems can be found in **Merrill Chemistry Problems and Solutions Manual.**

Answers to
Review and Practice

12. Electron configurations determine the properties of elements.
13. Electron configurations are also in periodic patterns.
14. increase
15. Atomic radii decrease, then increase at the noble gases. Across a row, the attraction of the nucleus increases with no increase in electron energy level. A noble gas is larger because of its stable electron configuration.

Chemistry Journal

Electron Affinities
Have students summarize in their journals the factors that affect the electron affinity of an atom.

PROGRAM RESOURCES

Vocabulary Review/Concept Review: Use the worksheets, p. 10 and pp. 53-54, to check students' understanding of the key terms and concepts of Chapter 10.

Evaluation: Use the Evaluation worksheets, pp. 37-40 to assess students' knowledge of Chapter 10.

Computer Test Bank: Use the IBM or Apple Test Bank and Question Selection Manual to prepare your own individualized test for Chapter 10.

15. How do atomic radii change from left to right across a horizontal row of the periodic table? What is the main reason for this pattern?

16. In general, how is the radius of an atom related to the atom's attraction for outer-level electrons?

17. How does the size of a negative ion compare with the size of the atom from which it formed?

18. How does the size of a positive ion compare with the size of its atom?

19. From each of the following pairs of particles, select the particle that is larger in radius.
   a. V, Nb
   b. Cl^-, Br^-
   c. Rn, Fr
   d. Cs, At
   e. Cs^+, At^-
   f. Co, Co^{2+}
   g. P, P^{3-}
   h. Sn^{2+}, Sn

20. Compare metals and nonmetals in terms of (a) position on the periodic table and (b) the way in which they form ions.

21. How do ions of metallic elements compare in size with the atoms from which they are derived? Give reasons for the difference.

22. How do ions of nonmetallic elements compare in size with the atoms from which they are derived? Give reasons for the difference.

23. When oxygen forms the oxide ion, O^{2-}, it achieves a noble gas configuration.
   a. What is a noble gas configuration?
   b. Why is a noble gas configuration stable?
   c. The oxide ion has the configuration of what noble gas?

24. Why can we say that density is a periodic property of elements?

25. Describe the trend in densities from left to right across a period of elements.

26. When atoms lose electrons, how is a noble gas configuration achieved? Which elements are most likely to do this?

27. Why would we predict an oxidation number of 2+ for the transition elements? Do all these elements, in fact, exhibit oxidation numbers of 2+? Give reasons for your answer.

28. Predict oxidation numbers for each of the following elements.
   a. aluminum
   b. samarium
   c. arsenic
   d. polonium
   e. mercury
   f. titanium
   g. barium
   h. rubidium
   i. silver
   j. nickel

29. What is ionization energy?

30. Describe the trend in first ionization energy moving from left to right across a period of elements in the periodic table.

31. Which atom in each of the following pairs of atoms would have the lower first ionization energy?
   a. Al, B
   b. B, Tl
   c. F, N
   d. Mg, Na
   e. K, Ca
   f. Br, Cl

32. Compare metals and nonmetals in terms of first ionization energies.

33. How do first ionization energies change from top to bottom in a group of elements? List factors that influence this change and tell how each factor contributes to the change in ionization energy.

34. Characterize metals and nonmetals in terms of electron affinities.

35. The following are the endings of the electron configurations for several elements. For each, predict the possible oxidation numbers.
   a. $3s^23p^3$
   b. $4s^23d^2$
   c. $5s^24d^{10}5p^1$
   d. $6s^24f^{14}5d^{10}6p^2$

36. Write the electron configurations for tin and lead. Using these configurations, explain why tin and lead show oxidation numbers of both 2+ and 4+.

37. For each of the following pairs, which ion would you expect to be larger? Give reasons for your answers.
   a. As^{3+} and As^{3-}
   b. Tl^{3+} and Tl^+

Answers to
Review and Practice (cont.)

16. The larger the atom, the less the attraction.

17. A negative ion is larger.

18. A positive ion is smaller.

19. a. Nb, **b.** Br^-, **c.** Fr, **d.** Cs, **e.** At^-, **f.** Co, **g.** P^{3-}, **h.** Sn

20. a. metals—left; nonmetals—right

b. Metals lose electrons; nonmetals gain electrons.

21. The ions are smaller; same number of protons attracting fewer electrons.

22. The ions are larger; same number of protons attracting more electrons.

23. a. octet in the outer energy level or the outer energy level is full, **b.** The outer level is full., **c.** Ne

24. The densities of the elements vary in a regular way when plotted against the atomic number of the element.

25. Density increases, then decreases.

26. Electrons are lost until there are 8 electrons in the outer energy level or the outer energy level is full; metals.

27. Most transition elements have 2 electrons in the outer energy level. No, some, such as Ag and Au, have more stable configurations than they would with two electrons in the outer energy level.

28. a. 3+, **b.** 2+(3+), **c.** 3–, 3+, 5+, **d.** 2–, 4+, 6+(2+), **e.** 1+, 2+, **f.** 2+, 3+, 4+, **g.** 2+, **h.** 1+, **i.** 1+ actual, 2+ predicted, **j.** 2+, 3+, 4+
Values in parentheses are from Appendix Table A-3 and are not predictable.

29. Ionization energy is the energy required to remove an electron from an atom.

30. increases

31. a. Al, **b.** Tl, **c.** N, **d.** Na, **e.** K, **f.** Br

32. Metals have lower first ionization energies.

33. First ionization energies decrease. Influencing factors are:

1. Nuclear charge; the larger the nuclear charge, the greater the ionization energy.

2. Shielding effect; the greater the shielding effect, the less the ionization energy.

3. Radius; the greater the distance from the nucleus to the outer electrons, the less the ionization energy.

34. Metals have lower electron affinities than do nonmetals.

35. a. 3-, 3+, 5+; **b.** 2+, 3+, 4+; **c.** 1+, 3+; **d.** 2+, 4+

36. Sn—$1s^22s^22p^63s^23p^64s^23d^{10}4p^65s^24d^{10}5p^2$

Pb—$1s^22p^22p^63s^23p^64s^23d^{10}4p^65s^24d^{10}5p^66s^24f^{14}5d^{10}6p^2$

Both atoms may lose either the two *p* electrons or both the two *p* and the two *s* electrons.

37. a. As^{3-} is larger because it has gained electrons while As^{3+} has lost electrons.

b. Tl^+ is larger; this ion results from the loss of its outer *p* electrons, while Tl^{3+} has lost the entire outer level of electrons.

38. The peaks represent the ionization energies of elements, such as Mg and P, that have electron configurations more stable than those of the rest of the elements in the period.

Chapter 10 REVIEW

38. What factors account for the small peaks and dips in ionization energy values moving from left to right across a period of elements? Use elements from the third period as examples in your answer.

39. Where on the periodic table are elements with the lowest densities? Which of the metal groups has the lowest densities?

40. Predict possible oxidation numbers for the following elements; state your reasoning. Check the actual oxidation numbers in Appendix Table A-3 and explain why your predictions may have deviated from those values.

a. argon
b. europium
c. gallium
d. uranium
e. silicon
f. antimony
g. bromine
h. cadmium
i. cerium
j. cobalt

41. Show how the octet rule would lead us to predict the formula Sr_3N_2 for the ionic compound strontium nitride. Use orbital notation to illustrate your answer.

Concept Mastery

42. What factors account for multiple oxidation numbers of a transition element? What factors tend to restrict the oxidation numbers an element can exhibit?

43. Using the data in Table 10.4, explain why the ionization energy changes so much for the third beryllium electron and for the fourth boron electron.

44. Explain the differences in the six ionization energies of carbon. See Table 10.4.

45. The electron affinities of magnesium and zinc are both negative. What structural features do their atoms possess that would lead to values less than zero for their electron affinities?

46. Both rubidium and silver have one electron in the outermost energy level (5*s*). Why do rubidium and silver differ in chemical properties?

47. The first ionization energies of hydrogen, helium, and lithium are 1312.0 kJ/mol, 2372.3 kJ/mol, and 520.2 kJ/mol, respectively. How do these first ionization values relate to the chemical properties of these three elements?

48. Why does Table 10.1 not show ionic radii for the noble gases?

49. Periodicity is not limited to a study of the elements. Many natural occurrences are periodic. Explain the concept of periodicity, using each of the following natural phenomena as examples.

a. ocean tides
b. the revolution of the moon around Earth
c. foliage on a maple tree

50. Magnesium and calcium both form ionic compounds that are present in hard water.

a. Draw an orbital notation diagram for both atoms and ions of magnesium and calcium.
b. Compare the atomic and ionic radii for both elements.
c. How are the principal quantum numbers of the outer electrons related to the atomic and ionic radii?

Application

51. In 1885, a rare metallic material, didymium, was found to consist of two metallic elements, praseodymium ($Z = 59$) and neodymium ($Z = 60$). Which of these elements would you expect to have a higher first ionization energy?

52. Selenium is named for the moon and has properties similar to those of tellurium, which was named for Earth. Which of these elements has the greater first ionization energy?

53. Graph the ionization energy against atomic radius for the elements lithium through chlorine. What conclusions can you make from the graph?

43. After Be loses two electrons, a full level is exposed. The same is true of B after losing 3 electrons.

44. The second is about double the first, because we are separating a 1- and a 2+. The third is more than three times the first because the third electron comes from the full *s* sublevel. The fourth is about 1/3 more than the third, as is to be expected (separating 1- from 4+). The sudden jump for the fifth is the result of breaking into the full, stable first level. The sixth is, as expected, 1/5 larger than the fifth.

45. Both elements have a stable electron configuration (filled sublevel). Gaining an electron would make the atom more unstable.

46. Ag has *d* electrons added after the *s* electron.

47. Li has the lowest ionization energy and readily loses an electron in a chemical reaction. H will lose an electron, but less readily. He has a stable electron configuration and does not easily react.

39. on the right; alkali metals

40. **a.** 0; stable electron configuration
b. 2+; lost 6*s* electrons
c. 1+, 3+; lose 4*p* or 4*s* and 4*p* electron(s)
d. 2+; loss of two 7*s* electrons
e. 4+; loss of two 3*p* and two 3*s* electrons
4-; gain of four electrons to fill outer level
2+; loss of two 3*p* electrons
f. 3+; loss of three 5*p* electrons
5+; loss of three 5*p* and two 5*s* electrons
3-; gain of three electrons to complete octet
g. 1-; gain of one electron to complete octet
h. 2+; loss of two 5*s* electrons
i. 2+; loss of two 6*s* electrons
j. 2+; loss of two 4*s* electrons

Predictions that differ from table. (There is no obvious explanation unless one is given):
b. 3+ also
d. 3+, 4+ 5+, 6+
e. 4- not included on table
f. 3- not included on table
i. 3+, 4+
j. 3+ also

41. Sr would lose two electrons, becoming Sr^{2+}, which now has an octet in its outer energy level. If N gains three electrons, it now has an octet in its outer energy level and is N^{3-}. Since Sr loses two electrons and N gains three electrons, three Sr atoms are needed to supply the electrons needed for two N atoms. For orbital notation, see *Merrill Chemistry Problems and Solutions Manual.*
The 5*s* electrons from three Sr atoms are lost to the second energy level of two N atoms, resulting in an octet in energy level 4 for Sr and energy level 2 for N.

Answers to
Concept Mastery

42. It is possible to lose not only the 2 electrons in the outer energy level, but also some lower level electrons, one at a time, until an element reaches a stable configuration.

CONCEPT DEVELOPMENT

- Emphasize to students that they are studying group behavior to accumulate knowledge. The characteristics and properties presented will become the foundation on which chemical bonding will be built in Chapter 12.

Misconceptions

Some students may think that any amount of NaCl, table salt, in the diet is harmful and will increase blood pressure. Too much can cause health problems, but so can too little.

Computer Demo: *Chemistry Games,* Order the elements on the periodic table. AP201, PC4604, Project SERAPHIM

✓ ASSESSMENT

Performance: Inform students that enzymes, which mediate chemical changes in all living organisms, are catalysts. Ask students to write a short description of enzymes based on the definition of the word *catalyst.*

GLENCOE TECHNOLOGY

Software

Mastering Concepts in Chemistry

Unit 8, Lesson 1

Periodic Properties

Unit 8, Lesson 2

Typical Elements

CD-ROM

Chemistry: Concepts and Applications

Interactive Simulation: *The Periodic Table*

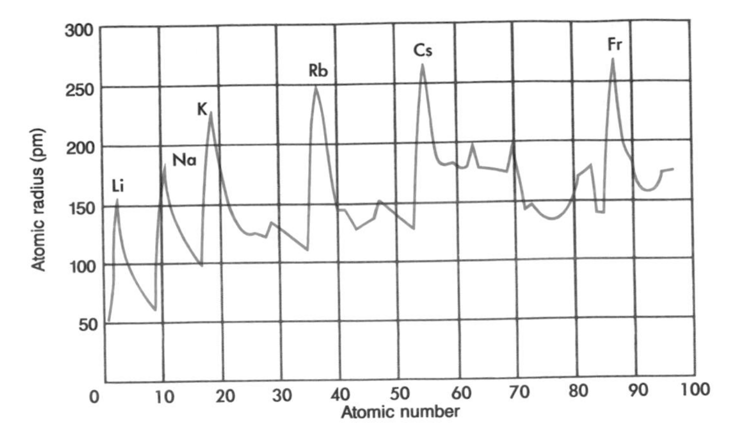

Figure 11.3 The alkali metal atoms increase in size with increasing atomic number. The peaks in the graph show that atomic radius is a periodic property.

table. All of the elements in this group are active enough to displace hydrogen from its compounds. For example, when potassium metal reacts with water, potassium hydroxide and hydrogen gas are formed.

Sodium compounds are among the most important in the chemical industry. Millions of tons of sodium hydroxide are used each year in producing other chemicals, paper, and petroleum products. Sodium carbonate is also produced in millions of tons and used in manufacturing glass and other chemicals. Sodium sulfate is another substance used in manufacturing glass as well as paper and detergents. Sodium silicate is widely used in making soaps, detergents, paper, and pigments in addition to its use as a catalyst. A **catalyst** is a substance that speeds up a reaction. Sodium tripolyphosphate ($Na_5P_3O_{10}$) is used as a food additive, in softening water, and in making detergents.

Sodium compounds are important to the human body. They supply sodium ions, which are essential to the transmission of nerve impulses. Sodium ions and potassium ions are found in

Figure 11.4 Sodium carbonate is used to control large and small chemical spills involving acids.

DEMONSTRATION

11-3 A Living Catalyst

Purpose: to show that yeast is a living organism whose enzymes speed up the rate of a chemical reaction by providing an alternate reaction pathway

Materials: clear glass disposable soda bottle, balloon, 1 Tbsp sugar, 2 Tbsp flour, 1 package yeast, 150 cm^3 water at 50°C, 150-cm^3 beaker, stirring rod

Procedure: 1. Mix together in the colorless glass, disposable soda bottle, the sugar, flour and 100 cm^3 hot water.
2. Cover the top of the bottle with a deflated balloon and observe for a few minutes.
3. Add the yeast to the remaining 50 cm^3 of hot water in a small beaker, stirring to mix. Remove the balloon and add the yeast-hot water mixture to the bottle with mixing. Replace the balloon, sealing the bottle. **CAUTION:** *Do not use the screw-on bottle cap to*

different concentrations in different parts of the body. Table 11.1 shows some of the concentrations of these ions in the body. Note that the relative amounts of these ions vary throughout the body. Sodium and potassium are necessary elements in the diet. Their compounds are found in many foods. One sodium compound, sodium hydrogen carbonate, is used in baked goods. This compound is commonly known as baking soda.

Table 11.1

Distribution of Sodium and Potassium Ions in the Body (mg/100 g)

	K^+	Na^+
Whole blood	200	160
Plasma	20	330
Cells	440	85
Muscle tissue	250-400	6-160
Nerve tissue	530	312

There are two other 1+ ions that, because of their size, behave in a fashion similar to the alkali metal ions. These ions are the ammonium ion (NH_4^+) and the thallium(I) ion (Tl^+). Their compounds follow much the same patterns as the alkali metal compounds.

Lithium

The reactions of the alkali metals involve mainly the formation of 1+ ions. In general, the reactivity increases with increasing atomic number, with one exception. Lithium reacts more vigorously with nitrogen than any other alkali metal. Lithium is exceptional in other ways, too. Its ion has the same charge as the other alkali metals, but the unusual behavior is due to its smaller size. The ratio of charge to radius is often a good indication of the behavior of an ion. The charge/radius ratio of lithium more closely resembles the magnesium ion (Mg^{2+}) of Group 2 (IIA). In its behavior, the lithium ion (Li^+) resembles Mg^{2+} more closely than it resembles sodium (Na^+), the next member of its own family. This diagonal relationship is not unusual among the lighter elements. One example of this relationship is that lithium burns in air to form the oxide, Li_2O, as does magnesium to form MgO. The other alkali metals burn in oxygen to form the peroxide, M_2O_2, or the superoxide, MO_2 (where M = Na, K, Rb, or Cs). Another example of this relationship concerns the solubilities of compounds. The solubility of lithium compounds is similar to that of magnesium compounds, but not to that of sodium compounds.

The lithium atom also differs from the other alkali metal atoms in some physical properties. Unlike the other alkali metals that dissolve in each other in any proportion, lithium is insoluble in all but sodium. It will dissolve in sodium only above 380°C. In other respects, lithium metal is like the other members of its family. For example, it is a soft, silvery metal with a low melting

Figure 11.5 Lithium metal is used to make batteries. The battery shown here is used in heart pacemakers.

CONCEPT DEVELOPMENT

Using The Table: Table 11.1 indicates to students that Na^+ and K^+ ions are important to the body's biochemistry. The Na ion from salt is used to produce compounds, such as $NaHCO_3$ and Na_2CO_3 that play a role in the blood's buffer system. The K^+ ion, obtained from orange juice, bananas, and milk, is important in nerve impulse transmission.

Enrichment: Lithium is often listed as being more reactive than sodium, contrary to predictions. Have a student find out why. Possible response: Although its ionization energy is higher, the Li^+ ion is so highly hydrated that the entire process

$Li(cr) \rightarrow Li(g) \rightarrow Li^+(g) \rightarrow Li^+(aq)$

is more exothermic than the similar process for Na.

$Na(cr) \rightarrow Na(g) \rightarrow Na^+(g) \rightarrow Na^+(aq)$

QUICK DEMO **A Quick Thaw**

Use ribbons to section the surface of a shallow pan of ice into four areas. Mass 10-g samples of rock salt, table salt, granular $CaCl_2$ and powdered $CaCl_2$. In the upper left corner, sprinkle rock salt sparingly on the surface. In the lower left corner, sprinkle granulated table salt on the ice. In the upper right corner, add large granules of calcium chloride. Add powdered $CaCl_2$ to the lower left. Which chemical and which form produced the best de-icing action? Use a chemical catalog to compare their costs. Point out that scientific knowledge can be used to make better consumer choices.

seal the bottle. *It will explode.* Have students observe the balloon every 10 minutes for the remainder of the class period. **Disposal:** A

Results: The sugar does not react with the flour or water. The balloon remains uninflated. After adding the yeast, the yeast enzymes oxidize the sugar, releasing CO_2 gas, which slowly inflates the balloon.

Questions: 1. What happens when the sugar, flour and water are mixed and sealed in the bottle? *No reaction is observed.*

2. What is observed after the yeast is added to the mixture? *The balloon slowly inflates, indicating that a chemical reaction occurs.*

3. Yeast contains enzymes. Why are these enzymes considered to be catalytic? *The enzymes increase the rate of a chemical reaction.*

✓ ASSESSMENT

Performance: Have interested s dents get permission to bake b at home.

Bring in a cookbook and go o recipe with the class. Contra cookbook recipe with the pro followed in the demonstratio students to describe in ch torms the process of alcoh mentation that the yeast carr

CONCEPT DEVELOPMENT

• Allow students to pull a piece of magnesium ribbon. They will observe that it is a very light, yet strong metal. Magnesium is alloyed with aluminum when making the frames of some jet aircraft. This can be a serious problem if there is a fire.

QUICK DEMO **Magnesium**

You can safely burn a short piece of magnesium ribbon for the class to see if you darken the room and use tongs to hold the burning ribbon inside a metal can or sink. The reflected light will not injure a student's eyes. **CAUTION:** *Don't look directly at the burning Mg; it causes serious retina burns.* Magnesium is used in parachute flares to illuminate large areas at night. These flares are also used to confuse heat-seeking anti-aircraft missiles.

QUICK DEMO **A Wet Comparison**

Calcium and magnesium metal react with water, liberating hydrogen gas bubbles. The reaction rate is much slower than that involving the alkali metals, Li, Na, and K. Place a small piece of each metal in a labeled test tube with water. Allow students to observe closely the reaction and compare the rates of Mg and Ca. **CAUTION:** *Do not stopper the test tubes or have them near a flame.*

GLENCOE TECHNOLOGY

Videodisc

Chemistry: Concepts and Applications

Properties of Alkali and Alkaline Earth Metals

Disc 1, Side 2, Ch. 7

Also available on CD-ROM.

Figure 11.6 Sodium vapor lights are used on highways because they can be seen through fog.

point, as are the other alkali metals. All of the alkali metals will dissolve in liquid ammonia to give faintly blue solutions. These solutions conduct electricity.

The alkali metals form binary compounds with almost all nonmetals. In these compounds, nonmetals are in the form of negative ions. In solution, the lithium ion, because of its high charge/radius ratio, attracts water molecules more strongly than any other alkali metal ion.

Alkaline Earth Metals

The alkaline earth metals, Group 2 (IIA), are quite similar to the alkali metals except that they form the 2+ ion. Also, the alkaline earth metals are not as reactive as the alkali metals. Most alkaline earth metal compounds are ionic. These compounds are soluble in water, except for some hydroxides, carbonates, and sulfates.

Figure 11.7 Lime can be added to lakes to counteract the acid conditions caused by acid rain. Very large quantities of lime must be added to a lake of this size.

Beryllium is used in making nonsparking tools, and magnesium is widely used in lightweight alloys. The other metals are too reactive to be used as free elements. The metals Ca, Sr, Ba, and Ra will displace hydrogen from water and other compounds.

Two calcium compounds find large markets. Lime (calcium oxide) is used to make steel, cement, and heat-resistant bricks. It is also applied to soils that are too acidic to farm without treatment. Some lakes that have become too acidic to support aquatic life (due to acid rain) have been treated with lime. Calcium chloride is used in a wide range of applications. One interesting use is in controlling road conditions through deicing in the winter and keeping down dust in the summer. It is also used in the paper and pulp industry.

CONTENT BACKGROUND

Sparkless Tools: Discuss with students what a sparkless tool is and why it is needed in industry. These expensive tools are made using beryllium. They are required when the atmosphere is explosive due to flammable vapors. For example, if an iron wrench is dropped on a cement floor, sparks may be produced. However, a beryllium containing wrench will not spark if dropped or struck against a surface.

Both calcium and magnesium ions are important biologically. The amount of calcium ion present in various tissues affects a very large number of biochemical processes. One of the most important of these processes is the coagulation of blood. There are at least 18 substances involved in the clotting of blood, and these substances undergo at least 13 different reactions. We say "at least" because the process is not yet entirely understood by biochemists, and we count only those substances and reactions known to be involved. Calcium ions are definitely known to be reactants in four of the reactions, and it is strongly suspected that they are required for a fifth reaction. In addition to their role in blood clotting, calcium ions are a major constituent of bone. They also have a significant effect upon the release and absorption of chemicals, called hormones, used by the body as "messengers" from one organ to another.

Bridge to

BOTANY

Magnesium in Plants

Plants produce their own food. In order to do so, they must have the necessary reactants and energy from sunlight. In addition, plants must contain chlorophyll, a complex compound that can capture the sun's energy and make it available to biochemical reactions. When plants make food, they are producing material for their own energy needs and for growth. Many plants are eaten by humans. In this process, we obtain the energy that was captured by the plant during photosynthesis. Humans also eat animals that eat plants. In this process, we are still obtaining energy from plants, although we do so indirectly.

At the center of the chlorophyll molecule is a magnesium ion. Thus, magnesium is an essential mineral for plants. If there is insufficient magnesium in the soil, plants cannot produce chlorophyll, and so they do not grow properly. Because chlorophyll is green, the lack of this compound can be detected visually. Plants that do not produce enough chlorophyll look pale or yellow. If the deficiency becomes severe, the plant may die because it cannot produce enough food to meet its energy requirements.

Exploring Further

1. From a drawing of the chlorophyll molecule on page 242, determine to what element(s) magnesium is attached in the chlorophyll molecule.
2. Use library materials to learn what properties of magnesium account for its being a part of a chlorophyll molecule.

BRIDGE TO BOTANY

Magnesium in Plants

Background: The sun is the energy source. Point out to students that they run on solar power. Use this statement to promote discussion that utilizes a student's knowledge of the food chain.

Teaching This Feature: The photosynthesis reaction discussed is referred to as the "light reaction." Ask a student to determine the "dark reaction" in photosynthesis. Possible response: *It is the reaction in which glucose is oxidized to produce CO_2 and H_2O with the release of energy.*

Connection to Pharmacology: Magnesium is considered to be an essential trace element. Have a student bring in a bottle of multivitamins with minerals. The label will list the minerals as essential trace elements. They may include Ca, Fe, P, Mg, Cu, Zn, Mn, K, Cr, Mo, Se, Ni, Sn, Si, V

Extension: Hydroponics, growing plants without soil, utilizes solutions that provide all of the essential nutrients. This method of growing plants can become a long-term class project for an interested student.

Answers to
Exploring Further

1. nitrogen

2. high charge-density, due to the small size of the ion

3 ASSESS

CHECK FOR UNDERSTANDING

Ask students to write the answers to Concept Review problems and Review and Practice Problems 14-18 from the end of the chapter.

RETEACHING

Put the names of the Group 1, 2, and 13 elements on individual cards in a box and have each student pick one. Have each student make a flash card of that element by writing its name and symbol on one side, and a property and use on the opposite side. Organize the students into teams and use competition to reteach the information.

Chemistry Journal

Many Uses for Many Metals

Ask students to look around their homes and list all the objects that contain or are made of metal. Have them make a guess as to the type of metal used in each object and what property or properties of the metal make it suitable for use in the object.

PROGRAM RESOURCES

Laboratory Manual: Use microlab 11, "Comparing Activities of Selected Metals" as enrichment for the concepts in Chapter 11.

EXTENSION

There are many types of batteries available for many different uses today. Find out the application to which the lithium battery is best suited. Possible response: *The Li battery has a long shelf life. It was invented to give heart pacemakers a longer implant life.*

MINUTE LAB

Airplane Skin

Purpose: to test the strength and corrosion resistance of aluminum

Materials: empty aluminum soft drink can, stiff L-shaped wire, 600-cm^3 beaker, 200 cm^3 6*M* HCl

Procedure Hints: The same acid can be used to treat cans for several classes. Plan on using two cans for each class because you will be asked to do it again.

Disposal: D

Results: The untreated can twists but does not tear. The acid corrosion causes the treated can to rip easily and cleanly in half.

✓ ASSESSMENT

Oral: Have a student read the list of ingredients on a soft drink can and indicate any acids.

Answers to
Concept Review

1. They have a similar arrangement of outer electrons.

2. Hydrogen can react in four ways: 1. by losing its electron to form a hydrogen ion, H^+, as in the Arrhenius theory of acids; 2. by sharing its electron with another atom, as in compounds such as H_2O and HCl; 3. by gaining an electron to form a hydride ion H^-, as in metal hydrides such as sodium hydride (NaH); 4. by forming bridges between atoms, as in the B_5H_9 molecule.

3. The shielding effect blocks the attraction of the nucleus for the outer electrons, which causes an alkali metal atom to lose its outer electron more readily and thus makes the atom more unstable.

4. hydroxides and hydrogen gas

5. Aluminum has three electrons in its outer level and tends to share electrons, while the elements in Groups 1 and 2 usually form ions.

Figure 11.8 The outer skin of airplanes is made of aluminum because this metal is light, strong, and flexible.

MINUTE LAB

Airplane Skin

The Federal Aviation Administration restricts the use of corrosive materials near the aluminum skins of airplanes. Check the strength of aluminum by trying to tear an empty aluminum soft drink can into two pieces. Grip the ends and twist. **CAUTION:** *Metal edges are sharp. Wear gloves, an apron, and goggles.* Using an L-shaped heavy wire, scratch the thin plastic coating on the inside of the can near the center. Place the can in an empty 600-cm^3 beaker. Add 6*M* HCl to the can so that the level is just above the scratch. **CAUTION:** *Acid is corrosive.* After a few minutes, the acid will have reacted with the can where the plastic liner was scratched. A thin black line will appear on the paint. Empty the acid into the beaker. Rinse the can with water to remove all traces of acid. Wearing gloves, try again to tear the can into two pieces. Explain your observations.

The Aluminum Group

Aluminum is the most plentiful metal in Earth's crust. Of the elements in Group 13 (IIIA), aluminum has the most practical uses. With three electrons in its outer level, aluminum is less metallic than the elements of Groups 1 (IA) and 2 (IIA). In forming compounds, it tends to share electrons rather than form ions. It is also less reactive than the metals of Groups 1 (IA) and 2 (IIA) metals. One aluminum compound, aluminum sulfate, is used in water purification, paper manufacture, and fabric dyeing. Large amounts of elemental aluminum are used to produce lightweight alloys for many items from soft drink cans to spacecraft.

When we think of electrical conductors, or wires, most of us think of copper. However, much long-distance wire is now made of aluminum. Although copper is a 50% better conductor than aluminum, it is also three times as dense as aluminum. As a result, an aluminum wire half the weight of a copper wire can conduct the same current. Many long-distance transmission lines from power plants are made of aluminum. Because aluminum's conductivity is dependent on its purity, only high quality aluminum can be used as a conductor. However, pure aluminum is a fairly soft metal. Consequently, long, thick aluminum wires would tend to sag under the force of their own weight. Engineers overcome this problem by wrapping the aluminum wire around a steel support cable. The combined weight and cost is still low enough to make aluminum wire competitive with copper.

11.1 CONCEPT REVIEW

1. Why do elements of a group have similar chemical properties?
2. List two of the four ways that hydrogen can form bonds. Give an example of each.
3. What effect does the shielding effect have on the stability of an alkali metal atom?
4. What substances are produced when the alkali metals react with water?
5. **Apply** Why is aluminum considered to be less metallic than the elements in Groups 1 and 2?

CONTENT BACKGROUND

More on Group 13(IIIA): Boron is the only nonmetal in Group 13(IIIA) and forms covalent bonds. However, it is "electron deficient" in most compounds with only three electrons to share. It is this characteristic that dominates its chemistry. Otherwise, it is much like silicon in its behavior. In addition to its hydrides (boranes) and borates, it forms many borides, as well as halides, carboranes (with carbon), B-N compounds and organoboron compounds.

The last three elements in Group 13(IIIA) are distinctly metallic. Gallium, like water, expands on freezing. All three metals (Ga, In, Tl) exhibit both the 1+ and the 3+ oxidation state, with 1+ being most common for thallium. In fact, there is a considerable similarity between the behavior of thallium(I) and the alkali metal ions.

11.2 Nonmetals

OBJECTIVES

- define catenation and explain how it affects the ability of carbon to form compounds.
- explain the importance of nitrogen and phosphorus compounds to living things.
- state the relationship between the activities of elements and their locations in the periodic table.
- list characteristics and give uses for representative main group nonmetals.

The elements in Groups 14 through 18 tend to share or gain electrons when they react. There are a few exceptions. Tin, lead, and bismuth tend to lose electrons because their outer electrons are so far from the nucleus and are so well shielded. As far as chemists are aware, helium, neon, and argon form no compounds. Arsenic, antimony, silicon, and germanium have some metallic and some nonmetallic properties. In this section we will take a closer look at some of these elements.

Group 14 (IVA)

Group 14 (IVA) is the carbon group. The elements of Group 14 (IVA) have atoms with four electrons in the outer level. These elements generally react by sharing electrons. However, the tendency to lose electrons increases as the atomic number of Group 14 (IVA) elements increases. There are a few compounds in which carbon in the form of a **carbide ion** (C^{4-}) exists. Silicon and germanium, the next members of this family, are metalloids. They do not form 4− ions under any conditions. However, there are compounds in which silicon and germanium exist as a 4+ ion.

You know from Chapter 3 that the major part of carbon chemistry is classed as organic chemistry. In most **organic compounds,** electrons are shared between a carbon atom and one or more other carbon atoms. This characteristic causes the formation of long, chainlike molecules. The tendency to form chains and rings of similar atoms is called **catenation** (kat uh NAY shuhn). Of all the elements, only carbon exhibits catenation to any great extent. In general, those compounds that do not contain carbon are called **inorganic compounds.** There are several exceptions to this rule. A few carbon-containing substances that are considered inorganic are carbon itself, carbonic acid and its salts, carbides, cyanides, and the oxides and sulfides of carbon.

catena: (L) chain
catenation—chain-forming

Figure 11.9 Diamond and graphite are two allotropes of carbon. Graphite is one component of pencil lead.

4 CLOSE

Activity: For each team of three students, cut the elements of groups 1, 2, and 13 from a periodic table and place then in 2 beakers. Have each team put the elements back into order on a blank outline of the periodic table. The team that completes the task first wins.

11.2 Nonmetals

PREPARE

Key Concepts

The major concepts presented in this section include the nonmetallic properties and unique chemistry of Groups 14, 15, 16, 17, and 18(IVA)-(VIIIA). Carbon, nitrogen, phosphorus, oxygen, fluorine, chlorine, and helium are discussed as detailed examples of these groups.

Key Terms

carbide ion
organic compound
catenation
inorganic compound
allotrope
silicate
amphoteric

1 FOCUS

Objectives

Preview with your students the objectives listed on the student page. Students can use them as a study guide for this major section.

Discussion

Review the properties of metals. Write a list on the chalkboard.

- luster
- malleable
- ductile
- high melting point
- good conductor

Point out to students that they already know the properties of the nonmetals. They are the opposites. Beside the first list, write these opposites.

- dull
- brittle
- not ductile
- low melting point
- insulator

FRONTIERS

Welding Gems

Background: In several earlier procedures, methane gas was heated and then passed over tiny diamond crystals. Carbon atoms crystallized onto the diamonds causing them to grow.

Teaching This Feature: Have students recall that in a diamond each carbon atom shares electrons with the four nearest carbon atoms, and in graphite each carbon atom shares electrons with the three nearest carbon atoms. Have students compare the physical properties of these two allotropes of carbon.

Connection to Dentistry: Today students can have their teeth sealed with a plastic coating. However, this coating wears off and must be reapplied. Diamond coatings would not.

Exploring Further: Have interested students investigate vapor deposition methods of producing crystals and their films.

allos: (GK) another
tropos: (GK) turn, or way

allotrope—found in another way (or form)

Elemental carbon is found in nature in two different molecular forms, diamond and graphite. Different forms of the same element are called **allotropes.** In diamond, each carbon atom shares electrons with the four nearest carbon atoms. In graphite, the sharing is to the nearest three carbon atoms. The difference in structure results in different properties for these two forms of carbon, as will be described in Chapter 16.

Industrially, carbon is used in a form called "carbon black" made by the incomplete burning of natural gas or other fuel. Carbon black is often referred to as soot when it has accumulated in a place where it is unwanted. It is actually a microcrystalline form of graphite, and is used as a black pigment and wear-resistant additive in rubber tires. Carbon dioxide gas is a by-product of several chemical processes. It is collected, compressed, and sold as a liquid in steel cylinders. Customers use it for refrigeration, carbonating beverages, and producing other chemicals. Solid carbon dioxide is called "dry ice." Small crystals of dry ice are sometimes used in cloud-seeding to induce rain.

FRONTIERS

Welding Gems

At extreme pressures and temperatures deep within Earth, carbon atoms slowly crystallize into diamonds. To replicate this geologic process, scientists use large hydraulic presses to make synthetic diamonds from graphite. Under pressures 10 000 times that of the atmosphere and at a temperature of 2000°C, the graphite slowly changes structure and forms small diamond crystals. Similar methods are now standard for manufacturing synthetic diamonds for industrial use. However, the costly manufacturing process makes the diamonds expensive. Now scientists are looking within the flame of a welder's torch and not the depths of Earth for a new and cheaper way of producing diamonds.

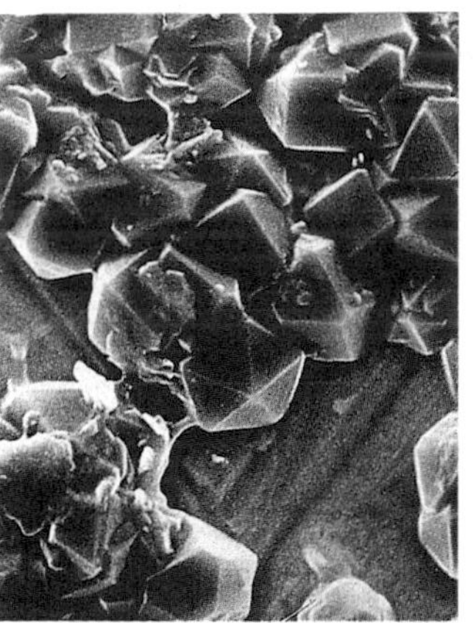

Researchers have found microscopic crystals of diamonds in the by-products formed from burning ethyne. Ethyne, commonly called acetylene, is an organic compound used to fuel oxyacetylene welding torches. To produce the diamonds, an oxyacetylene torch is used to heat the surface of a molybdenum disk. The disk is cooled from beneath to keep it from melting. After a few minutes of heating at normal atmospheric pressure, microscopic diamond crystals form on the surface of the disk. The micrograph shows similar crystals magnified 50 000 times.

How the crystals form is still not understood. Researchers guess that a series of reactions take place between free carbon atoms and hydrocarbons in the gases produced by the torch. The reactions produce ringlike molecules of carbon vapor. As the reactions continue, the rings grow until they become unstable. Then they break apart into more stable structures, namely, diamond crystals. The crystals that are deposited on the disk continue to grow as they capture carbon atoms from the hydrocarbons in the surrounding gases.

CONTENT BACKGROUND

Germanium: Germanium is a metalloid that expands on freezing. Tin and lead are both metals, and tin occurs in two allotropic forms. All three elements exhibit both the 2+ and 4+ oxidation states, with stability of the higher state decreasing from Ge to Pb where 2+ is the common form. The organic chemistry of tin and lead is exclusively the 4+ state, however.

Figure 11.10 The soot produced by burning wood in a fireplace must be removed periodically by a chimney sweep because it can become a fire hazard.

As you can see in Table 11.2, silicon is the second most plentiful element in Earth's crust after oxygen. Silicon is found in a large number of minerals such as quartz. Silicon shows only a slight tendency toward catenation, much less than carbon. The chemistry of silicon, like that of carbon, is characterized by electron sharing. In compounds called **silicates,** silicon is bound to oxygen, and each silicon atom is surrounded by four oxygen atoms.

Figure 11.11 Carbon compounds exhibit catenation, as shown by the structure of oleic acid (top). Silicon also exhibits this property, as shown by the structure of hexasilane (bottom).

$$CH_3-CH_2-CH_2-CH_2-CH_2-CH_2-CH_2-CH_2-CH=CH-CH_2-CH_2-CH_2-CH_2-CH_2-CH_2-CH_2-C(=O)-OH$$

Oleic Acid

$$H-SiH_2-SiH_2-SiH_2-SiH_2-SiH_2-SiH_2-H$$

Hexasilane

Silicon has many uses. Because it is a semiconductor, silicon is used in transistors, solar cells, and computer chips. Silicones are compounds of silicon. Some uses of silicones are in synthetic motor oils and in adhesives and automobile gaskets.

Tin and lead are distinctly metallic members of Group 14 (IVA). They are quite similar to each other except that for tin the 4+ state is more stable than the 2+, while the reverse is true for lead. These two metals are easily refined from their ores and have both been known since prehistoric times. In general, their uses are based on their lack of chemical reactivity. Tin and lead are common components of alloys, which are mixtures of metals. Examples of alloys are solder, which contains lead and tin, and bronze, which contains copper and tin.

Table 11.2

Abundances of Elements in Earth's Crust

Element	Percentage
Oxygen	45.5
Silicon	27.2
Aluminum	8.1
Iron	5.8
Calcium	4.66
Magnesium	2.76
Sodium	2.27
Potassium	1.84
Titanium	0.63
Hydrogen	0.152
Phosphorus	0.11
Manganese	0.1

2 TEACH

CONCEPT DEVELOPMENT

QUICK DEMO

Nonmetallic Properties Use a piece of roll sulfur to visually reinforce the properties of nonmetals. Show the class the dull surface. Cover it with a cloth and, using a hammer, break off a piece. Students will see that sulfur is brittle and powdery, not malleable. Use a conductivity checker to demonstrate that sulfur does not conduct an electric current. Place a very small piece in a spoon and heat it, demonstrating the low melting point that is characteristic of nonmetals.

MAKING CONNECTIONS

Music

There are two allotropic forms of tin. White tin is the metallic form. The gray tin allotrope forms when tin metal remains below 13°C. Because gray tin has no metallic properties, the tin organ pipes used in many of northern Europe's cathedrals disintegrated during the winter due to lack of heat.

✓ ASSESSMENT

Knowledge: Have students review the contents of Table 11.2. Then ask them to name the two most abundant metals and the two most abundant nonmetals in the Earth's crust. Students may have heard that iron is the most abundant element, but most of the iron is in the core of the planet.

PROGRAM RESOURCES

Study Guide: Use the Study Guide Worksheets, pp. 38-39, for independent practice with concepts in Section 11.2.

Transparency Master: Use the Transparency master and worksheet, pp. 49-50, for guided practice in the concept "Allotropes."

Color Transparency: Use Color Transparency 25 to focus on the concept of "Allotropes."

Chemistry Journal

Arsenic

Have students find out about this toxic metalloid and write brief reports. The reports should include the products in which arsenic is used, the ways in which a person might ingest arsenic, the biological effects of arsenic, and the treatment for arsenic poisoning.

interNET CONNECTION

A student hypertext on allotropes and buckminsterfullerene research is available for use.

World Wide Web
www.chem.wisc.edu/BDGtoc.html

CONCEPT DEVELOPMENT

Background: Arsenic and antimony are metalloids, while bismuth is a metal. All three elements exist in several allotropic forms, and antimony and bismuth both expand on freezing. The chemical trends in the group are those expected: increasing basicity and decreasing stability of the 5+ oxidation state with increasing atomic number. All three elements exhibit the 3+ and 5+ states and do not form 3- ions as such. There are arsenides, antimonides, and bismuthides, but these are intermetallic compounds, not ionic compounds.

HISTORY CONNECTION

Ancient Metallurgy in Peru

The Conquistadors reported that the streets of the New World were paved with gold. Speculate on their observation.

✓ ASSESSMENT

Knowledge: After reading this section, The Nitrogen and Phosphorus Group, ask students which element is present in most explosives. Point out that many nitrogen compounds are unstable and tend to decompose rapidly, forming large volumes of gas from small volumes of liquid or solid. Ask students how this property makes nitrogen compounds useful as explosives.

HISTORY CONNECTION

Ancient Metallurgy in Peru

Copper metallurgy was prevalent on the northern Peruvian coast as early as 500 B.C. Some burial grounds contain copper and gold-plated copper objects. In the early centuries A.D., the Mochicas made the first copper-arsenic alloys, although they were not produced in quantity until about 900 A.D. The Incas about 1450 A.D. attempted to use copper-tin bronzes as the state metal, but they allowed continued production of the copper-arsenic alloys. The invasion of Peru by Pizzaro in the middle 1500s led to the introduction of European metallurgy and wiped out the Peruvian metallurgical tradition.

The Nitrogen and Phosphorus Group

Nitrogen and phosphorus are found in Group 15 (VA). Both have five outer electrons, which can be shared to form compounds. However, nitrogen and phosphorus differ considerably for adjacent members of the same family. Nitrogen occurs in all oxidation states ranging from 3− through 5+; phosphorus shows only 3−, 0, 3+, and 5+. Most nitrogen compounds are relatively unstable, tending to decompose to N_2. Conventional high explosives, for example, trinitrotoluene (TNT) and dynamite, utilize nitrogen compounds. Elemental nitrogen, N_2 gas, is one of the most stable substances known. It makes up 78 percent of Earth's atmosphere. The unreactive gas is used to surround reactive materials that would otherwise react with oxygen in the air. Liquid nitrogen is used to maintain very low temperatures.

Nitrogen compounds are produced from atmospheric nitrogen by nitrogen-fixing bacteria. These bacteria naturally convert molecular nitrogen to nitrogen compounds that can be used readily by plants. The plants use the nitrogen compounds to make amino acids, the essential components of proteins.

Synthetic nitrogen compounds are usually produced from atmospheric nitrogen. Huge quantities of liquid N_2, obtained from the air, are used in the Haber Process to produce ammonia, as discussed in Chapter 22. The most common use of ammonia is as a fertilizer. Much of the ammonia not used directly as a fertilizer is converted to other nitrogen compounds that are themselves fertilizers. An example is ammonium nitrate, which is also used in explosives. Some ammonia is converted to nitric acid. Nitric acid is widely used in the manufacture of fertilizer and explosives. Large quantities of ammonium sulfate are obtained by the steel industry's conversion of coal to coke. The $(NH_4)_2SO_4$ by-product is then used as a fertilizer.

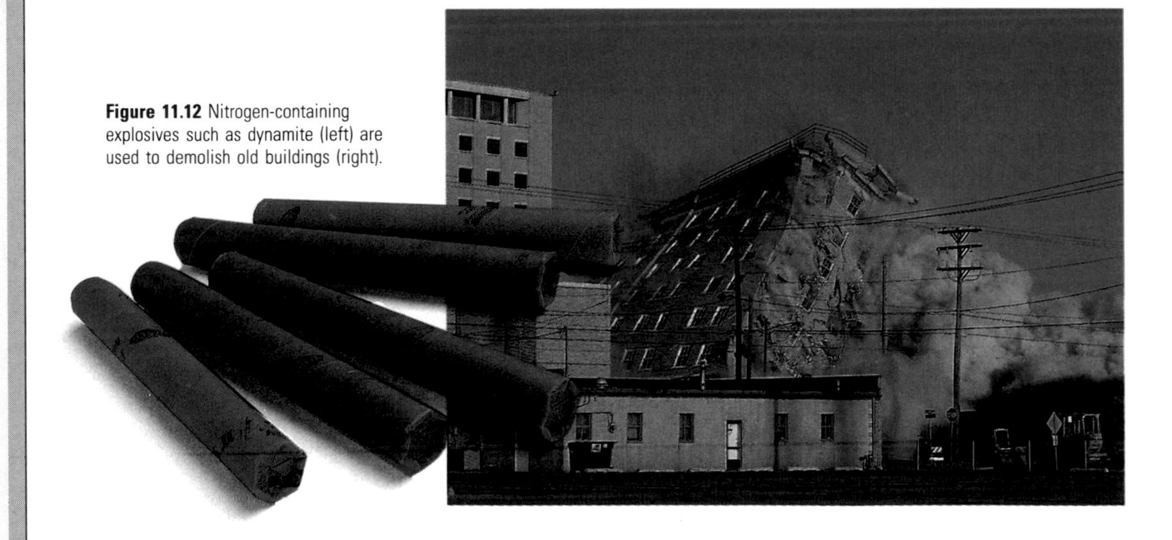

Figure 11.12 Nitrogen-containing explosives such as dynamite (left) are used to demolish old buildings (right).

CONTENT BACKGROUND

Natural Nitrogen Fertilizers: In 1857, the American chemist Evan Pugh proved that nitrogen fertilizers were important to plant growth. Guano, a substance composed mostly of bird excrement and used as fertilizer, was imported into the United States and Britain from the islands off Peru and Chile. Peru had a monopoly on this profitable business. The United States in 1856 declared 48 uninhabited islands to be part of America's domain. This action resulted in a skirmish between Peru's army and the U.S. Navy (Guano War). After 35 years, the natural supply was depleted because of poor conservation practices that resulted in the slaughter of the birds. Today, these islands supply fertilizer to Peruvian farmers because of enlightened conservation practices.

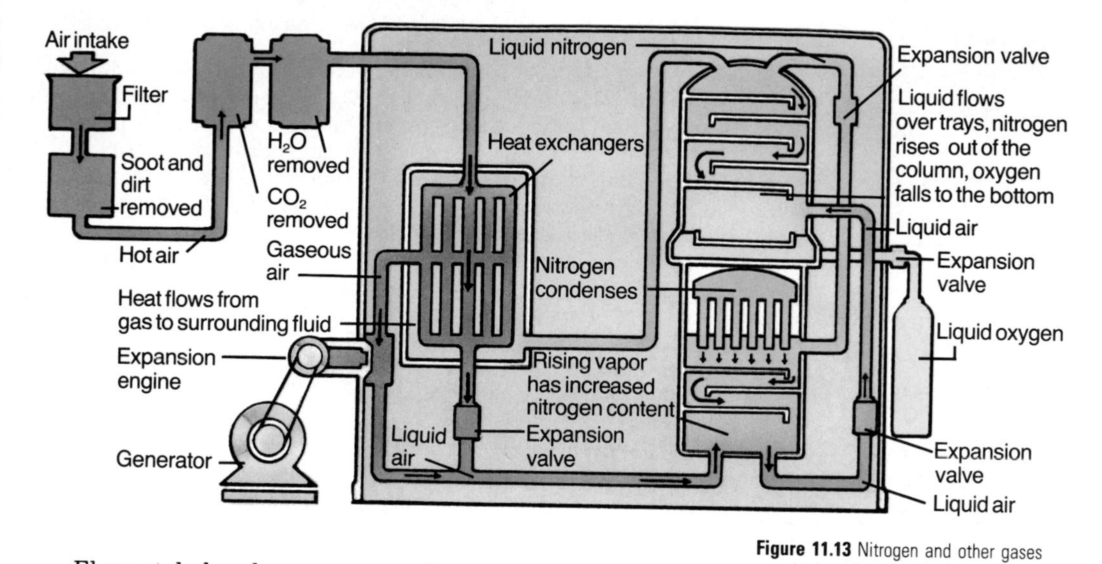

Figure 11.13 Nitrogen and other gases are obtained from liquid air. Air is liquefied through a complex process similar to that shown here.

Elemental phosphorus occurs as P_4 molecules and is solid at room temperature. The P_4 molecules "stack" in different ways to form several allotropes. One allotrope, white phosphorus, is so reactive that it ignites spontaneously on contacting air. Red phosphorus must be exposed to a flame to ignite. Black phosphorus, a third allotrope, is semiconducting. The principal source for phosphorus in nature is phosphate rock, $Ca_3(PO_4)_2$. Most phosphate rock is used in producing $Ca(H_2PO_4)_2$ and $CaHPO_4$ for use as fertilizer. Some phosphate rock is converted to phosphoric acid, which is used primarily in making fertilizer, but has other applications in industry.

Organic compounds containing both nitrogen and phosphorus are vital to living organisms. Utilization of energy by living systems involves a compound called adenosine triphosphate (ATP). The transfer of genetic information from generation to generation involves deoxyribonucleic acid (DNA). Each DNA molecule contains hundreds of nitrogen compounds attached to phosphate groups. Ribonucleic acid (RNA), used by cells in protein synthesis, also contains nitrogen compounds and phosphate groups.

Figure 11.14 Black and white phosphorus are allotropes. Note the difference in the geometric arrangement of the phosphorus atoms.

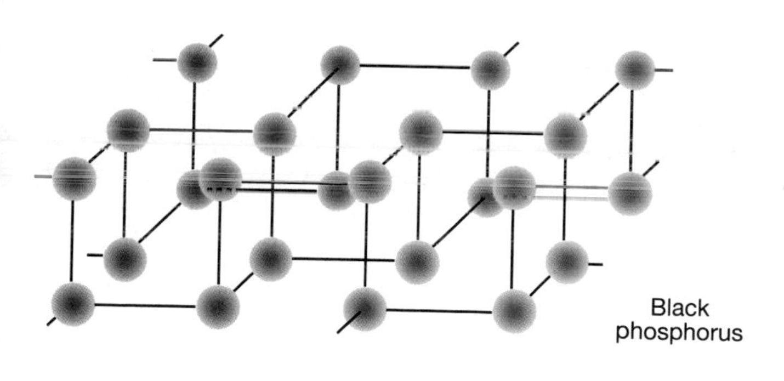

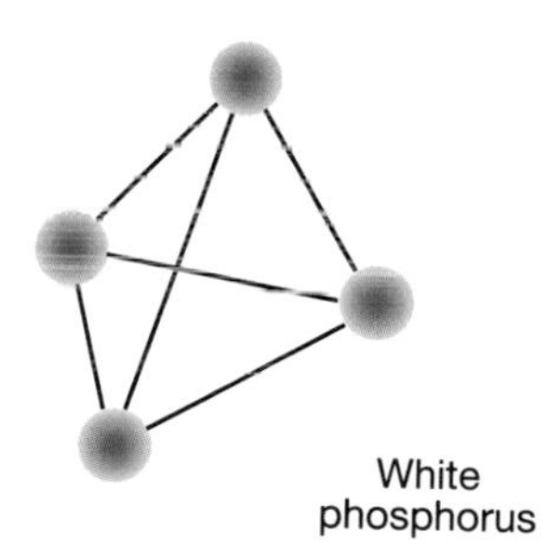

CULTURAL DIVERSITY

Peanut Shampoo

George Washington Carver was born to slave parents in 1861 in Missouri on the Carver Plantation. He was kidnapped as a baby and taken to Arkansas, but was later returned to the Carver Plantation. After the Civil War he stayed on the plantation and taught himself to read. At ten, he left to work odd jobs and attend school whenever possible. He finally finished high school and won a scholarship to Highland University. When he showed up to enroll, he was turned away because he was African-American. Carver did not let this stop him, however. He worked and saved and managed to attend Simpson College in Iowa and the Iowa Agricultural College. He stayed on to teach botany and chemistry there until he moved to Tuskegee in 1896 to carry out research at the personal request of Booker T. Washington.

Carver came up with more than 325 products to make from peanuts including coffee, face powder, ink, shampoo, and soap. He also found 118 uses for sweet potatoes including dyes, ink, and synthetic rubber.

✓ ASSESSMENT

Performance: Have students refer to Figure 11.14. Supply lab groups with small expanded foam spheres, pieces of packing "peanuts", or gumdrops and toothpicks. Have them construct models of the two allotropic forms of phosphorus shown in the figures.

CONTENT BACKGROUND

Nitrogen Explosives: Black gunpowder was the primary nitrogen-containing explosive until the beginning of the 20th century. In the United States, the du Ponts began to produce good quality gunpowder in 1802 along the Brandywine River in Delaware. They supplied much of America's need for explosives for 100 years. Highly nitrated cotton (guncotton) was produced in England by Schönbein and detonated by mercury fulminate. A smokeless powder was developed shortly after the Spanish-American War. Nitroglycerin was invented by an Italian, Ascanio Sobrero, in 1846. Nitrated aromatic compounds were introduced by the French military in 1885. The German military developed trinitrotoluene, TNT, which was first used in World War I.

PROGRAM RESOURCES

Reteaching: Use the Reteaching worksheet, pp. 16-17, to provide students another opportunity to understand the concept of "Understanding Allotropes."

CONCEPT DEVELOPMENT

Computer Demo: *Element Search,* Deductive reasoning is used to identify 10 elements. AP902, IIMC902, Project SERAPHIM

CAREERS IN CHEMISTRY
Forensic Chemist

Background: The course of study will include analytical chemistry, qualitative and quantitative analysis, instrumental analysis, and qualitative organic analysis. An internship with a police department is usually included as a requirement.

For More Information: Contact: The American Chemistry Society, 1155 Sixteenth Street, NW, Washington, DC 20036

GLENCOE TECHNOLOGY

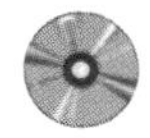

Videodisc

Chemistry: Concepts and Applications

Properties of Oxygen Gas
Disc 1, Side 1, Ch. 13

Show this video of Demonstration 11-4 to reinforce the concept of the properties of oxygen gas.

Also available on CD-ROM.

Word Origins

amphoteros: (GK) both
amphoteric—can be both acid and base

CAREERS IN CHEMISTRY

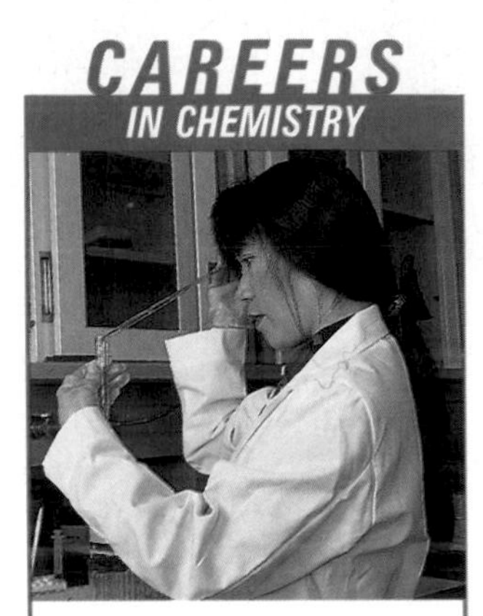

Forensic Chemist

As you know from numerous television shows, the laboratory plays a large part in crime detection and the analysis of evidence in preparation for a trial. Much of what takes place in a forensic laboratory is chemical in nature. Forensic chemists use modern instrumentation to analyze paint, blood, paper, hair, fabric, tissue, and other materials that could be used to solve a crime or as evidence in a trial.

Oxygen Family

Oxygen is the most plentiful element in Earth's crust and makes up about 21% of the atmosphere. Oxygen is a very reactive nonmetal, combining with all other elements except helium, neon, and argon. Since the oxygen atom has six electrons in its outer level, it can gain two electrons to achieve the octet configuration of neon. In so doing, it becomes the oxide ion, O^{2-}. With metals, which tend to lose electrons readily, oxygen forms ionic oxides. Oxygen can also react by sharing electrons with nonmetals. The behavior of oxides when dissolved in water depends on their structure. Ionic oxides generally react with water to produce basic solutions. However, oxides formed by the sharing of electrons tend to react with water to form acidic solutions. Some oxides can produce either acidic or basic solutions, depending on the other substances present. Such oxides are **amphoteric** (am foh TER ik) oxides.

Like carbon, oxygen has allotropes. Oxygen usually occurs in the form of diatomic oxygen molecules, O_2. The free oxygen you breathe from air is O_2. There is another allotrope of oxygen called ozone. Ozone is the triatomic form of oxygen, O_3, and is highly reactive. Ozone can be synthesized by subjecting O_2 to a silent electric discharge. In Europe, ozone is the main chemical used in water purification. Ozone is formed naturally in small amounts by lightning and, in the upper atmosphere, by ultraviolet radiation from the sun. It is also a major component of smog.

Pure oxygen is extracted from liquefied air, compressed, and sold in cylinders. Its largest uses are in the production of steel, artificial breathing atmospheres, rocket engines, and welding torches.

Figure 11.15 Ozone (O_3) is formed in the lower atmosphere by the reaction of lightning with oxygen. This ozone soon decomposes to ordinary oxygen (O_2).

DEMONSTRATION

11-4 Oxygen Gas

Purpose: to learn by direct observation that oxygen gas supports combustion but does not itself burn

Materials: 1 g MnO_2, 600-cm^3 beaker, 25 cm^3 30% H_2O_2, wood splints, matches, 10 g steel wool, crucible tongs, explosion shield, 150-cm^3 beaker, 25 cm^3 distilled water, laboratory burner

Procedure: **1.** Place 1 g of MnO_2 in a 600–cm^3 beaker.
2. In a separate beaker carefully dilute 30% hydrogen peroxide to 15% by adding 25 cm^3 of 30% H_2O_2 to 25 cm^3 of distilled water. **CAUTION:** *30% hydrogen peroxide is very hazardous, it causes severe burns. Wear rubber gloves, apron, and goggles.*
3. Slowly add the diluted H_2O_2 to the MnO_2 in the beaker. Oxygen gas will

The chemistry of sulfur is similar to that of oxygen, especially in the behavior of the 2− ion. The S^{2-} ion shows characteristics similar to the oxide ion in solubility and acid-base behavior. Unlike oxygen, which forms O_3 as its longest chain, sulfur can form long chains of atoms attached to the S^{2-} ion. These ions, for example S_6^{2-}, are called polysulfide ions. Sulfur exhibits two important positive oxidation states represented by the oxides SO_2 and SO_3. When sulfur or sulfur compounds are burned in an ample supply of air, SO_2 is produced. Using a catalyst, SO_2 can be converted to SO_3. When SO_2 is dissolved in water, sulfurous acid (H_2SO_3) is produced. If SO_3 is combined with water, sulfuric acid (H_2SO_4) is produced. When these processes occur in the atmosphere, acid rain is formed. In industry, the combination of SO_3 and water is achieved by dissolving the SO_3 in H_2SO_4 and then adding water. Sulfuric acid is produced in huge quantities. Twice as much H_2SO_4 (by mass) is produced as the next most common chemical. The principal consumers of sulfuric acid in the United States are fertilizer, petroleum refining, steel, synthetic fiber production, paint, and pigment industries.

The other members of the oxygen group are selenium, tellurium, and polonium. Selenium is a nonmetal. Tellurium and polonium are metalloids. Of these three elements, selenium is the most common.

Selenium is toxic, even in low doses. The Occupational Safety and Health Administration (OSHA) sets the permissible exposure limit at 200 μg Se/m^3. Yet, selenium is an essential part of our diet, in extremely small amounts. Red blood cells contain a vital compound that has four atoms of selenium per molecule. You will find, as you study chemistry and biology, that there are thousands of compounds with similar restrictions: a small amount of the substance is vital, but too much is toxic. The sixteenth century Swiss physician Paracelsus stated, "The dose makes the difference." When you read that a certain substance is harmful, you must know the amounts discussed.

Figure 11.16 When sulfur burns in sufficient air, it forms sulfur dioxide gas. It is this gas that is responsible for the formation of acid rain.

VITAMINS	QUANTITY	% OF U.S. RDA	MINERALS	QUANTITY	% OF U.S. RDA
VITAMIN A (AS BETA CAROTENE)	5000 I.U.	100	IRON (elemental)	18 mg	100
VITAMIN A (AS ACETATE)	1500 I.U.	30	CALCIUM	100 mg	12
VITAMIN C	60 mg	100	PHOSPHORUS	100 mg	10
THIAMINE (B-1)	1.5 mg	100	IODINE	150 mcg	100
RIBOFLAVIN (B-2)	1.7 mg	100	MAGNESIUM	100 mg	25
NIACIN	20 mg	100	COPPER	2 mg	100
VITAMIN D	400 I.U.	100	ZINC	15 mg	100
VITAMIN E	30 I.U.	100	CHROMIUM	10 mcg	*
VITAMIN B-6	2 mg	100	SELENIUM	10 mcg	*
FOLIC ACID	0.4 mg	100	MOLYBDENUM	10 mcg	*
VITAMIN B-12	6 mcg	100	MANGANESE	2.5 mg	*
BIOTIN	30 mcg	10	POTASSIUM	37.5 mg	*
PANTOTHENIC ACID	10 mg	100	CHLORIDE	34 mg	*

*No U.S. RDA established.

CONTAINS IRON, WHICH CAN BE HARMFUL IN LARGE DOSES. CLOSE TIGHTLY AND KEEP OUT OF REACH OF CHILDREN. IN CASE OF OVERDOSE, CONTACT A PHYSICIAN OR POISON CONTROL CENTER IMMEDIATELY.

MILES

Figure 11.17 Many elements are essential to proper body function, in very small quantities. Vitamin tablets with mineral supplements can be used as a dietary supplement for these elements.

CONCEPT DEVELOPMENT

Background: Sulfur is a very important industrial element. It has numerous allotropes with the rhombohedral crystals of S_8 molecules being the stable form at room temperature. Sulfur is quite plentiful in the solar system as evidenced by its occurrence in the atmosphere of Venus (along with considerable H_2SO_4) and on the Jovian satellite Io. Sulfur is very reactive, especially at elevated temperatures and forms an enormous number of compounds. It can form polysulfide chains and rings containing as many as 19 sulfur atoms. Many molecules and ions containing sulfur atoms are excellent ligands in the formation of complex ions, for instance, thiosulfate and thiocyanate.

The other elements of Group 16(VIA), selenium, tellurium, and polonium, are not very plentiful. All three elements, like oxygen and sulfur, are quite reactive. They are also highly toxic. Selenium is widely used in xerography.

PROGRAM RESOURCES

Applying Scientific Methods in Chemistry: Use the worksheet, "The Discovery of Oxygen," p. 27, to help students apply and extend their lab skills.

Enrichment: Use the worksheet "Sulfur," pp. 21-22, to challenge more able students.

be produced, filling the beaker by the upward displacement of air.

4. Lower a glowing splint into the beaker. Blow out the splint, and while the splint is still glowing, repeat the demonstration several times in rapid succession.

5. A small piece of steel wool can be heated in a burner flame until it glows. Using tongs, lower the glowing steel wool into the beaker of oxygen gas. **CAUTION:** *sparks will fly; use an explosion shield.* Have students observe the increased reaction. **Disposal:** E.

Results: The wood splint will burst into flame but the O_2 gas will not burn. The increased concentration of oxygen greatly increased the rate of the reaction of the steel wool to produce iron oxides.

Questions: 1. Are reactions with oxygen usually exothermic? *yes*

2. How many electrons does each oxygen atom gain to form the oxide ion? *2*

3. What happens to the rate of a combustion reaction as the concentration of oxygen is increased? *The rate increases.*

✓ ASSESSMENT

Oral: Ask students how breathing pure oxygen before a race would affect a track runner. Point out the deleterious results of such an action.

CONCEPT DEVELOPMENT

MAKING CONNECTIONS

Writing

Have several interested students write to different toothpaste companies asking for scientific test results that show the effectiveness of their fluoride treatment for teeth. Have the letter writers share the companies' responses with the class.

• Chemistry students should observe the gas and vapor forms of chlorine, bromine, and iodine and should be able to recognize their characteristic colors. Because these gases are toxic, they should be shown in sealed containers. See Demonstration 11-5.

Reinforce: Review and reinforce the s^2p^5 electronic configurations of the halogens. Discuss how the similar outer electron arrangements of the halogens explains their similar chemical behavior.

MAKING CONNECTIONS

Safety

Point out the tragic consequences that result when a toilet bowl cleaner is mixed with a chlorine bleach. The toxic fumes produced can kill quickly someone in a small, poorly-ventilated bathroom.

✓ ASSESSMENT

Knowledge: Have students refer to Figure 11.18. Ask students to name very reactive metals. Then ask them to name very reactive nonmetals. Finally, ask them to describe in their own words the pattern of reactivity of elements as it relates to the periodic table.

Mastering Concepts in Chemistry software: Use the lesson on Periodic Properties of Elements to introduce and reinforce the concept that patterns of reactivity correspond to the patterns of elements on the periodic table and, consequently, to the electronic properties of those elements.

The Halogens

Group 17 (VIIA) contains fluorine, chlorine, bromine, iodine, and astatine. The elements of this group are called the halogen (salt-forming) family. In many chemical reactions, halogen atoms gain one electron. They become negatively charged ions with a stable outer level of eight electrons. As in other families already discussed, three factors determine the reactivity of the halogens. They are (1) the distance between the nucleus and the outer electrons, (2) the shielding effect of inner level electrons, and (3) the size of the positive charge on the nucleus. Fluorine atoms contain fewer inner-level electrons than the other halogens, so the shielding effect is the least for this element. The distance between the fluorine nucleus and its outer electrons is less than in the other halogens. Thus, the fluorine atom has the greatest tendency to attract other electrons. This attraction makes fluorine the most reactive nonmetal.

The astatine nucleus has the largest number of protons and the largest positive charge of the halogen elements. However, the increased charge on the nucleus is not enough to offset the distance and shielding effects. Thus, of all halogen atoms, the astatine nucleus has the least attraction for outer electrons.

RULE OF THUMB

In general, *on the right side of the periodic table, the nonmetallic elements become more active as we move from the bottom to the top. On the left side of the table, the metals become more active as we move from the top to the bottom.* The most active elements are located toward the upper right-hand and at the lower left-hand corners of the periodic table. Notice that fluorine is active because the atoms of fluorine have a great tendency to gain one electron and become negative ions. At the other extreme, the alkali metals are active because they hold the single outer electron loosely and it is easily removed. The groups between 1 (IA), and 17 (VIIA) vary between these two extremes.

The halogens usually react by forming negative ions or by sharing electrons. Fluorine is the most reactive of all the chemical elements. It reacts with all other elements except helium, neon, and argon. Fluorine is obtained from the mineral fluorspar, CaF_2.

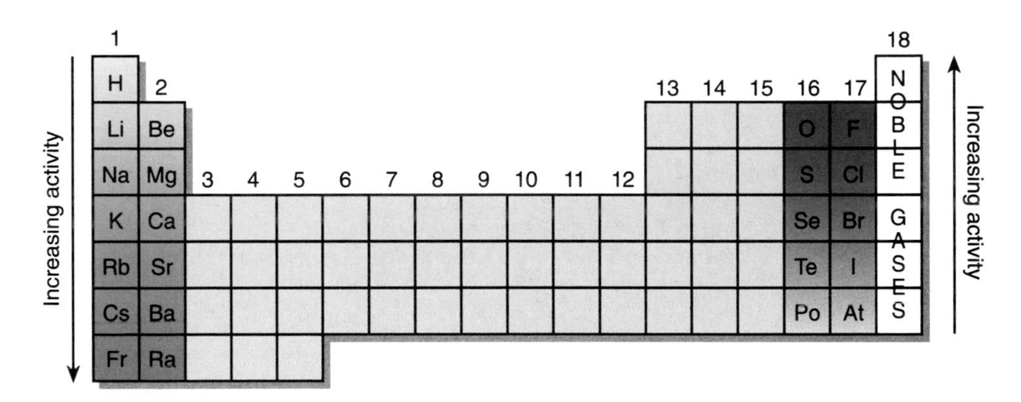

Figure 11.18 The activity of the elements on the two sides of the periodic table is indicated by the darkness of the tint in this illustration.

DEMONSTRATION

11-5 Chlorine, Bromine, and Iodine

Purpose: to observe the gas or vapor forms of chlorine, bromine, and iodine

Materials: 20 cm^3 6*M* HCl, 20 cm^3 chlorine bleach, three 500-cm^3 glass bottles with stoppers, few drops of liquid bromine, few crystals of iodine

Procedure: 1. Chlorine gas can be prepared by carefully adding a small amount of 6*M* HCl (10 cm^3 12*M* HCl to 10 cm^3 water) to 20 cm^3 of chlorine bleach. **CAUTION:** *Prepare the toxic chlorine gas in an operating fume hood.* The gas will collect above the liquid.

2. Securely stopper the 500-cm^3 glass bottle after the bubbling ceases so students can observe the pale yellow-green chlorine gas.

3. Add a few drops of liquid bromine to an empty glass bottle. Stopper it securely. **CAUTION:** *Use liquid bromine under a fume hood; wear rubber*

Like hydrogen, halogen atoms can form bridges between two other atoms. An example of such a compound is $BeCl_2$, shown in Figure 11.19. The halogens can also form compounds among themselves, as in ClF, ClF_5, BrF_5, IF_5, IF_7, BrCl, and ICl_3.

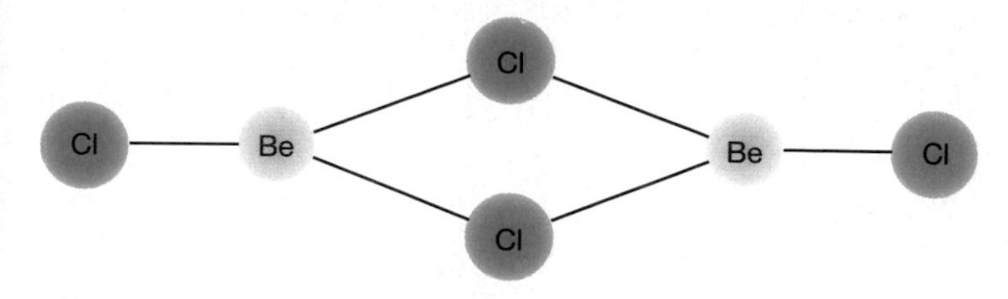

Figure 11.19 Beryllium chloride, $BeCl_2$, forms this molecule in which chlorine atoms form bridging bonds between two beryllium atoms.

Chlorine, though less abundant than fluorine in Earth's crust, is more commonly used in both the laboratory and industry. Chlorides of most elements are available commercially. These chlorides are quite often used in the laboratory as a source of positive metal ions bound with the chloride ions. The most familiar chloride is NaCl, common table salt.

Chlorine is produced primarily to produce other chemicals. Chlorine dioxide is used in water purification and as a bleaching agent in paper manufacturing. HCl is a by-product of many industrial processes. Its uses include steel manufacture, dye production, food processing, and oil well drilling.

The commercial preparation of chlorine led to one of the major water pollution crises of the early 1970s. Chlorine was produced by running an electric current through a solution of sodium chloride. Mercury, a toxic metal, was used to conduct current in the apparatus. The leakage of mercury into nearby water sources caused a public outcry. Industries had to remove the mercury from their waste before dumping it or switch to another method of producing chlorine. Chemists solved the problem by designing new types of electrochemical cells to produce chlorine.

Noble Gases

The noble gases, Group 18 (VIIIA), have very stable outer electron configurations. These elements are much less reactive than most other elements. For many years after their discovery, the noble gases were believed to be chemically unreactive and were commonly called inert. However, in 1962, the first "inert" gas compound was synthesized. Because these gases are not completely inert, they are usually referred to as the noble gases.

The first compound made with a noble gas was xenon hexafluoroplatinate ($XePtF_6$). Other compounds of xenon were soon produced. Once the techniques were known, the compounds could be made with increasing ease. Xenon difluoride (XeF_2) was first made by combining xenon and oxygen difluoride (OF_2) in a nickel tube at 300°C under pressure. The same compound can now be made directly from xenon and fluorine. In this process, fluorine

CONCEPT DEVELOPMENT

MAKING CONNECTIONS

Art

Ask students to bring in pictures of "neon" signs from magazines. The different colors observed are red-orange from Ne, yellow from He, blue from Xe, purple from Ar, and white from Kr. Have students arrange the pictures on the bulletin board.

gloves, goggles, and an apron. Have students note the red-brown color of Br_2.

4. Iodine vapors can be prepared by placing a few crystals of solid iodine in an empty glass bottle and stoppering it securely. The bottle can be placed in warm water to hasten the sublimation process. Point out the faint violet color of iodine vapors. **Disposal:** H then C

Results: Students will observe the pale yellow-green chlorine gas, the red-brown color of Br_2, the faint violet color of iodine vapors.

Questions: 1. What are the diatomic formulas of these three elements? *Cl_2, Br_2, and I_2*

2. How are the colors of the vapors of each substance described? *Cl_2, pale yellow-green; Br_2, red-brown; and I_2,* faint violet

✓ ASSESSMENT

Knowledge: Ask students if the odors of any of the gases remind them of a common household chemical. Chlorine bleach is often mentioned.

3 ASSESS

CHECK FOR UNDERSTANDING

Ask students to write the answers to Concept Review problems, and Review and Practice Problems 19-37 from the end of the chapter.

RETEACHING

Put the names of the Group 14-18 elements on individual cards in a box and have each student pick one and make a flash card of that element by writing its name and symbol on one side, and a property and use on the opposite side. Have student teams compete to learn the elements.

Cooperative Learning: Refer to pp. 28T-29T to select a teaching strategy to use with this activity.

EXTENSION

Demonstrate a neon gas discharge tube. Point out that in 1962 Bartlett and Chernick altered a well-established theory that the noble gases were inert. They do react and form compounds, some of which are unstable and detonate in ultraviolet light.

Answers to Concept Review

6. Catenation allows carbon to form a very large number of compounds.

7. Because the fluorine atom is small and has fewer inner-level electrons, it has a great tendency to attract electrons and is extremely reactive. Chlorine is a larger atom, has a greater shielding effect, and is less reactive than fluorine.

8. Nonmetals at the top of a group have a smaller shielding effect and a greater electron affinity than those at the bottom of the same group.

9. The size of the positive charge on the nucleus, the shielding effect of inner-level electrons, the distance of the outer electrons from the nucleus, and the configuration of the outer-level electrons.

10. Nitrogen compounds produced naturally from atmospheric nitrogen are used by plants to make amino acids, such as valine, which are essential components of proteins.

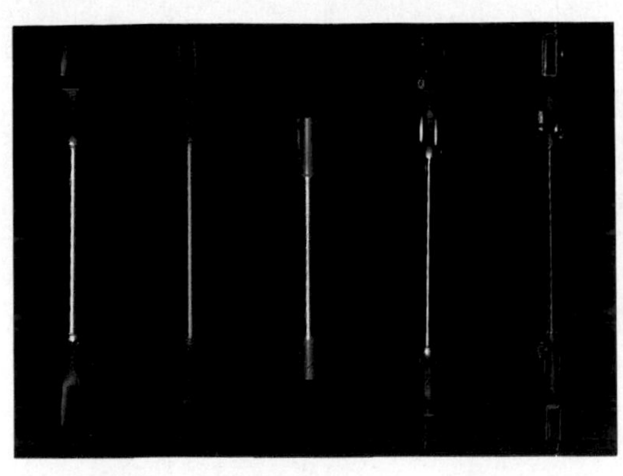

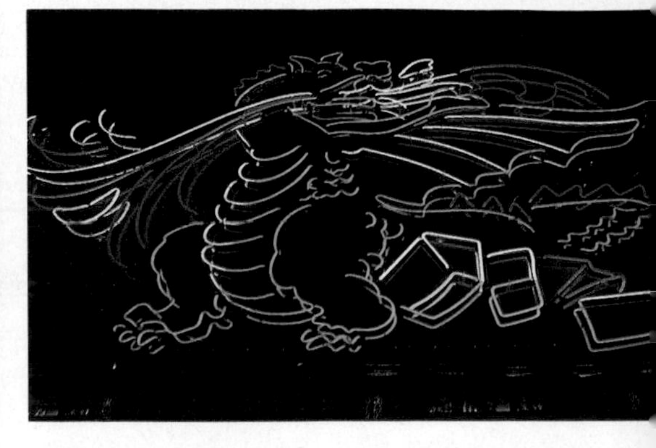

Figure 11.20 These gas discharge tubes (left) contain (from left to right) helium, neon, argon, krypton, and xenon. Because of the colored light they emit when a high voltage passes through them, the noble gases are used in "neon" lights for advertising and art (right).

and xenon are exposed to daylight in a glass container from which all the air has been removed. Compounds of xenon with oxygen and nitrogen have been synthesized. Krypton and radon compounds have also been produced.

Helium is the only element first discovered someplace other than on Earth. In 1868, the spectrum of an unknown element was discovered in sunlight. The element was then named helium after Helios, the Greek god of the sun. Helium is found in natural gas wells in the United States up to 7% by volume. This helium undoubtedly accumulated underground through the radioactive decay of certain elements emitting alpha particles. The boiling point of helium, −268.9°C, is the lowest of any known substance. Helium cannot be solidified at ordinary pressures, no matter how low the temperature. Liquid helium, because of its low boiling point, is used extensively in low temperature research. Helium is also used to fill balloons and in making artificial atmospheres such as those that deep-sea divers breathe.

Because they are inert, argon and helium are used to protect active metals during welding. Aluminum, for example, must be welded in an inactive atmosphere of argon or helium. If aluminum is welded in the air, the hot metal will catch on fire and burn in the air's oxygen. Argon is also used to fill light bulbs to protect the filament.

11.2 CONCEPT REVIEW

6. How does catenation affect the number of compounds carbon is able to form?
7. Fluorine differs from chlorine more than any other two adjacent halogens differ from each other. What explanation can you suggest for this phenomenon?
8. Why are nonmetallic elements at the top of a group more reactive than those at the bottom of the same group?
9. What three factors determine the reactivity of an element?
10. **Apply** Atmospheric nitrogen usually cannot be used by plants. Why, then, is it so important to plants?

CONTENT BACKGROUND

Noble Gas Compounds: Since 1962, research has been done on compounds of the noble gases. Ask a student to find an article about one of these projects and report on it to the class. Possible source of answer: *Science,* 138 (October 12, 1962) for an example of initial reports. XeF_4 is the compound most commonly reported.

CHEMISTRY AND
TECHNOLOGY

Recycling

Recycling has become an important environmental issue. There are two reasons for recycling. One is to reduce the amount of waste material that must be buried in landfills or incinerated. Landfills are unsightly, occupy space needed for more productive use, and may leak contaminants into water supplies. Incineration may introduce toxic gases and other pollutants into the atmosphere. The second reason for recycling is to slow down our consumption of resources that are nonrenewable or renewable only with difficulty. These resources include petroleum, natural gas, coal, ores, and radioactive minerals.

The principal materials being recycled at the present time are paper, glass, and aluminum. Paper is made primarily from trees and is the major component of trash. Trees, of course, are renewable, but replacing forests takes a long time. By recycling most of our paper (80-85%) we could greatly reduce the volume of trash entering landfills and save trees for other uses. Glass has been recycled for some time, because glass manufacturers have always introduced some broken glass into the glass-making process. Broken glass may make up from 10 to 80% of the glass furnace charge.

Aluminum is an ideal metal for making beverage cans. It is lightweight, and conducts heat rapidly so the food inside can be cooled quickly. Aluminum is produced from an ore called bauxite. Bauxite contains about 55% aluminum oxide, Al_2O_3. The bauxite is purified by dissolving the aluminum oxide in sodium hydroxide solution. The impurities are filtered off and the aluminum oxide precipitated by adding water. The aluminum oxide is then dissolved in cryolite, or sodium hexafluorosilicate, Na_3SiF_6. The solution is subjected to the passage of an electric current, which produces the aluminum metal. The production of aluminum from ore is a long, energy-consuming process.

One way to make the supply of aluminum ore last longer is to reuse the metal in objects that are no longer needed. Once a beverage can has been used, it cannot be refilled. However, it can be melted down and made into a new can. In addition to conserving the ore, recycling conserves energy. It takes only about one-third as much energy to make new cans from old as it does to make new cans from aluminum ore. Currently, 60% of aluminum cans are recycled in the United States.

Plastic products make up 9% of our trash by mass, but a whopping 30% by volume. With plastics, a new problem is encountered. There are many different kinds of plastics. Some of these are incompatible with others in a recycling procedure. Consequently, plastics must be separated by type before they can be reprocessed. Such a separation is labor intensive and, thus, expensive. Research into recycling plastics is currently being carried out in several locations in the United States.

Thinking Critically

1. Account for the difference in proportion of plastics in trash by mass (9%) and by volume (30%).
2. Find out why glass manufacturers introduce broken glass into the glass-making process. What properties of the finished product are determined by the amount of broken glass added?

4 CLOSE

Discussion: Most students drink chlorinated water. Ask why the water is treated with chlorine. Have a student report on the water treatment available in your community. You may want to send a video camera team to the local water treatment plant to video a tour that can be shared with the rest of the students.

CHEMISTRY AND TECHNOLOGY

Recycling

Background: Recycling did not become a major environmental issue until one state began shipping its trash to another state for burial. The people of the receiving state saw their landfills closing earlier than had been predicted.

Teaching This Feature: To make one aluminum can requires the same energy as is needed to make three cans from recycled metal. Show three candles burning, and then switch to just one.

Connection to Economics: See how much money the chemistry students can raise in one week by recycling aluminum cans. Establish a collection site, then transport them to the recycling center for sale. Have the students decide where to donate the money.

Answers to
Thinking Critically

1. Plastics are less dense than paper or metals.
2. Broken glass improves the finish of the glass product.

✓ ASSESSMENT

Portfolio: Ask students to write a report on personal recycling that details information they collected when surveying their own residence and that of two neighbors. Encourage students to prepare a list of questions about recycling before contacting the neighbors. They may choose to place this report, which should include information presented in tabular form, in their portfolio.

PROGRAM RESOURCES

Chemistry and Industry: Use the worksheets, pp. 15-18, "Glass: Is it really only melted sand?" as an extension to the concepts presented in Section 11.2.

11.3 Transition Metals

PREPARE

Key Concepts
The major concepts presented in this section include a presentation of typical transition metal chemistry. Chromium and zinc are discussed in detail. The inner transition elements are introduced and their chemistry is briefly described by studying neodymium, a lanthanoid, and curium, an actinoid.

Key Terms
galvanizing
inner transition element

1 FOCUS

Objectives
Preview with your students the objectives listed on the student page. Students can use them as a study guide for this major section.

Activity
Assemble a display of as many transition metals and their products as possible. Items may include coins, Cu, Ni, Ag, Zn; a spark plug, Mo, Pt; white paint, TiO_2; batteries, Ni, Cd, MnO_2; audio and videotapes, Fe and Cr oxides. Point out to students that the transition metal chemistry is very important to the quality of our daily lives. Attempt to establish a "need to know more attitude" in students regarding this area of chemistry.

✓ ASSESSMENT
Oral: Have students refer to Figure 11.21. Ask them which countries, if any, are entirely self-sufficient when considering natural resources.

OBJECTIVES
- define transition metals and list some of their uses.
- list representatives and some properties of lanthanoids and actinoids.

11.3 Transition Metals

Studying the transition metals is important for two reasons. First, these metals are the structural elements underlying much of our modern technology. For example, steel is an alloy of iron, a transition metal, and carbon. There are many specialty steels that include, in addition to iron, other transition metals such as cobalt, nickel, chromium, vanadium, tungsten and manganese.

The second reason for studying the transition metals is their fascinating chemistry. The involvement of the d sublevel electrons in their reactions leads to the formation of hundreds of thousands of compounds, many of which are important to biological systems.

Typical Transition Metals

The transition metals are those elements that have the highest energy electrons in d sublevels. In all groups except 12 (IIB), the d orbitals are only partially filled. These partially filled d orbitals make the chemical properties of the transition metals different from those of the main group metals. Remember that d electrons may be lost, one at a time, after the outer s electrons have been lost.

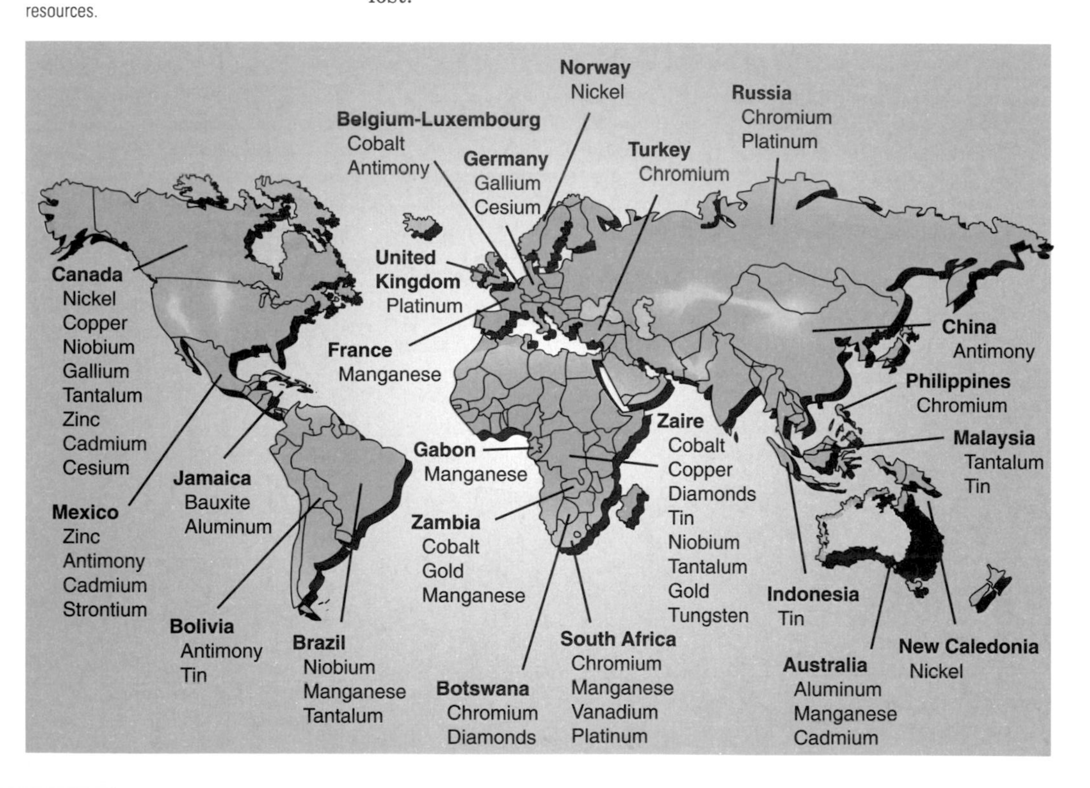

Figure 11.21 The map shows major sources of natural resources throughout the world. Note the number of transition metals that are included as important resources.

CONTENT BACKGROUND

Superconducting Ceramics: Many changes are occurring in the way high-temperature superconducting ceramics are being mixed. The transition metals are used to form these superconductors. $YBa_2Cu_3O_7$ is the formula of just one of these that superconducts at about 100 K. These elements are relatively common and inexpensive. In 1988, a thin film of a material containing a thallium-calcium-barium-copper oxide conducted an electric current without observable resistance at 97 K. This research is aimed at using these materials to make superconducting computer chips.

Figure 11.22 Chromium was used to protect steel parts of cars from corrosion (left). The colors of emeralds and rubies (right) are due to the presence of chromium impurities in the crystalline structure.

Transition metals have a wide variety of uses. Gold, silver, copper, and nickel are coinage metals. Catalytic converters contain platinum, palladium, and rhodium. Osmium is used to harden pen points. The radioactive isotope of cobalt, $^{60}_{27}Co$, is used in the treatment of cancer. Molybdenum is used in spark plugs. An important transition metal compound produced commercially is titanium(IV) oxide, TiO_2. The TiO_2 is used extensively as a white paint pigment. Other important oxides of transition metals include manganese dioxide, MnO_2, used in batteries, and iron and chromium oxides, used in audiotapes and videotapes.

Chromium

Chromium exhibits the typical properties of a transition metal. One property of chromium important to industry is its resistance to corrosion. Large quantities are used to make stainless steel and to "chromeplate" regular steel. In both cases, the chromium protects the iron in steel from corrosion.

Chromium usually reacts by losing three electrons to form the Cr^{3+} ion. Also of importance is the 6+ oxidation state. Recall that the outer electron configuration of chromium is $4s^13d^5$. This 6+ oxidation state is formed when chromium loses all five of its 3*d* electrons, as well as its outer level 4*s* electron. Two polyatomic ions are important examples of the 6+ state. These ions are the chromate ion, CrO_4^{2-}, and the dichromate ion, $Cr_2O_7^{2-}$.

Compounds containing chromium in the 6+ oxidation state are carcinogens; that is, they can cause cancer. Chromium also forms a Cr^{2+} ion. However, this ion is easily converted by oxygen in air to Cr^{3+}. Consequently, the 2+ state is not of great importance. The 4+ and 5+ states are very unstable, and so they have no practical uses. Many chromium compounds are highly colored. Small amounts of Cr^{3+} are found as impurities in the crystals of rubies and emeralds. These chromium ions cause the red color of rubies and the green color of emeralds.

ENVIRONMENT CONNECTION

Weeds Whack Wastes

Cadmium is a metal that contributes to toxic industrial waste along with copper and zinc. Now it has been discovered that the poisonous jimsonweed has the ability to absorb these metals from contaminated wastes. The weed manufactures a proteinlike material that absorbs and binds the metals within the plant cells in a nontoxic form. If enough of this proteinlike material can be produced synthetically, it could be used to remove metals from toxic wastes before they are thrown in a landfill.

2 TEACH

CONCEPT DEVELOPMENT

MAKING CONNECTIONS

International Studies

Transition metal chemistry is very important to high-technology industries. An uninterrupted supply of these metals is essential for the military-industrial complex to remain productive. No country has a supply of all of them. These metals must be imported and the prices can fluctuate greatly. Discuss how this interdependence makes the world a smaller place. Discuss how political issues and international relations can affect the supply of natural resources.

ENVIRONMENT CONNECTION

Weeds Whack Wastes

Find out what other methods have been used to extract metals from toxic wastes.

Computer Demo: *The Periodic Table as a Database,* Graphically display chemical and physical properties. AP207, Project SERAPHIM

Chemistry Journal

Transition Metals and Color

Have students research and write a report on how transition metals produce color. When white light shines on transition metals, *d* electrons of the metals absorb certain frequencies (colors) of the light and are excited to a higher energy state. The unabsorbed light is no longer white, but rather a color combination made up of the unabsorbed frequencies. These are transmitted or reflected and are what is perceived by the eye.

PROGRAM RESOURCES

Study Guide: Use the Study Guide Worksheets, p. 40, for independent practice with concepts in Section 11.3.

Critical Thinking/Problem Solving: Use the worksheet, p. 11, as enrichment for the concept of "Electron Configuration of Transition Metals and Their Ions."

3 ASSESS

CHECK FOR UNDERSTANDING

Ask students to write the answers to Concept Review problems, and Review and Practice Problems 38-44 from the end of the chapter.

RETEACHING

Have students bring in pictures of items made with the elements that have been studied in this section and make a bulletin board display. Use this display as a focus to reteach the properties and uses of these elements.

EXTENSION

Have a student investigate the lanthanide contraction (now called lanthanoid contraction) and see what effect it has on the behavior of the elements exhibiting it. Possible response: The addition of 14 protons to the nucleus without starting a new energy level causes a considerable contraction of the elements following the lanthanoids in the periodic table. Consequently, atoms of elements in the sixth row are smaller than we would expect. As a result, their chemical behavior is altered somewhat. As an example, because gold is less reactive than silver, gold will not tarnish in air, but silver will.

GLENCOE TECHNOLOGY

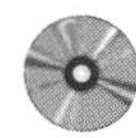

Videodisc

Chemistry: Concepts and Applications

Properties of Transition Metals
Disc 1, Side 2, Ch. 9

Also available on CD-ROM.

CD-ROM

Chemistry: Concepts and Applications

Exploration: *Organizing Elements in a Periodic Table*
Interactive Experiment: *Classify the Element*

Figure 11.23 Brass is an alloy of zinc and copper. One useful feature of brass is that it can be worked into thin-walled tubes. This property makes brass useful for producing musical instruments that require long lengths of tubing.

Zinc

Although zinc is classed as a transition element, its behavior differs slightly because of its full *d* sublevel. It exhibits only one oxidation state, 2+. The special stability of the full *d* sublevel leaves only the two outer 4*s* electrons available for reacting. Zinc is the second most important transition element (after iron) in biological systems. As an example, lack of zinc in the diet prevents the pancreas from producing some digestive enzymes. Over 25 zinc-containing proteins have been discovered.

Metallic zinc, like chromium, is corrosion resistant. It is used extensively as a coating to protect iron. The coating can be applied in three ways. When the iron is dipped in molten zinc, the process is called **galvanizing.** The coating is also applied electrically. The third method is to allow gaseous zinc to condense on the surface of iron. Another major use for metallic zinc is in the production of alloys. Especially important is its combination with copper to form brass.

Inner Transition Elements

The lanthanoids and actinoids are often called the **inner transition elements** because their highest energy electrons (*f* electrons) are inside the *d* sublevel and the outer level.

Neodymium, a Lanthanoid

The electron configuration of neodymium (nee oh DIHM ee um) ends $6s^2 4f^4$. From that configuration we would predict a 2+ oxidation state. However, Nd, as all the lanthanoids, shows 3+ as its most stable state. The 3+ ion is pale violet in solution. The metal itself is soft and quite reactive. It tarnishes when exposed to air. The metal can be mixed with other elements to form alloys with unusual conductivity and magnetic properties. The compound Nd_2O_3 is used in glass filters and in some lasers. Neodymium is

CONTENT BACKGROUND

The Not-So-Rare Earth Elements: Lanthanoids and actinoids were once called rare-earth elements. Canada, Australia, China, and the United States have reserves of these elements. More than 25 000 tons are sold annually. Metal ingots are available commercially for erbium, dysprosium, terbium, gadolinium, samarium, praseodymium, neodymium, cerium, and lanthanum. They are no longer rare. The prices range from ten cents to thirty dollars a gram for their oxides.

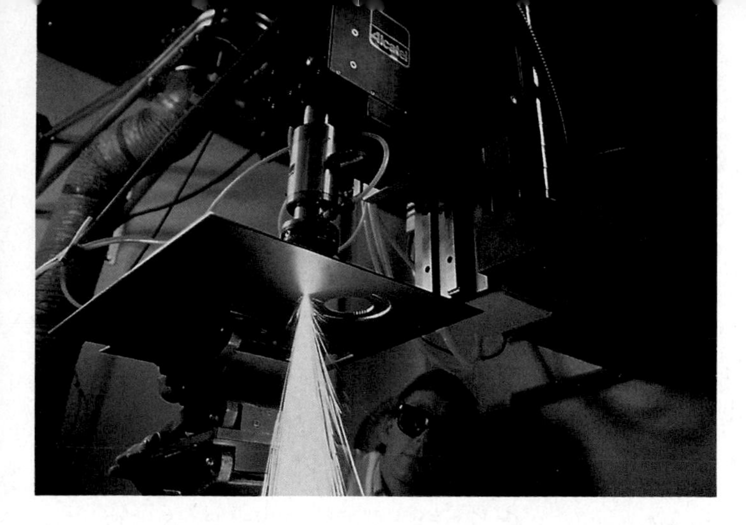

Figure 11.24 Neodymium is used in high-power pulsed lasers. These lasers are used to drill and weld metals and in experiments that attempt to produce nuclear fusion.

the second most abundant lanthanoid metal. Its compounds are separated from the other lanthanoids by a chromatographic technique, as explained further in Chapter 14. The metal is obtained from its fluoride by a single displacement reaction.

$$3Ca + 2NdF_3 \rightarrow 3CaF_2 + 2Nd$$

The lanthanoids have several interesting uses. Their compounds are used as the phosphors that give television picture tubes their colors. Lanthanoid compounds are also used to coat the insides of fluorescent light bulbs to alter the color given off.

Curium, an Actinoid

Curium's predicted electron configuration ends $7s^25f^8$. However, its actual configuration is $7s^25f^76d^1$. Plainly, the stability of the half-full f sublevel more than offsets the promotion of one electron to the $6d$. The element exhibits a 3+ oxidation state in its compounds and the ion is pale yellow in solution. The element itself is a silvery, hard metal of medium density and high melting point. It does not occur in nature. All multigram quantities of curium have been produced by the slow neutron bombardment of the artificial element plutonium. Curium is reactive and highly toxic to the human body. Curium has potential use as an energy source if the heat generated by its nuclear decay can be converted to electric energy. This metal may be used as the energy source in the nuclear generators in satellites.

HISTORY CONNECTION

A Town That Chemists Will Remember

In 1788, near the Swedish town of Ytterby, a new mineral was discovered. Over the course of 100 years, this mineral yielded 16 new elements, most of them in the lanthanoid series. Four of these elements were named for the town: yttrium, ytterbium, erbium, and terbium. John Gadolin, a Finnish chemist, found the first element, yttrium; gadolinium, another of the group, was named for him.

11.3 CONCEPT REVIEW

11. What property makes zinc useful in galvanizing?
12. The lanthanoids were once called the "rare earths." What is meant by "earth"?
13. **Apply** What properties are desirable in coinage metals? Which transition metals would make the best coins?

Answers to
Concept Review

11. Zinc is corrosion-resistant.
12. The terms *earths* refers to the oxide forms of these elements, which is the form in which chemists first isolated the lanthanoids.
13. Coinage metals should be relatively unreactive, resistant to corrosion, inexpensive, and able to form durable alloys. Silver, copper, and nickel are good coinage metals. Gold and platinum have also been used, but are generally too expensive.

HISTORY CONNECTION

A Town That Chemists Will Remember

Exploring Further: Have interested students look over the list of elements for names of countries, states, or cities. For example, Marie Curie named polonium after her native Poland. Others include elements with atomic numbers of 21, 31, 32, 49, 63, 84, 87, 95, 97, 98.

4 CLOSE

Discussion: Have students turn back to Chapter 6 in the textbook. Look at the Tables 6.7, 6.8, and 6.9, which show the electron configurations of the transition metals and the inner transition elements. Discuss how the Lewis electron dot diagram for each is the same.

PROGRAM RESOURCES

Vocabulary Review/Concept Review: Use the worksheets, p. 11 and pp. 55-56, to check students' understanding of the key terms and concepts of Chapter 11.

CHAPTER REVIEW 11

- Review Summary statements and Key Terms with your students.
- Complete solutions to Chapter Review Problems can be found in the **Merrill Chemistry Problems and Solutions Manual.**

Chapter 11

REVIEW

Summary

11.1 Hydrogen and Main Group Metals

1. Hydrogen is often considered a family by itself because of the unique ways in which it reacts.

2. The most active metals are listed toward the lower left-hand corner of the periodic table. The alkali metals, Group 1 (IA), are the most active metals. They react by losing one electron to form ions having a 1+ charge.

3. In large atoms, the shielding effect of inner-level electrons partially blocks the attraction of the nucleus for outer-level electrons. This effect, coupled with increased distance from the nucleus, results in increasing reactivity in larger alkali metal atoms.

4. Lithium behaves in many ways like other members of its family but, because of the small size of its ion, it acts in other ways like magnesium.

5. The alkaline earth metals, Group 2 (IIA), are second only to the alkali metals in reactivity. They form 2+ ions.

6. Aluminum, in Group 13 (IIIA), is less reactive than the alkaline earth metals, and forms 3+ ions. Aluminum is the most abundant metal in Earth's crust.

11.2 Nonmetals

7. The elements in Group 14 (IVA) have four electrons in the outer level. Carbon is a nonmetal; silicon and germanium are metalloids. Tin and lead, in the same family, are distinctly metallic.

8. An important property of carbon is its ability to form chains and rings of atoms. This ability is called catenation.

9. Nitrogen and phosphorus, in Group 15 (VA), form compounds by sharing electrons, but they differ from each other considerably.

10. An important use of nitrogen is in ammonia production. Compounds made from ammonia are used as fertilizers.

11. In Group 16 (VIA), oxygen is distinctly nonmetallic and reacts by gaining two electrons or by sharing electrons. Sulfur is similar to oxygen but also exhibits positive oxidation states.

12. Oxygen is the most plentiful element in Earth's crust. It forms compounds with all elements except certain noble gases. Many oxides react with water to form acids or bases.

13. The most active nonmetals are listed toward the upper right-hand corner of the periodic table. The halogens are the most active nonmetallic elements and usually react by gaining one electron.

14. In contrast to the alkali metals, increased shielding effect in the halogens makes the halogen elements less reactive moving downward through the family.

15. Fluorine is the most reactive of all elements. Chlorine is commonly produced by passing an electric current through a sodium chloride solution.

16. The noble gases, Group 18 (VIIIA), have very stable outer electron configurations and are much less reactive than most of the other elements.

11.3 Transition Metals

17. The transition elements are characterized by having partially filled *d* orbitals in all groups except 12 (IIB). These metals include major structural metals.

18. Chromium is a typical, corrosion-resistant transition metal, exhibiting more than one oxidation number.

19. Zinc is corrosion-resistant. Because of its full *d* sublevel, it does not exhibit more than one oxidation state.

20. Lanthanoids exhibit the 3+ oxidation state. Curium, an actinoid, also exhibits the 3+ oxidation state.

PROGRAM RESOURCES

Evaluation: Use the Evaluation worksheets, pp. 41-44, to assess students' knowledge of Chapter 11.

Computer Test Bank: Use the IBM or Apple Test Bank and Question Selection Manual to prepare your own individualized test for Chapter 11.

Chapter 11

REVIEW

Key Terms

hydrogen ion
hydride ion
catalyst
carbide ion
organic compound
catenation
inorganic compound
allotrope
silicate
amphoteric
galvanizing
inner transition element

Review and Practice

14. What can you generalize about the electron structure of the elements in any given family?

15. Describe the four ways in which hydrogen can react.

16. What is the difference between a hydrogen ion and a hydride ion?

17. Lithium is not a typical alkali metal.
- **a.** How does lithium differ chemically from other alkali metals?
- **b.** How does lithium differ physically from other alkali metals?
- **c.** What non-alkali metal does lithium resemble?
- **d.** How is lithium similar to other alkali metals?

18. Marble, used in sculpture and architecture, is composed of almost pure calcium carbonate. Calcium is an alkaline earth metal. List all the alkaline earth metals and give additional uses for at least three of the elements or their compounds.

19. Classify each of the following as either organic or inorganic.

a. Na_4C	**f.** C
b. CH_4	**g.** C_3H_7Br
c. CH_3COOH	**h.** CS_2
d. $(NH_4)_2CO_3$	**i.** H_2CO_3
e. CO_2	

20. Propane, C_3H_8, and butane, C_4H_{10}, are common organic fuels that are made up of chains of carbon atoms, as shown in the structures below.

```
    H   H   H            H   H   H   H
    |   |   |            |   |   |   |
H — C — C — C — H    H — C — C — C — C — H
    |   |   |            |   |   |   |
    H   H   H            H   H   H   H
     propane                 butane
```

- **a.** What is this tendency to form chains called?
- **b.** How does this tendency in carbon compare with the same tendency in silicon?
- **c.** Why would we call these compounds organic?

21. How do the structures of graphite and diamond differ? How do these materials illustrate allotropes?

22. What is the difference between silicon and silicates? What are three important products that use silicon?

23. What chemical property accounts for the uses of tin and lead, especially their uses in alloys?

24. Describe two ways that atmospheric nitrogen is changed to form useful compounds.

25. In 1669, Hennig Brandt noticed a new substance was produced when he heated urine and sand together in an attempt to make gold. Today we know the substance as phosphorus. In 1772, Daniel Rutherford noticed that a colorless gas—nitrogen—still remained after the oxygen and carbon dioxide were removed from an air sample. These two elements, phosphorus and nitrogen, have played an important part in chemistry.
- **a.** How do the properties of nitrogen and phosphorus differ?
- **b.** What properties do they both have?
- **c.** How are both nitrogen and phosphorus important to living things?

Answers to
Review and Practice

14. Different elements in the same family have the same outer level electron arrangement.

15. Four ways hydrogen can react:
1. loses an electron, becoming an H^+ ion
2. shares its electron
3. gains one electron, becoming a H^- ion
4. forms bridges between two atoms

16. A hydrogen ion is H^+; H atom has lost an electron.
A hydride ion is H^-; H atom has gained an electron.

17. a. Li reacts more vigorously with N, burns in air to form the oxide, and forms compounds with solubilities similar to those of Mg compounds.
b. Li is not soluble in alkali metals other than Na.
c. Li resembles Mg.
d. Li is a soft, silvery metal, has a low melting point and dissolves in liquid ammonia.

18. beryllium—nonsparking tools
magnesium—lightweight alloys
calcium—CaO in steel, cement, and heat-resistant bricks, lowers soil and pond acidity; $CaCl_2$ in deicing, paper, pulp. Ca in teeth, muscles, bones, and blood
strontium—radioisotope used as a tracer, fireworks
barium—$BaSO_4$ used to preserve statues, filler for paper.
radium—destroy malignant cells

19. a. inorganic, **b.** organic, **c.** organic, **d.** inorganic, **e.** inorganic, **f.** inorganic, **g.** organic, **h.** inorganic, **i.** inorganic

20. a. catenation
b. The tendency is much greater in carbon.
c. Electrons are shared between a carbon atom and one or more other carbon atoms.

21. In diamond, each C atom shares electrons with the four nearest C atoms. In graphite, each C atom shares electrons with the three nearest C atoms. They are different forms of the same element.

22. Silicon is an element.
Silicates are compounds formed from Si and O. Silicon is used in transistors, solar cells, computer chips, sandpaper abrasive.

23. Tin and lead are both relatively unreactive.

24. nitrogen—fixing bacteria, Haber process to make NH_3.

25. a. N and P differ in oxidation states, N is a gas, P is a solid, P forms allotropes, and they differ in stability of each element compared with the stability of its compounds.
b. N and P both have oxidation states of 3–, 0, 3+, and 5+.
c. N and P compounds are both used as fertilizers; N in amino acids, forming proteins, P in ATP, DNA, and RNA.
d. N is also used in explosives, ammonia, nitric acid, and protection of materials from air.
e. semiconductors

Answers to

Review and Practice (cont.)

26. Metals form ionic oxides; metals lose electrons to oxygen to form an octet in the outer level of oxygen. Nonmetals share electrons with oxygen.

27. O_2 is oxygen; O_3 is ozone.

28. a. Oxygen can accept electrons from metals to form ionic oxides or share electrons with nonmetals.

b. Ionic oxides generally form basic solutions. Other oxides form acidic solutions.

c. Amphoteric oxides may form either acidic or basic solutions, depending on what else is present.

29. Both oxygen and sulfur form 2− ions, which are similar in acid-base behavior, both combine with like atoms (e.g. O_2, S_8). S exhibits positive oxidation states also.

30. Salt-forming

31. Group 17 (VIIA)

32. The metal donates electrons to the halogens.

33. Fluorine is the most active; astatine is the least active. This difference in activity is a result of differences in atomic radius, shielding effect, and nuclear charge.

34. It increases with increasing atomic number on the left. It decreases with increasing atomic number on the right.

35. Bleaching agent, water purification, hydrochloric acid used in steel manufacture, dye production, food processing, oil well drilling

36. Chlorine is produced by running an electric current through a NaCl solution.

37. a. He, Ne, Ar, Kr, Xe, and Rn are the noble gases.

b. Noble gases are chemically less reactive than most elements and gases at room temperature.

c. Some form compounds.

d. Noble gases can be used to protect active metals during welding, fill light bulbs, and fill gas discharge tubes.

38. Transition elements are used in structures.

39. Chromium is resistant to corrosion.

40. Carcinogens are cancer-causing substances.

41. Zinc is resistant to corrosion and can be used to coat iron objects.

42. 3+

43. The actual configuration of curium is more stable with a half-full *f* sublevel.

44. All curium has been produced by slow neutron bombardment of plutonium.

45. Li is such a small atom that there is little shielding effect.

46. a. Co, **b.** Cu, **c.** Md, **d.** F, **e.** Ra, **f.** Zr, **g.** K, **h.** Co, **i.** Li, **j.** O

47. a. Ac, **b.** Am, **c.** Fm, **d.** Ga, **e.** Bk, **f.** Li **g.** Br, **h.** C, **i.** Rb, **j.**I

Chapter 11 REVIEW

d. List other uses for the element nitrogen and its compounds.

e. List other uses for phosphorus compounds.

26. How does the chemical behavior of oxygen differ when it combines with an active metal rather than a nonmetal?

27. What are the allotropes of oxygen?

28. Oxygen makes up one-fifth of the mass of the atmosphere, nine-tenths of the mass of Earth's water, and nearly half the mass of Earth's crust. Most of this oxygen is in the form of compounds resulting from the reactions between oxygen and other elements.

a. Explain the two ways oxygen can bond to form oxides.

b. If the resulting oxides are each dissolved in water, how do the solutions differ?

c. Compare the chemical behavior of amphoteric oxides with that of other oxides.

29. Compare the chemical properties of oxygen and sulfur.

30. What is the meaning of the word *halogen* in this chapter?

31. Where are the halogens found on the periodic table?

32. In electronic terms, what usually happens when halogens react with active metals?

33. Which halogen is the most active? Which is the least active? What characteristics account for this difference in activity?

34. How does chemical activity vary among the elements on the left side of the periodic table? How does it vary among elements on the right side of the periodic table?

35. List several uses of chlorine in the laboratory and in industry.

36. Describe a method by which chlorine is produced industrially.

37. The noble gas helium was discovered on the sun by spectroscopy before it was discovered on Earth.

a. List all the noble gases.

b. What are their properties?

c. Why are they no longer called inert gases?

d. List several practical uses for the noble gases.

38. What is the major practical use of the transition elements?

39. Why is chromium an important metal in plating iron or steel and in making alloys with iron?

40. What are carcinogens?

41. Why is zinc important in the production of materials made from iron?

42. What oxidation state is common to all of the lanthanoids?

43. Why does the electron configuration of curium vary from the predicted configuration?

44. What is the importance of plutonium to the production of curium?

45. What characteristics of the lithium atom account for the differences in behavior that it exhibits in relation to the other alkali metals?

46. Use the periodic table to predict which element in each of the following pairs would be more active.

a. cobalt, nickel
b. copper, gallium
c. mendelevium, nobelium
d. chlorine, fluorine
e. barium, radium
f. titanium, zirconium
g. sodium, potassium
h. arsenic, cobalt
i. beryllium, lithium
j. oxygen, phosphorus

47. Use the periodic table to predict which element in each of the following pairs would be more active.

a. actinium, thorium

Chapter 11 REVIEW

b. americium, europium
c. erbium, fermium
d. arsenic, gallium
e. berkelium, californium
f. carbon, lithium
g. bromine, iodine
h. boron, carbon
i. potassium, rubidium
j. iodine, tellurium

48. For each of the following pairs of elements, determine the family name, and state two chemical and two physical properties that we could expect these elements to have.
a. K, Na c. I, Br
b. Ar, Ne d. Ba, Ca

49. One important form of oxygen is ozone.
a. What is the formula for ozone?
b. How can ozone be produced?
c. What is a practical use for ozone?

Concept Mastery

50. Discuss the effects of shielding, atomic size, and nuclear charge on the relative reactivities of the alkali metals.

51. Lightning and the action of bacteria in the roots of legumes such as beans or peanuts are both said to be "nitrogen-fixing." What is meant by "nitrogen fixation?"

52. Why is phosphorus an essential nutrient in living systems?

53. How does the electronic structure of the transition elements differ from that in the main groups of the periodic table?

54. Many transition metals exhibit several oxidation states. Why does zinc exhibit only the 2+ oxidation state?

55. What is the shielding effect? How can it help you predict the difference in reactivity between calcium and barium?

56. Why are the alkali metals and most of the alkaline earth metals so little used in their uncombined forms?

57. Why do oxygen and lithium form the compound Li_2O when they combine, while calcium and oxygen form CaO?

58. Selenium is an example of a trace element that we must have in our diets. Why do we call it a trace element?

59. Why would you expect the noble gases to be relatively inert elements?

60. Which elements in the periodic table would you predict to have the least attraction for outer level electrons?

61. In some versions of the periodic table, hydrogen is placed at the top of both Group 1 (IA) and Group 17 (VIIA). Considering the chemistry of hydrogen, why might it fit in both places?

62. Write balanced equations for each of the following reactions.
a. formation of each of the two oxides of sulfur
b. reaction of each of these oxides with water

Application

63. Rubies and sapphires are two of the most valuable compounds of aluminum. They occur in nature and can also be synthesized. What are some other uses for aluminum or aluminum compounds?

64. Why does catenation make carbon compounds so important?

65. One property shared by most of the elements in Group 1 (IA) and Group 2 (IIA) is the ability to displace hydrogen from water. Write equations for the displacement of hydrogen from water by Li, Sr, K, and Ba.

66. Sodium compounds are used extensively in industry and in homes. Give at least one common use for each of the following compounds.
a. $NaHCO_3$ d. Na_2SO_4
b. $Na_5P_3O_{10}$ e. Na_2CO_3
c. NaOH f. Na_2SiO_3

48. a. alkali metals; loses one electron, displaces H from H_2O; soft, shiny
b. noble gases; relatively unreactive, may form compounds with O or F; gases at room temperature, colorless
c. halogens; diatomic, gain one electron when reacting; low melting and boiling points
d. alkaline earths; lose two electrons in a reaction, displace H from water; soft, shiny
49. a. O_3 is ozone.
b. subjecting O_2 to a silent electric discharge
c. water purification

Answers to
Concept Mastery
50. As atomic number increases, the nucleus and the outer level electron are separated by more electrons. The repulsive effects of this electron cloud make the outer level electron more easily lost. Thus, shielding increases reactivity. Even though the nuclear size and charge increase with each period, the shielding effect is greater.
51. Nitrogen fixation is changing elemental nitrogen to usable compounds.
52. Phosphorus is a component of DNA, ATP, and RNA.
53. The highest-energy electrons are in a *d* sublevel.
54. The *d* sublevel is full, leaving only the outer level electrons to react.
55. Shielding effect is the decrease in force between outer electrons and the nucleus due to the presence of other electrons between them. Ba has more shielding effect because it has more electrons between the nucleus and the outer energy level and is thus more reactive than Ca.
56. They so readily lose electrons that they are difficult to keep in their uncombined form.
57. Li has only one outer level electron, so two atoms of Li would be necessary to donate the two needed electrons for O. Ca has two outer level electrons.
58. It is present in very small amounts.
59. They have an octet in the outer energy level in a full outer level.
60. lower left
61. Hydrogen can either lose one electron (H^+) or gain one electron (H^-).
62. a. $S + O_2 \rightarrow SO_2$;
$2SO_2 + O_2 \rightarrow 2SO_3$
b. $SO_2 + H_2O \rightarrow H_2SO_3$;
$SO_3 + H_2O \rightarrow H_2SO_4$

Answers to
Application
63. $Al_2(SO_4)_3$—water purification, paper manufacture, fabric dying Al—alloys, wiring
64. Smaller molecules may combine into chains to form large molecules, so there are millions of organic compounds.
65. $2Li + 2H_2O \rightarrow 2LiOH + H_2$
$Sr + 2H_2O \rightarrow Sr(OH)_2 + H_2$
$2K + 2H_2O \rightarrow 2KOH + H_2$
$Ba + 2H_2O \rightarrow Ba(OH)_2 + H_2$
66. a. baking, **b.** food additive, **c.** paper production, **d.** glass manufacturing, **e.** glass manufacturing, **f.** catalyst

Answers to
Application (cont.)

67. working in high concentrations of O_2 or near flammable liquids or vapors

68. They are percent nitrogen and percent phosphorus. Answers will vary.

Answers to
Critical Thinking

69. The platinum speeds up the reaction. No, a catalyst speeds up or enables a reaction but is not permanently changed by the reaction.

70. H_2SO_4 is used in fertilizer, petroleum refining, steel, synthetic fiber production, paints, pigments, and many other industries.

71. sharing of electrons

72. a. Allotropes are different forms of the same element.

b. Graphite is in layers that can slip across one another; diamond has three-dimensional structure.

c. Diamond has a very definite structure, bonding is in all directions from any one C atom.

d. As the C atoms in graphite are forced closer together, they will bond layer to layer, not just within a layer.

67. One of beryllium's uses is in alloys from which nonsparking tools are made. Suggest a situation in which nonsparking tools would be of particular value.

68. The amounts of important nutritive ingredients in fertilizer are usually given on the package as a series of three numbers, for instance 5-10-5. The last number is the percent potassium. Find out what the first two numbers designate. Write a short, general statement regarding what proportions of nutrients are useful in what sorts of situations.

Critical Thinking

69. Finely divided platinum is used as a catalyst in many industrial processes. What is the function of the platinum? Do you think the platinum will be used up in the process? Explain your answer.

70. The production of sulfuric acid is considered to be one of a country's major economic indicators. Why do you think this is so?

71. Is the compound SO_3 produced by the sharing or transfer of electrons?

72. The structures of graphite and diamond are shown here. These are two allotropes of carbon.

a. What are allotropes?

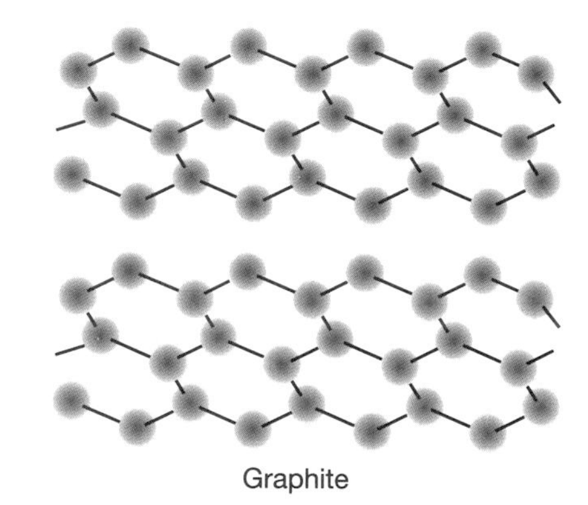

Graphite

Diamond

b. Study the structures and suggest reasons why graphite is slippery and is used as a lubricant, while diamond is not.

c. Study the structures and suggest reasons why diamond is harder than graphite and is used as an abrasive coating.

d. Use the structures to explain why graphite under great pressure turns into diamond.

Readings

Bachman, Peter K., and Russell Messier. "Diamond Thin Films." *Chemical and Engineering News* 67, no. 20 (May 15, 1989): 24-39.

Davenport, Derek A. "Going Against the Flow—The Isolation of Fluorine." *ChemMatters* 4, no. 4 (December 1986): 13-15.

Day, Richard. "Lead-Free Water." *Popular Science* 232, no. 1 (January 1988): 90-91, 116.

Dworetzky, Tom. "Gold Bugs." *Discover* 9, no. 3 (March 1988): 32.

Edwards, D. D. "Aluminum: A High Price for a Surrogate?" *Science News* 131, no. 16 (April 18, 1987): 245.

Raloff, Janet. "New Misgivings About Low Magnesium." *Science News* 133, no. 23 (June 4, 1988): 356.

Stolka, Milan. "Hard Copy Materials." *Chem-Tech* 19, no. 8 (August 1989): 487-495.

Cumulative Review

1. Complete and balance each of the following equations.
   a. $Zn + H_2SO_4 \rightarrow$
   b. $Mg + H_2O \rightarrow$
   c. $CuSO_4 + NaOH \rightarrow$
   d. $KI + Cl_2 \rightarrow$
2. How many grams of $AgNO_3$ are required to produce 38.7 grams of AgCNS according to the following reaction?
   $AgNO_3 + KCNS \rightarrow AgCNS + KNO_3$
3. Compute the atomic mass of bromine from the following data about its isotopes.

Isotopic mass	Abundance
78.918 332 u	50.54%
80.916 292 u	49.46%

4. Draw electron dot diagrams for yttrium, dysprosium, and indium.
5. Write balanced synthesis equations for the formation of each of the following.
   a. silver sulfide
   b. potassium bromide
   c. iron(III) oxide
   d. magnesium nitride
6. Look at the following orbital-filling diagrams, and decide which of them follow the arrow diagram and which do not. For those that are correct, write the electron configuration. For any others, correct the diagram. Do not change the total number of electrons present. Identify each element represented.

	1s	2s	2p
a.	⇅	⇅	↑ ○ ○
b.	↑	↑	○ ○ ○
c.	⇅	⇅	⇅ ○ ○
d.	⇅	⇅	↑ ↑ ↑
e.	⇅	↑	↑ ↑ ○

7. Explain why the second ionization energy for lithium is more than four times the second ionization energy for beryllium.
8. How do atomic radii change, moving left to right across a period of elements? What factors account for this change?
9. Nitrogen gas can be produced by the decomposition of ammonium nitrite to give nitrogen and water. Write the balanced equation for this reaction. Calculate the mass of ammonium nitrite needed to produce 0.100 mole of nitrogen.
10. The density of magnesium is 1.738 g/cm^3. A mole of magnesium atoms has a mass of 24.305 g.
    a. What is the volume of a mole of magnesium atoms?
    b. What is the space taken up by one atom of magnesium?

Answers to Cumulative Review

1. a. $Zn + H_2SO_4 \rightarrow ZnSO_4 + H_2$
   b. $Mg + 2H_2O \rightarrow Mg(OH)_2 + H_2$
   c. $CuSO_4 + 2NaOH \rightarrow Cu(OH)_2 + Na_2SO_4$
   d. $2KI + Cl_2 \rightarrow 2KCl + I_2$
2. 39.6 g $AgNO_3$
3. 79.91 u
4. Y:, Dy:, In:
5. a. $2Ag + S \rightarrow Ag_2S$
   b. $2K + Br_2 \rightarrow 2KBr$
   c. $4Fe + 3O_2 \rightarrow 2Fe_2O_3$
   d. $3Mg + N_2 \rightarrow Mg_3N_2$
6. a. $1s^2\ 2s^2\ 2p^1$; boron
   b. 1s
   ⇅; helium
   c. 1s 2s 2p
   ⇅ ⇅ ↑ ↑ ○; carbon
   d. $1s^2 2s^2 2p^3$; nitrogen
   e. 1s 2s 2p
   ⇅ ⇅ ↑ ○ ○; boron
7. Li has a stable electron configuration after losing one electron. Be is more stable after losing two electrons.
8. Decrease; there is no increase in the number of electron clouds but increased charge of the nucleus.
9. $NH_4NO_2 \rightarrow N_2 + 2H_2O$ 6.40 g NH_4NO_2
10. a. 13.98 cm^3/mol
    b. 2.321×10^{-23} $cm^3/atom$

CHAPTER 12

Chemical Bonding

CHAPTER ORGANIZER

CHAPTER OBJECTIVES	TEXT FEATURES	LABORATORY OPTIONS	TEACHER CLASSROOM RESOURCES
12.1 Bond Formation **1.** Identify the type of bonding between two elements given their electronegatives. **2.** List factors that influence electronegativity and recognize it as a periodic property of elements. **3.** Differentiate among properties of ionic, covalent, and metallic bonds. **4.** Explain the use of infrared and microwave spectroscopy to determine the structure of molecules.	**Careers in Chemistry:** Electrician, p. 305 **Home Economics Connection:** Food Frequencies, p. 307 **Bridge to Metallurgy:** Alloys, p. 310 **Frontiers:** Alien Alloys, p. 312	**Discovery Demo:** 12-1 Ionic and Covalent Bonds, p. 300 **Demonstration:** 12-2 Percent Ionic Character, p. 304 **Demonstration:** 12-3 Molecular Aerobics, p. 306 **Demonstration:** 12-4 The Alchemist's Gold, p. 310	**Lesson Plans:** pp. 25-26 **Study Guide:** pp. 41-42 **Enrichment:** Electronegativity, pp. 23-24 **Color Transparency 26 and Master:** Table of Electronegativities **Reteaching:** Types of Bonds, p. 18 **Applying Scientific Methods in Chemistry:** Electronegativity, pp. 28-29 **Critical Thinking/Problem Solving:** The Octet Rule and Bond Formation, p. 12 **Chemistry and Industry:** Dyes, pp. 19-20
Nat'l Science Stds: UCP.2, UCP.3, UCP.5, B.1, B.2, B.4, B.6, E.1, E.2			
12.2 Particle Sizes **5.** Differentiate among atomic radii, ionic radii, covalent radii, and van der Waals radii. **6.** Discuss factors that affect the values of ionic radii and covalent radii. **7.** Use covalent radii to calculate bond lengths.		**Minute Lab:** Two Bonds are Better Than One, p. 316 **ChemActivity 12:** A 3-D Periodic Table-microlab, p. 817 **Demonstration:** 12-5 van der Waals Radius, p. 314 **Laboratory Manual:** 12 Using Conductivity to Predict Bonding - microlab	**Study Guide:** p. 43 **Color Transparencies and Masters 27 & 28:** Ionic Bonding and Ionic Radii, Covalent Radii and van der Waals Radii **ChemActivity Master 12**
Nat'l Science Stds: UCP.2, UCP.3, UCP.5, B.1, B.2, B.3, B.4			
OTHER CHAPTER RESOURCES	**Vocabulary and Concept Review,** pp. 12, 57-58 **Evaluation,** pp. 45-48	**Videodisc Correlation Booklet** **Lab Partner Software**	**Test Bank** **Mastering Concepts in Chemistry—Software**

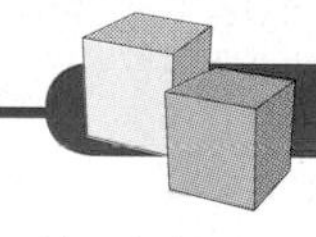

Block Schedule

For information on block scheduling, see the Lesson Plans booklet in the Teacher Resource Package.

GLENCOE TECHNOLOGY

Mastering Concepts in Chemistry Software

Electronegativity and Bond Types

Chemistry: Concepts and Applications Videodisc

Forming Ionic and Covalent Bonds
Electronegativity
Bonding
Ionic and Covalent Compounds

Chemistry: Concepts and Applications CD-ROM

The Periodic Table
Form an Ionic Bond
Forming Ionic and Covalent Bonds
Electronegativity
Bonding
Ionic and Covalent Compounds

Complete Glencoe Technology references are inserted within the appropriate lesson.

ASSIGNMENT GUIDE

Contents	Level	Practice Problems	Chapter Review	Solving Problems in Chemistry
12.1 Bond Formation				
Electronegativity	C	1-2	14, 24, 26, 27, 33, 34, 43	
Bond Character	C	3-4	15, 25, 28, 32	1-2
Ionic Bonds	C		16, 22, 31, 32, 35, 38, 40	
Covalent Bonds	C		17, 22, 23, 27, 29-31, 38	
Metallic Bonds	C		20, 21, 23, 36, 38, 41, 42	
Polyatomic Ions	C		22, 39	
12.2 Particle Sizes				
Ionic Radii	C		37	3-5
Covalent Radii	O	9	18, 19, 24, 37	
Van der Waals Radii	A		19, 37	
Summary of Radii	C			Chapter Review: 1-3

C=Core, A=Advanced, O=Optional

Laboratory Materials

Minute Labs	ChemActivities	Demonstrations
page 316 dropper glue, white plastic cup, disposable saturated sodium tetraborate solution stirring rod	**page 817** graph paper, large square grease pencil metric ruler 96-well microplate scissors soda straws (25 per group)	**pages 300, 304, 306, 310, 314** beaker, 400 cm^3 beaker tongs burner candle clear plastic lid from greeting card box copper or nichrome wire deflagrating spoon evaporating dish forceps glue, white granular zinc hot plate ice cubes magnesium ribbon marbles matches metal can, large and empty NaOH overhead projector paper towels pencil pennies, new and shiny plastic foam balls, 2-inch diameter protractor sulfur, roll tongs water

CHAPTER 12

Chemical Bonding

CHAPTER OVERVIEW

Main Ideas: Chapter 12 begins a series of chapters concerning the combination of atoms to form molecules (Chapters 12, 13, and 14). In Chapter 12, we discuss the underlying principles of bond formation, including chemical bonding and its effects on molecular structure and properties. The first section should develop naturally from the discussions of periodic properties in Chapter 10. Ionic and covalent bonding are presented in detail. The first section concludes with a look at metallic properties as explained by the metallic bond model. Time is spent discussing the different radii of bonded atoms in Section 12.2.

Theme Development: Stability and change as a theme of the textbook is described in terms of chemical stability. The relationship between an atom's electron structure and its stability is easily studied in the context of chemical bonding. A study of chemical bonding further develops the theme by relating the bond type to chemical and physical properties.

Tying To Previous Knowledge: Remind students of how the pieces of a jigsaw puzzle interlock to form a complete picture. In a similar way, atoms can bond together to form a compound. The atoms combine in a definite ratio and in a definite order in a way similar to the puzzle pieces.

ASSESSMENT PLANNER

Choose from the following assessment strategies.

Assess, pp. 311-312, 315-316
Assessment, pp. 304, 306, 309, 310, 312, 314, 316
Portfolio, p. 310
Minute Lab, p. 316
ChemActivity, pp. 817-818
Chapter Review, pp. 317-319

CHAPTER **Preview**

CONTENTS

DISCOVERY DEMO

12-1 Ionic and Covalent Bonds

Purpose: to demonstrate the formation of an ionic and a covalent bond

Materials: 5-cm long Mg ribbon; tongs; large, empty metal can; 2 g roll sulfur; deflagrating spoon; laboratory burner

Procedure: 1. Place a large can in the sink. Carefully ignite a 5-cm long piece of Mg ribbon. **CAUTION:** *Wear goggles when burning the Mg ribbon. Do not look directly at the burning Mg.* Hold the Mg with tongs while it burns inside the can to prevent students from looking directly at the burning Mg. If the room lights are turned off, the students can safely observe the reflected light.

2. In a darkened room, place a small piece of roll sulfur in a deflagrating spoon and briefly heat it in a burner flame. The sulfur will ignite and burn

Chemical Bonding

CHAPTER

12

Are we looking at an optical illusion? Is the picture turned sideways? No. The objects in the photograph are bonded with methyl cyanoacrylate, or Super Glue. How does Super Glue hold things together? And how is methyl cyanoacrylate itself held together?

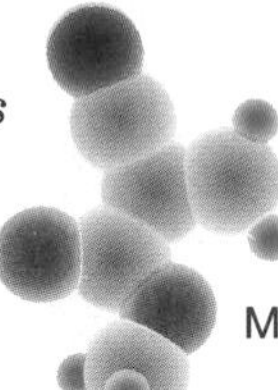

Methyl cyanoacrylate, $C_5H_5O_2N$

In past chapters, we have examined the structure and properties of atoms. This study has prepared us to learn how atoms attach together to form compounds. It is this attachment of atoms to each other that we refer to as a bond. In this chapter we will investigate three types of bonds, how they are formed, and how bonding affects the atoms that are bonded.

EXTENSIONS AND CONNECTIONS

Careers in Chemistry
Electrician, page 305

Home Economics Connection
Food Frequencies, page 307

Bridge to Metallurgy
Alloys: Metallic materials that aren't elements, page 310

Frontiers
Alien Alloys: A glimpse of high-tech alloys with remarkable properties, page 312

INTRODUCING THE CHAPTER

USING THE PHOTO

Super Glue® can be used in place of sutures in some types of surgery. It is also used by morticians to seal the eyes and lips of corpses. Super Glue® begins to polymerize when it is spread on a surface because the surface contains trace amounts of water and alcohols. The cyanide and carboxyl groups in methylcyanoacrylate, which are both attached to the same carbon, have high electron affinities. Because of this strong attraction for electrons, the carbon-carbon double bond can be easily broken. Super Glue® joins two surfaces by filling in any small cavities or cracks in them and by forming chemical bonds with surface molecules. Use this feature to start a discussion on why atoms bond. Point out that the interactions between particles of matter are not static.

REVEALING MISCONCEPTIONS

Students often think that a chemical bond physically links atoms just as a nail holds two pieces of wood together. Emphasize that the bond is an attractive force. The model kits used by chemists show a physical bond as a piece of wood, plastic or a metal spring. There is not a physical attachment that can be identified and called a chemical bond.

with a low blue flame. **CAUTION:** *Use a fume hood; the SO_2 gas is toxic.*
Disposal: E

Results: Magnesium burns with a very bright white hot flame. The stability of the MgO is evidenced by the large amount of heat and light given off.

$$2Mg(cr) + O_2(g) \rightarrow 2MgO(cr)$$
$$\Delta H_f = -602 \text{ kJ/mol}$$
$$3Mg(cr) + N_2(g) \rightarrow Mg_3N_2(cr)$$
$$\Delta H_f = -461 \text{ kJ/mol}$$

The sulfur burns with a low violet flame.

$$S(cr) + O_2(g) \rightarrow SO_2(g)$$
$$\Delta H_f = -297 \text{ kJ/mol}$$

Point out that the ionic bond in MgO is stronger, more stable, than the covalent bond in SO_2 because more energy is released when it formed.

Questions: 1. Both elements, Mg and S, react with the air by burning. With what element does the Mg and S react? *oxygen gas*

2. Write balanced chemical equations to represent the chemical reactions you observed. *See Results.*

3. Which reaction releases the most heat and light energy? *Mg*

4. If Mg forms an ionic bond, and S forms a covalent bond, which bond is the stronger bond? Why? *The ionic bond is stronger because more energy is released during its formation.*

12.1 Bond Formation

PREPARE

Key Concepts
The major concepts presented in this section include electronegativity, the ionic bond, the covalent bond, and the metallic bond along with their properties. The section ends with a discussion of the uniqueness of the bonding that is found in polyatomic ions.

Key Terms
electronegativity
ionic bond
covalent bond
molecule
bond axis
bond angle
bond length
metallic bond

Looking Ahead
In Demonstration 12-3, glue the models together one day before they will be used.

1 FOCUS

Objectives
Preview with students the objectives listed on the student page. Students can use them as a study guide for this major section.

Activity
Turn on the faucet to produce a thin stream of water. Rub an inflated balloon with wool, fur or silk. Bring the balloon near the stream of water and observe how the stream of polar water molecules is deflected. Point out to students that the electrons aren't always equally shared between atoms and as a result of unequal sharing, a chemical bond can have dipole properties.

GLENCOE TECHNOLOGY

Videodisc

Chemistry: Concepts and Applications

Forming Ionic and Covalent Bonds
Disc 1, Side 2, Ch. 10

Show this video of the Discovery Demo.

OBJECTIVES

- identify the type of bonding between two elements given their electronegativities.
- list factors that influence electronegativity and recognize it as a periodic property of elements.
- differentiate among properties of ionic, covalent, and metallic bonds.
- explain the use of infrared and microwave spectroscopy to determine the structure of molecules.

12.1 Bond Formation

We have experimental evidence that some atoms form bonds by transferring electrons. We also have evidence that atoms form bonds by sharing electrons. Therefore, we can expect that the bonds formed between atoms depend on the electron configurations of those atoms and on the attraction the atoms have for electrons. Because configuration and attraction are periodic properties, the bonding of atoms will vary in a systematic way.

Electronegativity

Both electron affinity and ionization energy are properties of isolated atoms. Chemists need a comparative scale relating the abilities of elements to attract electrons when their atoms are combined. The relative tendency of an atom to attract electrons to itself when it is bonded to another atom is called **electronegativity.** The elements are assigned electronegativities on the basis of many experimental tests.

Electronegativities of elements are influenced by the same factors that affect ionization energies and electron affinities. It is possible to construct an electronegativity scale using first ionization energies and electron affinities of the elements. Examine Table 12.1. Note that the trend in electronegativity follows the same trends as the ionization energies and electron affinities, as shown in Figure 12.2. *The most active metals* (lower left) *have the lowest electronegativities.* Fluorine, a nonmetal, has the highest electronegativity (4.10) of all the elements. Thus, in a bond involving fluorine, the bonding electrons are drawn closer to the fluorine atom than to the other atom. Electronegativities are not given for a number of the actinoids because not enough of these elements have been produced to measure their electronegativities accurately. Because only a few of the noble gases form any compounds and those few form only a small number of compounds, no electronegativities are given for the elements of Group 18 (VIIIA).

Figure 12.1 Potassium is a very active metal. In contact with water, it reacts vigorously, producing hydrogen.

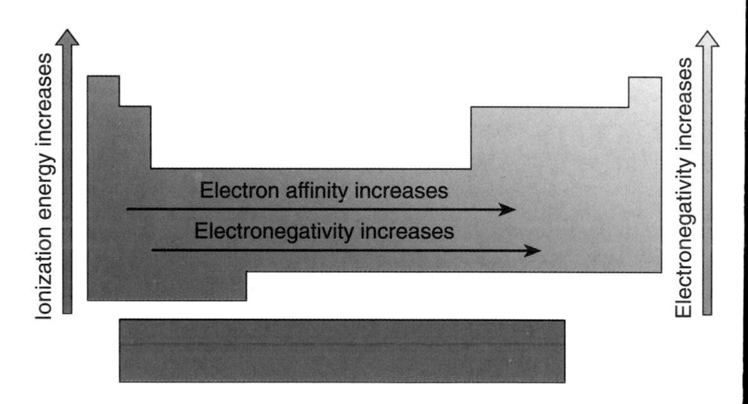

Figure 12.2 General trends for ionization energy, electron affinity, and electronegativity can be found in the periodic table.

CONTENT BACKGROUND

Calculating Electronegativity: There have been numerous equations used to calculate electronegativities. The most famous was proposed in 1930 by Pauling.

(1) Pauling's Equation:

$Q = 23.06\Sigma(X_A - X_B)^2 - 55.1N_N - 24.2N_O$

where $Q = \Delta H_f$ in kcal/mol

X_A = electronegativity of element A

X_B = electronegativity of element B

N_N = number of nitrogen atoms in compound

N_O = number of oxygen atoms in compound

(2) Mulliken's equation is an average of the first ionization energy and the electron affinity.

$1/2(I_A - E_A) - 1/2(I_B - E_B) = 2.78(X_A - X_B)$

where I_A and I_B = ionization energies of A and B

E_A and E_B = electron affinities of A

Table 12.1

Electronegativities

1	2	3	4	5	6	7	8	9	10	11	12	13	14	15	16	17	18
1 H 2.20																	2 He —
3 Li 0.97	4 Be 1.47											5 B 2.01	6 C 2.50	7 N 3.07	8 O 3.50	9 F 4.10	10 Ne —
11 Na 1.01	12 Mg 1.23											13 Al 1.47	14 Si 1.74	15 P 2.06	16 S 2.44	17 Cl 2.83	18 Ar —
19 K 0.91	20 Ca 1.04	21 Sc 1.20	22 Ti 1.32	23 V 1.45	24 Cr 1.56	25 Mn 1.60	26 Fe 1.64	27 Co 1.70	28 Ni 1.75	29 Cu 1.75	30 Zn 1.66	31 Ga 1.82	32 Ge 2.02	33 As 2.20	34 Se 2.48	35 Br 2.74	36 Kr —
37 Rb 0.89	38 Sr 0.99	39 Y 1.11	40 Zr 1.22	41 Nb 1.23	42 Mo 1.30	43 Tc 1.36	44 Ru 1.42	45 Rh 1.45	46 Pd 1.35	47 Ag 1.42	48 Cd 1.46	49 In 1.49	50 Sn 1.72	51 Sb 1.82	52 Te 2.01	53 I 2.21	54 Xe —
55 Cs 0.86	56 Ba 0.97	71 Lu 1.14	72 Hf 1.23	73 Ta 1.33	74 W 1.40	75 Re 1.46	76 Os 1.52	77 Ir 1.55	78 Pt 1.44	79 Au 1.42	80 Hg 1.44	81 Tl 1.44	82 Pb 1.55	83 Bi 1.67	84 Po 1.76	85 At 1.96	86 Rn —
87 Fr 0.86	88 Ra 0.97	103 Lr —	104 Unq —	105 Unp —	106 Unh —	107 Uns —	108 Uno —	109 Une —	110 Uun —	111 Uuu —	112 Uub —						
		57 La 1.08	58 Ce 1.08	59 Pr 1.07	60 Nd 1.07	61 Pm 1.07	62 Sm 1.07	63 Eu 1.01	64 Gd 1.11	65 Tb 1.10	66 Dy 1.10	67 Ho 1.10	68 Er 1.11	69 Tm 1.11	70 Yb 1.06		
		89 Ac 1.00	90 Th 1.01	91 Pa 1.14	92 U 1.30	93 Np 1.29	94 Pu 1.25	95 Am —	96 Cm —	97 Bk —	98 Cf —	99 Es —	100 Fm —	101 Md —	102 No —		

Many chemical properties of the elements can be organized in terms of electronegativities. For example, bond strength is a measure of the energy that is needed to break the bonds between atoms in molecules of a compound. As seen in Table 12.2, the greater the strength of the bond between two atoms, the greater the difference in electronegativities.

Consider a reaction between two elements. Their relative attraction for electrons determines how they react. We can use the electronegativity scale to determine this attraction. Because electronegativity represents a comparison of the same property for each element, it is a dimensionless number.

Table 12.2

Bonds between Hydrogen and Halogens

Bond	Bond Strength (kJ/mol)	Electronegativity Difference
H–F	568.1	1.90
H–Cl	431.951	0.63
H–Br	366.25	0.54
H–I	298.32	0.01

2 TEACH

CONCEPT DEVELOPMENT

Analogy: Ask students to imagine two of them becoming partners and buying a car together. Each one contributes an equal amount toward the purchase price. One student will drive it on M,W,F, and the other on T,Th, S and S. The students will immediately object that it's not equal sharing of the car. Point out that electronegativity is a measure of which bonded atom has the shared electron pair more of the time.

Background: Dr. Linus Pauling interpreted the bond energy of unsymmetrical molecules. Pauling's electronegativity scale is derived from fluorine as 4.0. Using this scale, you can decide if electrons favor one atom over another. For unequal attraction, the center of negative charge is displaced, resulting in a dipole. Then the bond will have some ionic character. Using the electronegativity table, students can predict bond stability (a large electronegativity difference) and dipole formation in the molecule (Chapter 14).

Using The Table: The values in Table 12.1 represent a modified Allred Rochow scale. Remind students that the actual attraction of an atom for electrons is greatly influenced by several factors, the chief of which are the element to which it is bonded and its oxidation state.

and B

X_A and X_B = electronegativities of A and B

(3) Allred and Rochow's equation developed a scale based on effective nuclear charges, the charge on the electron, and the covalent radii of the elements. The Z_{eff} values used are those of John C. Slater.

$X = 0.359\ Z_{eff} / r^2 + 0.744$

where X = electronegativity

Z_{eff} = effective nuclear charge

r = covalent radius

(4) Liu's equation:

$X = 0.313((n + 2.6)/r^{2/3})$

where X = electronegativity

n = number of valence electrons

r = covalent radius

(5) Gordy's equation:

$X = 0.31((n + 1)/r) + 0.50$

where X = electronegativity

n = number of valence electrons

r = covalent radius

PROGRAM RESOURCES

Lesson Plans: Use the Lesson Plan, pp. 25-26, to help you prepare for Chapter 12.

Study Guide: Use the Study Guide worksheets, pp. 41-42, for independent practice with concepts in Section 12.1.

Transparency Master: Use the Transparency master and worksheet, pp. 51-52, for guided practice in the concept "Table of Electronegativities."

Color Transparency: Use Color Transparency 26 to focus on the concept of "Table of Electronegativities."

CONCEPT DEVELOPMENT

Correcting Errors

The greatest difficulty students have is putting the elements in the correct increasing or decreasing order. Encourage them to read the question carefully and then think of a number line that increases or decreases as asked for in the question.

PRACTICE

Guided Practice: Have students work in class on Practice Problems 1-2.
Independent Practice: Your homework or classroom assignment can include Review and Practice 14 from the end of the chapter.

Answers to
Practice Problems: Have students refer to Appendix C for complete solutions to Practice Problems.
1. indium, antimony, selenium, fluorine
2. francium, zinc, gallium, germanium, phosphorus

 ASSESSMENT

Oral: Ask students to name and locate on the periodic table the elements that have the highest and the lowest electronegativities.

GLENCOE TECHNOLOGY

Videodisc

Chemistry: Concepts and Applications

Electronegativity
Disc 1, Side 2, Ch. 12

Also available on CD-ROM.

Bonding
Disc 1, Side 1, Ch. 15

Also available on CD-ROM.

PRACTICE PROBLEMS

Arrange the following elements in order of increasing attraction for electrons in a bond.

1. antimony, fluorine, indium, selenium
2. francium, gallium, germanium, phosphorus, zinc

Bond Character

Electrons are transferred between atoms when the difference in electronegativity between the atoms is quite high. If the electronegativity difference between two reacting atoms is small, we might expect a sharing of electrons. At what point in electronegativity difference does the changeover occur? The answer is not simple. For one thing, the electronegativity of an atom varies slightly depending upon the atom with which it is combining. Another factor is the number of other atoms with which the atom is combining. Therefore, a scale showing the percent of transfer of electrons (percent ionic character) has been constructed, Table 12.3. The amount of transfer depends on the electronegativity difference between two atoms.

Figure 12.3 When magnesium burns, it reacts with oxygen to form the ionic compound magnesium oxide.

When two atoms combine by a transfer of electrons, ions are produced. The attraction of the oppositely charged ions holds them together. When two elements combine by electron transfer, they form an ionic bond. If two elements combine by sharing electrons, they form a covalent bond.

From Table 12.3, we see that two atoms with an electronegativity difference of about 1.67 would form a bond that is 50% ionic and 50% covalent. For our purposes, we will consider an electronegativity difference of less than 1.67 as indicating a covalent bond. A difference of 1.67 or greater indicates an ionic bond.

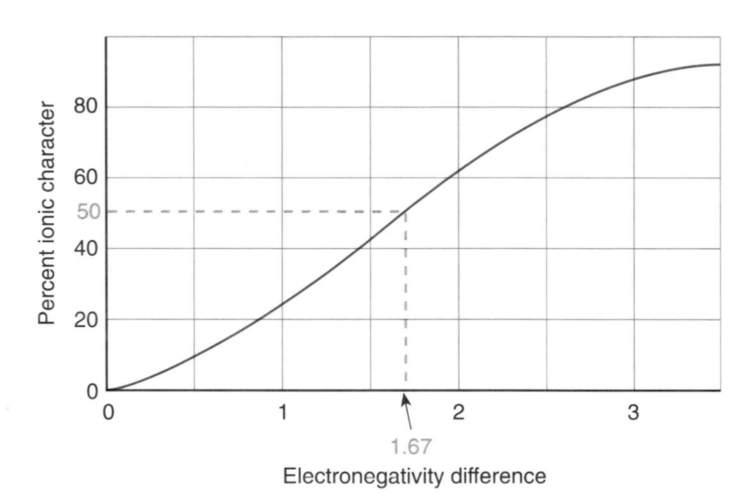

Figure 12.4 The percentage of ionic character of a bond between two atoms tends to increase as the difference in the electronegativities of the two elements increases.

DEMONSTRATION

12-2 Percent Ionic Character

Purpose: to demonstrate that reaction products can have a different percent of ionic character, which results in a difference in properties

Materials: candle, matches, 400-cm^3 beaker, ice cube, beaker tongs

Procedure: 1. Light a candle. **CAUTION:** *Fire risk.* As the candle burns, remind students that the gaseous products of the complete combustion of a hydrocarbon are CO_2 and H_2O.
2. Using beaker tongs, hold the inverted beaker over the candle flame. Place an ice cube on the bottom of the inverted beaker. After a few minutes have a student inspect the inside of the beaker and report the observations to the class.

Results: The water vapor condenses on the cool surface held near the

Table 12.3

Character of Bonds										
Electronegativity Difference	0.00	0.65	0.94	1.19	1.43	1.67	1.91	2.19	2.54	3.03
Percent Ionic Character	0%	10%	20%	30%	40%	50%	60%	70%	80%	90%
Percent Covalent Character	100%	90%	80%	70%	60%	50%	40%	30%	20%	10%

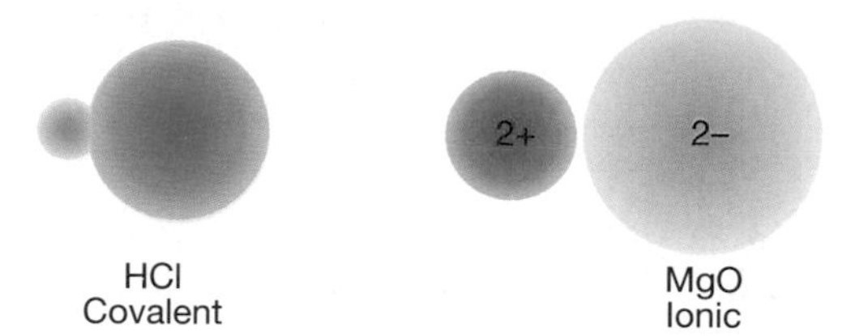

Figure 12.5 Covalent molecules, such as HCl, are held together by a mutual sharing of electrons as indicated by the overlap of the electron clouds. Ionic substances, such as MgO, are held together by an electrical force of attraction between oppositely charged ions.

Actually, unless the two atoms bonded are identical, all bonds have some ionic and covalent characteristics.

Think of magnesium reacting with oxygen. The difference in their electronegativities is 2.27 (| 3.50 − 1.23 |). We would predict the formation of an ionic bond between these two elements. On the other hand, consider boron and nitrogen. The difference in their electronegativities is 1.06 (| 3.07 − 2.01 |). We would predict that boron and nitrogen will form a covalent bond.

SAMPLE PROBLEM

Determining Bond Type

Decide whether an ionic bond or a covalent bond will form between the atoms for each of the following pairs of elements: B–P, Be–Si, C–Na, Li–O, Mg–N.

Solving Process:

The difference in electronegativity can be used to determine whether two atoms will form an ionic bond or a covalent bond. Therefore, use the values of the electronegativity found in Table 12.1 to find absolute value of the difference between the pairs of atoms listed above.

B–P: | 2.01 − 2.06 | = 0.05
Be–Si: | 1.47 − 1.74 | = 0.27
C–Na: | 2.50 − 1.01 | = 1.49
Li–O: | 0.97 − 3.50 | = 2.53
Mg–N: | 1.23 − 3.07 | = 1.84

From Table 12.3, we can see that differences in electronegativities of 1.67 or greater lead to the formation of ionic bonds. Differences in electronegativities less than 1.67 lead to the formation of covalent bonds. Therefore the element pairs B–P, Be–Si, and C–Na will form covalent bonds and the element pairs Li–O and Mg–N will form ionic bonds.

CAREERS IN CHEMISTRY

Electrician

An electrician uses knowledge of the movement of electrons through metal in order to wire houses and businesses for many purposes, including lighting and heating. An electrician must have a high school diploma, take additional work in apprentice classes, and serve an apprenticeship. Electricians in most states are required to pass competency tests in order to gain a license. In high school, future electricians should take physics, chemistry, and vocational courses.

flame, whereas the CO_2 remains a gas.

Questions: 1. What are the products of complete hydrocarbon combustion? *CO_2 and H_2O.*
2. Which product condenses on the inside of the beaker? *H_2O*
3. Use Table 12.3 and the electronegativities from Table 12.1 of C, O, and H, to determine the electronegativity differences for the H—O bond and the C—O bond. *C—O 1.00, H—O 1.30*
4. What is the percent of ionic character for each bond? *C—O about 22%, H—O about 36%.*
5. Based on these calculations, explain why water condenses and CO_2 does not. *The water molecules attract each other more and are easily condensed to a liquid.*

CONCEPT DEVELOPMENT

- Emphasize that the character of the bonds—ionic to covalent—is on a continuum. The use of 1.67 as the breaking point is simply a convenience for working with compounds. Most covalent bonds will have some ionic character and vice versa.

CAREERS IN CHEMISTRY
Electrician

Background: A knowledge of high school chemistry is useful to the electrician, but further college courses are not necessary.

For More Information: Contact the International Brotherhood of Electrical Workers, 1125 15th Street NW, Washington, DC, 20005.

Correcting Errors

To keep the introduction to bonding simple, emphasize that an electronegativity difference of 1.67 separates the ionic from the covalent bond type. The polar covalent bond will be introduced in Chapter 14. Encourage students to interpolate the numbers on Table 12.3 to predict closer percentages.

GLENCOE TECHNOLOGY

Software

Mastering Concepts in Chemistry
Unit 9, Lesson 1
Electronegativity and Bond Types

PROGRAM RESOURCES

Enrichment: Use the worksheet "Electronegativity," pp. 23-24, to challenge more able students.

Reteaching: Use the Reteaching worksheet, p. 18, to provide students another opportunity to understand the concept of "Types of Bonds."

Applying Scientific Methods in Chemistry: Use the worksheet, "Electronegativity," pp. 28-29, to help students apply and extend their lab skills.

PRACTICE

Guided Practice: Guide students through the sample problem, then have them work in class on Practice Problems 3-4.

Independent Practice: Your homework or classroom assignment can include Review and Practice 15 from the end of the chapter.

Answers to
Practice Problems: Have students refer to Appendix C for complete solutions to Practice Problems.

3. **a.** covalent **b.** ionic **c.** covalent **d.** covalent **e.** covalent **f.** covalent **g.** ionic **h.** covalent **i.** ionic

4. **a.** covalent **b.** covalent **c.** ionic **d.** covalent **e.** covalent **f.** ionic

CONCEPT DEVELOPMENT

Discussion: Discuss with students that high melting points are not exclusive to ionic substances. Metals have high melting points, as do macromolecular solids. In both cases the high melting points result from the three-dimensional attractions of each atom for its neighbors. The same factor occurs for ionic solids as well.

- Point out that infrared spectroscopy is one of many experimental processes that can be used to determine molecular structure. IR spectroscopy is very useful as an identification process.

✓ ASSESSMENT

Oral: Ask several students to contribute to a list of properties that characterize compounds with an ionic bond. Properties may include:

1. has a high melting point.
2. conducts electricity in molten state.
3. is soluble in water.
4. has sharply defined crystals.

PRACTICE PROBLEMS

3. Classify the bonds between the following pairs of atoms as principally ionic or covalent.

a. Al—Si **b.** Ba—O **c.** C—H **d.** Li—S **e.** Ca—P **f.** B—Na **g.** Ca—Cl **h.** F—S **i.** Br—Rb

4. For each atom pair listed below, decide whether an ionic or a covalent bond would form between the elements.

a. hydrogen-iodine **b.** astatine-beryllium **c.** cobalt-fluorine **d.** chlorine-tellurium **e.** bromine-cerium **f.** calcium-fluorine

Ionic Bonds

RULE OF THUMB

Sodium chloride is an excellent example of an ionic compound; that is, a compound with an ionic bond. *Ionic compounds are characterized by high melting points and the ability to conduct electricity in the molten state. They tend to be soluble in water and usually crystallize as sharply defined particles.*

Most of the properties of ionic compounds are best explained by assuming a complete transfer of electrons.

$$2Na^{\cdot} + :\ddot{\underset{..}{Cl}}:\ddot{\underset{..}{Cl}}: \rightarrow 2Na^{+} + 2:\ddot{\underset{..}{Cl}}:^{-}$$

If a chloride ion and a sodium ion are brought together, there will be an attractive force between them. If the ions are brought almost into contact, the force will be great enough to hold the two ions together. The electrostatic force that holds two ions together due to their differing charges is the **ionic bond.**

Elements can be assigned oxidation numbers for ionic bonding. Sulfur, for example, with six electrons in the outer level, will tend to gain two electrons. Thus, it attains the stable octet configuration. The oxidation number of sulfur for ionic bonding is 2−. The negative two is sulfur's electric charge after it gains two electrons.

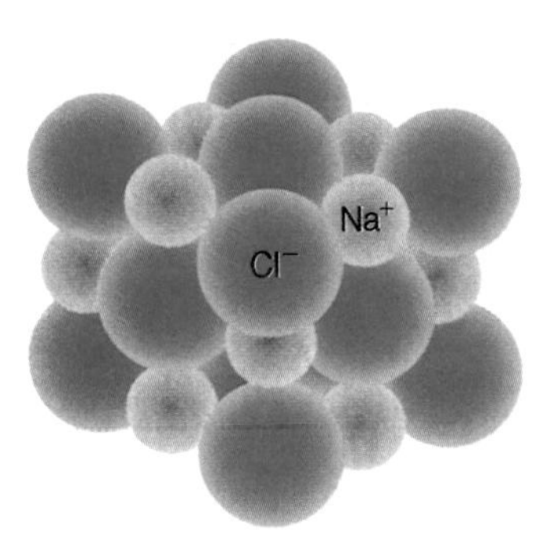

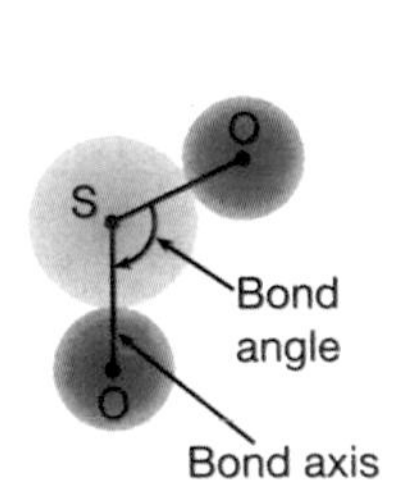

Figure 12.6 Sodium chloride exists as an array of sodium and chloride ions (left) while sulfur dioxide exists as molecules, each of which is composed of one sulfur and two oxygen atoms (right).

DEMONSTRATION

12-3 Molecular Aerobics

Purpose: to help students conceptualize the variety of motions that molecules exhibit as their bonds absorb energy

Materials: 1 m copper or nichrome wire, pencil, protractor, 3 2-inch diameter plastic foam balls, white glue

Procedure: **1.** A plastic foam model of water should be made a day before you plan to use it to allow the glue time to dry.
2. Make two metal springs by tightly winding copper or nichrome wire around a pencil. The springs should be about 3 cm long.
3. A good molecule to model is water. The bond angle is 104.5°. Glue the springs to the plastic balls where you have made slight indentations using the pencil eraser. Allow the model to set overnight.

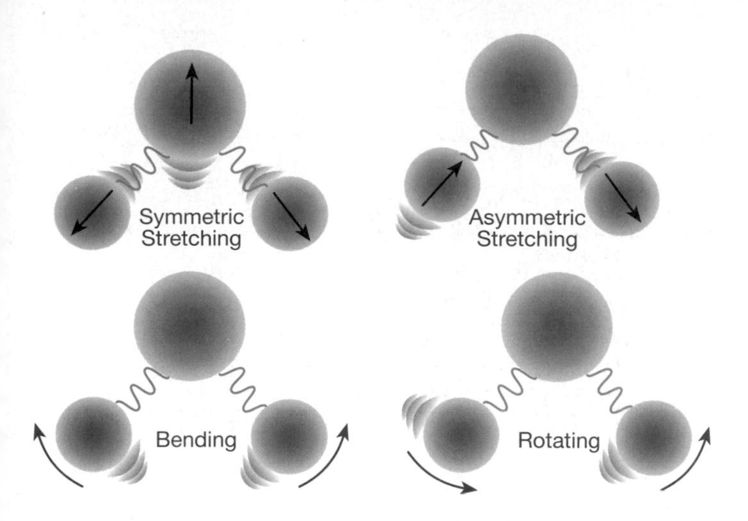

Figure 12.7 Atoms in molecules exhibit a variety of motions. Therefore, bond lengths and bond angles should be considered as average values.

Covalent Bonds

Atoms with the same or nearly the same electronegativities tend to react by sharing electrons. The shared pair or pairs of electrons constitute a **covalent bond.** Most covalent compounds are formed between atoms of nonmetals. *Covalent compounds typically have low melting points, do not conduct electricity, and are brittle.*

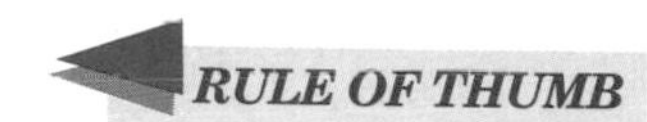

When two or more atoms bond covalently, the resulting particle is called a **molecule.** The line joining the nuclei of two bonded atoms in a molecule is called the **bond axis** as shown on the right of Figure 12.6. If one atom is bonded to each of two other atoms, the angle between the two bond axes is called the **bond angle.** The distance between nuclei along the bond axis is called the **bond length.** This length is not really fixed, because the bond acts much as if it were a stiff spring. The atoms vibrate as though the bond were alternately stretching and shrinking as shown in Figure 12.7.

Bonds also undergo bending, wagging, and rotational vibrations. These movements cause the bond angles and bond lengths to vary. The amplitudes of these vibrations are not large and the bond lengths and bond angles that we measure are, therefore, average values.

We have learned much of what we know about the structure of molecules from infrared and microwave spectroscopy. These methods of spectroscopy measure the amount of infrared or microwave radiation that a sample transmits or absorbs at different wavelengths. Infrared (IR) radiation has wavelengths of 700 nm to 50 000 nm and lies in a region of the electromagnetic spectrum between microwaves and visible light waves. The wavelengths of microwave radiation are typically between 50 000 nm and 30 cm, longer than infrared wavelengths.

HOME ECONOMICS CONNECTION

Food Frequencies

Microwave ovens operate by exposing food to frequencies (about 3 GHz) corresponding to the rotation of water molecules. Water molecules absorb this radiation and increase their rotational energy. The increased energy leads to a rise in the temperature of the food through molecular collisions.

CONCEPT DEVELOPMENT

QUICK DEMO **Melting Point Comparison**

Solid shortening is an example of a substance that is covalently bonded. Table salt has ionic bonds. Compare the melting points of these two substances by placing a small amount on the end of a metal spatula and placing it in the flame of a laboratory burner. The salt does not melt but the solidified vegetable oil does. Point out that ionic bonds give substances the property of high melting points.

Computer Demo: *Simulation of the Infrared Spectrum of HCl,* AP1303, Project SERAPHIM

Enrichment: Have a student investigate the methods used by chemists for determining bond lengths. Possible responses: X-ray diffraction of crystals, electron diffraction of gases, neutron diffraction of crystals, microwave spectroscopy are used by chemists to determine bond lengths. From the energy absorbed, the moment of inertia of the molecule can be calculated. Then, using known atomic masses, bond lengths can be calculated.

HOME ECONOMICS CONNECTION
Food Frequencies

Exploring Further: Ask students to share their favorite microwave recipes with the class. You may want to collect them and publish a chemistry class microwave cookbook of favorite recipes.

4. Use the model to simulate both symmetric stretching and asymmetric stretching as shown in Figure 12.7. Demonstrate the bending mode, and then the rotational mode by rotating about the model its center of mass. Show the translational mode by simply moving the model in a straight line. Show how the length of the bond changes as the atoms vibrate and angle of the bond changes as the atoms wag. **Disposal:** F

Results: The IR spectrum indicates energy changes in the bonding between particles of the molecule. At specific frequencies the atoms of the molecule stretch, twist, wag, and bend around the bonds joining them.

Questions: 1. Is the length of a covalent chemical bond constant? *No*

2. Is the angle constant between two chemical bonds in a molecule? *No*

3. If you answered *no* to either of the above questions, explain what is meant when a bond length or a bond angle is measured and reported. *Published bond angles or lengths are average values because the bond acts much as if it were a stiff spring.*

CULTURAL DIVERSITY

Ancient Steel of Tanzania

Two thousand years ago, Africans living on the Western shores of Lake Victoria were producing carbon steel using a method more sophisticated than any European technique in use before 1850. This process has been passed through oral tradition to the men of the Haya tribe, and was practiced by some of the older men as recently as 50 or 60 years ago. The Haya do not use the ancient process today because cheap, imported iron tools are readily available. Members of the tribe did reconstruct a traditional furnace in 1976 for a visiting anthropologist named Peter Schmidt.

The furnace looks like an inverted cone. It is lined with mud taken from a termite mound. Termites build their mounds to be water resistant, so the mud is full of alumina and silica. Blow pipes are used to supply preheated air into the base of the furnace, creating higher furnace temperatures and better fuel economy than the more recent cold-blast European smelting processes. A chemical reaction causes carbon from wood to penetrate the iron ore when the iron is added to the furnace. This causes large iron crystals to form—a unique process that is closer to semiconductor technology than traditional iron smelting techniques.

Archeological research has unearthed several similar furnaces dated to 2000 years ago, and tests of the slag, or fused refuse from smelting, shows it was formed at the extremely high temperatures (1350 - 1400°C) the Haya furnace produces.

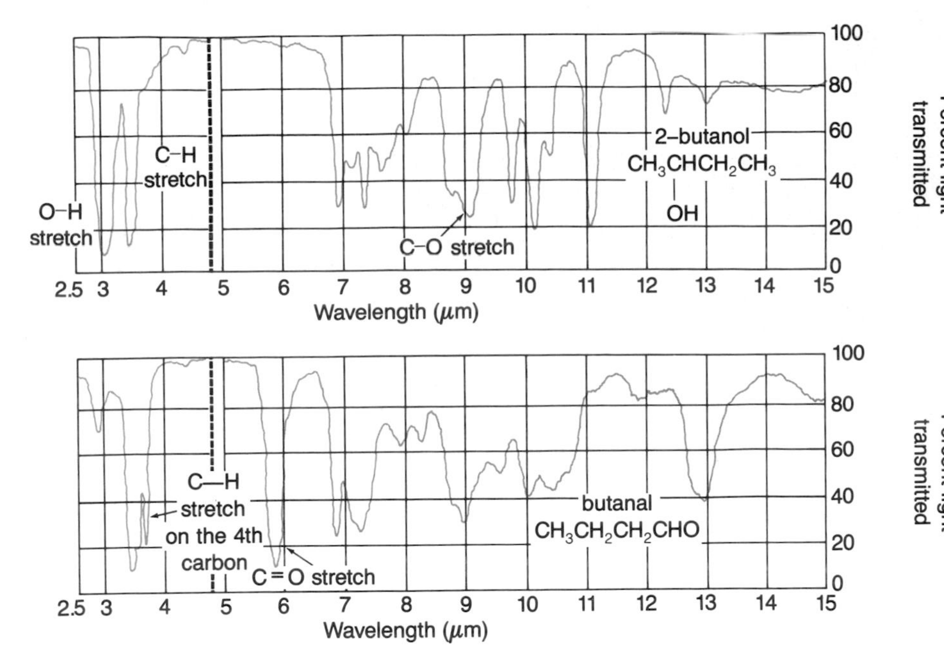

Figure 12.8 For the infrared spectra shown, note that the composition of both compounds is similar. However, the difference in the way the atoms are arranged causes different stretching motions in each molecule. Characteristic peaks in each spectrum can be used to identify the compound.

A molecular compound can often be identified by the infrared radiation it absorbs or transmits. Each molecular compound has a unique infrared spectrum. A sample of the compound is subjected to various wavelengths of IR radiation. The absorbance of IR radiation by the compound at each wavelength is measured and recorded graphically. Comparison with known spectra can be used to identify the compound, just as fingerprints can be used to reveal the identity of a person.

Look at the two spectra in Figure 12.8. The two compounds are not very different, yet the spectra are clearly distinguishable. As you can see, each compound absorbs IR radiation at specific wavelengths, and, therefore, at specific frequencies. The frequencies are determined by the structure of the molecule, the nature of the atoms bonded together in the molecule, and the effects of surrounding atoms. Radiation of wavelengths corresponding to those frequencies can be absorbed because the radiation has the same frequencies as the natural frequencies of vibration of the molecule. From measurements of these absorbances, we can compute bond strengths and determine a great deal about how and where particular atoms are bonded in the molecule.

Microwave radiation, on the other hand, affects the rotation of molecules. The way molecules rotate is determined principally by the mass of the molecule and the distribution of that mass. By substituting a different isotope of an atom in a molecule and seeing how that substitution changes its microwave spectrum, the location of that atom in the molecule can be determined.

A gallery of scanning tunneling microscope (STM) images of atoms is available on the Internet.

World Wide Web
http://www.almaden.ibm.com/vis/stm/gallery.html

CONTENT BACKGROUND

Metals: When metals form solid crystals, there is extensive overlap of the orbitals from adjacent atoms. Due to the Pauli exclusion principle, the interaction of all these orbitals causes a splitting of the outer energy level into an enormous number of extremely closely-spaced energy sublevels. The energy level is essentially altered into a band of energy. Because metals have relatively few electrons in their outer levels, electrons can easily flow anywhere in the crystal. Metals are, therefore, conductors of heat and electricity. In nonmetals, all the outer electrons are tied up in bonding, and they will conduct only if an electron is excited to the next higher energy level, a considerable increase in energy because a bond must be broken and the next level is much higher in energy. The difference in energy between the level occupied by the outer

Metallic Bonds

Metals are characterized by high electrical conductivity, luster (shininess), and malleability. What bonding arrangement could lead to these properties? Metals in the solid state form crystals in which each metal atom is surrounded by either eight or twelve neighboring metal atoms. Since metals have only 1, 2, or 3 outer-level electrons, they do not form normal covalent bonds with all their neighbors. Because all atoms of an element have the same attraction for their electrons, there is no tendency of metal atoms to form ions. Rather, metal crystals form when atoms crowd together and the outer-level orbitals from all those atoms overlap. The electrons can then move easily from one atom to the next. The electrons are said to be delocalized because they are not held in one "locality" as part of a specific ion or covalent bond. If an external electric field is applied, the electrons will flow through the metal, creating an electric current. Delocalized electrons interact readily with light, creating the luster of metals. When metals are pounded with a hammer, the atoms rearrange and the electrons rearrange into the orbitals of the atoms in the new positions. The delocalized electrons holding metallic atoms together constitute the **metallic bond.**

The properties of metals are determined by the number of outer electrons available. Group 1 (IA) metals have only one outer electron per atom. These metals are soft. Group 2 (IIA) metals have two outer electrons and are harder than Group 1 (IA) metals. In the transition elements, however, electrons from the partially filled *d* orbitals may take part in the metallic bond. Many of these metals are very hard.

Groups 3 (IIIB) through 6 (VIB) elements have three through six delocalized electrons. In the elements of Groups 7 (VIIB) through 10 (VIIIB), the number of delocalized electrons remains at six because not all of the *d* sublevel electrons of these elements are involved in the metallic bond. The number of delocalized electrons per atom begins to decrease with the metals of Groups 11

Figure 12.9 Some metals, such as iron, deform like putty during the forging process (left). Sodium, a Group 1 (IA) metal, is so soft it can be cut easily with a knife (right).

CONCEPT DEVELOPMENT

Background: Felix Bloch is credited with the band theory of the energy levels of solids. His theory provided an explanation of semiconductor behavior (Chapter 16).

Computer Demo: *IR Simulator,* Simulates IR spectrophotometer, MC1301, Project SERAPHIM

✓ ASSESSMENT

Oral: Ask several students to contribute to a list of properties that characterize a metallic bond. Properties may include:
1. high electrical conductivity
2. luster
3. malleability

GLENCOE TECHNOLOGY

CD-ROM

Chemistry: Concepts and Applications

Interactive Experiment: Form an Ionic Compound

Videodisc

Chemistry: Concepts and Applications

Ionic and Covalent Compounds
Disc 1, Side 1, Ch. 16

Also available on CD-ROM.

electrons and the level in which they conduct electricity is called the forbidden gap. For metals, there is no forbidden gap because the outer electrons are already in the conduction level. In the nonmetals, the forbidden gap is large. For semiconductors, there is a forbidden gap but it is smaller than that of nonmetals.

PROGRAM RESOURCES

Critical Thinking/Problem Solving: Use the worksheet, p. 12, as enrichment for the concept of "The Octet Rule and Bond Formation."

Chemistry and Industry: Use the worksheets, pp. 19-20, "Dyes: Not Just Blue Jeans and Red Sweaters," as an extension to the concepts presented in Section 12.1.

BRIDGE TO METALLURGY

Alloys

Background: Carbon, a nonmetal, is an essential element in steel. Nitrogen, phosphorus, oxygen, and sulfur may be present as impurities in alloys.

Teaching This Feature: Point out that there are other examples of solid solutions, such as Cu-Au, W-Mo, and Pt-Au. Alloys that are heterogeneous mixtures include Pb-Sn, Cu-Sn, and Ag-Cu.

Connection to Ceramics: Automobile engines today are made of different metal alloys. Research continues on the development of a ceramic engine that will be much lighter than today's automobile engine.

Extension: Have an interested student research what has been published about the Japanese ceramic car engine.

Answers to
Exploring Further

1. Alloys are classified according to solubility of one metal in another and in what proportion.

2. Pressure and temperature are plotted on the axes; the areas under the curve indicate solid, liquid, and vapor phases.

✓ ASSESSMENT

Portfolio: Ask each student to select a metal or alloy that they would like to know more about. Have them prepare a short report that can be presented to the class and then placed in their portfolio. Encourage students to include graphics such as maps, processes, and applications in their reports.

(IB) and 12 (IIB). Going from Groups 13 (IIIA) through 18 (VIIIA), the nonmetals, the metallic properties decrease more rapidly.

The strong metallic bond of our structural metals, such as iron, chromium, and nickel, makes them hard and strong. In general, the transition elements are the hardest and strongest elements. It is possible to strengthen some of the elements that have fewer delocalized electrons by combining them with other metals to form alloys. These alloys have properties different from those of pure elements.

Bridge to *METALLURGY*

Alloys

Many metallic materials are not pure elements. Brass, steel, and bronze are examples. These materials are alloys. An alloy is a metallic material that consists of two or more elements, usually metals.

Some pairs of metals are soluble in each other in all proportions. Alloys made from these pairs produce solid solutions, for example, copper-nickel. Some metal pairs will not dissolve completely in each other; thus, alloys of these pairs are heterogeneous mixtures, such as aluminum-silicon.

The solubility of one metal in another is determined mainly by the relative sizes of the atoms. Metals with atoms of similar size tend to be soluble in each other as are elements whose atoms are very much smaller than the other element.

Steel is an alloy of iron and a nonmetal, carbon, containing up to 2% carbon. Manufacturers often add other elements to steel to give it some special property. Iron is particularly subject to corrosion. By adding some chromium and nickel to the iron and carbon, we obtain stainless steel, which does not rust. Adding tungsten produces a steel that retains its hardness even at high temperatures. It is used to make cutting tools for metalworking. Manganese steels are very hard; they are used to make the jaws on rock crushing machinery and parts of bank vaults. Vanadium produces very tough steel that is used, among other things, to make the crankshafts in automobile engines.

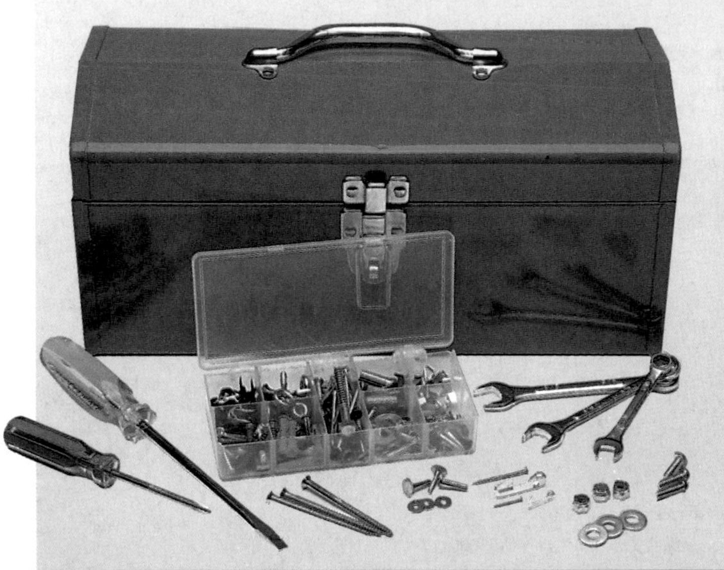

Exploring Further

1. How are alloys classified?
2. Examine a phase diagram for an alloy and learn how to interpret the various areas of the diagram.

DEMONSTRATION

12-4 The Alchemist's Gold

Purpose: to observe atomic motion and the formation of an alloy, brass

Materials: three new pennies, 1 g granular zinc, 5 g NaOH, 25 cm^3 water, porcelain evaporating dish, forceps, paper towels, hot plate

Procedure: 1. Dissolve 5 g NaOH in 25 cm^3 of water in an evaporating dish. **CAUTION:** *NaOH causes severe burns; avoid skin contact. Rinse spills with plenty of water. Wear goggles and apron.*

2. Add 1 g of granular zinc to the NaOH solution. Place the solution on the hot plate and heat to just below the boiling point.

3. Immerse two new, pennies in the NaOH-zinc plating bath. After a minute, use forceps to turn the pennies over. Keep the pennies immersed

Table 12.4

	Chemical Bond Summary				
	Bond Type	**Generally Formed between**	**Bond Formed by**	**Properties of Bond Type**	**Substances Utilizing Bond Type**
INTERATOMIC BONDS	**Covalent**	Atoms of nonmetallic elements of similar electronegativity	Sharing of electron pairs	Stable nonionizing molecules–not conductors of electricity in any phase.	OF_2, C_2H_6, $AsCl_3$, $GeCl_4$, C, SiC, Si
	Ionic	Atoms of metallic and nonmetallic elements of widely different electronegativities	Electrostatic attraction between ions resulting from transfer of electrons	Charged ions in gas, liquid, and solid. Solid is electrically nonconducting. Gas and liquid are conductors. High melting points.	NaCl, K_2O, BaS, LiH, CdF_2, $BaBr_2$, $ErCl_3$, CdO, Ca_3N_2
	Metallic	Atoms of metallic elements	Delocalized electron cloud around atoms of low electronegativity	Electrical conductors in all phases, lustrous, very high melting points.	Na, Au, Cu, Zn, Ac, Be, Gd, Fe, Dy

Polyatomic Ions

There are a large number of ionic compounds made of more than two elements. In these compounds, at least one of the ions consists of two or more atoms covalently bonded. However, the particle as a whole possesses an overall charge.

For example, consider the hydroxide ion (OH^-). $[:\ddot{O}:H]^-$ The oxygen atom is bonded covalently to the hydrogen atom. The hydrogen atom is stable with two electrons in its outer level. The hydrogen atom contributes only one electron to the octet of oxygen. The other electron required for oxygen to have a stable octet is the one that gives the 1– charge to the ion. Although the two atoms are bonded covalently, the combination still possesses charge. Such a group is called a polyatomic ion. Polyatomic ions form ionic bonds just as other ions do. Table 7.4 gives some of the more common polyatomic ions with their charges. Note that the charge of most polyatomic ions is negative. An important positive polyatomic ion is the ammonium ion, NH_4^+.

Figure 12.10 The sulfate ion contains S—O covalent bonds. The electron dot structure shows a stable octet for each atom. However, the particle as a whole has a negative charge.

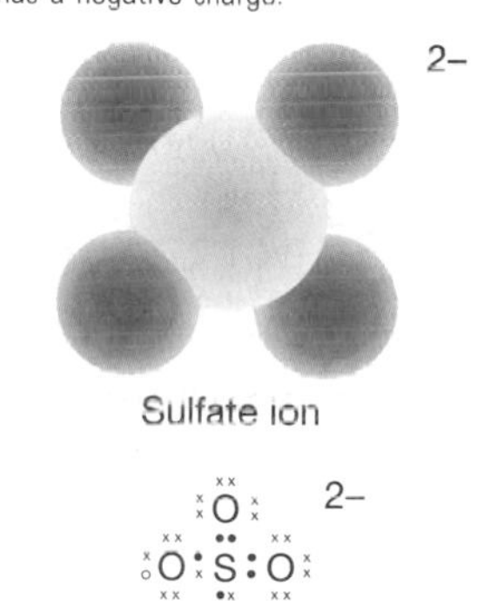

CONCEPT DEVELOPMENT

Using the Table: The Chemical Bond Summary Table 12.4 is a good review and study aid. Encourage students to develop their own summary tables for each chapter.

- Emphasize that although the atoms in a polyatomic ion are bonded covalently, the combination still possesses a charge. Thus, polyatomic ions form ionic bonds.

Background: The term *polyatomic ion* is used throughout this text in place of the term *radical,* which is reserved for the free radical of organic chemistry.

3 ASSESS

CHECK FOR UNDERSTANDING

Ask students to answer the Concept Review problems.

RETEACHING

Have students draw models of substances using dot diagrams and label them as having ionic or covalent bonds. Make magnetic "electrons" out of magnetic craft strips. Students can practice manipulating these on a metal chalkboard.

EXTENSION

Ask a student volunteer to prepare a report on the coordination number of the atom. Possible response: *In complexes, the coordination number is the number of places a ligand is bonded to the central ion. In crystals, it is the number of ions of opposite charge surrounding a given ion. In metals, it is the number of nearest neighbor atoms near the central atom.*

until silver in color.

4. Using forceps, remove the two coins from the hot bath, rinse with water and pat dry with paper towels.

5. Remove the evaporating dish from the hot plate. Place one of the plated pennies on the hot surface. Ask several students to observe and report their observations to the class. In a few seconds the plated coin will return to its original copper color. After a few more seconds the color of the coin appears golden. Remove it from the hot plate and allow it to cool. **CAUTION:** *Do not touch the hot penny.*

6. Use clear tape to attach the three pennies to a card. Pass the card around the room so each student can examine the alloy. **Disposal:** F

Questions: 1. What is an alloy? *A solid solution or heterogeneous mixture of two or more elements, usually metals.*

2. What two metals compose the alloy brass? *Cu and Zn*

3. Why is heat needed to cause the zinc coating to diffuse into the copper? *Heat increases the motion of the atoms, separates and expands the layers of atoms, allowing them to migrate easier.*

Extension: Give the card, with the pennies taped to it, as an award to a student who accomplishes something special in class.

Alien Alloys

Wing components of this aircraft are made of an aluminum-lithium alloy, a material that is stronger than aluminum, but less dense.

Mixing molten metals has been a standard method of manufacturing alloys for centuries. This traditional method has prohibited manufacture of alloys using the more active metals, such as lithium. However, the aluminum industry, realizing the potential uses of Al-Li alloys in the aerospace industry, has designed furnaces and techniques to produce these alloys safely and economically.

Many of the new techniques involve changes in the cooling of the alloy. In standard methods, alloys are cooled slowly, allowing atoms to form regular, crystalline structures within the alloy. What would happen if cooling is accelerated to a rate equivalent to a temperature drop of 100 000 C° in the blink of an eye? The atoms would be frozen in position—producing glasslike materials. This rapid solidification has led to some unusual alloys. One such alloy is made from iron, boron, and silicon and has the magnetic properties of iron, the rigidity of glass, and doesn't corrode. Its use in electrical transformers may save several billion dollars yearly by reducing the waste of electrical energy.

Another technique chemically combines metals to form materials called intermetallic compounds. Unlike alloys, intermetallic compounds have properties that differ significantly from the metals they contain. One such intermetallic compound is aluminum-titanium. Unlike most alloys, this alloy maintains its strength at extremely high temperatures. Your future car may be powered by an efficient and lightweight aluminum-titanium engine.

12.1 CONCEPT REVIEW

5. Describe the periodic nature of electronegativity values of the elements.
6. Compare and contrast ionic, covalent, and metallic bonds.
7. How does the behavior of molecules differ with the absorption of infrared radiation and microwave radiation?
8. **Apply** Is the attraction of the sulfur atom for one of the bonding pairs of electrons in a molecule of SO_3 greater than or less than that of the sulfur atom in a molecule of SO_2? Explain.

FRONTIERS

Alien Alloys

Background: Alloys and intermetallic compounds can be formed by another process called mechanical alloying. In this process, cold metals are pulverized and then mixed together to form the material.

Teaching This Feature: Point out that students probably have alloys in their mouths. Dental amalgams used to fill cavities are alloys of mercury, silver, and zinc. The material is prepared as a liquid but quickly hardens. You may wish to have students discuss the uses and compositions of other traditional alloys with which they may be familiar.

Exploring Further: Other examples of intermetallic compounds include $CuMg_2$, Au_2Bi, and Mg_3Sb_2, and KBi_2. Have students verify that these compounds do not follow the octet rule.

Answers to
Concept Review

5. Electronegativity within a group tends to decrease with increasing period and to increase from metals to nonmetals. The most active metals have the lowest electronegativities. Fluorine has the highest electronegativity of all the elements.

6. Ionic bonds are formed when there is a complete transfer of electrons between two atoms; covalent bonds are formed when electrons are shared by two atoms; metallic bonds are formed when atoms crowd together and the outer level orbitals of these atoms overlap, resulting in delocalized electrons.

7. Absorption of infrared radiation causes a molecule to vibrate more vigorously, while absorption of microwave radiation causes a molecule to rotate faster.

8. The attraction of an atom for bonding electrons decreases when the atom is bonded to two or more other atoms, because the attraction is divided among two or more electron pairs. Therefore, the attraction of the S atom for one of the bonding pairs is less in SO_3 than in SO_2.

✓ ASSESSMENT

Knowledge: Have students use the information provided in Table 12.4 to make a network tree concept map for chemical bonds. The concept map should include ionic, covalent, and metallic bonds. Students should draw linking words from the column heads.

Chemistry Journal

Alloys and Their Uses

Ask students to make a table in their journals with a column for each of several alloys such as stainless steel, bronze, brass, solder, and 14-karat gold. Have students list in each of the columns everyday objects made of that alloy and the way the object is used. Have students add to their tables whenever they observe a new alloy.

12.2 Particle Sizes

> **OBJECTIVES**
> - differentiate among atomic radii, ionic radii, covalent radii, and van der Waals radii.
> - discuss factors that affect the values of ionic radii and covalent radii.
> - use covalent radii to calculate bond lengths.

In trying to make predictions about the reactivity or the properties of a substance, chemists frequently need to know how large the particles are that compose the substance. In Chapter 10, you learned about the concept of atomic radius and its periodic nature. In that chapter you also learned about the comparative sizes of ions and the atoms from which they were formed. Now we need to take a closer look at the sizes of particles in compounds.

Ionic Radii

We saw in Chapter 10 that a sodium ion is smaller than a sodium atom. We also found that a chloride ion is larger than a chlorine atom. Values for the radii of many ions have been determined, and some are listed in Table 10.1. These values are found from a combination of experimental data and simplifying assumptions. By adding the radii of two ions in a compound, we may find their **internuclear distance** in a crystal. How can ionic radii be determined? Consider compounds such as LiI or LiBr in which the negative ions are very much larger than the positive ions. In crystals of these compounds, we can assume that the negative ions are in contact. Therefore, the ionic radii of the negative ions are one-half the internuclear distances of these compounds. Once such radii are determined, the remainder can be determined from internuclear distances of other compounds. Remember that the radii are not fixed values. One reason for their variability is the "fuzziness" of the electron cloud. Another reason is the effect each ion has on its neighboring ions.

Covalent Radii

It is possible, by experiment, to determine the internuclear distance between two bonded atoms. For example, consider iodine(I) chloride, ICl. What are the radii of the iodine and chlorine atoms

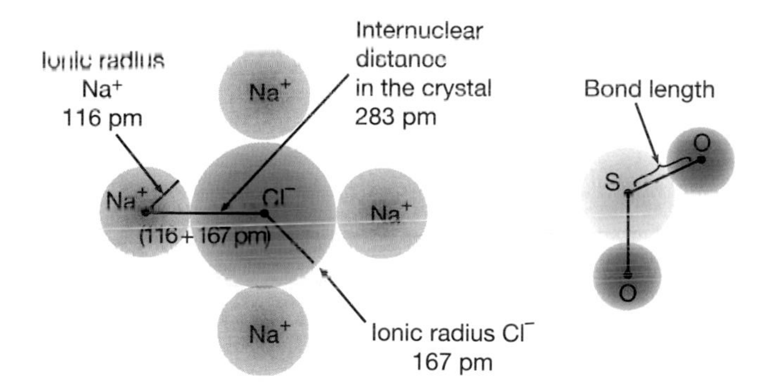

Figure 12.11 In ionic compounds such as sodium chloride (left), the internuclear distance is the sum of the radii of both ions. In a molecule such as sulfur dioxide (right), the bond length is the sum of the covalent radii of both atoms forming the bond.

4 CLOSE

Discussion: Elements that are close together on the periodic table often are covalently bonded; those far apart are ionically bonded. Ask student teams to prove or disprove this generalization with several examples.

Cooperative Learning: Refer to pp. 28T-29T to select a teaching strategy to use with this activity.

12.2 Particle Sizes

PREPARE

Key Concepts

The major concepts presented in this section include ionic and covalent radii, determination of bond length, and van der Waals radii.

Key Terms

internuclear distance
van der Waals radius

Looking Ahead

Obtain a clear plastic box lid for Demonstration 12-6; van der Waals radius.

1 FOCUS

Objectives

Preview with your students the objectives listed on the student page. Students can use them as a study guide for this major section.

PROGRAM RESOURCES

Study Guide: Use the Study Guide worksheet, p. 43, for independent practice with concepts in Section 12.2.

Transparency Master: Use the Transparency master and worksheet, pp. 53-54, for guided practice in the concept "Ionic Bonding and Ionic Radii."

Color Transparency: Use Color Transparency 27 to focus on the concept of "Ionic Bonding and Ionic Radii."

Discussion
Have students open their textbooks to Table 10.1, Atomic and Ionic Radii, in Chapter 10. Have students identify trends in ionic radii within a group and within a period. Review how the atomic and ionic radii compare for both nonmetals and metals.

2 TEACH

CONCEPT DEVELOPMENT

Using the Table: Students need help in conceptualizing that the atom is not a sphere having definite dimensions, but an electron cloud whose average dimensions have been measured. From Table 12.5 have students add the values for boron and chlorine, 82 pm + 99.5 pm = 181.5 pm. Compare this sum to the value obtained from Table 12.6 for BCl_3, 175 pm. Show that the two values differ by about 4%. Stress that the radius of an atom is not constant, but is influenced by the other atoms to which it is bonded.

✓ ASSESSMENT

Skill: Ask students to use Table 12.5 to determine the length of the bond formed between randomly selected pairs of atoms that you name.

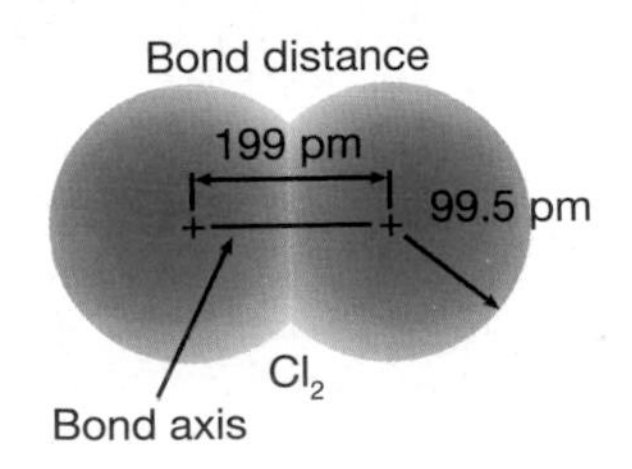

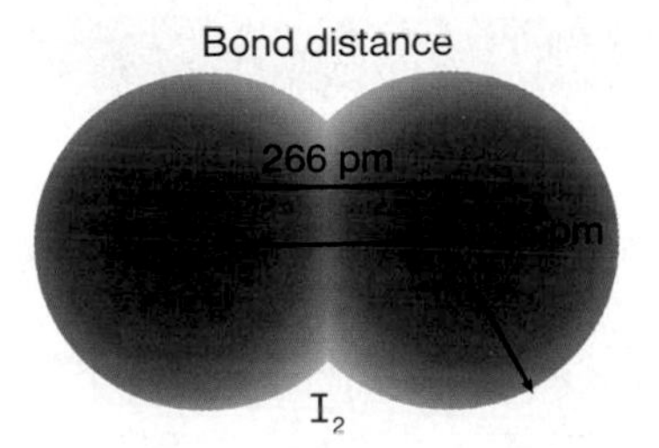

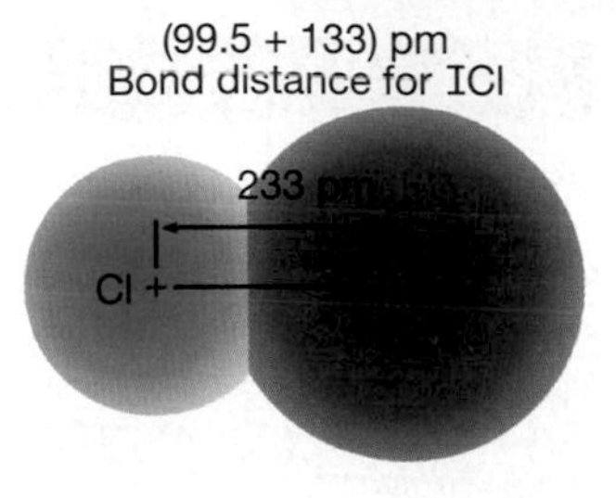

Figure 12.12 The internuclear distance is the sum of the covalent radii of each atom. Recall that the bond distances are not fixed values.

in this molecule? The internuclear distance in ICl is found to be 230 pm. The internuclear distance in Cl_2 is 199 pm, and in I_2 it is 266 pm as shown in Figure 12.12. One-half of each of these values might be taken as the radii of the chlorine and iodine atoms in a covalent bond: 99.5 pm and 133 pm. The sum of the iodine and chlorine radii would then be 233 pm. This sum is in good agreement with the observed bond distance in ICl. Covalent radii are only approximate. The value for hydrogen is less reliable than that for other atoms. Nevertheless, these radii are very useful in predicting bond lengths in molecules. Table 12.5 gives the covalent radii for some common atoms; Table 12.6 gives the bond lengths for some molecules. See for yourself how well the predicted bond lengths agree with the measured ones.

Remember that covalent radii are used to find the internuclear distance between atoms bonded to each other. Like electronegativities, covalent radii are average values. The radius of a particular atom is not constant. Its size is influenced by the other atom or atoms to which it is bonded.

Table 12.5

Covalent Radii (in picometers)*

Atom	Radius	Atom	Radius
Al	118	Li	134
As	120	Mg	130
B	82	N	70
Be	90	Na	154
Bi	150	O	73.5
Br	114	P	110
C	77.2	Pb	146
Ca	174	S	103
Cl	99.5	Sb	140
F	71.5	Sc	144
Ga	126	Se	119
Ge	122.3	Si	117.6
H	37.07	Sn	140.5
I	133	Te	142
K	196	Ti	152

*All digits given are significant.

Table 12.6

Experimental Bond Lengths (in picometers)

Molecule	Bond	Length
BCl_3	B—Cl	175
B_2H_6	B—H	132
$BeCl_2$	Be—Cl	177
Diamond	C—C	154.4
CH_4	C—H	109.4
CH_3I	C—H	109.6
	C—I	213.9
ClBr	Cl—Br	213.8
HF	H—F	91.7
H_2O	H—O	95.7
$LiCl_2(C_4H_8O_2)_2$	Li—O	195
NH_3	N—H	101.7
$OBe_4(CH_3COO)_6$	Be—O	166.6
OF_2	O—F	140.5
O_3	O—O	120.74
H_2SAlBr_3	S—Al	243
$(H_3Si)_2NN(SiH_3)_2$	Si—N	173

DEMONSTRATION

12-5 van der Waals Radius

Purpose: to demonstrate the van der Waals radius

Materials: clear plastic box lid, overhead projector, 3 or 4 marbles

Procedure: 1. Place the plastic box lid on an overhead projector and project its image onto a screen.
2. Add 3 or 4 marbles of different sizes to the lid. Agitate the lid so that the marbles collide. **Disposal:** F

Results: The marbles exhibit random motion and collisions. Point out that the radius of an atom's imaginary rigid shell is called the van der Waals radius. Contrast the van der Waals radius and the covalent radius, which is always shorter for a specific atom. For example, the van der Waals radius of oxygen is 152 pm, whereas the covalent radius is 73.5 pm.

SAMPLE PROBLEM

Bond Length

Predict the length of the bond formed between an atom of arsenic and an atom of sulfur.

Solving Process:
The length of the bond is the sum of the covalent radii of the two elements. From Table 12.5, the covalent radius of arsenic is 120 pm and that of sulfur is 103 pm. Therefore, the bond length is the sum of 120 pm and 103 pm, or 223 pm.

PRACTICE PROBLEMS

9. Predict the length of the bond formed between each of the following pairs of atoms.
a. Al—Cl
b. H—I
c. N—P
d. Se—S
e. B—F

Van der Waals Radii

CHEMISTRY IN DEPTH

A certain minimum distance is maintained between atoms that are not bonded to each other. This limitation exists because the electron cloud of one atom repels the electron cloud of other atoms.

In effect, colliding free atoms and molecules act as if they had a nearly rigid outer shell. This shell limits the closeness with which they may approach other atoms or molecules. As shown in Figure 12.13, bonded atoms are closer than atoms that are not bonded because the covalent bond consists of shared electrons. The radius of this imaginary rigid shell of a nonbonded atom is called the **van der Waals radius.** It is named for the Dutch physicist Johannes van der Waals. Chemists generally assume that an atom's rigid shell is at the point where the probability of finding electrons has dropped below 90 percent.

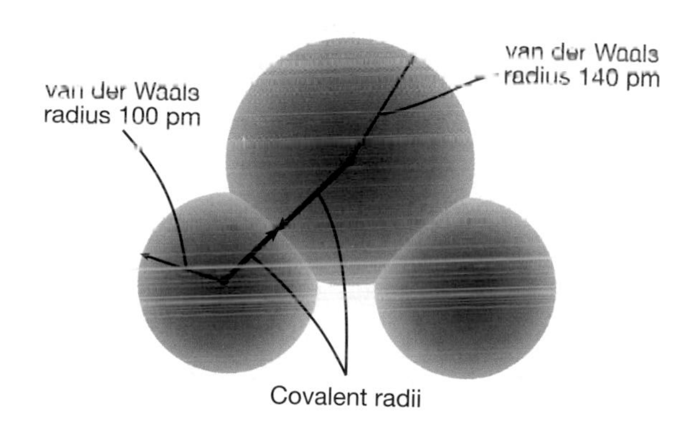

Figure 12.13 The van der Waals radius is the minimum distance between atoms on adjacent molecules. Note that the van der Waals radius will be larger than the covalent radius of a bonded atom.

Questions: 1. Assuming the marbles are atoms, could one atom penetrate the electron cloud of another atom? *No*
2. Do the atoms bond? *No*

Follow Up Activity: Repeat the demonstration using magnetic marbles to simulate atoms that do bond when they approach one another. For dramatic effect, don't let the students see you switch the marbles.

CHEMISTRY IN DEPTH

The section **Van der Waals Radii** on this page is recommended for more able students.

Correcting Errors

The most common error that students make is selecting the wrong element. Remind them that there is an alphabetic list of the elements with their symbols inside the back cover of the textbook. If there is a doubt, have them check to be certain they are using the radius of the correct element.

PRACTICE

Guided Practice: Guide students through the sample problem, then have them work in class on Practice Problem 9.
Independent Practice: Your homework or classroom assignment can include Review and Practice 18 from the end of the chapter.

Answers to
Practice Problem: Have students refer to Appendix C for complete solutions to Practice Problems.
9. a. 218 pm, **b.** 170 pm, **c.** 180 pm, **d.** 222 pm, **e.** 154 pm

3 ASSESS

CHECK FOR UNDERSTANDING

Use Concept Review 10 13 to assess students on mastery of the concepts presented in this section. Review and Practice 17-19 can also be used to check for understanding.

RETEACHING

Have students write their own definitions of ionic radius, covalent radius, and van der Waals radius. Then have them compare their definitions with the ones given in the textbook.

PROGRAM RESOURCES

Laboratory Manual: Use the microlab 12 "Using Conductivity to Predict Bonding" as enrichment for the concepts in Chapter 12.

Transparency Master: Use the Transparency master and worksheet, pp. 55-56, for guided practice in the concept "Covalent Radii and van der Waals Radii."

Color Transparency: Use Color Transparency 28 to focus on "Covalent Radii and van der Waals Radii."

MINUTE LAB

Two Bonds Are Better Than One

Purpose: to demonstrate cross-linking in polymers

Materials: 5 cm^3 white glue, 3-oz disposable plastic cup, 1 cm^3 saturated sodium tetraborate solution (add 4 g sodium tetraborate to 100 cm^3 water), stirring rod, medicine dropper

Procedure Hints: Have students do this lab individually.Make a model of polymer chains using paper clips as the monomers. Use other paper clips to cross-link the two chains together to form a ladder shape.

Results: A synthetic rubber-like ball forms. It will bounce, but the material with time will lose its shape when placed on a flat surface.

Answers to Questions: The glue is a thick liquid, but turns to a solid when cross-linked by the borate. The cross-linking prevents the easy movement of the polymer chains past each other that occurred in the liquid.

✓ ASSESSMENT

Knowledge: Have a student use a dictionary to define the term *vulcanization.*

Answers to
Concept Review

10. The atomic radius is the radius of an atom without regard to surrounding atoms. It is different from the ionic radius, because an ion is charged by the gain or loss of electrons. The atomic radius also differs from the covalent radius and van der Waals radius, which may vary when the subject atom bonds to different atoms.

11. Sum of van der Waals radii = 155 pm + 180 pm = 335 pm

12. Ionic radii are found from a combination of experimental data and simplifying assumptions. Also, there is an effect of each ion on its neighboring ions. Covalent radii depend on the atoms to which the subject atom bonds.

13. Bonding the atom to another atom with a higher electronegativity will decrease the van der Waals radius of the atom. Bonding the atom to another atom with a lower electronegativity will increase the radius.

MINUTE LAB

Two Bonds Are Better Than One

Put on an apron and goggles. Place 5 cm^3 of white glue in a small, disposable plastic cup. Stir in about 1 cm^3 of a saturated solution of sodium tetraborate, one drop at a time, until a sticky lump forms. Remove the lump from the cup and thoroughly rinse it under running tap water. Roll the lump into a ball. Can you bounce the ball? Set the ball on a flat surface and let it remain there undisturbed for 5 minutes. What happens to the ball?

White glue consists of long, chainlike molecules called polymers. When the glue and the sodium tetraborate solution are mixed, the borate ions cross-link the polymers in the glue like rungs on a ladder. The borate ion can form cross-links because it has two bonding sites, one at each end of the ion. Conjecture how cross-linking could account for the properties of the material you just made.

Table 12.7

van der Waals Radii (in picometers)			
Atom	**Radius**	**Atom**	**Radius**
As	185	O	152
Br	185	P	180
C	170	Pb	202
Cl	175	S	180
F	147	Se	190
Ga	187	Si	210
H*	120	Sn	217
I	196	Te	206
N	155		

*The van der Waals radius for hydrogen when it is hydrogen-bonded (Section 17.2) is 100 pm.

Summary of Radii

Thus far, we have studied four radii—atomic, ionic, covalent, and van der Waals. How do these various measurements differ? How are these measurements the same? Atomic radii are measured on atoms in crystals.

Ionic radii differ from atomic radii because of the loss or gain of electrons. This difference was discussed in Chapter 10. Since most ionic radii are determined from ionic crystals, their values are consistent with the data. Covalent radii and van der Waals radii are quite variable due to the wide range of atoms to which the subject atom may be bonded. We would expect covalent radii to be less than an atomic radius. However, if an atom is bonded to more than one other atom, its electron cloud may be distorted. The distortion may make its covalent radius larger than the atomic radius. The same situation occurs with van der Waals radii. For both of these radii, the data given in the tables in this chapter represent average values for the atom bonded to its usual number of neighboring atoms. In every case, we can use the radii to predict the internuclear distance between atoms.

12.2 *CONCEPT REVIEW*

10. Explain how an atomic radius differs from each of the following: an ionic radius; a covalent radius; a van der Waals radius.

11. What is the minimum internuclear distance for a collision between a nitrogen molecule and a phosphorus molecule?

12. Why are the values of ionic and covalent radii approximate values?

13. Apply Select an atom from those listed in Table 12.7. Then choose two elements that will each covalently bond with the atom that you selected. Choose the two elements so that one decreases the van der Waals radius of the atom that you selected and the other increases its van der Waals radius. Explain how you made your two choices.

Chemistry Journal

Analogy

Have students write an analogy in their journals to describe changes in atomic radii. For example, just as a building becomes higher as additional floors are added, the atoms of the elements in a group become larger as additional energy levels are added.

PROGRAM RESOURCES

Vocabulary Review/Concept Review: Use the worksheets, p. 12 and pp. 57-58, to check students' understanding of the key terms and concepts of Chapter 12.

Evaluation: Use the Evaluation worksheets, pp. 45-48, to assess students' knowledge of Chapter 12.

Computer Test Bank: Use the IBM or Apple Test Bank and Question Selection Manual to prepare your own individualized test for Chapter 12.

Chapter 12
REVIEW

Summary

12.1 Bond Formation

1. The relative tendency of a bonded atom to attract shared electrons to itself when it is bonded to another atom is called electronegativity.
2. Ionic bonds are formed by transfer of electrons between atoms with a large difference in electronegativity.
3. Ionic compounds are characterized by high melting points, solubility in water, and crystal formation.
4. Covalent bonds are formed by the sharing of electrons between atoms with either no difference or slight differences in electronegativity.
5. The bond axis is a line joining the nuclei of two bonded atoms. The length of the bond axis is called the bond length. The angle between two bond axes is called the bond angle.
6. A metallic bond is formed between atoms with few electrons in the outer level. These electrons circulate as delocalized electrons and allow metals to carry an electric current.
7. Polyatomic ions are composed of groups of atoms bonded covalently and possess an overall charge just as other ions.

12.2 Particle Sizes

8. The ionic radius is the best estimate chemists can make of the effective size of an ion.
9. Covalent radii can be used to predict the distance between bonded atoms.
10. Electron clouds repel each other strongly when two nonbonding atoms approach each other. The distance of closest approach for an atom is called the van der Waals radius of the atom.

Key Terms

electronegativity	bond angle
ionic bond	bond length
covalent bond	metallic bond
molecule	internuclear distance
bond axis	van der Waals radius

Review and Practice

14. Based on its location on the periodic table, predict which element in each of the following pairs has the greater electronegativity.
    - **a.** Rh — Ru
    - **b.** Ga — Ti
    - **c.** Cs — Sr
    - **d.** Al — Ti
    - **e.** Mo — Zr
    - **f.** Co — Re
15. Characterize the bond between the following pairs of elements as principally ionic or covalent.
    - **a.** Li and O
    - **b.** Se and F
    - **c.** Sr and Br
    - **d.** Ca and S
    - **e.** Cr and Cl
    - **f.** Mn and S
    - **g.** C and Br
    - **h.** Zn and I
16. What force holds ions together in ionic bonds? What causes this force?
17. Why is it difficult to give exact values for bond lengths and bond angles?
18. Using Table 12.5, predict the bond lengths indicated for the following substances.
    - **a.** F—F in F_2
    - **b.** C—Pb in $Pb(C_2H_5)_4$
    - **c.** Li—P in Li_3P
    - **d.** Rb—Si in Rb_4Si (Rb = 198 pm)
    - **e.** C—C in CH_3CH_3
    - **f.** N—O in N_2O_4
19. How does van der Waals radius differ from covalent radius?
20. List three characteristics of metals.
21. Why are the alkali metals and the alkaline earth metals very soft metals? How could they be strengthened?

4 CLOSE

Art Activity: After students have studied chemical bonding and the associated radii, have them convey their impressions of two atoms in each type of bond using color, form, and modeling to emphasize differences in bond type and radii. Have an art show in the chemistry room to display the students' works.

CHAPTER REVIEW 12

- Review Summary statements and Key Terms with your students.
- Complete solutions to Chapter Review Problems can be found in **Merrill Chemistry Problems and Solutions Manual.**

Answers to
Review and Practice

14. a. Rh, **b.** Ga, **c.** Sr, **d.** Al, **e.** Mo, **f.** Co

15. If the difference in electronegativities is 1.67 or greater, the bond is principally ionic; otherwise, it is principally covalent.
a. ionic, **b.** covalent, **c.** ionic, **d.** covalent, **e.** covalent, **f.** covalent, **g.** covalent, **h.** covalent

16. Ions are held together in an ionic bond by electrostatic force caused by differing charges on ions.

17. The values vary because of the constant motion of the atoms.

18. a. 71.5 pm + 71.5 pm = 143.0 pm, **b.** 223 pm, **c.** 244 pm, **d.** 316 pm, **e.** 154 pm, **f.** 144 pm

19. Van der Waals radius is the distance between the nuclei of nonbonded adjacent atoms, and covalent radius is the distance between the nuclei of bonded atoms.

20. Metals exhibit malleability, ductility, conductivity, luster, hardness, high density, and high melting point.

21. They have only 1 or 2 delocalized electrons per atom; they form alloys.

Answers to
Review and Practice

22. N and O are covalently bonded in the polyatomic ion, NO_3^-. An ionic bond exists between Na+ and NO_3^-.

23. See graph below.

24. $1s^22s^22p^4$ = O, 3.50, 73.5 pm
$1s^22s^22p^63s^1$ = Na, 1.01, 154 pm
$1s^22s^22p^63s^23p^5$ = Cl, 2.83, 99.5 pm

25. Electronegativity difference = 2.06 − 1.23 = 0.83, 16% ionic, 84% covalent

Answers to
Concept Mastery

26. Nonmetals with high electronegativities and metals with low electronegativities are the most active.

27. Similar electronegativities mean similar attractions for electrons. The electrons will be shared by the two atoms, not "captured" by one atom.

28. The difference in the electronegativities of Ca and Cl is ≥ 1.67, but the difference in eletronegativities for P and Cl is < 1.67. When a difference of electronegativity is less than 1.67, the bond is considered covalent.

29. The more delocalized electrons there are, the more enhanced are the metallic properties.

30. Delocalized electrons are free to flow under the influence of an external electric field.

31. Chlorine's attraction for electrons is so much greater than that of sodium, that chlorine "captures" an electron from sodium, and two ions are formed. Carbon and chlorine attract electrons similarly, and thus must share them.

32. By definition, a molecule contains covalent bonding.

33. Electronegativity is influenced by nuclear charge, distance between nucleus and outer electron level, shielding, and sublevel filling. Although the nuclear charge of K is greater than that of F, all other factors favor a greater electronegativity for F; F has smaller atomic radius, fewer electrons shielding the nucleus, and lacks one electron to fill a sublevel.

34. As distance from the nucleus is increased, the nucleus exerts less attraction for an electron. As other electrons are between the nucleus and outer electrons, their repulsion for another electron offsets the attraction of the nucleus.

35. They are the same. The absolute value of the ion charge equals the number of transferred electrons.

36. Since each atom of a metal has the same attraction for electrons as does any other atom of the metal, no ionic or covalent bond is formed. Atoms are free to move from their positions to other positions because of this loose bonding.

22. Explain how $NaNO_3$ is an example of two different types of bonding.

23. Construct a graph of the number of delocalized electrons versus the atomic number for elements Z = 21 through Z = 30.

24. Three elements have electron configurations of $1s^22s^22p^4$, $1s^22s^22p^63s^1$, and $1s^22s^22p^63s^23p^5$. Their electronegativities (not in the same order) are 1.01, 2.83, and 3.50. The covalent radii in picometers are 73.5, 99.5, and 154. Identify each element and match with the correct electronegativity and covalent radius.

25. Use Tables 12.1 and 12.3 to estimate the percent ionic character and the percent covalent character of a bond between a magnesium atom and a phosphorus atom.

Concept Mastery

26. How is the chemical activity of elements related to their electronegativities?

27. Use the definition of electronegativity to explain why small differences in electronegativity between two atoms result in the formation of a covalent rather than an ionic bond between the atoms.

28. Considering your study of bond character and its relationship to electronegativity, how should you interpret the statement that $CaCl_2$ is an ionic compound while PCl_3 is a covalent compound?

29. How does the number of delocalized electrons in a metal affect its properties?

30. Explain how the nature of the metallic bond accounts for electrical conductivity of metals.

31. Without using the term electronegativity, formulate a statement to explain why sodium and chlorine atoms form an ionic bond, while carbon and chlorine atoms form a covalent bond.

32. Why is the term *ionic molecule* incorrect?

33. What factors influence electronegativity? Use your answer to explain the difference in electronegativities for fluorine and potassium.

34. Examine the way electronegativity changes as you look down through Group 1 (IA), the alkali metals, in Table 12.1. Explain how distance from the nucleus and the shielding effect may contribute to this change in electronegativity.

35. How does the oxidation number of an element relate to its charge in an ionic bond? How does the ion charge relate to the number of electrons transferred when the atom becomes an ion?

36. Many metals will stand a great deal of distortion without breaking. They can be bent, folded, rolled thin into sheets of foil, drawn through progressively smaller dies into thin wire, and can be beaten into various shapes. How does the electron structure of metals account for these properties?

Application

37. Each of the following diagrams indicates a type of radius. Tell whether each radius indicated is atomic, covalent, ionic, or van der Waals.

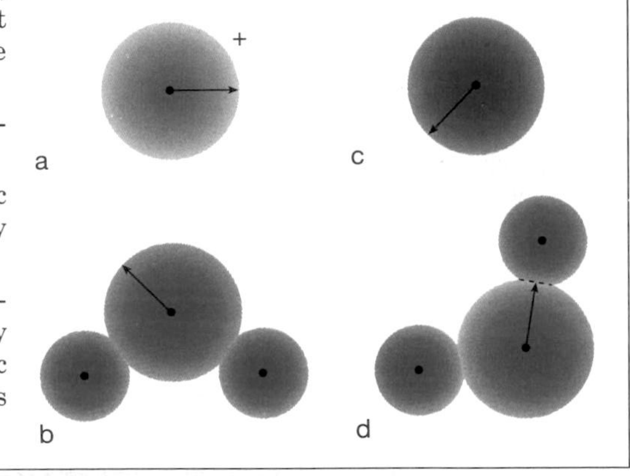

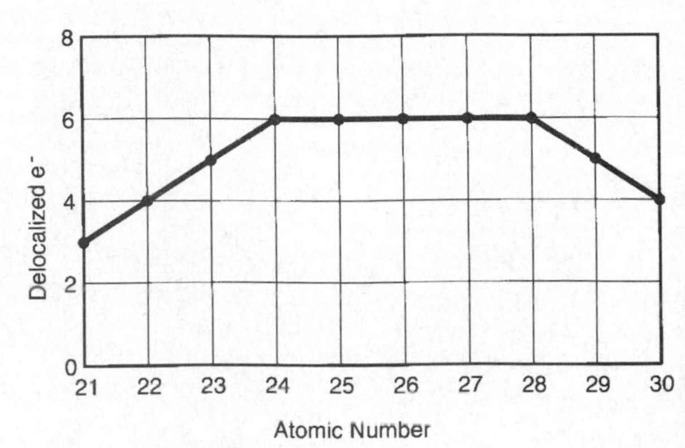

38. Think of ways different materials are used in building houses. For each of the following, give two examples of places where this type of substance, considering its properties, should be used? Why?
   a. a covalent compound
   b. an ionic compound
   c. a metal

39. Draw electron dot structures for the following polyatomic ions. Use dots to indicate the electrons from one type of atom, *x*'s to indicate electrons from the other type of atom, and *o*'s to indicate any electrons that are gained to stabilize the total electron configuration.
   a. NH_4^+ **b.** OH^- **c.** $C_2O_4^{2-}$ **d.** O_2^{2-}

42. Many pure metals are good conductors of electric current. However, the same metals are less conductive when in an unrefined or crude state. Explain this change in behavior.

43. The general trend in electronegativity from left to right across the periodic table is an increase. The trend down a column is a decrease. Explain each of the following observations.
   a. Zinc has an electronegativity of 1.66 and copper, which precedes it, a value of 1.75.
   b. Gallium has an electronegativity of 1.82 and aluminum, which precedes it, a value of 1.47.

Critical Thinking

40. Why are most ionic compounds brittle?

41. The metals copper, zinc, silver, cadmium, and gold are useful but are soft and relatively weak when pure—especially in comparison with metals that precede them in each period on the periodic table. Explain why this is fact. Suggest a way to add strength to these metals for practical use.

Readings

Borman, Stu. "New Tin-Based Propellane May Shed Light on Bonding Questions." *Chemical and Engineering News* 67, no. 32 (August 7, 1989): 29-31.

Imura, Toru. "Advanced Metallic Materials." *Chemtech* 19, no. 4 (April 1989): 234-237.

Cumulative Review

1. List ten elements having more than two possible oxidation states.
2. Briefly describe the development of classification for elements from Dobereiner to the modern periodic table.
3. Predict the oxidation number of the following elements, using only the periodic table as a guide.
   a. astatine, germanium, mercury, polonium, tin
   b. francium, hafnium, neodymium, rubidium, tellurium
4. Describe the structure of the modern periodic table.
5. What four factors affect the values obtained for ionization energies of an element?
6. Use the periodic table to predict oxidation numbers for the following elements.
   a. cesium **c.** niobium **e.** iodine
   b. zirconium **d.** selenium **f.** boron
7. Classify each of the elements in the previous problem as metal, metalloid, or nonmetal.
8. What are the family names of the elements in the following groups?
   a. 1 (IA) **c.** 17 (VIIA)
   b. 2 (IIA) **d.** 18 (VIIIA)

Answers to
Application

37. a. ionic, **b.** van der Waals, **c.** atomic, **d.** covalent

38. a. framing—flexible plastic plumbing—flexible and does not corrode, **b.** wallboard—does not burn, high melting point masonry—rigidity, **c.** wiring—conductivity, siding—resists corrosion

39. a. [H:N:H with H above and below N]⁺

b. [:Ö:H]⁻

c. [:Ö:C:C:Ö: with Ö above each C]²⁻

d. [:Ö:Ö:]²⁻

Answers to
Critical Thinking

40. Any attempt to move the ions in the compound places ions of like charge near each other. The resulting repulsive forces cause the crystal to break.

41. These metals have fewer delocalized electrons and thus weaker metallic bonds than do those metals preceding them in each period of the periodic table. Alloy them with stronger metals.

42. Electronegativities of the impurities will not be the same as that of the metal atoms, so there will be fewer delocalized electrons.

43. Having the 3*d* sublevel completed, zinc has a more stable electron configuration and thus less attraction for additional electrons. Gallium has only one more energy level than does aluminum but has many more protons. The increased nuclear charge increases electronegativity.

Answers to
Cumulative Review

1. See last column of Appendix Table A-3.

2. Dobereiner's triads; Newland's law of octaves; Mendeleev's and Meyer's atomic mass-based periodic tables; Moseley's atomic number-based periodic table

3. a. At, 1–; Ge, 2+ or 4+; Hg, 2+; Po, 2–; Sn, 2+ or 4+
b. Fr, 1+; Hf, 2+, 3+, or 4+; Nd, 2+; Rb 1+; Te, 2-

4. Each period (row) lists elements with the same principal quantum number for its highest energy electrons. Each group (column) lists elements with the same number of electrons in their highest energy sublevel.

5. Nuclear charge, distance from nucleus to electron, shielding, and electron sublevel affect the values of ionization energies of an element.

6. a. 1+, **b.** 2+, 3+, 4+, **c.** 2+, 3+, 4+, 5+, **d.** 2–, **e.** 1–, **f.** 3+

7. a. metal, **b.** metal, **c.** metal, **d.** nonmetal, **e.** nonmetal, **f.** metalloid

8. a. alkali metals, **b.** alkaline earth metals, **c.** halogens, **d.** noble gases

CHAPTER

13 Molecular Structure

CHAPTER ORGANIZER

Chapter Objectives	Text Features	Laboratory Options	Teacher Classroom Resources
13.1 Bonds in Space **1.** Use models to explain the structure of a given organic or inorganic molecule. **2.** Describe hybrid orbitals and use hybridization theory to explain the bond angles in compounds. **3.** Differentiate sigma and pi bonding and saturated and unsaturated carbon compounds. **Nat'l Science Stds: UCP.1, UCP.2, UCP.5, B.1, B.2, B.4**		**Minute Lab:** Like Charges Repel, p. 323 **Discovery Demo:** 13-1 Modeling the VSEPR Theory, p. 320 **Laboratory Manual:** 13-1 Polarity and Molecular Shape **ChemActivity 13:** Electron Clouds, p. 819 **Demonstration:** 13-2 A Sigma Overlap, p. 328 **Demonstration:** 13-3 Reactive Double Bonds, p. 330 **Demonstration:** 13-4 Reactive Triple Bonds, p. 332	**Study Guide:** pp. 45-46 **Lesson Plans:** pp. 27-28 **Color Transparency 29 and Master:** Geometry of the Methane Molecule **Critical Thinking/ Problem Solving:** The VSEPR Model and Molecular Structure, p. 13 **ChemActivity Master 13** **Reteaching:** Molecular Shapes, pp. 19-20 **Color Transparency 30 and Master:** Formation of Hybrid Bonds **Enrichment:** Molecular Orbital Theory, pp. 25-26 **Color Transparency 31 and Master:** Bonding in Ethene **Color Transparency 32 and Master:** Bonding in Ethyne **Chemistry and Industry:** Penicillin, pp. 21-22
13.2 Molecular Arrangements **4.** Name and write formulas for simple organic compounds. **5.** Define, explain, and give examples of isomerism. **Nat'l Science Stds: UCP.1, UCP.2, B.1, B.2, B.4, E.2**	**Chemists and Their Work:** Mary Lura Sherrill, p. 337 **Everyday Chemistry:** Gelatin, p. 338 **Frontiers:** Super Soot, p. 340 **Chemistry and Technology:** The Boranes, p. 342	**Laboratory Manual:** 13-2 Modeling Molecular Polarity	**Study Guide:** pp. 47-49 **Applying Scientific Methods in Chemistry:** Saturated vs. Unsaturated Compounds, p. 30
Other Chapter Resources	**Vocabulary and Concept Review,** pp. 13, 59-60 **Evaluation,** pp. 49-52	**Videodisc Correlation Booklet** **Lab Partner Software**	**Test Bank**

Block Schedule

For information on block scheduling, see the Lesson Plans booklet in the Teacher Resource Package.

GLENCOE TECHNOLOGY

Chemistry: Concepts and Applications Videodisc
Molecular Shapes
Modeling Hydrocarbons
Isomers

Chemistry: Concepts and Applications CD-ROM
Using Lewis Dot Structures
Molecular Shapes

Science and Technology Videodisc Series (STVS)
Chemistry
Images of Atoms

Complete Glencoe Technology references are inserted within the appropriate lesson.

ASSIGNMENT GUIDE

Contents	Level	Practice Problems	Chapter Review	Solving Problems in Chemistry
13.1 Bonds in Space				
Electron Distribution	C	1, 2	15, 16	1-3
Electron Pair Repulsion	C	3	17, 18, 26, 38, 42, 63	
Hybrid Orbitals	C		19, 28, 43	
Geometry of Carbon Compounds	C		44, 57	
Sigma and Pi Bonds	C		25, 27, 28, 45-48, 51, 56, 59, 61	4
Multiple Bond Molecular Shapes	C		20, 28, 49	
Benzene	A		29, 30, 50, 53, 60	
13.2 Molecular Arrangements				
Organic Names	C	9, 10	21-24, 37, 59, 65	
Isomers	C		31-35, 37, 52, 54, 55, 58, 62, 64	
Inorganic Compounds	C		36	
Bond Summary	C			Chapter Review: 1-18

C–Core, A=Advanced, O–Optional

LABORATORY MATERIALS

Minute Labs	ChemActivities	Demonstrations
page 323 balloons string wool or fur	**page 819** balloons, round and pear-shaped clear tape string	**pages 320, 328, 330, 332** balloons, 10 small, 6 large beaker, 600 cm^3 bromine water calcium carbide chunks colored acetate cottonseed oil crucible tongs cyclohexane cyclohexene dropper matches overhead projector permanent markers string test tubes with stoppers large test tubes with stoppers water

CHAPTER 13

Molecular Structure

CHAPTER OVERVIEW

Main Ideas: Chapter 13 presents a study of chemical bonds and their geometric arrangement. Different approaches to explain chemical bonding in molecules are presented in a sequence that proceeds from the electron pair repulsion theory (VSEPR) to the hybrid orbital theory. The chapter introduces the student to the structure of organic molecules, including those with multiple bonds. The unique structure of benzene and its resultant properties are described in detail. Students are introduced to the isomeric structures of organic molecules.

Theme Development: Scale and structure as a theme of the textbook is introduced through a presentation of bonding theories. The relationship that exists between the structure of molecules and the properties of those molecules are described and developed in detail.

Tying To Previous Knowledge: Ask students what is meant by the phrase, "form and function." Students realize that a kitchen gadget usually has a particular form that facilitates its function in preparing food. A racing car has a sleek design that enhances its speed and handling. Remind students that the properties of a material can be predicted when one knows its molecular structure.

CHAPTER Preview

CONTENTS

ASSESSMENT PLANNER

Choose from the following assessment strategies.

Assess, pp. 333-334, 341-342
Assessment, pp. 321, 323, 325, 326, 330, 331, 336, 338, 340
Portfolio, p. 340
Minute Lab, p. 323
ChemActivity, pp. 819-820
Chapter Review, pp. 344-347

DISCOVERY DEMO

13-1 Modeling the VSEPR Theory

Purpose: to demonstrate the geometry of shared and unshared electron pairs

Materials: 10 small and 6 large balloons, 16 20-cm lengths of string

Procedure: 1. Inflate and tie off the ends of all the balloons. Use a permanent felt marker to mark an *S* to represent a shared pair of electrons on each small balloon and a *U* to represent an unshared pair of electrons on each large balloon. Tie a 20-cm length of string to each balloon.

2. Form the following molecules by connecting the balloons: HF (3*U*, 1*S*); H_2O (2*U*, 2*S*); NH_3 (1*U*, 3*S*); CH_4 (4*S*). Discuss the shapes due to the interaction of the *S* and *U* electron clouds.

Disposal: A

Results: *HF linear, H_2O bent, bond*

Molecular Structure

CHAPTER 13

The structure shown at the right is a model of a molecule of buckminsterfullerene, C_{60}. It is named for Buckminster Fuller, an American engineer and architect who invented the geodesic dome shown in the larger photograph. The domes designed by Fuller proved to be especially stable and able to withstand strong environmental forces.

Like the geodesic dome, the structure of a C_{60} molecule is especially stable. In fact, there are a large number of even-numbered carbon atom clusters that are stable. These molecules are called, collectively, fullerenes. The C_{60} molecule is of special interest to scientists because it has many interesting and useful properties, such as magnetism and superconductivity, all of which are due to the structure of the molecule.

EXTENSIONS AND CONNECTIONS

Chemists and Their Work

Mary Lura Sherrill, page 337

Everyday Chemistry

Gelatin: How the properties of this molecule depend on its structure, **page 338**

Frontiers

Super Soot: A further look at buckminsterfullerene, **page 340**

Chemistry and Technology

The Boranes: A look at several of these boron-hydrogen compounds, **page 342**

321

INTRODUCING THE CHAPTER

USING THE PHOTO

Each carbon in the C_{60} molecule is connected to three other carbons. The carbons are sp^2 hybridized. The molecule is stable because of the strong carbon-carbon bonds and no loose "ends." Compare the structure of buckminsterfullerene to those of graphite and benzene. The carbons in both graphite and benzene are sp^3 hybridized.

The geodesic domes designed by R. Buckminster Fuller were the inspiration for the name buckminsterfullerene. The compound is one of a class of fullerenes, which are molecules that consist of carbons arranged in pentagons and hexagons joined in a cage structure. After studying this chapter students will be better able to understand a presentation on the structure of this molecule.

REVEALING MISCONCEPTIONS

There are numerous ways to represent atoms and their combinations in molecules. Thus, students should realize that the pictures in the text are simply models. Different approaches to bonding may dictate different representations. Do not attach too much importance to pictorial aspects of the orbital representations. Emphasize that these models are useful because they provide a basis for the prediction of structure and properties. Being able to predict molecular shapes avoids the necessity of memorizing hundreds of individual ones.

angle 104.5°, NH_3 trigonal pyramid, bond angle 107.3°, CH_4 tetrahedron, bond angle 109.5°

Questions: 1. Describe the molecular shapes of the four models that were constructed. *See Results section.*

2. Do you expect an unshared pair of electrons to exert a greater, equal, or lesser repulsion than a shared pair when near another shared pair of electrons? *greater repulsion*

✓ ASSESSMENT

Performance: Have students demonstrate the shapes that result when two, three, and four electron clouds (electron pairs) arrange themselves in space so as to minimize repulsion. Students should use five balloons of the same color and size. Make two sets by tying two balloons together. Remind your students that these are electron clouds formed by a pair of electrons. Students should be able to predict that if there are two pairs of electrons, the molecular arrangement will be linear with a 180° bond angle (beryllium). Three pairs (use one balloon and a pair tied together) will minimize repulsion if they are trigonal planar, 120° bond angle (boron). Four pairs will form a typical carbon tetrahedron.

13.1 Bonds in Space

PREPARE

Key Concepts

The major concepts presented in this section include electron-pair repulsion theory, hybrid orbital theory and the resulting molecular geometries, sigma and pi bonds, and the molecular shapes that result when the molecule contains double or triple bonds.

Key Terms

shared pair
unshared pair
hybridization
hybrid orbital
sigma bond
pi bond
double bond
triple bond
unsaturated compound
delocalization
conjugated system

1 FOCUS

Objectives

Preview with your students the objectives listed on the student page. Students can use them as a study guide for this major section.

Discussion

Splash some inexpensive cologne into a Petri dish before students enter the room. Ask students to classify the cologne as inexpensive or expensive. Point out that colognes usually contain synthetic esters to imitate the natural esters found in more expensive colognes. If the ester structures vary, the inexpensive cologne will not smell like the more expensive cologne.

OBJECTIVES

- use models to explain the structure of a given organic or inorganic molecule.
- describe hybrid orbitals and use hybridization theory to explain the bond angles in compounds.
- differentiate sigma and pi bonding and saturated and unsaturated carbon compounds.

13.1 Bonds in Space

A primary goal of studying chemistry is to learn how the macroscopic properties of matter are a consequence of its molecular structure. You have learned a great deal about the structure and properties of atoms. Their behavior is determined chiefly by their electron configurations. The behavior of molecules also depends on their structural characteristics. This section is devoted to examining the shapes of molecules, and what characteristics of their bonds produce those shapes.

Many consumer products are packaged in polystyrene foam, which is made of huge molecules of a substance named polystyrene. There are thousands of atoms in a molecule of polystyrene. How is it possible to study such a large molecule? Fortunately, atoms usually combine in large molecules in the same manner that they do in small molecules. By studying the relationship between structure and properties in small molecules, we can better understand the structure and properties of large molecules.

Electron Distribution

There are several ways of looking at the structure of molecules to account for their shape. We will consider two of these models. The first model takes into account the repulsive forces of electron pairs around an atom. The second model considers ways in which atomic orbitals can overlap to form orbitals around more than one nucleus. The electrons in these combined orbitals then serve to bind the atoms together.

In order to describe the shape of a molecule or polyatomic ion, it is useful to draw a Lewis electron dot diagram. In these diagrams, we arrange the outer electrons as dots around the atoms so that each atom ends up with a full outer level. *For all atoms that form covalent bonds, except hydrogen, eight electrons represent a full outer level.* Consider the water molecule. It is composed of one oxygen and two hydrogen atoms. The electron dot diagrams are ·Ö: and H× for these elements. All electrons are identical. We use different symbols for the electrons only to help us understand how we arrive at the final structure. By combining an oxygen and two hydrogens, we obtain the following electron dot diagram.

RULE OF THUMB

H:Ö:
 H

It is the only arrangement of electrons in which all three atoms can achieve a full outer level. Note that two pairs of electrons in the outer level of oxygen are involved in bonding the hydrogens. They are called **shared pairs.** The other two pairs of electrons are not involved in bonding. They are called **unshared pairs,** or lone pairs. Note that in counting electrons, a shared pair of electrons contributes to a full outer level for both atoms sharing the electron pair.

CONTENT BACKGROUND

Resonance: There are many instances in which it is not possible to draw a single Lewis structure for a species. A good example is the nitrate ion, NO_3^-. The theory of electron cloud spacing would predict 120° O—N—O bond angles and a trigonal planar ion. Experiments show this prediction is true. The most popular explanation of this structure is sp^2 hybridization on the atoms with delocalized orbitals formed from the unhybridized *p* orbitals to account for the remaining electrons. An older, but still popular, approach to such species is the concept of resonance. If an attempt is made to draw an electron dot diagram for the nitrate ion, nitrogen ends up with six electrons in the outer level.

:Ö: −
:Ö:N:Ö:

SAMPLE PROBLEM

Lewis Electron Dot Diagram

Draw the Lewis electron dot diagram for AsI_3.

Solving Process:

From the periodic table, we can see that arsenic has five electrons in its outer level and iodine has seven. Thus, arsenic tends to form three bonds and iodine one. The diagram is then

:I : As : I :
:I:

PROBLEM SOLVING HINT

The symbol used to represent electrons of a particular atom is arbitrary. However, each different element in a molecule should have a different electron symbol.

PRACTICE PROBLEMS

1. Draw Lewis electron dot diagrams for the following.
   a. H_2Te c. NI_3
   b. PF_3 d. CBr_4
2. How many unshared pairs of electrons are on the central atoms in 1c and 1d?

MINUTE LAB

Like Charges Repel

As you read about electron pair repulsion theory, a balloon model of H_2O is helpful. Tie a string 50 cm long to each of two inflated balloons. Keeping the balloons separated, rub each balloon with a piece of wool cloth. Bring the strings together and observe how the balloons behave. Draw the electron dot diagram of water. Let the two balloons represent the two unshared pairs of electrons on the oxygen atom. Will the two shared pairs between oxygen and hydrogen be in the same plane as the unshared pairs? What is the bond angle between the H—O bonds?

Electron Pair Repulsion

One way to account for the shape of molecules is to consider electron repulsion. Each bond and each unshared pair in the outer level of an atom form a charge cloud that repels all other charge clouds. In part, this repulsion is due to all electrons having the same charge. Another more important factor is the Pauli exclusion principle. Although two electrons of opposite spin may occupy the same orbital, electrons of the same spin may not do so. Repulsions resulting from the Pauli principle are much greater than electrostatic ones at small distances. Because of these repulsions, atoms cannot be compressed.

The repulsions between the charge clouds in the outer level of atoms determine the arrangement of the orbitals. The orbital arrangement, in turn, determines the shape of molecules. As a result, the following rule of thumb may be stated. *Electron pairs spread as far apart as possible to minimize repulsive forces.* Refer to Table 13.1 for diagrams of the bonding arrangement discussed here. If there are only two electron pairs in the outer level of the central atom, they will be on opposite sides of the nucleus. This arrangement is called linear. If there are three electron pairs, the axes of their charge clouds will be 120° apart. This arrangement is called trigonal planar and the electron pairs lie in the same plane as the nucleus of the central atom. If there are four electron pairs, the axes of the charge clouds will be farthest apart when they intersect at an angle of 109.5°. These axes will not all lie in the same plane but will form a tetrahedron. A tetrahedron is a figure having four faces, each of which is an equilateral triangle.

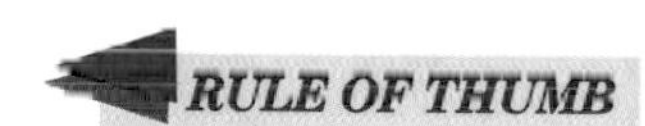

In order to satisfy the octet rule, one oxygen could share two pairs with nitrogen, forming a double bond.

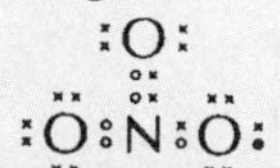

If that were actually the case, because double bonds are shorter than single, the oxygen doubly bonded to the nitrogen should be closer to the nitrogen than the other two oxygen atoms. However, research has shown that all three O—N bonds are identical. In the concept of resonance, there are three equivalent electron dot diagrams. The actual ion is said to be an average of the three structures. The true structure does NOT "resonate" among the three possibilities, but rather has one-third the character of each.

:Ö:N(:O:)(:Ö:)$^-$ ⟷ :Ö:N(:O:)(:Ö:)$^-$ ⟷ Ö::N(:O:)(:Ö:)$^-$

2 TEACH

Correcting Errors

Students often wonder which element is the central atom. Oversimplified, Groups 1,2 and 13-18 have bond capacities of 1, 2, 3, 4, 3, 2, 1, and 0 respectively. The element that typically forms the most bonds is placed in the center of a dot diagram.

PRACTICE

Guided Practice: Guide students through the Sample Problem, then have them work in class on Practice Problems 1 and 2.

Independent Practice: Your homework or classroom assignment can include Review and Practice 15-16 from the end of the chapter.

Answers to
Practice Problems:

1. Have students refer to Appendix C for solutions to practice problem 1.
2. 1c—one unshared; 1d—none

MINUTE LAB

Like Charges Repel

Purpose: To show that like charges repel

Materials: 2 large inflated balloons, 1-m length of string, piece of wool or fur.

Results: The like-charged balloons will swing away from each other when the two strings are held together in one hand. (Humid air makes it more difficult to get dramatic effects.)

Answers to Questions: 1. No, the four pairs are arranged in a distorted tetrahedron. **2.** 104.5°

✓ ASSESSMENT

Performance: Have students use two more balloons (shared pairs) to show the arrangement of the electron clouds around oxygen in water. See Figure 13.3.

CONCEPT DEVELOPMENT

Background: At one time the electron-pair repulsion theory was called the valence shell electron pair repulsion theory, VSEPR. The term *outer energy level* is now used instead of *valence shell.*

Computer Demo: *VSEPR,* Display and rotate molecular structures. AP301, Project SERAPHIM

QUICK DEMO **Modeling**
Use styrofoam balls and pipe cleaners to make a model of methane (bond angle of 109.5°) with the four bonds tetrahedrally arranged. Remove a hydrogen atom and leave a pipe cleaner to represent the unshared pair of electrons. Show the students how the remaining three bonds are pushed together. Point out that this model represents the shape of an ammonia molecule in which the bond angles are 107°. Remove another hydrogen atom and squeeze the two remaining bonds closer together. Point out that this model represents the shape of H_2O in which the bond angles are 104.5°.

PROGRAM RESOURCES

Study Guide: Use the Study Guide worksheets, pp. 45-46, for independent practice with concepts in Section 13.1.

Lesson Plans: Use the Lesson Plan, pp.27-28, to help you prepare for Chapter 13.

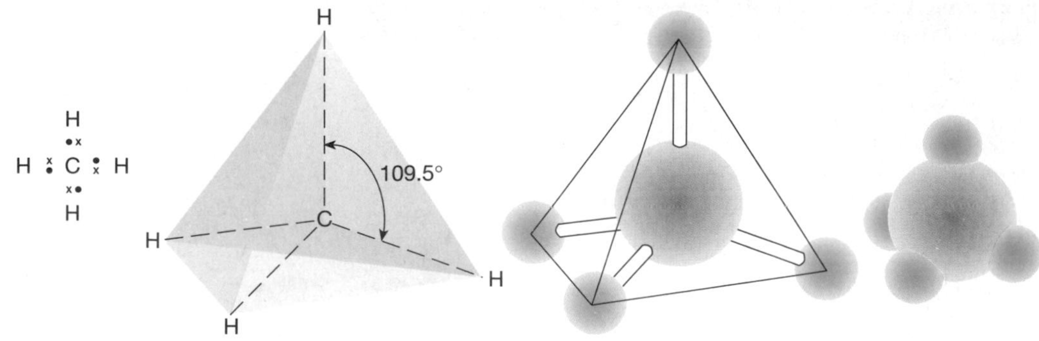

Figure 13.1 On the left, the Lewis dot diagram for methane shows a complete octet when four hydrogens bond to one carbon. The next diagram shows that each bond in methane points toward the vertex of a tetrahedron. The tetrahedral structure is also shown with a ball-and-stick model. The space-filling model on the right shows the relative volumes of the electron clouds.

The central atom is at the center and the axes extend out to the corners. See Figure 13.1.

The bonds and unshared electron pairs within a molecule determine the shape of the molecule. An unshared electron pair is acted upon by only one nucleus. Its charge cloud is shaped like a blunt pear with its stem end at the nucleus. A shared pair of electrons moves within the field of two nuclei. The cloud is therefore more slender because the attraction of the two nuclei restricts electron movement.

The electron pair repulsions in a molecule may not all be equal. The repulsion between two unshared pairs is greatest because they occupy the most space. The repulsion between two shared pairs is least because they occupy the least space. The repulsion between an unshared pair and a shared pair is an intermediate case.

unshared-unshared repulsion $>$ unshared-shared repulsion $>$ shared-shared repulsion

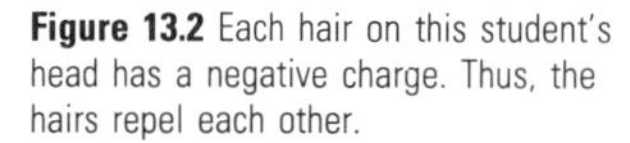

Figure 13.2 Each hair on this student's head has a negative charge. Thus, the hairs repel each other.

Let us consider the molecular shapes of the compounds CH_4, NH_3, and H_2O to illustrate this repulsion. See Figure 13.3. In each of these compounds, the central atom has four clouds around it. We expect the axes of all four charge clouds to point approximately in the direction of the corners of a tetrahedron.

In methane (CH_4) molecules, all clouds are shared pairs, so their sizes are equal and each bond angle is in fact 109.5°. The CH_4 molecule is therefore a perfect tetrahedron. In NH_3 molecules, there are one unshared pair and three shared pairs. The unshared pair occupies more space than any of the other three, so the bond clouds are pushed together and form an angle of 107° with each other. It is important to note that, although the electron clouds form a tetrahedron, one cloud is not involved in bonding. Therefore, the atoms composing the molecule form a trigonal pyramid. In H_2O molecules, two unshared pairs are present; both of these clouds are larger than the bond clouds. This additional cloud size results in a still greater reduction in the bond angle which is, in fact, 104.5°. Again, note that the electron clouds are tetrahedral but the molecule is "V" shaped, or bent. Note that in the three molecules discussed, each has four electron clouds. The differences in molecular shape result from the unequal space occupied by the unshared pairs and the bonds. The shapes for different numbers of clouds can be predicted in the same way and are listed in Table 13.1.

In most compounds, the outer level is considered full with four pairs or eight electrons. The outer level in some atoms can contain more than eight electrons (if the outer level is the third or higher level). A number of nonmetals, mainly the halogens, form compounds in which the outer level is expanded to 10, 12, or 14 electrons. Such an arrangement would also explain the formation of noble gas compounds. An example is xenon tetrafluoride, XeF_4. The structure of this compound is shown in Table 13.1. Xenon has eight electrons of its own in its outer level together with one electron from each of the four fluorine atoms.

SAMPLE PROBLEM

Bond Angles

Would the Cl—N—Cl bond angle in NCl_3 be greater than, less than, or equal to, 109.5°?

Solving Process:

First, draw an electron dot diagram for the molecule.

$$\begin{array}{c} :\ddot{Cl}: \\ :\ddot{Cl}:\ddot{N}:\ddot{Cl}: \end{array}$$

Note that there are four electron pairs in the outer level of the nitrogen atom. Because the four pairs would be expected to form a tetrahedron, the first approximation of a bond angle would be 109.5°. However, one of those pairs is unshared and has a greater repulsive effect. That force would push the three shared pairs closer together. Therefore, the Cl—N—Cl bond angle should be a little less than 109.5°.

PROBLEM SOLVING HINT

Drawing Lewis electron dot diagrams will help determine the number of electron pairs and whether these pairs are shared or unshared.

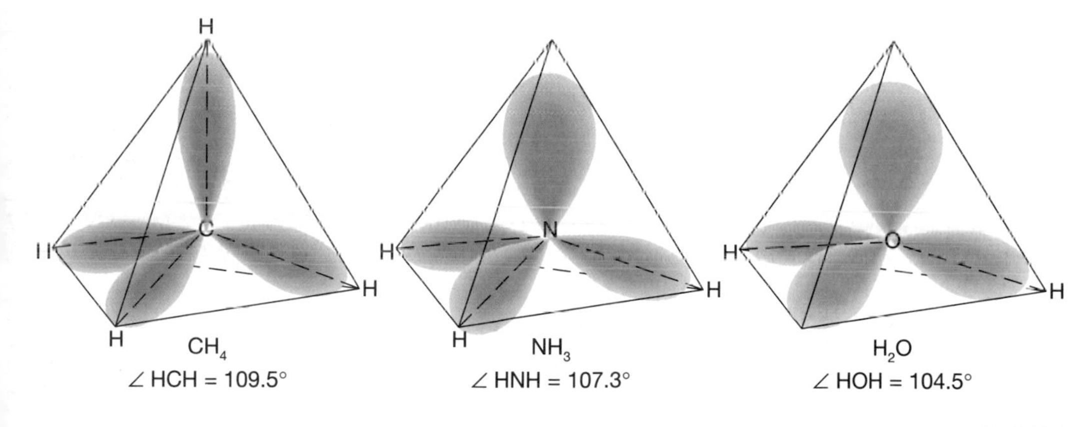

Figure 13.3 The central atom in each molecule is surrounded by four electron pairs. The influence of shared and unshared pairs accounts for the different shapes.

Correcting Errors

Remind students that the outer *s* and *p* electrons are shown in a dot diagram. The *s* electrons are on one side, the p_x, p_y, and p_z electrons occupy the remaining three sides. The negative cloud generated by an unshared pair is more negative than the cloud that is generated by a shared pair of electrons.

ASSESSMENT

Oral: Ask students to recall the Minute Lab on page 323 and visualize the structure of the shared and unshared pairs of electrons around the oxygen atom in the water molecule (also see Figure 13.3 and Table 13.1). Then give the students randomly selected repulsions and ask them which one is the stronger of the two. Example: Which is stronger, unshared-unshared (U-U) or the unshared-shared (U-S) repulsion? Students should respond that the unshared-unshared repulsion is the greater. The order of repulsion is U-U > U-S > S-S.

PROGRAM RESOURCES

Critical Thinking/Problem Solving: Use the worksheet, p.13, as enrichment for the concept of "The VSEPR Model and Molecular Structure."

Laboratory Manual: Use macrolab 13-1 "Polarity and Molecular Shape" as enrichment for the concepts in Chapter 13.

CONCEPT DEVELOPMENT

Using the Table: Provide students with 2-inch diameter plastic foam balls for atoms, dowel sticks for bonds, small balloons for unshared pairs of electrons, wire to connect balloons, and a 3 x 5-inch card for the label. Assign one molecular shape described in Table 13.1 to each student.

✓ ASSESSMENT

Knowledge: Ask students in cooperative learning groups to describe and record how many of each type and which types of electron cloud repulsions exist in each molecule shown in Table 13.1. For example, H_2O has one U-U, two U-S, and one S-S repulsion.

GLENCOE TECHNOLOGY

CD-ROM

Chemistry: Concepts and Applications

Exploration: *Using Lewis Dot Structures*

Videodisc

Chemistry: Concepts and Applications

Molecular Shapes
Disc 1, Side 2, Ch. 13

Also available on CD-ROM.

Videodisc

STVS: Chemistry

Disc 2, Side 1
Images of Atoms (Ch. 4)

Table 13.1

Molecular Shapes				
Molecule	**Total Number of Electron Pairs**	**Number of Shared Pairs**	**Number of Unshared Pairs**	**Molecular Shape**
BeF_2	2	2	0	Linear
GaF_3	3	3	0	Trigonal planar
O_3	3	2	1	Bent
CH_4	4	4	0	Tetrahedral
NH_3	4	3	1	Trigonal pyramidal
H_2O	4	2	2	Bent
$NbBr_5$	5	5	0	Trigonal bipyramidal

PROGRAM RESOURCES

Reteaching: Use the Reteaching worksheets, pp. 19-20, to provide students another opportunity to understand the concept of "Molecular Shapes."

Transparency Master: Use the Transparency master and worksheets, pp. 59-60, for guided practice in the concept "Formation of Hybrid Bonds."

Color Transparency: Use Color Transparency 30 to focus on the concept "Formation of Hybrid Bonds."

Table 13.1 *continued*

Molecular Shapes				
Molecule	Total Number of Electron Pairs	Number of Shared Pairs	Number of Unshared Pairs	Molecular Shape
SF_4	5	4	1	Irregular tetrahedron
BrF_3	5	3	2	T-shaped planar
XeF_2	5	2	3	Linear
SF_6	6	6	0	Octahedral
IF_5	6	5	1	Square pyramidal
XeF_4	6	4	2	Square planar

CULTURAL DIVERSITY

Electron Handball

Earl Shaw, an African American scientist at Bell Laboratory, explores changes in electronic behavior of molecules. Early lasers operated only on a fixed wavelength. This meant they had limited use—like a radio that can receive only one station. Scientists were interested in building tunable lasers, that is, lasers that can produce light of different wavelengths. The only way to do this is to change the electronic behavior of the molecules involved.

Shaw is the co-inventor of the spin-flip Raman laser that is used for air pollution and other molecular measurements. Shaw used mirrors to guide the light from a carbon dioxide laser to a synthetic indium antimonide crystal chilled below −250°C. The light beam emitted by the crystal could be adjusted, or tuned, using a magnetic field.

Shaw is also working on technology referred to as free-electron lasing. He is using an accelerator, like those used in cancer treatments, to eject electrons traveling almost at the speed of light into a magnet as long as a handball court. Mirrors feed the light that is emitted back and forth within the electron beam. This laser will allow scientists to better study solid materials.

interNET CONNECTION

Molecule models in MIME format can be downloaded.

World Wide Web
http://www.ch.ic.ac.uk/chemical_mime.html

PRACTICE

Guided Practice: Guide students through Practice Problem 3a and b, then have teams of students work in class on Practice Problems 3c-f that follow.
Cooperative Learning: Refer to pp. 28T-29T to select a teaching strategy to use with this activity.
Independent Practice: Your homework or classroom assignment can include Review and Practice 17-18, and 26, and Concept Mastery 39-42 from the end of the chapter.

Answers to
Practice Problems: Have students refer to Appendix C for complete solutions to Practice Problems.

3. For structures see Appendix C.

a. < 109.5°
b. < 109.5°
c. > 109.5°
d. equatorial-equatorial angles are > 109.5°, axial-equatorial angle is < 109.5°
e. The form is nearly tetrahedral, with some angles > 109.5°, and some angles < 109.5°
f. The angle is 109.5°

CONCEPT DEVELOPMENT

• The number of atomic orbitals that hybridize is conserved and results in the same number of hybrid orbitals being produced.

Enrichment Ask a student to determine the geometry of sp^3d^2 hybrid orbitals. See if the volunteer can discover some compounds or ions containing such a hybrid. Possible response: *octahedral;* SF_6, $CdCl_6^{4-}$, ICl_4^-, BrF_4^-, BrF_5, IF_5, $Co(NH_3)_6^{3+}$, $PdCl_6^{2-}$, $Fe(CN)_6^{3-}$

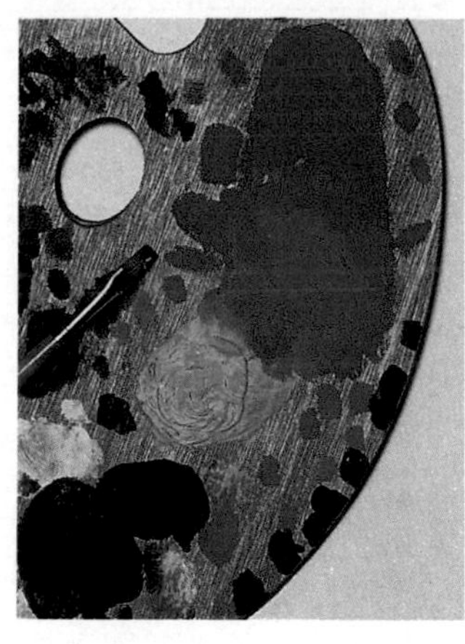

Figure 13.4 Forming hybrid orbitals is similar to mixing two colors of paint. The result is a combination of the two paints mixed and differs from each of them.

PRACTICE PROBLEMS

3. Predict whether the bond angle of each of the following is greater than, less than, or equal to 109.5°.

a. F—N—F in NF_3
b. F—O—F in OF_2
c. F—Be—F in BeF_2
d. F—As—F in AsF_5
e. F—Te—F in TeF_4
f. O—Xe—O in XeO_4

Hybrid Orbitals

Another model of molecular shape considers the different ways *s* and *p* orbitals can overlap when electrons are shared. This model is best illustrated using the element carbon. However, this model can also be applied to other atoms forming covalent bonds.

Consider the electron configuration of carbon, $2s^22p^2$. The electron dot diagram for carbon would be $\cdot\ddot{C}\cdot$. Because the 2*s* orbital is full, we might expect that only the two unpaired electrons would be used to form bonds with other atoms, thus, X:$\ddot{C}$:X. We would predict the electron pairs to be distributed in a trigonal planar manner and the molecule itself to be V-shaped. The predicted bond angle would be slightly less than 120° because of the unshared pair. Instead, carbon actually forms four bonds distributed tetrahedrally. How is this structure possible?

Suppose we wish to make a gallon of fruit drink to serve at a birthday party. On hand we have one quart of limeade and three quarts of lemonade. However, we do not want to worry about who gets one flavor and who gets the other. Serving will be simpler if everyone could be served the same flavor. By mixing the limeade and the lemonade, we can obtain a gallon of single-flavored fruit drink. Note that we put together four nonidentical quarts and will get four identical quarts of mixture.

An analogous process takes place in the outer level of the carbon atom. The *s* orbital and the three *p* orbitals merge to form four orbitals that are identical. This process is call **hybridi-**

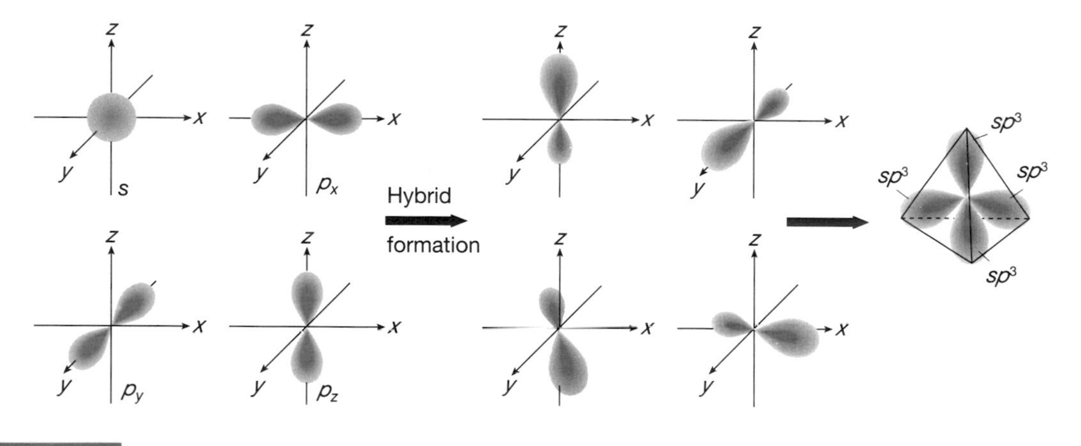

Figure 13.5 When carbon bonds, the *s* and three *p* orbitals merge to form four equivalent sp^3 hybrid orbitals. They are arranged in a tetrahedral shape.

DEMONSTRATION

13-2 A Sigma Overlap

Purpose: to demonstrate that a sigma bond is formed by the direct overlap of two half-filled orbitals

Materials: two sheets of colored acetate (transparency film with colored cellophane attached will work), permanent felt marker, overhead projector

Procedure: 1. You can demonstrate why the two nuclei, having the same charge, appear to be attracted to each other rather than being repelled. With a marker, place a large + sign on one end of each sheet of colored acetate.
2. Overlap the two sheets half way so that + signs are at opposite ends.
Disposal: F

Results: The color darkens in the area of overlap. Point out that this is a

zation and the orbitals formed are called **hybrid orbitals.** Note that four nonequivalent orbitals produced four equivalent orbitals.

Because one s and three p orbitals have merged, the hybrid orbitals are represented as sp^3 hybrids. These four orbitals are degenerate and contain one electron each. They are arranged in a regular tetrahedral shape. Each of these hybrid orbitals can then bond to another atom. If each of the hybrid orbitals bonds to an identical atom, the four bonds formed are equivalent.

Unhybridized Carbon			
	$1s$	$2s$	$2p$
carbon	⇅	⇅	↓ ↓ ○
			unhybridized orbitals

Hybridized Carbon		
	$1s$	$2sp^3$
carbon	⇅	↑ ↑ ↑ ↑
		4 hybrid orbitals

Geometry of Carbon Compounds

The bonding of four hydrogen atoms to one carbon atom forms methane. The bonds involve the overlap of the s orbital of each hydrogen atom with one of the sp^3 hybrid orbitals of a carbon atom. A three-dimensional representation of the formula of methane is shown in Figure 13.6. There is an angle of 109.5° between each carbon-hydrogen bond axis.

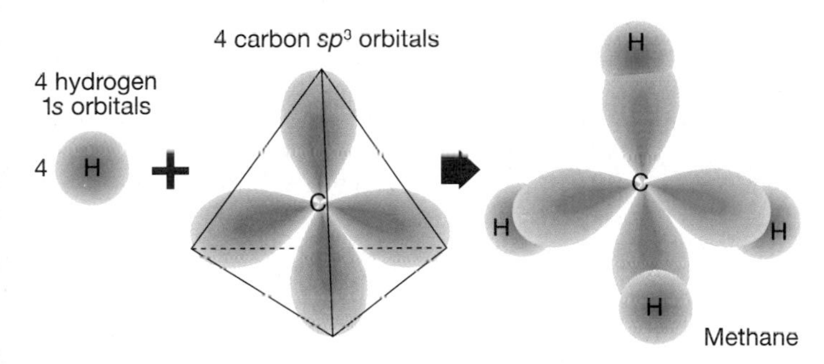

Figure 13.6 The methane structure can be explained by combining a hydrogen s orbital with each of the sp^3 hybrid orbitals of carbon.

Recall that carbon exhibits catenation. This structure occurs when two carbon atoms bond to each other by the overlap of an sp^3 orbital of one carbon with an sp^3 orbital of another carbon. The remaining three sp^3 orbitals of each carbon atom may bond with the s orbital of three hydrogen atoms. The compound formed in this way, C_2H_6, is called ethane. A three-dimensional structure of ethane is shown in Figure 13.7.

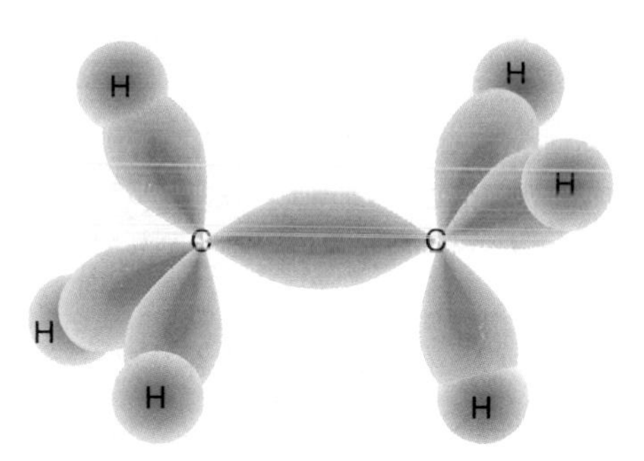

Figure 13.7 The shape of ethane is explained by combining two sp^3 hybrid carbon atoms. The single bond is a direct overlap of one hybrid orbital from each carbon atom.

CONCEPT DEVELOPMENT

QUICK DEMO **Tetrahedrons**

Give each student 2 drinking straws and have them cut the straws into thirds. Have students use a piece of 100-cm length of string to connect three of the straw pieces to make a triangle. Use other pieces of string to connect the other straw pieces to the triangle to form a tetrahedron. Remind students that carbon is in the center of the tetrahedron.

PROGRAM RESOURCES

Enrichment: Use the worksheet "Molecular Orbital Theory," pp. 25-26, to challenge more able students.

Transparency Master: Use the Transparency master and worksheets, pp. 61-62, for guided practice in the concept "Bonding in Ethane."

Color Transparency: Use Color Transparency 31 to focus on the concept "Bonding in Ethene."

Transparency Master: Use the Transparency master and worksheet, pp. 57-58, for guided practice in the concept "Geometry of the Methane Molecule."

Color Transparency: Use Color Transparency 29 to focus on the concept of "Geometry of the Methane Molecule."

model of how the electron clouds overlap. The electron probability thus becomes greater in the area between the two nuclei. As a result, the two nuclei, which repel one another, are attracted toward the denser electron cloud. The bonded distance between the two nuclei is always less than the sum of their respective van der Waals radii (Chapter 12).

Questions: 1. What do you observe happening to the charge density between the two nuclei as the orbitals overlap? *The darkening represents an increased charge density (electron probability).*

2. Will the two positive nuclei move toward the negative charge cloud between them or away from it? *toward it*

3. Is the bond distance shorter than, longer than, or the same as the nonbonded distance? *shorter*

Extension: Summarize by reminding students that sigma bonds can form by half-filled overlapping orbitals.

1. 2 s orbitals
2. s and p orbitals
3. 2 p orbitals end to end
4. 2 hybrid orbitals
5. hybrid orbital and s orbital
6. hybrid orbital and p orbital

CONCEPT DEVELOPMENT

Misconceptions

Students may interpret the diagram of a *p* orbital overlap above and below the sigma bond as two more bonds. Remind students that the parallel overlap of two half-filled *p* orbitals results in the formation of one pi bond.

✓ ASSESSMENT

Oral: Ask groups of students to make a list of the types of orbital overlaps that can occur between orbitals in a sigma bond. Then ask each group to add one type to a list that is being recorded on the chalkboard. The list should include the following.

1. two *s* orbitals
2. *s* and *p* orbitals
3. two *p* orbitals end to end
4. two hybrid orbitals
5. hybrid orbital and *s* orbital
6. hybrid orbital and *p* orbital

GLENCOE TECHNOLOGY

Videodisc

Chemistry: Concepts and Applications

Modeling Hydrocarbons

Disc 2, Side 2, Ch. 10

Show this video of Demonstration 29-2 on pages 740–741 to introduce the concept of modeling hydrocarbon molecules.

Also available on CD-ROM.

Sigma and Pi Bonds

A covalent bond is formed when an orbital of one atom overlaps an orbital of another atom and they share the electron pair in the bond. For example, a bond may be formed by the overlap of two half-filled *s* orbitals. When two orbitals form a bond that lies directly on the bond axis, the bond is called a **sigma bond,** and is designated σ. See Figure 13.8. A sigma bond can also be formed by the overlap of an *s* orbital with a *p* orbital, the overlap of two *p* orbitals, the overlap of two hybrid orbitals, or the overlap of a hybrid orbital with an *s* orbital. In Figure 13.7, the *s* orbitals of hydrogen atoms are shown overlapping hybrid sp^3 orbitals of carbon. Also in Figure 13.7, the overlap of two sp^3 hybrid orbitals of carbon is shown.

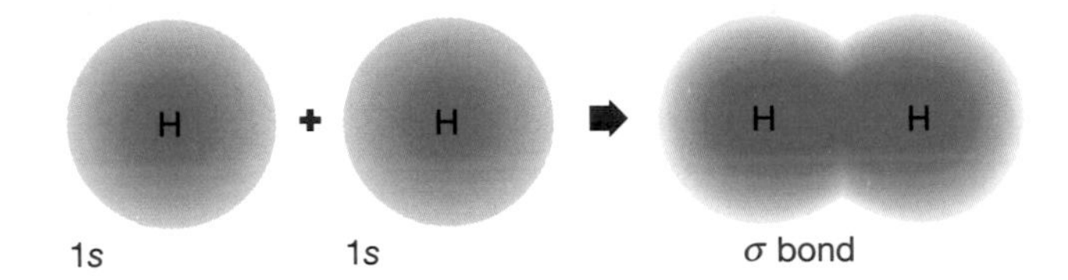

Figure 13.8 The overlap of two *s* orbitals is a sigma bond. The H-H bond in hydrogen gas is a sigma bond.

Because *p* orbitals are not spherical, when two half-filled *p* orbitals overlap, one of two types of bonds can form. If the two *p* orbitals overlap along an axis in an end-to-end fashion, a sigma bond forms. If the two *p* orbitals overlap sideways (parallel), they form a **pi bond,** designated π.

A molecule that shows both types of bonds is ethylene (ethene). In ethylene, one *s* orbital and two *p* orbitals of a carbon atom are considered to merge to form three sp^2 hybrid orbitals. These three sp^2 hybrid orbitals lie in the same plane at 120° bond angles. The third *p* orbital is not involved in this hybridization. It is perpendicular to the plane of the three sp^2 orbitals. See a model of this hybridization in Figure 13.9.

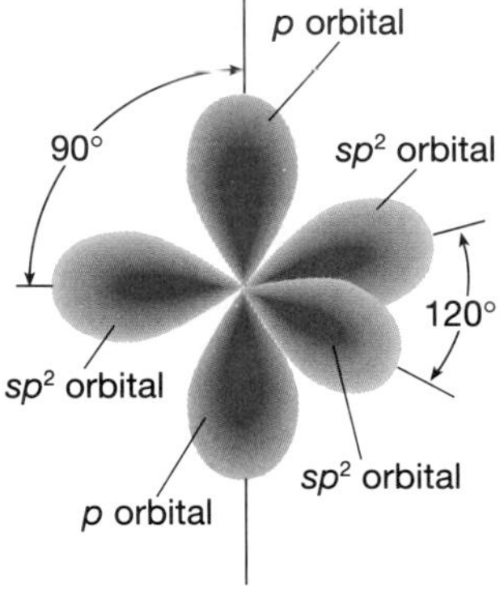

Figure 13.9 When one *s* orbital and two *p* orbitals merge to form three sp^2 orbitals, one *p* orbital is not hybridized. It is perpendicular to the sp^2 orbitals, which lie in the same plane at 120° angles from each other.

Now, if two sp^2 orbitals of adjacent carbon atoms overlap in an end-to-end fashion, a sigma bond is formed. Imagine that the two remaining sp^2 hybridized orbitals on each carbon atom overlap with an *s* orbital of two separate hydrogen atoms. Two other sigma bonds are formed, this time between sp^2 hybridized orbitals and *s* orbitals.

What happens to the remaining unhybridized *p* orbitals on the carbon atoms? These orbitals are now in position to overlap in a sideways fashion. They form a pi bond. Pi bonds are always formed by the sideways overlap of unhybridized *p* orbitals. The molecule formed by this series of bond formations, ethylene, is illustrated in Figure 13.10. Notice that ethylene contains one sigma bond and one pi bond between the carbon atoms. Thus, there is a double bond between the two carbon atoms, $H_2C{=}CH_2$. In a **double bond,** two pairs of electrons are shared between the bonding atoms. A double bond always consists of one sigma bond and one pi bond. The six atoms of ethylene lie in one plane.

DEMONSTRATION

13-3 Reactive Double Bonds

Purpose: to demonstrate that double bonds are reactive due to the presence of the strained parallel overlap found in the pi bond

Materials: 5 cm^3 bromine water, 5 cm^3 cyclohexene, 10 cm^3 cottonseed oil, 5 cm^3 cyclohexane, 4 test tubes with stoppers, medicine dropper

Procedure: 1. Place 5 cm^3 of cottonseed oil in a large test tube. Prepare a second test tube as a control. **CAUTION:** *Use a fume hood: hydrocarbons are both flammable and toxic.*
2. Use the medicine dropper to add bromine water to the cottonseed oil in small aliquots. Stopper the test tube and shake after each addition.
3. Compare the test tube that has had bromine added to the control.
Disposal: A

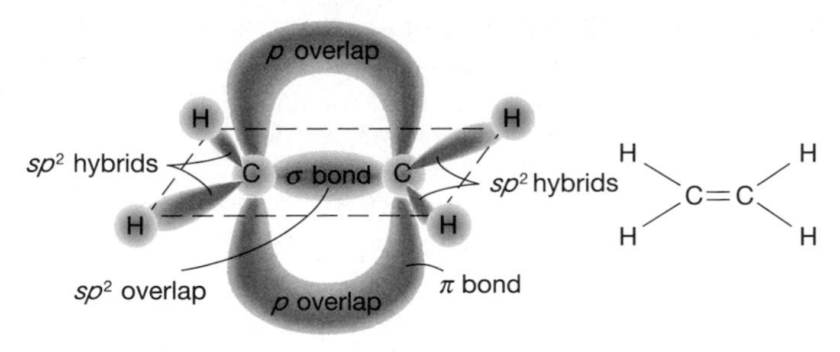

Figure 13.10 The shape of ethylene is explained by combining two sp^2 hybrid carbon atoms. The C-H sigma bonds all lie in the same plane. The unhybridized *p* orbitals of the carbon atoms combine to form the pi bond. The plane of the pi bond is perpendicular to the plane of the sigma bonds.

Now consider the bonding in the acetylene (ethyne) molecule, H—C≡C—H. In this molecule, the *s* orbital and one *p* orbital of each carbon atom form two *sp* hybrid orbitals. This particular hybridization leaves two *p* orbitals perpendicular to each other and to the *sp* hybrid orbitals. Two *sp* hybrid orbitals, one from each carbon, overlap to form one sigma bond. The two *p* orbitals from each atom overlap to form two pi bonds. Therefore, acetylene has a triple bond between its carbon atoms. The triple bond consists of one sigma bond and two pi bonds. In a **triple bond** three pairs of electrons are shared between the bonded atoms. Figure 13.11 summarizes the bonding that occurs in the acetylene molecule.

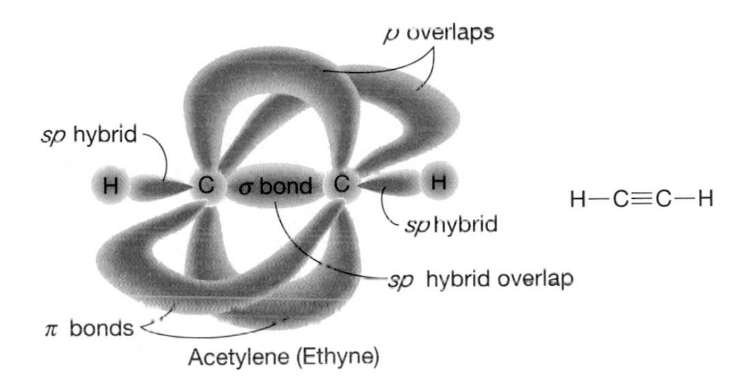

Figure 13.11 The shape of ethyne is explained by combining two *sp* hybrid carbon atoms. The two pi bonds are formed from the unhybridized *p* orbitals of both carbon atoms. The two pi bonds lie in planes that are perpendicular to each other.

Both double and triple bonds are less flexible than single bonds are. Also, a multiple bond between two atoms is shorter and therefore stronger than a single bond between the same two atoms. Pi bonds are more easily broken than sigma bonds are because the electrons forming pi bonds are farther from the nuclei of the two atoms. As a result, molecules containing multiple bonds are usually more reactive than are similar molecules containing only single bonds. Compounds that contain double or triple bonds between carbon atoms are called **unsaturated compounds.**

CONCEPT DEVELOPMENT

QUICK DEMO **Triple Bond = Triple Overlap**

Arrange 3 meter sticks to represent the 3 perpendicular half-filled *p* orbitals of a nitrogen atom. Give them to a student assistant to hold. Use three more meter sticks to similarly represent the p_x, p_y, and p_z orbitals of a second nitrogen atom. Move the one set of axes toward the other set of axes as when bonding occurs. The end-to-end overlap, sigma bond, along one axis occurs first. Students can observe that two parallel overlaps are possible along the other two axes. Show students the electron dot diagram for N_2 with three pairs of electrons between the two atoms, :N:::N: .

Analogy: The electron cloud overlap of the pi bond can be explained with the following analogy. The normally dumbbell-shaped *p* orbital is distorted when pi bonding. Point out to students that the two lobes of the p_x orbital are attracted to the nucleus of the second atom. They lean toward that atom's nucleus. At the same time the parallel p_x orbital of the second atom is attracted to the nucleus of the first atom. The shape of these two *p* orbitals is similar to two people sitting on opposite sides of a table and trying to touch heads and toes.

Results: The presence and reactivity of the double bond is demonstrated by the decolorization of the bromine water. The red-orange color disappears.

Questions: 1. What do you observe? *decolorization of bromine water*
2. The bromine in the water causes the pi bond to break. Do you think the sigma overlap in the double bond also broke? *No, it did not break.*
3. Which part of the double bond is more reactive, the sigma or pi bond? *The pi bond is more reactive.*

✓ ASSESSMENT

Performance: Have a student lab group repeat the demonstration procedures for the class using cyclohexane and then cyclohexene instead of cottonseed oil. Ask students which compound contains the double bond.

PROGRAM RESOURCES

Transparency Master: Use the Transparency master and worksheets, pp. 63-64, for guided practice in the concept "Bonding in Ethyne."

Color Transparency: Use Color Transparency 32 to focus on the concept "Bonding in Ethyne."

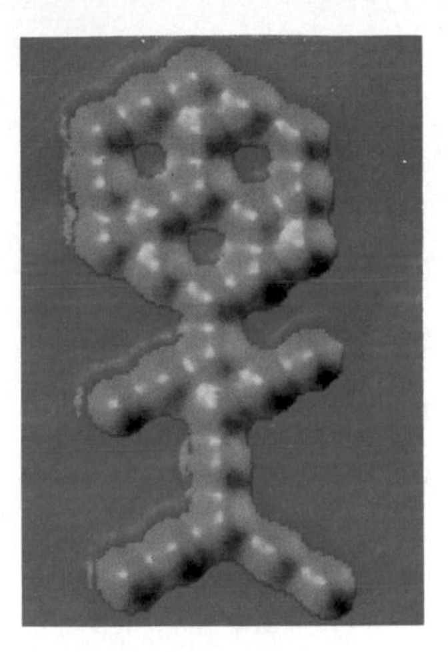

Figure 13.12 Twenty-eight carbon monoxide molecules on a platinum surface were used to create this figure, which is 5×10^3 pm tall.

Multiple Bond Molecular Shapes

The molecular formula for formaldehyde (methanal) is CH_2O. What is its electron dot diagram? If we allot the four electrons from carbon, one from each hydrogen, and six from the oxygen, the dot diagram ends up looking like this:

```
   H
H:C:Ö:
```

Each hydrogen atom has two electrons, so the hydrogen outer levels are full. Likewise, the oxygen has eight electrons, and its outer level is full. However, carbon's outer level contains only six electrons. How can the electrons be adjusted so that carbon has eight, yet none are taken away from other atoms? If atoms share more than one pair of electrons, all atoms in the molecule can have full outer levels. Remember that both atoms get full credit for all shared pairs. For formaldehyde, carbon and oxygen share two pairs of electrons and the diagram becomes

```
   H
H:C::Ö:
```

This diagram of formaldehyde is an example of a Lewis electron dot diagram of a molecule containing a double bond.

In the diatomic molecule, N_2, the two nitrogen atoms share three pairs of electrons. They are bound to each other by a triple bond. The electron dot diagram for the nitrogen molecule is

```
:N:::N:
```

How does the electron-pair repulsion theory predict the shapes of molecules containing multiple bonds? Recall that a double bond consists of four electrons occupying the space between the bonded atoms. The resulting cloud will occupy more space than a single bond. The triple bond occupies still more space than the double bond because six electrons occupy the space between the bonded atoms. How is molecular shape affected by the presence of multiple bonds? In the case of formaldehyde (methanal), there are three clouds around the carbon atom, two single bonds and one double bond. We anticipate that the clouds will assume a trigonal planar shape. The double bond will occupy somewhat more space. As a result, the H—C—H bond angle should be a little less than 120°, and the H—C=O bond angle a little more. Experiment shows a H—C—H bond angle of 116° and a H—C=O angle of 122°.

Figure 13.13 The methanal molecule contains a double bond. Thus, the H-C-H bond angle is smaller than predicted because a double bond occupies more space than a single bond does.

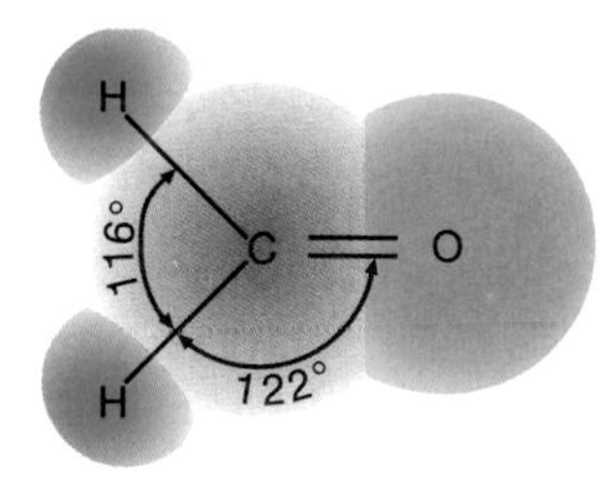

With two double bonds, each to a separate atom, we get a linear molecule. The organic compound, ketene, illustrates two double bonds as well as a double bond with two single bonds. The structure of the ketene molecule is

```
H
 \
  C=C=O
 /
H
```


CONCEPT DEVELOPMENT

Background: It is also possible to have three pairs of outer level electrons arranged as a double bond and a single bond with another atom. The three pairs form two bonds and the result is a linear molecule. This arrangement is found in some unstable organic molecules.

Enrichment: Ask a team of students to predict shapes for the following molecules: **a.** $H_2C=C=CH_2$ and **b.** $H_2C=C=C=CH_2$. Possible response: **a.** *$H_2C=C=CH_2$ is linear, and the entire molecule is planar, but one —CH_2 is in a plane perpendicular to the other —CH_2 group.* **b.** *$CH_2=C=C=CH_2$ is linear and the entire molecule is planar.*

- Summarize for students the following bond types and their resulting bond angles.

H—C—H	116°
H—C=O	122°
C=C=O	180°
H—C=O	120°
H—C≡N	180°

QUICK DEMO **Free Rotation**
Use plastic foam balls and toothpicks to make a model of an ethane molecule, C_2H_6, and demonstrate its free rotation about the single C—C bond. Construct ethene, C_2H_4, by inserting a second toothpick to form the C=C bond and removing two H. Point out that this double bond prevents ethene's free rotation. Discuss the trigonal planar structure associated with the one double bond, and two single bonds around a carbon atom. The students will be able to see this trigonal planar structure in the model of ethene. You will be able to refer back to this demonstration when you discuss *cis-trans* isomers in 13.2.
Disposal: F

DEMONSTRATION

13-4 Reactive Triple Bonds

Purpose: to demonstrate that triple bonds are reactive due to the presence of the two strained parallel overlaps found in the pi bonds

Materials: 3 calcium carbide chunks, large test tube with stopper, 600-cm^3 beaker, 500 cm^3 water, crucible tongs, matches

Procedure: **1.** Fill the test tube with water and stopper it.
2. Invert the test tube into a water-filled beaker. Remove the stopper while keeping the mouth of the test tube under water.
3. Place a chunk of calcium carbide in the water in the beaker and cover it with the inverted water-filled test tube. Collect the gas by water displacement. If gas from the first chunk doesn't completely fill the test tube,

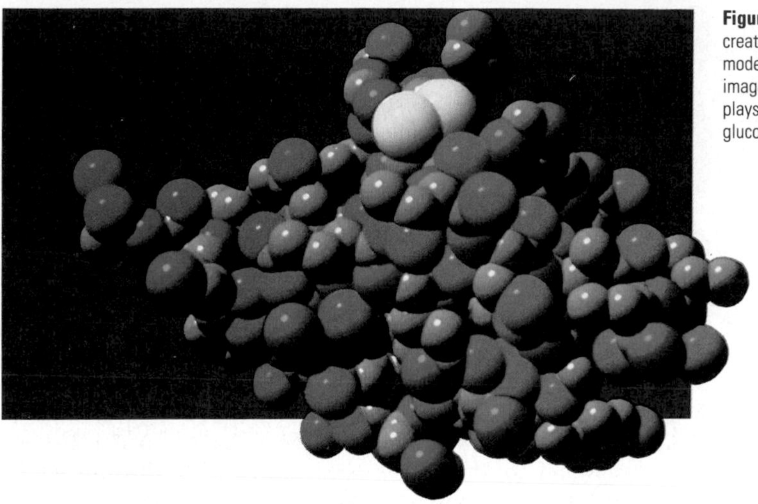

Figure 13.14 Computer simulations can create a picture of a three-dimensional model of a large molecule such as this image of the hormone, insulin, which plays a major role in regulating blood glucose in humans and other mammals.

The C=C=O bond angle is 180°, while the H—C=C bond angle is slightly greater than 120°.

The one remaining case is that of eight electrons shared as a triple bond and a single bond. As you might expect, such an arrangement gives rise to a linear shape. Hydrogen cyanide illustrates this arrangement. It has the structure

$$H-C\equiv N$$

The H—C≡N bond angle is 180°. When individual molecules are investigated in detail, the experimental bond angles are not always exactly as we would predict. Bond angles are influenced by other factors that we will not study in this course.

Benzene

CHEMISTRY IN DEPTH

Pages 333–334

One of the top 20 industrial chemicals in the United States is benzene, C_6H_6. It is used extensively in making drugs, dyes, and coatings as well as solvents. It is also highly toxic and can cause cancer. Each of the six carbon atoms in a benzene ring has three sp^2 hybrid orbitals and one p orbital. Sigma bonds are formed by the overlap of the sp^2 orbitals of six carbon atoms forming a ring of single bonds. The pi bonds of the benzene ring are formed by the sideways overlap of the p orbitals of the carbon atoms. However, note in Figure 13.15 that the unhybridized p orbitals can overlap to *both* sides. Because all six of these orbitals overlap to both sides, a circular orbital forms around the whole ring.

One of the characteristics of benzene is that the pi electrons can be shared all around the ring. Because the pi electrons are shared equally among all the carbon atoms and not confined to one atom or bond, they are delocalized. This **delocalization** of pi electrons among the carbon atoms in benzene results in greater stability of the compound.

CHEMISTRY IN DEPTH

The section **Benzene** on pages 333-334 is recommended for more able students.

CONCEPT DEVELOPMENT

Background: Point out to students that the area of organic chemistry that deals with benzene and its derivatives is called aromatics.

Misconceptions

Beginning students unfamiliar with organic chemistry may think that benzene is another name for a cyclohexane, or 1,3,5-cyclohexatriene. Contrast the structures of the two hydrocarbons.

3 ASSESS

CHECK FOR UNDERSTANDING

Have students answer the Concept Review problems and Review and Practice 15-19, and 26-30 from the end of the chapter.

RETEACHING

Have students use gum drops and tooth picks to make models that represent the molecular shapes and bond angles that are found in the compounds formed from the elements of Groups 1, 2, 13-18. As each model is made, discuss the bonding orbitals, bond angles, and molecular structure of each model.

Substance	Bonding Orbital	Bond Angle	Molecular Structure
LiF	*s-p*	none	diatomic
BeF_2	*sp-p*	180°	linear
BF_3	sp^2-p	120°	trigonal planar
CH_4	sp^3-s	109.5°	tetrahedron
NH_3	sp^3-s	107°	trigonal pyramid
H_2O	sp^3-p	104.5°	bent
F_2	*p-p*	none	diatomic
Ne	none	none	monatomic

drop in a second chunk. **CAUTION:** *Ethyne gas is flammable.*

4. Darken the room, then remove the test tube from the beaker, keeping the mouth of the test tube down. Use tongs to bring a burning match to the mouth of the test tube. **Disposal:** A

Results: The gas-filled test tube will burn around the mouth of the test tube with a yellow, sooty flame. When the room lights are turned back on, students will be able to see soot particles floating near the demonstration.

Questions: 1. The gas produced is called ethyne and contains a triple bond. Is it reactive? *Yes, it burns.*

2. Gases such as methane also burn and contain only sigma bonds. Which fuel would you expect to produce more energy when burned? *Fuels that contain double and triple bonds produce more energy.*

Extension: Car engines and furnaces both need regular tune-ups to operate at maximum efficiency with minimum pollution. When the ethyne gas burned without sufficient oxygen, the soot indicated that the reaction was not complete, and fuel was wasted. Point out that complete combustion of a hydrocarbon produces CO_2 and H_2O.

EXTENSION

Predicting the shape of cyclohexane, C_6H_{12}, can be assigned as an enrichment activity. Ask students to use references to determine the two shapes of cyclohexane. Have them make models of the two shapes to illustrate why one form is more stable than the other. *C_6H_{12} is a puckered ring structure. It has two shapes; the chair form and the boat form. The chair form is more stable because the H—H interference is much less than that found in the boat form.*

Answers to
Concept Review

4. For structure, see *Merrill Chemistry Problems and Solutions Manual.*

C shares its 4 outer electrons with the single atom of H and the 3 atoms of Cl.

5. The molecule is essentially tetrahedral, like that of CH_4. The bond angles are about 109.5°. Because carbon is bonded to two different types of atoms, chlorine and hydrogen, the molecule is probably not a perfectly symmetrical tetrahedron.

6. *sp^2* hybridization

7. For structure, see *Merrill Chemistry Problems and Solutions Manual.*

8. Molecular formula: $C_{10}H_8$
Its structure is that of two benzene rings sharing one of the C—C bonds.

4 CLOSE

Discussion: Students who have camped in a tent will recall that the bottom was probably a plastic called polypropylene. Most plastic wrap in the kitchen is polyethylene. Plastic milk jugs are polyvinyl chloride. Vinyl is used to cover notebooks. Discuss with students that these plastics have different uses because they have different properties resulting from different molecular structures.

CHEMISTRY IN DEPTH

Chemistry in Depth ends on this page.

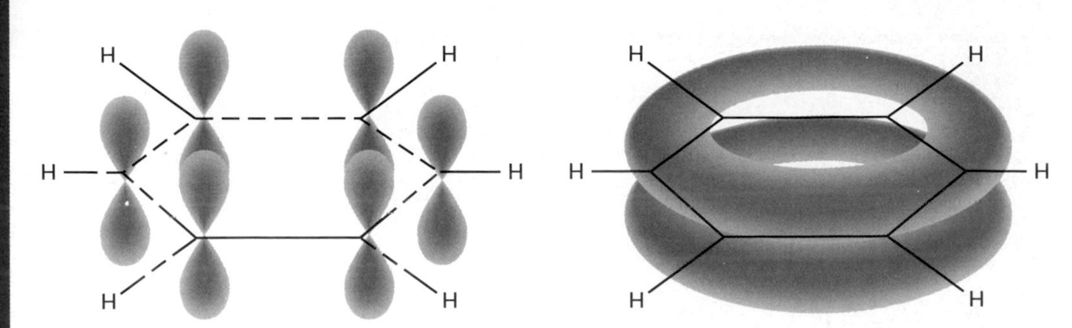

Figure 13.15 The benzene ring results from the overlap of six sp^2 hybrid carbon atoms. The unhybridized p orbitals form two pi clouds allowing electrons to be shared around the ring.

The following are representations of the benzene molecule. The solid lines in the drawing on the left represent sigma bonds formed between carbon sp^2 hybrid orbitals and hydrogen *s* orbitals, as well as between two adjacent carbon sp^2 hybrid orbitals. The circle represents the delocalized electrons in the pi bond formed from unhybridized carbon *p* orbitals. The drawing on the right is the simplified representation of the bonding in benzene. The vertices of the figure represent carbon atoms, and each straight line represents a bond formed between two carbon atoms.

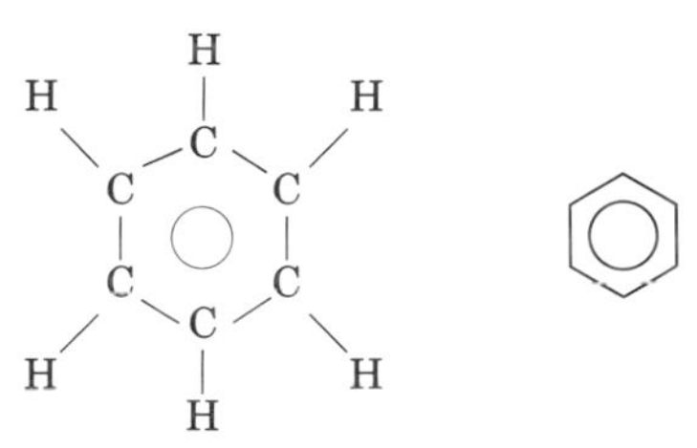

Whenever multiple *p* orbital overlap can occur, the molecule is said to contain a **conjugated system.** Conjugated systems can occur in chains as well as in rings of atoms. 1,3-butadiene, $CH_2{=}CH{-}CH{=}CH_2$, is an example. Again, the conjugated system imparts a special stability to the molecule.

13.1 CONCEPT REVIEW

4. Draw a Lewis electron dot diagram for $CHCl_3$. Explain what each chemical symbol represents and why the electrons have the indicated locations.

5. Predict the shape and bond angles for the molecule in Problem 4.

6. A molecule has a trigonal planar shape. What hybridization would you expect to find on its central atom?

7. Draw a structural diagram for a propyne molecule, C_3H_4. Label each sigma bond and each pi bond.

8. Apply The structure of naphthalene can be represented by the structural formula [naphthalene structure]. Write the molecular formula for naphthalene, and compare its structure to that of benzene.

PROGRAM RESOURCES

Chemistry and Industry: Use the worksheets, pp. 21-22, "Penicillin: From Fermentation to Medicine" as an extension to the concepts presented in Section 13.1

Study Guide: Use the Study Guide worksheets, pp. 47-49, for independent practice with concepts in Section 13.2.

Chemistry Journal

Hydrocarbon Series
Have students research the three hydrocarbon series that start with methane, ethene, and ethyne. Ask them to list, in their journals, the formulas and molecular shapes for the first four members of each series.

13.2 Molecular Arrangements

OBJECTIVES

- name and write formulas for simple organic compounds.
- define, explain, and give examples of isomerism.

Sometimes two or more compounds that are quite different have the same molecular formula. In order to clearly identify such compounds, we need to draw structural formulas and write distinctive names. The same principles apply to both organic and inorganic compounds. In this section we will deal with these variations in molecular shape.

Organic Names

In Chapter 7, it was mentioned that names for organic compounds carry a suffix describing how the atoms are bonded. The ending *-ane* was introduced at that time. You are now ready to expand your knowledge of organic compound names.

Compounds having only single bonds between carbon atoms are said to be **saturated compounds.** The ending *-ane* is used to name saturated compounds. For compounds containing a carbon-carbon double bond, the suffix *-ene* is used and for carbon-carbon triple bonds *-yne* is used. Recall that the name for $H_2C{=}CH_2$ is ethene, which indicates a double bond between the carbon atoms. The name for $H{-}C{\equiv}C{-}H$ is ethyne because there is a triple bond between carbon atoms. Ethene and ethyne are unsaturated compounds.

The compound known as propene, CH_3CHCH_2, has the following structure

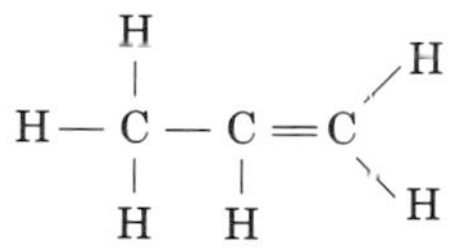

```
     H            H
     |          /
H — C — C = C
     |    |     \
     H    H       H
```

also written as $CH_3{-}CH{=}CH_2$.

Cyclopentene has the structure

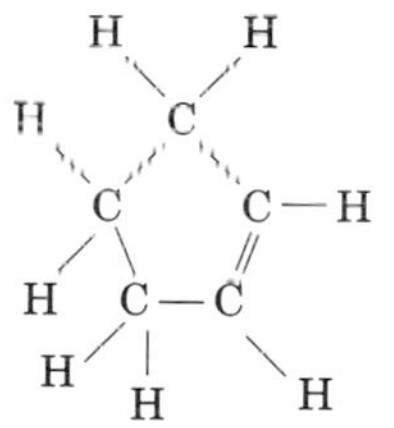

```
     H     H
       \  /
 H      C
   \  /   \
    C       C — H
  /  \     //
 H    C — C
    / |     \
   H  H      H
```

also written as ⬠ (cyclopentene ring).

Figure 13.16 shows other simplified diagrams used to represent cyclic compounds. It is understood that there is a carbon atom at the intersection of each pair of straight lines. We also know from Section 13.1 that carbon always forms four bonds. This bonding could be accomplished by four single bonds, a double and two single bonds, a single and a triple bond, or two double bonds. Therefore, when we use these simplified diagrams, we assume enough hydrogen atoms are bonded to each carbon to give that carbon four bonds

13.2 Molecular Arrangements

PREPARE

Key Concepts

The major concepts presented in this section include elementary nomenclature of alkanes, alkenes, alkynes, and cyclic organic molecules, four different types of isomers, hybridization in inorganic molecules, and a complete summary of bonding and the resultant structures.

Key Terms

saturated compound
isomerism
isomer
structural isomer
positional isomer
functional group
functional isomer
geometric isomer

1 FOCUS

Objectives

Preview with your students the objectives listed on the student page. Students can use them as a study guide for this major section.

Activity

A pair of balls, called "happy and sad balls," is available from science suppliers. As students enter the room, bounce the "happy ball". Without being observed, switch to the "sad ball," which looks the same, but doesn't bounce. Ask students to explain what happened to the ball. Point out that the two materials in the balls have different structures, and, therefore, different properties.

CONTENT BACKGROUND

Saturated/Unsaturated: The term *saturated* hydrocarbon is applied to organic molecules that have only single C—C bonds. Unsaturated hydrocarbons contain double or triple carbon bonds. If each carbon bonding position is filled, the atom is saturated. *Unsaturated* indicates that additional bonding can occur.

cyclopropane

cyclobutane

cyclopentane

cyclohexane

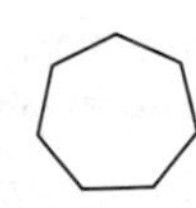
cycloheptane

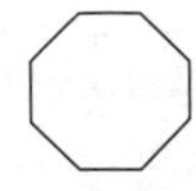
cyclooctane

Figure 13.16 These simplified diagrams are used to represent the first six cycloalkanes.

2 TEACH

CONCEPT DEVELOPMENT

Discussion: Ask students to turn to Table 7.12 to review the hydrocarbon stems and the endings that indicate single, double and triple bonds used in naming organic molecules.

Correcting Errors

Some memorization is necessary in chemistry. Encourage students to learn the ten hydrocarbon stems and the three suffixes used to indicate the types of bonds present. To help them learn this information, have them prepare a 3 x 5 card listing the ten stems and the *-ane, -ene,* and *-yne* endings.

PRACTICE

Guided Practice: Guide students through the Sample Problems, then have them work in class on Practice Problems 9 and 10.

Independent Practice: Your homework or classroom assignment can include Review and Practice 20-25 from the end of the chapter.

Answers to
Practice Problems

9. **a.** propyne
   b. cyclobutene
10. For structures, see Appendix C.

✓ ASSESSMENT

Skill: Review with students Table 7.12 on page 176 and the Sample Problem on page 177. Then, ask them to draw the straight chain model including hydrogen atoms for alkanes from Table 7.12 as you call the names of five alkanes at random.

PROBLEM SOLVING HINT

Prefixes are listed in Table 7.9, page 175.

PROBLEM SOLVING HINT

Note that the location of a multiple bond in a cyclic compound is arbitrary when only one multiple bond is present in the molecule.

SAMPLE PROBLEMS

1. Organic Nomenclature

Name the compound 

Solving Process:

There are three vertices in the figure; therefore, there are three carbon atoms in the compound and the stem of the name will be *prop.* The atoms are arranged in a ring, so the prefix *cyclo-* is used. Finally, there is a double bond, so the suffix is *-ene.* The name of the compound is *cyclopropene.*

2. Representing Organic Compounds

Draw cycloheptene.

Solving Process:

The stem of the name is *hept,* indicating seven carbon atoms. The prefix *cyclo-* means they are arranged in a ring. The suffix *-ene* means there is a double bond in the molecule.

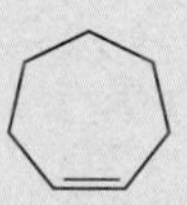

PRACTICE PROBLEMS

9. Name the following compounds.

   a. 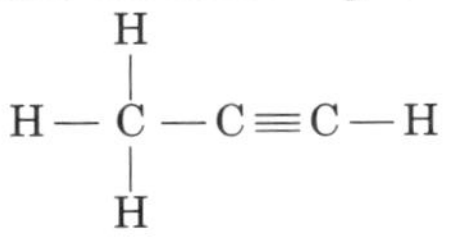 H—C(H)(H)—C≡C—H **b.**  [square with one double bond]

10. Draw structural formulas for the following compounds.
    a. cyclooctyne
    b. cyclohexene

Word Origins

isos: (GK) equal, same
meros: (GK) part
isomer—made of the same parts

Isomers

The existence of two or more substances with the same molecular formula, but different structures, is called **isomerism** (i SOM eh rihz uhm). These structures are **isomers** (I soh murs). Because isomerism is common in organic chemistry, we will study the isomers of carbon compounds.

Consider the compound with the formula C_4H_{10}. There are two structures that can be written for this formula. Butane and methylpropane are examples of **structural isomers** or skeleton isomers, since it is the carbon chain that is altered. Methylpropane was at one time called isobutane since it is made of the same atoms as butane.

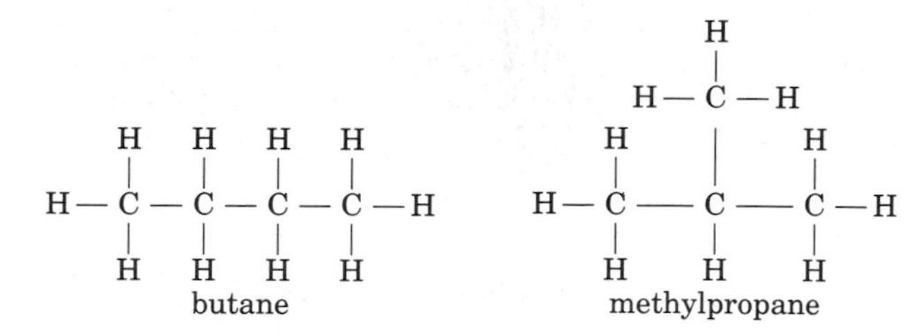

butane methylpropane

A second type of isomerism is positional isomerism. **Positional isomers** are two compounds that differ only in the position of something such as a double bond or an atom other than hydrogen and carbon. For example, consider the following two molecules.

$$CH_3-CH=CH-CH_3 \quad \text{and} \quad CH_2=CH-CH_2-CH_3$$

Both compounds are named butene, but plainly they are not the same compound. We name isomers such as these by specifying on which carbon atom the double bond begins. Number from the end, giving the double bond the lowest number. Thus, the compound $CH_3-CH=CH-CH_3$ is named 2-butene and the compound $CH_2=CH-CH_2-CH_3$ is named 1-butene.

When an atom or atoms different from hydrogen and carbon are introduced into an organic molecule, that part of the molecule is called a **functional group.** One of the most common functional groups is the alcohol group, $-OH$. For example, propanol is a three-carbon compound with single bonds and an alcohol group. However, the alcohol group could be placed on the end carbon or in the middle. The corresponding names reflect these locations.

1-propanol 2-propanol

Functional isomers have an element other than hydrogen or carbon being bonded in two different ways. In other words, the two isomers have the same atoms arranged in different functional groups. Ethanol is a liquid while methoxymethane is a gas at room temperature.

ethanol methoxymethane

CHEMISTS AND THEIR WORK

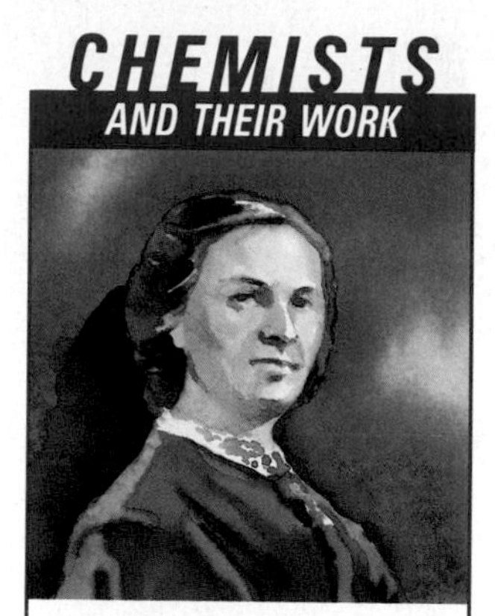

Mary Lura Sherrill
(1888-1968)

Leadership in educating young people in chemistry is an important quality for which some chemists have an exceptional talent. Mary Lura Sherrill was such a chemist. For 34 years, Sherrill taught chemistry at Mt. Holyoke College. Prior to that experience she had taught at Randolph-Macon Women's College, her alma mater, and the University of North Carolina. During World War II she also investigated antimalarial drugs for use by servicemen and women abroad. The American Chemical Society presented her with one of its top awards in 1947.

CHEMISTS AND THEIR WORK
Mary Lura Sherrill

Background: This year over one million people will die from malaria, most under the age of five years. As a chemist, Sherrill investigated ways to protect people from this mosquito-carried disease. Her research was done at the Edgewood Arsenal for the United States Army.

Exploring Further: Investigate Sherrill's research work involving antimalarial drugs.

Background: A 22-year-old, Dutch-born chemist, in his first publication proposed that the four carbon bonds were directed toward the corners of a regular tetrahedron. He used this three-dimensional representation to explain *cis-trans* isomerism. Jacob van't Hoff (1852-1911) received the first Nobel prize in chemistry.

QUICK DEMO — Isomers

Repeat the Focus Activity. "Happy and Unhappy Balls" are a pair of commercially available black spheres that appear to be identical. However, they exhibit very different physical properties. The "happy ball" is made of neoprene rubber and bounces normally. The "unhappy ball" is made of Norsorex and does not bounce at all. These are listed in science supply catalogs and can be used to give students a very dramatic impression of how two isomers might behave. Ask students to think of practical uses for both types of rubber. Possible response: The "dead" rubber absorbs shocks and would make a good running shoe sole or car bumper.

CONTENT BACKGROUND

Isomer Separation: You will want to emphasize that mixtures of isomers can be separated because each isomer has its own chemical and physical properties. Three-dimensional models help students visualize this concept. You may want to give extra credit to students who volunteer to build models for your use in class.

CONCEPT DEVELOPMENT

Computer Demo: *PEEKS—1984* (revised), Student inputs molecular formula, then program produces isotope pattern graphically. AP205, Project SERAPHIM

EVERYDAY CHEMISTRY

Gelatin

Background: Gels are liquid colloids that have been cooled into a semisolid state. Discuss them in relation to sols (see Chapter 20). The enzyme in pineapple is bromelin.

Teaching This Feature: Show a bowl of gelled gelatin and ask students if they recognize it. Allow each student to feel a small cube.

Connection to Food Science: Make students aware of this field of study in universities. Food scientists provide quality assurance during production, as well as research and development of new food items that are nutritious, inexpensive and tasty.

Extension: Interested students may want to read about colloids in Chapter 20 of this textbook.

Answers to

Exploring Further

1. Apples and pears are two fruits. The pH should be about 5.

2. Sols are colloids formed by solids dispersed in a medium. Gels are coagulated sols.

3. Gelatin is often used in products where a semisolid product is desired, such as pie and pudding mixes.

✓ ASSESSMENT

Knowledge: Use molecular models, the overhead, or the chalkboard to show students structural diagrams like those on pages 337 and 338 that are examples of the four types of isomerization they have studied. Ask them to identify and record the type of isomerization as the models or diagrams are shown. Try to use (or have students make) molecular models to demonstrate geometric isomers.

EVERYDAY CHEMISTRY

Gelatin

Nearly everyone has had the opportunity to see gelatin dissolve in water, cool in the refrigerator, and thicken. This process happens because of the structure and properties of the proteins that make up gelatin.

Gelatin is collagen that has undergone a structural change. Collagen, the protein found in the connective tissue of animals, is composed of three separate chains of amino acids intertwined. When heated in water, the weak bonds forming collagen are broken and the chains of proteins are unbraided.

Gelatin is generally dissolved in hot water. The dissolved gelatin is then added to other liquids and refrigerated. The weak bonds begin to form again, but they randomly form a large network that captures the liquid ingredients. The semisolid formed is called a gel. The more the gel is cooled, the more solid it becomes, because more weak bonds have had the opportunity to form.

Sometimes gels do not form or they are not firm enough. The best conditions for this particular gel formation are a pH of about 5 and addition of a small amount of sugar. Therefore, if fruits are added, those that are slightly acidic have a better chance of jelling.

Using pineapple in a gel illustrates the instability of gels. An enzyme in fresh pineapple will break the protein chains into small pieces that will not gel. Cooked or canned pineapple does not have this effect because the enzyme that destroys gelatin is itself destroyed in cooking.

Exploring Further

1. What are some fruits that are acidic enough to set up well in gelatin?
2. Look up sols and gels in a science handbook. How are they alike? How are they different?
3. Besides gelatin dessert, gelatin is also used to govern the consistency of many other products of the kitchen. What are some of these products?

A fourth type of isomerism can be seen in the two structures shown for butene. The formation of a π bond prevents the atoms on each end of the bond from rotating with respect to each other. Some compounds containing double bonds exhibit a kind of isomerism called geometric isomerism. **Geometric isomers** are composed of the same atoms bonded in the same order, but with a different arrangement of atoms around a double bond. We will use 2-butene (C_4H_8) to illustrate geometric isomers. Note that in the *cis* form of 2-butene, the $—CH_3$ groups and the hydrogen atoms are on the same side of the double bond. In the *trans* form, the $—CH_3$ groups and the hydrogen atoms are on opposite sides.

```
CH3        CH3          CH3        H
   \      /                \      /
    C = C                   C = C
   /      \                /      \
  H        H              H        CH3
 cis-2-butene            trans-2-butene
```

cis-2-butene   *trans*-2-butene

PROGRAM RESOURCES

Applying Scientific Methods in Chemistry: Use the worksheet "Saturated vs. Unsaturated Compounds," p. 30, to help students apply and extend their lab skills.

Laboratory Manual: Use macrolab 13-2 "Modeling Molecular Polarity," as enrichment for the concepts in Chapter 13.

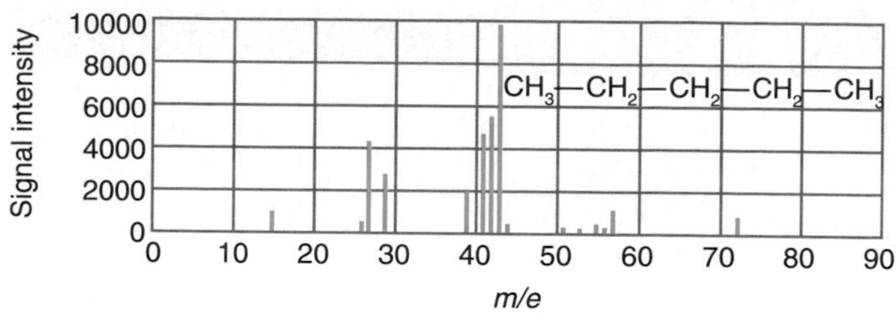

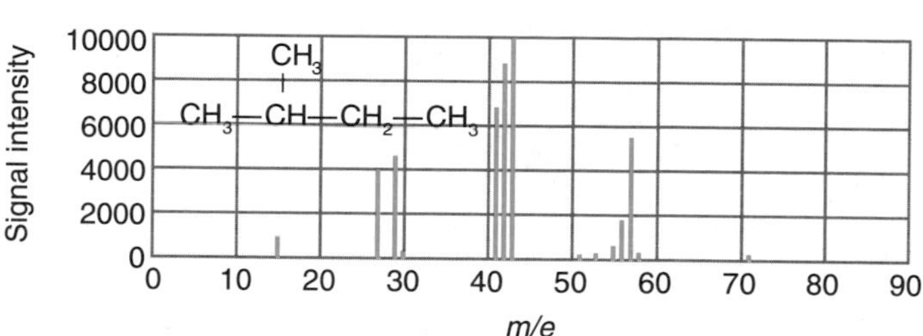

Figure 13.17 The mass spectrograms for the isomers pentane (top) and 2-methylbutane (bottom) are similar, but the difference in values for each peak can be used to identify each compound.

The mass spectrometer can be used to distinguish between isomers that have very similar properties. During the analysis, the sample being studied is ionized and divided into ion fragments. Each fragment has an *m/e* value (mass to charge ratio). In Figure 13.17, the mass spectrogram for pentane shows peaks at 15 different *m/e* values. Thus, the pentane molecule can be broken into 15 fragments that each have a different mass. Note the spectrogram for 2-methylbutane shows similar fragments. However, you can see that the relative intensities of the two isomers are significantly different.

Inorganic Compounds

The ground state electron configuration of carbon ends in $2s^22p^2$. Each of the four electrons was placed in a separate orbital. We treated the carbon atom as if it were in the state $2s^12p^3$, the configuration in which hybridization of orbitals could occur.

What about other atoms? For example, beryllium has two outer electrons, $2s^2$. We would expect *sp* hybridization, leading to linear bonding. Analysis bears out this prediction for molecules having a central atom ending in an s^2 configuration. For atoms with three outer electrons, we would predict sp^2 hybridization. Again, analysis confirms trigonal planar bond arrangements.

H — Be — H

beryllium hydride

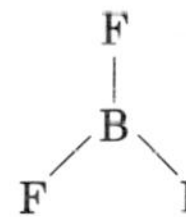

boron trifluoride

We can also apply other principles of molecular geometry and isomerism to inorganic compounds.

CONCEPT DEVELOPMENT

Computer Demo: *Conformational Analysis,* Newman projections of methane, ethane, butane. AP706, Project SERAPHIM.

- Point out to students that the electron dot diagrams for beryllium and boron compounds do not follow the octet rule. Beryllium is stable with four outer electrons and boron with six outer electrons.

MAKING CONNECTIONS

Daily Life

You may want to send a team of students to a local carpet store to interview a salesperson about the different fibers used in carpet manufacture. Have them determine the advantages and disadvantages of the various carpet fibers. Most synthetic fibers used have double bonds present in their structures.

GLENCOE TECHNOLOGY

Videodisc

Chemistry: Concepts and Applications

Isomers

Disc 2, Side 2, Ch. 11

Also available on CD-ROM.

Chemistry Journal

Geometric Isomers

Have students practice drawing geometric isomers by asking them to draw in their journals the *cis* and *trans* isomers for a hydrocarbon molecule with five carbon atoms and one double bond. The molecule should have a methyl group on one carbon and an ethyl group on the other carbon involved in the double bond.

CONCEPT DEVELOPMENT

Using the Table: Take time to discuss Table 13.2 in detail. It summarizes much of the information presented in this chapter. You may wish to have copies of it available to students during the test on this chapter.

FRONTIERS

Super Soot

Background: Samples of C_{60}, buckminsterfullerene, are produced by passing a large current through two carbon electrodes. The vaporized graphite is deposited as soot and then dissolved in an organic solvent and centrifuged.

Teaching this Feature: Have students look at the photograph of the model of a C_{60} molecule and identify single and double bonds. Discuss how the bonds indicate conjugated systems within the molecule. Have students recall that conjugated systems impart special stability to molecules and relate this to the stability of buckminsterfullerene.

Connection to Architecture: Have interested students research the design and structural dynamics of R. Buckminster Fuller's geodesic domes. They may want to build a small model using drinking straws and thin wire.

Exploring Further: Have interested students construct three-dimensional models of C_{60} molecules. Students will each need to construct an equilateral pentagon and an equilateral hexagon on pieces of card stock to use as patterns. The length of one side of the pentagon must be equal to the length of one side of the hexagon. Students will need to cut out 12 pentagons and 20 hexagons from heavy construction paper for each molecule. To assemble the molecules, students must realize that each of the 60 carbon atoms in a C_{60} molecule is at a vertex formed by the junction of a pentagon and two hexagons. By taping the polygons together in such a manner, the students will be able to construct their own buckyballs.

FRONTIERS

Super Soot

What has 32 sides, is hollow, and is the roundest molecule that can possibly exist? The answer, introduced in the chapter opener, is buckminsterfullerene, C_{60}, the best known member of a class of chemical substances called fullerenes. The structure of this molecule is so stable that experiments have revealed molecules of C_{60} bouncing back to shape after striking a steel plate at a speed of 7000 m/s. Because of the soccer ball-like shape and behavior of buckminsterfullerene molecules, tongue-tied and tired researchers have named them "buckyballs!"

Because of the hollow structure of the buckyball, substances may be trapped within the molecule. Theorists predict that C_{60} will not react with substances trapped within its molecular cage and envision buckyballs as molecular reaction chambers. Substances entrapped within the molecules would be forced to react with each other as the molecule is compressed. After decompression, the products would be extracted from the intact C_{60} molecule.

On the surface of a C_{60} molecule, things are different. Researchers have reacted buckminsterfullerene with fluorine and isolated $C_{60}F_{60}$. With the possibility of 60 bonding sites on each molecule, chemists anticipate sticking groups of atoms at various locations on the surface of the molecule. Such compounds could be designed to have specific chemical and physical properties.

Bond Summary

The substances in Table 13.2 illustrate the principles of molecular geometry. The angular values given are the actual values determined experimentally. Boron trichloride, BCl_3, has the shape we would expect from the electron-pair repulsion model. This compound is used in metallurgy to produce high purity metals including some used in the manufacture of computer chips.

Carbon dioxide, CO_2, is already familiar to you as a product of combustion and of respiration in living cells. It is also produced commercially for use in carbonated beverages, fire extinguishers, as the refrigerant dry ice, and as a propellant in aerosol cans containing such things as whipped cream.

Cyclohexane, C_6H_{12}, is a widely used solvent in many industries. It is also used by the chemical industry as a reagent in the synthesis of other chemicals.

Carbon tetrachloride, CCl_4, is no longer used extensively. At one time it was a common solvent and dry-cleaning fluid. However, it has been found that prolonged exposure to its vapors causes severe liver damage and may cause cancer. Carbon tetrachloride was used as a fire extinguishing liquid until it was discovered that, at high temperatures, it is converted to a very toxic gas called phosgene. Carbon tetrachloride is used in the chemical industry as a reagent in the synthesis of other chemicals.

✓ ASSESSMENT

Portfolio: Ask students to design a network tree concept map that shows how hydrocarbons are classified as saturated and unsaturated. Have students map the type of bond (single, double, or triple), and the nomenclature suffix that is used with each type of bond. Finally, ask students to map the approximate bond angle for each type of bond and to draw a 3-carbon structure illustrating that type of bond. Remind students that this concept map can be included as a sample in their portfolio.

Oxygen difluoride, OF_2, is also quite toxic. It is of interest mainly because it is one of the very few compounds in which oxygen exhibits a positive oxidation state.

Ammonia, NH_3, is produced commercially in huge quantities. It is used to make nitric acid, explosives, synthetic fibers, and hundreds of other materials. In addition, ammonia is used extensively in the production of fertilizers and as a refrigerant.

Ozone, O_3, is found in our atmosphere at an altitude of about 25 km. It protects us from the sun's harmful ultraviolet rays. Ozone is also a major component of smog. Industrially it is used as a disinfectant and as a bleaching agent. Small amounts of ozone are also used in the synthesis of organic compounds.

Methane, CH_4, is the principal constituent of natural gas and is therefore one of our major fuels. It is also widely used in the chemical industry in the synthesis of other organic compounds.

Acetylene (ethyne), C_2H_2, is another fuel that is important for welding and use in cutting torches. Acetylene is also a raw material for plastics and synthetic fibers.

The bond angles (96°) of trimethylarsine, $(CH_3)_3As$, seem to indicate that the bonding orbitals of the arsenic atom are *p* orbitals rather than the sp^3 hybrid orbitals we might have expected. Generally we find that the bonding orbitals of higher atomic mass elements are hybridized much less than in lighter elements. The reason may be that the heavier atoms can accommodate more bonded atoms around them because they are larger.

Table 13.2

Experimental Molecular Shapes		
Cl, B, Cl, Cl 120°	Cl, C, Cl, Cl, Cl 109.5°	H, C, H, H, H 109.5°
O=C=O 180°	O, F, F 103°	H—C≡C—H 180°
H, H, H, C, C, C, C, C, H, H, H, H, H, H, H, H 109.5°	O, H, H 104.5°	As, H_3C, CH_3, CH_3 96°
	N, H, H, H 107°	[Cl, Cl, Cl, Cl, Re—Re, Cl, Cl, Cl, Cl]$^{2-}$ 87°, 104°
	O, O, O 117°	

3 ASSESS

CHECK FOR UNDERSTANDING

Have students answer the Concept Review problems, and Review and Practice 20-25, 27, 31-35, and 37 from the end of the chapter.

RETEACHING

By assigning individual problems to individual students, and then presenting the answers in class,everyone can think about the answers to a large number of questions without having a heavy homework load. Assign each student one Review and Practice and one Concept Mastery question to research and present the answers.

EXTENSION

Ask a student to research and report on the historical development of molecular structure theories. Possible response: *Alfred Werner → Langmuir → G.N. Lewis → Sidwick → Born → Pauling → Ingold*

CHEMISTRY AND TECHNOLOGY

The Boranes

Background: Magnesium boride and hydrochloric acid were reacted in 1879 by F. Jones to produce the first boranes. The gas produced by the reaction was condensed with liquid air.

Teaching This Feature: Point out to students that we believe that boron has three sp^2 hybrid orbitals. Two of the sp^2 hybrid orbitals are used to sigma bond to end hydrogen atoms. The other sp^2 orbital is used to bond to the "bridge" hydrogen.

Connection to Aerospace: Chemists continue to develop new fuels and fuel mixtures to propel the next generation of spacecraft into orbit. The diboranes are often used as a fuel component.

Answers to

Thinking Critically

1. The structure of borazine is identical to that of benzene except alternating N and B atoms replace the six C atoms that are in benzene.

2. Each pair of electrons connecting to a bridging hydrogen is shared among the two boron atoms and the hydrogen atom.

Exploring Further: Investigate A.E. Stock's studies of the boranes.

CHEMISTRY AND
TECHNOLOGY

The Boranes

Boron and hydrogen form several compounds with unusual bonding arrangements. From the electron configurations of boron and hydrogen, we would predict boron hydride, BH_3, to be a planar molecule with 120° H—B—H bond angles. Instead, the two elements form a whole series of compounds called boranes. The simplest compound in the series is diborane(6), B_2H_6.

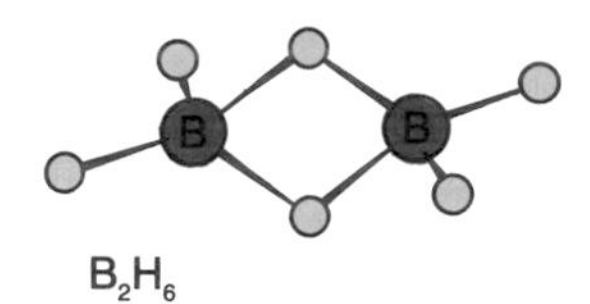

B_2H_6

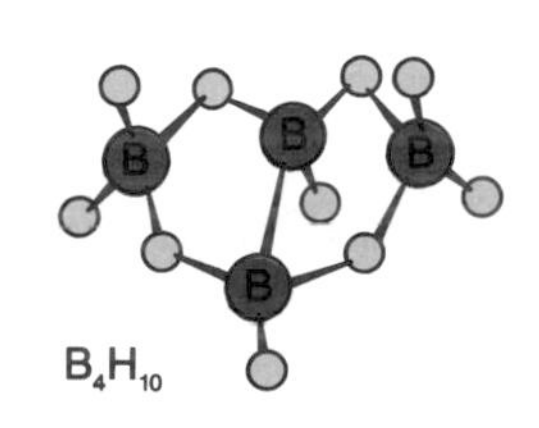

B_4H_{10}

All of these borane compounds are said to be electron deficient. That is, there are not enough electrons in the outer levels of the atoms to form any electron dot diagram to satisfy the octet rule. As a result, chemists have developed a three-center, two-electron bond model. In this model, a pair of electrons occupies an orbital spread over three atoms, so that the middle atom acts as a bridge between the other two atoms.

In the structural diagrams shown here, note that each contains bridging hydrogen atoms in three-center, two-electron bonds. In addition, all boranes having five or more boron atoms have at least one boron atom acting as a bridge between other boron atoms.

Most boranes are unstable. Several, including diborane(6), are spontaneously flammable. Diborane(6) is used as a reactant in making other chemicals. Diborane(6) has also been used on an experimental basis as a rocket fuel. The best rocket fuels are composed of low atomic mass elements. Lighter atoms travel faster than heavy atoms at the same temperature. The combustion of boranes produces higher-velocity gases in the rocket motor nozzle. Also, these compounds produce more energy than similar molecular mass organic fuels. For example, the energy produced by burning a mole of diborane(6), B_2H_6, is 2170 kJ, while one mole of ethane, C_2H_6, produces only 1560 kJ.

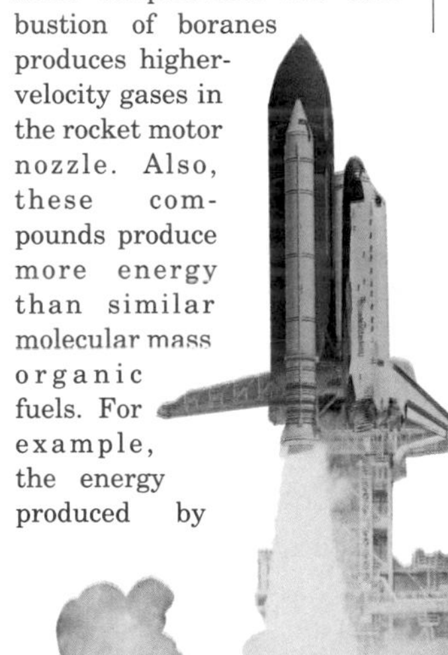

Carboranes are compounds in which one or more boron atoms have been replaced by carbon atoms. Their chemistry has been studied extensively in the hope that they too can be used as rocket fuels. Carborane polyesters have been used as ablative materials. An ablative material has the ability to carry away heat by gradually peeling away in layers. Space capsules use ablative materials in their heat shields. The peeling occurs when the capsule reenters Earth's atmosphere.

Thinking Critically

1. What is the structure of the borazine molecule? How does its electronic structure compare with benzene's?
2. Draw a Lewis electron dot diagram for diborane(6), and use this diagram to explain three-center, two-electron bonds.

Figure 13.18 As was mentioned on page 341, ammonia is used extensively to produce many useful commercial products such as these synthetic fabrics.

The octachlorodirhenate(III) ion, $[Re_2Cl_8]^{2-}$, is shown in Table 13.2 as an illustration of a metal-metal covalent bond, something that is not very common. In this ion, the chlorine atoms occupy the eight vertices of a cube.

In studying atomic structure chemists treat the electron as a particle, wave, and negative cloud. Several different approaches to bonding have also been studied and used to explain what can be observed. It is plain that chemists do not have a complete understanding of all factors in bonding. Therefore, more than one explanation is often needed to account for observations.

When faced with multiple explanations, scientists follow a basic rule. That rule is to try the simplest explanation first. If that method does not suffice, then the more complex ideas are applied until one is found to fit. The models of molecular structure presented in this chapter are arranged in order of increasing complexity. Electron-pair repulsion is the simplest model and should be applied first when attempting to explain the structure of a molecule.

13.2 *CONCEPT REVIEW*

11. Name these compounds.
  a. $CH_2{=}CH{-}CH_2{-}CH_2{-}CH_3$
  b. $CH_3{-}C{\equiv}C{-}CH_3$

12. Draw the structure of each of the following compounds.
  a. cyclopentene **b.** nonane

13. Draw structures for two structural isomers, two positional isomers, and two geometric isomers of a molecule with the formula, C_5H_{10}.

14. Apply Are these structures examples of structural isomers? Justify your answer.

```
    H   H   H   H   H
    |   |   |   |   |
H — C — C — C — C — C — H
    |   |   |   |   |
    H   H   H   H   H
```

```
                 H
                 |
    H      HH —  C — H
    |       |    |
H — C   —   C —  C — H
    |       |    |
H — C — HH       H
    |
    H
```


Answers to
Concept Review
11. a. 1-pentene
  b. 2-butyne
12. For structures, see *Merrill Chemistry Problems and Solutions Manual.*
13. For structures, see *Merrill Chemistry Problems and Solutions Manual.*
14. No, both structures are unbranched pentane molecules, $CH_3CH_2CH_2CH_2CH_3$.

4 CLOSE

Discussion: Vitamin C is important to our health. It can be a part of our diet through the foods we eat, or it can be obtained as a supplement in the form of a vitamin pill. Ask students if the naturally-obtained vitamin C has a different structure than the laboratory-produced vitamin C. Direct the discussion so that students realize that if two molecules, one from orange juice, and one from a pill, have the same structure, they will have the same properties and chemical behavior.

PROGRAM RESOURCES

Vocabulary Review/Concept Review: Use the worksheets, p. 13 and pp. 59-60, to check students' understanding of the key terms and concepts of Chapter 13.

CHAPTER REVIEW 13

- Review Summary statements and Key Terms with your students.
- Complete solutions to Chapter Review problems can be found in **Merrill Chemistry Problems and Solutions Manual.**

Answers to
Review and Practice

15. For structures, see *Merrill Chemistry Problems and Solutions Manual.*
16. For structures, see *Merrill Chemistry Problems and Solutions Manual.* The electron pairs represented by "x·" are shared pairs because one electron comes from N and one comes from H, and they are shared by the two atoms. The N has one unshared pair, "··" that is not shared by any other atom.
17. a. tetrahedral, **b.** trigonal pyramidal, **c.** octahedral, **d.** tetrahedral, **e.** tetrahedral, **f.** square planar
18. a. trigonal planar, **b.** bent, **c.** trigonal planar, **d.** octahedral, **e.** zigzag, **f.** irregular tetrahedron
19. Hybrid means it has the average characteristics of the orbitals merged to form it.
20. For structures, see *Merrill Chemistry Problems and Solutions Manual.*

Chapter 13
REVIEW

Summary

13.1 Bonds in Space

1. The structure of a molecule or a polyatomic ion can be represented by a Lewis electron dot diagram, which shows the pattern of shared and unshared pairs of electrons.
2. The shape of a molecule can be predicted by taking into account the mutual repulsion of electron pairs due to the same electrostatic charge and the Pauli exclusion principle. Electron pairs spread as far apart as possible.
3. The shape of a molecule containing three or more atoms is determined by the number and type of electron clouds in outer levels of the atoms.
4. The actual bond angles may vary from predicted angles. Unshared electron pairs occupy the most space. Shared pairs occupy the least space.
5. The *s* and *p* orbitals of the same atom can combine to form hybrid orbitals. One *s* and three *p* orbitals can merge to form four tetrahedral sp^3 hybrid orbitals. One *s* and two *p* orbitals can merge to form three planar sp^2 hybrid orbitals. One *s* and one *p* orbital can merge to form two linear *sp* hybrid orbitals.
6. If two *s* orbitals or an *s* and a *p* orbital overlap, a sigma (σ) bond is formed. If two *p* orbitals overlap end to end, a sigma bond is formed. If two *p* orbitals overlap sideways, a pi (π) bond is formed. Pi electrons are farther from the nucleus.
7. Two atoms sometimes share more than one pair of electrons, forming double and triple bonds. This possibility must be considered when predicting geometry of a molecule.
8. Delocalization of pi electrons in conjugated chain and ring compounds such as benzene increases the stability of the structure.

13.2 Molecular Arrangements

9. Names of organic compounds carry a suffix to indicate how the atoms are bonded. The ending *-ane* is used for compounds in which all bonds are single bonds. The ending *-ene* is used for compounds with a carbon-carbon double bond. The ending *-yne* is used for compounds with a carbon-carbon triple bond.
10. Cyclic compounds may be represented using skeletal outlines with double and triple lines representing double and triple bonds respectively.
11. Isomers are compounds with the same molecular formula but with different arrangements of atoms and bonds.
12. Compounds may have several types of isomers including structural, geometric, positional, and functional.
13. Electron-pair repulsion is the simplest model used in predicting the structure of a molecule. Therefore, it should be applied before more complex explanations are tried.

Key Terms

shared pair	delocalization
unshared pair	conjugated system
hybridization	saturated compound
hybrid orbital	isomerism
sigma bond	isomer
pi bond	structural isomer
double bond	positional isomer
triple bond	functional group
unsaturated compound	functional isomer
	geometric isomer

Review and Practice

15. Draw Lewis electron dot structures for the following substances.
a. H_2S **b.** Br_2 **c.** HBr **d.** CH_2Cl_2 **e.** NH_3 **f.** $AlBr_3$

21. a. cyclohexene, **b.** butyne, **c.** propene, **d.** butane
22. a. cycloheptane, **b.** cyclooctyne, **c.** cyclooctene, **d.** cyclopentene
23. a. **b.** **c.** **d.**

PROGRAM RESOURCES

Evaluation: Use the Evaluation worksheets, pp. 49-52, to assess students' knowledge of Chapter 13.

Computer Test Bank: Use the IBM or Apple Test Bank and Question Selection Manual to prepare your own individualized test for Chapter 13.

16. Using the ammonia molecule as an example, distinguish between shared pairs and unshared pairs of electrons.

17. Predict shapes for the following ions.
a. IO_4^- **c.** SiF_6^{2-} **e.** PO_4^{3-}
b. ClO_3^- **d.** SO_4^{2-} **f.** ClF_4^-

18. Predict the shapes of the following molecules. See Table 13.1.
a. H_2CO **c.** BF_3 **e.** S_2Cl_2
b. SeO_2 **d.** SF_6 **f.** SF_4

19. What is meant by hybrid in the term hybrid orbital?

20. Sketch a molecular diagram of butyne and one of pentyne, showing all bonds.

21. Name the following compounds.

a.

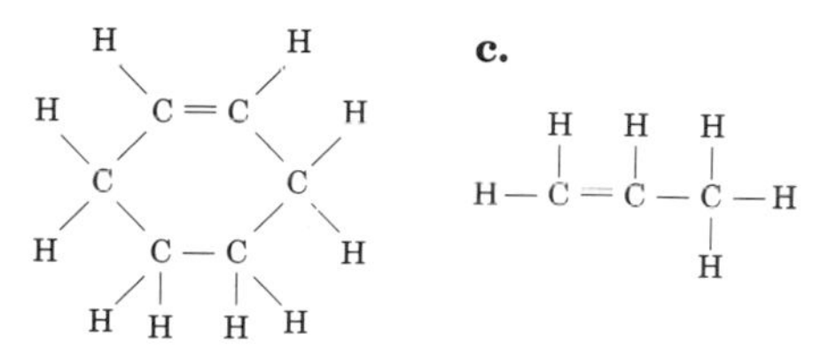

c.

H H H
| | |
H—C=C—C—H
|
H

b. **d.**

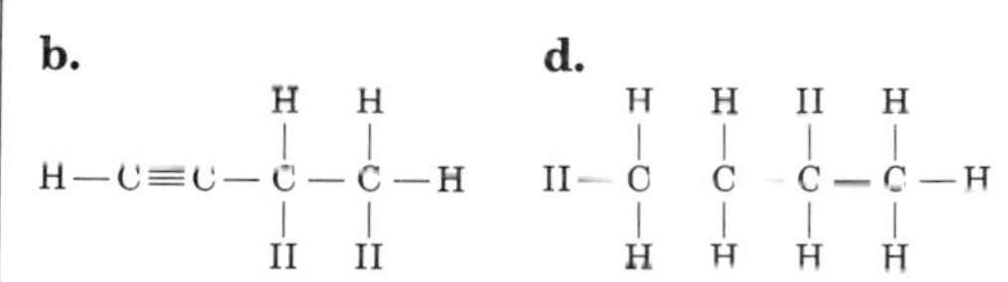

22. Name the following compounds.

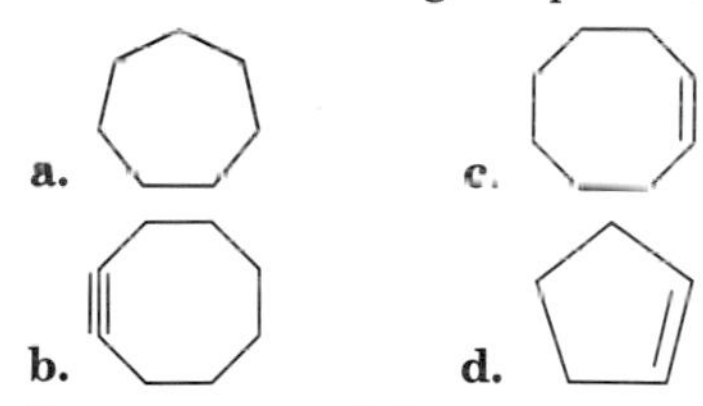

a. **b.** **c.** **d.**

23. Draw structural formulas for each of the following hydrocarbons.
a. cycloheptene **c.** cyclobutene
b. cyclobutane **d.** cyclononyne

24. Draw structural formulas for each of the following hydrocarbons.
a. propene **c.** heptane
b. octyne **d.** ethyne

25. Sketch a molecular diagram of butene, showing its orbitals and bonds.

26. Predict the bond angles indicated in the following compounds. Use Table 13.2.
a. H—Se—H in H_2Se
b. H—P—H in PH_3
c. C—N—C in $N(CH_3)_3$
d. Cl—P—Cl in PCl_3
e. F—C—F in CF_4
f. C—Pb—C in $Pb(C_2H_5)_4$

27. How many pairs of electrons are shared by two atoms with each of the following bonds between them?
a. single **b.** double **c.** triple

28. In the HCN molecule, the bonds are H—C and C≡N. Predict the shape of the molecule, the hybridization of the orbitals on the carbon atom, and the type (σ, π) of each bond.

29. A conjugated system stabilizes a molecule. Define "conjugated system" and give an example of one.

30. Why is benzene a particularly stable compound?

31. Define isomerism. Draw examples of two isomers of hydrocarbons containing five carbon atoms each.

32. Draw the structural isomers of the compound with the formula C_6H_{14}.

33. How do positional and functional isomers differ?

34. In what kinds of molecules do geometrical isomers occur? Draw examples using hydrocarbon molecules with five carbon atoms and one double bond.

35. What are functional isomers? Draw an example and include names.

36. Phosphorus pentachloride occurs as PCl_5 molecules in the gaseous state. As a solid, it is an ionic compound, PCl_4^+ PCl_6^-. Describe the shape and bonding in each of these species. See Table 13.1.

37. Draw two possible structural formulas for decyne and one for cyclononene.

24. For structures, see *Merrill Chemistry Problems and Solutions Manual.*

25. 2-butene

H H H H
$H\overset{\sigma}{-}C\underset{\sigma}{-}C\underset{\pi}{\overset{\sigma}{=}}C\underset{\sigma}{-}C\overset{\sigma}{-}H$
H H

Sigma and pi bonds are indicated. Carbons 1 and 4 have 4 sp^3 hybrid orbitals each. Carbons 2 and 3 have three sp^2 and one p orbitals each. The p orbitals form the π bond.

26. **a.** 104.5°, **b.** 107°, **c.** 96°, **d.** 107°, **e.** 109.5°, **f.** 109.5°

27. **a.** 1, **b.** 2, **c.** 3

28. linear; *sp*; H—C is sigma, C≡N is one sigma and two pi bonds.

29. Multiple *p* overlap can occur; benzene, 1, 3-butadiene.

30. The delocalized electrons stabilize the structure.

31. Isomerism is the existence of two or more substances with the same molecular formula but different structures. For structures, see *Merrill Chemistry Problems and Solutions Manual.* (Also included are isomers of cyclopentane, cyclopentene, cyclopentyne, pentene, and pentyne.)

32. For structures, see *Merrill Chemistry Problems and Solutions Manual.*

33. Positional isomers are formed when an added particle may occupy two or more positions in the molecule. Functional isomers are formed when a new element is bonded in different ways in the molecule.

34. Geometrical isomers occur in hydrocarbons with double bonds.

CH_3 C_2H_5 / C = C / H H — *cis*

CH_3 H / C - C / H C_2H_5 — *trans*

35. Functional isomers have a new element being bonded in two different ways. For example:

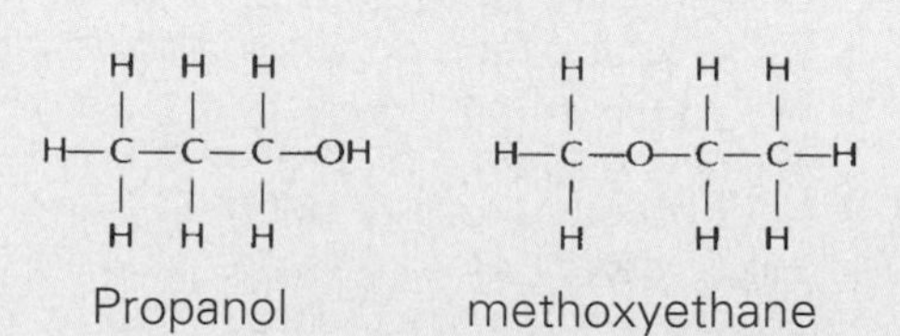

36. PCl_5—trigonal bipyramidal; 5σ bonds
PCl_4^+—tetrahedral; 4σ bonds
PCl_6^-—octahedral; 6σ bonds
All are covalently bonded within each unit, PCl_4^+ and PCl_6^- are ionically bonded.

37. decyne: For structures, see *Merrill Chemistry Problems and Solutions Manual.*

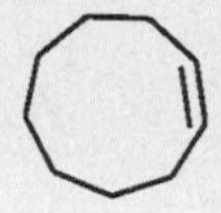
cyclononene

Concept Mastery

38. How do forces resulting from the Pauli exclusion principle affect the shape of molecules? How do these forces compare with the electrostatic repulsions in a molecule? Taking both kinds of forces into account, state the general rule that applies to the shape of molecules when considering electron-pair repulsions.

39. What accounts for the difference in shape between the electron clouds of shared pairs and unshared pairs of electrons?

40. Why is the bond angle in NH_3 only 107° when the bond angle for BF_3 is 120°?

41. Explain why water has a bond angle of 104.5° instead of the predicted 109.5°.

42. In general, what effect does the presence of unshared pairs in an atom have on the orientation of bonds to the same atom?

43. Explain why the carbon atom forms four equal bonds instead of just two bonds involving electrons from p orbitals.

44. One might expect the bond angle for each C—H bond in methane to be 90°. Why is this prediction incorrect?

45. How are covalent bonds formed?

46. What is the major difference between sigma and pi bonds?

47. What kinds of orbital arrangements contribute to the bonding in ethene, $H_2C{=}CH_2$?

48. In what characteristics do multiple bonds differ from single bonds?

49. In general, how does the presence of double or triple bonds affect the geometry of adjacent single bonds?

50. What is meant by electrons being delocalized? In the case of benzene, describe how electrons are delocalized.

51. What is an unsaturated compound? Draw structural formulas of a saturated and an unsaturated organic hydrocarbon, each containing five carbon atoms.

52. How can isomers with very similar properties be distinguished from each other?

53. Explain the significance of the circles in the structural formula for naphthalene, $C_{10}H_8$.

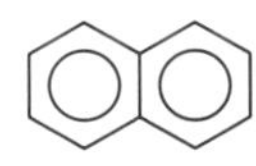

54. Butanol is an alcohol. One of its positional isomers is shown. Draw another positional isomer of butanol. Are there still other positional isomers? Explain your answer.

```
    H   H   H   H
    |   |   |   |
H — C — C — C — C — OH
    |   |   |   |
    H   H   H   H
```

55. Draw functional isomers of the following compounds.

a.

```
    H  O  H
    |  ‖  |
H — C — C — C — H
    |     |
    H     H
```

acetone

b.

```
H       H   H   H
 \      |   |   |
  N  —  C — C — C — H
 /      |   |   |
H       H   H   H
```

1-aminopropane

Application

56. Acetylene, ethyne, is widely used as fuel for welding torches and as a raw material for synthetic fibers and rubber. Sketch a molecular diagram of an ethyne molecule showing its bonds. Describe how these bonds are formed.

57. The largest moon of Saturn, Titan, has an atmosphere containing methane, ethane, ethene, propyne, hydrogen cyanide, and the compound diacetylene, C_4H_2, which contains two triple bonds. Draw the structural formula of each compound.

58. Fumaric acid, $C_2H_2(COOH)_2$, is used commercially in beverages. It occurs naturally in the metabolism of glucose in animal and plant cells. The structural formula of the *trans* isomer is shown.

Answers to Concept Mastery

38. Although electrons of opposite spin may occupy the same volume of space, electrons of the same spin may not do so. This results in electron repulsions that are much stronger than the electrostatic repulsions. Electron pairs spread as far apart as possible to minimize repulsive forces.

39. An unshared pair of electrons is attracted to only one nucleus. A shared pair is attracted to two different nuclei.

40. NH_3 has an unshared pair of electrons; BF_3 does not.

41. Repulsion of the two unshared pairs of electrons is greater than repulsion of the shared pairs, and the shared pairs are pushed closer together.

42. Repulsion causes bond angles to be less.

43. The 2s and 2p orbitals combine to form four equivalent hybrid sp^3 orbitals.

44. For four equal bonds in a three-dimensional figure, the bonds are farthest apart in a tetrahedron, not a square.

45. An orbital of one atom overlaps an orbital of another atom and the two atoms share the resulting electron pair.

46. Sigma is end-to-end overlap of orbitals. Pi is sideways overlap of p orbitals.

47. One s orbital and two p orbitals of each C atom hybridize to form three sp^2 orbitals. The third p orbital is perpendicular to the plane of the three sp^2 orbitals in each atom. The double bond between the carbon atoms is formed by a sigma bond formed from overlap of two sp^2 orbitals and a pi bond formed from the two non-hybrid p orbitals. H atoms form sigma bonds with the unused sp^2 orbitals.

48. Multiple bonds are less flexible, shorter, stronger, and more reactive.

49. Bond angle is less than predicted; multiple bonds occupy more space than do single bonds.

50. Delocalized electrons are shared equally and are not confined to any atom or bond. In benzene, the pi electrons are shared equally among the 6 carbon atoms in the ring. (The electrons in the sigma bonds are not delocalized.)

51. An unsaturated compound contains one or more double or triple bonds between carbon atoms. For structures, see *Merrill Chemistry Problems and Solutions Manual.*

52. mass spectrometer

53. These represent delocalized electrons in the pi bond formed from unhybridized carbon p orbitals.

54.

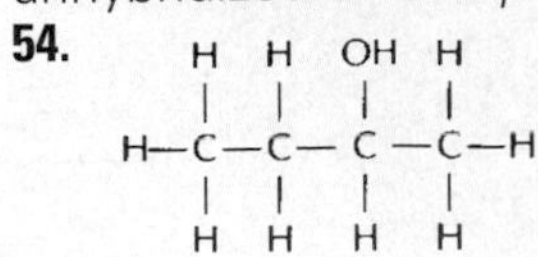

No, the OH^- attached to either end carbon is identical to the isomer in the problem, and the OH^- attached to either middle carbon is identical to the isomer in the answer.

55. a.

```
H           H
 \          |
  C = C—O—C—H
 /    |     |
H     H     H
```

b.

```
  H   H   H   H
  |   |   |   |
H—C—N—C—C—H
  |       |   |
  H       H   H
```

```
        O
        ‖
  H     C       H
   \   / \     /
    O     C = C     O
         /     \   / \
        H       C     H
                ‖
                O
```

Sketch the structural formula of the *cis* isomer, known as maleic acid.

59. In nutrition, scientists refer to fats as being saturated, unsaturated, or polyunsaturated. Considering that fats have long hydrocarbon chains, what do these terms mean when applied to fats?

64. Do the following two structures represent isomers, or are they representations of the same substance? Explain your reasoning.

```
      Cl                 Cl
      |                  |
      C                  C
    / | \              / | \
   H  Cl  H           H  H  Cl
```

65. Explain why the general formula for a non-cyclic alkane such as pentane, C_5H_{12}, or hexane, C_6H_{14}, is C_nH_{2n+2}, where n is the number of carbon atoms in the chain. What are the general formulas for a cyclic alkane, a noncyclic alkene, and a noncyclic alkyne?

Critical Thinking

60. How does the conjugated bonding arrangement in benzene compare with the bonding in metals?

61. Why are there no compounds named methene or methyne?

62. Why are there no isomers for propane or ethane?

63. Explain why some noble gases such as Xe will form compounds and some such as Ne will not.

Readings

Al-Mousawi, Saleh M. "Molecular Shape Prediction and the Lone-Pair Electrons on the Central Atom." *Journal of Chemical Education* 67, no. 10 (October 1990): 861.

Curl, Robert F., and Richard E. Smalley. "Fullerenes." *Scientific American* 265, no. 4 (October 1991): 54–63.

Kluger, Jeffrey. "A Dream Come True." *Discover* 10, no. 1 (January 1989): 56.

Cumulative Review

1. Complete and balance the following equations for synthesis reactions.

a. $Al + O_2 \rightarrow$ **c.** $Na + I_2 \rightarrow$
b. $Mg + S_8 \rightarrow$ **d.** $Na + P_4 \rightarrow$

2. Classify the bonds between the following pairs of atoms as principally ionic or principally covalent.

a. barium and fluorine
b. bromine and rubidium
c. cesium and oxygen
d. iodine and antimony
e. nitrogen and sulfur
f. silicon and carbon

3. Explain why metals conduct electricity.

4. Sodium amide is used in preparing the dye indigo. This dye is used to color blue jeans. If the percentage composition of the compound is 58.9% sodium, 35.9% nitrogen, and 5.17% hydrogen, what is the empirical formula of the compound?

5. Arrange the following elements in order of increasing attractive force between the nucleus and the outer electrons.

a. boron **c.** manganese
b. francium **d.** zinc

Answers to
Application

56. H—C≡C—H; for each carbon atom, one *s* orbital and one *p* orbital hybridize to form two *sp* orbitals. For each carbon atom this leaves two *p* orbitals perpendicular to each other and to the *sp* hybrid orbitals. The triple bond between the carbon atoms is formed by a sigma bond formed by one *sp* orbital from each atom and two pi bonds formed from sideways overlap of the unhybridized *p* orbitals. H atoms form sigma bonds with th unused *sp* orbital in each carbon atom.

57. For structures, see *Merrill Chemistry Problems and Solutions Manual.*

58. For structures, see *Merrill Chemistry Problems and Solutions Manual.*

59. Saturated has no multiple bonds. Unsaturated contains at least one multiple bond. Polyunsaturated contains more than one multiple bond.

Answers to
Critical Thinking

60. Both contain delocalized electrons.

61. They would each contain only one carbon atom and thus cannot contain a double or triple bond.

62. With chains of only two and three carbon atoms and no multiple bonds, it is impossible to have isomers because their structures cannot vary. For structures, see *Merrill Chemistry Problems and Solutions Manual.*

63. The shielding effect and large atomic radius decreases the attraction of the nucleus for the outer electrons in a large atom such as that of Xe. For noble gas atoms from Argon up, the outer level has a capacity for more than eight electrons ($2n^2$).

64. Because all bonds are identical sp^3 bonds, and they are all attached to the same carbon atom, the compounds are not isomers.

65. Each carbon atom has 4 bonds to neighboring atoms in a molecule. Along the carbon chain, each carbon atom is bonded to two other carbon atoms, with the other two bonds being to two hydrogen atoms. At the ends of the chain, the C atom is bonded to only one other C atom, and another H is necessary for the fourth bond. Thus, each C has two H plus one at each end of the molecule. The general formulas are cyclic alkane: C_nH_{2n}, noncyclic alkene: C_nH_{2n}, and noncyclic alkyne: C_nH_{2n-2}.

Answers to
Cumulative Review

1. a. $4Al + 3O_2 \rightarrow 2Al_2O_3$
b. $8Mg + S_8 \rightarrow 8MgS$
c. $2Na + I_2 \rightarrow 2NaI$
d. $12Na + P_4 \rightarrow 4Na_3P$

2. a. ionic, **b.** ionic, **c.** ionic, **d.** covalent, **e.** covalent, **f.** covalent

3. delocalized electrons are free to move

4. $NaNH_2$. For a complete solution, see *Merrill Chemistry Problems and Solutions Manual.*

5. b, c, d, a

CHAPTER

14 Polar Molecules

CHAPTER ORGANIZER

Chapter Objectives	Text Features	Laboratory Options	Teacher Classroom Resources
14.1 Molecular Attraction **1.** Distinguish between polar and nonpolar covalent bonds. **2.** Use electronegativities to predict the comparative polarities of bonds. **3.** Define dipole and compare the strengths of intermolecular forces based on dipole moments. **4.** Define and describe the types of van der Waals forces and list the three factors contributing to them.	**Frontiers:** Ion Trapping, p. 355	**ChemActivity 14-2:** Miscibility of Liquids - microlab, p. 822 **Discovery Demo:** 14-1 All Charged Up, p. 348 **Demonstration:** 14-2 Induced Dipoles, p. 352 **Demonstration:** 14-3 Dipole Attractions, p. 354	**Study Guide:** pp. 51-52 **Lesson Plans:** pp. 29-31 **ChemActivity Master 14-2** **Color Transparency 33 and Master:** Structure of Polar Molecules **Reteaching:** Molecular Forces, p. 21 **Chemistry and Industry:** Soaps and Detergents, pp. 23-26
	Nat'l Science Stds: UCP.2, B.1, B.2, B.4, B.6, E.1, E.2		
14.2 Coordination Chemistry **5.** Define complex ion, ligand, coordination number, and coordination compound. **6.** Name complex ions given their formulas, and write formulas for complex ions given their names. **7.** Name coordination compounds given their formulas, and write formulas for coordination compounds given their names.	**Math Connection:** Platonic Solids, p. 358 **Bridge to Medicine:** Removing Toxic Metals from the Body, p. 359 **Chemists and Their Work:** Jokichi Takamine, p. 364 **Printing Connection:** Ink Composition, p. 365	**Laboratory Manual:** 14 Separating Metal Ions - microlab	**Study Guide:** p. 53 **Enrichment:** Capacitors, pp. 27-28 **Critical Thinking/ Problem Solving:** Coordination Chemistry - A Closer Look, p. 14
Nat'l Science Stds: UCP.1, UCP.2, UCP.5, B.1, B.2, B.4, B.6, F.1, F.5			
14.3 Chromatography **8.** Define chromatography, mobile phase, and stationary phase. **9.** Define, describe and name uses for the different types of chromatography	**Everyday Chemistry:** New Tips on the Laundry, p. 367 **Chemistry and Technology:** Drug Detection, p. 370 **Home Economics Connection:** Hard Water, p. 371	**Minute Lab:** The Ink Blot Test, p. 368 **ChemActivity 14-1:** Paper Chromatography - microlab, p. 820 **Demonstration:** 14-4 Column Chromatography, p. 368 **Demonstration:** 14-5 Paper Chromatography, p. 370	**Study Guide:** p. 54 **ChemActivity Master 14-1** **Applying Scientific Methods in Chemistry:** Chromatography, pp. 31-32
Nat'l Science Stds: UCP.1, A.2, B.6, E.2, G.2			
Other Chapter Resources	**Vocabulary and Concept Review,** pp. 14, 61-62 **Evaluation,** pp. 53-56	**Videodisc Correlation Booklet** **Lab Partner Software**	**Test Bank**

Block Schedule

For information on block scheduling, see the Lesson Plans booklet in the Teacher Resource Package.

GLENCOE TECHNOLOGY

Mastering Concepts in Chemistry Software

Chemistry: Concepts and Applications Videodisc

Chemistry: Concepts and Applications CD-ROM

Complete Glencoe Technology references are inserted within the appropriate lesson.

ASSIGNMENT GUIDE

Contents	Level	Practice Problems	Chapter Review	Solving Problems in Chemistry
14.1 Molecular Attraction				
Polarity	C	1, 2	24, 27, 30, 53, 55, 60, 65, 71	1-2
Weak Forces	C		25, 26, 28-30, 54, 73, 75-77	3-4
14.2 Coordination Chemistry				
Ligands	C		31-37, 49, 50, 70, 78	
Names and Formulas of Complex Ions	A	7-12	38-42, 52, 61, 67, 68	5-6
Coordination Compounds	A	13,14	43, 44, 51, 52, 66, 69	
Bonding in Complexes	A		56, 62	
14.3 Chromatography				
Fractionation	O		47, 57	
Chromatography	O		48, 58, 59, 63, 64, 72, 74	Chapter Review: 1-17

C=Core, A=Advanced, O=Optional

Laboratory Materials

Minute Labs

page 368
beaker, 400-cm^3
coffee filter paper
ink pens and felt markers
scissors
water

ChemActivities

page 820
black, water soluble marker
distilled water
ethanol
filter paper, 1 x 9 cm strips
heavy plastic bag
microplate, 24-well
pipet, plastic microtip

page 822
construction paper, white
distilled water
food coloring, blue
liquids to be tested:
1-butanol
2-butanol
tert-butanol
hexane
1-propanol
2-propanol
96-well microplate
Microplate Data Form
plastic microtip pipets (8 per group)
small beaker or plastic cup
TTE (trichlorotrifluoroethane)
toothpicks

Demonstrations

pages 348, 352, 354, 368, 370
Al_2O_3 packing
aspirator
beakers, 100, 250 cm^3
buret and stands
$CoCl_2 \cdot 6H_2O$
capillary tube
copper wire, 22 guage or finer
cyclohexane or TTE
distilled water
$FeCl_3 \cdot 6H_2O$
filter paper
food coloring
forceps
glass tubing
HCl
2-propanol
lucite rod
$NiCl_2 \cdot 6H_2O$
overhead projector
permanent bar magnets
petri dish
silk, fur or wool
straight pins
large test tubes with cork stoppers
thumbtack
water

CHAPTER 14

Polar Molecules

CHAPTER OVERVIEW

Main Ideas: Properties of molecules which are a consequence of their structure are discussed. Polar molecules are introduced. Students will investigate how some properties of similar compounds differ because of varying molecular shapes and structures. Polarity shows the relationship between properties and electric dipoles in molecules. Students will learn that the presence of *d* orbitals gives no new bonding types, only some new shapes. A study of the weak van der Waals forces between molecules is important if students are to have a complete understanding of the relationship between structure and properties. Finally, the knowledge of polarity is applied in learning how to separate the components of a mixture by chromatography. Several types of chromatography are discussed.

Theme Development: Chemical stability and the relationship between molecular structure and interactions are developed in this chapter.

Tying To Previous Knowledge: Have students recall how clothes containing synthetic fiber cling together as they are removed from a dryer. Point out that the clothes attract each other because of static electric charges. Polar molecules attract and "cling" to one another in an analogous manner.

ASSESSMENT PLANNER

Choose from the following assessment strategies.

Assess, pp. 354-355, 365, 371
Assessment, pp. 351, 353, 358, 360, 363, 368
Portfolio, p. 360
Minute Lab, p. 368
ChemActivity, pp. 820-824
Chapter Review, pp. 372-375

CHAPTER **Preview**

CONTENTS

DISCOVERY DEMO

14-1 All Charged Up

Purpose: to demonstrate that one end of a polar molecule can be attracted to an opposite charge

Materials: 2 burets and buret stands; 50 cm^3 water; 50 cm^3 cyclohexane or TTE(1,1,2 trichlorotrifluoroethane); piece of silk, fur, or wool; a lucite rod; 2 250-cm^3 beakers

Procedure: **1.** Place two burets in a buret clamp and fill one with water and one with cyclohexane or TTE. **CAUTION:** *Cyclohexane vapors are poisonous and flammable. Perform the demonstration in a fume hood.*
2. Charge the lucite rod by rubbing it with silk cloth, wool, or fur. Open the cyclohexane or TTE buret and allow a thin stream to run into the beaker.
3. Bring the charged rod near the stream. Have students observe that

CHAPTER

14

Polar Molecules

The word polar *applies to bears, regions of Earth, magnets, and some molecules. These molecules are so named not because they are found exclusively in polar areas, but because they have two poles, just as Earth does.*

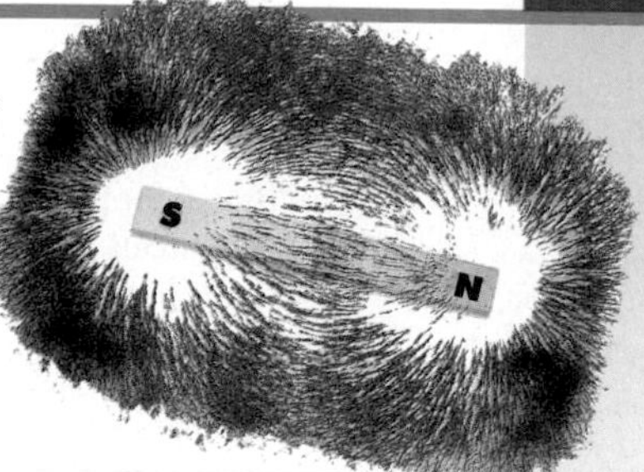

Most of the substances we encounter in our everyday life are made up of polar molecules. Water is, perhaps, the most important of all polar molecules. Because water molecules are polar, water is attracted by a charged rod, ice floats on liquid water, and water dissolves many substances. What does the word polar mean? What features cause a water molecule to be polar? What are the consequences of some molecules being polar? In this chapter we will investigate the structural features that cause a molecule to be polar and cause the behavior of polar molecules.

EXTENSIONS AND CONNECTIONS

Frontiers
Ion Trapping: Researchers measure minute amounts of chemicals, **page 355**

Math Connection
Platonic Solids, **page 358**

Bridge to Medicine
Removing Toxic Metals from the Body: Using ligands to remove transition metals, **page 359**

Chemists and Their Work
Jokichi Takamine, **page 364**

Printing Connection
Ink Composition, **page 365**

Everyday Chemistry
New Tips on the Laundry: How polar molecules make clothing dirty—and get it clean, **page 367**

Chemistry and Technology
Drug Detection: A view of the use of chromatography to detect drug abuse, **page 370**

Home Economics Connection
Hard Water, **page 371**

INTRODUCING THE CHAPTER

USING THE PHOTO

When the polar bonds in a molecule are not symmetrically arranged, a polar molecule is produced. Suggest that students use the idea of a polar bear on ice to help them remember that water is very polar. Encourage students to discuss how the world would be different if water were nonpolar. Possible responses: *Water would not be a liquid. The water content of the oceans would be in the atmosphere.*

REVEALING MISCONCEPTIONS

Students can confuse a polar covalent bond and a polar attractive force such as a dipole-dipole attraction. The polar covalent chemical bond involves the unequal sharing of electrons between two nuclei due to the overlap of half-filled orbitals. The polar attractions are not nearly as strong as a chemical bond. You can erase this misconception by boiling a beaker of water. Remind students that the water molecules are separating because the dipole-dipole attractions are being broken. The molecules of water are still intact, but just exist as a vapor instead of a liquid.

the stream is not deflected.
4. Repeat steps 2-3 using the water-filled buret. Have students observe the deflection of the stream.
Disposal: F

Results: The cyclohexane or TTE stream is not deflected, but the water stream is.

Questions: 1. Which of the two liquids has a polar nature? *water*
2. What is the molecular geometry and bond angle of the water molecule? *bent, 104.5°*
3. Would you expect a symmetrical molecule to display a polar nature? *no*
4. Would you expect a molecule of cyclohexane or TTE to be symmetrical? *yes*

Extension: Draw the structures of both molecules for the students. Even though every C—H bond is polar in cyclohexane, the molecule is nonpolar because of its symmetry. Both H—O bonds in the water are polar; the oxygen end has a partial negative charge, and the hydrogen end has a partial positive charge. Because the polar bonds are not symmetrically arranged, the water molecule is polar.

14.1 Molecular Attraction

PREPARE

Key Concepts

The major concepts presented in this section include polarity, van der Waals forces, intramolecular and intermolecular forces. Dipole-dipole, dipole-induced dipole, and dispersion forces are described in detail.

Key Terms

polar covalent
dipole
dipole moment
van der Waals force
intramolecular force
intermolecular force
dipole-dipole force
induced dipole
dipole-induced dipole force
temporary dipole
dispersion force

Looking Ahead

Prepare the straight pin for Demonstration 14-3: Dipole Attraction the day before you plan to use the demonstration.

1 FOCUS

Objectives

Preview with your students the objectives listed on the student page. Students can use them as a study guide for this major section.

Activity

Use tape to attach a length of rubber tubing to a ring stand above the sink. Attach the other end to the faucet. Turn on the water so as to produce a thin stream. Rub a balloon with wool or fur and bring it near the thin stream of water as students enter the classroom. Have a student comb his or her hair and bring the comb near the stream of water as the class observes. Encourage students to read the chapter to explain their observations.

OBJECTIVES

- distinguish between polar and nonpolar covalent bonds.
- use electronegativities to predict the comparative polarities of bonds.
- define *dipole* and compare the strengths of intermolecular forces based on dipole moments.
- define and describe the types of van der Waals forces and list the three factors contributing to them.

14.1 Molecular Attraction

Think about table salt and sugar. They look alike. They both dissolve in water. Yet they are quite different. Salt is made of ions. The sodium ions and chloride ions are oppositely charged and attract each other. Knowing the properties of ionic substances, we understand why salt is a solid. Sugar, on the other hand, is made of molecules. Each molecule of sugar contains twelve atoms of carbon, twenty-two atoms of hydrogen, and eleven atoms of oxygen bonded covalently. A sugar molecule is neutral, so what holds one sugar molecule to another? Why don't the sugar molecules just float away from each other and become a gas?

Substances composed of molecules exhibit a wide range of melting and boiling points. There must be, therefore, a wide range in the strength of forces holding molecules to each other. These forces are determined by the internal structure of a molecule. In this section, we will look at the way that internal structure affects the forces holding molecules to each other.

Polarity

Recall that electronegativity is an atom's ability to attract the electrons involved in bonding. Because no two elements have exactly the same electronegativities, in a covalent bond between different elements, one of the atoms attracts the shared pair more strongly than does the other. The resulting bond is said to be **polar covalent.** In the bond, the atom with higher electronegativity attracts the electrons more strongly, and that end of the bond will have a partial negative charge. The atom at the other end of the bond will have a partial positive charge.

Partial charges within a molecule are indicated by δ (delta). A water molecule, therefore, would be represented as follows:

$$\delta^{+}\mathrm{H}:\ddot{\underset{\cdot\cdot}{\mathrm{O}}}:\delta^{-} \quad \text{bond angle } 104.5^\circ$$
$$\mathrm{H}\ \ \delta^{+}$$

Figure 14.1 CH_3Cl, HF, NH_3, and H_2O are polar molecules because the arrangement of the polar bonds is not symmetrical. The CCl_4 and CO_2 molecules are nonpolar because the polar bonds are symmetrically arranged. The arrows point toward the more electronegative atom.

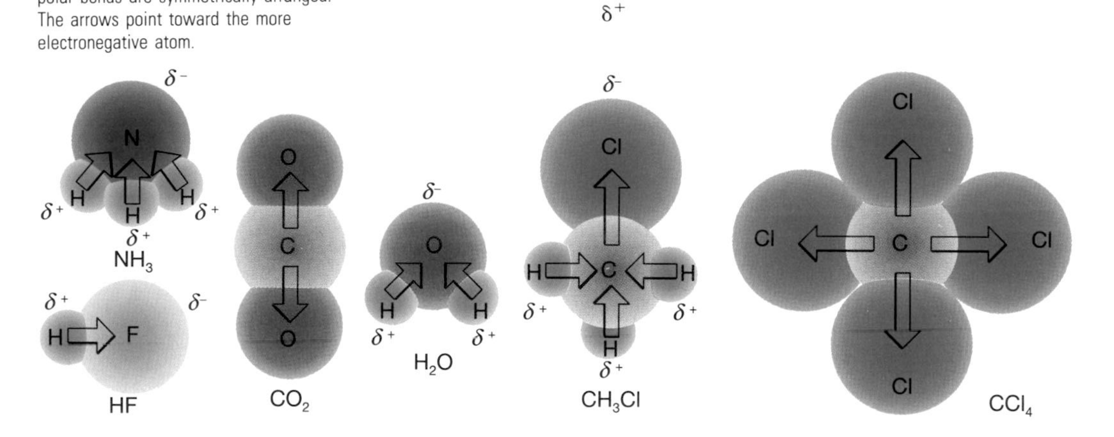

Chemistry Journal

Properties of Water

Ask students to describe in their journals how and why the properties of water would change if the molecules were linear. Ask them to add to the journal entry as various types of molecular attractions are discussed in this section of the chapter.

Figure 14.2 On the left, the polar molecules of water are attracted to a charged rod. The nonpolar molecules of carbon tetrachloride on the right are not affected by the rod.

Polar bonds may produce polar molecules. For example, look at Figure 14.1. The arrows show the direction in which electrons are attracted in each bond. In the ammonia and hydrogen fluoride molecules there is a concentration of negative charge near the nitrogen and fluorine atoms and positive charge near the hydrogen atoms. Consequently, ammonia and hydrogen fluoride are polar molecules. Now consider the carbon dioxide molecule. Each carbon-oxygen bond in this molecule is polar because oxygen has a greater electronegativity than carbon does. However, the polarities of the two bonds are in exactly opposite directions and therefore cancel each other. Even though the bonds are polar, the carbon dioxide molecule is nonpolar due to its linear geometry.

What about the water molecule? Is it polar or nonpolar? Both of the oxygen-hydrogen bonds are polar. In this case, however, the polar bonds do not cancel each other. Recall that the water molecule has an angular geometry. Therefore, there is a net concentration of negative charge near the oxygen atom and a slight positive charge near each hydrogen atom. For this reason, water is a polar molecule. In carbon tetrachloride, the four polar carbon-chlorine bonds are symmetrically distributed and therefore cancel each other. Carbon tetrachloride is not a polar molecule. In chloromethane, there is an asymmetrical distribution of charge. Therefore, chloromethane is a polar molecule. Polar bonds are a necessary but not a sufficient condition for polar molecules. In a polar molecule, the polar bonds cannot be symmetrically arranged.

Because it has both a positive and a negative pole, a polar molecule, such as water, is also said to be a **dipole,** or to have a dipole moment. A **dipole moment** is a measure of the strength of the dipole and is a property that results from the asymmetrical charge distribution in a polar molecule.

The dipole moment depends upon the size of the partial charges and the distance between them. In symbols,

$$\mu = Qd$$

2 TEACH

CONCEPT DEVELOPMENT

- You may wish to introduce vectors and vector addition to show why an unsymmetrical molecule has a dipole moment.
- Remind students that Table 12.1 Electronegativities is useful for determining the relative amount of unequal sharing of electrons. The greater the electronegativity difference, the greater the polarity of the chemical bond between the two atoms.

MAKING CONNECTIONS

Construction

Concrete becomes hard because the water and cement chemically combine to form new compounds. Concrete does not dry; it cures. This chemical process continues for months. The conditions under which it cures determine the concrete's final strength.

✓ ASSESSMENT

Skill: Have students refer to Table 12.1 on page 303. Give them two element pairs and ask them to determine the electronegativity differences for each pair and tell you which bond is the stronger dipole. Example: In the C-Cl bond the difference is 0.33 and in the H-O bond it is 1.3. The H-O bond is the stronger dipole.

CONTENT BACKGROUND

Asphalt: Researchers in Wyoming and Pennsylvania describe asphalt as a closely packed, 3-D network of polar molecules that extends throughout a neutral hydrocarbon medium. This new understanding of asphalt chemistry will lead to better roads.

PROGRAM RESOURCES

Study Guide: Use the Study Guide worksheets, pp. 51-52, for independent practice with concepts in Section 14.1.

Lesson Plans: Use the Lesson Plan, pp.29-31, to help you prepare for Chapter 14.

Chemistry and Industry: Use the worksheets, pp.23-26, "Soaps and Detergents: A Very 'Clean' Industry" as an extension to the concepts presented in Section 14.1.

PRACTICE

Guided Practice: Have students work in class on Practice Problems 1 and 2.

Independent Practice: Your homework or classroom assignment can include Review and Practice 24, and Concept Mastery 53 from the end of the chapter.

Answers to

Practice Problems: Have students refer to Appendix C for complete solutions to Practice Problems.

1. In order of decreasing bond polarity: f, c, a, e, b, d

2. a. HBr is polar because of the concentration of negative charge near the Br atom and of positive charge near the H atom.

b. SO_2 is polar because oxygen has a greater electronegativity than sulfur and the molecule is asymmetric.

GLENCOE TECHNOLOGY

Software

Mastering Concepts in Chemistry
Unit 9, Lesson 1
Electronegativity and Bond Types

Videodisc

Chemistry: Concepts and Applications
Electronegativity
Disc 1, Side 2, Ch. 12

Also available on CD-ROM.

Molecular Shapes
Disc 1, Side 2, Ch. 13

Also available on CD-ROM.

where μ is the dipole moment, Q is the size of the partial charge in coulombs, and d is the distance in meters between the partial charges. Dipole moment is then expressed in coulomb·meters.* The higher the dipole moment, the stronger the intermolecular forces; and, consequently, the higher the melting point and boiling point for molecules of similar mass. Table 14.1 lists a number of common solvents in order of increasing polarity. The first three solvents are nonpolar.

*Older tables of dipole moments may be expressed in debyes (D) where 1 D = 3.338×10^{-30} C·m.

Table 14.1

Polarity of Solvents

Name	Formula	$\mu \times 10^{30}$ C·m
Cyclohexane	C_6H_{12}	0
Carbon tetrachloride	CCl_4	0
Benzene	C_6H_6	0
Toluene	$C_6H_5CH_3$	1.5
Ethoxyethane	$CH_3CH_2OCH_2CH_3$	4.07
Ammonia	NH_3	4.90
1-Butanol	$CH_3CH_2CH_2CH_2OH$	5.54
1-Propanol	$CH_3CH_2CH_2OH$	5.57
Ethanol	CH_3CH_2OH	5.64
Methanol	CH_3OH	5.64
Ethyl acetate	$CH_3CH_2OOCCH_3$	6.14
Water	HOH	6.14
Propanone	CH_3COCH_3	9.25

Figure 14.3 Liquids tend to form drops because their molecules have intermolecular attraction.

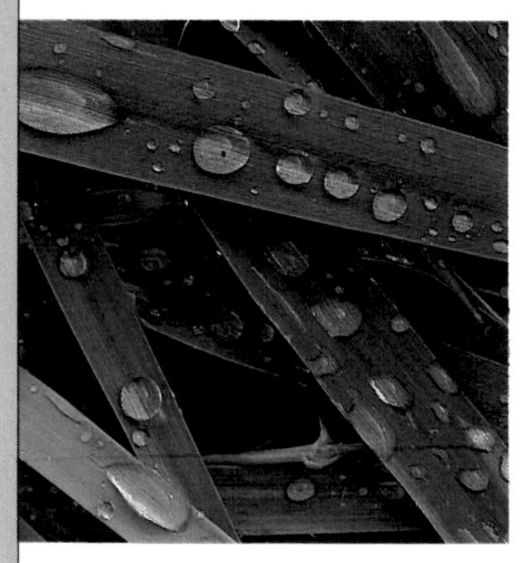

PRACTICE PROBLEMS

1. The following pairs of atoms are all covalently bonded. Arrange the pairs in order of decreasing polarity of the bonds. (See Table 12.1, page 303.)
   a. aluminum and phosphorus
   b. chlorine and carbon
   c. molybdenum and tellurium
   d. hydrogen and sulfur
   e. phosphorus and sulfur
   f. chlorine and silicon
2. Which of the following molecules are polar?
   a. HBr
   b. SO_2
   c. N_2
   d. BF_3

Weak Forces

Covalent compounds show a melting point range of over 3000 C°. How can we account for such a wide variation? The forces involved in some of these cases are called **van der Waals forces.** Johannes van der Waals was the first to account for these forces in calculations concerning gases. These forces are sometimes

DEMONSTRATION

14-2 Induced Dipoles

Purpose: to demonstrate that polar water molecules can induce a dipole in other substances

Materials: plastic petri dish, forceps, 100 cm^3 distilled water, 2 1-cm lengths of 22 or higher gauge copper wire

Procedure: 1. Place a plastic petri dish on the stage of an overhead projector and fill it with distilled water.
2. Use forceps to place the 2 pieces of wire in the center of the petri dish so that they float end-to-end about 1 cm apart.

Results: The ends of the wires either attract or repel. Point out that a dipole is induced in the copper wire by the polar arrangement of the water molecules.

Questions: 1. What types of attractive

referred to as weak forces because they are much weaker than chemical bonds. Weak forces involve the attraction of the electrons of one atom for the protons of another.

It is important to note the difference between intramolecular forces and intermolecular forces. **Intramolecular forces** are forces within a molecule that hold atoms together, that is, covalent bonds. **Intermolecular forces** are forces between molecules that hold molecules to each other, that is, van der Waals forces.

The first van der Waals force that we will consider is the dipole-dipole force. With **dipole-dipole forces,** two molecules of the same or different substances that are both permanent dipoles, will be attracted to each other, Figure 14.4. Such would also be the case between two trichloromethane molecules, $CHCl_3$, or between a trichloromethane and an ammonia molecule.

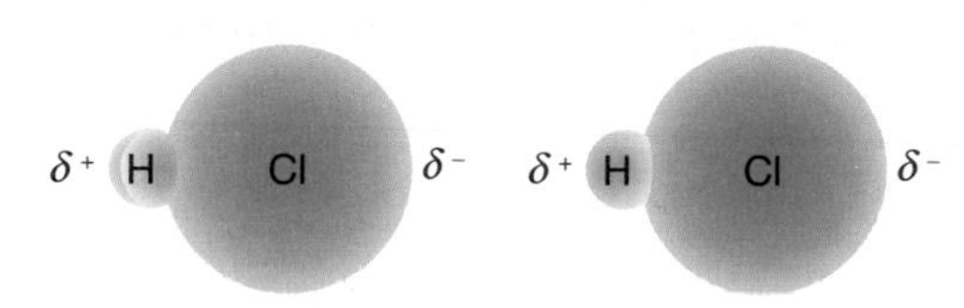

Figure 14.4 Dipole-dipole attraction exists between molecules that are permanent dipoles.

A dipole can also attract a molecule that is ordinarily not a dipole. When a dipole approaches a nonpolar molecule, its partial charge either attracts or repels the electrons of the other particle. For instance, if the negative end of the dipole approaches a nonpolar molecule, the electrons of the nonpolar molecule are repelled by the negative charge. The electron cloud of the nonpolar molecule is distorted by bulging away from the approaching dipole as shown in Figure 14.5. As a result, the nonpolar molecule is itself transformed into a dipole. We say it has become an **induced dipole.** Since it is now a dipole, it can be attracted to the permanent dipole. Interactions such as these are called **dipole-induced dipole forces.** An example of this force also occurs in a water solution of iodine. The I_2 molecules are nonpolar while the water molecules are highly polar.

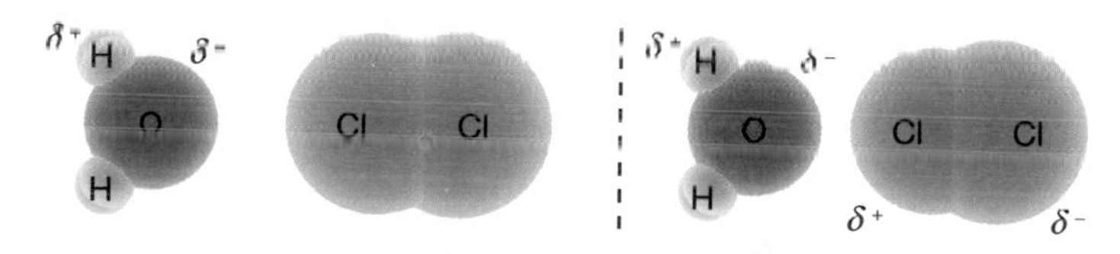

Figure 14.5 Dipole-induced dipole attraction occurs when a molecule that is a dipole causes a nonpolar molecule to become an induced dipole.

The attraction of two nonpolar molecules for each other must also be taken into account. For instance, there must be some force between hydrogen molecules; otherwise it would be impossible to form liquid hydrogen. Consider a hydrogen molecule. We usually assume that the electrons of the hydrogen molecule move uniformly about the nuclei of the two atoms. For an instant, both electrons may be at the same end of the molecule. Thus, the hydrogen molecule develops a temporary asymmetrical electron

force exists between water molecules? *dipole-dipole force*

2. What type of force exists between the water and the copper? *dipole-induced dipole.*

3. Which force is stronger, dipole-dipole force or dipole-induced dipole force? *dipole-dipole force*

CONCEPT DEVELOPMENT

Background: Chemical bond energies range from about 200 to 800 kJ/mol, but weaker dipole-dipole forces are less than 125 kJ/mol.

QUICK DEMO **Magnetic Marbles** Magnetic marbles can be purchased at national chain toy stores. On an overhead projector, they simulate the action of polar molecules as they link up to form chains and aggregates of molecules. Show regular marbles moving past each other in a glass or clear plastic tray to simulate nonpolar molecules. Introduce students to the vocabulary of this section as they observe the marbles.

✓ ASSESSMENT

Performance: Ask students to use a magnet to induce a magnetic dipole in a small piece of metal such as a small nail. Have them show that bringing the magnet near (but not touching) one end of this nail enables the other end to attract iron filings. Two students with two magnets and two nails on strings can show attraction and repulsion due to the induced magnetic field on the nail. Avoid permanently magnetizing the nail by keeping it as far away from the magnet as possible. Have students discuss this demonstration as an analogy for dipole-induced dipole forces.

PROGRAM RESOURCES

Transparency Master: Use the Transparency master and worksheet, pp. 65-66, for guided practice in the concept "Structure of Polar Molecules."

Color Transparency: Use Color Transparency 33 to focus on the concept of "Structure of Polar Molecules."

Reteaching: Use the Reteaching worksheet, p. 21, to provide students another opportunity to understand the concept of "Molecular Forces."

CONCEPT DEVELOPMENT

Using the Table: Take the time to carefully review Table 14.2, which summarizes this section.

3 ASSESS

CHECK FOR UNDERSTANDING

Ask students to answer the Concept Review problems, and Review and Practice 24-30 from the end of the chapter.

RETEACHING

Ask why water is a liquid at room temperature when hydrogen sulfide, H_2S, is a gas. Direct the discussion to include boiling points and relate these to polarity and the resulting weak attractive forces.

EXTENSION

Many polar compounds are used in the qualitative analysis of inorganic ions because they form complex ions that have a distinctive, identifying color. Ask a volunteer to investigate the substances used in detecting nickel, aluminum and zirconium ions by such a method. Possible response: *Ni — dimethylglyoxime, Al — ammonium aurin tricarboxylate, Zr — sodium alizarin sulfonate.*

Answers to **Concept Review**

3. Dipole-dipole forces exist between two molecules that are both permanent dipoles. If a dipole approaches a nonpolar molecule, its partial charge distorts the electron cloud of the nonpolar molecule, causing it to become an induced dipole. Dispersion forces exist between nonpolar molecules and are the result of temporary dipoles. All three forces are attractive and act over very short distances. Generally, dipole-dipole forces are stronger than dipole-induced dipole forces, which are in turn stronger than dispersion forces.

4. A dipole is a combination of two electrically charged particles of opposite sign which are separated by a very small distance. "Dipole" is more descriptive than "polar" because it expresses the idea of two charges.

5. A polar molecule has an unsymmetrical charge distribution,

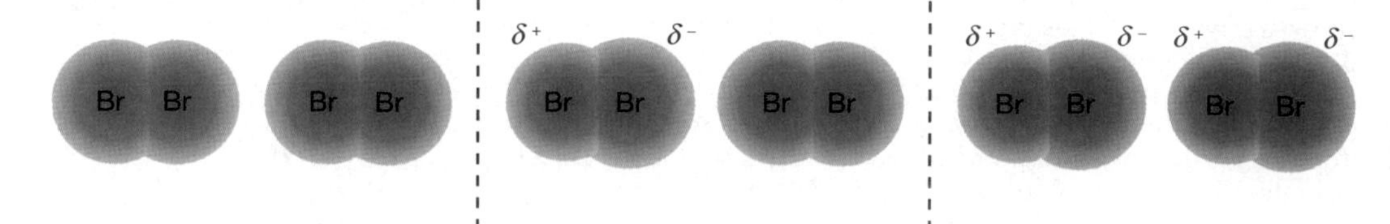

Figure 14.6 Dispersion forces exist between nonpolar molecules and are the result of the formation of temporary dipoles.

distribution. The end of the molecule where the electrons are concentrated develops a partial negative charge. The opposite end of the molecule develops a partial positive charge. In this way, a temporary dipole is set up. A **temporary dipole** can induce a dipole in the molecule next to it and an attractive force results. Another example is shown in Figure 14.6. The forces generated in this way are called **dispersion forces** or London forces after Fritz W. London, the physicist who first investigated them.

Of the factors contributing to van der Waals forces, dispersion forces are the most important. They are the only attractive forces that exist between nonpolar molecules. Even for most polar molecules, dispersion forces account for 85% or more of the van der Waals forces. Only in some special cases, such as those involving NH_3 or H_2O, do dipole-dipole interactions become more important than dispersion forces. We will examine these special cases in Chapter 17.

Many molecules will exhibit both dipole and dispersion interactions. However, we are only interested in the net result. The liquid and solid states of many compounds exist because of these intermolecular forces. Substances composed of nonpolar molecules are generally gases at room temperature or low-boiling liquids. Substances composed of polar molecules generally have

Table 14.2

Weak Forces Summary		
Type of Force	Dispersion forces	Dipole
Substances exhibiting force	Nonpolar molecules	Polar covalent molecules
Source of the force	Weak electric fluctuations which destroy spherical symmetry of electronic fields about atoms	Electric attraction between dipoles resulting from polar bonds
Properties due to the force	Low melting and boiling points	Higher boiling and melting points than nonpolar molecules of similar size
Examples	Ne (20 u; m.p. = −249°C) O_2 (32 u; m.p. = −219°C) F_2 (38 u; m.p. = −220°C) C_9H_{20} (128 u; m.p. = −54°C) Br_2 (160 u; m.p. = −7°C)	HF (20 u; m.p. = −84°C) CH_3OH (32 u; m.p. = −98°C) HCl (36.5 u; m.p. = −114°C) SeO_3 (127 u; m.p. = 118°C) ICl (162 u; m.p. = 27°C)

DEMONSTRATION

14-3 Dipole Attractions

Purpose: to simulate a polar molecule and its interactions with other molecules

Materials: 2 permanent bar magnets, overhead projector, 3 straight pins

Procedure: 1. Attach a straight pin to a magnet overnight so that the pin becomes magnetized.

2. Place two magnets on an overhead projector. Bring two bar magnets together so that they align. Have students observe that the magnets either attract or repel each other depending upon their orientation. Point out that polar molecules behave similarly except that the force is electric, not magnetic.

3. Place two nonmagnetized straight pins on the overhead projector and show that they do not attract each other.

4. Substitute the magnetized pin for

higher boiling points than nonpolar compounds have. Many substances composed of polar molecules are solids under normal conditions.

Weak intermolecular forces are effective only over short distances. They vary inversely as the sixth power of distance, $1/d^6$. Thus, if distance is doubled, the attractive force is 1/64 as large.

Another weak force is the attractive force between an ion and the oppositely charged end of a dipole that comes near the ion. These ion-dipole forces are somewhat stronger than van der Waals forces, yet are not nearly as strong as chemical bonds.

14.1 *CONCEPT REVIEW*

3. Compare and contrast dipole-dipole forces, dipole-induced dipole forces, and dispersion forces.
4. What does the term *dipole* mean? Explain why *dipole* is a more descriptive term than *polar* to describe a polar molecule.
5. How does a polar molecule differ from a nonpolar molecule?
6. **Apply** Using Table 14.1, which solvent would be predicted to have a lower melting point, ethanol or toluene? Water or methanol?

FRONTIERS

Ion Trapping

The illustration is not showing the motion of a wobbly ice-skater but that of a single aluminum ion whizzing around in the electric field of an ion trap. Ion traps were once exotic experimental tools used to study the behavior of isolated ions. However, ion trap mass spectrometers may soon become standard tools used to measure minute amounts of chemicals.

The chamber of an ion trap confines individual ions in a quadripole electric field, which is the electric field associated with four point charges. The ion trap's electric field is generated by applying a high frequency potential difference to electrodes within the trap. In operation, gaseous chemical samples are ionized by reagents within the chamber. By applying a potential difference of precise frequency, ions of a specific charge-to-mass ratio are confined to certain trajectories in the electric field. The ions are literally trapped in different regions of the chamber. By slightly changing the frequency, these ions move into unstable trajectories, exit the trap, and are counted by an external detector. Because ions of a specific charge-to-mass ratio are ejected at a precise frequency, the number of ions of a particular molecular mass can be correlated to the frequency of the applied potential difference.

Researchers are using ion trap mass spectrometers to detect trace amounts of carcinogenic agricultural pesticides. Researchers have accurately measured concentrations of pesticide residue to as low as 50 parts per billion. These spectrometers may be used to analyze the safety of food crops.

while a nonpolar molecule has a symmetrical charge distribution.
6. Ethanol and toluene have quite different molecular masses (46 vs. 92), and although ethanol has a higher polarity, it has a lower melting point. Methanol has a lower melting point than water.

4 CLOSE

Discussion: Ask students how the top leaves of trees get water. Point out that pressure cannot force water that high. Explain that polarity and the resultant dipole-dipole attractions are responsible for water transport in trees. As a molecule of water is lost through transpiration, it pulls the next water molecule into position like the cars in a railroad train.

FRONTIERS

Ion Trapping

Background: The ion trap was invented at the University of Bonn in Germany by Wolfgang Pauli and co-workers in the late 1950s. Hans Dehmelt of the University of Washington used ion traps to make exacting measurements of the energy of electron transitions. Both shared the 1989 Nobel Prize in Physics for their related work.

Teaching this Feature: Point out that ion trap mass spectrometers can measure ions of very large mass-to-charge ratios making them useful for studying proteins and nucleic acids.

one of the other pins. Have students observe that the pins will be attracted to each other. Explain that the substituted pin was magnetized and that it attracted the nonmagnetized pin because it induced an opposite magnetic field in the other pin. Point out that a polar molecule can induce an electric dipole in another molecule. Use the pins as a model of dispersion forces.

Disposal: F

Results: The two magnets attract each other as did the magnetized pin and the nonmagnetized pin.

Questions: 1. The two magnets simulate which type of attractive force? *dipole-dipole forces*

2. The magnetized and the nonmagnetized straight pin simulate which type of force? *dipole-induced dipole force*

3. Would you expect the boiling point of a polar substance to be higher or lower than that of a nonpolar substance having a similar molecular mass? *higher*

14.2 Coordination Chemistry

PREPARE

Key Concepts

The major concepts presented in this section on coordination chemistry include ligands, coordination numbers, nomenclature and writing formulas for complex ions and coordination compounds, and bonding in complexes.

Key Terms

complex ion
ligand
coordination number
coordinate covalent bond

1 FOCUS

Objectives

Preview with your students the objectives listed on the student page. Students can use them as a study guide for this major section.

Discussion

From the stockroom, display samples of some very colorful compounds in clear glass or plastic containers with closures. Point out to students that complex ions result in colorful coordination compounds. Inform them that they will learn to name and write the formulas of some of these compounds in this section.

OBJECTIVES

- define *complex ion, ligand, coordination number,* and *coordination compound.*
- name complex ions given their formulas, and write formulas for complex ions given their names.
- name coordination compounds given their formulas, and write formulas for coordination compounds given their names.

14.2 Coordination Chemistry

One important property of polar molecules is their behavior toward ions in solution. As an ionic crystal dissolves in water, surface ions become surrounded by polar H_2O molecules that adhere to the surface. The water molecule-ion clusters formed enter solution. The stability of these clusters is greatest when they have at their center a small ion of high charge. Because positive ions are usually smaller than negative ions, the clusters with which we are concerned have a positive ion at the center.

Ligands

When polar molecules or negative ions cluster around a central positive ion, a **complex ion** is formed. These polar molecules or negative ions are known as **ligands.** The number of points of attachment of the ligands around a central positive ion in a complex is called the **coordination number.** The most common coordination number found in complexes is six. These complex ions are described as octahedral. The ligands may be thought of as lying at the vertices of a regular octahedron with the central positive ion in the middle as shown in Figure 14.8. $[PtCl_6]^{2-}$ is an octahedral complex. Complex ions are widely used in analytical chemistry and as catalysts in industry.

Word Origins

ligare: (L) to bind

ligand—binds to the central ion in a complex

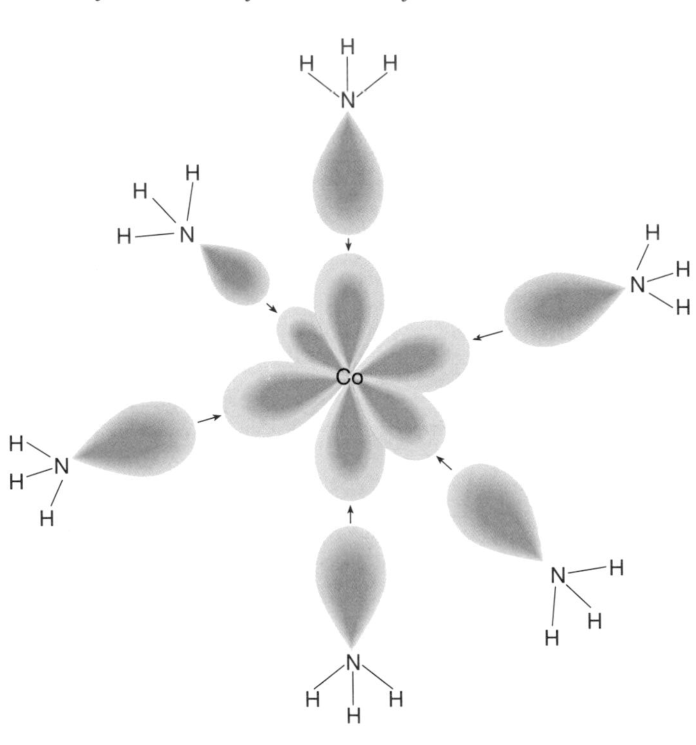

Figure 14.7 Unpaired electrons in NH_3 ligands are accepted by d^2sp^3 hybrid orbitals in a Co^{3+} ion.

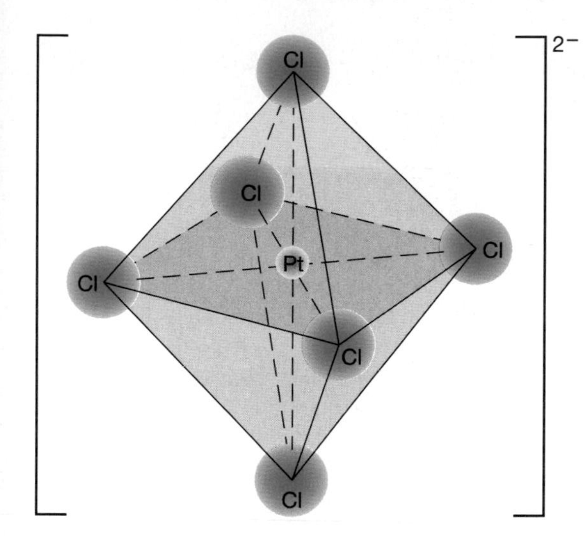

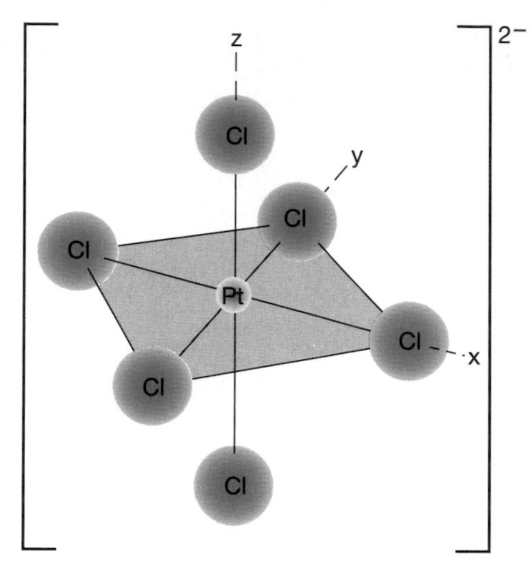

Figure 14.8 $[PtCl_6]^{2-}$ is a complex ion having an octahedral shape. Chloride ions are the ligands. On the right, a simpler representation of the octahedral shape is shown.

The coordination number four is also common. These complexes may be square-planar, with the ligands at the corners of a square and the central positive ion in the center. Others may be tetrahedral, with the ligands at the vertices of a regular tetrahedron and the central positive ion in the middle. Complexes of Ni^{2+}, Pd^{2+}, and Pt^{2+} are usually planar if the coordination number is four. An example of a square planar complex is $[NiBr_4]^{2-}$. A typical tetrahedral complex is $[CoI_4]^{2-}$, Figure 14.9.

The coordination number two is found in complexes of Ag^+, Au^+, and Hg^{2+}. Complex ions with coordination number two are always linear. The ligands are always located at the ends and the positive ion in the middle of a straight line. An example is $[Ag(CN)_2]^-$.

$$[NC—Ag—CN]^-$$

Ligands can be either molecules or negative ions. Molecular ligands are always polar and always have an unshared pair of electrons that is shared with the central ion. The most common

Figure 14.9 $[NiBr_4]^{2-}$ (left) forms a square planar complex, and $[CoI_4]^{2-}$ (right) exists as a tetrahedron. Both Ni^{2+} and Co^{2+} have a coordination number of four in these complexes.

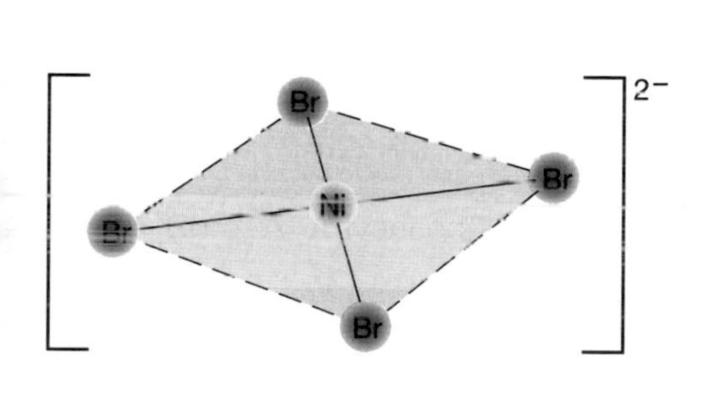

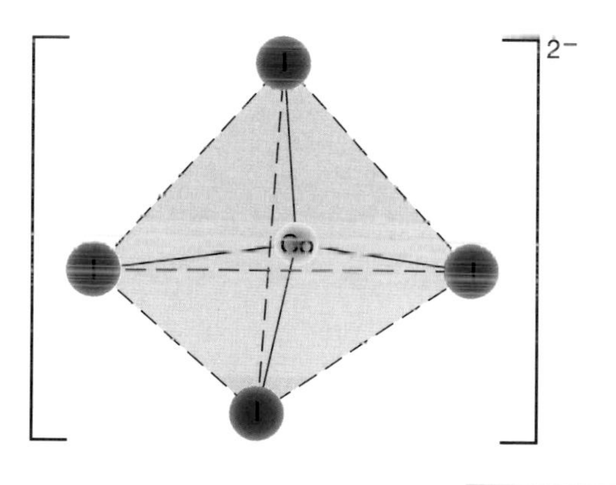

2 TEACH

CONCEPT DEVELOPMENT

- Ball and stick models will be helpful to students as they attempt to visualize the coordination number of the central positive ion in a complex. Point out that the structures of complexes are usually either linear (*sp* bonding), tetrahedral (sp^3), square planar (dsp^2), or octahedral (d^2sp^3).

QUICK DEMO **Complex Models** If you do not have ball and stick models, use plastic soda straws and thin wire to build three-dimensional models of the complexes. Use one color straw to represent the chemical bonds and another color of straw to show the outside edges of the geometric figure. Use spray paint or permanent markers to change the color of the plastic straws.

Enrichment: Certain materials called chelating agents are used for medical and industrial purposes. Have a student research chelating agents and report on their relationship to complexes. Possible response: *The formation of a heterocyclic ring with a metal atom in the ring is the process of chelation.*

disodium ethylenediamminetetraacetate copper chelate

PROGRAM RESOURCES

Study Guide: Use the Study Guide worksheet, p.53, for independent practice with concepts in Section 14.2.

Enrichment: Use the worksheet "Capacitors," pp. 27-28, to challenge more able students.

MATH CONNECTION
Platonic Solids

Exploring Further: Students can use card stock to draw and then cut out the regular geometric shapes necessary to construct the Platonic solids listed. Instruct them to leave small tabs along the edges of the shapes to use to glue the shapes together.

Extension: Encourage students to use geometry textbooks to determine why these solids are called Platonic solids and how Euler's Theorem pertains to them.

✓ ASSESSMENT

Skill: Have students use the glossary on pp. 884-894 to look up and record the definitions of the terms complex ion, ligand, and coordination number. Ask students to give an example of each one using a diagram on the same sheet of paper on which they recorded the definitions.

GLENCOE TECHNOLOGY

Videodisc

STVS: Chemistry

Disc 2, Side 1

Images of Atoms (Ch. 4)

Videodisc

Chemistry: Concepts and Applications

Properties of Transition Metals

Disc 1, Side 2, Ch. 9

Also available on CD-ROM.

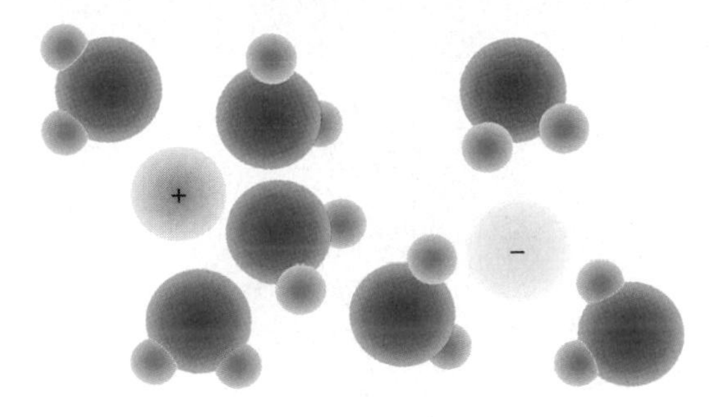

Figure 14.10 When an ionic substance dissolves in water, the water molecules cluster around the ions, forming complex ions.

ligand is water. The hydrated compounds mentioned in Section 8.2 are composed of positive ions surrounded by water ligands and the negative ions. Ammonia, NH_3, is also a common ligand. Many negative ions can also act as ligands in complexes. Some of the most important are the following: fluoride, F^-; chloride, Cl^-; bromide, Br^-; iodide, I^-; cyanide, CN^-; thiocyanate, SCN^-; and oxalate, $C_2O_4^{2-}$.

The oxalate ion has two of its oxygen atoms attached to the positive ion. Such a ligand is shown in Figure 14.11, and is called didentate ("two-toothed"). A didentate ligand attaches at two points. Therefore, two didentate ligands can form a tetrahedral complex. Three didentate ligands can form an octahedral complex. Tridentate and tetradentate ligands are also known.

One widely used hexadentate ligand is the ethylenediaminetetraacetate ion (EDTA), usually produced in the form of one of its salts. The calcium disodium salt, called calcium sodium edetate, is used in medicine. EDTA is also found as a cleansing agent in detergents and shampoos, as a preservative in foods, and as a cleansing agent in metal polishes. It is used in purifying vegetable oil. EDTA has even been used to clean surfaces contaminated by radioactive materials.

MATH CONNECTION

Platonic Solids

The regular octahedron made of eight equilateral triangles is one of only five Platonic solids. The Platonic solids are those solids whose faces are congruent regular polygons. Others that are common in coordination chemistry and in crystals are the hexahedron (cube) with six square faces and the tetrahedron with four equilateral triangles. The remaining two solids are the dodecahedron with twelve pentagons as faces and the icosahedron with twenty equilateral triangular faces.

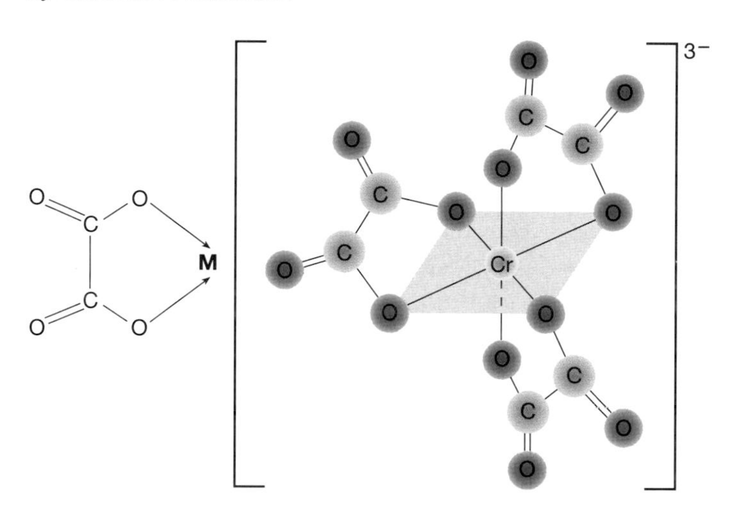

Figure 14.11 The oxalate ion (left) is a didentate ligand. In $[Cr(C_2O_4)_3]^{3-}$ (right), three oxalate ions form an octahedral complex with Cr^{3+}.

Chemistry Journal

EDTA

Have students choose one use of EDTA listed on this page, research its use, and write a journal report.

Bridge to MEDICINE

Removing Toxic Metals from the Body

The human body does not have built-in processes for eliminating most transition metal atoms. Small amounts of several of these metals, such as chromium, copper, iron, manganese, and molybdenum, are essential to health. However, most transition metals, and large amounts of all of them, are toxic. In cases where a human has ingested large amounts of a transition metal, physicians use polydentate ligands to form complexes with the metal atoms. These complexes prevent the metal atoms from interfering with bodily processes. In addition, in most cases the body is able to eliminate the complex, thereby getting rid of the toxic atom. For example, calcium sodium edetate is used to complex lead. Dimercaprol complexes arsenic, gold, and mercury. Penicillamine complexes copper.

There are some problems that may arise from using some of these complexing agents. For example, the lanthanoid metals are more toxic when complexed by organic ligands than they are as the free metals.

Exploring Further

1. If most metals are toxic, what metals can be used to make artificial body parts?
2. What problems might result from using silver and mercury in dental amalgams?
3. Find out why lead poisoning is a particular problem for young children. Why is lead poisoning less of a problem than it was thirty years ago?

Names and Formulas of Complex Ions

CHEMISTRY IN DEPTH

Pages 359–365

In naming a complex ion, the ligands are named first, followed by the name of the central ion. The names used to identify the ligands are listed in Table 14.3. The name for each ligand is modified by using a prefix indicating how many of that ligand appear in the formula. The prefixes used are *di-*, *tri-*, *tetra-*, *penta-*, and *hexa-*. If no prefix is used, it is assumed that the ligand appears only once in the formula. If more than one kind of ligand appears in the formula, the names are listed alphabetically without regard to numerical prefixes. For example, pentaammine would precede chloro because we disregard the *penta,* and the *a* of ammine comes before the *c* of chloro.

The central ion is named in the usual manner if the whole complex ion has a positive charge. If, however, the whole complex ion has a negative charge, the ending of the central ion is changed to *-ate*. In the case of the elements listed in Table 14.4, the Latin stems are used in negatively charged complex ions. If the central ion has more than one possible oxidation number, a Roman numeral indicating the correct oxidation number is included in parentheses following the name of the complex ion.

BRIDGE TO MEDICINE

Removing Toxic Metals from the Body

Background: As the world becomes increasingly industrialized, more metals are finding their way into the air we breathe and the water we drink.

Teaching this Feature: Point out that selenium, another trace metal that is essential for immune systems to operate efficiently, is very toxic in slightly larger amounts.

Connection to Economics: Have a team of students contact the water treatment department of your city to determine the techniques and costs of removing metals from water for human consumption.

Extension: Have interested students use a water analysis kit to test for metals in water. If such a kit is unavailable have students obtain a qualitative chemical analysis book and identify procedures that are used to test for the presence of metal ions in water.

Answers to **Exploring Further**

1. stainless steel, titanium
2. Small amounts of these metals may enter the body and both are toxic.
3. Lead poisoning causes blindness, mental retardation, and death. Lead oxide is no longer used as a paint pigment.

CONCEPT DEVELOPMENT

- Students will need practice to master the nomenclature rules of complex ions. It is recommended that students not be required to memorize the table of ligands or their Latin stems. Both of these tables of information should be provided to students during evaluation of this material.

PROGRAM RESOURCES

Laboratory Manual: Use microlab 14 "Separating Metal Ions" as enrichment for the concepts in Chapter 14.

Critical Thinking/Problem Solving: Use the worksheet, p.14, as enrichment for the concept of "Coordination Chemistry — A Closer Look."

CHEMISTRY IN DEPTH

The sections **Names and Formulas of Complex Ions, Coordination Compounds,** and **Bonding in Complexes** on pages 359-365 are recommended for more able students.

CULTURAL DIVERSITY

Emerald City

Gemstones have had a fascination for humans since prehistoric times. They are found in graves as decorations and have served as currency at various times. Emeralds, highly prized stones, are beryl, colored by traces of chromic oxide or iron and vanadium. The earliest known emerald mines are from the Zubara region of Egypt near the Red Sea. There are also mines in the Ural Mountains of Siberia, Western Australia, India, Brazil, Transvaal, and Zimbabwe. The finest emeralds today come from Colombia, from one of two major mines. The Chivor mine produces blue-green emeralds in pockets of crystals, and the Muzo mine produces rich green emeralds found with quartz in limestones and shales.

The Incas had vast quantities of emeralds and when the Spaniards invaded they seized many of the gems. Some, they took back to Europe. Others, they took to Southeast Asia across the Pacific. Thousands of Colombian emeralds, five of which are over 300 carats in weight, decorate the crown jewels of Iran.

✓ ASSESSMENT

Portfolio: Have students select a gemstone they want to know more about, and write a two paragraph summary of what they learned as a result of their research on that gemstone. They should emphasize chemical structure and composition.

Figure 14.12 From left to right are solutions containing the colorless $[Cu(CN)_2]^-$ ion and colorful $[Cu(H_2O)_4]^{2+}$, $[Cu(en)_2]^{2+}$, and $[CuBr_4]^{2-}$ ions. The abbreviation *en* represents the ethylenediamine ligand.

Table 14.3

Some Common Ligands			
Neutral Molecules		**Anions**	
Ligand	**Name**	**Ligand**	**Name**
CO	carbonyl	Br^-	bromo
NH_3	ammine	CN^-	cyano
NO	nitrosyl	Cl^-	chloro
H_2O	aqua	F^-	fluoro
		I^-	iodo
		OH^-	hydroxo
		$C_2O_4^{2-}$	(oxalato)*
		O^{2-}	oxo
		S^{2-}	thio
		SCN^-	thiocyanato
		$S_2O_3^{2-}$	thiosulfato

*Names of organic ligands are always shown in parentheses.

Table 14.4

Latin Derived Names Used in Metal Complexes	
Metal	**Latin Form**
copper	cuprate
gold	aurate
iron	ferrate
lead	plumbate
silver	argentate
tin	stannate

PROBLEM SOLVING HINT

In writing the formula of a complex ion, the brackets do not include the charge on the ion.

SAMPLE PROBLEMS

1. Naming a Positive Complex Ion

Name the complex ion $[CrCl(NH_3)_5]^{2+}$.

Solving Process:

The two ligands present in this complex ion are *chloro* (the chloride ion) and *ammine* (the ammonia molecule). Ammine precedes chloro alphabetically. Because there are five ammonia molecules, ammine is prefixed with *penta-*. The ligands are, then, pentaamminechloro. The central ion, chromium, is named with its ending unchanged because the complex ion, as a whole, is positive. However, chromium has more than one oxidation number. We must, therefore, follow the name with a Roman numeral, in this case (III). The complete name is written as one word and followed by the word *ion*.

pentaamminechlorochromium(III) ion

*inter*NET CONNECTION

Basic chemistry software models are available from the *Journal of Chemical Education.*

World Wide Web
http://jchemed.chem.wisc/edu

2. Naming a Negative Complex Ion

Name the complex ion, $[IrCl_6]^{3-}$.

Solving Process:

The ligands are six chloride ions named *hexachloro*. The central ion is iridium. The complex ion is negative, so the name iridium is changed to *iridate*. Iridium here is in the oxidation state (III). The full name of the complex is

hexachloroiridate(III) ion.

Figure 14.13 The colors of many gemstones are due to small quantities of transition metal ions. For example, Cr in the mineral beryl produces an emerald, and Fe in beryl produces aquamarine. Cr in corundum produces a ruby, and corundum containing Fe or Ti is a sapphire.

PRACTICE PROBLEMS

7. Name the following complex ions.
   a. $[Cd(NH_3)_2]^{2+}$
   b. $[Cr(H_2O)_4]^{2+}$
   c. $[Co(NH_3)_6]^{2+}$
   d. $[Cu(NH_3)_4]^{2+}$
   e. $[PtBr(NH_3)_3]^{+}$
8. Name the following complex ions.
   a. $[PtI_6]^{2-}$
   b. $[GeF_6]^{2-}$
   c. $[Sb(OH)_6]^{-}$
   d. $[PdCl_4]^{2-}$
   e. $[Fe(CN)_6]^{4-}$
9. Name the following complex ions.
   a. $[SbF_6]^{-}$
   b. $[BF_4]$
   c. $[AlF_6]^{3-}$
   d. $[CoCl(H_2O)(NH_3)_4]^{2+}$
   e. $[Cd(NH_3)_4]^{2+}$

As you may have already noticed, formulas for complex ions are written in an order different from that of the names. In the formulas, the symbol for the central ion is placed first, followed by the symbols or formulas for the ligands. Negative ligands are listed first, followed by neutral molecules. Within those two groups, the ligands are listed in alphabetical order of their symbols. Formulas for complex ions are always enclosed in brackets. Polyatomic ions or molecules acting as ligands are always enclosed in parentheses even if only occurring once.

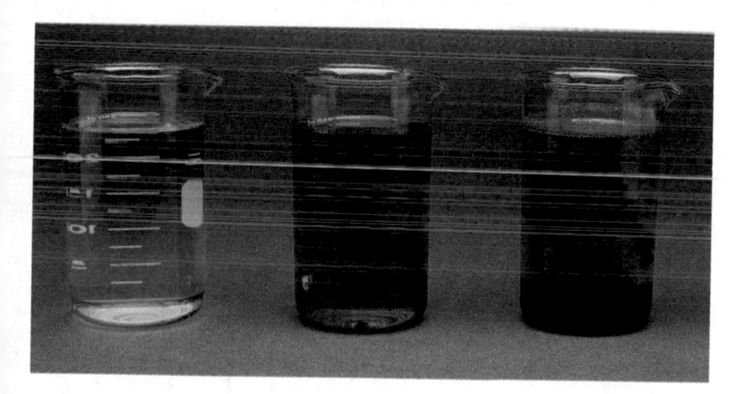

Figure 14.14 Nickel forms a variety of complex ions. Left to right are $[Ni(H_2O)_6]^{2+}$, $[Ni(NH_3)_6]^{2+}$, and a strawberry red precipitate of a nickel and dimethylglyoxime complex.

Correcting Errors

Encourage students to learn the pattern that is used by chemists when naming complex ions.
1. ligands first, in alphabetical order with prefixes, if needed
2. central ion
3. *-ate* ending, if the complex ion is negative
4. Roman numeral, if needed

Have students recite the prefixes *di-*, *tri-*, *tetra-*, *penta-*, and *hexa-* before they begin the Practice Problems.

PRACTICE

Guided Practice: Guide students through the Sample Problems, then have them work in groups on Practice Problems 7 and 8.

Cooperative Learning: Refer to pp. 28T-29T to select a teaching strategy to use with this activity.

Independent Practice: Your homework or classroom assignment can include the remaining Practice Problems, and Review and Practice 38-41 from the end of the chapter.

Answers to Practice Problems

7. a. $[Cd(NH_3)_2]^{2+}$ diamminecadmium ion
   b. $[Cr(H_2O)_4]^{2+}$ tetraaquachromium(II) ion
   c. $[Co(NH_3)_6]^{2+}$ hexaamminecobalt(II) ion
   d. $[Cu(NH_3)_4]^{2+}$ tetraamminecopper(II) ion
   e. $[PtBr(NH_3)_3]^{+}$ triamminebromoplatinum(II) ion
8. a. hexaiodoplatinate(IV) ion
   b. hexafluorogermanate(IV) ion
   c. hexahydroxoantimonate(V) ion
   d. tetrachloropalladate(II) ion
   e. hexacyanoferrate(II) ion
9. a. hexafluoroantimonate(V) ion
   b. tetrafluoroborate ion
   c. hexafluoroaluminate ion
   d. tetraammineaquachlorocobalt(III) ion
   e. tetraamminecadmium ion

CONCEPT DEVELOPMENT

- Emphasize that the ligands in the formulas of complex ions are written in a different order from ligands in the names.

Correcting Errors

Encourage students to learn the pattern that is used by chemists when writing the formula of a complex ion.

1. Symbol of the central ion is placed first.
2. Negative ligands are next in alphabetical order of their symbols.
3. Place neutral molecules acting as ligands in alphabetical order of their symbols.
4. Enclose polyatomic and molecular ligands in parentheses.
5. Enclose formula of complex ion in brackets with the overall charge of the complex ion outside the brackets.

PRACTICE

Guided Practice: Guide students through the Sample Problems, then have them work in groups on Practice Problems 10 and 11.
Cooperative Learning: Refer to pp. 28T-29T to select a strategy to use with this activity.

Independent Practice: Your homework or classroom assignment can include the remaining Practice Problem, and Review and Practice 42 from the end of the chapter.

Answers to
Practice Problems

10. a. $[Pt(NH_3)_4]^{2+}$
b. $[Co(NH_3)_6]^{3+}$
c. $[Ir(H_2O)_6]^{3+}$
d. $[Pd(NH_3)_4]^{2+}$
11. a. $[PdCl_6]^{2-}$
b. $[PtCl_3(NH_3)]^-$
c. $[AuI_4]^-$
d. $[Au(CN)_4]^-$
12. a. $[W(CN)_8]^{3-}$
b. $[Sb(OH)_6]^-$
c. $[Fe(CO)_5(NO)]^{2+}$
d. $[Co(CO)_4]^+$

Below are listed several other complex ions and their names. Use these examples as a check on your understanding of names and formulas for complex ions.

$[SiF_6]^{2-}$	hexafluorosilicate(IV) ion
$[Zn(NH_3)_2]^{2+}$	diamminezinc ion
$[Ir(H_2O)(NH_3)_5]^{3+}$	pentaammineaquairidium(III) ion

SAMPLE PROBLEMS

1. Writing a Positive Complex Ion Formula
Write the formula for the diamminepalladium(II) ion.

Solving Process:
The central ion is palladium, Pd. The ligands are two ammonia molecules, NH_3. The charge on the palladium ion is its oxidation number, 2+. The ammonia molecules are neutral, so they do not contribute to the charge of the complex ion. The complex ion, then, has a charge of 2+.

$$[Pd(NH_3)_2]^{2+}$$

2. Writing a Negative Complex Ion Formula
Write the formula for the carbonylpentacyanoferrate(II) ion.

Solving Process:
The central ion is iron. See Table 14.4. The two ligands are CO and CN^-. Because cyano is charged and carbonyl is not, cyano will precede carbonyl in the formula. The charge on the iron ion is 2+, on the carbonyl it is 0, and on the five cyanides, 5− (1− each). The formula, then, is

$$[Fe(CN)_5(CO)]^{3-}$$

Figure 14.15 The color of these iron-containing solutions depends on the ligand present. At the left, the yellow solution contains Fe^{3+} complexed with H_2O and OH^-. The deep red color in the flask on the right is produced as aqueous KSCN is added and the complex $[Fe(SCN)(H_2O)_5]^{2+}$ forms.

PRACTICE PROBLEMS

10. Write formulas for the following complex ions.
a. tetraammineplatinum(II) ion
b. hexaamminecobalt(III) ion
c. hexaaquairidium(III) ion
d. tetraamminepalladium(II) ion

11. Write formulas for the following complex ions.
a. hexachloropalladate(IV) ion
b. amminetrichloroplatinate(II) ion
c. tetraiodoaurate(III) ion
d. tetracyanoaurate(III) ion

12. Write formulas for the following complex ions.
a. octacyanotungstate(V) ion
b. hexahydroxoantimonate(V) ion
c. pentacarbonylnitrosyliron(II) ion
d. tetracarbonylcobalt(I) ion

CONTENT BACKGROUND

Isomerism in Complex Ions: Most of the types of isomerism found in complex ions and coordination compounds were discovered by Alfred Werner in the late nineteenth and early twentieth centuries. Square planar complexes can have geometric isomers if the complex is of the type *Mabcd* (three isomers) or MA_2B_2 (two isomers).

Octahedral complexes can also exhibit geometrical isomerism. Consider the complex $[PtCl_2(NH_3)_4]^{2+}$.

Optical isomerism arises in the case of didentate ligands in octahedral complexes. For instance, the complex $[Co(en)_3]^{3+}$, where en = ethylenediamine, has two enantiomers.

Ionization isomerism occurs where there is an exchange of ions between a ligand and an associated ion outside the complex, such as is seen in $[CoCl(NH_3)_5]SO_4$ and its isomer,

Figure 14.16 Colorless complex ions are formed by ions with empty or full *d* orbitals. The colored complex ions are formed when *d* orbitals are partially filled. Shown here from left to right are solutions containing the ions Sc^{3+}, Cr^{3+}, Co^{2+}, Ni^{2+}, Cu^{2+}, and Zn^{2+}. All ions are complexed with six water molecules.

Coordination Compounds

Sometimes, the charge of the central ion in a complex ion is just matched by the charges of the ligands. A neutral compound has then been formed. Occasionally, a complex will form in which neither the ligands nor the central atom has a charge. In that case, the name of the central atom is followed by zero (0). Complex ions can also form ionic compounds just as any monatomic ion can. The charges of all particles present must, of course, add up to zero. Coordination compounds are compounds formed in any of the ways just listed.

SAMPLE PROBLEMS

1. Naming a Coordination Compound

Name the following coordination compound:

$$[Ni(NH_3)_6]Br_2$$

Solving Process:

Determine the name of the complex ion. The complex ion is hexaamminenickel(II). Then name the compound as you would any other ionic compound. The complete name is hexaammine-nickel(II) bromide.

2. Writing a Formula for a Coordination Compound

Write the formula for the coordination compound, hexacarbonylchromium(0).

Solving Process:

The central atom is chromium without a charge. The ligands are carbonyl groups, also without charge. Thus, the formula is $[Cr(CO)_6]$.

Correcting Errors

Ask students to name the compound whose formula is NaBr. Remind students that a complex ion can be substituted for either or both of the ions in sodium bromide. The compound is named as you would any other ionic compound. The positive ion is named first, then the negative ion.

✓ ASSESSMENT

Knowledge: Have students work in cooperative learning groups to name and write formulas for the coordination compounds listed in the Review and Practice questions 41-44 on page 373.

$[CoSO_4(NH_3)_5]Cl$. Hydration isomers follow a similar pattern. For example, $[Cr(H_2O)_6]Cl_3$, $[CrCl(H_2O)_5]Cl_2 \cdot H_2O$, and $[CrCl_2(H_2O)_4]Cl \cdot 2H_2O$. Coordination isomerism is a similar kind of isomerism in which both anion and cation are complexes. Two examples are $[Co(NH_3)_6][Cr(CN)_6]$ and its isomer $[Cr(NH_3)_6][Co(CN)_6]$. Linkage isomerism occurs with ligands that contain more than one atom that can donate an electron pair. Such ligands are often referred to as ambidentate ligands. Examples are NO_2^-, SCN^-, CN^-, and $S_2O_3^{2-}$.

PRACTICE

Guided Practice: Guide students through the Sample Problems, then have them work in class on Practice Problems 13 and 14.

Independent Practice: Your homework or classroom assignment can include Review and Practice 43-44 from the end of the chapter.

Answers to
Practice Problems

13. **a.** tetraamminediaquanickel(II) nitrate
b. pentaamminechlorocobalt(III) chloride
c. potassium hexacyanoferrate(III)
d. potassium hexahydroxostannate(IV)
e. diamminesilverperrhenate

14. **a.** $[Co(H_2O)_6]I_2$
b. $[Co(NH_3)_6]Cl_2$
c. $K_2[Ni(CN)_4]$
d. $PdCl_2(NH_3)_2$
e. $Ru(CO)_5$

PRINTING CONNECTION
Ink Composition

Exploring Further: Ask a student to contact the local newspaper to determine if they are using an anti-smudge ink that doesn't rub off on hands or clothing. Determine the cost difference between regular newspaper ink and the newer, smudge-less type.

CHEMISTS AND THEIR WORK
Jokichi Takamine

Background: Many of the chemists in Asia were educated in the United States. As a tribute both to his native country and his adopted country Takamine convinced the mayor of Tokyo to send cherry trees as a gift to the United States. Each year when these trees bloom along the Potomac River in Washington, D.C. they reflect the social consciousness of a chemist who loved both his countries.

CHEMISTS AND THEIR WORK

Jokichi Takamine
(1854-1922)

It is not often that industrial chemists receive the same fame that research chemists do. However, Jokichi Takamine, a Japanese industrial chemist, had many "firsts" in the field of chemistry. He was the first to make a successful separation of the starch-hydrolyzing enzyme called diastase. As a result of this accomplishment, he made enzyme production commercially feasible. At the age of 30, while in the United States, he learned of the use of superphosphate as a fertilizer. He returned to his native country and soon started an artificial fertilizer company to use the process in Japan. In 1901, he was the first to isolate pure adrenalin.

PRACTICE PROBLEMS

13. Name the following coordination compounds.
a. $[Ni(H_2O)_2(NH_3)_4](NO_3)_2$
b. $[CoCl(NH_3)_5]Cl_2$
c. $K_3[Fe(CN)_6]$
d. $K_2[Sn(OH)_6]$
e. $[Ag(NH_3)_2]ReO_4$

14. Write formulas for the following coordination compounds.
a. hexaaquacobalt(II) iodide
b. hexaamminecobalt(II) chloride
c. potassium tetracyanonickelate(II)
d. diamminedichloropalladium(II)
e. pentacarbonylruthenium(0)

Below are several more examples of coordination compounds with their names. Use these examples as a check on your understanding of naming and writing formulas for coordination compounds.

$Na_2[Sn(OH)_6]$	sodium hexahydroxostannate(IV)
$Ag_3[Fe(CN)_6]$	silver hexacyanoferrate(III)
$[V(CO)_6]$	hexacarbonylvanadium(0)

Bonding in Complexes

Almost any positive ion might be expected to form complexes. In practice, the complexes of the metals of the first two groups in the periodic table have little stability. By far the most important and most interesting complexes are those of the transition metals. The transition metals have partially filled *d* orbitals that can become involved in bonding. These positive ions are small

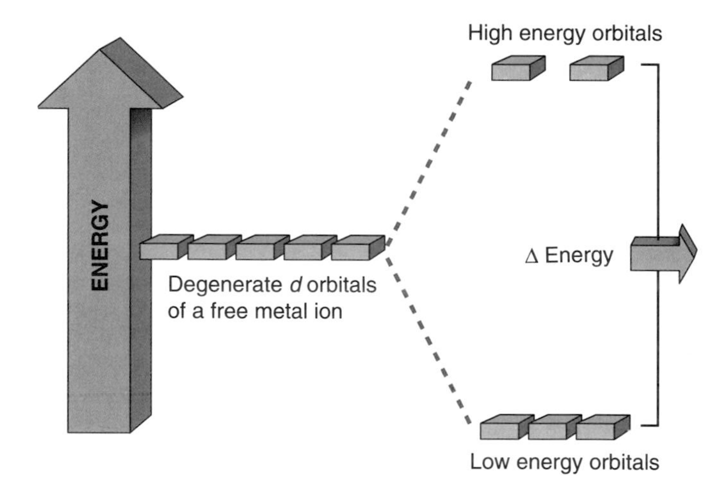

Figure 14.17 The intense colors of many complex ions are the result of the splitting of the *d* orbitals. Energy is absorbed as electrons move from the lower to higher energy levels.

CONTENT BACKGROUND

Gemstones: The colors of gemstones come from trace amounts of transition metal ions in the minerals beryl or corundum.

Mineral	Gemstone	Transition Metal
Beryl	emerald aquamarine	Cr Fe
Corundum	ruby sapphire	Cr Fe or Ti

Exploring Further: An interested student can do library research and prepare a report on the chemical composition of other common gemstones.

because of the strong attractive force of the nucleus for the inner electrons. They also have high oxidation numbers. The combination of small size and high charge results in a high charge density on the central ion. Such a high charge density is particularly favorable to complex ion formation.

In an isolated ion of one of the metals of the first transition series, the five 3*d* orbitals are degenerate. That is, they have the same energy. No energy is absorbed if an electron is transferred from one of the 3*d* orbitals to another. However, in the complex, the ligands affect the energies of the different 3*d* orbitals. In an octahedral complex, the *d* orbitals are split into a higher energy group of two orbitals and a lower energy group of three orbitals. Many complex ions have intense colors. The blue color that is typical of solutions of copper(II) compounds is an example. This blue color is due to the $[Cu(H_2O)_6]^{2+}$ ion. The colors of complexes arise from electron transitions between the split *d* orbitals. The split represents only a small energy gap. As the electrons move from the lower to the higher energy group, they absorb light energy. The phthalocyanine blue and green dyes and paint pigments are tetradentate ligand complexes of copper. Their intense color is due to the splitting of the *d* orbitals of the copper ion. These dyes find wide use in inks.

Ligands are either negative ions or polar molecules, while the central atom is a positive ion. These facts suggest that the bonding forces in a complex are electrostatic, just as they are in a salt crystal. In fact, there are strong similarities between the structures of salts and the structures of complex ions. On the other hand, the ligands have an unshared pair of electrons that they are capable of donating. The central ion always has unoccupied orbitals into which electron pairs might be placed. This combination suggests that the bonds are covalent. The name **coordinate covalent bond** has been given to covalent bonds in which both of the electrons in the shared pair come from the same atom. The chemistry of coordinate covalent bonds is exactly like that of any other covalent bond. The bonds of most complex ions have characteristics of both covalent and ionic bonding types. The covalent character dominates.

PRINTING CONNECTION

Ink Composition

Some printing inks consist of a pigment (color) dispersed in a vehicle such as linseed oil or kerosene. Other oils and solvents are also used as vehicles depending on the particular kind of printing machinery to be used. Additives are also often blended in the vehicle. They may stabilize the pigment if it is subject to decomposition. They may also alter the finish of the ink to give a glossy appearance or a matte effect.

14.2 CONCEPT REVIEW

15. If an ion has a positive charge, would you expect to find it in the central position in a complex, as a ligand, or either one?

16. How is it possible for a ligand to be a neutral molecule?

17. Name the following.
- **a.** $[MoI(CO)_5]^-$
- **b.** $[Fe(C_2O_4)_3]^{3-}$
- **c.** $[TiCl(H_2O)_5]^{2+}$

18. Write formulas for the following.
- **a.** hexacarbonylmolybdenum(0)
- **b.** dicyanocuprate(I) ion

19. Apply In terms of energy levels of electrons, explain the deep red color of the complex ion $[Fe(SCN)(H_2O)_5]^{2+}$.

3 ASSESS

CHECK FOR UNDERSTANDING

Ask students to answer the Concept Review problems and Review and Practice 31-46 from the end of the chapter.

RETEACHING

Ask students to recite how complex ion formulas are written when the name is given and the order in which a complex ion is named when the formula is given.

EXTENSION

Ask students to determine the central ion of hemoglobin and its coordination number. What is the coordinated group, and what is its spatial orientation about the central ion? *in hemoglobin: Fe, 4, porphyrin, planar*

Answers to
Concept Review

15. central position

16. Molecular ligands are always polar and always have an unshared pair of electrons that can be shared with the central ion.

17. a. pentacarbonyliodomolybdenate(0) ion
b. tri(oxalato)ferrate(III) ion
c. pentaaquachlorotitanium (III) ion

18. a. $[Mo(CO)_6]^-$, **b.** $[Cu(CN)_2]$

19. The color of the complex arises from electron transitions between split *d* orbitals. As the electrons move from the lower to the higher energy group, they absorb light energy. The deep red color of this complex indicates a relatively small energy gap between the lower and higher energy groups, since red light has a lower energy than other visible colors.

4 CLOSE

Discussion: Remind students that one polar molecule can attract another polar molecule, which in turn, can attract another. Dip the corner of a paper towel in colored water. As the water is wicked up the paper towel by capillary action, have the students discuss how relative polarities might be used to separate components of a solution.

CHEMISTRY IN DEPTH

Chemistry in Depth ends on this page.

14.3 Chromatography

PREPARE

Key Concepts

The major concepts presented in this section include fractionation, column chromatography, high performance liquid chromatography, ion exchange chromatography, paper chromatography, thin-layer chromatography, and gas chromatography.

Key Terms

fractionation
chromatography
mobile phase
stationary phase
column chromatography
high performance liquid chromatography (HPLC)
ion chromatography
paper chromatography
thin layer chromatography
gas chromatography

Looking Ahead

Obtain a raw egg to use in the Focus activity of this section.

1 FOCUS

Objectives

Preview with your students the objectives listed on the student page. Students can use them as a study guide for this major section.

OBJECTIVES

- define *chromatography, mobile phase,* and *stationary phase.*
- define, describe, and name uses for the different types of chromatography.

14.3 Chromatography

One of the routine tasks facing analytical chemists and scientists engaged in chemical research is the separation of different substances from mixtures. You have already been introduced to two such operations, distillation and crystallization. Another separation method is chromatography, which will be described in this section.

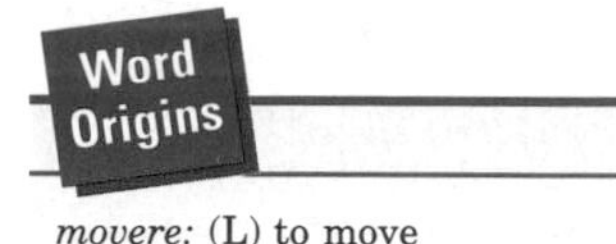

movere: (L) to move
mobile—moves easily

Fractionation

Substances separated from a mixture are called fractions because they are parts of the whole. The overall separation of parts from a whole by any process may be called **fractionation. Chromatography** is a method of fractionation based on polarity. We will examine several different chromatographic techniques.

In chromatography, a mobile phase containing a mixture of substances is allowed to pass over a stationary phase that has an attraction for polar materials. The **mobile phase** consists of the mixture to be separated dissolved in a liquid or gas. The **stationary phase** consists of either a solid or a liquid adhering to the surface of a solid. The substances in the mixture will travel at different rates due to varying polarity. There are several polarity factors determining the rate at which each component of the mixture migrates. A polar substance will have attraction for the solvent as well as an attraction for the stationary phase.

The stationary phase will attract some substances more strongly than others. The slowest migrating substance will be the one with the greatest attraction for the stationary phase. The fastest migrating substance will be the one with the least attraction for the stationary phase. Thus, substances can be separated.

Figure 14.18 The packing components for a typical chromatography column are shown.

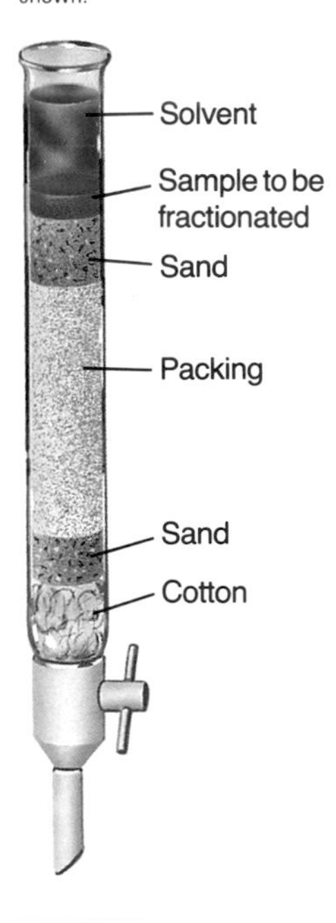

Chromatography

Column chromatography is used for extremely delicate separations. Complex substances such as vitamins, proteins, and hormones, not easily separated by other methods, can be separated by column chromatography. In column chromatography, a glass or plastic column, typically 1 cm in diameter and 50 cm long, is used to carry out the separation. The column is packed with a stationary phase such as aluminum oxide. Other packings used are calcium carbonate, magnesium carbonate, sodium carbonate, activated charcoal, clays, gels, and many organic compounds. The size of the particles of the stationary phase will be in the range of 150–200 μm.

The mixture to be fractionated is dissolved in a solvent. This solution is the mobile phase, which is added to the top of the column. The various components of the mixture are attracted to the stationary phase at the top of the tube. Then fresh solvent is

CONTENT BACKGROUND

R_f Values: In preparing paper chromatograms, identification of unknowns is aided by the measurement of R_f values. These values concern the distance an unknown has advanced compared to the distance the solvent front has advanced. R_f values must be found for a particular system with standards of known concentration. Once that particular piece of apparatus has been standardized for a substance, an unknown containing that substance can be analyzed. The volume of the chromatographic chamber, for instance, has a large effect on R_f values. Values from one chamber cannot be applied to another chamber of different geometry. Other factors affecting R_f values are the state of subdivision of the adsorbent, the temperature, the purity of the solvent, the separation of a multi-component solvent, and evaporation. Solvents and/or

poured onto the top of the column and allowed to percolate through the column. The solvent may be water, ammonia, an acid, an alcohol, or some other organic or inorganic solution.

Each substance in the mobile phase migrates down the tube at a different rate. The rate of travel of a particular substance depends upon the attraction of the substance for the stationary phase and its attraction for the solvent. If the substance has a high attraction for the stationary phase it will move down the column very slowly. A substance with less attraction for the stationary phase moves faster. The greater the concentration of solvent the faster a substance will move through the column. However, it is usually the case that very high concentrations of the solvent do not result in efficient separations. Solvent concentration, then, is usually a compromise between speed and effective separation.

After separation, it may be necessary to recover the material in each separate zone for identification. One method involves forcing the zones out the bottom of the tube, one at a time. The substance in each zone is dissolved and then recovered by evaporation. A second method uses solvents of increasing polarity until each zone comes out the bottom along with the solvent. Identification is then made by any number of methods.

chroma: (GK) color
graphein: (GK) to write
chromatography – writing with color

EVERYDAY CHEMISTRY

New Tips on the Laundry

Sebaceous glands are found near the hair follicles and secrete fats and oils, cellular debris, and keratin. These fats and oils rub off on clothing and cause dirt to adhere more readily. Researchers have some high-technology tools such as neutron bombardment, radiotracer analysis, and thin-layer chromotography to help them find out what happens after you put soiled clothing into the laundry basket.

Apparently, a series of reactions occurs. First, oxygen reactions make the oils more polar and thus more easily dissolved by soaps and detergents. However, as the oils stay in the clothing, more reactions take place which bind them even more closely.

One of the oils, squalene, gradually forms long chains that bond with fibers and turn them yellow. Polyesters sometimes resist the yellowing effects of the oil, but cottons are very difficult to lighten.

If you launder often, not allowing the oils to get to the second stage, you are more likely to be happy with the results of your laundry.

Exploring Further

1. Explain the action of polar molecules in helping detergents or soaps do their work.
2. Find out the names of the other oils that are secreted by the sebaceous glands. In your search, find out what sebum is.

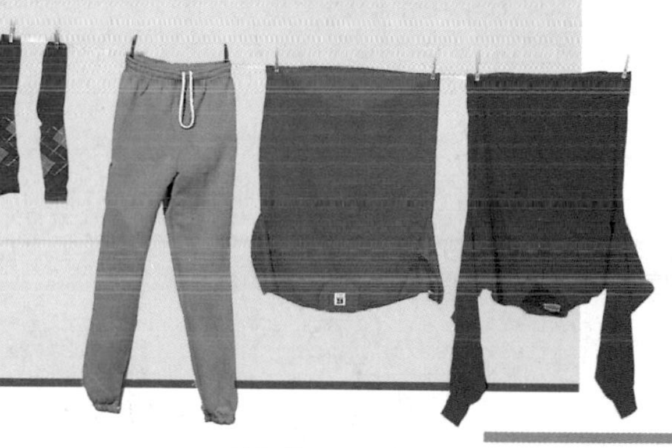

adsorbents from the same manufacturer but from different batches will give different R_f values. However, the relative values are quite useful in analysis.

Discussion

In front of students, break a raw egg into a clear glass container. Ask students which amino acids are present in the egg albumin. Possible response: *glycine, alanine, valine, and leucine.* Ask students how these amino acids could be separated for identification and/or purification. Point out that different types of chromatography are especially well suited to effect this type of separation.

2 TEACH

EVERYDAY CHEMISTRY

New Tips on the Laundry

Background: Today's detergents contain water softeners, surfactants, wetting agents, color brighteners, bleach, fragrance, coloring, and more. The research mentioned took place at Cornell University.

Teaching this Feature: Ask students to bring in samples of home laundry detergents. Determine the most widely used brand. The sample can be used in the Extension activity below.

Connection to Marketing: Have students look at home laundry detergent containers and note the part of the label or container that they first notice. Compare responses in class to determine if a pattern is used by the industry to market its products.

Extension: Simulate industrial product testing by placing a drop each of mustard, catsup, and ink on several small squares of white cloth. Give a square to each lab team with a sample of a different detergent. Have the class design and conduct a controlled experiment to determine which detergent cleans best. Afterwards discuss how polar forces helped in the cleaning.

Answers to
Exploring Further

1. Detergents have a polar end and a nonpolar end. The polar end attaches to water and the nonpolar end attaches to oils.

2. Sebum is a mixture of free cholesterol, waxes, triglyceride, free fatty acids, and squalene.

PROGRAM RESOURCES

Study Guide: Use the Study Guide worksheet, p.54, for independent practice with concepts in Section 14.3.

MINUTE LAB

The Ink Blot Test

Purpose: to separate the components of ink using chromatography

Materials: coffee-maker filter paper, scissors, assorted ink pens and felt markers, 400-cm^3 beaker, 100 cm^3 water, paper towel

Procedure Hints: Because four pieces of paper can fit in one large beaker, have students work in teams of four with each student preparing one chromatogram.

Results: Some inks are not mixtures. Others are nonpolar and will not separate using the polar solvent water. Other inks will produce brightly colored chromatograms.

Answers to Questions: 1. Mixtures of different dyes will separate. **2.** There is no diffusion of the insoluble ink. **3.** nonpolar **4.** no **5.** Ink that moves is polar. **6.** Paper is the stationary phase. The ink is the mobile phase.

✓ ASSESSMENT

Performance: Ask interested students to test other colored materials that they find at home. Have them bring their chromatograms to class to show. The student should be prepared to describe in detail how they made their chromatogram. This should include the solvent used, the time required, and the source of the colored material that was separated.

Chemistry Journal

Chromatography

Have students design a laboratory method by which they could obtain pure samples of the components of a mixture that has been separated by using paper chromatography. Students should write their reports in their journals.

MINUTE LAB

The Ink Blot Test

Put on an apron and goggles. Flatten a piece of filter paper of the type used in coffee makers. Use scissors to cut four long strips, each 2 cm wide, from the circle. Using a different pen or marker for each, draw a line 1.5 cm from the bottom of each of the four paper strips. Add water to a large beaker until it is about 1 cm deep. Arrange the four filter paper strips around the edge of the beaker so the bottom edge is just immersed in the water. The ink line should be above the water level. Fold the top of the paper strip over the edge of the beaker to hold the strip in place.

After five minutes, remove the paper strips and hang them up to dry. How do you know that some of the inks are a mixture of different dyes? What evidence do you have to prove that some of the inks are insoluble in water? Water is a polar solvent. What type of substance, polar or nonpolar, would water be unable to separate? Did the dye of any ink go as far up the filter paper as the water traveled? What would you guess about the ink being polar or nonpolar? Identify the ink and the paper as being either the mobile or stationary phase.

High performance liquid chromatography (HPLC) is a technique designed to overcome the speed limitations of conventional column chromatography. If the particle size in the stationary phase is reduced to the range 5-10 μm, the surface area of the particles in the stationary phase is increased, and much better separations can be made. However, the small particle size severely restricts the rate of flow of the mobile phase. Consequently, high pressures are required to force the mobile phase through the stationary phase at a reasonable rate. Typically, pressures as much as several hundred times normal atmospheric pressure are employed. The apparatus needed for HPLC is considerably more expensive than that for conventional column chromatography. The pumps used must not only achieve high pressures, but must produce a stream of mobile phase at a very steady rate without any pulsing. Columns may be 1 to 5 mm in diameter and 10 to 30 cm long.

Ion chromatography is another specialized application of column chromatography. In this case, the column is packed with a material called an ion exchange resin. The resin is a complex material made of extremely large molecules called polymers (Chapter 30). Attached to the polymers are ions that have a weak attraction for the resin. When the mobile phase passes over the resin, ions in the mixture have a greater attraction for the resin, and they displace the weakly held ions. The ions with the strongest attraction for the resin "stick" first. Ions with less attraction stick farther down the column and a separation is thus achieved. Different resins are available, so that by using an appropriate resin, either positive ions or negative ions may be separated.

Paper chromatography is a form of chromatography in which the separations are carried out on paper rather than in glass columns. Strips of paper are placed in a closed container in which the atmosphere is saturated with water vapor or solvent vapor. A drop of the solution to be separated is placed at the top of

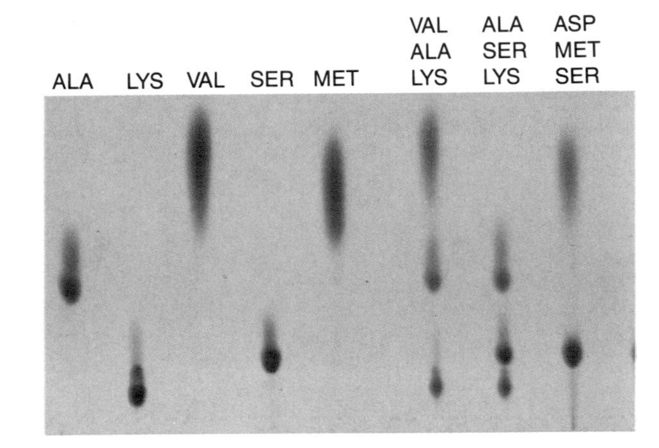

Figure 14.19 Thin layer chromatography may be used to separate amino acids. The letters at the top are abbreviations of different amino acids. Note that each amino acid separates out at a certain level on the chromatogram.

DEMONSTRATION

14-4 Column Chromatography

Purpose: to demonstrate the techniques to be used to separate the components in a mixture using column chromatography

Materials: 1.4 g $FeCl_3 \cdot 6H_2O$, 1.3 g $NiCl_2 \cdot 6H_2O$, 1.3 g $CoCl_2 \cdot 6H_2O$, 65 cm^3 9*M* HCl, 20-cm length of 10-mm inside diameter glass tubing, 20 cm^3 5*M* HCl, 25 cm^3 1*M* HCl, 40 g Al_2O_3 packing, aspirator, 3 100-cm^3 beakers

Procedure: 1. You can purchase commercial columns, many of which are shipped slightly moist. **CAUTION:** *Do not breathe any of the dust if you use a dry packing material.* The easiest way to make sure no air becomes trapped in your column is to prepare a slurry of the packing material, Al_2O_3. A flow rate that works well is 2-3 cm^3 per minute. If the rate is too slow, use either air pressure on top (rubber

the paper. The paper is then overlapped into a tray of solvent at the top of the container. The solvent moves down the paper by capillary action, separating the constituents of the drop. In an alternate method, the paper may be placed in solvent at the bottom of the box. In this case, the drop of the solution would be placed at the bottom of the paper and the solvent would ascend the paper. In either case, any separations are seen as a series of colored spots on the paper strip. If the separated fractions are colorless, they can be sprayed with solutions that will produce colored compounds. Some of these compounds may fluoresce under ultraviolet light.

Paper chromatography is simple, fast, and has a high resolving power. One great difficulty with using paper chromatography separation arises out of its extremely small scale. Also, quantitative determinations are difficult. In addition, a control is needed to determine which spots on the paper belong to which compound. Even with these difficulties, the method is still extremely useful.

Thin layer chromatography combines some of the techniques and principles used in both column and paper chromatography. A glass or plastic plate is coated with a very thin layer of stationary phase, as is used in column chromatography. A spot of an unknown mixture is applied to the plate as a spot is applied to the paper in paper chromatography. The glass plate is then placed in an atmosphere of solvent vapor and solvent, as is done in paper chromatography. The procedure from this point is just the same as that for paper chromatography. The thin layer technique is used frequently in separating mixtures of different biological materials.

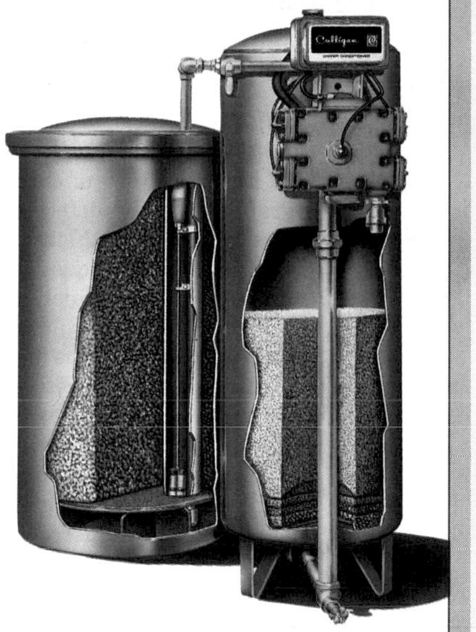

Figure 14.20 Ion exchange resins are often used in home water softeners.

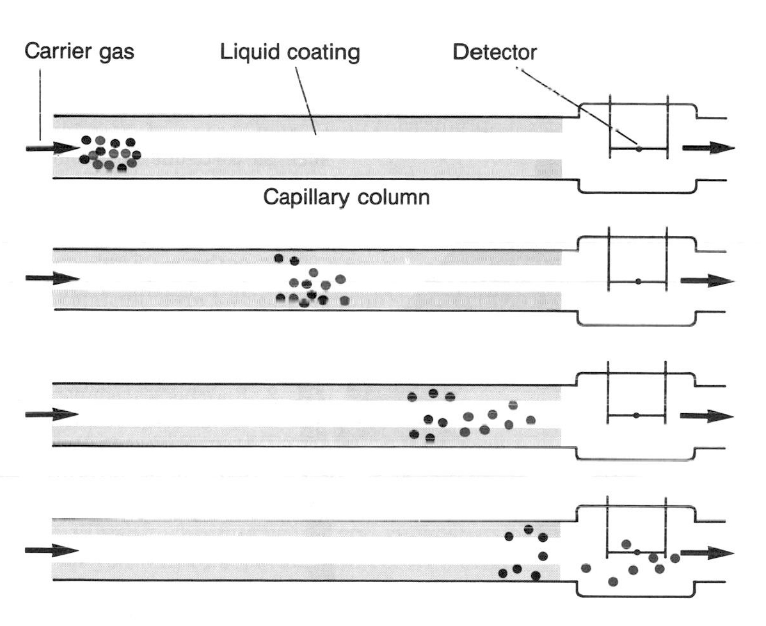

Figure 14.21 In gas chromatography a mixture is driven through a column by an unreactive carrier gas. The column is packed with a solid coated with a liquid that gradually adsorbs various components of the mixture to different degrees, thus separating them. Each component then passes over the detector and its presence is recorded.

CONCEPT DEVELOPMENT

MAKING CONNECTIONS

Food Science

Chemists have identified 21 different substances that contribute to the flavor of fresh orange juice using gas chromatography. The citrus industry will use this research to improve the flavor of processed orange juice.

Computer Demo: *GC,* Simulation of gas chromatograph. PC3901, Project SERAPHIM

- You may want to demonstrate column chromatography, which may be new to your students. If you do not have a commercial column, a list of suggested adsorbents include aluminum oxide, Fuller's earth, calcium oxide, calcium carbonate, calcium phosphate, talc, and starch. A list of suggested solvents, beginning with the least polar, are petroleum ether, cyclohexane, ether, acetone, chloroform, alcohol, water, pyridine, and acid. Columns can be kept from year to year if kept moist and sealed. Old columns can be reconditioned by percolating 2*M* HCl through them.

PROGRAM RESOURCES

Applying Scientific Methods in Chemistry: Use the worksheets, "Chromatography," pp.31-32, to help students apply and extend their lab skills.

Vocabulary Review/Concept Review: Use the worksheets, p.14, and pp.61-62, to check students' understanding of key terms and concepts of Chapter 14.

squeeze bulb) or an aspirator on the bottom of the tube.

2. A good solution for separation can be prepared by dissolving 1.4 g $FeCl_3 \cdot 6H_2O$, 1.3 g $NiCl_2 \cdot 6H_2O$, and 1.3 g $CoCl_2 \cdot 6H_2O$ in 50 cm^3 of 9*M* HCl. Use 2 cm^3 of this solution for a column 10 mm x 20 cm.

3. To follow a common manner of separation, begin with three aliquots of 5 cm^3 each of 9*M* HCl, then 4 portions of 5 cm^3 of 5*M* HCl then 5 portions of 5 cm^3 of 1*M* HCl. **CAUTION:** *Acid is corrosive.* After the solution to be separated has entered the column, keep the top of the packing material covered with acid.

4. Collect each of the chemicals to be separated in different beakers as they move through the column. **Disposal:** D for solutions, F for columns

Results: Students will observe three bands of color move down the column through the packing. The separate bands of color can be collected in beakers, thus effectively separating the components of the mixture.

Questions: 1. Is the column packing material the stationary or mobile phase? *stationary phase*

2. After collecting the sample that contains the separated substance, how could the pure substance be separated from the solvent? *Evaporate the solvent.*

CHAPTER
16 Solids

CHAPTER ORGANIZER

CHAPTER OBJECTIVES	TEXT FEATURES	LABORATORY OPTIONS	TEACHER CLASSROOM RESOURCES
16.1 Crystal Structure **1.** Describe characteristics of all solid substances. **2.** Distinguish among cubic, body-centered cubic, and face-centered cubic cells. **3.** Explain the relationship of melting point to bonding type and to crystal type. **4.** Distinguish isomorphous and polymorphous crystals. **Nat'l Science Stds: UCP.1, UCP.2, UCP.5, B.2, B.6**	**Industrial Arts Connection:** Shop Crystals, p. 402 **Frontiers:** Quirky Quasicrystals, p. 405	**Minute Lab:** Hard as Diamonds, p. 404 **ChemActivity 16:** Crystals and Their Structure - microlab, p. 825 **Discovery Demo:** 16-1 Blow Up a Crystal, p. 394 **Demonstration:** 16-2 Crystal Systems in 3-D, p. 396 **Laboratory Manual:** 16-1 Crystal Shapes **Demonstration:** 16-3 Compound Unit Cell - NaCl, p. 400 **Demonstration:** 16-4 Hexagonal Closest Packing, p. 402 **Demonstration:** 16-5 Heat-Treated Metal Crystals, p. 404	**Study Guide:** pp. 57-59 **Lesson Plans:** pp. 34-35 **Applying Scientific Methods in Chemistry:** Crystal Formation, p. 35 **Color Transparency 36 and Master:** Crystal Structure of Sodium Chloride **Critical Thinking/ Problem Solving:** Unit Cell Dimensions, p. 16 **Enrichment:** X-Ray Diffraction, pp. 31-32 **ChemActivity Master 16**
16.2 Special Structures **5.** Identify and explain the types of crystal defects. **6.** Describe the chemistry of semiconductors. **7.** Distinguish between hydrated ions and anhydrous substances. **8.** Describe the structure and properties of crystals, liquid crystals, and amorphous solids. **Nat'l Science Stds: UCP.1, UCP.2, UCP.5, B.2, B.6, E.1, E.2**	**Chemistry and Society:** Space and Defense Spin-offs, p. 410 **Everyday Chemistry:** Using Anhydrous Materials as Desiccants, p. 412 **Everyday Chemistry:** Liquid Crystal Thermometers, p. 414	**Demonstration:** 16-6 Crystal Defects, p. 408 **Laboratory Manual:** 16-2 Enthalpy of Hydration **Demonstration:** 16-7 Glass Flows, p. 414	**Study Guide:** pp. 60-61 **Reteaching:** Crystals, pp. 23-24 **Chemistry and Industry:** Artificial Gems, pp. 27-30
OTHER CHAPTER RESOURCES	**Vocabulary and Concept Review,** pp. 16, 65-66 **Evaluation,** pp. 61-64	**Videodisc Correlation Booklet** **Lab Partner Software**	**Test Bank**

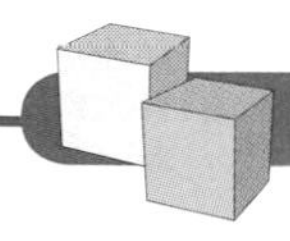

Block Schedule

For information on block scheduling, see the Lesson Plans booklet in the Teacher Resource Package.

GLENCOE TECHNOLOGY

Chemistry: Concepts and Applications Videodisc
Shape-Memory Metals
A Deliquescent Compound

Chemistry: Concepts and Applications CD-ROM
Shape-Memory Metals
A Deliquescent Compound

Science and Technology Videodisc Series (STVS)
Chemistry
Images of Atoms
Snowflakes
Composite Materials
Advanced Composites
High-Tech Ceramics
Glass Making for Science

Earth and Space
Frazil Ice
Fibers from Rocks

Physics
Making Integrated Circuits

Complete Glencoe Technology references are inserted within the appropriate lesson.

Solids

CHAPTER 16

The photo on the left is the dazzling interior of a geode — a crystal rock garden in a stone. What was once a liquid-filled cavity such as the shell of a clam has changed slowly into an array of shimmering mineral crystals. These minerals have many of the properties that characterize solids. As you will see, this is because all solids are composed of crystals.

The study of crystals is an extension of the study of kinetic theory. As you recall, solids are composed of particles that seemingly vibrate about fixed positions. What determines the arrangement of these positions? Does the arrangement account for the properties of a solid? As you know, different solids have different properties. Do these solids differ in the arrangement of their particles? To answer these questions we must look at solids on the atomic level.

EXTENSIONS AND CONNECTIONS

Industrial Arts Connection
Shop Crystals, page 402

Frontier
Quirky Quasicrystals: Strange solids that may consist of more than one type of unit cell, **page 405**

Chemistry and Society
Space and Defense Spin-offs: Products developed from the national space and defense programs, **page 410**

Everyday Chemistry
Using Anhydrous Materials as Desiccants: A look at how anhydrous materials help to keep things dry, **page 412**
Liquid Crystal Thermometers: Temperature-dependent properties of liquid crystals make these thermometers possible, **page 414**

the arrangement of atoms in a solid. The orientation of the bubbles conforms to an energy minimum. The coins and washers will also be in a hexagonal close pack arrangement. The hexagonal closest packing, HCP, arrangement is the most efficient of solid packings. Point out that the bubbles naturally achieved this arrangement without external manipulation.

Questions: 1. Do you see a repeating pattern in the bubbles and coins? *Yes*
2. What is the simplest repeating unit in a crystal called? *unit cell*
3. What holds a crystal together? *chemical bonds*

Follow Up Activity: Encourage students to repeat this demonstration at home so they can show their parents and other children in the family how unit cells arrange to form a crystal.

INTRODUCING THE CHAPTER

USING THE PHOTO

Have students identify general properties of solids, such as shape and rigidity, that can be accounted for by crystal structure. Have students make inferences about the possible arrangements and the motion of particles in a solid in contrast to those of a liquid or a gas. In comparing the pictures, students should suggest that the shape of a crystal is an extension of the way its particles are packed at the atomic level. Have students think of materials that appear to be solids but do not seem to have crystals. Point out that most of these materials either are not true solids or have crystals that are too small to be evident.

REVEALING MISCONCEPTIONS

Show students a glass bottle and a paraffin candle. Ask them if these two items are true solids. Students may respond that they aren't liquid, and they aren't gases, so they must be solids. Point out that there are many substances that appear to be solids, but are not crystalline. Materials having a disordered arrangement are said to be amorphous materials or rigid bodies without crystalline form.

16.1 Crystal Structure

PREPARE

Key Concepts

The major concepts presented in this section include crystal unit cells, compound unit cells, closest packing, elementary crystal types, network crystals, isomorphism and polymorphism.

Key Terms

crystal
unit cell
simple cubic
body-centered cubic
face-centered cubic
space lattice
hexagonal closest packing
cubic closest packing
network crystal
macromolecule
isomorphous
polymorphous

Looking Ahead

Paint the twelve 0.75-inch foam balls several days before you plan to conduct Demonstration 16-3; Compound Unit Cell-NaCl. Prepare the overhead transparency several days before you plan to conduct Demonstration 16-4.

1 FOCUS

Objectives

Preview with your students the objectives listed on the student page. Students can use them as a study guide for this major section.

Discussion

Point out to students the similarity between the arrangement of desks in a typical classroom and the regular, repeating, geometric array of a crystal space lattice. You can refer to the arrangement of desks when you discuss crystal defects and the doping of crystals.

OBJECTIVES

- describe characteristics of all solid substances.
- distinguish among cubic, body-centered cubic, and face-centered cubic cells.
- explain the relationship of melting point to bonding type and to crystal type.
- distinguish isomorphous and polymorphous crystals.

16.1 Crystal Structure

Have you ever examined table salt under a magnifying glass? If so, you have seen that the crystals appear to be little cubes. The lengths of the edges may vary, but the angles between the surfaces are always exactly 90°.

The systematic study of crystals began with Nicolaus Steno, in 1669. He observed that corresponding angles between faces on different crystals of the same substance were always the same. This fact was true regardless of the size or source of the crystals.

Steno's observation has been extended to all intensive properties (density, melting point, face angles, and other similar properties). For example, a single crystal of the mineral beryl, $Be_3Al_2Si_6O_{18}$, with a mass of approximately 40 metric tons was unearthed in New Hampshire. This huge crystal was found to be identical in intensive properties to all other beryl crystals. The size or mass of the crystal is not important. It can now be stated that *the extensive properties of crystals vary while the intensive properties remain the same.* As you recall, extensive properties include mass, volume, and length.

Crystals

The study of the solid state is really a study of crystals. All true solid substances are crystalline. Apparent exceptions to this statement can be explained in either of two ways. In some cases, substances we think of as solids are not solids at all. In other cases, the crystals are so small that the solid does not appear crystalline to the unaided eye. The relationship of properties to structure is important to chemists in the study of crystals as well as in other aspects of chemistry. Once the relationship is established, the chemist can use that knowledge to predict the properties of new substances.

Figure 16.1 All edges of a salt crystal meet at angles of 90° whether the crystal is large or small.

DEMONSTRATION

16-2 Crystal Systems in 3-D

Purpose: to show students how to construct a set of crystal systems

Materials: 4 plastic or paper soda straws, thin wire (22 or 26 gauge)

Procedure: 1. To make the inner crystal structure visible to your students you may wish to purchase a set of commercial crystal models. As an alternative to the commercial crystals, show students how to make their own crystal systems. Cut 4 soda straws into thirds. Wire the twelve pieces into a cubic form.

2. Construct another cubic structure and then deform it into the rhombohedral form.

3. To make the hexagonal crystal, construct another cubic form and flatten it on a table. Have students observe that the six edges form the shape of a

Figure 16.2 Two minerals that form cubic crystals are fluorite, CaF_2 (left); and galena, PbS (right).

All crystals of a certain substance are made of similar small units. These units are then repeated over and over again as the crystal grows. The units that compose a crystal are too small to be seen. Yet before any methods existed for studying crystal structure, scientists suggested that crystals form by repetition of identical units. Consider the patterns on wallpaper or drapery fabrics. These patterns are applied by rollers that repeat the design with each rotation. In a crystal, the forces of chemical bonding play the same role that the roller does in printing. They cause the basic pattern to be repeated. However, the "design" of a crystal must be composed of atoms, molecules, or ions instead of ink. Unlike the two-dimensional units of wallpaper, the units in crystals are three-dimensional.

Figure 16.3 These minerals represent different crystal systems. Molybdenite, MoS_2 (left), has a hexagonal structure; manganite, MnO(OH) (center), has a monoclinic structure; and amazonite, $KAlSi_3O_8$ (right) is triclinic.

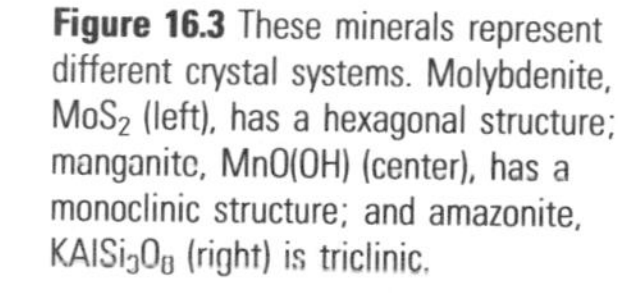

2 TEACH

CONCEPT DEVELOPMENT

- You may wish to emphasize the definition of a crystal as given in the text. Crystals have well-defined melting points. They have flat faces that meet at definite angles, resulting in sharp edges.

PROGRAM RESOURCES

Study Guide: Use the Study Guide worksheets, pp. 57-59, for independent practice with concepts in Section 16.1.

Lesson Plan: Use the Lesson Plans, pp.34-35, to help you prepare for Chapter 16.

Applying Scientific Methods in Chemistry: Use the worksheet, "Crystal Formation," p. 35, to help students apply and extend their lab skills.

hexagon and that the three axes are of equal length and meet at 60°.

4. To form the tetragonal form, cut 4 straws in half. Then, take 4 of the halves and cut each of them in half again. Wire these 8 shorter pieces to form the square ends, and wire the 4 long straws to complete the structure.

5. To make the orthorhombic form, take 4 straws and cut them in half. Take 4 of the halves and cut off 1/3 of the length. You now should have 4 long pieces of equal length, 4 medium pieces of equal length, and 4 short pieces of equal length. Wire these pieces together so that each side is a rectangle.

6. To make the monoclinic and the triclinic form, construct two orthorhombic forms. Deform one into the monoclinic form and the other into the triclinic form. **Disposal:** F

Results: After students have constructed these crystal forms have them discuss the length of the axes and angle between the axes. You may wish to display these models in the room.

Questions: 1. Which crystal system has three axes that are the same length and meet at 90° angles? *cubic*

2. Which system has three axes that are the same length and has no angles that are 90°? *rhombohedral*

Follow Up Activity: Have seven volunteers each make one large model of a crystal system using full-length straws.

CONCEPT DEVELOPMENT

Using the Table: Use the results of Demonstration 16-2 to discuss the seven crystal systems shown in Table 16.1.

Misconceptions

Do not leave students with the misconception that a crystal system is a unit cell and that there are only seven different unit cells found in nature. Use Table 16.2 to contrast crystal systems and unit cells.

✓ ASSESSMENT

Performance: Assign each student lab group one of the seven crystal systems shown in Table 16.1. Have them construct a model of the system using soda straws and thin wire or string. Have each group then show their model to the class, name the system, and describe the length of the unit cell axes and the angles between the axes. See Demonstration 16-2 on page 396 for more detailed instructions.

GLENCOE TECHNOLOGY

Videodisc

STVS: Chemistry

Disc 2, Side 1

Images of Atoms (Ch. 4)

Snowflakes (Ch. 3)

STVS: Earth & Space

Disc 3, Side 2

Frazil Ice (Ch. 16)

A **crystal** is defined as a rigid body in which the particles are arranged in units which form a repeating pattern. The arrangement of these units is determined by the bonds between the particles. Therefore, the bonding in the crystal partially determines the properties of the crystal.

There is a relationship between the repeating units and the external shape of the crystal. Long ago, crystallographers (krihs tuh LAHG ruh furs) classified crystals on the basis of their external shapes into seven "crystal systems," Table 16.1.

Table 16.1

Seven Crystal Systems

Lengths of the Unit Cell Axes	Angles between the Unit Cell Axes	Crystal System
all equal	all = 90°	cubic (Figure 16.2)
2 equal 1 unequal	all = 90°	tetragonal (Figure 16.10, left)
3 equal 1 unequal	1 = 90° 3 = 60°	hexagonal (Figure 16.3, left)
all equal	all ≠ 90°	rhombohedral (Figure 16.15)
all unequal	all = 90°	orthorhombic (Figure 16.13)
all unequal	2 = 90°, 1 ≠ 90°	monoclinic (Figure 16.3, center)
all unequal	all ≠ 90°	triclinic (Figure 16.3, right)

CONTENT BACKGROUND

Unit Cells: The unit cells that are not discussed in detail can be pictured as analogous to the three cubic unit cells. There is one new type introduced, the single face-centered, sometimes called end-centered cell. One particle is added to each of two opposite faces of a primitive cell to obtain the single-face centered unit cell.

interNET CONNECTION

The Smithsonian Gem and Mineral Exhibit on the World Wide Web has examples of common, crystalline minerals.

World Wide Web
http://galaxy.einet.net/images/gems/gems–icons.html

Unit Cells

Each substance that crystallizes does so according to a particular geometric arrangement. The simplest repeating unit in this arrangement is called the **unit cell.** It is possible to have more than one kind of unit cell with the same shape. For example, a unit cell in a tetragonal crystal may be simple tetragonal or it may be body-centered.

Consider the different kinds of three-dimensional unit cells. Fourteen such cells are possible, Table 16.2.

Table 16.2

Unit Cells

Crystal System	Unit Cell
Cubic	simple, body-centered, face-centered
Tetragonal	simple, body-centered
Orthorhombic	simple, single face-centered, body-centered, face-centered
Monoclinic	simple, single face-centered
Triclinic	simple
Rhombohedral	simple
Hexagonal	simple

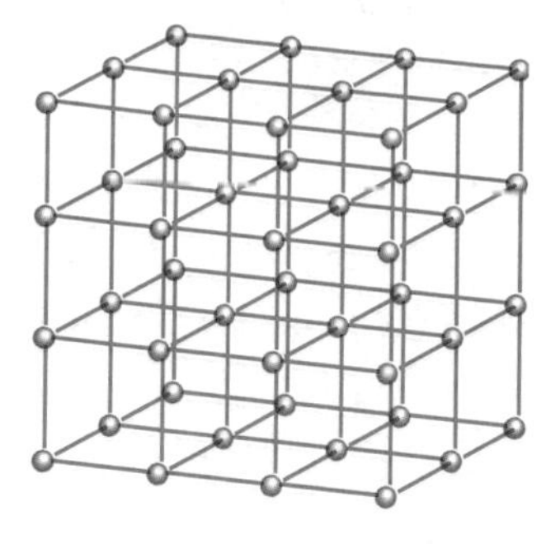

Figure 16.4 A simple cubic cell (top) is shown expanded into its space lattice (bottom). Unit cells and space lattices are mental models.

Three of the simplest unit cells are **simple cubic, body-centered cubic (bcc),** and **face-centered cubic (fcc).** These unit cells are illustrated in Figure 16.5. Note the packing arrangement of the particles in each unit cell. In the simple cubic cell, each particle has six immediate neighbors. In the body-centered cell, each particle has eight immediate neighbors; and in the face-centered unit cell, each has twelve. The three-dimensional arrangement of unit cells repeated over and over in a definite geometric arrangement is called a **space lattice.**

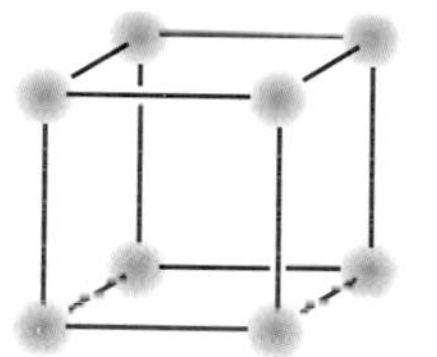

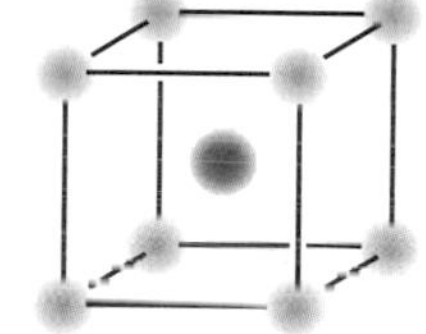

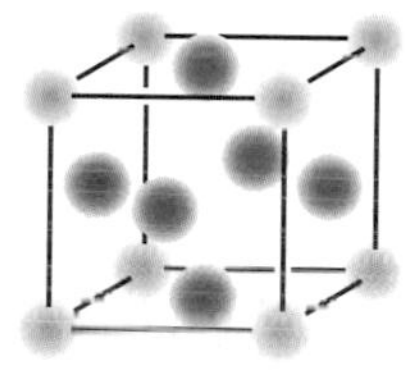

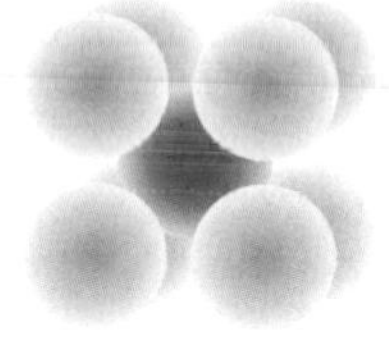

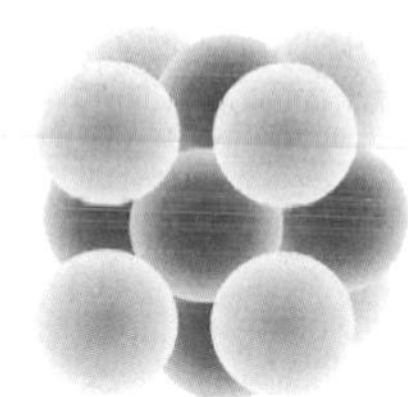

Figure 16.5 The three unit cells of the cubic system shown at the top have different packing arrangements within the same shape. The same arrangements shown in the bottom row indicate that the particles in crystals are much closer than those used to illustrate unit cells.

CULTURAL DIVERSITY

Letters from the Sky

Snowflakes are six-sided crystals. The Chinese correctly described their shapes as early as the second century A.D., yet as late as 1550 it seemed to be a well-kept secret.

Snowflakes form when frozen water molecules attach to a microscopic particle, usually dust. More water molecules attach to the edges in a hexagonal pattern.

Ukichuo Nakaya, a nuclear physicist, became the greatest snowflake scientist of the twentieth century through circumstance. When he took a position in physics in Hokkaido, Japan, he found no facilities for nuclear physics. Instead, he turned his studies to the snowflakes he found in abundance. He became the first scientist to produce snowflakes in the lab. He stressed that each flake reflected the unique way that it had been formed. Dry air at −15°C causes plate-shaped flakes to form; at −25°C, a solid column forms; and the feathery, six-arm flakes form in very wet air around −14°C. As the growing flake passes through air of varying temperature and humidity, it alters the pattern it is building while keeping what has already formed. By reading a flake's form, one can trace its travels from sky to Earth. For this reason, Nakaya wrote "a snowflake is a letter to us from the sky."

Chemistry Journal

Snowflakes

Have students write a journal entry explaining why snowflakes always have six sides.

PROGRAM RESOURCES

Transparency Master: Use the Transparency master and worksheet, pp. 71-72, for guided practice in the concept "Crystal Structure of Sodium Chloride."

Color Transparency: Use Color Transparency 36 to focus on the concept "Crystal Structure of Sodium Chloride."

Laboratory Manual: Use macrolab 16-1 "Crystal Shapes" as enrichment for the concepts in Chapter 16.

CONCEPT DEVELOPMENT

Background: One of the most important crystal structures found in compounds is exemplified by the compound ZnS. Zinc sulfide is found in nature in two minerals having different structures: wurtzite and zincblende. In both structures, each zinc atom is surrounded tetrahedrally by four sulfur atoms and each sulfur atom is surrounded tetrahedrally by four zinc atoms. In viewing both structures, the atoms are seen lying in puckered sheets. In zincblende each sheet is the same, but in wurtzite every other sheet is reversed by 180° from the sheets on either side. The following list gives some structures for simple AB-type compounds.

Cesium chloride structures: CsCl, CsBr, CsI, NH_4Cl, NH_4Br

Sodium chloride structures: lithium, sodium, potassium and rubidium halides, CsF, AgF, AgCl, AgBr, NH_4I, MgO, CaO, SrO, BaO, VO, MnO, FeO, CoO, NiO, MgS, CaS, SrS, BaS, MnS

Zinc sulfide structures: copper(I) halides, NH_4F, AgI, BeO, BeS, CdS, HgS, ZnO, MnS, AlN, GaN, InN, SiC

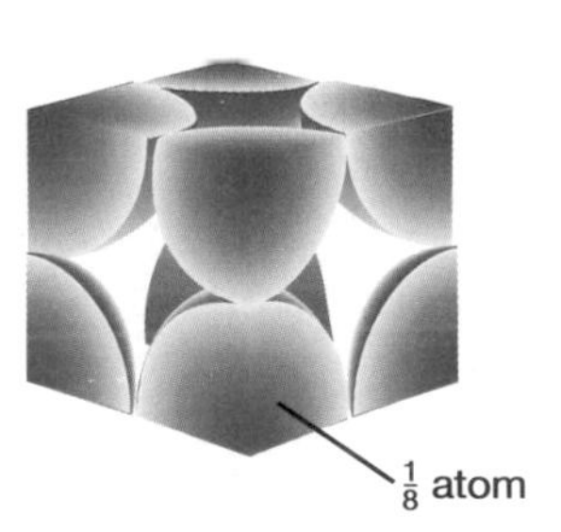

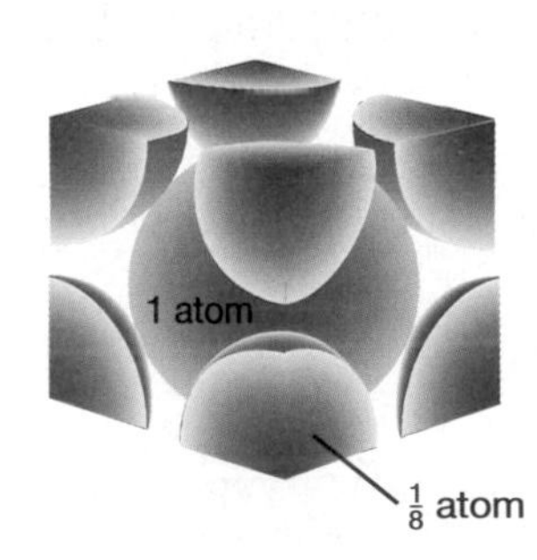

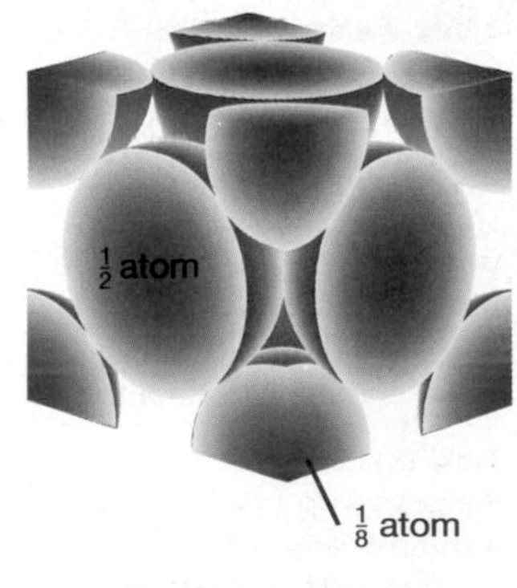

Figure 16.6 This representation of the cubic unit cells shows how atoms are shared between unit cells. The contents of the unit cells shown are 1, 2, and 4 atoms, respectively.

It should also be pointed out that space lattices and unit cells have no real physical existence. They are mental models that help demonstrate crystal structures.

Compound Unit Cells

Sodium chloride crystallizes in a structure similar to that pictured in Figure 16.7. How would you classify this unit cell? It apparently is some form of cubic, but which one? Look more closely at the arrangment of the ions. Each Cl^- ion is surrounded by six Na^+ ions. Each Na^+ ion in turn is surrounded by six Cl^- ions. If you consider either arrangment alone, you can see that the unit of repetition is face-centered cubic. Thus, the unit cell can be considered face-centered cubic, even though more than one kind of particle is present.

The particular crystal structure of an ionic compound such as sodium chloride is determined principally by the ratio of the radii of the ions. Since negative ions are generally larger than positive ions, the positive ions must be large enough to keep the negative

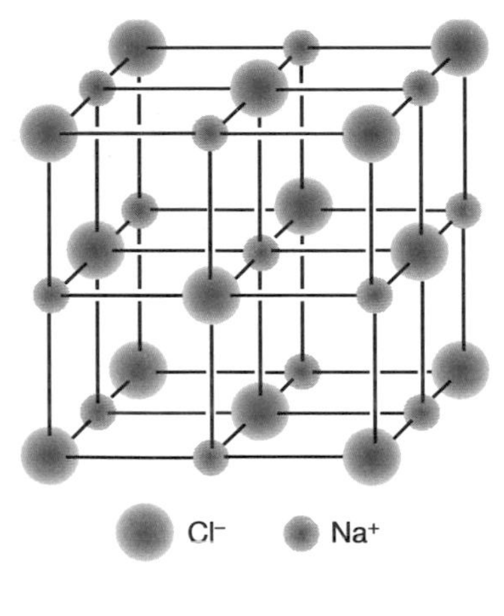

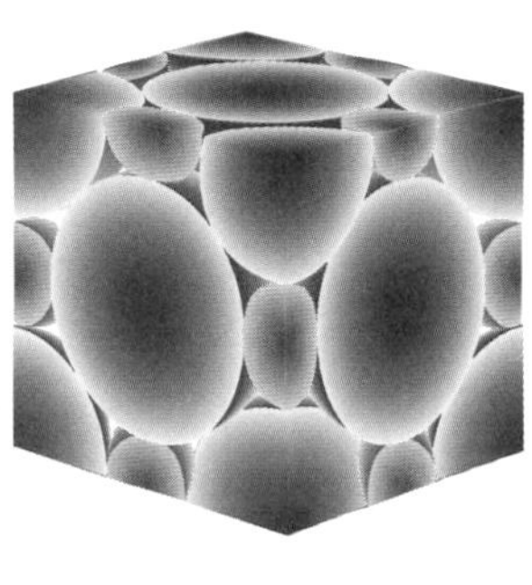

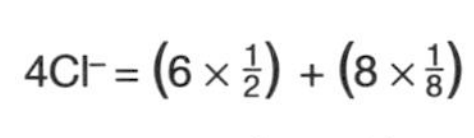

Figure 16.7 The sodium chloride unit cell is face-centered cubic, as all three representations show. The center representation shows how the ions are shared by adjacent unit cells.

DEMONSTRATION

16-3 Compound Unit Cell—NaCl

Purpose: to construct a model of the compound unit cell of NaCl

Materials: 14 1.5-inch plastic foam balls, 12 0.75-inch similarly-colored plastic foam balls, pipe cleaners, white glue

Procedure: 1. Use the 1.5-inch balls to make three layers. Two layers each use 5 balls. For each layer use pipe cleaners to arrange 4 balls around a center ball to form the corners of a square. The balls should touch. Glue the balls in position. A third layer is made by gluing 4 balls tangentially together to form a square.
2. To each of the three layers, use pipe cleaners and glue to attach 4 of the 0.75-inch latex painted balls. Fit them into the spaces around the outer perimeter of the layer, one on each edge.

ions from coming into contact with each other, but small enough not to come into contact with neighboring positive ions. The radius ratio of Na^+ to Cl^- is 0.69 and that of Cs^+ to Cl^- is 1.08. This difference is the reason NaCl has a face-centered cubic structure and CsCl is simple cubic.

Simple salts are those formed by the chemical combination of the elements of Group 1 (IA) (the alkali metals) and the elements of Group 17 (VIIA) (the halogens). These salts, except for those of Cs, always have structures based on the face-centered cubic lattice. Salts of cesium differ because of the large size of the Cs^+ ion. Sodium chloride is typical of this class of compounds. The same model would serve for other members of the group and also for many other binary compounds, like MgO and CaO.

Closest Packing

The crystal structures of elements are usually simple. Imagine placing a group of spheres as close as possible on a table and holding them so that they cannot roll apart. Then place another layer on top of the first one in an equally close arrangement. If we continue with more layers, a close-packed structure results. This structure is the kind found in the majority of metals. It is difficult to visualize the lattice in such a structure, since the lattice is really only a system of imaginary lines. However, the experienced eye will detect that the lattice is either hexagonal or face-centered cubic in the close-packed arrangement. They are often called **hexagonal closest packing** (hcp) and **cubic closest packing**

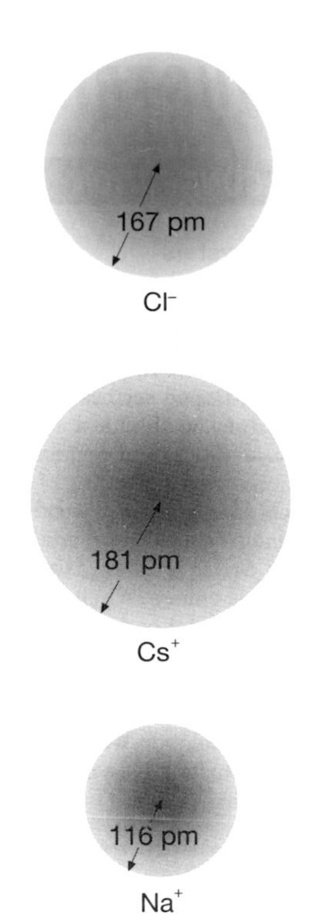

Figure 16.8 The size difference between the sodium ion and the cesium ion accounts for the difference in unit cell arrangements in NaCl and CsCl.

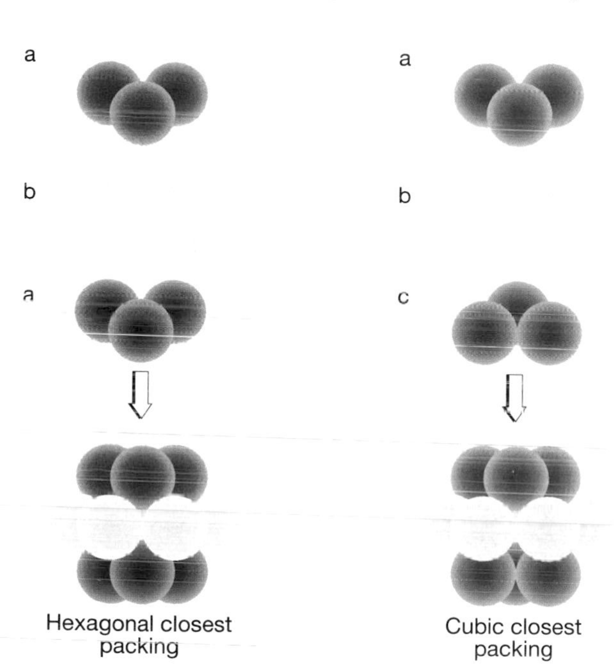

Figure 16.9 When spheres are closest packed so that spheres in the third layer are directly over those in the first layer *(aba)*, the arrangement is hexagonal closest packed (left). When the spheres are packed in the *abc* arrangement, the arrangement of spheres in the third layer is rotated 60° with respect to those in the first layer. The result is a cubic-closest packed structure (right).

CONCEPT DEVELOPMENT

QUICK DEMO **Hexagonal Closest Packing**

Demonstrate that hexagonal closest packing form has a coordination number of 12 by placing 7 pennies on the stage of an overhead projector. Show students that the next layer has only three particles, the next seven, the next three, etc.

Background: Iron has a body-centered cubic structure at ordinary temperatures. Above 910°C it has a face-centered cubic structure. The amount of carbon in the iron also affects its structure and properties as will be shown in Demonstration 16-5.

✓ ASSESSMENT

Performance: Have students refer to Figure 16.9 as they construct the three layers shown in the figure using expanded foam spheres and pipe cleaners or tooth picks. Have student lab groups exhibit to the class the *aba* (hexagonal-closest packing) and the *abc* (cubic-closest packing) arrangement of spheres and refer to the structures using their correct names.

3. Stack the layers on top of each other to form the compound unit cell that is found in NaCl. So you can show students the inner structure of the middle layer, do not glue the layers together.

Results: Use the model to discuss compound unit cells.

Questions: 1. Is the NaCl unit cell a simple cubic, body-centered cubic, or face-centered cubic structure? *face-centered cubic*

2. How many larger chloride ions are arranged around the smaller sodium ion? *6*

3. How many smaller sodium ions are arranged around a larger chloride ion? *6*

4. The coordination number of a face centered crystal is 12. Because NaCl is a compound unit cell, does this number agree with your answers to questions 2 and 3? *Yes, 6+6=12.*

PROGRAM RESOURCES

Critical Thinking/Problem Solving: Use the worksheet, p. 16, as enrichment for the concept of "Unit Cell Dimensions."

17.1 Changes of State

PREPARE

Key Concepts
The major concepts presented in this section include melting and freezing, vapor equilibrium, Le Chatelier's principle, solids changing state, liquefaction of gases, phase diagrams, and problem solving involving the enthalpies of fusion and vaporization.

Key Terms
vapor
equilibrium
dynamic equilibrium
saturated
reversible change
melting point
sublimation
normal boiling point
volatile
liquefaction
phase diagram
triple point
enthalpy of fusion
enthalpy of vaporization

Looking Ahead
Place a large container of water in the freezer overnight for the Focus Activity.

1 FOCUS

Objectives
Preview with your students the objectives listed on the student page. Students can use them as a study guide for this major section.

Activity
Pour hot water over the outside of the container to free a large block of ice. Place a thin wire over the block of ice. Attach heavy masses to each end of the wire. Have students observe that the wire will pass through without cutting the ice in half. Emphasize that students will need to study the properties of liquids to explain this observation.

OBJECTIVES
- explain the properties of liquids and changes of state in terms of the kinetic theory.
- use Le Chatelier's principle to explain reversible changes of state in a closed system.
- determine the relationship between energy and change of state and perform related calculations.

17.1 Changes of State

Water exists in three states: solid, liquid, and gas. Each of these states is stable under certain conditions. If large changes of the conditions occur, the motion of water molecules is altered. This causes the state to change. The same principles apply to all other substances. To understand the role water and other liquids play, we need to investigate the processes of change of state.

Melting and Freezing

Almost all solids and liquids expand when the temperature is raised. According to the kinetic theory, when the temperature of a solid is raised, the velocity of the particles increases. As the temperature increases, the particles collide with each other more often and with a greater force. Thus, they move farther apart. If the temperature of a solid is raised sufficiently, the particles will move far enough apart for some of them to slip over one another. The ordered arrangement of the solid state breaks down. When such a change takes place, we say the solid has melted.

In the liquid state, there will be a temperature (and pressure) at which particles travel so slowly that they can no longer slip past one another. The particles settle into an ordered arrangement and form a solid. Then, we say the liquid freezes. All pure liquids have a definite freezing point and all pure solids have a definite melting point. Melting and freezing take place at the same temperature for a pure substance.

Vapor Equilibrium

The average kinetic energy of atoms or molecules in a gas is a constant for all substances at a given temperature. This average kinetic energy can be calculated for any particular temperature. If we were to measure the kinetic energy of individual particles in a gas, we would find that few had exactly the predicted kinetic energy. Some particles would have more and some would have less kinetic energy than the average, as shown in Figure 17.1.

Figure 17.1 At a given temperature, the particles of a substance have a range of kinetic energies. At a higher temperature (green curve), the average kinetic energy of the particles is greater than at a lower temperature (red curve). More particles can escape the surface at the higher temperature, as seen by the steam condensing over the beaker.

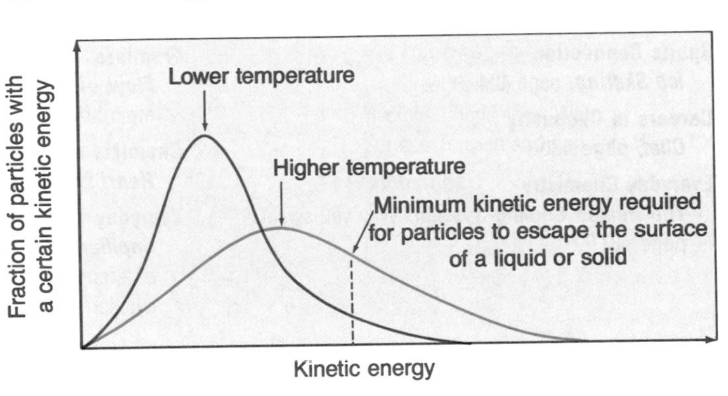

CONTENT BACKGROUND

Vapor Pressure: The equilibrium vapor pressure of a liquid has been assumed to depend solely on temperature. That assumption is not strictly true. The vapor pressure, *p*, depends upon the total pressure in the gas phase, *P*, according to the equation $(p/P)T = V_1/V_2$ where V_1 and V_2 are the molar volumes, at temperature *T*, of the liquid and vapor at the pressures *P* and *p*, respectively.

For many liquids, the vapor pressure variation with temperature follows an empirical relationship of $\log p = A - B/(T+C)$. The constants *A*, *B*, and *C*, for many liquids are available in handbooks. A few are given at the right for *T* in degrees Celsius and *p* in torr.
1 torr = 1 mm Hg.

Equilibrium in a closed container of fluid

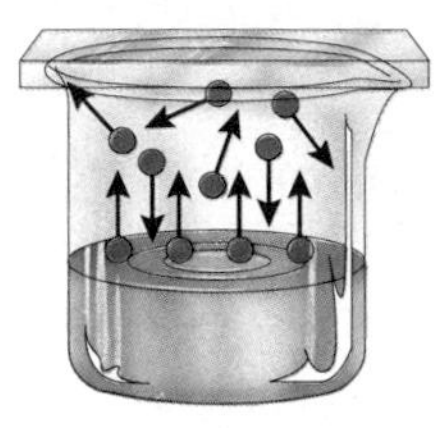

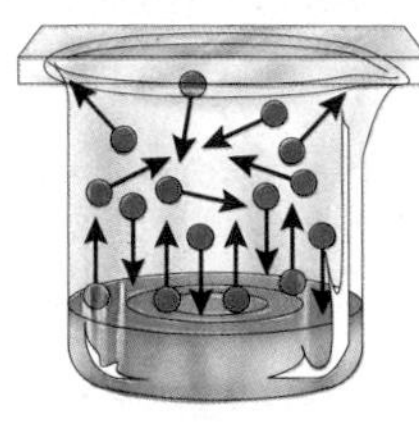

Figure 17.2 When the container is sealed initially, the particle movement is away from the liquid surface. The system reaches dynamic equilibrium when the number of molecules leaving the liquid surface equals the number reentering the liquid.

Most would have a kinetic energy close to the calculated amount. However, we would sometimes find a molecule with a kinetic energy considerably above or below the average.

All that has been said about the collisions of particles in a gas is also true of particles in a solid or a liquid. A molecule in a liquid, because of several rapid collisions with other molecules, might gain kinetic energy considerably above the average value. Imagine that molecule on the surface of the liquid. If it gains enough kinetic energy to overcome the attractive force of nearby molecules, it may escape from the liquid surface. The same process may also occur at the surface of a solid. The molecules that escape from the surface of a solid or a liquid form a vapor. This vapor is made of molecules or atoms of the substance in the gaseous state. A gas and a vapor are the same. We usually use the word *gas* for those substances that are gaseous at room temperature. The word **vapor** is used for the gaseous state of substances that are liquids or solids at room temperature.

A molecule of a solid or liquid that has escaped the surface behaves as a gaseous molecule. It is possible for this molecule to collide with the surface of the liquid it left. If its kinetic energy is sufficiently low at the time of such a collision, the molecule may be captured and again be a part of the liquid. However, if the container holding the liquid is open, there is little chance that the molecule will return to the surface it left.

If the solid or liquid is in a closed container, then there is an increased chance that the molecule will return to the surface. In fact, a point will be reached where just as many molecules return to the surface as leave the surface, as shown in Figure 17.2. There will be a constant number of molecules in the solid or liquid phase, and a constant number of molecules in the vapor phase. Such a situation in which there is no net change is known as an **equilibrium** condition. The situation we just discussed is a dynamic equilibrium because molecules are continuously escaping from and returning to the surface. In a **dynamic equilibrium,** two opposing changes are occurring at the same rate. Thus, the overall result remains constant. When a substance is in equilibrium with its vapor, the gaseous phase of the system is said to be **saturated** with the vapor of that substance. A gaseous phase that is saturated holds all the vapor it can at that temperature and pressure.

2 TEACH

CONCEPT DEVELOPMENT

QUICK DEMO **Melting and Freezing** As you discuss melting and freezing, melt a large test tube full of crystals such as 1,4 dichlorobenzene (moth crystals) or stearic acid in a beaker of hot water. **CAUTION:** Toxic vapors. Carry out in a well-ventilated place or under a fume hood. Do not overheat. Ask the students what is happening. Review with them the vocabulary used in Chapter 16. Inform them that in Chapter 17 they will build on their understanding of the solid phase as they study the liquid phase. **Disposal:** F

QUICK DEMO **Vapor Pressure Differences** Demonstrate differences in vapor pressure by placing ethanol, water, and acetone in three separate spray bottles or wash bottles. Squirt a stream of each onto the chalk board and have students discuss their observations. The students will observe that substance with the highest vapor pressure will be the most volatile because the intermolecular attractive forces are the weakest. **Disposal:** F

✓ ASSESSMENT

Oral: Ask students to define equilibrium and dynamic equilibrium and distinguish between the two definitions.

Chemistry Journal

Water Molecules

Ask each student to imagine that he or she is a water molecule and to describe his/her positions and movements as the temperature changes from −10°C to +110°C. Ask students to write the account in their journals.

Liquid	*A*	*B*	*C*
NH_3	7.36050	926.132	240.17
Ar	6.61651	304.227	267.32
Br_2	6.87780	1119.68	221.38
CO	6.69422	291.743	267.99
CO_2	9.81066	1347.786	273.00
Cl_2	6.93790	861.34	246.33
F_2	6.76588	304.35	266.54
I_2(cr)	9.8109	2901.0	256.00
I_2(l)	7.0181	1610.9	205.0

PROGRAM RESOURCES

Study Guide: Use the Study Guide worksheets, pp. 63-65, for independent practice with concepts in Section 17.1

Lesson Plans: Use the Lesson Plans, pp. 36-37, to help you prepare for Chapter 17.

SPORTS CONNECTION
Ice Skating

Exploring Further: The phase diagram of water has a negative slope to the line that separates the solid and liquid areas. Use the phase diagram to show how an increase in pressure, with no increase in temperature can result in the solid's becoming a liquid. Note, however, that the pressure required to melt ice is much greater than that exerted by a human skater. For more information on this topic, see James D. White, "The Role of Surface Melting in Ice Skating," *The Physics Teacher*, November 1992, pp. 495-497.

Background: Le Chatelier's principle applies to any equilibrium system. It is introduced here in relation to a liquid-solid equilibrium system. Le Chatelier's principle is discussed again in Chapter 22 in relation to chemical equilibria.

PROGRAM RESOURCES

Applying Scientific Methods in Chemistry: Use the worksheet "Changes of State," p. 36, to help students apply and extend their lab skills.

Reteaching: Use the Reteaching worksheet, p. 25, to provide students another opportunity to understand the concept of "Interpreting Vapor Pressure Graphs."

SPORTS CONNECTION

Ice Skating

Why does ice have a slick surface that makes skating (and automobile accidents) possible? Recent research indicates that every solid has a thin layer of liquid on its surface. As the temperature of the solid increases, the liquid layer increases in thickness, causing the solid's surface to become slicker. If you live in areas where outdoor temperatures fall significantly below zero, you know that it becomes increasingly difficult to skate on very cold ice. This difficulty occurs because the liquid layer becomes so thin that it no longer reduces the friction between the ice and the skate to any great extent.

The physical change from liquid to vapor is represented in equation form as

$$X_{(l)} \rightarrow X_{(g)}$$

where X represents any vaporizable substance, such as water. The opposite process can be represented by

$$X_{(l)} \leftarrow X_{(g)}$$

The two equations can be combined

$$X_{(l)} \leftrightarrows X_{(g)}$$

This kind of change is called a **reversible change.** A reversible process has reached equilibrium when the changes are occurring at the same rate in both directions.

Le Chatelier's Principle

In a liquid-vapor system, the vapor phase exerts a pressure that depends on the temperature of the system. In a closed container, a liquid and its vapor will reach equilibrium at a specific pressure for a particular temperature. If the temperature is increased, a new equilibrium will be established in which the vapor exerts a greater pressure. Thus, the equilibrium shifts in a way that accommodates the change in temperature. This shifting equilibrium was observed and described in 1884 by a French scientist, Henri Louis Le Chatelier. Le Chatelier's principle is expressed as follows: *If stress is applied to a system at equilibrium, the system will tend to readjust so that the stress is reduced.* The stress may be a change in temperature, pressure, concentration, or other external factor. Le Chatelier's principle applies to any system in equilibrium.

Let's look at an example of Le Chatelier's principle. A thermos is an insulated container that is known in science as a Dewar flask. A Dewar flask insulates its contents so that little or no energy can enter or leave. Suppose you have a Dewar flask nearly filled with a mixture of ice and liquid water at equilibrium. This equilibrium can be expressed by the equation

$$H_2O(l) \rightleftarrows H_2O(cr)$$

The temperature inside the flask will be 0°C, the melting point of ice. What will happen if you place a warmer or colder object into the system at equilibrium? The system will no longer be at equilibrium when an object of different temperature is introduced.

Suppose that you heat a copper penny to 100°C in boiling water and immediately drop it into the flask. Heat from the penny places a stress on the water-ice system. Gradually, the penny will lose heat to the system until it reaches the temperature of the system, 0°C. The heat absorbed by the system will be the same as the heat lost by the penny. What happens to the heat absorbed by the system? The heat melts a small quantity of the

DEMONSTRATION

17-2 Equilibrium Vapor Pressure

Purpose: to demonstrate vapor pressure and equilibrium

Materials: 500-cm^3 Florence flask, 3-hole stopper to fit flask, mercury manometer, rubber tubing and a clamp, dropping funnel or buret equipped with a stopcock, 5 cm^3 of acetone, 2 5-cm lengths of glass tubing

Procedure: Fit a 500-cm^3 Florence flask with a 3-hole stopper. Place the pieces of glass tubing in two of the stopper holes. Connect one of the tubes to a manometer. Fit the second tube with rubber tubing and a clamp. In the third hole, insert a dropping funnel or buret equipped with a stopcock. Partially evacuate the flask by attaching a vacuum pump to the rubber tubing that is clamped. A difference of 150 mm Hg in the manometer levels

ice and the equilibrium is restored. However, the system now has a little more liquid water and a little less ice.

$$\underset{\text{More liquid}}{H_2O(l)} \rightleftarrows \underset{\text{Less ice}}{H_2O(cr)}$$

Suppose that instead of heating the penny, you cool it to −10°C in a freezer and add it to the Dewar flask. Because the penny is colder than the system, it will absorb heat from the system until it reaches 0°C. As a result, liquid water will freeze. When equilibrium is restored, there is more ice and less water, but the temperature is still 0°C.

There are many chemical reactions in which reactants do not change completely into products. Instead, a reaction of this type reaches an equilibrium between reactants and products. Many of the products of these types of reactions are often commercially important. Chemical engineers apply Le Chatelier's principle to find ways to shift the equilibrium to increase the yield of products. Engineers may manipulate factors such as temperature, pressure, concentration of reactants, and time of reaction.

Measuring Vapor Pressure

Many techniques are available for measuring vapor pressure. Figure 17.3 shows two methods of finding the vapor pressures of substances. The apparatus in Figure 17.3a is especially useful for finding the vapor pressures of solids at elevated temperatures. A carefully controlled constant temperature bath is used to maintain the temperature of a substance. When the substance and its vapor reach equilibrium, the vapor pressure and temperature can be measured with a manometer and thermometer, respectively. Often, the vapor pressure will be read at temperatures over a wide range. Using these data, the changing behavior of a substance with temperature is obtained. Such information is valuable in industrial processes involving the substance tested.

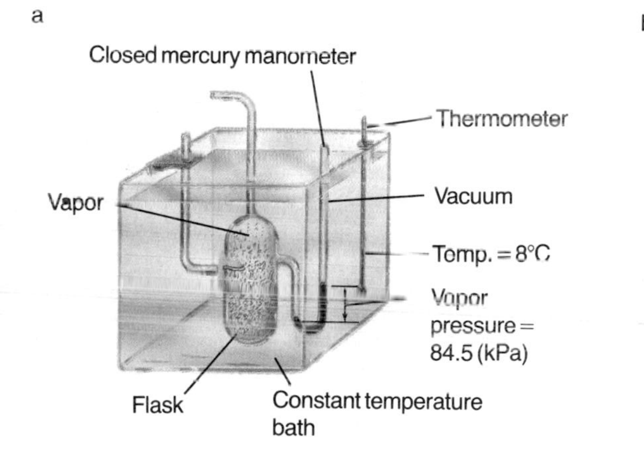

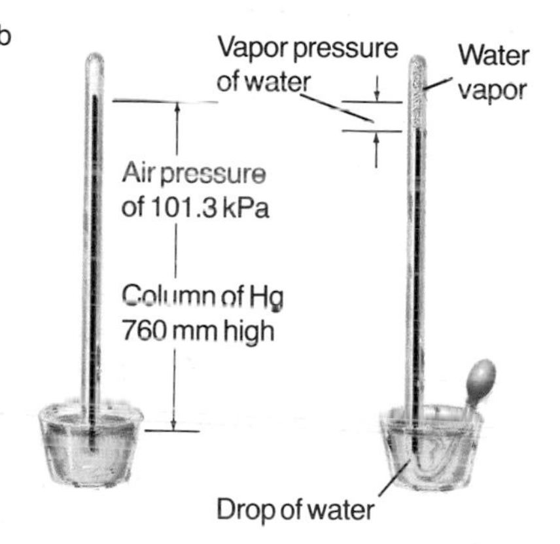

Figure 17.3 Two devices used for measuring vapor pressure are shown. In each, the vaporized liquid exerts a force (pressure) on the mercury in the tube.

CONCEPT DEVELOPMENT

QUICK DEMO **Measuring Vapor Pressure**

Refer students to Figure 17.3b. Use water tinted with food coloring to simulate a mercury barometer. Introduce some air into a gas-measuring tube that was filled with colored water and then inverted into a beaker of water. The space above the liquid simulates a vacuum. Point out how the air pressure forces the liquid up into the tube. The vapor pressure of the introduced substance would cause some of the mercury (H_2O) to come out of the gas measuring tube.

Discussion: Discuss the ice skating example presented in the text in detail. Stress that the expansion of water in forming the ice crystals is highly unusual. Remind students that water and ice expand and contract as expected except during the temperature range of 0°C to 4°C. This discrepancy can be related to the negative slope of the solid-liquid line when you present the phase diagram. Because of the negative slope, as the pressure on the ice is increased by the skater's blade, the solid-liquid line is crossed.

Using the Illustration: Take time to discuss the text's explanation of the two methods used to determine the vapor pressure of a substance. Explain each part of Figure 17.3a and b.

is satisfactory. Add a few cubic centimeters of acetone to the dropping funnel. Open the stopcock and allow a few drops of acetone to enter the Florence flask. Have a student read the mercury level every minute for 8-10 minutes. Then, add a few more drops of acetone to establish vapor equilibrium. Have 2 students ascertain that the flask is cooler than room temperature. The vapor pressure of acetone at 15°C is approximately 20 kPa.

Disposal: H

Results: Students will observe the vapor pressure of the acetone increase, then level off and remain constant as vapor pressure equilibrium is established.

Questions: 1. Why do we use the term vapor pressure instead of gas pressure for the vaporized acetone? *Vapor is used for the gaseous state of substances that are liquids or solids at room temperature.*

2. After the second addition of acetone, does the vapor pressure remain constant because the molecules of liquid are no longer being vaporized? *No, the rate of vaporization is equal to the rate of condensation.*

CONCEPT DEVELOPMENT

Analogy: If a 45 kg (100 lb) person stepped on your foot in shoes it would hurt, but you can stand the pain. If that person puts on ice skates and steps on your foot the pressure has changed dramatically because of the reduced surface area of the blade. This pressure change causes a shift in the equilibrium of the ice-water system.

- Guide students' thinking in the following pattern. (a) What is the stress? (b) How can the stress be relieved? (c) What changes occur in the system to relieve the stress and regain equilibrium?

Enrichment: Ask a student to explain the observation of the Focus Activity. Possible response: *The pressure exerted by the wire causes an equilibrium shift directly under the wire, the ice melts and the wire moves downward. The water remaining above the wire is at 0°C and is no longer under pressure, so it refreezes. As the wire passes through the block of ice, the equilibrium shifts first from solid to liquid, then from liquid to solid.*

MAKING CONNECTIONS

Daily Life

Point out that winter blends of gasolines have a higher vapor pressure. Discuss how this vapor pressure difference affects fuel economy and ease of starting on a cold morning. You get poorer fuel economy in winter than in summer, but the winter blend aids quicker starting of a cold engine. Also discuss how vapor pressure difference creates vapor lock in the fuel lines on a hot summer day. Summer blend is designed to prevent the formation of vapor in the fuel line by having a lower vapor pressure.

Table 17.1

Vapor Pressures of Some Substances at 25°C

Substance	Vapor Pressure (kPa)	Substance	Vapor Pressure (kPa)
Mercury (Hg)	0.000 246 0	Bromine (Br_2)	28.720
Turpentine ($C_{10}H_{16}$)	0.588 4	Acetone (CH_3COCH_3)	30.786
Water (H_2O)	3.167 2	Carbon disulfide (CS_2)	48.113
Carbon tetrachloride (CCl_4)	15.250	Sulfur dioxide (SO_2)	392.23

The left side of Figure 17.3b shows an ordinary barometer. The space above the mercury column in the closed tube is essentially a vacuum because mercury has a very low vapor pressure. Thus, there are no molecules bombarding the surface of the mercury. The pressure on the enclosed column is zero. If we introduce a drop of a liquid to the column, as shown on the right side of Figure 17.3b, the liquid will float to the top of the mercury because the mercury is more dense. When the liquid reaches the vacuum above the mercury, evaporation takes place. Eventually, an equilibrium will be established between the liquid and its vapor. Now there are gas molecules bombarding the top of the mercury column, which will push it down. The pressure exerted on the mercury column from inside the tube is the vapor pressure of the liquid and is equal to the change in level of the mercury column. This method is accurate but not very practical. For one thing, the barometer is now contaminated. For another, it is difficult to change or control the temperature in such an apparatus.

RULE OF THUMB

Table 17.1 gives the vapor pressures of some substances near room temperature. *Substances with low vapor pressures have strong intermolecular forces. Those with high vapor pressures have weak intermolecular forces. Ionic compounds do not exert a significant vapor pressure because the interionic forces are too strong to be overcome.*

Solids Changing State

Melting Point

In a mixture of solid and liquid states in a closed container, there will be a dynamic equilibrium between the molecules of the solid and the liquid. Remember that each state is also in equilibrium with its vapor. Because there is only one vapor, the solid and liquid have the same vapor pressure, Figure 17.4. In fact, **melting point** is defined as the temperature at which the vapor pressure of the solid and the vapor pressure of the liquid are equal.

RULE OF THUMB

The melting point temperature of a substance depends on the intermolecular forces in the substance. *Substances with weak intermolecular forces have lower melting points than substances with strong intermolecular forces.* Thus, nonpolar substances with low molar masses have lower melting points than polar substances with similar molar masses.

DEMONSTRATION

17-3 Vapor Phase Combustion

Purpose: to make the combustion of vapors visible

Materials: 6-gallon transparent plastic bottle, 40 cm^3 ethanol, matches, tongs

Procedure: 1. A large 6-gallon transparent plastic bottle can be obtained for a small fee from a company that sells or rents bottled-water dispensing machines. Place 40 cm^3 of ethanol in the empty bottle. Rotate the bottle on its side so that a mixture of ethanol vapor and air fills the bottle. Pour out from the bottle any excess ethanol and wash it down the drain.

2. Set the bottle upright. Turn off the room lights. Holding a burning match with tongs, ignite the vapors at the mouth of the bottle. **CAUTION:** *Keep bottle away from flammable materials*

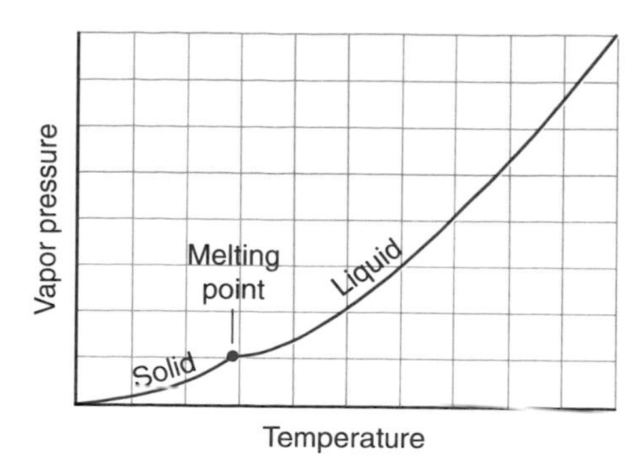

Figure 17.4 The graph shows the melting point of a substance to be that temperature where the vapor pressures of the solid and liquid phases are equal.

Sublimation

Some solids have a vapor pressure high enough at room temperature to vaporize rapidly if not kept in a closed container. Such substances will change directly from a solid to a gas, without passing through the liquid state. This process is known as **sublimation.** Dry ice (solid CO_2) and moth crystals are two examples of substances that sublime.

Word Origins

sublimare: (L) to uplift, elevate, refine

sublime—to change directly from a solid to a gas

Boiling Point

A liquid and its vapor can be at equilibrium only in a closed container. There is little chance that the molecules leaving the surface of a liquid will return to the liquid phase if the liquid container is open to the air. When a liquid is exposed to the air, it may gradually disappear. The disappearance is due to the constant escape of molecules from its surface. The liquid is said to evaporate.

As the temperature of a liquid is increased, the vapor pressure of that liquid increases because the kinetic energy of the molecules increases. Eventually, the kinetic energy of the molecules becomes large enough to overcome the internal pressure of the liquid. The internal pressure is caused by the pressure of the atmosphere on the liquid surface. When this pressure is overcome, the molecules are colliding violently enough to push each

Figure 17.5 Dry ice (left) and iodine (right) are two substances that sublime.

because the burning vapors will roar out the neck of the bottle. **Disposal:** F

Results: The bottle will not be damaged by the burning vapors. The bottle will be warm to the touch.

Questions: 1. What burned in the bottle? *Ethanol vaporized in the bottle due to its relatively high vapor pressure. The vapor burned.*

2. Does the vapor burn evenly throughout the bottle? *No, the flame is seen in different areas of the bottle.*

3. Does this alcohol have a higher or lower vapor pressure than water? *higher*

CONCEPT DEVELOPMENT

Discussion: Have students explain the second Rule of Thumb on page 426. Possible response: *Students usually suggest that less energy is required to overcome a weak attractive force resulting in a lower melting point.*

QUICK DEMO **Sublimation**

Demonstrate that some solids have a vapor pressure large enough to vaporize at room temperature. Place a few crystals of solid iodine in a stoppered glass flask. Have students observe the pale violet vapor that forms. **Disposal:** F or G.

✓ ASSESSMENT

Knowledge: Give the students the names of two of the substances from Table 17.1. Ask students to determine which one has the stronger intermolecular forces. In general, strong forces correspond to low vapor pressure.

GLENCOE TECHNOLOGY

Videodisc

Chemistry: Concepts and Applications

Changes of State
Disc 1, Side 2, Ch. 17

Also available on CD-ROM.

Combustion of Ethanol
Disc 1, Side 2, Ch. 11

Show this video of Demonstration 17-3 to demonstrate the combustion of vapors.
Also available on CD-ROM.

PROGRAM RESOURCES

Laboratory Manual: Use microlab 17 "Determining Melting Points" as enrichment for the concepts in Chapter 17.

EVERYDAY CHEMISTRY

Capillary Action and Types of Paper

Background: In 105 A.D. the first paper made from wood fibers came from China. A paper industry existed in Baghdad in 795 A.D. Crusaders brought paper back to Europe about 1200 A.D. The invention of the printing press increased the demand for paper. Germantown, PA was the site of the first paper mill in America in 1690.

Teaching This Feature: There are over 7000 different kinds of papers. Ask students to bring in samples and arrange them on a bulletin board.

Connection to Recycling: 75% of the paper made in the U.S. is manufactured from wood pulp. The high cost of building new paper mills, the scarcity of forest land, and increased demand have encouraged chemists to research and develop processes that use recycled paper.

Extension: Have interested students prepare reports on paper making.

Answers to
Exploring Further:

1. The water mark is impressed into the wet paper before it goes through the final drying and pressing stages of paper making.

2. The capillary action of chromatography paper allows the solvent to travel through the paper carrying the components at different rates which causes their separation.

3 ASSESS

CHECK FOR UNDERSTANDING

Have students answer the Concept Review problems, and Review and Practice 49-54 from the end of the chapter.

GLENCOE TECHNOLOGY

Videodisc

Chemistry: Concepts and Applications

Surface Tension

Disc 2, Side 1, Ch. 9

Video of Demo 17-10.

Word Origins

capillus: (L) hair

capillary—a tube that has a hair-thin internal diameter

The unbalanced force also accounts for the phenomenon known as capillary rise. **Capillary rise,** or capillary action, is the rise of a liquid in a tube of small diameter. If there is an attractive force between a liquid and the solid wall of the capillary tube, the liquid will rise in the tube. Capillary rise for water is shown in Figure 17.21 (left). The attractive force between the water molecules and the glass relieves the unbalanced force on the surface water molecules. The water will continue to rise in the capillary tube until the various forces (attraction between glass and water, attraction between water molecules, and the force of gravity on the water column) are balanced. Capillary rise is one method used to measure surface tension. The amount of surface tension is related to the density of the liquid, the radius of the capillary tube, the force of gravity on the liquid column, and the height to which the liquid rises.

EVERYDAY CHEMISTRY

Capillary Action and Types of Paper

Paper is manufactured almost entirely from vegetable fibers. These fibers may come from rags, grass, straw, or waste paper, but primarily they come from wood. The fibers are prepared as a pulp in water. The pulp is then formed into sheets, and the water is drained away, leaving a mat of fibers.

So that the paper will take ink without it running, the surface of the paper may then be treated to produce various specialty papers, such as the paper this book is printed on or your notebook paper. Untreated papers are used for paper bags and paper boxes.

Another use for untreated paper is in the making of paper towels. If you dip the corner of a paper towel in water, you can watch the water gradually move up the paper. This movement is a good example of capillary action. The spaces between the individual fibers act just like a capillary tube. The attraction of the water for the fibers causes it to move up the spaces between fibers.

Try writing with a felt-tip marker on a paper towel. Note how the ink spreads through capillary action on the fibers. Many inks will spread so much that the writing becomes illegible. If you try the same thing with a coated paper, such as notebook paper, you will note little or no capillary action. The reason no movement takes place is that the coating materials have filled the spaces between the fibers.

Exploring Further

1. Determine how watermarks are put into paper.
2. Learn how the technique of paper chromatography depends on capillary action to separate components of a mixture.

DEMONSTRATION

17-10 Surface Tension

Purpose: to demonstrate surface tension

Materials: 1-pint size canning jar and with ring lid, wire screen, 1-dm^3 water with food coloring added, piece of card stock

Procedure: 1. Cut a circular piece of wire window screen to fit inside the ring lid of a 1-pint canning jar.

2. Fill the jar with colored water to overflowing. Screw on the lid with the wire screen center.

3. Have students predict what will happen when the jar is inverted.

4. Cover the top of the jar with a piece of card stock.

5. Repeat the question in step 3. Turn the jar upside down. Remove the card stock by sliding it off to the side.

6. Have a student volunteer look up

Figure 17.20 Surface tension holds a water strider on the surface of water (left). Falling drops of a liquid are spherical because particles at the surface of the drop have a net force inward (right).

The surface tension of both water and mercury is high at room temperature, though that of mercury is higher. Water will rise quite readily in a capillary tube while mercury is depressed, as shown in Figure 17.21 (center). Mercury will not "wet" the glass of the tube. That is, there is not enough attractive force between the mercury and the glass to overcome the attractive forces that exist between the atoms of mercury. Compare the meniscus of water with that of mercury in Figure 17.21 (right). Can you think of an explanation for the difference in behavior?

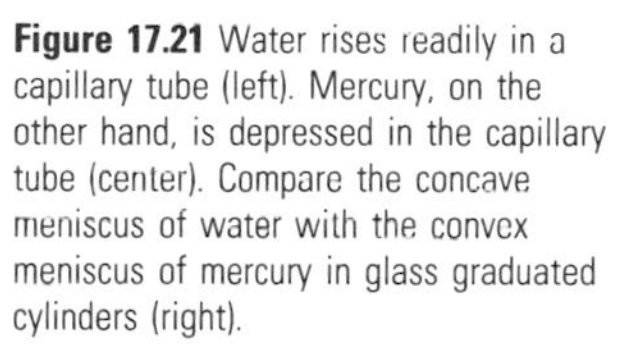

Figure 17.21 Water rises readily in a capillary tube (left). Mercury, on the other hand, is depressed in the capillary tube (center). Compare the concave meniscus of water with the convex meniscus of mercury in glass graduated cylinders (right).

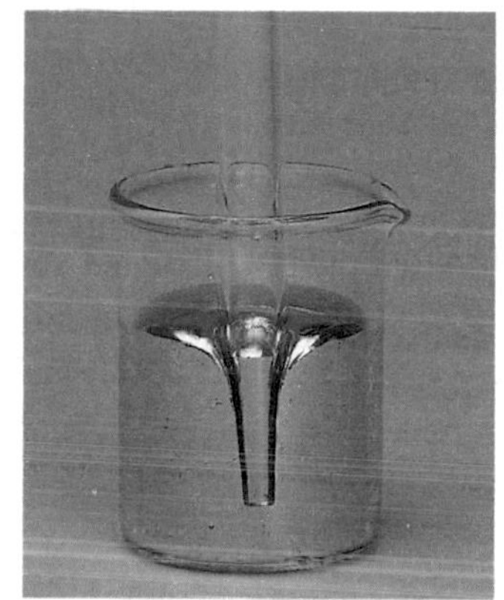

17.2 CONCEPT REVIEW

16. Why is the partial charge on a hydrogen atom bonded to a highly electronegative element more concentrated than for any other atom bonded to that same element?
17. Explain why water expands when it freezes.
18. Water has a greater capillary rise than does ethanol under identical circumstances. Explain why this is true.
19. **Apply** Why do paper towels absorb water throughout when only one corner is immersed in the water?

into the jar and report the observation.
7. While holding the jar over a sink, slowly tip the inverted jar sideways until the water flows into the sink.
Disposal: F

Results: The surface tension is sufficient to form a film that holds the water in the jar when inverted.

Questions: 1. Why does the water remain in the inverted jar? *Surface tension of water forms a film over the little square holes in the window screen.*
2. Why does water fall from the jar when it is tilted? *Air enters the jar and pushes the water into the sink.*

✓ ASSESSMENT

Performance: Have students design and conduct experiments to determine if the results of this demonstration depend on the size of the jar.

RETEACHING

Ask students if they have ever used a parachute to exercise in physical education class. Have them recall the shape of the parachute when the students holding the edge of the parachute all move quickly toward its center. A canopy is formed which resembles the apparent skin that forms on the surface of a water drop. The skin is due to the surface tension.

EXTENSION

A transparent polymer coating can be applied to glass surfaces to make them hydrophobic. One such coating is Rain X®, which can be purchased in the automotive section of many stores. Have a student research this surface-tension altering product. *The coating consists of isomeric polymers that yields hydrophobic, monomolecular, optically clear films that are chemically and physically bonded to the surface.*

Answers to
Concept Review

16. The nucleus of the hydrogen atom has no other electrons to act as a shield.
17. When water is frozen, bonding of the molecules forms an open crystalline structure that takes more space than when the water is in the liquid state.
18. Water rises higher because the water-to-glass attraction is large in comparison to the water-to-water attraction. The ethanol-to-glass attraction is less strong.
19. Paper towels are made mostly of vegetable fibers. When the corner of a towel is dipped in water, the spaces between the individual fibers act as capillary tubes. The attraction of water for the fibers causes it to move up the spaces between fibers, and hence up the paper towel.

4 CLOSE

Discussion: In deep ocean vents, hot lava erupts from cracks in Earth's crust, producing boiling water. Discuss why steam bubbles are never observed at the ocean's surface. *The steam loses heat and condenses as it rises.*

Answers to
Review and Practice (cont.)

42. *bp* 249 K, *mp* 133 K, *tp* 124 K, T_c 255 K, P_c 120 kPa

43. a. liquid, **b.** solid, **c.** solid, **d.** liquid, **e.** vapor

44. The temperature (average kinetic energy) increases. The substance melts, but the potential energy increases and the temperature (average kinetic energy) remains constant.

45. Enthalpy of fusion is the energy required to change one gram of a substance from solid to liquid at its melting point. Enthalpy of vaporization is the energy required to change one gram of a substance from liquid to gaseous state at its boiling point.

46. 6930 J

47. 1 367 000 J

48. 0.0°C

49. Hydrogen must be covalently bonded to a highly electronegative atom (N, F, or O) in the molecule.

50. The hydrogen that is bonded to the highly electronegative atom is positively charged and attracts the negative portion of other molecules. This force is stronger than that in other polar molecules because the hydrogen has no other electrons to shield the nucleus.

51. Hydrogen is the only reactive atom that has no inner electron clouds to shield its nucleus when the outer cloud is pulled away.

52. Molecules move apart to form the very open crystal lattice of ice.

53. Surface molecules are subject to unbalanced forces.

54. Water has enough attraction for the walls of the tube to overcome surface tension, and it will rise in the tube. Mercury does not have enough attraction for the tube to overcome the attraction Hg atoms have for each other, and it will not rise in the tube.

55. It changes from a solid to a vapor; sublimation.

56. The kinetic energy of some molecules on the surface of the solid is sufficient for them to break the attractive forces holding them in the crystal lattice.

57. 12.0 g

58. 32 000 J

59. See graph at the right.

42. Using the phase diagram shown, determine the boiling point, melting point, triple point, and the critical temperature and pressure for substance X.

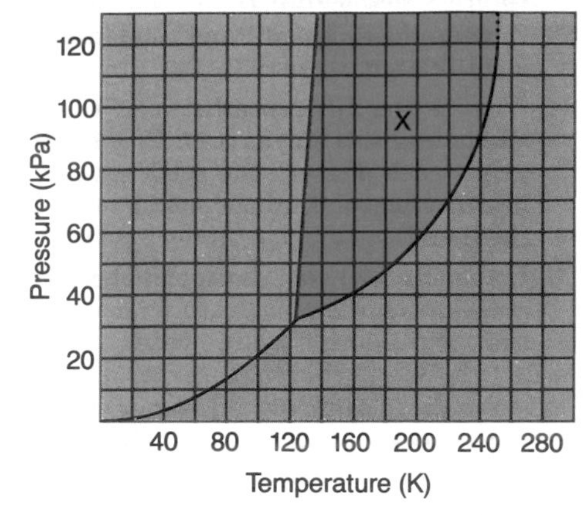

43. Using Figure 17.10, determine the state of matter that exists for hydrogen under the following conditions.
- **a.** 20 K and 750 kPa
- **b.** 7 K and 255 kPa
- **c.** 16 K and 1250 kPa
- **d.** 15 K and 250 kPa
- **e.** 22 K and standard pressure

44. What happens as heat is added to a solid substance below its melting point? What happens as the melting point is reached? For both questions, answer in terms of temperature and give a kinetic description as well.

45. Define enthalpy of fusion and enthalpy of vaporization for a substance.

46. Calculate the energy in joules required to melt 86.4 g of gallium at its melting point. Use information from Appendix Table A-3.

47. Calculate how much energy it takes to convert 400.0 g of ice at −25°C to steam at 275°C.

48. What is the final temperature when 750 J of energy is added to 9.0 g of ice at 0.0°C?

49. What conditions must be met in order for hydrogen bonding to take place in a collection of molecules?

50. Explain hydrogen bonding in terms of polarity.

51. Why do molecules with hydrogen bonded to nitrogen, oxygen, or fluorine have unusual properties when compared with other molecules having atoms with similar electronegativity differences?

52. Why does water decrease in density when it changes from a liquid to a solid?

53. Why do liquids exhibit surface tension?

54. Explain differences in behavior between water and mercury in a capillary tube.

55. At room temperature, what happens to a solid that has a high vapor pressure? What is this process called?

56. Describe how it is possible for a molecule to leave the surface of a solid. (How do solids evaporate?)

57. An ice cube at 0.0°C is heated until it melts and the water reaches 20.0°C. The amount of heat needed for this is 5.000×10^3 J. What is the mass of the ice cube?

58. From the following data and that of Appendix Table A-3, calculate the heat required to raise 45.0 g of cesium metal from 24.0°C to 880.0°C. The specific heat of solid Cs is 0.2421 J/g·C°; specific heat of liquid Cs is 0.2349 J/g·C°; specific heat of gaseous Cs is 0.1564 J/g·C°.

59. Bromine, Br_2, has a critical temperature of 315°C and a critical pressure of 1.03×10^4 kPa. Its triple point is at −7.3°C and 5.87 kPa. Its normal boiling point is 59.35°C, and its normal melting point is −7.25°C. Draw a phase diagram.

Concept Mastery

60. What is Le Chatelier's principle? Describe the equilibrium that exists in an inflated balloon. How does pressing on a

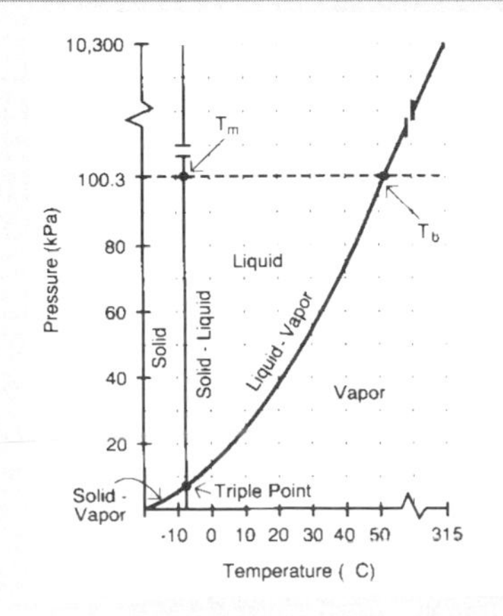

Chapter 17 REVIEW

balloon to decrease its size demonstrate Le Chatelier's principle?

61. If an H_2O sample is at 100°C, is the sample liquid water or steam? Explain.

62. Ethyl acetate is a sweet-smelling substance often used in organic chemistry as a solvent or a reactant. When it is in an open container, its odor soon is evident at some distance. From this description, what can you infer about its volatility, boiling point, evaporation rate, critical temperature, and vapor pressure?

63. At constant pressure, how does the quantity of heat necessary to change a liquid to a gas compare with the quantity of heat that must be removed in order to condense that gas?

64. Describe what happens to the crystal lattice as ice melts.

65. Why would you expect the boiling point of HF to be higher than that of HBr?

66. Why can many substances be identified by their melting or boiling points?

67. At any given temperature, the molecules in an object can have a large range of kinetic energies. How, then, is it possible to measure the temperature of the object? What does that temperature represent?

68. If there were a 20-g sample of water in a container at 80°C, how could the water be made to boil without raising the temperature?

69. Place the following hydrogen halides in order of expected vapor pressure, lowest to highest, at room temperature.
   a. HBr b. HI c. HF d. HCl

70. On the basis of molecular mass, where would HF be expected to fit in the list in the previous problem? However, HF does not follow the pattern expected. How can you explain this?

71. What happens to a boiling substance when the heat source is turned up? Will the temperature increase? Explain.

72. A cup contains 10 cm^3 of a liquid. Explain how each of the following factors affects its rate of evaporation.
   a. temperature
   b. surface area
   c. intermolecular forces

73. Figure 17.10 shows the phase diagram for hydrogen. What effect will increasing the pressure have on the freezing point of hydrogen? The boiling point?

74. A flask contains a mixture of acetone and water. How would knowledge of the boiling points of the liquids enable them to be separated?

75. A well-stirred mixture of 100 grams of ice and 100 grams of water is at equilibrium in an insulated container. Fifteen grams of water at 40°C are added. Will the temperature of the water change? Explain your reasoning.

76. A student is cooling a liquid to its freezing point and below. She records the temperature every minute. The readings in Celsius degrees are 169, 151, 140, 128, 118, 113, 113, 113, 113, 109, 100, 92, and 87. Explain the pattern of these temperature readings.

77. Use a dictionary to look up the general meaning of the word *saturated*. Use this definition to characterize the vapor phase of a liquid-vapor equilibrium.

78. Why would it be unlikely that the liquid-vapor equilibrium line in a phase diagram for a substance would ever have a negative slope?

Application

79. Explain why wet laundry hung on an outside clothesline when the temperature is less than 0°C will freeze, but then dry.

80. Explain the "freezer burn" that occurs on food stored in a freezer for long periods. How can it be prevented?

Answers to
Concept Mastery

60. If a stress is applied to a system in equilibrium, the system shifts so as to relieve the stress. The equilibrium in an inflated balloon consists of pressure inside the balloon and elasticity of the balloon equalling pressure outside the balloon. Pressing on a balloon increases the pressure inside the balloon. The balloon relieves this stress by expanding in some other direction until pressures are again equal.

61. The sample could be either liquid water or stream. Boiling takes place at 100°C with a change in potential, not kinetic energy. Thus, both states can exist at 100°C.

62. volatile, low boiling point, evaporates readily, low critical temperature, high vapor pressure

63. same

64. The lattice collapses and the water molecules move closer together.

65. HF exhibits hydrogen bonding; HBr does not. Thus, the intermolecular forces for HF are much higher.

66. Pure substances have sharp, characteristic melting and boiling points.

67. The temperature measured is an average temperature of all molecules in the object.

68. Lower the pressure.

69. c, b, a, d

70. last; hydrogen bonding

71. It boils faster; no, boiling is a change in potential energy.

72. **a.** Increasing temperature increases evaporation rate.
b. Increasing surface area increases evaporation rate.
c. Increased intermolecular forces decrease evaporation rate.

73. Both are raised.

74. The liquid with the lower boiling point would boil off first. The vapors could be collected, condensed, and thus separated.

75. No, the added energy will melt some of the ice, but the temperature stays at 0°C.

76. The readings from 169°C to 113°C indicate that the liquid is being cooled to its freezing point. The readings at 113°C are the time the energy loss is in the form of potential, not kinetic energy, as the crystal lattice is formed. When the lattice is completely formed, energy loss is again indicated by a drop in temperature (kinetic energy).

77. Saturated means "filled to capacity." The vapor phase of a liquid-vapor equilibrium is "filled to capacity" with vapor particles. For every new particle that vaporizes, one must condense.

78. Increasing the temperature always increases the fraction of molecules possessing sufficient energy to escape the liquid.

Answers to
Application

79. The ice sublimes.

80. "Freezer burn" results from sublimation of moisture from the food. Wrap the food better to prevent moisture loss.

81. Many boxed baking mixes give high altitude directions. Why are these directions necessary? Would you expect the high altitude baking temperature to be greater than, less than, or equal to the baking temperature at sea level?

82. Freeze drying is a process in which food products and other items are dried after being frozen. Often, this is done in a "vacuum" chamber in which a very low air pressure can be maintained. Explain why the vacuum chamber might be required.

83. People frequently run dehumidifiers in their homes in the summer and humidifiers in the winter. How can the action of these devices help keep us comfortable? Why might these devices reduce fuel bills for air conditioning and heating?

84. In a weather report, if the relative humidity is 100%, the air is saturated with water vapor. Explain this saturated system in terms of dynamic equilibrium.

85. Would the boiling point of water at each of the following locations be greater than, less than, or equal to 100°C? Why?
   a. top of Mt. Whitney
   b. on a New York City sidewalk
   c. Death Valley
   d. Miami Beach

86. The human body has its own cooling system. In terms of the kinetic energy of the molecules involved, explain why the evaporation of perspiration cools the body. If the day is warm, why are you more uncomfortable if the humidity is high?

87. Before a person donates blood, a small sample of the blood is taken for testing. This sample is usually obtained by pricking a finger to get a drop of blood. A small glass tube is touched to the drop, and the blood goes up into the tube. Explain why the blood rises in the tube.

88. Describe the effect of applying a large pressure to ice that is at or very near its melting point. To answer this question, notice the negative slope of the solid-liquid line on the phase diagram of water in Figure 17.9 on page 432. Recall also that water expands when it freezes. What will happen to the melting point of ice that is under great pressure?

89. In past centuries, tall structures called shot towers were used to manufacture lead shot. Kettles of molten lead were poured slowly down through the center of the tower from a platform at the top. The shot cooled as it fell. It was gathered at the bottom and graded into sizes by passing it through a series of sieves. Explain why shot towers were able to produce uniformly shaped shot.

Critical Thinking

90. Use information from Appendix Table A-3 to determine whether attractive forces are stronger in magnesium or in aluminum. On what information did you base your judgment?

91. Why might a burn from steam be more severe than a burn from boiling water?

92. Many mixtures of organic compounds can be separated by distillation. Often, the process of vacuum distillation is used. In this process, the distillation apparatus is attached to a vacuum pump, which maintains the system inside at a very low pressure. Vacuum distillation has two main applications. The first is separating mixtures of components having high boiling points. The second is the separation of compounds that decompose at high temperatures. Explain how vacuum distillation could be beneficial in each of these situations.

93. A student needs 245 grams of ice for an experiment involving heat transfer. She decides to make the ice by measuring 245 grams of liquid water into a shallow bowl placed in a freezer. A week later, she is

Answers to
Application (cont.)

81. A change in altitude means a change in atmospheric pressure and thus a change in boiling points of liquids present in the prepared mixes; baking temperature is increased.

82. Water molecules in ice will change directly to vapor at 0°C if atmospheric pressure is lowered below the triple point pressure.

83. Increased humidity in the air in the winter reduces the rate of evaporation of perspiration. Decreased humidity in the air in the summer increases the rate of evaporation of perspiration. Because evaporation is a cooling process, it is desirable in the summer but not in the winter. Comfort can be maintained with a higher room temperature in the summer and a lower room temperature in the winter.

84. For every water molecule that condenses, one molecule evaporates.

85. **a.** less—lower atmospheric pressure, **b.** equal—sea level, **c.** greater—low elevation, high atmospheric pressure, **d.** equal—sea level

86. Only high-energy molecules have enough energy to escape. When they leave, the average kinetic energy is less. When the humidity is high, the air around the body becomes saturated, and an equilibrium is established between the air and the moisture on the skin.

87. Capillary action of the water in the blood causes the blood to rise in the tube.

88. The pressure will tend to lower the melting point of ice causing it to change to a liquid. This is logical when you consider that liquid water takes up less space than the equivalent amount of ice. Pressure, then, would favor the formation of liquid water.

89. The net force acting on a surface particle is pulling the surface molecules inward. Because a sphere has the least surface area, spheres of lead are formed. They maintain this shape as they cool in the tower.

Answers to
Critical Thinking

90. Aluminum, it has a higher melting point, enthalpy of fusion, boiling point, and enthalpy of vaporization.

91. Steam at 100°C has more energy than water at 100°C. Above 100°C, it would have even more energy.

92. In both cases, lowering the pressure lowers the boiling points. Less energy is used in the first case, and no decomposition occurs in the second case.

93. The ice sublimed, which it would not have done in a sealed container.

ready to conduct her experiment. Before beginning, she decides to check the mass of the ice. To her surprise, she finds the mass to be only 239 grams. Can you explain what happened to the rest of the water and suggest a way the student could avoid this problem in the future?

94. Hot molten rock wells up at fissures deep in the oceans. Boiling has been observed at these sites. Would the boiling occur at higher, lower, or the same temperature as boiling at the ocean's surface? Give reasons for your answer. No bubbles of steam ever make it to the ocean surface to give evidence of the boiling far below. How can you account for this fact?

95. For any substance, the enthalpy of vaporization is larger than the enthalpy of fusion. Suggest an explanation for this observation.

Readings

Aubert, James H., *et al.* "Aqueous Foams." *Scientific American* 254, no. 5 (May 1986): 74-82.

Berry, R. Stephen. "When the Melting and Freezing Points are Not the Same." *Scientific American* 263, no. 2 (August 1990): 68-74.

Davis, Gode. "Self-Cooling Cans." *Popular Science* 230, no. 4 (April 1987): 53.

Morris, Daniel Luzon. "Cooking With Steam." *ChemMatters* 5, no. 1 (February 1987): 17-19.

Ring, Terry A. "Making Powders." *CHEMTECH* 18, no. 1 (January 1988): 60-64.

Rowell, Charles F. "Flash Point!" *ChemMatters* 4, no. 4 (December 1986): 10-11.

Salem, Lionel. *Marvels of the Molecule.* New York: VCH Publishers, 1987.

Cumulative Review

1. What is an exothermic reaction? Why is energy often required to start an exothermic reaction? What is this energy called?

2. Balance the equation below, then answer the following questions.

$$As + Cl_2 \rightarrow AsCl_3$$

a. What type of reaction is this?
b. How much chlorine will be needed to react with 3.44 g of arsenic?

3. How do atomic radii and ionic radii differ for (a) the alkali metals, and (b) the halogens? Explain differences between the two families.

4. Describe the nature of ionic bonds. List two pairs of elements that are likely to form ionic bonds.

5. How is it possible for a molecule to have polar bonds but not be a dipole? Give an example of a substance that fits this description.

6. What kinds of forces exist among molecules of nonpolar substances? How do these forces affect the melting points and boiling points of these substances?

7. What is a crystal? Why do we study crystals in order to learn about solids?

8. How many particles are found in a simple cubic unit cell?

9. What are the most common crystal structures found in metals?

10. How does doping affect the physical properties of silicon and germanium?

11. How do gases exert pressure?

12. Compare the distance between molecules in a gas at room temperature with the size of the molecules themselves.

13. An open manometer shows a mercury level 97.1 mm higher in the arm connected to the confined gas. The air pressure is 98.5 kPa. What is the pressure of the confined gas?

94. Higher, because pressures in the ocean raise the boiling point; rising bubbles cool to condensing point before they reach the surface.
95. Molecules are being moved much farther apart.

Answers to
Cumulative Review

1. An exothermic reaction emits energy. Activation energy is needed to form an activated complex.
2. a. synthesis
b. 4.89 g Cl_2
3. a. Ionic < atomic; they have lost the electron in the outer energy level and thus lost that energy level.
b. Ionic > atomic; they have gained an electron and the size increases because more electrons are being held by the same number of protons.
4. Ionic bonds are electrostatic attraction of opposite charges, as in the ions of Na and Cl, and K and F.
5. There is symmetrical arrangement of polar bonds; CO_2.
6. dispersion forces; low boiling and melting points
7. A crystal is a solid in which the particles are arranged in a regular, repeating pattern; all true solids are crystalline.
8. one
9. bcc, fcc, hcp
10. Doping increases their conductivity.
11. Gas particles have collisions with the wall of their container.
12. Size of particles is negligible compared with the distance between particles.
13. 85.6 kPa

✓ ASSESSMENT

Portfolio: Application questions 79-89, and Critical Thinking questions 90-95 are very practical questions. Consider assigning them to be completed in cooperative learning groups. A copy of each question with its answer would be a good homework sample to include in a student's portfolio.

CHAPTER

18 Gases

CHAPTER ORGANIZER

Chapter Objectives	Text Features	Laboratory Options	Teacher Classroom Resources
18.1 Variable Conditions **1.** Explain the concept of an ideal gas. **2.** Describe the conditions of STP. **3.** Relate the laws of Boyle, Dalton, and Charles and perform calculations using these laws.	**Careers in Chemistry:** Meteorologist, p. 455 **Sports Connection:** Ballooning and Charles's Law, p. 458 **Frontiers:** Stirling Shiver, p. 462	**Minute Lab:** Air Pressure vs Gravity, p. 454 **Minute Lab:** Charles's Law in a Freezer, p. 461 **Discovery Demo:** 18-1 Can Crushing, p. 450 **Demonstration:** 18-2 Pressure-Volume Relationships, p. 452 **Demonstration:** 18-3 Dalton's Law and Water Vapor, p. 458 **Laboratory Manual:** 18-1 Using Charles's Law to Determine Absolute Zero - microlab **Demonstration:** 18-4 Gas Against Fire, p. 462	**Study Guide:** pp. 69-70 **Lesson Plans:** pp. 38-39 **Critical Thinking/ Problem Solving:** The Kinetic Molecular Theory, p. 18 **Reteaching:** Boyle's and Charles's Laws, pp. 26-27 **Color Transparency 39 and Master:** Boyle's Law **Applying Scientific Methods in Chemistry:** Charles's Law, p. 37
Nat'l Science Stds: UCP.1, UCP.2, UCP.3, B.2, B.4, B.5, B.6, E.1, E.2			
18.2 Additional Considerations of Gases **4.** Solve problems involving the change of more than one condition for gases. **5.** Explain Graham's law and solve problems using it. **6.** Differentiate between an ideal gas and a real gas.	**Chemists and Their Work:** Jacques Alexandre César Charles, p. 466 **Everyday Chemistry:** Compressed Gas and Conveniences, p. 468 **Chemistry and Society:** Losing the Ozone Layer, p. 470	**ChemActivity 18:** Graham's Law of Diffusion - microlab, p. 832 **Laboratory Manual:** 18-2 Rates of Diffusion of Gases **Demonstration:** 18-5 Joule-Thomson Effect, p. 468	**Study Guide:** p. 71 **ChemActivity Master 18** **Enrichment:** Effusion, Diffusion, and Flow Rate, pp. 35-36
Nat'l Science Stds: UCP.1, UCP.2, UCP.3, B.2, B.4, B.5, B.6, E.1, E.2, F.6			
Other Chapter Resources	**Vocabulary and Concept Review,** pp. 18, 69-72 **Evaluation,** pp. 69-72	**Videodisc Correlation Booklet** **Lab Partner Software**	**Test Bank** **Mastering Concepts in Chemistry—Software**

Block Schedule

For information on block scheduling, see the Lesson Plans booklet in the Teacher Resource Package

GLENCOE TECHNOLOGY

Mastering Concepts in Chemistry Software

Pressure
General Concepts, Boyle's Law, Dalton's Law
Charles's Law, Combined Gas Law

Chemistry: Concepts and Applications Videodisc

Pumping Gas
Demonstrating Boyle's Law
Charles's Law
Gas Against Fire

Chemistry: Concepts and Applications CD-ROM

Relating Particle Motion and Temperature
Using the Gas Laws
Pumping Gas
Demonstrating Boyle's Law
Charles's Law
Gas Against Fire

Science and Technology Videodisc Series (STVS)

Chemistry

Hole in the Ozone

Complete Glencoe Technology references are inserted within the appropriate lesson.

ASSIGNMENT GUIDE

Contents	Level	Practice Problems	Chapter Review	Solving Problems in Chemistry
18.1 Variable Conditions				
Kinetic Theory of Gases	C		25, 27, 28, 67-70	
Boyle's Law	C		29, 58	1-6
Applying Boyle's Law	C	1-4	30-34, 59, 72, 76, 82	
Dalton's Law of Partial Pressure	C	5, 6	26, 35-38, 60, 65, 66	7-10
Charles's Law	C		39, 64, 75	
Applying Charles's Law	C	7-9	40-42, 58, 77, 78, 84	11-13
18.2 Additional Considerations of Gases				
Combined Gas Law	C	14, 15	43-47, 58, 62, 63, 65, 66, 76, 79-81, 83	14-19
Diffusion and Graham's Law	O	16-20	48-52, 71, 74	20-24
Deviations from Ideal Behavior	A		53-57, 61, 73	
				Chapter Review: 1-10

C-Core, A=Advanced, O=Optional

Laboratory Materials

Minute Labs	ChemActivities	Demonstrations
page 454 card, 3x5 inches small fruit juice glass water **page 461** balloon freezer marker ruler string tape	**page 832** concentrated ammonia solution, NH_3 (aq) concentrated hydrochloric acid, HCl (aq) 96-well microplate plastic microtip pipets, 3 plastic sandwich bag with zipper seal soda straws universal indicator	**pages 450, 452, 458, 462, 468** aquarium baking soda balloon, large beaker, 600 cm^3 candles, 4 dropper hot pad or beaker tongs hot plate ice innertube filled with air jar, small pie pan side-arm vacuum flask, 500 cm^3 soda can, empty vinegar water, hot, ice-cold, and room temperature

CHAPTER 18

Gases

CHAPTER OVERVIEW

Main Ideas: The gas laws are developed with the aid of the kinetic theory model. The chapter begins with the introduction of the characteristics of an ideal gas. As each gas law is introduced students are given the opportunity to develop their problem-solving abilities. The chapter closes with applications of the kinetic theory model so students will better understand the properties of gases.

Theme Development: The theme of this chapter is systems and interactions. A volume of gas can be considered a system as it interacts with its environment.

Tying To Previous Knowledge: Have students recall and discuss the pain or "popping" sensation of their ears as they drove over mountains or rode in airplanes. Have students discuss how air temperature and humidity can affect people with breathing problems.

GLENCOE TECHNOLOGY

Software

Mastering Concepts in Chemistry
Unit 10, Lesson 1
Pressure

Videodisc

Chemistry: Concepts and Applications
Pumping Gas
Disc 2, Side 1, Ch. 2

Also available on CD-ROM.

CD-ROM

Chemistry: Concepts and Applications
Exploration: *Relating Particle Motion and Temperature*

CHAPTER **Preview**

CONTENTS

450

DISCOVERY DEMO

18-1 Can Crushing

Purpose: to demonstrate the relationships between temperature, pressure, volume, and the number of moles of a confined gas

Materials: empty soda can, 15 cm^3 H_2O, hot plate, hot pad or beaker tongs, 600-cm^3 beaker filled with ice water

Procedure: 1. Place a small amount of water in an empty soda can. Heat the can until the water boils and steam comes from the opening in the top.
2. Using a hot pad or beaker tongs, remove the can from the heat and immediately invert the open top into a large container of ice water. **Disposal:** A

Results: The can will be crushed quickly and with some noise. The can was sealed by inverting it into the ice

Gases

CHAPTER 18

*The scuba (**s**elf-**c**ontained **u**nderwater **b**reathing **a**pparatus) allows divers to descend to depths of as much as 100 meters. How are divers able to breathe underwater?*

In the atmosphere, the pressure of the air entering the lungs is the same as the pressure on the chest. When a diver is beneath the water's surface, the water exerts an additional pressure on the chest. Below the 0.5 meter dive mark, the pressure on the chest is too great to allow breathing of atmospheric air. A scuba removes the pressure differential problem by supplying a breathing mixture at an elevated pressure. Using the pressure regulator shown here, a diver adjusts the pressure of his or her breathing gas to equal that of the surrounding water. In this chapter, you will study the behavior of gases and how various conditions change the characteristics of the gases.

EXTENSIONS AND CONNECTIONS

Careers In Chemistry
Meteorologist, page 455

Sports Connection
Ballooning and Charles's Law, page 458

Frontiers
Stirling Shiver: Engines that use helium, **page 462**

Chemists and Their Work
Jacques Alexandre César Charles, page 466

Everyday Chemistry
Compressed Gas and Conveniences: How refrigerators and air conditioners work, **page 468**

Chemistry and Society
Losing the Ozone Layer: Earth's atmosphere is in trouble, **page 470**

451

INTRODUCING THE CHAPTER

USING THE PHOTO

The weight of the water and the air pressure above the water exert a large force on the diver. You can demonstrate that air pressure exerts a force by doing the Discovery Demo 18-1: Can Crushing. Point out that a diver's depth gauge is a straightforward Boyle's law device. Air trapped within the tube of the gauge is compressed by the surrounding water. As the diver goes deeper, the added water pressure compresses the gas more and more. One can easily see the meniscus of the water in the tube.

REVEALING MISCONCEPTIONS

Students may think that a real gas behaves as an ideal gas. Remind students that real gases do exert attractive forces between molecules. Have students discuss how these attraction forces might affect the pressure. Possible response: *Measured pressure is less than predicted pressure at room temperature, because real gases have a volume and are not point masses.*

ASSESSMENT PLANNER

Choose from the following assessment strategies.

Assess, pp. 462-463, 468-469
Assessment, pp. 451, 454, 457, 459, 461, 464, 466, 470
Portfolio, p. 466
Minute Lab, pp. 454, 461
ChemActivity, pp. 832-833
Chapter Review, pp. 471-475

water. Because the volume of the water vapor decreased as it condensed and its temperature decreased, the pressure inside the can decreases. The pressure of the atmosphere is constant and sufficient to crush the can symmetrically.

Questions: 1. What happens to the water vapor inside the can when it is inverted in the ice water? *It condenses into a liquid.*
2. Why is the can crushed on all sides symmetrically? *The air molecules strike the can equally on all sides.*
3. Why is the can full of water at the conclusion of the activity? *A partial vacuum occurs when the water vapor condenses. The liquid enters the can to replace the water vapor's volume.*

✓ ASSESSMENT

Performance: If a bell jar and vacuum pump are available, have a student place a marshmallow in the bell jar, seal the jar, and evacuate the jar using a vacuum pump. The marshmallow expands as the air dispersed within it expands. When air is allowed to re-enter the bell jar, the marshmallow shrinks back to its original size. Ask students to explain their observations.

18.1 Variable Conditions

PREPARE

Key Concepts

The major concepts presented in this section include the kinetic theory of gases, Boyle's law, Dalton's law of partial pressures, and Charles's law. Problem solving skills are introduced with the laws.

Key Terms

point mass
ideal gas
standard atmospheric pressure
standard temperature
STP
Boyle's law
Dalton's law
Charles's law

Looking Ahead

If each student or lab team is to make a Cartesian diver (Quick Demo), inform the students in advance to bring to class an empty, clear plastic 2-Liter soda bottle.

1 FOCUS

Objectives

Preview with your students the objectives listed on the student page. Students can use them as a study guide for this major section.

Activity

Tie ribbon streamers to the grill of an electric fan and have the fan operating as students enter the room. Have students observe the streamers and then discuss what they can imply about the makeup of gas from these observations.

2 TEACH

CONCEPT DEVELOPMENT

Background: Point out to your students that gases can be compressed, whereas liquids and solids are considered to be noncompressible. Gas molecules at STP have about ten diameters of space between each other. Emphasize that gases are unique in the way temperature, volume, and pressure are related.

OBJECTIVES

- explain the concept of an ideal gas.
- describe the conditions of STP.
- relate the laws of Boyle, Dalton, and Charles and perform calculations using these laws.

18.1 Variable Conditions

Look at the photograph of the scuba diver again. Notice particularly the size of the tank containing the breathing mixture. It has a volume of about 30 dm^3. Normal human respiration requires roughly 0.5 dm^3 of breathing mixture for each breath. How can the tank hold enough mixture for the diver to stay down for an hour or more? The answer is that the mixture in the tank is under pressure. Why does pressure ensure a sufficient supply of mixture? This section of Chapter 18 is devoted to examining how various conditions affect gases.

Kinetic Theory of Gases

You already know many characteristics of gases because air is composed of gases. When we speak of a cubic centimeter of a solid or a hundred cubic centimeters of a liquid, we are referring to a definite amount of matter at a given temperature and pressure. Both solids and liquids expand and contract with pressure and temperature changes. However, the change is usually small enough to be ignored. This statement is not true for gases. When the temperature of a gas is raised or lowered, the change in its volume is significant. A gas has no particular volume. Instead, a given amount of gas occupies the entire volume of its container. According to the kinetic theory, a gas is made of very small particles that are in constant random motion. Gas particles of a substance are not held in a fixed position as they are in the solid form of the substance. Also, the gas particles are not held close together by van der Waals forces as they are in the liquid form of the substance. Instead, gas particles are free to spread far apart from each other.

Gas molecules are much smaller than the distances between molecules. They can be treated as **point masses,** which have no volume or diameter because they are so small and so far apart. A

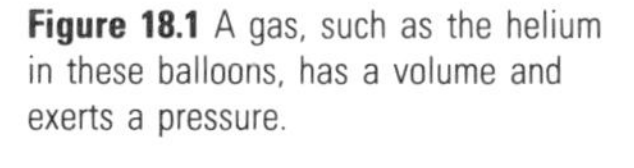

Figure 18.1 A gas, such as the helium in these balloons, has a volume and exerts a pressure.

DEMONSTRATION

18-2 Pressure–Volume Relationships

Purpose: to generate a discussion of how pressure and volume are related

Materials: 500-cm^3 side-arm vacuum flask, large balloon, 25 cm^3 water, hot plate, medicine dropper

Procedure: **1.** Place a small amount of water in a side-arm vacuum flask.
2. Stretch a large balloon over the mouth of the flask.
3. Heat the flask on a hot plate until the water boils and the flask vents steam through the side-arm.
4. Use the rubber bulb from a medicine dropper to seal the side-arm. Promptly remove the flask from the heat.
5. Cool the flask by placing it into a beaker of cold water. **Disposal:** A

Results: The steam in the flask will condense and the surrounding air will

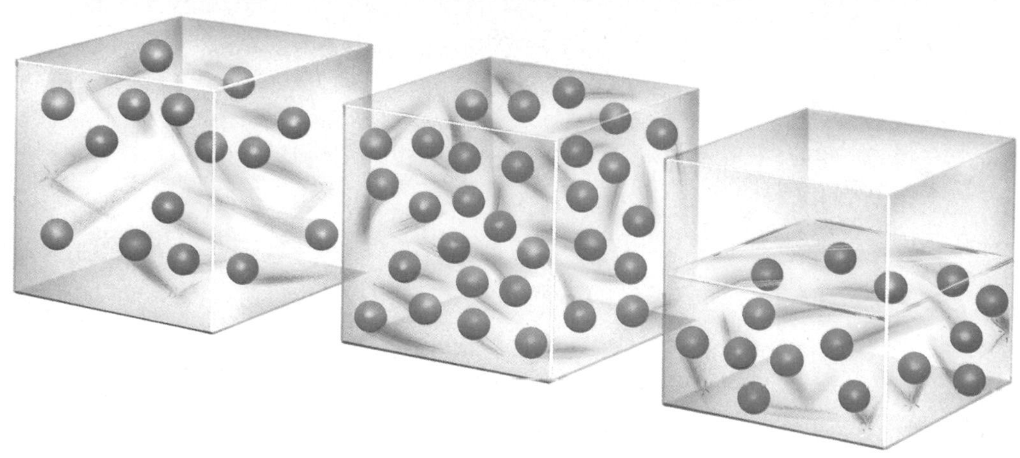

Figure 18.2 The force of the collisions between the gas particles and the sides of the container represents the pressure of the gas. The pressure in a given container (left) is greater if the number of particles increases (center) or if the volume decreases (right).

gas composed of point masses does not actually exist, but it is a useful way of thinking about gases. An **ideal gas** is composed of molecules with mass but with no volume and no mutual attraction between the molecules. In the latter part of this chapter, you will study more about real gases and how they differ in behavior from ideal gases.

The volume of a gas, the number of gas particles in that volume, the pressure of the gas, and the temperature of the gas are variables that depend on one another. Therefore, in discussing quantities of gases, it is necessary to specify not only the volume but also the pressure and the temperature. **Standard atmospheric pressure** is 101.325 kilopascals and **standard temperature** is 0°C. We indicate that a gas has been measured at standard conditions by the capital letters **STP** (standard temperature and standard atmospheric pressure). In scientific literature, gas volumes are reported as so many m^3, dm^3, or cm^3 at STP. The system is convenient, but what can we do if we must actually measure the volume of a gas in the laboratory when the pressure is 98.7 kilopascals and the temperature is 22°C? In this chapter, we will describe how a measured gas volume can be adjusted mathematically to the volume the gas will occupy at STP.

We have seen that a gas exerts pressure on the walls of its container because gas molecules collide with the walls. The pressure exerted by a gas then depends on three factors. The three factors are the number of molecules, the volume they are in, and the average kinetic energy of the molecules. A change in any of these factors changes the pressure exerted by the gas. If the number of molecules in a constant volume increases, the pressure increases. If the number of molecules and the volume remain constant but the kinetic energy of the molecules increases, the pressure increases. The kinetic energy depends on the temperature. If the temperature increases, the average kinetic energy will also increase.

CONCEPT DEVELOPMENT

Computer Demo: *Molecular Speed Distribution,* Graphical Display of molecular speed vs. temperature. AP603, PC2603, Project SERAPHIM

CULTURAL DIVERSITY

Clearing the Air

Garrett Morgan was born in Kentucky in 1875 and moved to Cincinnati, Ohio at the age of 14. He had received only six years of schooling, but hired his own tutor to help him with grammar. In 1895 he moved to Cleveland, Ohio where he worked as a sewing machine adjuster, eventually opening his own sewing machine repair business.

In 1913 Morgan was working on a polish for sewing machine needles and discovered a process to straighten hair. This product made him very wealthy. He also invented the three-way traffic light—a patent he sold to General Electric Company for $40,000.

One of his major contributions was a "breathing device" or gas mask. In 1914, he won grand prize at a New York Safety and Sanitation fair for the device, but it was largely ignored. On July 25, 1916, however, an explosion trapped 30 workers 250 feet below Lake Erie in a tunnel full of poisonous gases. Someone remembered Morgan's device and contacted him for help. Garrett and his brother Frank rescued the trapped workers using the breathing device. Fire departments immediately became interested in the gas masks, but prejudice against African Americans was a problem. In the south, Morgan had to have a white man demonstrate the mask while he posed as an "Indian" assistant. Because of his experiences, Morgan founded the Cleveland Call to give African Americans their own news voice.

push the balloon into the flask.

Questions: 1. What happens to the steam inside the flask when the flask is cooled? *It condenses.*

2. Why is the balloon pushed into the flask? *The outside air pressure is greater than that inside.*

Extension: Ask students how to remove the balloon from the flask. Possible response: *Heat the sealed flask and the balloon will pop out of the flask.*

PROGRAM RESOURCES

Study Guide: Use the Study Guide worksheets, pp. 69-70, for independent practice with concepts in Section 18.1.

Lesson Plans: Use the Lesson Plans, pp. 38-39, to help you prepare for Chapter 18.

Critical Thinking/Problem Solving: Use the worksheet, p. 18, as enrichment for the concept of "The Kinetic Molecular Theory."

CONCEPT DEVELOPMENT

Cartesian Diver

To make Boyle's law visible, remove the label from an empty, colorless 2-liter soft drink bottle. Add water up to 4 cm from the top. Half fill a medicine dropper with water that has been colored with vegetable dye. Drop the medicine dropper into the 2-liter bottle, and screw the cap on tight. Squeeze the sides of the bottle firmly until the medicine dropper descends. The students will see that the level of the colored water in the dropper changes. The dropper has more mass and descends when the air in the dropper is compressed and is displaced by water. **Disposal:** A

MINUTE LAB

Air Pressure vs. Gravity

Purpose: to show that air pressure is exerted in all directions

Materials: 3" × 5" card, small fruit juice glass, water

Procedure Hints: Ask students to predict what they believe will happen when your hand is removed. Use a glass rather than a plastic or foam cup. **Disposal:** A

Results: The air pressure will continue to hold the card in place, and the water will remain in the inverted glass. Students will be better able to conceptualize the idea of molecules colliding and creating pressure (force/unit area).

Answers To Questions: 1. See Results. **2.** air pressure **3.** 101.3 kPa

✓ ASSESSMENT

Performance: Have students think of ways to vary the lab. Some may use glasses of different diameters. Other students may use a cover that varies in composition or thickness from notebook paper to cardboard. Some students will vary the amount of water in the glass, or its temperature. Ask them to explain their observations.

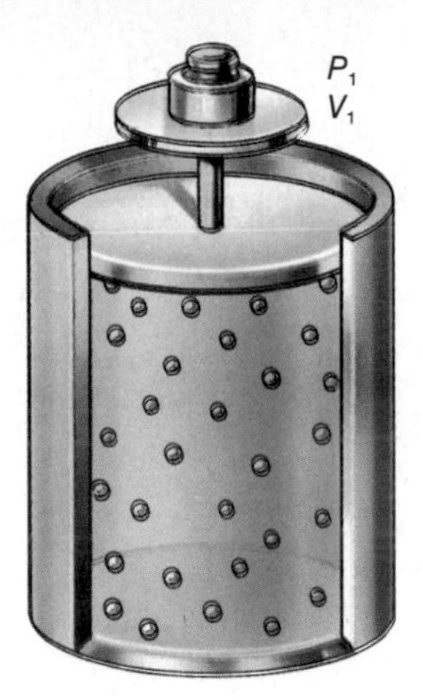

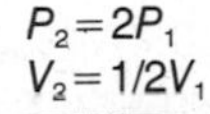

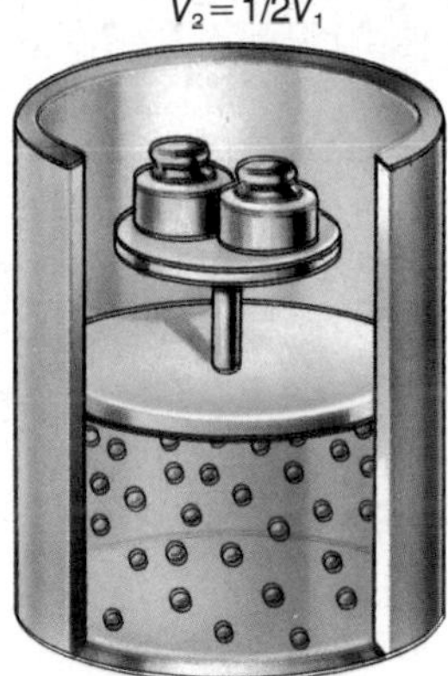

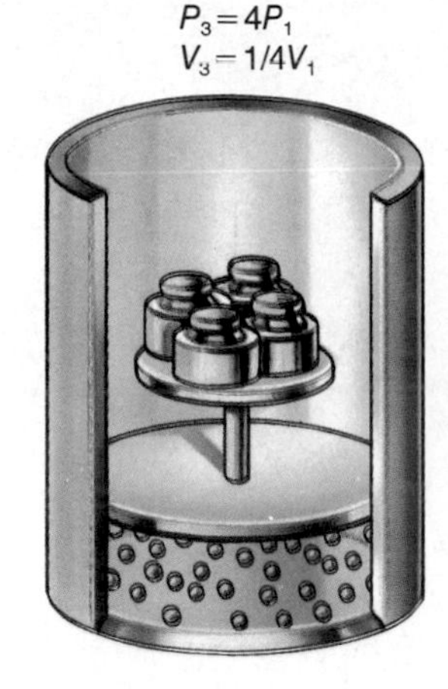

Figure 18.3 At constant temperature, an increase in pressure on a gas decreases the volume of that gas. The number of molecules in each cylinder remains constant.

Boyle's Law

Let us consider first the relationship between the pressure and volume of a gas when both the number of molecules and their average kinetic energy are constant; that is, they do not change when the pressure and volume change. Consider the container of gas with a movable piston in the top as shown at the left in Figure 18.3. If the piston is lowered until it is half the original distance from the bottom, there is only half as much space as before. The same number of molecules occupy half the volume. The molecules hit the walls of the container twice as often and with the same force per collision. So, the pressure is twice as high when the volume is half as much. When the volume is one fourth the original volume, the pressure is four times as high. We conclude that, at constant temperature, pressure increases as volume decreases. More precisely, the pressure varies inversely with volume. The product of pressure and volume is then a constant.

The British chemist Robert Boyle arrived at this principle by experiment 300 years ago. The relationship is called Boyle's law. **Boyle's law** states: *If the amount and the temperature of a gas remain constant, the pressure exerted by the gas varies inversely as the volume.* When we put the relationship into mathematical form, we obtain

$$P \propto \frac{1}{V}$$

where the symbol $\propto$ is a sign that means "varies with."

Then $$P = \frac{k}{V} \quad \text{or} \quad PV = k$$

In this equation P is the pressure and V is the volume. The k is a constant that takes into account the number of molecules and the temperature. Thus, pressure varies inversely with a change in volume.

Air Pressure vs. Gravity

Put on goggles and an apron. Pour water into a small fruit-juice glass until it overflows. Place a 3" × 5" index card over the mouth of the glass. Hold the card across the top of the glass while you invert the glass over a sink. Remove your hand from the card. What did you observe? Explain. Which force appears to be stronger, gravity or air pressure? What is the pressure of one standard atmosphere?

CONTENT BACKGROUND

Applying Boyle's law: Use the following discussion to introduce the ratio method of solving ideal gas problems.

Consider the equation for finding the area of a rectangle, $A = lw$. If we have a rectangle that is 3 cm wide and 4 cm long, the area equals 12 cm^2. What happens to the area when we double the length? The new area is 24 cm^2. In this case, it was easy to calculate the new area from the original equation. However, let us look at the problem differently. By using the basic equation, we get A = 12 cm^2, l = 4 cm, and w = 3 cm. Now use the equation $A' = l'w$, where A' represents the new area and l' represents the new length. w remains the same because the width did not change. If we divide the first equation by the second we get

$$A/A' = l/l'$$

When we solve the final equality for

Applying Boyle's Law

If we know the volume a gas occupies at one pressure, we can use Boyle's law to calculate the volume it will occupy at a different pressure. For example, it is difficult to experiment with gases under standard conditions. Experiments are often carried out at room temperature and pressure. Since the temperature and atmospheric pressure vary from day to day, experimental results cannot be compared easily. It is desirable, therefore, to adjust all results mathematically to standard conditions. If P_1 is measured pressure and V_1 is measured volume, P_2 is standard atmospheric pressure and V_2 is volume at standard atmospheric pressure, then

$$V_1 = \frac{k}{P_1} \quad \text{and} \quad V_2 = \frac{k}{P_2}$$

In the first equation, $k = V_1P_1$. Since k is a constant, we may substitute V_1P_1 for k in the second equation:

$$V_2 = \frac{V_1P_1}{P_2} \quad \text{or} \quad V_2 = \frac{V_1}{} \left| \frac{P_1}{P_2} \right.$$

We have derived this relationship by using Boyle's law. Note that the original volume is simply multiplied by the ratio of the two pressures to find the new volume. In dealing with changes in pressure, there will always be possible two pressure ratios, one of which will be greater than one and the other less than one. Which ratio to use is decided by applying Boyle's law to the problem. For instance, if a gas is changing from 87.1 kPa to 101.3 kPa, the pressure is increasing. Boyle's law enables you to say immediately that the volume will decrease. The new volume after the pressure change will be less than the volume before the change. Since the new volume is less, you must multiply the old volume by a ratio whose value is less than one. The two possible ratios are 87.1/101.3 and 101.3/87.1. You would choose 87.1/101.3 because multiplying the old volume by a ratio whose value is less than one will produce a new volume smaller than the original volume.

The same process can be used to change the volume of a gas to correspond to a pressure other than standard. For instance, if we wish to change the volume of a gas measured at 16.0 kPa to the volume it would occupy at 8.8 kPa, the mathematical operations would be as follows. Correcting for a pressure change from 16.0 kPa to 8.8 kPa is equivalent to expanding the gas. The new volume would be greater, and the ratio of pressures must be greater than 1. The proper ratio is 16.0 kPa/8.8 kPa.

In a similar manner, you can compute a new pressure for a gas if you know its original volume, original pressure, and new volume. In this case, you would multiply the original pressure by a ratio of volumes. The same decision concerning which ratio to use would still apply. If the volume increased, for instance, then Boyle's law could be used to predict that the pressure decreased. If the pressure decreased, you would multiply the old pressure by

CAREERS IN CHEMISTRY

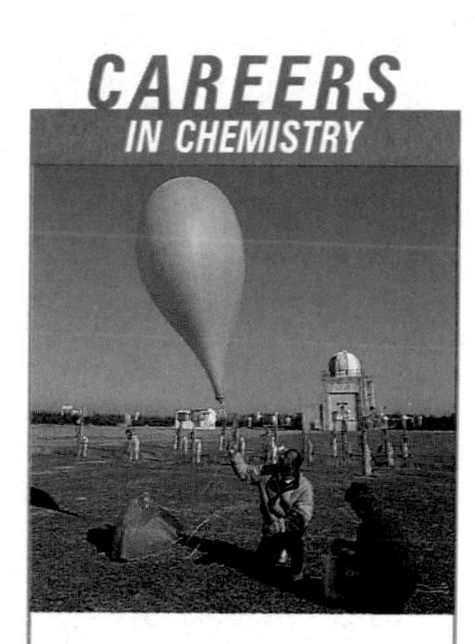

Meteorologist

Most people think of meteorologists as weather forecasters. While forecasting is certainly one of the important functions of meteorologists, their work actually covers a much broader field. Meteorologists study all aspects of Earth's atmosphere. In doing so, they also become involved in the study of the atmospheres of other components of the solar system. The advent of spacecraft that can send us pictures of atmospheres on other bodies has helped meteorologists understand our atmosphere much better. In studying atmospheric phenomena, the meteorologist must understand the chemical reactions that take place and how those changes affect weather and climate. The problems we are experiencing with the ozone layer illustrate how important chemistry is to the study of atmospheric behavior.

the new area, A', we get

$$A' = Al'/l$$

In other words, we find the new area by multiplying the old area by the ratio of the two lengths. Should we then multiply the old area (12 cm^2) by 4/8 or 8/4? Because the new length is greater than the old, we can expect to get a new area that is greater than the old. Multiplying the old area by a fraction less than one (4/8) would give a new area less than the old. Multiplying by a fraction greater than one (8/4) will give us an area larger than the original. Therefore, we choose the fraction that is greater than one.

In applying the relationships between volume, pressure, and temperature, the same kind of reasoning allows students to easily determine the solution for gas-related problems.

CONCEPT DEVELOPMENT

QUICK DEMO **Measuring Pressure** Challenge students to demonstrate Boyle's law using a bicycle tire or a football inflating pump and an air pressure gauge, such as a tire gauge. Remind the students that the air in the pump already is at atmospheric pressure before the plunger is depressed. Student results will vary but should approximate $PV = k$, Boyle's law. You may wish to have students graph their data. For example, the volume of the air in the pump, represented by $1/n$, where n is the number of compression of the pump handle, can be plotted against the pressure readings.

Computer Demo: *Boyle's Law Simulation,* Laboratory simulation of Boyle's law experiment. AP401, PC2601 Project SERAPHIM

CAREERS IN CHEMISTRY

Meteorologist

Background: Stratospheric chemistry involves more than 160 known chemical reactions. There are many more that are not yet known. The effects of these chemicals on our weather and health are unknown. Computer models of our atmosphere give predictions. According to the EPA there could be 40 million cases of skin cancer resulting in 800,000 deaths through 2075 due to ozone thinning.

For More Information: Contact: National Oceanic and Atmospheric Administration, NOAA, Washington, DC

PROGRAM RESOURCES

Reteaching: Use the Reteaching worksheets, pp. 26-27, to provide students another opportunity to understand the concept of "Boyle's and Charles's Laws."

Transparency Master: Use the Transparency master and worksheet, pp. 77-78, for guided practice in the concept "Boyle's Law."

Color Transparency: Use Color Transparency 39 to focus on the concept "Boyle's Law."

Correcting Errors

To help students organize the information in a gas law problem, write a data table on the chalkboard that shows both the initial and final conditions.

Observation	Volume	Pressure	Temperature
initial	V_1	P_1	T_1
final	V_2	P_2	T_2

As the Sample Problem is read, complete the data table.

Observation	Volume	Pressure	Temperature
initial	242 cm^3	87.6 kPa	T_1
final	?	101.3 kPa	T_2

Students can more easily see that the pressure is increasing. Therefore, according to Boyle's law, the volume will decrease because the temperature is constant. Thus, the ratio must be less than one to decrease the volume.

PRACTICE

Guided Practice: Guide students through the Sample Problem, then have them work in class on Practice Problems 1 and 2 that follow.

Independent Practice: Your homework or classroom assignment can include the remaining Practice Problems 3-4 and Review and Practice 29-34.

Answers to

Practice Problems: Have students refer to Appendix C for complete solutions to Practice Problems.

1. 4.10 dm^3

2. 51 kPa

3. **a.** 821 cm^3 **c.** 103 m^3
b. 1.60×10^2 cm^3 **d.** 80.5 m^3

4. **a.** 2.80×10^2 cm^3 **c.** 37.4 cm^3
b 0.805m^3 **d.** 552 cm^3

GLENCOE TECHNOLOGY

Videodisc

Chemistry: Concepts and Applications

Demonstrating Boyle's Law
Disc 2, Side 1, Ch. 3

a ratio whose value was less than one. Do not fall into the habit of "plugging" numbers into equations. Visualize the change to be made in the volume or pressure, and then multiply by the appropriate ratio.

PROBLEM SOLVING HINT

Determine whether the gas must be expanded or compressed to STP. This change determines which ratio to use.

SAMPLE PROBLEM

Pressure Correction

A gas is collected in a 242-cm^3 container. The pressure of the gas in the container is measured and determined to be 87.6 kPa. What is the volume of this gas at standard atmospheric pressure? (Assume that the temperature remains constant.)

Solving Process:

Standard atmospheric pressure is 101.325 kPa. A change to standard atmospheric pressure would compress the gas. Therefore, the gas would occupy a smaller volume. If the volume is to decrease, then the ratio of pressures by which the original volume is to be multiplied must be less than 1. The two possible ratios by which the original volume could be multiplied are

$$\frac{101.3 \text{ kPa}}{87.6 \text{ kPa}} \quad \text{and} \quad \frac{87.6 \text{ kPa}}{101.3 \text{ kPa}}$$

The latter value is the proper one in this case because it is less than 1 and decreases the volume. The corrected volume is

$$\frac{242 \text{ cm}^3 \mid 87.6 \cancel{\text{kPa}}}{\mid 101.3 \cancel{\text{kPa}}} = 209 \text{ cm}^3$$

PRACTICE PROBLEMS

1. If 4.41 dm^3 of gas are collected at a pressure of 94.2 kPa, what volume will the same gas occupy at standard atmospheric pressure, assuming the temperature remains the same?
2. If some oxygen gas at 101 kPa and 25°C is allowed to expand from 5.0 dm^3 to 10.0 dm^3 without changing the temperature, what pressure will the oxygen gas exert?
3. Correct the following volumes of gas from the indicated pressures to standard atmospheric pressure. (Use 101.3 kPa.)
   a. 844 cm^3 at 98.5 kPa **c.** 116 m^3 at 90.0 kPa
   b. 273 cm^3 at 59.4 kPa **d.** 77.0 m^3 at 105.9 kPa
4. Make the indicated corrections in the following gas volumes. Assume constant temperature.
   a. 338 cm^3 at 86.1 kPa to 104.0 kPa
   b. 0.873 m^3 at 94.3 kPa to 102.3 kPa
   c. 31.5 cm^3 at 97.8 kPa to 82.3 kPa
   d. 524 cm^3 at 110.0 kPa to 104.5 kPa

Chemistry Journal

The Vacuum Cleaner and the Gas Laws

Ask students to write a journal entry in which they evaluate the validity of the statement "A vacuum cleaner sucks up dirt." Entries should be supported by references to gas behavior.

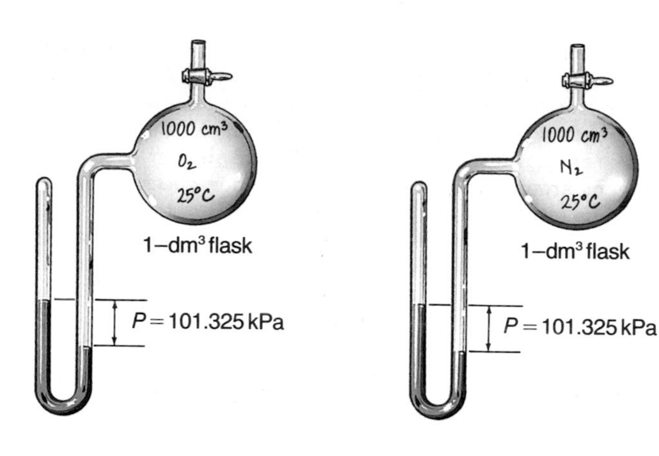

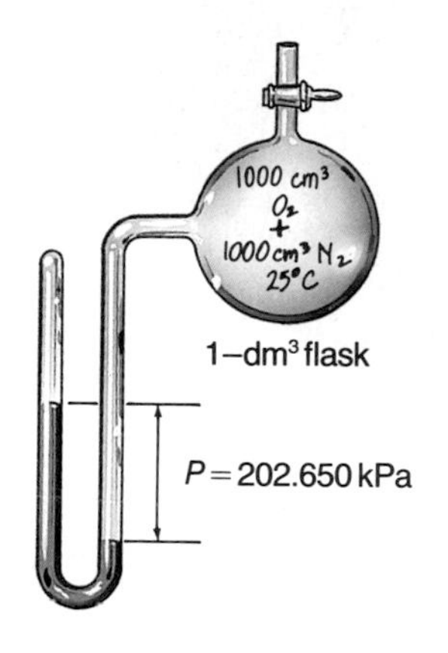

Figure 18.4 When 1000 cm^3 O_2 (left) and 1000 cm^3 N_2 (center) are mixed (right), the pressure of the mixture is the sum of the individual pressures. Note that all three flasks have the same volume and temperature.

Dalton's Law of Partial Pressure

How much pressure is exerted by a particular gas in a mixture of gases? Because more than one gas is occupying the container, each gas contributes to the total pressure. We say that each gas in the mixture exerts its partial pressure. John Dalton was the first to form a hypothesis about partial pressures. After experimenting with gases, he concluded that each gas exerts the same pressure it would if it were present alone at the same temperature. When a gas is one of a mixture, the pressure it exerts is called its partial pressure. **Dalton's law** of partial pressure states: *The total pressure in a container is the sum of the partial pressures of all the gases in the container.*

$$P_{total} = P_1 + P_2 + \ldots + P_n$$

P_1, P_2, and P_n are partial pressures. Gases in a single container are all at the same temperature and have the same volume. Therefore, the difference in their partial pressures is due only to the difference in the numbers of molecules present.

We can mix 1000 cm^3 of O_2 and 1000 cm^3 of N_2, each at room temperature and 101.325 kilopascals. The volume of the mixture is then adjusted to 1000 cm^3 with no change in temperature. The

Figure 18.5 At very high altitudes, mountain climbers must use a supplementary oxygen supply because the partial pressure of oxygen is too low for normal breathing.

CONCEPT DEVELOPMENT

Enrichment: Have a student interested in aviation prepare a research report on how a wing develops lift. Possible response: *The air under the wing is at a higher pressure than the air above the wing because of the difference in motion of air across the top and bottom surfaces of the wing. Planes fly because of a difference in air pressures.*

Discussion Question: Introduce Dalton's law of partial pressure by having students recall inflating a flat tire. Point out that the air pump is putting both nitrogen (80%) and oxygen (20%) molecules into the tire. Ask if the final tire pressure would be different if you could inflate the tire with both gases separately. *No.* Students readily understand that the total pressure in the tire is the sum of the partial pressures.

Using The Illustration: If you choose to use the discussion question to introduce Dalton's law, Figure 18.4 can be used to aid the visual learner. The illustration shows that when two gases are mixed, their partial pressures add to equal the total pressure, just as the partial pressures of nitrogen and oxygen gases do in an automobile tire.

✓ ASSESSMENT

Knowledge: Have students study Figure 18.4 for a few minutes, then give students blank diagrams that are similar to Figure 18.4. Have students fill in the diagrams with numbers that are different from the book's numbers, but which make sense. Ask students to explain to the class why they selected the numbers shown on their diagrams.

Mastering Concepts in Chemistry software: Use the lesson on Applying the Gas Laws to introduce and reinforce Boyle's law, Charles' law, Dalton's law, and the combined gas law.

CONTENT BACKGROUND

History: It is interesting to note the gases used by the men for whom the gas laws are named. Boyle worked with air in establishing the pressure-volume relationship. Dalton worked with air in developing the pressure-number law, although he also worked with other gases. Dalton's interest in gases principally was not chemical, but meteorological. Charles principally worked with air and hydrogen, although like Dalton, he had experimented with other gases. Graham worked with nitrogen, oxygen, carbon monoxide, methane, hydrogen, ethene, hydrogen sulfide, nitrogen(I) oxide, carbon dioxide, and sulfur dioxide. It also should be noted that Gay-Lussac discovered the temperature-volume relationship independently of Charles.

volume that the dry gas would occupy at standard atmospheric pressure and the indicated temperature. Assume that the temperature remains constant.

a. 888 cm^3 at 14°C and 93.3 kPa
b. 30.0 cm^3 at 16°C and 77.5 kPa
c. 34.0 m^3 at 18°C and 82.4 kPa
d. 384 cm^3 at 12°C and 78.3 kPa
e. 8.23 m^3 at 27°C and 87.3 kPa

Charles's Law

Jacques Charles, a French physicist, noticed a simple relationship between the volume of a gas and the temperature. He found that the volume of any gas doubled when the temperature increased from 0°C to 273°C (at constant pressure). For each Celsius degree increase in temperature, the volume of the gas increased by 1/273 of its volume at 0°C. Similarly, Charles found that a gas will decrease by 1/273 of its 0°C volume for each Celsius degree decrease in temperature. This finding would suggest that at −273°C, a gas would have no volume, or would disappear. This temperature is called absolute zero. However, all gases become liquid before they are cooled to this low temperature, and Charles's relationship does not hold for liquids or solids.

Charles's experimental information led to the formation of the absolute or Kelvin temperature scale. Thus far, we have always defined the Kelvin scale in terms of the Celsius scale. Now we are in a position to define the Kelvin scale directly. The zero point of the Kelvin scale is absolute zero. The other reference point in defining the Kelvin scale is the triple point of water, 0.01°C, which is defined as 273.16 K.

We can now describe the volume-temperature relationship that Charles found in terms of the Kelvin scale. *The volume of a quantity of gas, held at a constant pressure, varies directly with the Kelvin temperature.* This is called **Charles's law.** In calculations involving gases, the temperatures given in Celsius degrees must be converted to Kelvins where K = °C + 273.15.

Figure 18.8 Charles's law can be demonstrated when air-filled balloons are placed in liquid nitrogen (at 77 K). The volume of the air is greatly reduced at this temperature. When the balloons are poured out of the nitrogen and warmed to room temperature, they reinflate to their original volume.

Applying Charles's Law

Charles's law states that the volume of a definite amount of gas varies directly as the absolute temperature: $V = k'T$. For a volume of gas V_1, $V_1 = k'T_1$. After a temperature change with pressure constant, the new volume would be $V_2 = k'T_2$. Since $k' = V_1/T_1$, substituting gives

$$V_2 = \frac{V_1}{T_1}T_2 \quad \text{or} \quad V_2 = \frac{V_1}{} \left| \frac{T_2}{T_1} \right.$$

To correct the volume for a change in temperature, you must multiply the original volume by a ratio. For temperature changes, the ratio must be expressed in kelvins.

CONCEPT DEVELOPMENT

- Emphasize to students that a temperature increase of 1C° results in a 1/273 increase in volume.

QUICK DEMO **A Shriveled Balloon** If you have access to liquid nitrogen, inflate a small balloon and place it in a large beaker. Slowly pour liquid nitrogen over the balloon. **CAUTION:** *Liquid nitrogen boils at 77 K and causes tissue damage.* Have students observe that the balloon shrinks and shrivels dramatically. Allow the balloon to warm to room temperature. Have students observe that the balloon returns to normal size. **Disposal:** A

Using The Graph: Remind students that the extrapolation is for an ideal gas. Real gases will condense and freeze before reaching absolute zero.

Correcting Errors

To help students organize the information in a gas law problem, write a data table on the chalkboard that shows both the initial and final conditions. This table should be similar to that used for Boyle's law on p. 456.

Students can more easily see that the temperature is decreasing. Therefore, according to Charles's law, the volume will decrease because the pressure is constant. Thus, the ratio must be less than one to decrease the volume.

GLENCOE TECHNOLOGY

Videodisc

Chemistry: Concepts and Applications
Charles's Law
Disc 2, Side 1, Ch. 4

Software

Mastering Concepts in Chemistry
Unit 11, Lesson 2
Charles's Law, Combined Gas Law

PROGRAM RESOURCES

Applying Scientific Methods in Chemistry: Use the worksheet, "Charles's Law," p. 37, to help students apply and extend their lab skills.

Laboratory Manual: Use microlab 18-1 "Using Charles's Law to Determine Absolute Zero" as enrichment for the concepts in Chapter 18.

CONTENT BACKGROUND

Zero Point Energy: You will have noted that absolute zero has been defined in the text as the temperature at which all molecular motion should cease. We have not said "ceases" because the expression for the energy levels of a molecule (or ion or atom) includes the expression $E_n = (n + 1/2)h\omega$ where E_n is the energy of level n, h = Planck's constant, and $\omega = (k/m)^{1/2}$. In the expression for ω, k is the force

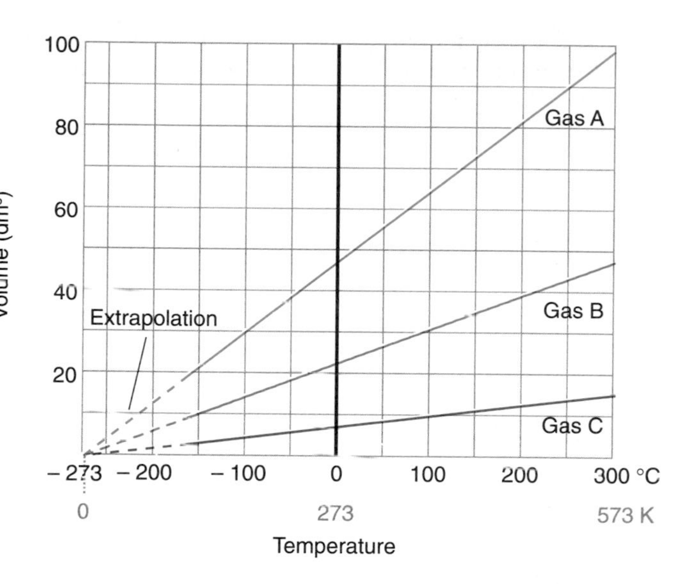

Figure 18.9 At constant pressure, the volume of a gas varies directly with temperature. Extrapolation of the volume-temperature line shows that, at absolute zero, the volume of a gas is theoretically zero. Note that the two temperature scales are comparable.

SAMPLE PROBLEM

Temperature Correction

A 225-cm^3 volume of gas is collected at 58°C. What volume would this sample of gas occupy at standard temperature? Assume a constant pressure.

Solving Process:

The temperature decreases and the pressure remains constant. Charles's law states that if the temperature of a gas decreases at constant pressure, the volume will decrease. Therefore, the original volume must be multiplied by a fraction less than 1. Convert both the initial (58°C) and the final (0°C) temperatures to the Kelvin scale (331 K and 273 K). The two possible temperature ratios are the following.

$$\frac{331 \text{ K}}{273 \text{ K}} \quad \text{and} \quad \frac{273 \text{ K}}{331 \text{ K}}$$

The correct ratio is 273 K/331 K because it is less than 1. The corrected gas volume is

$$\frac{225 \text{ cm}^3 \mid 273 \text{ K}}{\quad \mid 331 \text{ K}} = 186 \text{ cm}^3$$

PRACTICE PROBLEMS

7. A gas occupies a volume of 60.0 cm^3 at 36°C. What volume will the same gas occupy at standard temperature if the pressure remains constant?

MINUTE  *LAB*

Charles's Law in a Freezer

At home, inflate a small balloon and tie it closed. Tape one end of a piece of string to the side of the balloon at its widest point. Wrap the string around the balloon tightly but without stretching the string or deforming the balloon. Tape the string to the balloon where it meets the previously taped end. Use a marker to mark the string where the two ends meet so you can later measure the circumference. Use the marker to mark where the string lies on the balloon also. Place the balloon in a freezer and wait fifteen minutes. When you check the cold balloon, be prepared to pull the string tight and mark it where the two ends meet so you can later determine the circumference of the cold balloon. What did you observe about the balloon's volume when you opened the freezer? State a relationship between temperature and volume when the pressure remains constant. Is this a direct or an inverse mathematical relationship?

constant of the bond under consideration and m is the mass of the atom concerned. In the ground state, $n = 0$, and there is still remaining the energy $(1/2)\hbar\omega$. This energy is called the zero point energy of the molecule. Although the expression falls out naturally from the quantum mechanical description of the molecule, there is the additional requirement of the Heisenberg Uncertainty Principle. If, in the ground state, the molecule had no energy, it would not be moving. We would, therefore, know exactly its momentum (0), and we would be able to locate it exactly. Such a situation is clearly a violation of the Uncertainty Principle.

PRACTICE

Guided Practice: Guide students through the Sample Problem, then have them work in class on Practice Problems 7 and 8a that follow on page 462.

Independent Practice: Your homework or classroom assignment can include the remaining Practice Problems 8b-8d, 9a-e, and Review and Practice 39-42.

Answers to
Practice Problems: Have students refer to Appendix C for complete solutions to Practice Problems.
7. 53.0 cm^3

MINUTE LAB

Charles's Law In A Freezer

Purpose: to show that gas volume is directly related to absolute temperature

Materials: balloon, string, measuring ruler, marker, freezer

Procedure Hints: Remind students not to stretch the string when they are measuring the circumference of the balloon. **Disposal:** A

Results: The balloon will be smaller after being in the freezer; sample data: circumference changed from 71 cm to 67 cm. Students will expect a much greater volume change. Point out that the change is proportionate to the absolute temperature change.

Answers To Questions: 1. The balloon is smaller when cold. **2.** When the temperature decreases, the volume of a gas decreases. **3.** direct relationship

✓ ASSESSMENT

Oral: Ask students what happens to the density of the gas when the volume decreases due to decreasing temperature. *The gas becomes more dense.* This increased density is one reason that doctors recommend air conditioning for people with respiratory ailments, such as emphysema.

Answers to
Practice Problems (continued)

8. a. 597 cm^3 **c.** 872 cm^3
b. 467 cm^3 **d.** 1.97 m^3
9. a. 3.80 m^3 **c.** 486 cm^3
b. 13.1 m^3 **d.** 4.97 m^3

FRONTIERS

Stirling Shiver

Teaching this Feature: Have students compare and contrast refrigerators and cryogenic coolers.

Exploring Further: Have students research recent developments in the use of Stirling engines.

3 ASSESS

CHECK FOR UNDERSTANDING

Ask students to answer the Concept Review problems, and Review and Practice 25-42 from the end of the chapter.

RETEACHING

Ask why Air Force fighter pilots wear oxygen masks. Explain that at high altitude the air pressure is very low. Therefore, the partial pressure of oxygen is too low for normal breathing (Dalton's law).

EXTENSION

Have a local scuba diver talk to the class about the temperature and pressure changes that are experienced by a diver and how these changes affect their air supply. Point out that a one atmosphere pressure is added for every 10 meters of depth.

GLENCOE TECHNOLOGY

Videodisc

Chemistry: Concepts and Applications

Gas Against Fire

Disc 2, Side 1, Ch. 1

Show this video of Demo 18-4 to demonstrate how a carbon dioxide fire extinguisher works.

Also available on CD-ROM.

8. Correct the following volumes of gases for a change from the temperature indicated to standard temperature (pressure is constant).
   a. 617 cm^3 at 9°C **c.** 942 cm^3 at 22°C
   b. 609 cm^3 at 83°C **d.** 7.12 m^3 at 988 K
9. Correct the following volumes of gases for the temperature changes indicated (P is constant).
   a. 2.90 m^3 at 226 K to 23°C
   b. 7.91 m^3 at 52°C to 538 K
   c. 667 cm^3 at 431 K to 41°C
   d. 4.82 m^3 at 22°C to 31°C

18.1 *CONCEPT REVIEW*

10. Which of these variables are held constant in Boyle's law: number of particles, volume, temperature, pressure?
11. What are the properties of an ideal gas?
12. What is meant by STP? Why is it important when discussing quantities of gases?
13. **Apply** If a scuba diver is to remain submerged for 1 hour, what pressure must be applied to force sufficient air into the tank to be used? Assume 0.500 dm^3 of air per breath at standard atmospheric pressure, a respiration rate of 38 breaths per minute, and a tank capacity of 30 dm^3.

FRONTIERS

Stirling Shiver

On the hottest days, the coolest room in any office building will be the computer room. Computers work best with cool computer chips. The noisiest room is the one in which the air conditioners are located. Thanks to an invention from an eighteenth century Scottish clergyman, there may be a quiet change coming.

In 1816, Robert Stirling was granted a patent for an external combustion engine—the Stirling engine. In an internal combustion engine, the heat source is the rapidly-burning gasoline or diesel fuel inside a cylinder of the engine. The heat causes the entrapped gases to expand and push down on a piston. In the Stirling engine, the heat source is external to the engine. The heat is transferred through a plate to an inert gas. The heat causes the entrapped gas to expand and push down on a piston. The Stirling engine can become a cooling mechanism by reversing the process; that is, by *pulling* the piston upward. The trapped gas beneath the piston will expand and cool rapidly. The cooled gas can then be used as a refrigerant.

Researchers are presently designing cryogenic coolers based on the Stirling engine. These coolers, which use helium as the entrapped gas, have been used to reach temperatures close to −200°C. Helium is environmentally safer than the chlorofluorocarbons presently used as refrigerants.

DEMONSTRATION

18-4 Gas Against Fire

Purpose: to show how a carbon dioxide fire extinguisher works

Materials: aquarium; 4 candles of different heights; 100 g baking soda, $NaHCO_3$; 200 cm^3 vinegar

Procedure: 1. Place four candles of increasing heights in an empty aquarium. Cover the bottom of the aquarium with baking soda.
2. Light the candles with a match. Have a student predict what will happen when vinegar is added.
3. Lower the room lights and add the vinegar to the aquarium. **CAUTION:** *Acetic acid is corrosive.* **Disposal:** C for chemicals.

Results: The carbon dioxide gas produced by the chemical reaction will remain in the bottom of the aquarium.

18.2 Additional Considerations of Gases

OBJECTIVES

- solve problems involving the change of more than one condition for gases.
- explain Graham's law and solve problems using it.
- differentiate between an ideal gas and a real gas.

You now know how pressure, volume, temperature, and number of particles affect gas characteristics. However, you have dealt with these variables only two at a time, with the other variables held constant. What would happen to the volume of a gas if both pressure and temperature changed? What happens when a gas is released from its container? Thus far, you have worked only with the ideal gas. How do real gases behave? In this section, we will look into the answers to these questions.

Combined Gas Law

As you have seen, the volume of a gas is affected by both temperature and pressure. It is possible to correct a volume to a new set of conditions. For example, laboratory experiments are almost always made at temperatures and pressures other than standard. It is then necessary to correct the laboratory volumes of gases for both temperature and pressure. The correction is made by multiplying the original volume by two ratios, one for temperature and the other for pressure. Because multiplication is commutative, we can use the ratios in either order. We may think of the process as correcting the volume for a pressure change while the temperature is held constant. Then, we correct for the temperature change while the pressure is held constant. The two changes do not have any effect on each other.

SAMPLE PROBLEM

Volume Correction to STP

The volume of a gas measured at 75.6 kPa pressure and 60.0°C is to be corrected to correspond to the volume it would occupy at STP. The measured volume of the gas is 10.0 cm^3.

Solving Process:

The pressure must be increased from 75.6 kPa to 101.3 kPa. Thus, the volume must decrease, which means the pressure ratio must be less than one. The correct pressure ratio is 75.6 kPa/101.3 kPa. The temperature must be decreased from 333 K to 273 K. This change will also decrease the volume. Therefore, the correct temperature ratio is 273 K/333 K. The problem then becomes

$$\frac{10.0 \text{ cm}^3}{} \left| \frac{75.6 \text{ kPa}}{101.3 \text{ kPa}} \right| \frac{273 \text{ K}}{333 \text{ K}} = 6.12 \text{ cm}^3 \text{ at STP.}$$

PROBLEM SOLVING HINT

Remember that STP = 101.325 kPa and 0°C and that temperature must be in kelvins, where K = °C + 273.

PRACTICE PROBLEMS

14. Freon-12, dichlorodifluoromethane (CCl_2F_2), was a widely used refrigerant. Consider a 2.23-dm^3 sample of gaseous CCl_2F_2 at a pressure of 4.85 kPa and a temperature of

As the CO_2 rises and fills the aquarium, it will extinguish the candles one at a time beginning with the shortest.

Questions: 1. Why does the gas produced not extinguish the candles at the same time? *The gas is more dense than air, and it filled the aquarium by rising, therefore encountering the shortest candle first.*

2. How does a carbon dioxide fire extinguisher work? *It displaces oxygen, necessary for combustion.*

Follow Up Activities: Discuss the various classes of fire extinguishers. Have students use a science supply catalog to determine which type of fire extinguisher would be best suited for use in the basement, garage, kitchen, and computer room. *The ABC type of fire extinguisher is generally recommended for use in most locations except the computer room.*

Answers to
Concept Review

10. Numbers of particles and temperature are constant.

11. An ideal gas is composed of molecules with mass but no volume and no mutual attraction between the molecules.

12. Standard temperature and standard atmospheric pressure (STP) are 0°C and 101.325 kilopascals. In discussing quantities of gases, it is necessary to specify not only the volume but also the pressure and temperature. Gas volumes are often reported in m^3, dm^3, or cm^3 at STP.

13. For a complete solution refer to *Merrill Chemistry Problems and Solutions Manual.* 3.8 x 10^3 kPa

4 CLOSE

Discussion: Inform students that 60 years ago automobile tires were not as strong as they are today. Have students consider why then people waited until dark to drive across the desert. Possible response: *Because of high temperature and lack of tire strength, tire blowouts were common during the heat of the day. Cars didn't have air conditioning.*

18.2 Additional Considerations of Gases

PREPARE

Key Concepts

The major concepts presented in this section include the combined gas law, diffusion and Graham's law, and deviations of real gases from ideal behavior.

Key Terms

diffusion	real gas
Graham's law	adiabatic system

1 FOCUS

Objectives

Preview with your students the objectives listed on the student page. Students can use them as a study guide for this major section.

BRIDGE TO ENGINEERING

Heat Engines

Teaching This Feature: Demonstrate a model steam engine to the class.

Connection to History: The development of the American frontier west of the Mississippi River was hastened by the steam engine. River and rail traffic moved people west and the steam threshing machine enabled the great plains farmer to increase his productivity.

Extension: Have an interested student research and report on the steam turbine and its use in generating electricity in modern power plants.

Answers to

Exploring Further

1. Boyle's law; pressure and volume are inversely related.
2. Method differs for each type of engine. A gasoline engine's horsepower rating is generally obtained from compression data. Diesel engine horsepower rating is shaft horsepower. Steam engine is brake horsepower.

Correcting Errors

Guide the students through the algebraic substitution of *m/M* for *n* in the ideal gas equation. Review with students how to calculate the number of moles using a given mass and the formula mass. Example: 22 g CO_2/44 g/mol CO_2 = 0.50 moles.

PRACTICE

Guided Practice: Guide students through the Sample Problems, then have them work in class on Practice Problems 6-7.

✓ ASSESSMENT

Knowledge: Ask students to write down the ideal gas equation from memory. Then have them write the units for each of the variables.

Bridge to *ENGINEERING*

Heat Engines

Consider a gas in a cylinder containing a piston, as shown in the figure. If the internal pressure exceeds the external pressure (P_{ex}), the gas expands, doing work. Work is represented by w, a positive quantity, which is a symbol for work done *on* the gas. If the gas *does work*, w will be a negative quantity.

The actual amount of work done by a gas depends on the conditions under which expansion takes place. Specifically, w depends on whether the volume change occurred in an adiabatic system or in an isothermal (constant temperature) system. It also depends upon the difference between the external and internal pressures during the change.

Chemists define a reversible expansion of a gas as an ideal change. Reversible changes do not occur in practice, but they are convenient to use to find the maximum amount of work in a change. In actual practice we deal with irreversible changes that always involve less than the ideal maximum work. The most practical change is the irreversible change against a constant pressure. This change is the type we find in an engine. In this case

$$w = -P_{ex}\Delta V$$

Exploring Further

1. In the mathematical expressions for the work done in an isothermal reversible change, one form uses V_2/V_1 and another uses P_1/P_2. Why are initial and final conditions reversed in the two expressions?
2. Find out how horsepower rating is obtained for one type of heat engine.

Molecular Mass Determination

The ideal gas equation can be used to solve a variety of problems. One type, which we will illustrate here, is the calculation of the molecular mass of a gas from laboratory measurements. Such calculations are of importance to the chemist in determining the formulas and structures of unknown compounds.

The number of moles n of a substance is equal to mass m divided by the molecular mass M; $n = \frac{m}{M}$. Therefore, the ideal gas equation may be written

$$PV = \frac{mRT}{M}, \text{ or } M = \frac{mRT}{PV}$$

This modified form of the ideal gas equation may be used in many other types of problems. For example, the equation can be modified to solve for the density of a gas. Because $D = m/V$, then

$$D = \frac{m}{V} = \frac{PM}{RT}$$

GLENCOE TECHNOLOGY

Videodisc

Chemistry: Concepts and Applications

Molar Mass

Disc 2, Side 1, Ch. 6

Also available on CD-ROM.

SAMPLE PROBLEM

Molecular Mass from Gas Measurements

Suppose we measure the mass of the vapor of an unknown compound contained in a 273-cm^3 gas bulb. We find that the bulb contains 0.750 g of gas at 97.2 kPa pressure and 61°C. What is the molecular mass of the gas?

Solving Process:

Since the unknown quantity is the molecular mass, we will use the modified ideal gas equation

$$M = \frac{mRT}{PV}$$

Before we can substitute the known values into the ideal gas equation, °C must be converted to K. We get the following expression

$$\underset{m}{\frac{0.750\ \text{g}}{97.2\ \cancel{\text{kPa}}}}\ \Big|\ \underset{R}{\frac{8.31\ \cancel{\text{dm}^3}\cdot\cancel{\text{kPa}}}{\text{mol}\cdot\cancel{\text{K}}}}\ \Big|\ \underset{T}{\frac{334\ \cancel{\text{K}}}{273\ \cancel{\text{cm}^3}}}\ \Big|\ \frac{1000\ \cancel{\text{cm}^3}}{1\ \cancel{\text{dm}^3}}$$

(P under 97.2 kPa; V under 273 cm^3; conversion to dm^3)

$$= 78.4\ \text{g/mol}$$

The result is, therefore, 78.4 g/mol. Note that all other units in the problem divide out.

PROBLEM SOLVING HINT

The units remaining at the end of the problem serve as a check on the problem setup. Answers other than g/mol indicate an error in the setup.

PRACTICE PROBLEMS

6. What is the molecular mass of a gas if 150.0 cm^3 have a mass of 0.922 g at 99°C and 107.0 kPa?
7. What is the molecular mass of a gas if 3.59 g of it occupy 4.34 dm^3 at 99.2 kPa and 31°C?
8. What is the molecular mass of a gas if 0.858 g of it occupies 150.0 cm^3 at 106.3 kPa and 2°C?
9. What is the molecular mass of a gas if 8.11 g of it occupy 2.38 dm^3 at 109.1 kPa and 10.0°C?

19.1 CONCEPT REVIEW

10. State Avogadro's principle.
11. What is meant by the term *molar volume*? What are its numerical value and units?
12. State the ideal gas equation.
13. Find the number of moles of an ideal gas contained in a 4.70-dm^3 tank at 4°C and 91.5 kPa.
14. **Apply** A chemist was asked to find the molecular mass of a gas. She massed the gas and found that 2.73 g of the gas occupied 315 cm^3 at 24°C and 94.2 kPa. What is the molecular mass of the gas?

Independent Practice: Your homework or classroom assignment can include the remaining Practice Problems 8-9, and Review and Practice 43-44 from the end of the chapter.

Answers to
Practice Problems: Have students refer to Appendix C for complete solutions to Practice Problems.

6. 178 g/mol
7. 21.1 g/mol
8. 123 g/mol
9. 73.5 g/mol

3 ASSESS

CHECK FOR UNDERSTANDING

Have students answer the Concept Review problems, and Review and Practice 33-44 from the end of the chapter.

RETEACHING

Use **Solving Problems in Chemistry,** Chapter 19, Sections 1 and 2 to reteach this section. Problems can be assigned to provide review and reinforcement of the needed problem-solving skills.

EXTENSION

Molecular masses of substances are often found today by using a mass spectrograph. Have an interested student research "time-of-flight" spectrometers. Possible response: *In a time-of-flight spectrometer, a burst or pulse of ionized particles is fired through a series of charged grids. Ions of different masses require varying times to transit the path; by proper interpretation of the electrode currents, the sample may be analyzed for e/m (charge to mass ratio) values.*

Answers to
Concept Review

10. At equal temperatures and equal pressures, equal volumes of gases contain the same numbers of molecules.
11. The molar volume is the volume occupied by one mole of a substance. For a gas at STP, this volume is 22.4 dm^3.
12. $PV = nRT$
13. 0.187 mol
14. 227 g/mol

PROGRAM RESOURCES

Enrichment: Use the worksheets "Behavior of Real Gases," pp. 37-38, to challenge more able students.

Laboratory Manual: Use the macrolab 19-2 "Molar Volume of a Gas" as enrichment for the concepts in Chapter 19.

Laboratory Manual: Use the microlab 19-3 "Molar Volume of a Gas" as enrichment for the concepts in Chapter 19.

4 CLOSE

Discussion: The ideal gas equation gives close approximations for real gases under normal conditions. Review with students the laboratory data that is needed to solve for the volume of the gas. Have students use similar reasoning processes to determine the data that is needed to calculate the molecular mass of a gas.

CHEMISTRY AND TECHNOLOGY

Gases From Air

Teaching This Feature: Obtain a small amount of liquid nitrogen from a neighboring college or university. Demonstrate how liquid nitrogen affects a banana, hot dog, rubber ball, and fresh flower. **CAUTION:** *Use tongs and wear gloves.*

Connection to Medicine: Liquid nitrogen probes are used to remove external warts. These probes can also be inserted into the nerve sheath to freeze a nerve thus giving a patient relief from chronic pain.

Answers to

Thinking Critically

1. Force molecules close enough together for van der Waals forces to take effect.

2. Air is filtered, compressed, cooled, and stripped of H_2O. Air is then fed to a molecular sieve where only O_2 is adsorbed. When the adsorption vessel is saturated, stream is switched to a new vessel. The saturated vessel is stripped by depressurizing and purging.

CHEMISTRY AND
TECHNOLOGY

Gases from Air

Several of the gases comprising air have extensive industrial and commercial uses. Oxygen, nitrogen, neon, argon, krypton, and xenon are these useful gases, and they are all produced by the fractional distillation of liquid air. Before air can be liquefied, it is filtered to remove soot and other small dirt particles that would clog the equipment used later in the process. The air is then compressed to five or six times normal atmospheric pressure and cooled. Some of the water and carbon dioxide solidify during the cooling process. The remainder of the water and carbon dioxide, as well as impurities such as hydrocarbons, are removed by a special filter. The filtered air is then compressed to about 10 megapascals (MPa). As a result of being compressed, the temperature of the air rises. This energy is removed in a device called a heat exchanger. A heat exchanger consists of a series of tubes passing through a cylindrical vessel. One fluid passes through the tubes. Another cooler fluid flows through the vessel surrounding the tubes. Heat flows from the hotter fluid to the cooler fluid. The compressed air is cooled and then further compressed to about 15 MPa.

Some of the cold gas is used to run one or more compressors. In the process of doing that work, the gas uses some of its internal energy. Thus, its temperature drops even more. Most of the cold, compressed gas is allowed to expand through a valve. A large part of the gas is liquefied through the Joule-Thomson effect.

The liquid and extremely cold gases then enter a two-stage distillation column. In the column, the air is separated into high-purity nitrogen, oxygen, and other fractions. The high-purity products are sold and low-purity nitrogen gas is used as a coolant in the heat exchangers. The other liquid fraction passes to a specialized distillation unit where neon, argon, krypton, and xenon are produced. During distillation, each component is removed at its boiling point.

Oxygen is the most important gas obtained from liquid air, with nitrogen ranking second. The major industrial use for oxygen is in the steelmaking industry, where it is used to burn off impurities in open-hearth or basic oxygen furnaces. Oxygen is also consumed by the chemical industry in the production of artificial atmospheres, for the oxidizing of rocket fuels, and for processing wastewater. Nitrogen is used mostly for the production of ammonia. Nitrogen is also used in creating an inert atmosphere where the presence of oxygen might cause problems. It is also used as a refrigerant and for quick-freezing foods. Argon is another air-derived gas used to produce inert atmospheres. In the welding of very active metals, for instance, nitrogen may be reactive and argon is used to protect the metal during welding. Aluminum, titanium, and zirconium are often protected in this manner. Argon is also used extensively to fill light bulbs. The major use for neon, krypton, and xenon is the filling of advertising signs, although new applications are being found, especially in the field of lighting. It should be noted that helium is not recovered from the air but is obtained from natural gas wells where it makes up as much as 2% of the natural gas produced.

Thinking Critically

1. What is the purpose of the compression in the liquefying process?
2. A small amount of liquid oxygen is produced by the pressure-swing adsorption system. How does that system work?

484

CONTENT BACKGROUND

Equations of State: As noted in Chapter 18, the ideal gas equation, or equation of state, can be modified to take into account the behavior of real gases. For example, the most famous modification is that of van der Waals.

$$(P + an^2/V^2)(V - nb) = nRT$$

The quantities *a* and *b* are called van der Waals constants for the gases. The values of some gases appear in the following table.

Gas	*a* ($dm^6 \cdot kPa/mol^2$)	*b* (dm^3/mol)
He	3.457	0.02370
Ne	21.35	0.01709
Ar	136.3	0.03219
N_2	140.8	0.03913
H_2O	553.6	0.03049
C_6H_6	1820	0.1154

19.2 Gas Stoichiometry

OBJECTIVES

- solve gas volume-mass, mass-gas volume, and volume-volume problems.
- identify the limiting reactant and be able to solve problems based upon it.

One of the most frequent calculations performed by chemists and chemical engineers concerns the quantitative relationships between and among reactants and products in chemical reactions. In earlier work you learned about mass-mass relationships. Working with gases, however, we frequently encounter quantities of gas expressed in volume units. In this section you will learn how to deal with problems involving gas volumes in quantitative chemical calculations.

Mass-Gas Volume Relationships

In Section 9.2, we discussed a method of finding the mass of one substance from a specific mass of another substance. It is usually awkward to measure the mass of a gas. It is easier to measure the volume under existing conditions and convert this measurement to the volume under standard conditions. One mole of gas molecules occupies 22.4 dm^3 (STP). This knowledge enables us to determine the volume of gas in a reaction by using the balanced equation for the reaction.

All mass-gas volume problems in this book can be solved in a manner similar to that shown in the following sample problem. Try to keep in mind the following four steps:

Step 1. *Write a balanced equation.*
Step 2. *Find the number of moles of the given substance.*
Step 3. *Use the ratio of the moles of the given substance to the moles of required substance to find the moles of gas.*
Step 4. *Express moles of gas in terms of volume of gas.*

Remember that 1 mole of gas occupies 22.4 dm^3 at STP.

SAMPLE PROBLEM

Mass-Gas Volume

What volume of hydrogen at STP can be produced when 6.54 g of zinc reacts with hydrochloric acid?

Solving Process:

(a) Write a balanced equation for the reaction.

$$2HCl(aq) + Zn(cr) \rightarrow H_2(g) + ZnCl_2(aq)$$

(b) Express the mass (6.54 g) of zinc in moles.

$$\frac{6.54 \text{ g Zn}}{} \left| \frac{1 \text{ mol Zn}}{65.4 \text{ g Zn}} \right. \cdots$$

(c) From the balanced equation, determine the mole ratio. Note that 1 mole of zinc yields 1 mole of hydrogen gas.

$$2HCl(aq) + Zn(cr) \rightarrow H_2(g) + ZnCl_2(aq)$$
$$1 \text{ mole Zn} \rightarrow 1 \text{ mole } H_2$$

Figure 19.4 The reaction of zinc metal with hydrochloric acid produces hydrogen gas.

19.2 Gas Stoichiometry

PREPARE

Key Concepts

The major concepts presented in this section include mass-gas volume, gas volume-mass, volume-volume, and limiting reactant stoichiometric relationships and appropriate problem-solving strategies.

Key Terms

limiting reactant
excess reactant

1 FOCUS

Objectives

Preview with your students the objectives listed on the student page. Students can use them as a study guide for this major section.

Demonstration

To focus student thoughts on gases and the mole, place a broken Alka Seltzer tablet inside a large balloon. Place 50 cm^3 of warm water in a 250-cm^3 Erlenmeyer flask. Fit the balloon over the mouth of the flask. Then, spill the tablet pieces from the balloon into the water. Introduce the section's vocabulary by focusing the discussion on the amount of solid substance reacting, the volume of gas produced, and the temperature and pressure of the gas collected inside the balloon. The reaction is
$HCO_3^- + H^+ \rightarrow H_2O + CO_2$.
Disposal: C

Additional values can be calculated from the relationships

$a = 27R^2T_c^2/64P_c$ and $b = RT_c/8P_c$

where T_c and P_c are the critical temperature and the critical pressure of the gas.

Other equations of state are listed below. Note that the meaning of V_m is molar volume and a and b are not the same as those in the van der Waals equation.

Berthelot: $P = RT/(V_m - b) - a/TV_m^2$
Dieterici: $P = [RT/(V_m - b)]e^{(-a/RTV_m)}$
Beattie-Bridgeman: $P = (1 - \gamma)RT(V_m + \beta)/V_m^2 - \alpha/V_m^2$

PROGRAM RESOURCES

Study Guide: Use the Study Guide worksheets, pp. 75-76, for independent practice with concepts in Section 19.2.

SPORTS CONNECTION
Humid Homers

Exploring Further: Ask students what other sport performances might be affected by increasing or decreasing humidity in the air. Possible response: *The air's relative humidity has an effect on the ability of the body to cool itself through the evaporation of water. Breathing dry air can cause the athlete's throat and mouth to be dry and sore. Goggles may be necessary to prevent the eyes from being dried by less humid air.*

2 TEACH

CONCEPT DEVELOPMENT

• You may want to review mass-mass stoichiometry problems from Chapter 9 before introducing mass-volume and volume-volume problems. Emphasize the factor-label method. Attempt to get students to ask themselves, "Is this answer reasonable?"

Correcting Errors

Remind students that the elements hydrogen, nitrogen, oxygen, fluorine, chlorine, bromine, and iodine exist as diatomic molecules, H_2, N_2, O_2, F_2, Cl_2, Br_2, and I_2. Students may fail to remember these forms when they write and balance equations. Encourage students to use cancellation lines to double-check their factor labels.

PRACTICE

Guided Practice: Guide students through the Sample Problem, then have them work in class on Practice Problems 15-16.

Independent Practice: Your homework or classroom assignment can include Review and Practice 46-47 from the end of the chapter.

Answers to

Practice Problems: Have students refer to Appendix C for complete solutions to Practice Problems.

15. 22 400 cm^3 NH_3

16. 9590 cm^3 H_2

SPORTS CONNECTION

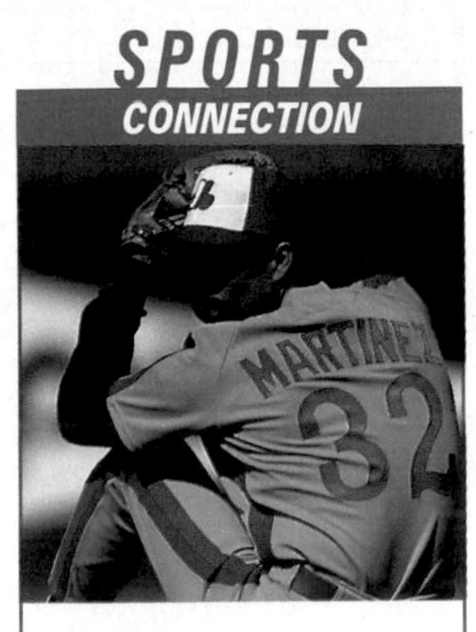

Humid Homers

Are you more likely to hit a home run in humid weather or in dry weather? By measuring the mass of 22.4 dm^3 of air, you can determine the "average molecular mass" of the mixture of gases composing air. Such a measurement will show that the average molecular mass of air is about 28. The molecular mass of water vapor is 18. Therefore, humid air is less dense than dry air. As a result, the lighter water molecules provide less resistance to the passage of a ball through the air. The difference, however, is extremely small. A well-hit ball will travel only about 2 to 3 cm farther in humid air.

Find the moles of hydrogen produced.

$$6.54 \text{ g Zn} \times \frac{1 \text{ mol Zn}}{65.4 \text{ g Zn}} \times \frac{1 \text{ mol } H_2}{1 \text{ mol Zn}} \cdots$$

(d) Use molar volume to express the volume of hydrogen in terms of dm^3 of hydrogen.

$$6.54 \cancel{\text{g Zn}} \times \frac{1 \cancel{\text{mol Zn}}}{65.4 \cancel{\text{g Zn}}} \times \frac{1 \cancel{\text{mol}} H_2}{1 \cancel{\text{mol Zn}}} \times \frac{22.4 \text{ dm}^3}{1 \cancel{\text{mol}}} = 2.24 \text{ dm}^3 \, H_2$$

We conclude that 2.24 dm^3 of hydrogen at STP are produced when 6.54 g of zinc react completely with hydrochloric acid.

PRACTICE PROBLEMS

15. An excess of hydrogen reacts with 14.0 grams of nitrogen. How many cm^3 of ammonia are produced at STP?

16. How many cm^3 of hydrogen at STP are produced from 28.0 grams of zinc reacting with an excess of sulfuric acid?

Gas Volume-Mass Relationships

It is also possible to determine the mass of a substance in a reaction when the volume of a gaseous substance is known. We use a procedure similar to that used in mass-gas volume relationships.

We begin with gas volume and find the mass of another substance. Previously, we started with the mass of a substance and found the volume of gas. However, we are still concerned with the mole relationships. The procedure in the sample problem below is as follows.

Step 1. *Write a balanced equation.*
Step 2. *Change volume of gas to moles of gas.*

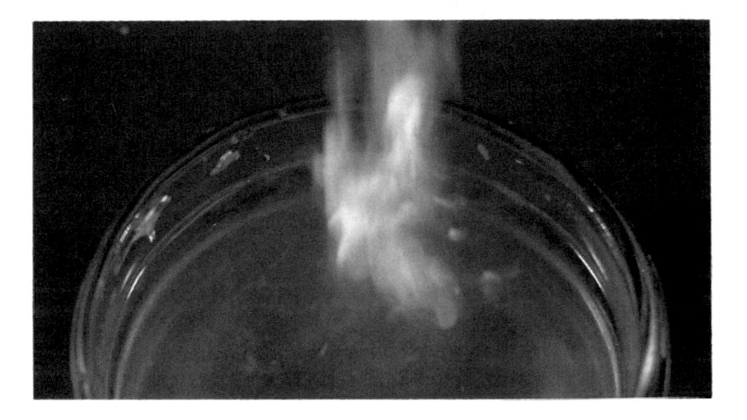

Figure 19.5 If you could measure the volume of hydrogen given off when potassium reacts with excess water, you could determine the mass of potassium that reacted. Unfortunately, the heat given off in this reaction usually causes the hydrogen to catch fire in air.

DEMONSTRATION

19-4 Air Bags

Purpose: to observe that an expanding gas does work

Materials: 4 small, plastic sandwich bags; 4 one-hole rubber stoppers; 4 drinking straws; 4 rubber bands; cafeteria tray

Procedure: 1. Use a rubber band to tightly attach a plastic sandwich bag to a one-hole stopper. Tightly fit the straw into the hole in the stopper. When air is blown into the straw the plastic bag should inflate. Prepare a total of four bags in a similar way.

2. Place the four deflated bags under the corners of a cafeteria tray. Ask four volunteers to inflate the bags and lift the tray by blowing through the straws.

3. Deflate the bags. Then place a large mass on the tray and have the

Step 3. *Determine the ratio of moles of the given substance to moles of required substance.*

Step 4. *Express moles of required substance as grams of required substance.*

After you have worked enough problems to become familiar with these procedures, you should be able to vary your approach to suit the problem.

Figure 19.6 When powdered antimony is placed in an atmosphere of chlorine, antimony (III) chloride is produced in a vigorous reaction.

SAMPLE PROBLEM

Gas Volume-Mass

How many grams of NaCl can be produced by the reaction of 112 cm^3 of chlorine at STP with an excess of sodium?

Solving Process:

(a) Determine the balanced equation.

$$2Na(cr) + Cl_2(g) \rightarrow 2NaCl(cr)$$

(b) Express the volume of chlorine as moles of chlorine at STP.

$$\frac{112\ cm^3\ Cl_2}{} \left| \frac{1\ dm^3}{1000\ cm^3} \right| \frac{1\ mol}{22.4\ dm^3} \ldots$$

(c) Determine the mole ratio.

$$2Na(cr) + Cl_2\ (g) \rightarrow 2NaCl(cr)$$
$$1\ mole\ Cl_2 \rightarrow 2\ moles\ NaCl$$

$$\frac{112\ cm^3\ Cl_2}{} \left| \frac{1\ dm^3}{1000\ cm^3} \right| \frac{1\ mol}{22.4\ dm^3} \left| \frac{2\ mol\ NaCl}{1\ mol\ Cl_2} \right. \ldots$$

(d) Convert moles of NaCl to grams of NaCl.

Na	$1 \times 23.0\ g = 23.0\ g$
Cl	$1 \times 35.5\ g = 35.5\ g$
molar mass of NaCl	$= 58.5\ g$

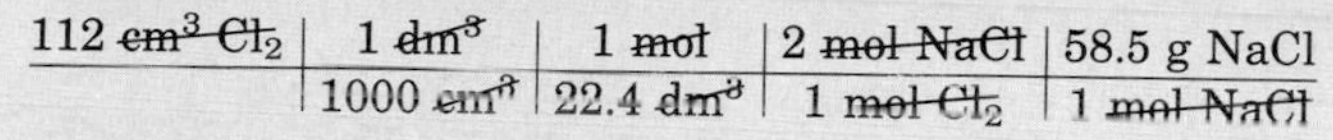

$$\frac{112\ \not{cm^3}\ \not{Cl_2}}{} \left| \frac{1\ \not{dm^3}}{1000\ \not{cm^3}} \right| \frac{1\ \not{mol}}{22.4\ \not{dm^3}} \left| \frac{2\ \not{mol\ NaCl}}{1\ \not{mol\ Cl_2}} \right| \frac{58.5\ g\ NaCl}{1\ \not{mol\ NaCl}}$$

$$= 0.585\ g\ NaCl$$

We conclude that 112 cm^3 of Cl_2, plus enough sodium to react completely with the Cl_2, will yield 0.585 g of NaCl.

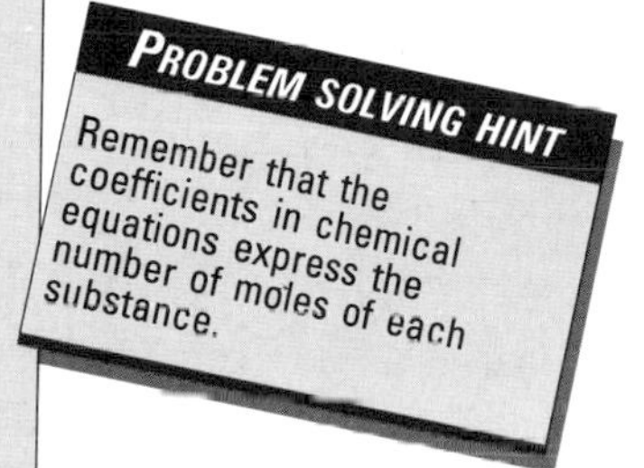

PROBLEM SOLVING HINT
Remember that the coefficients in chemical equations express the number of moles of each substance.

PRACTICE PROBLEMS

17. Bromine reacts with 5.60×10^3 cm^3 of hydrogen to yield what mass of hydrogen bromide at STP?

18. How many grams of solid antimony(III) chloride can be produced from 3570 cm^3 of chlorine gas at STP reacting with an excess of antimony?

Correcting Errors

Point out that the problems in this chapter are solved in the same way as those in Section 9.2. The only difference is that the conversion factor 22.4 dm^3/1 mol is now used. Remind students that there are 1000 cm^3 in 1 dm^3 by showing them a 1-dm^3 graduated cylinder.

PRACTICE

Guided Practice: Guide students through the Sample Problem, then have them work in class on Practice Problems 17-18.

Independent Practice: Your homework or classroom assignment can include Review and Practice 55-56 from the end of the chapter.

Answers to

Practice Problems: Have students refer to Appendix C for complete solutions to Practice Problems.

17. 40.5 g HBr

18. 25.4 g $SbCl_3$

CONCEPT DEVELOPMENT

Background: Large air bags are used to upright semi tractor-trailers that have flipped over on freeways. The air bags are placed under the side of a trailer and inflated with exhaust gases from the wrecker trucks. When the trailer is lifted nearly upright, the wreckers pull the trailer upright. Fire rescue trucks also use air bags to quickly lift heavy objects off people.

✓ ASSESSMENT

Skill: Ask students to solve Review and Practice questions 46 and 47 on page 494.

students inflate the bags again to lift it. **Disposal:** F

Results: Students should realize that their breath acts as an exhaust gas just like the wrecker truck's fuel that is burned in the engine to produce the exhaust gases to inflate the large air bags that are placed under the flipped tractor trailer to upright it.

✓ ASSESSMENT

Skill: Have students calculate how long it would take to inflate a 1- m^3 air bag with CO_2 if a wrecker truck engine burns 22.8 g of octane each minute according to the following equation.

$$2C_8H_{18} + 25O_2 \rightarrow 16CO_2 + 18H_2O$$

It would take 28 minutes assuming STP and all the H_2O was a liquid.

MAKING CONNECTIONS

Daily Life

Have students cite how compressed gases are used to do work. Possible responses: *air-impact wrench, which is used to change tires; jack hammer; or a paint sprayer*

Correcting Errors

Allow students to work volume-volume problems completely, showing all the steps needed in the solutions. With experience they will gain confidence and drop out the two self-canceling steps as shown in the sample problem.

PRACTICE

Guided Practice: Guide students through the Sample Problem, then have them work in class on Practice Problems 19-23.

Independent Practice: Your homework or classroom assignment can include Review and Practice 48-49 from the end of the chapter.

✓ ASSESSMENT

Skill: Ask cooperative learning groups to solve Review and Practice questions 68 and 69 on page 496.

Volume-Volume Relationships

The equation for the complete burning of methane is

$$CH_4(g) + 2O_2(g) \rightarrow CO_2(g) + 2H_2O(g)$$

Notice that all reactants and all products of the reaction are gases. Gas is more easily measured by volume than by mass. Therefore, we will solve problems involving gases by using Avogadro's principle instead of converting to moles. According to Avogadro's principle, equal numbers of molecules of a gas occupy equal volumes at the same temperature and pressure. Therefore, the ratio of the combining volumes of gases is the same as the ratio of the combining moles. Thus, you can use the coefficients of the balanced equation as the ratios of the combining volumes.

SAMPLE PROBLEM

Volume-Volume

How many dm^3 of oxygen are required to burn 1.00 dm^3 of methane? (All of these substances are gases measured at the same temperature and pressure.)

Solving Process:

(a) Write a balanced equation.

$$CH_4(g) + 2O_2(g) \rightarrow CO_2(g) + 2H_2O(g)$$

(b) Determine the ratio of moles from the equation.

$$CH_4 + 2O_2 \rightarrow CO_2 + 2H_2O$$

We see that 1 mole CH_4 reacts with 2 moles O_2.

(c) This mole ratio is the same as the ratio of volumes, so one dm^3 of CH_4 reacts with 2 dm^3 of O_2.

$$\frac{1.00\ \cancel{dm^3\ CH_4}}{} \left| \frac{2\ dm^3\ O_2}{1\ \cancel{dm^3\ CH_4}} \right. = 2.00\ dm^3\ O_2$$

We conclude that 1.00 dm^3 of methane will be completely burned by 2.00 dm^3 of oxygen.

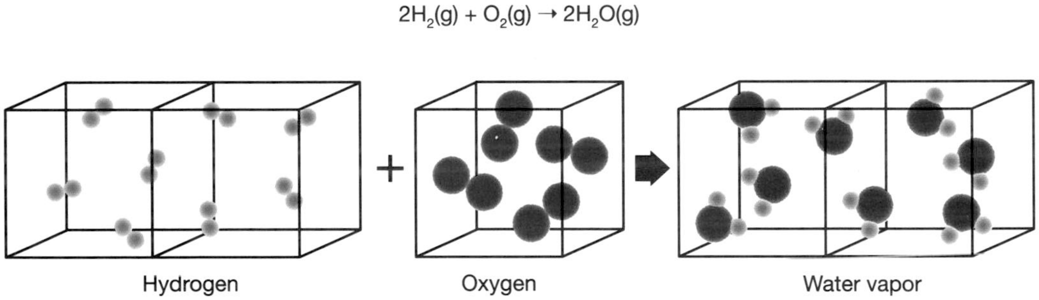

Figure 19.7 Just as two molecules of hydrogen react with one molecule of oxygen to form two molecules of water, so two volumes of hydrogen react with one volume of oxygen to produce two volumes of water vapor.

$$2H_2(g) + O_2(g) \rightarrow 2H_2O(g)$$

Hydrogen

Oxygen

Water vapor

DEMONSTRATION

19-5 Limiting Reactants

Purpose: to demonstrate the effects of limiting a reactant

Materials: 10 test tubes, cafeteria tray, 4 150-cm^3 beakers, 20 rubber stoppers that fit the test tubes

Procedure: 1. Place ten empty test tubes on a cafeteria tray. In each of four small beakers place 5 rubber stoppers that fit the test tubes.

2. Show students that reactant A (test tube) plus reactant B (stopper) will react to form product AB (a stoppered test tube). This procedure is a model of a synthesis reaction (Chapter 9).

3. Ask students how much product will form when the first beaker containing 5 stoppers is added to the cafeteria tray containing the 10 test tubes. *5*

4. Add another beaker full of stoppers

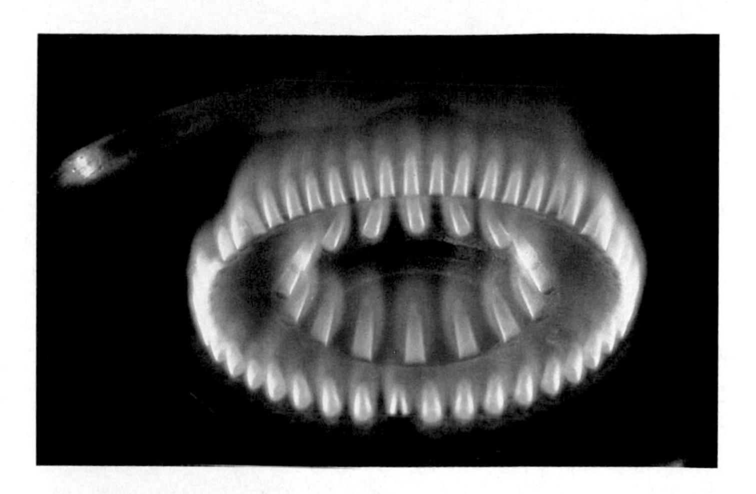

Figure 19.8 The methane in natural gas burns as it reacts with oxygen in air to produce carbon dioxide and water vapor.

PRACTICE PROBLEMS

19. What volume of oxygen is required to burn 401 cm^3 of butane, C_4H_{10}, completely? (All substances are gases measured at the same temperature and pressure.)

20. What volume of bromine gas is produced if 75.2 dm^3 of Cl_2 react with excess HBr? (All substances are gases measured at the same temperature and pressure.)

$$Cl_2(g) + 2HBr(g) \rightarrow Br_2(g) + 2HCl(g)$$

21. Heptane, C_7H_{16}, is a typical component of gasoline. When gasoline is injected into the cylinder of an automobile engine, it is vaporized. What volume of oxygen is required to burn 917 cm^3 of heptane vapor when both gases are at the same temperature and pressure? What volume of air (20.9% O_2) will supply the necessary oxygen?

22. What volume of O_2 is required to oxidize 499 cm^3 of NO to NO_2? (Assume STP.)

23. What volume of hydrogen is produced when 941 m^3 of C_6H_6 are produced? (All substances are gases measured at the same temperature and pressure.)

$$C_6H_{14}(g) \rightarrow C_6H_6(g) + 4H_2(g)$$

Limiting Reactants

Suppose 4.00 dm^3 of hydrogen and 1.00 dm^3 of oxygen are placed in a container and ignited by means of a spark. An explosion occurs and water is formed.

$$2H_2(g) + O_2(g) \rightarrow 2H_2O(g)$$

We know that two volumes of hydrogen are all that can combine with one volume of oxygen, so 2.00 dm^3 of hydrogen are left

Answers to
Practice Problems: Have students refer to Appendix C for complete solutions to Practice Problems.
19. 2610 cm^3 O_2
20. 75.2 dm^3 Br_2
21. 10 100 cm^3 O_2
48 300 cm^3 air
22. 2.50×10^2 cm^3 O_2
23. 3770 m^3 H_2

GLENCOE TECHNOLOGY

Software

Mastering Concepts in Chemistry
Unit 6, Lesson 3
Analytical Calculations Using Moles
Unit 7, Lesson 1
Chemical Equations
Unit 7, Lesson 2
Mass Stoichiometry
Unit 12, Lesson 2
Gas Stoichiometry

CD-ROM

Chemistry: Concepts and Applications
Exploration: *Balancing an Equation*
Interactive Experiment: *How much oxygen is available?*
Exploration: *Predicting Mass of Product*

and repeat the question. *10*
5. Add the third beaker and then the fourth beaker, each time asking how much product will be produced. *10 each time* Ask students which reactant was in excess *(stoppers)* and which one was the limiting reactant *(test tubes)* for this step. **Disposal:** F

Results: The students will observe that the model demonstrates the effects of limiting reactants.

Questions: 1. For a school bake sale, you volunteer to bake several cakes. You find there are three boxes of cake mix in the pantry and 10 eggs in the refrigerator. How many cakes will be baked if each mix requires two eggs? *three cakes*
2. Which of the items is the limiting reactant? *cake mixes*
3. How much of the excess reactant remains after the cakes are baked? *Four eggs remain.*

PROGRAM RESOURCES

Reteaching: Use the Reteaching worksheet, pp. 28-29, to provide students another opportunity to understand the concept of "Working with Gas Masses and Volumes."

CULTURAL DIVERSITY

Blowing Off Steam

Iceland, an island country in the North Atlantic, is 500 miles from its nearest European neighbor, Scotland. Although it has many glaciers—the largest glacier, Vatnajökull, is equal in area to all of the glaciers on the entire European continent—it is also one of the most active volcanic regions of the globe. Because of this subterranean thermal activity, Iceland has large numbers of hot springs and hot gas vents. These form a vast energy resource. Geothermal heat from the hot springs heats 90 percent of the island's capital Reykjavik and provides steam for industrial energy and commercial vegetable farming in greenhouses. The total power output of the largest hot gas vent region, Torfajökull, is estimated to equal 1 500 000 000 watts.

✓ ASSESSMENT

Knowledge: Ask students to construct an events chain concept map that details the method used to solve a mass-gas volume problem. If students need help, they can look at page 485.

Figure 19.9 Limiting factors are common in everyday life. For example, when the student runs out of either dough or candies, no more cookies of this type can be made.

unreacted. To take another example, suppose we drop nine moles of sodium into a vessel containing four moles of chlorine. If we warm the container slightly, the sodium will burn with a bright yellow flame, and crystals of sodium chloride will be formed.

$$2Na(cr) + Cl_2(g) \rightarrow 2NaCl(cr)$$

We know that one mole of Cl_2 will react completely with two moles of Na, so one mole of sodium will remain unreacted. For these two reactions we say that the hydrogen and sodium are in "excess" and that the oxygen and chlorine are "limiting reactants." In a chemical reaction, the **limiting reactant** is the one that is completely consumed in the reaction. It is not present in sufficient quantity to react with all of any other reactant. The reactants that are left are said to be in excess and are called **excess reactants.** The amount of product is therefore determined by the limiting reactant.

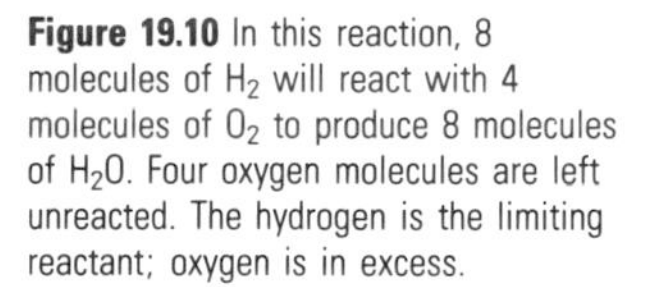

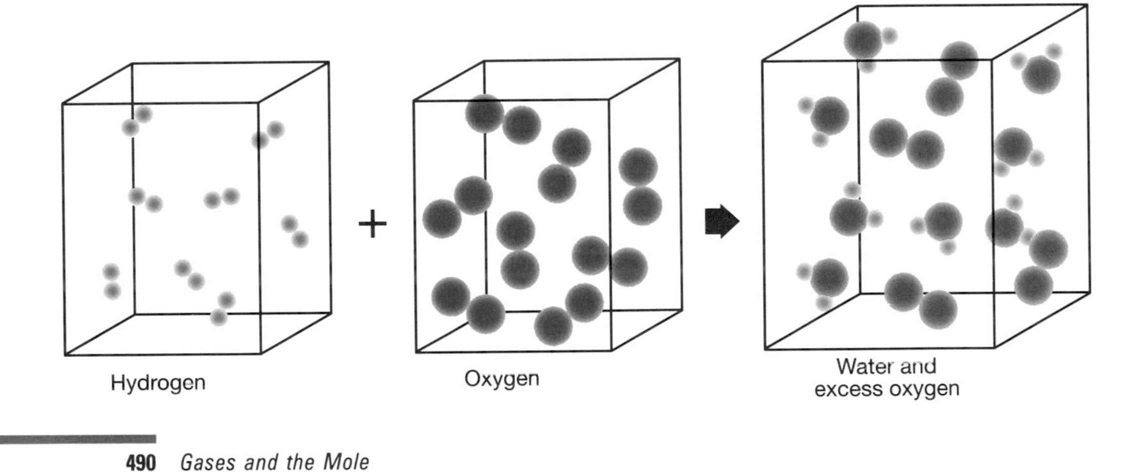

Figure 19.10 In this reaction, 8 molecules of H_2 will react with 4 molecules of O_2 to produce 8 molecules of H_2O. Four oxygen molecules are left unreacted. The hydrogen is the limiting reactant; oxygen is in excess.

Chemistry Journal

Limiting Reactant

Ask students to use models to simulate the reaction $CH_4 + 2O_2 \rightarrow CO_2 + 2H_2O$, starting with $2CH_4$ molecules and $3O_2$ molecules. Students should write up a description of their models with drawings and determine which is the limiting reactant.

*inter*NET CONNECTION

Illustrated demonstrations of heat engines are posted on the School of Physics, University of Sidney web page.

World Wide Web
http://www.physics.usyd.edu.au/teach/thermal/engine.html

SAMPLE PROBLEMS

1. Limiting Reactants

How many grams of CO_2 are formed if 10.0 g of carbon are burned in 20.0 dm^3 of oxygen? (Assume STP.)

Solving Process:

(a) Write a balanced equation.

$$C(amor) + O_2(g) \rightarrow CO_2(g)$$

(b) Change both quantities to moles.

$$\frac{10.0\ \cancel{g\ C}}{} \left| \frac{1\ mol\ C}{12.0\ \cancel{g\ C}} \right. = 0.833\ mol\ C$$

$$\frac{20.0\ \cancel{dm^3}\ O_2}{} \left| \frac{1\ mol}{22.4\ \cancel{dm^3}} \right. = 0.893\ mol\ O_2$$

(c) Compare the amount of CO_2 produced by complete reaction of each of the reactants. The equation indicates that

$$1\ mole\ C + 1\ mole\ O_2 \rightarrow 1\ mole\ CO_2$$

Therefore, 0.833 mol C would produce 0.833 mol CO_2, and 0.893 mol O_2 would produce 0.893 mol CO_2. Because carbon would produce fewer moles of carbon dioxide, the carbon limits the reaction. Some oxygen (0.060 mole) is left unreacted. We call carbon the limiting reactant.

(d) Complete the problem on the basis of the limiting reactant.

$$\frac{0.833\ \cancel{mol\ C}}{} \left| \frac{1\ \cancel{mol\ CO_2}}{1\ \cancel{mol\ C}} \right| \frac{44.0\ g\ CO_2}{1\ \cancel{mol\ CO_2}} = 36.7\ g\ CO_2$$

We conclude that 10.0 g of carbon will react with excess O_2 to form 36.7 g of CO_2.

Figure 19.11 A fire is a chemical reaction between oxygen and combustible substances in fuel. In a forest fire, a firebreak makes the fuel the limiting reactant. When a campfire is covered with dirt, oxygen becomes the limiting reactant.

Correcting Errors

Point out to students that these stoichiometry problems are different in that they have two quantities of information (numbers). In all of the previous stoichiometry problems there was only one number quantity. Some students prefer to work two complete problems, one for each given quantity, and then select the smaller answer as the correct one for the problem.

✓ ASSESSMENT

Skill: Ask students to answer Concept Mastering question 62 on page 496 and to give the reasoning behind their answers.

PRACTICE

Guided Practice: Guide students through the Sample Problems, then have them work in class on Practice Problems 24-27.

Independent Practice: Your homework or classroom assignment can include Review and Practice 50-53 from the end of the chapter.

Answers to

Practice Problems

Have students refer to Appendix C for complete solutions to Practice Problems.

24. 1.63 g Br_2

25. 0.747 dm^3 CO

26. 0.456 g S

27. 58.0 cm^3 Cl_2O

CHEMISTS AND THEIR WORK

James Prescott Joule

Background: Joule suffered ill health as a child, but because he came from a wealthy family, he was educated at home. He did have some brief tutoring from John Dalton. Due to his father's illness, he had to take part in the family business (a brewery) beginning at the age of 15. Nevertheless, he continued his investigation of heat.

Exploring Further: Have an interested volunteer investigate the Joule-Thomson effect and prepare a report for the class.

✓ ASSESSMENT

Portfolio: Encourage students to include in their portfolios examples of each of the following types of problems they have learned to solve in this chapter.

1. ideal gas equation
2. mass-gas volume
3. gas volume-mass
4. volume-volume
5. limiting reactants

CHEMISTS AND THEIR WORK

James Prescott Joule (1818-1889)

James Prescott Joule was an ardent measurer of natural phenomena, and most of his measurements had to do with heat. Even as a teenager, he published papers on the heat generated by electric motors. Eventually, he was able to produce specific relationships between an electric current in a circuit or mechanical work performed and the heat generated. His work was not immediately accepted because he was not in an academic position and was almost entirely self-educated. It was largely through the effort of William Thomson (Lord Kelvin) that Joule's work finally received the recognition it deserved. The unit of energy (and work) was named for him. His work led directly to the formulation of the law of conservation of energy. Together with Lord Kelvin, Joule formulated the Joule-Thomson effect. In addition, he made other scientific discoveries and eventually was to receive many honors from the scientific community.

SAMPLE PROBLEMS (continued)

2. Limiting Reactants—Another Example

How many grams of aluminum sulfide can form from the reaction of 9.00 g of aluminum with 8.00 g of sulfur?

Solving Process:

(a) $2Al(cr) + 3S(l) \rightarrow Al_2S_3(cr)$

(b) $\frac{9.00 \text{ g Al}}{} \Big| \frac{1 \text{ mol Al}}{27.0 \text{ g Al}} = 0.333 \text{ mol Al}$

$\frac{8.00 \text{ g S}}{} \Big| \frac{1 \text{ mol S}}{32.1 \text{ g S}} = 0.249 \text{ mol S}$

(c) 2 moles Al yield 1 mole Al_2S_3.

$\frac{0.333 \text{ mol Al}}{} \Big| \frac{1 \text{ mol } Al_2S_3}{2 \text{ mol Al}} = 0.167 \text{ mol } Al_2S_3$

3 moles S yield 1 mole Al_2S_3

$\frac{0.249 \text{ mol S}}{} \Big| \frac{1 \text{ mol } Al_2S_3}{3 \text{ mol S}} = 0.0830 \text{ mol } Al_2S_3$

(d) We know that 0.0830 mole of Al_2S_3 is less than 0.167 mole of Al_2S_3, so sulfur is the limiting reactant. Complete the problem on the basis of the limiting reactant.

$\frac{0.0830 \text{ mol } Al_2S_3}{} \Big| \frac{150.0 \text{ g } Al_2S_3}{1 \text{ mol } Al_2S_3} = 12.5 \text{ g } Al_2S_3$

Thus, using sulfur as the limiting reactant, 0.0830 times the formula mass of aluminum sulfide will give the mass of 12.5 g of aluminum sulfide produced from 8.00 g sulfur and an excess of aluminum. An alternate method for determining the limiting reactant is as follows.

3. Limiting Reactants—An Alternate Method

How many grams of aluminum sulfide can form from the reaction of 9.00 g of aluminum with 8.00 g of sulfur?

Solving Process:

(a) $2Al(cr) + 3S(l) \rightarrow Al_2S_3(cr)$

(b) According to the equation, 2 moles Al require 3 moles S. Thus, 9.00 g of Al, converted to moles, require

$\frac{9.00 \text{ g Al}}{} \Big| \frac{1 \text{ mol Al}}{27.0 \text{ g Al}} \Big| \frac{3 \text{ mol S}}{2 \text{ mol Al}} = 0.500 \text{ mol S}$

Since we have only

$\frac{8.00 \text{ g S}}{} \Big| \frac{1 \text{ mol S}}{32.1 \text{ g S}} = 0.249 \text{ mol S}$

sulfur is the limiting reactant. If we had more than 0.500 mole of sulfur, aluminum would have been the limiting reactant.

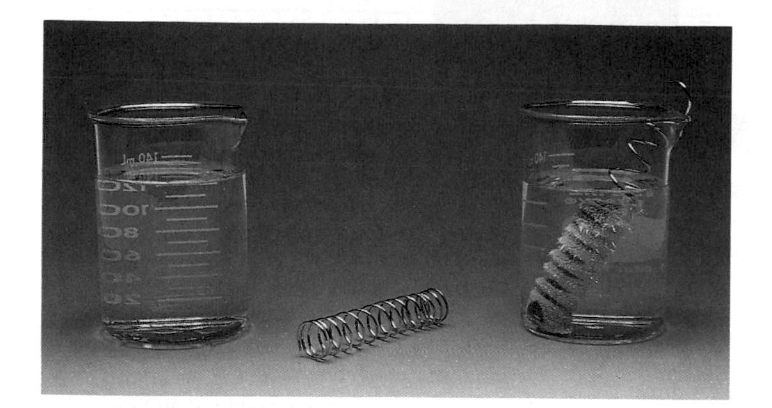

Figure 19.12 This copper wire has been placed in a silver nitrate solution. After the reaction has stopped, some copper wire remains. Silver nitrate was the limiting reactant.

PRACTICE PROBLEMS

24. $2NaBr(aq) + 2H_2SO_4(aq) + MnO_2(cr) \rightarrow Br_2(l) + MnSO_4(aq) + 2H_2O(l) + Na_2SO_4(aq)$
What mass of bromine could be produced from 2.10 g of NaBr and 9.42 g of H_2SO_4?

25. $2Ca_3(PO_4)_2(cr) + 6SiO_2(cr) + 10C(amor) \rightarrow P_4(cr) + 6CaSiO_3(cr) + 10CO(g)$
What volume, in cubic decimeters, of carbon monoxide gas at STP is produced from 4.14 g $Ca_3(PO_4)_2$ and 1.20 g SiO_2?

26. $H_2S(g) + I_2(aq) \rightarrow 2HI(aq) + S(cr)$
What mass of sulfur is produced by 4.11 g of I_2 and 317 cm^3 of H_2S at STP?

27. $2Cl_2(g) + HgO(cr) \rightarrow HgCl_2(cr) + Cl_2O(g)$
What volume, in cubic centimeters, of Cl_2O gas can be produced from 116 cm^3 of Cl_2 gas at STP and 7.62 g of solid HgO?

19.2 CONCEPT REVIEW

28. $2K_2MnO_4(aq) + Cl_2(g) \rightarrow 2KMnO_4(aq) + 2KCl(aq)$
What volume of chlorine gas at STP is required to completely react with 4.80 g of K_2MnO_4?

29. $Ca(cr) + 2HCl(aq) \rightarrow CaCl_2(aq) + H_2(g)$
What mass of calcium is required to produce 8.32 dm^3 of hydrogen by this reaction?

30. $H_2(g) + Cl_2(g) \rightarrow 2HCl(g)$
What volume of hydrogen chloride can be produced from 1.60 dm^3 of hydrogen?

31. $Cr(OH)_3(cr) + NaOH(aq) \rightarrow NaCrO_2(aq) + 2H_2O(l)$
What mass of $NaCrO_2$ can be obtained from the reaction of 7.40 g $Cr(OH)_3$ with 7.60 g NaOH?

32. Apply Why do we not have to convert to moles in the solution of a volume-volume problem?

3 ASSESS

CHECK FOR UNDERSTANDING

Have students answer the Concept Review problems and Review and Practice 45-57 from the end of the Chapter.

RETEACHING

Use **Solving Problems in Chemistry** Chapter 19, sections 3-5 to reteach this section. Problems can be assigned to provide review and reinforcement of the needed problem solving skills.

EXTENSION

Remind students that when gas volumes are not at standard conditions they cannot be compared to the molar volume of 22.4 dm^3. If the nonstandard volume is given, correct to STP before doing the stoichiometry problem. If the volume is the calculated answer, calculate it, then correct to nonstandard conditions as in Chapter 18. Have students solve Review and Practice 57-60.

Answers to
Concept Review
For complete solutions, refer to *Merrill Chemistry Problems and Solutions Manual.*

28. 0.273 dm^3 Cl_2
29. 14.9 g Ca
30. 3.20 dm^3 HCl
31. 7.68 g $NaCrO_2$
32. Equal volumes of gases at the same conditions of temperature and pressure contain equal numbers of molecules.

4 CLOSE

Activity: Give the students an ungraded pre-test that has one of each type of problem on it, mass-gas volume, gas volume-mass, volume-volume, and limiting reactant. Have students work with their lab partners to solve the problems. Allow the first teams that finish to go to the chalkboard and explain their problem-solving strategies to the other students.
Cooperative Learning: Refer to pp. 28T-29T to select a teaching strategy to use with this activity.

PROGRAM RESOURCES

Vocabulary Review/Concept Review: Use the worksheets, p. 19 and pp. 73-74, to check students' understanding of the key terms and concepts of Chapter 19.

20.1 Solutions

PREPARE

Key Concepts

The major concepts presented in this section include dissociation and solvation, various solute-solvent combinations, solution equilibrium, precipitation reactions, enthalpy of solution, concentration units such as molarity, molality, mole fraction, and mass percent, and dilution calculations.

Key Terms

solvent
solute
dissociation
solvation
hydration
miscibility
solution equilibrium
saturated solution
unsaturated solution
solubility
supersaturated solution
enthalpy of solution
molality
mole fraction

1 FOCUS

Objectives

Preview with students the objectives listed on the student page. Students can use them as a study guide for this major section.

Demonstration

Partially fill a petri dish with water and place it onto an overhead projector. Place a crystal of $KMnO_4$ into the dish. Have the students note the purple streamers as the water is very gently stirred. Demonstrate the effect of temperature on rate of solution by using two petri dishes side by side on the overhead. Place ice water in one, and hot water in the second dish. Put a crystal of $KMnO_4$ in each dish. Introduce students to the equilibrium nature of dissolving by having them note how the color density changes as the distance from the crystal increases. **Disposal:** C

OBJECTIVES

- describe and explain the processes of solvation, dissociation, and dissolving.
- discuss factors affecting the solubility of one substance in another.
- relate enthalpy of solution to endothermic and exothermic dissolving processes.
- differentiate among and solve problems involving molarity, molality, mole fraction, and mass percent.

20.1 Solutions

We will begin our study of solutions by looking at the physical process of dissolving. One important aspect of solutions is the relative amounts of the substances present in the solution. These amounts can be described by the concept of concentration.

The Dissolving Process

What do we mean by solution? In Chapter 3, we referred to homogeneous matter as being the same throughout. It is often made of only one substance—a compound or an element—but may also be a mixture of several substances. Such a homogeneous mixture is called a solution. Most solutions consist of a solid dissolved in a liquid. The particles in a true solution are molecules, atoms, or ions that will pass easily through the pores of filter paper. Solutions cannot be separated into their components by filtration.

The substance that occurs to the greater extent in a solution is said to do the dissolving and is called the **solvent.** The less abundant substance is said to be dissolved and is called the **solute.** The most common solvent is water. Using water as a typical solvent, let us look at the mechanism of solvent action. Water molecules are very polar. Because they are polar, they are attracted to other polar molecules and to ions. Table salt, NaCl, is an ionic compound made of sodium and chloride ions. If a salt crystal is put into water, the polar water molecules are attracted to ions on the crystal surfaces. The water molecules gradually surround and isolate the surface ions. The ions become hydrated, Figure 20.1. The attraction between the hydrated sodium and chloride ions and the remaining crystal ions becomes so small that the hydrated ions are no longer held by the crystal. They

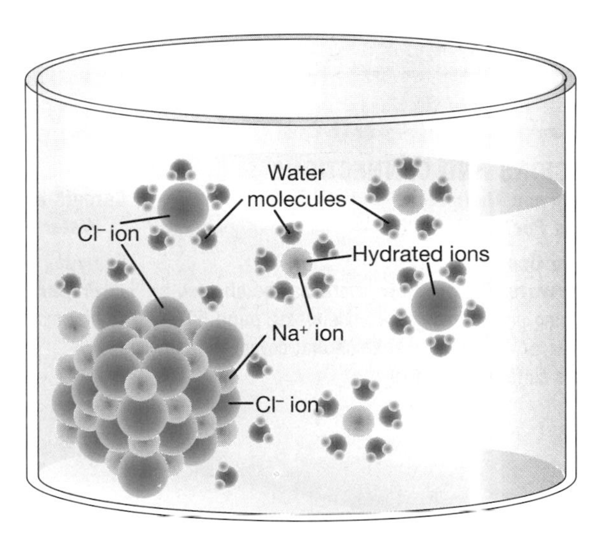

Figure 20.1 Table salt dissolves in water because polar water molecules gradually surround and isolate sodium and chloride ions.

DEMONSTRATION

20-2 Likes Dissolve Likes

Purpose: to demonstrate that nonpolar solutes are better dissolved by nonpolar solvents

Materials: 1 g iodine crystals, 100-cm³ graduated cylinder, 30 cm³ water, 30 cm³ dichloromethane (methylene chloride) or TTE (trichlorotrifluoroethane), 30 cm³ hexane, deflagrating spoon

Procedure: 1. Place the dichloromethane or TTE in the graduated cylinder.
2. Slowly pour 30 cm³ of water down the side to avoid mixing.
3. Slowly add 30 cm³ of hexane to form the third distinct layer.
4. Place several large I_2 crystals in a deflagrating spoon and slowly lower the spoon into the cylinder. **CAUTION:** *Avoid contact with the solvents. Do not breathe the vapors. Use a fume hood for the demonstration.*

gradually move away from the crystal into solution. This separation of ions from each other is called **dissociation.** The surrounding of solute particles by solvent particles is called **solvation.**

Whenever you are considering a solution of an ionic compound, it is important to keep the following in mind. When the ions are dissociated, each ion species in the solution acts as though it were present alone. Thus, a solution of sodium chloride acts as a solution of sodium ions and chloride ions. There is no characteristic behavior of "sodium chloride" in solution because there really is no sodium chloride in solution. There is simply a solution containing both sodium ions and chloride ions uniformly mixed with water molecules.

Solvent-Solute Combinations

Four simple solution situations can be considered. They are listed in Table 20.1. Not all possible combinations of substances will fit into these four rigid categories. We will include ionic substances with polar substances in our discussion.

Table 20.1

Solvent-Solute Combinations

Solvent Type	Solute Type	Is Solution Likely?
Polar	Polar	yes
Polar	Nonpolar	no
Nonpolar	Polar	no
Nonpolar	Nonpolar	yes

(1) Polar solvent-polar solute.
The mechanism of solution involving a polar solvent and a polar solute is the one we have already described for salt and water. The polar solvent particles solvate the polar solute particles. If the solvent is water, this process is called **hydration.** The solvent particles attach themselves due to the polar attraction. The intracrystalline forces are reduced so much that the surface particles are carried away by the solvent particles.

(2) Polar solvent-nonpolar solute.
Because the solvent particles are polar, they are attracted to each other. However, the solute particles in this case are nonpolar and have little attraction for particles of the solvent. Thus, solvation does not occur, and solution to any extent is unlikely, as we see if we try to dissolve wax in water.

(3) Nonpolar solvent-polar solute.
Reasoning similar to that of the second case applies here. The solvent particles are nonpolar and thus have little attraction for the solute particles. In addition, the solute particles in this case are polar and are attracted to each other. Again, because there is no solvation, solution to any extent is unlikely as we see if we try to dissolve salt in hexane.

CHEMISTS AND THEIR WORK

Agnes Pockels (1862-1935)
Agnes Pockels was the daughter of an Austrian army officer. Although she attended high school for some time, she was largely self-taught. In her late teens, she made her first observations on the changes in surface tension of water contaminated by oil. Her first paper was printed nearly ten years later. Most of her experimentation was done at home with homemade apparatus. Among other makeshift devices, she developed the Pockels' trough, which was used to measure surface tension.

At 69, Pockels won the Laura-Leonard prize for her work on the properties of surface layers and films. An extension of her work in surface chemistry led to the Nobel Prize for Irving Langmuir in 1932.

2 TEACH

CONCEPT DEVELOPMENT

Misconceptions

Make sure that students do not confuse the terms *solvolysis* and *hydrolysis,* which imply chemical reactions, with the terms *solvation* and *hydration,* which imply no chemical change.

CHEMISTS AND THEIR WORK
Agnes Pockels

Background: Agnes Pockels, using makeshift equipment, observed phenomenon that scientists in laboratories overlooked.

Exploring Further: Use forceps to lower a needle onto the surface of a glass of water. With the needle floating on the water's surface, add a drop of oil. Have students explain what happened to the surface tension.

MAKING CONNECTIONS

Medical Research

Chemists have been involved in sickle-cell anemia research since the late 1940s. In 1949, Dr. Linus Pauling, a two time Nobel prize winner, reported that normal and abnormal hemoglobin are large molecules with molecular masses of about 64 000 u. They differ in structure in one place, making the abnormal hemoglobin nonpolar. The abnormal hemoglobin is not as soluble in water, hence it produces a fibrous precipitate. This hemoglobin cannot carry oxygen and the sickle-shaped cells clog capillaries.

5. Begin at the top and slowly move the spoon through each layer. Stir each layer with the spoon, allowing time for some of the I_2 to dissolve.
6. Slowly remove the deflagrating spoon. Then allow the solvents to evaporate in an operating fume hood. Collect and reuse the I_2. **Disposal:** B

Results: The nonpolar iodine dissolves in the top and bottom layer, but not in the water layer.

Questions: 1. Which type of solvent, polar or nonpolar, dissolves more iodine? *nonpolar*
2. What is meant by the phrase, "likes dissolve likes"? *Nonpolar solutes are more soluble in nonpolar solvents and polar solutes are more soluble in polar solvents.*

Extension: Explain that the person responsible for pre-treating spots at the dry cleaners needs to know the nature of the spot. Different solvents are used to remove different stains.

Key Terms
crystalloid
colloid
suspension
colloid chemistry
Tyndall effect
Brownian motion
adsorption
electrophoresis

Looking Ahead
Obtain a can of aerosol shaving cream for the Quick Demo: *Aerosols.*

1 FOCUS

Objectives
Preview with your students the objectives listed on the student page. Students can use them as a study guide for this major section.

Discussion
Have on display examples of different colloids, such as a marshmallow, can of shaving foam, mayonnaise, jar of jelly, can of paint, and cheese. Have students discuss what these display items have in common. Inform them that they are all colloids. Students will realize that they use colloids daily.

2 TEACH

CONCEPT DEVELOPMENT
Enrichment: An ultracentrifuge can be used to separate colloids. Have a student prepare a report for the class on the use of this apparatus in characterizing colloids. Possible response: The ultracentrifuge was developed by the Swedish scientist, Svedberg. It can, through centrifugal force at 50 000 rpm, produce an effective gravitational field about 240 000 times that of Earth's. Because dissolved particles of large molecular mass settle out quickly, mass values can be determined by this sedimentation rate.

Using the Table: Point out the particle size of solution, suspensions and colloids, using Table 20.4. Students are usually surprised that colloids, like solutions, are a permanent mixture, unlike a suspension which will separate with time.

foam; marshmallows are solid foam. Emulsions are colloids in which liquids are dispersed in other liquids or in solids. Mayonnaise is a liquid emulsion; cheese is a solid emulsion. Sols are dispersions of solids in liquids or other solids. Jellies and paints are sols; pearls and opals are solid sols.

Colloidal Sizes

If a finely ground substance is placed in water, one of three things will happen. First, it may form a true solution that is simply a dispersion of atoms, molecules, or ions of the substance into a solvent. Particles in a solution do not exceed ~ 1 nm in size.

Second, the particles may remain larger than ~ 100 nm. These particles are large enough to be seen with a microscope and gradually fall to the bottom of the container. Because the particles are temporarily suspended and settle out upon standing, this mixture is called a **suspension.**

RULE OF THUMB

Particles from ~ 1 to ~ 100 nm in size usually remain dispersed throughout the medium. Such a mixture is called a colloid. Colloids represent a transition between solutions and suspensions. However, they are considered heterogeneous, with the medium as the continuous phase and the dispersed substance as a separate phase.

Actually, substances show unusual properties even when only one of the three dimensions of the particles is less than 100 nm. **Colloid chemistry** is the study of the properties of matter whose particles are colloidal in size in at least one dimension.

Table 20.4

Comparison of Solutions, Suspensions, and Colloids

Type	Particle size	Permanence
Solution	<1 nm	permanent
Colloid	>1 nm but < 100 nm	permanent
Suspension	>100 nm	settle out

Figure 20.10 A laboratory centrifuge (right) can be used to separate some colloids into components. Milk (left) is a colloid that can be separated into cream and skim milk.

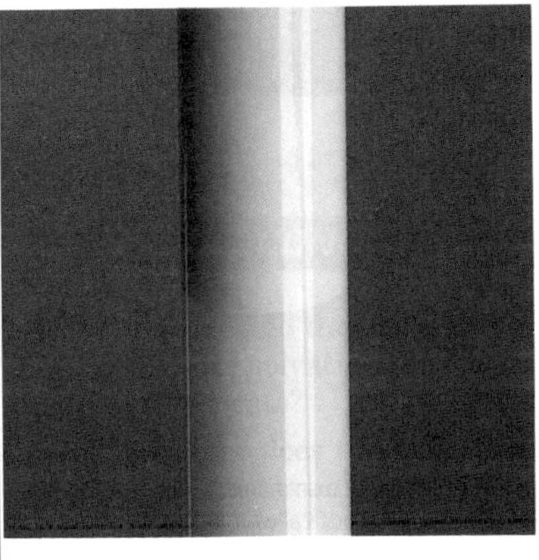

CONTENT BACKGROUND
History of Colloid Chemistry: Colloids were known by medical alchemists. They called colloidal gold "purple of Cassius" and "aurum potabile." The difference between solutions and colloids was not understood until 1844 when Francesco Selmi, an Italian, referred to colloids as "pseudosolutions." When Graham developed the use of membranes to separate crystalloids and colloids in 1861, he first used the term "dialysis" to represent such separation. In 1892 Lindner and Picton demonstrated that colloids carry an electrical charge and move toward electrodes when placed in an electrical field. Electrophoresis became a valuable technique in the separation and purification of proteins. Einstein was interested in Brownian motion and developed the kinetic theory of molecular bombardment.

Figure 20.11 A light beam passes through a solution, but a colloid scatters the light (left). The Tyndall effect can also be seen in air (right) when suspended dust and water droplets scatter light.

Properties of Colloids

If a beam of light is allowed to pass through a true solution, some of the light will be absorbed, and some will be transmitted. The particles in solution are not large enough to scatter the light. However, if light is passed through a colloid, the light is scattered by the larger, colloidal particles. The beam becomes visible from the side. This effect is called the **Tyndall effect.** You may have seen this effect in the beam of a searchlight in the night air. In this situation, the light is scattered by suspended water droplets in air. You may also have observed this effect as the sunbeam coming through a hole in the blinds. Here, suspended dust particles in the air scatter the sunlight.

As you know, colloidal particles are too small to be seen with an ordinary microscope. In 1912, Richard Zsigmondy, a German professor of chemistry, designed the ultramicroscope. Through ultramicroscopy, in which the Tyndall effect is used, it is possible to "see" colloidal particles.

If you observe colloids under an ultramicroscope, you will notice that they have another interesting property. The particles of a colloid are in continuous random motion. This motion is called **Brownian motion.** The constant bombardment of the colloid particles by the smaller molecules of the medium is the source of the motion. The motion is the result of the collision of

Table 20.5

Properties of Solutions, Colloids, and Suspensions		
Solutions	**Colloids**	**Suspensions**
Do not settle out	Do not settle out	Settle out on standing
Pass unchanged through ordinary filter paper	Pass unchanged through ordinary filter paper	Separated by ordinary filter paper
Pass unchanged through membrane	Separated by a membrane	Separated by a membrane
Do not scatter light	Scatter light	Scatter light
Affect colligative properties	Do not affect colligative properties	Do not affect colligative properties

MAKING CONNECTIONS

Clean Air

Ask a student to research what methods are now being used industrially to remove colloidal dust from stack gases and reduce air pollution. Possible response: *(1) Cyclone separators: Stack gas is forced into a spiral path; dust particles strike the sides of the stack where they settle. (2) Cottrell precipitators: Charged, colloidal dust particles are electrostatically precipitated. (3) Scrubbing towers: Gas is bubbled through water, or other wash liquid, which strips out dust as well as noxious gases.*

CONCEPT DEVELOPMENT

Using the Table: Have students consider ways that can be used to separate the mixtures listed in Table 20.5 into their components.

QUICK DEMO

Brownian Motion

Hit two chalkboard erasers together near a window and have students observe and record the motion of the particles. Ask them to record what they have observed. Have several students read their observations to the class.

QUICK DEMO

Aerosols

To emphasize that colloids are a part of our daily life, bring an aerosol can of shaving soap to class. Have a student read the definition of aerosol from a dictionary. Most definitions will state that an aerosol is colloidal particles dispersed in a gas. Release some of the shaving foam onto a watch glass. Allow the class to observe the shaving cream for a minute and relate their observations to the definition. Discuss the difference in particle size of a solution, colloid, and suspension. **Disposal:** C

✓ ASSESSMENT

Portfolio: Tables that summarize information are valuable learning tools. Provide students with a copy of Table 20.5 that has some material missing. Ask students to write in the missing information. The completed table should be considered for inclusion in their portfolios.

EVERYDAY CHEMISTRY

Preparing Preserves with Pectin

Background: Ripe fruits contain the highest concentrations of pectin. Unripe fruit contains protopectin, which does not gel, but which will change to pectin as the fruit ripens. Overripe fruit contains pectic acid which does not gel.

Answers to
Exploring Further

1. Jams are made using the entire fruit while jellies are made from the fruit juice. **2.** Add both pectin and lemon juice. **3.** Water is polar and is attracted to the bridging oxygen between the fructose and glucose molecules.

3 ASSESS

CHECK FOR UNDERSTANDING

Have students answer the Concept Review problems, and Review and Practice 54-61 from the end of the chapter.

RETEACHING

Ask students how gelatin can be used as a protective colloid and an emulsifier. Possible response: *Gelatin forms the most useful protective colloids. It is used with silver bromide emulsions in photography; to prevent ice crystals in ice cream; and as an emulsifying agent in many mixtures.*

EXTENSION

Have students determine why the sky is blue. Possible response: *Light from the sun is scattered by the air molecules in our atmosphere. The sky is blue because of shorter wavelength blue waves. Shorter blue waves are scattered more than longer wavelength red light.*

many molecules with the particle. It is as if a large crust of bread in a pond is moved back and forth as first one fish and then another hits the bread from opposite sides. Brownian motion is named in honor of Robert Brown, a biologist. Brown first noticed it while observing the motion of particles in a suspension of pollen grains in water.

Word Origins

ad: (L) to
sorbere: (L) to soak up
adsorption—the soaking up of substances

As a result of electric forces, solid and liquid surfaces tend to attract and hold substances with which they come into contact. This phenomenon is called **adsorption.** The stationary phase in chromatography is an adsorbing material. Colloidal particles, because of their small size, have an extremely large ratio of surface to mass. Consider a cube that measures 1 cm on an edge. It has a volume of 1 cm^3 and a surface area of 6 cm^2. If that cube is cut into 1000 smaller cubes of equal size, the surface area of that same amount of matter becomes 60 cm^2. If we subdivide these cubes even further, until they measure 10 nm on edge, the surface area increases to 6 000 000 cm^2. Such a large surface area makes colloidal particles excellent adsorbing materials, or adsorbents. Dispersed particles have the property of adsorbing charge on the surface.

elektron: (GK) electricity
phoreus: (GK) carrier
electrophoresis—carrying by electric charges

If a colloid is subjected to an electric field, a migration of the particles can be observed. The positive particles are attracted to the cathode. The negative particles are attracted to the anode. This migration, called **electrophoresis,** is evidence that colloidal particles are charged. The separation of amino acids and peptides obtained in protein analysis is accomplished rapidly using electrophoresis. The process is also common in nucleic acid research.

Semipermeable membranes can be used to separate ions and colloidal particles. The ions pass through the membrane but the colloidal particles, because they are larger, do not. This type of separation is utilized in kidney dialysis.

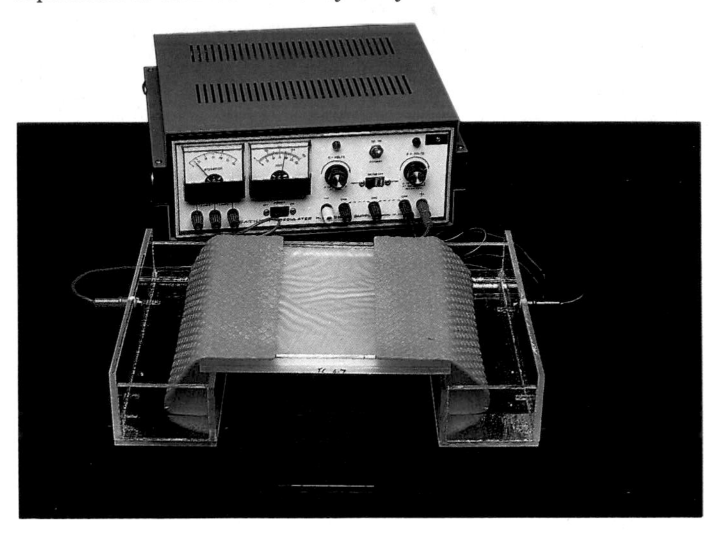

Figure 20.12 An electrophoresis apparatus is used in separating mixtures containing charged particles. Negatively-charged particles move toward the anode. Positively-charged particles move toward the cathode.

DEMONSTRATION

20-6 Tyndall Effect

Purpose: to demonstrate the Tyndall effect

Materials: 1 g sodium thiosulfate, 120 cm^3 distilled water, 150-cm^3 beaker, overhead projector, opaque paper, 1 cm^3 12*M* HCl, medicine dropper

Procedure: 1. Dissolve the sodium thiosulfate in the distilled water in the beaker. Cover the top of an overhead projector with opaque paper.
2. Cut a hole in the paper slightly smaller than the beaker. Turn on the projector and turn off the room lights.
3. Place the beaker containing the $Na_2S_2O_3$ over the hole in the paper, and focus the image on the screen.
4. Add the 12*M* HCl drop by drop to the beaker, stirring occasionally.

Disposal: C.

Results: The presence of colloidal sulfur tends to scatter light of shorter wavelengths (blue). When viewed from the side, the beaker appears blue as the light is scattered (Tyndall effect), whereas the transmitted light on the screen becomes dimmer and red, because the longer wavelength red light does not scatter as much. Have students note that the color resembles that of the setting sun. Students will notice that the liquid becomes whiter as the colloidal sulfur

EVERYDAY CHEMISTRY

Preparing Preserves with Pectin

If you enjoy preserves such as jams and jellies, you can thank a long molecule named pectin for your pleasure. Pectin is found naturally in grapes, berries, green apples, and citrus fruits. It is a long, chainlike molecule made up of smaller structures similar to those of sugars. Pectin is what causes preserves to gel.

To make preserves from fruits that contain natural pectin, the fruits are first boiled to release the pectin from the walls of the cells. Pectin is added to fruits that do not have it naturally. Within the boiling juice, the negatively-charged pectin molecules repel each other. The charge on the molecules also causes them to be surrounded by water molecules. Fruits like grapes or green apples are acidic and neutralize the charge on the pectin molecule. Neutralization of fruits that are not naturally acidic is achieved by the addition of lemon juice. Sugar is added and the solution is boiled until the sugar concentration reaches 60-65%. The concentrated sugar solution attracts water molecules from the pectin. The pectin molecules then bond to each other, trapping the other ingredients to form a gel similar to gelatin.

The amount of pectin added to preserves is important. Too much pectin causes preserves to be rubbery; too little pectin makes them runny. Similarly, if the fruit lacks sufficient acid or too little sugar is added, no gel will form. The result will be syrup.

Making preserves used to be an annual ritual in many homes, and allowed fruits to be enjoyed throughout the year. Getting the right balance of the critical ingredients was an art. The chemistry of the process may not have been understood, but the products were delicious!

Exploring Further

1. Find out the difference between jams and jellies.
2. Peaches and apricots are not acidic and do not contain pectin. How can preserves be made from them?
3. Concentrated sugar solutions are hygroscopic. What is the mechanism by which sugar attracts water?

20.2 CONCEPT REVIEW

20. Distinguish between the dispersed phase and the continuous phase in a colloid.
21. How do colloids differ from solutions and suspensions?
22. How is the Tyndall effect dependent upon particle size?
23. **Apply** List the products found in your kitchen that you can classify as aerosols, foams, emulsions, gels, or sols.

continues to form and more of the light is reflected. The chemical reactions are:

$Na_2S_2O_3(aq) + 2HCl(aq) \rightarrow H_2S_2O_3(aq) + 2NaCl(aq)$

$H_2S_2O_3(aq) \rightarrow H_2O(l) + SO_2(g) + S(cr)$

Questions: 1. Describe the solution before any acid was added. *clear and colorless*

2. As colloidal sulfur forms in the solution, what do you observe? *See results.*

3. What name is used to describe the scattering of light by colloidal particles? *Tyndall effect*

Extension: The property of a colloid to scatter light can be observed at night. Students have seen the sweeping search lights used to attract people to grand openings. They have observed the Tyndall effect. The light beam is visible because the colloidal-sized particles of moisture and pollutants scatter the light.

Answers to
Concept Review

20. The dispersed phase is composed of very small particles that are distributed throughout the continuous phase. The dispersed phase is "solutelike," while the continuous phase is "solventlike."

21. Particles in colloids are larger than those in solutions and smaller than those in suspensions. Colloid particles range between 1 and 100 nm in at least one dimension. The result is that colloid properties are intermediate compared with the properties of solutions and suspensions.

22. The Tyndall effect is the scattering of light by colloidal particles. Colloidal particles are large enough to scatter light, but particles in solution are not.

23. Answers may vary. Possible answers include: aerosol, such as air freshener spray; foam, such as marshmallows, whipped cream; emulsion, such as mayonnaise, cheese; gel, such as pearl; and sol, such as paint.

4 CLOSE

Discussion: Ask students why the tail lights on an automobile are red. Students will probably reply that red means danger. Point out that red light has a longer wavelength and is scattered less than the other colors. Therefore, red light can be transmitted farther through fog, rain, snow, or colloidal pollutants.

PROGRAM RESOURCES

Vocabulary Review/Concept Review: Use the worksheets, p. 20 and pp. 75-76, to check students' understanding of the key terms and concepts of Chapter 20.

CHAPTER REVIEW 20

- Review Summary statements and Key Terms with your students.
- Complete solutions to Chapter Review Problems can be found in **Merrill Chemistry Problems and Solutions Manual**.

Answers to
Review and Practice

24. The less abundant substance in a solution is the solute. The more abundant substance is the solvent.
25. Water is highly polar.
26. The polar water molecules are attracted to ions on the crystal surfaces. The water molecules surround, isolate, and hydrate the surface ions that now gradually move away from the crystal and into solution. Entropy is increased.
27. Polar-polar attractions help to separate the solute particles and promote the formation of a solution. Polar-nonpolar interactions involve small attractive forces, which are usually not strong enough to form a solution.

Chapter 20 REVIEW

Summary

20.1 Solutions

1. A true solution is a homogeneous mixture of molecules, atoms, or ions that cannot be separated by filtration.
2. When ionic compounds are dissolved in water, the ions become hydrated and dissociate.
3. Polar solvents tend to dissolve polar solutes; nonpolar solvents tend to dissolve nonpolar solutes.
4. Two liquids that are mutually soluble in all proportions are completely miscible.
5. A solution has reached solution equilibrium when the rates of particles leaving and returning to the solution are equal. When solution equilibrium is reached, the solution is said to be saturated.
6. The rate of solution is increased by increasing the surface area of crystal exposed, stirring to prevent the buildup of saturated solution around solute crystals, and increasing the kinetic energy of solute and solvent.
7. Enthalpy of solution is the energy change that occurs when one substance is dissolved in another.
8. Henry's law: The mass of a gas that will dissolve in a liquid at a given temperature varies directly with the partial pressure of that gas.
9. Molarity, molality, mole fraction, and mass percent are common concentration units.

20.2 Colloids

10. A colloid is composed of two phases, the dispersed phase and the continuous phase.
11. In terms of particle size, colloids are intermediate between solutions and suspensions. Colloidal particles range between ~ 1 and ~ 100 nm in at least one dimension.
12. Colloids can scatter light (Tyndall effect), undergo constant random motion (Brownian motion) as a result of bombardments by the particles of the suspending medium, and act as excellent adsorbing materials.

Key Terms

solvent
solute
dissociation
solvation
hydration
miscibility
solution equilibrium
saturated solution
unsaturated solution
solubility
supersaturated solution
enthalpy of solution
molality
mole fraction
crystalloid
colloid
suspension
colloid chemistry
Tyndall effect
Brownian motion
adsorption
electrophoresis

Review and Practice

24. Define solute and solvent.
25. What is the most common solvent? Why is it the most common?
26. Describe the process by which water dissolves an ionic compound.
27. Why is a solution likely to form when a polar solute and a polar solvent are combined? Why is a solution unlikely to form when a polar solute and a nonpolar solvent are combined or when a nonpolar solute and a polar solvent are combined?
28. What factors determine the solubility of a nonpolar solid in a nonpolar liquid?
29. "Oil and water do not mix" is an old adage. What term describes two liquids that are not mutually soluble?
30. What is a saturated solution? Describe the equilibrium process that takes place in a saturated solution.
31. How is solubility specified?

28. Melting point and enthalpy of fusion of the solid are indicators of how strong the attractive forces are within the solid.
29. Immiscible
30. A saturated solution contains an undissolved substance in equilibrium with the dissolved substance. The solute is dissolving at the same rate it is crystallizing.
31. g solute/100 g solvent
32. **a.** Cl^- will precipitate Ag^+ but not Na^+
b. OH^- will precipitate Fe^{3+} but not Na^+
c. SO_4^{2-} will precipitate Ba^{2+} but not K^+
d. OH^- will precipitate Cu^{2+} but not K^+
33. **a.** $FeCO_3(cr) + Na_2SO_4(aq)$
b. $BaCrO_4(cr) + KNO_3(aq)$
c. $CaC_2O_4(cr) + NaCl(aq)$
d. $Zn_3(PO_4)_2(cr) + KCH_3COO(aq)$

Chapter 20
REVIEW

32. Using Appendix Table A-7, select a reagent that will precipitate only one of the positive ions in each of the following mixtures.
a. $AgNO_3$ and $NaNO_3$
b. $FeCl_3$ and NaCl
c. $BaCl_2$ and KNO_3
d. KCl and $CuSO_4$

33. Predict the results of the following reactions.
a. Na_2CO_3 (aq) + $FeSO_4$ (aq) →
b. K_2CrO_4 (aq) + $Ba(NO_3)_2$ (aq) →
c. $Na_2C_2O_4$ (aq) + $CaCl_2$ (aq) →
d. K_3PO_4 (aq) + $Zn(CH_3COO)_2$ (aq) →

34. Discuss three actions that can increase the rate of solution of a solute in a solvent. Explain how each of these actions works to increase rate of solution.

35. Explain why ΔH_{sol} is negative for some solids, such as sodium hydroxide, dissolved in water.

36. Relate temperature conditions to the solubilities of solids and gases in liquids.

37. State Henry's law. Explain how Henry's law accounts for what happens when a bottle of soda is opened.

38. What effect would an increase in pressure have on the solubility of a solid in a liquid? What effect would it have on the solubility of a gas in a liquid?

39. Describe the content of a 1*M* solution of a molecular substance such as glucose, $C_6H_{12}O_6$.

40. Calculate the molarity of the following for the given volume of solution.
a. 31.1 g $Al_2(SO_4)_3$ in 1.00 × 10^3 cm^3
b. 48.4 g $CaCl_2$ in 1.00 × 10^2 cm^3
c. 313.5 g $LiClO_3$ in 2.50 × 10^2 cm^3

41. Determine the concentration of each type of ion in 0.100*M* solutions of each of the following compounds.
a. KBr **c.** $CuCl_2$
b. Rb_2S **d.** $Mg(NO_3)_2$

42. Calculate the concentration of Ca^{2+} ions in a solution of 21.0 g of $CaCl_2$ in 5.00 × 10^2 cm^3 of solution. Determine the concentration of Cl^- ions in the same solution.

43. How many moles of CH_3COO^- ions are in 46.0 cm^3 of a 0.250*M* solution of $Pb(CH_3COO)_2$?

44. Calculate the molality of the following solutions.
a. 20.0 g of NH_4Cl in 4.00 × 10^2 g of water
b. 145 g of CH_3COCH_3 in 0.320 kg of water

45. Calculate the molality of the following solutions.
a. 98.0 g RbBr in 824 g water
b. 85.2 g $SnBr_2$ in 1.40 × 10^2 g water
c. 10.0 g $AgClO_3$ in 201 g water
d. 0.059 g KF in 0.272 g water

46. Calculate the mole fraction for each component in the following solutions.
a. 67.4 g of C_9H_7N in 2.00 × 10^2 g C_2H_6O
b. 5.48 g of $C_5H_{10}O_5$ and 3.15 g of CH_6N_4O in 21.2 g H_2O

47. The mole fraction of benzene in a solution with cyclohexane is 0.125. What does this statement tell you about the solution? How much benzene would have to be combined with 2.00 mol of cyclohexane to give this mole fraction?

48. Find the mass percent concentration of 12.0 g of glucose, $C_6H_{12}O_6$, in 2.50 × 10^2 g of water.

49. Compute the masses of solute needed to make the following solutions.
a. 1.00 dm^3 of 0.780*M* $Sc(NO_3)_3$
b. 2.00 dm^3 of 0.179*M* $Er_2(SO_4)_3$
c. 1.00 × 10^2 cm^3 of 0.626*M* VBr_3
d. 2.50 × 10^2 cm^3 of 0.0965*M* $DyCl_3$

50. Consider the mixture of sodium nitrate and barium chloride mentioned on page 504. The mass of the mixture was measured, then the mixture was dissolved in water. Excess sulfuric acid was added. The precipitate was dried and its mass was measured.

PROGRAM RESOURCES

Evaluation: Use the Evaluation worksheets, pp. 77-80, to assess students' knowledge of Chapter 20.

Computer Test Bank: Use the IBM or Apple Test Bank and Question Selection Manual to prepare your own individualized test for Chapter 20.

34. Solubility is increased by crushing the solute (increases the surface area in contact with solvent), heating (increases movement of solvated particles away from the surface of the solute), and stirring (moves saturated solvent away from the surface of the solute).
35. The degree of hydration is so high that more order exists in the solution than exists in the separate solute and solvent.
36. For most solids, solubility increases with increasing temperature. For all gases, solubility decreases with increasing temperature.
37. The mass of a gas that will dissolve in a specific amount of liquid varies directly with the pressure. As the bottle is opened, pressure decreases, so solubility of the gas also decreases. CO_2 leaves the solution.
38. Pressure has negligible effect on the solubility of a solid in a liquid, but a direct effect on solubility of a gas in a liquid.
39. A 1*M* solution contains one mole of the molecular substance in 1 dm^3 of solution.
40. a. 0.0909*M* $Al_2(SO_4)_3$
b. 4.36*M* $CaCl_2$
c. 13.9*M* $LiClO_3$
41. a. 0.100*M* K^+; 0.100*M* Br^-
b. 0.200*M* Rb^+; 0.100*M* S^{2-}
c. 0.100*M* Cu^{2+}; 0.200*M* Cl^-
d. 0.100*M* Mg^{2+}; 0.200*M* NO_3^-
42. 0.378*M* Ca^{2+}; 0.757*M* Cl^-
43. 0.230 mol CH_3COO^-
44. a. 0.935*m* NH_4Cl
b. 7.80*m* CH_3COCH_3
45. a. 0.719*m* RbBr
b. 2.18*m* $SnBr_2$
c. 0.260*m* $AgClO_3$
d. 3.7*m* KF
46. a. 0.107 mole fraction C_9H_7N; 0.893 mole fraction C_2H_6O
b. 0.0292 mole fraction $C_5H_{10}O_5$; 0.0280 mole fraction CH_6N_4O; 0.944 mole fraction H_2O
47. The solution is composed of 0.125 mol benzene for every 0.875 mol cyclohexane.
= 0.286 mol benzene
48. 4.58% glucose
49. a. 1.80 x 10^2 g $Sc(NO_3)_3$
b. 223 g $Er_2(SO_4)_3$
c. 18.2 g VBr_3
d. 6.49 g $DyCl_3$
50. a. $BaSO_4$
b. 22.6 g $BaCl_2$
c. 40.2% $BaCl_2$

CHAPTER 22

Reaction Rate and Chemical Equilibrium

CHAPTER ORGANIZER

Chapter Objectives	Text Features	Laboratory Options	Teacher Classroom Resources
22.1 Reaction Rates **1.** Distinguish between thermodynamic stability and kinetic stability. **2.** List and describe the factors that influence the rate of reaction. **3.** Distinguish among heterogeneous catalyst, homogeneous catalyst, and inhibitor. **4.** Describe and determine reaction mechanisms for simple reactions. **Nat'l Science Stds: UCP.1, UCP.3, UCP.4, B.1, B.2, B.3, B.5, B.6, C.5, E.2**	**Careers in Chemistry:** Photographer, p. 547 **Frontiers:** Laser-Zapped Ions, p. 549 **Math Connection:** Increasing Surface Area, p. 552 **Bridge to Biology:** Enzymes in the Body, p. 559	**Minute Lab:** More is Better, p. 550 **Minute Lab:** Faster and Faster, p. 558 **ChemActivity 22:** Preparation and Properties of Oxygen - microlab, p. 838 **Discovery Demo:** 22-1 Oscillating Clock Reaction, p. 542 **Demonstration:** 22-2 A Reversible Reaction, p. 544 **Demonstration:** 22-3 Nature of Reactants, p. 546 **Laboratory Manual:** 22-1 Measuring the Rate of a Chemical Reaction - microlab; 22-2 Determining the Order of a Chemical Reaction - microlab **Demonstration:** 22-4 Temperature Affects Reaction Rate, p. 552 **Demonstration:** 22-5 Reaction Mechanism & Rate Determining Step, p. 556	**Study Guide:** pp. 85-87 **Lesson Plans:** pp. 46-47 **ChemActivity Master 22** **Reteaching:** Reaction Equilibrium, p. 32 **Critical Thinking/Problem Solving:** Chemical Kinetics—Graphically Speaking, p. 22 **Applying Scientific Methods in Chemistry:** Reaction Rates, p. 42 **Enrichment:** Enzyme Action and Inhibitors, pp. 43-44 **Color Transparency 42 and Master:** Activation Energy **Color Transparency 43 and Master:** Catalysis
22.2 Chemical Equilibrium **5.** Determine an equilibrium constant expression for a system at equilibrium. **6.** Use Le Chatelier's principle to explain the effects of changes in concentration, pressure, and temperature on an equilibrium system. **7.** Relate relative amounts of product and reactant to the equilibrium constant. **8.** Calculate equilibrium constants and concentrations of reactants or products for a reaction. **Nat'l Science Stds: UCP.1, UCP.3, UCP.4, B.2, B.3, B.5, B.6, E.2**	**Math Connection:** Calculators and Roots, p. 563 **Chemists and Their Work:** Fritz Haber, p. 565 **Bridge to Agriculture:** Ammonia as a Fertilizer, p. 567	**Demonstration:** 22-6 Equilibrium is Not 50:50, p. 562 **Demonstration:** 22-7 Le Chatelier's Principle Made Visible, p. 564 **Laboratory Manual:** 22-3 Chemical Equilibrium **Laboratory Manual:** 22-4 Effects of Concentration on Chemical Equilibrium - microlab	**Study Guide:** pp. 88-89 **Chemistry and Industry:** Ammonia, pp. 41-42
Other Chapter Resources	**Vocabulary and Concept Review,** pp. 22, 79-80 **Evaluation,** pp. 85-88	**Videodisc Correlation Booklet** **Lab Partner Software**	**Test Bank** **Mastering Concepts in Chemistry—Software**

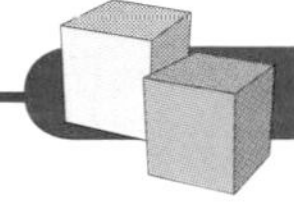

Block Schedule

For information on block scheduling, see the Lesson Plans booklet in the Teacher Resource Package.

GLENCOE TECHNOLOGY

Mastering Concepts in Chemistry Software
Reaction Rates
Chemical Equilibrium

Chemistry: Concepts and Applications Videodisc
Activation Energy

Chemistry: Concepts and Applications CD-ROM
Speed Up the Reaction

Complete Glencoe Technology references are inserted within the appropriate lesson.

ASSIGNMENT GUIDE

Contents	Level	Practice Problems	Chapter Review	Solving Problems in Chemistry
22.1 Reaction Rates				
Stability of Compounds	C		43, 46, 51	
Reversible Reactions and Equilibrium	C		40, 44, 52, 56	
Reaction Rate	C		20, 23, 35	1-3
Nature of Reactants	C		25, 39, 45, 53, 55, 57, 59	
Concentration	C		21, 22, 24, 36, 48, 49, 54, 58	
Temperature	C		25, 36, 48, 49	
Catalysis	C		26-28, 38	
Reaction Mechanism	A	1-4	29-31, 37, 47, 60	
22.2 Chemical Equilibrium				
Equilibrium Constant	C	10-15	32-34, 41, 42	5-9
Le Chatelier's Principle	C		49, 50, 52	4 Chapter Review: 1-7

C=Core, A=Advanced, O=Optional

LABORATORY MATERIALS

Minute Labs	ChemActivities	Demonstrations		
page 550 beaker, 250-cm^3 burner glass plate 6% H_2O_2 MnO_2 tongs wood splint **page 558** beakers, 100, 250-cm^3 $CoCl_2 \cdot 6H_2O$ hot plate 6% H_2O_2 potassium sodium tartrate thermometer water	**page 838** household bleach, 5% NaOCl 0.1*M* cobalt (II) nitrate, $Co(NO_3)_2(aq)$ matches 24-well microplate microtip pipet thin stem micropipet toothpicks	**pages 542, 544, 546, 552, 556, 562, 564** acetone $AgNO_3$ Alka-Seltzer tablets beakers, 400 cm^3 $CoCl_2 \cdot 6H_2O$ dextrose distilled water Erlenmeyer flask, 125, 250, 500, 1000 cm^3 with stopper $FeSO_4 \cdot 7H_2O$ food coloring funnels with 3 flow rates	glass tubing graduated cylinders, 100 cm^3 12*M* HCl $H_2C_2O_4 \cdot 2H_2O$ 6% H_2O_2 18*M* H_2SO_4 KIO_3 $KMnO_4$ $MnSO_4 \cdot H_2O$ malonic acid 1% methylene blue indicator solution NaOH or KOH overhead projector plastic sandwich bags	ring stand and 3 rings rubber bands, small soluble starch water, hot, ice-cold, and room temperature

CHAPTER 22
Reaction Rate and Chemical Equilibrium

CHAPTER OVERVIEW

Main Ideas: Chapter 22 is used to introduce both the qualitative and quantitative aspects of the factors affecting rates of chemical reactions. The quantitative study of reversible reactions at equilibrium is based on the study of reaction rates. Using Le Chatelier's principle, students may predict how changes in concentration, pressure, and temperature affect equilibrium concentrations.

Theme Development: The theme of stability and change is developed through a presentation of how chemical reaction rates and chemical systems at equilibrium can be altered.

Tying to Previous Knowledge: Ask students how they can shorten the cooking time of some foods. Most suggest that one can increase the temperature. Point out that they will learn ways in which chemists increase the rates of chemical reactions.

ASSESSMENT PLANNER

Choose from the following assessment strategies.

Assess, pp. 559, 566
Assessment, pp. 547, 548, 550, 551, 554, 558, 563, 565, 567
Portfolio, p. 567
Minute Lab, pp. 550, 558
ChemActivity, pp. 838-840
Chapter Review, pp. 568-571

PROGRAM RESOURCES

Study Guide: Use the Study Guide worksheets, pp. 85-87, for independent practice with concepts in Section 22.1.

Lesson Plans: Use the Lesson Plans, pp. 46-47, to help you prepare for Chapter 22.

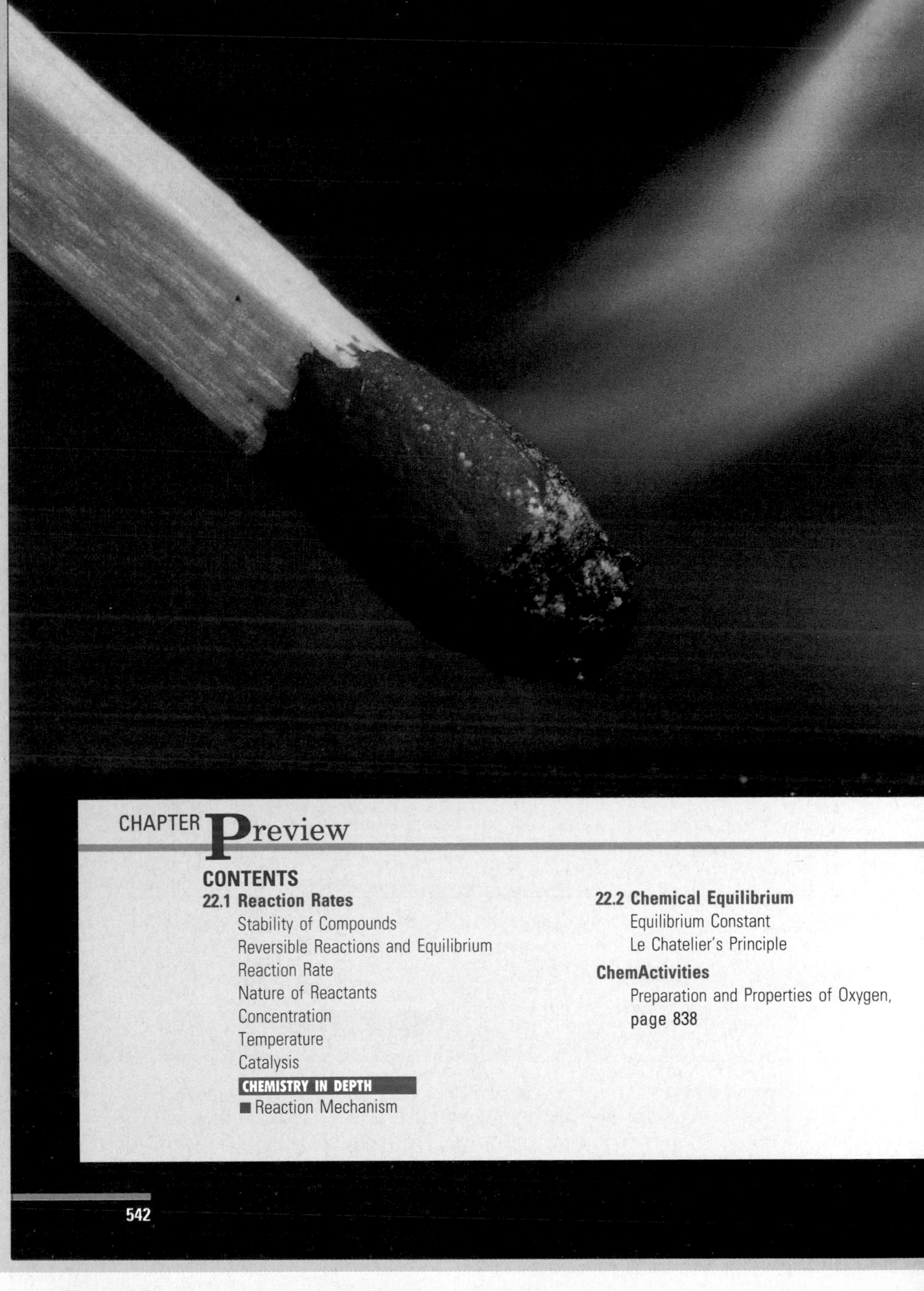

CHAPTER Preview

CONTENTS

542

DISCOVERY DEMO

22-1 Oscillating Clock Reaction

Purpose: to introduce the concepts of reversibility, equilibrium, and multistep reaction mechanism by an oscillating clock reaction

Materials: 2.0 cm^3 18*M* H_2SO_4, 850 cm^3 distilled H_2O, 11.7 g KIO_3, 250 cm^3 6% H_2O_2, 1 g $MnSO_4 \cdot H_2O$, 4 g malonic acid, 4 g soluble starch, overhead projector, 500-cm^3 Erlenmeyer flask

Procedure: 1. Prepare each solution in advance. Solution *A*: Very slowly add 2.0 cm^3 18*M* H_2SO_4 to 250 cm^3 distilled H_2O. **CAUTION:** *Sulfuric acid is corrosive.* Dissolve 11.7 g of KIO_3 in the acidified water (15 minutes with a magnetic stirrer).
2. Solution *B*: 250 cm^3 of 6% H_2O_2.
3. Solution *C*: Add 1 g $MnSO_4 \cdot H_2O$ and 4 g malonic acid to 250 cm^3 distilled H_2O.
4. Solution *D*: Make a paste using 4 g

Reaction Rate and Chemical Equilibrium

CHAPTER

22

Striking a kitchen match isn't a very mysterious way of starting a chemical reaction. As you probably guessed, the heat generated by the friction of the match head in contact with a rough surface started the combustion reaction. You now know that heat was the activation energy of this reaction. Why is activation energy necessary for a reaction? What does activation energy activate?

Some reactions take place until one of the reactants is used up. Other reactions run for a while and then appear to stop even though unused reactants remain. These reactions have reached equilibrium. In this chapter, you will learn about reaction rates and systems that reach equilibrium.

EXTENSIONS AND CONNECTIONS

Careers in Chemistry

Photographer, page 547

Frontiers

Laser-Zapped Ions: Controlling chemical reactions with lasers, page 549

Math Connection

Increasing Surface Area, page 552

Calculators and Roots, page 563

Bridge to Biology

Enzymes in the Body: A look at how enzymes control the processes of life, page 559

Chemists and Their Work

Fritz Haber, page 565

Bridge to Agriculture

Ammonia as a Fertilizer: Improving food production through agrichemicals, page 567

INTRODUCING THE CHAPTER

USING THE PHOTO

The tips of many old-fashioned wooden kitchen matches are coated with the compound P_4S_3, which is highly flammable. It will not spontaneously ignite until activation energy is supplied from the heat produced by friction. Have students identify that the light produced by the burning match provided the activation energy to produce the photograph.

REVEALING MISCONCEPTIONS

Ask students what is formed when carbon burns in air. Possible response: *Carbon dioxide gas is produced.* Have students analyze this response by considering the statement that they may have read on labels of cigarette packages that warns that cigarette smoke contains carbon monoxide. Point out that the production of CO occurs as a two-step mechanism with CO_2 forming first; then the CO_2 reacts with additional C to form CO. Emphasize that very few chemical reactions occur in one step.

of soluble starch and a little H_2O. Add the paste to 100 cm^3 of boiling H_2O. (You can use a can of spray starch as a quicker substitute. Spray it into a beaker of water while stirring.) Use 25 cm^3 of the starch solution *D* per demonstration. Use 125 cm^3 each of solutions *A*, *B*, and *C*.

5. Place the 500-cm^3 Erlenmeyer flask on the stage of an overhead projector and darken the room.

6. Add solutions A, C and D to the flask. Swirl together, and add B. Observe the color changes.

Disposal: G

Results: The color changes from blue-black to yellow and back again half a dozen times. Introduce the terms *reversibility, mechanisms,* and *rate* in discussing the students' observations.

Questions: 1. Is this reaction due to a single or multistep reaction mechanism? *multistep reaction mechanism*

2. Do both the blue-black product and the yellow product disappear at the same rate? *No, the yellow product disappears sooner.*

3. What substance when present turns starch blue-black? *iodine*

Extension: Have students recall oscillating biochemical reactions in their bodies. Possible response: *The conversion of ADP to ATP to ADP as the cell stores and uses energy is one such oscillating biochemical reaction.*

CHAPTER 23
Acids, Bases, and Salts

CHAPTER OVERVIEW

Main Ideas: The three classes of electrolytes—acids, bases, and salts—are discussed in Chapter 23 with an emphasis on their behavior in water solution. The chapter begins with an introduction to three acid-base theories. After the students know the definition of an acid, they learn how to name them. The properties and behavior of acids, bases, and salts are discussed in a quantitative manner. Students apply the knowledge gained in Chapter 22 on equilibrium to solve problems involving the ionization constant, percent of ionization, and common ion effect.

Theme Development: The theme of the chapter, systems and interactions, is developed through a presentation of the Arrhenius, Brønsted-Lowry, and Lewis systems of acids and bases. The interactions that occur between an acid and base to produce a salt are studied quantitatively.

Tying to Previous Knowledge: Have students recall taking medications to relieve stomach discomfort. Point out that these stomach remedies contain a base, such as magnesium or calcium hydroxide, which neutralizes hydrochloric acid, HCl, in the stomach. Remind students that the substance that they sprinkle on their lunch is the product of an acid-base reaction.

CHAPTER Preview

CONTENTS

572

DISCOVERY DEMO

23-1 When Is An Acid Strong?

Purpose: to demonstrate relative ion concentrations by conductivity

Materials: 60 cm^3 12*M* HCl, 41 cm^3 glacial acetic acid, 8 100-cm^3 beakers, conductivity tester with a light bulb, 1 dm^3 distilled water

Procedure: 1. Prepare the following solutions as indicated. **CAUTION:** *Both acids are corrosive.* 6.0*M* HCl: Add 50.0 cm^3 12*M* HCl to 50.0 cm^3 H_2O. 1.0*M* HCl: Add 8.3 cm^3 12*M* HCl to 91.7 cm^3 H_2O. 0.1*M* HCl: Add 10.0 cm^3 1.0*M* HCl to 90.0 cm^3 H_2O. 0.01*M* HCl: Add 10.0 cm^3 0.1*M* HCl to 90.0 cm^3 H_2O. 6.0*M* acetic acid: Add 34.5 cm^3 glacial acid acetic acid to 65.5 cm^3 H_2O. 1.0*M* acetic acid: Add 5.8 cm^3 glacial acetic acid to 94.2 cm^3 H_2O. 0.1*M* acetic acid: Add 10.0 cm^3 1.0*M* acetic acid to 90.0 cm^3 H_2O. 0.01*M* acetic acid: Add 10.0 cm^3 0.1*M* acetic acid to 90.0 cm^3 H_2O.

2. Dim the room lights and test each solution with the conductivity tester. Have students note the brightness of the bulb for each test. **Disposal:** D

Results: The HCl solutions conduct electricity. However, the bulb becomes dimmer as the lower concentrations are tested. Using the type of conductivity apparatus described

Acids, Bases, and Salts

CHAPTER 23

Have you ever noticed that window cleaner has a slick feel? The cleaning solution contains ammonia, which creates a basic solution, and bases feel slick. If you were to taste soap, you would find that it tastes bitter, another property of a base. Lemonade, however, tastes sour because it contains an acid. Sour taste is a property of acids.

Of the top 100 chemical products produced in the United States, seven are acids, three are bases, and twelve are a type of compound known as salts. Many of these substances are consumed in quantities of billions of pounds per year. Most of that consumption is by industry in the production of consumer products. The role played by these substances makes them important groups of compounds to understand. In this chapter, you will learn what these three groups are, their relationship to each other, how they react, and how they may be used.

EXTENSIONS AND CONNECTIONS

Chemistry and Society

Chemical Economics. A view of how chemical processes affect the economics of production, **page 585**

Everyday Chemistry

Baking: An Acid-Base Reaction: How this reaction makes batter rise, **page 589**

Chemists and Their Work

Mary Engle Pennington, **page 591**

INTRODUCING THE CHAPTER

USING THE PHOTO

Have students name foods, such as lemon, apple, grapefruit, grape, rhubarb, and vinegar, that taste sour. Point out to them that the Latin word *acidus* means "sour." Have students recall the taste of soap suds and that of table salt. Point out that soap is a base and table salt is the product of an acid-base reaction. Remind students that they will be studying acids, bases, and salts in this chapter.

REVEALING MISCONCEPTIONS

Have students cite a common property of an acid. Most will state that an acid is corrosive because they have the misconception that all acids are very strong. Bring in some common items that contain acids, such as vitamin C (ascorbic acid), eye wash (boric acid), drink mixes (citric acid), and soft drinks (phosphoric acid). Point out that most acids are weak acids.

ASSESSMENT PLANNER

Choose from the following assessment strategies.

Assess, pp. 583-585, 595-596
Assessment, pp. 575, 577, 584, 586, 588, 589, 592, 594
Portfolio, p. 589
Minute Lab, p. 584
ChemActivity, pp. 840-841
Chapter Review, pp. 597-599

below, the 0.01*M* HCl does not light the bulb. The bulb does not light for any of the acetic acid solutions.

Questions: 1. As the concentration of the HCl decreases, what happens to the conductivity as measured by the light bulb's brightness? *The conductivity decreases.*

2. Does the acetic acid conduct an electric current? *no*

Follow Up Activity: A safer conductivity apparatus can be easily made using an AC adaptor (battery eliminator). Remove the special plug from the end of the DC wire. Attach alligator clips to the ends of the two wires. Wire in series a flashlight bulb socket into one side of the wire. Place a bulb in the socket. The alligator clips can be attached to straightened paper clips (electrodes) that have been pushed through cardboard. The conductivity apparatus can now be used with a small beaker.

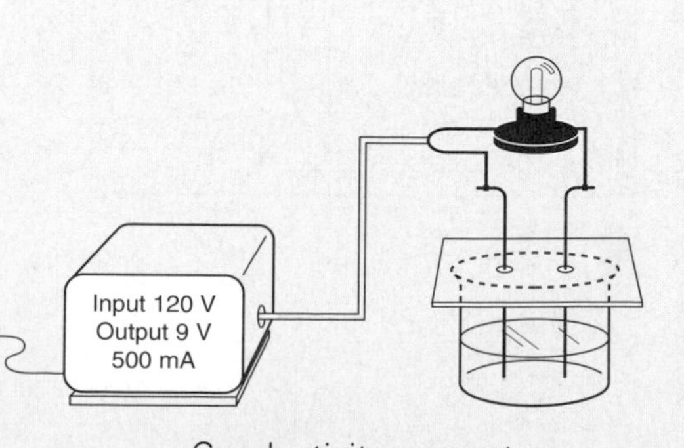

Conductivity apparatus

23.1 Acids and Bases

PREPARE

Key Concepts

The major concepts presented in this section include the Arrhenius, Brønsted-Lowry, and Lewis theories, binary, ternary, and organic acid nomenclature, acid-base behavior, anhydrides, and acid-base strengths.

Key Terms

electrolyte	anhydrous
hydronium ion	acidic anhydride
conjugate base	basic anhydride
conjugate acid	strong acid
binary acid	strong base
ternary acid	weak acid
amphoteric	weak base

Looking Ahead

Obtain a 5-pint PVC coated bottle for Demonstration 23-3.

1 FOCUS

Objectives

Preview with your students the objectives listed on the student page. Students can use them as a study guide for this major section.

Activity

Use a medicine dropper to place a dilute solution of hydrochloric acid, HCl, on pieces of limestone or marble chips. As the students observe the CO_2 gas bubbles and the rock disintegrating, have them consider the effects of acid rain on a statue made of marble, which is mostly calcite, $CaCO_3$.

PROGRAM RESOURCES

Study Guide: Use the Study Guide worksheets, pp. 91-93, for independent practice with concepts in Section 23.1.

Lesson Plans: Use the Lesson Plans, pp. 48-49, to help you prepare for Chapter 23.

OBJECTIVES

- distinguish the definitions of acids and bases as outlined in the theories of Arrhenius, Brønsted-Lowry, and Lewis.
- name acids and bases.
- define acidic and basic anhydrides and write formulas for them.
- define and give examples of strong and weak acids and bases.

23.1 Acids and Bases

Acids and bases have a particular relationship to one another. Therefore, we will look at these two different groups of compounds together. There are a number of ways of defining acids and bases. In this section, three such definitions will be presented. The definitions of acids and bases have changed over the years because of the changing ways chemists have dealt with substances. Originally, most chemistry was carried out in aqueous solution. Consequently, the first definitions of acids and bases included their behavior in water. As chemists began using other solvents, the definitions were broadened.

Arrhenius Theory

It was discovered long ago that most acids, bases, and salts, when dissolved in water, conduct an electric current. Because solutions of these substances conduct a current, these substances are called **electrolytes.**

In 1887, the Swedish chemist Svante Arrhenius published a paper that discussed acids and bases as electrolytes. He knew that aqueous solutions containing acids or bases conducted an electric current. Arrhenius concluded that these substances released charged particles when dissolved. He called these charged particles ions. His theory was that *acids were substances that ionized in water solution to produce hydrogen ions, H^+, or free protons.* He also believed that *bases were substances that ionized to produce hydroxide ions, OH^-, in water solution.* The following are examples.

$$HCl(g) \rightarrow H^+(aq) + Cl^-(aq)$$
$$NaOH(cr) \rightarrow Na^+(aq) + OH^-(aq)$$

We now know that in the case of NaOH, the ions already exist in the solid and merely dissociate in solution. Note the difference between dissociation and ionization. In dissociation, ions separate. In ionization, neutral molecules react with water to form charged ions.

Figure 23.1 Some sea slugs defend themselves by ejecting a stream of sulfuric acid, H_2SO_4. Sulfuric acid is a strong acid.

DEMONSTRATION

23-2 Arrhenius Acids and Bases

Purpose: to demonstrate a typical Arrhenius acid-base reaction

Materials: 0.6 cm^3 18*M* H_2SO_4, 0.32 g $Ba(OH)_2 \cdot 8H_2O$, conductivity tester with light bulb, 2 250-cm^3 beakers, magnetic stirrer

Procedure: 1. Prepare 100 cm^3 of 0.1*M* H_2SO_4 by adding 0.6 cm^3 18*M* H_2SO_4 to 99.4 cm^3 water. **CAUTION:** *Sulfuric acid is very corrosive.* Add 10 cm^3 of the 0.1*M* H_2SO_4 to 90 cm^3 water to prepare 100 cm^3 of 0.01*M* H_2SO_4.

2. Separately prepare 100 cm^3 of 0.01*M* $Ba(OH)_2 \cdot 8H_2O$ by adding 0.32 g to 100 cm^3 water. **CAUTION:** *The solution is toxic.*

3. Add the 0.01*M* H_2SO_4 to the beaker of a conductivity apparatus so that the beaker is half filled with acid. Place the electrodes in the beaker.

Brønsted-Lowry Theory

As the knowledge of catalysts and nonaqueous solutions increased, it became necessary to redefine the terms acid and base. In 1923, an English scientist, T. M. Lowry, and a Danish scientist, J. N. Brønsted, independently proposed new definitions. They stated that *in a chemical reaction, any substance that donates a proton is an acid* and *any substance that accepts a proton is a base.* Remember that H^+ is just a proton. For example, when hydrogen chloride gas is dissolved in water, ions are formed.

$$\underset{\text{acid}}{HCl(g)} + \underset{\text{base}}{H_2O(l)} \rightarrow H_3O^+(aq) + Cl^-(aq)$$

In this reaction, hydrogen chloride is an acid, and water is a base. Notice that the hydrogen ion (H^+) from the acid has combined with a water molecule to form the polyatomic ion H_3O^+, which is called the **hydronium** (hi DROH nee uhm) **ion.** There is strong evidence that the hydrogen ion is never found free as H^+. The bare proton is so strongly attracted by the electrons of surrounding water molecules that H_3O^+ forms immediately. Consider the opposite reaction.

$$\underset{\text{acid}}{H_3O^+(aq)} + \underset{\text{base}}{Cl^-(aq)} \rightarrow HCl(g) + H_2O(l)$$

In this reaction, the H_3O^+ ion is an acid. It acts as an acid because it donates a proton to the chloride ion, which is a base. The hydronium ion is said to be the conjugate acid of the base, water. The chloride ion is called the conjugate base of the acid, hydrochloric acid. In general, any acid-base reaction is described as:

acid + base → conjugate base + conjugate acid

The **conjugate base** of an acid is the particle that remains after a proton has been released by the acid. The **conjugate acid** of a base is formed when the base acquires a proton from the acid. Table 23.1 contains some bases and their conjugate acids.

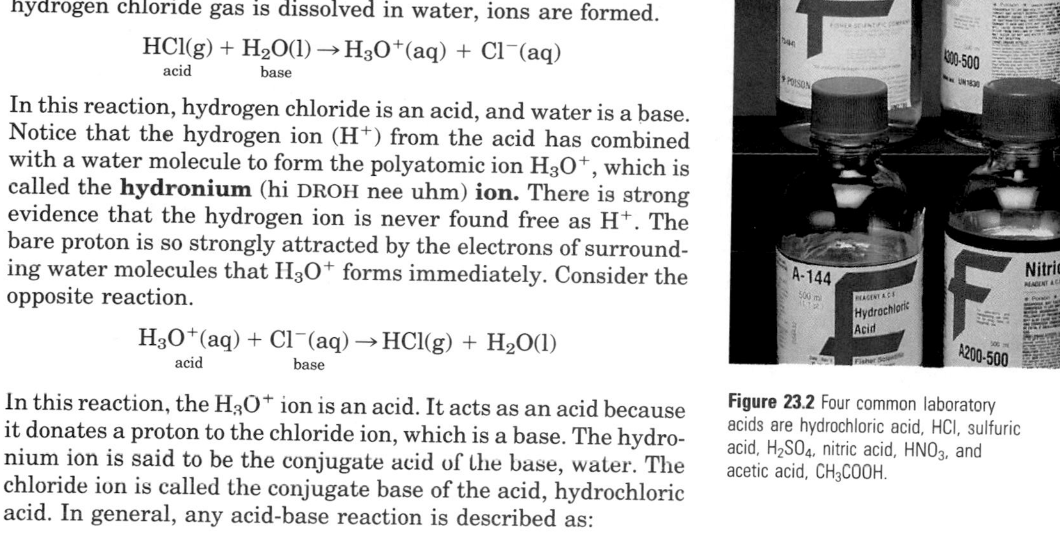

Figure 23.2 Four common laboratory acids are hydrochloric acid, HCl, sulfuric acid, H_2SO_4, nitric acid, HNO_3, and acetic acid, CH_3COOH.

Table 23.1

Bases and Their Conjugate Acids

Base	Name	Conjugate acid	Name
CH_3COO^-	acetate ion	CH_3COOH	acetic acid
NH_3	ammonia	NH_4^+	ammonium ion
CO_3^{2-}	carbonate ion	HCO_3^-	hydrogen carbonate ion
CN^-	cyanide ion	HCN	hydrocyanic acid
$H_2PO_4^-$	dihydrogen phosphate ion	H_3PO_4	phosphoric acid
H_2NNH_2	hydrazine	$H_2NNH_3^+$	hydrazinium ion
HSO_4^-	hydrogen sulfate ion	H_2SO_4	sulfuric acid
OH^-	hydroxide ion	H_2O	water
NO_3^-	nitrate ion	HNO_3	nitric acid
ClO_4^-	perchlorate ion	$HClO_4$	perchloric acid
S^{2-}	sulfide ion	HS^-	hydrogen sulfide ion
H_2O	water	H_3O^+	hydronium ion

2 TEACH

CONCEPT DEVELOPMENT

- Point out to students that the first section of the chapter includes a discussion of the three popular theories regarding the definition of acids and bases. Point out to students that each definition has two parts—descriptive and theoretical. All three theories describe acid properties (sour taste, electrolyte, affect indicators, and react with metals to produce H_2). What makes each theory unique is the theoretical description of behavior (ionizable hydrogen, proton donor, or electron pair acceptor). Students may ask, "Which one is the right definition?" Point out that they all are. As with molecular structure, different observations can be explained by different theories. Arrhenius's theory applies only to water solutions. The Brønsted-Lowry theory can be used with all solvents that will gain or lose a proton, including water. As a result, this theory is more widely applicable.

QUICK DEMO **Slippery Bases**
Have students wash their hands with soap and feel how slippery it is when wet. Test a wet bar of soap with litmus paper to show students that it is basic.

Using the Table: Use several examples from Table 23.1 to show students that the conjugate acid of a base is what forms when an acid accepts a proton from a base.

4. While stirring, add the 0.01*M* $Ba(OH)_2$ solution by drops. Have students observe both the formation of the salt, $BaSO_4$, and the conductivity change indicated by the light bulb. Have students explain their observations using the Arrhenius theory.
Disposal: E

Results: $Ba^{2+} + 2OH^- + 2H^+ + SO_4^{2-} \rightarrow 2H_2O + BaSO_4$
The light bulb dims as the amount of precipitate increases.

Questions: 1. What are the name and formula of the precipitate that forms? *barium sulfate, $BaSO_4$*
2. As the amount of precipitate produced increases, what happens to the number of ions remaining in solution? *The number decreases.*
3. Is the quantitative relationship between the number of ions in solution and the electrical conductivity a direct or inverse mathematical relationship? *direct relationship*

✓ ASSESSMENT

Oral: Ask students for evidence that blood and other body fluids contain ions. Possible responses: *The fact that people get electric shocks and are electrocuted indicates that the human body conducts electric current. Most of the body is fluid. Therefore, it seems appropriate to say that body fluids contain ions.*

CONCEPT DEVELOPMENT

• It is helpful for students to know that the terms *acid* and *base* are applied to reactants, that is, those substances on the left of the →, while the terms *conjugate acid* and *conjugate base* are used to describe products, that is, those substances on the right of the →.

CHEMISTRY IN DEPTH

The section **Lewis Theory** on pages 576-578 is recommended for more able students.

PRACTICE

Guided Practice: Have students work in class on Practice Problems 1 and 2.

Answers to
Practice Problems

1. a. acid—HNO_3
base—NaOH
conjugate acid—H_2O
conjugate base—NO_3^- in $NaNO_3$

b. acid—HCl
base—$NaHCO_3$
conjugate acid—H_2CO_3
conjugate base—Cl^- in NaCl

2. a. hydrogen sulfite ion
b. hydrogen carbonate ion
c. amide ion
d. fluoride ion

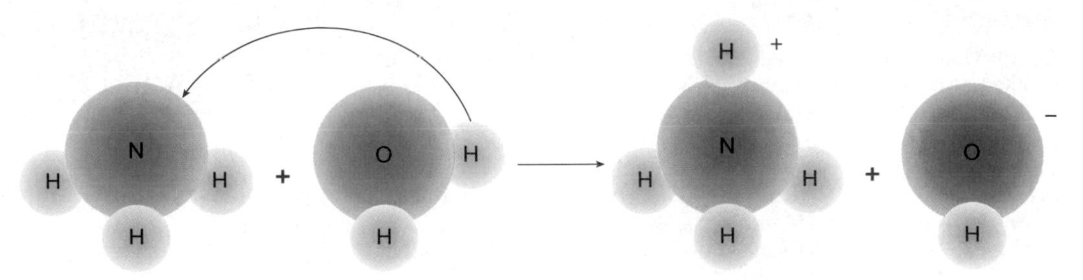

Figure 23.3 When ammonia is added to water, a water molecule donates a proton to the ammonia molecule. Water acts as an acid, and ammonia acts as a base.

Consider what happens when ammonia is added to water.

$$\underset{\text{base}}{NH_3(g)} + \underset{\text{acid}}{H_2O(l)} \rightarrow \underset{\text{conjugate acid}}{NH_4^+(aq)} + \underset{\text{conjugate base}}{OH^-(aq)}$$

In this reaction, water acts as an acid because it donates a proton to the ammonia molecule. The ammonium ion is the conjugate acid of ammonia, a base, which receives a proton from water. The hydroxide ion is the conjugate base of the acid, water.

PRACTICE PROBLEMS

1. Identify the acid, base, conjugate acid, and conjugate base in the following reactions.
   a. $HNO_3(aq) + NaOH(aq) \rightarrow H_2O(l) + NaNO_3(aq)$
   b. $NaHCO_3(aq) + HCl(aq) \rightarrow NaCl(aq) + H_2CO_3(aq)$
2. What is the conjugate base of each of the following?
   a. H_2SO_3 **c.** NH_3
   b. H_2CO_3 **d.** HF

CHEMISTRY IN DEPTH

Pages 576–578

Lewis Theory

In 1923, the same year that Brønsted and Lowry proposed their acid-base theory, another idea appeared. Gilbert Newton Lewis, an American chemist, proposed even broader definitions of acids and bases. However, Lewis focused on electron behavior rather than proton transfer. He defined *an acid as an electron-pair acceptor and a base as an electron-pair donor*. These definitions are more general than Brønsted's. They apply to solutions and reactions that do not even involve hydrogen or hydrogen ions.

Consider the reaction that occurs between ammonia and boron trifluoride.

$$BF_3(g) + NH_3(g) \rightarrow F_3BNH_3(g)$$

The electronic structures of boron trifluoride and ammonia are

$$\begin{matrix} & F & \\ F: & \ddot{B} & \\ & \underset{\cdot\cdot}{F} & \end{matrix} \quad \text{and} \quad \begin{matrix} & H & \\ : & \ddot{N} & :H \\ & \underset{\cdot\cdot}{H} & \end{matrix}$$

DEMONSTRATION

23-3 NH_3 Fountain

Purpose: to demonstrate that NH_3 gas is very soluble in water

Materials: 250-cm^3 Erlenmeyer flask, dropping funnel or buret, drying tube filled with $CaCl_2$, 25 g NaOH pellets, 50 cm^3 14.8*M* $NH_3(aq)$, 5-pint PVC-coated bottle with 2-hole stoppers, medicine dropper, length of glass tubing, 2-dm^3 beaker, 1 cm^3 universal indicator, ring stand with ring

Procedure: 1. Set up an ammonia gas generator in the fume hood. Use a 250-cm^3 Erlenmeyer flask fitted with a dropping funnel or buret and a delivery tube. The delivery tube must be connected to a drying tube filled with $CaCl_2$. Place 25 g of NaOH pellets in the flask. **CAUTION:** *NaOH is caustic.* Begin to generate NH_3 gas by slowly dropping the 14.8*M* $NH_3(aq)$ onto the NaOH pellets. Fill the dry, 5-pint bottle by the downward displacement of air. Tightly stopper the bottle until ready to use. **CAUTION:** *Do not breathe ammonia gas.*

2. Set up the NH_3 fountain. The long glass tube should have one end drawn into a tapered tip which should reach three quarters of the way up into the bottle. The large bottle is then unstoppered and quickly fitted onto the stopper fitted with a large medicine dropper full of H_2O and the

Note that boron has an empty orbital and can accept two more electrons in its outer level. Since boron trifluoride can accept an electron pair, it is a Lewis acid. Now consider the structure of ammonia. Note that the nitrogen atom has an unshared electron pair that can be donated to the boron atom. Ammonia is a Lewis base because it can donate an electron pair. If we use dots to represent the electrons involved in the reaction, the equation can be written

$$\underset{\text{Lewis base}}{H_3N:} + \underset{\text{Lewis acid}}{BF_3} \rightarrow \underset{\text{Addition product}}{H_3N:BF_3}$$

Consider again the reaction of ammonia gas and water.

$$\underset{\text{Lewis base}}{H:\overset{H}{\underset{H}{\ddot{N}}}:} + \underset{\text{Lewis acid}}{H-O-H} \rightarrow \left[H:\overset{H}{\underset{H}{\ddot{N}}}:H\right]^+ + OH^-$$

The ammonia donates an electron pair and is the Lewis base. The hydrogen atom attached to the oxygen of the water molecule acts as the Lewis acid. Notice that ammonia is a base in the Arrhenius theory, the Brønsted-Lowry theory, and the Lewis theory.

The formation of complex ions can be viewed in terms of the Lewis acid-base theory. Recall from Chapter 14 that central ions have empty orbitals and can therefore act as electron pair acceptors. These ions are Lewis acids. Ligands, on the other hand, have unshared electron pairs that they can donate. Ligands are Lewis bases. For example, the aluminum ion, Al^{3+}, has all outer orbitals empty, and water,

$$:\ddot{O}-H$$
$$|$$
$$H$$

has unshared electron pairs. It is to be expected, then, that the reaction between Al^{3+} and water where aluminum has coordination number 6,

$$Al^{3+} + 6H_2O \Rightarrow [Al(H_2O)_6]^{3+},$$

proceeds vigorously to the right.

Al3+

Figure 23.4 The Al^{3+} ion acts as a Lewis acid by accepting electron pairs from water and forms the octahedral complex, $[Al(H_2O)_6]^{3+}$.

Table 23.2

Summary of Acid-Base Theories

Theory	Acid Definition	Base Definition
Arrhenius	Any substance that releases H^+ ion in water solution	Any substance that releases OH^- ions in water solution
Brønsted-Lowry	Any substance that donates a proton	Any substance that accepts a proton
Lewis	Any substance that can accept an electron pair	Any substance that can donate an electron pair

CONCEPT DEVELOPMENT

- Inform students that the Arrhenius theory is the traditional approach and is adequate for introductory chemistry. The Brønsted-Lowry theory is the most prevalent of the three theories in inorganic chemistry. The Lewis theory is the most prevalent in organic chemistry.
- Students should note that the Lewis theory is applicable to reactions carried out in the absence of solvents.

Misconceptions

Students tend to think of substances that are classified as Lewis acids and bases as separate, unrelated cases that do not fit the traditional definition of acids and bases. Dispel this misconception with an example. HCl contains an ionizable hydrogen; therefore, it is an acid according to the Arrhenius theory. The H^+ ion of HCl acts as an electron pair acceptor; therefore, HCl is also a Lewis acid.

Using the Table: Point out to students that a summary table such as Table 23.2 is an important and useful study aid when preparing for a test or quiz. Encourage students to make their own table.

✓ ASSESSMENT

Knowledge: Give students a copy of Table 23.2 from page 577 that has some information missing. Ask students to complete the table.

long glass tube.

3. The open end of the long glass tube is placed in a 2-dm^3 beaker filled with water by positioning the inverted bottle in the ring attached to the ring stand.

4. Add 10–15 drops of universal indicator to the water in the large beaker. To start the NH_3 fountain, squeeze the dropper hard to squirt water into the bottle. **CAUTION:** *Bottles may implode. Use only PVC coated bottles and an explosion shield.* Have students observe the fountain and the color produced in the bottle.

Disposal: C

Results: Water is forced up the tube by the atmospheric pressure because as the gas dissolves in the water, the pressure in the bottle drops. Discuss the colors produced by the universal indicator in the two bottles.

Question: Explain the following reaction, which you saw demonstrated, using the Brønsted-Lowry theory. Identify the conjugate acid and conjugate base.

$$\underset{\text{base}}{NH_3} + \underset{\text{acid}}{H_2O} \rightarrow \underset{\textit{conjugate acid}}{NH_4^+} + \underset{\textit{conjugate base}}{OH^-}$$

The water donates a proton to the unshared pair of electrons on the nitrogen in ammonia to form the ammonium ion.

CONCEPT DEVELOPMENT

GLENCOE TECHNOLOGY

Software

Mastering Concepts in Chemistry
Unit 14, Lesson 1
Background and Theories
Unit 14, Lesson 2
Names and Formulas of Common Acids and Bases

PRACTICE

Guided Practice: Have students work in class on Practice Problems 3-8.

Answers to
Practice Problems

3. **a.** Lewis base
   b. Lewis base
   c. Lewis acid
   d. Lewis base
4. $Ag^+ + 2NH_3 \rightarrow [Ag(NH_3)_2]^+$
5. $Cu^{2+} + 4NH_3 \rightarrow [Cu(NH_3)_4]^{2+}$
6. $Co^{3+} + 6NH_3 \rightarrow [Co(NH_3)_6]^{3+}$
7. $Fe^{3+} + 6CN^- \rightarrow [Fe(CN)_6]^{3-}$
8. $Zn^{2+} + 4OH^- \rightarrow [Zn(OH)_4]^{2-}$

CHEMISTRY IN DEPTH

Chemistry in Depth ends on this page.

MAKING CONNECTIONS

Medicine

Ask students how they would classify magnesium or calcium hydroxide, which is present in some antacids. Possible response: *Each is a base because it neutralizes excess stomach acid.* Confirm the response by crushing a tablet, dissolving it in water, and testing the solution with red litmus paper.

PRACTICE PROBLEMS

3. Classify each of the following substances as either a Lewis acid or a Lewis base.
   a. Cl^- **c.** Na^+
   b. CO_3^{2-} **d.** Br^-

Write equations for the formation of complex ions in the following reactions:

4. $Ag^+ + NH_3 \rightarrow$ (coordination number = 2)
5. $Cu^{2+} + NH_3 \rightarrow$ (coordination number = 4)
6. $Co^{3+} + NH_3 \rightarrow$ (coordination number = 6)
7. $Fe^{3+} + CN^- \rightarrow$ (coordination number = 6)
8. $Zn^{2+} + OH^- \rightarrow$ (coordination number = 4)

Naming Binary Acids

Binary acids are acids containing only two elements. If you look at Table 23.3, you will notice that the prefix is *hydro-* and the suffix is *-ic* in the names of binary acids.

To name a binary acid, we determine what stem to use with the prefix *hydro-* and the suffix *-ic* by finding the stem of the name of the element that combines with hydrogen to form the acid. For instance, in HCl, chlorine will have the stem *-chlor-*, and, in HF, fluorine will have the stem *-fluor-*. To this stem, the prefix *hydro-* and the suffix *-ic* are added. One example of a non-binary acid named according to these rules is hydrocyanic acid, HCN. An exception to the naming system for a binary acid is seen in the naming of hydroazoic acid, HN_3, where the root *−azo−* is used for nitrogen.

Table 23.3

Naming Binary Acids

Binary Compound + Water	Prefix	Stem	Suffix	Name
Hydrogen chloride gas dissolved in water	Hydro-	-chlor-	-ic	Hydrochloric acid
Hydrogen iodide gas dissolved in water	Hydro-	-iod-	-ic	Hydroiodic acid
Hydrogen sulfide gas dissolved in water	Hydro-	-sulfur-	-ic	Hydrosulfuric acid

Naming Ternary Acids and Bases

Ternary acids are acids that contain three elements. The ternary acids we will be working with have oxygen as the third element. We find the stem by determining what element is combined with oxygen and hydrogen in the acid molecule. We determine the prefix, if there is one, and the suffix by the number of oxygen atoms in each molecule.

DEMONSTRATION

23-4 A Visible Lewis Acid-Base Reaction

Purpose: to demonstrate that the reaction of an acidic anhydride, CO_2, with a basic anhydride, CaO, is an example of a Lewis acid - base reaction

Materials: 20 g powdered soda lime (a mixture of NaOH and CaO), 500-cm^3 side-arm vacuum flask with stopper, 200 cm^3 vinegar, 75 g $NaHCO_3$, empty litmus paper vial with stopper, 12-inch balloon, scissors, 250-cm^3 beaker, rubber tubing

Procedure: 1. Place 20 g freshly powdered soda lime, NaOH and CaO, into an empty litmus paper vial. **CAUTION:** *Soda lime is caustic.* Loosely stopper the vial and place it into a 250-cm^3 beaker.
2. Use vinegar and $NaHCO_3$ to generate CO_2 gas in a gas generator (stop-

Generally, the most common form of a ternary acid is given the suffix *-ic*. No prefix is used. Examples of common ternary acids are sulfur*ic* acid, H_2SO_4, chlor*ic* acid, $HClO_3$, and nitr*ic* acid, HNO_3.

If a second acid is formed containing the same three elements, but having one fewer oxygen atom, this acid is given the suffix *-ous*. Again, there is no prefix. Examples of these acids are sulfu*rous* acid, H_2SO_3, chlor*ous* acid, $HClO_2$, and nitr*ous* acid, HNO_2.

If a third acid containing still fewer oxygen atoms is formed, it is given the prefix *hypo-* and the suffix *-ous*. An example is *hypo*chlor*ous* acid, $HClO$.

Acids containing one more oxygen atom than the common form contains are named by adding the prefix *per-* to the common name; for example, $HClO_4$ is *per*chloric acid. See the examples in Table 23.4.

Table 23.4

Naming Ternary Acids

Compound	Number of Oxygen Atoms	Prefix	Stem	Suffix	Name of Acid
H_2SO_4	4	no prefix	sulfur-	-ic	sulfuric
H_2SO_3	3	no prefix	sulfur-	-ous	sulfurous
$HClO_4$	4	per-	-chlor-	-ic	perchloric
$HClO_3$	3	no prefix	chlor-	-ic	chloric
$HClO_2$	2	no prefix	chlor-	-ous	chlorous
$HClO$	1	hypo-	-chlor-	-ous	hypochlorous

It is not possible, without previous knowledge, to know which form of an acid is most common. If the name of one form is known, the other ternary acids containing the same elements can be named. Bromine forms only two acids with hydrogen and oxygen: $HBrO$ and $HBrO_3$. Instead of being named bromous and bromic acids, they are named hypobromous and bromic acids. The exception occurs because they contain the same number of oxygen atoms as hypochlorous and chloric acids. The pattern of the chlorine acids is followed for the bromine acids because both elements are in the same group in the periodic table. The same pattern is followed in naming the ternary acids of the other halogens. In addition to those acids previously mentioned, the most common acid containing phosphorus, oxygen, and hydrogen is H_3PO_4. The most common inorganic acid containing carbon, oxygen, and hydrogen is H_2CO_3. As you can see, it is useful to memorize the names of common acids.

Arrhenius bases are composed of metallic, or positively charged, ions and the negatively charged hydroxide ion. These bases are named by adding the word *hydroxide* to the name of the positive ion. Examples are sodium hydroxide, $NaOH$, and calcium hydroxide, $Ca(OH)_2$.

Figure 23.5 Nitric acid, HNO_3, reacts with copper, producing nitrogen dioxide gas, NO_2.

CONCEPT DEVELOPMENT

- Summarize the ternary acid naming for students with the following list.

per-STEM-ic = more oxygen
STEM-ic = most common
STEM-ous = less oxygen
hypo-STEM-ous = still less oxygen

Using the Table: Emphasize that in Table 23.4 the number of oxygen atoms in the *-ic* acid must be memorized; it cannot be predicted. However, once this acid is known, the other members of the series can be predicted, although they may not exist.

MAKING CONNECTIONS

Language Arts

The archaic name for sulfuric acid is oil of vitriol. Have a student use a dictionary to find the meanings of the words *vitriol* and *vitriolic.* Have students generate sentences containing the word *vitriolic.*

PROGRAM RESOURCES

Reteaching: Use the Reteaching worksheets, pp. 33-34, to provide students another opportunity to understand the concept of "The Brønsted-Lowry Theory."

Chemistry and Industry: Use the worksheets, pp. 43-46, "Sulfuric Acid: The Mark of a Healthy Economy" as an extension to the concepts presented in Section 23.1.

pered side-arm vacuum flask). Fill the beaker containing the vial with CO_2 by the upward displacement of air.

3. Cut the bottom third off a large (12-inch) balloon. Use the top of the balloon to cover and seal the beaker containing the CO_2 and vial.

4. Stretch the balloon as you reach through it to remove the vial's stopper. Shake the balloon-covered beaker to spill the soda lime from the vial.

Results: An exothermic reaction occurs as a solid reacts with a gas to produce a solid. Air pressure forces the balloon into the beaker in a few seconds.

Questions: 1. Use electron dot diagrams to show that the reaction of the OH^- ion from NaOH with CO_2 is a Lewis acid-base reaction.

(from NaOH)

:Ö:H⁻ + :Ö::C::Ö: → :Ö:C::O: (HCO_3^-) with :Ö:H⁻ above C

e⁻ pair donor; e⁻ pair acceptor; 1 of 2 resonance structures

2. Use electron dot diagrams to show that the reaction of the O^{2-} ion from CaO with CO_2 is a Lewis acid-base reaction.

:Ö:²⁻ + :Ö::C::Ö: → :Ö::C:Ö: with :Ö:²⁻ above C

Lewis base; Lewis acid; 1 of 3 resonance structures

CONCEPT DEVELOPMENT

Enrichment: Have students find and explain the names of the following acids: H_2SO_5, $H_2S_2O_7$, $H_2S_2O_8$, HNCO, HSCN.

H_2SO_5, peroxysulfuric acid (contains -O-O- peroxy group)
$H_2S_2O_7$, disulfuric acid (contains 2 sulfuric acid molecules condensed)
$H_2S_2O_8$, peroxydisulfuric acid (both of above reasons)
HNCO, isocyanic acid (HOCN is cyanic acid)
HSCN, thiocyanic acid (S substituted for O in cyanic acid)

MAKING CONNECTIONS

Cellular Biology

Remind students that carboxylic acids and amines are weak acids and bases when compared to strong inorganic acids and bases. Ask students why this would be a logical assumption. Possible response: *Organic acids and bases, which occur in living cells, would destroy cells if they were very strong.*

PRACTICE

Guided Practice: Have students work in class on Practice Problems 9-13.

Answers to
Practice Problems

9. a. hydrobromic acid
b. hydrofluoric acid
10. a. selenous acid
b. nitrous acid
c. phosphoric acid
d. arsenous acid
e. iodic acid
11. a. H_2CO_3
b. HNO_3
c. H_3AsO_4
d. H_2SeO_4
e. HIO
12. a. calcium hydroxide
b. potassium hydroxide
c. aluminum hydroxide
d. methanamine
e. rubidium hydroxide
13. a. CsOH
b. $CH_3CH_2CH_2NH_2$
c. $CH_3CH_2CH_2CH_2NH_2$
d. LiOH

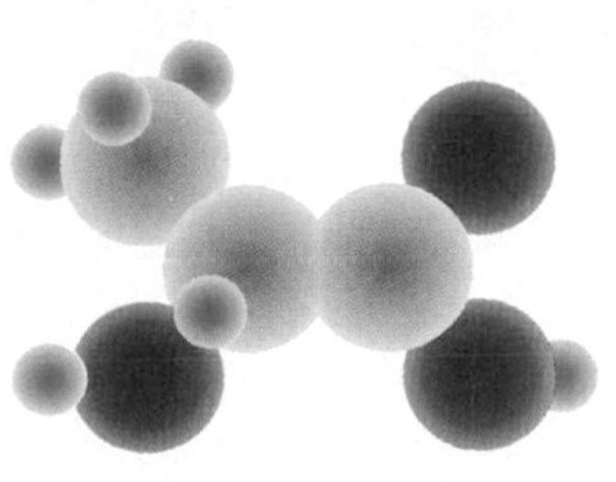

Figure 23.6 Lactic acid is a carboxylic acid. It is a solid at room temperature and is widespread in nature. The sour taste of yogurt, the tartness of sauerkraut, and the acid taste of perspiration are all due to lactic acid.

Naming Organic Acids and Bases

There are several kinds of organic acids. The most important type is the carboxylic acid, characterized by the carboxyl group

$$-\overset{\overset{\displaystyle O}{\|}}{C}-OH \quad \text{or} \quad -COOH$$

Carboxylic acids are named by adding the ending *-oic acid* to the name of the hydrocarbon from which the acid is derived. For example, HCOOH is methanoic acid, which is also known as formic acid. Similarly, $CH_3CH_2CH_2CH_2CH_2COOH$ is hexanoic acid and CH_3CH_2COOH is propanoic acid.

Organic bases contain nitrogen with its unshared pair of electrons. The most common organic bases contain the amine group

$$-\overset{\overset{\displaystyle H}{|}}{N}-H \quad \text{or} \quad -NH_2$$

Amines are named by adding the ending *-amine* to the name of the hydrocarbon from which they were derived. For example, $CH_3CH_2NH_2$ is ethanamine.

PRACTICE PROBLEMS

9. Name each of the following binary acids.
a. HBr(aq)
b. HF(aq)

10. Name each of the following ternary acids.
a. H_2SeO_3 **c.** H_3PO_4 **e.** HIO_3
b. HNO_2 **d.** H_3AsO_3

11. Write formulas for each of the following acids.
a. carbonic acid **c.** arsenic acid **e.** hypoiodous acid
b. nitric acid **d.** selenic acid

12. Name each of the following bases.
a. $Ca(OH)_2$ **c.** $Al(OH)_3$ **e.** RbOH
b. KOH **d.** CH_3NH_2

13. Write formulas for each of the following bases.
a. cesium hydroxide **c.** butanamine
b. propanamine **d.** lithium hydroxide

Acid-Base Behavior

Consider a compound having the formula HO*X*. If the element *X* is highly electronegative, it will have a strong attraction for the electrons it is sharing with the oxygen. As these electrons are pulled toward *X*, the oxygen, in turn, will pull strongly on the electrons it is sharing with the hydrogen. The hydrogen ion, or

CONTENT BACKGROUND

Organic Acids Discovered: Only four organic acids (acetic, formic, succinic, and benzoic) were known before Carl Wilhelm Scheele (1742-1786), a Swedish pharmacist and chemist, discovered ways to separate organic acids based on their salt's solubility. Scheele was able to separate, purify, and identify such water-soluble acids as tartaric, lactic, citric, oxalic, uric, malic, gallic, pyrogallic, mucic, and pyromucic acids from natural sources. A widespread belief, called vitalism, held that organic compounds could not be produced from the elements in the laboratory. They could be produced only in living plants and animals through the agency of a vital force. Friedrich Wohler in 1828 synthesized urea. The final destruction of vitalism was accomplished when Kolbe in 1844 synthesized acetic acid.

proton, will then be easily lost. In this case, HO*X* is behaving as an acid. For example, consider hypochlorous acid, HOCl. The combination of oxygen and chlorine, both of which have a high electronegativity, creates a greater attraction for the electrons shared with hydrogen than oxygen would alone. The hydrogen is then easily lost.

$$H \rightarrow O - Cl$$

If the element *X* has a relatively low electronegativity, the oxygen will tend to pull the shared electrons away from *X*. The hydrogen will remain joined to the oxygen. Since in this case the formation of the hydroxide ion, OH^-, is likely, HO*X* is behaving as a base. For example, in lithium hydroxide, oxygen has such a high electronegativity it has captured the outer electron of lithium, which has a relatively low electronegativity. Oxygen, then, has a lesser attraction for the electrons shared with the hydrogen. As a result, the lithium and hydroxide ions dissociate to produce a base in water.

$$Li \rightarrow O - H$$

We know that nonmetals have high electronegativities and metals have low electronegativities. We can conclude, then, that nonmetals will tend to form acids, and metals will tend to form bases.

Some substances can react as either an acid or a base. If one of these substances is in the presence of a proton donor, then it reacts as a base. In the presence of a proton acceptor, it acts as an acid. Such a substance is said to be **amphoteric.** Water is the most common amphoteric substance.

$$\underset{\text{proton donor}}{HCl} + \underset{\text{base}}{H_2O} \rightarrow H_3O^+ + Cl^-$$

$$\underset{\text{proton acceptor}}{NH_3} + \underset{\text{acid}}{H_2O} \rightarrow NH_4^+ + OH^-$$

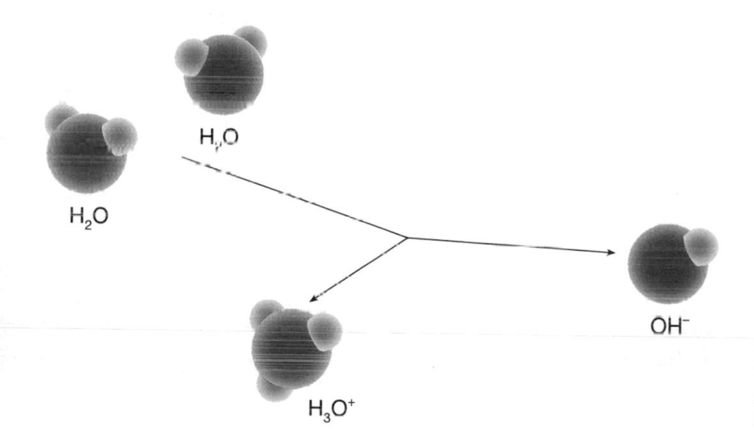

Figure 23.7 The reaction between two water molecules shows the amphoteric nature of water. The reaction produces a hydronium ion and a hydroxide ion.

CONCEPT DEVELOPMENT

- Have students recall that in the Lewis theory, as in the Brønsted-Lowry theory, many substances may act as acids or bases. For example,
base: $NH_3 + H^+ \rightarrow NH_4^+$
acid: $NH_3 \rightarrow H^+ + NH_2^-$ (amide)
Water is the most common amphoteric substance. In the presence of a strong base, it will act as an acid, and vice-versa.

CULTURAL DIVERSITY

Tung Oil and Lacquer

Lacquer is an important Chinese contribution from ancient times. It is anti-corrosive, resistant to acid and alkali, durable, lightweight, and attractive. The Chinese devoted a great deal of time and effort to perfecting its chemical makeup through empirical research. Lacquer has been used on vehicles, handles, tables, trays, musical instruments, and even to make houses more durable.

Raw lacquer, the sap from the Chinese rhus tree (a member of the cashew family) is watery. When it is stirred in the sunshine, the excess moisture evaporates and a black, viscous liquid is formed. The Chinese found that by adding tung oil as a drying agent, the lacquer was made thinner, easier to apply, and brown. Tung oil is extracted from the seeds of the tung tree and has as its chief ingredient eleostearic acid ($C_{17}H_{29}COOH$). Mercury(II) sulfide was added to make the lacquer red.

Although the production of lacquerware has been explained chemically only quite recently, the Chinese people long ago applied scientific methods to develop the involved chemical processes.

interNET CONNECTION

The EPA's acid rain control efforts are described on the World Wide Web.

World Wide Web
http://www.epa.gov/acidrain/ardhome.html

PROGRAM RESOURCES

Critical Thinking/Problem Solving: Use the worksheet, p. 23, as enrichment for the concept of "Understanding Acids and Bases."

Enrichment: Use the worksheets "Ionization of Weak Diprotic Acids," pp. 45-46, to challenge more able students.

Laboratory Manual: Use microlab 23-1 "K_a of a Chemical Compound" as enrichment for the concepts in Chapter 23.

CONCEPT DEVELOPMENT

• Students are helped when told that some metallic oxides are considered to be basic anhydrides, and acidic anhydrides are often nonmetallic oxides.

QUICK DEMO **Basic Anhydride**
Before class, burn a few centimeters of magnesium ribbon in a crucible. **CAUTION:** *Do not look directly at the bright light.* Use a stirring rod to powder the residue, MgO. In class, add 1 cm^3 of distilled water and heat to near boiling. Test the resulting $Mg(OH)_2$ with red litmus paper to show that metal oxides react with water to form a basic solution.

Correcting Errors

To help students determine the formula of the anhydride have them use the following steps.
1. Check to be certain the formula is copied correctly. **2.** Determine if the central element is a metal or nonmetal. **3.** Recall that metals tend to form bases and nonmetals tend to form acids.

PRACTICE

Guided Practice: Guide students through the Sample Problem, then have them work in groups on Practice Problems 14 and 15.
Cooperative Learning: Refer to pp. 28T-29T to select a teaching strategy to use with this activity.

Independent Practice: Your homework or classroom assignment can include Review and Practice 50-52 from the end of the chapter.

Answers to
Practice Problems:

14. a. basic
b. basic
c. acidic
d. basic
e. acidic

15. a. BaO
b. I_2O_7
c. TeO_3
d. Al_2O_3
e. ZnO

Acidic and Basic Anhydrides

Anhydrous means without water, so anhydrides are acids or bases that have had water removed. When sulfur dioxide is dissolved in water, sulfurous acid is formed. Any oxide that will produce an acid when dissolved in water is called an **acidic anhydride**.

$$\underset{\text{acidic anhydride}}{SO_2(g)} + \underset{\text{water}}{H_2O(l)} \rightarrow \underset{\text{acid}}{H_2SO_3(aq)}$$

When sulfur-containing fuels such as coal and petroleum are burned, SO_2 is introduced into the atmosphere. Then, when water vapor condenses and rain falls, the SO_2 reacts with the water to produce sulfurous acid.

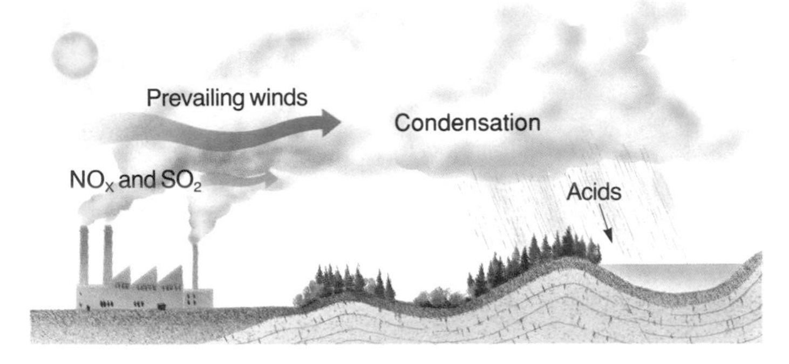

Figure 23.8 Acid rain is produced when sulfur oxides and nitrogen oxides in the air dissolve in water.

If sodium oxide is added to water, sodium hydroxide, a base, is formed. Any oxide that will produce a base when dissolved in water is called a **basic anhydride.**

$$\underset{\text{basic anhydride}}{Na_2O(cr)} + \underset{\text{water}}{H_2O(l)} \rightarrow \underset{\text{base}}{2NaOH(aq)}$$

Figure 23.9 Many valuable documents are disintegrating because acidic anhydrides were used as filler by paper manufacturers.

DEMONSTRATION

23-5 Etching Glass

Purpose: to show that hydrofluoric acid can etch glass

Materials: 12 g CaF_2, 100 cm^3 18*M* H_2SO_4, plastic dish with cover, glass plate, 500 g paraffin, hot plate, empty metal coffee can

Procedure: 1. Carefully warm a metal pan containing paraffin wax. **CAUTION:** *Do not leave unattended; there is a risk of fire.* Using tongs, dip a glass plate into the wax, thoroughly coating both sides.
2. Use a needle probe from a biology dissection kit to remove the wax in a design. Place 12 g of CaF_2 in a plastic dish. Add enough 18*M* H_2SO_4 to the dish to cover the glass plate. **CAUTION:** *Acid is very corrosive. Wear rubber gloves when using CaF_2 and H_2SO_4. Work in a fume hood.*

Figure 23.10 The effects of acid rain caused by the dissolving of acidic anhydrides present in air can cause serious damage to structures made from materials such as limestone.

Organic acids may also form anhydrides. An example is acetic anhydride. It is formed by removing a water molecule from two acetic acid molecules.

$$CH_3-\overset{\overset{O}{\|}}{C}-O-H + H-O-\overset{\overset{O}{\|}}{C}-CH_3 \rightarrow CH_3-\overset{\overset{O}{\|}}{C}-O-\overset{\overset{O}{\|}}{C}-CH_3 + H_2O$$

SAMPLE PROBLEM

Determining the Formula of an Anhydride

Write the formula for the anhydride of nitric acid, HNO_3.

Solving Process:

The word anhydride means *without water*, so we need to remove H_2O from HNO_3. Because there is only one H atom in each HNO_3 molecule, we need to use two molecules.

$$\left.\begin{matrix} HNO_3 \\ HNO_3 \end{matrix}\right\} - H_2O \text{ gives } N_2O_5$$

Thus, removing water from HNO_3 leaves the anhydride N_2O_5.

PRACTICE PROBLEMS

14. Predict the acidic or basic nature of the following anhydrides.
    a. Li_2O
    b. MgO
    c. CO_2
    d. K_2O
    e. SeO_2
15. Write formulas for the anhydrides of the following.
    a. $Ba(OH)_2$
    b. HIO_4
    c. H_6TeO_6
    d. $Al(OH)_3$
    e. $Zn(OH)_2$

3 ASSESS

CHECK FOR UNDERSTANDING

Have students answer the Concept Review problems and Review and Practice 42-54 from the end of the chapter.

RETEACHING

Have each student look at the food labels in his or her pantry or refrigerator to find items that contain weak acids. Have students record the name of each item and that of the acid it contains. Have students generate a master list of the names of the food items and acids.

EXTENSION

Have the class determine what is meant by the "leveling effect" of water on very strong acids and very strong bases. Possible response: *All strong acids ionize completely in water because water is a strong enough base to remove all protons from an acid. However, in less basic solvents, these acids (for example, $HClO_4$, HCl, HNO_3) would not be completely ionized, and an order of relative strength can be set up. Because water can act as a base, it "levels" the effect of these acids. Water can also act as an acid in the presence of very strong bases.*

PROGRAM RESOURCES

Laboratory Manual: Use microlab 23-2 "Determining Percent Acetic Acid in Vinegar" as enrichment for the concepts in Chapter 23.

3. Place the glass plate into the dish and cover with a plastic lid. After 20 minutes, use tongs to remove the glass plate from the acid bath.
4. Rinse the acid off the plate with water. Use a large beaker of boiling water to remove the wax. **Disposal:** D

Results: The HF fumes will etch the design onto the glass plate.

Questions: 1. What is the name of the binary acid, HF? *hydrofluoric acid*
2. Why is hydrofluoric acid commonly prepared in the laboratory instead of being stored for use? *It is very reactive and cannot be stored in a glass container.*

Extension: You may want to extend the discussion of acid-base behavior into what determines the strength of an acid or base. You can use the etched glass plate as a prize for the student who contributes the most to the day's discussion.

CHEMISTRY AND SOCIETY

Chemical Economics

Teaching This Feature: Have teams of lab partners develop written outlines of the feature. Ask teams to describe how the business decisions of chemical companies affect team members.

Answers to
Analyzing the Issue

1. cost of equipment, amount of electricity produced, price for electricity; other answers possible

2. Possible responses might be that old equipment should be used if design changes in the product might be imminent, and new equipment should be used when there is a strong, long-term demand for the product produced in its present form.

MINUTE LAB

Let's Bag Air Pollution

Purpose: to simulate the formation of acid rain around a smokestack

Materials: petri dish, zip-close plastic bag, 3 medicine droppers, 1 cm^3 0.04% bromocresol green indicator, 1 drop 0.5M KNO_2, 2 drops 2.0M H_2SO_4

Procedure Hints: Disposal: C

Results: The blue drops change to yellow as the gas dissolves in the water.

Answer to Question: nitric acid and nitrous acid

✓ ASSESSMENT

Performance: Have students, using a similar procedure, introduce other gases into the sealed plastic bag through a small slit in the top that can be sealed with a piece of tape. Have students suggest what other gases they have available or could generate to test with the indicator. For example, they could place a crystal of Na_2SO_3 in the dish and then add the H_2SO_4 solution to generate SO_2 gas. Also, using a straw to exhale into the bag will produce an increased level of CO_2.

MINUTE LAB

Let's Bag Air Pollution

Put on an apron and goggles. You can see how gases responsible for acid rain diffuse around and from their source. Place drops of varying sizes of 0.04% bromocresol green indicator in the bottom half of a plastic petri dish so that the drops are about 1 cm apart. Leave the center of the petri dish empty. Place a zipper-close plastic bag on a piece of white paper. Carefully place the bottom half of the petri dish, with the drops of indicator, inside the plastic bag. In the center of the petri dish, place one drop of 0.5 M KNO_2 and two drops of 2.0 M H_2SO_4. **CAUTION:** *NO_2 gas is corrosive; KNO_2 and H_2SO_4 are skin irritants.* Seal the bag. Observe, comparing the various color changes that take place in large and small drops, and drops near to and far from the source. Record your observations every 15 seconds. To clean up, open the bag inside an operating fume hood. Rinse out the petri dish. The NO_2 gas produced is soluble in the water of the indicator drops and reacts according to the equation: $2NO_2 + H_2O \rightarrow HNO_3 + HNO_2$. Name the two acids that form.

Strengths of Acids and Bases

Not all acids and bases are completely ionized in water solution. An acid such as hydrochloric that is considered to ionize completely into positive and negative ions is called a **strong acid**. A base such as sodium hydroxide that is completely dissociated into positive and negative ions is called a **strong base**.

Some acids and bases ionize only slightly in solution and are called **weak acids** and **weak bases.** The most important base of this kind is ammonia. In water solution, this base ionizes only partially into NH_4^+ and OH^-. The major portion of the ammonia molecules remains unreacted. Ammonia is a weak base. Acetic acid also ionizes only slightly in water solution. Acetic acid is a weak acid.

Do not equate the solubility of an acid or base with its strength. For example, some bases, such as $Ca(OH)_2$ or $Mg(OH)_2$, are not considered soluble in water, but what small amounts of the bases that do dissolve dissociate completely.

Table 23.5

Relative Strengths of Some Acids and Bases

Compound	Formula	Relative Strength
Hydrochloric acid	HCl	strong acid ↑
Phosphorous acid	H_3PO_3	
Phosphoric acid	H_3PO_4	
Hydrofluoric acid	HF	
Hydroselenic acid	H_2Se	
Acetic acid	CH_3COOH	
Carbonic acid	H_2CO_3	
Hydrosulfuric acid	H_2S	neutral solution
Hypochlorite ion	ClO^-	
Cyanide ion	CN^-	
Ammonia	NH_3	
Carbonate ion	CO_3^{2-}	
Aluminum hydroxide	$Al(OH)_3$	
Phosphate ion	PO_4^{3-}	
Silicate ion	SiO_3^{2-}	
Hydroxide ion	OH^-	↓ strong base

23.1 CONCEPT REVIEW

16. Describe the differences in the three acid-base definitions of this chapter.

17. What are conjugate acids and conjugate bases?

18. What are acidic and basic anhydrides?

19. Name the following substances:
a. HCl(aq) **b.** H_2SO_4 **c.** KOH

20. Apply What kind of element would you expect to find in position X of the compound HOX if the compound is determined to be amphoteric?

Chemistry Journal

Acids in Foods

Ask students to look at the labels of food items in their home and to list the acids that are present in the foods. They should also evaluate the strengths of the acids and explain why strong acids are not used in foods. (Citric, phosphoric, acetic, and carbonic acids are weak acids and can be found in processed foods. Strong acids would probably destroy the foods and would make them unsafe for consumption.) Have students include the lists and other information in their journals.

CHEMISTRY AND
SOCIETY

Chemical Economics

Just as a chemist must balance equations, so must a chemical company balance its account books if it is to stay in business. Chemical firms take in raw materials and produce a product with more monetary value than the raw materials have. The increase in value is due to the cost of the process, including the company's investment and the company profit.

The two main factors that determine the price of a product are cost of materials and cost of processing. Gold is expensive because the raw material, gold-bearing rock, is quite scarce. Other elements that are expensive because of their scarcity include bismuth, iridium, osmium, palladium, platinum, rhenium, rhodium, ruthenium, and tellurium. In addition, most of the radioactive elements beyond bismuth in the periodic table are unavailable from Earth's crust and must be produced synthetically. Titanium ore is plentiful, yet titanium metal is expensive because the process used to produce it is costly. The most expensive materials are those that are both scarce and costly to produce.

Chemists and chemical engineers can do nothing about the scarcity of raw materials, but they can, and do, work continuously to create new processing methods that will lower the cost of making a product. One interesting method of cutting costs that has become important in recent years is cogeneration. Many of the processes in the chemical industry produce heat. In the past, this heat has been transferred to cooling water. Thus, that energy was essentially wasted. Now that energy is used to operate a plant in which electricity is generated. The electricity can then be sold to a local electric utility. The income from the sale of the electricity is used to offset some of the cost of manufacturing the product.

Some economic factors affecting the chemical industry are those common to all industries: equipment wears out, competition becomes keener, and labor-management conflicts must be resolved. In recent years, the United States chemical industry has had to meet increased competition from foreign firms, both at home and abroad. Research is one way chemists meet competition. Chemical engineers contribute in a number of ways. Optimizing the usage of equipment is one such way. As the technology of a particular manufacturing process is improved, a company must make a decision about replacement. That is, which path is better: continuing to use the old equipment in an expensive process until it wears out, or spending the money to replace the old equipment with the latest design now in order to recognize an immediate saving in the process?

Analyzing the Issue

1. If you were a plant engineer, what factors would you consider when deciding whether or not to build a cogeneration unit?
2. A plant manager must decide whether to keep old equipment, which results in an expensive process, or replace the old equipment with expensive new equipment. The new equipment costs much less to run. When might it be a wise decision to use the old equipment? When might it be more economical to use the new equipment?

585

Answers to
Concept Review

16. *The Arrhenius Theory* described acids and bases as substances that release charged particles when dissolved.

The Brønsted-Lowry Theory stated that in a chemical reaction, any substance that donates a proton is an acid and any substance that accepts a proton is a base.

The Lewis Theory defined an acid as an electron-pair acceptor, and a base as an electron-pair donor.

17. The conjugate base of an acid is the particle remaining after a proton has been released by the acid. The conjugate acid of a base is formed when the base accepts a proton from the acid.

18. Any oxide that will produce an acid when dissolved in water is called an acidic anhydride (SO_2), whereas a basic anhydride is any oxide that will produce a base when dissolved in water.

19. **a.** hydrochloric acid
b. sulfuric acid
c. potassium hydroxide

20. An amphoteric compound can act as either an acid or a base. For HOX to be amphoteric, X would have to be an element with a moderate electronegativity. If X were highly electronegative the HOX molecule would behave as an acid due to the electron attractions of X and oxygen. If X had a low electronegativity the HOX molecule would behave as a base, because the oxygen would tend to pull electrons away from X, and hydrogen would remain joined with oxygen.

4 CLOSE

Discussion: Ask students to discuss if it is better to treat marble statues with preservatives like polyurethane coatings and lakes with basic compounds or to remove the offending oxides from exhausts by using catalytic converters and smokestack precipitators or scrubbers. Have students discuss the role that economics plays in determining their solutions.

GLENCOE TECHNOLOGY

Videodisc

Chemistry: Concepts and Applications
The Actions of Acids and Bases
Disc 2, Side 1, Ch. 15

Also available on CD-ROM.

Forming a Basic Anhydride
Disc 2, Side 1, Ch. 14

Show this video of the Quick Demo on page 582 to reinforce the concept of basic anhydrides.
Also available on CD-ROM.

23.2 Salts and Solutions

PREPARE

Key Concepts

The major concepts presented in this section include salts, net ionic equations, the ionization constant, percent of ionization, common ion effect, and polyprotic acids.

Key Terms

salt
neutralization reaction
spectator ion
polyprotic acid
ionization constant
percent of ionization
common ion effect

Looking Ahead

Obtain several samples of liquid drain cleaners for the CLOSE activity.

1 FOCUS

Objectives

Preview with your students the objectives listed on the student page. Students can use them as a study guide for this major section.

QUICK DEMO **A Salt**
Use a medicine dropper to place 20 drops of 6*M* HCl and 20 drops of 6*M* NaOH in a small beaker. Evaporate the water using a hot plate. Show the students the white NaCl salt crystals.

GLENCOE TECHNOLOGY

Videodisc

Chemistry: Concepts and Applications

Forming a Salt

Disc 2, Side 1, Ch. 18

Show this video of the Quick Demo on this page.

Also available on CD-ROM.

OBJECTIVES

- explain the concept of neutralization and the composition of a salt and be able to name salts.
- write net ionic equations.
- derive and use ionization constants.
- compute the percent ionization of a weak electrolyte.

23.2 Salts and Solutions

Now that you have seen the relationship of acids and bases, you are ready to examine the third group of inorganic substances, the salts. When you see the word "salt," you most likely think of common table salt, NaCl. There are, however, thousands of salts. You may have already worked with many of them in the laboratory and in the questions and problems in this book. Salts bear a close relationship to both acids and bases, as you will learn in this section.

Definition of a Salt

An Arrhenius acid is composed of positive hydrogen ions combined with negative nonmetallic ions. Metallic bases are composed of negative hydroxide ions combined with positive metallic ions. An Arrhenius acid reacts with an Arrhenius base to form a salt and water. A **salt** is a crystalline compound composed of the negative ion of an acid and the positive ion of a base. The water is formed from the hydrogen ion of the acid and the hydroxide ion of the base. If the water is evaporated, the negative ions of the acid will unite with the positive ions of the base to form another compound called a salt. If such a reaction results in the removal of all hydrogen and hydroxide ions in solution, the resulting salt will be neither an acid nor a base. We say that the acid and base have neutralized each other. The reaction of an acid and a base is called a **neutralization reaction**. For example, if equivalent amounts of chloric acid and sodium hydroxide react, the salt sodium chlorate and water are formed.

$$\underset{\text{acid}}{HClO_3(aq)} + \underset{\text{base}}{NaOH(aq)} \rightarrow \underset{\text{water}}{H_2O(l)} + \underset{\text{salt}}{NaClO_3(aq)}$$

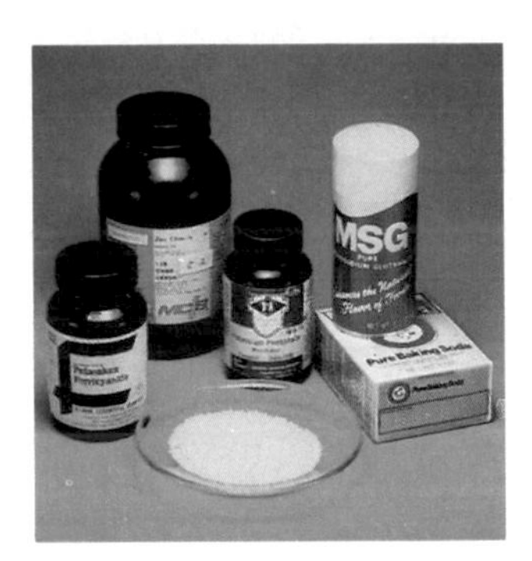

Figure 23.11 A wide variety of salts are used in chemical laboratories and in everyday life.

Figure 23.12 Basic minerals, such as limestone, react with acidic solutions, such as rainwater, to form solutions of dissolved salts. These cave formations result when water evaporates from the salt solutions.

DEMONSTRATION

23-6 Amphoteric Neutralization

Purpose: to demonstrate that an amphoteric compound can react with both an acid and a base

Materials: 0.5 g $Al_2(SO_4)_3$, 90 cm^3 distilled water, few drops universal indicator, petri dish, overhead projector, 2 medicine droppers, stirring rod, 5 cm^3 12*M* HCl, 1.8 g NaOH

Procedure: 1. Dissolve 0.5 g of $Al_2(SO_4)_3$ in 60 cm^3 of distilled water.
2. Place a few cm^3 of this solution in a petri dish on an overhead projector. Add a few drops of universal indicator. Have students note the color of the solution.
3. Add 3*M* (1.8 g NaOH in 15 cm^3 water) sodium hydroxide solution by drops to the petri dish while stirring until a clear, violet solution is produced. Have students note the color

Salts may also result from the reactions of acidic or basic anhydrides with a corresponding base, acid, or anhydride. Note that water is not a product of a reaction between two corresponding anhydrides.

$Na_2O + H_2SO_4 \rightarrow Na_2SO_4 + H_2O$ (basic anhydride + acid)
$2NaOH + SO_3 \rightarrow Na_2SO_4 + H_2O$ (base + acidic anhydride)
$Na_2O + SO_3 \rightarrow Na_2SO_4$ (basic anhydride + acidic anhydride)

Although a salt is formed by neutralization, solutions of some salts in water are not neutral. It is possible to obtain salts that are acidic or basic. For example, if sodium hydroxide reacts with sulfuric acid in a 1:1 mole ratio, the product, sodium hydrogen sulfate, is called an acidic salt.

$$H_2SO_4(aq) + NaOH(aq) \rightarrow H_2O(l) + NaHSO_4(aq)$$

$NaHSO_4$ still contains an ionizable hydrogen atom. In a similar manner, partially neutralized bases form basic salts. Acidic salts such as $NaHSO_4$ and basic salts such as $Cu_2(OH)_2CO_3$ are not neutral in solution.

There is a relationship between the name of an acid and the name of the salt it forms. Binary acids (prefix *hydro-*, and suffix *-ic*) form salts ending in *-ide*. As an example, hydrochloric acid forms chloride salts. Ternary acids form salts in which *-ic* acids form *-ate* salts and *-ous* acids form *-ite* salts. Prefixes from ternary acid names remain in the salt names.

In naming acidic and basic salts, each ion is named separately. Hydrogen is generally named immediately before the names of any negative ions and hydroxide immediately after the names of any positive ions. Thus, $NaHC_2O_4$ is sodium hydrogen oxalate and $Pb_2(OH)_2CO_3$ is lead(II) hydroxide carbonate. Table 23.6 lists names and formulas of several acids and their negative ions.

Table 23.6

Acid and Ion Names

Acid		Ion
HF hydrofluoric acid	→	F^- fluoride
$HMnO_4$ permanganic acid	→	MnO_4^- permanganate
H_2SO_4 sulfuric acid	→	SO_4^{2-} sulfate
HNO_2 nitrous acid	→	NO_2^- nitrite
HClO hypochlorous acid	→	ClO^- hypochlorite
H_2CO_3 carbonic acid	→	HCO_3^- hydrogen carbonate

2 TEACH

CONCEPT DEVELOPMENT

- Review with students the four types of neutralization reactions discussed.
  - acid + base
  - acid + basic anhydride
  - acidic anhydride + base
  - acidic anhydride + basic anhydride

Misconceptions

Students think that the salt produced in a neutralization reaction should form neutral solutions. Point out to students that acidic or basic salts may be formed by neutralization reactions, and their solutions are not neutral.

Using the Table: Remind students of the ternary nomenclature relationship that is shown in Table 23.6.

Acid	Ion
per-STEM-ic	per-STEM-ate
STEM-ic	STEM-ate
STEM-ous	STEM-ite
hypo-STEM-ous	hypo-STEM-ite

PROGRAM RESOURCES

Study Guide: Use the Study Guide worksheets, pp. 94-95, for independent practice with concepts in Section 23.2.

Applying Scientific Methods in Chemistry: Use the worksheet, "Formation of Salts," p. 43, to help students apply and extend their lab skills.

of the solution.

4. Add 3*M* HCl (5 cm^3 12*M* HCl in 15 cm^3 water) solution dropwise to the petri dish while stirring until a clear red solution results. Have students note the color of the solution.

Disposal: D

Results: The entire range of the universal indicator will appear as the acid and base react with the aluminum sulfate. Discuss the following reactions in order. The salt is acidic in water. The aluminum ion forms the hydrated complex which then hydrolyzes.

$Al^{3+} + 6H_2O \rightarrow [Al(H_2O)_6]^{3+}$

$[Al(H_2O)_6]^{3+} + H_2O \rightarrow [Al(H_2O)_5OH]^{2+} + H_3O^+$

When a base is added, a gelatinous precipitate forms.

$Al^{3+} + 3OH^- \rightarrow Al(OH)_3$

In excess base the aluminate complex ion forms.

$Al^{3+} + 4OH^- \rightarrow [Al(OH)_4]^-$

In acid the $Al(OH)_3$ dissolves.

$Al(OH)_3 + 3H^+ \rightarrow Al^{3+} + 3H_2O$

Questions: 1. What evidence is there that the aluminum ion from the salt acts as an acid? *It reacts with a base.* **2.** Does the aluminum solution behave as a base and react with an acid? *yes* **3.** What term is used to describe a substance that can act as either an acid or a base? *amphoteric*

CHAPTER 24
Solutions of Electrolytes

CHAPTER OVERVIEW

Main Ideas: Chapter 24 presents the practical side of acid, base, and salt behavior to students. Solubility product constant is described both qualitatively and quantitatively. Common ion effect is used mathematically to predict precipitation. The sequence is for students to progress from the ionization of water, to pH, and then to the hydrolysis of salts. The concepts of hydrolysis and the common ion effect are linked in a practical study of buffers. Students are prepared for laboratory work with a study of indicators followed by titration calculations.

Theme Development: The theme of the chapter, systems and interactions, is developed through a presentation of electrolyte interactions such as precipitation, hydrolysis, buffer systems, and titration.

Tying to Previous Knowledge: Have students recall that Arrhenius acids and bases contain ionizable hydrogen and hydroxide ions, respectively. Remind students that these ions contribute to the acid's or base's conductivity.

ASSESSMENT PLANNER

Choose from the following assessment strategies.

Assess, pp. 613-614, 619
Assessment, pp. 603, 605, 606, 609, 611, 612, 615, 616, 618
Portfolio, p. 618
Minute Lab, p. 612
ChemActivity, pp. 842-843
Chapter Review, pp. 620-623

PROGRAM RESOURCES

Study Guide: Use the Study Guide worksheets, pp. 97-99, for independent practice with concepts in Section 24.1

Lesson Plans: Use the Lesson Plans, pp. 50-51, to help you prepare for Chapter 24.

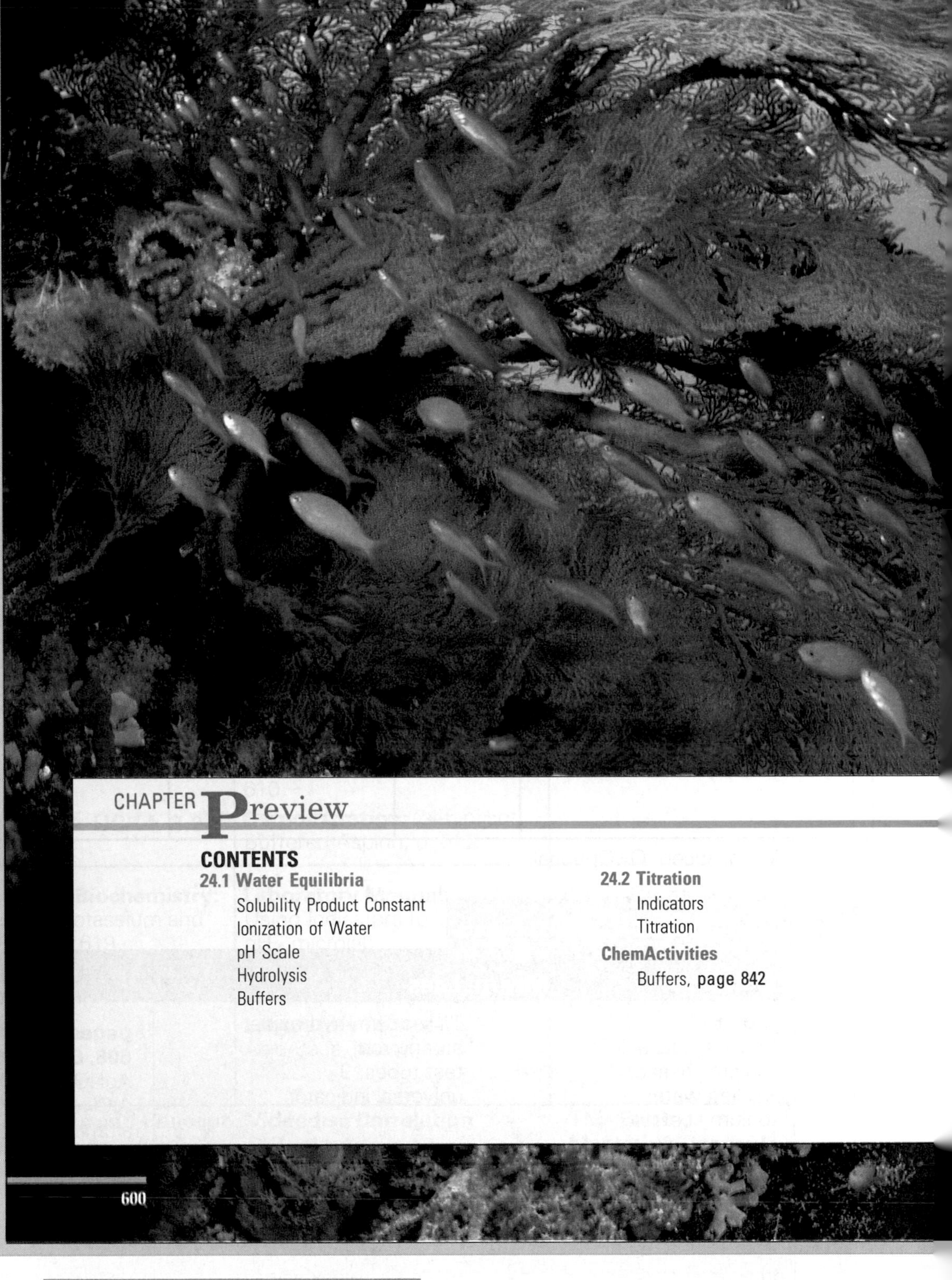

CHAPTER **Preview**

CONTENTS

600

DISCOVERY DEMO

24-1 K_{sp} Made Visible

Purpose: to demonstrate that an apparently insoluble precipitate is still very slightly dissolved and in equilibrium

Materials: 3 150-cm^3 beakers, 400-cm^3 beaker, 250 cm^3 water, 8 g $Pb(NO_3)_2$, 7 g Na_2SO_4 (anhydrous), 12 g Na_2S

Procedure: 1. Prepare 3 solutions in advance. Solution *A:* Dissolve 8 g of $Pb(NO_3)_2$ in 50 cm^3 of distilled H_2O. **CAUTION:** *Toxic chemical.* Solution *B:* Dissolve 7 g of anhydrous Na_2SO_4 in 100 cm^3 H_2O. Solution *C:* Dissolve 12 g Na_2S in 100 cm^3 H_2O.

2. Mix solutions *A* and *B* together in a 400-cm^3 beaker. Allow the white precipitate, $PbSO_4$, to settle. (K_{sp} = 1.9 x 10^{-8}). Have students observe the precipitate and agree that it appears to be insoluble.

3. Write the equation for the equilibri-

Solutions of Electrolytes

CHAPTER
24

The world's seas contain many unusual creatures. One group, the tunicates, squirts jets of water when disturbed—hence, their common name: sea squirts. Sea squirts, shown here, take up vanadium from sea water, where it occurs to the slight extent of 0.3 to 3 parts per million. The vanadium is found in blood cells of the sea squirt at concentrations of about 3700 parts per million. In other words, the organism can concentrate vanadium by a factor of over 1000.

Tunicates are so named because they build an envelope, or tunic, around themselves. Vanadium acts as a catalyst in the synthesis of tiny rods called microfibrils. The tunic is then spread over these microfibrils. Vanadium is found in sea water as ions. Recall that ions in solution are electrolytes. In this chapter, you will learn about the behavior of many electrolytes in solution.

EXTENSIONS AND CONNECTIONS

Health Connection
Keeping the pH Right, page 606

Careers in Chemistry
Swimming Pool Technician, page 610

Chemists and Their Work
Mae C. Jemison, page 611

Bridge to Medicine
Buffering and Hyperventilation: How hyperventilation upsets blood pH, **page 613**

Chemistry and Society
Acid Rain: A serious threat to our environment, page 614

Bridge to Biochemistry
Sodium, Potassium, and Nerves: The chemistry of nerve response, **page 619**

601

INTRODUCING THE CHAPTER

USING THE PHOTO

The sea squirt is able to concentrate vanadium by moving the ions against the normal osmotic pressure gradient from an area of low concentration to an area of high concentration by active transport. The cells use energy to accomplish this task. Ask students where this active transport occurs in their bodies. The catalytic vanadium allows the animal to synthesize cellulose like a plant. The tough cellulose is used for protection. Remind students that they synthesize a protein called skin for the same purpose. This synthesis is also catalytically controlled using enzymes. Point out to students that the blood of the sea squirt contains electrolytes just as human blood contains electrolytes.

REVEALING MISCONCEPTIONS

When an Arrhenius acid reacts with a base, a salt and water are the products. Because the term *neutralization* is applied to the reaction of an acid with a base, many students think that the resulting salt solution always has a neutral pH of 7. This is true when both acid and base are strong or both are weak. When a strong acid neutralizes a weak base, the salt solution has a pH less than 7. When a strong base neutralizes a weak acid, the resulting salt solution has a pH greater than 7.

um reaction on the chalkboard. $PbSO_4(cr) \rightleftarrows Pb^{2+}(aq) + SO_4^{2-}(aq)$ Briefly review with students Le Chatelier's principle. Add solution *C* to the precipitate in the beaker and have students observe the color change of the precipitate. **Disposal:** E

Results: The color of the precipitate changes from white to black. Write the equation for the reaction on the chalkboard. $Pb^{2+}(aq) + S_2^{-}(aq) \rightleftarrows PbS(cr)$ Point out that in solution, Pb^{2+} reacts with S^{2-} to form the more insoluble PbS ($K_{sp} = 8.4 \times 10^{-28}$). The equilibrium shifts to the right and the $PbSO_4$ dissolves.

Questions: 1. Using the solubility rules table in the Appendix, predict the name and formula of the white precipitate. *lead(II) sulfate, $PbSO_4$*
2. If the white precipitate is insoluble, where do the lead ions come from to react with the sulfide ions to form the black PbS precipitate? *The lead ions are in equilibrium with the white solid. As the few dissolved lead ions react with the S^{2-} ions, according to Le Chatelier's principle, the white solid dissolves and supplies the lead ions.*

Extension: Point out to students that industries use similar techniques to remove hazardous chemicals from waste water before returning it to the water cycle.

24.1 Water Equilibria

PREPARE

Key Concepts
The major concepts presented in this section include solubility product constant, ionization of water, pH scale, hydrolysis, and buffers.

Key Terms
solubility product constant
ion product constant of water
pH scale
pH
hydrolysis
buffer system

Looking Ahead
Have students bring in household products for pH testing for Demonstration 24-3. Have students begin collecting samples of rainwater at home for the CLOSE activity.

1 FOCUS

Objectives
Preview with your students the objectives listed on the student page. Students can use them as a study guide for this major section.

Discussion
Students often have difficulty comparing relative values of numbers expressed in scientific notation, especially when the exponent is a negative number. Review scientific notation with students before K_{sp} and pH problem solving is attempted.

GLENCOE TECHNOLOGY

Software

Mastering Concepts in Chemistry
Unit 14, Lesson 4
Ionization Constants
Unit 15, Lesson 1
Water Equilibria

OBJECTIVES
- explain the concept of solubility product and solve problems using the solubility product constant.
- discuss the auto-ionization of water and solve problems using the ion product constant for water.
- explain how the pH scale is used for measuring solution acidity.
- describe the processes of hydrolysis and buffering.

24.1 Water Equilibria

Have you ever seen a swimming pool attendant check the pH of the pool water? A balance must be found between conditions that will prevent the spread of bacteria in the water and conditions that are safe for human occupancy. In this section we will look at the meaning of pH as well as other characteristics of water-based solutions.

Solubility Product Constant

The pH is a measure of the concentration of hydrogen ions in solution. It may also be important to know about other ion concentrations. For example, it is possible to have many Na^+ and Cl^- ions in the same volume of solution. However, some pairs of ions form insoluble compounds, so it is not possible to have a high concentration of both of these ions in solution. In that case, their solubility product is used to calculate the amount of each ion that can exist in solution.

Silver bromide is an ionic compound that is only slightly soluble in water. When an ionic compound is in a solution so concentrated that the solid is in equilibrium with its ions, the solution is saturated. The equilibrium equation for a saturated solution of silver bromide is

$$AgBr(cr) \rightleftarrows Ag^+(aq) + Br^-(aq)$$

The equilibrium constant (K_{eq}) for this equilibrium system is

$$K_{eq} = \frac{[Ag^+][Br^-]}{[AgBr]}$$

However, because we are dealing with a solid substance, the AgBr concentration is constant. Both sides of the equation can be multiplied by [AgBr], giving a new expression.

$$K_{eq}[AgBr] = \frac{[Ag^+][Br^-][AgBr]}{[AgBr]}$$

$$K_{eq}[AgBr] = [Ag^+][Br^-]$$

The term $K_{eq}[AgBr]$ is a constant. This new constant is called the **solubility product constant** K_{sp}.

$$K_{sp} = [Ag^+][Br^-]$$

At room temperature, 25°C, K_{sp} of silver bromide is 5.01×10^{-13}. Consider some silver bromide allowed to stand in water. The silver bromide dissolves until an equilibrium exists between the undissolved solid and the ions in solution. The product of the Ag^+ and Br^- ion concentrations will then be 5.01×10^{-13}.

$$[Ag^+][Br^-] = 5.01 \times 10^{-13}$$

Using the following equation, we see that for every silver ion there is one bromide ion.

DEMONSTRATION

24-2 Predicting Precipitation

Purpose: to demonstrate the predictability of precipitation when concentrations and the K_{sp} value are known

Materials: 0.06 g NaCl, 0.09 g $AgNO_3$, 3 150-cm^3 beakers, 100 cm^3 distilled water, 100-cm^3 graduated cylinder

Procedure: 1. Prepare a 0.020*M* NaCl solution by dissolving 0.06 g NaCl in 50 cm^3 distilled water in a 150-cm^3 beaker.
2. Prepare a 0.010*M* $AgNO_3$ solution by dissolving 0.17 g $AgNO_3$ in 50 cm^3 distilled water in a 150-cm^3 beaker.
3. Mix the contents of the two beakers together and have students observe that a precipitate forms.

Disposal: E

Results: The mixture will result in the formation of a white precipitate.

$$AgBr(cr) \rightleftharpoons Ag^+(aq) + Br^-(aq)$$

$$[Ag^+] = [Br^-], \text{ or } [Ag^+][Ag^+] = 5.01 \times 10^{-13}$$

therefore,

$$[Ag^+]^2 = 5.01 \times 10^{-13}$$

$$[Ag^+] = 7.08 \times 10^{-7}M$$

In a saturated solution containing only silver bromide and water, the concentration of silver ions is 7.08×10^{-7} mol/dm^3. The concentration of bromide ions is also 7.08×10^{-7} mol/dm^3. Suppose, however, that some potassium bromide solution is added. What will happen? KBr dissociates completely to K^+ and Br^- ions. The increased number of bromide ions collide more frequently with silver ions. The equilibrium is thus shifted toward the solid silver bromide. When equilibrium is again established, the concentration of silver ion, $[Ag^+]$, has decreased. At the same time, the concentration of bromide ion, $[Br^-]$, has increased. Solid silver bromide has precipitated out, and the K_{sp}, 5.01×10^{-13}, has remained the same. This reaction is an example of the common ion effect. The addition of a common ion removes silver ion from solution and causes the equilibrium to shift toward the solid silver bromide. We could also say the addition of a common ion decreases the solubility of a substance in solution.

SAMPLE PROBLEMS

1. Common Ion

What will be the silver ion concentration in a saturated solution of silver bromide if 0.100 mol of KBr is added to 1.00 dm^3 of the solution? Assume no increase in the solution volume.

Solving Process:

Potassium bromide is a soluble salt. Thus, it will contribute a bromide ion concentration of 0.100M. The total bromide ion

Figure 24.1 The addition of the common ion Br^- to the AgBr equilibrium system decreases the Ag^+ concentration as more AgBr precipitates.

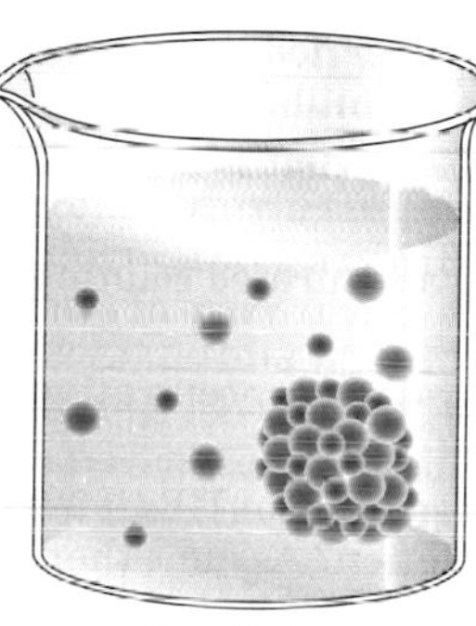

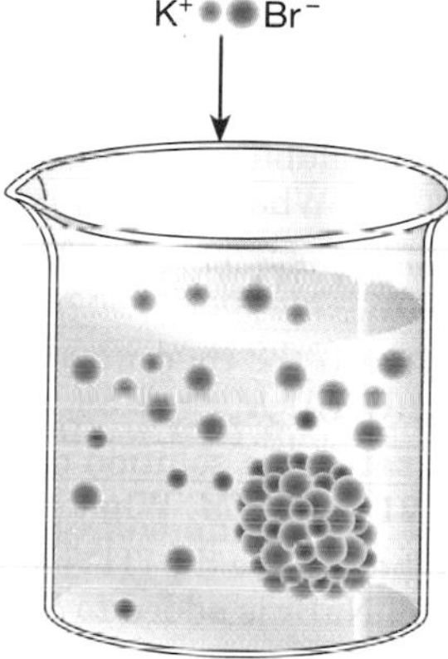

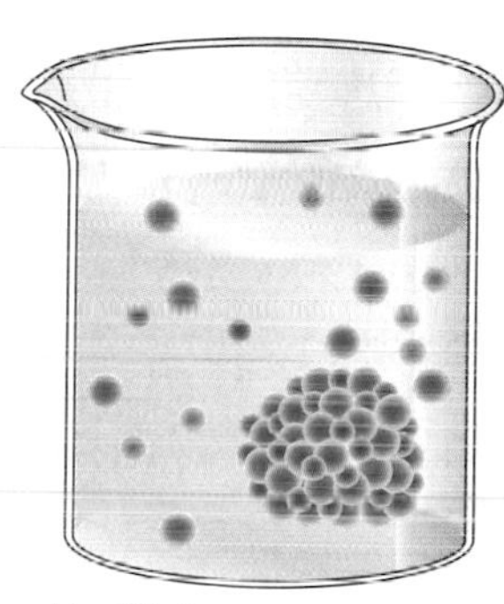

2 TEACH

CONCEPT DEVELOPMENT

- Remind students that ions in a saturated solution are in equilibrium with the undissolved solid.
- Point out to students that the concentration of a solid such as AgBr is a constant because the equilibrium between the solid and the liquid is established at the surface of the solid. The number of ions per square centimeter of solid surface is dependent on the crystal structure, which does not vary.
- Emphasize that just like K_a, the solubility product constant, K_{sp}, is derived from a special case of the equilibrium constant expression. Solids do not appear in the solubility product constant expression.

Mastering Concepts in Chemistry software: Use the lesson on Solubility, pH, and Titrations to introduce and reinforce the concept of solubility in water, the relationship of pH to the ionization of water, and the use of titrations to relate quantities of acids and bases in solutions.

Computer Demo: *Precipitation Game,* Form precipitates from ions. AP502, PC3401, Project SERAPHIM

Correcting Errors

Emphasize to students that the undissolved solid is always written to the left of the arrow and the ions are always written to the right as products. The solubility product constant is the product of the molar concentrations of the ions in the saturated solution, each raised to the power of the coefficients from the balanced chemical equation.

Show students how to calculate the solubility product constant from the concentration (K_{sp} = $[Ag^+][Cl^-]$ = $[0.0050][0.010] = 5.0 \times 10^{-5}$). Point out that because this K_{sp} is greater than that of AgCl (1.78×10^{-5}) a precipitate can be predicted.

Questions: 1. What color is the precipitate? *white*

2. Why was the original concentration of each ion halved? *Because each solution was diluted by a factor of 1/2 when added together.*

3. What is the calculated value for the experimental ion product? *5.0×10^{-5}*

4. How does the calculated value compare to that of AgCl? *The calculated value is larger.*

✓ ASSESSMENT

Oral: Have students compare the calculation made in this demonstration to those in the Sample Problem of the same title on page 605.

PROGRAM RESOURCES

Applying Scientific Methods in Chemistry: Use the worksheet, "Solubility in Water," p. 44 to help students apply and extend their lab skills.

Laboratory Manual: Use the microlab 24-1 "K_{sp} of Salts" as enrichment for the concepts in Chapter 24.

CHEMISTS AND THEIR WORK

Jane Haldimond Marcet

Background: Jane Marcet was the wife of a London physician. She wrote the small book on chemistry to help her niece with the subject. The book served as a textbook in both England and America.

Exploring Further: Michael Faraday at age 13 was apprenticed to a bookbinder in London. There he read Marcet's book, which stimulated his scientific curiosity. Have students research some of Faraday's contributions to science. Possible response: *In 1821 he produced mechanical motion by means of a permanent magnet and an electric current—the electric motor. In electrochemistry, Faraday coined the terms electrode, anode, ion, cathode, anion, cation, electrolyte, and electrolysis.*

CONCEPT DEVELOPMENT

Using the Table: Review with the students each of the oxidation number rules listed in Table 25.3.

- Point out to students that a reaction in which the oxidation number of any element changes is an oxidation-reduction (redox) reaction.
- Remind students that oxidation numbers are used as an aid in understanding atomic structure. However, point out that oxidation numbers do not necessarily indicate the actual relationships existing between atoms and electrons in compounds.

MAKING CONNECTIONS

Health

Chromium ions will have varying effects on body systems depending on their oxidation number. A Cr^{3+} deficiency produces symptoms similar to diabetes. However, $Cr_2O_7^{2-}$ is toxic and carcinogenic, primarily because it is such a strong oxidizing agent.

CHEMISTS AND THEIR WORK

Jane Haldimond Marcet (1769-1845)
Science writer Jane Haldimond Marcet had a profound effect on the chemists of her day and those that followed. Her book, *Conversations in Chemistry,* was first published in the early 1800s and was continually revised through sixteen editions to reflect new scientific breakthroughs. By the 1860s, nearly 200 000 copies had been sold in England and the United States. The book, written in dialogue form, provided experiments to illustrate chemical concepts—a method that today would be labeled a discovery approach. Marcet's work provided the chemical foundations for a young bookbindery clerk named Michael Faraday. Later, as a famous scientist, Faraday referred to *Conversations in Chemistry* as "an anchor in chemical knowledge."

each have 1+. Therefore, (6+) + 4 (2−) + 2 (1+) = 0, and the apparent charge of the compound is zero.

In sulfur dichloride, SCl_2, sulfur has a different oxidation state. Consider the electron dot structure for SCl_2.

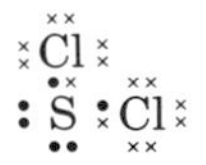

```
 ××
×Cl×
 •×  ××
:S ×Cl×
 ••  ××
```

The sulfur atom shares only four electrons with chlorine. Chlorine is more electronegative than sulfur. Thus the shared electrons are assigned to the chlorine. This assignment gives sulfur an oxidation state of 2+ (14 electrons, 16 protons). Each chlorine then has an oxidation state of 1− (18 electrons, 17 protons).

The oxidation number of an atom may change from compound to compound. Therefore, an electron dot structure must be made for each new compound and the oxidation number determined from this electronic structure. Drawing electron dot structures takes time, however. Fortunately, there is an easier way. General rules have been made to enable you to determine oxidation numbers more easily. These rules are summarized in Table 25.3.

Table 25.3

Oxidation Number Rules
Rule 1. *The oxidation number of any free element is 0.* This statement is true for all atomic and molecular structures: monatomic, diatomic, or polyatomic.
Rule 2. *The oxidation number of a monatomic ion* (Na^+, Ca^{2+}, Al^{3+}, Cl^-) *is equal to the charge on the ion.* Some atoms have several possible oxidation numbers. For example, iron can be either 2+ or 3+; tin, 2+ or 4+.
Rule 3. *The oxidation number of each hydrogen atom in most of its compounds is 1+.* There are some exceptions. In compounds such as lithium hydride (LiH), hydrogen, being the more electronegative atom, has an oxidation number of 1−.
Rule 4. *The oxidation number of each oxygen atom in most of its compounds is 2−* (H_2O). In peroxides, (Na_2O_2, H_2O_2), each oxygen is assigned an oxidation number of 1−.
Rule 5. *The sum of the oxidation numbers of all the atoms in a particle must equal the apparent charge of that particle.* In SO_4^{2-}, sulfur has an oxidation number of 6+ and each oxygen has an oxidation number of 2−, for a net charge of 2−.
Rule 6. *In compounds, the elements of Group 1 (IA), Group 2 (IIA), and aluminum have positive oxidation numbers of 1+, 2+, and 3+, respectively.*

DEMONSTRATION

25-5 Using Indicators to Identify Redox Reactions

Purpose: to demonstrate simultaneous oxidation and reduction

Materials: 4-cm × 4-cm iron sheet, 25 cm^3 acetone, steel wool, 0.1 g $K_3Fe(CN)_6$, 3 g NaCl, 1 cm^3 1% phenolphthalein indicator, 300 cm^3 distilled water, 15 g $FeSO_4{\cdot}7H_2O$, 5 drops 18*M* H_2SO_4, 0.4 g NaOH, medicine dropper, 3 150-cm^3 beakers, 2 test tubes

Procedure: 1. Clean the iron sheet with acetone and steel wool. **CAUTION:** *Acetone is flammable.* **2.** Prepare three solutions. Solution *A*, the Fe^{2+} indicator solution: Add the $K_3Fe(CN)_6$, NaCl, and phenolphthalein indicator solution to 100 cm^3 distilled water in a beaker. **3.** Solution *B*: Add $FeSO_4{\cdot}7H_2O$ and H_2SO_4 to 100 cm^3 water in a beaker. **CAUTION:** *Sulfuric acid is corro-*

SAMPLE PROBLEM

Oxidation Numbers

What are the oxidation numbers of the elements in Na_2SO_4?

Solving Process:

According to Rule 6, the oxidation number of sodium is 1+. According to Rule 4, the oxidation number of oxygen is 2−. According to Rule 5, the total of all oxidation numbers in the formula unit is 0. Letting x = oxidation number of sulfur, we have

$$2(1+) + x + 4(2-) = 0$$
$$x = 6+$$

PROBLEM SOLVING HINT

When you solve an equation, be sure to add the terms to *both* sides so the equation remains balanced.

PRACTICE PROBLEMS

1. Give the oxidation number for each indicated atom.
   - a. S in Na_2SO_3
   - b. Mn in $KMnO_4$
   - c. N in $Ca(NO_3)_2$
   - d. C in Na_2CO_3
   - e. N in NO_2
   - f. S in HSO_4^-
   - g. S in $H_2S_2O_7$
   - h. S in Al_2S_3
   - i. Mn in $MnCl_2$
   - j. C in $C_{12}H_{22}O_{11}$

Identifying Oxidation-Reduction Reactions

Oxidation numbers can be used to determine whether oxidation and reduction (electron transfer) occur in a specific reaction. Even the simplest reaction may be a redox reaction. Let us see how it is possible to determine whether a reaction is actually a redox reaction.

The direct combination of sodium and chlorine to produce sodium chloride is a simple example.

$$2Na(cr) + Cl_2(g) \rightarrow 2NaCl(cr)$$

As a reactant, each sodium atom has an oxidation number of 0. In the product, the oxidation number of each sodium atom is 1+. Similarly, each chlorine atom as a reactant has an oxidation number of 0. As a product, each chlorine atom has an oxidation number of 1−. Since a change of oxidation number has occurred, an oxidation-reduction reaction has taken place.

$$2Na(cr) + Cl_2(g) \rightarrow 2Na^+Cl^-(cr)$$

The change in oxidation number can result only from a shift of electrons between atoms. This shift of electrons alters the apparent charge, that is, the oxidation number.

A gain of electrons means the substance is reduced. It also means that the oxidation number is algebraically lowered. In contrast, a loss of electrons means the substance is oxidized. When an atom is oxidized, its oxidation number increases.

Figure 25.5 Sodium metal reacts vigorously in an atmosphere of chlorine. Sodium is oxidized; chlorine is reduced.

Correcting Errors

Have students practice adding signed numbers before you go through the Sample Problem with them.

3+ + 7− = ? *(4−)*

5+ + ? = 2+ *(3−)*

Point out to students that the difference between the absolute values of the two oxidation numbers yields the magnitude of the change. The sign is that of the oxidation number of the greater absolute value.

PRACTICE

Guided Practice: Guide students through the Sample Problem, then have them work in groups on Practice Problems 1a-j that follow.

Cooperative Learning: Refer to pp. 28T-29T to select a strategy to use with this activity.

Independent Practice: Your homework or classroom assignment can include Review and Practice 45-46, 52, and Concept Mastery 56 from the end of the chapter.

Answers to

Practice Problems: Have students refer to Appendix C for complete solutions to Practice Problems.

1. a. 4+
   b. 7+
   c. 5+
   d. 4+
   e. 4+
   f. 6+
   g. 6+
   h. 2−
   i. 2+
   j. 0

sive. Label the beaker *0.1M* $FeSO_4$ **4.** Solution *C*: Add 0.4 g NaOH to 100 cm^3 water in a beaker and label the beaker *0.1M NaOH.* **CAUTION:** *Sodium hydroxide is caustic.* **5.** Empty 1 full dropper Solution *A* onto the center of the clean iron sheet. Allow to stand for 5 minutes. **6.** While waiting, mix equal amounts Solutions *B* and *A* in a small test tube as a control to indicate change that occurs when Fe^{2+} is formed on the iron. **7.** Mix equal amounts of Solutions *C* and *A* in another small test tube as a control to indicate what happens when O_2 from the air reacts to form OH^- in the presence of the indicator. Have students observe the drop on the iron and each of the controls. **Disposal:** A

Results: The edge of the drop of liquid on the iron sheet is pink and its center is blue. In the presence of $K_3Fe(CN)_6$, production of a blue color indicates oxidation of iron metal to Fe^{2+} producing the deep blue compound iron(II) hexacyanoferrate(III), $Fe_3[Fe(CN)_6]_2$. In the presence of phenolphthalein, production of a pink color indicates that reduction of O_2 (from the air) to produce OH^- occurs.

Questions: 1. What evidence is there that indicates Fe was oxidized and O_2 was reduced simultaneously? *See results.*

2. Which is more likely to rust, wet or dry iron? *wet*

Correcting Errors

Use this mnemonic device to help students remember the definitions of oxidation and reduction. LEO the lion goes GER. Lose Electrons Oxidize; Gain Electrons Reduce.

Determine the oxidation number of each element in the equation. Then, determine which elements' oxidation numbers changed. If the number increased, the element was oxidized. If it decreased, the element was reduced.

PRACTICE

Guided Practice: Guide students through the Sample Problem, then have them work in class on Practice Problems 2-10.

Independent Practice: Your homework or classroom assignment can include the remaining Practice Problems 11-17, Review and Practice 40-41, 43-44, 47, and Application 61 from the end of the chapter.

Answers to

Practice Problems: Have students refer to Appendix C for complete solutions to Practice Problems.

2. Oxidizing agent Reducing Agent

	Oxidizing agent	Reducing Agent
a.	Cu^{2+}	Mg
b.	Sn^{4+}	Fe
c.	S	Na

3. a. Mg; 2 electrons
b. Fe; 2 electrons
c. Na; 1 electron per atom

4. Yes, H is oxidized, H_2 is the reducing agent, N is reduced, and N_2 is the oxidizing agent.

5. Yes, C is oxidized and the reducing agent, H is reduced, and H_2O is the oxidizing agent.

6. not a redox reaction

7. not a redox reaction

8. not a redox reaction

9. Yes, O is reduced, S is oxidized, H_2O_2 is the oxidizing agent, and PbS is the reducing agent.

10. not a redox reaction

11. Yes, N is reduced, P is oxidized, HNO_3 is oxidizing agent, and H_3PO_3 is reducing agent.

The equation for a reaction can be used to determine whether the reaction is a redox reaction. It can also be used to find the substance oxidized, the substance reduced, and the oxidizing and reducing agents. Since the oxidation number of sodium in the equation

$$2Na(cr) + Cl_2(g) \rightarrow 2NaCl(cr)$$

changed from 0 to 1+, sodium is oxidized, making sodium the reducing agent. A reducing agent always loses electrons to another substance and is, therefore, always oxidized as it reduces the other substance.

Solving problems involving reduced or oxidized elements and reducing and oxidizing agents is much easier if we write the equation in net ionic form. Redox reactions are also easier to balance in net ionic form. Before proceeding, review the rules in Chapter 23, pages 588–589, for writing net ionic equations.

PROBLEM SOLVING HINT

Use this mnemonic device to help you remember the definitions of oxidation and reduction: LEO goes GER. Lose Electrons Oxidize; Gain Electrons Reduce.

Figure 25.6 Liquid oxygen and liquid hydrogen mix in a space shuttle's main tank. They react to produce some of the energy that launches the space vehicle.

SAMPLE PROBLEM

Identifying Oxidizing and Reducing Agents

Which substance is the reducing agent and which is the oxidizing agent in the following reaction?

$$16H^+(aq) + 2MnO_4^-(aq) + 5C_2O_4^{2-}(aq) \rightarrow 2Mn^{2+}(aq) + 8H_2O(l) + 10CO_2(g)$$

Solving Process:

In this reaction, manganese is converted from 7+ to 2+. This conversion is accomplished by the gain of five electrons. Therefore, MnO_4^- is the oxidizing agent.

$$\underset{\text{oxidizing agent}}{MnO_4^-} \rightarrow Mn^{2+} \quad \text{reduction}$$

The carbon loses one electron, going from 3+ to 4+, and is converted to carbon dioxide. Therefore, carbon is oxidized, and $C_2O_4^{2-}$ is the reducing agent.

$$\underset{\text{reducing agent}}{C_2O_4^{2-}} \rightarrow CO_2 \quad \text{oxidation}$$

PRACTICE PROBLEMS

2. Identify the oxidizing and reducing agents in each of the following reactions.
a. $Mg(cr) + Cu(NO_3)_2(aq) \rightarrow Mg(NO_3)_2(aq) + Cu(cr)$
b. $SnCl_4(l) + Fe(cr) \rightarrow SnCl_2(cr) + FeCl_2(cr)$
c. $2Na(cr) + S(cr) \rightarrow Na_2S(cr)$

3. What element is oxidized in each of the reactions in Problem 2? How many electrons does each of these elements lose?

CONTENT BACKGROUND

Multiple Oxidation Numbers: Nitrogen exhibits nine oxidation numbers including zero. Have students determine the oxidation number of nitrogen in each of the following formulas.

ammonia	NH_3	3–
hydrazine	N_2H_4	2–
hydroxylamine	NH_2OH	1–
the free element	N_2	0
nitrogen(I) oxide	N_2O	1+
nitrogen(II) oxide	NO	2+
nitrogen(III) oxide	N_2O_3	3+
nitrogen(IV) oxide	NO_2	4+
nitrogen(V) oxide	N_2O_5	5+

Remind students that the oxidation numbers do not necessarily represent the actual number of electrons gained or lost in a chemical reaction. Oxidation numbers are a bookkeeping tool for the chemist.

SAMPLE PROBLEM

Redox

For the following reaction, tell what is oxidized, what is reduced, and identify the oxidizing and reducing agents.

$$3As_2S_3(cr) + 10NO_3^-(aq) + 10H^+(aq) + 4H_2O(l) \rightarrow 6H_3AsO_4(aq) + 9S(cr) + 10NO(g)$$

Solving Process:

Nitrogen is reduced ($5+ \rightarrow 2+$) and sulfur is oxidized ($2- \rightarrow 0$). The nitrate ion (NO_3^-) is the oxidizing agent because it contains nitrogen, which is reduced. The arsenic(III) sulfide (As_2S_3) is the reducing agent; it contains sulfur, which is oxidized. The oxidation number of arsenic does not change.

PRACTICE PROBLEMS

Some of the following unbalanced reactions are oxidation-reduction reactions, and some are not. In each case: **(a)** *Is the reaction redox?* **(b)** *If yes, name the element reduced, the element oxidized, the oxidizing agent, and the reducing agent.*

4. $H_2(g) + N_2(g) \rightarrow NH_3(g)$
5. $C(cr) + H_2O(g) \rightarrow CO(g) + H_2(g)$
6. $AgNO_3(aq) + FeCl_3(aq) \rightarrow AgCl(cr) + Fe(NO_3)_3(aq)$
7. $H_2CO_3(aq) \rightarrow H_2O(l) + CO_2(g)$
8. $MgSO_4(aq) + Ca(OH)_2(aq) \rightarrow Mg(OH)_2(aq) + CaSO_4(cr)$
9. $H_2O_2(aq) + PbS(cr) \rightarrow PbSO_4(cr) + H_2O(l)$
10. $KCl(cr) + H_2SO_4(aq) \rightarrow KHSO_4(aq) + HCl(g)$
11. $HNO_3(aq) + H_3PO_3(aq) \rightarrow NO(g) + H_3PO_4(aq) + H_2O(l)$
12. $HNO_3(aq) + I_2(cr) \rightarrow HIO_3(aq) + NO_2(g) + H_2O(l)$
13. $Na_2S(aq) + AgNO_3(aq) \rightarrow Ag_2S(cr) + NaNO_3(aq)$
14. $H^+(aq) + NO_3^-(aq) + Fe^{2+}(aq) \rightarrow H_2O(l) + NO(g) + Fe^{3+}(aq)$
15. $FeBr_2(aq) + Br_2(l) \rightarrow FeBr_3(aq)$
16. $S_2O_3^{2-}(aq) + I_2(cr) \rightarrow S_4O_6^{2-}(aq) + I^-(aq)$
17. $H_2O_2(aq) + MnO_4^-(aq) \rightarrow O_2(g) + Mn^{2+}(aq)$

25.1 CONCEPT REVIEW

18. How do oxidation and reduction differ?
19. What is an oxidizing agent? a reducing agent?
20. How are shared electrons assigned to the atoms in molecules and polyatomic ions?
21. How can you identify a redox reaction?
22. **Apply** Some of the energy needed to launch a space shuttle is provided by a redox reaction. Liquid oxygen and liquid hydrogen react, when combined, to form water vapor. Write an equation for this reaction. Circle the oxidizing agent and underline the reducing agent.

Answers to
Practice Problems: (continued)

12. Yes, N is reduced, I is oxidized, HNO_3 is the oxidizing agent, and I_2 is the reducing agent.
13. not a redox reaction
14. Yes, N is reduced, Fe^{2+} is oxidized and the reducing agent, and NO_3^- is the oxidizing agent.
15. Yes, Br_2 is reduced and the oxidizing agent, Fe^{2+} is oxidized, and $FeBr_2$ is the reducing agent.
16. Yes, I is reduced, S is oxidized, $S_2O_3^{2-}$ is the reducing agent, and I_2 is the oxidizing agent.
17. Yes, Mn is reduced, O is oxidized, MnO_4^- is the oxidizing agent, and H_2O_2 is the reducing agent.

3 ASSESS

CHECK FOR UNDERSTANDING

Have students answer Concept Review problems and Review and Practice 38-47 from the end of the chapter.

RETEACHING

Strike a match, and as it burns, ask the students to explain the chemical reaction using the key words from this section.

EXTENSION

Have a student use a chemical handbook to answer the following questions about Fe_3O_4. What color is it? What are the oxidation numbers of the elements composing it? How is it formed? What are its properties? *black; oxide is 2–, iron(II) is 2+, iron(III) is 3+; Fe_2O_3 + FeO; magnetic*

Answers to Concept Review

18. Oxidation is the part of a redox reaction in which electrons are removed or apparently removed from an atom or ion. Reduction is the part of a redox reaction in which electrons are added to or apparently added to atoms or ions.
19. The substance in a redox reaction that gives up electrons is called the reducing agent (it is oxidized). And the substance in a redox reaction that gains electrons is the oxidizing agent (it is reduced).
20. When electrons are shared, they are assigned to the more electronegative element. If two identical atoms share electrons,

Chemistry Journal

Common Redox Reactions

Ask students to list oxidation-reduction reactions that occur on a daily basis and to describe the ways in which each reaction is beneficial and useful or detrimental and harmful. Ask students to include the lists and descriptions in their journals.

interNET CONNECTION

A discussion of chemiluminescence and HPLC can be found on the Web.

World Wide Web
http://www.shsu.edu/~chm_tgc/chemilumdir/chemiluminescence2.html

CONCEPT DEVELOPMENT

- Point out to students that by convention the oxidation half-cell is written first. The electrode in this half-cell is the anode. The anode undergoes oxidation and loses mass as it changes from a metal into ions in solution. The two vertical lines represent the salt bridge or porous barrier. The reduction half-cell contains the cathode that gains mass because of metal plating that occurs on the electrode.

Summarize the information on the voltaic cell in a table.

Electrode	Anode (+)	Cathode (-)
Reaction	oxidation	reduction
Δe^-	–	+
Δm	–	+

3 ASSESS

CHECK FOR UNDERSTANDING

Have students answer Concept Review problems, and Review and Practice 25-37 from the end of the chapter.

RETEACHING

Draw the outline of a voltaic cell consisting of two beakers and a salt bridge on the chalkboard. Have students take turns adding to the drawing. The additions can include electrodes, voltmeter, ions in the solution, the directions of electron flow in the external circuit and the ion migration in the internal circuit, the anode, cathode, various anions and cations, and calculation of the theoretical reading of the voltmeter using values listed in Table 26.1.

EXTENSION

Have a student investigate fuel cells as a source of electric energy from chemical reactions and prepare a report for the class. *Fuels are combined with oxygen to release energy directly in the form of electric energy rather than as heat. Information on the hydrogen-oxygen fuel cells used on the shuttle spacecraft is available from NASA.*

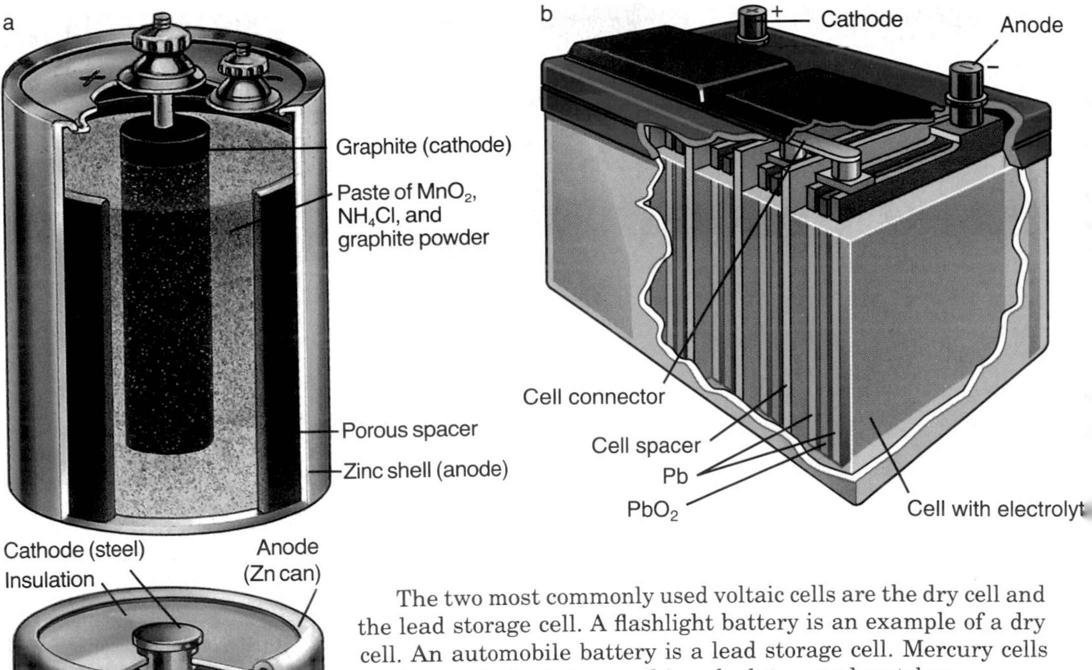

Figure 26.10 A dry cell (a), an automobile battery (b), and a mercury battery (c) are all voltaic cells. The automobile battery contains an electrolyte (H_2SO_4) in solution. The dry cell and mercury cell each have an electrolyte in paste form.

The two most commonly used voltaic cells are the dry cell and the lead storage cell. A flashlight battery is an example of a dry cell. An automobile battery is a lead storage cell. Mercury cells are increasingly being used in calculators and watches.

Chemists have a shorthand method of representing cell reactions. The oxidation half-cell is written first. The reduced and oxidized species are separated by a vertical line. The zinc half-cell from the zinc-copper cell would be written

$$Zn|Zn^{2+}$$

The reduction half-cell is written in the reverse order.

$$Cu^{2+}|Cu$$

The two half-cells are separated by two vertical lines representing the salt bridge. The entire zinc-copper cell is thus shown as

$$Zn|Zn^{2+}||Cu^{2+}|Cu$$

Fuel Cells

In the most common method of generating electric current, a fuel is burned in air and the heat produced is used to boil water. The steam is then passed into a turbine, which is used to turn a generator. Each step in this series has an efficiency of less than 100%. As a consequence, only about 20% of the potential energy in the fuel is finally realized as electric energy. If the intermediate steps between fuel and electricity could be eliminated, the efficiency would be greatly enhanced. A device that accomplishes just that end is the fuel cell. A fuel cell acts much like a battery. However, fuel cells do not have the very limited life that batteries

CONTENT BACKGROUND

Fuel Cells: Fuel cells are voltaic cells that have a continuous supply of reactants. A hydrogen-oxygen fuel cell was used in the Apollo space missions. However, the size of this 225-kg fuel cell is impractical for general use. New York City and Tokyo have permanent power sources that rely on hydrogen-oxygen fuel cells as a temporary backup.

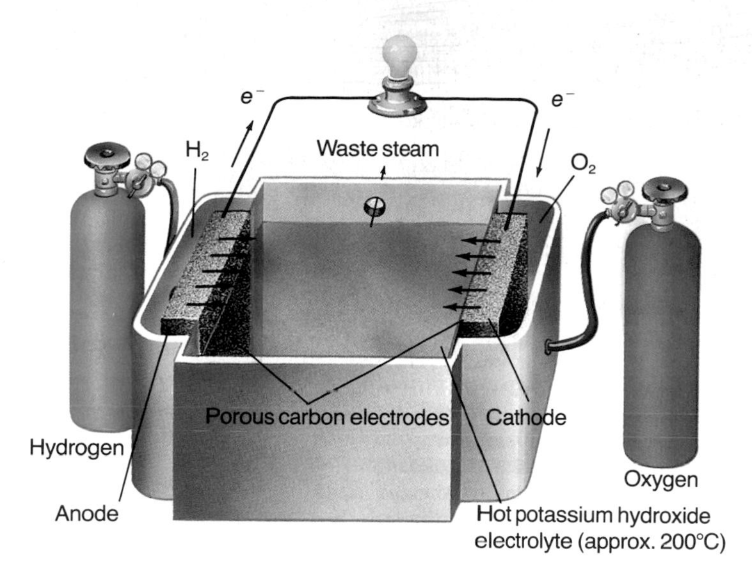

Figure 26.11 A hydrogen-oxygen fuel cell like this one was used in Apollo spacecraft. Instead of producing toxic wastes, as combustion engines do, this fuel cell produces drinkable water when the waste steam is condensed.

have, because there is a continuous feed of fuel and oxygen or air to the cell. The fuel is fed to an electrode on one side of an electrolyte solution, and the oxygen is fed to an electrode on the other side. The electrolyte must be capable of transferring electrons between the electrodes. In addition to the cell itself, an installation must have pumps, fuel reservoirs, and control circuits. The main feature of the fuel cell process is that it separates the oxidation of fuel molecules from the reduction of oxygen. Fuel cells are used in demanding installations such as manned spacecraft, but at present are too expensive for common applications.

26.1 CONCEPT REVIEW

1. Describe the difference between metallic conduction and electrolytic conduction.
2. Define cathode and anode in terms of oxidation and reduction.
3. What is the difference between an electrolytic cell and a voltaic cell?
4. The two types of electrodes in a lead storage battery are made of porous lead and compressed insoluble lead(IV) oxide. These electrodes are immersed in concentrated sulfuric acid. The equation for the overall reaction in this cell is

$$Pb(cr) + PbO_2(cr) + 2H_2SO_4(aq) \rightarrow 2PbSO_4(cr) + 2H_2O(l)$$

   a. Identify the reactions taking place at the anode and the cathode and write the shorthand representation for this cell.
   b. What substance loses electrons in the reaction? What substance gains electrons in the reaction?
5. **Apply** Describe what would happen if a sample of molten magnesium bromide were subjected to an electric current.

Answers to
Concept Review

1. Metallic conduction takes place by the movement of electrons through a metal, whereas electrolytic conduction takes place by the migration of ions in solution.
2. The cathode is the electrode at which reduction (gain of electrons) occurs; the anode is the electrode at which oxidation (loss of electrons) occurs.
3. In an electrolytic cell, passage of an electric current causes a chemical reaction. In a voltaic cell, a chemical reaction generates an electric current.
4. a. anode: $Pb + HSO_4^- \rightarrow PbSO_4 + 2e^- + H^+$
cathode: $PbO_2 + 3H^+ + HSO_4^- + 2e^- \rightarrow PbSO_4 + 2H_2O$
$Pb \mid Pb^{2+} \parallel Pb^{4+} \mid Pb^{2+}$
b. Pb; Pb^{4+}
5. Electrolysis of the magnesium bromide would occur. At the anode, bromide ions lose electrons and bromine gas is formed. At the cathode, magnesium ions gain electrons and magnesium metal is formed.

4 CLOSE

QUICK DEMO

9-Volt Battery
Obtain a used, 9-V battery. Use a small screwdriver and a pair of pliers to open the case. Show students that the battery consists of six voltaic cells connected in series. **CAUTION:** *Encourage students not to open the case of alkaline or mercury batteries at home.* Mercury cells contain a zinc anode and a mercury(II) oxide, HgO, cathode. HgO is reduced to give elemental mercury during the cell reaction. Students will be able to observe a separator that is needed in order to allow negative ions to pass while preventing mixing of the half-cell components. The insulator prevents the anode and cathode from touching and shorting out the battery. Point out to students that these batteries can and should be recycled to avoid mercury contamination in the environment. **Disposal:** F

Chemistry Journal

Everyday Electrochemistry
Ask students to list all the uses or occurrences of electrochemical reactions that they encounter in a typical day and to describe them in their journals.

*inter***NET CONNECTION**

Details of current fuel-cell research are available at the Morgantown Energy Technology Center's website.

World Wide Web
http://www.metc.doe.gov/research/power/fc.html

CONCEPT DEVELOPMENT

- To help students differentiate between the enthalpy of reaction and the enthalpy of formation, remind them that ΔH_f° is the change in enthalpy when one mole of a compound is produced from free elements in their standard states.
- Emphasize to students that thermodynamically stable compounds have large negative enthalpies of formation.
- Remind students that if the sign convention (exothermic $-\Delta H$) is to be observed, the equation used to solve for ΔH will have the sum of the enthalpies of the products minus the sum of the enthalpies of the reactants.

Misconceptions

Students may think that the thermodynamic standard pressure given in the text is a misprint. In 1988, the International Union of Pure and Applied Chemistry, IUPAC, changed the standard to 100.000 kPa. This pressure is equal to 750 mm Hg.

PROGRAM RESOURCES

Laboratory Manual: Use microlab 27 "Measuring the Enthalpy of Reaction" as enrichment for the concepts in Chapter 27.

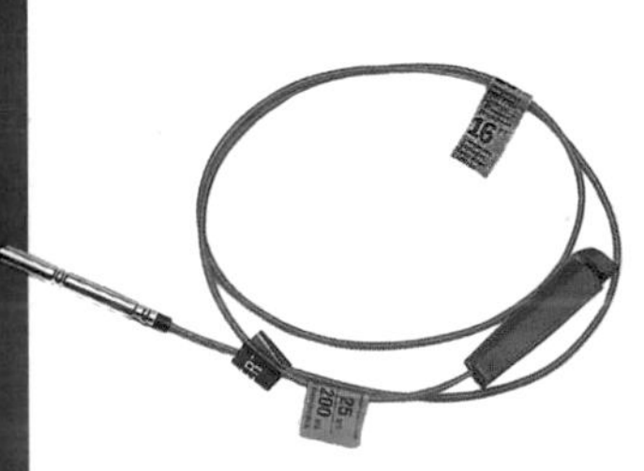

Figure 27.5 Mercury fulminate, used in blasting caps (left) is thermodynamically unstable. It detonates easily, producing large amounts of energy. Mercury fulminate is too unstable to be used in large quantities, but is used to supply activation energy for the detonation of dynamite (right).

Enthalpy of Formation

In reaction of carbon and oxygen, we know that carbon and oxygen are assigned enthalpies of zero because they are free elements. Therefore, the standard molar enthalpy for the formation of carbon dioxide must be −393.5 kJ/mol. The change in enthalpy when one mole of a compound is produced from the free elements in their standard states is known as the standard **enthalpy of formation.** This quantity is expressed in units of kilojoules per mole. A negative sign represents an exothermic reaction. Thus the compound has less enthalpy than the elements from which it was formed. By the same reasoning, a compound produced by an endothermic reaction would have a positive enthalpy of formation. Compounds like CO_2 with large negative enthalpies of formation are thermodynamically stable. **Thermodynamic stability** depends on the amount of energy that would be required to decompose the compound. One mole of CO_2 would require 393.5 kJ of energy to decompose it to the elements carbon and oxygen. On the other hand, mercury fulminate, $Hg(OCN)_2$, produces 268 kJ when one mole decomposes. It is explosive and is used in making detonator caps. Thus, CO_2 has higher thermodynamic stability than $Hg(OCN)_2$.

Appendix Table A-6 lists the enthalpies of formation for some compounds. The symbol used for enthalpy of formation is ΔH_f°. The superscript "°" is used to indicate that the values given are those at 100.000 kPa and 25°C. The subscript "f" designates the value as enthalpy of formation.

Calculation of Enthalpy of Reaction

Let us apply the law of conservation of energy to a reaction. To do so, we will make use of the mathematical symbol, Σ. This Greek uppercase sigma is used to represent a sum. Thus, ΣQ is the sum of all values of Q. Now, consider the reaction *reactants → products.* The enthalpy of the products, $\Sigma \Delta H_{f\ (products)}^\circ$, must equal the enthalpy of the reactants, $\Sigma \Delta H_{f\ (reactants)}^\circ$, plus any change in enthalpy (ΔH_r°) during the reaction.

DEMONSTRATION

27-4 Thermodynamically Unstable

Purpose: to illustrate that a substance can be both kinetically stable and thermodynamically unstable

Materials: 10 cm^3 30% hydrogen peroxide, 3 100-cm^3 graduated cylinders, 10 cm^3 liquid dishwashing detergent, 0.5 g KI crystals, 0.5 g MnO_2

Procedure: **1.** In each of the three graduated cylinders, place 3 cm^3 of 30% hydrogen peroxide, and 3 cm^3 of liquid soap solution. **CAUTION:** *Hydrogen peroxide causes severe burns to the skin. Wear rubber gloves.*
2. Have students note that the first cylinder, the control, is not bubbling.
3. To the second cylinder add a few crystals of KI.
4. To the third cylinder, add a small amount of MnO_2. Have students compare the relative amounts of bubbling

$$\Sigma \Delta H_{f\,(products)}^{\circ} = \Sigma \Delta H_{f\,(reactants)}^{\circ} + \Delta H_r^{\circ}$$

Solving for ΔH_r° we get

$$\Delta H_r^{\circ} = \Sigma \Delta H_{f\,(products)}^{\circ} - \Sigma \Delta H_{f\,(reactants)}^{\circ}$$

If the enthalpy of formation of each reactant and product is known, we can calculate the amount of energy produced or absorbed. We can then predict whether a reaction will be exothermic or endothermic.

SAMPLE PROBLEM

Enthalpy Change

Calculate the enthalpy change in the following reaction.

carbon monoxide + oxygen → carbon dioxide

Solving Process:

First, write a balanced equation. Include all the reactants and products.

$$2CO(g) + O_2(g) \rightarrow 2CO_2(g)$$

Each formula unit represents one mole. Remember that free elements have zero enthalpy by definition. Using the table of enthalpies of formation, Appendix Table A-6, the total enthalpy of the reactants is

$$\Sigma \Delta H_{f\,(reactants)}^{\circ} = \frac{2\ \cancel{mol\ CO}}{} \left| \frac{-110.5\ kJ}{\cancel{mol\ CO}} \right. + 0\ kJ = -221.0\ kJ$$

The total enthalpy of the product ($2CO_2$) is

$$\Sigma \Delta H_{f\,(products)}^{\circ} = \frac{2\ \cancel{mol\ CO_2}}{} \left| \frac{-393.5\ kJ}{\cancel{mol\ CO_2}} \right. = -787.0\ kJ$$

The difference between the enthalpy of the reactants and the enthalpy of the product is

$$\Delta H_r^{\circ} = \Sigma \Delta H_{f\,(products)}^{\circ} - \Sigma \Delta H_{f\,(reactants)}^{\circ}$$
$$\Delta H_r^{\circ} = -787.0\ kJ - (-221.0\ kJ) = -566.0\ kJ$$

This difference between the enthalpy of the products and the reactants (−566.0 kJ) is released as the enthalpy of reaction.

$$2CO(g) + O_2(g) \rightarrow 2CO_2(g) + \textit{enthalpy of reaction}$$

PROBLEM SOLVING HINT

Be careful with signs. Most enthalpies of formation are negative. Likewise, enthalpies of reaction for most spontaneous processes are negative.

PRACTICE PROBLEMS

5. Compute ΔH_r° for the following reaction.
$2NO(g) + O_2(g) \rightarrow 2NO_2(g)$
6. Compute ΔH_r° for the following reaction.
$4FeO(cr) + O_2(g) \rightarrow 2Fe_2O_3(cr)$

Correcting Errors

Have students turn to Table A-6 in the Appendix, Thermodynamic Properties. Have students determine that the ΔH_f column contains the information needed to solve the Sample Problem. Have students look at the top of that column to determine that the unit of ΔH_f is kJ/mol. Remind students to multiply this value by the number of moles found in the balanced chemical equation.

PRACTICE

Guided Practice: Guide students through the Sample Problem, then have them work in class with a partner on Practice Problems 5 and 6 that follow.

Cooperative Learning: Refer to pp. 28T-29T to select a teaching strategy to use with this activity.

Independent Practice: Your homework or classroom assignment can include Review and Practice 34 from the end of the chapter.

Answers to **Practice Problems:** Have students refer to Appendix C for complete solutions to Practice Problems.

5. −144.14 kJ
6. −560.4 kJ

in the last two cylinders. **Disposal:** C

Results: The first cylinder demonstrates that H_2O_2 is kinetically stable even though the reaction shows that the decomposition of H_2O_2 is spontaneous. The last two cylinders indicate that H_2O_2 is thermodynamically unstable and quickly decomposes into water and oxygen. The last two cylinders also demonstrate the effect of two different catalysts that use different pathways for the decomposition. The MnO_2 pathway has a lower activation energy than does the KI pathway.

Questions: 1. Is H_2O_2 kinetically stable? *Yes, it is very slow to decompose unless catalyzed.*
2. Write the balanced equation for the decomposition of H_2O_2. $2H_2O_2(l) \rightarrow O_2(g) + 2H_2O(l)$
3. Is the action of a catalyst pathway dependent? *Yes*
4. Which catalyst provides a pathway that has a lower activation energy? *MnO_2*

✓ ASSESSMENT

Skill: Ask students to use Table A-6 in the Appendix to determine the enthalpy of formation of $H_2O(g)$. $\Delta H_f^{\circ} = -242\ kJ/mol$

PRACTICE

Guided Practice: Have students work in class on Practice Problem 7.

Answers to **Practice Problems:** Have students refer to Appendix C for the complete solution to the Practice Problem.

7. $\Delta H_r° = -135.2$ kJ

3 ASSESS

CHECK FOR UNDERSTANDING

Have students answer Concept Review problems, and Review and Practice 26-35 from the end of the chapter.

RETEACHING

QUICK DEMO **Light a Candle**

Dim the room lights and light a candle with a match. Ask students to observe what is happening, then turn on the lights and group the students so the teams can discuss and write an explanation of their observations using as many of the section's Key Terms as possible.

EXTENSION

Ask a student to report to the class about the difference between specific heat and the enthalpy change. *Specific heat is the energy required to raise the temperature of a certain amount of substance a specified number of degrees, usually one degree. Enthalpy change is the energy gained or lost by the system during an isothermal reaction.*

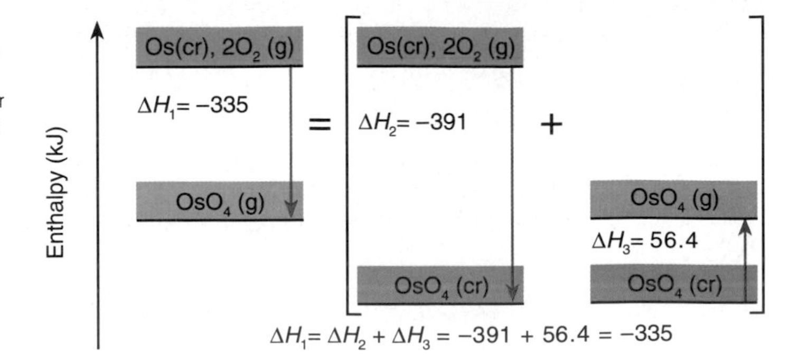

Figure 27.6 Shown here is an illustration of Hess's law. When Os and O_2 react to form gaseous OsO_4, the same change in enthalpy occurs whether the reaction goes in one step (left) or in two steps (right).

Hess's Law

Consider a reaction A → C that can be broken into two steps, A → B and B → C.

$$\Delta H_{r(1)}° = \Delta H_f°B - \Delta H_f°A \qquad \Delta H_{r(2)}° = \Delta H_f°C - \Delta H_f°B$$

The enthalpy change for the overall change of A to C is $\Delta H° = \Delta H_{r(1)}° + \Delta H_{r(2)}°$ because enthalpy is a state function. The principle just illustrated is known as **Hess's law:** *the enthalpy change for a reaction is the sum of the enthalpy changes for a series of reactions that add up to the overall reaction.*

For the reaction in the previous sample problem, we can make the steps

$$C + O_2 \rightarrow CO_2 \qquad \Delta H = -393.5 \text{ kJ}$$
$$2CO \rightarrow CO_2 + C \qquad \Delta H = -172.5 \text{ kJ}$$

Adding, we get $2CO + O_2 \rightarrow 2CO_2$, $\Delta H = -566.0$ kJ, which is the same result as before.

PRACTICE PROBLEMS

7. Barium oxide reacts with sulfuric acid as follows.

$$BaO(cr) + H_2SO_4(l) \rightarrow BaSO_4(cr) + H_2O(l)$$

Calculate the enthalpy of the reaction from these data:

$$SO_3(g) + H_2O(l) \rightarrow H_2SO_4(l) \qquad \Delta H_r° = -78.2 \text{ kJ}$$
$$BaO(cr) + SO_3(g) \rightarrow BaSO_4(cr) \qquad \Delta H_r° = -213.4 \text{ kJ}$$

27.1 CONCEPT REVIEW

8. List two ways to change the internal energy of a system.

9. In a reaction whose enthalpy change is positive, compare the energy involved in breaking the bonds of reactants with that involved in forming bonds of products.

10. State two reasons that reactions occur.

11. Apply Does the burning of propane in an outdoor grill have a positive or negative enthalpy of reaction? Explain.

GLENCOE TECHNOLOGY

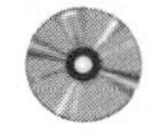

CD-ROM

Chemistry: Concepts and Applications

Exploration: *Burning Calories*

Videodisc

STVS: Ecology

Disc 6, Side 2

Lake Erie Recovery (Ch. 22)

NATIONAL GEOGRAPHIC SOCIETY

Videodisc

STV: Water

Water Quality

Unit 1, Side 1 *Water Quality*

00392-17833

interNET CONNECTION

A hypertext stack for reviewing basic concepts of thermodynamics is available for download.

Anonymous FTP
ftp://archive.umich.edu/mac/misc/chemistry/thermochemistry/sit.hqx

CHEMISTRY AND
SOCIETY

Environmental Quality

The news media today are full of articles expressing concern about the quality of our environment. In some localities the air is polluted. In other areas, waterways are contaminated. The very existence of some species of plants and animals is threatened. Yet, there are points of encouragement.

As an example of what has been done and what still remains to be done, consider the Great Lakes region. This natural drainage system was in serious trouble when both U.S. and Canadian clean-up efforts began in 1972. A bilateral agreement between the United States and Canada established goals for reducing pollutants and a mechanism for the joint monitoring of compliance. The first target was phosphorus pollution, which causes algal blooms and subsequent depletion of the oxygen in the water when the algae die. Sewage treatment, low-phosphorus detergents, and restriction of phosphorus discharge from industry have reduced the phosphorus by 80 percent.

After the initial cleanup of the lakes, some less-obvious problems emerged. One was farm runoff of fertilizer, wastes, and pesticides. However, urban runoff can be worse because lawns and golf courses often get a more intensive fertilizer and pesticide treatment than do farmers' fields. The agricultural runoff can be controlled through no-till planting, careful use of fertilizer, and animal waste management.

Only in the last few years has it been recognized that many contaminants reach the lakes through the air. Automobile exhaust contributes nitrogen in the form of NO_x, which represents a mixture of nitrogen oxides. Contaminated sewage sludge that is spread on fields as fertilizer or burned in municipal incinerators also contributes toxic substances to the atmosphere. Treatment of toxic waste and contaminated groundwater by aeration results in airborne contaminants. Contaminated groundwater discharging into the lakes contributes pollutants. The sediments in the lakes have accumulated pollutants and can give up these pollutants over a long period of time, thus complicating cleanup efforts.

Enormous progress has been made in cleaning up the whole Great Lakes region. At one time, Lake Erie was considered "dead." Today, commercial fishing once again goes on. Even so, the Great Lakes cleanup has a long way to go.

Analyzing the Issue.

1. What is thermal pollution? What kinds of processes contribute to thermal pollution? What steps do industries take to avoid thermal pollution?
2. Find out how phosphorus pollution depletes the oxygen in bodies of water.
3. Is air or water pollution a problem in your community? If so, research the problem, its cause, and any measures being taken to clean up the pollution. Organize a debate on the economic costs of the cleanup versus the health benefits obtained.

CHEMISTRY AND SOCIETY

Environmental Quality

Teaching This Feature: Have students discuss if it is less expensive to cleanup the waste from the environment or at its source before it gets into the air, water, or soil.

Answers to
Analyzing the Issue

1. Thermal pollution results when hot water is released into a body of water. Power plants use a great deal of cooling water. Cooling towers are used to release the heat into the air.
2. Phosphorus is a fertilizer and causes algal blooms. These die and their decomposition requires large quantities of dissolved oxygen.
3. Answers will vary.

Answers to
Concept Review

8. The internal energy of a system may be changed by heating the system or by doing work on it. A system may also transfer energy to its surroundings by giving off heat or by doing work on the surroundings.
9. The energy required to form bonds in the products is greater than the energy released by breaking the bonds of the reactants.
10. Systems in nature tend to go from one state of higher energy to one of lower energy, thereby creating spontaneous reactions to release energy. Also, natural processes tend to go from an orderly state to a less orderly one, which also causes reactions to occur.
11. When propane is burned, the reactants are releasing large quantities of heat energy when being transformed to the products, so the reaction is exothermic with a negative enthalpy. In an exothermic reaction, the products have less enthalpy than the reactants, so $\Delta H < 0$.

4 CLOSE

Discussion: Point out that in the first section, students studied that a system tends to go from a state of higher energy to one of lower energy. Explain that in the second section, they will study processes that tend to go from an orderly to a disorderly state.

CONCEPT DEVELOPMENT

Discussion Question: Ask students why some nuclei are stable and why others are unstable and undergo radioactive decay. A stable isotope is one whose nuclei will not decay spontaneously. The nuclear mass defect of such an isotope is high. This mass is converted into binding energy.

- Students are familiar with Einstein's famous equation, $E = mc^2$, but some students do not know what each letter represents. Review this with those students; E = energy, m = mass, c = speed of light, a constant. An atomic mass unit must be converted to kilograms using the conversion factor $1.660\ 40 \times 10^{-27}$ kg/u.
- To help students with the units of this problem, point out that a newton is 1 kg·m/s^2. One joule is a newton-meter or kg·m^2/s^2.

CHEMISTS AND THEIR WORK

James Chadwick

Background: In 1930 the German scientists Bothe and Becker observed a very penetrating radiation associated with alpha particle bombardment of Be, B and Li. They thought it was from very high-energy gamma rays. James Chadwick repeated the experiments in 1932 and demonstrated that the results could better be explained if the radiation were produced by a neutral particle having the mass of a proton.

Exploring Further: Ask students how the discovery of a neutral particle helped scientists explain atomic structure. Possible response: *Chemists could account for the difference between atomic number and atomic mass. The difference between isotopes could be explained as merely a difference in the number of neutrons.*

CHEMISTRY IN DEPTH

The section **Stability of Nuclides** on pages 720-723 is recommended for more able students.

CHEMISTRY IN DEPTH — **Stability of Nuclides**

Pages 720–723

Not all isotopes of an element are equally stable. What do we mean by a stable isotope? A stable isotope is one whose nucleus will not spontaneously decay. An unstable nucleus is one that undergoes spontaneous change of some kind. Some nuclei, such as $^{99m}_{43}$Tc, decay by giving off a quantum of gamma radiation. Various other unstable nuclei decay by emitting a particle with or without an accompanying quantum. There is also the possibility of the nucleus capturing an electron from the innermost energy level. This process is called *K*-capture because an old name for the innermost energy level was the *K*-level. It is possible to estimate which nuclides will be the most stable by applying the three following rules.

RULE OF THUMB

Rule 1. *The greater the binding energy per nucleon the more stable the nucleus.* The binding energy is the energy needed to separate the nucleus into individual protons and neutrons.

Consider the $^{16}_{8}$O nuclide. It contains eight protons, eight electrons, and eight neutrons. We can think of it as eight hydrogen (protium) atoms, each of which contains one proton and one electron, and eight neutrons. Each protium atom has a mass of 1.007 825 2 u. Each neutron has a mass of 1.008 665 2 u. Thus, the total mass of an $^{16}_{8}$O atom should be 16.131 923 2 u. However, the actual mass of the $^{16}_{8}$O atom is 15.994 915 0 u. The difference between the calculated mass and the actual mass is called the **mass defect.**

$$\text{mass of 8 } ^{1}_{1}\text{H atoms} + \text{mass of 8 } ^{1}_{0}n = \text{expected mass of } ^{16}_{8}\text{O atom}$$

$$\begin{aligned} 8(1.007\ 825\ 2\ \text{u}) + 8(1.008\ 665\ 2\ \text{u}) &= 16.131\ 923\ 2\ \text{u} \\ \text{actual mass of } ^{16}_{8}\text{O atom} &= \underline{15.994\ 915\ 0\ \text{u}} \\ \text{mass defect} &= 0.137\ 008\ 2\ \text{u} \end{aligned}$$

For an $^{16}_{8}$O atom, the mass defect is 0.137 008 2 u. This mass has been converted to energy and released in the formation of the oxygen nucleus. Thus, it is also the energy that must be put back into the nucleus to separate the nucleons.

We can convert this mass defect of 0.137 008 2 u into its energy equivalent using the equation

$$E = mc^2$$

$$E = \frac{0.137\ 008\ 2\ \text{u}}{} \left| \frac{1.660\ 40 \times 10^{-27}\ \text{kg}}{1\ \text{u}} \right| \frac{(2.997\ 93 \times 10^{8}\ \text{m})^2}{(\text{s})^2}$$

$$= 2.044\ 57 \times 10^{-11}\ \frac{\text{kg·m}^2}{\text{s}^2}$$

Recall that one joule is the energy required to maintain a force of one newton through a distance of one meter. A newton is equivalent to kg·m/s^2. Thus,

$$E = 2.044\ 57 \times 10^{-11}\ \text{J}$$

CHEMISTS AND THEIR WORK

James Chadwick (1891-1974)

Capitalizing on the research and theories of Rutherford, Bothe, and Joliot-Curie, James Chadwick produced experimental evidence of the existence of the neutron. The discovery of the neutron explained a number of puzzling features about atomic structure, including the existence of isotopes and the masses of atoms. Chadwick spent some of the time during World War II in the United States, working on a combined British and U.S. effort to develop a nuclear weapon.

CONTENT BACKGROUND

Natural Radioactivity: Radioactivity is a natural occurrence. Students are aware that nuclear bombs and reactors produce large amounts of heat. Radium undergoes natural radioactive decay that produces heat—552 J/h for each gram. Thus 1.0 gram of radium produces 1.00×10^{10} J of heat in its lifetime. This energy is equivalent to the heat produced by half a ton of coal when it burns. The heat from natural radioactivity can be detected deep within Earth. Nuclear energy and geothermal energy are related. Solar energy is from a great nuclear furnace also.

The energy described by this equation is called the **binding energy.** If we divide the total binding energy by the total number of nucleons in the oxygen atom, we obtain the binding energy per nucleon.

Total nucleons = 8 protons + 8 neutrons = 16 nucleons

$$\text{Energy per nucleon} = \frac{2.044\,57 \times 10^{-11}\ \text{J}}{1.6 \times 10^{1}\ \text{nucleons}}$$

$$= 1.277\,85 \times 10^{-12}\ \frac{\text{J}}{\text{nucleon}}$$

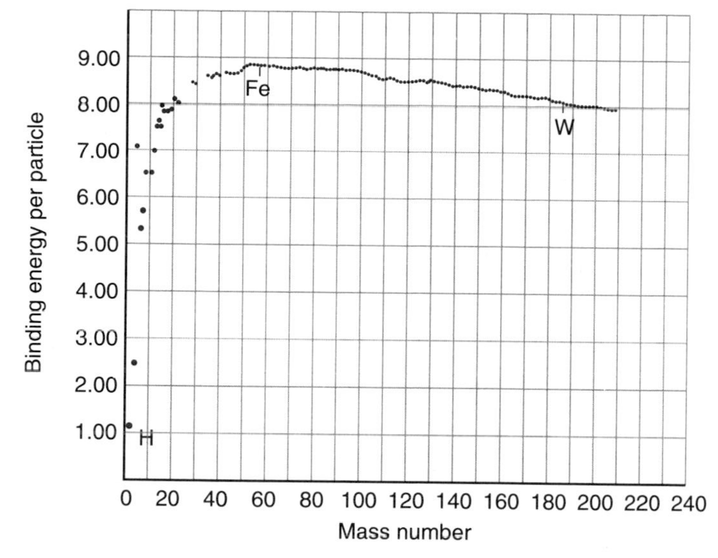

Figure 28.8 The curve of the graph shows binding energy per particle with increasing mass number. The curve reaches a maximum at iron (Fe). All other nuclei, both smaller and larger, are less stable than iron nuclei.

The greater the binding energy per nucleon, the greater the stability of the nucleus. In Figure 28.8, the binding energy per nucleon is graphed against the mass number of known nuclides. Note that energy will be released in two reaction types involving the nucleus. It is released when one large nucleus splits to form two medium-sized nuclei as in a nuclear reactor. It is also released when two small nuclei join to form a medium-sized nucleus. In both cases, the medium-sized nuclei have greater binding energies per nucleon than the nuclei from which they were produced.

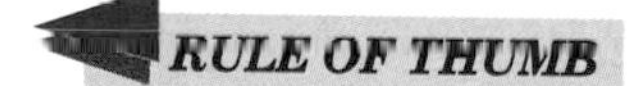

Rule 2. *Nuclei of low atomic numbers with a 1:1 neutron-proton ratio are very stable.* In Figure 28.9, the ratio of neutrons to protons is plotted for the known stable nuclei. For low atomic numbers, the ratio has a value very close to one. However, as the atomic number increases, the value of the neutron-proton ratio steadily increases. The closer the value of the neutron-proton ratio of a nuclide is to the broken line in this figure, the more stable the nuclide is.

CONCEPT DEVELOPMENT

- Emphasize that the rules are comparative rules and not absolute rules.
- Inform students that two kinds of nuclear change lead to nuclei with greater binding energy: the breakup of large nuclei (fission) and the combination of light nuclei (fusion).

Using the Illustration: In Figure 28.8 the total binding energy divided by the number of nucleons gives the binding energy per particle. This energy is plotted against the mass number. Remind students that the binding energy is the energy needed to separate the nucleus into individual particles.

Discussion Question: Ask students to apply Rule 2 and use the periodic table to give examples of stable, light nuclei (Z= 20 or less) with a 1:1 neutron-proton ratio. Possible response: *helium-4, carbon-12, nitrogen-14, oxygen-16, neon-20, magnesium-24, silicon-28, sulfur-32*

CONCEPT DEVELOPMENT

- Summarize Rule 3 for students by informing them that if the nucleus has both an even number of protons and neutrons it is stable, meaning not radioactive. If either the number of protons or the number of neutrons is even, but not both, then the nucleus is slightly less stable. If both the number of protons and neutrons are odd, there is a 98.5% chance that a given nucleus will be radioactive.

Misconceptions

Ask students how much material is left after one half-life. The typical answer is one-half. The correct answer is all of the material is left. If you start with 10 g of uranium, you still have 10 g at the end of one half-life. Half of the uranium (5 g) remains and half (5 g) has been transmutated into another element, thorium.

MAKING CONNECTIONS

Nuclear Medicine

Almost 80% of the radioactive waste generated comes from hospitals. Most radioactive isotopes used in medicine have a half-life of a few days or less. Some isotopes are used in diagnostic testing as markers or tracers. Others are used in the treatment of cancer. Have interested students ask a local university for information about a program in allied medicine with a specialty in nuclear medicine. After five years of schooling, the board certified radiologic technologist can expect to start at $40,000 per year. Write the American Society of Radiologic Technologists, 15000 Central Ave. S.E., Albuquerque, NM 87123.

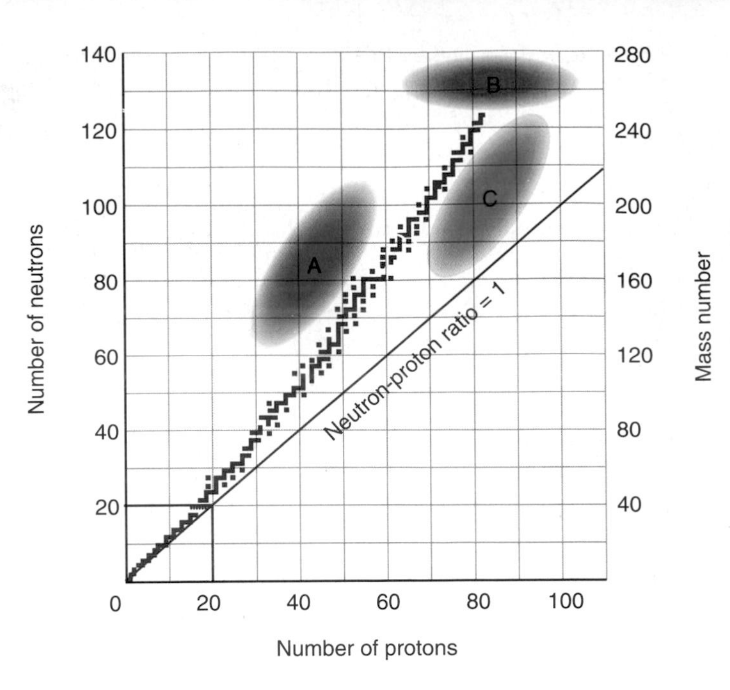

Figure 28.9 Unstable nuclei are represented by the shaded areas A, B, and C. The nuclei in region A emit neutrons or beta particles. The nuclei in region B emit alpha particles. Those nuclei in C emit positrons or capture electrons.

RULE OF THUMB

Rule 3. *The most stable nuclei tend to contain an even number of both protons and neutrons.* Of the known stable nuclei, 57.8 percent have an even number of protons and an even number of neutrons. Those nuclei with an even number of one kind of nucleon but an odd number of the other are slightly less stable. Thus, 19.8 percent of stable nuclei have an even number of neutrons but an odd number of protons; and 20.9 percent of stable nuclei have an even number of protons and an odd number of neutrons. Only 1.5 percent of stable nuclei have both an odd number of neutrons and an odd number of protons.

In Figure 28.9, nuclei falling within the regions *A*, *B*, and *C* are all unstable. Those nuclei lying in region *A* have excess neutrons and become more stable either by emitting neutrons or, more commonly, beta particles. For example,

$$^{131}_{53}\text{I} \rightarrow {}^{131}_{54}\text{Xe} + {}^{0}_{-1}e$$

Nuclei falling within the *B* region are too large for stability and are usually alpha emitters.

$$^{227}_{92}\text{U} \rightarrow {}^{223}_{90}\text{Th} + {}^{4}_{2}\text{He}$$

When a nucleus in the *B* region emits an alpha particle, its composition moves parallel to the 1:1 ratio line toward the *A* region. A nucleus in the *C* region has excess protons and can become stable by positron emission or by *K*-electron capture. Either the loss of a positron (β^+) or the capture of an electron (β^-) results in a new atom. This atom has the same mass number as the original atom but has an atomic number with a value one unit lower.

DEMONSTRATION

28-4 Simulating a Nuclear Scaler

Purpose: to simulate a nuclear scaler that records the number of disintegrations per second during a radioactive decay and use the data to determine the element's half-life

Materials: A free printout in Basic is available that can be adapted for PCs and MACs. An Apple II version is available on disk. Write to Richard G. Smith, 1416 Walker Dr. N.W., Lancaster, OH 43130. He will send you a copy of the program at minimum cost.

Procedure: The computer program written in Basic for Chapter 28 can be used to simulate a nuclear scaler measuring a radioactive source. Students can use the program to collect data that has some built-in experimental error. They can graph this data to get a curved line. When students graph

$$^{129}_{55}Cs + ^{0}_{-1}e \rightarrow ^{129}_{54}Xe$$
$$^{145}_{64}Gd \rightarrow ^{145}_{63}Eu + ^{0}_{1}e$$

Note that in the heavy-element radioactivity series, several alpha emissions are always followed by beta emission. The beta emission moves the nucleus back toward the stable region.

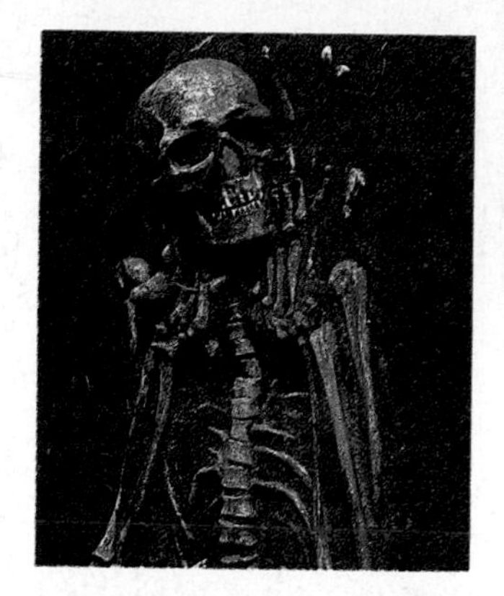

Figure 28.10 Carbon-14, an isotope of carbon with a half-life of 5730 years, can be used to date previously living artifacts that are less than 20 000 years old.

Half-Life

The rate at which many radioactive nuclides decay has been determined experimentally. The rate of disintegrations is measured in a unit called a becquerel, Bq. One becquerel is one disintegration per second. An older unit for rate was the curie, Ci, equal to 3.7×10^{10} Bq.

The number of atoms that disintegrate in a unit of time varies directly as the number of atoms present. The length of time it takes for one-half of the atoms to disintegrate has been chosen as a standard for comparison purposes. This time interval is called **half-life.** For example, the half-life of $^{131}_{56}Ba$ is 12 days. If we start with a given number n of atoms of $^{131}_{56}Ba$, then at the end of 12 days, $n/2$ atoms will have changed into another element or isotope. At that time, we will have $n/2$ $^{131}_{56}Ba$ atoms left. At the end of the next 12 days, half of the remaining atoms will have disintegrated and we will have $n/4$ $^{131}_{56}Ba$ atoms left. In 12 more days, half of these atoms will have disintegrated and $n/8$ atoms of $^{131}_{56}Ba$ will remain. How many atoms will remain at the end of another 12 days?

Table 28.1

Half-Life and Decay Mode of Selected Nuclides

Nuclide	Half-Life	Decay Mode	Nuclide	Half-Life	Decay Mode
$^{3}_{1}H$	12.26 years	β^-	$^{100}_{46}Pd$	4.0 days	K-capture and γ
$^{6}_{2}He$	0.797 second	β^-	$^{129}_{55}Cs$	32.1 hours	K-capture and γ
$^{14}_{6}C$	5730 years	β^-	$^{149}_{61}Pm$	53.1 hours	β^- and γ
$^{19}_{8}O$	29.1 seconds	β^- and γ	$^{145}_{64}Gd$	25 minutes	β^+ and γ
$^{20}_{9}F$	11.56 seconds	β^- and γ	$^{183}_{76}Os$	12.0 hours	K-capture and γ
$^{26}_{14}Si$	2.1 seconds	β^+ and γ	$^{212}_{82}Pb$	10.64 hours	β^- and γ
$^{39}_{17}Cl$	55.5 minutes	β^- and γ	$^{194}_{84}Po$	0.5 second	α
$^{49}_{21}Sc$	57.5 minutes	β^- and γ	$^{210}_{84}Po$	138.40 days	α and γ
$^{60}_{27}Co$	52 seconds	β^-	$^{226}_{88}Ra$	1602 years	α and γ
$^{71}_{30}Zn$	2.4 minutes	β^- and γ	$^{227}_{92}U$	1.3 minutes	α
$^{69}_{32}Ge$	36 hours	β^+ and γ	$^{235}_{92}U$	7.1×10^8 years	α and γ
$^{81}_{33}As$	33 seconds	β^-	$^{238}_{92}U$	4.51×10^9 years	α
$^{82}_{35}Br$	35.34 hours	β^- and γ	$^{236}_{94}Pu$	2.85 years	α and γ
$^{87}_{37}Rb$	4.8×10^{10} years	β^-	$^{242}_{94}Pu$	3.79×10^5 years	α
$^{91}_{42}Mo$	15.49 minutes	β^+ and γ	$^{250}_{96}Cm$	1.7×10^4 years	Spontaneous fission

CHEMISTRY IN DEPTH

Chemistry in Depth ends on this page.

CONCEPT DEVELOPMENT

Computer Demo: *Nuke 4,* Lab simulation of decay. AP1001, Project SERAPHIM.

Using the Table: You can use Table 28.1 as a source of information when writing additional nuclear equations for students to balance. The information on Table 28.1 can also be used when writing half-life problems for additional drill and practice.

3 ASSESS

CHECK FOR UNDERSTANDING

Have students answer Concept Review problems, and Review and Practice 19-50 from the end of the chapter.

RETEACHING

Use examples from **Merrill Chemistry Solving Problems in Chemistry,** Chapter 28 to reteach the essential skills and concepts.

EXTENSION

Ask a student to investigate the dating of minerals by the use of naturally occurring radioactive nuclides.

the log of the data, the line is straight.

Disposal: F

Results: Students can use the graphs to determine the half-life of the element.

Questions: These are provided with the printed material that accompanies the program or disk.

PROGRAM RESOURCES

Critical Thinking/Problem Solving: Use the worksheet, p. 29, as enrichment for the concept of "Nuclear Decay Series."

Enrichment: Use the worksheets "Decay Rate," pp. 55-56, to challenge more able students.

Applying Scientific Methods in Chemistry: Use the worksheet, "Radioactive Decay," p. 48, to help students apply and extend their lab skills.

ARCHEOLOGY CONNECTION
Carbon-14 Dating

Exploring Further: Ask a student to research how carbon-14 is formed. Possible response: *In the upper atmosphere, neutrons that are produced by cosmic rays collide with atmospheric nitrogen.*

$^{14}_{7}N$ + neutron → $^{14}_{6}C$ + proton

The radioactive carbon is then oxidized to carbon-14 dioxide, which is dispersed throughout the atmosphere.

PRACTICE

Guided Practice: Guide students through the Sample Problem, then have them work in class on Practice Problems 6-8.

Independent Practice: Your homework or classroom assignment can include Review and Practice 41-43 and 65 from the end of the chapter.

Answers to
Practice Problems: Have students refer to Appendix C for complete solutions to Practice Problems.

6. 1/1024 of the original sample remains after 15 days
7. 13.4 days
8. 2.27 x 10^{26} atoms

Answers to
Concept Review

9. The structure and properties of the nucleus are investigated by bombarding nuclei with high-energy particles and observing the results. The high-energy particles may be charged particles from an accelerator or neutrons from a nuclear reactor. In a linear accelerator, charged particles are accelerated to very high velocities by electromagnetic waves.

10. 1) The greater the binding energy per nucleon, the more stable the nucleus.

2) Nuclei of low atomic numbers with a 1:1 neutron-proton ratio are very stable.

3) The most stable nuclei tend to contain an even number of both protons and neutrons.

Because of its large mass number, the binding energy per nucleon for $^{212}_{82}Pb$ is relatively small. This tends to make the nuclide unstable. The $^{212}_{82}Pb$ nuclide has 82 protons and 130 neutrons, both even numbers. This may

ARCHEOLOGY CONNECTION

Carbon-14 Dating

The radioactive nuclide $^{14}_{6}C$ is formed at a constant rate in the upper atmosphere. The $^{14}_{6}C$ is also decaying at a steady rate. Therefore, its concentration in the atmosphere remains constant. All $^{14}_{6}C$ is converted to CO_2 immediately after formation. This CO_2 is used in the photosynthetic process and becomes a part of the plant. Animals eat the plant or plant-eating animals and take in $^{14}_{6}C$. Once an organism dies, however, no more $^{14}_{6}C$ is taken in, and that $^{14}_{6}C$ already present continues to decay, with a half-life of 5730 years. Thus, by comparing the $^{14}_{6}C$ content of the once-living archeological specimen with its concentration in living organisms, a date of death for the artifact can be found. After four half-lives, the $^{14}_{6}C$ activity is so low that it becomes impossible to separate from background radiation. Consequently, $^{14}_{6}C$ dating is only good for specimens up to 20 000 years old.

These half-life figures are determined experimentally for a large number of atoms of an individual nuclide. They predict the behavior of large numbers of atoms. At present, it is not possible to predict the exact instant when an individual atom will decay.

SAMPLE PROBLEM

Half-Life

If you start with 2.97 × 10^{22} atoms of $^{91}_{42}Mo$, how many atoms will remain after 62.0 minutes? The half-life of $^{91}_{42}Mo$ is 15.49 minutes.

Solving Process:
Divide 62.0 min by 15.49 min to find the number of half-lives.

$$\text{Number of half-lives} = \frac{62.0 \not{\text{min}}}{15.49 \not{\text{min}}} = 4.00$$

The $^{91}_{42}Mo$ will go through four half-life decay cycles in 62.0 minutes.

$$(1/2)^4 = 1/16$$

One-sixteenth of the atoms will remain. Multiplying 1/16 by the number of atoms you started with gives

$$2.97 \times 10^{22} \text{ atoms } (1/16) = 1.86 \times 10^{21} \text{ atoms}$$

Thus, after 62.0 minutes, 1.86 × 10^{21} atoms of $^{91}_{42}Mo$ remain.

PRACTICE PROBLEMS

6. What part of a sample of $^{69}_{32}Ge$ will remain after 15 days?

7. If there are 5.32 × 10^9 atoms of $^{129}_{55}Cs$, how much time will pass before the amount remaining is 5.20 × 10^6 atoms?

8. If you start with 5.80 × 10^{28} atoms of $^{242}_{94}Pu$, how many will remain after 3.03 × 10^6 years?

28.1 CONCEPT REVIEW

9. What are the ways in which scientists investigate the structure and properties of the nucleus? Describe the action of a linear particle accelerator as it is used in this investigation.

10. What three rules allow us to estimate relative nuclide stability? Would $^{212}_{82}Pb$ be predicted to be stable or unstable?

11. Explain the functions of the moderator and the containment vessel in a nuclear reactor.

12. Complete the following nuclear reaction: $^{231}_{91}Pa \rightarrow ^{227}_{89}Ac + ?$

13. Apply A sample of $^{129}_{55}Cs$ contains 2.54 × 10^{18} atoms. How many of these atoms will remain after 96.3 hours?

GLENCOE TECHNOLOGY

Videodisc

Chemistry: Concepts and Applications
Half-Life
Disc 3, Side 1, Ch. 6

Also available on CD-ROM.

Videodisc

STVS: Chemistry
Disc 2, Side 2
Carbon-14 Dating (Ch. 4)

28.2 Nuclear Applications

There is more to nuclear chemistry than huge accelerators and fission reactors. There are practical applications of radioactive nuclides in the chemistry laboratory, the hospital, the geologist's laboratory, and the archeologist's "dig." We must also be aware that radioactive nuclides, improperly handled, can pose a threat to health. In this section you will learn about some of the applications and dangers of radioactive nuclides.

OBJECTIVES

- explain how synthetic elements are produced.
- list the biological effects of radiation and the units used to measure them.
- describe the uses for radioactive nuclides.
- describe fusion reactions and the ongoing research being devoted to controlling the fusion process.

Synthetic Elements

Elements with atomic numbers greater than 92 are called the transuranium elements. All of the synthesized transuranium elements have been produced by converting a lighter element into a heavier one. Such a change requires an increase in the number of protons in the nucleus. One of the processes of synthetic transmutation occurs as follows. A nuclear reactor produces a high concentration of neutrons that are "packed" into the nucleus of $^{239}_{94}Pu$. As the mass number builds, a beta particle is emitted. When beta emission occurs, a neutron is lost, and a proton is gained. This process produces an element that has an atomic number greater than that of the original element. The process can be written

$$^{239}_{94}\text{Pu} + {}^{1}_{0}n \rightarrow {}^{240}_{94}\text{Pu}$$

$$^{240}_{94}\text{Pu} + {}^{1}_{0}n \rightarrow {}^{241}_{94}\text{Pu}$$

$$^{241}_{94}\text{Pu} \rightarrow {}^{241}_{95}\text{Am} + {}^{0}_{-1}e$$

The nuclide $^{241}_{95}Am$ in turn can be used as a target to produce another element with a higher atomic number. The element with the highest atomic number reached in this manner is $^{256}_{100}Fm$.

A second method of synthesizing transuranium elements makes use of nuclear explosions that produce vast numbers of neutrons. Some of these neutrons are captured by uranium

Figure 28.11 Ionization-type smoke detectors use $^{241}_{95}Am$ to ionize the air. The ions conduct an electric current, indicated by the arrows in the diagram on the right. The current is monitored by a current sensor that triggers the alarm when the current is interrupted by particles in smoke.

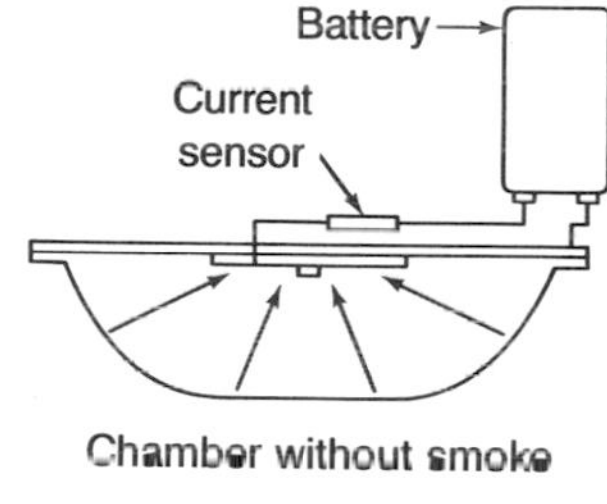

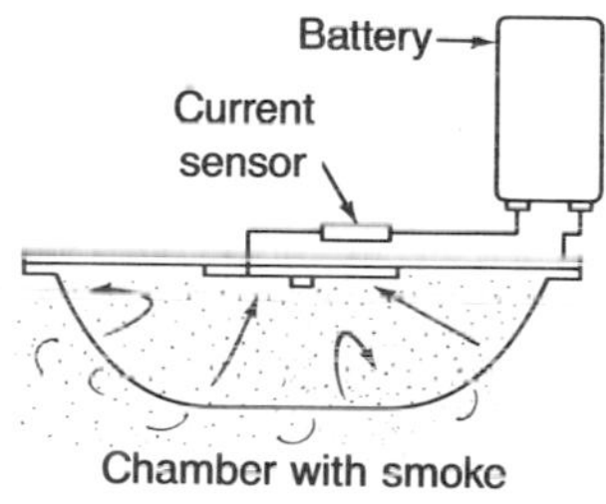

GLENCOE TECHNOLOGY

CD-ROM

Chemistry: Concepts and Applications
Exploration: *Dating Artifacts*

Videodisc

Chemistry: Concepts and Applications
Detecting Radiation

Disc 3, Side 1, Ch. 5

Show this video of Demonstration 28-5 to reinforce the concept of the effectiveness of materials in blocking radiation.

Also available on CD-ROM.

contribute to stability of the nuclide. Considering all the evidence, one expects $^{212}_{82}Pb$ to be fairly unstable.

11. The moderator slows down neutrons through collisions so that they initiate fission efficiently. The containment vessel keeps radioactive materials from escaping into the environment.

12. $^{231}_{91}\text{Pa} \rightarrow {}^{227}_{89}\text{Ac} + {}^{4}_{2}\text{He}$

13. 3.18×10^{17} atoms

4 CLOSE

Discussion: Ask students if they think the work of nuclear chemists has been beneficial to people. In Section 28.2 students will learn how the discoveries in nuclear chemistry have been applied in an attempt to make life better.

28.2 Nuclear Applications

PREPARE

Key Concepts

The major concepts presented in this section include synthetic elements, biological effects of radiation, uses of radioactive nuclides, fusion, and fusion reactors.

Key Terms

gray	tracer
sievert	fusion reaction

1 FOCUS

Objectives

Preview with your students the objectives listed on the student page. Students can use them as a study guide for this major section.

QUICK DEMO Using a Synthetic Element

Bring a battery-operated smoke detector to class. Open the case to show students the location of the americium-241 that is used to ionize the air. This specific use of a transuranium element will help focus attention as you begin the discussion of nuclear applications.

2 TEACH

Misconceptions

Many people think that only radioactive substances can be used as tracers. Point out to students that both radioactive and nonradioactive nuclides can be used as tracers in industry and in medicine.

MINUTE LAB

Too Close for Comfort

Purpose: to allow students to discover the inverse square law

Materials: flashlight with batteries, piece of newspaper, felt-tip marker, calculator, meter stick

Procedure Hints: A light meter can be used to replace the newspaper. Each decrease in *f*-stop reading represents half as much light as did the previous reading.

Disposal: F

Results: The lighted area gets larger, and the intensity of the illumination decreases as the distance increases. The inverse square law is represented by the formula $E \propto 1/d^2$, where E is energy and d is distance. Point out that the intensity of nuclear radiation, like that of light, varies as the inverse of the square of the distance from the source.

Answers to Questions: 1. less intense **2.** Use $A = \pi r^2$; student answers will vary. **3.** The area increases by 4 and then by a factor of 9. **4.** At 10 cm = 1, 20 cm = 1/4, 30 cm= 1/9. (Student measurements vary ± 30%) **5.** X-ray radiation at 4 m is 1/16 of that at 1 m.

✓ ASSESSMENT

Performance: See Assessment in the Demonstration on page 727.

MINUTE LAB

Too Close for Comfort

A color TV emits X rays. Investigate what happens to the intensity of the radiation you absorb when you move farther from the source. Cover a tabletop with newspaper. Cover the reflector of a flashlight with dark paper or tempera paint. In a darkened room, shine the flashlight on the tabletop from a height of 10 cm. Mark the limits of the lighted area. Repeat this procedure from 20 cm and then 30 cm above the table. What do you notice about the intensity of the illuminated area as the light moves farther from the paper's surface? Calculate the area illuminated during each of the three trials. Compute a ratio of areas to the nearest whole number. The same amount of energy continues to fall on the paper as the light moves farther away, but the energy is spread over a larger area. If the intensity was 1 unit at 10 cm, what was it at 20 cm? At 30 cm?

Predict how much X-ray energy you would receive if you were 4 meters from the TV instead of 1 meter. Test your hypothesis using the flashlight 40 cm above the table.

atoms. Successive electron emissions produce new elements. The element with the highest atomic number produced in this way is also $^{256}_{100}Fm$.

Elements with atomic numbers greater than 100 have been produced using other elements to bombard target elements. The nuclide $^{256}_{101}Md$ was created by bombarding $^{254}_{99}Es$ with alpha particles.

$$^{254}_{99}Es + ^{4}_{2}He \rightarrow ^{256}_{101}Md + 2^{1}_{0}n$$

Nobelium, atomic number 102, was created by using carbon nuclei and curium.

$$^{12}_{6}C + ^{244}_{96}Cm \rightarrow ^{254}_{102}No + 2^{1}_{0}n$$

The production of lawrencium, atomic number 103, can use neon and californium or einsteinium.

$$^{22}_{10}Ne + ^{254}_{99}Es \rightarrow ^{262}_{103}Lr + ^{12}_{6}C + 2^{1}_{0}n$$

One way that element 104 can be produced is by the bombardment of plutonium by neon. An alternate method uses carbon and californium.

$$^{12}_{6}C + ^{249}_{98}Cf \rightarrow ^{257}_{104}Unq + 4^{1}_{0}n$$

Element 105 can be produced by the bombardment of californium with nitrogen.

$$^{15}_{7}N + ^{249}_{98}Cf \rightarrow ^{260}_{105}Unp + 4^{1}_{0}n$$

Other nuclides have also been produced.

$$^{209}_{83}Bi + ^{58}_{26}Fe \rightarrow ^{267}_{109}Une$$

A heavy-ion accelerator in California is being prepared to accelerate particles as heavy as bromine nuclei.

It is highly possible that elements with even greater atomic numbers can be produced. The elements produced thus far are characterized by low yields and extremely short half-lives. Only a few atoms of elements 103–109 were first prepared, and these had half-lives of seconds. However, the latest lawrencium isotope produced, $^{262}_{103}Lr$, has a half-life of 216 minutes. Nuclear scientists believe that elements with atomic numbers as high as 126 might be produced.

Biological Effects of Radiation

All radiation (particles and electromagnetic waves) has an effect on living organisms. If the radiation has enough energy, it can penetrate living cells and disrupt their function. The disruption is particularly dangerous if a nucleic acid molecule is affected. Nucleic acids make up the genetic material of cells.

The amount of radiation, called kerma, is measured in a unit called the **gray,** Gy. One gray is equivalent to the transfer of exactly one joule of energy to one kilogram of living tissue. An older unit used to measure the same quantity is the rad, which is

DEMONSTRATION

28-5 Using a Geiger Counter

Purpose: to demonstrate shielding materials and their effectiveness, and to demonstrate the inverse square law

Materials: Geiger counter with meter, packaged radioactive source, small pieces of cloth, aluminum foil, glass plates, roofing shingles, paper, wood, meter stick

Procedure: 1. If the Geiger counter has a meter, you may want to take quantitative data. Mount the Geiger tube about 5 cm from the packaged radioactive source. Record the meter reading. Place pieces of the shielding material being tested between the source and the geiger tube. Add pieces one at a time and take readings after each piece. You may want to graph the data. You can have the students bring in materials that they

equal to 0.01 gray. The damage done biologically is better indicated by the absorbed dose, which is measured in the unit **sievert,** Sv. A sievert is equal to a gray multiplied by factors that determine how much of the energy transferred was actually absorbed by the tissue. An older measure of absorbed dose is the rem, which is equal to 0.01 sievert.

We are always exposed to some radiation from rocks containing radioactive elements, cosmic rays, and radioactive atoms naturally present in foods and water. These sources expose the average person to about one millisievert each year. About 0.02 mSv per year has been added to the exposure by testing nuclear weapons and operating nuclear power plants.

The radiation from radioactive sources varies in its effect on humans. All three types of radiation (α, β, and γ) occur in a range of energies. However, we may make the following general observations. Alpha particles are the least penetrating; a thin cotton garment will stop them. However, if an alpha particle gets inside the body, for example, when we breathe or eat contaminated material, it will do the most damage. Remember that an alpha particle has a large mass and charge compared with the other types of radiation. Gamma rays are the most penetrating.

Radiation exposure must sometimes be balanced against any other factors in a value judgment. For example, an X ray of a leg would produce about 0.2 mSv. However, if your leg is broken, you may decide to have your leg x-rayed. The danger of an incorrectly set bone outweighs the slight additional exposure to radiation.

Uses of Radioactive Nuclides

Because radioactive elements are easily detected by their radiation, they can be used as **tracers.** Tracers have a number of practical applications in chemistry. In quantitative analysis, a small amount of a radioactive nuclide of the element sought is introduced into the sample. The proportion of the unstable nuclide recovered in the analytical process is measured. That ratio, along with the actual total amount of substance recovered, can be used to compute the original quantity of unknown.

CAREERS IN CHEMISTRY

Science Writer

Science writers explore and report on major scientific and technological developments, explaining science discoveries and breakthroughs in everyday language. They call attention to issues that have important implications for society and spread information outside the scientific community. Sometimes, science writers are scientists, but more often they are people with excellent writing skills and a strong interest in science. Science writers are usually employed by the media. After a writer develops a strong reputation, he or she may be offered a contract for a book.

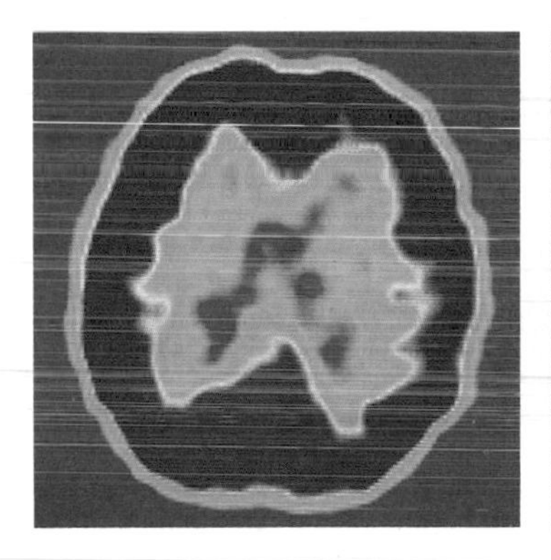

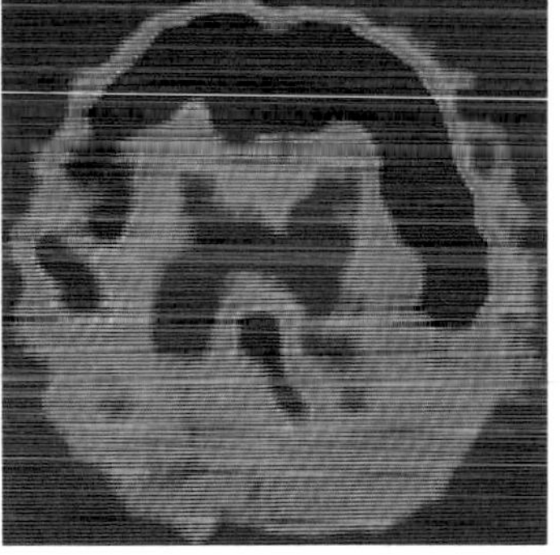

Figure 28.12 Radioisotopes can be used to find brain problems. The left scan shows a healthy brain. The right scan shows the brain of a patient with Alzheimer's disease.

CONCEPT DEVELOPMENT

- You may want to have a student contact a doctor of radiation medicine and then make a report to the class on the use of radiation in the treatment of cancer. If you can get a doctor to come and talk to one of your classes, you will want to videotape the presentation to show future classes.
- You may want to inform students that on a relative scale, if the radiation is internal, the biological damage of beta particles or gamma rays =1, neutrons = 5, and alpha particles = 10.
- Point out to students that nuclear reactor workers are permitted to receive up to 0.05 Sv/y of radiation. 1 Sv/h results in radiation poisoning. 3-5 Sv/h can result in death in 50% of the cases. Background radiation is about 0.001 Sv/y.

CAREERS IN CHEMISTRY
Science Writer

Background: Technical writers may serve as the link between the engineers who design equipment and the people who will install, use, and service it. Students interested in such a position will need a science and engineering background in addition to good writing skills.

For More Information: Have interested students write the Society for Technical Communications, 901 N. Stuart Street, Arlington, VA 22203-1821.

PROGRAM RESOURCES

Study Guide: Use the Study Guide worksheet, p.118, for independent practice with concepts in Section 28.2.

want tested, such as cloth, aluminum foil, glass plates, roofing shingles, and wood.

2. Use tongs to move the radioactive source along a meter stick. Record the Geiger counter's meter reading as the distance between the tube and the source increases. **Disposal:** F

Results: The relative effectiveness for equal thicknesses of paper is 1, masonry is 16, iron is 50, and lead is 100. Figures will vary with type of paper and type of masonry. As the distance between the radioactive source and the geiger tube is doubled, the amount of radiation received is only one-fourth as much ($E = 1/d^2$)

Questions: 1. Which material provides the best shielding? *lead*

2. What happens to the radiation level as the distance from the source increases? *It falls dramatically.*

ASSESSMENT

Performance: Have students simulate the inverse square law by varying the distance of a thermometer from a heat lamp and recording the thermometer's reading. The bulb of the thermometer should be wrapped with black paper.

CONCEPT DEVELOPMENT

- Use the diagram in the text's narrative to point out to students that double bonds can be found in cyclic compounds. Have them look at cyclohexene.
- Use the example in the text, 1,3-butadiene, to reinforce with students that a molecule whose name ends in *-diene* contains two double bonds.

CULTURAL DIVERSITY

Amino Acid Killer

Seemingly healthy Amish babies who experienced infections involving fever and diarrhea lapsed into comas. Many died within 48 hours. Others lived, but suffered progressive paralysis. In one family, five of seven children became ill, but no one could figure out why.

Dr. Holmes Morton, on a research fellowship at The Children's Hospital of Philadelphia, noticed an odd peak on a graph analyzing the blood of four-year-old Danny Lapp and found the key to understanding the genetic disease, *glutaric aciduria*, which affects one out of every 200 Amish children in the United States. When affected children are stressed by infection, their bodies are unable to break down two of the 20 amino acids that make up protein, and a toxic byproduct, glutaric acid, builds up in the bloodstream and muscle tissues. The acid attacks the liver, nervous system, and brain, causing death or brain damage and paralysis.

Dr. Morton and Rebecca Huyard, an Old Order Amish woman who taught school for 15 years, are currently screening over 65% of Amish newborns for glutaric aciduria. Infants testing positive are placed on a low protein diet, given riboflavin to break down glutaric acid, and kept well hydrated. This home therapy is inexpensive and effective.

Figure 29.4 Ethene (ethylene) is a plant hormone essential to the growth and maturation of fruits and vegetables. Tomatoes treated with ethene ripen at the same time for harvesting.

Two alkenes are commercially important, ethene (common name *ethylene*) and propene (common name *propylene*). In fact, ethene is the number one organic chemical in industry. More than 17 million tons are produced annually! It is obtained chiefly as one product of the refining of petroleum (Chapter 30). It is used to produce plastics, antifreeze, synthetic fibers, and solvents. Propene is also a by-product of petroleum refining and is used to manufacture plastics and synthetic fibers. Double bonds may also be found in cyclic compounds. A typical example of a cycloalkene is cyclohexene.

```
H         H   H
 \        |   |
  C == C — C — H
 /            |
H             H
```

propene

cyclohexene

A single molecule may contain more than one double bond. In that case the *-ene* ending must be preceded by a prefix indicating the number of double bonds in the molecule. Thus, the compound $CH_2{=}CH{-}CH{=}CH_2$ is named 1,3-buta*di*ene. 1,3-butadiene is produced from petroleum in large quantities to be used in the production of synthetic rubber.

Alkynes

A third homologous series of hydrocarbons consisting of molecules containing triple bonds between carbon atoms is called **alkynes** (AL kyns).

H:C:::C:H

carbon-carbon triple bond

DEMONSTRATION

29-6 An Alkyne Named Acetylene

Purpose: to produce ethyne, and to observe complete and incomplete combustion

Materials: 3 chunks of calcium carbide, 3 test tubes with stoppers, 3 400-cm^3 beakers, 3 wood splints, laboratory burner

Procedure: 1. Fill a 400-cm^3 beaker with water. Fill a test tube completely with water. Place a chunk of calcium carbide in the water in the beaker and cover with the inverted, water-filled test tube. Collect the gas by H_2O displacement.

2. Collect two additional tubes of gas, one 50% gas (start with a tube half-filled with water and half air), and one 10% gas (start with a tube 10% filled with water, 90% air).

3. Keeping the mouth of the test tubes down, ignite each gas mixture by placing a burning splint in the mouth of each test tube. **CAUTION:** *Fire risk.*

Alkynes constitute a homologous series with the general formula C_nH_{2n-2}. They are important raw materials for industries producing synthetic materials such as plastics and fibers. Chemically, alkynes are very reactive. The alkynes are named just as the alkenes, except the ending *-yne* replaces *-ene*. *Acetylene* is the common name for ethyne, the first member of this series. Ethyne is commercially the most important member of the alkyne family. The first two members of the alkyne family are ethyne, C_2H_2, and propyne, C_3H_4.

$$\mathrm{H{-}C{\equiv}C{-}H} \qquad \mathrm{H{-}C{\equiv}C{-}\underset{\displaystyle H}{\overset{\displaystyle H}{C}}{-}H}$$

ethyne (acetylene) — propyne

In naming alkynes, the numbering system for location of the triple bond and the substituent groups follows the same pattern as was used for naming the alkenes. For example, the name of the following compound is 4,4-dimethyl-2-pentyne.

$$\overset{1}{\mathrm{CH_3}}{-}\overset{2}{\mathrm{C}}{\equiv}\overset{3}{\mathrm{C}}{-}\underset{\displaystyle CH_3}{\overset{\displaystyle CH_3}{\overset{4}{\mathrm{C}}}}{-}\overset{5}{\mathrm{CH_3}}$$

4,4-dimethyl-2-pentyne

If a compound contains both double and triple bonds, the double bonds take precedence in numbering; that is, the double bond is given the lower number and is named first.

Aromatic Hydrocarbons

To an organic chemist, one of the most important organic compounds is benzene, a cyclic hydrocarbon.

The benzene ring is diagrammed as ⌬ .

In this structural representation of C_6H_6, it is assumed that there is a carbon atom at each corner with one hydrogen atom attached.

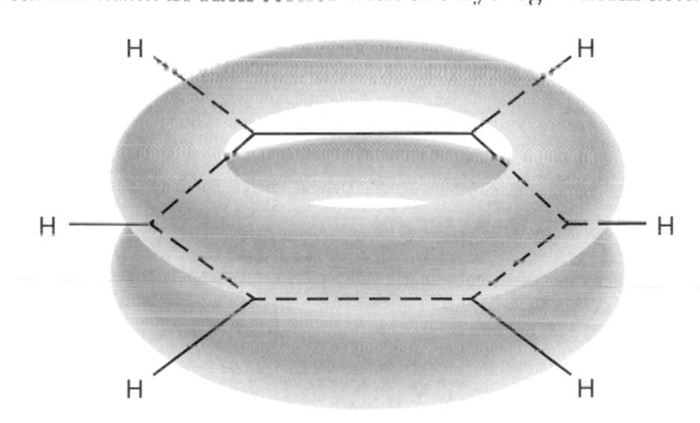

Figure 29.5 Aromatic hydrocarbons have a benzene ring or similar structure. The common characteristics of these compounds arise from the delocalization of the π electrons as shown here.

CONCEPT DEVELOPMENT

- Students will find that learning the general formula for an alkyne, C_nH_{2n-2}, is helpful if you plan to ask them to write formulas.
- Emphasize to students that because of the importance the triple bond plays in the compound's reactivity, the longest carbon chain that is numbered must contain the triple bond. The parent chain is numbered so the lowest position number is assigned to the first triple-bonded carbon atom.
- Advanced students may want to know that compounds with two or more triple bonds belong to groups called alkadiynes, alkatriynes, etc. The second triple bond has little effect on the first unless they are relatively close in the chain.
- Ask students to recall that in Chapter 13 they learned that benzene contains a conjugated system of bonds. Point out that aromatic compounds are generally prepared from benzene or benzene derivatives in the laboratory.

Misconceptions

Explain to students that it is not necessary for a compound to have a distinctive odor for it to be an aromatic compound. It is only necessary that the compound include a benzene ring.

Disposal: C

Results: The tube that was 100% ethyne gas burned at the mouth of the tube only and produced soot. The tube that is 50% gas produced large amounts of soot. The 10% gas tube burned with no soot being produced. It made a sound as the hot gases left the tube.

Questions: 1. Why does the test tube that was 100% gas only burn at the mouth? *Oxygen is not inside the tube, and is required for combustion.*

2. Why does the tube that was 50% gas produce soot? *There is not enough air for efficient and complete combustion. The soot is unburned carbon.*

3. Which tube is the cleanest after combustion? Why? *In the 10% gas tube the fuel is efficiently burned, releasing the energy that is stored in its chemical bonds.*

Extension: Discuss the pollution caused by an improperly tuned automobile engine.

PROGRAM RESOURCES

Chemistry and Industry: Use the worksheets, pp. 53-56, "Benzene: The Making of an Organic Chemical" as an extension to the concepts presented in Section 29.1.

Note from Chapter 13 that the benzene ring has a system of delocalized electrons. Therefore, it possesses great stability. The benzene ring diagram should not be confused with the symbol for cyclohexane. Cyclohexane, C_6H_{12}, is an alkane composed of six carbon atoms bonded only by single bonds.

Thousands of compounds are derived from benzene, and their study constitutes a whole branch of organic chemistry. Most of these compounds have rather distinctive odors. Thus, they are called aromatic compounds. Aromatic compounds are organic compounds that have a benzene ring or a similar structure. These compounds are normally named as derivatives of benzene. Aromatic compounds occur in small quantities in some petroleum reserves. They occur to a large extent in coal tar obtained from the distillation of coal. Some compounds consist of a fused system of several rings. These compounds have properties similar to benzene. An example of a fused ring compound is naphthalene.

naphthalene

The radical formed by removing a hydrogen atom from a benzene ring is called the phenyl radical. The symbol for the phenyl radical is

phenyl radical

Some examples of benzene compounds are

CH_3

methylbenzene (toluene)

CH_2 CH_3

ethylbenzene

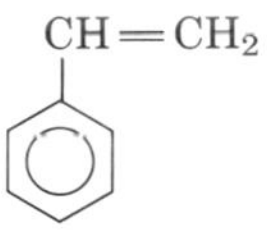

phenylethene (styrene)

$CH_3-CH-CH_3$

2-phenylpropane (cumene)

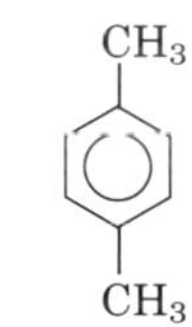

1,4-dimethylbenzene (*p*-xylene)

Benzene, toluene, and xylene are synthesized from petroleum. Ethylbenzene and cumene are made from benzene, while styrene is made from ethylbenzene. All of these compounds are important to our economy. Benzene is used in making plastics and fibers. Toluene is used to improve the quality of gasoline, as well as in the production of explosives and other chemicals. The explosive TNT is 2,4,6-trinitrotoluene. Ethylbenzene is converted almost entirely to styrene. Styrene is a vital component of synthetic rubber, plastics, and paints. The compound *p*-xylene (p = *para*) is a raw material for polyester fibers. Cumene is used in making plastics. A very large part of our synthetic materials industry is based upon aromatic compounds containing a benzene ring.

BRIDGE TO MEDICINE

Nuclear Magnetic Resonance

Background: The first measurement of nuclear magnetic moment in gases was done in 1933 by German scientists Otto Stern and I. Estermann.

Teaching This Feature: Point out that MRI does not involve X rays as a CAT scan does. The technique is especially useful in examining the brain, because bone does not give a response and the skull doesn't obscure the brain as in an X ray. Unlike surgery, MRI is noninvasive.

Answers to
Exploring Further

1. NMR can be used to differentiate between isomers, to provide evidence of hydrogen bonding, to follow the kinetics (rate) of a reaction, and perform quantitative analysis.

2. Hydrogen-1, carbon-13, and phosphorus-31 are naturally magnetic and are present in compounds that are of interest to the biochemist.

3 ASSESS

CHECK FOR UNDERSTANDING

Have students answer Concept Review problems, and Review and Practice 16-34 from the end of the chapter.

RETEACHING

On a handout, draw several structural formulas of different hydrocarbons. Have students decide if they are alkanes, alkenes, or alkynes. Review how the compounds are named.

GLENCOE TECHNOLOGY

Videodisc

Chemistry: Concepts and Applications

Properties of Ethyne: A Covalent Compound
Disc 1, Side 1, Ch. 12

Show this video of the Minute Lab on page 154.
Also available on CD-ROM.

CONTENT BACKGROUND

Stereo Isomerism: Stereoisomers can be divided into two major groups. One group, the diastereoisomers, are not mirror images of each other. A good example of diastereoisomerism is the geometric isomerism exhibited by cis-2-butene and trans-2-butene. Enantiomers are isomers that are mirror images of each other. Enantiomers can exist only for chiral molecules. Chiral comes from the Greek word for hand. Your hands are mirror images of each other.

Chiral molecules contain a chiral center, or a carbon atom to which four different groups are attached. The substance 2-butanol is a good example of a chiral molecule. Carbon number 2 is attached to one hydrogen atom, one hydroxyl group, one methyl group, and one ethyl group. It exists in two enantiomeric forms, a left-handed version and a right-handed version.

Bridge to *MEDICINE*

Nuclear Magnetic Resonance

Both protons and neutrons have the property of spin and that motion creates a magnetic field. Just as electrons pair in orbitals, like nucleons also pair. If two neutrons pair, their spins will be opposite and their magnetic fields will cancel each other. However, for nuclei with unpaired nucleons, the nucleus as a whole should possess a magnetic field.

If a nucleus with a magnetic field is placed in an external magnetic field, two energy states are possible for the nucleus. The two fields are either aligned or opposed. The opposed fields are a higher energy state. The energy required to "flip" the nucleus from one state to the other is affected by nearby atoms in the molecule. The energy can be measured by a process called nuclear magnetic resonance, NMR, spectroscopy.

About two-thirds of naturally occurring nuclei have magnetic fields. An NMR spectrum of an organic molecule shows several energy peaks. These peaks correspond to hydrogen atoms (protons) in different locations in a molecule.

NMR spectroscopy has been adapted to a medical diagnostic technique called magnetic resonance imaging, MRI. By having the detector scan the body while varying the magnetic field, a three-dimensional image can be obtained. MRI can be used to detect tumors, to monitor chemical processes in the body, and to locate malformations in organs.

Exploring Further

1. What are other applications of NMR spectroscopy?
2. What natural nuclides could be used by biochemists in NMR spectroscopy?

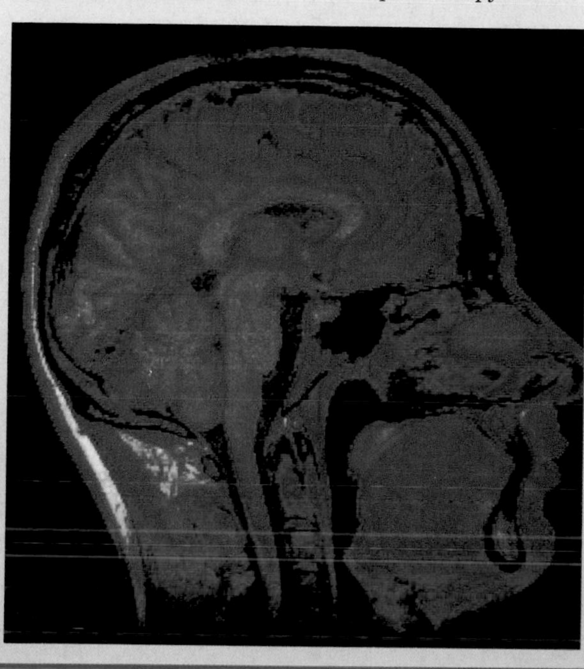

29.1 *CONCEPT REVIEW*

8. How do the following hydrocarbons differ?
   a. aromatic and aliphatic
   b. chain and cyclic
   c. saturated and unsaturated
9. How is the numbering of the parent chain determined in naming a branched hydrocarbon?
10. What are the general formulas of the alkane, alkene, and alkyne homologous groups? Name the compound in each series with the lowest molecular mass.
11. **Apply** 1,2- and 1,3-dimethylbenzenes are xylenes used in producing industrial chemicals, plastics, and in enriching gasoline. Write their structural formulas.

EXTENSION

Unbranched isomers have higher melting and boiling points than do highly branched isomers. Ask a student to verify this statement by using a chemistry handbook.

Answers to
Concept Review

8. a. Aromatic hydrocarbons contain one or more benzene rings; aliphatic hydrocarbons do not contain such a structure.

b. Cyclic hydrocarbons contain carbon atoms bonded in a closed ring; chain hydrocarbons do not contain a ring structure.

c. In saturated hydrocarbons, each carbon-carbon bond is a single bond; unsaturated hydrocarbons contain one or more multiple bonds.

9. Carbons of the parent chain are numbered so that the lowest possible position number is given to the first branch.

10. alkane C_nH_{2n+2}; methane CH_4; alkene C_nH_{2n}; ethene C_2H_4; alkyne C_nH_{2n-2}; ethyne or acetylene C_2H_2

11. For structures, see *Merrill Chemistry Problems and Solutions Manual.*

4 CLOSE

Discussion: Organic chemistry was once defined as the chemistry of compounds found in living or once-living tissue. It was thought that a "vital force" was required to produce these compounds, and they could never be produced in a laboratory. Discuss the work of Friedrich Wöhler, who synthesized urea. Ask students what our world would be like today if chemists had adhered to the "vital force" theory.

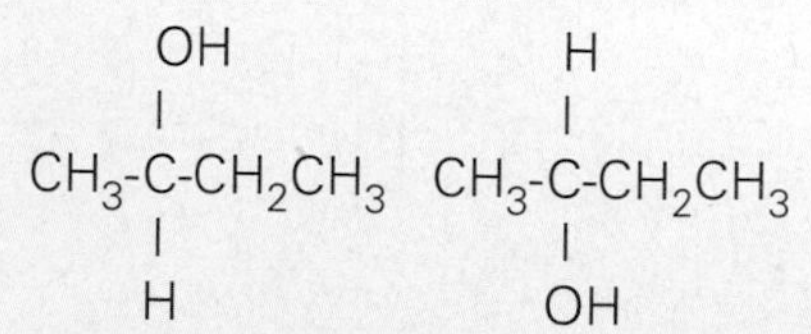

Stereoisomers have the property of rotating the plane of plane-polarized light. Such substances are called optically active, and the amount of rotation is a characteristic physical property of the molecule. Optical activity is measured in a polarimeter.

In an equimolar mixture of both enantiomers, polarized light will be unaffected, since one enantiomer would rotate the light in one direction and the other would rotate it in the other direction. In living systems, however, usually only one form is metabolically important. In fact, in a few instances, the enantiomer of the naturally occurring form is toxic to a living organism.

29.2 Other Organic Compounds

PREPARE

Key Concepts

The major concepts presented in this section include halogen-substituted hydrocarbon derivatives, organic oxygen compounds including alcohols, ethers, aldehydes, ketones, acids, and esters, and organic nitrogen compounds.

Key Terms

alcohol
phenol
inductive effect
amine
amide

1 FOCUS

Objectives

Preview with your students the objectives listed on the student page. Students can use them as a study guide for this major section.

QUICK DEMO **Organic Diversity**

Obtain samples of organic compounds that contain halogens, oxygen, and nitrogen from the chemistry and biology stockrooms and exhibit them to students. Point out to students that 9 million of the 10 million compounds known to chemists are organic compounds. These chemical compounds are a part of our daily lives. Reinforce the idea that students should know more about them.

OBJECTIVES

- identify and explain the inductive effect.
- differentiate among the functional groups of substituted hydrocarbons, organic oxygen compounds, and organic nitrogen compounds.
- recognize, name, and write structural formulas of representative halogen substituted hydrocarbons, organic oxygen compounds, and organic nitrogen compounds.

29.2 Other Organic Compounds

Atoms other than carbon and hydrogen can be substituted for part of a hydrocarbon molecule. When this substitution occurs, the chemical reactivity of the hydrocarbon is generally increased. The nonhydrocarbon part of the molecule is called a functional group. Most of the chemical reactivity of the substituted hydrocarbon is due to the functional group.

Halogen Derivatives

One family of substituted hydrocarbon molecules has a halogen atom substituted for a hydrogen atom. For example, if we substitute a bromine atom for a hydrogen atom on methane, we obtain CH_3Br. The name of this compound is bromomethane. It is also possible to replace more than one hydrogen atom by halogen atoms. In the compound $CHCl_3$, three chlorine atoms have been substituted for three of the hydrogen atoms in a methane molecule. The name of this compound is trichloromethane. You may know this compound by its common name, *chloroform*. Chloroform has been widely used as a solvent. It was once used as an anesthetic. In the compound CCl_4, four chlorine atoms have been substituted for the four hydrogen atoms in a methane molecule. The common name of this tetrachloromethane compound is carbon tetrachloride. These compounds are named as derivatives of the hydrocarbons.

```
     H                H                Cl
     |                |                |
Cl — C — Cl      H — C — Br      Cl — C — Cl
     |                |                |
     Cl               H                Cl
```

trichloromethane — bromomethane — tetrachloromethane

In large chains, we number the carbon atoms to avoid any confusion in naming the compounds. Suppose we have a chain that contains both a double bond and a halogen. In this case, begin the numbering at the end closer to the double bond. Thus, the name of the following compound is 1,4-dichloro-1-butene.

```
     H    H
     |    |
H — C⁴ — C³ — C² = C¹ — H
     |    |    |    |
     Cl   H    H    Cl
```

In aromatic compounds, it is necessary to indicate the relative positions of the various substituent groups on the ring. If two or more substituents are attached to the benzene ring, it is necessary to assign position numbers to the carbon atoms of the ring. The atoms in the benzene ring are numbered to give the smallest position numbers to the substituents. For example, the name of the following compound is 1,3-dibromobenzene, not 1,5-dibromobenzene.

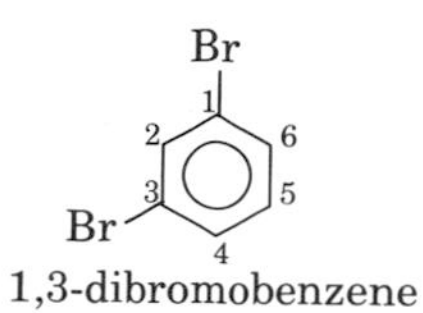

There are four 1-positions possible in each molecule of naphthalene. The 1-position that gives the lowest numbers to substituents is always used. The 1-position is next to the atom without a hydrogen atom attached. The numbering system for naphthalene is as follows.

naphthalene 2-chloronaphthalene 1,3-dibromonaphthalene

Several more examples that illustrate the naming of substituted hydrocarbons are

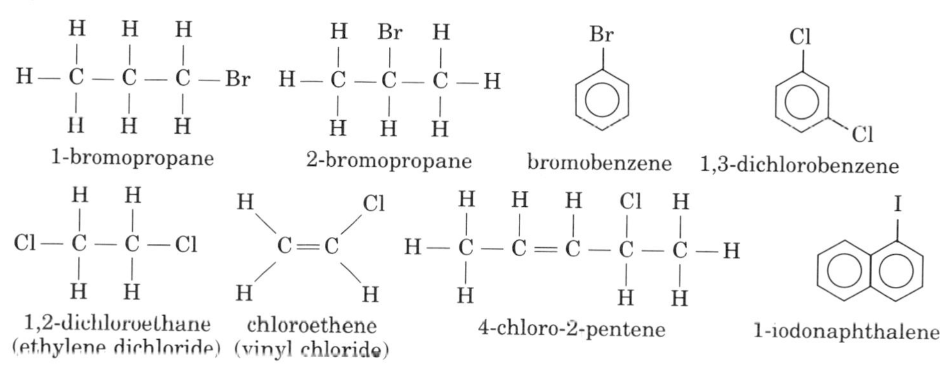

1,2-dichloroethane (common name *ethylene dichloride*) is manufactured in large quantities from ethene. In turn, it is converted to chloroethene (common name *vinyl chloride*). The vinyl chloride is used to make a plastic, polyvinyl chloride (PVC). PVC has numerous applications, such as water and waste piping.

Figure 29.6 Plumbing pipes are manufactured from polyvinyl chloride (PVC), a compound that is produced from halogen derivatives of ethene.

2 TEACH

CONCEPT DEVELOPMENT

- Point out to students that substitutions in hydrocarbons generally cause the resulting compound to be more reactive than the unsubstituted hydrocarbon.
- Use models to help students visualize that a halogen derivative has a halogen atom substituted for a hydrogen atom.
- Remind students that when numbering the parent chain of substituted hydrocarbons, it is numbered from the end that gives the lowest position number to the substituent. Note that a double bond has precedence over the substituent when numbering the parent chain.

MAKING CONNECTIONS

Medicine

The halogen-substituted hydrocarbon, ethyl chloride, is often used as a local anesthetic. Iodoform is used as an antiseptic.

PROGRAM RESOURCES

Study Guide: Use the Study Guide worksheets, pp. 122-123, for independent practice with concepts in Section 29.2.

Transparency Master: Use the Transparency master and worksheet, pp. 99-100, for guided practice in the concept "Aliphatic and Aromatic Hydrocarbons."

Color Transparency: Use Color Transparency 50 to focus on the concept of "Aliphatic and Aromatic Hydrocarbons."

CHAPTER 30

Organic Reactions and Biochemistry

CHAPTER ORGANIZER

Chapter Objectives	Text Features	Laboratory Options	Teacher Classroom Resources
30.1 Organic Reactions and Products **1.** Describe substitution, addition, elimination, esterification, saponification, addition polymerization, and condensation polymerization reactions. **2.** Describe the processing of petroleum and the octane rating of gasolines. **3.** Explain how synthetic fibers, plastics, and elastomers are produced industrially. **Nat'l Science Stds: UCP.1, UCP.2, B.2, B.3, E.1, E.2, F.3, G.1, G.2**	**Chemists and Their Work:** Bertram O. Fraser-Reid, p. 771 **Home Ec Connection:** Soaps and Detergents, p. 772	**Minute Lab:** Time to Be Inventive, p. 773 **ChemActivity 30:** Production of Aspirin - microlab, p. 852 **Discovery Demo:** 30-1 Making Nylon, p. 768 **Demonstration:** 30-2 Monomers→Polymers p. 772 **Demonstration:** 30-3 The Superabsorbent, p. 774 **Demonstration:** 30-4 A Cross-Linked Polymer, p. 776	**Study Guide:** pp. 125-126 **Lesson Plan,** pp. 62-63 **Critical Thinking/Problem Solving:** Everyday Organic Chemistry, pp. 31-32 **Applying Scientific Methods in Chemistry:** Synthesizing an Organic Compound, p. 50 **Chemistry and Industry:** Nylon, pp. 57-60 **ChemActivity Master 30**
30.2 Biochemistry **4.** State the structure and function of the four main biomolecules: proteins, carbohydrates, lipids, and nucleic acids. **5.** Describe the progress in and problems associated with the biomaterials industry. **Nat'l Science Stds: UCP.1, UCP.5, B.2, B.3, C.1, C.5, E.2, F.6, G.1, G.2**	**History Connection:** The First Synthetic Fiber, p. 786 **Frontiers:** DNA Fingerprinting, p. 787 **Chemistry and Society:** Genetic Engineering, p. 790	**Minute Lab:** Sugar Fermentation—It's a Gas, p. 783 **Demonstration:** 30-5 What's in a Protein?, p. 782 **Demonstration:** 30-6 Friction-Produced Luminescence in Sugar, p. 784 **Demonstration:** 30-7 Chromatographic Separation of Amino Acids, p. 786 **Laboratory Manual:** 30-1 Biochemical Reactions; 30-2 Qualitative Analysis of Food - microlab; 30-3 An Enzyme-Catalyzed Reaction - microlab	**Study Guide:** pp. 127-128 **Reteaching:** Proteins, pp. 44-45 **Enrichment:** Steroids, pp. 59-60
Other Chapter Resources	**Vocabulary and Concept Review,** pp. 30, 101-102 **Evaluation,** pp. 117-120	**Videodisc Correlation Booklet** **Lab Partner Software**	**Test Bank**

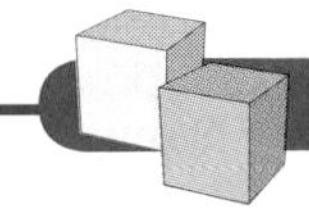

Block Schedule

For information on block scheduling, see the Lesson Plans booklet in the Teacher Resource Package.

GLENCOE TECHNOLOGY

Chemistry: Concepts and Applications Videodisc
Making Soap
Manufacturing Soap
Polymers

Chemistry: Concepts and Applications CD-ROM
Speed Up the Reaction
Making Soap

Science and Technology Videodisc Series (STVS)
Plants and Simple Organisms
Oil from Wood

Complete Glencoe Technology references are inserted within the appropriate lesson.

ASSIGNMENT GUIDE

Contents	Level	Practice Problems	Chapter Review	Solving Problems in Chemistry
30.1 Organic Reactions and Products				
Substitution Reactions	O	1, 2, 3, 4, 5	21, 27	3
Addition Reactions	O	6, 7, 8	22, 27, 56, 57, 60, 66, 88	2
Elimination Reactions	O	9, 10, 11	28, 60, 67	4
Esterification and Saponification Reactions	O		23, 29-31, 58, 60, 61, 65, 79	5
Polymerization	O		32, 33, 35, 59, 62, 68, 69, 89, 91, 92	
Petroleum	O		24, 26, 34, 70, 86, 90	
Elastomers	O		62, 71	
Plastics	O		35	
Synthetic Fibers	O		36, 80, 83, 94	
30.2 Biochemistry				
Proteins	O		37-44, 63, 64, 72, 73, 80, 81	
Carbohydrates	O		45-49, 74, 84	
Lipids	O		50-52, 75, 82, 85, 93	
Nucleic Acids	O		38, 53-55, 76, 87	
Biomaterials	O		77, 78	Chapter Review: 1-2

C=Core, A=Advanced, O=Optional

Laboratory Materials

Minute Labs	ChemActivities	Demonstrations		
pages 773, 783 beakers cheese cloth ethanol evaporating dish sodium chloride sodium hydroxide vegetable oil, solid ammonia, household balloon, large bromothymol blue drinking straw sucrose test tubes vinegar dry yeast	**page 852** acetic anhydride 100-cm^3 glass beakers, 2 ethanol 10-cm^3 graduated cylinder hot plate ice paper towel plastic micropipets, 2 plastic cup salicylic acid scissors sulfuric acid (concentrated) thermometer toothpick	**pages 768, 772, 774, 776, 782, 784, 786** amino acids anhydrous sodium carbonate beakers bottle chromatography paper clothespins cobalt chloride test paper 1,6-diaminohexane food coloring	graduated cylinder 0.05*M* HCl hot plate hydrolyzed polyvinyl alcohol lead acetate test paper magnetic stirrer NaCl 2% NH_3 paper clips paper cups peanuts pliers	2-propanol red litmus paper sebacoyl chloride sodium tetraborate sodium polyacylate stirring rod sugar cube test tube toothpick TTE UV light wintergreen candy

30.1 Organic Reactions and Products

PREPARE

Key Concepts

The major concepts presented in this section include substitution reactions, addition reactions, elimination reactions, esterification and saponification reactions, polymerization, petroleum, elastomers, plastics, and synthetic fibers.

Key Terms

substitution reaction
addition reaction
elimination reaction
esterification
saponification
polymer
addition polymerization
condensation polymerization
octane rating
elastomer

1 FOCUS

Objectives

Preview with your students the objectives listed on the student page. Students can use them as a study guide for this major section.

Discussion

Ask students if they have heard of starch, glycogen, protein, and nucleic acids. In biology students may have learned that these are biological polymers. The synthesis of these polymers is controlled by another polymer, DNA. Encourage students with the fact that in this section they will learn about polymers and the reactions that form them.

OBJECTIVES

- describe substitution, addition, elimination, esterification, saponification, addition polymerization, and condensation polymerization reactions.
- describe the processing of petroleum and the octane rating of gasolines.
- explain how synthetic fibers, plastics, and elastomers are produced industrially.

30.1 Organic Reactions and Products

In addition to being organic, what do garbage bags, antifreeze, laundry detergent, garden hose, and aspirin have in common? They are all manufactured, at least in part, from ethene, or ethylene ($CH_2{=}CH_2$). How can so great a variety of substances be produced from a simple raw material? Organic compounds undergo an enormous variety of reactions. However, there are a few common types of reactions that are frequently found in industrial processes. We will investigate some of these reactions in this section. In addition, we will look at some of the more common consumer products composed of organic substances.

Substitution Reactions

A reaction in which either a hydrogen atom of a hydrocarbon is replaced by a functional group or a functional group is replaced by another functional group is called a **substitution reaction**. Alkanes react with chlorine in sunlight to produce chloro-substituted compounds. The product is a mixture of different isomers with very similar properties. A number of aromatic substitution reactions can be controlled to produce specific products. For example, benzene reacts with nitric acid in the presence of concentrated sulfuric acid to form nitrobenzene.

$$C_6H_6(l) + HNO_3(l) \xrightarrow[H_2SO_4]{conc.} C_6H_5NO_2(l) + H_2O(l)$$

benzene　nitric acid　nitrobenzene　water

Alkyl groups and halogen atoms can also be substituted easily onto a benzene ring.

In the second type of substitution reaction, one functional group replaces another. Alcohols undergo substitution reactions with hydrogen halides to form alkyl halides. For example, 2-propanol reacts with hydrogen iodide to form 2-iodopropane. The reaction is reversible and reaches equilibrium. By introducing a substance to absorb the water, the reaction can be forced to the right, producing more product.

$$CH_3{-}CH(OH){-}CH_3(l) + HI(aq) \rightarrow CH_3{-}CH(I){-}CH_3(l) + H_2O(l)$$

2-propanol　hydrogen iodide　2-iodopropane　water

Alkyl halides, in turn, react with ammonia to produce amines. For example, bromoethane reacts with ammonia to produce ethylamine.

$$CH_3{-}CH_2{-}Br(l) + NH_3(aq) \rightarrow CH_3{-}CH_2{-}NH_2(aq) + HBr(aq)$$

bromoethane　ammonia　ethylamine　hydrogen bromide

CONTENT BACKGROUND

Grignard Reagents: In 1901, the French chemist Victor Grignard made the important observation that alkyl halides dissolved in ethoxyethane (diethyl ether, $C_2H_5{-}O{-}C_2H_5$) will react with metallic magnesium to form ether-soluble alkylmagnesium halides. These halides are now called Grignard reagents. The reaction producing them will be illustrated here by the formation of methylmagnesium iodide from methyl iodide. The diethyl ether solvent must be very pure, and the apparatus and reagents must be moisture free.

$$CH_3I + Mg \rightarrow CH_3MgI$$

Grignard reagents react with water to produce alkanes. Since the step leading to the formation of the Grignard reagent and the hydrolysis process both produce a good yield, this reaction sequence is an attractive one for the conversion of alkyl halides

PRACTICE PROBLEMS

Complete and balance each substitution reaction.

1. $HOH + (CH_3)_3CI \rightarrow$
2. $CH_3CH_2CH_3 + Cl_2 \xrightarrow{sunlight}$
3. (benzene ring) $+ Br_2 \rightarrow$
4. $CH_3CH_2OH + HF \rightarrow$
5. $CH_3Cl + NH_3 \rightarrow$

Addition Reactions

In a double bond between two carbon atoms, each carbon atom contributes two electrons to the bond. Suppose one bond is broken and the other remains intact. Each carbon atom then has one electron available to bond with some other atom. A number of substances will cause one bond of a double bond to break by adding on at the double bond. This type of reaction is called an **addition reaction.** An example is the addition of bromine to the double bond of ethene. The product of this reaction is 1,2-dibromoethane.

$$H_2C{=}CH_2(g) + Br_2(l) \rightarrow BrH_2C{-}CH_2Br(l)$$

ethene bromine 1,2-dibromoethane

Atoms of many substances can be added at the double bond of an alkene. Some common addition agents are the halogens (except fluorine), the hydrogen halides, and sulfuric acid. The double bonds in the benzene ring of aromatic compounds are so stabilized by delocalization that addition reactions do not occur readily in these compounds.

PRACTICE PROBLEMS

Complete and balance each addition reaction.

6. $CH_3CH_2CH{=}CH_2 + Br_2 \rightarrow$
7. $CH_2{=}CHCH_3 + HCl \rightarrow$
8. $CH_2{=}CH_2 + H_2SO_4 \rightarrow$

Elimination Reactions

We have seen that under certain circumstances, atoms can be "added on to" a double bond. It is also possible to remove certain atoms from a molecule to create a double bond. Such a reaction is known as an **elimination reaction.** In the most common elimination reactions, a water molecule is removed from an alcohol. A hydrogen atom is removed from one carbon atom, and a hydroxyl

CHEMISTS AND THEIR WORK

Bertram O. Fraser-Reid (1934-)

Bertram Fraser-Reid is a biochemist and a former professor of chemistry at Duke University. His major research efforts over the years have been in organic chemistry, notably sugars. He has worked on biological regulators for immune systems and has developed methods for producing pheromones, insect sex attractants, from glucose. These pheromones were used in place of DDT to control timber-destroying insects by disrupting their mating patterns. Fraser-Reid's work in sugars has led him to believe that many products now made from petroleum can also be made from sugar—a renewable resource.

to the corresponding alkanes. The hydrolysis of methylmagnesium iodide produces methane in the following manner:

$$CH_3MgI + HOH \rightarrow CH_4 + MgI(OH)$$

This type of reaction sequence may be generalized as:

$$R{-}X + Mg \rightarrow RMgX$$
$$R{-}MgX + HO{-}H \rightarrow R{-}H + MgX(OH)$$

2 TEACH

PRACTICE

Guided Practice: Guide students through the examples in the text, then have them work in groups on Practice Problems 1-5 and 6-8 that follow.

Cooperative Learning: Refer to pp. 28T-29T to select a teaching strategy to use with this activity.

Independent Practice: Your homework or classroom assignment can include Review and Practice 21-22, and 27 from the end of the chapter.

Answers to

Practice Problems: Have students refer to Appendix C for complete solutions to Practice Problems.

1. $H_2O + (CH_3)_3CI \rightarrow HI + (CH_3)_3COH$
2. $CH_3CH_2CH_3 + Cl_2 \rightarrow CH_3CH_2CH_2Cl + HCl$ or any of 15 other products through perchloropropane, $CCl_3{-}CCl_2{-}CCl_3$
3. (benzene) $+ Br_2 \rightarrow$ (bromobenzene, Br) $+ HBr$
4. $CH_3CH_2OH + HF \rightarrow CH_3CH_2F + H_2O$
5. $CH_3Cl + NH_3 \rightarrow CH_3NH_2 + HCl$
6. $CH_3CH_2CH{=}CH_2 + Br_2 \rightarrow CH_3CH_2CHBrCH_2Br$
7. $CH_2{=}CHCH_3 + HCl \rightarrow CH_3CHClCH_3$
8. $CH_2{=}CH_2 + H_2SO_4 \rightarrow CH_3CH_2OSO_3H$

CHEMISTS AND THEIR WORK
Bertram O. Fraser-Reid

Background: A native of Jamaica, Dr. Frasier-Reid has written over 200 papers. He is also an accomplished organist. The regulators of the immune system are complex sugars known as oligisaccharides.

Exploring Further: Find out how the pheromones serve as a substitute for insecticides. *The sex attractant causes the male insect to fall into a trap.* Ask students if they have used Japanese beetle traps or gypsy moth traps. Both of these depend on pheromones to attract the animal into the trap.

2 TEACH

CONCEPT DEVELOPMENT

Using the Table: Point out to students that essential amino acids (listed below) are not synthesized by the body. Therefore, when students plan their diets they must be sure to include foods that contain sufficient amounts of these amino acids. Eggs, dairy products, kidneys, and liver contain all of the essential amino acids. Corn, wheat, gelatin, soybeans, peanuts, potatoes, poultry, fish, and red meats can be combined in ways that provide all of the essential amino acids.

Essential Amino Acids

isoleucine	leucine
threonine	lysine
methionine	phenylalanine
tryptophan	valine

- Point out that arginine and histidine are nonessential for adults but are essential for growing children because they do not synthesize sufficient amounts of these amino acids.

GLENCOE TECHNOLOGY

CD-ROM

Chemistry: Concepts and Applications

Interactive Experiment: *Speed up the Reaction*

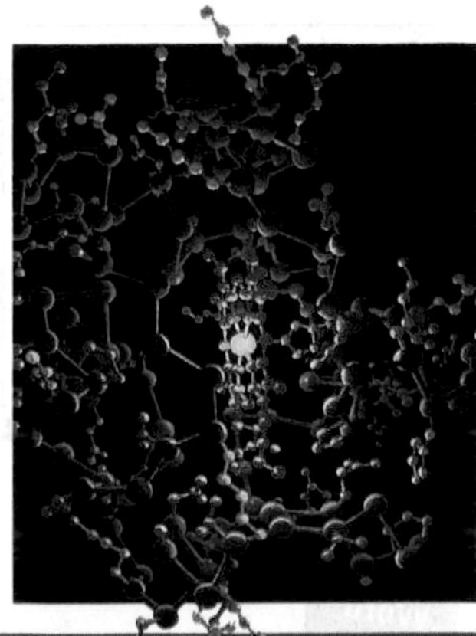

Figure 30.11 This model of the protein cytochrome c (top) shows that protein chains are folded and coiled into a three-dimensional shape. Cytochrome c is present in muscle tissue such as this steak (bottom).

Table 30.1

Amino Acids

Amino acids have the form

$$H-\underset{\displaystyle G}{\overset{\displaystyle COOH}{\underset{|}{\overset{|}{C}}}}-NH_2$$

In the table, only the composition of *G* is represented.

Name	G	Symbol
Glycine	$H-$	Gly
Alanine	CH_3-	Ala
Valine	CH_3-CH- | CH_3	Val
Leucine	$CH_3-CH-CH_2-$ | CH_3	Leu
Isoleucine	CH_3-CH_2-CH- | CH_3	Ile
Tryptophan	(indole ring)$-CH_2-$, N–H	Trp
Lysine	$H_2N-CH_2-CH_2-CH_2-$	Lys
Arginine	$H_2N-C-NH-CH_2-CH_2-CH_2-$ ‖ NH	Arg
Phenylalanine	(benzene ring)$-CH_2-$	Phe
Histidine	(imidazole ring: N, NH)$-CH_2-$	His
Asparagine	$O{=}C-CH_2-$ | NH_2	Asn
Glutamine	$O{=}C-CH_2-CH_2-$ | NH_2	Gln
Serine	$HO-CH_2-$	Ser
Threonine	CH_3-CH- | OH	Thr
Aspartic acid	$HOOC-CH_2-$	Asp
Glutamic acid	$HOOC-CH_2-CH_2-$	Glu
Tyrosine	$HO-$(benzene ring)$-CH_2-$	Tyr
Methionine	$CH_3-S-CH_2-CH_2-$	Met
Cysteine	$HS-CH_2-$	Cys
Proline (an exception to the general formula)	H_2C-CH_2, H_2C, $CH-COOH$, N, H (ring)	Pro

DEMONSTRATION

30-5 What's in a Protein?

Purpose: to perform a destructive distillation of a protein

Materials: a few peanuts, test tube, red litmus paper, lead acetate test paper, cobalt chloride test paper

Procedure: 1. In a way similar to the destructive distillation of wood, a protein can be placed in a test tube and heated strongly. Place a few peanuts in a test tube and heat strongly.
2. Place a piece of moist red litmus paper over the mouth of the tube while heating.
3. In a similar way, use lead acetate paper to test for hydrogen sulfide.
4. Use cobalt chloride test paper to indicate the presence of water.

Disposal: A

Results: The protein will blacken when heated. The ammonia will turn the red

Two different dipeptides can be made from two amino acids, depending on which ends of the molecules react. For example,

$$H_2N-CH_2-COOH + H_2N-CH(CH_3)-COOH \rightarrow H_2N-CH_2-CO-NH-CH(CH_3)-COOH + H_2O$$

glycine + alanine → glycylalanine dipeptide

$$H_2N-CH(CH_3)-COOH + H_2N-CH_2-COOH \rightarrow H_2N-CH(CH_3)-CO-NH-CH_2-COOH + H_2O$$

alanine + glycine → alanylglycine dipeptide

Reactions utilizing catalysts consume or produce energy. Some of the energy produced appears as heat in the cell. However, most of the energy must be converted to forms other than heat or the cell will die. In cells, this excess energy is used to produce a product whose synthesis requires the input of energy. This compound is usually adenosine-5′-triphosphate, or ATP for short. All organisms, from single-celled bacteria to humans, use ATP as their energy transfer molecule. When an enzymatic reaction requiring energy occurs, some ATP can be decomposed to provide that energy. For example, glucose is an important energy source for organisms. The first step in the use of glucose is the formation of glucose-6-phosphate. The ΔG for glucose reacting with phosphate ion is +12 600 kJ/mol. However, the ΔG for glucose reacting with ATP to form glucose-6-phosphate and ADP (adenosine diphosphate) is −21 400 kJ/mol. The structure of ATP is

triphosphate group: $HO-P(=O)(OH)-O-P(=O)(OH)-O-P(=O)(OH)-O-$; adenine (NH_2); ribose (5′ CH_2, 4′, 3′, 2′, 1′; OH OH)

Carbohydrates

Carbohydrates are also important to living systems. These compounds contain the elements carbon, hydrogen, and oxygen. The word *carbohydrate* literally means "hydrate of carbon." Almost all **carbohydrates** are either simple sugars or condensation polymers of sugars. The most common simple sugar is glucose. Another common simple sugar is fructose.

MINUTE LAB

Sugar Fermentation — It's a Gas

In a beaker, dissolve 5 teaspoons of sucrose in 250 cm^3 of warm water. Place 1 teaspoon of dry yeast and 3 teaspoons of warm water in a 500-cm^3 soda bottle. Add the sucrose solution. Fit a large, flat balloon over the mouth of the bottle. Place 3 cm^3 of distilled water in a small test tube, 3 cm^3 of vinegar in a second tube, and 3 cm^3 of household ammonia in a third tube. Add 1 drop of 0.1% bromothymol blue to each tube. Observe the color in neutral, acidic, and basic solutions. Fill a large test tube with distilled water. Add bromothymol blue until the solution is pale green. Pinch the balloon closed and remove it from the bottle. Place a plastic straw in the large tube. Hold the open end of the balloon tightly around the straw, allowing the CO_2 in the balloon to bubble through the solution. Is the solution acidic, basic, or neutral? Write an equation for the reaction.

CONCEPT DEVELOPMENT

• Emphasize to students that the complexity and diversity of proteins is due to the variety possible in the order of amino acids, the way in which the chain is coiled, folded, or twisted, and in the type of bonding that holds the polymer in its particular shape.

MAKING CONNECTIONS

Biological Pesticide

Nematodes, which are microscopic roundworms, do $5 billion damage to U.S. crops annually. Chitin, a protein present in marine organisms' shells, stimulates soil microorganisms to produce an enzyme that destroys the nematodes' skin and eggs. Millions of tons of crab shells are being converted into this safe biological pesticide to be used instead of the faster-acting chemicals.

MINUTE LAB

Sugar Fermentation—It's a Gas

Purpose: to demonstrate that when yeast oxidize sugar for energy, they produce CO_2 gas

Materials: 5 teaspoons sugar, 1 teaspoon dry yeast, 500-cm^3 soda bottle, large balloon, a few drops 0.1% bromothymol blue indicator, 3 test tubes, 3 cm^3 vinegar, 3 cm^3 household ammonia, drinking straw

Procedure Hints: Place the soda bottle in a water bath at 37°C. Adjust the indicator color in distilled water to green by adding dilute acid or dilute base.

Results: The balloon inflates. The gas causes the indicator solution to change from green to yellow. This indicates that the solution is now acidic.

Answers to Questions: 1. acidic
2. $CO_2 + H_2O \rightarrow H_2CO_3$

✓ ASSESSMENT

Knowledge: Have students research the process by which corn is fermented to make ethanol. This ethanol is mixed with gasoline to produce a cleaner burning fuel. Some students might want to examine the economics of using corn to produce fuel.

litmus paper to blue. The lead acetate paper turns black for a positive test for H_2S. The cobalt chloride test paper indicated water was present.

Questions: 1. What does the darkening of the protein indicate is in a protein? *carbon*

2. What other elements are shown to be present by the test papers? *S, H, O, N*

PROGRAM RESOURCES

Enrichment: Use the worksheets, "Steroids," pp. 59-60, to challenge more able students.

Laboratory Manual: Use microlab 30-2 "Qualitative Analysis of Food" and microlab 30-3 "An Enzyme-Catalyzed Reaction" as enrichment for the concepts in Chapter 30.

CHEMACTIVITY 11-1

Preparation and Properties of Hydrogen

Hydrogen is the most common element in the universe. It is one of the more common elements on the surface of the Earth when combined with oxygen to form water. In this activity, students will generate hydrogen and will also practice following a detailed set of laboratory directions.

Process Skills

Observing, measuring, predicting, interpreting data, inferring, communicating

Procedure Hints

Before proceeding, be sure students have read and understood the purpose, procedure, and safety precautions for this ChemActivity.

- One liter of 1*M* HCl may be prepared by adding 86 cm^3 of commercial concentrated HCl to 914 cm^3 of water. **CAUTION:** *Concentrated acid should be added to the water.*
- You may save some class time by preparing the two types of micropipets, the generator (G) pipets and collector (C) pipets in advance.
- Students may hold the plastic pipet when they test for the presence of hydrogen provided that they are wearing goggles and aprons.
- **Troubleshooting:** Magnesium ribbon that is old or has been stored in a non-airtight container may not react since the surface of the metal could be coated with a layer of magnesium oxide. The surface can be cleaned by rubbing gently with fine sandpaper or emery cloth.
- A hydrochloric acid solution more concentrated than 1*M* makes the reaction occur too quickly.
- When students are testing for hydrogen, be sure that the water plug does not put out the flaming toothpick.

Data and Observations

1. Hydrogen is a clear, colorless gas. Hydrogen does not appear to dissolve in water.

Micro

Preparation and Properties of Hydrogen

Hydrogen is the most common element in the universe, and water is the most common hydrogen compound on the surface of Earth. Hydrogen is a colorless, odorless, tasteless gas. It is less dense than air and is the least dense of all the elements.

The reaction of hydrogen with oxygen releases a large amount of energy. For this reason, hydrogen can be used as a fuel. Liquid hydrogen is the fuel of choice for rockets like the space shuttle. The purpose of this activity is to demonstrate how hydrogen gas can be prepared and tested for.

Objectives

- **Set up** apparatus for generating and collecting hydrogen gas.
- **Prepare** hydrogen gas by a displacement reaction.
- **Demonstrate** a chemical test for hydrogen gas.

Materials

thin stem pipet	toothpicks
micropipets (3)	magnesium ribbon (1 cm)
clear tape	1*M* hydrochloric acid (HCl)
matches	distilled water
24-well microplate	$250\text{-}cm^3$ beaker
scissors	
matches	

Procedure

1. Cut a small slit in the middle of the bulb of the thin-stem pipet as illustrated in Figure A. This will be the generator pipet, pipet G.
2. Insert a piece of magnesium into pipet G through the slit.
3. Cover the slit in the pipet with clear tape to seal the magnesium inside the pipet bulb.
4. Trim the stem of pipet G to 2.5 cm.

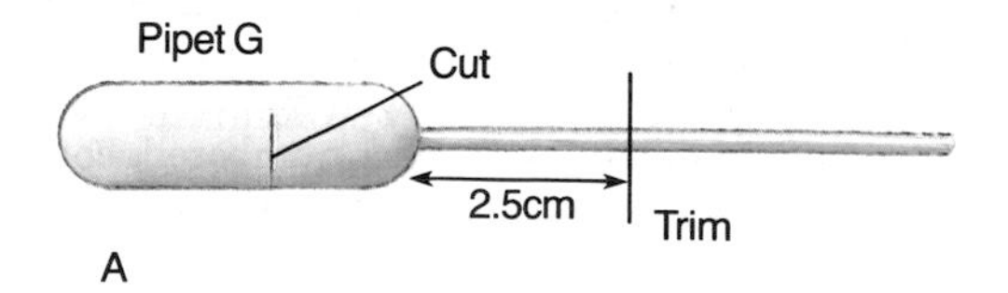

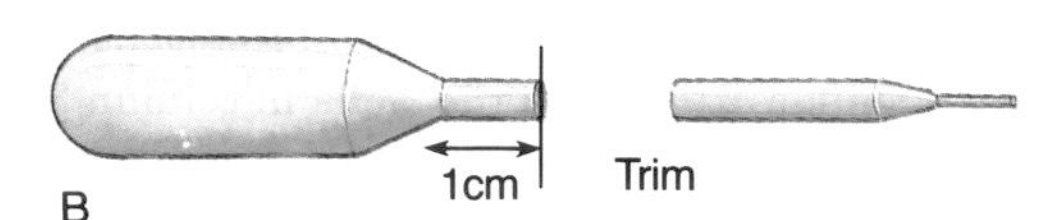

5. Cut the stems of the two micropipets to a length of 1 cm as in Figure B. These will be the collector pipets, C1 and C2.
6. Submerge pipets C1 and C2 under water with the stems pointing up in a $250\text{-}cm^3$ beaker. Squeeze the bulbs repeatedly until no more air is expelled. The pipets should be completely filled with water.
7. Using a new micropipet, place about 1/2 pipet of 1*M* HCl in well C3 in the 24-well microplate.
8. Turn pipet G so the tip is pointing down. Squeeze the air out of the pipet.
9. Place the stem of pipet G in well C3 and draw up the acid in the well.
10. Immediately, turn the pipet over and insert the stem of pipet G into the stem of water-filled pipet C1. Be sure to place the stem of pipet G into pipet C1 as far as it will go. Stand pipets G and C1 in the microplate as shown in Figure C.
11. Replace pipet C1 with C2 when C1 is almost full of hydrogen gas. Allow a "plug" of water to remain in the neck of the pipet before removing it from pipet G.
12. Stand the gas-filled pipets in wells of the microplate as shown in Figure D.
13. Light a toothpick with a match.
14. Quickly invert pipet C1 to dislodge the plug of water from the neck of the pipet.

PROGRAM RESOURCES

ChemActivity Masters: Use worksheets for ChemActivity 11-1 & 11-2 to provide students with a copy of procedures and a data table for use in the lab.

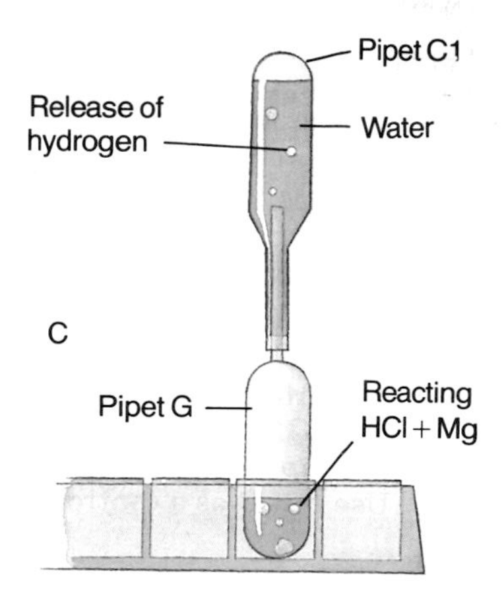

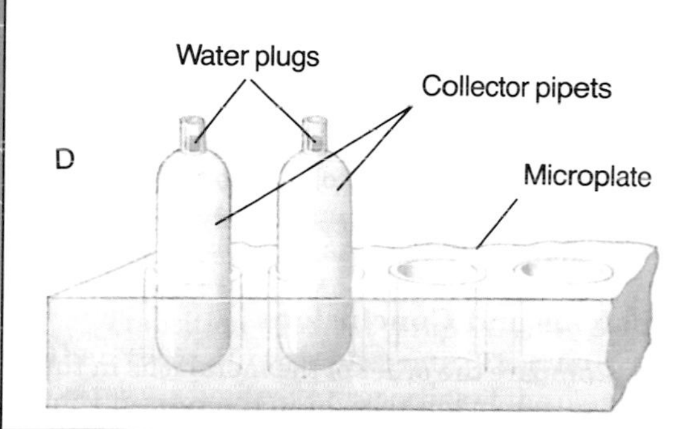

15. Place the flaming end of the toothpick close to the mouth of pipet C1.
16. A small pop indicates the presence of hydrogen gas. Repeat the test for hydrogen with pipet C2.

Data and Observations

1. What are the physical properties of hydrogen gas that you observed?
2. What happened to hydrogen gas when it was exposed to the flame?

Analysis and Conclusions

1. What compound do you think will be left in pipet G after the reaction of magnesium metal and hydrochloric acid? How could you isolate this product?
2. What are the products when hydrogen gas is burned in air?

Extension and Application

Commercial drain cleaners sometimes use aluminum metal and sodium hydroxide as active agents. What is the purpose of both the metal and the sodium hydroxide? What makes commercial drain cleaners flammable?

Chapter 11 ChemActivity 2

Micro

Transition Metals

The elements with atomic numbers 21 through 30 exhibit many properties different from those of other elements in the same period. The elements directly below in the next two periods, however, have similar properties. Together, these elements in Groups 3 to 12 are known as the transition metals. The purpose of this activity is to compare the chemical reactions of some transition metal ions with those of non-transition metals from the same period.

Objectives

- **Observe** physical and chemical properties of transition metal ions in aqueous solution.
- **Observe** results of mixing three different chemicals (NH_3, KSCN, HCl) with metal ions.
- **Compare** chemical reactions of transition metal ions with those of other metal ions.

- Substitution for the compounds shown in the Materials list should be avoided since many of the transition elements cited have multiple oxidation states and the results will vary.
- Waste material from this laboratory experience should be handled as heavy metal waste.
- If available, the lanthanide ions (*f* sublevel fillers) provide an interesting contrast to the reaction of elements whose ions have incomplete *d* sublevels.

2. When hydrogen gas is ignited, it burns violently or explodes giving a pop or "hydrogen bark."

Analysis and Conclusions

1. The reaction of magnesium metal and hydrochloric acid produces magnesium chloride and hydrogen gas. Magnesium chloride remains in solution. This compound may be isolated by evaporating the water.

2. When pure hydrogen is burned in air the only product formed is water.

Extension and Application

The hydrogen gas bubbles inside the drain to agitate the contents of the drain in an attempt to clear it. Sodium hydroxide is also known as lye. Hydrogen gas is generated when aluminum reacts with sodium hydroxide.

CHEMACTIVITY 11-2

Transition Metals

Most of the salts of the transition elements are brightly colored, and most transition metal ions in water solution have a distinctive color. The color of the solution is due to the interaction of the metal ions with water molecules and the formation of hydrated ions. Similar complex ions with characteristic colors are also formed by transition metal ions with a variety of reagents, including NH_3, SCN^-, and Cl^-.

Process Skills

Observing, measuring, classifying, predicting, recognizing cause and effect, interpreting data, comparing and contrasting

Procedure Hints

Before proceeding, be sure students have read and understood the purpose, procedure, and safety precautions for this ChemActivity.

- Care should be taken when handling hydrochloric acid and ammonia.

✓ ASSESSMENT

Performance: Assess student performance for this ChemActivity by having students complete copies of the masters from the *ChemActivity Masters* booklet.

CHEMACTIVITY 26

Lemon Battery

Metals differ in activity, that is, in the tendency to give up electrons. When two different metals are placed in an electrolyte solution, electrons flow from the more active metal to the less active metal, and an electromagnetic force (EMF) is generated. EMF is measured in volts and is commonly called voltage. The greater the difference in the activity of the metals, the higher the voltage produced.

Process Skills
Observing, measuring, classifying, predicting, interpreting data, sequencing, comparing and contrasting, inferring, communicating

Procedure Hints
The lemon must be fresh.
- The metal strips must be clean and free from corrosion. Make sure they are rinsed and dried immediately after use.
- A high impedance or vacuum tube voltmeter with a low range must be used.

Data and Observations
See data in the table below.

Analysis and Conclusions
1. The combination of magnesium and copper gave the highest voltage reading.
2. The combination of copper and carbon gave the lowest reading.

Extension and Application
The electrolyte in a commercial battery is a paste instead of a liquid. Commercial batteries are of three general types: carbon-zinc batteries, alkaline batteries, and rechargeable batteries, either lead-acid or nickel-cadmium. Commercial batteries provide stronger current and last longer than a lemon battery.

Micro

Chapter 26 ChemActivity

Lemon Battery

A battery is a device in which chemical energy is converted to electrical energy. The energy obtained from a battery is produced by a difference in activity of two different metals. When two metals are placed in an electrolyte and are connected by a conductor, electrons flow from the more active metal through the conductor to the less active metal. The flow of electrons is an electrical current and can be made to do work. The purpose of this activity is to investigate the activity of different metals in various combinations in a simple battery. Note that carbon is a conductor and will be considered a metal in this activity.

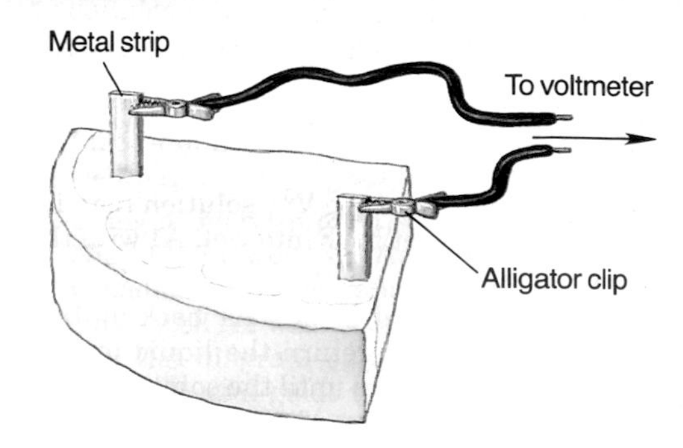

Objectives

- **Make** a simple battery, using a lemon.
- **Compare** the activity of metals used in different combinations in the battery.

Materials

lemon (1/4 per group)
carbon rod (pencil lead will do)
voltmeter or multimeter
short lengths of connecting wire
alligator clips
small knife
chemical scoop
small strips of the following metals:
magnesium
lead
zinc
copper

Procedure

1. Using a chemical scoop, pierce the flesh of a piece of lemon in two places approximately 1 cm apart.
2. Select a strip of two different metals.
3. Insert each strip of metal into a different slit in the flesh of the lemon, as shown in the Figure.
4. With connecting wire and alligator clips, connect each metal strip to the voltmeter. Observe the needle and scale of the voltmeter as you complete the connection. If the needle does not move or you get a negative reading, reverse the connections to the metals. Note the polarity of the metals, that is, whether the metal is positive or negative. Record the voltmeter reading and the polarity (+ or −) of each metal in a data table like the one shown.
5. Remove the metal strips from the lemon and rinse them in tap water.
6. Repeat steps 2–5 until you have tested each of the ten possible combinations.

Data and Observations

Polarity of Metals				
Pb	C	Zn	Mg	
				Cu
			Mg	
		Zn		
	C			

Analysis and Conclusions

1. Which pair of metals gave the highest reading on the voltmeter?
2. Which pair of metals gave the lowest reading on the voltmeter?

Extension and Application

Compare your battery with a commercially produced battery. In what ways are the two batteries similar? How do they differ?

PROGRAM RESOURCES

ChemActivity Masters: Use worksheets for ChemActivity 26 & 27 to provide students with a copy of procedures and a data table for use in the lab.

Data and Observations

Polarity of Metals				
Pb	C	Zn	Mg	
Pb-Cu+ 0.35	C-Cu+ 0.18	Zn-Cu+ 1.0	Mg-Cu+ 1.9	Cu
Pb+Mg- 0.8	C+Mg- 1.7	Zn+Mg- 0.4	Mg	
Pb+Zn- 1.2	C+Zn- 1.3	Zn		
Pb-C+ 0.45	C			

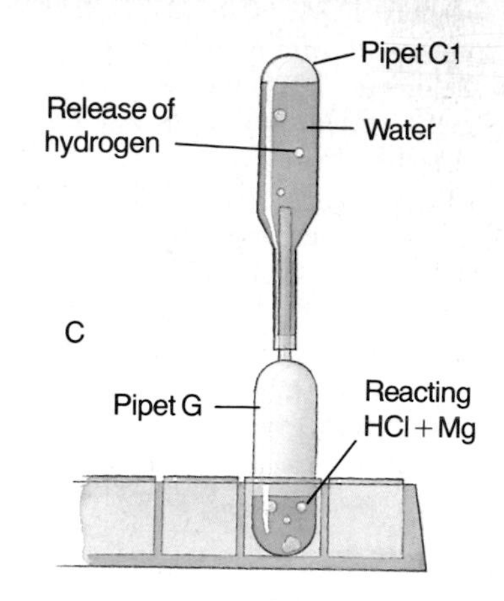

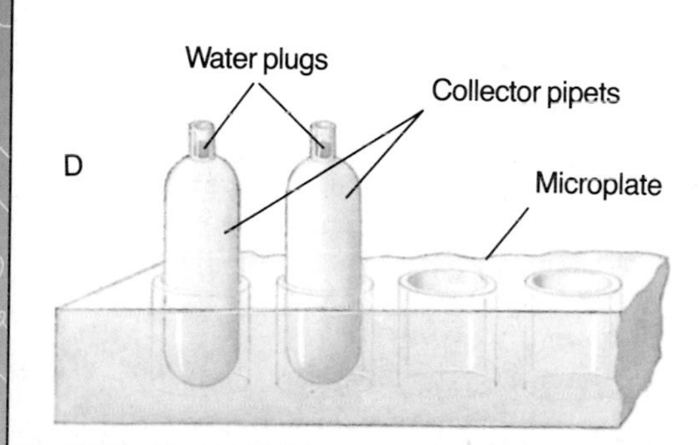

15. Place the flaming end of the toothpick close to the mouth of pipet C1.
16. A small pop indicates the presence of hydrogen gas. Repeat the test for hydrogen with pipet C2.

Data and Observations

1. What are the physical properties of hydrogen gas that you observed?
2. What happened to hydrogen gas when it was exposed to the flame?

Analysis and Conclusions

1. What compound do you think will be left in pipet G after the reaction of magnesium metal and hydrochloric acid? How could you isolate this product?
2. What are the products when hydrogen gas is burned in air?

Extension and Application

Commercial drain cleaners sometimes use aluminum metal and sodium hydroxide as active agents. What is the purpose of both the metal and the sodium hydroxide? What makes commercial drain cleaners flammable?

Micro

Transition Metals

The elements with atomic numbers 21 through 30 exhibit many properties different from those of other elements in the same period. The elements directly below in the next two periods, however, have similar properties. Together, these elements in Groups 3 to 12 are known as the transition metals. The purpose of this activity is to compare the chemical reactions of some transition metal ions with those of non-transition metals from the same period.

Objectives

- **Observe** physical and chemical properties of transition metal ions in aqueous solution.
- **Observe** results of mixing three different chemicals (NH_3, KSCN, HCl) with metal ions.
- **Compare** chemical reactions of transition metal ions with those of other metal ions.

2. When hydrogen gas is ignited, it burns violently or explodes giving a pop or "hydrogen bark."

Analysis and Conclusions

1. The reaction of magnesium metal and hydrochloric acid produces magnesium chloride and hydrogen gas. Magnesium chloride remains in solution. This compound may be isolated by evaporating the water.
2. When pure hydrogen is burned in air the only product formed is water.

Extension and Application

The hydrogen gas bubbles inside the drain to agitate the contents of the drain in an attempt to clear it. Sodium hydroxide is also known as lye. Hydrogen gas is generated when aluminum reacts with sodium hydroxide.

CHEMACTIVITY 11-2

Transition Metals

Most of the salts of the transition elements are brightly colored, and most transition metal ions in water solution have a distinctive color. The color of the solution is due to the interaction of the metal ions with water molecules and the formation of hydrated ions. Similar complex ions with characteristic colors are also formed by transition metal ions with a variety of reagents, including NH_3, SCN^-, and Cl^-.

Process Skills

Observing, measuring, classifying, predicting, recognizing cause and effect, interpreting data, comparing and contrasting

Procedure Hints

Before proceeding, be sure students have read and understood the purpose, procedure, and safety precautions for this ChemActivity.

- Care should be taken when handling hydrochloric acid and ammonia.
- Substitution for the compounds shown in the Materials list should be avoided since many of the transition elements cited have multiple oxidation states and the results will vary.
- Waste material from this laboratory experience should be handled as heavy metal waste.
- If available, the lanthanide ions (*f* sublevel fillers) provide an interesting contrast to the reaction of elements whose ions have incomplete *d* sublevels.

✓ ASSESSMENT

Performance: Assess student performance for this ChemActivity by having students complete copies of the masters from the *ChemActivity Masters* booklet.

Misconceptions

Students tend to focus their attention on the reactivity differences and similarities in the periodic table groups of elements, particularly the main-group elements. This activity emphasizes differences across the table and the properties of the transition metals.

- **Troubleshooting:** If the concentrations of HCl and NH_3 are too small, the desired effects are not noted.
- In particular, the formation of precipitates or complex ions with ammonia can vary with concentrations or oxidation states of the metals.

Data and Observations
Provide students with a copy of the Microplate Data Form provided in your ChemActivity Teacher Resource Book.

Analysis and Conclusions
1. The ions in solution are in the same order as the elements in the periodic table. The elements scandium and titanium are omitted, however, since the salts of these elements are not usually found in the high school chemistry stockroom. The ions are those of the first two non-transition metals and all the transition metals in period 4.
2. There are no visible reactions of these ions. The ions in these columns are K^+, Ca^{2+}, and Zn^{2+}. The first two are not transition metals. Zinc, although it is a transition metal, is more like the non-transition metals in its properties because it has a full *d* sublevel.
3. All of the observable changes indicating possible chemical reactions occurred in these columns. All of the ions in these columns are those of transition metals. Two types of changes—color changes and formation of precipitates—indicate that two different types of reactions have occurred. Color changes without precipitation are due to the formation of complex ions, a type of reaction common for transition metal ions and unusual for main-group metals. Precipitation indicates formation of hydroxides with ammonia solution.

Materials

96-well microplate
Microplate Data Form
plastic micropipets
toothpicks
$0.1M$ solutions of compounds containing period 4 metals: KNO_3, $Ca(NO_3)_2$, NH_4VO_3, $Cr(NO_3)_3$, $Mn(NO_3)_2$, $Co(NO_3)_2$, $Fe(NO_3)_3$, $Ni(NO_3)_2$, $Cu(NO_3)_2$, $Zn(NO_3)_2$
$6.0M$ ammonia, NH_3
$1M$ potassium thiocyanate, KSCN
$6M$ hydrochloric acid, HCl
white paper, one sheet

Procedure

1. Review the list of metals that you will be testing. **Hypothesize** which three are likely to have properties different from the others.
2. Use a sheet of white paper to make a microplate template by copying the diagram below. Place a 96-well microplate on your template with the numbered columns at the top and the lettered rows to the left.
3. Place 5 drops of KNO_3 in each of wells A1, B1, C1, and D1.
4. Place 5 drops of $Ca(NO_3)_2$ in each of wells A2, B2, C2, and D2.
5. Continue to repeat this process of placing 5 drops of solutions of metal compounds in subsequent columns of the microplate. Use chemicals in the order in which they are given in the materials list. The last drops will be in column 10.
6. Add 5 drops of $6M$ NH_3 to each well in row A. Mix well with a toothpick. **CAUTION:** *Ammonia solution is caustic.*
7. Add 5 drops of KSCN solution to each well in row B. Mix well with a toothpick.
8. Finally, add 5 drops of HCl to each well in row C. **CAUTION:** *Do not mix HCl and KSCN. Clean up all spills immediately.* Mix well. Use row D as a control for comparing with the other rows.

Data and Observations

Label the rows and columns of your Microplate Data Form the same way as your template.

1. Observe row D of your microplate. Record your observations in the Data Form.
2. Compare the solutions in the wells in rows A, B, and C in each column to the solution in row D in that column. Where there was a change, record what you observe.

Analysis and Conclusions

1. Compare the order of the metal ions in the wells from left to right in the rows of your microplate. Note the relationship of your metal ions to the positions of the metals in the periodic table.

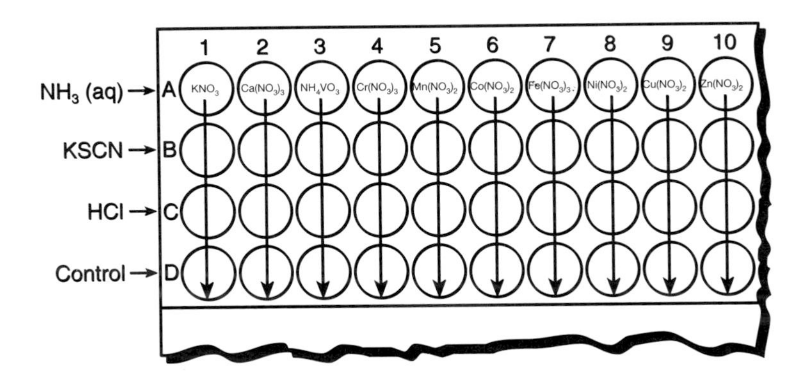

	1	2	3	4	5	6	7	8	9	10
A	NR	NR	Orange cpx	Green ppt.	White ppt.	Green ppt.	Brown ppt.	Pale Blue cpx	Blue cpx	NR
B	NR	NR	Pale green cpx	Gray cpx	NR	NR	Red cpx	Pale green cpx	Green cpx	NR
C	NR	NR	NR	Gray cpx	NR	Blue cpx	Orange cpx	Light green cpx	Green cpx	NR
D	Colorless ←→		Yellow soln	Blue gray soln	Color-less soln	Pink soln	Dark yellow soln.	Emerald green soln	Blue soln	Color-less soln

NR = No reaction ppt = Precipitate cpx = Complex soln = solution

2. What did you observe in columns 1, 2, and 10 in your microplate? What do the metal ions in these columns have in common?
3. Review what you observed in columns 3 through 9 of your microplate. What do the results in these columns have in common? What do the metal ions in these columns have in common? Is there any evidence for the occurrence of more than one type of chemical reaction?
4. Was your hypothesis correct? Why or why not?

Extension and Application

1. What physical or chemical properties would help to identify a salt as containing a transition metal ion?
2. Modern atomic theory provides an explanation for the similarities of the transition elements. Describe the basis for this explanation. How are the properties of zinc accounted for?

Chapter 12
ChemActivity

Micro

A 3-D Periodic Table

The periodic table is the most valuable tool in chemistry. The organization of the elements according to atomic number and electron configuration gives insights into the chemical and physical properties of the elements and the compounds that they form.

Transition element properties vary less than those of other elements. Therefore, this activity is limited to properties of the elements in Groups 1, 2, and 13-18 (Groups IA to VIIIA), which are known as the main-group elements. The purpose of this activity is to show periodic relationships among properties of the main-group elements.

Objectives

- **Assemble** a three-dimensional scale for a property of the elements.
- **Examine** properties of the elements according to their periodic relationships.
- **Identify** properties with similar periodic relationships.

Materials

96-well microplate
soda straws (25 each group)
metric ruler
large-square graph paper
scissors
grease pencil

Procedure

1. Your teacher will select a property for you to study from the list below:

density	boiling point
atomic radius	melting point
ionization energy	electronegativity

2. Locate the data for your property of elements in Groups IA through VIIIA in tables in Chapter 10, Chapter 12, or Appendix Table A-3 in your textbook.
3. Measure the length of a soda straw; determine one-half the length of the straw, and record the results in a data table like the one shown.
4. Select an appropriate scale so that the largest value of your property can be represented by one-half the length of the soda straw. For example, suppose the highest value of your property is 35 units. If the straw is 24 cm long, half of which is 12 cm, the scale for your property would be

$$\frac{12\text{ cm}}{35\text{ units}} = 0.34\text{ cm/unit}$$

4. Students' hypotheses will vary. If they said K^+, Ca^{2+}, and Zn^{2+} they were correct.

Extension and Application

1. Properties that would indicate the presence of a transition metal ion are a brightly colored solid, a colored water solution, and changes in color with NH_3, KSCN, HCl or other reagents that form complex ions.

2. The transition elements are those elements in which a *d* sublevel is being filled with electrons across each period. Such similarities in electron configurations lead to similarities in properties. Zinc, as the final transition element in period 4, has a filled *d* sublevel and two electrons in the outer *s* sublevel, making it more like a main-group element in electron configuration.

CHEMACTIVITY 12

A 3-D Periodic Table

The periodic table is the most useful tool in chemistry. Students should be told that they will always have access to the table for tests, quizzes, and exams. This will serve to motivate students in this activity.

Process Skills

Measuring, classifying, sequencing, comparing and contrasting, inferring, using space/time relationships, formulating models, communicating

Misconceptions

Periodic relationships are generalities. Students should be reminded that for all physical and chemical properties there are exceptions to the trend.

PROGRAM RESOURCES

ChemActivity Masters: Use worksheets for ChemActivity 12 to provide students with a copy of procedures and a data table for use in the lab.

✓ ASSESSMENT

Performance: Assess student performance for this ChemActivity by having students complete copies of the masters from the *ChemActivity Masters* booklet.

Procedure Hints

Before proceeding, be sure students have read and understood the purpose, procedure, and safety precautions for this ChemActivity.

- Supply the same brand or kind of straw for all parts.
- Refer students to Appendix Table A-3 (boiling point, melting point, density) in their textbooks.
- Students may question the omission of the transition elements. The transition elements have very similar chemical and physical properties. The periodic variations in properties among these elements are small.

Data and Observations

Sample data for electronegativity:
Length of soda straw, 24 cm
Half the straw length, 12 cm
Largest value of property to be investigated, 4.10 EN units
Scale of length vs. property,
12 cm/4.10 EN value = 2.9 cm/EN unit

Summaries of trends (Student answers will vary.)

Density—Generally decreases down families; across rows, densities are higher for metals than nonmetals.

Boiling point—Generally decreases down families at the left in the table and increases down families at the right. Values increase across rows up to Group 14 (IVA).

Atomic radius—Decreases across the rows and increases down the families. Cesium and francium have the largest radii.

Melting point—Similar to boiling point, but less regular.

Ionization energy—Increases across the rows and decreases down the families. Helium has the highest ionization energy.

Electronegativity—Increases across the rows and decreases down the families.

Analysis and Conclusions

1. Answers will differ depending on the properties chosen. Electronegativity and ionization energy vary in the same way. Atomic radii vary opposite from electronegativity and ionization energy.

2. Ionization energy and electronegativity are similar, as noted above. Density, melting point, and boiling point are for the most part similar in their periodic relationships.

Extension and Application

1. Students should be able to replace the straw based on the location of the removed straw. They are hypothesizing the probable value of a property of the element based on periodic trends.

2. Results should be the same as in the first trial.

3. Results should be the same as in the first trial.

4. Generalities about individual elements or series of individual elements can be deduced based on their positions in the periodic table.

The length of the straw representing the element with the value of 35 units would be 0.34 cm/unit × 35 units = 12 cm. Enter the scale for your property in your data table.

5. Cut soda straws to the proper lengths to represent your assigned property for each of the main-group elements. Mark each straw near the top with the symbol of the element.

6. By comparing your 96-well microplate with a periodic table, locate the well that represents each main-group element. Place the cut straw for that element in the well. Be sure to preserve the location of the elements relative to each other as shown in the diagram.

7. Using a grease pencil, label the outside of your plate with the name of the physical or chemical property represented by your 3-D table.

8. Record in your data table how your property varies across the rows and down the families.

9. Observe the microplates of other students in your class. In one sentence, summarize in your data table your observations about the properties studied.

Data and Observations

Length of soda straw (cm)	
Half the straw length (cm)	
Largest value of property to be investigated	
Scale for straw length vs. property (cm): ½ straw length ÷ largest value of property	
Observation notes on trend for property:	

Analysis and Conclusions

1. Choose two pairs of properties. For each pair, draw conclusions about the relationship between the trends. For example, what seems to be the relationship between electronegativity and ionization energy?
2. Can several of the properties be grouped according to similar periodic trends?

Extension and Application

1. While you are not looking, have your lab partner remove one of the straws from the set you have constructed. Cut another soda straw so that it will fit into the general trend of your 3-D table. Predict and describe to your partner the property value for the unknown element. Compare your description to the actual value.
2. Reverse roles and repeat the exercise.
3. Switch plates with another group and repeat Extension 1.
4. How is the periodic table useful in identifying chemical elements and their properties?

Electron Clouds

Electrons occupy space about the nucleus. The exact path or location of electrons in an atom cannot be known. Instead, we speak about the probability of locating an electron in an electron cloud—a volume determined by the energy of the electron. Balloons allow us to model the electron clouds of bonding electron pairs whose arrangement determines the shapes of molecules.

Round balloons will be used to represent the spherical electron cloud occupied by one or two *s* electrons. Pear-shaped balloons will be used to represent the electron clouds occupied by pairs of bonding electrons. The purpose of this activity is to provide you with a model of how atoms join together to form molecules.

Objectives

- **Use** balloons to model the arrangement of electron pairs around the nucleus of an atom.
- **Observe** a model of bonding in three dimensions.

Materials

round and pear-shaped balloons
clear tape
string

Procedure

1. Blow up five round balloons to the same size. Tie their ends closed.
2. Make a loop of tape on each of the balloons to represent an electron. It takes a pair of loops to make a chemical bond.
3. Stick two of the balloons together at the two tapes to form a model of a covalent bond as shown in Figure A. This is a model of two hydrogen atoms and an example of linear bonding.
4. Blow up four pear-shaped balloons. Tie three together at their knotted ends as shown in Figure B.

A

Linear bonding

B

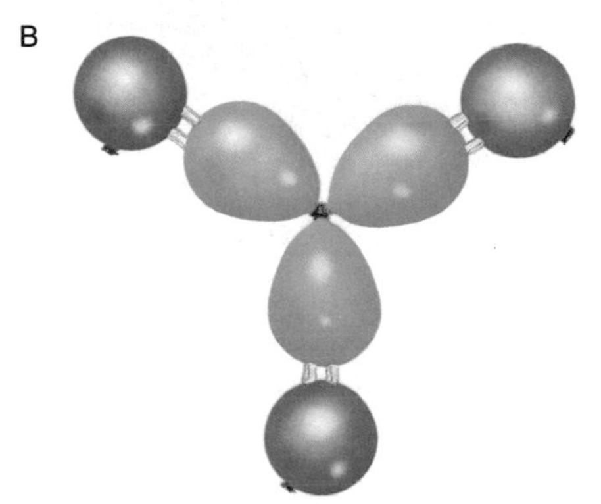

Planar bonding

5. Place a loop of tape at the end of each balloon opposite where they are knotted together. Attach a round balloon by its loop of tape on the end of each pear-shaped balloon. This is an example of planar bonding, as shown in Figure B.
6. Add the fourth pear-shaped balloon to the other three pear-shaped balloons by tying it where the others are knotted together.
7. Place a loop of tape at the end of the fourth pear-shaped balloon.
8. Place a round balloon at the end of the fourth pear-shaped balloon, as shown in Figure C. This is a model of CH_4, and an example of tetrahedral bonding.

Data and Observations

1. Compare your three models and picture the location of the bonded atoms. Using atomic symbols and straight lines for the bonds, diagram each type of molecule.

CHEMACTIVITY 13

Electron Clouds

The repulsion of electron pairs is a simple model that shows the arrangement of electron pairs in molecules. This activity focuses on the shapes of molecules in which there are no unshared electron pairs. It also demonstrates how the linear, trigonal, and tetrahedral arrangements of bonds arise naturally by electron-pair repulsion.

Process Skills

Observing, classifying, predicting, recognizing cause and effect, comparing and contrasting, inferring, using space/time relationships, formulating models

Procedure Hints

Before proceeding, be sure students have read and understood the purpose and procedure for this ChemActivity.

- Use only round or pear-shaped balloons in this activity.
- Do not allow students to use double-sided tape to connect the balloons. Double-sided tape sticks to the balloons without any give. This leads to a large number of broken balloons.

Misconceptions

- Most students picture a molecule as flat or planar based on the pictures they see in textbooks.
- Students often picture electrons as standing still or having limited motion.

Data and Observations

1. See structures below.
2. In the planar molecule BCl_3 the B-Cl bonds are at 120° from each other. In the CH_4 tetrahedral molecule, the C-H bonds are at 109.5° from each other.

Analysis and Conclusions

1. Electron pairs in molecules stay as far away from each other as possible because their negative charges repel each other. The tied, pear-shaped balloons natural-

✓ ASSESSMENT

Performance: Assess student performance for this ChemActivity by having students complete copies of the masters from the *ChemActivity Masters* booklet.

PROGRAM RESOURCES

ChemActivity Masters: Use worksheets for ChemActivity 13 to provide students with a copy of procedures and a data table for use in the lab.

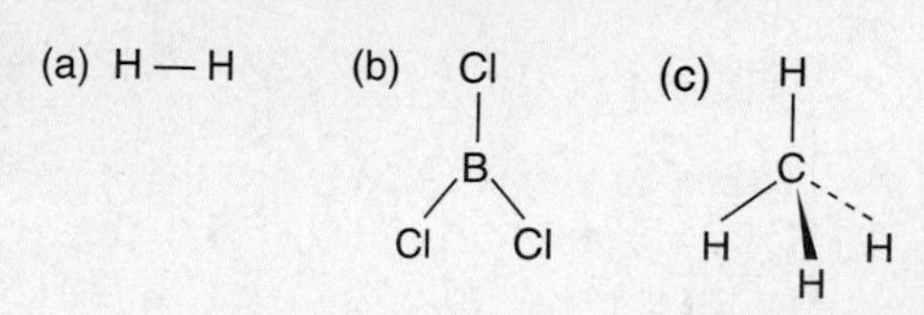

ly assume the same geometry in which they are as far apart as possible.

2. (a) tetrahedral
(b) linear
(c) tetrahedral
(d) linear
(e) planar

Extension and Application

1. The $BeCl_2$ molecule is linear, with a 180° angle between the bonds.

2. The larger balloons create a strain on the tetrahedral arrangement of the pear-shaped balloons. The bond angle between the balloons supporting the large round balloons becomes larger than 109.5° and the other bond angles become smaller. This happens when there is an unshared electron pair on the central atom in a molecule.

3. Each carbon atom in acetylene has two bonds that stay away from each other in a flat, linear arrangement. The bond angles will be 180°. Each carbon atom in ethylene has three bonds that stay away from each other in a flat, triangular arrangement. The molecule will be planar with 120° bond angles.

CHEMACTIVITY 14-1

Paper Chromatography

Separation of the components of a mixture is important in the laboratory and in applications of chemistry. Chromatography is one of the most useful tools for carrying out separations. It has numerous applications in medicine, biology, and chemical research. Paper chromatography of marker pen ink is a simple experiment that illustrates the principles of chromatography.

Process Skills

Observing, measuring, using numbers, interpreting data, comparing and contrasting, inferring, using space/time relationships, communicating

Procedure Hints

Before proceeding, be sure students have read and understood

2. What are the angles between the bonds in your models of BCl_3 and CH_4?

Tetrahedral bonding

Analysis and Conclusions

1. Explain in your own words how the balloons represent the forces that shape molecules.

2. Which of the molecules listed below will be linear? planar? tetrahedral?
(a) CH_3Cl **(b)** Cl_2 **(c)** CCl_4
(d) HCl **(e)** BF_3

Extension and Application

1. Use your round and pear-shaped balloons to make a model of $BeCl_2$. What are the bond angle and the molecular shape of this molecule?

2. Blow up two round balloons four times larger than before. Replace two round balloons in the tetrahedral molecule. What happens? When does something similar happen in a real molecule?

3. In determining molecular shape by the repulsion of electron pairs, double or triple bonds have the same effect as single bonds. Are ethylene, $H_2C{=}CH_2$, and acetylene, $H{-}C{\equiv}C{-}H$, planar or three-dimensional? What will be the bond angles of the carbon-carbon-hydrogen bonds in each molecule?

Chapter 14
ChemActivity 1

Paper Chromatography

Chromatography is a technique of separation. There are many forms of chromatography, but in this activity you will be using paper chromatography. The separation of compounds in paper chromatography depends on how strongly the components of a mixture are attracted to a solvent and to the paper.

In paper chromatography, the *stationary phase* is a piece of paper. The *mobile phase* is the mixture to be separated, often a solvent in which a sample of unknown composition is dissolved. Your unknown sample will be the black ink from a marking pen. The purpose of this activity is to observe the effect of solvent polarity on the separation of a mixture by paper chromatography.

Objectives

- **Set up** a paper chromatography experiment.
- **Observe** the separation of the dyes in an ink sample.
- **Examine** the relationship between separation and solvent polarity.

Materials

1 × 9 cm strips of filter paper (3)
plastic microtip pipet
24-well microplate
black, water-soluble marker
heavy plastic bag
distilled water
ethanol

PROGRAM RESOURCES

ChemActivity Masters: Use worksheets for ChemActivity 14-1 to provide students with a copy of procedures and a data table for use in the lab.

Procedure

1. Place ¼ pipet of water and ¼ pipet of ethanol (EtOH) in well A1 of a 24-well microplate.
2. Place ½ pipet of water in well B1.
3. Place ½ pipet of ethanol (EtOH) in well C1.
4. One by one place each strip of filter paper in an empty well and draw a pencil line on it even with the top of the well.
5. Remove each strip and make a small circle on the pencil line with the black marker as shown in Figure A.

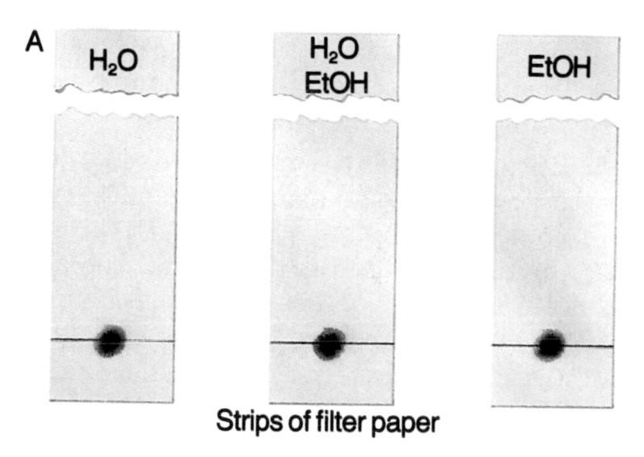

Strips of filter paper

6. Allow the mark to dry and mark the spot again.
7. In pencil, label the tops of the paper strips with the solvents to be used: A1, the water-ethanol mixture (H_2O-EtOH); B1, water (H_2O); C1, ethanol (EtOH). See Figure A.
8. Place the marked strips in the appropriate wells, A1, B1, and C1.
9. Fold over the marked strips and place the entire assembly in a plastic bag as shown in Figure B.

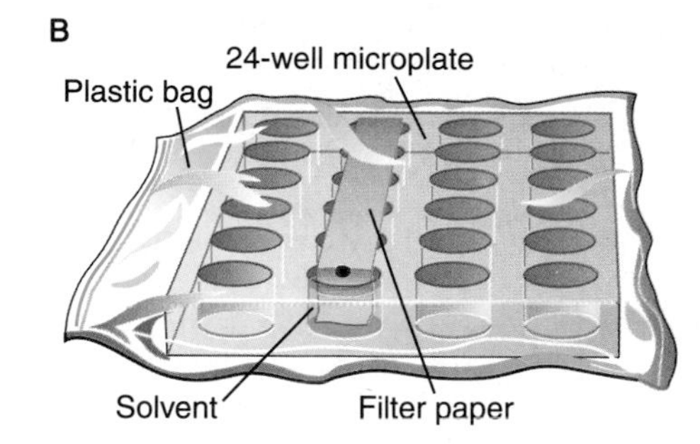

10. Allow the chromatogram to "develop" until the solvent comes close to the top of the paper.
11. Remove the paper chromatograms.
12. Choose spots of the same color on each chromatogram for your analysis. Measure the distance that the spot of this color moved from the original spot on each chromatogram, as in Figure C. Enter this distance in a data table like the one shown.

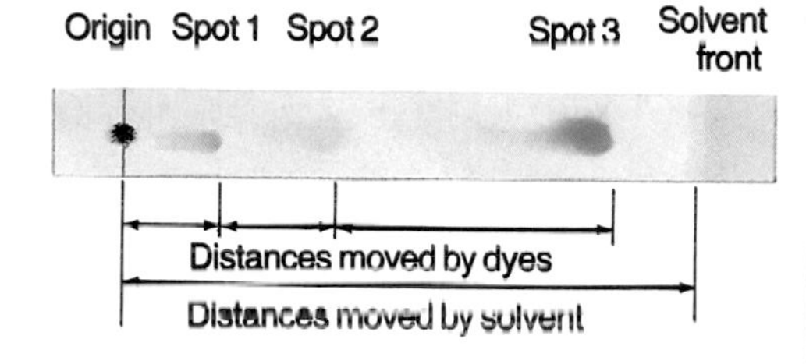

Data and Observations

Well	Solvent	Distance to solvent front	to spot 1	to spot 2	R_f Value
A1	H_2O + EtOH				
B1	H_2O				
C1	EtOH				

Misconceptions

- Students may think that a marker contains only one color of ink. The chromatography shows that a single color like black may be a mixture of four or five colors.
- Most students think that the separation will be the same even if the solvent is different.

the purpose, procedure, and safety precautions for this ChemActivity.

- Be sure to use a water-soluble marker in this activity.
- The paper strips must be about 6 inches long.
- Use the same color marker for all students. Only three or four markers will be necessary for an entire class.
- Use markers from the same manufacturer. Different manufacturers use different combinations of dyes for each color.
- Students usually want to continue this activity. Have them repeat the experiment with several different colored markers. Ask them to discover whether the same color dye is used in more than one color marker. Manufacturers simply mix the same basic color dyes in different proportions to make different colored markers.

Data and Observations

See table below.

1. The mixture consisted of different colored dyes. The number and colors of the dyes depend on the make of the marker used. Each dye is an organic compound used in the marker for its characteristic color.
2. All the chromatograms will show some degree of separation of the dye mixture. Depending on the formulation of the mixture, the dyes will migrate at different rates in different solvents.
3. The solvent in which the dye spots moved the farthest gives the best separation.

Data and Observations **Sample data**

Well	Solvent	Distance to: solvent front	to spot 1	R_f Value
A1	H_2O + EtOH	10.5 cm	8.3 cm	0.79
B1	H_2O	10.0 cm	5.7 cm	0.57
C1	EtOH	10.2 cm	9.6 cm	0.94

✓ ASSESSMENT

Performance: Assess student performance for this ChemActivity by having students complete copies of the masters from the *ChemActivity Masters* booklet.

Analysis and Conclusions

1. See table on previous page.
2. No two dye spots will have the same R_f value since the solvents used in each chromatogram are different.
3. If the dye spot moved farthest in water, it was a highly polar dye. If the dye spot moved farthest in ethanol, it was a less polar dye. Student answers will depend upon the spot chosen and the type of marker used.

Extension and Application

1. The separation of the mixture depends on the dyes being more attracted to the solvent than they are to the solid paper. A more polar solvent will attract polar dyes more strongly than the paper.
2. The individual dye spots could be cut out of the paper, the dyes separately redissolved from the paper, and the chromatography repeated with the same solvent for the dye from each spot. If the dyes are identical, the resulting R_f values of the dye spots will be the same.
3. Many answers are possible. In a chemical research lab, for example, chromatography is used to separate unknown substances from a new kind of reaction or to separate natural products during their identification. In a forensic lab, chromatography is useful in the identification of illegal drugs. In a chemical manufacturing plant, it can be used to monitor the composition of a product mixture or identify an impurity in a product.

CHEMACTIVITY 14-2

Miscibility of Liquids

Liquid solubility gives some indication of the molecular makeup of the molecules in the solution. Polar molecules can interact with polar molecules to form a stronger association than that they can form with nonpolar molecules. Nonpolar molecules exhibit the same tendencies with nonpolar molecules.

Process Skills

Observing, classifying, predicting, recognizing cause and

13. Measure the distance that the solvent moved from the origin to the solvent front on each chromatogram. Enter this distance in your data table.

Data and Observations

1. How many colored spots do you have? What are the components of the mixture?
2. How are the three chromatograms similar? How are they different?
3. The degree of separation is shown by the distance between the dye spots on the paper. Which solvent was best at separating the different dyes?

Analysis and Conclusions

1. Determine the R_f values for each of your dye colors by using the formula

$$R_f = \frac{\text{Distance dye moved}}{\text{Distance to solvent front}}$$

Enter your calculated R_f values in your data table.

2. Compare the R_f values for your spots. Do any two of these spots have the same R_f value? Explain your answer.
3. Water is more polar than ethanol and the mixture is intermediate between the two in polarity. The most polar molecules will move farthest in the most polar solvent. What can you conclude about the polarity of the dye in the spots you analyzed?

Extension and Application

1. Explain how a polar solvent works in the separation of polar dyes.
2. How could you prove that two dye spots of the same color but with different R_f values in different solvents are the same compound?
3. Chromatography of many kinds is widely used to isolate and identify chemical compounds and mixtures. How might chromatography be used in a chemical research lab? In a police forensic lab? In a chemical manufacturing plant?

Chapter 14
ChemActivity 2

Miscibility of Liquids

The solubility of a substance depends on several factors. One is the polarity of both the substance and the solvent. A general guide is that "like dissolves like." A solvent that is polar will dissolve polar substances. A solvent that is nonpolar will dissolve nonpolar substances.

Water is a polar liquid and TTE (trichlorotrifluoroethane) is a nonpolar liquid. The purpose of this activity is to classify liquids as more polar or less polar based on their solubilities in water and TTE.

Objectives

- **Observe** the solubility of liquids in water and TTE.
- **Classify** liquids as more or less polar.

Materials

96-well microplate (3)
plastic microtip pipets (8 per group)
small beaker or plastic cup
toothpicks
construction paper, white
distilled water
TTE (trichlorotrifluoroethane)

PROGRAM RESOURCES

ChemActivity Masters: Use worksheets for ChemActivity 14-2 to provide students with a copy of procedures and a data table for use in the lab.

liquids to be tested:

1-butanol	1-propanol
2-butanol	2-propanol
tert-butanol	hexane

food coloring, blue
Microplate Data Form

Procedure

1. Place a 96-well microplate with the numbered columns away from you and the lettered rows to the left.
2. Place 5 drops of 1-butanol in each of wells A1 to A10.
3. With a clean microtip pipet, place 5 drops of 2-butanol in each of wells H1 to H10.
4. Using a clean microtip pipet for each liquid, repeat steps 2 and 3 with pairs of remaining test liquids and two more microplates. Use only rows A and H in each microplate so that you can observe the test mixture from the sides of the microplates.
5. Place 20 cm^3 of distilled water in a small plastic cup or beaker. Add 3 drops of blue food coloring.
6. In the first microplate, using a clean microtip pipet, add 1 drop of the colored water to wells A1 and H1. Add 2 drops of colored water to wells A2 and H2. Continue adding more colored water until you reach wells A5 and H5. Mix the contents of each row with a separate toothpick.
7. Next, with another clean microtip pipet, add 1 drop of TTE to wells A10 and H10 of the first microplate. Add 2 drops of TTE to wells A9 and H9. Continue adding more water to each successive well until you reach wells A6 and H6. Using clean toothpicks, mix the contents of each row of wells with a separate toothpick.
8. Hold the plate over a piece of white paper. Observe the wells from the side of the plate. Note your observations in a Microplate Data Form like the one shown.
9. Repeat steps 6–8 for the four liquids in the other two microplates.

Data and Observations

Relabel your Microplate Data Form so that you can record the results from all six liquids on the same form:

Test Liquids	H_2O					TTE				
	1	2	3	4	5	6	7	8	9	10
1-butanol A										
2-butanol H										
tert-butanol A										
1-propanol H										
2-propanol A										
Hexane H										

1. Two liquids that mix together in all proportions are said to be *miscible,* meaning they are soluble in each other. The formation of two distinct layers means that two liquids are not miscible. Which combinations of liquids formed two layers?
2. The formation of a white cloudy mixture that does not separate means that the two liquids are partially miscible. Which combinations of liquids formed cloudy mixtures?
3. Which combinations of liquids formed clear solutions, indicating that they are miscible?
4. Were any of the liquids tested miscible with both water and TTE?

Analysis and Conclusions

1. Which were the most polar liquids tested? Which were the least polar?
2. What do you think would be the result of mixing colored water and TTE? Try it.

effect, interpreting data, comparing and contrasting, inferring, communicating

Procedure Hints

Before proceeding, be sure students have read and understood the purpose, procedure, and safety precautions for this ChemActivity.

- If you cannot supply three 96-well microplates per lab group, have the students carry out the activity with two liquids at a time. They should rinse the microplate and shake it as dry as possible before each trial.
- The water is tinted blue by food coloring to increase the visibility of different layers when they are formed.
- Dispense liquids in small containers to facilitate student access.
- Do not allow students to return solvents to stock bottles.
- No flames should be present in the room when this activity is in progress.
- Adequate ventilation is necessary since some students may be sensitive to organic fumes.
- Waste organic solvents should be collected, stored, and discarded in an approved manner. Organic liquids should not be discarded down the drain.

Data and Observations

See the data table below.

1. Water with 1-butanol, 2-butanol, hexane.
2. Water with higher proportions of 1-butanol and 2-butanol; TTE with a high proportion of 2-butanol.
3. Water with *tert*-butanol, 1-propanol, 2-propanol; TTE with 1-butanol, 2-butanol, *tert*-butanol, 1-propanol, 2-propanol, hexane.
4. Yes; *tert*-butanol, 1-propanol, 2-propanol.

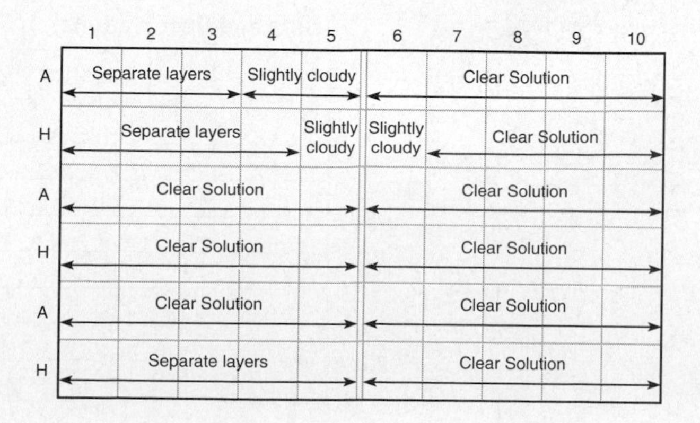

1 2 3 4 5 6 7 8 9 10
A Separate layers | Slightly cloudy | Clear Solution
H Separate layers | Slightly cloudy | Slightly cloudy | Clear Solution
A Clear Solution | Clear Solution
H Clear Solution | Clear Solution
A Clear Solution | Clear Solution
H Separate layers | Clear Solution

Analysis and Conclusions

1. The most polar liquids are *tert*-butanol, 1-propanol, and 2-propanol. The least polar liquids

✓ ASSESSMENT

Performance: Assess student performance for this ChemActivity by having students complete copies of the masters from the *ChemActivity Masters* booklet.

are 1-butanol, 2-butanol, and hexane.

2. Water and TTE are not miscible.

Extension and Application

1. 1-Butanol and hexane were not polar enough to be miscible with water, while 1-propanol was miscible with water. It looks like more carbon atoms means less polar and less soluble in water. Hexane was the substance least soluble in water. Since like dissolves like, the fact that hexane has no -OH group like water might be important.

2. Since methanol and ethanol both have fewer carbon atoms than 1-propanol, which was miscible with water, they should also be miscible with water. Since octane has more carbon atoms than hexane, it should be immiscible with water.

CHEMACTIVITY 15

Gas Pressure

The volume of a confined gas varies inversely with the pressure on the gas. In this activity the confined gas is trapped in the stem of the plastic pipet. The gas volume is defined as the volume of a cylinder.

$V = \pi \times r^2 \times h$

Since the radius (r) of the tube does not change and π is a constant, the volume of the confined gas can be expressed as a function of the length of the air column.

Process Skills

Observing, measuring, using numbers, predicting, recognizing cause and effect, interpreting data, inferring, communicating

Extension and Application

1. The formulas of three of the liquids in your experiment are shown below:

   $CH_3CH_2CH_2CH_2OH$ 1-butanol $CH_3CH_2CH_2OH$ 1-propanol

   $CH_3CH_2CH_2CH_2CH_2CH_3$ hexane

   What do you see in the structures that might explain the results of your experiment with these substances?

2. Examine the three structures below and predict which of these substances would be miscible or immiscible with water.

   CH_3OH methanol CH_3CH_2OH ethanol

   $CH_3CH_2CH_2CH_2CH_2CH_2CH_2CH_3$ octane

Chapter 15 ChemActivity

Gas Pressure

The volume of a gas depends on three factors. The first is the number of moles of the gas, the second is the temperature at which the volume is measured, and the third is the pressure. For measurements at constant temperature, there is a simple mathematical relationship between the volume of a gas and the pressure. This relationship is known as Boyle's law in honor of Sir Robert Boyle, the British chemist who first recognized it about 300 years ago. The purpose of this activity is to discover the relationship between the pressure and the volume of a gas.

Objectives

- **Observe** the effect of increased pressure on the volume of a confined gas.
- **Graph** the pressure to volume relationship.
- **Describe** the pressure to volume relationship in mathematical terms.

Materials

thin stem pipet	matches
plastic cup	glass rod
graph paper	metric ruler
textbooks (8 of equal size and mass)	water
	methylene blue

Procedure

1. **Form a hypothesis** to explain what would happen to the volume of a gas as the pressure is increased or decreased.
2. Fill a plastic cup with 20 cm^3 of water.
3. Add a few drops of methylene blue solution to the water.
4. Fill only the bulb of a thin-stem plastic pipet with the tinted water.
5. Soften the tip of the filled pipet by slightly heating it in a match flame. **CAUTION:** *Be careful not to heat the tip of the pipet more than necessary to soften it; it could catch fire.*
6. Seal the tip by pressing down on it with a glass rod.
7. Allow the pipet tip to cool.
8. Place two equal-sized books on the bulb of the pipet. The stem of the pipet should be visible as shown in the diagram.
9. Measure the column length of the trapped air in the stem of the pipet in millimeters and enter this measurement in a data table like the one shown.
10. Place another book on top of the first two, and measure and record the length of the air column again.

Procedure Hints

Before proceeding, be sure students have read and understood the purpose, procedure, and safety precautions for this ChemActivity.

- Caution the students not to overheat the end of the pipet since the plastic can catch fire.
- Caution the students not to pinch the end of the plastic pipet with their fingers while is it hot.

PROGRAM RESOURCES

ChemActivity Masters: Use worksheets for ChemActivity 15 & 16 to provide students with a copy of procedures and a data table for use in the lab.

Data and Observations

Pressure (books)	Length of air column (mm)	1/Pressure
2	80	0.5
3	72	0.33
4	50	0.25
5	45	0.20
6	38	0.17
7	35	0.14
8	30	0.13

11. Continue to add the remaining books to the stack one-by-one. Measure and record each new air-column length.

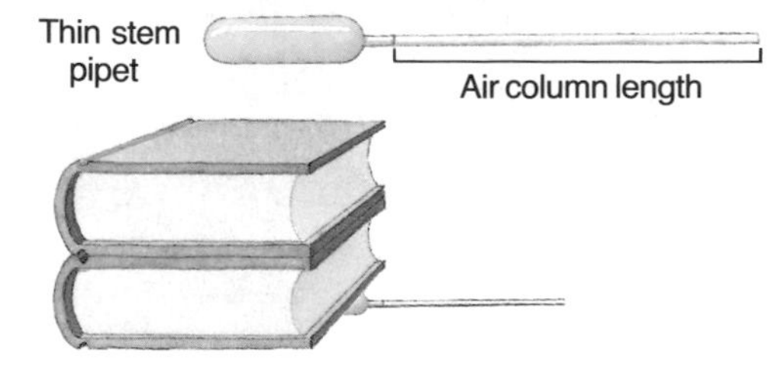

Data and Observations

Pressure (books)	Length of air column (mm)	1/Pressure
2		
3		
4		

Analysis and Conclusions

1. Plot a graph of your data. Use the pressure in books as the x variable and the length of the air column as the y variable. Is the relationship between gas pressure and volume linear?
2. Was your hypothesis correct? Explain why or why not.
3. Calculate the inverse of the pressure, 1/pressure, in books. Enter your calculations in your data table.
4. Plot a second graph. Use the inverse, 1/pressure, as the x variable and the length of the air column as the y variable. Is the relationship between 1/pressure and the volume of the gas linear?
5. What generalization can you make about the effect of increased book pressure on the volume of the confined gas?

Extension and Application

1. Write a mathematical equation for the relationship between the pressure and volume of a gas at constant temperature. Use P for pressure, V for volume, and k for any necessary constant.
2. What can happen to a gas under extremely high pressures or extremely low temperatures?
3. If a pressurized gas is released from a vessel, ice can be seen forming on the outside of the vessel. Why does this happen?

Chapter 16 ChemActivity

Micro

Crystals and their Structure

Crystals are classified according to their unit cells, the simplest repeating units of atoms, ions, or molecules. A crystal that is large enough to see is a collection of microscopic unit cells.

Crystals can be many sizes. The shape of a large crystal gives a hint to the characteristic shape of the unit cell. The purpose of this activity is to study the possible shapes of crystals by constructing unit-cell models and comparing the models with real crystals.

Objectives

- **Construct** some unit-cell models.
- **Observe** the process of crystallization.
- **Observe** and **classify** the unit cells of some common crystals.

Materials

gumdrops (3 colors)
toothpicks (16)
protractor
plastic microtip pipet
24-well microplate
chemical scoop
glass microscope slides (4)

- Food coloring may be used instead of methylene blue for tinting purposes.
- If you wish, increase the number of books used to 10 to provide more data points.

Data and Observations
See data table below.

Analysis and Conclusions
1. A sample student plot is shown below. The relationship between pressure and volume is not linear.
2. Answers will vary.
3. See sample data below.
4. A sample student plot is shown below. The relationship between the inverse of the pressure and the volume is linear.
5. There is a directly proportional relationship between the inverse of pressure and the volume of a confined gas at constant temperature.

Extension and Application
1. P = k/V or PV = k
2. A gas will liquefy under appropriate combinations of high pressure and low temperature.
3. When a pressurized gas is released, the molecules absorb thermal energy from the surroundings. This cools the outside of the vessel. The outside atmosphere contains water vapor that condenses and freezes on the vessel's surface.

CHEMACTIVITY 16

Crystals and Their Structure
The study of crystals and crystallography is complex. The crystals examined here are only a representative sampling of the wide variety of crystal types.

Process Skills
Observing, measuring, using numbers, classifying, predicting, comparing and contrasting, using space/time relationships, formulating models, communicating

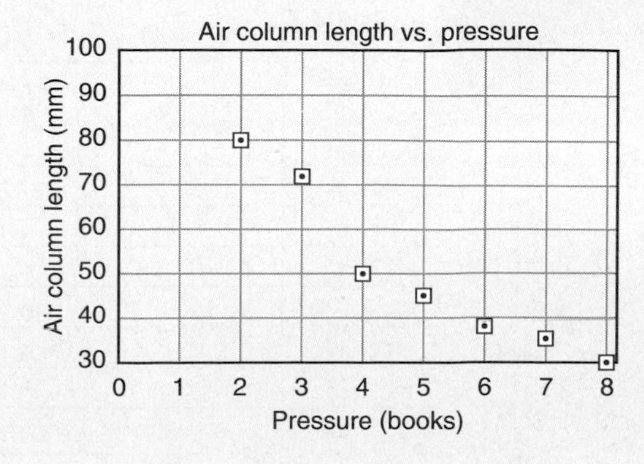

Air column length vs. 1/pressure
Air column length (mm)
1/pressure (books)

✓ ASSESSMENT
Performance: Assess student performance for this ChemActivity by having students complete copies of the masters from the *ChemActivity Masters* booklet.

Misconceptions

- Students should be reminded that not all substances form crystals. Some substances are amorphous.
- Glass is not a crystal. It is a supercooled liquid.
- Because it is hard to observe the crystals in a metal, students may not recognize that metals are crystalline solids.

Procedure Hints

Before proceeding, be sure students have read and understood the purpose, procedure, and safety precautions for this ChemActivity.

- Use small gumdrops, also known as spice drops.
- The unit cells that are built in this activity are the simplest ones. A student may wish to build others. These may be kept as models for other students for many years.
- Do not use marshmallows because they tend to crystallize over a long period of time.
- **Troubleshooting:** Students tend to eat the gumdrop "crystals" before the end of the class. Be aware of this problem.
- Warn students not to push the toothpick into the gumdrop too firmly. The toothpick may go through the gumdrop and pierce a finger or hand.
- Students may have trouble focusing a microscope. Caution them to use low power at all times for this activity.

Materials (continued)

- hand lens or low-power microscope
- sodium chloride $NaCl(cr)$
- sodium nitrate, $NaNO_3(cr)$
- benzoic acid ($C_7H_6CO_2$)
- aspirin (acetylsalicyclic acid, $C_9H_8O_4$) (1 tablet)
- ethanol (C_2H_5OH)

Procedure

Unit-cell models

1. Obtain 14 gumdrops from your teacher. Take 4 of one color, 4 of another color, and 6 of a third color.
2. Place one gumdrop of the first color on your lab table. Using a protractor to measure angles, gently push three toothpicks into the gumdrop at right angles to each other as shown in Figure A.

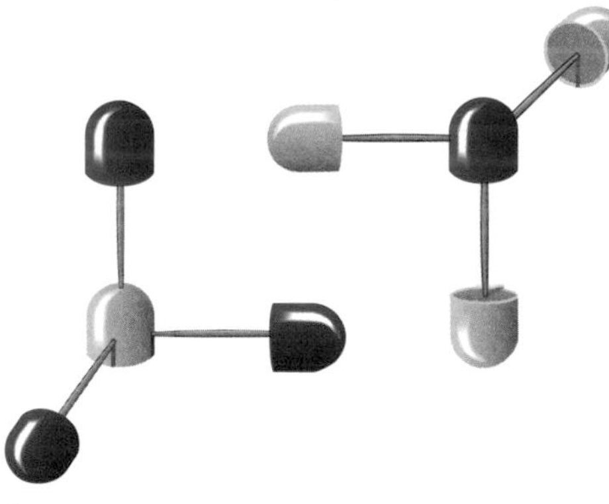

A

3. Place a gumdrop of the second color at the end of each of the three toothpicks.
4. Repeat steps 2 and 3 using the opposite color scheme.
5. Place the two models around a common center so that the central gumdrops of each model are situated at opposite corners of a cube. Join opposite colored gumdrops to each other with toothpicks to form a cube as shown in Figure B.

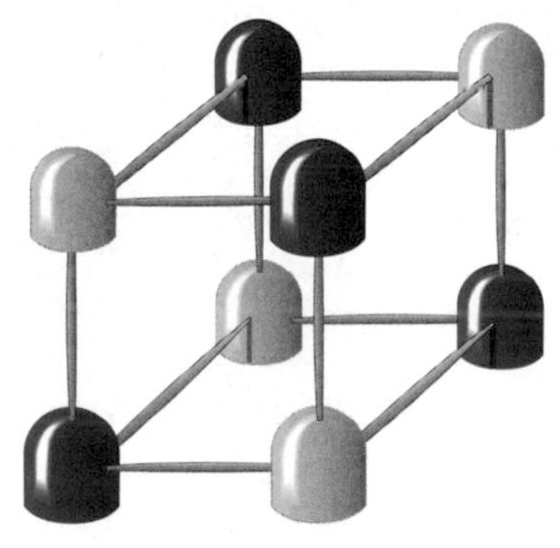

B Simple cubic unit cell

6. In a data table like the one below, enter the value of the three angles of this simple cubic unit cell indicated in the table.
7. Place a gumdrop of the third color in the center of your cubic unit cell and secure it with the toothpicks to four of the corners of the cube as shown. This is a model of a body-centered cubic unit, Figure C.

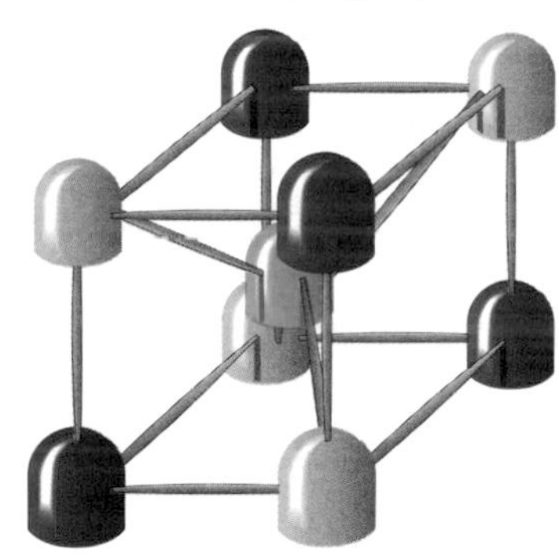

C Body-centered cubic unit cell

8. In your data table, enter the values of the three angles for a body-centered cubic unit.
9. Remove the center gumdrop.
10. Relocate the center gumdrop in the center of one of the faces of the cube. Place other gumdrops of the same color at the center of each of the other faces to make a model of a face-centered cubic unit cell like that shown in Figure D.

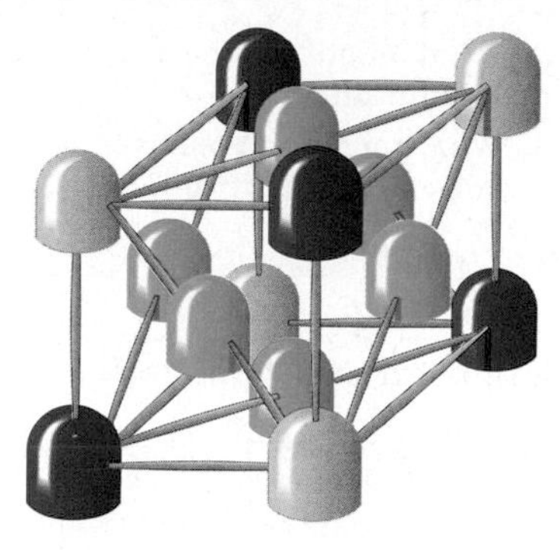

D Face-centered cubic unit cell

11. Record the indicated three angles of this face-centered cubic unit cell in your data table.
12. Obtain eight more gumdrops of two colors and assemble a simple cubic unit cell as you did in steps 2–5.
13. Make a rhombohedral unit cell from your simple cubic unit cell. Hold the bottom of the new cubic unit cell and push the top four gumdrops to one side about 3 cm, Figure E. Measure the three angles and record them in your data table.

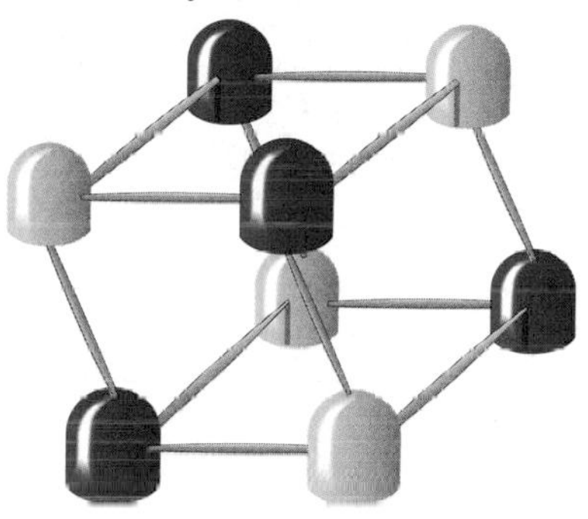

E Rhombohedral unit cell

Crystals

14. Place ½ pipet of ethanol in each of wells A1 and B1 of a 24-well microplate.
15. Add 0.2 g of benzoic acid to well A1. Add 1 aspirin to well B1.
16. Stir the contents of each well with a separate toothpick. The aspirin will seem not to dissolve. The insoluble material is starch, not the aspirin itself.
17. Allow any solids to settle to the bottom of each well.
18. Place a clean glass slide on a microscope or white piece of paper.
19. Using a plastic pipet, transfer three drops of liquid from well A1 to the slide.
20. Observe the slide under a microscope or with a hand lens. Describe the shapes of crystals that you observe in a data table like the one shown.
21. Repeat steps 18–20 using the liquid in well B1.
22. Use a chemical scoop to place a few crystals of sodium chloride on another glass slide. Observe the slide under a microscope or with a hand lens. Describe the crystal shapes you observe.
23. In the same way, observe crystals of sodium nitrate. Describe the crystal shapes in your data table.

Data and Observations

Unit-cell models

3 D Unit Cell	Angle 1	Angle 2	Angle 3
Simple Cubic			
Body-centered cubic			
Face-centered cubic			
Rhombo-hedral			

Analysis and Conclusions

1. The sodium chloride and sodium nitrate crystals were clearly three-dimensional and box-like while the benzoic acid and aspirin crystals were flat and pointed.
2. Sodium chloride could be simple cubic, body-centered cubic, or face-centered cubic.
3. Sodium nitrate was not cubic. It appeared to be rhombohedral.
4. Based on the crystal shape, the unit cell would not be cubic, but would be narrow with angles that are not 90°.

✓ ASSESSMENT

Performance: Assess student performance for this ChemActivity by having students complete copies of the masters from the *ChemActivity Masters* booklet.

Data and Observations

Unit-cell models			
Unit-cell type	Angle 1	Angle 2	Angle 3
Simple cubic	90°	90°	90°
Body-centered cubic	90°	90°	90°
Face-centered cubic	90°	90°	90°
Rhombo-hedral	<90°	<90°	<90°

Extension and Application

1. By stacking layers of unit cells on top of one another, a basic unit cell model could be built up to a model of a macroscopic crystal that would mirror the shape of the unit cell.

2. If impurities were added to the crystal, the shape of the crystal might be altered. The impurities might also be fixed within the crystal as flaws or faults in the crystal structure.

3. Because X rays are diffracted by crystals, they can be used to study crystal structure. The diffraction pattern of the X ray is determined by the location of atoms and molecules in crystals. Thus, x-ray diffraction can be used to determine the dimensions and arrangement of particles in unit cells.

CHEMACTIVITY 17-1

Enthalpy of Fusion of Ice

A change of state requires the addition or removal of energy from matter. For example, ice at its melting point changes to liquid water when enough energy is added for the water molecules to pull away from each other. As long as any solid remains in the ice/water mixture, addition of energy serves only to change more solid into liquid without changing the temperature of the mixture.

Process Skills

Observing, measuring, using numbers, predicting, interpreting data, comparing and contrasting, inferring, communicating

Crystals

Compound	Description of Crystals
Benzoic acid	
Aspirin	
Sodium chloride	
Sodium nitrate	

Analysis and Conclusions

1. How did the benzoic acid and aspirin crystals compare with the sodium chloride and sodium nitrate crystals?
2. What might be the type of unit cell for sodium chloride?
3. What might be the type of unit cell for sodium nitrate?
4. What might be the shape and angles of a benzoic acid unit cell?

Extension and Application

1. How could a unit-cell model be expanded into a macroscopic crystal model?
2. What would be the physical effects on a crystal if impurities were added?
3. The study of crystals has been aided by the development of X-ray crystallography. How do scientists study crystals using X rays? What information do they obtain by these studies?

Micro

Chapter 17 ChemActivity 1

Enthalpy of Fusion of Ice

When energy is added to ice at its melting point of 0°C, the temperature of the ice does not increase. Instead, the added energy allows the water molecules to pull away from each other and enter the liquid state. The energy required to melt one gram of a substance at its melting point is known as the enthalpy of fusion of that substance. The purpose of this activity is to determine the enthalpy of fusion of ice at atmospheric pressure.

Objectives

- **Determine** the enthalpy of fusion of ice.
- **Compare** an experimental value for enthalpy of fusion of ice with the actual value.

Materials

50-cm^3 graduated cylinder	balance
plastic ketchup cups with two tops (4)	labels
rubber band	thermometer
	crushed ice
	distilled water

Procedure

1. A calorimeter, an instrument used to measure energy changes in chemical systems, must exchange as little heat with the air around it as possible. **Hypothesize** how accurate your measurements are likely to be using a simple plastic calorimeter.
2. Construct a plastic calorimeter by placing a rubber band around one ketchup cup, and then placing this cup inside the second cup as shown here. With a paper punch, make one hole in a ketchup-cup top.

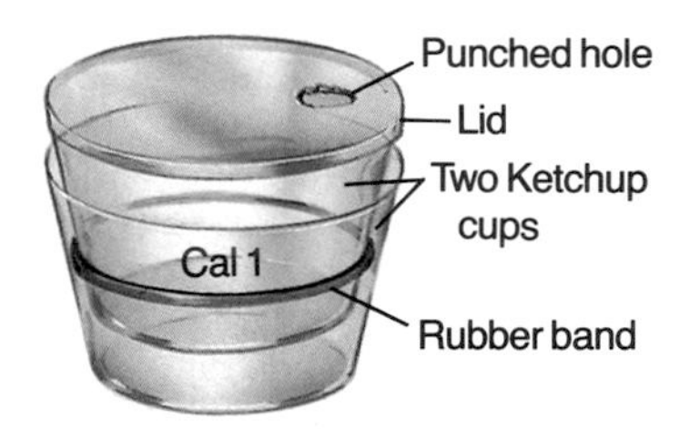

Procedure Hints

Before proceeding, be sure students have read and understood the purpose, procedure, and safety precautions for this ChemActivity.

- Plastic cups for this activity can be obtained from any restaurant or party store that has portion control cups. Ketchup, soy sauce, and mayonnaise are often supplied in this kind of cup.
- To give accurate results, ice must be at 0°C.

Data and Observations

	Crystals
Compound	Description of crystals
Benzoic acid	The crystals are narrow and look like needles.
Aspirin	The crystals are needles, but have irregular, not smooth edges.
Sodium chloride	The crystals are shaped like cubes
Sodium nitrate	The crystals are like cubes pushed to one side so that the angles are not 90°

3. Repeat step 2 to make a second plastic calorimeter.
4. Label the two plastic calorimeters *Cal 1* and *Cal 2*.
5. Measure the mass of Cal 1 with its lid and record the mass in a data table like the one shown.
6. Place 20 cm^3 of distilled water in Cal 1. Cover it and measure the total mass of Cal 1 with the water. Record your result.
7. Insert a thermometer into Cal 1 and record this initial water temperature.
8. Fill Cal 2 with crushed ice and thoroughly drain off any water.
9. Remove the thermometer from Cal 1, and add the water from Cal 1 to the ice in Cal 2.
10. Swirl the water/ice mixture for 1 minute. Pour all the water from Cal 2 into Cal 1.
11. Insert the thermometer and measure the temperature of the water in Cal 1. Record this final temperature.
12. Replace the lid on Cal 1, and measure and record the total mass of Cal 1, which now contains the original amount of water plus the melted ice.
13. Discard the water and ice from both calorimeters. Dry the two calorimeters, and repeat the activity starting from Step 4 in order to get another set of data so that you can average the results of two runs.

Analysis and Conclusions

1. Complete the data table by finding the values of C, F, and H for each run. Disregard the minus sign in data F because only the amount of temperature change is of concern here.
2. The heat that melted the ice came from the original water sample, causing a drop in water temperature. This quantity of heat is found from the equation

$$q\text{, in joules} = m(\Delta T)C_p$$

where m is the initial mass of water (data C), ΔT is the change in temperature of this water (data F), and C_p is the specific heat of water, which is 4.18 J/g·C°. Calculate q for each of your runs and record the values for data I in your data table.
3. The enthalpy of fusion is the heat needed, in joules, to melt one gram of ice. To find the experimental values for your two runs, divide the results of analysis step 2 (data I) by the mass of ice melted (data H). Record the data in your data table.
4. Find the average enthalpy of fusion from your two runs (data K). Compare the result of your experiment with the actual value for the enthalpy of fusion of water, which is 334 J/g. Suggest some reasons for the differences. Was your hypothesis supported by your data?

Data and Observations

Observations	Run 1	Run 2
A. Mass of empty Cal 1 (g)		
B. Mass of Cal 1 + water (g)		
C. Mass of water (B − A) (g)		
D. Initial temperature of water in Cal 1 (°C)		
E. Final temperature of water in Cal 1 (°C)		
F. Change in temperature of water (D − E) (°C)		
G. Mass of Cal 1, water, and melted ice (g)		
H. Mass of ice melted (G − B) (g)		
I. Heat to melt ice (J)		
J. Enthalpy of fusion (I/H) (J/g)		
K. Average enthalpy of fusion (J/g)		

- **Troubleshooting:** It is important to drain all liquid from Cal 2 before adding the water from Cal 1.
- The energy required to melt the ice in Cal 2 is drawn from the water poured from Cal 1, causing the temperature of the water to drop.
- Each gram of ice melted appears as an increase in the mass of water in Cal 1.

Data and Observations

Sample student data for one run:

A. 2.52 g	F. 9.9 °C
B. 23.72 g	G. 26.56 g
C. 21.20 g	H. 2.84 g
D. 22.1 °C	I. 880 J
E. 12.2 °C	J. 310 J/g

Analysis and Conclusions

1. See sample data for C, F, and H above.
2. q, in joules = $m(\Delta T)C_p$
 = (C) x (F) x (4.19 J/g·°C)
 = 21.20 g x 9.9°C x (4.19 J/g·°C)
 − 880 J
3. Enthalpy of fusion = $\frac{q\text{, in joules}}{\text{grams of ice melted}}$
 = 880 J/2.84 g = 310 J/g
4. Some possible reasons for differences between the experimental value and the actual value for the enthalpy of fusion are gain of heat from the surroundings as well as the water, limit due to accuracy of thermometer, or failure to get rid of all water before ice was transferred to Cal 1. Students' answers will depend on their hypotheses.

✓ ASSESSMENT

Performance: Assess student performance for this ChemActivity by having students complete copies of the masters from the *ChemActivity Masters* booklet.

PROGRAM RESOURCES

ChemActivity Masters: Use worksheets for ChemActivity 17-1 to provide students with a copy of procedures and a data table for use in the lab.

Extension and Application

1. An ice/water mixture remains at 0°C at atmospheric pressure as long as some ice remains in the mixture.

2. Dry ice is solid carbon dioxide. The temperature of dry ice at atmospheric pressure is much lower than that of water ice (-79.5°C). Dry ice does not form a liquid as it melts, but changes directly to gas at atmospheric pressure. A solid that directly changes to a gas is said to sublime.

3. Water ice is relatively inexpensive and easily available. Although dry ice provides lower temperatures, it is more difficult to store for long periods of time and is more difficult to handle because it is so cold that it "burns" the skin.

4. Because freezing is the reverse of melting, it releases the same amount of heat, 334 J/g.

CHEMACTIVITY 17-2

Enthalpy of Vaporization of Water

The enthalpy of vaporization of any liquid gives some indication of the intermolecular forces that hold molecules of the substance in the liquid phase. The higher the enthalpy of vaporization, the stronger the intermolecular forces that keep the molecules in the liquid state.

Process Skills

Observing, measuring, using numbers, interpreting data, comparing and contrasting, inferring, communicating

Procedure Hints

Before proceeding, be sure students have read and understood the purpose, procedure, and safety precautions for this ChemActivity.

- The students need not be concerned with the differences in sign between enthalpy of vaporization and enthalpy of condensation.
- **CAUTION:** *This activity should not be repeated with any other liquid.*
- You may wish students to carry out multiple runs.

Extension and Application

1. When is ice always 0°C?
2. What is dry ice? How is dry ice different from water ice?
3. Ice is used as a coolant. What are the advantages and disadvantages in the use of water ice as a coolant instead of dry ice?
4. In the equilibrium between ice and water as shown by the equation

$$H_2O(cr) \rightleftharpoons H_2O(l)$$

the change in the forward direction is melting and the reverse change is freezing or crystallization. How much heat do you think is released when 1 gram of liquid water freezes?

Chapter 17
ChemActivity 2

Enthalpy of Vaporization of Water

When energy is added to water at its boiling point of 100°C, the temperature of the water does not increase. Instead, the added energy allows the water molecules to pull away from each other completely and enter the gaseous state. The energy required to vaporize one gram of a substance at its boiling point is known as the enthalpy of vaporization of that substance. An equal amount of energy is released when 1 g of the vapor condenses to a liquid at its boiling point. The purpose of this activity is to determine the enthalpy of vaporization of water at atmospheric pressure by measuring the heat released in the condensation of steam.

Objectives

- **Determine** the enthalpy of vaporization of water.
- **Compare** the enthalpy of vaporization of water with the enthalpies of vaporization of other liquids.

Materials

plastic ketchup cups with one lid (2)
rubber band
125-cm^3 Erlenmeyer flask
one-hole rubber stopper
glass bend
hot plate
50-cm^3 graduated cylinder
thermometer
distilled water

Procedure

1. Construct a plastic calorimeter from two ketchup cups as described in ChemActivity 17-1, procedure Step 2.
2. Measure the mass of the calorimeter and its lid. Enter this mass (A) in a data table like the one shown.
3. Half-fill the 125-cm^3 Erlenmeyer flask with water.
4. Fit the flask with a glass bend and rubber stopper as shown here.
5. Heat the flask on a hot plate as shown.
6. Place 20 cm^3 of distilled water in the calorimeter. Measure and record the total mass of the calorimeter and water (B).

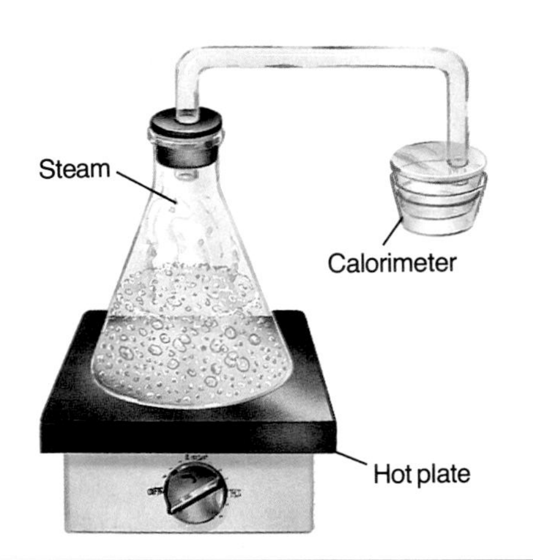

- **Troubleshooting:** Students should avoid having condensed water drip from the glass tube into the calorimeter. The increase in mass in the calorimeter must be from condensed water vapor only.
- The energy released by the condensing steam causes the temperature of the water in the calorimeter to rise.
- Each gram of condensed steam appears as an increase in the mass of water in the calorimeter.

Data and Observations

Sample Data

A. 4.15 g
B. 24.25 g
C. 20.10 g
D. 22.5 °C
E. 46.5 °C
F. 24.0 °C
G. 25.20 g
H. 0.95 g

Analysis and Conclusions

1. See sample data for C, F, and H.

2. q, in joules = $m(\Delta T)C_p$
= (C) x (F) x (4.19 J/g·°C)
= 20.10 g x 24.0 °C
x 4.19 J/g·°C – 2020 J

7. Place a thermometer in the calorimeter and record this initial temperature of the water (D).
8. When vapor is seen coming out of the glass bend, insert the tip of the glass bend through the hole in the lid and down into the water in the calorimeter. Hold it in this position for 30-45 seconds.
   CAUTION: *Wear a thermal glove to avoid scalds from the water vapor.*
9. Remove the calorimeter from under the glass bend. Place the thermometer in the water in the calorimeter and record this final temperature (E).
10. Measure the mass of the calorimeter with its lid, water, and condensed water vapor and record the mass (G).

Data and Observations

A. Mass of the empty calorimeter (g)	
B. Mass of the calorimeter and water (g)	
C. Mass of water (B − A) (g)	
D. Initial temperature of water in the calorimeter (°C)	
E. Final temperature of water in the calorimeter (°C)	
F. Change in temperature of water (E − D) (C°)	
G. Mass of the calorimeter, water, and condensed water vapor (g)	
H. Mass of condensed water vapor (G − B) (g)	
I. Heat (q) released in condensation (J)	
J. Enthalpy of vaporization (J/g)	

Analysis and Conclusions

1. Complete your data table by finding the values of C, F, and H.
2. The heat released when the steam condensed was added to the water in the calorimeter, causing its temperature to increase. This quantity of heat is found from the equation

$$q\text{, in joules} = m(\Delta T)C_p$$

   where m is the initial mass of water (data C), ΔT is the change in temperature of this water (data F), and C_p is the specific heat of water, which is 4.18 J/g·C°. Calculate q for your experiment.
3. The enthalpy of vaporization is the heat needed, in joules, to vaporize one gram of water. It is equal in value to the heat released per gram of water in condensation. Divide the results of analysis step 2 (data I) by the mass of steam condensed (data H). Record the results.
4. The enthalpy of vaporization of water is 2260 J/g. How does your experimental enthalpy of vaporization of water compare with the actual value? How does it compare with the enthalpy of fusion of water (334 J/g)? Explain the difference.
5. Use a chemistry handbook to look up the value for the enthalpies of vaporization of isopropyl alcohol and benzene. How do these values compare with the enthalpy of vaporization of water? How would you explain the difference?

Extension and Application

1. What is superheated steam? How is this substance used for practical purposes?
2. How is the enthalpy of vaporization related to the boiling point of a liquid and the strength of intermolecular forces in the liquid?
3. A burn caused by exposure to steam is more harmful than a burn caused by exposure to boiling water. Why do you think this is so?

3. Enthalpy of vaporization =

$$\frac{q\text{, in joules}}{\text{grams of vapor condensed}}$$

= 2020 J/0.95 g
= 2130 J/g

4. The enthalpy of vaporization of water is about six times larger than the enthalpy of fusion. A larger enthalpy of vaporization is to be expected because it takes more energy to separate molecules completely to form a gas than just to free them to slip around each other in a liquid.

5. The enthalpy of vaporization of isopropyl alcohol is 900 J/g and that of benzene is 395 J/g. These values are significantly lower than for water, as can be expected because of the strong hydrogen bonding in water. The enthalpy of vaporization of isopropyl alcohol is higher than that of benzene because there is hydrogen bonding in the alcohol and none in benzene.

Extension and Application

1. Superheated steam is steam generated at higher temperature and higher pressures than steam at one atmosphere and 100°C. Because it contains a larger amount of thermal energy, superheated steam can do more work than ordinary steam.

2. The higher the enthalpy of vaporization, the higher the boiling point. This is a direct relationship since the enthalpy of vaporization reflects the strength of the intermolecular forces that hold molecules in the liquid state.

3. When the body comes in contact with boiling water, it absorbs the heat released as the water cools down. In contact with steam, the body also absorbs the large amount of heat released when steam condenses.

PROGRAM RESOURCES

ChemActivity Masters: Use worksheets for ChemActivity 17-2 to provide students with a copy of procedures and a data table for use in the lab.

✓ ASSESSMENT

Performance: Assess student performance for this ChemActivity by having students complete copies of the masters from the *ChemActivity Masters* booklet.

CHEMACTIVITY 18

Graham's Law of Diffusion
Molecules of different gases at the same temperature have the same kinetic energy, which for each gas is given by $1/2\ mv^2$. Mathematically, then, for gases 1 and 2 $KE_1 = 1/2\ m_1v_1^2$ and $KE_2 = 1/2\ m_2v_2^2$ Since $KE_1 = KE_2$, then

$$1/2\ m_1v_1^2 = 1/2\ m_2v_2^2$$
$$m_1v_1^2 = m_2v_2^2$$

or

$$\frac{v_2}{v_1} = \frac{\sqrt{m_1}}{\sqrt{m_2}}$$

This relationship shows that the velocity of a molecule is inversely proportional to the square root of its mass. Simply put, the more massive the molecule, the more slowly it moves.

Process Skills
Observing, measuring, using numbers, predicting, interpreting data, comparing and contrasting, inferring, using space/time relationships, communicating

Misconceptions
Students may think that there will be no difference in the rates of diffusion of gases. One mole of any gas contains the same number of molecules and occupies the same relative volume. Students may extend the mole concept of sameness in number and volume to the process of diffusion.

Procedure Hints
Before proceeding, be sure students have read and understood the purpose, procedure, and safety precautions for this ChemActivity.
- **CAUTION:** *Both ammonia and hydrogen chloride solutions are toxic by inhalation or ingestion and are corrosive. Dispense these solutions in a fume hood. Do not allow these solutions to be handled in open containers. Be sure to have eyewash accessible. Note that universal indicator is flammable.*
- Do not allow one student to

Micro

Chapter 18 ChemActivity

Graham's Law of Diffusion

Molecules are constantly in motion, and their kinetic energy depends on their mass and velocity. At the same temperature and pressure, the average kinetic energy of the molecules in any gas is the same. For this to be true, more massive molecules must move more slowly. One way to observe this difference is in the rate of diffusion of gases. A gas of lower molar mass diffuses faster than a gas of higher molar mass. The purpose of this activity is to discover the relationship between the mass and velocity of two different gases.

Objectives

- **Observe** the color change of universal acid-base indicator exposed to hydrochloric acid and ammonia (a base).
- **Determine** the relative velocities of ammonia and hydrogen chloride molecules.
- **Identify** the mathematical relationship between the mass and velocity of two different gases.

Materials

96-well microplate
plastic microtip pipets (3)
soda straws
plastic sandwich bag with zipper seal
concentrated ammonia solution, $NH_3(aq)$
concentrated hydrochloric acid, HCl (aq)
universal indicator

Procedure

1. **Hypothesize** which will move faster, NH_3 molecules or HCl molecules.
2. Fill all the wells in a 96-well microplate with water by placing the plate under a tap of slowly running water.
3. Place the filled microplate on a flat surface with the numbered columns away from you and the lettered rows to the left.
4. Remove the water from wells D1 and D12 with a plastic pipet.
5. Add one drop of universal indicator to each well in the microplate except wells D1 and D12. Note the color of the indicator in a data table like the one shown.
6. Cut 6 pieces of soda straw, each approximately 3 cm long.
7. Place the soda straws in wells D1, D12, A1, A12, H1 and H12, as shown in the diagram. The straws will prevent a plastic bag from coming in contact with the surface of the liquid in the plate.
8. Hold the plate horizontally, slip it into a sandwich bag, and seal the bag.
9. Puncture the plastic bag with a pencil or pen point just above wells D1 and D12.
10. One partner should obtain a few drops of ammonia solution in a plastic pipet. **CAUTION:** *Do not inhale fumes of ammonia or hydrochloric acid. Avoid contact with skin or eyes. Be sure no ammonia or hydrochloric acid remains in the stem of the pipet before you carry it back to your work area.*
11. The other partner will obtain a few drops of hydrochloric acid in a separate plastic pipet.
12. While one partner places the ammonia pipet into the hole above well D1, the other partner should place the hydrochloric acid pipet into the hole above well D12.
13. Simultaneously add the acid and ammonia to wells D1 and D12. Wait for 30 seconds.

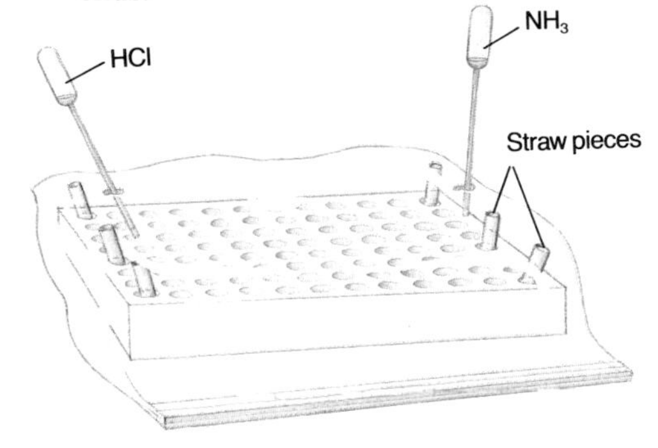

PROGRAM RESOURCES

ChemActivity Masters: Use worksheets for ChemActivity 18 to provide students with a copy of procedures and a data table for use in the lab.

14. Remove the plate from the plastic bag.
15. Record the number of wells that changed to red or yellow due to the reaction of HCl with the indicator (B in your data table).
16. Record the number of wells that changed to blue or violet due to the reaction of NH_3 with the indicator (C).
17. Dispose of the solutions in your microplate according to the instructions given by your teacher.

Data and Observations

A. Initial color of universal indicator in water	
B. Wells with red or yellow color from exposure to HCl	
C. Wells with blue or violet color from exposure to NH_3	
D. NH_3/HCl velocity ratio (C/B)	
E. Molar mass of NH_3 (g/mol)	
F. Molar mass of HCl (g/mol)	

Analysis and Conclusions

1. Find the ratio of blue or violet wells to red or yellow wells. This is the velocity ratio:

$$\frac{\text{Blue or violet wells}}{\text{Red or yellow wells}} = \frac{NH_3 \text{ velocity}}{HCl \text{ velocity}}$$

Which gas moved faster? Did your results support your hypothesis?

2. Calculate the molar masses of ammonia (E) and hydrogen chloride (F) and record them in your data table.
3. Estimate the NH_3/HCl velocity ratio (D) and enter this in your data table.
4. Calculate the four molar mass ratios for NH_3 and HCl listed below:

$$\text{Mass ratio 1} = \frac{\text{molar mass } NH_3}{\text{molar mass HCl}}$$

$$\text{Mass ratio 2} = \frac{\text{molar mass HCl}}{\text{molar mass } NH_3}$$

$$\text{Mass ratio 3} = \sqrt{\frac{\text{molar mass } NH_3}{\text{molar mass HCl}}}$$

$$\text{Mass ratio 4} = \sqrt{\frac{\text{molar mass HCl}}{\text{molar mass } NH_3}}$$

5. Compare each of the molar mass ratios with the velocity ratio. Which mass ratio is most like the estimated velocity ratio? Write the equation for the relationship of the molar masses and molecular velocities of two different gases.

Extension and Application

1. Suggest a method for determining the molar mass of a gas based on Graham's law of diffusion. Use your data to apply this method, assuming that the molar mass of HCl is unknown.
2. Uranium has several isotopes. Only one of these isotopes is useful in nuclear fission. How does Graham's law of diffusion aid in the separation of these isotopes?
3. Oxygen and nitrogen gases are common in Earth's atmosphere. Hydrogen is not. Explain this difference using the kinetic molecular theory and Graham's law of diffusion.

take samples of hydrochloric acid and ammonia at the same time. The two gases released may cause a cloud of ammonium chloride to be generated as the student walks from the supply table to the work area.

• When students are finished with the experiment, submerge microplates in a beaker filled with water to wash out the hydrochloric acid and ammonia solutions.

Data and Observations
A. Green
B. 36
C. 54
D. 1.5
E. 17 g/mol
F. 36.5 g/mol

Analysis and Conclusions
1. Ammonia molecules moved faster.
2. See data E and F.
3. See data D.
4. Mass ratio 1 = 0.47
Mass ratio 2 = 2.1
Mass ratio 3 = 0.68
Mass ratio 4 = 1.5
5. The value of mass ratio 4 is most like the estimated velocity ratio.

$$\sqrt{\frac{\text{molar mass 1}}{\text{molar mass 2}}} = \frac{\text{velocity 2}}{\text{velocity 1}}$$

Extension and Application
1. The molar mass of one gas could be found from the Graham's law relationship if the velocity ratio and the molar mass of the other gas were known.

$$\frac{\sqrt{\text{molar mass HCl}}}{\sqrt{\text{molar mass } NH_3}} = \frac{\text{velocity } NH_3}{\text{velocity HCl}}$$

$$\sqrt{\text{molar mass HCl}} = 1.5 \times \sqrt{17}$$

= 6.2
molar mass HCl = 38

2. Diffusion chambers allow the two isotopes of uranium to migrate at the same temperature and pressure. Since the useful uranium-235 is lighter than the more abundant uranium-238, the lighter element diffuses farther in the same amount of time. This is the way that fissionable material was first separated.

3. The molecules of all atoms are in motion. The lighter the molecules of a gas, the more rapid the motion of these molecules. Since hydrogen is such a light gas, it can reach a velocity high enough to escape from the gravitational pull of Earth. If hydrogen is generated on the surface of Earth and remains unreacted, the gas will escape into space.

✓ ASSESSMENT

Performance: Assess student performance for this ChemActivity by having students complete copies of the masters from the *ChemActivity Masters* booklet.

CHEMACTIVITY 19

Molar Mass of Butane

An essential step in identifying an unknown substance is determination of its molar mass. For a gas, this can be done by measuring the mass and volume of a gas sample at known temperature and pressure and then using the ideal gas equation. This activity introduces the students to the manipulation and measurements of a gas, and the calculation of molar mass from experimental data. Students should not, however, expect highly accurate results.

Process Skills

Observing, measuring, using numbers, interpreting data, inferring, communicating

Misconceptions

Students often do not believe that a gas has mass. This misconception is easy to understand since air, the most common gas in student experience, is invisible and we are not conscious of its mass. You may wish to explain that the pressure of the atmosphere is due to the mass of the gases above the surface of Earth.

Procedure Hints

Before proceeding, be sure students have read and understood the purpose, procedure, and safety precautions for this ChemActivity.

- **CAUTION:** *It is important that no flames are used in the laboratory, for any reason, during this activity.*
- The plastic bags should be new to minimize leakage.
- The water bath used to immerse the plastic bag must be close to room temperature.
- The plastic bag should not be overfilled, since doing so might rupture it or create a leak in the plastic.
- **Troubleshooting:** Remind students that the volumes of both the empty plastic bag and lighter must be subtracted from the total volume of the water displaced.

Molar Mass of Butane

The ideal gas equation gives the relationships among the pressure, volume, temperature, and number of moles of a gas sample.

$$PV = nRT$$

Because n, the moles of gas, is equal to the mass of the gas divided by its molar mass, the ideal gas equation provides a method for finding molar mass. It is necessary to measure the mass and volume of a gas sample and the temperature and pressure at which the volume was measured. The only unknown remaining in the equation is then the molar mass of the gas. The purpose of this activity is to determine the molar mass of butane using the ideal gas equation.

Objectives

- **Measure** the mass and volume of a sample of butane gas.
- **Measure** the temperature and pressure at which the volume was measured.
- **Determine** the molar mass of butane gas using the ideal gas equation.

Materials

thermometer
barometer
100-cm^3 graduated cylinder
plastic sandwich bag with zipper seal
empty coffee can (2- or 3-lb size)
disposable butane lighter

Procedure

1. Measure the mass of the butane lighter. Record the mass (A) in a data table like the one shown.
2. Put the lighter into a plastic bag. Place the lighter and bag on a flat surface, squeeze as much air out of the bag as possible, and seal the bag.
3. Set a large coffee can in the bottom of a sink and fill it to the top with water.
4. Find the release valve on the lighter and allow butane to fill the bag.
5. Place the filled bag into the can of water. Allow water to spill over the top of the can until the entire bag is below the surface.
6. Remove the bag from the can.
7. Determine the volume of water that spilled out of the coffee can. To do this, use a graduated cylinder to measure the necessary volume of water, and refill the can to the top. Enter this volume (D) in your data table.
8. Open and empty the bag of gas in the fume hood or near an open window.
9. Remove the lighter from the bag. Measure the mass of the lighter again and record the value (B).
10. Measure out 50 cm^3 of water in the graduated cylinder. Lower the lighter into the graduated cylinder and record the new volume (F). Remove the lighter.

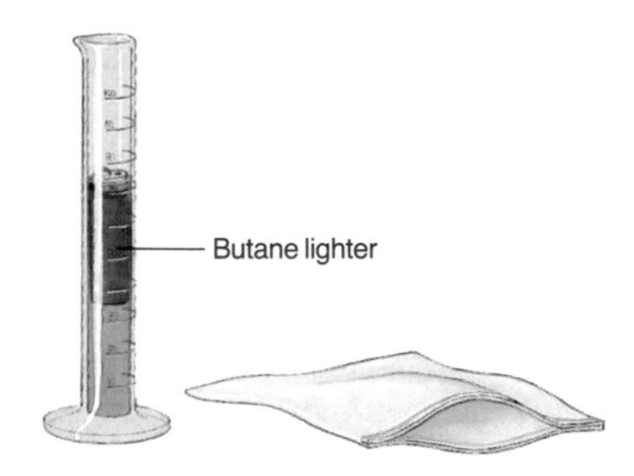

11. Refill the graduated cylinder with water up to the 80-cm^3 mark.
12. Flatten your plastic bag. Fold the bag into a convenient shape to fit into the graduated cylinder.
13. Immerse the bag completely and record the new volume (I).
14. Use your thermometer to measure and record the temperature of the room (L).
15. Read the atmospheric pressure in the room from a barometer and record the pressure in millimeters of mercury (M).

PROGRAM RESOURCES

ChemActivity Masters: Use worksheets for ChemActivity 19 to provide students with a copy of procedures and a data table for use in the lab.

Data and Observations

A.	Mass of butane lighter (g)	
B.	Mass of lighter after releasing gas (g)	
C.	Mass of butane used (A − B) (g)	
D.	Volume of water displaced (volume of butane, uncorrected) (cm^3)	
E.	Volume of water in graduated cylinder	50 cm^3
F.	Volume of water and lighter	
G.	Volume of lighter (Volume − 50 cm^3)	
H.	Volume of water in graduated cylinder	80 cm^3
I.	Volume of water and plastic bag (cm^3)	
J.	Volume of bag (Volume − 80 cm^3)	
K.	Volume of gas in bag (D − (G + J)) (cm^3)	(___ dm^3)
L.	Temperature of the room (°C)	
M.	Barometric pressure (mm Hg)	(___ kPa)
N.	No. of moles of gas (mol)	
O.	Molar mass of gas (g/mol)	

Analysis and Conclusions

1. Find the values of data table entries C, G, J, and K.
2. The data must be converted to appropriate units for use in the ideal gas equation with R = 8.31 dm^3·kPa/mol·K. Convert the pressure (M) in mm Hg to pressure in kPa (1 kPa = 7.50 mm Hg).
3. Convert the volume of the gas sample (K) to dm^3 (1 dm^3 = 1000 cm^3).
4. Convert the temperature (L) to Kelvin (C° + 273 = Kelvin).
5. Using the values of P, V, and T from analysis steps 2, 3, and 4 in the ideal gas equation, calculate the number of moles of gas in your sample and record the value (N).
6. Use the mass of gas in your sample (C) and the number of moles of gas (N) to calculate the molar mass of butane (O).

Extension and Application

1. Write the form of the ideal gas equation used to find the density of a gas. At the same temperature and pressure, which of the gases—hydrogen, hydrogen chloride, helium, or oxygen—will be most dense? Which will be least dense?
2. Gases are usually stored and transported under pressure as liquids. Butane is sold in small, light containers (cigarette lighters) but helium for balloons is sold in heavy metal tanks. Why?
3. At the same temperature and pressure, which weighs more, 1 dm^3 of ammonia or 1 dm^3 of hydrogen chloride gas?
4. At the same temperature and pressure, which will occupy a larger volume, 100 g of helium or 100 g of oxygen?

Data and Observations

A. 17.83 g
B. 17.61 g
C. 0.22 g
D. 149 cm^3
E. 50 cm^3
F. 71 cm^3
G. 21 cm^3
H. 80 cm^3
I. 92 cm^3
J. 12 cm^3
K. 116 cm^3 (0.116 dm^3)
L. 22°C (295 K)
M. 748 mm Hg (99.7 kPa)
N. 0.00472 mol
O. 47 g/mol

Analysis and Conclusions

1.–4. See sample data C, G, J, K.

5. $PV = nRT$

$$n = \frac{0.116\ dm^3 \times 99.7\ kPa}{8.31\ dm^3 \cdot kPa/mol \cdot K \times 295K} = 0.00472\ mol$$

6.

$$\frac{\text{mass of gas sample}}{\text{moles of gas sample}} = \frac{0.22\ g}{0.00472\ mol} = 47\ g/mol$$

Extension and Application

1. $$D = \frac{M}{V} = \frac{PM}{RT}$$

where M is the molar mass of the gas. The equation shows that as M increases, the density will increase. Therefore, hydrogen chloride, which has the highest molar mass, will be the most dense gas, and hydrogen (H_2), which has the lowest molar mass, will be the least dense.

2. Helium is much less dense than butane. Therefore, it takes higher pressure and a much stronger container to hold it in storage as a liquid.

3. hydrogen chloride, because it has a higher molar mass ($m = VMP/RT$)

4. Helium, because it has a lower molar mass ($V = mRT/MP$)

✓ ASSESSMENT

Performance: Assess student performance for this ChemActivity by having students complete copies of the masters from the *ChemActivity Masters* booklet.

CHEMACTIVITY 20

Colloids

The differences among solutions, suspensions, and colloids are not immediately obvious. By producing suspensions and a colloid and then observing them over time, it can be demonstrated that colloidal solid particles do not settle out of solution.

Process skills

Observing, classifying, interpreting data, comparing and contrasting, inferring, communicating

Misconceptions

Students will probably be unfamiliar with the large number of colloids encountered in everyday life, such as smoke, blood, gelatin desserts, shaving cream, clay, and cheese.

Procedure Hints

Before proceeding, be sure students have read and understood the purpose, procedure, and safety precautions for this ChemActivity.

- Caution students to handle sodium hydroxide solution and hydrochloric acid with care.
- Prepare 0.1*M* $Fe(NO_3)_3$ solution by dissolving 4.04 g of $Fe(NO_3)_3 \cdot 9H_2O$ in 100 cm^3 of water.
- Prepare 0.1*M* $Al(NO_3)_3$ solution by dissolving 3.75 g of $Al(NO_3)_3 \cdot 9H_2O$ in 100 cm^3 of water.
- Prepare 0.3*M* NaOH by dissolving 1.2 g of NaOH pellets or flakes in 100 cm^3 of water.
- Prepare 0.15*M* $Na_2S_2O_3$ by dissolving 3.72 g of $Na_2S_2O_3 \cdot 5H_2O$ in 100 cm^3 of water.

Data and Observations

See sample data in table below.

1. $Fe(NO_3)_3(aq) + 3NaOH(aq) \longrightarrow Fe(OH)_3\ (cr) + 3NaNO_3(aq)$
$Al(NO_3)_3(aq) + 3NaOH(aq) \longrightarrow Al(OH)_3\ (cr) + 3NaNO_3(aq)$
$Na_2S_2O_3(aq) + 2HCl(aq) \longrightarrow S(cr) + SO_2(g) + H_2O + 2\ NaCl(aq)$

Analysis and Conclusions

1. A solid formed in each well.
2. The reactions in wells A1 and A2 produced suspensions as shown by settling out of solids.

Chapter 20 ChemActivity

Colloids

Colloids are intermediate between solutions and suspensions. Solutions are homogeneous mixtures of solvent and solute. Suspensions are mixtures of a solvent and insoluble particles so large that they do not remain in suspension but eventually settle out. Colloids are mixtures of a continuous phase (like the solvent) and a dispersed phase composed of insoluble particles so small that they do not settle out. The purpose of this activity is to compare suspensions with colloids.

Objectives

- **Use** chemical reactions to produce suspensions or colloids.
- **Compare** the properties of suspensions and colloids.

Materials

plastic microtip pipets (5)
24-well microplate
toothpicks
hand lens or low-power microscope
0.1*M* iron(III) nitrate, $Fe(NO_3)_3$
0.1*M* aluminum nitrate, $Al(NO_3)_3$
0.3*M* sodium hydroxide, NaOH
0.15*M* sodium thiosulfate, $Na_2S_2O_3$
6*M* hydrochloric acid, HCl(aq)
Microplate Data Form

Procedure

1. Place a 24-well microplate on a flat surface with the numbered columns away from you and the lettered rows to the left.
2. Use a microtip pipet to place 30 drops of 0.1*M* iron(III) nitrate solution in well A1.
3. Use a clean pipet to place 30 drops of 0.1*M* aluminum nitrate solution in well A2.
4. Place 30 drops of 0.15*M* sodium thiosulfate solution in well A3.
5. Add 30 drops of 0.3*M* sodium hydroxide solution to well A1 and 30 drops to well A2.
6. Stir the contents of wells A1 and A2 with different toothpicks.
7. Observe the contents of wells A1 and A2 with a hand lens or low-power microscope. Record your observations in a Microplate Data Form like the one shown here.

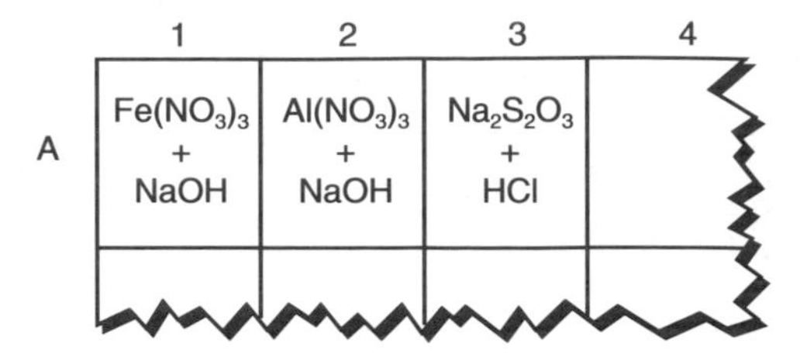

	1	2	3	4
A	$Fe(NO_3)_3$ + NaOH	$Al(NO_3)_3$ + NaOH	$Na_2S_2O_3$ + HCl	

8. Add 30 drops of 6*M* hydrochloric acid to well A3. Stir with a different toothpick.
9. Observe the contents of well A3 with a hand lens or low-power microscope. Record your observations.
10. After 10 minutes, reexamine the contents of all three wells and record your observations. Make certain that you also observe the mixtures from the side of the plate.

Data and Observations

1. Complete and balance the following equations:
Reaction in well A1
$Fe(NO_3)_3(aq) + NaOH\ (aq) \longrightarrow ?$
Reaction in well A2
$Al(NO_3)_3(aq) + NaOH(aq) \longrightarrow ?$
Reaction in well A3
$Na_2S_2O_3(aq) + HCl(aq) \longrightarrow S(cr) + SO_2(g) + ?$

Analysis and Conclusions

1. How could you tell that a chemical reaction occurred in each well?
2. In which well(s) did the reaction produce a suspension? Explain why you reached this conclusion.
3. In which well(s) did the reaction produce a colloid? Explain why you reached this conclusion.

PROGRAM RESOURCES

ChemActivity Masters: Use worksheets for ChemActivity 20 & 21 to provide students with a copy of procedures and a data table for use in the lab.

	1	2	3
A	Brown, precipitate that sinks to bottom	White, gel-like precipitate that settles to bottom	Granular, yellow precipitate that stays suspended

4. Identify the continuous and dispersed phases in your colloid(s).

Extension and Application

1. Describe two more tests with larger samples that you could use to distinguish between your suspensions and colloids. What would be the results of each with your three mixtures?
2. Mayonnaise is a type of colloid known as an emulsion. Look up the definition of an emulsion. Of the three main ingredients in mayonnaise—oil, water, and egg yolk—identify the emulsifying agent, the continuous phase, and the dispersed phase.

Chapter 21 ChemActivity

Super Steam

Pure water boils at 100°C at one atmosphere of pressure. The addition of soluble inorganic or organic compounds changes the boiling point of water. This phenomenon is an example of a colligative property of a solvent. The purpose of this activity is to observe the effects of a solute on the boiling point of water.

Objectives

- **Observe** boiling point elevation.
- **Determine** the effect of solute concentration on the boiling point of a solvent.
- **Determine** experimentally the value of the molal boiling point constant for water.

Materials

notebook paper
100-cm^3 beaker
hot plate
thermometer
stirring rod
graph paper
sodium chloride

Procedure

1. **Hypothesize** what will happen to the boiling point of water as the concentration of added sodium chloride is increased.
2. Prepare five 0.58-g samples of sodium chloride on individual pieces of paper.
3. Measure the mass of a 100-cm^3 beaker and record the mass in a data table like the one shown.
4. Place 50 cm^3 of water in the 100-cm^3 beaker. Measure the mass of the beaker and water. Record this new mass.
5. Heat the water in the beaker on a hot plate until the water boils.
6. Measure and record the temperature of the boiling water to the nearest 0.1°C.
7. Add one of the prepared samples of NaCl. Stir until the NaCl is completely dissolved.
8. Bring the solution to a boil on the hot plate and record the temperature.
9. Repeat steps 7 and 8 until all the NaCl samples have been used.

Data and Observations

	Trial 1	Trial 2	Trial 3
NaCl mass (grams)	0	0.58	1.16
Temp. boiling water/NaCl mixture (°C)			
Moles of NaCl			
Moles of Ions			
Moles of ions/kilogram of water			

Data and Observations

	Trial 1	Trial 2	Trial 3	Trial 4	Trial 5	Trial 6
NaCl mass (grams)	0	0.58	1.16	1.74	2.32	2.90
Temp. boiling water/NaCl mixture (°C)	99.7	100.0	100.8	101.0	101.5	102.0
Moles of NaCl	0	0.010	0.020	0.030	0.040	0.050
Moles of ions	0	0.020	0.040	0.060	0.080	0.10
Moles of ions/kilogram of water	0	0.40	0.80	1.2	1.6	2.0

3. The reaction in well A3 produced a colloid as shown by the fact that the solid did not settle out.
4. In the colloid in well A3 the continuous phase is water and the dispersed phase is sulfur.

Extension and Application

1. First test — The three mixtures could be passed through filter paper. In wells A1 and A2 the solid would remain behind on the filter paper, and in well A3 the solid would pass through.
Second test — A light beam could be passed through the mixtures. The beam would not be visible as it passed through the mixtures from wells A1 and A2, but the beam would be visible in the mixture from well A3.
2. An emulsion is a colloid in which one liquid is dispersed in another with which it is immiscible. The droplets are held in suspension by an emulsifying agent that coats their surface. In mayonnaise, water is the continuous phase, oil is the dispersed phase, and egg yolk is the emulsifying agent.

CHEMACTIVITY 21

Super Steam

The boiling point of a liquid is constant at a constant pressure only if there are no impurities present. The addition of a solute to a pure liquid changes the boiling point of the liquid by a constant amount per mole of particles added. The constant that defines the amount of change in boiling point of a pure liquid is called the molal boiling point constant (K_{bp}). The value of this constant is different for each solvent. For water, the molal boiling point constant is 0.515°C/molal solution.

Process Skills

Observing, measuring, using numbers, predicting, recognizing cause and effect, interpreting

ASSESSMENT

Performance: Assess student performance for this ChemActivity by having students complete copies of the masters from the *ChemActivity Masters* booklet.

data, comparing and contrasting, inferring, using space/time relationships, communicating

Procedure Hints

- Before proceeding, be sure students have read and understood the purpose, procedure, and safety precautions for this ChemActivity.
- Caution students not to reach directly over the beaker of water while the water is boiling.
- Use only the solute suggested. **CAUTION:** *Do not use ethyl or methyl alcohol.*
- Do not allow students to continue to raise the boiling point beyond the range of the thermometer.
- **Troubleshooting:** Be sure students read the thermometer to the nearest 0.1 C°.

Data and Observations

A. 61.53 g
B. 111.64 g
C. 50.11 g
(0.05011 kg)

See data table on page 837.

1. As more NaCl is added to the water, the temperature at which the solution boiled increased.

2. When additional NaCl is added to the boiling water, the boiling stops for a moment, the temperature of the liquid increases and boiling resumes.

Analysis and Conclusions

1.–4. See sample data.

5.–6. See sample plot below.

7. The slope of the straight line obtained using sample data is 1.16. Some students may obtain values that are closer to the true molal boiling point elevation, 0.515 C°/*m*. The units for the slope of the straight line are C°/moles of ions per kilogram, or C°/molal unit.

Extension and Application

1. Acetone is lower boiling than water. Therefore, it would boil away, leaving behind water at its normal boiling point of 100°C.

2. An azeotrope is a solution of two or more liquids that cannot be separated by distillation. An azeotrope has a distinct boiling point for a specific percentage mixture of substances and behaves as if it were a pure substance.

A. Mass of beaker (g)	
B. Mass of beaker and water (g)	
C. Mass of water (B – A) (g)	

1. What happened to the temperature of the water/NaCl mixture as more NaCl was added?
2. What happened to the boiling water when more NaCl was added?

Analysis and Conclusions

1. For each trial, convert grams of NaCl to the number of moles of NaCl and record the results.
2. For each trial, convert the number of moles of NaCl to the number of moles of ions (moles NaCl × 2 ions/mole). Record the results.
3. How many kilograms of water did you use (Data C)(1 kg/1000 g)?
4. Complete the data table entries for moles of ions per kilogram of water.
5. Construct a graph of your results. Plot total moles of ions/kg of water on the *x*-axis and the temperature of the boiling water on the *y*-axis.
6. Draw the best-fitting straight line through your points.
7. Find the slope of this line. What are the units of the slope of the line? The slope is the molal boiling point constant. Compare your value with the actual value of 0.515 C°/*m*.

Extension and Application

1. How would the results differ if you repeated this activity with acetone instead of sodium chloride?
2. Look up the meaning of the term *azeotrope*, a special kind of solution of two liquids. How is an azeotrope different from other solutions?

Chapter 22 ChemActivity

Preparation and Properties of Oxygen

Oxygen gas, O_2, can be prepared in the laboratory by releasing the element from a compound that contains oxygen. Decomposition reactions usually require the addition of energy as heat or electrical current in order to take place. In some cases, addition of a catalyst will cause decomposition to occur. Sodium hypochlorite decomposes to give oxygen and sodium chloride in a reaction catalyzed by the cobalt ion, Co^{2+}:

$$2NaOCl\ (aq) \xrightarrow{Co^{2+}} 2NaCl\ (aq) + O_2\ (g)$$

Sodium hypochlorite is found as a 5 percent solution in household bleach. The purpose of this activity is to prepare oxygen gas from sodium hypochlorite.

Objectives

- **Prepare** oxygen from a compound containing oxygen.
- **Observe** the action of a catalyst.
- **Investigate** the properties of oxygen.

Materials

- thin stem micropipet
- microtip pipets (2)
- 24-well microplate
- household bleach (5% NaOCl)
- 0.1*M* Cobalt(II) nitrate, $Co(NO_3)_2$ (aq)
- toothpicks
- matches

Procedure

1. Use a microtip pipet to place 30 drops of household bleach in well A1 of a 24-well microplate. Wash the pipet with water.

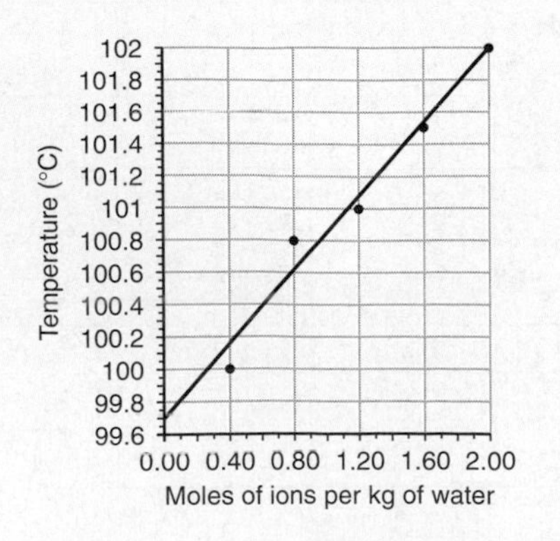

2. Trim the stem of a thin-stem pipet to a length of about 2.5 cm. Label this pipet G; it is the generator pipet.
3. Draw up the 30 drops of household bleach into the bulb of pipet G. Stand the pipet up in well A6 until you are ready to use it, as shown in Figure A. Do not allow any bleach to remain in the stem of the pipet.
4. Using another microtip pipet, place 10 drops of cobalt nitrate solution in well A2. Wash the pipet with water.
5. Cut the stems of the two microtip pipets to a length of 1 cm. These will be the collector pipets, C1 and C2.
6. Place pipets C1 and C2 under water with the stems pointing up in a large beaker. Squeeze the bulbs repeatedly until no more air is expelled. The pipets should be completely filled with water. Stand the two pipets up in wells C1 and C2 as shown in Figure A.

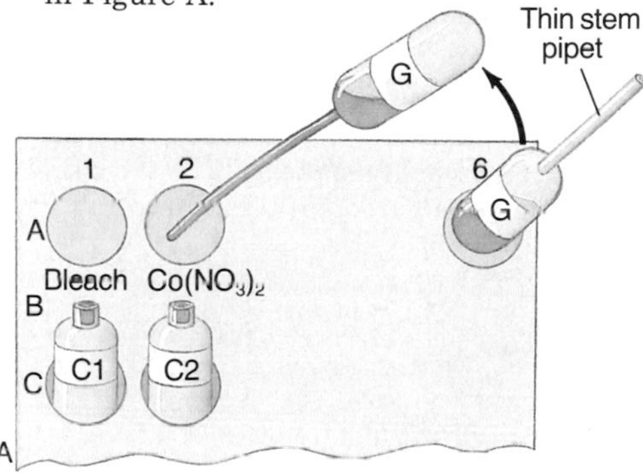

7. Hold pipet G, containing the NaOCl solution, with the stem pointing up. Carefully squeeze the air out of pipet G and insert the stem into well A2. Draw up the cobalt nitrate solution from the well.
8. Immediately, remove pipet G, insert the short tube of pipet G into the stem of pipet C1, and stand pipet G with C1 in well C3 as shown in Figure B. As the reaction continues and oxygen gas collects in pipet C1, the displaced water will collect in microplate well C3.

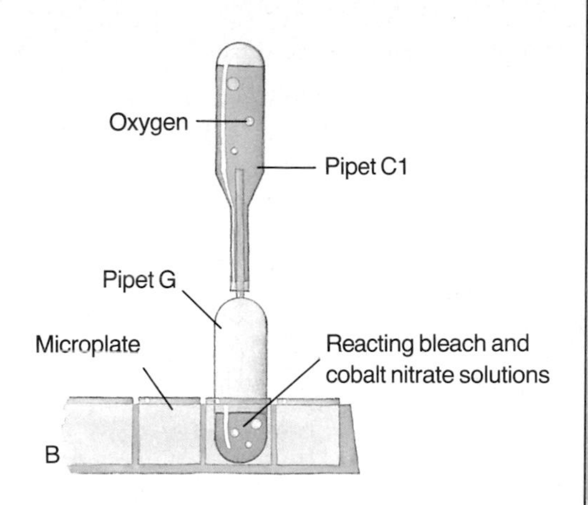

9. Replace pipet C1 with C2 when C1 is almost filled with oxygen gas.
10. Allow a small "plug" of water to be trapped in the neck of the pipet as shown in Figure C. Stand the filled pipet in an empty well of the microplate.

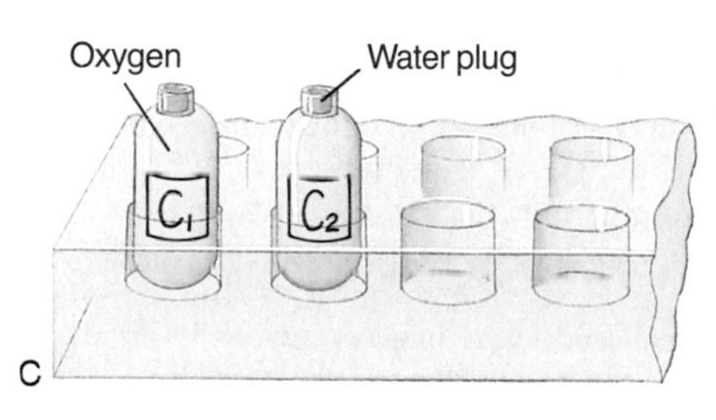

11. Light a toothpick with a match. Gently blow out the flame so that the tip of the toothpick remains glowing.
12. Hold the stem of the gas-filled pipet close to the toothpick and squeeze the bulb of the pipet to force the oxygen gas over the glowing end.

Data and Observations

1. What physical properties of oxygen did you observe?
2. What happened when the glowing toothpick was placed in pure oxygen?

CHEMACTIVITY 22

Preparation and Properties of Oxygen

Most of the life forms on Earth require oxygen gas, which makes up about one-fifth of Earth's atmosphere. Oxygen is the most common element, by mass, on the surface of Earth. Most of the metallic elements are found combined with oxygen.

Process Skills

Observing, interpreting data, inferring, communicating

Procedure Hints

Before proceeding, be sure students have read and understood the purpose, procedure, and safety precautions for this ChemActivity.

Data and Observations

1. Oxygen is a clear colorless gas. Oxygen does not appear to dissolve in water.
2. It burst into flame.

Analysis and Conclusions

1. Oxygen is slightly soluble in water. The students should be able to conclude that oxygen must be water-soluble, because it is oxygen in this form that is used by aquatic animals.
2. Oxygen gas does not burn. It makes burning possible because burning is the combination of substances with oxygen.
3. By proving that cobalt ion remains in the solution when the reaction is over, it could be proved that the ion was a catalyst.
4. The end of the glowing toothpick was burning slowly in air, which is one-fifth oxygen. The reaction rate increased in pure oxygen, an example of an increase in reaction rate with an increase in concentration of a reactant.

✓ ASSESSMENT

Performance: Assess student performance for this ChemActivity by having students complete copies of the masters from the *ChemActivity Masters* booklet.

PROGRAM RESOURCES

ChemActivity Masters: Use worksheets for ChemActivity 22 to provide students with a copy of procedures and a data table for use in the lab.

Extension and Application

1. Almost any compound that contains oxygen can be used for the production of oxygen. One method of producing oxygen is the electrolysis of water in the presence of dilute acid. Another method that students may find referred to is the laboratory preparation from potassium chlorate in the presence of manganese dioxide catalyst.

2. Vigorously exercising athletes are burning food to provide energy at a rapid rate and this requires more oxygen than normal. During vigorous exercise, lactic acid can build up in muscles. An increased supply of oxygen allows the body to burn the lactic acid to carbon dioxide and water. Otherwise, the buildup of lactic acid, called oxygen debt, creates fatigue.
3. The oxygen used in rockets is stored in liquid form (often called LOX), making it possible to store a large quantity in less space than if it were stored as a gas.

CHEMACTIVITY 23

Acidic and Basic Anhydrides

When the oxide of a nonmetal or a metal is placed in water, the water reacts with the oxide to form an acidic solution or a basic solution, respectively.

Process Skills

Observing, measuring, classifying, predicting, recognizing cause and effect, interpreting data, comparing and contrasting, inferring, communicating

Misconceptions

- Students often assume that salt solutions always have a neutral pH.
- Students seldom realize that water reacts with a salt to form an acid and a base in solution.

Procedure Hints

Before preceeding, be sure students have read and understood the purpose, procedure, and safety precautions for this ChemActivity.

- Small pieces of marble are desirable. Do not use powdered calcium carbonate in the micropipet.
- The solution of carbonic acid will not stay acidic for a long time. Advise students to use the solution from well C2 as soon as possible.
- The water in well C4 is used for purposes of comparison. Since carbon dioxide in the air can change the pH and the indicator

Analysis and Conclusions

1. Is oxygen gas soluble in water? What evidence can you offer to support your answer?
2. Does oxygen gas burn?
3. How would it be possible to demonstrate that the cobalt ion was a catalyst in this reaction?
4. Explain what happened to the reaction rate when the glowing toothpick was exposed to oxygen and explain why it happened.

Extension and Application

1. What other compounds could be used to isolate oxygen? Suggest another method of preparing oxygen, using a reference source if necessary.
2. Oxygen gas is often administered to athletes who have been playing at a rapid pace. Why is this done?
3. Great quantities of oxygen gas are required for missiles such as the Saturn rocket, which propelled astronauts to the moon. How is this large amount of oxygen stored in a rocket?

Micro

Chapter 23 ChemActivity

Acidic and Basic Anhydrides

When dissolved in water, the oxide of a metal forms a basic solution. The oxide of a nonmetal, when dissolved in water, forms an acidic solution. Unless water is added to them, the oxides of both metals and nonmetals are anhydrides, that is, compounds without water. Since a metallic oxide in water forms a basic solution, a metallic oxide is called a basic anhydride. The purpose of this activity is to form acidic and basic solutions using an acidic anhydride and a basic anhydride.

Objectives

- **Use** the oxide of a metal to form a basic solution.
- **Use** the oxide of a nonmetal to form an acidic solution.
- **React** an acid and a base to form a salt.

Materials

thin stem micropipet	tap water
24-well microplate	$2M$ HCl
universal indicator	scissors
calcium oxide (CaO)	toothpick
marble chips	clear tape
Microplate Data Form	

Procedure

1. **Form a hypothesis** about whether the solutions formed by dissolving a calcium oxide in tap water and a carbon oxide in tap water will be acidic or basic.
2. Use scissors to make a slit in the bulb of a thin stem micropipet, as shown in Figure A.

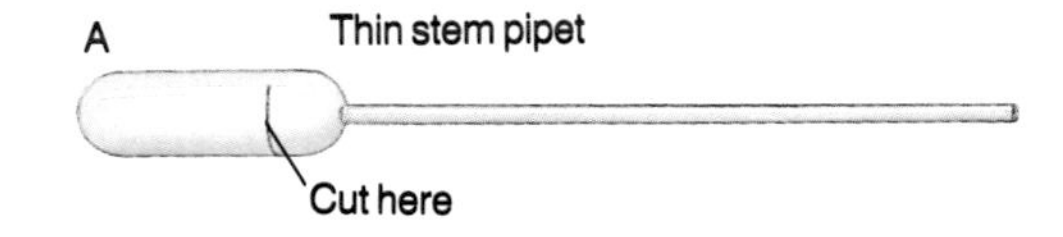

3. Push a small piece of marble through the slit into the bulb.
4. Seal the slit with a piece of clear tape.
5. Use Figure B as a guide to the following procedures.
6. Using a clean micropipet, place 1/2-pipet of tap water in wells C2, C3, and C4 of the microplate.
7. Add 2 drops of universal indicator to wells C2, C3, and C4. Note the color of the universal indicator and the approximate pH in your Microplate Data Form.

PROGRAM RESOURCES

ChemActivity Masters: Use worksheets for ChemActivity 23 to provide students with a copy of procedures and a data table for use in the lab.

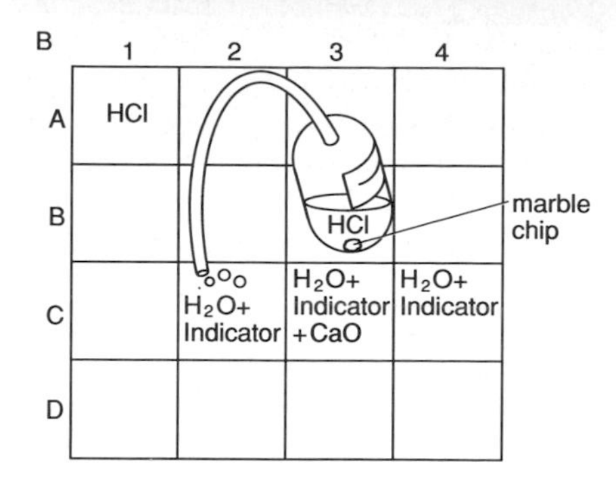

8. Using a clean pipet, place 1/2-pipet of HCl in well A1 of the microplate.
9. Squeeze the bulb of the micropipet containing the piece of marble to eject the air from the pipet. Continue to hold the bulb so that air cannot enter the pipet.
10. Invert the micropipet and place the stem into well A1. Release the bulb of the micropipet and draw up approximately 1/4-pipet of HCl into the bulb.
11. Place the bulb of the micropipet in well B3 and direct the stem into well C2.
12. Allow the gas produced to bubble through well C2. Note the color of the universal indicator-water solution in your Microplate Data Form.
13. Using a chemical scoop, place a few crystals of calcium oxide into well C3. Note the color of the universal indicator-water solution in your Microplate Data Form.
14. Allow the solid to settle out in well C3. Remove approximately 1/8 pipet of the liquid from well C3.
15. Place 10 drops of liquid from well C3 into well D1.
16. Using a clean micropipet, remove approximately 1/8-pipet of liquid from well C2. Place 10 drops of the liquid from well C2 into well D1.
17. Mix the liquid in D1 thoroughly with a toothpick. Note the color of the universal indicator-water solution in your Microplate Data Form.

Data and Observations

1. Label a Microplate Data Form as indicated by Figure B and use it to record your observations.
2. What color was the universal indicator in tap water alone?
3. What was the color of the universal indicator when the gas was bubbled through the water in well C2?
4. What was the color of the universal indicator when calcium oxide was added to the water in well C3?
5. What happened to the universal indicator when the two solutions were mixed in well D1?

Analysis and Conclusions

1. Balance this equation, which describes the reaction in the micropipet containing the marble chip.

$$CaCO_3 + HCl \rightarrow CO_2 + H_2O + CaCl_2$$

2. Balance this equation, which describes what happened to the water in well C2.

$$CO_2 + H_2O \rightarrow H_2CO_3$$

Look at your Microplate Data Form and determine whether CO_2 is an acid anhydride or a basic anhydride.

3. Balance this equation, which describes what happened when calcium oxide was added to the water in well C3.

$$CaO + H_2O \rightarrow Ca(OH)_2$$

Determine whether CaO is an acid anhydride or a basic anhydride.

4. Balance this equation, which describes the reaction in well D1.

$$Ca(OH)_2 + H_2CO_3 \rightarrow CaCO_3 + H_2O$$

Extension and Application

When a salt is dissolved in water, the salt solution almost never has a neutral pH. Why? What are the products when a salt is dissolved in water? How is the pH of a solution changed by a salt?

color in a short time, do not save the solution in well C4 for another class.

- 2*M* HCl is prepared by dissolving 17.2 cm^3 of concentrated HCl in 50 cm^3 of distilled water. Add water to 100 cm^3 to complete the solution.

Data and Observations

1. See sample data below.
2. The universal indicator in water alone should show a color indicating a relatively neutral pH. The color of the indicator is usually green or yellow-green depending on the pH of the local water supply.
3. When carbon dioxide is bubbled through water, the weak acid carbonic acid is formed and the universal indicator turns yellow or red.
4. When calcium oxide is added to water, the weak base calcium hydroxide is formed and the universal indicator turns blue.
5. When the carbonic acid solution from well C2 is mixed with the calcium hydroxide solution from well C3, the acid and the base react to form a salt and water.

Analysis and Conclusions

1. $CaCO_3 + 2HCl \rightarrow CO_2 + H_2O + CaCl_2$
2. $CO_2 + H_2O \rightarrow H_2CO_3$
Carbon dioxide is an acidic anhydride.
3. $CaO + H_2O \rightarrow Ca(OH)_2$
Calcium oxide is a basic anhydride.
4. $Ca(OH)_2 + H_2CO_3 \rightarrow CaCO_3 + 2H_2O$

Extension and Application

When a salt dissolves in water, the ions that make up the salt react with water molecules in a process called hydrolysis. The ions in the salt alter the balance of H^+ and OH^- in the solution.

✓ ASSESSMENT

Performance: Assess student performance for this ChemActivity by having students complete copies of the masters from the *ChemActivity Masters* booklet.

	1	2	3	4
C		Before CO_2 green pH≈7 / After CO_2 yellow $pH<7$	Before CaO pH≈7 green / After CaO blue $pH>7$	Before CONTROL pH≈7 / After CONTROL pH≈7
D	Mixture of C_2 and C_3			

CHEMACTIVITY 24

Buffers

The titration curves of acids with bases show that buffer systems tend to maintain a relatively constant pH over a wide range. A buffer system utilizing a weak acid depends on the slight potential for ionization of the acid, which acts as a source of H^+ ions, and the presence of a negative ion of the salt, which acts as a receptor of H^+ ions.

Process Skills

Observing, measuring, using numbers, interpreting data, comparing and contrasting, inferring, communicating

Misconceptions

- Most students think that when an acid reacts stoichiometrically with a base, the resulting solution is always neutral. In fact, the endpoint of a titration is not always at pH7.
- Students might think that all salts in solutions are neutral.

Data and Observations

See sample data in the table below.

1. When a base is present in sufficient quantity, it will neutralize any acid.

2. The same number of drops of base was required to reach stoichiometric equality. The stoichiometric equivalent of an acid and a base does not necessarily result in a solution at pH7.

Drops of Base	pH of Solution A	B	C	D
0	1	1	2	2
2	2	2	3	4
4	2	2	3	6
6	4	2	4	8
8	4	3	5	8
10	7	5	7	10
12	10	7	8	10
14	12	7	8	12
16	12	10	9	12
18	12	10	10	12

Chapter 24
ChemActivity

Buffers

When an acid reacts with a base, a salt and water are formed. A solution that contains either a weak acid and the salt of a weak acid or a weak base and the salt of a weak base is called a buffer system. Buffer systems have special properties that are of great importance in certain biological and chemical processes. The purpose of this activity is to compare the reactions of strong acid and base combinations with the reactions of buffer systems.

Objectives

- **Develop** a buffer system.
- **Investigate** the properties of a buffer system.

Materials

- 24-well microplate
- 96-well microplate
- microtip pipets (4)
- drinking straws
- scissors
- distilled water
- universal indicator (with color scale)
- toothpick
- labels
- 0.1*M* solutions of the following:
  - HCl (strong acid)
  - NaOH (strong base)
  - $HC_2H_3O_2$ (weak acid)
  - NH_3 (aq) (weak base)

Procedure

Part A

1. With drinking straws, construct a ringstand, following the instructions in the diagram. Set up the ringstand in well A1 in the 96-well microplate.
2. Label four microtip pipets HCl, NaOH, $HC_2H_3O_2$, and NH_3 (aq).
3. Fill the labeled pipets with HCl, NaOH, $HC_2H_3O_2$, or NH_3 (aq) solution. Then store the pipets, tip up, in row D of the 24-well microplate.
4. Place 20 drops of distilled water in well A1 of the 24-well plate.
5. Add 10 drops of HCl to the distilled water in well A1.
6. Place the NaOH pipet in the straw ringstand, as shown in the diagram.
7. Add 2 drops of universal indicator to well A1 of the 24-well plate.
8. Record in the data table the pH of the solution in well A1.
9. Add 2 drops of NaOH solution to well A1. Stir with a toothpick.
10. Record in the data table the pH of the solution in A1.
11. Repeat steps 9 and 10 until a pH of 12 or greater is reached.

Part B

Repeat the procedure in steps 4–11 of Part A, using HCl in well A2 and NH_3 (aq) in the pipet.

Part C

Repeat the procedure in steps 4–11 of Part A, using $HC_2H_3O_2$ in well A3 and NH_3 (aq) in the pipet.

Part D

Repeat the procedure in steps 4–11 of Part A, using $HC_2H_3O_2$ in well A4 and NaOH in the pipet.

Data and Observations

Drops of Base	pH of Solution PART A	B	C	D
0				
2				
4				
6				

1. What is the general effect of a base on an acid?
2. Was the number of drops of base required to neutralize the acid the same, regardless of which acid and which base were involved?

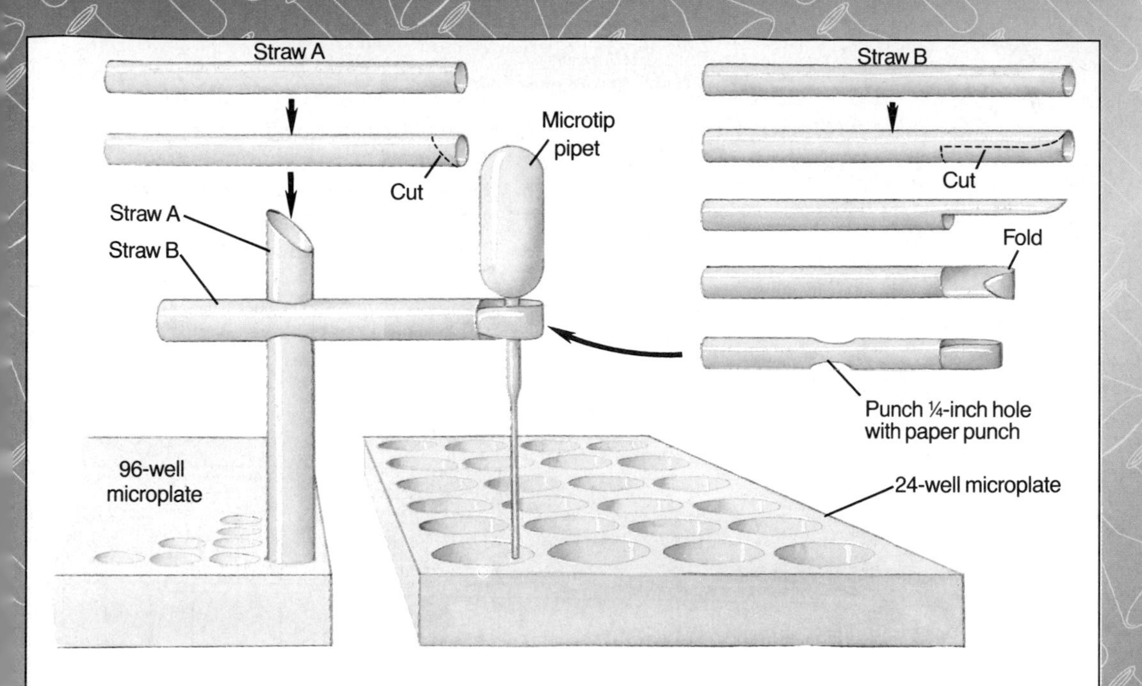

Analysis and Conclusions

1. Make a graph of your data. Make the x-axis the number of drops of base used and the y-axis the pH of the solution. Graph each set of data on the same piece of graph paper.
2. How does neutralization of strong and weak acids and strong and weak bases compare?
3. What data supports your conclusions?
4. Look at your graphs. Which combinations of acids and bases produced a buffer system?
5. What is the special property of a buffer system?

Extension and Application

1. Why might buffer systems be important in living things?
2. What other combinations of acids and bases would produce a buffer system?

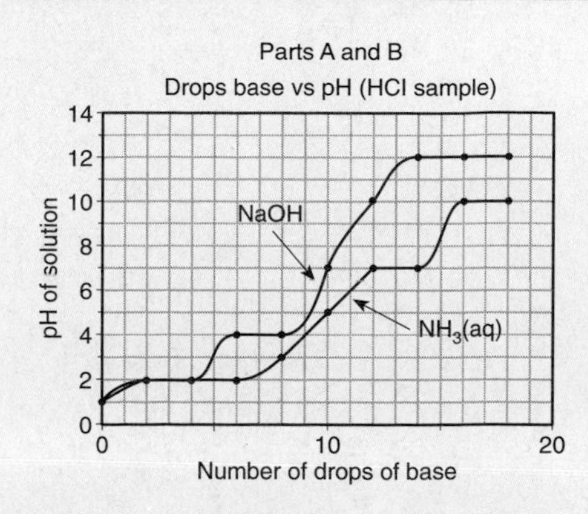

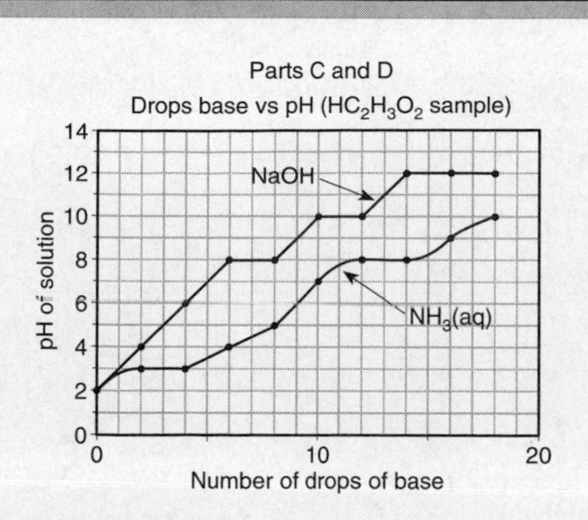

Analysis and Conclusions

1. See sample graphs below.
2. Weak acids form buffer systems when they are titrated with strong bases, and weak bases form buffer systems when they are titrated with strong acids.
3. The flat parts of the slopes of the graphs of the weak acid-strong base and the weak base-strong acid combinations show the buffering effect.
4. $HC_2H_3O_2$ titrated with NaOH, and HCl titrated with NH_3(aq) produced buffer systems.
5. A buffer system tends to resist change in pH.

Extension and Application

1. Nearly all chemical reactions in living things take place at a pH level close to 7. Without buffering, most reactions in a living world would slow down and life processes would be seriously disrupted.
2. Any combination of a weak acid and a strong base will produce a buffer system. For example, formic acid (HCOOH) forms a buffer system when it is combined with NaOH to form sodium formate salt (NaHCOO). Excess formic acid added to the salt solution forms a buffer system.

✓ ASSESSMENT

Performance: Assess student performance for this ChemActivity by having students complete copies of the masters from the *ChemActivity Masters* booklet.

PROGRAM RESOURCES

ChemActivity Masters: Use worksheets for ChemActivity 24 to provide students with a copy of procedures and a data table for use in the lab.

CHEMACTIVITY 25

Oxidation/Reduction of Vanadium

Students will cause the same element to undergo six changes in oxidation state. The reaction responsible for the reduction of V^{n+} is $V^{n+} + Zn^0 \rightarrow Zn^{2+} + V^{(n+)-1}$

The reaction responsible for the formation of the oxidized vanadium is

$V^{n+} + 2OCl^- + 2H^+ \rightarrow V^{n+(+1)} + 2Cl^0 + 2OH^-$

Process Skills

Observing, classifying, predicting, recognizing cause and effect, interpreting data, sequencing, comparing and contrasting, inferring, communicating

Procedure Hints

Prepare 0.1*M* NH_4VO_3 by dissolving 0.4 g of NaOH in 50 cm^3 of distilled water. Stirring constantly, add 1.0 g of NH_4VO_3. When the NH_4VO_3 has dissolved completely, add 1 cm^3 of concentrated H_2SO_4. Then add distilled water to make a total volume of 100 cm^3.

- 40-mesh zinc metal provides the best combinaton of surface area and size.
- Prepare diluted bleach by adding 50 cm^3 of household bleach to 50 cm^3 of distilled water.
- It is important that students use a microtip pipet to dispense the vanadium solution over the zinc metal.
- It is essential that no zinc metal be carried over to any well from well A1.
- The first color change that students will notice is from yellow to green. This is not a change in the oxidation state of vanadium. The color results from a mixture of both yellow V^{5+} ions and blue V^{4+} ions.

Data and Observations

The numbers of passes between pipet and well will differ because color changes are gradual and it is a matter of judgment as to when the color changes are complete. Students will observe these colors:

Yellow	V^{5+}
Blue	V^{4+}
Green	V^{3+}
Violet	V^{2+}

Micro

Oxidation/Reduction of Vanadium

Vanadium, element 23 on the periodic table, is one of the transition elements of period 4. One characteristic of transition elements is that they have variable oxidation states. This means that elements 21 through 29 and the elements directly below them on the periodic table have several stable oxidation states. For example, the vanadium atom in an ionic form can have oxidation states of 5+, 4+, 3+ and 2+. The purpose of this activity is to prepare ions of an element that are in different oxidation states.

Objectives

- **Prepare** ions of the same element that have different oxidation states.
- **Compare** some properties of the element in different oxidation states.

Materials

24-well microplate	0.1*M* NH_4VO_3
microtip pipets (2)	household bleach (diluted 1:1)
scissors	white paper
zinc granules	
micropipet	

Procedure

Part A Reduction of Vanadium

1. **Form a hypothesis** about whether vanadium ions in different oxidation states will have different physical and chemical properties.
2. Place a 24-well microplate on a piece of white paper with the numbered columns away from you and the lettered rows to the left.

Micropipet

Cut bulb to form microscoop

A

3. Make a microscoop by cutting the bulb of of a micropipet, as shown in Figure A.
4. Place a half microscoop of zinc metal into well A1.
5. Using a microtip pipet, place ½-pipet of NH_4VO_3 in well A2. Ammonium metavanadate contains vanadium in the 5+ state. Well A2 will serve as a control well. Use the diagram of your microplate in Figure B as a guide to the steps that follow.

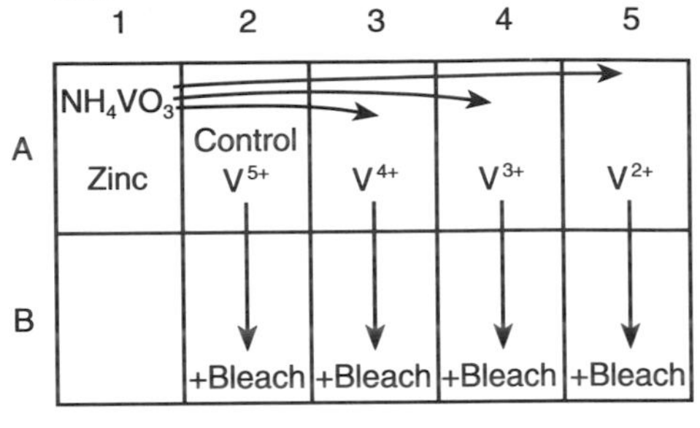

B

6. Three quarters fill a microtip pipet with NH_4VO_3 solution. Place the NH_4VO_3 in well A1 with the zinc metal.
7. Draw the liquid in well A1 back up into the microtip pipet. Then return the liquid to well A1. This is one pass. Repeat this process for as many passes as necessary until the solution turns color. The solution of vanadium is now in a V^{4+} state. Record the number of times you must return the V^{5+} solution to the zinc for a color change in a data table like the one shown.
8. Place 20 drops of the V^{4+} solution into well A3.
9. Place the rest of the V^{4+} solution remaining in the pipet back into well A1 with the zinc metal.
10. Draw the liquid in well A1 back up into the pipet. Then return the liquid to well A1. Repeat passes until the solution turns color. The solution of vanadium in A1 is now in the V^{3+} state. Record the number

PROGRAM RESOURCES

ChemActivity Masters: Use worksheets for ChemActivity 25 to provide students with a copy of procedures and a data table for use in the lab.

of times you must return the solution to the zinc for a color change in your data table.

11. Place 20 drops of the V^{3+} solution into well A4.
12. Place the rest of the V^{3+} solution remaining in the pipet back into well A1 with the zinc metal.
13. Draw the liquid in well A1 back up into the pipet. Then return the liquid to well A1. Repeat passes until the solution turns color. The solution of vanadium in A1 is now in the V^{2+} state. Record the number of times you must return the solution to the zinc for a color change in your data table.
14. Place 20 drops of the V^{2+} solution in well A5. Return any remaining solution to well A1.

Part B Oxidation of Vanadium

1. Rinse your micropipet and transfer 10 drops of the contents of well A5 to well B5. Rinse your micropipet again.
2. Transfer 10 drops of the contents of well A4 to well B4. Rinse your micropipet.
3. Transfer 10 drops of the contents of well A3 to well B3.
4. Fill a clean micropipet with diluted household bleach.
5. Drop by drop, add the diluted bleach to well B5 until a color change occurs. Record your observations and the number of drops required in the column labeled "Color with Bleach" in your data table.
6. Following the procedure in step 5, add diluted bleach to wells B4, B3, and B2. Record your observations.

Data and Observations

Well	Number of Passes	Color	Oxidation State	Color with Bleach
A2				
A3				
A4				
A5				

Analysis and Conclusions

1. How many changes in oxidation state did vanadium go through in Parts A and B?
2. Which change in oxidation state required the greatest number of passes through the zinc metal?
3. What happened to the solutions in the A row when the bleach was added?
4. If V^{2+} is allowed to stand for any length of time, it reverts to V^{5+}. Explain why you think this happens.

Extension and Application

Iron as Fe^{2+} is an essential nutrient, but iron as Fe^{3+} has no value as a nutrient. Some iron-enriched cereals contain iron filings. What is the advantage of using iron in this form?

Analysis and Conclusions

1. Vanadium goes through three changes in oxidation states from V^{5+} to V^{2+} with zinc and three more changes from V^{2+} to V^{5+} with bleach.
2. The change from V^{3+} to V^{2+} required the greatest number of passes through the zinc metal. Reduction becomes more difficult as more electrons are added to the metal ion.
3. The vanadium solutions changed colors indicating a change in oxidation states from V^{2+} to V^{3+} to V^{4+} to V^{5+}.
4. The fully reduced V^{2+} ion is reoxidized by oxygen in the air.

Extension and Application

Iron filings are used in enriched products because the metal is less easily oxidized than the Fe^{2+} ion. If iron were added as Fe^{2+} (as iron(II) sulfate, for example), the compound would be easily oxidized by oxygen in the air to Fe^{3+}. The metal Fe is converted to Fe^{2+} by the hydrochloric acid in the stomach.

✓ ASSESSMENT

Performance: Assess student performance for this ChemActivity by having students complete copies of the masters from the *ChemActivity Masters* booklet.

CHEMACTIVITY 26

Lemon Battery

Metals differ in activity, that is, in the tendency to give up electrons. When two different metals are placed in an electrolyte solution, electrons flow from the more active metal to the less active metal, and an electromagnetic force (EMF) is generated. EMF is measured in volts and is commonly called voltage. The greater the difference in the activity of the metals, the higher the voltage produced.

Process Skills

Observing, measuring, classifying, predicting, interpreting data, sequencing, comparing and contrasting, inferring, communicating

Procedure Hints

The lemon must be fresh.

- The metal strips must be clean and free from corrosion. Make sure they are rinsed and dried immediately after use.
- A high impedance or vacuum tube voltmeter with a low range must be used.

Data and Observations

See data in the table below.

Analysis and Conclusions

1. The combination of magnesium and copper gave the highest voltage reading.
2. The combination of copper and carbon gave the lowest reading.

Extension and Application

The electrolyte in a commercial battery is a paste instead of a liquid. Commercial batteries are of three general types: carbon-zinc batteries, alkaline batteries, and rechargeable batteries, either lead-acid or nickel-cadmium. Commercial batteries provide stronger current and last longer than a lemon battery.

Micro

Chapter 26 ChemActivity

Lemon Battery

A battery is a device in which chemical energy is converted to electrical energy. The energy obtained from a battery is produced by a difference in activity of two different metals. When two metals are placed in an electrolyte and are connected by a conductor, electrons flow from the more active metal through the conductor to the less active metal. The flow of electrons is an electrical current and can be made to do work. The purpose of this activity is to investigate the activity of different metals in various combinations in a simple battery. Note that carbon is a conductor and will be considered a metal in this activity.

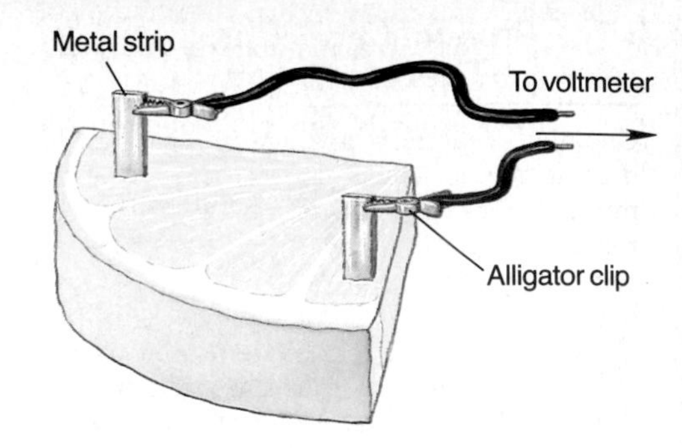

Objectives

- **Make** a simple battery, using a lemon.
- **Compare** the activity of metals used in different combinations in the battery.

Materials

lemon (1/4 per group)
carbon rod (pencil lead will do)
voltmeter or multimeter
short lengths of connecting wire
alligator clips
small knife
chemical scoop
small strips of the following metals:
magnesium
lead
zinc
copper

Procedure

1. Using a chemical scoop, pierce the flesh of a piece of lemon in two places approximately 1 cm apart.
2. Select a strip of two different metals.
3. Insert each strip of metal into a different slit in the flesh of the lemon, as shown in the Figure.
4. With connecting wire and alligator clips, connect each metal strip to the voltmeter. Observe the needle and scale of the voltmeter as you complete the connection. If the needle does not move or you get a negative reading, reverse the connections to the metals. Note the polarity of the metals, that is, whether the metal is positive or negative. Record the voltmeter reading and the polarity (+ or −) of each metal in a data table like the one shown.
5. Remove the metal strips from the lemon and rinse them in tap water.
6. Repeat steps 2–5 until you have tested each of the ten possible combinations.

Data and Observations

Polarity of Metals				
Pb	C	Zn	Mg	
				Cu
			Mg	
		Zn		
	C			

Analysis and Conclusions

1. Which pair of metals gave the highest reading on the voltmeter?
2. Which pair of metals gave the lowest reading on the voltmeter?

Extension and Application

Compare your battery with a commercially produced battery. In what ways are the two batteries similar? How do they differ?

PROGRAM RESOURCES

ChemActivity Masters: Use worksheets for ChemActivity 26 & 27 to provide students with a copy of procedures and a data table for use in the lab.

Data and Observations

Polarity of Metals				
Pb	C	Zn	Mg	
Pb-Cu+ 0.35	C-Cu+ 0.18	Zn-Cu+ 1.0	Mg-Cu+ 1.9	Cu
Pb+Mg- 0.8	C+Mg- 1.7	Zn+Mg- 0.4	Mg	
Pb+Zn- 1.2	C+Zn- 1.3	Zn		
Pb-C+ 0.45	C			

Physical and Chemical Thermodynamics

Chemical reactions are accompanied by two driving forces—the tendency to reach minimum energy (enthalpy) and the tendency to reach maximum disorder (entropy). Although some reactions tend toward more order or higher energy, these reactions are the exceptions. All chemical reactions eventually cease when maximum entropy and minimum enthalpy are reached. The purpose of this activity is to investigate the flow of energy and the effects of energy on physical and chemical systems.

Objectives

- **Explore** enthalpy and entropy, using a physical system as a model for chemical reactions.
- **Classify** reactions as exothermic or endothermic and as entropic or more orderly.

Materials

thick rubber band
hot plate
250-cm^3 beakers (2)
1- or 2-kg mass
ringstand and iron ring
24-well microplate
scissors
ruler
ammonium chloride (solid)
sodium hydrogen carbonate (solid)
hydrochloric acid (6*M*)
thermometer
thin stem micropipets (2)
toothpicks
Microplate Data Form

Procedure

Part A

1. Half-fill a 250 cm^3 beaker with water and place the beaker on a hot plate to heat.
2. While the water is heating, put a thick rubber band around both of your index fingers and hold the fingers apart so that the rubber band is at its full length, *but is not stretched.* Briefly touch the rubber band to your upper lip and sense the "temperature" of the band.
3. Move your index fingers apart to stretch the rubber band as much as you can. Briefly touch the stretched rubber band to your lip. Notice whether you feel a change in the "temperature" from your earlier observation. Record your observations in a data table like the one shown.
4. Slip the rubber band onto the ring attached to the ringstand and let it dangle. Hang a mass weighing 1 to 2 kg from the end of the rubber band, as shown in the diagram.
5. With a ruler, measure the length of the rubber band, and enter the length in your data table.
6. Place an empty 250-cm^3 beaker below the mass and the rubber band.
7. Wearing thermal mitts, carefully pour hot water from the other beaker over the rubber band, collecting the water in the beaker beneath. Measure the length of the rubber band and record this new length.

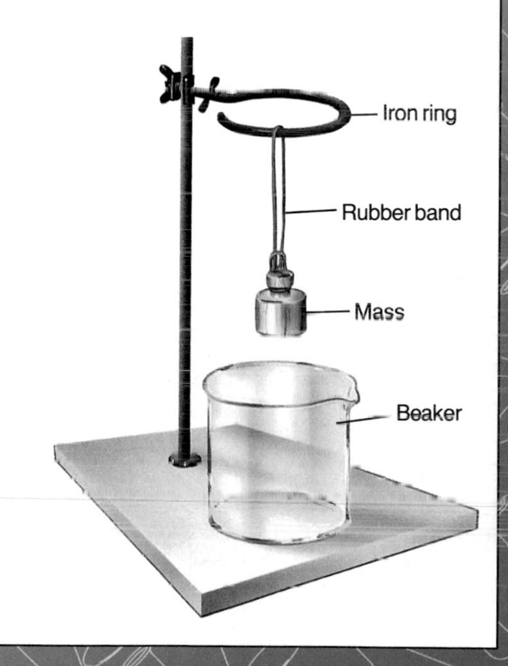

CHEMACTIVITY 27

Physical and Chemical Thermodynamics

The physical and chemical systems in this activity have been chosen to provide experiences with a variety of physical and chemical thermodynamic systems. The physical system in this activity is a rubber band under stress. When unstretched, the rubber band is in the state of higher enthalphy and lower entropy. When the rubber band is stretched, the molecules of rubber become more orderly along the long axis of the band. Entropy is decreased, so ΔS is –. Heat is given off (The band feels slightly warm when touched to the lip), so ΔH is –. The stretching of the band is not spontaneous, so the sign of ΔG is +.

Some of the chemical reactions in Part B are exothermic; others are endothermic. All products of these reactions are more disordered than the reactants, so all the ΔS values are +.

Process Skills

Observing, measuring, classifying, interpreting data, comparing and contrasting, inferring, formulating models, communicating

Misconceptions

Students may have difficulty visualizing physical and chemical systems that are spontaneous, exothermic, endothermic, and more or less orderly. This activity will aid their comprehension of these properties.

Procedure Hints

- Explain that not all chemical reactions are spontaneous, exothermic, nor produce products with higher entropy.
- **Troubleshooting:** To make it

✓ ASSESSMENT

Performance: Assess student performance for this ChemActivity by having students complete copies of the masters from the *ChemActivity Masters* booklet.

easier to attach the mass to the rubber band, choose objects that have a loop or other projection. A slip knot made in the dangling end of the rubber band can be slipped over the projection of the mass.

Data and Observations

Part A

Sample data
A. warmer
B. 38.1 cm
C. 34.2 cm

Part B

See sample data in the table below.

Analysis and Conclusions

1. Unstretched band $\rightarrow$ stretched band
$\Delta H-$ $\Delta S-$ $\Delta G+$
2. $NH_4Cl(cr) \rightarrow NH_4^+(aq) + Cl^-(aq)$
$\Delta H+$ $\Delta S+$ $\Delta G-$
3. $NaHCO_3(cr) \rightarrow Na^+(aq) + HCO_3^-(aq)$
$\Delta H-$ $\Delta S+$ $\Delta G-$
4. $NaHCO_3(cr) + HCl \rightarrow Na^+(aq) + Cl^-(aq) + CO_2(g) + H_2O$
$\Delta H+$ $\Delta S+$ $\Delta G-$
5. Reactions 2, 3, and 4 occur spontaneously. ΔG is –.
6. stretched $\rightarrow$ unstretched
$\Delta H+$ $\Delta S+$ $\Delta G-$
It would occur spontaneously.

Extension and Application

a. photosynthesis
$\Delta H+$ $\Delta S-$ $\Delta G+$
b. rusting of a car
$\Delta H-$ $\Delta S+$ $\Delta G-$
c. formation of a diamond (at 1 atmosphere)
$\Delta H+$ $\Delta S-$ $\Delta G+$

	1	2
A	Before: 23° C After: 17° C	Before: 23° C After: 26° C
B	Before: 23° C After: 18° C	Before: 23° C After: Bubbles 18° C

Part B

1. Make a chemical microscoop by cutting the end off the bulb of a micropipet. (See ChemActivity 25, Procedure Step 3.)
2. Place 1/2-microscoop of solid ammonium chloride in well A1 and well B1 of a 24-well microplate. Add 1/2-microscoop of solid sodium hydrogen carbonate to well A2 and well B2.
3. With the thermometer, measure the temperature of the solid chemicals in wells A1, A2, B1, and B2. Record these data in your Microplate Data Form.
4. Using the thin stem pipet, place 1/2-pipet of water in wells A1 and A2. Stir with a toothpick. Enter any physical changes you observe in your Microplate Data Form.
5. With the thin stem pipet, add HCl to wells B1 and B2. Stir with a toothpick. Enter any physical changes you observe in your Microplate Data Form.
6. Use the thermometer to measure the temperature of the chemicals as they dissolve in wells A1, A2, B1, and B2. Be sure to rinse your thermometer in cold water between each reading. Record these data in your Microplate Data Form.

Data and Observations

Part A

A	The stretched rubber band felt warmer/cooler than the unstretched rubber band.	
B	Length of stretched rubber band at room temperature (cm)	
C	Length of stretched and heated rubber band (cm)	

Part B

Record your observations in the appropriate boxes of your Microplate Data Form as indicated here.

	1	2
A	NH_4Cl + H_2O	$NaHCO_3$ + H_2O
B	NH_4Cl + HCl	$NaHCO_3$ + HCl

Analysis and Conclusions

Use your data and the symbols in the table below to analyze the reactions in Part A and Part B. Write the reaction and replace each question mark by the appropriate sign for ΔH, ΔS, and ΔG.

$\Delta H+$	=	endothermic (cool)
$\Delta H-$	=	exothermic (warm)
$\Delta S+$	=	more disorder
$\Delta S-$	=	less disorder
$\Delta G+$	=	not spontaneous
$\Delta G-$	=	spontaneous

1. unstretched band $\rightarrow$ stretched band
ΔH? ΔS? ΔG?
2. $NH_4Cl(cr) \rightarrow NH_4^+(aq) + Cl^-(aq)$
ΔH? ΔS? ΔG?
3. $NaHCO_3(cr) \rightarrow Na^+(aq) + HCO_3^-(aq)$
ΔH? ΔS? ΔG?
4. $NaHCO_3(cr) + HCl \rightarrow Na^+(aq) + Cl^-(aq) + CO_2(g) + H_2O$
ΔH? ΔS? ΔG?
5. Which reaction took place spontaneously?
6. What would be the reverse reaction with the rubber band?

Extension and Application

1. Find out about the familiar reactions named below. What are the signs (+ or −) for ΔG, ΔH, and ΔS for each system?
   a. photosynthesis
   b. rusting of a car
   c. formation of a diamond

PROGRAM RESOURCES

ChemActivity Masters: Use worksheets for ChemActivity 28 to provide students with a copy of procedures and a data table for use in the lab.

A Half-Life Model

Radioactive isotopes are unstable atoms that decompose spontaneously to the atoms of a different element. The breakdown of atoms takes place at a set rate, called half-life, which differs for each radioactive isotope. Half-life is the amount of time it takes for one-half the atoms in a sample of a radioactive isotope to decay to the atoms of a different element. Because it is not practical for you to study the half-life of a real radioactive isotope, you will use a model of such an element in this activity. The purpose of this activity is to explore the phenomenon of half-life.

Objectives

- **Use a model** to study half-life.
- **Construct** a decay curve of the model atoms.
- **Use a model** to evaluate the effect of a catalyst on half-life.

Materials

split peas (100 per group)	250-cm^3 beakers (2 per group)
lima beans (10 per group)	clock or watch
	graph paper

Procedure

Part A

1. Label the beakers *Decayed Atoms* and *Undecayed Atoms*. Make a table like the one shown.
2. Count out 100 split peas and place them in the beaker labeled Undecayed Atoms. Each pea will represent one atom of the imaginary element peanium.
3. Note the time in your data table.
4. Shake the Undecayed Atoms beaker and dump the split peas out onto the table.
5. Note that some peas landed curved side down and the others landed curved side up. Separate the peas into two groups, according to the position, up or down, of the curved side. The group with the curved side down will represent the peanium atoms that have decayed to the atoms of a different element. The group with the curved side up will represent the undecayed peanium atoms.
6. Count the undecayed peanium atoms and record the number in your data table.
7. Place the decayed peanium atoms in the Decayed Atoms beaker. Return the undecayed atoms to the Undecayed Atoms beaker.
8. Repeat steps 4–7 until there are no more atoms in the Undecayed Atoms beaker. Note the time in your data table.

Part B

Repeat Part A, but this time beans will represent beanium atoms. Add 10 lima beans to the peanium atoms in step 2. Since the sides of the lima beans do not differ significantly, always sort the beanium atoms into the group of undecayed peanium atoms.

Data and Observations

	Part A	Part B
Time: start end total		
Number of runs		

Number of Undecayed Atoms		
Run #	Part A	Part B
1 (Start)		
2		
3		
4		
5		

CHEMACTIVITY 28

A Half-Life Model
Radioactive elements decompose to the atoms of a different element. Radioactive decay takes place at a set rate called half-life. Half-life is the time it takes for one half of a radioactive sample to decay. Half-lives of the radioisotopes range from milliseconds to billions of years.

Process Skills
Measuring, using numbers, classifying, interpreting data, inferring, formulating models

Procedure Hints
Any objects that have two distinct sides can be used as models of radioactive atoms.
- Students will be able to perform the necessary runs in a very short time. They will have ample time to use their data to construct graphs.

Data and Observations
See sample data in the tables below.

Analysis and Conclusions
1. See graphs below.

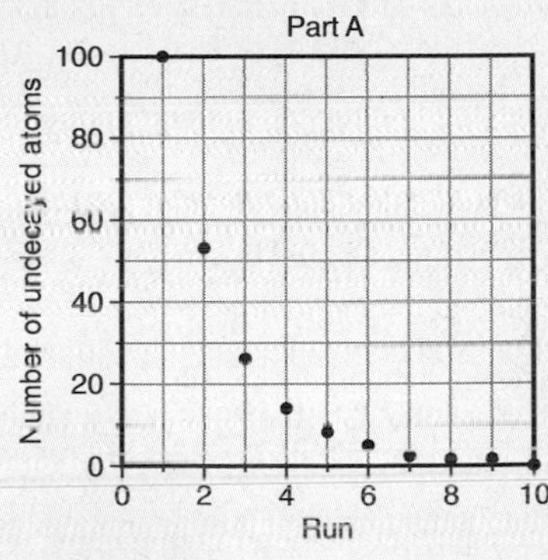

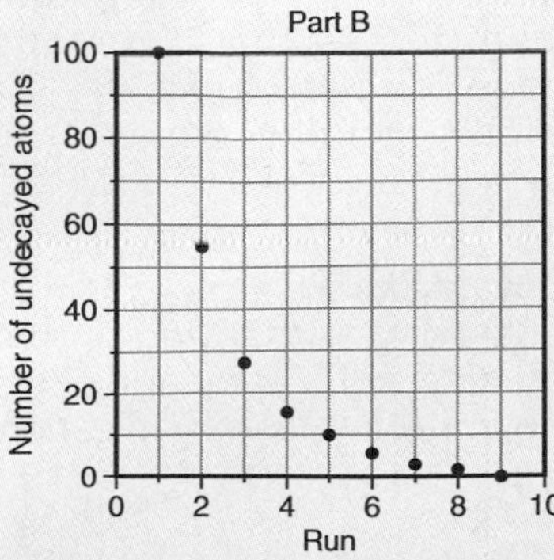

Data and Observations

	Part A	Part B
Time: start end total	10:30 am 10:54 am 24 min.	11:00 am 11:20 am 20 min.

Number of Undecayed Atoms		
Run #	Part A	Part B
1 (Start)	100	100
2	53	55
3	27	28
4	14	16
5	9	10
6	5	6
7	2	3
8	1	1
9	1	0

✓ ASSESSMENT

Performance: Assess student performance for this ChemActivity by having students complete copies of the masters from the *ChemActivity Masters* booklet.

2. See graphs below.
3. The graph of the natural log of the remaining atoms versus time is linear. The other graphs are not linear.
4. The natural log of the number of peanium atoms varies indirectly with time; that is, the number of stable peanium atoms decreases logarithmically with time.
5. Beanium does not change the rate at which peanium decomposes.
6. If the model is valid, there are no catalysts for the decomposition of radioactive isotopes.

Extension and Application
1. Because radioactive isotopes decay at a constant rate logarithmically, scientists can calculate the estimated age of any object that contains radioactive elements. They first determine the proportion of undecayed radioisotope to the stable element formed by the isotope's decay.
2. Under normal conditions of heat and pressure, half-life is not altered.
3. Topics of radioactive waste management that students could investigate include: waste transportation, waste storage, and waste recycling.

CHEMACTIVITY 29

Formation of Ethyl and Butyl Acetates

Students generally enjoy doing this activity. Preparing chemicals that have familiar odors intrigues and delights students and helps them to see that chemistry is, indeed, a part of real life. If time allows, students may prepare another ester, using butyl alcohol and acetic acid.

Process Skills
Observing, classifying, interpreting data, comparing and contrasting, inferring, communicating

Analysis and Conclusions

1. Make a graph of your results. Place time (the run number), on the x-axis and the number of remaining undecayed peanium atoms on the y-axis.
2. Regraph your data using time on the x-axis and the natural logarithm of the number of remaining peanium atoms on the y-axis. Remember, the natural logarithm of a number is 2.3 times the logarithm of the number ($\ln x = 2.3 \log x$).
3. Which of the graphs is linear?
4. What is the relationship between time and the number of remaining peanium atoms?
5. In your experiment, how did beanium affect the rate of decomposition of peanium?
6. Assuming that lima beans are a valid model for the atoms of any catalyst, what statement can you make about the role of catalysts in the decay of radioisotopes?

Extension and Application

1. The rate of decay of every radioactive element is a constant logarithmically. How do scientists use this constant to determine the age of natural objects, such as rocks and bones, and materials made by humans, such as cloth and baskets?
2. Find out if any physical processes, such as heat or pressure, can change the half-life of an element.
3. Nuclear power plants use radioactive materials to produce power. Some of the waste products from these plants are radioactive. Research the topic of radioactive waste management and report to the class on the problems associated with nuclear waste.

Chapter 29
ChemActivity

Formation of Ethyl and Butyl Acetates

When an organic acid reacts with an alcohol, an organic compound called an ester is formed. Esters are noted for their aromas and for their often distinctive flavors. Esters are responsible for the odors and flavors of many fruits. The artificial flavors and fragrances that are added to many food and toiletry products are also usually esters. The general reaction for the formation of an ester is:

$$R—CH_2OH + HOOC—R' \rightarrow R—CH_2OOC—R' + H_2O$$

The symbols R and R′ represent chains of carbon atoms. Different combinations of acids and alcohols produce different esters. Some are given in the following table

Acid	Alcohol	Ester	Aroma
butyric	ethyl	ethyl butyrate	pineapple
acetic	amyl	amyl acetate	banana
acetic	ethyl	ethyl acetate	nail polish remover

The purpose of this activity is to prepare an ester.

Objectives

- **Prepare** an organic compound, an ester.
- **Compare** the properties of the ester with those of the compounds from which the ester was prepared.

850 *ChemActivities*

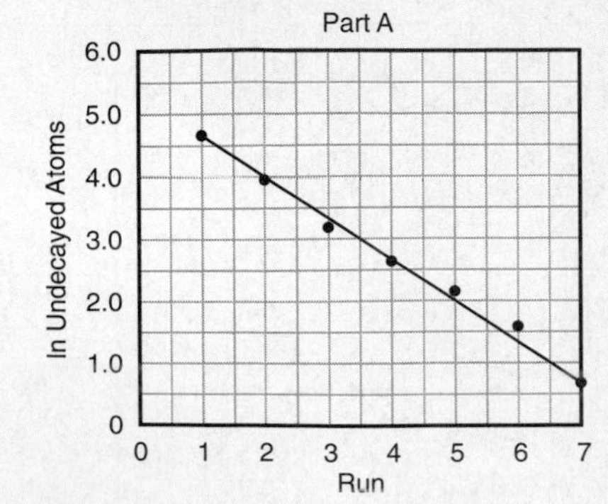

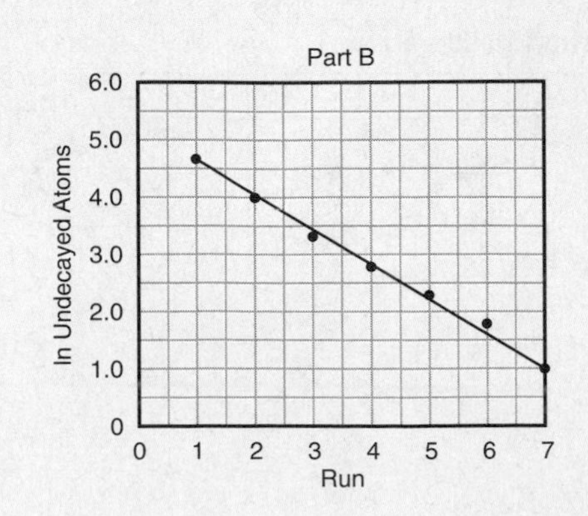

Materials

100-cm^3 glass beaker
thermometer
10-cm^3 graduated cylinder
ethanol
acetic acid
sulfuric acid (concentrated)
hot plate
anhydrous sodium sulfate
plastic micropipet
plastic cup
water
small test tube
paper towel

CAUTION: *Do not use a laboratory burner for this experiment. Do not allow vapors from this experiment to come into contact with an open flame from any source.*

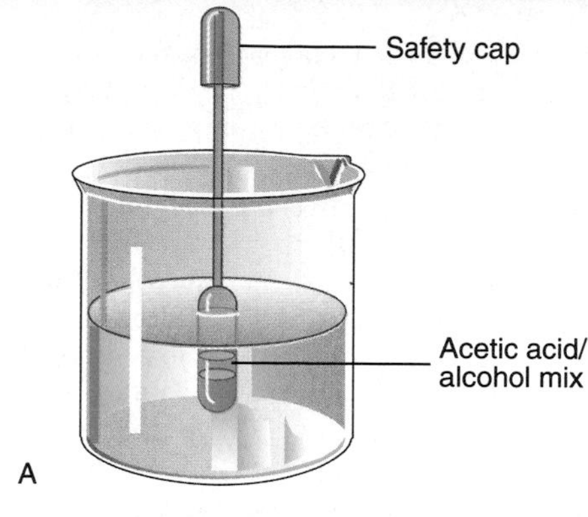

Procedure

1. Half-fill the 100-cm^3 beaker with water.
2. Begin heating the water in the beaker to 50°C while you proceed with the ChemActivity.
3. Place 1.0 cm^3 each of ethanol and acetic acid in a plastic cup. Note the odor of both the alcohol and the acid as you work with these compounds. Record your observations in a data table like the one shown on the next page.
4. Add 5 drops of sulfuric acid to the mixture in the plastic cup. **CAUTION:** *Handle sulfuric acid with great care. It can cause severe burns if it touches the skin and will damage clothing if it comes into contact with it.*
5. Cut through the bulb of a pipet at about the midpoint. Using a second pipet, draw up the mixture from the plastic cup. Invert the cutoff bulb of the first pipet and place it over the end of the stem of the second pipet to form a safety cap.
6. Rinse the plastic cup in tap water and dry with a paper towel.
7. Place the pipet with the stem pointing upward into the heated water, as shown in Figure A.
8. Maintain the temperature of the water bath at 45–50°C for 10 minutes.
9. At the end of the 10 minutes, remove the pipet from the warm water bath and place it stem upward in the plastic cup.
10. Replace the warm water with cold water.
11. Holding the pipet with the stem pointing upward, squeeze the bulb to force air out of the bulb and the stem. Then invert the pipet and place its stem in the beaker of cold water. Release the bulb so that cold water will be drawn into the pipet.
12. Holding the pipet by the bulb, swirl the pipet to mix its contents. The mixture in the pipet now has an aqueous layer at the bottom and a crude ester layer above the aqueous layer, as shown in Figure B.

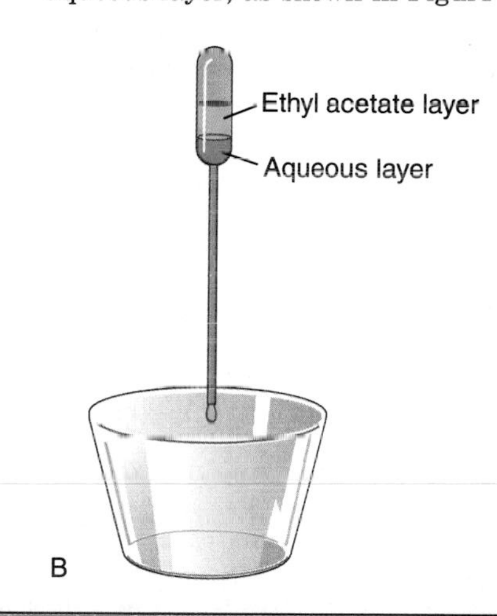

Misconceptions

Students usually are unaware that some flavors and fragrances are manufactured.

Procedure Hints

Do not use laboratory burners or open flames in this activity.

• Demonstrate the safe way to sample the odor of a compound. With open hand, waft the fumes from the compound past your nose. Remind students to use this method whenever they investigate odors in a laboratory.

• **CAUTION:** *Handle sulfuric acid with great care. Sulfuric acid can cause severe skin burns and can destroy clothing by contact.*

✓ ASSESSMENT

Performance: Assess student performance for this ChemActivity by having students complete copies of the masters from the *ChemActivity Masters* booklet.

PROGRAM RESOURCES

ChemActivity Masters: Use worksheets for ChemActivity 29 to provide students with a copy of procedures and a data table for use in the lab.

Data and Observations
See sample data table below.

Analysis and Conclusions

1. The ester did not have an odor similar to either the acid or the alcohol. Since the ester is a different compound, its physical and chemical properties are different from the compounds from which it was prepared.

2. See data table.

3. The ester, ethyl acetate, is not soluble in water. Both acetic acid and ethanol are soluble in water. Ethyl acetate appears to be oily while ethanol and acetic acid have the viscosity of water.

Extension and Application
Some of the products grouped with spices in the supermarket contain artificial flavors. Artificial flavors are often among the ingredients listed in the labels of baked goods and other food products. Toiletries, such as toothpaste, also use artificial flavors and fragrances. Often, several esters are mixed in a product to give a blending of flavors and fragrances.

CHEMACTIVITY 30

Production of Aspirin
An aspirin tablet is formed from a mixture of starch, acetylsalicylic acid, and, in some cases, special buffering and coating agents. The active compound, acetylsalicylic acid, has been in use for nearly 100 years to relieve pain and reduce fever and inflammation. Other chemicals have been produced that have the same physiological effects, but the aspirin tablet remains a widely used and prescribed over-the-counter medicine.

Process Skills
Observing, measuring, using numbers, interpreting data, inferring, communicating

Misconceptions
Students are likely to think that an aspirin tablet is formed from a pure chemical compound. Explain that most of an aspirin is starch.

13. Separate the aqueous layer from the ester layer by holding the pipet stem downward over the beaker of water. Squeeze the bulb gently to expel only the layer of water.
14. Place 2 grams of anhydrous sodium sulfate in the plastic cup.
15. Holding the pipet stem downward over the plastic cup, squeeze the bulb to expel the ester layer and add it to the sodium sulfate. Swirl the plastic cup until the ester is clear.
16. Carefully sniff the odor of the ester and record it in your data table. Also record any other physical properties of the ester, the acid, and the alcohol.

Analysis and Conclusions

1. Did the ester formed have an odor similar to either the alcohol or the acid?
2. Of what common substance did the odor of acetic acid remind you? Ethanol? The ester, ethyl acetate?
3. What physical property showed that a new chemical compound had been formed when ethanol reacted with acetic acid?

Data and Observations

Odor and Physical Properties		
	Odor	Physical Properties
Ethanol		
Acetic acid		
Ethyl acetate		

Extension and Application

In a supermarket, look at products containing artificial flavors. If the names of the chemicals that produced the flavors are given, make a list of them. Also check reference books and textbooks to find the chemical names for artificial flavors. Choose one name that includes *ester* and find out from which acid and alcohol the ester is made.

Micro

Chapter 30
ChemActivity

Production of Aspirin

Aspirin is one of the most commonly used medicines in the world. It is effective in relieving pain such as headache and can reduce fever and inflammation. Its chemical name is acetylsalicylic acid. The purpose of this activity is to produce aspirin and to isolate it.

Objectives

- **Prepare** and **isolate** an organic compound.
- **Observe** the formation of crystals from a chemical reaction.

Materials

- 100-cm^3 glass beakers (2)
- hot plate
- ice
- thermometer
- 10-cm^3 graduated cylinder
- acetic anhydride
- salicylic acid
- sulfuric acid (concentrated)
- ethanol
- plastic micropipets (2)
- plastic cup
- toothpick
- scissors
- microscale
- paper towel

852 *ChemActivities*

Data and Observations

Odor and Physical Properties		
	Odor	Physical Properties
Ethanol	reminiscent of doctor or dentist's office	soluble in water; viscosity similar to water's
Acetic acid	vinegar	soluble in water; viscosity similar to water's
Ethyl acetate	nail polish remover	not soluble in water; oily

Procedure

1. Half-fill a 100-cm^3 beaker with water.
2. Begin heating the water to 60°C while you proceed with the activity.
3. Place 2.0 cm^3 of acetic anhydride in a plastic cup.
4. Add 5 drops of sulfuric acid to the acetic acid in the plastic cup. **CAUTION:** *Handle sulfuric acid with great care. It can cause severe burns if it touches the skin and will also damage clothing.*
5. Place 1.0 g of salicylic acid in the plastic cup. Mix the chemicals with a toothpick.
6. Cut through the bulb of a pipet at about the midpoint. Using a second pipet, draw up the mixture from the plastic cup. Invert the cutoff bulb of the first pipet and place it over the end of the stem of the second pipet to form a safety cap.
7. Rinse the plastic cup in tap water and dry.
8. Place the pipet with its stem pointing upward into the heated water.
9. Maintain the temperature of the water bath at 55°–60°C for 25 minutes.
10. At the end of the 25 minutes, remove the pipet from the warm water bath and place it stem upward in the plastic cup.
11. Do not discard the warm tap water.
12. Fill another 100-cm^3 beaker with cold tap water and ice.
13. Holding the pipet with the stem upward, place the bulb in the ice-cold water.
14. Swirl the bulb to cool the contents of the pipet. You may begin to notice white crystals forming. These crystals are aspirin.
15. Separate the aqueous layer (the reaction mixture) from the aspirin by holding the pipet downward over the beaker of water and squeezing the bulb gently.
16. Draw approximately ½ pipet of ethanol into a clean pipet. Place the pipet stem upward in the beaker containing warm water.
17. Heat the ethanol for 5 minutes.
18. When the ethanol is warm, eject it from the pipet into the plastic cup.
19. Draw up the warm ethanol from the plastic cup into the pipet containing the aspirin crystals. Mix well.
20. Holding the pipet with the stem pointing upward, squeeze the air out of the pipet.
21. Invert the pipet and place the stem in the ice-cold water. Draw up approximately ½-pipet of water. Mix well.
22. Allow the pipet to cool.
23. Cut open the pipet to reveal the crystals of aspirin.
24. Mass the crystals of aspirin and enter the mass in a data table like the one shown.

Data and Observations

Compound	Mass (g)	Moles
Acetic anhydride		$\frac{\text{mass}}{102\text{g/mole}} =$
Salicylic acid		$\frac{\text{mass}}{138\text{g/mole}} =$
Aspirin		$\frac{\text{mass}}{180\text{g/mole}} =$

Analysis and Conclusions

1. Complete the data table. First, enter the mass of each compound. Then calculate the moles of the compounds you used to produce the aspirin and the moles of aspirin you obtained.
2. What compounds reacted to form aspirin?
3. Why do you think sulfuric acid was added to the reacting compounds?
4. What was the purpose of the ice-water bath?

Extension and Application

Find out when aspirin first came into use as a medicine. What ancient folk remedy is aspirin related to?

Procedure Hints

Chemicals should be dispensed in small plastic cups with lids to avoid spillage and contamination.

- **CAUTION:** *Concentrated sulfuric acid should be handled with great care.* Consider adding the sulfuric acid to the student's reaction mixtures yourself rather than having students handle the acid.
- Students should be cautioned about breathing the vapors that may be produced as they heat the reactant mixture. Acetic anhydride produces a strong odor of vinegar.
- All acetylsalicylic acid should be collected. Students should not be allowed to keep their product.

Data and Observations

See table below.

Analysis and Conclusions

1. See data table below. Sample mass is given for aspirin.
2. Acetic anhydride and salicylic acid reacted to form the product, acetylsalicylic acid (aspirin).
3. The sulfuric acid was a catalyst for the reaction.
4. The ice-water bath brought about rapid cooling and promoted the formation of crystals.

Extensions and Applications

Acetylsalicylic acid came into use as a medicine about 1899, when it was found that the compound can relieve pain and reduce fever. For centuries, people had used the bark of the willow tree to accomplish the same medicinal purposes. The bark contains a compound that the human body converts to a salicylate.

✓ ASSESSMENT

Performance: Assess student performance for this ChemActivity by having students complete copies of the masters from the *ChemActivity Masters* booklet.

PROGRAM RESOURCES

ChemActivity Masters: Use worksheets for ChemActivity 30 to provide students with a copy of procedures and a data table for use in the lab.

Data and Observations

Compound	Mass	Moles
Acetic anhydride	(cm^3 × 1.08 g/cm^3) 216 g	0.0212 mole
Salicylic acid	1.0 g	0.0072 mole
Aspirin	0.71 g	0.0039 mole

APPENDIX A

Table A-1

Definitions of Standards

1 ampere is the constant current which, if maintained in two straight parallel conductors of infinite length, of negligible circular cross-section, and placed 1 meter apart in a vacuum, would produce a force of 2×10^{-7} newton per meter of length between these conductors.

1 candela is the luminous intensity, in the perpendicular direction, of a surface of 1/600 000 m^2 of a blackbody at the temperature of freezing platinum at a pressure of 101 325 pascals.

1 cubic decimeter is equal to 1 liter.

1 kelvin is 1/273.16 of the thermodynamic temperature of the triple point of water.

1 kilogram is the mass of the international prototype kilogram.

1 meter is the distance light travels in 1/299 792 458 of a second.

1 mole is the amount of substance containing as many elementary entities as there are atoms in 0.012 kilogram of carbon-12.

1 second is equal to 9 192 631 770 periods of the natural electromagnetic oscillation during that transition of ground state $^2S_{1/2}$ of cesium-133 which is designated $(F = 4, M = 0) \leftrightarrow (F = 3, M = 0)$.

Avogadro constant = $6.022\ 136\ 7 \times 10^{23}$

1 electronvolt = $1.602\ 177\ 33 \times 10^{-19}$ J

Faraday constant = 96 485.309 C/mole e^-

Ideal gas constant = 8.314 471 J/mol · K = 8.314 471 dm^3 · kPa/mol · K

Molar gas volume at STP = 22.414 10 dm^3

Planck's constant = $6.626\ 075 \times 10^{-34}$ J · s

Speed of light = $2.997\ 924\ 58 \times 10^8$ m/s

Table A-2

SI Prefixes

Prefix	Symbol	Meaning	Multiplier (Numerical)	Multiplier (Exponential)
		Greater than 1		
exa	E	quintillion	*1 000 000 000 000 000 000	1×10^{18}
peta	P	quadrillion	1 000 000 000 000 000	1×10^{15}
tera	T	trillion	1 000 000 000 000	1×10^{12}
giga	G	billion	1 000 000 000	1×10^{9}
mega	M	million	1 000 000	1×10^{6}
kilo	k	thousand	1 000	1×10^{3}
hecto	h	hundred	100	1×10^{2}
deka	da	ten	10	1×10^{1}
		Less than 1		
deci	d	tenth	0.1	1×10^{-1}
centi	c	hundredth	0.01	1×10^{-2}
milli	m	thousandth	0.001	1×10^{-3}
micro	μ	millionth	0.000 001	1×10^{-6}
nano	n	billionth	0.000 000 001	1×10^{-9}
pico	p	trillionth	0.000 000 000 001	1×10^{-12}
femto	f	quadrillionth	0.000 000 000 000 001	1×10^{-15}
atto	a	quintillionth	0.000 000 000 000 000 001	1×10^{-18}

*Spaces are used to group digits in long numbers. In some countries, a comma indicates a decimal point. Therefore, commas will not be used.

Table A-3

Some Properties of the Elements

Element	Symbol	Atomic Number (Z)	Atomic Mass* (u)	Melting Point (°C)	Boiling Point (°C)	Density (g/cm^3) (gases measured at STP)	Atomic Radius (pm)	First Ionization Energy (kJ/mol)	Standard Reduction Potential (V) (for elements from or to oxidation state indicated)	Enthalpy of Fusion (kJ/mol)	Specific Heat (J/g · C°)	Enthalpy of Vaporization (kJ/mol)	Abundance in Earth's Crust (%)	Major Oxidation States
Actinium	Ac	89	[227.0278]	1050	3300	10.07	203	666	(3+)−2.13	14.3	0.120	293	trace	3+
Aluminum	Al	13	26.981539	660.37	2517.6	2.699	143	577.5	(3+)−1.67	10.71	0.9025	290.8	8.1	3+
Americium	Am	95	[243.0614]	994	2600	13.67	183	579	(3+)−2.07	10	—	238.5	—	2+, 3+, 4+
Antimony	Sb	51	121.760	630.7	1635	6.697	161	834	(3+)+0.15	19.5	0.2072	193	2×10^{-5}	3+, 5+
Argon	Ar	18	39.948	−189.37	−185.86	0.001784	191	1521	—	1.18	0.52033	6.52	4×10^{-6}	
Arsenic	As	33	74.92159	816 (2840 kPa)	615 (sublimes)	5.778	121	947	(3+)+0.24	27.7	0.3289	— (sublimes)	1.9×10^{-4}	3+, 5+
Astatine	At	85	[209.98037]	300	350	—	—	916	(1−)+0.2	23.8	—	90.3	trace	1−, 5+
Barium	Ba	56	137.327	726.9	1845	3.62	222	502.9	(2+)−2.92	8.012	0.2044	140	0.039	2+
Berkelium	Bk	97	[247.0703]	986	—	14.78	170	601	(3+)−2.01	—	—	—	—	3+, 4+
Beryllium	Be	4	9.012182	1287	2468	1.848	112	899.5	(2+)−1.97	7.895	1.824	297.6	2×10^{-4}	2+
Bismuth	Bi	83	208.98037	271.4	1564	9.808	151	703	(3+)+0.317	10.9	0.1221	179	8×10^{-7}	3+, 5+
Boron	B	5	10.811	2080	3865	2.46	85	800.6	(3+)−0.89	50.2	1.026	504.5	9×10^{-4}	3+
Bromine	Br	35	79.904	−7.25	59.35	3.1028	119	1139.9	(1−)+1.065	10.571	0.47362	29.56	2.5×10^{-4}	1−, 1+, 3+, 5+
Cadmium	Cd	48	112.411	320.8	770	8.65	151	867.7	(2+)−0.4025	6.19	0.2311	100	1.6×10^{-5}	2+
Calcium	Ca	20	40.078	841.5	1500.5	1.55	197	589.8	(2+)−2.84	8.54	0.6315	155	4.66	2+
Californium	Cf	98	[251.0796]	900	—	—	186	608	(3+)−2	—	—	—	—	3+, 4+
Carbon	C	6	12.011	3620	4200	2.266	77	1086.5	(4−)+0.132	104.6	0.7099	711	0.018	4−, 2+, 4+
Cerium	Ce	58	140.115	804	3470	6.773	181.8	541	(3+)−2.34	5.2	0.1923	313	0.007	3+, 4+
Cesium	Cs	55	132.90543	28.4	674.8	1.9	262	375.7	(1+)−2.923	2.087	0.2421	67	2.6×10^{-4}	1+
Chlorine	Cl	17	35.4527	−101	−34	0.003214	91	1255.5	(1−)+1.3583	6.41	0.47820	20.41	0.013	1−, 1+, 3+, 5+
Chromium	Cr	24	51.9961	1860	2679	7.2	128	652.8	(3+)−0.74	20.5	0.4491	339	0.01	2+, 3+, 6+
Cobalt	Co	27	58.9332	1495	2912	8.9	125	758.8	(2+)−0.277	16.192	0.4210	382	0.0028	2+, 3+
Copper	Cu	29	63.546	1085	2570	8.92	128	745.5	(2+)+0.34	13.38	0.38452	304	0.0058	1+, 2+
Curium	Cm	96	[247.0703]	1340	3540	13.51	174	581	(3+)−2.06	—	—	—	—	3+, 4+
Dysprosium	Dy	66	162.5	1407	2600	8.536	178.1	572	(3+)−2.29	10.4	0.1733	250	6×10^{-4}	2+, 3+
Einsteinium	Es	99	[252.0828]	860	—	—	186	619	(3+)−2	—	—	—	—	3+
Erbium	Er	68	167.26	1497	2900	9.045	176.1	589	(3+)−2.32	17.2	0.1681	293	3.5×10^{-4}	3+
Europium	Eu	63	151.965	826	1439	5.245	208.4	547	(3+)−1.99	10.5	0.1820	176	2.1×10^{-3}	2+, 3+
Fermium	Fm	100	[257.0951]	—	—	—	—	627	(3+)−1.96	—	—	—	—	2+, 3+
Fluorine	F	9	18.9984032	−219.7	−188.2	0.001696	69	1681	(1−)+2.87	0.51	0.8238	6.54	0.0544	1−
Francium	Fr	87	[223.0197]	24	650	—	280	375	—	2	—	63.6	trace	1+
Gadolinium	Gd	64	157.25	1312	3000	7.886	180.4	592	(3+)−2.29	15.5	0.2355	311.7	6.3×10^{-4}	3+
Gallium	Ga	31	69.723	29.77	2203	5.904	134	573.8	(3+)−0.529	5.59	0.3709	256	0.0018	1+, 3+
Germanium	Ge	32	72.61	945	2350	5.323	123	761.2	(4+)+0.124	31.8	0.3215	334.3	1.5×10^{-4}	2+, 4+

*[ ] indicates mass of longest-lived isotope

Element	Symbol	Atomic Number (Z)	Atomic Mass* (u)	Melting Point (°C)	Boiling Point (°C)	Density (g/cm³) (gases measured at STP)	Atomic Radius (pm)	First Ionization Energy (kJ/mol)	Standard Reduction Potential (V) (for elements from or to oxidation state indicated)	Enthalpy of Fusion (kJ/mol)	Specific Heat (J/g · C°)	Enthalpy of Vaporization (kJ/mol)	Abundance in Earth's Crust (%)	Major Oxidation States
Gold	Au	79	196.96654	1064	2808	19.32	144	889.9	(3+)+1.52	12.4	0.12905	324.4	3×10^{-7}	1+, 3+
Hafnium	Hf	72	178.49	2227	4691	13.28	159	654.4	(4+)−1.56	29.288	0.1442	661	3×10^{-4}	4+
Helium	He	2	4.002602	−269.7 (2536 kPa)	−268.93	0.00017847	122	2372	—	0.02	5.1931	0.084	—	
Holmium	Ho	67	164.9032	1461	2600	8.78	176.2	581	(3+)−2.33	17.1	0.1646	251	1.5×10^{-4}	3+
Hydrogen	H	1	1.00794	−259.19	−252.76	0.0000899	78	1312	(1+) 0.0000	0.117	14.298	0.904	—	1−, 1+
Indium	In	49	114.82	156.61	2080	7.29	167	558.2	(3+)−0.3382	3.26	0.2407	231.8	2×10^{-5}	1+, 3+
Iodine	I	53	126.90447	113.6	184.5	4.93	138	1008.4	(1−)+0.5355	15.517	0.21448	41.95	4.6×10^{-5}	1−, 1+, 5+, 7+
Iridium	Ir	77	192.217	2447	4550	22.65	135.5	880	(4+)+0.926	26.4	0.1306	563.6	1×10^{-7}	3+, 4+, 5+
Iron	Fe	26	55.845	1536	2860	7.874	126	759.4	(3+)−0.4	13.807	0.4494	350	5.8	2+, 3+
Krypton	Kr	36	83.8	−157.2	−153.35	0.0037493	201	1351	—	1.64	0.2480	9.03	—	
Lanthanum	La	57	138.9055	920	3420	6.17	187	538	(3+)−2.37	8.5	0.1952	402	0.0035	3+
Lawrencium	Lr	103	[260.1054]	—	—	—	—	—	(3+)−2.06	—	—	—	—	3+
Lead	Pb	82	207.2	327	1746	11.342	175	715.6	(2+)−0.1251	4.77	0.1276	178	0.0013	2+, 4+
Lithium	Li	3	6.941	180.5	1347	0.534	156	520.2	(1+)−3.045	3	3.569	148	0.002	1+
Lutetium	Lu	71	174.967	1652	3327	9.84	173.8	524	(3+)−2.3	11.9	0.1535	414	8×10^{-5}	3+
Magnesium	Mg	12	24.305	650	1105	1.738	160	737.8	(2+)−2.356	8.477	1.024	127.4	2.76	2+
Manganese	Mn	25	54.93805	1246	2061	7.43	127	717.5	(2+)−1.18	12.058	0.4791	219.7	0.1	2+, 3+, 4+, 6+, 7+
Mendelevium	Md	101	[258.0986]	—	—	—	—	635	—	—	—	—	—	2+, 3+
Mercury	Hg	80	200.59	−38.9	357	13.534	151	1007	(2+)+0.8535	2.2953	0.13950	59.1	2×10^{-6}	1+, 2+
Molybdenum	Mo	42	95.94	2623	4679	10.28	139	685	(6+) 0.114	36	0.2508	590	1.2×10^{-4}	4+, 5+, 6+
Neodymium	Nd	60	144.24	1024	3111	7.003	181.4	530	(3+)−2.32	7.13	0.1903	283.7	0.004	2+, 3+
Neon	Ne	10	20.1797	−248.61	−246.05	0.0008999	131	2081	—	0.34	1.0301	1.77	—	
Neptunium	Np	93	237.0482	640	3900	20.45	155	597	(5+)−0.91	9.46	—	336	—	2+, 3+, 4+, 5+, 6+
Nickel	Ni	28	58.6934	1455	2883	8.908	124	736.7	(2+)−0.257	17.15	0.4442	375	0.0075	2+, 3+, 4+
Niobium	Nb	41	92.90638	2477	4858	8.57	146	664.1	(5+)−0.65	26.9	0.2648	690	0.002	4+, 5+
Nitrogen	N	7	14.00674	−210	−195.8	0.0012409	71	1402	(3−)−0.092	0.72	1.0397	5.58	0.002	3−, 2−, 1−, 1+, 2+, 3+, 4+, 5+
Nobelium	No	102	[259.1009]	—	—	—	—	642	(2+)−2.5	—	—	—	—	2+, 3+
Osmium	Os	76	190.2	3045	5025	22.57	135	840	(4+)+0.687	31.7	0.130	627.6	2×10^{-7}	4+, 6+, 8+
Oxygen	O	8	15.9994	−218.8	−183	0.001429	60	1313.9	(2−) 0.815	0.44	0.91738	6.82	45.5	2−, 1−
Palladium	Pd	46	106.42	1552	2940	11.99	137	805	(2+) 0.915	17.6	0.2441	362	3×10^{-7}	2+, 4+
Phosphorus	P	15	30.973762	44.2	280.5	1.823	109	1012	(3−)−0.063	0.659	0.76968	49.8	0.11	3−, 3+, 5+
Platinum	Pt	78	195.08	1769	3824	21.41	138.5	863	(4+)+1.15	19.7	0.1326	510.4	1×10^{-6}	2+, 4+
Plutonium	Pu	94	[244.0642]	640	3230	19.86	162	585	(4+)−1.25	2.8	0.138	343.5	—	3+, 4+, 5+, 6+
Polonium	Po	84	[208.9824]	254	962	9.4	164	813	(4+)+0.73	3.81	0.125	103	—	2−, 2+, 4+, 6+
Potassium	K	19	39.0983	63.2	766.4	0.862	231	418.8	(1+)−2.925	2.334	0.7566	76.9	1.84	1+
Praseodymium	Pr	59	140.90765	935	3343	6.782	182.4	522	(3+)−2.35	11.3	0.1930	332.6	9.1×10^{-4}	3+, 4+
Promethium	Pm	61	[144.9128]	1168	2460	7.2	183.4	536	(3+)−2.29	8.17	—	293	trace	3+

*[ ] indicates mass of longest-lived isotope

Element	Symbol	Atomic Number (Z)	Atomic Mass* (u)	Melting Point (°C)	Boiling Point (°C)	Density (g/cm³) (gases measured at STP)	Atomic Radius (pm)	First Ionization Energy (kJ/mol)	Standard Reduction Potential (V) (for elements from or to oxidation state indicated)	Enthalpy of Fusion (kJ/mol)	Specific Heat (J/g · C°)	Enthalpy of Vaporization (kJ/mol)	Abundance in Earth's Crust (%)	Major Oxidation States
Protactinium	Pa	91	231.03588	1552	4227	15.37	163	568	(5+)−1.19	14.6	—	481	trace	3+, 4+, 5+
Radium	Ra	88	226.0254	700	1630	5	228	509.1	(2+)−2.916	8.36	—	136.8	—	2+
Radon	Rn	86	[222.0176]	−71	−62	0.00973	232	1037	—	16.4	—	16.4	—	
Rhenium	Re	75	186.207	3180	5650	21.232	137	760	(7+)+0.34	33.4	0.1368	707	1 × 10⁻⁷	3+, 4+, 6+, 7+
Rhodium	Rh	45	102.9055	1960	3727	12.39	134	720	(3+)+0.76	21.6	0.2427	494	1 × 10⁻⁷	3+, 4+, 5+
Rubidium	Rb	37	85.4678	39.5	697	1.532	248	403	(1+)−2.925	2.19	0.36344	69.2	0.0078	1+
Ruthenium	Ru	44	101.07	2310	4119	12.41	134	711	(4+)+0.68	25.5	0.2381	567.8	—	2+, 3+, 4+, 5+
Samarium	Sm	62	150.36	1072	1800	7.536	180.4	542	(3+)−2.3	8.9	0.1965	191	7 × 10⁻⁴	2+, 3+
Scandium	Sc	21	44.95591	1539	2831	3	162	631	(3+)−2.03	15.77	0.5677	304.8	0.0022	3+
Selenium	Se	34	78.96	221	685	4.79	117	940.7	(2−)−0.67	5.43	0.3212	26.3	5 × 10⁻⁶	2−, 2+, 4+, 6+
Silicon	Si	14	28.0855	1411	3231	2.336	118	786.5	(4−)−0.143	50.2	0.7121	359	27.2	2+, 4+
Silver	Ag	47	107.8682	961	2195	10.49	144	730.8	(1+)+0.7991	11.65	0.23502	255	8 × 10⁻⁶	1+
Sodium	Na	11	22.989768	97.83	897.4	0.968	186	495.9	(1+)−2.714	2.602	1.228	97.4	2.27	1+
Strontium	Sr	38	87.62	776.9	1412	2.6	215	549.5	(2+)−2.89	7.4308	0.301	137	0.0384	2+
Sulfur	S	16	32.066	115.2	444.7	2.08	103	999.6	(2−)−0.45	1.7272	0.7060	9.62	0.03	2−, 4+, 6+
Tantalum	Ta	73	180.9479	2980	5505	16.65	146	760.3	(5+)−0.81	36.57	0.1402	737	2 × 10⁻⁴	4+, 5+
Technetium	Tc	43	97.9072	2200	4567	11.5	136	702	(6+)+0.83	23.0	—	577	—	2+, 4+, 6+, 7+
Tellurium	Te	52	127.6	450	990	6.25	138	869	(2−)−1.14	17.4	0.2016	50.6	2 × 10⁻⁷	2−, 2+, 4+, 6+
Terbium	Tb	65	158.92534	1356	2800	8.272	177.3	564	(3+)−2.31	10.3	0.1819	293	1 × 10⁻⁴	3+, 4+
Thallium	Tl	81	204.3833	303.5	1457	11.85	170	589.1	(1+)−0.3363	4.27	0.1288	162	7 × 10⁻⁵	1+, 3+
Thorium	Th	90	232.0381	1750	4787	11.78	179	587	(4+)−1.83	16.11	0.1177	543.9	8.1 × 10⁻⁴	4+
Thulium	Tm	69	168.93421	1545	1727	9.318	175.9	596	(3+)−2.32	18.4	0.1600	213	5 × 10⁻⁵	—
Tin	Sn	50	118.71	232	2623	7.265	141	708.4	(4+)+0.064	7.07	0.2274	296	2.1 × 10⁻⁴	2+, 4+
Titanium	Ti	22	47.867	1660	3358	4.5	147	658.1	(4+)−0.86	14.146	0.5226	425	0.63	2+, 3+, 4+
Tungsten	W	74	183.85	3680	6000	19.3	139	770.4	(6+)−0.09	35.4	0.1320	806	1.2 × 10⁻⁴	4+, 5+, 6+
Unnilennium	Une	109	[266]	—	—	—	—	—	—	—	—	—	—	—
Unnilhexium	Unh	106	[263]	—	—	—	—	—	—	—	—	—	—	—
Unniloctium	Uno	108	[265]	—	—	—	—	—	—	—	—	—	—	—
Unnilpentium	Unp	105	[262]	—	—	—	—	—	—	—	—	—	—	—
Unnilquadium	Unq	104	[261]	—	—	—	—	—	—	—	—	—	—	—
Unnilseptium	Uns	107	[262]	—	—	—	—	—	—	—	—	—	—	—
Uranium	U	92	238.0289	1130	3930	19.05	156	584	(6+)−0.83	12.6	0.11618	423	2.3 × 10⁻⁴	3+, 4+, 5+, 6+
Vanadium	V	23	50.9415	1917	3417	6.11	134	650.3	(4+)−0.54	22.84	0.4886	459.7	0.0136	2+, 3+, 4+, 5+
Xenon	Xe	54	131.29	−111.8	−108.09	0.0058971	218	1170	—	2.29	0.15832	12.64	—	
Ytterbium	Yb	70	173.04	824	1427	6.973	193.3	603	(3+)−2.22	7.66	0.1545	155	3.4 × 10⁻⁴	2+, 3+
Yttrium	Y	39	88.90585	1530	3264	4.5	180	616	(3+)−2.37	17.15	0.2984	393	0.0035	3+
Zinc	Zn	30	65.39	419.6	907	7.14	134	906.4	(2+)−0.7626	7.322	0.3884	115	0.0076	2+
Zirconium	Zr	40	91.224	1852	4400	6.51	160	659.7	(4+)−1.7	20.92	0.2780	590.5	0.0162	4+

*[ ] indicates mass of longest-lived isotope

Table A-4

Names and Charges of Polyatomic Ions			
1−	**2−**	**3−**	**4−**
Acetate, CH_3COO^-	Carbonate, CO_3^{2-}	Arsenate, AsO_4^{3-}	Hexacyanoferrate(II), $Fe(CN)_6^{4-}$
Amide, NH_2^-	Chromate, CrO_4^{2-}	Arsenite, AsO_3^{3-}	Orthosilicate, SiO_4^{4-}
Astatate, AtO_3^-	Dichromate, $Cr_2O_7^{2-}$	Borate, BO_3^{3-}	Diphosphate, $P_2O_7^{4-}$
Azide, N_3^-	Hexachloroplatinate, $PtCl_6^{2-}$	Citrate, $C_6H_5O_7^{3-}$	
Benzoate, $C_6H_5COO^-$	Hexafluorosilicate, SiF_6^{2-}	Hexacyanoferrate(III), $Fe(CN)_6^{3-}$	
Bismuthate, BiO_3^-	Molybdate, MoO_4^{2-}	Phosphate, PO_4^{3-}	
Bromate, BrO_3^-	Oxalate, $C_2O_4^{2-}$	**1+**	**2+**
Chlorate, ClO_3^-	Peroxide, O_2^{2-}	Ammonium, NH_4^+	Mercury(I), Hg_2^{2+}
Chlorite, ClO_2^-	Peroxydisulfate, $S_2O_8^{2-}$	Neptunyl(V), NpO_2^+	Neptunyl(VI), NpO_2^{2+}
Cyanide, CN^-	Phosphite, HPO_3^{2-}	Plutonyl(V), PuO_2^+	Plutonyl(VI), PuO_2^{2+}
Formate, $HCOO^-$	Ruthenate, RuO_4^{2-}	Uranyl(V), UO_2^+	Uranyl(VI), UO_2^{2+}
Hydroxide, OH^-	Selenate, SeO_4^{2-}	Vanadyl(V), VO_2^+	Vanadyl(IV), VO^{2+}
Hypobromite, BrO^-	Selenite, SeO_3^{2-}		
Hypochlorite, ClO^-	Silicate, SiO_3^{2-}		
Hypophosphite, $H_2PO_2^-$	Sulfate, SO_4^{2-}		
Iodate, IO_3^-	Sulfite, SO_3^{2-}		
Nitrate, NO_3^-	Tartrate, $C_4H_4O_6^{2-}$		
Nitrite, NO_2^-	Tellurate, TeO_4^{2-}		
Perbromate, BrO_4^-	Tellurite, TeO_3^{2-}		
Perchlorate, ClO_4^-	Tetraborate, $B_4O_7^{2-}$		
Periodate, IO_4^-	Thiosulfate, $S_2O_3^{2-}$		
Permanganate, MnO_4^-	Tungstate, WO_4^{2-}		
Perrhenate, ReO_4^-			
Thiocyanate, SCN^-			
Vanadate, VO_3^-			

Table A-5

Specific Heat Values (J/g · K)					
Substance	C_p	**Substance**	C_p	**Substance**	C_p
AlF_3	0.8948	Fe_3C	0.5898	$NaVO_3$	1.540
$BaTiO_3$	0.79418	$FeWO_4$	0.37735	$Ni(CO)_4$	1.198
BeO	1.020	HI	0.22795	PbI_2	0.1678
CaC_2	0.9785	K_2CO_3	0.82797	SF_6	0.6660
$CaSO_4$	0.7320	$MgCO_3$	0.8957	SiC	0.6699
CCl_4	0.85651	$Mg(OH)_2$	1.321	SiO_2	0.7395
CH_3OH	2.55	$MgSO_4$	0.8015	$SrCl_2$	0.4769
CH_2OHCH_2OH	2.413	MnS	0.5742	Tb_2O_3	0.3168
CH_3CH_2OH	2.4194	Na_2CO_3	1.0595	$TiCl_4$	0.76535
CdO	0.3382	NaF	1.116	Y_2O_3	0.45397
$CuSO_4 \cdot 5H_2O$	1.12				

Table A-6

Thermodynamic Properties (at 25°C and 100.000 kPa)

$\Delta H_f°$ (kJ/mol) $\Delta G_f°$ (kJ/mol) $S°$ (J/mol · K)
(concentration of aqueous solutions is 1*M*)

Substance	$\Delta H_f°$	$\Delta G_f°$	$S°$
Ag(cr)	0	0	42.55
AgCl(cr)	−127.068	−109.789	96.2
AgCN(cr)	146.0	156.9	107.19
Al(cr)	0	0	28.33
Al_2O_3(cr)	−1675.7	−1582.3	50.92
$BaCl_2$(aq)	−871.95	−823.21	122.6
$BaSO_4$(cr)	−1473.2	−1362.2	132.2
Be(cr)	0	0	9.50
BeO(cr)	−609.6	−580.3	—
Bi(cr)	0	0	56.74
$BiCl_3$(cr)	−379.1	−315.0	177.0
Bi_2S_3(cr)	−143.1	−140.6	200.4
Br_2(l)	0	0	152.231
CH_4(g)	−74.81	−50.72	186.264
C_2H_2(g)	+226.73	+209.20	200.94
C_2H_4(g)	+52.26	+68.15	219.56
C_2H_6(g)	−84.68	−32.82	229.60
CO(g)	−110.525	−137.168	197.674
CO_2(g)	−393.509	−394.359	213.74
CS_2(l)	+89.70	+65.27	151.34
Ca(cr)	0	0	41.42
$Ca(OH)_2$(cr)	−986.09	−898.49	—
Cl_2(g)	0	0	223.066
Co_3O_4(cr)	−891	−774	—
CoO(cr)	−237.94	−214.20	52.97
Cr_2O_3(cr)	−1139.7	−1058.1	81.2
CsCl(cr)	−443.04	−414.53	101.17
Cs_2SO_4(cr)	−1443.02	−1323.58	211.92
CuI(cr)	−67.8	−69.5	96.7
CuS(cr)	−53.1	−53.6	66.5
Cu_2S(cr)	−79.5	−86.2	120.9
$CuSO_4$(cr)	−771.36	−661.8	109
F_2(g)	0	0	202.78
$FeCl_3$(cr)	−399.49	—	—
FeO(cr)	−272.0	—	—
Fe_2O_3(cr)	−824.2	−742.2	87.40
Fe_3O_4(cr)	−1118.4	−1015.4	146.4
H(g)	+217.965	—	114.713
H_2(g)	0	0	130.684
HBr(g)	−36.40	−53.45	198.695
HCl(g)	−92.307	−95.299	186.908
HCl(aq)	−167.159	−131.228	56.5
HCN(aq)	+150.6	+172.4	94.1
HCHO(g)	−108.57	−102.53	218.77
HCOOH(l)	−424.72	[illegible]	128.95
HF(g)	−271.1	−273.2	173.779
HI(g)	+26.48	+1.70	206.594
H_2O(l)	−285.830	−237.129	69.91
H_2O(g)	−241.818	−228.572	188.825
H_2O_2(l)	—	−120.35	109.6
H_3PO_2(l)	−595.4	—	—
H_3PO_3(aq)	−964.4	—	—
H_3PO_4(aq)	−1279.0	−1119.1	110.50
H_2S(g)	−20.63	−33.56	205.79
H_2SO_3(aq)	−608.81	−537.81	232.2
H_2SO_4(aq)	−909.27	−744.53	20.1
$HgCl_2$(cr)	−224.3	−178.6	—
Hg_2Cl_2(cr)	−265.22	−210.745	192.5
Hg_2SO_4(cr)	−743.12	−625.815	200.66
I_2(cr)	0	0	116.135
K(cr)	0	0	64.18
KBr(cr)	−393.798	−380.66	95.90
$KMnO_4$(cr)	−837.2	−737.6	171.71
KOH(cr)	−424.764	—	—
LiBr(cr)	−351.213	—	—
LiOH(cr)	−484.93	−438.95	42.80
Mn(cr)	0	0	32.01
$MnCl_2$(aq)	−555.05	−490.8	38.9
$Mn(NO_3)_2$(aq)	−635.5	−450.9	218
MnO_2(cr)	−520.03	−465.14	53.05
MnS(cr)	−214.2	—	—
N_2(g)	0	0	191.61
NH_3(g)	−46.11	−16.45	192.45
NH_4Br(cr)	−270.83	−175.2	113
NO(g)	+90.25	86.55	210.761
NO_2(g)	+33.18	+51.31	240.06
N_2O(g)	+82.05	+104.20	219.85
Na(cr)	0	0	51.21
NaBr(cr)	−361.062	—	—
NaCl(cr)	−411.153	−384.138	72.13
$NaNO_3$(aq)	−447.48		—
NaOH(cr)	−425.609	—	—
Na_2S(aq)	−447.3	—	—
Na_2SO_4(cr)	−1387.08	−1270.16	149.58
O_2(g)	0	0	205.138
P_4O_6(cr)	−1640.1	—	—
P_4O_{10}(cr)	−2984.0	−2697.7	228.86
$PbBr_2$(cr)	270.7	−261.92	161.5
$PbCl_2$(cr)	−359.41	−314.10	136.0
S(cr)	0	0	31.80
SO_2(g)	−296.830	−300.194	[illegible]
SO_3(g)	−454.51	374.21	70.7
SrO(cr)	−592.0	−561.9	54.4
Ti(cr)	0	0	30.63
TiO_2(cr)	939.7	−884.5	49.92
TlI(cr)	−123.8	−125.39	127.6
UCl_4(cr)	−1019.2	−930.0	197.1
UCl_5(cr)	−1059	−950	242.7
Zn(cr)	0	0	41.63
$ZnCl_2$(aq)	−488.19	−409.50	0.8
ZnO(cr)	−348.28	−318.30	43.64
$ZnSO_4$(aq)	−1063.15	−891.59	−92.0

Table A-7

Solubility Rules

You will be working with water solutions, and it is helpful to have a few rules concerning what substances are soluble* in water. The more common rules are listed below.

1. All common salts of the Group 1(IA) elements and ammonium ion are soluble.
2. All common acetates and nitrates are soluble.
3. All binary compounds of Group 17(VIIA) elements (other than F) with metals are soluble except those of silver, mercury(I), and lead.
4. All sulfates are soluble except those of barium, strontium, lead, calcium, silver, and mercury(I).
5. Except for those in Rule 1, carbonates, hydroxides, oxides, sulfides, and phosphates are insoluble.

*A substance is considered soluble if more than 3 grams of the substance dissolve in 100 mL of water.

Table A-8

Molal Freezing–Point Depression and Boiling–Point Elevation Constants

Substance	K_{fp} (C° kg/mol)	Freezing Point (°C)	K_{bp} (C° kg/mol)	Boiling Point (°C)
Acetic Acid	3.90	16.66	2.530	117.90
Benzene	5.12	5.533	2.53	80.100
Camphor	37.7	178.75	5.611	207.42
Cyclohexane	20.0	6.54	2.75	80.725
Cyclohexanol	39.3	25.15	—	—
Nitrobenzene	6.852	5.76	5.24	210.8
Phenol	7.40	40.90	3.60	181.839
Water	1.853	0.000	0.515	100.000

Table A-9

Ionization Constants

Substance	Ionization Constant	Substance	Ionization Constant	Substance	Ionization Constant
$HCOOH$	1.77×10^{-4}	HBO_3^{-2}	1.58×10^{-14}	HS^-	1.26×10^{-13}
CH_3COOH	1.75×10^{-5}	H_2CO_3	4.37×10^{-7}	HSO_4^-	1.02×10^{-2}
$CH_2ClCOOH$	1.36×10^{-3}	HCO_3^-	4.68×10^{-11}	H_2SO_3	1.29×10^{-2}
$CHCl_2COOH$	5.50×10^{-2}	HCN	6.17×10^{-10}	HSO_3^-	6.17×10^{-8}
CCl_3COOH	3.02×10^{-1}	HF	6.61×10^{-4}	$HSeO_4^-$	2.19×10^{-2}
$HOOCCOOH$	5.36×10^{-2}	HNO_2	7.24×10^{-4}	H_2SeO_3	2.29×10^{-3}
$HOOCCOO^-$	5.35×10^{-5}	H_3PO_4	7.08×10^{-3}	$HSeO_3^-$	5.37×10^{-9}
CH_3CH_2COOH	1.34×10^{-5}	$H_2PO_4^-$	6.31×10^{-8}	$HBrO$	2.51×10^{-9}
C_6H_5COOH	6.25×10^{-5}	HPO_4^{2-}	4.17×10^{-13}	$HClO$	2.88×10^{-8}
H_3AsO_4	6.03×10^{-3}	H_3PO_3	6.31×10^{-2}	HIO	2.29×10^{-11}
$H_2AsO_4^-$	1.05×10^{-7}	$H_2PO_3^-$	2.00×10^{-7}	NH_3	1.74×10^{-5}
H_3BO_3	5.75×10^{-10}	H_3PO_2	5.89×10^{-2}	H_2NNH_2	8.71×10^{-7}
$H_2BO_3^-$	1.82×10^{-13}	H_2S	1.07×10^{-7}	H_2NOH	8.91×10^{-9}

Table A-10

Solubility Product Constants (at 25°C)					
Substance	K_{sp}	Substance	K_{sp}	Substance	K_{sp}
AgBr	5.01×10^{-13}	$BaSO_4$	1.10×10^{-10}	Li_2CO_3	2.51×10^{-2}
$AgBrO_3$	5.25×10^{-5}	$CaCO_3$	2.88×10^{-9}	$MgCO_3$	3.47×10^{-8}
Ag_2CO_3	8.13×10^{-12}	$CaSO_4$	9.12×10^{-6}	$MnCO_3$	1.82×10^{-11}
AgCl	1.78×10^{-10}	CdS	7.94×10^{-27}	$NiCO_3$	6.61×10^{-9}
Ag_2CrO_4	1.12×10^{-12}	$Cu(IO_3)_2$	7.41×10^{-8}	$PbCl_2$	1.62×10^{-5}
$Ag_2Cr_2O_7$	2.00×10^{-7}	CuC_2O_4	2.29×10^{-8}	PbI_2	7.08×10^{-9}
AgI	8.32×10^{-17}	$Cu(OH)_2$	2.19×10^{-20}	$Pb(IO_3)_2$	3.24×10^{-13}
AgSCN	1.00×10^{-12}	CuS	6.31×10^{-36}	$SrCO_3$	1.10×10^{-10}
$Al(OH)_3$	1.26×10^{-33}	FeC_2O_4	3.16×10^{-7}	$SrSO_4$	3.24×10^{-7}
Al_2S_3	2.00×10^{-7}	$Fe(OH)_3$	3.98×10^{-38}	TlBr	3.39×10^{-6}
$BaCO_3$	5.13×10^{-9}	FeS	6.31×10^{-18}	$ZnCO_3$	1.45×10^{-11}
$BaCrO_4$	1.17×10^{-10}	Hg_2SO_4	7.41×10^{-7}	ZnS	1.58×10^{-24}

Table A-11

Acid-Base Indicators			
Indicator	Lower Color	Range	Upper Color
Methyl violet	yellow-green	0.0–2.5	violet
Malachite green HCl	yellow	0.5–2.0	blue
Thymol blue	red	1.0–2.8	yellow
Naphthol yellow S	colorless	1.5–2.6	yellow
p-Phenylazoaniline	orange	2.1–2.8	yellow
Methyl orange	red	2.5–4.4	yellow
Bromophenol blue	orange-yellow	3.0–4.7	violet
Gallein	orange	3.5–6.3	red
2,5-Dinitrophenol	colorless	4.0–5.8	yellow
Ethyl orange	salmon	4.2–4.6	orange
Propyl red	pink	5.1–6.5	yellow
Bromocresol purple	green-yellow	5.4–6.8	violet
Bromoxylenol blue	orange-yellow	6.0–7.6	blue
Phenol red	yellow	6.4–8.2	red-violet
Cresol red	yellow	7.1–8.8	violet
m-Cresol purple	yellow	7.5–9.0	violet
Thymol blue	yellow	8.1–9.5	blue
Phenolphthalein	colorless	8.3–10.0	dark pink
o-Cresolphthalein	colorless	8.6–9.8	pink
Thymolphthalein	colorless	9.5–10.4	blue
Alizarin yellow R	yellow	9.9–11.8	dark orange
Methyl blue	blue	10.6–13.4	pale violet
Acid fuchsin	red	11.1–12.8	colorless
2,4,6-Trinitrotoluene	colorless	11.7–12.8	orange

Table A-12

Symbols and Abbreviations

α	= rays from radioactive materials, helium nuclei
β	= rays from radioactive materials, electrons
γ	= rays from radioactive materials, high-energy quanta
Δ	= change in
λ	= wavelength
ν	= frequency
Π	= osmotic pressure
A	= ampere (*electric current*)
Bq	= becquerel (*nuclear disintegration*)
°C	= Celsius degree (*temperature*)
C	= coulomb (*quantity of electricity*)
c	= speed of light
cd	= candela (*luminous intensity*)
C_p	= specific heat
D	= density
E	= energy, electromotive force
F	= force, Faraday
G	= free energy
g	= gram (*mass*)
Gy	= gray (*radiation*)
H	= enthalpy
Hz	= hertz (*frequency*)
h	= Planck's constant
h	= hour (*time*)
J	= joule (*energy*)
K	= kelvin (*temperature*)
K_a	= ionization constant (acid)
K_b	= ionization constant (base)
K_{eq}	= equilibrium constant
K_{sp}	= solubility product constant
kg	= kilogram
M	= molarity
m	= mass, molality
m	= meter (*length*)
mol	= mole (*amount*)
min	= minute (*time*)
N	= newton (*force*)
N_A	= Avogadro's number
n	= number of moles
P	= pressure, power
Pa	= pascal (*pressure*)
$\boldsymbol{p}$	= momentum
q	= heat
R	= gas constant
S	= entropy
s	= second (*time*)
Sv	= sievert (*absorbed radiation*)
T	= temperature
U	= internal energy
u	= atomic mass unit
V	= volume
V	= volt (*electromotive force*)
v	= velocity
W	= watt (*power*)
w	= work
x	= mole fraction

APPENDIX B

LOGARITHMS

A logarithm or log is an exponent. We will work with exponents given in terms of base 10.

$$N = b^x$$

$$\text{number} = \text{base}^{\text{exponent or logarithm}}$$

$$100 = 10^{2.0000}$$

For the log 2.000, the part of the numeral to the left of the decimal point is the characteristic. The part to the right of the decimal point is the mantissa.

$$\text{Log } 100 = \underset{\text{characteristic}}{2}.\underset{\text{mantissa}}{000}$$

SAMPLE PROBLEM

How to Find a Logarithm

Find the log of 657.

(a) Write the number in scientific notation, 6.57×10^2.

(b) Look in the table under the column (N). Find the first two digits, (65).

(c) Look to the right and find the mantissa that is in the vertical column under the third digit of the number (7). It is .8176.

(d) From the scientific notation, write the power of ten as the characteristic, to the left of the decimal point.

(e) Write the four digits from the table as the mantissa to the right of the characteristic and the decimal point, 2.8176.

$$\text{thus } 657 = 10^{2.8176} \text{ or } \log 657 = 2.8176$$

When given a logarithm and asked to find the number it represents, we use the table to find the first three digits for the number. We use the characteristic to determine where to locate the decimal point with respect to these digits.

SAMPLE PROBLEM

How to Find the Antilogarithm

Given the logarithm 2.8176, find the number it represents (antilog).

(a) In the log table, find the mantissa that is closest to .8176.

(b) We find by looking under the column (N) that this mantissa corresponds to 65. The third digit is found at the top of the column in which the mantissa appears, 7. (657).

(c) Write the three digits (657) in scientific notation, 6.57×10^x.

(d) The characteristic will be the power of ten.

$$\text{antilog } 2.8176 = 6.57 \times 10^2 \text{ or } 657.$$

SAMPLE PROBLEMS

1. Logarithms of Numbers Less Than 1

Find the log 0.00657.

(a) Write the number in scientific notation, 6.57×10^{-3}

(b) Look in the table under the column *N* for the first two digits, 6.5, and to the right in the column under the third digit, 7, for the mantissa. Note that the mantissa is always a positive number, .8176

(c) From the scientific notation, we get the negative characteristic, −3.

(d) Add the negative characteristic and the positive mantissa $(-3.0000) + (+.8176) = -2.1824$. This value is more commonly represented as 7.8176-10. However, the negative logarithm −2.1824 is more useful in pH calculations.

2. Antilog of a Negative Logarithm

Find the antilog of −2.1824.

(a) We ask ourselves what number would we add to the next, lesser integer, −3., to get the log −2.1824. It would be 0.8176.

$$\begin{array}{rr} & -3.0000 \\ \text{subtract} & -2.1824 \\ \hline & 0.8176 \end{array}$$

We know that logarithm tables do not give mantissas for negative numbers. So, we have changed the −2.1824 into the sum of a negative characteristic and a positive mantissa. The characteristic is always the next negative number. The positive mantissa was determined by asking ourselves what positive number would we add to the negative characteristic to get −2.1824.

$$-2.1824 = -3. + 0.8176$$

(b) Antilog $-2.1824 = \text{antilog } -3. \times \text{antilog } 0.8176$
We know the antilog of −3. is 10^{-3}. From the table, we find that the antilog $0.8176 = 6.57$. Therefore the antilog of $-2.1824 = 6.57 \times 10^{-3}$.

SIGNIFICANT DIGITS AND LOGARITHMS

The characteristic of a logarithm simply tells us how many decimal places the number has. The mantissa, on the other hand, represents the actual value of the number. Thus, there may be only as many significant digits in the mantissa as there are significant digits in the number whose logarithm is being obtained. Thus,

$$\log 2.34 \times 10^7 = 7.369$$

but

$$\log 2.34 = 0.369$$

When taking antilogarithms of numbers, the same principle applies. The antilog contains as many significant digits as the mantissa. Thus,

$$\text{antilog } 4.357 = 22\,800$$

and

$$\text{antilog } 0.357 = 2.28$$

Table B-1

Logarithms of Numbers										
N	**0**	**1**	**2**	**3**	**4**	**5**	**6**	**7**	**8**	**9**
10	0000	0043	0086	0128	0170	0212	0253	0294	0334	0374
11	0414	0453	0492	0531	0569	0607	0645	0682	0719	0775
12	0792	0828	0864	0899	0934	0969	1004	1038	1072	1106
13	1139	1173	1206	1239	1271	1303	1335	1367	1399	1430
14	1461	1492	1523	1553	1584	1614	1644	1673	1703	1732
15	1761	1790	1818	1847	1875	1903	1931	1959	1987	2014
16	2041	2068	2095	2122	2148	2175	2201	2227	2253	2279
17	2304	2330	2355	2380	2405	2430	2455	2480	2504	2529
18	2553	2577	2601	2625	2648	2672	2695	2718	2742	2765
19	2788	2810	2833	2856	2878	2900	2923	2945	2967	2989
20	3010	3032	3054	3075	3096	3118	3139	3160	3181	3201
21	3222	3243	3263	3284	3304	3324	3345	3365	3385	3404
22	3424	3444	3464	3483	3502	3522	3541	3560	3579	3598
23	3617	3636	3655	3674	3692	3711	3729	3747	3766	3784
24	3802	3820	3838	3856	3874	3892	3909	3927	3945	3962
25	3979	3997	4014	4031	4048	4065	4082	4099	4116	4133
26	4150	4166	4183	4200	4216	4232	4249	4265	4281	4298
27	4314	4330	4346	4362	4378	4393	4409	4425	4440	4456
28	4472	4487	4502	4518	4533	4548	4564	4579	4594	4606
29	4624	4639	4654	4669	4683	4698	4713	4728	4742	4757
30	4771	4786	4800	4814	4829	4843	4857	4871	4886	4900
31	4914	4928	4942	4955	4969	4983	4997	5011	5024	5038
32	5051	5065	5079	5092	5105	5119	5132	5145	5159	5172
33	5185	5198	5211	5224	5237	5250	5263	5276	5289	5302
34	5315	5328	5340	5353	5366	5378	5391	5403	5416	5428
35	5441	5453	5465	5478	5490	5502	5514	5527	5539	5551
36	5563	5575	5587	5599	5611	5623	5635	5647	5658	5670
37	5682	5694	5705	5717	5729	5740	5752	5763	5775	5786
38	5798	5809	5821	5832	5843	5855	5866	5877	5888	5899
39	5911	5922	5933	5944	5955	5966	5977	5988	5999	6010
40	6021	6031	6042	6053	6064	6075	6085	6096	6107	6117
41	6128	6138	6149	6160	6170	6180	6191	6201	6212	6222
42	6232	6243	6253	6263	6274	6284	6294	6304	6314	6325
43	6335	6345	6355	6365	6375	6385	6395	6405	6415	6425
44	6435	6444	6454	6464	6474	6484	6493	6503	6513	6522
45	6532	6542	6551	6561	6571	6580	6590	6599	6609	6618
46	6628	6637	6646	6656	6665	6675	6684	6693	6702	6712
47	6721	6730	6739	6749	6758	6767	6776	6785	6794	6803
48	6812	6821	6830	6839	6848	6857	6866	6875	6884	6893
49	6902	6911	6920	6928	6937	6946	6955	6964	6972	6981
50	6990	6998	7007	7016	7024	7033	7042	7050	7059	7067
51	7076	7084	7093	7101	7110	7118	7126	7135	7143	7152
52	7160	7168	7177	7185	7193	7202	7210	7218	7226	7235
53	7243	7251	7259	7267	7275	7284	7292	7300	7308	7316
54	7324	7332	7340	7348	7356	7364	7372	7380	7388	7396

N	**0**	**1**	**2**	**3**	**4**	**5**	**6**	**7**	**8**	**9**
55	7404	7412	7419	7427	7435	7443	7451	7459	7466	7474
56	7482	7490	7497	7505	7513	7520	7528	7536	7543	7551
57	7559	7566	7574	7582	7589	7597	7604	7612	7619	7627
58	7634	7642	7649	7657	7664	7672	7679	7686	7694	7701
59	7709	7716	7723	7731	7738	7745	7752	7760	7767	7774
60	7782	7789	7796	7803	7810	7818	7825	7832	7839	7846
61	7853	7860	7868	7875	7882	7889	7896	7903	7910	7917
62	7924	7931	7938	7945	7952	7959	7966	7973	7980	7987
63	7993	8000	8007	8014	8021	8028	8035	8041	8048	8055
64	8062	8069	8075	8082	8089	8096	8102	8109	8116	8122
65	8129	8136	8142	8149	8156	8162	8169	8176	8182	8189
66	8195	8202	8209	8215	8222	8228	8235	8241	8248	8254
67	8261	8267	8274	8280	8287	8293	8299	8306	8312	8319
68	8325	8331	8338	8344	8351	8357	8363	8370	8376	8382
69	8388	8395	8401	8407	8414	8420	8426	8432	8439	8445
70	8451	8457	8463	8470	8476	8482	8488	8494	8500	8506
71	8513	8519	8525	8531	8537	8543	8549	8555	8561	8567
72	8573	8579	8585	8591	8597	8603	8609	8615	8621	8627
73	8633	8639	8645	8651	8657	8663	8669	8675	8681	8686
74	8692	8698	8704	8710	8716	8722	8727	8733	8739	8745
75	8751	8756	8762	8768	8774	8779	8785	8791	8797	8802
76	8808	8814	8820	8825	8831	8837	8842	8848	8854	8859
77	8865	8871	8876	8882	8887	8893	8899	8904	8910	8915
78	8921	8927	8932	8938	8943	8949	8954	8960	8965	8971
79	8976	8982	8987	8993	8998	9004	9009	9015	9020	9025
80	9031	9036	9042	9047	9053	9058	9063	9069	9074	9079
81	9085	9090	9096	9101	9106	9112	9117	9122	9128	9133
82	9138	9143	9149	9154	9159	9165	9170	9175	9180	9186
83	9191	9196	9201	9206	9212	9217	9222	9227	9232	9238
84	9243	9248	9253	9258	9263	9269	9274	9279	9284	9289
85	9294	9299	9304	9309	9315	9320	9325	9330	9335	9340
86	9345	9350	9355	9360	9365	9370	9375	9380	9385	9390
87	9395	9400	9405	9410	9415	9420	9425	9430	9435	9440
88	9445	9450	9455	9460	9465	9469	9474	9479	9484	9489
89	9494	9499	9504	9509	9513	9518	9523	9528	9533	9538
90	9542	9547	9552	9557	9562	9566	9571	9576	9581	9586
91	9590	9595	9600	9605	9609	9614	9619	9624	9628	9633
92	9638	9643	9647	9652	9657	9661	9666	9671	9675	9690
93	9685	9689	9694	9699	9703	9708	9713	9717	9722	9727
94	9731	9736	9741	9745	9750	9754	9759	9763	9768	9773
95	9777	9782	9786	9791	9795	9800	9805	9809	9814	9818
96	9823	9827	9832	9836	9841	9845	9850	9854	9859	9863
97	9868	9872	9877	9881	9886	9890	9894	9899	9903	9908
98	9912	9917	9921	9926	9930	9934	9939	9943	9948	9952
99	9956	9961	9965	9969	9974	9978	9983	9987	9991	9996

APPENDIX C

SOLUTIONS TO IN-CHAPTER PRACTICE PROBLEMS

Chapter 2

4. **a.** centimeter **b.** micrometer **c.** kilogram **d.** deciliter
5. **a.** millimeter **b.** microsecond **c.** centigram **d.** picosecond
6. Student 2 has the precise data because the largest difference in values is 0.03 g while Student 1 has a 0.59-g difference.
7. No, the balance error would subtract out.
8. $\% \text{ error} = \dfrac{11.342 \text{ g} - 10.95 \text{ g}}{11.342 \text{ g}}(100) = 3.5\%$
9. $\% \text{ error} = \dfrac{59.35°\text{C} - 40.6°\text{C}}{59.35°\text{C}}(100) = 31.6\%$
10. **a.** 1 **b.** 2 **c.** 2 **d.** 2 **e.** 3 **f.** 1 **g.** 2 **h.** 3 **i.** 4 **j.** infinite
11. $D = \dfrac{m}{V} = \dfrac{8.76 \text{ g}}{3.07 \text{ cm}^3} = 2.85 \text{ g/cm}^3$
12. $D = \dfrac{m}{V} = \dfrac{26.8 \text{ g}}{14.5 \text{ cm}^3} = 1.85 \text{ g/cm}^3$
13. $D = \dfrac{m}{V} = \dfrac{0.61 \text{ g}}{0.26 \text{ cm}^3} = 2.3 \text{ g/cm}^3$
14. $m = DV = \left(\dfrac{2.72 \text{ g}}{\cancel{\text{cm}^3}}\right) \times (24.0 \cancel{\text{cm}^3}) = 67.7 \text{ g}$
15. $V = \dfrac{m}{D} = (7.91 \text{ g}) \times \left(\dfrac{\text{cm}^3}{2.50 \text{ g}}\right) = 3.16 \text{ cm}^3$
16. $m = DV = \left(\dfrac{1.84 \text{ g}}{\cancel{\text{cm}^3}}\right) \times (7.62 \cancel{\text{cm}^3}) = 14.0 \text{ g}$
17. $(795 \cancel{\text{kg}})\left(\dfrac{1000 \text{ g}}{1 \cancel{\text{kg}}}\right)\left(\dfrac{\cancel{\text{cm}^3}}{0.788 \text{ g}}\right)\left(\dfrac{\text{m}^3}{10^6 \cancel{\text{cm}^3}}\right) = 1.01 \text{ m}^3$
18. $D = \dfrac{m}{V} = \dfrac{2580 \text{ g}}{(4.05 \text{ cm})(8.85 \text{ cm})(164 \text{ cm})}$ $= 0.439 \text{ g/cm}^3$
19. $D = \dfrac{m}{V} = \dfrac{m}{(\pi d^2/4)(L)} = \dfrac{51.6 \text{ g}}{\pi(0.622^2 \text{ cm}^2/4)(22.1 \text{ cm})}$ $= 7.68 \text{ g/cm}^3$
20. The first balance permits measurement within 0.1 g; the second balance permits measurement within 1 g.
21. **a.** 5000 mg
    b. 0.15 m
    c. 0.2 s
    d. 5 cm
    e. 0.06 kg
    f. 2000 μm
22. 21°C, 3 m of tape, 1.4 kg, 90 s

Chapter 3

5. **a.** 38 g/100 g H_2O
   b. 230 g/100 g H_2O
   c. 46 g/100 g H_2O
   d. 95 g/100 g H_2O
10. **a.** $(1980 \cancel{\text{J}})\left(\dfrac{1}{4.184 \cancel{\text{J}}/\text{cal}}\right) = 473 \text{ cal}$
    b. $(1.11 \cancel{\text{Cal}})(1000 \cancel{\text{cal}}/\cancel{\text{Cal}})(4.184 \text{ J}/\cancel{\text{cal}}) = 4640 \text{ J}$
    c. $(800 \cancel{\text{cal}})\left(\dfrac{1}{1000 \cancel{\text{cal}}/\text{Cal}}\right) = 0.8 \text{ Cal}$
    d. $(3.40 \cancel{\text{J}})\dfrac{1}{(4.184 \cancel{\text{J}}/\cancel{\text{cal}})(1000 \cancel{\text{cal}}/\text{Cal})}$ $= 0.000\,813 \text{ Cal or } 8.13 \times 10^{-4} \text{ Cal}$
    e. $(47.0 \cancel{\text{cal}})(4.184 \text{ J}/\cancel{\text{cal}}) = 197 \text{ J}$
11. $q = m(\Delta T)C_p$
    $q = (854 \cancel{\text{g}})(61.5 \cancel{\text{C°}})\left(\dfrac{4.18 \text{ J}}{\cancel{\text{g}}\cdot\cancel{\text{C°}}}\right)$
    $= 220\,000 \text{ J or } 2.20 \times 10^5 \text{ J}$
12. $q = m(\Delta T)C_p$
    $q = (96.7 \cancel{\text{g}})(37.5 \cancel{\text{C°}})\left(\dfrac{0.874 \text{ J}}{\cancel{\text{g}}\cdot\cancel{\text{C°}}}\right) = 3170 \text{ J or } 3.17 \times 10^3 \text{ J}$
13. $q = m(\Delta T)C_p$
    $q = (10.35 \cancel{\text{g}})(24.3 \cancel{\text{C°}})\left(\dfrac{0.856 \text{ J}}{\cancel{\text{g}}\cdot\cancel{\text{C°}}}\right) = 215 \text{ J}$
14. $q_{\text{lost}} = q_{\text{gained}}$
    $m(\Delta T)C_p = m(\Delta T)C_p$
    $(3.90 \cancel{\text{g}})(99.3°\text{C} - T_f)\left(\dfrac{0.903 \cancel{\text{J}}}{\cancel{\text{g}}\cdot\cancel{\text{C°}}}\right) =$
    $(10 \cancel{\text{g}})(T_f - 22.6°\text{C})\left(\dfrac{4.18 \cancel{\text{J}}}{\cancel{\text{g}}\cdot\cancel{\text{C°}}}\right)$
    $T_f = 28.6°\text{C}$
15. $q_{\text{lost}} = q_{\text{gained}}$
    $m(\Delta T)C_p = m(\Delta T)C_p$
    $(65.6 \cancel{\text{g}})(100.0 \text{ °C} - T_f)\left(\dfrac{0.231 \cancel{\text{J}}}{\cancel{\text{g}}\cdot\cancel{\text{C°}}}\right) =$
    $(25.0 \cancel{\text{g}})(T_f - 23.0 \text{ °C})\left(\dfrac{4.18 \cancel{\text{J}}}{\cancel{\text{g}}\cdot\cancel{\text{C°}}}\right)$
    $T_f = 32.7 \text{ °C}$
16. $q_{\text{lost}} = q_{\text{gained}}$
    $m(\Delta T)C_p = m(\Delta T)C_p$
    $(23.8 \cancel{\text{g}})(67.50 \cancel{\text{C°}})(C_p) = (50.0 \cancel{\text{g}})(8.5 \cancel{\text{C°}})\left(\dfrac{4.18 \text{ J}}{\text{g}\cdot\text{C°}}\right)$
    $C_p = 1.1 \text{ J/g}\cdot\text{C°}$

Chapter 4

1. 23 protons
2. 47
3. 11
4. rubidium
5. barium
6. **a.** 89 protons, 89 electrons, 132 neutrons
   b. 12 protons, 12 electrons, 13 neutrons
   c. 45 protons, 45 electrons, 60 neutrons
   d. 57 protons, 57 electrons, 76 neutrons

12. $\dfrac{5(176) + 19(177) + 27(178) + 14(179) + 35(180.0)}{100}$

– 179 u

13. $\dfrac{(92.21 \times 27.977) + (4.70 \times 28.976) + (3.09 \times 29.974)}{100}$

= 28.09 u

Chapter 5

6. The greatest number of electrons in a given energy level can be determined by finding the value of $2n^2$ where n is the energy level involved; for
   $n = 2,\ 2n^2 = 2(2)^2 = 8$
   $n = 3,\ 2n^2 = 2(3)^2 = 18$
   $n = 5,\ 2n^2 = 2(5)^2 = 50$
   $n = 7,\ 2n^2 = 2(7)^2 = 98$
7. **a.** 1 **b.** 3 **c.** 4
8. **a.** An *s* sublevel contains 1 orbital.
   b. A *p* sublevel contains 3 orbitals.
   c. A *d* sublevel contains 5 orbitals.
   d. An *f* sublevel contains 7 orbitals.
14.

1	$1s^1$	11	$1s^22s^22p^63s^1$
2	$1s^2$	12	$1s^22s^22p^63s^2$
3	$1s^22s^1$	13	$1s^22s^22p^63s^23p^1$
4	$1s^22s^2$	14	$1s^22s^22p^63s^23p^2$
5	$1s^22s^22p^1$	15	$1s^22s^22p^63s^23p^3$
6	$1s^22s^22p^2$	16	$1s^22s^22p^63s^23p^4$
7	$1s^22s^22p^3$	17	$1s^22s^22p^63s^23p^5$
8	$1s^22s^22p^4$	18	$1s^22s^22p^63s^23p^6$
9	$1s^22s^22p^5$	19	$1s^22s^22p^63s^23p^64s^1$
10	$1s^22s^22p^6$	20	$1s^22s^22p^63s^23p^64s^2$

15. **a.** $1s^22s^22p^63s^23p^64s^23d^8$ Ni:
    b. $1s^22s^22p^63s^23p^6$:Ar:
    c. $1s^22s^22p^63s^23p^4$ ·S:
    d. $1s^22s^22p^63s^23p^64s^23d^{10}4p^65s^24d^9$ Ag: (actually $4d^{10}5s^1$ Ag·)
    e. $1s^22s^22p^63s^23p^64s^1$ K·
    f. $1s^22s^22p^63s^23p^64s^23d^{10}4p^2$ ·Ge:

Chapter 6

6. **a.** nonmetal **b.** metal **c.** metalloid **d.** metal **e.** metal **f.** metal **g.** metal

Chapter 7

1. **a.** $CaCl_2$ **b.** $NaCN$ **c.** MgO **d.** BaO **e.** NaF **f.** $Al(NO_3)_3$ **g.** ZnI_2 **h.** $CoCO_3$
2. **a.** AgF **b.** NiS **c.** $CrBr_3$ **d.** $Pb_3(PO_4)_2$ **e.** $(NH_4)_2C_2O_4$ **f.** SrI_2 **g.** Li_2O
3. **a.** KH **b.** $Hg(CN)_2$ **c.** $ZnC_4H_4O_6$ **d.** $CdSiO_3$ **e.** $(NH_4)_2Cr_2O_7$ **f.** $Pb(NO_3)_2$ **g.** $Cu(ClO_4)_2$ **h.** $Na_2B_4O_7$
8. **a.** barium sulfide **b.** bismuth(III) iodide **c.** magnesium nitride **d.** lead(II) bromide **e.** zinc fluoride
9. **a.** calcium hydride **b.** sodium phosphide **c.** calcium sulfide **d.** thallium(I) iodide **e.** cobalt(II) bromide
10. **a.** barium chloride **b.** sodium bromide **c.** aluminum fluoride **d.** lithium carbonate **e.** potassium chloride **f.** mercury(II) iodide **g.** zinc nitrate **h.** barium hydroxide
11. **a.** diphosphorus pentoxide **b.** phosphorus pentachloride **c.** sulfur hexafluoride **d.** phosphorus trichloride
12. **a.** ammonium nitrate **b.** acetic acid or hydrogen acetate **c.** sodium phosphate **d.** hydrochloric acid or hydrogen chloride **e.** nitric acid or hydrogen nitrate **f.** copper(II) acetate **g.** potassium oxide **h.** sulfuric acid or hydrogen sulfate
13. **a.** 4 **b.** 7 **c.** 3 **d.** 9 **e.** 8 **f.** 5
14. **a.** hexane **b.** cyclobutane
15. **a.** NO_2 **b.** $C_3H_4O_3$ **c.** CH_3 **d.** CH_4 **e.** HgI **f.** C_4H_9
16. **a.** 1 **b.** 3 **c.** 2 **d.** 4 **e.** 6 **f.** 5 **g.** 3 **h.** 3

Chapter 8

1. $(0.143\ \text{h})\left(\dfrac{60\ \text{min}}{1\ \text{h}}\right)\left(\dfrac{60\ \text{s}}{1\ \text{min}}\right) = 515\ s$
2. $(0.84\ \text{m})\left(\dfrac{100\ \text{cm}}{1\ \text{m}}\right) = 84\ \text{cm}$
3. $(31.5\ \text{cg})\left(\dfrac{1\ \text{g}}{100\ \text{cg}}\right)\left(\dfrac{1000\ \text{mg}}{1\ \text{g}}\right) = 315\ \text{mg}$
4. $(65.22\ \text{mg})\left(\dfrac{1\ \text{g}}{1000\ \text{mg}}\right) = 0.065\ 22\ \text{g}$
5. $(531\ \text{cm}^3)\left(\dfrac{1\ \text{m}^3}{100^3\ \text{cm}^3}\right)\left(\dfrac{10^3\ \text{dm}^3}{1\ \text{m}^3}\right) = 0.531\ \text{dm}^3$
6. $(718\ \text{nm})\left(\dfrac{1\ \text{m}}{10^9\ \text{nm}}\right)\left(\dfrac{100\ \text{cm}}{1\ \text{m}}\right) = 7.18 \times 10^{-5}\ \text{cm}$
7. $\begin{array}{c|c} 0.032\ \text{g} & 1000\ \text{mg} \\ \hline & 1\ \text{g} \end{array} = 32\ \text{mg}$
8. $\begin{array}{c|c} 0.436\ \text{m}^3 & 100^3\ \text{cm}^3 \\ \hline & 1\ \text{m}^3 \end{array} = 436\ 000\ \text{cm}^3 \text{ or } 4.36 \times 10^5\ \text{cm}^3$
9. $\begin{array}{c|c} 302.1\ \text{mL} & 1\ \text{cm}^3 \\ \hline & 1\ \text{mL} \end{array} = 302.1\ \text{cm}^3$
10. $\begin{array}{c|c|c} 0.693\ \text{dm}^3 & 1\ \text{m}^3 & 100^3\ \text{cm}^3 \\ \hline & 10^3\ \text{dm}^3 & 1\ \text{m}^3 \end{array} = 693\ \text{cm}^3$
11. $\begin{array}{c|c|c} 9.06\ \text{km} & 1000\ \text{m} & 1\ \text{h} \\ \hline \text{h} & 1\ \text{km} & 60\ \text{min} \end{array} = 151\ \text{m/min}$
12. $\begin{array}{c|c} 0.307\ \text{mg} & 1\ \text{g} \\ \hline \text{cm}^3 & 1000\ \text{mg} \end{array} = 3.07 \times 10^{-4}\ \text{g/cm}^3$
13. $\begin{array}{c|c|c} 822\ \text{dm}^3 & 1\ \text{L} & 60\ \text{s} \\ \hline \text{s} & 1\ \text{dm}^3 & 1\ \text{min} \end{array} = 49\ 300\ \text{L/min}$
    or 4.93×10^4 L/min
14. $\begin{array}{c|c|c|c} 0.78\ \text{L} & 1000\ \text{mL} & 1\ \text{cm}^3 & 1\ \text{min} \\ \hline \text{min} & 1\ \text{L} & 1\ \text{mL} & 60\ \text{s} \end{array} = 13\ \text{cm}^3/\text{s}$
15. $\begin{array}{c|c|c|c|c} 0.848\ \text{kg} & 1000\ \text{g} & 1000\ \text{mg} & 1\ \text{L} & 1\ \text{mL} \\ \hline \text{L} & 1\ \text{kg} & 1\ \text{g} & 1000\ \text{mL} & 1\ \text{cm}^3 \end{array}$
    $= 848\ \text{mg/cm}^3$
16. $\begin{array}{c|c|c|c} 81.42\ \text{nm} & 1\ \text{m} & 100\ \text{cm} & 60\ \text{s} \\ \hline \text{s} & 10^9\ \text{nm} & 1\ \text{m} & 1\ \text{min} \end{array}$
    $= 4.885 \times 10^{-4}\ \text{cm/min}$
17. $\begin{array}{c|c|c|c} 7.56\ \text{mm}^3 & 1\ \text{m}^3 & 10^3\ \text{dm}^3 & 60\ \text{s} \\ \hline \text{s} & 1000^3\ \text{mm}^3 & 1\ \text{m}^3 & 1\ \text{min} \end{array}$
    $= 4.54 \times 10^{-4}\ \text{dm}^3/\text{min}$
18. $\begin{array}{c|c|c|c|c} 0.03\ \text{cm} & 1\ \text{m} & 1\ \text{km} & 60\ \text{s} & 60\ \text{min} \\ \hline \text{s} & 100\ \text{cm} & 1000\ \text{m} & 1\ \text{min} & 1\ \text{h} \end{array}$
    $= 0.001\ \text{km/h}$

19. $\frac{0.0775\ \cancel{cg}}{\cancel{cm^3}}\ \Big|\ \frac{1\ g}{100\ \cancel{cg}}\ \Big|\ \frac{100^3\ \cancel{cm^3}}{1\ m^3} = 775\ g/m^3$

20. $\frac{0.95\ \cancel{kg}}{\cancel{cm^3}}\ \Big|\ \frac{1000\ \cancel{g}}{1\ \cancel{kg}}\ \Big|\ \frac{1000\ mg}{1\ \cancel{g}}\ \Big|\ \frac{100^3\ \cancel{cm^3}}{1\ \cancel{m^3}}\ \Big|\ \frac{1\ \cancel{m^3}}{1000^3\ mm^3}$
$= 950\ mg/mm^3$ or $9.5 \times 10^2\ mg/mm^3$

21. $0.0753\ m^2$

22. $1.1 \times 10^6\ mm^2$

23. $7.4 \times 10^2\ g/dm^3$

24. 763 g

25. 9.40 kg

26. $1.33 \times 10^{-5}\ cm^2$

27. 3.59×10^{12} cm

32. a.

1 C atom	1 × 12.0	= 12.0 u
4 H atoms	4 × 1.01	= 4.04 u
1 O atom	1 × 16.0	= 16.0 u
		32.0 u

b.

2 C atoms	2 × 12.0	= 24.0 u
6 H atoms	6 × 1.01	= 6.06 u
		30.1 u

c.

12 C atoms	12 × 12.0	= 144 u
22 H atoms	22 × 1.01	= 22.2 u
11 O atoms	11 × 16.0	= 176 u
		342 u

d.

3 C atoms	3 × 12.0	= 36.0 u
8 H atoms	8 × 1.01	= 8.08 u
1 O atom	1 × 16.0	= 16.0 u
		60.1 u

33. a.

1 Ta atom	1 × 181	= 181 u
1 C atom	1 × 12.0	= 12.0 u
		193 u

b.

1 Al atom	1 × 27.0	= 27.0 u
1 N atom	1 × 14.0	= 14.0 u
		41.0 u

c.

4 P atoms	4 × 31.0	= 124 u
3 S atoms	3 × 32.1	= 96.3 u
		2.20×10^2 u

d.

3 Ca atoms	3 × 40.0	= 120.0 u
2 P atoms	2 × 31.0	= 62.0 u
8 O atoms	8 × 16.0	= 128 u
		3.10×10^2 u

e.

1 Ba atom	1 × 137	= 137 u
2 Cl atoms	2 × 35.5	= 71.0 u
2 O atoms	2 × 16.0	= 32.0 u
		2.40×10^2 u

34. $\frac{0.638\ \cancel{mol\ Ba(CN)_2}}{}\ \Big|\ \frac{189\ g\ Ba(CN)_2}{1\ \cancel{mol\ Ba(CN)_2}} = 121\ g\ Ba(CN)_2$

35. $\frac{50.4\ \cancel{g\ CaBr_2}}{}\ \Big|\ \frac{1\ mol\ CaBr_2}{2.00 \times 10^2\ \cancel{g\ CaBr_2}} = 0.252\ mol\ CaBr_2$

36. $\frac{1.26\ \cancel{mol\ NbI_5}}{}\ \Big|\ \frac{727.4\ g\ NbI_5}{1\ \cancel{mol\ NbI_5}} = 917\ g\ NbI_5$

37. $\frac{86.2\ \cancel{g\ C_2H_4}}{}\ \Big|\ \frac{1\ mol\ C_2H_4}{28.0\ \cancel{g\ C_2H_4}} = 3.08\ mol\ C_2H_4$

38. $\frac{0.943\ \cancel{mol}\ H_2O}{}\ \Big|\ \frac{6.02 \times 10^{23}\ molecules}{1\ \cancel{mol}}$
$= 5.68 \times 10^{23}$ molecules H_2O

39. $\frac{7.74 \times 10^{26}\ \cancel{formula\ units}\ Al_2O_3}{}\ \Big|\ \frac{1\ mol}{6.02 \times 10^{23}\ \cancel{formula\ units}} = 1.29 \times 10^3\ mol\ Al_2O_3$

40. $\frac{91.9\ \cancel{g\ NH_4IO_3}}{}\ \Big|\ \frac{1\ \cancel{mol}\ NH_4IO_3}{193\ \cancel{g\ NH_4IO_3}}\ \Big|\ \frac{6.02 \times 10^{23}\ formula\ units}{1\ \cancel{mol}}$
$= 2.87 \times 10^{23}$ formula units NH_4IO_3

41. $\frac{6.63 \times 10^{23}\ \cancel{molecules}\ C_6H_{12}O_6}{}\ \Big|\ \frac{1\ mol}{6.02 \times 10^{23}\ \cancel{molecules}}$
$= 1.10\ mol\ C_6H_{12}O_6$

42. $\frac{5.23\ \cancel{g\ Fe(NO_3)_2}}{100.00\ \cancel{cm^3}\ soln}\ \Big|\ \frac{1\ mol\ Fe(NO_3)_2}{1.80 \times 10^2\ \cancel{g\ Fe(NO_3)_2}}\ \Big|\ \frac{1000\ \cancel{cm^3}}{1\ dm^3}$
$= 0.291M\ Fe(NO_3)_2$

43. $\frac{8.55\ \cancel{g\ NH_4I}}{50.0\ \cancel{cm^3}\ soln}\ \Big|\ \frac{1\ mol\ NH_4I}{145\ \cancel{g\ NH_4I}}\ \Big|\ \frac{1000\ \cancel{cm^3}}{1\ dm^3}$
$= 1.18M\ NH_4I$

44. $\frac{9.94\ \cancel{g\ CoSO_4}}{2.50 \times 10^2\ \cancel{cm^3}\ soln}\ \Big|\ \frac{1\ mol\ CoSO_4}{155\ \cancel{g\ CoSO_4}}\ \Big|\ \frac{1000\ \cancel{cm^3}}{1\ dm^3}$
$= 0.257M\ CoSO_4$

45. $\frac{44.3\ \cancel{g\ Pb(ClO_4)_2}}{250.0\ \cancel{cm^3}\ soln}\ \Big|\ \frac{1\ mol\ Pb(ClO_4)_2}{406\ \cancel{g\ Pb(ClO_4)_2}}\ \Big|\ \frac{1000\ \cancel{cm^3}}{1\ dm^3}$
$= 0.436M\ Pb(ClO_4)_2$

46. $\frac{1.00\ \cancel{dm^3\ soln}}{}\ \Big|\ \frac{3.00\ \cancel{mol\ NiCl_2}}{1\ \cancel{dm^3\ soln}}\ \Big|\ \frac{1.30 \times 10^2\ g\ NiCl_2}{1\ \cancel{mol\ NiCl_2}}$
$= 3.90 \times 10^2\ g\ NiCl_2$

Dissolve 3.90×10^2 g $NiCl_2$ in enough water to make 1.00 dm^3 of solution.

47. $\frac{2.50 \times 10^2\ \cancel{cm^3\ soln}}{}\ \Big|\ \frac{4.00\ \cancel{mol\ CoCl_2}}{1\ \cancel{dm^3\ soln}}\ \Big|\ \frac{1.30 \times 10^2\ g\ CoCl_2}{1\ \cancel{mol\ CoCl_2}}\ \Big|\ \frac{1\ \cancel{dm^3}}{1000\ \cancel{cm^3}} = 1.30 \times 10^2\ g\ CoCl_2$

Dissolve 1.30×10^2 g $CoCl_2$ in enough water to make $2.50 \times 10^2\ cm^3$ of solution.

48. $\frac{0.500\ \cancel{dm^3\ soln}}{}\ \Big|\ \frac{1.50\ \cancel{mol\ AgF}}{1\ \cancel{dm^3\ soln}}\ \Big|\ \frac{127\ g\ AgF}{1\ \cancel{mol\ AgF}} = 95.3\ g\ AgF$

Dissolve 95.3 g AgF in enough water to make 0.500 dm^3 of solution.

49. $\frac{2.50 \times 10^2\ \cancel{cm^3\ soln}}{}\ \Big|\ \frac{0.002\ 00\ \cancel{mol\ Cd(IO_3)_2}}{1\ \cancel{dm^3\ soln}}\ \Big|\ \frac{462\ g\ Cd(IO_3)_2}{1\ \cancel{mol\ Cd(IO_3)_2}}\ \Big|\ \frac{1\ \cancel{dm^3}}{1000\ \cancel{cm^3}} = 0.231\ g\ Cd(IO_3)_2$

Dissolve 0.231 g $Cd(IO_3)_2$ in enough water to make $2.50 \times 10^2\ cm^3$ of solution.

50. NH_3:

1 N atom	1 × 14.0 u =	14.0 u
3 H atoms	3 × 1.01 u =	3.03 u
		17.0 u

Percentage of N in $NH_3 = \frac{mass\ N}{mass\ NH_3} \times 100$

$= \frac{14.0\ \cancel{u}}{17.0\ \cancel{u}} \times 100 = 82.4\%$

$CO(NH_2)_2$:

1 C atom	1×12.0 u =	12.0 u
1 O atom	1×16.0 u =	16.0 u
2 N atoms	2×14.0 u =	28.0 u
4 H atoms	4×1.01 u =	4.04 u
		60.0 u

$$\text{Percentage of N in } CO(NH_2)_2 = \frac{\text{mass 2N}}{\text{mass } CO(NH_2)_2} \times 100$$

$$= \frac{28.0\ \cancel{u}}{60.0\ \cancel{u}} \times 100 = 46.7\%$$

51.

2 Al atoms	2×27.0 u =	54.0 u
3 S atoms	3×32.1 u =	96.3 u
		1.50×10^2 u

$$\text{Percentage of Al} = \frac{\text{mass 2Al}}{\text{mass } Al_2S_3} \times 100$$

$$= \frac{54.0\ \cancel{u}}{1.50 \times 10^2\ \cancel{u}} \times 100 = 36.0\%$$

$$\text{Percentage of S} = \frac{\text{mass 3S}}{\text{mass } Al_2S_3} \times 100$$

$$= \frac{96.3\ \cancel{u}}{1.50 \times 10^2\ \cancel{u}} \times 100 = 64.2\%$$

52.

1 Ni atom	1×58.7 u =	58.7 u
2 I atoms	2×127 u =	254 u
		313 u

$$\text{Percentage of Ni} = \frac{\text{mass Ni}}{\text{mass } NiI_2} \times 100$$

$$= \frac{58.7\ \cancel{u}}{313\ \cancel{u}} \times 100 = 18.8\%$$

$$\text{Percentage of I} = \frac{\text{mass 2I}}{\text{mass } NiI_2} \times 100$$

$$= \frac{254\ \cancel{u}}{313\ \cancel{u}} \times 100 = 81.2\%$$

53.

1 Ca atom	1×40.1 u =	40.1 u
2 C atoms	2×12.0 u =	24.0 u
2 N atoms	2×14.0 u =	28.0 u
		92.1 u

$$\text{Percentage of Ca} = \frac{\text{mass Ca}}{\text{mass } Ca(CN)_2} \times 100$$

$$= \frac{40.1\ \cancel{u}}{92.1\ \cancel{u}} \times 100 = 43.5\%$$

$$\text{Percentage of C} = \frac{\text{mass 2C}}{\text{mass } Ca(CN)_2} \times 100$$

$$= \frac{24.0\ \cancel{u}}{92.1\ \cancel{u}} \times 100 = 26.1\%$$

$$\text{Percentage of N} = \frac{\text{mass 2N}}{\text{mass } Ca(CN)_2} \times 100$$

$$= \frac{28.0\ \cancel{u}}{92.1\ \cancel{u}} \times 100 = 30.4\%$$

54.

1 Cu atom	1×63.5 u =	63.5 u
2 Cl atoms	2×35.5 u =	71.0 u
8 O atoms	8×16.0 u =	128 u
		263 u

$$\text{Percentage of Cu} = \frac{\text{mass Cu}}{\text{mass } Cu(ClO_4)_2} \times 100$$

$$= \frac{63.5\ \cancel{u}}{263\ \cancel{u}} \times 100 = 24.1\%$$

$$\text{Percentage of Cl} = \frac{\text{mass 2Cl}}{\text{mass } Cu(ClO_4)_2} \times 100$$

$$= \frac{71.0\ \cancel{u}}{263\ \cancel{u}} \times 100 = 27.0\%$$

$$\text{Percentage of O} = \frac{\text{mass 8O}}{\text{mass } Cu(ClO_4)_2} \times 100$$

$$= \frac{128\ \cancel{u}}{263\ \cancel{u}} \times 100 = 48.7\%$$

55.

2 N atoms	2×14.0 u =	28.0 u
8 H atoms	8×1.01 u =	8.08 u
2 C atoms	2×12.0 u =	24.0 u
4 O atoms	4×16.0 u =	64.0 u
		124 u

$$\text{Percentage of N} = \frac{\text{mass 2N}}{\text{mass } (NH_4)_2C_2O_4} \times 100$$

$$= \frac{28.0\ \cancel{u}}{124\ \cancel{u}} \times 100 = 22.6\%$$

$$\text{Percentage of H} = \frac{\text{mass 8H}}{\text{mass } (NH_4)_2C_2O_4} \times 100$$

$$= \frac{8.08\ \cancel{u}}{124\ \cancel{u}} \times 100 = 6.52\%$$

$$\text{Percentage of C} = \frac{\text{mass 2C}}{\text{mass } (NH_4)_2C_2O_4} \times 100$$

$$= \frac{24.0\ \cancel{u}}{124\ \cancel{u}} \times 100 = 19.4\%$$

$$\text{Percentage of O} = \frac{\text{mass 4O}}{\text{mass } (NH_4)_2C_2O_4} \times 100$$

$$= \frac{64.0\ \cancel{u}}{124\ \cancel{u}} \times 100 = 51.6\%$$

56. $$\frac{1.67\ \cancel{\text{g Ce}}}{} \Bigg| \frac{1\ \text{mol Ce}}{1.40 \times 10^2\ \cancel{\text{g Ce}}} = 0.0119\ \text{mol Ce}$$

$$\frac{4.54\ \cancel{\text{g I}}}{} \Bigg| \frac{1\ \text{mol I}}{127\ \cancel{\text{g I}}} = 0.0357\ \text{mol I}$$

$0.0357/0.0119 \approx 3$; 1:3 ratio; $\therefore CeI_3$

57. $$\frac{0.556\ \cancel{\text{g C}}}{} \Bigg| \frac{1\ \text{mol C}}{12.0\ \cancel{\text{g C}}} = 0.0463\ \text{mol C}$$

$$\frac{0.0933\ \cancel{\text{g H}}}{} \Bigg| \frac{1\ \text{mol H}}{1.01\ \cancel{\text{g H}}} = 0.0924\ \text{mol H}$$

$0.0924/0.0463 \approx 2$; 1:2 ratio; $\therefore CH_2$

58. Assume 100.0 g sample.

$$\frac{68.8\ \cancel{\text{g C}}}{} \Bigg| \frac{1\ \text{mol C}}{12.0\ \cancel{\text{g C}}} = 5.73\ \text{mol C}$$

$$\frac{4.95\ \cancel{\text{g H}}}{} \Bigg| \frac{1\ \text{mol H}}{1.01\ \cancel{\text{g H}}} = 4.90\ \text{mol H}$$

$$\frac{26.2\ \cancel{\text{g O}}}{} \Bigg| \frac{1\ \text{mol O}}{16.0\ \cancel{\text{g O}}} = 1.64\ \text{mol O}$$

$5.73/1.64 \approx 3.5$; $4.90/1.64 \approx 3$; 7:6:2 ratio; $\therefore C_7H_6O_2$

59. Assume 100.0 g sample

$$\frac{9.93\ \cancel{g\ C}}{} \left| \frac{1\ mol\ C}{12.0\ \cancel{g\ C}} \right. = 0.828\ mol\ C$$

$$\frac{58.6\ \cancel{g\ Cl}}{} \left| \frac{1\ mol\ Cl}{35.5\ \cancel{g\ Cl}} \right. = 1.65\ mol\ Cl$$

$$\frac{31.4\ \cancel{g\ F}}{} \left| \frac{1\ mol\ F}{19.0\ \cancel{g\ F}} \right. = 1.65\ mol\ F$$

$1.65/0.828 \approx 2$; 1:2:2 ratio; $\therefore CCl_2F_2$

60. $CH = 12.0 + 1.01 = 13.0\ u$

$$\frac{78\ \cancel{u}}{13.0\ \cancel{u}} = 6.0,\ 6(CH) = C_6H_6$$

61. $CHOCl = 12.0 + 1.01 + 16.0 + 35.5 = 64.5\ u$

$$\frac{129\ \cancel{u}}{64.5\ \cancel{u}} = 2.00$$

$2(CHOCl) = C_2H_2O_2Cl_2$ or $HCCl_2COOH$

62. $CClN = 12.0 + 35.5 + 14.0 = 61.5\ u$

$$\frac{184\ \cancel{u}}{61.5\ \cancel{u}} = 3.00,\ 3(CClN) = C_3Cl_3N_3$$

63. Assume 100.0 g sample.

$$\frac{40.9\ \cancel{g\ C}}{} \left| \frac{1\ mol\ C}{12\ \cancel{g\ C}} \right. = 3.41\ mol\ C$$

$$\frac{4.58\ \cancel{g\ H}}{} \left| \frac{1\ mol\ H}{1.01\ \cancel{g\ H}} \right. = 4.53\ mol\ H$$

$$\frac{54.5\ \cancel{g\ O}}{} \left| \frac{1\ mol\ O}{16.0\ \cancel{g\ O}} \right. = 3.41\ mol\ O$$

$4.53/3.41 = 1.33$, 3:4:3 ratio
$\therefore C_3H_4O_3$ is the empirical formula.
$C_3H_4O_3 = 36.0 + 4.04 + 48.0 = 88.0\ u$
$176\ u/88.0\ u = 2$
$\therefore C_6H_8O_6$ is the molecular formula.

64. Assume 100.0 g sample.

$$\frac{60.0\ g}{} \left| \frac{1\ mol}{12.0\ g} \right. = 5.00\ mol\ C$$

$$\frac{4.48\ g}{} \left| \frac{1\ mol}{1.01\ g} \right. = 4.44\ mol\ H$$

$$\frac{35.5\ g}{} \left| \frac{1\ mol}{16.0\ g} \right. = 2.22\ mol\ O$$

$$\frac{5.00}{2.22} = 2.25,\ \frac{4.44}{2.22} = 2.00,\ \frac{2.22}{2.22} = 1.00$$

multiply by 4 to get $C_9H_8O_4$; formula mass $= 1.80 \times 10^2\ u$
$\therefore C_9H_8O_4$ is also the molecular formula.

65. **a.** sodium thiosulfate pentahydrate
b. calcium sulfate dihydrate

66. $Na_2B_4O_7 \cdot 10H_2O$ (borax)

67. Li_2SiF_6: molecular mass $= 2(6.94) + 28.1 + 6(19.0) = 156\ u$

$$\frac{0.391\ \cancel{g\ Li_2SiF_6}}{} \left| \frac{1\ mol}{156\ \cancel{g\ Li_2SiF_6}} \right. = 0.002\ 51\ mol$$

H_2O: molecular mass $= 2(1.01) + 16.0 = 18.0\ u$

$$\frac{0.0903\ \cancel{g\ H_2O}}{} \left| \frac{1\ mol}{18.0\ \cancel{g\ H_2O}} \right. = 0.005\ 02\ mol$$

$0.005\ 02/0.002\ 51 = 2.00$; 1:2 ratio
$\therefore Li_2SiF_6 \cdot 2H_2O$

68.

$$\frac{0.737\ \cancel{g\ MgSO_3}}{} \left| \frac{1\ mol}{104\ \cancel{g\ MgSO_3}} \right. = 0.007\ 09\ mol$$

$$\frac{0.763\ \cancel{g\ H_2O}}{} \left| \frac{1\ mol\ H_2O}{18.0\ \cancel{g\ H_2O}} \right. = 0.0424\ mol\ H_2O$$

$0.0424/0.007\ 09 \approx 6.00$; 1:6 ratio
$\therefore MgSO_3 \cdot 6H_2O$

69. Assume 100.0 g of sample.
76.9 g $CaSO_3$, 23.1 g H_2O

$$\frac{76.9\ \cancel{g\ CaSO_3}}{} \left| \frac{1\ mol}{120\ \cancel{g\ CaSO_3}} \right. = 0.641\ mol$$

$$\frac{23.1\ g}{} \left| \frac{1\ mol\ \cancel{H_2O}}{18.0\ \cancel{g\ H_2O}} \right. = 1.28\ mol$$

$1.28/0.641 \approx 2.00$; 1:2 ratio
$\therefore CaSO_3 \cdot 2H_2O$

70. Assume 100.0 g of material.
$\therefore$ 89.2 g $BaBr_2$, 10.8 g H_2O

$$\frac{89.2\ \cancel{g\ BaBr_2}}{} \left| \frac{1\ mol}{297\ \cancel{g\ BaBr_2}} \right. = 0.300\ mol$$

$$\frac{10.8\ \cancel{g\ H_2O}}{} \left| \frac{1\ mol}{18.0\ \cancel{g\ H_2O}} \right. = 0.600\ mol$$

$0.600/0.300 = 2.00$; 1:2 ratio
$\therefore BaBr_2 \cdot 2H_2O$

Chapter 9

1. balanced
2. $Al(NO_3)_3 + 3NaOH \rightarrow Al(OH)_3 + 3NaNO_3$
3. $2KNO_3 \rightarrow 2KNO_2 + O_2$
4. $2Fe + 3H_2SO_4 \rightarrow Fe_2(SO_4)_3 + 3H_2$
5. $3O_2 + CS_2 \rightarrow CO_2 + 2SO_2$
6. balanced
7. $3Mg + N_2 \rightarrow Mg_3N_2$
8. $CuCO_3 \rightarrow CuO + CO_2$
9. $2Na + 2H_2O \rightarrow 2NaOH + H_2$
10. $2Cu + S \rightarrow Cu_2S$
11. $2AgNO_3 + H_2SO_4 \rightarrow Ag_2SO_4 + 2HNO_3$
12. $2C_2H_6 + 7O_2 \rightarrow 4CO_2 + 6H_2O$
13. single displacement
14. decomposition
15. synthesis
16. combustion
17. single displacement
18. double displacement
19. $2Al + 3Cu(NO_3)_2 \rightarrow 3\underline{Cu} + 2\underline{Al(NO_3)_3}$
20. $2Hg + O_2 \rightarrow 2\underline{HgO}$
21. $H_2SO_4 + 2KOH \rightarrow \underline{K_2SO_4} + 2\underline{H_2O}$
22. $2C_5H_{10} + 15O_2 \rightarrow 10\underline{CO_2} + 10\underline{H_2O}$

28.

$$\frac{12.5\ \cancel{g\ C_6H_{12}O_6}}{} \left| \frac{1\ \cancel{mol\ C_6H_{12}O_6}}{1.80 \times 10^2\ \cancel{g\ C_6H_{12}O_6}} \right| \frac{6\ \cancel{mol\ O_2}}{1\ \cancel{mol\ C_6H_{12}O_6}} \left| \frac{32.0\ g\ O_2}{1\ \cancel{mol\ O_2}} \right. = 13.3\ g\ O_2$$

29. $\frac{3.41\ \cancel{g\ H_2}}{} \times \frac{1\ \cancel{mol\ H_2}}{2.02\ \cancel{g\ H_2}} \times \frac{2\ \cancel{mol\ NH_3}}{3\ \cancel{mol\ H_2}} \times \frac{17.0\ g\ NH_3}{1\ \cancel{mol\ NH_3}}$
$= 19.1\ g\ NH_3$

30. $3KClO_3 \rightarrow 2KCl + 3O_2$

$\frac{80.5\ \cancel{g\ O_2}}{} \times \frac{1\ \cancel{mol\ O_2}}{32.0\ \cancel{g\ O_2}} \times \frac{2\ \cancel{mol\ KCl}}{3\ \cancel{mol\ O_2}} \times \frac{74.6\ g\ KCl}{1\ \cancel{mol\ KCl}}$
$= 125\ g\ KCl$

31. $2Al + 6HCl \rightarrow 2AlCl_3 + 3H_2$

$\frac{9.23\ \cancel{g\ Al}}{} \times \frac{1\ \cancel{mol\ Al}}{27.0\ \cancel{g\ Al}} \times \frac{3\ \cancel{mol\ H_2}}{2\ \cancel{mol\ Al}} \times \frac{2.02\ g\ H_2}{1\ \cancel{mol\ H_2}} = 1.04\ g\ H_2$

32. a. $\frac{2.50\ \cancel{g\ K_2PtCl_4}}{} \times \frac{1\ \cancel{mol\ K_2PtCl_4}}{415\ \cancel{g\ K_2PtCl_4}} \times \frac{1\ \cancel{mol\ PtCl_2(NH_3)_2}}{1\ \cancel{mol\ K_2PtCl_4}} \times \frac{3.00 \times 10^2\ g\ PtCl_2(NH_3)_2}{1\ \cancel{mol\ PtCl_2(NH_3)_2}} = 1.81\ g\ Pt(NH_3)_2Cl_2$

b. $\frac{2.50\ \cancel{g\ K_2PtCl_4}}{} \times \frac{1\ \cancel{mol\ K_2PtCl_4}}{415\ \cancel{g\ K_2PtCl_4}} \times \frac{2\ \cancel{mol\ NH_3}}{1\ \cancel{mol\ K_2PtCl_4}} \times \frac{17.0\ g\ NH_3}{1\ \cancel{mol\ NH_3}} = 0.205\ g\ NH_3$

33. percentage yield $= \frac{39.7\ g}{65.6\ g} \times \frac{100}{} = 60.5\%$

34. $N_2(g) + 3H_2(g) \rightarrow 2NH_3(g)$

$\frac{5.50\ \cancel{g\ H_2}}{} \times \frac{1\ \cancel{mol\ H_2}}{2.02\ \cancel{g\ H_2}} \times \frac{2\ \cancel{mol\ NH_3}}{3\ \cancel{mol\ H_2}} \times \frac{17.0\ g\ NH_3}{1\ \cancel{mol\ NH_3}}$
$= 30.9\ g\ NH_3$, theoretical yield

percentage yield $= \frac{20.4\ \cancel{g\ NH_3}}{30.9\ \cancel{g\ NH_3}} \times \frac{100}{} = 66.0\%$

35. $NH_3 + HBr \rightarrow NH_4Br + 188.32\ kJ$

$\frac{193\ \cancel{g\ NH_4Br}}{} \times \frac{1\ \cancel{mol\ NH_4Br}}{97.9\ \cancel{g\ NH_4Br}} \times \frac{188.32\ kJ}{1\ \cancel{mol\ NH_4Br}} = 371\ kJ$

$q_r = -371\ kJ$

36. $\frac{0.772\ \cancel{g\ CoCO_3}}{} \times \frac{1\ \cancel{mol\ CoCO_3}}{119\ \cancel{g\ CoCO_3}} \times \frac{81.6\ kJ}{1\ \cancel{mol\ CoCO_3}} = 0.529\ kJ$

$q_r = 0.529\ kJ$, or 529 J

37. $\frac{0.0663\ \cancel{g\ Br_2}}{} \times \frac{1\ \cancel{mol\ Br_2}}{160\ \cancel{g\ Br_2}} \times \frac{100.18\ kJ}{1\ \cancel{mol\ Br_2}} = 0.0415\ kJ$

$q_r = -0.0415\ kJ$, or $-41.5\ J$

38. $\frac{6.18\ \cancel{kJ}}{} \times \frac{1\ \cancel{mol\ CrO_3}}{5.4\ \cancel{kJ}} \times \frac{99.99\ g\ CrO_3}{1\ \cancel{mol\ CrO_3}} = 110\ g\ CrO_3$

Chapter 10

1. **a.** Ar **b.** B **c.** P **d.** Cl **e.** Ca
2. **a.** O **b.** Mg^{2+} **c.** Te **d.** Ti^{4+}

Chapter 12

1. Electronegativity values:
Sb: 1.82, F: 4.10, In: 1.49, Se: 2.48
∴ In, Sb, Se, F
indium, antimony, selenium, fluorine

2. Electronegativity values:
Fr: 0.86, Ga: 1.82, Ge: 2.02, P: 2.06, Zn: 1.66
∴ Fr, Zn, Ga, Ge, P
francium, zinc, gallium, germanium, phosphorus

3. **a.** $|1.47 - 1.74| = 0.27$ covalent
b. $|0.97 - 3.50| = 2.53$ ionic
c. $|2.50 - 2.20| = 0.30$ covalent
d. $|0.97 - 2.44| = 1.47$ covalent
e. $|1.04 - 2.06| = 1.02$ covalent
f. $|2.01 - 1.01| = 1.00$ covalent
g. $|1.04 - 2.83| = 1.79$ ionic
h. $|4.10 - 2.44| = 1.66$ covalent
i. $|2.74 - 0.89| = 1.85$ ionic

4. **a.** $|2.20 - 2.21| = 0.01$ covalent
b. $|1.96 - 1.47| = 0.49$ covalent
c. $|1.70 - 4.10| = 2.40$ ionic
d. $|2.83 - 2.01| = 0.82$ covalent
e. $|2.74 - 1.08| = 1.66$ covalent
f. $|1.04 - 4.10| = 3.06$ ionic

9. **a.** sum of covalent radii = 118 pm + 99.5 pm = 218 pm
b. 37.1 pm + 133 pm = 170 pm
c. 70 pm + 110 pm = 180 pm
d. 119 pm + 103 pm = 222 pm
e. 82 pm + 71.5 pm = 154 pm

Chapter 13

1. **a.** H:Te: with H below **b.** :F:P:F: with F below **c.** :I:N:I: with I below **d.** :Br:C:Br: with Br above and below

2. 1c—one unshared; 1d—none

3. **a.** one unshared pair; < 109.5°
b. two unshared pairs; < 109.5°
c. no unshared pairs; > 109.5°
d. Form is trigonal bipyramidal: As in center, one F atom at each end of polar axis, and three F atoms on equator.
Five shared pairs; no unshared pairs; equatorial-equatorial angles are > 109.5°, axial-equatorial angle is < 109.5°.
e. There are four shared pairs of electrons. The outer shell of Te is expanded to 10 electrons. The form is nearly tetrahedral, with some angles > 109.5°, and some angles < 109.5°.
f. The Xe atom shares all eight of its outer electrons with the four O atoms. The angle is 109.5°.

9. **a.** propyne
b. cyclobutene

10. **a.** **b.**

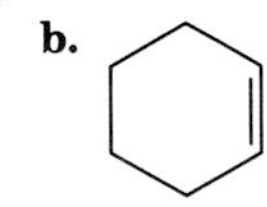

Chapter 14

1.

	Bond	*Electronegativity difference*
a.	Al—P	$|1.47 - 2.06| = 0.59$
b.	Cl—C	$|2.83 - 2.50| = 0.33$
c.	Mo—Te	$|1.30 - 2.01| = 0.71$
d.	H—S	$|2.20 - 2.44| = 0.24$
e.	P—S	$|2.06 - 2.44| = 0.38$
f.	Cl—Si	$|2.83 - 1.74| = 1.09$

In order of decreasing bond polarity: f, c, a, e, b, d

2. a. HBr is polar because of the concentration of negative charge near the Br atom and of positive charge near the H atom.
   b. SO_2 is polar because oxygen has a greater electronegativity than sulfur and the molecule is asymmetric.
7. a. $[Cd(NH_3)_2]^{2+}$ diamminecadmium ion
   b. $[Cr(H_2O)_4]^{2+}$ tetraaquachromium(II) ion
   c. $[Co(NH_3)_6]^{2+}$ hexaamminecobalt(II) ion
   d. $[Cu(NH_3)_4]^{2+}$ tetraamminecopper(II) ion
   e. $[PtBr(NH_3)_3]^{+}$ triamminebromoplatinum(II) ion
8. a. hexaiodoplatinate(IV) ion
   b. hexafluorogermanate(IV) ion
   c. hexahydroxoantimonate(V) ion
   d. tetrachloropalladate(II) ion
   e. hexacyanoferrate(II) ion
9. a. hexafluoroantimonate(V) ion
   b. tetrafluoroborate ion
   c. hexafluoroaluminate ion
   d. tetraammineaquachlorocobalt(III) ion
   e. tetraamminecadmium ion
10. a. $[Pt(NH_3)_4]^{2+}$
    b. $[Co(NH_3)_6]^{3+}$
    c. $[Ir(H_2O)_6]^{3+}$
    d. $[Pd(NH_3)_4]^{2+}$
11. a. $[PdCl_6]^{2-}$
    b. $[PtCl_3(NH_3)]^{-}$
    c. $[AuI_4]^{-}$
    d. $[Au(CN)_4]^{-}$
12. a. $[W(CN)_8]^{3-}$
    b. $[Sb(OH)_6]^{-}$
    c. $[Fe(CO)_5(NO)]^{2+}$
    d. $[Co(CO)_4]^{+}$
13. a. tetraamminediaquanickel(II) nitrate
    b. pentaamminechlorocobalt(III) chloride
    c. potassium hexacyanoferrate(III)
    d. potassium hexahydroxostannate(IV)
    e. diamminesilverperrhenate
14. a. $[Co(H_2O)_6]I_2$
    b. $[Co(NH_3)_6]Cl_2$
    c. $K_2[Ni(CN)_4]$
    d. $PdCl_2(NH_3)_2$
    e. $Ru(CO)_5$

Chapter 15

1. $62\ \text{mm} \times \frac{1\ \text{kPa}}{7.50\ \text{mm}} = 8.3\ \text{kPa}$

   $97.7\ \text{kPa} - 8.3\ \text{kPa} = 89.4\ \text{kPa}$

2. $691\ \text{mm} \times \frac{1\ \text{kPa}}{7.50\ \text{mm}} = 92.1\ \text{kPa}$

3. $38\ \text{mm} \times \frac{1\ \text{kPa}}{7.50\ \text{mm}} = 5.1\ \text{kPa}$

   $96.3\ \text{kPa} + 5.1\ \text{kPa} = 101.4\ \text{kPa}$

4. $86.0\ \text{mm} \times \frac{1\ \text{kPa}}{7.50\ \text{mm}} = 11.5\ \text{kPa}$

9. a. $86\ \text{K} - 273 = -187°\text{C}$
   b. $191\ \text{K} - 273 = -82°\text{C}$
   c. $533\ \text{K} - 273 = 2.60 \times 10^{2}°\text{C}$
   d. $321\ \text{K} - 273 = 48°\text{C}$
   e. $894\ \text{K} - 273 = 621°\text{C}$
10. a. $23°\text{C} + 273 = 296\ \text{K}$
    b. $58°\text{C} + 273 = 331\ \text{K}$
    c. $-90°\text{C} + 273 = 183\ \text{K}$
    d. $18°\text{C} + 273 = 291\ \text{K}$
    e. $25°\text{C} + 273 = 298\ \text{K}$
11. a. $872\ \text{K} - 273 = 599°\text{C}$
    b. $690\ \text{K} - 273 = 417°\text{C}$
    c. $384\ \text{K} - 273 = 111°\text{C}$
    d. $20\ \text{K} - 273 = -253°\text{C}$
    e. $60\ \text{K} - 273 = -213°\text{C}$
12. a. N_2, because it has the lowest molecular mass.

Chapter 17

1. a. 50°C b. 99°C c. 70°C d. 48°C
2. $CHCl_3$
3. a. $T_c = 33\ \text{K}$
   b. $P_c = 1290\ \text{kPa}$
   c. $T_t = 14\ \text{K}$
   d. $P_t = 5\ \text{kPa}$
   e. $T_m = 14\ \text{K}$
   f. $T_b = 18\ \text{K}$
4. a. gas b. solid c. liquid d. liquid
5. a. $46.0\ \text{g} \times 58.0\ \text{C°} \times \frac{2.06\ \text{J}}{\text{g·C°}} = 5.50 \times 10^{3}\ \text{J}$

   b. $46.0\ \text{g} \times \frac{334\ \text{J}}{\text{g}} = 15\ 400\ \text{J}$

   c. $46.0\ \text{g} \times 100\ \text{C°} \times \frac{4.18\ \text{J}}{\text{g·C°}} = 19\ 200\ \text{J}$

   d. $46.0\ \text{g} \times \frac{2260\ \text{J}}{\text{g}} = 104\ 000\ \text{J}$

   e. $46.0\ \text{g} \times 14\ \text{C°} \times \frac{2.02\ \text{J}}{\text{g·C°}} = 1300\ \text{J}$

6. $25.4\ \text{g} \times \frac{61.7\ \text{J}}{\text{g}} = 1570\ \text{J}$

7. $4.24\ \text{g} \times \frac{162\ \text{J}}{\text{g}} = 687\ \text{J}$

8. $5.58\ \text{kg} \times \frac{1000\ \text{g}}{1\ \text{kg}} \times \frac{0.4494\ \text{J}}{\text{g·C°}} \times 980.0\ \text{C°} \times \frac{1\ \text{kJ}}{1000\ \text{J}} = 2460\ \text{kJ}$

9. $70.0\ \text{g} \times 64\ \text{C°} \times \frac{3.06\ \text{J}}{\text{g·C°}} = 9200\ \text{J}$

   $70.0\ \text{g} \times \frac{334\ \text{J}}{\text{g}} = 23\ 400\ \text{J}$

   $70.0\ \text{g} \times 100.0\ \text{C°} \times \frac{4.18\ \text{J}}{\text{g·C°}} = 29\ 300\ \text{J}$

   $70.0\ \text{g} \times \frac{2260\ \text{J}}{\text{g}} = 158\ 000\ \text{J}$

   $70.0\ \text{g} \times 422\ \text{C°} \times \frac{2.02\ \text{J}}{\text{g·C°}} = 59\ 700\ \text{J}$

   $9200\ \text{J} + 23\ 400\ \text{J} + 29\ 300\ \text{J} + 158\ 000\ \text{J} + 59\ 700\ \text{J} = 2.80 \times 10^{5}\ \text{J}$

10. $28.9\ \text{g} \times (1085 - 25)\ \text{C°} \times \frac{0.384\ 52\ \text{J}}{\text{g·C°}} = 11\ 800\ \text{J} = 1.18 \times 10^{4}\ \text{J}$

Chapter 18

1. $\dfrac{4.41\text{ dm}^3 \mid 94.2\text{ kPa}}{\mid 101.3\text{ kPa}} = 4.10\text{ dm}^3$

2. $\dfrac{101\text{ kPa} \mid 5.0\text{ dm}^3}{\mid 10.0\text{ dm}^3} = 51\text{ kPa}$

3. a. $\dfrac{844\text{ cm}^3 \mid 98.5\text{ kPa}}{\mid 101.3\text{ kPa}} = 821\text{ cm}^3$

   b. $\dfrac{273\text{ cm}^3 \mid 59.4\text{ kPa}}{\mid 101.3\text{ kPa}} = 1.60 \times 10^2\text{ cm}^3$

   c. $\dfrac{116\text{ m}^3 \mid 90.0\text{ kPa}}{\mid 101.3\text{ kPa}} = 103\text{ m}^3$

   d. $\dfrac{77.0\text{ m}^3 \mid 105.9\text{ kPa}}{\mid 101.3\text{ kPa}} = 80.5\text{ m}^3$

4. a. $\dfrac{338\text{ cm}^3 \mid 86.1\text{ kPa}}{\mid 104.0\text{ kPa}} = 2.80 \times 10^2\text{ cm}^3$

   b. $\dfrac{0.873\text{ m}^3 \mid 94.3\text{ kPa}}{\mid 102.3\text{ kPa}} = 0.805\text{ m}^3$

   c. $\dfrac{31.5\text{ cm}^3 \mid 97.8\text{ kPa}}{\mid 82.3\text{ kPa}} = 37.4\text{ cm}^3$

   d. $\dfrac{524\text{ cm}^3 \mid 110.0\text{ kPa}}{\mid 104.5\text{ kPa}} = 552\text{ cm}^3$

5. $P_{gas} = 101.1\text{ kPa} - 8.6\text{ kPa} = 92.5\text{ kPa}$

   $\dfrac{596\text{ cm}^3 \mid 92.5\text{ kPa}}{\mid 101.3\text{ kPa}} = 544\text{ cm}^3$

6. a. $P_{gas} = 93.3\text{ kPa} - 1.6\text{ kPa} = 91.7\text{ kPa}$

   $\dfrac{888\text{ cm}^3 \mid 91.7\text{ kPa}}{\mid 101.3\text{ kPa}} = 804\text{ cm}^3$

   b. $77.5\text{ kPa} - 1.8\text{ kPa} = 75.7\text{ kPa}$

   $\dfrac{30.0\text{ cm}^3 \mid 75.7\text{ kPa}}{\mid 101.3\text{ kPa}} = 22.4\text{ cm}^3$

   c. $82.4\text{ kPa} - 2.1\text{ kPa} = 80.3\text{ kPa}$

   $\dfrac{34.0\text{ m}^3 \mid 80.3\text{ kPa}}{\mid 101.3\text{ kPa}} = 27.0\text{ m}^3$

   d. $78.3\text{ kPa} - 1.4\text{ kPa} = 76.9\text{ kPa}$

   $\dfrac{384\text{ cm}^3 \mid 76.9\text{ kPa}}{\mid 101.3\text{ kPa}} = 292\text{ cm}^3$

   e. $87.3\text{ kPa} - 3.6\text{ kPa} = 83.7\text{ kPa}$

   $\dfrac{8.23\text{ m}^3 \mid 83.7\text{ kPa}}{\mid 101.3\text{ kPa}} = 6.80\text{ m}^3$

7. $\dfrac{60.0\text{ cm}^3 \mid 273\text{ K}}{\mid 309\text{ K}} = 53.0\text{ cm}^3$

8. a. $\dfrac{617\text{ cm}^3 \mid 273\text{ K}}{\mid 282\text{ K}} = 597\text{ cm}^3$

   b. $\dfrac{609\text{ cm}^3 \mid 273\text{ K}}{\mid 356\text{ K}} = 467\text{ cm}^3$

   c. $\dfrac{942\text{ cm}^3 \mid 273\text{ K}}{\mid 295\text{ K}} = 872\text{ cm}^3$

   d. $\dfrac{7.12\text{ m}^3 \mid 273\text{ K}}{\mid 988\text{ K}} = 1.97\text{ m}^3$

9. a. $\dfrac{2.90\text{ m}^3 \mid 296\text{ K}}{\mid 226\text{ K}} = 3.80\text{ m}^3$

   b. $\dfrac{7.91\text{ m}^3 \mid 538\text{ K}}{\mid 325\text{ K}} = 13.1\text{ m}^3$

   c. $\dfrac{667\text{ cm}^3 \mid 314\text{ K}}{\mid 431\text{ K}} = 486\text{ cm}^3$

   d. $\dfrac{4.82\text{ m}^3 \mid 304\text{ K}}{\mid 295\text{ K}} = 4.97\text{ m}^3$

14. $\dfrac{2.23\text{ dm}^3 \mid 4.85\text{ kPa} \mid 278.5\text{ K}}{\mid 1.38\text{ kPa} \mid 271.6\text{ K}} = 8.04\text{ dm}^3$

15. a. $\dfrac{7.51\text{ m}^3 \mid 273\text{ K} \mid 59.9\text{ kPa}}{\mid 278\text{ K} \mid 101.3\text{ kPa}} = 4.36\text{ m}^3$

    b. $\dfrac{351\text{ cm}^3 \mid 309\text{ K} \mid 82.5\text{ kPa}}{\mid 292\text{ K} \mid 94.5\text{ kPa}} = 324\text{ cm}^3$

    c. $\dfrac{7.03\text{ m}^3 \mid 273\text{ K} \mid 111\text{ kPa}}{\mid 304\text{ K} \mid 101.3\text{ kPa}} = 6.92\text{ m}^3$

    d. $\dfrac{955\text{ cm}^3 \mid 349\text{ K} \mid 108.0\text{ kPa}}{\mid 331\text{ K} \mid 123.0\text{ kPa}} = 884\text{ cm}^3$

    e. $\dfrac{960.0\text{ cm}^3 \mid 286\text{ K} \mid 107.2\text{ kPa}}{\mid 344\text{ K} \mid 59.3\text{ kPa}} = 1440\text{ cm}^3$

16. $\dfrac{v_{He}}{v_{Ar}} = \sqrt{\dfrac{m_{Ar}}{m_{He}}} = \sqrt{\dfrac{39.9}{4.00}} = 3.16$

17. $\dfrac{v_{Ar}}{v_{Rn}} = \sqrt{\dfrac{m_{Rn}}{m_{Ar}}} = \sqrt{\dfrac{222}{39.9}} = 2.36$

18. $\dfrac{v_{hydrogen}}{v_{oxygen}} = \dfrac{\sqrt{m_{O_2}}}{\sqrt{m_{H_2}}} = \sqrt{\dfrac{32.0}{2.02}} = 3.98$

19. $\dfrac{v_{He}}{v_{Rn}} = \dfrac{\sqrt{m_{Rn}}}{\sqrt{m_{He}}} = \dfrac{\sqrt{222\text{ g}}}{\sqrt{4.00\text{ g}}} = 7.45$

20. $\dfrac{v_{He}}{v_{O_2}} = \sqrt{\dfrac{m_{O_2}}{m_{He}}}$

    $\sqrt{\dfrac{32.0\text{ g}}{4.00\text{ g}}} = 2.83$

    $v_{He} = 2.83 \times 0.0760\text{ m/s} = 0.215\text{ m/s}$

Chapter 19

1. $P = \dfrac{nRT}{V} = \dfrac{0.622\text{ mol} \mid 8.31\text{ dm}^3\cdot\text{kPa} \mid 289\text{ K}}{9.22\text{ dm}^3 \mid \text{mol}\cdot\text{K} \mid} = 162\text{ kPa}$

2. $n = \dfrac{PV}{RT} = \dfrac{66.7\text{ kPa} \mid 0.486\text{ dm}^3 \mid \text{mol}\cdot\text{K}}{\mid 284\text{ K} \mid 8.31\text{ dm}^3\cdot\text{kPa}}$

   $= 0.0137\text{ mol}$

3. $V = \dfrac{nRT}{P} = \dfrac{0.684\text{ mol} \mid 8.31\text{ dm}^3\cdot\text{kPa} \mid 282\text{ K}}{99.1\text{ kPa} \mid \text{mol}\cdot\text{K} \mid} = 16.2\text{ dm}^3$

4. $T = \dfrac{PV}{nR} = \dfrac{100.4\text{ kPa} \mid 604\text{ cm}^3 \mid \text{mol}\cdot\text{K} \mid}{0.0851\text{ mol} \mid \mid 8.31\text{ dm}^3\cdot\text{kPa} \mid}$

   $\dfrac{\mid 1\text{ dm}^3}{\mid 1000\text{ cm}^3} = 85.8\text{ K} = -187°\text{C}$

5. $P = \frac{nRT}{V} = \frac{0.003\,06\text{ mol}}{25.9\text{ cm}^3} \cdot \frac{8.31\text{ dm}^3\cdot\text{kPa}}{\text{mol}\cdot\text{K}} \cdot \frac{282\text{ K}}{} \cdot \frac{1000\text{ cm}^3}{1\text{ dm}^3} = 277\text{ kPa}$

6. $M = \frac{mRT}{PV} = \frac{0.922\text{ g}}{107.0\text{ kPa}} \cdot \frac{8.31\text{ dm}^3\cdot\text{kPa}}{\text{mol}\cdot\text{K}} \cdot \frac{372\text{ K}}{0.1500\text{ dm}^3}$
$= 178\text{ g/mol}$

7. $M = \frac{mRT}{PV} = \frac{3.59\text{ g}}{99.2\text{ kPa}} \cdot \frac{8.31\text{ dm}^3\cdot\text{kPa}}{\text{mol}\cdot\text{K}} \cdot \frac{304\text{ K}}{4.34\text{ dm}^3}$
$= 21.1\text{ g/mol}$

8. $M = \frac{mRT}{PV} = \frac{0.858\text{ g}}{106.3\text{ kPa}} \cdot \frac{8.31\text{ dm}^3\cdot\text{kPa}}{\text{mol}\cdot\text{K}} \cdot \frac{275\text{ K}}{150.0\text{ cm}^3} \cdot \frac{1000\text{ cm}^3}{1\text{ dm}^3} = 123\text{ g/mol}$

9. $M = \frac{mRT}{PV} = \frac{8.11\text{ g}}{109.1\text{ kPa}} \cdot \frac{8.31\text{ dm}^3\cdot\text{kPa}}{\text{mol}\cdot\text{K}} \cdot \frac{283\text{ K}}{2.38\text{ dm}^3}$
$= 73.5\text{ g/mol}$

15. $N_2(g) + 3H_2(g) \rightarrow 2NH_3(g)$

$\frac{14.0\text{ g }N_2}{} \cdot \frac{1\text{ mol }N_2}{28.0\text{ g }N_2} \cdot \frac{2\text{ mol }NH_3}{1\text{ mol }N_2} \cdot \frac{22.4\text{ dm}^3}{1\text{ mol}} \cdot \frac{1000\text{ cm}^3}{1\text{ dm}^3} = 22\,400\text{ cm}^3\text{ }NH_3$

$= 22\,400\text{ cm}^3\text{ }NH_3$

16. $Zn(cr) + H_2SO_4(aq) \rightarrow ZnSO_4(aq) + H_2(g)$

$\frac{28.0\text{ g Zn}}{} \cdot \frac{1\text{ mol Zn}}{65.4\text{ g Zn}} \cdot \frac{1\text{ mol }H_2}{1\text{ mol Zn}} \cdot \frac{22.4\text{ dm}^3}{1\text{ mol}} \cdot \frac{1000\text{ cm}^3}{1\text{ dm}^3}$

$= 9590\text{ cm}^3\text{ }H_2$

17. $H_2(g) + Br_2(g) \rightarrow 2HBr(g)$

$\frac{5600\text{ cm}^3\text{ }H_2}{} \cdot \frac{1\text{ dm}^3}{1000\text{ cm}^3} \cdot \frac{1\text{ mol}}{22.4\text{ dm}^3} \cdot \frac{2\text{ mol HBr}}{1\text{ mol }H_2} \cdot \frac{80.9\text{ g HBr}}{1\text{ mol HBr}} = 40.5\text{ g HBr}$

18. $2Sb(cr) + 3Cl_2(g) \rightarrow 2SbCl_3(cr)$

$\frac{3570\text{ cm}^3\text{ }Cl_2}{} \cdot \frac{1\text{ dm}^3}{1000\text{ cm}^3} \cdot \frac{1\text{ mol}}{22.4\text{ dm}^3} \cdot \frac{2\text{ mol }SbCl_3}{3\text{ mol }Cl_2} \cdot \frac{228\text{ g }SbCl_3}{1\text{ mol }SbCl_3} = 25.4\text{ g }SbCl_3$

19. $2C_4H_{10}(g) + 13O_2(g) \rightarrow 8CO_2(g) + 10H_2O(g)$

$\frac{401\text{ cm}^3\text{ }C_4H_{10}}{} \cdot \frac{13\text{ cm}^3\text{ }O_2}{2\text{ cm}^3\text{ }C_4H_{10}} = 2610\text{ cm}^3\text{ }O_2$

20. $\frac{75.2\text{ dm}^3\text{ }Cl_2}{} \cdot \frac{1\text{ dm}^3\text{ }Br_2}{1\text{ dm}^3\text{ }Cl_2} = 75.2\text{ dm}^3\text{ }Br_2$

21. $C_7H_{16}(g) + 11O_2(g) \rightarrow 7CO_2(g) + 8H_2O(g)$

$\frac{917\text{ cm}^3\text{ }C_7H_{16}}{} \cdot \frac{11\text{ cm}^3\text{ }O_2}{1\text{ cm}^3\text{ }C_7H_{16}} = 10\,100\text{ cm}^3\text{ }O_2$

volume of air = 10 100/0.209 = 48 300 cm^3 air

22. $2NO(g) + O_2(g) \rightarrow 2NO_2(g)$

$\frac{499\text{ cm}^3\text{ NO}}{} \cdot \frac{1\text{ cm}^3\text{ }O_2}{2\text{ cm}^3\text{ NO}} = 2.50 \times 10^2\text{ cm}^3\text{ }O_2$

23. $\frac{941\text{ m}^3\text{ }C_6H_6}{} \cdot \frac{4\text{ m}^3\text{ }H_2}{1\text{ m}^3\text{ }C_6H_6} = 3770\text{ m}^3\text{ }H_2$

24. $2NaBr(aq) + 2H_2SO_4(aq) + MnO_2(cr) \rightarrow Br_2(l) + MnSO_4(aq) + 2H_2O(l) + Na_2SO_4(aq)$

$\frac{2.10\text{ g NaBr}}{} \cdot \frac{1\text{ mol NaBr}}{103\text{ g NaBr}} = 0.0204\text{ mol NaBr}$

$\frac{9.42\text{ g }H_2SO_4}{} \cdot \frac{1\text{ mol }H_2SO_4}{98.1\text{ g }H_2SO_4} = 0.0960\text{ mol }H_2SO_4$

∴ NaBr is the limiting reactant.

$\frac{0.0204\text{ mol NaBr}}{} \cdot \frac{1\text{ mol }Br_2}{2\text{ mol NaBr}} \cdot \frac{159.8\text{ g }Br_2}{1\text{ mol }Br_2}$

$= 1.63\text{ g }Br_2$

25. $2Ca_3(PO_4)_2(cr) + 6SiO_2(cr) + 10C(amor) \rightarrow P_4(cr) + 6CaSiO_3(cr) + 10CO(g)$

$\frac{4.14\text{ g }Ca_3(PO_4)_2}{} \cdot \frac{1\text{ mol }Ca_3(PO_4)_2}{310\text{ g }Ca_3(PO_4)_2}$

$= 0.0134\text{ mol }Ca_3(PO_4)_2$

$\frac{1.20\text{ g }SiO_2}{} \cdot \frac{1\text{ mol }SiO_2}{60.1\text{ g }SiO_2} = 0.0200\text{ mol }SiO_2$

SiO_2 is the limiting reactant, since the ratio is 3:1 in the equation.

$\frac{0.0200\text{ mol }SiO_2}{} \cdot \frac{10\text{ mol CO}}{6\text{ mol }SiO_2} \cdot \frac{22.4\text{ dm}^3}{1\text{ mol}} = 0.747\text{ dm}^3\text{ CO}$

26. $H_2S(g) + I_2(aq) \rightarrow 2HI(aq) + S(cr)$

$\frac{4.11\text{ g }I_2}{} \cdot \frac{1\text{ mol }I_2}{254\text{ g }I_2} = 0.0162\text{ mol }I_2$

$\frac{317\text{ cm}^3\text{ }H_2S}{} \cdot \frac{1\text{ dm}^3}{1000\text{ cm}^3} \cdot \frac{1\text{ mol}}{22.4\text{ dm}^3} = 0.0142\text{ mol }H_2S$

H_2S is the limiting reactant.

$\frac{0.0142\text{ mol }H_2S}{} \cdot \frac{1\text{ mol S}}{1\text{ mol }H_2S} \cdot \frac{32.1\text{ g S}}{1\text{ mol S}} = 0.456\text{ g S}$

27. $2Cl_2(g) + HgO(cr) \rightarrow HgCl_2(cr) + Cl_2O(g)$

$\frac{116\text{ cm}^3\text{ }Cl_2}{} \cdot \frac{1\text{ dm}^3}{1000\text{ cm}^3} \cdot \frac{1\text{ mol}}{22.4\text{ dm}^3} = 0.005\,18\text{ mol }Cl_2$

$\frac{7.62\text{ g HgO}}{} \cdot \frac{1\text{ mol HgO}}{217\text{ g HgO}} = 0.0351\text{ mol HgO}$

Cl_2 is the limiting reactant.

$\frac{116\text{ cm}^3\text{ }Cl_2}{} \cdot \frac{1\text{ cm}^3\text{ }Cl_2O}{2\text{ cm}^3\text{ }Cl_2} = 58.0\text{ cm}^3\text{ }Cl_2O$

Chapter 20

1. Suggested reagents include:
   - a. H_3PO_4
   - b. H_2SO_4
   - c. H_2CO_3
   - d. NH_4OH
2. Suggested reagents include:
   - a. $Pb(CH_3COO)_2$
   - b. $Hg_2(NO_3)_2$
   - c. $SrCl_2$
   - d. $AgNO_3$
3. a. $2HF(aq) + Hg_2(NO_3)_2(aq) \rightarrow Hg_2F_2(cr) + 2HNO_3(aq)$
   b. $LiI(aq) + AgNO_3(aq) \rightarrow AgI(cr) + LiNO_3(aq)$
   c. $K_2S(aq) + Co(CH_3COO)_2(aq) \rightarrow CoS(cr) + 2KCH_3COO(aq)$
   d. $6NaOH(aq) + Cr_2(SO_4)_3(aq) \rightarrow 3Na_2SO_4(aq) + 2Cr(OH)_3(cr)$

4. $MgBr_2(s) \rightarrow Mg^{2+} + 2Br^-$

$$\frac{193\text{ g }MgBr_2}{5.00 \times 10^2\text{ cm}^3\text{ soln}} \Big| \frac{1\text{ mol }MgBr_2}{184\text{ g }MgBr_2} \Big| \frac{1000\text{ cm}^3}{1\text{ dm}^3} \Big| \frac{2\text{ mol }Br^-}{1\text{ mol }MgBr_2} = 4.20M\ Br^-$$

5. $Ca(C_5H_9O_2)_2(s) \rightarrow Ca^{2+} + 2C_5H_9O_2^-$

$$\frac{8.28\text{ g }Ca(C_5H_9O_2)_2}{2.50 \times 10^2\text{ cm}^3\text{ soln}} \Big| \frac{1\text{ mol }Ca(C_5H_9O_2)_2}{242\text{ g }Ca(C_5H_9O_2)_2} \Big| \frac{1000\text{ cm}^3}{1\text{ dm}^3} \Big| \frac{1\text{ mol }Ca^{2+}}{1\text{ mol }Ca(C_5H_9O_2)_2} = 0.137M\ Ca^{2+}$$

6. $CaCl_2(s) \rightarrow Ca^{2+} + 2Cl^-$

$$\frac{0.523\text{ mol }CaCl_2}{1\text{ dm}^3} \Big| \frac{2.00\text{ dm}^3}{} \Big| \frac{1\text{ mol }Ca^{2+}}{1\text{ mol }CaCl_2}$$
$= 1.05\text{ mol }Ca^{2+}$

7. $$\frac{199\text{ g }NiBr_2}{5.00 \times 10^2\text{ g }H_2O} \Big| \frac{1\text{ mol }NiBr_2}{219\text{ g }NiBr_2} \Big| \frac{1000\text{ g }H_2O}{1\text{ kg }H_2O}$$
$= 1.82m\ NiBr_2$ (mol/kg = m)

8. $$\frac{92.3\text{ g KF}}{1.00 \times 10^3\text{ g }H_2O} \Big| \frac{1\text{ mol KF}}{58.1\text{ g KF}} \Big| \frac{1000\text{ g }H_2O}{1\text{ kg }H_2O}$$
$= 1.59m$ KF

9. $$\frac{0.059\text{ g KF}}{0.272\text{ g }H_2O} \Big| \frac{1\text{ mol KF}}{58.1\text{ g KF}} \Big| \frac{1000\text{ g }H_2O}{1\text{ kg }H_2O} = 3.7m\text{ KF}$$

10. $$\frac{12.3\text{ g }C_4H_4O}{} \Big| \frac{1\text{ mol }C_4H_4O}{68.0\text{ g }C_4H_4O} = 0.181\text{ mol }C_4H_4O$$

$$\frac{1.00 \times 10^2\text{ g }C_2H_6O}{} \Big| \frac{1\text{ mol }C_2H_6O}{46.1\text{ g }C_2H_6O} = \frac{2.17\text{ mol }C_2H_6O}{2.35\text{ mol soln}}$$

$$\frac{0.181}{2.35} = 0.0760\text{ mole fraction }C_4H_4O$$

$$\frac{2.17}{2.35} = 0.923\text{ mole fraction }C_2H_6O$$

11. $$\frac{156\text{ g }C_{12}H_{22}O_{11}}{} \Big| \frac{1\text{ mol }C_{12}H_{22}O_{11}}{342\text{ g }C_{12}H_{22}O_{11}}$$
$= 0.456\text{ mol }C_{12}H_{22}O_{11}$

$$\frac{3.00 \times 10^2\text{ g }H_2O}{} \Big| \frac{1\text{ mol }H_2O}{18.0\text{ g }H_2O} = \frac{16.7\text{ mol }H_2O}{17.2\text{ mol soln}}$$

$$\frac{0.456}{17.2} = 0.0265\text{ mole fraction }C_{12}H_{22}O_{11}$$

$$\frac{16.7}{17.2} = 0.971\text{ mole fraction }H_2O$$

12. $$\frac{75.6\text{ g }C_{10}H_8}{} \Big| \frac{1\text{ mol }C_{10}H_8}{128\text{ g }C_{10}H_8} = 0.591\text{ mol }C_{10}H_8$$

$$\frac{6.00 \times 10^2\text{ g }C_4H_{10}O}{} \Big| \frac{1\text{ mol }C_4H_{10}O}{74.0\text{ g }C_4H_{10}O}$$
$$= \frac{8.11\text{ mol }C_4H_{10}O}{8.70\text{ mol soln}}$$

$$\frac{0.591}{8.70} = 0.0679\text{ mole fraction }C_{10}H_8$$

$$\frac{8.11}{8.70} = 0.932\text{ mole fraction }C_4H_{10}O$$

13. $$\frac{199\text{ g}}{199\text{ g} + (5.00 \times 10^2)\text{ g}} \times 100 = 28.5\%\ NiBr_2$$

$$\frac{92.3\text{ g}}{92.3\text{ g} + (1.00 \times 10^3)\text{ g}} \times 100 = 8.45\%\text{ KF}$$

14. $$\frac{0.059\text{ g}}{0.059\text{ g} + 0.272\text{ g}} \times 100 = 18\%\text{ KF}$$

$$\frac{12.3\text{ g}}{12.3\text{ g} + (1.00 \times 10^2)\text{ g}} \times 100 = 10.9\%\ C_4H_4O$$

15. $$\frac{156\text{ g}}{156\text{ g} + (3.00 \times 10^2)\text{ g}} \times 100 = 34.2\%\ C_{12}H_{22}O_{11}$$

$$\frac{75.6\text{ g}}{75.6\text{ g} + (6.00 \times 10^2)\text{ g}} \times 100 = 11.2\%\ C_{10}H_8$$

Chapter 21

1. v.p. H_2O at 25°C = 3.2 kPa
   v.p. soln = (3.2 kPa)(1.000 − 0.163) = 2.7 kPa
2. v.p. H_2O = (101.325 kPa)(0.900) = 91.2 kPa
3. v.p. ethanal = (86.3 kPa)(0.300) = 25.9 kPa
   v.p. methanol = (11.6 kPa)(0.700) = 8.12 kPa
   v.p. soln = 25.9 kPa + 8.12 kPa = 34.0 kPa

8. $$\frac{25.5\text{ g }C_7H_{11}NO_7S}{1.00 \times 10^2\text{ g }H_2O} \Big| \frac{1\text{ mol }C_7H_{11}NO_7S}{253\text{ g }C_7H_{11}NO_7S} \Big| \frac{1000\text{ g}}{1\text{ kg}}$$
$= 1.01m\ C_7H_{11}NO_7S$

$$\Delta T_{fp} = mK_{fp} = \frac{1.01m}{} \Big| \frac{1.853\text{ C}^\circ}{m} = 1.87\text{ C}^\circ$$

$$\Delta T_{bp} = \frac{1.01m}{} \Big| \frac{0.515\text{ C}^\circ}{m} = 0.520\text{ C}^\circ$$

Freezing point = 0°C − 1.87 C° = −1.87°C
Boiling point = 100°C + 0.520 C° = 100.520°C

9. $$\frac{1.00 \times 10^2\text{ g }C_{10}H_8O_6S_2}{1.00 \times 10^2\text{ g }H_2O} \Big| \frac{1\text{ mol }C_{10}H_8O_6S_2}{288\text{ g }C_{10}H_8O_6S_2} \Big| \frac{1000\text{ g}}{1\text{ kg}}$$
$= 3.47m\ C_{10}H_8O_6S_2$

$$\Delta T_{fp} = \frac{3.47m}{} \Big| \frac{1.853\text{ C}^\circ}{m} = 6.43\text{ C}^\circ$$

$$\Delta T_{bp} = \frac{3.47m}{} \Big| \frac{0.515\text{ C}^\circ}{m} = 1.79\text{ C}^\circ$$

Freezing point = 0°C − 6.43 C° = −6.43°C
Boiling point = 100°C + 1.79 C° = 101.79°C

10. $NiSO_4 \rightarrow Ni^{2+} + SO_4^{2-}$

$$\frac{21.6\text{ g }NiSO_4}{1.00 \times 10^2\text{ g }H_2O} \Big| \frac{1\text{ mol }NiSO_4}{155\text{ g }NiSO_4} \Big| \frac{1000\text{ g}}{1\text{ kg}} \Big| \frac{2\text{ mol ions}}{1\text{ mol }NiSO_4}$$
$= 2.79m$ ions

$$\Delta T_{fp} = \frac{2.79m}{} \Big| \frac{1.853\text{ C}^\circ}{m} = 5.17\text{ C}^\circ$$

$$\Delta T_{bp} = \frac{2.79m}{} \Big| \frac{0.515\text{ C}^\circ}{m} = 1.44\text{ C}^\circ$$

Freezing point = 0°C − 5.17 C° = −5.17°C
Boiling point = 100°C + 1.44 C° = 101.44°C

11. $Mg(ClO_4)_2 \rightarrow Mg^{2+} + 2ClO_4^-$

$$\frac{77.0\text{ g }Mg(ClO_4)_2}{2.00 \times 10^2\text{ g }H_2O} \Big| \frac{1\text{ mol }Mg(ClO_4)_2}{223\text{ g }Mg(ClO_4)_2} \Big| \frac{1000\text{ g}}{1\text{ kg}} \Big| \frac{3\text{ mol ions}}{1\text{ mol }Mg(ClO_4)_2} = 5.18m\text{ ions}$$

$$\Delta T_{fp} = \frac{5.18m}{} \Big| \frac{1.853\text{ C}^\circ}{m} = 9.60\text{ C}^\circ$$

$$\Delta T_{bp} = \frac{5.18m \mid 0.515\ \text{C}^\circ}{\mid m} = 2.67\ \text{C}^\circ$$

Freezing point = 0°C − 9.60 C° = −9.60°C
Boiling point = 100°C + 2.67 C° = 102.67°C

12. $$\frac{41.3\ \text{g } C_{15}H_9NO_4 \mid 1\ \text{mol } C_{15}H_9NO_4 \mid 1000\ \text{g}}{1.00 \times 10^2\ \text{g } C_6H_5NO_2 \mid 267\ \text{g } C_{15}H_9NO_4 \mid 1\ \text{kg}}$$
$= 1.55m\ C_{15}H_9NO_4$

$$\Delta T_{fp} = \frac{1.55m \mid 6.852\ \text{C}^\circ}{\mid m} = 10.6\ \text{C}^\circ$$

$$\Delta T_{bp} = \frac{1.55m \mid 5.24\ \text{C}^\circ}{\mid m} = 8.12\ \text{C}^\circ$$

Freezing point = 5.76°C − 10.6 C° = −4.8°C
Boiling point = 210.8°C + 8.12 C° = 218.9°C

13. $$\frac{8.02\ \text{g solute} \mid 1000\ \text{g}}{861\ \text{g } H_2O \mid 1\ \text{kg}} = 9.31\ \text{g solute/kg } H_2O$$

$$m = \frac{\Delta T_{fp}}{K_{fp}} = \frac{0.430\ \text{C}^\circ \mid m}{\mid 1.853\ \text{C}^\circ}$$
$= 0.232$ mol solute/kg H_2O

$$\frac{9.31\ \text{g solute} \mid \text{kg } H_2O}{\text{kg } H_2O \mid 0.232\ \text{mol solute}} = 40.1\ \text{g/mol}$$

14. $$\frac{64.3\ \text{g solute} \mid 1000\ \text{g}}{3.90 \times 10^2\ \text{g } H_2O \mid 1\ \text{kg}} = 165\ \text{g solute/kg } H_2O$$

$$m = \frac{0.680\ \text{C}^\circ \mid m}{\mid 0.515\ \text{C}^\circ} = 1.32\ \text{mol solute/kg } H_2O$$

$$\frac{165\ \text{g solute} \mid \text{kg } H_2O}{\text{kg } H_2O \mid 1.32\ \text{mol solute}} = 125\ \text{g/mol}$$

15. $$\frac{20.8\ \text{g solute} \mid 1000\ \text{g}}{128\ \text{g } CH_3COOH \mid 1\ \text{kg}} = 163\ \text{g solute/kg } CH_3COOH$$

$\Delta T_{fp} = 16.66°\text{C} - 13.5°\text{C} = 3.2\ \text{C}°$

$$m = \frac{3.2\ \text{C}^\circ \mid m}{\mid 3.90\ \text{C}^\circ} = 0.82\ \text{mol solute/kg } CH_3COOH$$

$$\frac{163\ \text{g solute} \mid \text{kg } CH_3COOH}{\text{kg } CH_3COOH \mid 0.82\ \text{mol solute}} = 2.0 \times 10^2\ \text{g/mol}$$

16. $$\frac{10.4\ \text{g solute} \mid 1000\ \text{g}}{164\ \text{g phenol} \mid 1\ \text{kg}} = 63.4\ \text{g solute/kg phenol}$$

$\Delta T_{fp} = 40.90°\text{C} - 36.3°\text{C} - 4.6\ \text{C}°$

$$m = \frac{4.6\ \text{C}^\circ \mid m}{\mid 7.40\ \text{C}^\circ} = 0.62\ \text{mol solute/kg phenol}$$

$$\frac{63.4\ \text{g solute} \mid \text{kg phenol}}{\text{kg phenol} \mid 0.62\ \text{mol solute}} = 1.0 \times 10^2\ \text{g/mol}$$

17. $$\frac{2.53\ \text{g solute} \mid 1000\ \text{g}}{63.5\ \text{g nitrobenzene} \mid 1\ \text{kg}}$$
$= 39.8$ g solute/kg nitrobenzene

$\Delta T_{fp} = 5.76°\text{C} - 3.40°\text{C} = 2.36\ \text{C}°$

$$m = \frac{2.36\ \text{C}^\circ \mid m}{\mid 6.852\ \text{C}^\circ}$$
$= 0.344$ mol solute/kg nitrobenzene

$$\frac{39.8\ \text{g solute} \mid \text{kg nitrobenzene}}{\text{kg nitrobenzene} \mid 0.344\ \text{mol solute}} = 116\ \text{g/mol}$$

18. $$M = \frac{\Pi}{RT} = \frac{780\ \text{kPa} \mid \text{mol}\cdot\text{K}}{298\ \text{K} \mid 8.31\ \text{dm}^3\cdot\text{kPa}}$$
$= 0.31M\ C_6H_{12}O_6$

19. $$\Pi = MRT = \frac{mRT}{MV}; \quad M = \frac{mRT}{\Pi V}$$

$$M = \frac{0.300\ \text{g hemoglobin} \mid 1000\ \text{cm}^3 \mid 8.31\ \text{dm}^3\cdot\text{kPa} \mid 298\ \text{K}}{30.0\ \text{cm}^3 \mid 1\ \text{dm}^3 \mid \text{mol}\cdot\text{K} \mid 0.40\ \text{kPa}} = 62\ 000\ \text{g/mol}$$

20. $$\Pi = \frac{0.490\ \text{mol} \mid 8.31\ \text{dm}^3\cdot\text{kPa} \mid 298\ \text{K}}{1.00\ \text{dm}^3 \mid \text{mol}\cdot\text{K} \mid}$$
$= 1.21 \times 10^3$ kPa

Chapter 22

1. **a.** doubled $[2H_2] = 2[H_2]$
   b. eight times faster $[2NO]^2[2H_2] = 8[NO]^2[H_2]$
   c. slows down
2. **a.** Rate = $k[H_2][I_2]$
   b. $0.2 = k(1)(1)$
   $k = 0.2\ \text{dm}^3/\text{mol}\cdot\text{s}$
   $0.8 = k(2)(2)$
   $k = 0.2\ \text{dm}^3/\text{mol}\cdot\text{s}$
   Rate = 0.2(0.5)(0.5)
   Rate = $0.05(\text{mol/dm}^3)/\text{s}$
3. rate = $k[CH_3COCH_3]$ (Iodine has no effect)
4. $H^+ + CH_3-C(=O)-CH_3 \rightarrow CH_3-C^{\oplus}(OH)-CH_3$ slow

   Mechanisms for other steps:

   $CH_3-C^{\oplus}(OH)-CH_3 \rightarrow CH_3-C(OH)=CH_2 + H^+$ fast

   $CH_3-C(OH)=CH_2 + I_2 \rightarrow CH_3-C^{\oplus}(OH)-CH_2I + I^-$ fast

   $CH_3-C^{\oplus}(OH)-CH_2I + I^- \rightarrow CH_3-C(=O)-CH_2I + HI$ fast

10. **a.** $$K_{eq} = \frac{[CO_2][NH_3]^2}{[NH_2COONH_4]}$$
    b. $$K_{eq} = \frac{[Cl_2]^2[H_2O]^2}{[HCl]^4[O_2]}$$
    c. $$K_{eq} = \frac{[NH_3][H_2S]}{[NH_4HS]}$$
    d. $$K_{eq} = \frac{[CuSO_4][H_2O]^5}{[CuSO_4\cdot 5H_2O]}$$
11. $$K_{eq} = \frac{[H_2][CO_2]}{[CO][H_2O]} = \frac{(0.32)(0.42)}{(0.200)(0.500)} = 1.3$$
12. $$K_{eq} = \frac{[H_2]^2[S_2]}{[H_2S]^2}$$
    $$= \frac{(2.22 \times 10^{-3})^2(1.11 \times 10^{-3})}{(7.06 \times 10^{-3})^2} = 1.10 \times 10^{-4}$$

13. $[HI]^2 = \frac{[H_2][I_2]}{K_{eq}}$

$[HI]^2 = \frac{(2.00 \times 10^{-4})(2.00 \times 10^{-4})}{1.40 \times 10^{-2}} = 2.86 \times 10^{-6}$

$[HI] = 1.69 \times 10^{-3} M$

14. $[CO] = \frac{[H_2][CO_2]}{K_{eq}[H_2O]}$

$[CO] = \frac{(0.32)(0.42)}{(2.40)(0.500)} = 0.11M$

15. $K_{eq} = \frac{[NO_2]^2}{[N_2O_4]}$

$[NO_2]^2 = K_{eq}[N_2O_4]$
$= (8.75 \times 10^{-2})(1.72 \times 10^{-2})$
$= 1.51 \times 10^{-3}$
$[NO_2] = 3.88 \times 10^{-2} M$

Chapter 23

1. a. acid — HNO_3
   base — NaOH
   conjugate acid — H_2O
   conjugate base — NO_3^- in $NaNO_3$
   b. acid — HCl
   base — $NaHCO_3$
   conjugate acid — H_2CO_3
   conjugate base — Cl^- in NaCl
2. a. hydrogen sulfite ion
   b. hydrogen carbonate ion
   c. amide ion
   d. fluoride ion
3. a. Lewis base
   b. Lewis base
   c. Lewis acid
   d. Lewis base
4. $Ag^+ + 2NH_3 \rightarrow Ag(NH_3)_2^+$
5. $Cu^{2+} + 4NH_3 \rightarrow Cu(NH_3)_4^{2+}$
6. $Co^{3+} + 6NH_3 \rightarrow Co(NH_3)_6^{3+}$
7. $Fe^{3+} + 6CN^- \rightarrow Fe(CN)_6^{3-}$
8. $Zn^{2+} + 4OH^- \rightarrow Zn(OH)_4^{2-}$
9. a. hydrobromic acid
   b. hydrofluoric acid
10. a. selenous acid
    b. nitrous acid
    c. phosphoric acid
    d. arsenous acid
    e. iodic acid
11. a. H_2CO_3
    b. HNO_3
    c. H_3AsO_4
    d. H_2SeO_4
    e. HIO
12. a. calcium hydroxide
    b. potassium hydroxide
    c. aluminum hydroxide
    d. methanamine
    e. rubidium hydroxide
13. a. CsOH
    b. $CH_3CH_2CH_2NH_2$
    c. $CH_3CH_2CH_2CH_2NH_2$
    d. LiOH
14. a. basic
    b. basic
    c. acidic
    d. basic
    e. acidic
15. a. BaO
    b. I_2O_7
    c. TeO_3
    d. Al_2O_3
    e. ZnO
21. a. sodium hydrogen sulfate
    b. potassium hydrogen tartrate
    c. sodium dihydrogen phosphate
    d. sodium hydrogen sulfide
    e. aluminum hydroxide silicate
22. a. $NaHCO_3$
    b. Na_2HPO_4
    c. NH_4HS
    d. $KHSO_4$
    e. $Sn(OH)NO_3$
23. $2Ag^+(aq) + HSO_4^-(aq) \rightarrow Ag_2SO_4(cr) + H^+(aq)$
24. $H_4SiO_4(aq) + 4OH^-(aq) \rightarrow SiO_4^{4-}(aq) + 4H_2O(l)$
25. $2Cl^-(aq) + 2Cr^{2+}(aq) + 2Hg^{2+}(aq) \rightarrow 2Cr^{3+}(aq) + Hg_2Cl_2(cr)$
26. $2Mn^{2+}(aq) + 5NaBiO_3(cr) + 14H^+(aq) \rightarrow 2MnO_4^-(aq) + 5Bi^{3+}(aq) + 7H_2O(l) + 5Na^+(aq)$
27. $2Cu^{2+}(aq) + SO_4^{2-}(aq) + 2CNS^-(aq) + H_2SO_3(aq) + H_2O(l) \rightarrow 2CuCNS(cr) + 2HSO_4^-(aq) + 2H^+(aq)$
28. $HClO + H_2O \rightleftarrows H_3O^+ + ClO^-$

$K_a = \frac{[H_3O^+][ClO^-]}{[HClO]} = 2.88 \times 10^{-8} = \frac{x^2}{0.0200 - \otimes}$ ← neglect

$x = 2.40 \times 10^{-5} M$

29. $N_2H_4 + H_2O \rightleftarrows N_2H_5^+ + OH^-$

$K_b = \frac{[N_2H_5^+][OH^-]}{[N_2H_4]} = \frac{(1.23 \times 10^{-3})^2}{0.499}$

$K_b = 3.03 \times 10^{-6}$

30. $K_a = \frac{[CH_3COO^-][H_3O^+]}{[CH_3COOH]}$

$1.75 \times 10^{-5} = \frac{x^2}{0.200 - \otimes}$ ← neglect

$x^2 = 3.50 \times 10^{-6}$
$x = 1.87 \times 10^{-3}$

$\% = \frac{1.87 \times 10^{-3}}{0.200} \left| \frac{100}{1} \right. = 0.935\%$

31. $K_a = \frac{[A^-][H_3O^+]}{[HA]} = \frac{(0.0200)^2}{0.980} = 4.08 \times 10^{-4}$

32. $HNO_2 + H_2O \rightleftarrows H_3O^+ + NO_2^-$

$K_a = \frac{[H_3O^+][NO_2^-]}{[HNO_2]}$

$7.24 \times 10^{-4} = \frac{x(\otimes + 0.03)}{(0.150 - \otimes)}$ ← neglect

$x = 3.26 \times 10^{-3} M$

33. $HBrO + H_2O \rightleftarrows H_3O^+ + BrO^-$

$$K_a = \frac{[H_3O^+][BrO^-]}{[HBrO]}$$

$$2.51 \times 10^{-9} = \frac{x(\otimes + 0.000\ 195)}{0.403 - \otimes} \leftarrow \text{neglect}$$

$x = 5.19 \times 10^{-6}M$

34. Add either HCO_3^- or CO_3^{2-} in the form of soluble salts; add H_3O^+.

35. b, d, e, f

36. oxalic acid $\frac{5.36 \times 10^{-2}}{5.35 \times 10^{-5}} = 1.00 \times 10^3$

arsenic acid $\frac{6.03 \times 10^{-3}}{1.05 \times 10^{-7}} = 5.74 \times 10^4$

boric acid $\frac{5.75 \times 10^{-10}}{1.82 \times 10^{-13}} = 3.16 \times 10^3$

carbonic acid $\frac{4.37 \times 10^{-7}}{4.68 \times 10^{-11}} = 9.34 \times 10^3$

phosphoric acid $\frac{7.08 \times 10^{-3}}{6.31 \times 10^{-8}} = 1.12 \times 10^5$

phosphorus acid $\frac{6.31 \times 10^{-2}}{2.00 \times 10^{-7}} = 3.16 \times 10^5$

hydrogen sulfide $\frac{1.07 \times 10^{-7}}{1.26 \times 10^{-13}} = 8.49 \times 10^5$

sulfurous acid $\frac{1.29 \times 10^{-2}}{6.17 \times 10^{-8}} = 2.09 \times 10^5$

selenous acid $\frac{2.29 \times 10^{-3}}{5.37 \times 10^{-9}} = 4.26 \times 10^5$

The first ionization constant is considerably larger than the second ionization constant.

Chapter 24

1. a. $K_{sp} = [Pb^{2+}][I^-]^2$
   b. $K_{sp} = [Cu^{2+}]^3[PO_4^{3-}]^2$

2. $AgI \rightleftarrows Ag^+ + I^-$
   $K_{sp} = [Ag^+][I^-]$
   $[Ag^+] = [I^-] = x$
   $x^2 = 8.32 \times 10^{-17}$
   $x = 9.12 \times 10^{-9}M = [Ag^+]$

3. $D_2A \rightleftarrows 2D^+ + A^{2-}$
   $K_{sp} = [D^+]^2[A^{2-}]$
   $2[A^{2-}] = [D^+] = 1.00 \times 10^{-5}$
   $[A^{2-}] = 2.00 \times 10^{-5}$
   $K_{sp} = (2.00 \times 10^{-5})^2(1.00 \times 10^{-5})$
   $K_{sp} = 4.00 \times 10^{-15}$

4. $Be(OH)_2 \rightleftarrows Be^{2+} + 2OH^-$
   $K_{sp} = [Be^{2+}][OH^-]^2$
   $1.58 \times 10^{-22} = x(2x)^2$
   $x = 3.41 \times 10^{-8}M$

5. $PbI_2 \rightleftarrows Pb^2 + 2I^-$
   $K_{sp} = [Pb^{2+}][I^-]^2$
   $K_{sp} = (1.21 \times 10^{-3})(2.42 \times 10^{-3})^2$
   $K_{sp} = 7.09 \times 10^{-9}$

6. $TlBr \rightleftarrows Tl^+ + Br^-$
   $K_{sp} = [Tl^+][Br^-]$

$$[Br^-] = \left.\frac{0.050\ mol}{500.0\ \cancel{cm^3}}\right|\left.\frac{10^3\ \cancel{cm^3}}{1\ \cancel{m^3}}\right|\frac{1\ \cancel{m^3}}{10^3\ dm^3} = 0.100M$$

$3.39 \times 10^{-6} = [Tl^+](0.100)$
$[Tl^+] = 3.39 \times 10^{-5}M$

7. $$[Fe^{3+}] = \left.\frac{0.100\ mol}{1\ dm^3}\right|\frac{0.250\ \cancel{dm^3}}{0.500\ \cancel{dm^3}} = 5.00 \times 10^{-2}M$$

$$[OH^-] = \left.\frac{0.010\ mol}{1\ dm^3}\right|\frac{0.250\ \cancel{dm^3}}{0.500\ \cancel{dm^3}} = 5.0 \times 10^{-3}M$$

$[Fe^{3+}][OH^-]^3 = (5.00 \times 10^{-2})(5.0 \times 10^{-3})^3 = 6.3 \times 10^{-9}$
Since the K_{sp} of $Fe(OH)_3$ is much smaller than this ($K_{sp} = 3.98 \times 10^{-38}$), a precipitate will form.

8. $$[Li^+] = \left.\frac{0.100\ mol}{1\ dm^3}\right|\frac{0.500\ \cancel{dm^3}}{1.000\ \cancel{dm^3}} = 5.00 \times 10^{-2}M$$

$$[CO_3^{2-}] = \left.\frac{0.100\ mol}{1\ dm^3}\right|\frac{0.500\ \cancel{dm^3}}{1.000\ \cancel{dm^3}} = 5.00 \times 10^{-2}M$$

$[Li^+]^2[CO_3^{2-}] = (5.00 \times 10^{-2})^2(5.00 \times 10^{-2})$
$= 1.25 \times 10^{-4}$
Since the K_{sp} of Li_2CO_3 is larger than this ($K_{sp} = 2.51 \times 10^{-2}$), no precipitate will form.

9. $H_2O + H_2O \rightleftarrows H_3O^+ + OH^-$
   $K_w = [H_3O^+][OH^-]$
   $1.00 \times 10^{-14} = x(6.80 \times 10^{-10})$
   $x = 1.47 \times 10^{-5}M$

10. $H_2O + H_2O \rightleftarrows H_3O^+ + OH^-$
    $K_w = [H_3O^+][OH^-]$
    $1.00 \times 10^{-14} = x(5.67 \times 10^{-3})$
    $x = 1.76 \times 10^{-12}M$

11. a. $-\log(1.00 \times 10^{-3}) = 3.000$
    b. $-\log(1.00 \times 10^{-6}) = 6.000$
    c. $-\log(6.59 \times 10^{-10}) = 9.181$
    d. $-\log(7.01 \times 10^{-6}) = 5.154$
    e. $-\log(9.47 \times 10^{-8}) = 7.024$
    f. $-\log(6.89 \times 10^{-14}) = 13.162$

12. a. $\text{antilog}(-3.000) = 1.00 \times 10^{-3}M$
    b. $\text{antilog}(-10.000) = 1.00 \times 10^{-10}M$
    c. $\text{antilog}(-6.607) = 2.47 \times 10^{-7}M$
    d. $\text{antilog}(-2.523) = 3.00 \times 10^{-3}M$
    e. $\text{antilog}(-6.149) = 7.10 \times 10^{-7}M$
    f. $\text{antilog}(-7.622) = 2.18 \times 10^{-8}M$

13. a. $pH = 14.00 - pOH = 14.00 - 2.00 = 12.00$
    b. $pH = 14.00 - pOH = 14.00 - 7.00 = 7.00$
    c. $pH = 14.00 - pOH = 14.00 - 1.263 = 12.737$
    d. $pH = 14.00 - pOH = 14.00 - 4.976 = 9.024$
    e. $pH = 14.00 - pOH = 14.00 - 9.714 = 4.286$
    f. $pH = 14.00 - pOH = 14.00 - 3.004 = 10.996$

14. a. $[H_3O^+] = \frac{1.00 \times 10^{-14}}{1.00 \times 10^{-4}} = 1.00 \times 10^{-10}M$;
    $pH = -\log(1.00 \times 10^{-10}) = 10.000$
    b. $[H_3O^+] = \frac{1.00 \times 10^{-14}}{1.00 \times 10^{-6}} = 1.00 \times 10^{-8}M$;
    $pH = -\log(1.00 \times 10^{-8}) = 8.000$
    c. $[H_3O^+] = \frac{1.00 \times 10^{-14}}{2.64 \times 10^{-13}} = 3.79 \times 10^{-2}M$;
    $pH = -\log(3.79 \times 10^{-2}) = 1.422$
    d. $[H_3O^+] = \frac{1.00 \times 10^{-14}}{3.45 \times 10^{-8}} = 2.90 \times 10^{-7}M$;
    $pH = -\log(2.90 \times 10^{-7}) = 6.538$
    e. $[H_3O^+] = \frac{1.00 \times 10^{-14}}{4.97 \times 10^{-10}} = 2.01 \times 10^{-5}M$;
    $pH = -\log(2.01 \times 10^{-5}) = 4.696$

f. $[H_3O^+] = \dfrac{1.00 \times 10^{-14}}{2.93 \times 10^{-2}} = 3.41 \times 10^{-13}M$;

$pH = -\log(3.41 \times 10^{-13}) = 12.467$

15. $[H_3O^+] = \text{antilog}(-pH)$
- **a.** antilog $(-12.00) = 1.00 \times 10^{-12}M$
- **b.** antilog $(-7.00) = 1.00 \times 10^{-7}M$
- **c.** antilog $(-12.737) = 1.83 \times 10^{-13}M$
- **d.** antilog $(-9.024) = 9.46 \times 10^{-10}M$
- **e.** antilog $(-4.286) = 5.18 \times 10^{-5}M$
- **f.** antilog $(-10.966) = 1.01 \times 10^{-11}M$

16. **a.** neutral **b.** basic **c.** acidic

17. **a.** acidic **d.** acidic
b. acidic **e.** basic
c. acidic **f.** basic

24. $\dfrac{21.4\ \cancel{\text{cm}^3\ \text{HCl}}}{} \left| \dfrac{0.106\ \cancel{\text{mol HCl}}}{1000\ \cancel{\text{cm}^3\ \text{HCl}}} \right| \dfrac{1\ \cancel{\text{mol NaOH}}}{1\ \cancel{\text{mol HCl}}} \left| \dfrac{1000\ \text{cm}^3\ \text{NaOH}}{0.0947\ \cancel{\text{mol NaOH}}} \right. = 24.0\ \text{cm}^3\ \text{NaOH}$

25. $\dfrac{21.7\ \cancel{\text{cm}^3\ \text{soln}}}{} \left| \dfrac{0.500\ \cancel{\text{mol HBr}}}{1000\ \cancel{\text{cm}^3\ \text{soln}}} \right| \dfrac{1\ \text{mol LiOH}}{1\ \cancel{\text{mol HBr}}} \left| \dfrac{}{26.4\ \cancel{\text{cm}^3}\ \text{soln}} \right| \dfrac{1000\ \cancel{\text{cm}^3}}{1\ \text{dm}^3} = 0.411M\ \text{LiOH}$

26. $\dfrac{23.4\ \cancel{\text{cm}^3\ \text{soln}}}{} \left| \dfrac{0.551\ \cancel{\text{mol NaOH}}}{1000\ \cancel{\text{cm}^3\ \text{soln}}} \right| \dfrac{1\ \text{mol HCl}}{1\ \cancel{\text{mol NaOH}}} \left| \dfrac{}{50\ \cancel{\text{cm}^3}\ \text{soln}} \right| \dfrac{1000\ \cancel{\text{cm}^3}}{1\ \text{dm}^3} = 0.258M\ \text{HCl}$

Chapter 25

1.
- **a.** $2(1+) + x + 3(2-) = 0; x = 4+$
- **b.** $(1+) + x + 4(2-) = 0; x = 7+$
- **c.** $(2+) + 2x + 6(-2) = 0; x = 5+$
- **d.** $2(1+) + x + 3(2-) = 0; x = 4+$
- **e.** $x + 2(2-) = 0; x = 4+$
- **f.** $(1+) + x + 4(2-) = 1-; x = 6+$
- **g.** $2(1+) + 2x + 7(2-) = 0; x = 6+$
- **h.** $2(3+) + 3x = 0; x = 2-$
- **i.** $x + 2(1-) = 0; x = 2+$
- **j.** $12x + 22(1+) + 11(2-) = 0; x = 0$

2.

	Oxidizing agent	Reducing Agent
a.	Cu^{2+}	Mg
b.	Sn^{4+}	Fe
c.	S	Na

3. **a.** Mg; 2 electrons
b. Fe; 2 electrons
c. Na; 1 electron per atom

4. yes
H is oxidized and H_2 is the reducing agent.
N is reduced and N_2 is the oxidizing agent.

5. yes
C is oxidized and the reducing agent
H is reduced
H_2O is oxidizing agent

6. not a redox reaction

7. not a redox reaction

8. not a redox reaction

9. yes
O is reduced
S is oxidized
H_2O_2 is oxidizing agent
PbS is reducing agent

10. not a redox reaction

11. yes
N is reduced
P is oxidized
HNO_3 is oxidizing agent
H_3PO_3 is reducing agent

12. yes
N is reduced
I is oxidized
HNO_3 is oxidizing agent
I_2 is the reducing agent

13. not a redox reaction

14. yes
N is reduced
Fe^{2+} is oxidized and the reducing agent
NO_3^- is oxidizing agent

15. yes
Br_2 is reduced and the oxidizing agent
Fe^{2+} is oxidized
$FeBr_2$ is reducing agent

16. yes
I is reduced
S is oxidized
$S_2O_3^{2-}$ is the reducing agent
I_2 is the oxidizing agent

17. yes
Mn is reduced
O is oxidized
MnO_4^- is oxidizing agent
H_2O_2 is reducing agent

23. **a.** $K_2Cr_2O_7$ is the oxidizing agent
Na_2SO_3 is the reducing agent
b. MnO_2 is the oxidizing agent
KI is the reducing agent
c. H_2O_2 is the reducing agent
MnO_2 is the oxidizing agent
d. K_2MnO_4 is the oxidizing agent
HCl is the reducing agent
e. Al is the reducing agent
HCl is the oxidizing agent

24. **a.** Reduction half-reaction:

Skeleton	$Ag^+ \rightarrow Ag$
Electrons	$Ag^+ + e^- \rightarrow Ag$
Check Charge	$0 = 0$

Oxidation half-reaction:

Skeleton	$Cu \rightarrow Cu^{2+}$
Electrons	$Cu \rightarrow Cu^{2+} + 2e^-$
Check Charge	$0 = 0$

Multiply reduction half-reaction by 2, add, and simplify:

$$2Ag^+ + 2e^- \rightarrow 2Ag$$
$$Cu \rightarrow Cu^{2+} + 2e^-$$
$$\overline{2Ag^+ + Cu \rightarrow 2Ag + Cu^{2+}}$$

b. Reduction half-reaction:

Skeleton	$NO_3^- \rightarrow NO$
Electrons	$NO_3^- + 3e^- \rightarrow NO$
Oxygen	$NO_3^- + 3e^- \rightarrow NO + 2H_2O$
Hydrogen	$NO_3^- + 3e^- + 4H^+ \rightarrow NO + 2H_2O$
Check Charge	$0 = 0$

Oxidation half-reaction:

Skeleton	$H_2S \rightarrow S$
Electrons	$H_2S \rightarrow S + 2e^-$
Hydrogen	$H_2S \rightarrow S + 2e^- + 2H^+$
Check Charge	$0 = 0$

Multiply to balance electrons, add, and simplify:

$$2NO_3^- + 6e^- + 8H^+ \rightarrow 2NO + 4H_2O$$
$$3H_2S \rightarrow 3S + 6e^- + 6H^+$$
$$\overline{2H^+ + 2NO_3^- + 3H_2S \rightarrow 3S + 2NO + 4H_2O}$$

c. Reduction half-reaction:

Skeleton	$PbO_2 + HSO_4^- \rightarrow PbSO_4$
Electrons	$PbO_2 + HSO_4^- + 2e^- \rightarrow PbSO_4$
Oxygen	$PbO_2 + HSO_4^- + 2e^- \rightarrow PbSO_4 + 2H_2O$
Hydrogen	$PbO_2 + HSO_4^- + 2e^- + 3H^+ \rightarrow PbSO_4 + 2H_2O$
Check Charge	$0 = 0$

Oxidation half-reaction:

Skeleton	$Pb + HSO_4^- \rightarrow PbSO_4$
Electrons	$Pb + HSO_4^- \rightarrow PbSO_4 + 2e^-$
Hydrogen	$Pb + HSO_4^- \rightarrow PbSO_4 + 2e^- + H^+$
Check Charge	$1- = 1-$

Add and simplify:

$$PbO_2 + HSO_4^- + 2e^- + 3H^+ \rightarrow PbSO_4 + 2H_2O$$
$$Pb + HSO_4^- \rightarrow PbSO_4 + 2e^- + H^+$$
$$\overline{Pb + PbO_2 + 2HSO_4^- + 2H^+ \rightarrow 2PbSO_4 + 2H_2O}$$

25. $(8H^+ + 5e^- + MnO_4^- \rightarrow Mn^{2+} + 4H_2O) \times 2$
$(H_2O + H_2SO_3 \rightarrow HSO_4^- + 2e^- + 3H^+) \times 5$
$16H^+ + 10e^- + 2MnO_4^- \rightarrow 2Mn^{2+} + 8H_2O$
$5H_2O + 5H_2SO_3 \rightarrow 5HSO_4^- + 10e^- + 15H^+$
$\overline{H^+ + 2MnO_4^- + 5H_2SO_3 \rightarrow 2Mn^{2+} + 5HSO_4^- + 3H_2O}$

26. $(14H^+ + 6e^- + Cr_2O_7^{2-} \rightarrow 2Cr^{3+} + 7H_2O)$
$(2I^- \rightarrow I_2 + 2e^-) \times 3$
$14H^+ + 6e^- + Cr_2O_7^{2-} \rightarrow 2Cr^{3+} + 7H_2O$
$6I^- \rightarrow 3I_2 + 6e^-$
$\overline{Cr_2O_7^{2-} + 14H^+ + 6I^- \rightarrow 2Cr^{3+} + 3I_2 + 7H_2O}$

27. $(5OH^- + NH_3 \rightarrow NO + 5e^- + 4H_2O) \times 4$
$4H_2O + 4e^- + O_2 \rightarrow 2H_2O + 4OH^-$
Simplifying the 2nd equation gives:
$2H_2O + 4e^- + O_2 \rightarrow 4OH^-$
$(2H_2O + 4e^- + O_2 \rightarrow 4OH^-) \times 5$
$20OH^- + 4NH_3 \rightarrow 4NO + 20e^- + 16H_2O$
$10H_2O + 20e^- + 5O_2 \rightarrow 20OH^-$
$\overline{4NH_3 + 5O_2 \rightarrow 4NO + 6H_2O \text{ (basic)}}$

28. $(5H_2O + As_2O_3 \rightarrow 2H_3AsO_4 + 4e^- + 4H^+) \times 3$
$(4H^+ + 3e^- + NO_3^- \rightarrow NO + 2H_2O) \times 4$
$15H_2O + 3As_2O_3 \rightarrow 6H_3AsO_4 + 12e^- + 12H^+$
$16H^+ + 12e^- + 4NO_3^- \rightarrow 4NO + 8H_2O$
$\overline{3As_2O_3 + 4H^+ + 4NO_3^- + 7H_2O \rightarrow 6H_3AsO_4 + 4NO}$

29. $(2e^- + I_2 \rightarrow 2I^-) \times 1$
$(H_2O + H_2SO_3 \rightarrow HSO_4^- + 2e^- + 3H^+) \times 1$
$\overline{I_2 + H_2SO_3 + H_2O \rightarrow 2I^- + HSO_4^- + 3H^+}$

30. $(8H^+ + 8e^- + H_3AsO_4 \rightarrow AsH_3 + 4H_2O) \times 1$
$(Zn \rightarrow Zn^{2+} + 2e^-) \times 4$
$8H^+ + 8e^- + H_3AsO_4 \rightarrow AsH_3 + 4H_2O$
$4Zn \rightarrow 4Zn^{2+} + 8e^-$
$\overline{H_3AsO_4 + 8H^+ + 4Zn \rightarrow AsH_3 + 4H_2O + 4Zn^{2+}}$

31. $(MnO_4^{2-} \rightarrow MnO_4^- + e^-) \times 2$
$(4H^+ + 2e^- + MnO_4^{2-} \rightarrow MnO_2 + 2H_2O) \times 1$
$2MnO_4^{2-} \rightarrow 2MnO_4^- + 2e^-$
$\overline{3MnO_4^{2-} + 4H^+ \rightarrow 2MnO_4^- + MnO_2 + 2H_2O}$

32. $(8H^+ + 5e^- + MnO_4^- \rightarrow Mn^{2+} + 4H_2O) \times 2$
$(2H_2O + SO_2 \rightarrow SO_4^{2-} + 2e^- + 4H^+) \times 5$
$16H^+ + 10e^- + 2MnO_4^- \rightarrow 2Mn^{2+} + 8H_2O$
$10H_2O + 5SO_2 \rightarrow 5SO_4^{2-} + 10e^- + 20H^+$
$\overline{2MnO_4^- + 5SO_2 + 2H_2O \rightarrow 2Mn^{2+} + 5SO_4^{2-} + 4H^+}$

33. $e^- + NO_2 \rightarrow NO_2^-$
$2OH^- + NO_2 \rightarrow NO_3^- + e^- + H_2O$
$\overline{2NO_2 + 2OH^- \rightarrow NO_2^- + NO_3^- + H_2O \text{ (basic)}}$

34. $(4Cl^- + HgS \rightarrow S + 2e^- + HgCl_4^{2-}) \times 3$
$(4H^+ + 3e^- + NO_3^- \rightarrow NO + 2H_2O) \times 2$
$12Cl^- + 3HgS \rightarrow 3S + 6e^- + 3HgCl_4^{2-}$
$8H^+ + 6e^- + 2NO_3^- \rightarrow 2NO + 4H_2O$
$\overline{8H^+ + 3HgS + 12Cl^- + 2NO_3^- \rightarrow 3HgCl_4^{2-} + 3S + 2NO + 4H_2O}$

Chapter 26

6. -1.676 V + 0.257 V = -1.419 V; no reaction
7. 0.7991 V + 0.277 V = 1.076 V; reaction occurs
8. -0.154 V + 0.1251 V = -0.029 V; no reaction
9. 1.52 V + 0.44 V = 1.96 V; reaction occurs

10.

a.	$Zn \rightarrow Zn^{2+} + 2e^-$	0.7626 V
	$Fe^{2+} + 2e^- \rightarrow Fe$	−0.44 V
		0.32 V
b.	$Mn \rightarrow Mn^{2+} + 2e^-$	1.18 V
	$Br_2 + 2e^- \rightarrow 2Br^-$	1.0652 V
		2.25 V
c.	$H_2C_2O_4 \rightarrow 2CO_2 + 2H^+ + 2e^-$	0.475 V
	$MnO_4^- + 8H^+ + 5e^- \rightarrow Mn^{2+} + 4H_2O$	1.51 V
		1.99 V
d.	$Ni \rightarrow Ni^{2+} + 2e^-$	0.257 V
	$Hg_2^{2+} + 2e^- \rightarrow 2Hg$	0.7960 V
		1.053 V
e.	$Cu \rightarrow Cu^{2+} + 2e^-$	−0.340 V
	$Ag^+ + e^- \rightarrow Ag$	0.7991 V
		0.459 V
f.	$Pb \rightarrow Pb^{2+} + 2e^-$	0.1251 V
	$Cl_2 + 2e^- \rightarrow 2Cl^-$	1.35828 V
		1.4834 V

11. a. H_2 and O_2—Hydrogen reduction requires less energy than that of sodium; sulfate cannot be oxidized.
b. Cu and O_2—Copper reduction requires less energy than that of hydrogen; oxygen oxidation requires less energy than that of chlorine.
c. Co and O_2—Cobalt reduction requires less energy than that of hydrogen; oxygen oxidation requires less energy than that of fluorine.
d. Pb and O_2—Lead reduction requires less energy than that of hydrogen; nitrate cannot be oxidized.
e. H_2 and O_2—Hydrogen reduction requires less energy than that of lithium; oxygen oxidation requires less energy than that of bromine.
f. H_2 and O_2—Hydrogen reduction requires less energy than that of sodium; oxygen oxidation requires less energy than that of chlorine.

12. $Pb + Sn^{4+} \rightarrow Pb^{2+} + Sn^{2+}$

$$E = E° - \frac{0.059\ 16}{n}\log K = E° - \frac{0.059\ 16}{n}\log\frac{[Pb^{2+}][Sn^{2+}]}{[Sn^{4+}]}$$

$$E = 0.279 - \frac{0.059\ 16}{2}\log\frac{(0.458)(0.346)}{(0.652)} = 0.296 \text{ V}$$

13. $Ni + Cu^{2+} \rightarrow Ni^{2+} + Cu$

$$E = E° - \frac{0.059\ 16}{n}\log K = E° - \frac{0.059\ 16}{n}\log\frac{[Ni^{2+}]}{[Cu^{2+}]}$$

$$E = 0.597 - \frac{0.059\ 16}{2}\log\frac{(1.00)}{(0.0100)} = 0.538 \text{ V}$$

14. $Fe + 2Ag^+ \rightarrow Fe^{2+} + 2Ag$

$$E = E^\circ - \frac{0.059\ 16}{n}\log K = E^\circ - \frac{0.059\ 16}{n}\log\frac{[Fe^{2+}]}{[Ag^+]^2}$$

$$E = 1.24 - \frac{0.059\ 16}{2}\log\frac{(0.720)}{(0.785)^2} = 1.24\ V$$

15. $Ag^+(aq) + e^- \rightarrow Ag(cr)$

$$\frac{1.00\ A}{} \Big| \frac{9650\ s}{} \Big| \frac{1\ mol\ e^-}{96\ 485\ A{\cdot}s} \Big| \frac{1\ mol\ Ag}{1\ mol\ e^-} \Big| \frac{108\ g\ Ag}{1\ mol\ Ag}$$

$= 10.8$ g Ag

16. $Ag^+(aq) + e^- \rightarrow Ag(cr)$

$$\frac{5.00\ A}{} \Big| \frac{10.0\ min}{} \Big| \frac{60\ s}{1\ min} \Big| \frac{1\ mol\ e^-}{96\ 485\ A{\cdot}s} \Big| \frac{1\ mol\ Ag}{1\ mol\ e^-} \Big| \frac{108\ g\ Ag}{1\ mol\ Ag} = 3.36\ g\ Ag$$

17. $Cu^{2+}(aq) + 2e^- \rightarrow Cu(cr)$

$$\frac{1.75\ mol\ Cu}{6.24\ min} \Big| \frac{2\ mol\ e^-}{1\ mol\ Cu} \Big| \frac{96\ 485\ A{\cdot}s}{1\ mol\ e^-} \Big| \frac{1\ min}{60\ s} = 902\ A$$

18. $2Cl^-(aq) \rightarrow Cl_2(g) + 2e^-$

$$\frac{1.00\ mol\ Cl_2}{5.00\ A} \Big| \frac{2\ mol\ e^-}{1\ mol\ Cl_2} \Big| \frac{96\ 485\ A{\cdot}s}{1\ mol\ e^-} = 3.86 \times 10^4\ s$$

$= 643$ min $= 10.7$ h

19. $Ca^{2+}(aq) + 2e^- \rightarrow Ca(cr)$

$$\frac{0.375\ g\ Ca}{3.93\ A} \Big| \frac{1\ mol\ Ca}{40.1\ g\ Ca} \Big| \frac{2\ mol\ e^-}{1\ mol\ Ca} \Big| \frac{96\ 485\ A{\cdot}s}{1\ mol\ e^-} \Big| \frac{1\ min}{60\ s}$$

$= 7.65$ min

Chapter 27

1. $\Delta U = q + w = 350 + 230 = 580$ J

2. $w = \Delta U - q = -2.37 - 0.65 = -3.02$ kJ

3. $\Delta U = q + w = 419 + (-389) = 3.0 \times 10^1$ kJ

4. $\Delta U = q + w = -196 + (+4.20 \times 10^2) = 224$ kJ

5. $2NO(g) + O_2(g) \rightarrow 2NO_2(g)$

$\Delta H_r^\circ = [2(33.18)] - [2(90.25) + 0]$

$\Delta H_r^\circ = -144.14$ kJ

6. $4FeO(cr) + O_2(g) \rightarrow 2Fe_2O_3(cr)$

$\Delta H_r^\circ = [2(-824.2)] - [4(-272.0) + 0]$

$\Delta H_r^\circ = -560.4$ kJ

7. $H_2SO_4(l) \rightarrow SO_3(g) + H_2O(l)$

$BaO(cr) + SO_3(g) \rightarrow BaSO_4(cr)$

$BaO(cr) + SO_3(g) + H_2SO_4(l) \rightarrow BaSO_4(cr) + SO_3(g) + H_2O(l)$

$\Delta H_r^\circ = 78.2$ kJ

$\Delta H_r^\circ = -213.4$ kJ

$\Delta H_r^\circ = -135.2$ kJ

12. **a.** positive
b. negative
c. negative
d. positive

13. **a.** positive
b. negative

14. $\Delta G_r^\circ = \Sigma\Delta G_{f\ (products)}^\circ - \Sigma\Delta G_{f\ (reactants)}^\circ$

$$= \left[\frac{1\ mol\ BaSO_4}{} \Big| \frac{(-1362.2)\ kJ}{1\ mol\ BaSO_4}\right] + \left[\frac{2\ mol\ HCl(aq)}{} \Big| \frac{(-131.2)\ kJ}{1\ mol\ HCl(aq)}\right] -$$

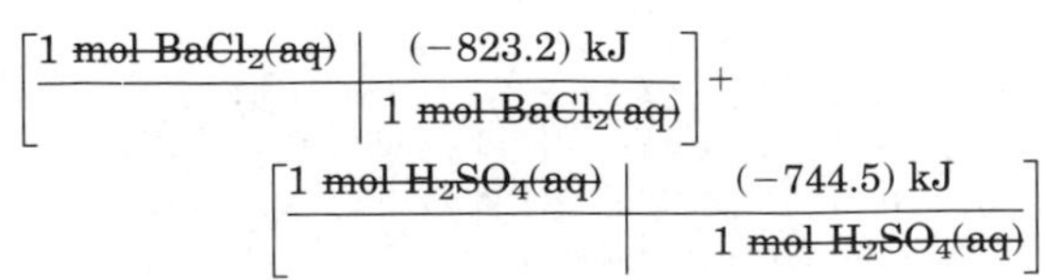

$$\left[\frac{1\ mol\ BaCl_2(aq)}{} \Big| \frac{(-823.2)\ kJ}{1\ mol\ BaCl_2(aq)}\right] + \left[\frac{1\ mol\ H_2SO_4(aq)}{} \Big| \frac{(-744.5)\ kJ}{1\ mol\ H_2SO_4(aq)}\right]$$

$= -56.9$ kJ; Yes, the reaction is spontaneous.

15. $\Delta H_r^\circ = \Sigma\Delta H_{f\ (products)}^\circ - \Sigma\Delta H_{f\ (reactants)}^\circ$

$$= \left[\frac{1\ mol\ Ca(OH)_2}{} \Big| \frac{(-986.09)\ kJ}{1\ mol\ Ca(OH)_2}\right] + \left[\frac{1\ mol\ H_2}{} \Big| \frac{0\ kJ}{1\ mol\ H_2}\right] - \left[\frac{1\ mol\ Ca}{} \Big| \frac{0\ kJ}{1\ mol\ Ca} + \frac{2\ mol\ H_2O}{} \Big| \frac{(-285.830)\ kJ}{1\ mol\ H_2O}\right]$$

$= -414.43$ kJ

$\Delta G_r^\circ = \Sigma\Delta G_{f\ (products)}^\circ - \Sigma\Delta G_{f\ (reactants)}^\circ$

$$= \left[\frac{1\ mol\ Ca(OH)_2}{} \Big| \frac{(-898.49)\ kJ}{1\ mol\ Ca(OH)_2}\right] + \left[\frac{1\ mol\ H_2}{} \Big| \frac{0\ kJ}{1\ mol\ H_2}\right] - \left[\frac{1\ mol\ Ca}{} \Big| \frac{0\ kJ}{1\ mol\ Ca} + \frac{2\ mol\ H_2O}{} \Big| \frac{(-237.129)\ kJ}{1\ mol\ H_2O}\right]$$

$= -424.23$ kJ

$$\Delta G_r^\circ = \Delta H_r^\circ - T\Delta S_r^\circ;\ \Delta S_r^\circ = \frac{\Delta H_r^\circ - \Delta G_r^\circ}{T}$$

$$\Delta S_r^\circ = \frac{-414.43\ kJ - (-424.23)\ kJ}{298\ K}$$

$= 32.9$ J/K

16. $\Delta G_r^\circ = \Sigma\Delta G_{f\ (products)}^\circ - \Sigma\Delta G_{f\ (reactants)}^\circ$

$$= \left[\frac{1\ mol\ H_2}{} \Big| \frac{0\ kJ}{1\ mol\ H_2} + \frac{1\ mol\ F_2}{} \Big| \frac{0\ kJ}{1\ mol\ F_2}\right] - \left[\frac{2\ mol\ HF}{} \Big| \frac{(-273.2)\ kJ}{1\ mol\ HF}\right]$$

$= +546.4$ kJ, not spontaneous

17. $\Delta S_r^\circ = \Sigma S^\circ_{(products)} - \Sigma S^\circ_{(reactants)}$

$$\Delta S_r^\circ = 463.6\ \frac{J}{mol{\cdot}K} - \left(462.8\ \frac{J}{mol{\cdot}K} + 130.7\ \frac{J}{mol{\cdot}K}\right)$$

$$= -129.9\ \frac{J}{mol{\cdot}K}$$

18. $$\log K_{eq} = -\frac{\Delta G}{2.30RT} = -\frac{28.95}{(2.30)(8.31 \times 10^{-3})(298)}$$

$= -5.0828$

$\text{antilog}(-5.0828) = 8.26 \times 10^{-6}$

19. $\Delta G = -2.30RT(\log K_{eq})$

$= (-2.30)(0.008\ 31)(900)(\log 3.16 \times 10^{-4})$

$= 60.2$ kJ/mol

20. $\Delta G = -2.30RT(\log K_{eq})$

$= (-2.30)(0.008\ 31)(473)(\log 0.457)$

$= 3.07$ kJ/mol

21. $E^\circ = (0.52 + 0.8535)V = 1.37$ V

$\Delta G = -nFE^\circ = (-2)(96\ 485)(1.37) = -264$ kJ

22. $E^\circ = (0.9110 + 0.7626)V = 1.6736$ V

$\Delta G = -nFE^\circ = (-2)(96\ 485)(1.6736) = -322.95$ kJ

Chapter 28

1. $^{27}_{12}Mg \rightarrow {}^{0}_{-1}e + {}^{27}_{13}Al$
2. $^{49}_{24}Cr \rightarrow {}^{0}_{+1}e + {}^{49}_{23}V$
3. $^{76}_{36}Kr + {}^{0}_{-1}e \rightarrow {}^{76}_{35}Br$
4. $^{213}_{88}Ra \rightarrow {}^{4}_{2}He + {}^{209}_{86}Rn$
5. $^{231}_{90}Th \rightarrow {}^{4}_{2}He + {}^{227}_{88}Ra$
6. From Table 28.1, the half-life of $^{69}_{32}Ge$ is 36 hours.

$$\text{Number of half-lives} = \frac{15\ \cancel{\text{days}}}{36\ \cancel{\text{h}}} \times \frac{24\ \cancel{\text{h}}}{1\ \cancel{\text{day}}} = 10$$

$\left(\frac{1}{2}\right)^{10} = \frac{1}{1024}$ of the original sample remains after 15 days.

7. $\dfrac{5.32 \times 10^{9}}{5.20 \times 10^{6}} = 1020 = 2^{10} = 10$ half-lives

$10(32.1\ \text{h}) = 321\ \text{h} = 13.4$ days

8. $\dfrac{3.03 \times 10^{6}}{3.79 \times 10^{5}} = 7.99 = 8$ half-lives

$2^{8} = 256; \dfrac{5.80 \times 10^{28}}{256} = 2.27 \times 10^{26}$ atoms

Chapter 29

1. c. 2,3-dimethylpentane
2. a. 2,2-dimethylbutane
   b. 3-methylpentane
   c. hexane
   d. 3-ethyl-2-methylhexane
3. a.

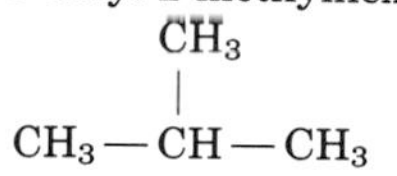

```
         CH3
          |
    CH3 — CH — CH3
```

b.

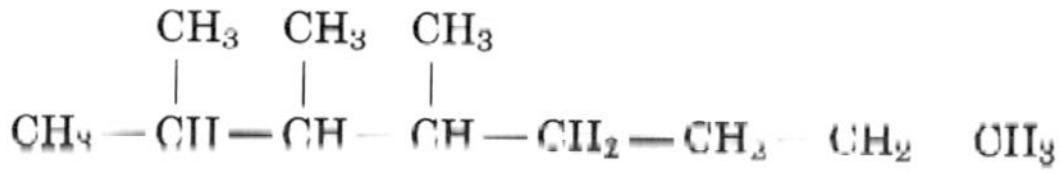

```
         CH3   CH3   CH3
          |     |     |
    CH3 — CH — CH — CH — CH2 — CH2 — CH2 — CH3
```

c.

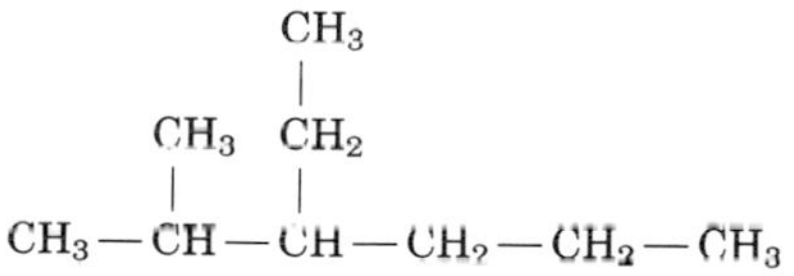

```
                CH3
                 |
          CH3   CH2
           |     |
    CH3 — CH — CH — CH2 — CH2 — CH3
```

d.

```
                          CH2CH3
                            |
                           CH2
                            |
    CH3—CH2—CH2—CH—CH2—CH2—CH2—CH3
```

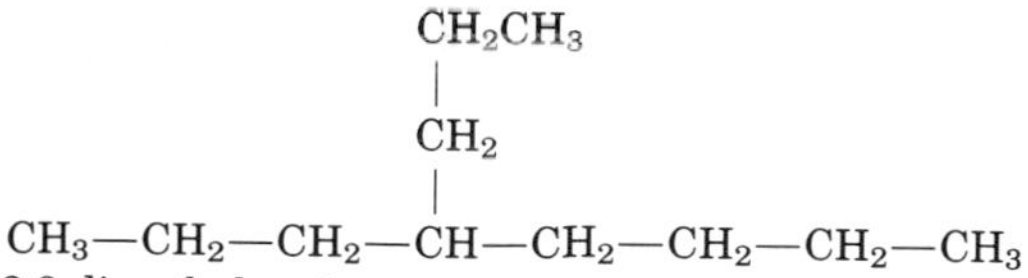

4. a. 2,3-dimethylpentane
   b. octane
   c. 2,2,3-trimethylheptane
   d. 2,3,4-trimethylhexane
   e. 2,2-dimethylbutane
5. a.

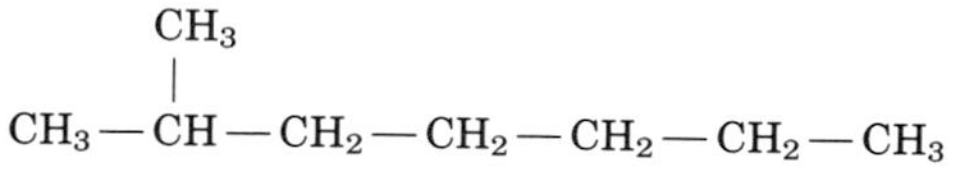

```
          CH3
           |
    CH3 — CH — CH2 — CH2 — CH2 — CH2 — CH3
```

b.

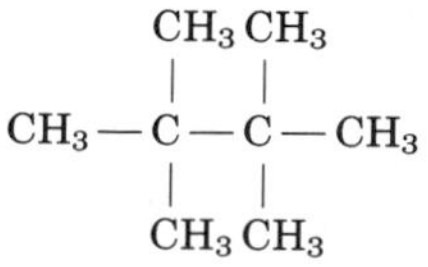

```
          CH3 CH3
           |   |
    CH3 — C — C — CH3
           |   |
          CH3 CH3
```

c.

```
          CH3         CH3
           |           |
    CH3 — C — CH2 — CH — CH3
           |
          CH3
```

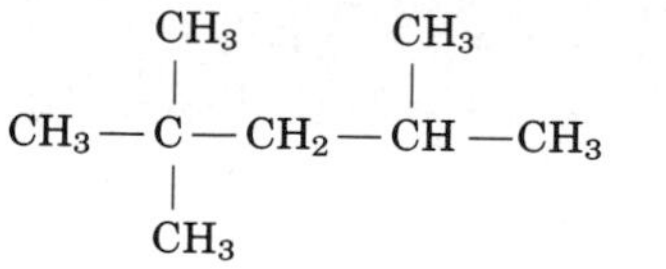

d.

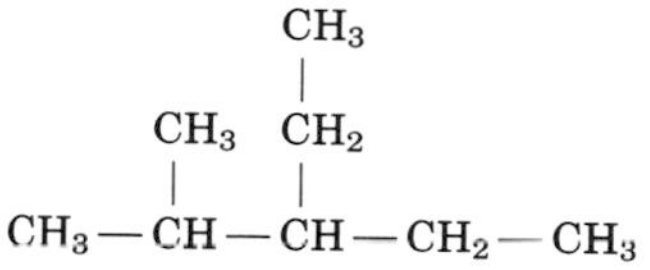

```
                CH3
                 |
          CH3   CH2
           |     |
    CH3 — CH — CH — CH2 — CH3
```

e.

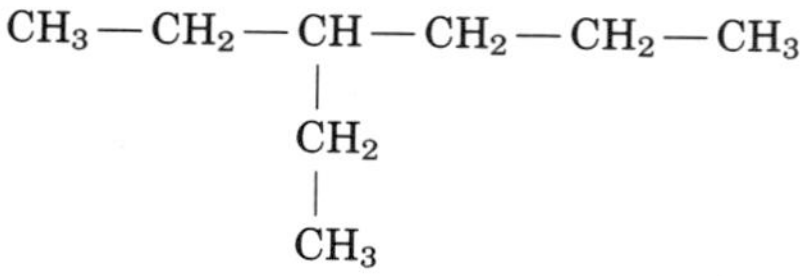

```
    CH3 — CH2 — CH — CH2 — CH2 — CH3
                 |
                CH2
                 |
                CH3
```

6.

```
    C — C — C — C — C — C            C
                                      |
        C                       C — C — C — C
        |                             |
    C — C — C — C — C                 C
            C                         C   C
            |                         |   |
    C — C — C — C — C           C — C — C — C
```

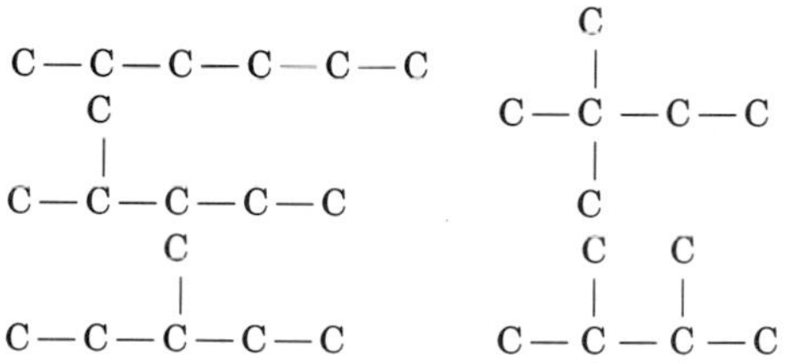

(Note: carbon forms four bonds. Hydrogen atoms are not shown because of space limitations.)

7. hexane
   2-methylpentane
   3-methylpentane
   2,2-dimethylbutane
   2,3-dimethylbutane

Chapter 30

1. $H_2O + (CH_3)_3Cl \rightarrow HI + (CH_3)_3COH$
2. $CH_3CH_2CH_3 + Cl_2 \rightarrow CH_3CH_2CH_2Cl + HCl$ or any of fifteen other products through perchloropropane, $CCl_3-CCl_2-CCl_3$
3. 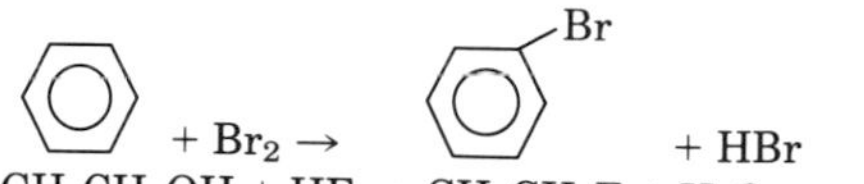 $C_6H_6 + Br_2 \rightarrow C_6H_5Br + HBr$ (benzene ring structures)
4. $CH_3CH_2OH + HF \rightarrow CH_3CH_2F + H_2O$
5. $CH_3Cl + NH_3 \rightarrow CH_3NH_2 + HCl$
6. $CH_3CH_2CH{=}CH_2 + Br_2 \rightarrow CH_3CH_2CHBrCH_2Br$
7. $CH_2{=}CHCH_3 + HCl \rightarrow CH_3CHClCH_3$
8. $CH_2{=}CH_2 + H_2SO_4 \rightarrow CH_3CH_2OSO_3H$
9. $CH_3CH_2CH_2CH_2OH \xrightarrow{H_2SO_4} CH_3CH_2CH{=}CH_2 + H_2O$
10. $CH_3CH_2OH + CH_3CH_2CH_2COOH \rightarrow CH_3CH_2OCCH_2CH_2CH_3 + H_2O$ (C=O on the carbonyl carbon)
11. $CH_3OH + HCOOH \rightarrow CH_3OCH + H_2O$ (C=O on the CH carbon)

GLOSSARY

A

absolute zero: the temperature at which all molecular motion should cease
accelerator: a device that is used to accelerate charged particles to high speeds
accuracy: the relationship between the graduations on a measuring device and the actual standard for the quantity being measured
acid: a substance that produces hydrogen ions in water solution; a proton donor
acidic anhydride: a nonmetallic oxide that can react with water to form an acid
actinoid series: fourteen elements beginning with actinium in which the arrow diagram predicts the highest energy electrons to be in the 5*f* sublevel
activated complex: an assembly of atoms in an excited state between reactants and products in a chemical reaction
activation energy: the energy required to form the activated complex
activity (ion): the effective concentration of a species
addition polymerization: the formation of a polymer through addition reactions
addition reaction: the combining of two or more molecules through adding on at the double or triple bond of an unsaturated organic compound
adiabatic system: system in which heat neither leaves nor enters
adsorption: the process of one substance being attracted and held to the surface of another substance
alcohol: one of a class of organic compounds characterized by the presence of the hydroxyl group, $-OH$
aldehyde: one of the class of organic compounds characterized by the presence of the carbonyl group ($\rangle C=O$) on the end of the carbon chain ($RCHO$)
aliphatic: hydrocarbons consisting of chains or nonaromatic rings
alkali metal: an element in Group 1(IA)
alkaline earth metal: an element in Group 2(IIA)
alkane: an aliphatic hydrocarbon having only single carbon-carbon bonds
alkene: an aliphatic hydrocarbon having one or more double bonds
alkyne: an aliphatic hydrocarbon having one or more triple bonds
allotrope: form of an element differing in crystal or molecular structure
alloy: a mixture of a metal and one or more other elements, usually metals
alpha particle: a helium nucleus
amide: an organic compound containing the $-CO-NH_2$ group; an inorganic compound containing NH_2^-
amine: an organic compound derived from ammonia by replacement of one or more hydrogen atoms by hydrocarbon radicals
amino acid: an organic compound characterized by the presence of an amino group and a carboxyl group on the same carbon atom
amorphous: a noncrystalline material that appears solid but without long-range order; supercooled liquid
ampere: the unit of electric current equal to one coulomb per second
amphoteric: having the ability to act as either an acid or a base
amplitude: the maximum value attained by a wave
anhydrous: without water
anion: a negative ion
anode: the positive electrode (general); the electrode at which oxidation occurs (electrochemical)
antiparticle: a particle identical to a second particle in all respects except for opposite charge and magnetic moment
aromatic compound: an organic ring compound containing one or more benzene rings
arrow diagram: a system for predicting the order of filling energy sublevels with electrons
atom: the smallest particle of an element
atomic mass: the mass of an atom in atomic mass units; the average mass of the atoms of an element
atomic mass unit: one-twelfth the mass of the $^{12}_{6}C$ atom
atomic number: the number of protons in the nucleus of an atom
atomic radius: the distance from the center of an atom to the 90% probability surface of the electron cloud
atomic theory: the body of knowledge concerning the existence of atoms and their characteristic structure
Avogadro constant: the number of objects in a mole; $6.022\ 136\ 7 \times 10^{23}$
Avogadro's principle: Equal volumes of gases at the same temperature and pressure contain the same number of molecules.

B

balance: an instrument used to measure mass
barometer: a manometer used to measure atmospheric pressure
baryon: a subatomic particle classified as a heavy hadron
base: a substance that produces hydroxide ions in water solution; a proton acceptor
basic anhydride: a metallic oxide that will react with water to form a base
beta particle: an electron (−) or positron (+)
binary acid: an acid containing only hydrogen and one other element
binary compound: a compound composed of only two elements

binding energy: the energy required to split the nucleus into separate nucleons
biochemistry: the study of the substances and reactions involved in life processes
body-centered cubic: having a unit cell with a particle at each vertex of a cube and a particle in the center of the cube
Bohr atom: planetary atom model
boiling point: the temperature at which the vapor pressure of a liquid equals the atmospheric pressure
bond: the force holding atoms together in a compound or molecule
bond angle: the angle between two bond axes extending from the same atom
bond axis: the imaginary line connecting the nuclei of two bonded atoms
bond character: the relative ionic or covalent nature of a chemical bond
bond length: the distance between the nuclei of bonded atoms
bond strength: the energy required to break a bond
Boyle's law: The volume of a specific amount of gas varies inversely as the pressure if the temperature remains constant.
branch: a carbon group, named as a radical, that is attached to the main carbon chain in an organic compound
Brownian motion: the random motion of colloidal particles due to their bombardment by the molecules of the dispersing medium
buffer system: a solution that can receive moderate amounts of either acid or base without significant change in its pH

C

calorimeter: a device for measuring the transfer of heat during a chemical or physical change
capillary rise: the tendency of a liquid to rise in a tube of small diameter due to the surface tension of the liquid
carbide ion: a carbon atom that has gained four electrons; C^{4-}
carbohydrate: compounds of carbon, hydrogen, and oxygen that are mostly simple sugars or condensation polymers of sugars
carboxylic acid: the class of organic compounds characterized by the carboxyl group ($-COOH$)
catalysis: the speeding up of chemical reactions by the presence of a substance that is unchanged after the reaction
catalyst: a substance that speeds a chemical reaction without being permanently changed itself
catenation: the joining of like atoms in chains
cathode: the negative electrode (general); the electrode at which reduction occurs (electrochemical)
cathode rays: the beam of electrons in a gas discharge tube
cation: a positive ion
cell potential: the voltage obtained from a voltaic cell
cellulose: a polymer of glucose
Celsius scale: the temperature scale based on the boiling point of water as 100 degrees and the freezing point of water as 0 degrees
chain reaction: a reaction in which the product from each step acts as a reactant for the next step
chalcogen: an element in Group 16(VIA)
Charles's law: The volume of a specific amount of gas varies directly as the absolute temperature if the pressure remains constant.
chemical change: a rearrangement of atoms and/or molecules to produce one or more new substances with new properties
chemical formula: the notation using symbols and numerals to represent the composition of substances
chemical property: a property characteristic of a substance when it is involved in a chemical change
chemical reaction: a change in which one or more substances are changed into one or more new substances
chemical symbol: a notation using one to three letters to represent an element
chemistry: the study of the structure and properties of matter
chromatography: the separation of a mixture using a technique based upon differential adsorption between a stationary phase and a mobile phase
closest packing: the crystal structure in which space between particles is minimized
coefficient: a numeral, representing the number of formula units of the substance, placed before a formula
colligative properties: the properties of solutions that depend only on the number of particles present without regard to type
colloid: a dispersion of particles from 1 nm to 100 nm in at least one dimension in a continuous medium
colloid chemistry: the study of colloids, especially of surfaces
column chromatography: chromatography in which the stationary phase is held in a column
combustion: burning, or reaction with oxygen producing heat and usually light
common ion effect: an equilibrium phenomenon in which an ion common to two or more substances in a solution shifts the point of equilibrium away from itself
complex ion: a central positive ion surrounded by bonded ligands
compound: a substance composed of atoms of two or more elements linked by chemical bonds
concentrated solution: a solution with a high ratio of solute to solvent
concentration: the ratio of the amount of solute to the amount of solvent or solution
condensation polymer: the formation of a polymer and a small molecule, usually water, from monomers
condensed state: solid or liquid
conductivity: the relative ability to conduct heat and electricity
conjugate acid: the particle obtained after a base has gained a proton
conjugate base: the particle remaining after an acid has donated a proton
conjugated system: a group of four or more adjacent atoms in a molecule with an extended π-bonding system
consumer product: an item for sale to the general public

contact catalyst: a catalyst that functions by adsorbing one of the reactants on its surface; a heterogeneous catalyst
containment vessel: a reinforced concrete and steel structure designed to contain any leakage from a nuclear reactor
continuous phase: the dispersing medium in a colloid
control rod: the neutron-absorbing substance used to control the rate of reaction in a nuclear reactor
coordinate covalent bond: a covalent bond in which both electrons of the shared pair come from the same atom
coordination number: the number of points at which ligands are attached to the central atom or ion in a complex ion or coordination compound
corrosion: the chemical interaction of a metal with its environment
coulomb: the quantity of electricity equal to the flow of one ampere for one second
counting number: natural number; any cardinal number except zero
covalent bond: a bond characterized by the sharing of one or more pairs of electrons between two atoms
covalent radius: the radius of an atom along the bond axis
critical pressure: the pressure needed to liquefy a gas at its critical temperature
critical temperature: the highest temperature at which the vapor and liquid states of a substance can exist in equilibrium
crystal: a solid in which the particles are arranged in a regular, repeating pattern
crystal defect: an imperfection in a crystal lattice
crystallization: the forming of crystals by evaporation or cooling
crystalloid: a substance that can penetrate a semipermeable membrane
cubic closest packing: face-centered cubic
cyclic: consisting of atoms bonded in a closed ring
cycloalkanes: hydrocarbons in which the carbon atoms are bonded in a ring and all bonds are single bonds

D

Dalton's law: In a mixture of gases, the total pressure of the mixture is the sum of the partial pressures of each component gas.
de Broglie's hypothesis: Particles may have the properties of waves.
decomposition: a reaction in which a compound breaks into two or more simpler substances
degenerate: having the same energy
dehydrating agent: a substance that can absorb water from other substances
deliquescent: the property of a solid to absorb sufficient water from the air to form a liquid solution
delocalization: the concept in which bonding electrons are not confined to the region between two atoms, but may be spread over several atoms or a whole piece of metal
density: mass per unit volume
desiccant: a drying agent
didentate: a ligand that attaches to the central ion in a complex in two places
diffusion: the spontaneous spreading of particles throughout a given volume until they are uniformly distributed
dilute solution: a solution with a low ratio of solute to solvent
dipeptide: two amino acids linked by an amide bond
dipole: a polar molecule
dipole-dipole force: an attraction between dipoles; component of van der Waals forces
dipole-induced dipole force: an attraction between a dipole and a nonpolar molecule that has been induced to become a dipole; component of van der Waals forces
dipole moment: the strength of a dipole expressed as charge multiplied by distance
dislocation: a crystal defect
dispersed phase: colloidal particles distributed throughout the continuous phase
dispersion forces: the forces between particles that are not permanent dipoles; component of van der Waals forces
dissociation: the separation of ions in solution
distillation: a separation method based on the evaporation of a liquid and the condensation of its vapor
doping: the addition of impurities to a semiconductor to increase electrical conductivity
double bond: a covalent bond in which two atoms share two pairs of electrons
double displacement: a chemical reaction in which the positive part of one compound combines with the negative part of another compound, and vice versa
drift tube: an uncharged tube through which particles being accelerated travel while the decelerating part of an electromagnetic wave passes
ductility: the ability of a substance to be drawn out into a thin wire
dynamic equilibrium: the state in which two opposite changes take place simultaneously and at the same rate so that there is no overall change in the system

E

edge dislocation: crystal defect in which an extra layer of atoms is found between unit cells
effusion: the movement of gas through a small opening
elastic: describing collisions in which kinetic energy is conserved
elastomer: a substance that can be deformed under the influence of an outside force but will return to its original shape once the force is removed
electric current: the flow of charged particles
electrochemistry: the study of the interaction of electric current and chemical reactions
electrode potential: the potential of a reduction half-cell compared to that of the standard hydrogen half-cell
electrolysis: a chemical change caused by an electric current
electrolyte: a substance whose aqueous solution conducts electricity

electrolytic cell: an electric cell in which passage of an electric current causes a chemical reaction
electrolytic conduction: the migration of ions in solution
electromagnetic energy: radiant energy; energy transferred by electromagnetic waves
electron: an elementary particle with unit negative charge
electron affinity: the attraction of an atom for an electron
electron cloud: the space effectively occupied by an electron in an atom
electron configuration: a description of the arrangement of the electrons in an atom
electronegativity: the relative attraction of an atom for a shared pair of electrons
electronic conduction: the flow of electrons in a metal
electrophoresis: the migration of colloidal particles in an electric field
element: a substance whose atoms all have the same number of protons in the nucleus
elimination reaction: an organic reaction in which a small molecule is removed from a larger molecule leaving a double bond in the larger molecule
empirical formula: the formula giving the simplest ratio between the atoms of the elements present in a compound
endergonic: a process having an increase in Gibbs free energy
endothermic: a change that takes place with the absorption of heat
endpoint: the point in a titration where equivalent amounts of reactants are present
energy: a property of matter that can be converted to work under the proper circumstances
energy level: a specific energy or group of energies that may be possessed by electrons in an atom
energy sublevel: a specific energy that may be possessed by electrons within an energy level in an atom
enthalpy: that part of the energy of a substance that is due to the motion of its particles added to the product of its volume and pressure
enthalpy of formation: the net amount of energy produced or consumed when a mole of a compound is formed from its elements
enthalpy of fusion: the energy required to change 1 gram of a substance from solid to liquid
enthalpy of reaction: the change in enthalpy accompanying a chemical reaction
enthalpy of solution: the change in enthalpy when one substance is dissolved in another
enthalpy of vaporization: the energy needed to change 1 gram of a substance from liquid to gas
entropy: the degree of disorder in a system
enzyme: a biological catalyst
equation: a symbolic expression representing a chemical change
equilibrium: a state in which no net change takes place in a system
equilibrium constant: a mathematical expression giving the ratio of the product of the concentrations of the products to the product of the concentrations of reactants in a chemical reaction
ester: an organic compound characterized by the functional group $R-CO-O-R'$
esterification: the production of an ester by the reaction of an alcohol with a carboxylic acid
ether: an organic compound characterized by the functional group $R-O-R'$
evaporation: the process by which surface particles of liquids escape into the vapor state
excess reactant: reactant remaining when all of some other reactant has been consumed
exergonic: a process having a decrease in Gibbs free energy
exothermic: a change that produces heat
experiment: a test of a hypothesis under controlled conditions
extensive property: a property dependent on the amount of matter present

F

face-centered cubic: having a cubic unit cell with the addition of a particle in the center of each face
factor-label method: a problem-solving method in which units (labels) are treated as factors
family: the elements composing a vertical column of the periodic table
fat: a biological ester of glycerol and a fatty acid
first ionization energy: the energy required to remove the most loosely held electron from an atom
fission: the splitting of an atomic nucleus into two approximately equal parts
fluid: a material that flows (liquid or gas)
formula: the symbolic representation of a chemical compound
formula mass: the sum of the atomic masses of the atoms in a formula
formula unit: the amount of a substance represented by its formula
fractional crystallization: a separation method based on the difference in the solubility of substances
fractional distillation: a separation method based on the difference in the boiling points of substances
fractionation: separating a whole into its parts, a mixture into its components
free electrons: the delocalized electrons that are in a metal
freezing point: the temperature equal to the melting point of a pure substance
frequency: the number of complete wave cycles per unit of time
functional group: an atom other than hydrogen or carbon introduced into an organic molecule
functional isomers: organic compounds with the same formula, but with the nonhydrocarbon part of the molecule bonded in different ways
fusion reaction: nuclear reaction in which small nuclei are combined to make a larger nucleus

G

galvanizing: coating iron with a protective layer of zinc
galvanometer: an instrument used to detect an electric current

gamma ray: a quantum of energy of very high frequency and very short wavelength

gas: the state of matter in which particles are far apart and moving randomly

gas chromatography: a chromatographic method in which a carrier gas (inert) distributes the vapor being analyzed in a packed column

geometric isomers: compounds with the same formula but different arrangement of substituents around a double bond

Gibbs free energy: the chemical reaction potential of a substance or system

gluon: a theoretical massless particle exchanged by quarks

glycogen: a biological polymer of glucose

Graham's law: The ratio of the relative rates of diffusion of gases is equal to the square root of the inverse ratio of their molecular masses.

gray: the unit of absorbed dose of radiation equivalent to 1 J/kg of living tissue

ground state: the state of lowest energy of a system

group: the elements of a vertical column in the periodic table

H

hadrons: a class of heavy subatomic particles

half-cell: the part of an electrochemical cell in which either the oxidation or reduction reaction is taking place; single electrode in contact with the solution of an electrolyte

half-life: the length of time necessary for one-half an amount of a radioactive nuclide to disintegrate

half-reaction: either the oxidation or the reduction part of a redox reaction

halogen: an element in Group 17(VIIA)

heat: energy transferred due to differences in temperature

Heisenberg uncertainty principle: It is impossible to know exactly both the position and momentum of an electron at the same instant.

Henry's law: The mass of gas that will dissolve in a specific amount of a liquid varies directly with the pressure.

hertz: the unit of frequency equal to one cycle per second

Hess's law: The enthalpy change for an overall reaction is equal to the sum of the enthalpy changes for all steps of the reaction.

heterogeneous: composed of more than one phase

heterogeneous catalyst: a catalyst in a phase different from that of the reactants

heterogeneous mixture: a combination of two or more substances that are not uniformly dispersed

heterogeneous reaction: a reaction in which not all reactants are in the same phase

hexagonal closest packing: having a crystal structure in which space between particles is minimized; found in most metals

high performance liquid chromatography (HPLC): a type of column chromatography in which the surface area of the particles in the stationary phase is increased

homogeneous: uniform throughout

homogeneous catalyst: a catalyst in the same phase as the reactants

homogeneous reaction: a reaction in which the reactants are in the same phase

homologous series: compounds that differ from each other by a specific structural unit

hybrid orbitals: equivalent orbitals formed from orbitals of different energies

hybridization: the merging of two or more unlike orbitals to form an equal number of identical orbitals in an atom

hydrate: a compound (crystalline) in which the ions are attached to one or more water molecules

hydrated ion: complex ion in which the ligands are water molecules

hydration: the adhering of water molecules to dissolved ions

hydride ion: a hydrogen atom that has gained an electron; H^-

hydrocarbon: compound containing only the elements hydrogen and carbon

hydrogen bonding: a very strong dipole-dipole interaction involving molecules in which hydrogen is bonded to a highly electronegative element (N, O, F)

hydrogen ion: a hydrogen atom that has lost its electron; H^+; a proton

hydrolysis: the reaction of a salt with water to form a weak acid or weak base, or both

hydronium ion: H_3O^+

hygroscopic: absorbing water from the air

hypothesize: to propose an explanation based on observations

I

ideal gas: a model in which gas particles are mass points and exert no attraction for each other

ideal gas equation: $PV = nRT$

ideal solution: a solution in which all intermolecular forces are roughly equal

immiscible: a property of two liquids that will not dissolve in each other at all

indicator: a weak organic acid whose color differs from that of its conjugate; used to indicate the pH of a solution

induced dipole: a nonpolar molecule that is transformed into a dipole by an electric field

inductive effect: the influence of one functional group on another

inertia: the tendency of an object to resist any change in its velocity

infrared spectroscopy: the study of the behavior of matter when it is exposed to infrared radiation

inhibitor: a substance that stops or retards a chemical reaction by forming a complex with a reactant

inner transition elements: those elements that fall between numbers 57 and 70 (the lanthanoids) and between numbers 89 and 102 (the actinoids) of the periodic table

inorganic compound: a molecular compound that does not contain carbon

inorganic substance: a substance that is not a hydrocarbon or a derivative of a hydrocarbon
insulator: a material that does not conduct heat or electricity
intensive property: a property of a substance that is independent of the amount of matter present
interface: the area of contact between two phases
intermediate: the material that is produced from raw materials and processed further to produce some consumer products
intermolecular force: the force holding molecules to each other
internal energy: that energy of a system that is altered by the absorption or release of heat and by doing work or having work done on it; energy of a system due to the energy of its constituent particles, but excluding the kinetic and potential energy of the system as a whole
internuclear distance: the distance between the nuclei of two atoms or ions
intramolecular force: the force holding atoms together in a molecule
ion: an atom or molecule that has gained or lost one or more electrons
ion chromatography: a type of column chromatography in which the column is packed with an ion exchange resin
ion product constant of water: the product of the hydronium and hydroxide ion concentrations in water solutions, equal to 1.00×10^{-14} at 25°C
ionic bond: the electrostatic attraction between ions of opposite charge
ionic compound: a compound that is formed by ionic bonds
ionic radius: the radius of an ion
ionization constant: the equilibrium constant for the ionization of a weak electrolyte
ionization energy: the energy required to remove an electron from an atom
irreversible thermodynamic change: change in volume or pressure in which some energy is lost to an entropy change
isobaric process: a process taking place at constant pressure
isomer: a substance that has the same molecular formula as another substance, but differs in structure; a substance that exhibits isomerism with another substance
isomerism: the property of having more than one structure for the same formula
isomorphism: condition of two or more compounds having the same crystalline structure
isothermal process: a process taking place at constant temperature
isotope: one of two or more atoms having the same number of protons but different numbers of neutrons

J

joule: the SI unit of energy; 1 kg·m^2/s^2
Joule-Thomson effect: the cooling effect observed when a compressed gas is allowed to expand rapidly through a small opening

kelvin: the SI unit of temperature; 1/273.16 of the interval between absolute zero and the triple point of water
Kelvin scale: the temperature scale with 0 equal to absolute zero and 273.16 equal to the triple point of water
ketone: an organic compound characterized by the functional group R—CO—R′
kilogram: the SI unit of mass
kinetic energy: the energy of an object due to its motion
kinetic theory: the group of ideas explaining the interaction of matter and energy due to particle motion
kinetically stable: property of a compound for which the activation energy for decomposition is so high that reaction proceeds too slowly for a change to be detected

L

lanthanoid series: fourteen elements beginning with lanthanum in which the arrow diagram predicts the highest energy electrons to be in the 4*f* sublevel
law of conservation of energy: Energy is conserved in all nonnuclear changes; it cannot be created or destroyed.
law of conservation of mass: Mass is conserved in all nonnuclear changes; it cannot be created or destroyed.
law of conservation of mass-energy: Although they can be interconverted, the total amount of mass and energy in the universe is constant.
law of definite proportions: The elements composing a compound are always found in the same ratio by mass.
law of multiple proportions: The masses of one element that combine with a fixed amount of another element to form more than one compound are in the ratio of small whole numbers.
law of octaves: The same properties appear every eighth element when the elements are listed in order of their atomic masses.
Le Chatelier's principle: If a system at equilibrium is subjected to a stress, the system will adjust so as to relieve the stress.
length: the distance between two points
leptons: light subatomic particles
Lewis electron dot diagram: the representation of an atom, ion, or molecule in which an element symbol stands for the nucleus and all inner level electrons while dots stand for outer level electrons
ligand: a negative ion or polar molecule attached to a central ion in a complex
limiting reactant: the reactant that is consumed completely in a chemical reaction
linear accelerator: a device for accelerating particles in a straight line
lipid: a biological molecule that is soluble in nonpolar solvents
liquefaction: condensing a gas to a liquid

liquid: the state of matter characterized by its constituent particles appearing to vibrate about moving points
liquid crystal: a substance that has order in the arrangement of its particles in only one or two dimensions
liter: one cubic decimeter
London forces: dispersion forces

M

macromolecule: a crystal composed of a single molecule with all atoms covalently bonded in a network fashion
magnetohydrodynamics: the study of the behavior of plasmas in magnetic fields
malleability: the property of a substance that allows it to be beaten into thin sheets
manometer: a device for measuring gas pressure
mass: measure of the amount of matter
mass defect: the difference between the mass of an atom and the sum of the masses of the particles composing it
mass-energy problem: a problem in which the amount of energy absorbed or released during a reaction can be calculated from the mass of materials
mass-mass problem: a problem in which the mass of one substance is provided and the mass of another substance must be calculated
mass number: the total number of protons and neutrons in an atom
mass spectrometry: an analysis of substances on the basis of the behavior of their ionized forms in magnetic and electric fields
material: a specific kind of matter
matter: anything that exhibits the property of inertia
mean free path: the average distance a particle travels between collisions
melting point: the temperature at which the vapor pressures of the solid and liquid phases of a substance are equal
meson: a subatomic particle classified as a hadron
metal: an element that tends to lose electrons in chemical reactions
metallic bond: a force holding metal atoms together and characterized by free or delocalized electrons
metallic conduction: electronic conduction; flow of electrons
metalloid: an element that has properties characteristic of a metal and a nonmetal
metastable: the state in which no change will occur unless acted upon by an outside force, but not the most stable state
meter: the SI unit of length
miscibility: the ability of two liquids to dissolve in each other in all proportions
mixture: a material consisting of two or more substances
mobile phase: the fluid containing the mixture to be fractionated in chromatography
model: an arrangement analogous to, and useful for, understanding a system in nature, but existing only in one's mind
moderator: a substance used to slow neutrons in a nuclear reactor
molal boiling point constant: the change of the boiling point of a solvent in a one-molal solution
molal freezing point constant: the change of the freezing point of a solvent in a one-molal solution
molality: a unit of concentration equal to the number of moles of solute per kilogram of solvent
molar heat capacity: the energy necessary to raise the temperature of one mole of a substance by one Celsius degree
molar mass: the mass in grams of one mole of a substance
molar volume: the volume occupied by one mole of a substance; equal to 22.414 10 dm^3 for a gas at standard temperature and standard atmospheric pressure
molarity: a unit of concentration equal to the number of moles of solute in a cubic decimeter of solution
mole: the Avogadro constant number of objects
mole fraction: a unit of concentration equal to the number of moles of component per mole of solution
molecular formula: a formula indicating the actual number of atoms of each element making up a molecule
molecular mass: the mass found by adding the atomic masses of the atoms comprising the molecule
molecule: a neutral group of atoms held together by covalent bonds
momentum: the product of mass and velocity

N

nematic substance: a liquid crystal with one dimension of order
net ionic equation: a chemical equation with spectator ions eliminated
network crystal: a crystal in which each atom is covalently bonded to all its nearest neighbors, so that the entire crystal is one molecule
neutral: neither acidic nor basic (electrolytes); neither positive nor negative (electricity)
neutralization: combining equivalent amounts of acid and base
neutralization reaction: the double displacement reaction between an acid and a base to produce salt and water
neutrino: a neutral particle associated with leptons
neutron: a neutral subatomic particle; a hadron
Newtonian mechanics: the laws of mechanics applicable in the macroscopic world
nitrile: an organic compound characterized by the functional group —CN
nitro: the functional group $—NO_2$
noble gas: an element in Group 18(VIIIA)
noble gas configuration: an arrangement of eight electrons in the outer energy level, except for helium with two electrons in the outer level
nonmetal: an element that tends to gain electrons in chemical reactions
nonvolatile: does not evaporate easily

normal boiling point: the temperature at which the vapor pressure of a liquid is equal to standard atmospheric pressure
nuclear force: the force holding nucleons together in a nucleus
nuclear magnetic resonance spectroscopy (NMR): the analysis of the structure of a substance by the behavior of its nuclei in a magnetic field
nuclear reactor: a device engineered to run a controlled nuclear reaction
nucleic acid: an organic compound containing nitrogenous bases, sugars, and phosphate groups; a compound that either transfers genetic information (DNA) or synthesizes biomolecules (RNA)
nucleon: a particle found in the nucleus of an atom; a proton or a neutron
nucleotide: a substance containing a nitrogenous base, a sugar, and a phosphate group
nuclide: an atom of a specific energy with a specified number of protons and a specified number of neutrons in its nucleus

O

observe: to note with the senses, aided or unaided
octahedral: the shape in which six objects are equally spaced about a central object
octane rating: a system of rating gasoline based upon the proportions of heptane and 2,2,4-trimethylpentane in the mixture
octet: an especially stable arrangement of four pairs of electrons in the outer energy level of an atom
octet rule: the tendency of atoms to gain or lose electrons so that they acquire eight electrons in their outer level
ohm: the SI unit of electrical resistance; one volt per ampere
olefin: an alkene
optimum conditions: the conditions maximizing the product of an equilibrium reaction
orbital: the space that can be occupied by 0, 1, or 2 electrons with the same energy level, energy sublevel, and spacial orientation
organic: pertaining to carbon compounds
organic chemistry: the chemistry of the compounds of carbon
organic compound: a compound containing carbon, with a very few exceptions
organic oxidation reaction: the conversion of an organic compound to carbon dioxide, water, and other appropriate oxides
organic substance: a compound that contains the element carbon; a few carbon compounds are considered inorganic
osmotic pressure: the pressure developed across a semipermeable membrane by differential diffusion through the membrane
oxidation: the loss of electrons
oxidation number: the apparent charge on an atom if the electrons in a compound are assigned according to established rules
oxidation-reduction reaction: a chemical reaction in which electrons are transferred
oxidizing agent: a substance that tends to gain electrons

P

packing: the adsorbent in a chromatographic column
pair repulsion: a model used to predict molecular shape based on the mutual repulsion of electron clouds
paper chromatography: a chromatographic method that uses paper as the stationary phase; the mobile phase moves by capillary action
parent chain: the longest continuous chain of carbon atoms in an organic compound
partially miscible: property of two liquids that dissolve in each other to some extent, but not completely
pascal: the SI unit of pressure; 1 N/m^2
Pauli exclusion principle: No two electrons in an atom can have the same set of quantum numbers.
peptide bond: an amide link; $-CO-NH-$
percent of ionization: the amount ionized divided by the original amount, multiplied by 100
percentage composition: the mass of an element in a compound divided by the mass of the compound, multiplied by 100
percentage yield: the mass of product actually obtained from a chemical reaction divided by the amount of product expected from a mass-mass calculation, multiplied by 100
period: a horizontal row of the periodic table
periodic law: The properties of the elements are a periodic function of their atomic numbers.
periodic property: a property of elements that appears periodically when the elements are arranged in order of their atomic numbers
periodic table: a pictorial arrangement of the elements based upon their atomic numbers and electron configurations
petroleum: a raw material consisting chiefly of a complex mixture of hydrocarbons
pH: $-\log[H_3O^+]$
pH meter: an electronic device for the determination of pH values in solutions
pH scale: a logarithmic scale expressing degree of acidity or basicity
phase: a physically distinct section of matter with uniform properties set off from the surrounding matter by physical boundaries
phase diagram: a graphical representation of the equilibrium relationships of the phases of a substance
phenol: C_6H_5OH; any compound having a hydroxyl group attached to a benzene ring
photoelectric effect: ejection of electrons from a surface exposed to light
photon: quantum of radiant energy
physical change: a change in which the same substance is present before and after the change
physical property: a property that can be observed without a change of substance
pi bond: a bond formed by the sideways overlap of *p* orbitals
planetary model: the model of the atom in which the sun represents the nucleus and the planets represent the electrons
plasma: a state composed of electrons and positive ions that have been knocked apart by collisions at very high temperatures

pOH: $-\log[OH^-]$
point mass: an ideal gas particle with mass but no dimensions
polar covalent: a bond formed by a shared pair of electrons that are more strongly attracted to one atom than to the other
polarity: property of a molecule caused by an unsymmetrical charge distribution
polyatomic ion: a group of atoms covalently bonded but possessing an overall charge
polymer: a very large molecule made from the same simple units repeated many times
polymerization: the formation of a polymer from monomers
polymorphism: the property of a substance whereby it exists in more than one crystalline form
polyprotic acid: an acid with more than one ionizable hydrogen atom
positional isomers: two or more molecules having the same formula but having a functional group in different positions on the parent chain
positron: the antiparticle of the electron
potential difference: the difference in electric potential
potential energy: the energy of an object due to its position
precipitate: a solid, produced by a reaction, that separates from a solution
precision: the measure of the reproducibility of measurements within a set
pressure: force per unit area
principal quantum number: the quantum number designating energy level and electron cloud size
probability: mathematical expression of "chance" or "odds"
product: a substance produced as the result of a chemical change
protein: a biological polymer of amino acids linked by amide groups
proton: positive nucleon

Q

qualitative: concerning the kinds of matter present
quantitative: concerning the amounts of matter present
quantum: a discrete "packet" of energy (*plural:* quanta)
quantum mechanics: the laws of mechanics concerning the interaction of matter and radiation at the atomic and subatomic level
quantum number: a number describing a property of an electron in an atom
quantum theory: the concept that energy is transferred in discrete units
quark: a theoretical particle believed to be elementary and a constituent of a hadron

R

rad: 0.01 gray
radiant energy: energy being transferred between objects by electromagnetic waves
radical: a fragment of a molecule; neutral, yet at least one atom lacking its octet of electrons
radioactivity: spontaneous nuclear decay
Raoult's law: The vapor pressure of a solution of a nonvolatile solute is the product of the vapor pressure of the pure solvent and the mole fraction of the solvent.
rate determining step: the slowest step in a reaction mechanism
raw material: a crude, unprocessed material found in nature and used to make intermediates or consumer products
reactant: a starting substance in a chemical reaction
reaction mechanism: the series of steps through which the reactants pass in being converted to the products in a chemical reaction
reaction rate: the rate of disappearance of a reactant or the rate of appearance of a product
real gas: a gas with particles of finite volume and van der Waals forces between particles
redox reaction: an oxidation-reduction reaction
reducing agent: a substance that tends to give up electrons
reduction: the gain of electrons
rem: 0.01 sievert
reversible change: a change that can also go in the opposite direction
reversible reaction: a reaction in which the products may react to produce the original reactants
reversible thermodynamic change: an ideal change in which the difference in pressure is infinitesimal
Rutherford-Bohr atom: the planetary atom model

S

salt: a compound formed from a positive ion other than hydrogen and a negative ion other than hydroxide
salt bridge: an ionic solution used to complete an electric circuit in a voltaic cell
saponification: the reaction of an ester with a strong aqueous base to form a soap and glycerol
saturated: (vapor) the gaseous phase of a system with equilibrium between a substance and its vapor
saturated compound: a compound having only single bonds between carbon atoms
saturated hydrocarbon: a hydrocarbon in which all carbon-carbon bonds are single bonds
saturated solution: a solution in which undissolved solute is in equilibrium with dissolved solute
science: the systematic investigation of nature
scientific notation: the expression of numbers in the form $M \times 10^n$ where $1.00 \leq M < 10$ and n is an integer
screw dislocation: a crystal defect due to improperly aligned unit cells
second: the SI unit of time
semiconductor: a substance that conducts electricity, but poorly
semipermeable membrane: a barrier allowing the passage of small ions and molecules but blocking passage of large particles

shared pair: a pair of electrons bonding two atoms together by being shared by the two atoms
shielding effect: the decrease in the attraction between outer electrons and the nucleus due to the presence of other electrons between them
SI units: the internationally accepted set of standards for measurements
side chain: a branch on the parent chain of an organic molecule
sievert: the SI unit used to measure the absorbed dose of radiation; ionizing radiation equal to 100 rem
sigma bond: a bond formed by the direct or end-to-end overlap of atomic orbitals
significant digits: the reliable digits in a measurement based on the accuracy of the measuring instrument
silicates: compounds containing silicon and oxygen
simple cubic: having a unit cell with one particle centered on each vertex of a cube
single displacement: a reaction in which one element replaces another in a compound
smectic substance: a liquid crystal having two dimensions of order
solid: the state of matter characterized by particles that appear to vibrate about fixed points
solubility: the quantity of solute that will dissolve in a specified amount of solvent at a specific temperature
solubility product constant: the equilibrium constant for the dissolving of a slightly soluble salt
solute: the substance present in lesser quantity in a solution
solution: a homogeneous mixture composed of solute and solvent
solution equilibrium: the state in which solute is dissolving at the same rate that solute is coming out of solution
solvation: the attaching of solvent particles to solute particles
solvent: the substance present in the greater amount in a solution
space lattice: the arrangement pattern of the unit cells in a crystal
specific heat: the amount of energy required to raise the temperature of one gram of a substance by one Celsius degree
specific rate constant: a constant relating the rate of a reaction to reactant concentrations
spectator ion: an ion present in a solution but not taking part in a chemical reaction
spectroscopy: the study of the interaction of matter and radiant energy
spectrum: a unique set of wavelengths absorbed or emitted by a substance
spin: a property of subatomic particles corresponding to rotation on an axis
spontaneous: occurring without outside influence
square planar: the arrangement in which four objects are at the corners of a square around a fifth object in the center
stability: the ability of a substance to remain undecomposed
standard atmospheric pressure: 101.325 kPa
standard solution: a solution whose concentration is known with a high degree of accuracy
standard state: thermodynamic reference conditions, 25°C, 100 kPa, 1*M*
standard temperature: 0°C for gases; 25°C for thermodynamics
starch: a biological polymer of glucose
state: the particle arrangement in a phase as solid, liquid, gas, or plasma
state function: a thermodynamic quantity that is determined solely by the conditions, not the method of arriving at those conditions
stationary phase: the adsorbent in chromatography
stoichiometry: mass and volume relationships in chemical changes
STP: standard temperature and atmospheric pressure (273 K and 101.325 kPa)
strong (acid or base): a completely ionized electrolyte
structural isomers: two or more compounds with the same formula but differing arrangements of the parent carbon chain
subatomic particle: a particle smaller than an atom
sublevel: energy subdivision of an energy level
sublimation: the change directly from solid to gas
substance: a material with a constant composition
substituent: a hydrocarbon branch or functional group attached to the parent chain of an organic compound
substitution reaction: a reaction of organic compounds in which a hydrogen atom or functional group is replaced by another functional group
supercooled liquid: a liquid cooled below its normal freezing point without having changed state to the solid form
supersaturated solution: a solution containing more solute than would a saturated solution at the same temperature
surface tension: the apparent "skin" effect on the surface of a liquid or solid due to unbalanced forces on the surface particles
suspension: a dispersion of particles $>$ 100 nm in a continuous medium
synchrotron: a device for accelerating particles in a circular path
synthesis: the formation of a compound from two or more substances
synthetic element: an element not occurring in nature
system: that part of the universe under consideration

T

technology: the practical applications of scientific discoveries
temperature: a measure of the average kinetic energy of the particles composing a material
temporary dipole: a dipole formed from a nonpolar molecule for a brief period due to the presence of an electric field
ternary: composed of three elements
ternary acid: an acid containing hydrogen, usually oxygen, and one other element
tetradentate: describing a ligand that attaches to the central ion in four locations

tetrahedral: four objects equally spaced in three dimensions around a fifth object
theory: an explanation of a phenomenon
thermodynamic stability: the stability of a substance due to a positive change in the Gibbs free energy for the decomposition of the substance
thermodynamics: the study of the flow of energy in systems
thermometer: a device for measuring temperature
thin layer chromatography: a method of chromatography utilizing an adsorbent spread over a flat surface in a thin layer
time: the interval between two occurrences
titration: a laboratory technique for measuring the relative concentrations of solutions
tracer: a radioactive nuclide used to follow the progress of a reaction or a process
transistor: an electronic device made from a doped semiconductor
transition element: an element whose highest energy electron is in a *d* sublevel
transmutation: the conversion of one element into another
transuranium element: an element with an atomic number greater than that of uranium
triad: a group of three elements with similar properties
tridentate: a ligand that attaches to the central ion in three locations
tripeptide: three amino acids joined by amide links
triple bond: a bond in which two atoms share three pairs of electrons
triple point: the temperature and pressure at which all three states of a substance are in equilibrium
Tyndall effect: the scattering of light by colloidal or suspended particles

U

ultraviolet spectroscopy: the study of the interaction of matter and ultraviolet radiation
unit cell: the simplest unit of repetition in a crystal
unsaturated compound: an organic compound containing one or more multiple bonds
unsaturated hydrocarbon: hydrocarbon containing one or more multiple bonds
unsaturated solution: a solution containing less than the saturated amount of solute
unshared pair: a pair of electrons in an orbital belonging to one atom

V

van der Waals forces: weak forces of attraction between molecules
van der Waals radius: radius of closest approach of a nonbonded atom
vapor: the gaseous state of a substance that is liquid or solid at room temperature and pressure
vapor equilibrium: the equilibrium state between a liquid and its vapor
vapor pressure: the pressure exerted by a vapor in equilibrium with its liquid
velocity: the speed and direction of motion
viscosity: the resistance of a liquid to flow
visible spectroscopy: the study of the interaction of matter with visible radiation
vitamin: a group of biochemicals that are necessary for some enzymatic reactions to take place
volatile: an easily evaporated liquid
volt: the SI unit of electric potential difference
voltaic cell: a cell in which a chemical reaction generates an electric current

W

wave: a periodic disturbance in a medium
wave equation: the equation describing the behavior of the electron as a wave
wavelength: the distance between two successive crests of a wave
wave-particle duality of nature: All particles have wave properties and all waves have particle properties.
weak (acids and bases): a slightly ionized electrolyte
weak forces: an attraction of molecules for each other through the action of dipoles
weight: the gravitational attraction of Earth or a celestial body for matter
work: a force moving through a distance

INDEX

A

B

Boron trifluoride, 339; reaction with ammonia, 576–577

D

E

F

G

H

I

J

K

L

M

N

O

P

Q

R

S

T

U

V

W

Y

Z

Cover, Erich Schrempp/Photo Researchers; iii, Aaron Haupt; iv, (t) First Image, (bl) Les Losego/Crown Studios, (bc) MAK-1 Photodesign, (br) E.R. Degginger; v, First Image; vi, (l) Doug Martin, (r) image by Tom Palmer, Visualization Group, NC Supercomputing Center/based on work by Jerzy Bernholc, Qiming Zhang, Jae-Yel Yi, Charles Brabec, NC State University; vii, (t) Spencer Swanger/Tom Stack & Associates, (bl) simulations and image: Jerzy Bernholc, Qiming Zhang, Jae-Yel Yi, Charles Brabec, NC State University/Tom Palmer, Cray Research, NC Supercomputing Center, (br) Tom Van Sant/Eyes On Earth; viii, (t) Robert Mathena/Fundamental Photographs, (b) F. Stuart Westmorland/Tom Stack & Associates; ix, (l, c) Matt Meadows, (r) T.J. Florian from Rainbow; x, (l) R. Feldman from Rainbow, (r) Julie Houck/Stock Boston; xi, (l) GE Research & Technology Center, (r) Science Source/Photo Researchers; xii, courtesy Tecktronics, Inc.; xiii, William McCoy from Rainbow; xiv, courtesy AT&T Bell Laboratories; 1, Aaron Haupt; 2, Tom Sanders/Adventure Photo; 3, ©G. Kramer/Helga Lade Fotoagentur/Peter Arnold, Inc.; 4, Doug Martin; 5, 6, Bob Daemmrich; 7, Los Alamos National Laboratory; 8, (t) Les Lesego/Crown Studios, courtesy Bath & Brass Emporium, (b) Bob Thomason/Tony Stone Images; 9, Matt Meadows; 10, Paul Silverman/Fundamental Photographs; 11, John Durham/Science Photo Library/Photo Researchers; 12, (l) courtesy Dr. Bruce N. Ames, (c) Andy Sacks/Tony Stone Images, (r) D. Wilder/Tom Stack & Associates; 13, Fred Ward/Black Star; 14, (t) Dr. Jeremy Burgess/Science Photo Library/Photo Researchers, (b) Dr. Gerardo B. Staal, Sandoz Crop Protection Corp., Zoecon Research Institute, Palo Alto CA; 15, 16, First Image; 17, Matt Meadows,19, Julie Houck/Stock Boston; 20, Stacey Pick/Stock Boston; 21, First Image; 22, Hank Morgan from Rainbow; 23, Matt Meadows; 24, Doug Martin; 25, Bettmann Archive; 26, Bob Daemmrich/Stock Boston; 27, Bill Everitt/Tom Stack & Associates; 30, (l) National Bureau of Weights and Measures, (r) First Image; 31, Chip Clark; 32, Skip Comer Photography; 33, Matt Meadows; 34, Stephen Frisch/Stock Boston; 36, Doug Martin; 37, Matt Meadows; 39, Chip Clark; 40, Skip Comer Photography; 41, (l) Chip Clark, (r) Les Losego/Crown Studios; 46, Doug Martin, courtesy Ohio Historical Society; 47, MAK-1 Photodesign; 48, (tl) Aaron Haupt, (bl) Doug Martin, (r) MAK-1 Photodesign; 49,

Eastcott-Momatiuk/Woodfin Car
Associates;51, Chip Clark; 52, G
Research & Technology Center;
MAK-1 Photodesign; 54-55, fro
Periodic Systems of the Element y;
P. Menzel. © by Ernst Klett Sch
Verlag GmbH, Stuttgart, Germa
available from Science Imports,
465, Sillery, Quebec Canada G1
56, Rod Joslin, courtesy Columb k;
Cultural Arts Center; 57, ©Zev R
58, Manfred Kage/Peter Arnold
Chip Clark; 60, (l) Chip Clark, (
D. Franke/Peter Arnold, Inc.; 61
Clark; 66, NASA; 67, Wesley
Bocxe/Photo Researchers; 70, W
McCoy from Rainbow; 76, KS S
77, Doug Martin; 79, Chip Clar
Comer Photography; 86, Stuart
Anderson/Caltech; 88, Chip Cla
©1978 Dan Guravich/Photo Res s;
93, Bud Fowle; 98, Dan McCoy
Rainbow; 99, Les Losego/Crow
100, Sven-Olof Lindblad/Photo
Researchers; 102, Ron Kimball
Stephen Webster; 111, (l) IBM
Corporation Research Division E
Research Center, (r) Doug Mar
William E. Ferguson; 116, Ski
Photography; 118, Sepp Dietri
Stone Images; 125, ©1980 Ang s;
Australian Telescope Board, p
David F.Malin; 126, courtesy C
Arthur Freeman, Northwestern
130, IBM Corporation Researc
Almaden Research Center; 13
Jerry Watcher Photography Lt
Researchers; 137, Aaron H
140, courtesy National Fo
for History of Chemistry; 1
Chip Clark; 150, First Ima
Doug Martin, courtesy All
Coin Shop, Westerville, O
Argonne National Lab; 15
Burroughs Wellcome Co.;
Meadows; 157, StudiOhio; 16
Martin, 162, Andy Sacks/Ton
Images; 164, Matt Meadows;
Koons; 170, Matt Meadows;
Clark; 172, MAK-1 Photodes
Chip Clark; 174, (t) Matt Me
Ted Rice; 176, KS Studios;17
Fowle; 178, courtesy Burroug
Wellcome Co.; 179, (l) Bud F
Matt Meadows; 181, (l) Dan
Rainbow, (r) courtesy John P
184, Panographics/L.S. Stepa
Doug Martin; 188, MAK-1 P
189, Matt Meadows; 191, Le
Losego/Crown Studios; 192,
Johnston/Tony Stone Images 7,
Martin; 194, (l) courtesy IB
Corporation, (r) file photo; 1
Losego/Crown Studios; 197,

International Atomic Masses

Element	Symbol	Atomic number	Atomic mass	Element	Symbol	Atomic number	Atomic mass
Actinium	Ac	89	227.027 8*	Neon	Ne	10	20.179 7
Aluminum	Al	13	26.981 539	Neptunium	Np	93	237.048 2
Americium	Am	95	243.061 4*	Nickel	Ni	28	58.6934
Antimony	Sb	51	121.757	Niobium	Nb	41	92.906 38
Argon	Ar	18	39.948	Nitrogen	N	7	14.006 74
Arsenic	As	33	74.921 59	Nobelium	No	102	259.100 9*
Astatine	At	85	209.987 1*	Osmium	Os	76	190.2
Barium	Ba	56	137.327	Oxygen	O	8	15.999 4
Berkelium	Bk	97	247.070 3*	Palladium	Pd	46	106.42
Beryllium	Be	4	9.012 182	Phosphorus	P	15	30.973 762
Bismuth	Bi	83	208.980 37	Platinum	Pt	78	195.08
Boron	B	5	10.811	Plutonium	Pu	94	244.064 2*
Bromine	Br	35	79.904	Polonium	Po	84	208.982 4*
Cadmium	Cd	48	112.411	Potassium	K	19	39.098 3
Calcium	Ca	20	40.078	Praseodymium	Pr	59	140.907 65
Californium	Cf	98	251.079 6*	Promethium	Pm	61	144.912 8*
Carbon	C	6	12.011	Protactinium	Pa	91	231.035 88
Cerium	Ce	58	140.115	Radium	Ra	88	226.025 4
Cesium	Cs	55	132.905 43	Radon	Rn	86	222.017 6*
Chlorine	Cl	17	35.452 7	Rhenium	Re	75	186.207
Chromium	Cr	24	51.996 1	Rhodium	Rh	45	102.905 50
Cobalt	Co	27	58.933 20	Rubidium	Rb	37	85.467 8
Copper	Cu	29	63.546	Ruthenium	Ru	44	101.07
Curium	Cm	96	247.070 3*	Samarium	Sm	62	150.36
Dysprosium	Dy	66	162.50	Scandium	Sc	21	44.955 910
Einsteinium	Es	99	252.082 8*	Selenium	Se	34	78.96
Erbium	Er	68	167.26	Silicon	Si	14	28.085 5
Europium	Eu	63	151.965	Silver	Ag	47	107.868 2
Fermium	Fm	100	257.095 1*	Sodium	Na	11	22.989 768
Fluorine	F	9	18.998 403 2	Strontium	Sr	38	87.62
Francium	Fr	87	223.019 7*	Sulfur	S	16	32.066
Gadolinium	Gd	64	157.25	Tantalum	Ta	73	180.947 9
Gallium	Ga	31	69.723	Technetium	Tc	43	97.907 2*
Germanium	Ge	32	72.61	Tellurium	Te	52	127.60
Gold	Au	79	196.966 54	Terbium	Tb	65	158.925 34
Hafnium	Hf	72	178.49	Thallium	Tl	81	204.383 3
Helium	He	2	4.002 602	Thorium	Th	90	232.038 1
Holmium	Ho	67	164.930 32	Thulium	Tm	69	168.934 21
Hydrogen	H	1	1.007 94	Tin	Sn	50	118.710
Indium	In	49	114.82	Titanium	Ti	22	47.88
Iodine	I	53	126.904 47	Tungsten	W	74	183.85
Iridium	Ir	77	192.22	Unnilennium†	Une	109	266*
Iron	Fe	26	55.847	Unnilhexium†	Unh	106	263*
Krypton	Kr	36	83.80	Unniloctium†	Uno	108	265*
Lanthanum	La	57	138.905 5	Unnilpentium†	Unp	105	262*
Lawrencium	Lr	103	260.105 4*	Unnilquadium†	Unq	104	261*
Lead	Pb	82	207.2	Unnilseptium†	Uns	107	262*
Lithium	Li	3	6.941	Uranium	U	92	238.028 9
Lutetium	Lu	71	174.967	Vanadium	V	23	50.941 5
Magnesium	Mg	12	24.305 0	Xenon	Xe	54	131.29
Manganese	Mn	25	54.938 05	Ytterbium	Yb	70	173.04
Mendelevium	Md	101	258.098 6*	Yttrium	Y	39	88.905 85
Mercury	Hg	80	200.59	Zinc	Zn	30	65.39
Molybdenum	Mo	42	95.94	Zirconium	Zr	40	91.224
Neodymium	Nd	60	144.24				

*The mass of the isotope with the longest known half-life.

†Names for elements 104-109 have been approved for temporary use by the IUPAC. The USSR has proposed Kurchatovium (Ku) for element 104, and Bohrium (Bh) for element 105. The United States has proposed Rutherfordium (Rf) for element 104, and Hahnium (Ha) for element 105.